SOYBEANS: Improvement, Production, and Uses

Second Edition

AGRONOMY

A Series of Monographs

The American Society of Agronomy and Academic Press published the first six books in this series. The General Editor of Monographs 1 to 6 was A. G. Norman. They are available through Academic Press, Inc., 111 Fifth Avenue, New York NY 10003.

1. C. EDMUND MARSHALL: The Colloid Chemical of the Silicate Minerals, 1949
2. BYRON T. SHAW, *Editor*: Soil Physical Conditions and Plant Growth, 1952
3. K. D. JACOB, *Editor*: Fertilizer Technology and Resources in the United States, 1953
4. W. H. PIERRE and A. G. NORMAN, *Editors*: Soil and Fertilizer Phosphate in Crop Nutrition, 1953
5. GEORGE F. SPRAGUE, *Editor*: Corn and Corn Improvement, 1955
6. J. LEVITT: The Hardiness of Plants, 1956

The Monographs published since 1957 are available from the American Society of Agronomy, 677 S. Segoe Road, Madison, WI 53711.

7. JAMES N. LUTHIN, *Editor*: Drainage of Agricultural Lands, 1957 *General Editor,* D. E. Gregg
8. FRANKLIN A. COFFMAN, *Editor*: Oats and Oat Improvement, 1961
   Managing Editor, H. L. Hamilton
9. A. KLUTE, *Editor*:Methods of Soil Analysis, 1986
   Part 1—Physical and Mineralogical Methods, Second Edition *Managing Editor,* R. C. Dinauer
   A. L. PAGE, *Editor*: Methods of Soil Analysis, 1982
   Part 2—Chemical and Microbiological Properties, Second Edition *Managing Editor,* R. C. Dinauer
10. W. V. BARTHOLOMEW and F. E. CLARK, *Editors*: Soil Nitrogen, 1965
    (Out of print; replaced by no. 22) *Managing Editor,* H. L. Hamilton
11. R. M. HAGAN, H. R. HAISE, and T. W. EDMINSTER, *Editors*: Irrigation of Agricultural Lands, 1967 *Managing Editor,* R. C. Dinauer
12. FRED ADAMS, *Editor*: Soil Acidity and Liming, Second Edition, 1984
    Managing Editor, R. C. Dinauer
13. K. S. QUISENBERRY and L. P. REITZ, *Editors*: Wheat and Wheat Improvement, 1967
    Managing Editor, H. L. Hamilton
14. A. A. HANSON and F. V. JUSKA, *Editors*: Turfgrass Science, 1969
    Managing Editor, H. L. Hamilton
15. CLARENCE H. HANSON, *Editor*: Alfalfa Science and Technology, 1972
    Managing Editor, H. L. Hamilton
16. J. R. WILCOX, *Editor*: Soybeans: Improvement, Production, and Uses, Second Edition, 1987
    Senior Managing Editor, D. A. Fuccillo
17. JAN VAN SCHILFGAARDE, *Editor*: Drainage for Agriculture, 1974
    Managing Editor, R. C. Dinauer
18. GEORGE F. SPRAGUE, *Editor*: Corn and Corn Improvement, 1977
    Managing Editor, D. A. Fuccillo
19. JACK F. CARTER, *Editor*: Sunflower Science and Technology, 1978
    Managing Editor, D. A. Fuccillo
20. ROBERT C. BUCKNER and L. P. BUSH, *Editors*: Tall Fescue, 1979
    Managing Editor, D. A. Fuccillo
21. M. T. BEATTY, G. W. PETERSEN, and L. D. SWINDALE, *Editors*: Planning the Uses and Management of Land, 1979 *Managing Editor,* R. C. Dinauer
22. F. J. STEVENSON, *Editor*: Nitrogen in Agricultural Soils, 1982
    Managing Editor, R. C. Dinauer
23. H. E. DREGNE and W. O. WILLIS, *Editors*: Dryland Agriculture, 1983
    Managing Editor, D. A. Fuccillo
24. R. J. KOHEL and C. F. LEWIS, *Editors*: Cotton, 1984
    Managing Editor, D. A. Fuccillo
25. N. L. TAYLOR, *Editor*: Clover Science and Technology, 1985
    Managing Editor, D. A. Fuccillo
26. D. C. RASMUSSON, *Editor*: Barley, 1985
    Managing Editor, D. A. Fuccillo
27. M. A. TABATABAI, *Editor*: Sulfur in Agriculture, 1986
    Managing Editor, R. C. Dinauer

SOYBEANS: Improvement, Production, and Uses

Second Edition

J. R. Wilcox, ***editor***

Editorial Committee
H. R. Boerma
E. J. Kamprath
L. E. Schrader

Number 16 in the series

AGRONOMY

American Society of Agronomy, Inc.
Crop Science Society of America, Inc.
Soil Science Society of America, Inc.
Publishers
Madison, Wisconsin, USA

1987

American Society of Agronomy, Inc.
Crop Science Society of America, Inc.
Soil Science Society of America, Inc.
677 South Segoe Road, Madison, WI 53711 USA

Library of Congress Cataloging-in-Publication Data

Soybeans: Improvement, Production, and Uses

(Agronomy ; no. 16)
Includes bibliographies and index.
1. Soybean. I. Wilcox, J. R. II. Series.
SB205.S7S56 1987 633.3′4 86-28757
ISBN 0-89118-090-7

Printed in the United States of America

Top left, soybean flower; *top right,* developing soybean pods; *bottom,* raceme of mature soybean pods. Courtesy, Michiel A. Smit, Republic of South Africa.

Contents

Foreword

The first edition of *Soybeans: Improvement, Production, and Uses* was published in 1973, when demand for soybean exceeded supplies. World production was continuing to expand and prices had reached unprecedented highs. Fourteen years later, the U.S. cropland planted to soybean has stabilized at levels somewhat lower than the record high in 1982. Ample world supplies of soybean has resulted from U.S. production and greatly increased production in South America, primarily in Brazil and Argentina.

The high demand for soybean at the time the initial edition of this monograph was being prepared stimulated increased research. This research focused on production practices that would have an immediate effect on increasing soybean supplies. Additional research efforts were directed toward protecting the crop from losses due to pathogens, insects, and weeds. Research efforts also concentrated on improving its productivity (yield) through breeding and genetics. Increased research was also directed toward increasing our knowledge of physiological processes of the plant that would open the way for further improvements in soybean production efficiency and in quality of soybean products.

The increased research effort on soybean rapidly expanded our knowledge of this crop. This edition of the soybean monograph is a reflection of that increased research effort. This edition summarizes our current knowledge on the soybean as a plant, as a crop, and on the utilization of soybean products. It will be a useful reference on the state of our current knowledge and should serve as a basis for continuing research on this economically important crop plant.

D. N. Moss, *president,* 1986
American Society of Agronomy

J. B. Beard, *president,* 1986
Crop Science Society of America

John Pesek, *president,* 1986
Soil Science Society of America

Preface

Soybean, *Glycine max* (L.) Merr., has become the major source of edible vegetable oils and of high protein feed supplements for livestock in the world. A native of Eastern Asia, the soybean was introduced into the USA where it has become a major agricultural crop and a significant export commodity.

Research on soybean has been in proportion to production of the crop. As demand for soybean increased and area planted to the crop expanded, research support for the crop was also increasing. The increased support for soybean research has resulted in a rapid increase in the breadth and depth of our knowledge about this major crop. The second edition of *Soybeans: Improvement, Production, and Uses,* presents the current status of our knowledge at a time when this knowledge base is expanding rapidly.

This edition has been organized to reflect those areas where recent research has had the greatest impact. In all chapters, authors have reviewed our knowledge in specific areas but emphasis has been on research progress during the past 14 yrs.

Chapter 1 documents recent changes in world distribution and production of soybean and the economic and political reasons for those changes. Chapter 2 reports changes in the taxonomic history of the Phaseoleae with respect to the genus *Glycine* and our current knowledge of relationships among species in this genus.

There are two chapters on morphology, one each on vegetative and reproductive morphology. These were included in this edition of the monograph because information on anatomy and morphology are basic to our understanding of all plant processes. These chapters are a consolidation of this information where it will be readily accessible to those involved in soybean research and technology.

Information on genetics, cytogenetics, and breeding methods is included in chapters 5 through 7. There has been a rapid increase in our knowledge of the genetics of this crop, including the identification of over 200 genes and several linkage groups. The rate of cultivar development has increased rapidly during the past 14 yrs with the research efforts of commercial plant breeders. These changes have had a major impact on cultivar development and on the wide choice of cultivars available to soybean producers.

Various aspects of soybean production are reviewed in chapters 8 through 12. Seed production and technology is an important aspect of soybean production that is covered for the first time in this edition of the monograph. Chapter 9 on crop management reflects the rapid changes in production practices that have occurred during the past 14 yrs. Edaphic factors, including fertility, liming, tillage practices to minimize soil erosion, and effective water management are reviewed in these chapters of the monograph.

Perhaps our greatest advances in knowledge about the soybean has been in plant metabolism. This is reflected in individual chapters 13, 14, and 16 on N, C, and seed metabolisms. Included with these is chapter 15 on stress physiology since the soybean plant may be subjected to stress throughout its development.

Chapter 17 documents the recent advances in our knowledge of fungal disease of soybean. Although no new serious fungal diseases of soybean have been identified, many new races of specific pathogens, such as *Phytophthora megasperma* f. sp. *glycinea* have been identified. Locating genes for resistance to these races and incorporating the genes into new cultivars has been an integral part of most soybean improvement programs. Chapter 18 documents information about viral and bacterial diseases that are naturally occurring on soybean.

Nematodes, major pathogens of soybean in the South, have become a significant factor affecting soybean production in the Midwest. Chapter 19 reviews the progressive increase in the area infested by the soybean cyst nematode and strategies to limit losses associated with these pathogens.

Losses due to many soybean pests are limited by integrated systems of pest management. This recently developed control strategy is illustrated in chapter 20 on controlling insect pests of soybean.

Chapter 21 is on processing and utilization. This chapter describes in detail how soybean crops are processed, utilized, and require special handling. The final use of the crop will affect research to expand our knowledge of the crop and to improve production efficiency.

Mention of a trade name, proprietary product, or specific equipment does not constitute a guarantee or warranty by the U.S. Department of Agriculture or the American Society of Agronomy and does not imply its approval to the exclusion of other products that may be suitable.

My appreciation and gratitude is expressed to the Editorial Committee of this edition of the soybean monograph, Drs. H. R. Boerma, E. J. Kamprath, and L. E. Schrader. They represent different disciplines in soybean research and as members of the Editorial Committee have been fully involved in the development of chapter subjects, selection of authors, and in reviewing completed manuscripts. My appreciation is also extended to the many scientists who assisted with chapter reviews. I am especially grateful to Domenic Fuccillo and the staff at the American Society of Agronomy Headquarters who have been responsible for the myriad of details involved in the final editing of the manuscripts and the printing of this monograph.

James R. Wilcox, *editor*

Contributors

Kirk L. Athow — Professor (now Professor Emeritus), Department of Botany and Plant Pathology, Purdue University, West Lafayette, Indiana

H. Roger Boerma — Professor, Department of Agronomy, University of Georgia, Athens, Georgia

J. W. Burton — Research Agronomist, USDA-ARS, North Carolina State University, Raleigh, North Carolina

John B. Carlson — Professor, Department of Biology, University of Minnesota, Duluth, Minnesota

H. D. Coble — Professor of Crop Science, Department of Crop Science, North Carolina State University, Raleigh, North Carolina

Dennis B. Egli — Professor of Agronomy, Department of Agronomy, University of Kentucky, Lexington, Kentucky

Walter R. Fehr — Professor of Agronomy, Department of Agronomy, Iowa State University, Ames, Iowa

Duane Merlin Ford — Assistant Professor of Agronomy, Northeast Missouri State University, Kirksville, Maryland

J. E. Harper — Plant Physiologist and Professor, USDA and Department of Agronomy, University of Illinois, Urbana, Illinois

Wipada Huyser — Associate, International Monetary Fund, Washington, DC

T. Hymowitz — Professor of Plant Genetics, Department of Agronomy, University of Illinois, Urbana, Illinois

Richard R. Johnson — Senior Scientist, Deere & Company Technical Center, Moline, Illinois

T. N. Jordan — Professor of Weed Science, Department of Botany and Plant Pathology, Purdue University, West Lafayette, Indiana

Eugene J. Kamprath — Professor, Department of Soil Science, North Carolina State University, Raleigh, North Carolina

Thomas C. Kilen — Research Geneticist, USDA-ARS Delta Branch Experiment Station, Stoneville, Mississippi

Marcos Kogan — Professor Entomology, University of Illinois and Illinois Natural History Survey, Champaign, Illinois

Paul J. Kramer — James B. Duke Professor of Botany, Emeritus, Department of Botany, Duke University, Durham, North Carolina

Nels R. Lersten — Professor, Department of Botany, Iowa State University, Ames, Iowa

W. H. Martinez — Staff Scientist, National Program Staff, USDA-ARS, Beltsville, Maryland

David B. Mengel — Professor of Agronomy, Department of Agronomy, Purdue University, West Lafayette, Indiana

T. L. Mounts — Supervisory Research Chemist, Vegetable Oil Research, Northern Regional Research Center, USDA-ARS, Peoria, Illinois

Reid G. Palmer — Research Geneticist and Professor, USDA-ARS and Department of Agronomy, Iowa State University, Ames, Iowa

C. David Raper, Jr. — Professor of Soil Science, Department of Soil Science, North Carolina State University, Raleigh, North Carolina

George W. Rehm — Extension Specialist–Soil Fertility, Department of Soil Science, University of Minnesota, St. Paul, Minnesota

D. C. Reicosky — Soil Scientist, USDA-ARS North Central Soil Conservation Research Laboratory, Morris, Minnesota

R. D. Riggs — Professor, Department of Plant Pathology, University of Arkansas, Fayetteville, Arkansas

J. P. Ross — Research Plant Pathologist, USDA-ARS and Department of Plant Pathology, North Carolina State University, Raleigh, North Carolina

D. P. Schmitt — Associate Professor, Department of Plant Pathology, North Carolina State University, Raleigh, North Carolina

L. E. Schrader — Professor and Head, Department of Agronomy, University of Illinois, Urbana, Illinois

Jacob Secor — Plant Physiologist, Dow Chemical USA, Walnut Creek, California

William Segars — Extension Agronomist-Soils and Fertilizers, University of Georgia, Athens, Georgia

Richard Shibles — Professor, Department of Agronomy, Iowa State University, Ames, Iowa

R. J. Singh — Agronomist (Cytogenetics), Department of Agronomy, University of Illinois, Urbana, Illinois

Keith J. Smith — Staff Vice President, Research and Utilization, American Soybean Association, St. Louis, Missouri

Dennis M. TeKrony — Professor of Agronomy, Department of Agronomy, University of Kentucky, Lexington, Kentucky

Sam G. Turnipseed — Professor of Entomology, Clemson University, Edisto Research and Education Center, Blacksville, South Carolina

D. M. Van Doren, Jr. — Professor of Agronomy (now Professor Emeritus), Ohio Agricultural Research and Development Center and Ohio State University, Wooster, Ohio

L. M. Wax — Research Agronomist and Professor, USDA-ARS and Department of Agronomy, University of Illinois, Urbana, Illinois

Gerald M. White — Professor of Agricultural Engineering, Department of Agricultural Engineering, University of Kentucky, Lexington, Kentucky

J. R. Wilcox — Supervisory Research Geneticist and Professor, USDA-ARS and Department of Agronomy, Purdue University, West Lafayette, Indiana

R. F. Wilson — Supervisory Plant Physiologist and Professor of Crop Science, USDA-ARS-SAA and Crop Science Department, North Carolina State University, Raleigh, North Carolina

W. J. Wolf — Research Chemist, Northern Regional Research Center, Peoria, Illinois

Conversion Factors for SI and non-SI Units

To convert Column 1 into Column 2, multiply by	Column 1 SI Unit	Column 2 non-SI Unit	To convert Column 2 into Column 1 multiply by
	Length		
0.621	kilometer, km (10^3 m)	mile, mi	1.609
1.094	meter, m	yard, yd	0.914
3.28	meter, m	foot, ft	0.304
1.0	micrometer, μm (10^{-6} m)	micron, μ	1.0
3.94×10^{-2}	millimeter, mm (10^{-3} m)	inch, in	25.4
10	nanometer, nm (10^{-9} m)	Angstrom, Å	0.1
	Area		
2.47	hectare, ha	acre	0.405
247	square kilometer, km^2 (10^3 m)^2	acre	4.05×10^{-3}
0.386	square kilometer, km^2 (10^3 m)^2	square mile, mi^2	2.590
2.47×10^{-4}	square meter, m^2	acre	4.05×10^3
10.76	square meter, m^2	square foot, ft^2	9.29×10^{-2}
1.55×10^{-3}	square millimeter, mm^2 (10^{-6} m)^2	square inch, in^2	645
	Volume		
6.10×10^4	cubic meter, m^3	cubic inch, in^3	1.64×10^{-5}
2.84×10^{-2}	liter, L (10^{-3} m^3)	bushel, bu	35.24
1.057	liter, L (10^{-3} m^3)	quart (liquid), qt	0.946
3.53×10^{-2}	liter, L (10^{-3} m^3)	cubic foot, ft^3	28.3
0.265	liter, L (10^{-3} m^3)	gallon	3.78
33.78	liter, L (10^{-3} m^3)	ounce (fluid), oz	2.96×10^{-2}
2.11	liter, L (10^{-3} m^3)	pint (fluid), pt	0.473
9.73×10^{-3}	$meter^3$, m^3	acre-inch	102.8
35.3	$meter^3$, m^3	cubic foot, ft^3	2.83×10^{-2}

continued on next page

Conversion Factors for SI and non-SI Units

To convert Column 1 into Column 2, multiply by	Column 1 SI Unit	Column 2 non-SI Unit	To convert Column 2 into Column 1 multiply by
	Mass		
2.20×10^{-3}	gram, g (10^{-3} kg)	pound, lb	454
3.52×10^{-2}	gram, g	ounce (avdp), oz	28.4
2.205	kilogram, kg	pound, lb	0.454
10^{-2}	kilogram, kg	quintal (metric), q	10^{2}
1.10×10^{-3}	kilogram, kg	ton (2000 lb), ton	907
1.102	megagram, Mg (tonne)	ton (U.S.), ton	0.907
	Yield and Rate		
0.893	kilogram per hectare, kg ha^{-1}	pound per acre, lb $acre^{-1}$	1.12
7.77×10^{-2}	kilogram per cubic meter, kg m^{-3}	pound per bushel, lb bu^{-1}	12.87
1.49×10^{-2}	kilogram per hectare, kg ha^{-1}	bushel per acre, 60 lb	67.19
1.59×10^{-2}	kilogram per hectare, kg ha^{-1}	bushel per acre, 56 lb	62.71
1.86×10^{-2}	kilogram per hectare, kg ha^{-1}	bushel per acre, 48 lb	53.75
0.107	liter per hectare, L ha^{-1}	gallon per acre	9.35
893	megagram per hectare, Mg ha^{-1}	pound per acre, lb $acre^{-1}$	1.12×10^{-3}
0.446	megagram per hectare, Mg ha^{-1}	ton (2000 lb) per acre, ton $acre^{-1}$	2.24
2.24	meter per second, m s^{-1}	mile per hour	0.447
	Specific Surface		
10	square meter per kilogram, m^2 kg^{-1}	square centimeter per gram, cm^2 g^{-1}	0.1
10^{3}	square meter per kilogram, m^2 kg^{-1}	square millimeter per gram, mm^2 g^{-1}	10^{-3}

Pressure			
9.90	megapascal, MPa (10^6 Pa)	atmosphere	0.101
10	megapascal, MPa (10^6 Pa)	bar	0.1
1.00	megagram per cubic meter, Mg m^{-3}	gram per cubic centimeter, g cm^{-3}	1.00
2.09×10^{-2}	pascal, Pa	pound per square foot, lb ft^{-2}	47.9
1.45×10^{-4}	pascal, Pa	pound per square inch, lb in^{-2}	6.90×10^{3}
Temperature			
1.00 (K − 273)	Kelvin, K	Celsius, °C	1.00 (°C + 273)
(9/5 °C) + 32	Celsius, °C	Fahrenheit, °F	5/9 (°F −32)
Energy, Work, Quantity of Heat			
9.52×10^{-4}	joule, J	British thermal unit, Btu	1.05×10^{3}
0.239	joule, J	calorie, cal	4.19
10^{7}	joule, J	erg	10^{-7}
0.735	joule, J	foot-pound	1.36
2.387×10^{-5}	joule per square meter, J m^{-2}	calorie per square centimeter (langley)	4.19×10^{4}
10^{5}	newton, N	dyne	10^{-5}
1.43×10^{-3}	watt per square meter, W m^{-2}	calorie per square centimeter minute (irradiance), cal cm^{-2} min^{-1}	698
Transpiration and Photosynthesis			
3.60×10^{-2}	milligram per square meter second, mg m^{-2} s^{-1}	gram per square decimeter hour, g dm^{-2} h^{-1}	27.8
5.56×10^{-3}	milligram (H_2O) per square meter second, mg m^{-2} s^{-1}	micromole (H_2O) per square centimeter second, μmol cm^{-2} s^{-1}	180
10^{-4}	milligram per square meter second, mg m^{-2} s^{-1}	milligram per square centimeter second, mg cm^{-2} s^{-1}	10^{4}
35.97	milligram per square meter second, mg m^{-2} s^{-1}	milligram per square decimeter hour, mg dm^{-2} h^{-1}	2.78×10^{-2}
Angle			
57.3	radian, rad	degrees (angle), °	1.75×10^{-2}

continued on next page

Conversion Factors for SI and non-SI Units

To convert Column 1 into Column 2, multiply by	Column 1 SI Unit	Column 2 non-SI Unit	To convert Column 2 into Column 1 multiply by
	Electrical Conductivity		
10	siemen per meter, S m^{-1}	millimho per centimeter, mmho cm^{-1}	0.1
	Water Measurement		
9.73×10^{-3}	cubic meter, m^3	acre-inches, acre-in	102.8
9.81×10^{-3}	cubic meter per hour, m^3 h^{-1}	cubic feet per second, ft^3 s^{-1}	101.9
4.40	cubic meter per hour, m^3 h^{-1}	U.S. gallons per minute, gal min^{-1}	0.227
8.11	hectare-meters, ha-m	acre-feet, acre-ft	0.123
97.28	hectare-meters, ha-m	acre-inches, acre-in	1.03×10^{-2}
8.1×10^{-2}	hectare-centimeters, ha-cm	acre-feet, acre-ft	12.33
	Concentrations		
1	centimole per kilogram, cmol kg^{-1} (ion exchange capacity)	milliequivalents per 100 grams, meq 100 g^{-1}	1
0.1	gram per kilogram, g kg^{-1}	percent, %	10
1	megagram per cubic meter, Mg m^{-3}	gram per cubic centimeter, g cm^{-3}	1
1	milligram per kilogram, mg kg^{-1}	parts per million, ppm	1
	Plant Nutrient Conversion		
	Elemental	***Oxide***	
2.29	P	P_2O_5	0.437
1.20	K	K_2O	0.830
1.39	Ca	CaO	0.715
1.66	Mg	MgO	0.602

23 April 1986

1 World Distribution and Significance of Soybean

Keith J. Smith
American Soybean Association
St. Louis, Missouri

Wipada Huyser
Development Planning and Research Associates, Inc.
Manhattan, Kansas

The fascinating early history of the soybean [*Glycine max* (L.) Merr.] has been thoroughly reviewed in several texts. Probst and Judd (1973) presented an extensive review of the origin and early history of this crop with highlighted references to soybean in books written over about 4500 yrs. The early Chinese history is particularly interesting.

The history of soybean in the USA has been recently expanded by Hymowitz and Harlan (1983). They have evidence that Henry Yonge first planted soybean on his farm in Thunderbolt, GA in 1765. Samuel Bowen, a former seaman employed by the East India Company, brought soybean to the USA from China via London.

From 1804 to 1890 in the USA, occasional references to soybean were made in experiment station publications and scientific literature. Virtually without exception these references were positive. The soybean was highly praised for its high productivity and yield, ability to grow in many different climates and soils, and quality as a silage and forage crop. After 1890, soybean research intensified. The U.S. Department of Agriculture (USDA) published three early bulletins devoted solely to soybean (Ball, 1907; Morse, 1918; Piper and Nielsen, 1909).

In the USA, soybean was primarily a forage crop and grown for hay and silage with cowpea (*Vigna sinensis* L.), millet (*Panicum* spp.), or sorghum [*Sorghum bicolor* (L.) Moench]. Soybean was frequently grown with corn (*Zea mays* L.) to increase soil N and to improve the quality of silage.

Successful use of soybean as an oilseed in Europe from about 1900 to 1910 promoted interest in its use in the USA. Limited U.S. production necessitated imports of oil and meal from the Orient for processing in

Soybeans: Improvement, Production, and Uses, 2nd ed.–Agronomy Monograph no. 16.

Table 1–1. World production of major oilseeds (Tg). Source: USDA Foreign Agricultural Service (1983).

Year	Soybean	Cottonseed	Peanut	Sunflowerseed	Rapeseed	Flaxseed	Copra	Kernel	Total
1964–1965	28.3	22.1	16.5	8.6	4.4	3.4	3.4	0.9	87.7
1965–1966	32.4	22.5	15.7	8.1	4.8	3.4	3.6	0.9	91.4
1966–1967	35.0	20.7	16.4	9.4	4.7	3.0	3.3	0.8	93.3
1967–1968	36.4	20.7	16.9	9.7	5.8	2.5	3.4	0.8	96.4
1968–1969	40.2	22.6	15.8	9.8	5.5	2.9	3.5	0.8	101.2
1969–1970	40.9	21.8	16.8	10.2	5.2	3.5	3.5	0.9	103.1
1970–1971	42.4	21.9	17.6	9.6	7.4	4.0	4.0	0.9	108.1
1971–1972	46.4	24.5	18.3	9.8	7.9	2.8	4.6	0.9	115.5
1972–1973	51.4	25.4	15.4	9.6	7.7	2.3	3.9	0.9	116.8
1973–1974	63.9	26.1	16.5	12.1	7.5	2.4	3.6	1.0	133.3
1974–1975	56.6	26.6	17.0	10.6	8.0	2.4	4.6	1.1	127.0
1975–1976	67.9	22.5	18.9	9.9	8.8	2.4	5.3	1.1	136.9
1976–1977	61.3	23.2	17.2	10.1	7.5	2.2	4.7	1.2	127.4
1977–1978	74.6	25.2	16.9	12.9	8.1	2.9	4.9	1.2	146.9
1978–1979	77.4	23.7	17.6	12.8	10.7	2.5	4.4	1.3	150.4
1979–1980	93.7	25.1	17.2	15.3	10.1	2.7	4.6	1.5	170.1
1980–1981	80.8	25.6	16.1	13.1	11.1	2.1	5.0	1.5	155.3
1981–1982	86.3	28.2	19.9	14.7	12.4	2.1	4.7	1.9	170.2
1982–1983	93.9	27.3	17.5	16.5	15.1	2.6	4.5	1.8	179.3
1983–1984	80.3	27.5	18.8	15.7	14.6	2.2	4.3	2.0	165.4

the USA. In 1911, the first oil and meal was processed in the USA from Manchurian soybean by a crushing plant in Seattle, WA. Not until 1915 was oil processed from domestically grown soybean by a cottonseed (*Gossypium hirsutum* L.) oil mill in Elizabeth City, NC. Nearly 30 000 t were processed the 1st yr, owing to the combination of surplus soybean seed and high cottonseed prices.

Processors reverted to using Manchurian soybean during the 1916 to 1917 processing season because of limited availability of domestically grown seed. Southern cottonseed oil mills saw a potential for soybean as an oilseed and contracted with producers for their 1917 crop. These contracts led to increased plantings. Another factor promoting increased soybean acreage was damage to the cotton crop associated with the spread of the boll weevil (*Anthonomus grandis* Boh.) through the South. Soybean was an alternative crop to cotton.

Not until 1920 was domestic soybean processed in the Midwest, by a linseed mill in Chicago Heights, IL. In 1928, several farm groups and an Illinois processor agreed to take production of 20 000 ha to guarantee the availability of enough soybean to operate a processing plant. This agreement helped convince Midwest farmers that the soybean industry wanted a dependable supply of soybean, thus facilitating expansion of soybean production in the Midwest.

Even though interest in soybean production was on the rise during the 1920s and 1930s, most planted soybean acres were used for forage. Large quantities of meal were imported in the 1920s for use as fertilizer. During the mid-1920s soybean meal became an accepted ingredient of livestock and poultry feed rations. Significant quantities of soybean, oil, and meal continued to be imported until the mid-1920s when domestic meal production finally exceeded meal imports. Soybean acreage and yields have consistently increased, making soybean a major U.S. crop.

1–1 WORLD SOYBEAN PRODUCTION

There are eight major oilseeds traded in international markets: soybean, cottonseed, peanut (*Arachis hypogaea* L.), sunflower (*Helianthus annuus* L.), rapeseed (*Brassica napus* L.), flaxseed (*Linum usitatissimum* L.), copra (*Cocus nucifera* L.), and palm kernel (*Elaeis guineensis* L.). They account for over 97% of all world oilseed production. Table 1–1 shows the world production of these oilseeds since 1965.

Soybean production has dominated world oilseed production. It is followed by cottonseed, peanut, and sunflower. Since 1970, soybean production has been at least double that of any other oilseed. The share of soybean in world oilseed production has increased from 32% in 1965 to over 50% in the 1980s. Peanuts' share of production, by contrast, has experienced a reduction of world oilseed share from 18% to only 11% during the same period. Soybean will probably maintain its dominant role for the foreseeable future.

Relatively few countries produce soybean, thus giving soybean-producing countries significant economic power in world oilseed trade. Table 1–2 presents world soybean production by country. Until 10 yrs ago, the USA and the People's Republic of China had long been the only major producers and world suppliers of soybean.

Remarkable increases of soy production in Brazil since 1970 have changed the soybean market structure tremendously. Brazil is now a major producer, having surpassed the People's Republic of China in 1974. At present, Brazil is the second largest soybean producer and world supplier.

Argentina recently became the newest major soybean producer. Soybean production in Argentina has continued to increase rapidly since 1975. Their pattern of production expansion resembles that of Brazil.

The four major producers, USA, Brazil, China, and Argentina together account for 90 to 95% of the world production. Figure 1–1 shows soybean production of these countries since 1965. Prior to the 1970s, the USA and China were the only major soybean producers. The USA's production share climbed yearly at the expense of China's share, because the USA increased soybean production at a much faster rate. The U.S. share of world soybean production increased from 60% in 1960 to its peak share of 75% in 1969. China's production share had decreased from 32 to 16% over the same period.

Recently, from 1980 to 1983, the USA has produced 63% of world soybean. Brazil and China, the second and third largest producers, have shares of 16 and 9%, respectively, of world production. Argentina continues the persistent strong increase of soybean production annually with a 6% share.

1–1.1 United States Soybean Production

World demand for cooking and salad oil and red meat increased substantially during and immediately after World War II. These demands stimulated the rapid expansion of soybean production, and made soybean second only to corn in generating farm income.

The expansion of soybean acreage has come primarily from the diversion of land from other cultivated crops and the development of new cropland. Soybean acreage has increased from less than 5.6 million ha (Mha) in the early 1950s to over 17 Mha in 1971 to 1972, and a record 28.5 Mha in 1979 to 1980. The USA is now producing, consuming, and exporting more soybean than any other country.

There are five major soybean production regions according to the USDA: Western Corn Belt, Eastern Corn Belt, Southeast, Delta, and Atlantic states (Hazera and Fryar, 1981). Soybean production areas are illustrated in Fig. 1–2. One can see production is concentrated in the Midwest and Mississippi Valley.

Corn Belt states, with their rich agricultural land, have the highest soybean yields and also produce the most soybean in the nation. Soybean

Table 1-2. World soybean production. Source: USDA-Foreign Agricultural Service, 1983.

Continent and country	Area 1980-1981	Area 1981-1982	Area 1982-1983	Production 1980-1981	Production 1981-1982	Production 1982-1983
	1 000 ha			1 000 t		
North America						
Canada	283	279	364	713	607	857
Mexico	150	370	390	280	680	550
USA	27 461	26 858	28 645	48 772	54 435	61 969
South America						
Argentina	1 740	1 985	2 000	3 500	4 000	3 400
Bolivia	27	45	45	40	78	79
Brazil	8 534	8 293	8 300	15 200	12 800	14 600
Chile	1	1	1	1	1	1
Columbia	44	50	56	89	100	110
Ecuador	21	21	21	31	33	20
Paraguay	400	420	370	600	600	550
Peru	10	12	14	8	10	11
Uruguay	35	17	20	45	20	25
Venezuela	0	0	0	0	0	0
Africa						
Egypt	35	46	62	32	130	175
Morocco	1	1	0	1	1	0
Nigeria	165	165	165	56	60	65
South Africa	22	22	25	26	21	20
Cimbabwe	38	60	70	65	86	100
Europe						
Bulgaria	94	94	90	107	113	120
Czechoslovakia	5	5	5	5	5	5
France	8	10	9	14	18	20
Hungary	20	25	25	39	45	50
Romania	364	310	315	448	268	325
Spain	7	4	5	14	7	9
Yugoslavia	17	47	77	34	93	180
Soviet Union	854	864	800	525	450	480
Asia						
China						
Mainland	7 226	8 024	8 300	7 940	9 330	8 700
Taiwan	10	10	10	16	15	15
India	500	500	550	450	500	650
Indonesia	811	725	780	687	609	620
Iran	34	50	50	50	80	80
Japan	142	149	147	174	212	226
North Korea	300	300	300	330	330	330
Republic of Korea	188	202	183	216	257	233
Philippines	10	11	12	10	10	12
Thailand	128	130	112	131	117	110
Turkey	3	17	25	3	15	50
Oceania						
Australia	46	46	45	73	79	69
World total	49 734	50 168	52 417	80 784	86 215	94 843

acreage has increased at a remarkable rate in Corn Belt states. The acreage has doubled nearly every 20 yrs since the early 1920s. Over half of the soybean acreage increase in the Corn Belt came from acreage reductions

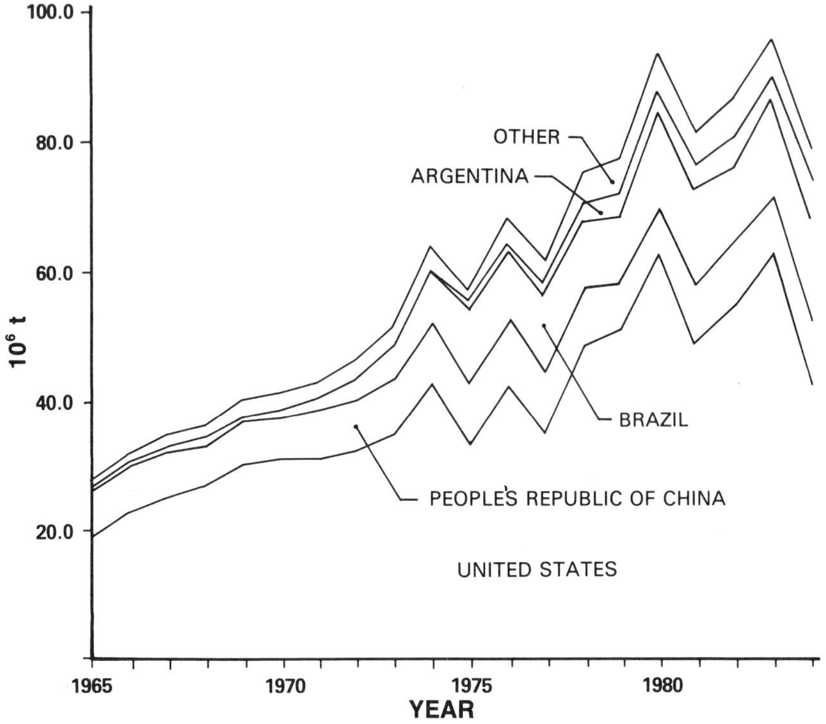

Fig. 1–1. World soybean production, 1965–1984.

in oat (*Avena sativa* L.), hay, and barley (*Hordeum vulgare* L.). While there have been acreage controls on wheat and corn production, there has been no such restriction on soybean acreage, a situation that has allowed soybean expansion to new cultivated land as well as to land diverted from other crops.

Western Corn Belt states have produced about 35% of the U.S. soybean during each of the past three decades. Soybean account for 25% of the total cropland in the region, while corn makes up approximately 40%. Past acreage increases have been made at the expense of oat. This displacement appears completed, so any future acreage increases must come from new cropland or from corn acreages.

Delta States, which include Arkansas, Louisiana, and Mississippi, have expanded soybean harvested acreage for beans at an impressive rate. Soybean acreage has more than doubled every 10 yrs, with increases occurring at the expense of acreage reductions for cotton, corn, small grains, and roughage. Double cropping soybean and wheat (*Triticum aestivum* L.) also contributed to acreage increases. Further expansion of soybean acreage in this region depends upon the government programs for cotton, which competes with soybean for land.

Atlantic states—Virginia, Maryland, Delaware, and North and South Carolina—also expanded their soybean acreage steadily, though at a slower

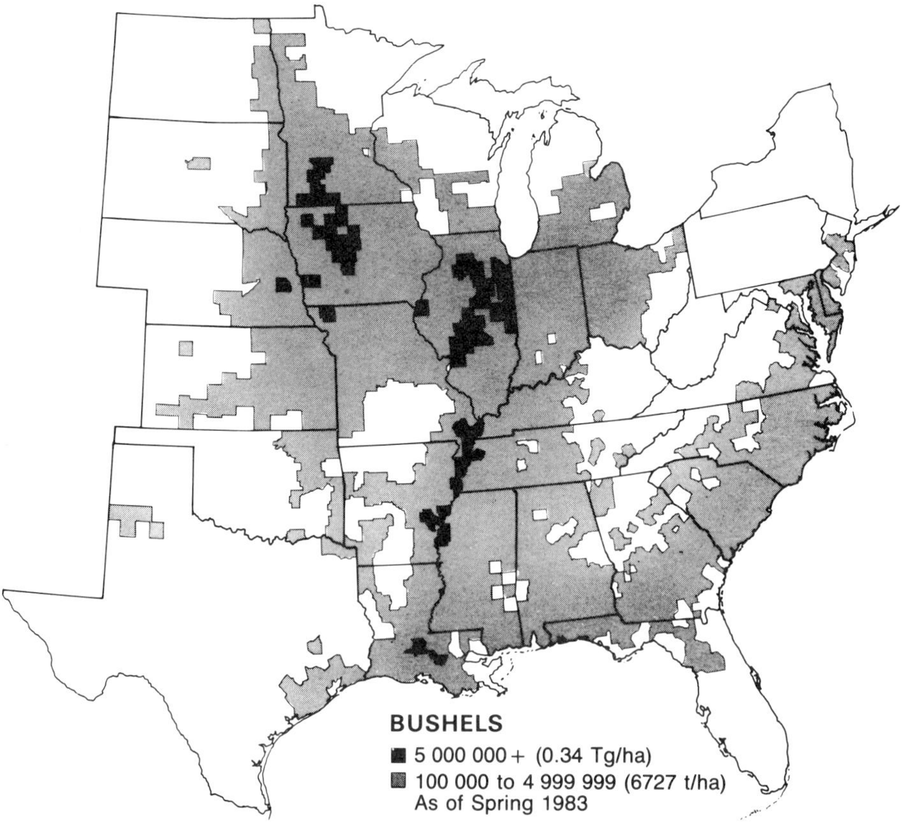

Fig. 1–2. Soybean production in the USA. Courtesy American Soybean Association.

rate than the other regions. New acreage has come from reduction of small grains and roughage production as well as new cultivated land. Access to port facilities on the Atlantic gives incentive for soybean production to satisfy the increasing export demand.

Soybean production in the Southeastern region has expanded rapidly. Acreage increases came from new lands being put into production, from lands previously planted to cotton, corn, and small grains, and from expansion of double-cropped soybean and wheat.

Soybean yields increased rapidly during the period of 1920s to 1940s, and at a slower rate since the 1960s. Yields continue to climb upward as a result of new and better varieties in all regions, improved weed, insect, and disease control, and improved management practices by growers. The trendline for yields may reach about a 20 kg ha^{-1} annual increase based on historical yields. There has been no major genetic breakthrough like hybrid corn for soybean, and no major technological change is expected in the foreseeable future.

The Corn Belt region has the highest soybean yields in the country, between 2.02 to 2.35 t ha^{-1}. States with the highest annual yields are Illinois, Iowa, and Indiana.

Soybean production has continued to increase year after year at an impressive rate. Table 1–3 presents U.S. soybean production, acreage, and utilization data.

Table 1–3. Soybean production and disappearance (local market years). Source: USDA, Foreign Agricultural Service (1983).

Crop year	Harvested area	Yield	Production	Market year export	Crush†
	Mha	kg ha^{-1}	Tg		
		USA‡			
1963–1964	12.46	1531	19.08	5.77	13.04
1965–1966	13.94	1651	23.01	6.82	14.63
1966–1967	14.79	1709	25.27	7.12	15.23
1967–1968	16.11	1650	26.58	7.26	15.69
1968–1969	16.75	1800	30.13	7.81	16.49
1969–1970	16.73	1843	30.84	11.77	20.07
1970–1971	17.10	1794	30.68	11.81	20.69
1971–1972	17.28	1852	32.01	11.34	19.61
1972–1973	18.49	1870	34.58	13.05	19.65
1973–1974	22.53	1870	42.12	14.67	22.35
1974–1975	20.78	1593	33.10	11.45	19.09
1975–1976	21.70	1942	42.14	15.11	23.55
1976–1977	19.99	1754	35.07	15.35	21.51
1977–1978	23.40	2055	48.10	19.06	25.22
1978–1979	25.76	1974	50.86	20.12	27.70
1979–1980	28.48	2189	61.54	23.82	30.57
1980–1981	27.45	1807	48.94	19.71	27.17
1981–1983	26.79	2053	54.14	25.28	28.03
1982–1983	28.11	2148	59.60	24.63	30.16
1983–1984	25.03	1725	42.65	20.69	26.40
		Brazil§			
1965	0.43	1211	0.52	0.08	0.28
1966	0.49	1212	0.59	0.12	0.40
1967	0.61	1170	0.72	0.31	0.42
1968	0.72	906	0.65	0.07	0.47
1969	0.91	1167	1.06	0.31	0.61
1970	1.32	1144	1.51	0.29	0.93
1971	1.72	1210	2.08	0.23	1.70
1972	2.84	1291	3.67	1.01	2.13
1973	3.62	1386	5.01	1.79	2.71
1974	5.14	1531	7.88	2.86	4.30
1975	5.82	1698	9.89	3.52	5.52
1976	6.42	1750	11.23	3.33	6.37
1977	7.07	1770	12.51	2.58	8.67
1978	7.78	1226	9.53	0.66	8.88
1979	8.26	1240	10.24	0.64	9.09
1980	8.77	1727	15.16	1.53	13.01
1981	8.49	1791	15.20	1.50	13.80
1982	8.20	1565	12.84	0.81	12.73
1983	8.23	1793	14.75	1.32	12.87

(continued on next page)

Table 1–3. Continued.

Crop year	Harvested area	Yield	Production	Market year export	Crush†
	Mha	kg ha^{-1}	Tg		
1984	9.25	1643	15.20	1.50	12.90
		Argentina#			
1965	0.02	1063	0.02	–	0.02
1966	0.02	1125	0.02	–	0.02
1967	0.02	1176	0.02	–	0.02
1968	0.03	1100	0.02	–	0.01
1969	0.03	1143	0.03	–	0.02
1970	0.03	1038	0.03	–	0.02
1971	0.04	1639	0.06	–	0.04
1972	0.07	1147	0.08	–	0.05
1973	0.16	1732	0.27	0.05	0.20
1974	0.34	1442	0.50	0.08	0.28
1975	0.36	1362	0.48	–	0.53
1976	0.43	1601	0.70	0.11	0.50
1977	0.66	2121	1.40	0.62	0.59
1978	1.25	2160	2.70	1.98	0.68
1979	1.60	2313	3.70	2.83	0.64
1980	2.03	1773	3.60	2.79	0.72
1981	1.74	2011	3.50	2.19	1.00
1982	1.98	2091	4.15	2.15	1.91
1983	2.12	1688	3.57	1.35	2.05
1984	2.58	2252	5.80	2.45	3.10

† Includes meal exported. ‡ Crop year (September-August). § Crop year (February-January). # Crop year (April-March).

1–1.2 Soybean Production in Brazil

Japanese immigrant farmers introduced soybean to Brazil over 65 yrs ago, but soybean production did not become an important crop in Brazil until the late 1960s. Since the mid-1960s, Brazil has remarkably increased soybean production and become an important world soybean and soybean meal producer and supplier. Four main reasons explain the boom in soybean production and trade in Brazil (Broadbent and Dixon, 1976).

First, high soybean prices during the late 1960s to the late 1970s made soybean a more profitable crop than alternatives such as corn, cotton, rice (*Oryza sativa* L.), pasture, and beef production on the basis of profit per hectare. Therefore, land previously used for these operations was shifted to soybean production.

Second, the U.S. embargo on soybean exports in 1973 forced soybean importers to seek alternative sources for soybean, an action that stimulated Brazilian production.

Third, Brazilian soybean production has been encouraged and assisted through many government policies and a favorable market environment. The Brazilian government offered liberal credit at interest rates lower than current inflation rates for acquisition of machinery and production inputs such as fertilizer. These loans make it easier for farmers to start or expand soybean production to take advantage of high soybean prices and cheaper production costs.

Fourth, multinational firms that saw economic opportunity assisted soybean farmers by transferring personnel, proven industry technology, capital, and management to develop a new agricultural industry in Brazil.

These motivating forces successfully pushed up Brazilian soybean production from an insignificant 1.8% share of world soybean production in 1965 to more than 15% in 1985. Table 1–3 shows Brazilian soybean production. A dramatic increase in soybean production during the 15 yrs from 1965 to 1980 resulted from vast soybean acreage expansion rather than yield increases. Brazil planted only 432 000 ha of soybeans in 1965 compared to 8.774 Mha in 1980, thus producing 0.52 Tg compared to 15.2 Tg, respectively. The annual rate of production growth has been impressive, especially during the late 1960s to the early 1970s. On the average, soybean production increased more than 25% per year during this past 20 yrs.

Soybean yields in Brazil have not changed much over the past 20 yrs (Huyser, 1983). Some increase in yields in the early 1970s resulted from improved cultural practices, a shift to more fertile virgin soils, and introduction of soybean varieties adaptable to local conditions. Brazilian yields on the average are lower than U.S. yields. In 1975, Brazilian yield was 1.48 t ha^{-1} (22 bushels $acre^{-1}$) compared to 1.88 t ha^{-1} (28 bushels $acre^{-1}$) for the USA.

Soybean yields in Brazil also vary considerably from area to area. Average yields reported during the 1975 to 1977 period were as follows; Parana 2.10 t ha^{-1}; Sao Paulo, 1.78 t ha^{-1}; Rio Grande do Sul, 1.57 t ha^{-1}; Mato Grasso, 1.50 t ha^{-1}; Santa Catarina, 1.28 t ha^{-1}; and others 1.23 t ha^{-1} (Williams, 1981). The major factors affecting these yield differences are the soil quality and climate conditions.

Weather has, in the past, played an important role in production. For example, the reduction of soybean production in 1978 and 1979 was due to severe drought in most southern states which reduced yield considerably and, in turn, drastically reduced soybean production.

Most of the Brazilian soybean supply is domestically crushed for meal and oil. Table 1–3 provides data on Brazilian soybean production and disappearance. Soybean crush demand has increased continuously from only 0.282 Tg in 1965 to 5.516 Tg in 1975 and 13.300 Tg in 1981. The crush demand is a dominant soybean use and export is a residual demand. During a year of low soybean supply, soybean exports decrease while soybean crush maintains its increasing trend level. The Brazilian government has policies to encourage the domestic crushing industry and exports of soybean meal and soybean oil, rather than exports of soybean.

1–1.3 Soybean Production in Argentina

Soybean, a relatively new crop in Argentina, has been produced in significant quantity only since 1961 to 1962 as a result of government encouragement (Huyser, 1983). Since 1962, soybean production has steadily expanded every year. (Table 1–3) Production and acreage began to increase rapidly during the 1970s, as illustrated in a comparison of acreage: 0.03 Mha in 1970, 0.16 Mha in 1973, 0.43 Mha in 1976, and 2.12 Mha in 1983. Production also increased dramatically over this period from only 27 000 t in 1970 to 5.80 Tg in 1984.

Soybean yield, which had been about 1.1 t ha^{-1} during the 1960s, improved during the early 1970s. By 1979 Argentina's average yields reached a record of 2.313 t ha^{-1}.

Soybean crops are mainly produced in the provinces of Misiones, the southern part of Santa Fe, the northern part of Buenos Aires, Tucuman, and the eastern part of Cordoba. These areas contain about 96% of the total soybean land. During 1964 to 1965 to 1967 to 1968, Misiones soybean land averaged 48% of the total soybean production area. This area of soybean production is located in the corn-wheat-soybean belt of Argentina. Most soybean land was switched from grains and pasture.

Soybean crops are frequently double-cropped with wheat, especially in the major provinces. Soybean also competes with corn production depending upon weather conditions and soil moisture levels. Future expansion of soybean acreage in Argentina is more limited than in Brazil because of land constraints. Expanding soybean acreage means less land is available for other crops. The government does not exercise any supporting policy for wheat-soybean production as in Brazil. Thus, the future of soybean expansion depends upon the soybean-corn price relationship.

Traditionally, all soybean supplies were for domestic consumption to satisfy crush demand. Less than 50 000 t were crushed annually prior to 1973. Since 1973, soybean crush demand has increased from around 200 000 t in 1973 to 3.1 t in 1984. In 1985, there is sufficient crushing capacity in Argentina for domestic supply of soybean meal and oil. Most existing firms crush several types of seed, mainly sunflower and other oilseeds. The soybean crushing industry in Argentina has the disadvantage of higher operating costs than in other countries, especially in the major producing provinces of Santa Fe and Buenos Aires.

At times, these costs are partly offset by financial incentives such as tax rebates on exports of oilseed products. Since 1980, Argentina has expanded its soybean crushing rapidly to domestically capture the industry value added. The 1984 soybean crushing level is more than four times its 1980 level. The expansion has been encouraged by recent unusually high vegetable oil prices. Current indications are that Argentina will continue to expand its domestic crushing industry rather than export whole beans as in the 1970s.

1–2 WORLD TRADE IN SOYBEAN

Twenty-five percent of total world soybean production enters the international trade market in the form of whole beans. The major soybean exporters are the USA, Brazil, and Argentina. Numerous countries import soybeans both for human food products and for processing into soybean meal and oil. The major traditional soybean importers include European Economic Community (EEC): West Germany, United Kingdom, France, Netherlands, Belgium, Luxemburg, Italy, Denmark, Ireland, and Greece; Japan; Eastern European countries (Poland, Bulgaria, Czechoslovakia, East Germany, Hungary, Romania, and Yugoslavia); and Spain. Tables 1–4 and 1–5 present the amount of soybean trade by country.

After World War II, the USA and the Mainland China were the major soybean exporters with the USA having a much larger share of the export market—around 70% of the trade. The Chinese population explosion and a leveling of soybean production forced China to turn nearly all domestic production to fill demand at home. China reduced exports and by 1965 the export share was only 10% of the world trade. In 1974 for the first time, China became a soybean importer and has continued to import soybeans to meet domestic needs.

The USA has been the largest soybean exporter since the end of Word War II. United States soybean exports have increased rapidly from 1.89 Tg in 1955 to 5.8 t in 1965, 11.8 Tg in 1970, and 23.8 Tg in 1980. The USA dominated the soybean export market with a share of 90% or more until 1972 when the share started to drop because of the increase in Brazilian soybean exports. Table 1–4 shows the soybean export share of major soybean exporters.

Major importers of U.S. soybean are the EEC, Japan, and Spain. The EEC is the USA's biggest customer. The EEC soybean imports ac-

Table 1–4. Soybean exports by major trading countries (Tg). Source: USDA-Foreign Agricultural Service (1983).

Year†	USA	Brazil	People's Republic of China‡	Argentina
1964–1965	5.77	0.08	0.58	0.00
1965–1966	6.82	0.12	0.55	0.00
1966–1967	7.12	0.30	0.56	0.00
1967–1968	7.26	0.07	0.57	0.00
1968–1969	7.80	0.31	0.49	0.00
1969–1970	11.77	0.29	0.37	0.00
1970–1971	11.81	0.23	0.46	0.00
1971–1972	11.34	1.20	0.37	0.00
1972–1973	13.05	1.78	0.06	0.05
1973–1974	14.67	2.86	−0.28	0.08
1974–1975	11.45	3.52	0.29	0.00
1975–1976	15.11	3.33	0.15	0.11
1976–1977	15.35	2.58	−0.13	0.62
1977–1978	19.06	0.57	−0.10	1.98
1978–1979	20.12	0.38	0.00	2.83
1979–1980	23.82	1.15	−0.81	2.31
1980–1981	19.71	1.80	−0.54	2.70
1981–1982	25.28	0.86	0.00	1.88
1982–1983	24.63	1.32	0.00	1.42
1983–1984	20.69	1.50	0.00	2.80

† Crop year (September-August).
‡ Negative numbers indicate import of soybean.

Table 1–5. Soybean imports by major trading blocks (Tg). Source: USDA-Foreign Agricultural Service (1983).

Year†	European Economic Community	Japan	Spain	Eastern Europe	Other	Total
1964–1965	3.07	1.85	0.34	0.12	1.05	6.43
1965–1966	3.53	2.17	0.64	0.08	1.07	7.49
1966–1967	3.72	2.17	0.81	0.11	1.18	7.99
1967–1968	3.62	2.42	0.92	0.07	0.86	7.89
1968–1969	3.98	2.59	1.03	0.21	0.79	8.60
1969–1970	5.69	3.24	1.23	0.16	2.10	12.42
1970–1971	5.79	3.21	1.31	0.20	1.99	12.50
1971–1972	6.53	3.40	1.43	0.16	1.21	12.73
1972–1973	7.12	3.64	0.83	0.18	3.17	14.94
1973–1974	9.12	3.24	1.59	0.24	3.41	17.60
1974–1975	8.25	3.33	1.74	0.13	1.81	15.26
1975–1976	9.27	3.55	1.94	0.35	3.59	18.70
1976–1977	9.20	3.60	1.84	0.25	3.67	18.56
1977–1978	11.20	4.26	2.18	0.63	3.34	21.61
1978–1979	12.17	4.13	2.24	0.56	4.23	23.33
1979–1980	12.28	4.17	3.10	0.85	6.71	27.11
1980–1981	10.70	4.21	2.79	0.54	8.75	26.99
1981–1982	12.26	4.47	3.20	0.57	8.82	29.32
1982–1983	11.54	4.87	3.04	0.76	7.89	28.10
1983–1984	10.18	4.70	2.80	0.75	7.86	26.29

† Crop year (September-August).

count for about 40 to 45% of U.S. soybean exports. Japan and Spain import significant amounts of soybean from the USA, accounting for 20 and 8%, respectively. It is anticipated Spain and Eastern Europe will increase their percentage share of soybean imports from the USA.

The U.S. soybean export embargo during 1973 motivated importing countries to look for soybean from alternative sources. This action was a major factor in expanding the Brazilian export market.

Brazil's share of soybean in the world market fluctuates considerably since its actual exports fluctuate from year to year. Since 1965, its export share in the world soybean market varied from less than 1 to 23%. Soybean exports increased from around 300 000 t in 1969 to 1 Tg in 1972 and peaked in 1975 at 3.5 Tg. Due to a drought during 1978 to 1979, soybean exports were restricted to about 650 000 t. Major Brazilian soybean customers include the EEC, Japan, Spain, and more recently, the USSR, the People's Republic of China, and some Eastern European countries.

Brazilian soybean exports enjoy a seasonal advantage of high world prices since its harvest season, March to May, differs from the U.S. major soybean export season of September and October. The annual world price fluctuation has been influenced mainly by the USA because of its domination of world supplies. The normal world soybean price reaches its peak in August, just before the U.S. harvest, and decreases until February when the price starts increasing again. To take advantage of these seasonal prices, Brazil exports soybean slowly in April and accelerates exports from July to August.

The Brazilian government exercises a number of trade policies to control the quantity of soybean exports, in addition to some domestic policies discussed previously. The objectives of these policies are to encourage the domestic crushing industry, to control domestic soybean meal and soybean oil prices, and to take advantage of seasonal price fluctuations. The trade policies include export quota and licenses, export taxes and subsidies, export embargoes, and bilateral trade agreements.

Argentina is the newest major soybean exporter. Argentina did not export any soybean until 1973 when the first 50 000 t were exported. Soybean exports have increased dramatically since 1976 from only 111 000 t to nearly 3 Tg in 1981. The marked increase in soybean production in the late 1970s was primarily for export purposes. The major markets for Argentine soybean exports are the EEC, especially the Netherlands, the USSR, Brazil, and to some extent, Mexico. The USSR recently became a major soybean importer from Argentina. In 1981, the USSR made a trade agreement with Argentina in 1981 to purchase a minimum of 500 000 t of soybean annually until 1985—about 25% of total Argentine soybean exports. Most soybean is exported during May to July each year.

Argentina has exercised a variety of trade policies to manipulate the quantity of soybean and soybean meal exports. These policies include export quotas, issuance of export licenses and export taxes (Huyser, 1983).

Major soybean importers are EEC, Japan, Spain, and Eastern Europe. Table 1–5 presents soybean import shares of major importers. The EEC has always imported the most soybean, since these countries are heavily dependent upon oilseed imports for supplying needed protein for the feed industry. Despite the steady increase in EEC oilseed production, mainly rapeseed and olive (*Olea europaea* L.) the EEC demand for oilseed protein has increased much faster, thus their self-sufficiency in oilseeds has not changed. The EEC is only a little more than 8% self-sufficient in oilseeds. Currently, soybean imports account for nearly 90% of all EEC oilseed imports, which include peanuts, rapeseed, copra, and palm kernels. The USA had long been the only soybean exporter to this region until recently when Brazil and Argentina captured significant market shares in the EEC.

The European soybean-processing industry has increased continuously from 2.5 Tg in 1960, 5.3 Tg in 1970 to 10.2 Tg in 1978 (Williams, 1981). Currently, EEC produces more oil than domestic demand, so EEC exports some soybean oil competing with the USA, Brazil, and Argentina.

The rapid increase of high protein feeds and the low oilseed self-sufficiency of the EEC have led to a heavy reliance on the USA as the major supplier of soybean and soybean meal.

The U.S. soybean embargo in 1973 made the EEC aware of its dependence upon a single supplier. As a result, the EEC attempts to decrease its vulnerability by encouraging greater domestic protein meal production, and substituting other forms of domestically produced high protein feeds for imported protein.

Soybean have also been included under Common Agricultural Policy to encourage domestic oilseed output. The EEC has also tried to diversify import sources by using bilateral trading agreements. Agreements with India now provide for the supply of peanut meal, and agreements with Brazil and Argentina provide for soybean and soybean meal. Finally, some attempt has been made to improve the community stocking policies and to participate in negotiations in international forums which aim to improve the overall performance of world commodity markets.

The second largest soybean importer is Japan. Like the EEC, Japan relies heavily upon imported soybean to satisfy domestic consumption. Japan has continuously decreased agricultural land and concentrated more on industrialization. The area of agricultural land has declined from a peak of 6.1 Mha in 1961 to about 5.4 Mha in 1978. The acreage for oilseeds has decreased even more because more existing agricultural land is devoted to orchards, permanent plantations, and arable grasslands. This decrease was due to the change of consumption preference in Japan from cereal products to meats, fruits, vegetables, and vegetable oil as personal income increased. The area planted to soybean has declined from 385 000 ha in 1955 to 1956 to 79 000 ha in 1977 to 1978.

Japan has increasingly relied upon soybean imports. Soybean imports were around 1 Tg in the late 1950s and early 1960s, reached 2.2 Tg in 1966, 3.2 Tg in 1970. Imports have totaled more than 4 Tg annually since 1978. At present, soybean imports account for 90 to 95% of the total soybean supply available to Japanese oilseed crushers. Japan also imports some soybean from Brazil and Argentina in an attempt to diversify import sources to reduce dependence upon the U.S. soybean imports.

Traditionally Japanese soybean imports were primarily from the USA, supplemented by imports from the People's Republic of China. While U.S. soybean are mainly crushed for soybean meal, Japanese and Chinese soybean imports are processed for human consumption. Chinese soybean exports have declined due to tremendous increases in domestic demand, which make the People's Republic of China a net soybean importer.

Spain is the third largest soybean importer. Soybean importers account for over 99% of domestic soybean consumption. Since 1965, Spain has increased soybean imports at an impressive average rate of 15% per year. Soybean imports have increased from only 340 000 t in 1965 to 1.2 Tg in 1970 and over 3 Tg in 1980. Peak soybean imports were reported in 1982 at 3.2 Tg. Currently, imports have decreased to an estimated 2.5 Tg for 1983. The increased soybean meal demand by the developing poultry industry have created a growing Spanish market for soybean. The Spanish government has encouraged development of the poultry industry as a means to increase per capita meat consumption. The USA has been a major soybean supplier to Spain.

Another fast-growing soybean importer is Eastern Europe. These countries have imported some soybean for crushing in order to make optimum use of existing crushing capacity in the years when sunflower seed or rapeseed production (domestic oilseeds) is low. Table 1–5 shows soybean imports. Recently, soybean imports have increased substantially, ranging from 120 000 t in 1965 to over 800 000 t in 1981. Eastern Europe has attempted to increase its crushing industry. The soybean imports are still relatively small compared to soybean meal equivalent consumption. The USA has been the traditional soybean exporter to Eastern Europe. Brazil has recently expanded its soybean export market to Eastern Europe.

A number of high income developing countries, like the Republic of Korea and Taiwan, have increased their soybean imports rapidly in recent years. Taiwan imported 161 000 t of soybean in 1978. The peak Taiwanese import of 1.111 Tg was in 1979. The Republic of Korea started importing soybean in 1967 and its imports have gradually increased. In 1976 its imports were 129 000 t and this reached 500 000 t in 1981. These countries can be strong potential soybean importer markets as their economies improve.

1-3 IMPORTANCE OF SOYBEAN MEAL AND OIL

The demand for soybean is derived mainly from the oil and meal products, and to only a small extent from whole bean products. Houck et al (1972) have summarized soybean utilization (Fig. 1-3). Soybean meal and oil are the most produced, traded, and utilized meal and oil in the world. Table 1-6 compares utilization of soybean meal and oil to other major meals and oils. Soybean meal is 60 to 65% of the total. Soybean meal accounts for an even higher share in the high protein

Table 1-6. World meal and oil utilization (Tg).
Source: USDA-Foreign Agricultural Service (1983).

	1979-1980	1980-1981	1981-1982	1982-1983
	Protein meal consumption			
Soybean	58.24	56.92	59.58	61.01
Cottonseed	9.19	9.70	10.24	9.96
Rapeseed	5.32	6.02	7.22	8.54
Sunflower	5.77	5.48	5.78	6.46
Fish	4.61	4.35	4.84	4.76
Peanut	4.22	3.86	4.96	4.35
Copra	1.54	1.61	1.59	1.43
Palm kernel	0.69	0.75	0.89	0.98
Total	89.58	88.70	95.10	97.48
	Major vegetable and marine oil consumption			
Soybean	12.42	12.62	13.09	13.55
Palm	4.43	4.95	5.47	5.99
Sunflower	4.66	4.51	5.06	5.42
Rapeseed	3.28	3.36	4.55	5.30
Cottonseed	3.15	3.26	3.36	3.33
Peanut	3.10	2.92	3.40	3.34
Coconut	2.59	2.99	2.88	2.61
Olive	1.66	1.69	1.61	1.68
Fish	1.07	1.11	1.18	1.07
Palm kernel	0.06	0.06	0.77	0.80
Total	36.39	38.62	41.60	42.99

Table 1-7. Soybean meal: World production trade and consumption (Tg). Source: USDA-Foreign Agricultural Service (1983).

	1979–1980	1980–1981	1981–1982	1982–1983
	Production			
USA	24.59	22.06	22.36	24.00
Brazil	8.12	10.62	9.64	10.60
Argentina	0.56	0.72	1.03	1.69
European Economic Community	9.31	8.22	9.40	8.71
Portugal	0.18	0.22	0.41	0.52
Mexico	1.02	1.21	1.19	1.12
Japan	2.70	2.70	2.78	3.00
Spain	2.44	2.25	2.53	2.40
China				
Mainland	2.84	2.92	3.43	3.08
Taiwan	0.67	0.68	0.80	0.81
Soviet Union	0.94	0.99	1.30	1.16
Eastern Europe	1.13	0.87	0.79	1.08
Other	3.13	3.22	3.47	3.74
Total	57.64	56.68	59.45	61.12
	Gross exports			
USA	7.20	6.15	6.27	6.45
Brazil	5.44	7.74	8.35	8.24
Argentina	0.26	0.41	0.74	1.50
European Economic Community	3.57	3.91	4.45	5.32
Other	0.70	0.72	1.10	1.42
Total	17.16	18.93	20.90	22.92
	Gross imports			
European Economic Community 10	9.42	10.56	11.46	11.95
Portugal	0.27	0.29	0.03	0.00
Mexico	0.16	0.15	0.04	0.18
Japan	0.26	0.29	0.10	0.18
Spain	0.09	0.11	0.12	0.47
Soviet Union	0.44	1.00	1.38	2.63
Eastern Europe	3.92	4.16	3.19	2.75
Other	3.17	3.53	4.13	4.28
Total	17.73	20.08	20.45	22.43
	Consumption			
USA	17.43	15.96	16.08	17.52
Brazil	2.46	2.55	2.05	2.29
Argentina	0.32	0.29	0.21	0.19
European Economic Community	15.30	14.82	16.38	15.31
Eastern Europe	5.08	5.03	3.98	3.82
Portugal	0.45	0.50	0.45	0.39
Mexico	1.13	1.22	1.37	1.17
Japan	2.99	2.93	2.93	3.18
Spain	2.50	2.30	2.40	2.30
China				
Mainland	2.78	2.75	3.23	2.85
Taiwan	0.67	0.68	0.77	0.81
Soviet Union	1.38	1.99	2.68	3.79
Other	5.67	6.28	7.03	7.41
Total	58.15	57.30	59.57	61.02

†Preliminary data.

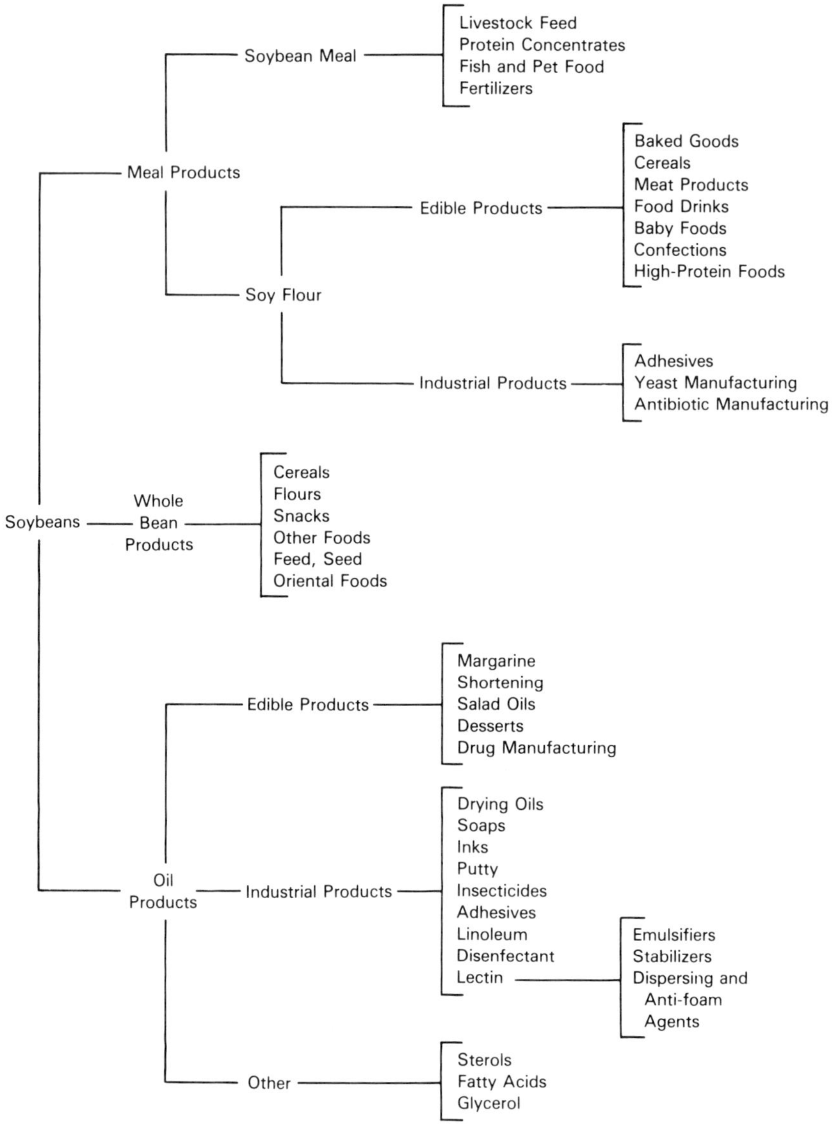

Fig. 1–3. Soybean utilization. Source: Hauck et al. (1972).

international trade, i.e., between 70 and 75%. Major soybean meal competitors include cottonseed meal, fish meal, sunflower meal, and rapeseed meal.

Soybean meal is used mainly in developed countries and some higher income developing countries. Table 1–7 presents soybean meal production, trade, and utilization by country. The USA, EEC, and Brazil are the major soybean meal producers, while the major soybean meal consumers are the USA, EEC, Eastern Europe, Japan, and the People's Republic of China. Brazil produces soybean meal mainly for exports. Currently, Brazil has about the same market share as the USA in the soybean meal export market. The EEC re-exports significant amounts of soybean meal, and some soybean meal export goes to Eastern Europe.

Soybean meal is a major protein component in livestock feed. Livestock and poultry feeds containing soybean meal as their major protein ingredient are used worldwide. The relative importance of soybean meal in the international market has increased over the years as seen by comparing percentage of soybean meal net trade in the international market to total soybean meal production.

Soybean oil accounts for 20 to 25% of total world fats and oils production and 30 to 35% of total edible vegetable oil production (Table 1–6). Tallow and grease, butter, sunflower, and palm oils are the next most produced fats and oils. The production of these competing oils is much smaller than soybean oil. However, the production and trade of palm oil has increased dramatically during recent years. Palm oil's competition in the markets is anticipated to become more intense in future years.

The USA, EEC, and Brazil produce the most soybean oil accounting for 43, 16, and 13% of total world soybean oil production. Table 1–8 presents soybean oil production, trade, and consumption by country. Major soybean oil exporters are the USA, Brazil, EEC, and Spain.

The EEC and Spain normally import soybean to crush for soybean meal for domestic consumption, thus producing more soybean oil than can be absorbed domestically. Major soybean oil importers include India, Middle Eastern and African countries, and Pakistan.

Historically, soybean meal accounts for approximately 60 to 70% of the value of the soybean with the balance coming from soybean oil. Soybean oil has many more substitute products such as vegetable oil, palm oil, and animal fats, whereas soybean meal dominates the feed protein market. These competitive factors imply that soybean is relatively more dependent upon meal markets than upon oil markets.

1–4 WORLD PRODUCTION TRENDS

World soybean production will increase at a slower rate over the next 20 yrs than it has in the past (Sharpe, 1983). Elanco Product Company and the American Soybean Association (1983) have conducted an extensive study of the prospect of soybean production and markets involving over 200 experts in various relevant fields. The study projects a 4% annual increase in soybean production and demand of soybean over the next 20 yrs. World soybean production and demand are estimated to be about 190.5 Mt (7×10^9 bushels) by 2002.

Table 1–8. Soybean meal: World production trade and consumption.
Source: USDA-Foreign Agricultural Service (1983).

	1979–1980	1980–1981	1981–1982	1982–1983†
	Tg			
	Production			
USA	5.49	5.11	4.98	5.46
Brazil	2.00	2.60	2.41	2.56
Argentina	0.12	0.16	0.22	0.35
European Economic Community	2.04	1.81	2.02	1.90
Mexico	0.24	0.27	0.26	0.26
Japan	0.62	0.63	0.63	0.68
Spain	0.53	0.48	0.54	0.52
China				
Mainland	0.40	0.41	0.48	0.43
Taiwan	0.15	0.15	0.17	0.17
Soviet Union	0.22	0.23	0.28	0.25
Eastern Europe	0.25	0.20	0.18	0.24
Other	0.73	0.75	0.86	0.95
Total	12.78	12.79	13.04	13.78
	Gross exports			
USA	1.22	0.74	0.94	0.82
Brazil	0.53	1.15	0.85	1.02
Argentina	0.11	0.06	0.12	0.28
European Economic Community	0.90	0.89	0.96	0.94
Portugal	0.02	0.03	0.05	0.04
Spain	0.37	0.41	0.48	0.42
Other	0.09	0.09	0.10	0.13
Total	3.24	3.38	3.61	3.73
	Gross imports			
European Economic Community	0.49	0.48	0.52	0.51
Mexico	0.03	0.03	0.08	0.03
Brazil	0.13	0.00	0.01	0.01
India	0.64	0.60	0.42	0.40
Pakistan	0.22	0.26	0.32	0.24
China (Mainland)	0.10	0.07	0.03	0.04
Soviet Union	0.05	0.08	0.20	0.23
Eastern Europe	0.09	0.20	0.12	0.23
Mid-East/N. Africa	0.68	0.73	0.79	0.85
Latin America	0.34	0.46	0.47	0.48
Other	0.44	0.40	0.45	0.50
Total	3.21	3.31	3.42	3.53
	Consumption			
USA	4.07	4.13	4.32	4.47
Brazil	1.42	1.53	1.50	1.65
Argentina	0.01	0.08	0.80	0.07
Latin America	0.44	0.52	0.58	0.61
European Economic Community 10	1.60	1.47	1.49	1.35
Mexico	0.25	0.32	0.32	0.29
Japan	0.62	0.63	0.69	0.69
Spain	0.11	0.10	0.10	0.90
China				

(continued on next page)

Table 1–8. Continued

	1979–1980	1980–1981	1981–1982	1982–1983†
	Tg			
Mainland	0.50	0.48	0.51	0.47
Taiwan	0.14	0.14	0.18	0.17
Soviet Union	0.27	0.30	0.48	0.48
Eastern Europe	0.34	0.39	0.29	0.47
India	0.75	0.66	0.49	0.48
Pakistan	0.18	0.28	0.32	0.25
Mid-East/N. Africa	0.81	0.86	0.92	1.01
Other	0.78	0.74	0.79	0.91
Total	12.31	12.65	13.07	13.62

†Preliminary data.

This projection of production growth is based on several factors, including an estimated world economic growth rate of 3% per year in the future, which is lower than the post-World War boom of 1950 to 1972. The U.S. soybean production share will probably continue to decrease as production in Brazil, Argentina, and other countries increases.

Slower world economic growth will slow growth in consumption of soybean meal and high protein meals. The same study estimates growth in soybean meal consumption to be 4.6% per year during the next 20 yrs. This amount is substantially lower than the past 20-yr rate of 7.5% per year.

Soybean acreage could be altered significantly by government controls, trade restrictions or sanctions, soybean reserve policies, world trade protectionism, and other policies under administrative control. The actions could regulate both the supply and demand for soybean and soybean products.

Soybean yields will reflect research advancements and will be one of the determining factors in the future of soybean. Main contributors to higher yields will be new cultural practices; more productive varieties; better herbicides, insecticides and nematicides; new plant growth regulators; and more effective grower education programs.

The impact of biotechnology, genetic engineering, and molecular biology studies on basic plant processes is an unknown factor in the future of soybean. These basic tools may open a new chapter in soybean expansion. They may greatly improve soybean production through improved insect and disease resistance. Biotechnology may also be important in improving quality characteristics of soybean. These scientific advances could improve both the efficiency of soybean production and the utilization of soybean products.

REFERENCES

American Soybean Association. 1983. Soybean bluebook. American Soybean Association, St. Louis.

Ball, C.R. 1907. Soybean varieties. USDA Bureau of Plant Industries Bull. 98. U.S. Government Printing Office, Washington, DC.

Broadbent, E.E., and F.P. Dixon. 1976. Exploratory study of Brazil soybean marketing. Ill. Agric. Econ. Exp. Stn. Pub. AERR 144.

Foreign Agricultural Service, U.S. Department of Agriculture. 1983. Various issues. U.S. Government Printing Office, Washington, DC.

Hazera, Jorge, and Ed Fryar. 1981. Regional soybean production since 1960 and the outlook for the 1980's. Economic Res. Serv., USDA FOS-305. U.S. Government Printing Office, Washington, DC.

Hauck, J.P., M.E. Ryan, and A. Subotnik. 1972. Soybeans and their products. University of Minnesota Press, Minneapolis.

Huyser, W.S. 1983. A regional analysis of trade policies affecting the soybean and soymeal market. Ph.D. thesis. Iowa State Univ., Ames (Diss. Abstr. DEP 83-16149).

Hymowitz, T., and J.R. Harlan. 1983. Introduction of soybeans to North America by Samuel Bowen in 1765. Econ. Bot. 37(4):371-379.

Morse, W.J. 1918. Soy bean: Its culture and uses. USDA Farmers Bull. 973. U.S. Government Printing Office, Washington, DC.

Piper, C.V., and H.T. Nielsen. 1909. Soy beans. USDA. Farmers Bull. 372. U.S. Government Printing Office, Washington, DC.

Probst, A.H., and R.W. Judd. 1973. Origin, U.S. history and development, and world distribution. *In* Soybeans: Improvement, production, and uses. Agronomy 16:1-15.

Sharpe, D. 1983. What's your future in soybeans: Project 2002. Soybean Dig. 1983 (November):29.

Williams, G.W. 1981. The U.S. and world oilseeds and derivative markets: economic structure and policy interventions. Ph.D. diss. Purdue Univ., West Lafayette, IN (Diss. Abstr. DDJ81-23724).

2 Taxonomy and Speciation

T. Hymowitz and R. J. Singh
University of Illinois
Urbana, Illinois

The genus *Glycine* Willd. is a member of the family Leguminosae, subfamily Papilionoideae, and tribe Phaseoleae. The Phaseoleae is the most economically important tribe of the Leguminosae. It contains members that have considerable importance as sources of food and feed, e.g., *Glycine max* (L.) Merr.—soybean; *Cajanus cajan* (L.) Millsp.—pigeon pea; *Lablab purpureus* (L.) Sweet-hyacinth bean; *Phaseolus* spp.—common bean, lima bean, tepary bean; *Psophocarpus tetragonolobus* (L.) DC.—winged bean; and *Vigna* spp.—cowpea, mung bean, urd, black gram, adzuki bean, Bambarra groundnut. Within the tribe Phaseoleae, *Glycine* is a member of the subtribe Glycininae along with allied genera *Eminia* Taub., *Pseudeminia* Verdc., *Pseudovigna* (Harms) Verdc., *Nogra* Merrill, *Sinodolichos* Verdc., and *Pueraria* DC. *Nogra*, *Sinodolichos*, and *Pueraria* are essentially of Asian origin, *Eminia*, *Pseudeminia*, and *Pseudovigna* are African, and *Glycine* is of Asian and Australian origin (Lackey, 1981).

2-1 TAXONOMIC HISTORY

Glycine has a confused taxonomic history which dates back to the time of its first inception. The name *Glycine* was originally introduced by Linnaeus in the first edition of his *Genera Plantarum* (Linnaeus, 1737) and based on *Apios* of Boerhaave (Linnaeus, 1754). *Glycine* is derived from the Greek *glykys* (sweet) and probably refers to the sweetness of the edible tubers produced by *G. apios* L. (Henderson, 1881), now *Apios americana* Medik. In the *Species Plantarum* of 1753, Linnaeus listed eight *Glycine* spp. (Table 2-1). All of these were subsequently moved to other genera, although *G. javanica* remained as the lectotype for the genus until 1966 (Hitchcock and Green, 1947). Thus, when *G. apios* became *A. americana*, the original justification for the name *Glycine* was removed from the genus. Therefore, the Greek *glykys* does not refer to any of the current *Glycine* spp.

The soybean was described by Linnaeus (1753, p. 725) as both *Phaseolus max* based on specimens that he saw, and *Dolichos soja* (1753, p.

Soybeans: Improvement, Production, and Uses, 2nd ed.—Agronomy Monograph no. 16.

Table 2–1. The species of *Glycine* according to Linnaeus (1753), and their subsequent classification.

Glycine species		
apios	=	*Apios*
frutescens	=	*Wisteria*
abrus	=	*Abrus*
tomentosa	=	*Rhynchosia*
comosa	=	*Amphicarpa*
javanica	=	
bracteata	=	*Amphicarpa*
bituminosa	=	*Fagelia*

727) based upon descriptions of other writers. Linnaeus apparently intended the name *P. max* to apply to the mung bean of India (Piper, 1914; Piper and Morse, 1923). Several years later, he obtained seed of *D. soja* and grew the plants at Uppsala. It was then that he realized that *P. max* and *D. soja* were the same plant and that the mung bean was still without a name. Therefore, in the *Mantissa Plantarum* published in 1767, Linnaeus described the mung bean for the first time under *P. mungo.* Since then, the correct nomenclature for the soybean has been the subject of much debate (Lawrence, 1949; Paclt, 1949; Piper, 1914; Piper and Morse, 1923; Ricker and Morse, 1948). Currently, the combination *G. max* proposed by Merrill in 1917 is widely accepted as the valid designation for the soybean.

According to Bentham, by the time of De Candolle's *Prodromus* in 1825, "the genera *Glycine* and *Dolichos* had become the receptacle for all the Phaseoleae which had no striking character to distinguish them" (Bentham, 1865). This led to an enormous proliferation of species attributed to *Glycine,* such that 286 species were eventually listed in *Index Kewensis,* with additional subspecies and taxonomic varieties bringing the total to 323 (Hermann, 1962). Bentham arranged the genus into three sections containing 11 species (Bentham, 1864, 1865): *Leptolobium* which comprised six species of Australian origin; *Johnia* which included *G. javanica,* the sole remaining Linnaean species of African and Asian origin; and *Soja* which included the cultivated soybean (Table 2–2).

Hermann (1962) published a revision of the genus *Glycine* and its allies (Table 2–3). He brought together the pertinent literature on *Glycine* nomenclature and listed those species that had been published as *Glycine* in the past but later excluded from the genus. According to his classification, *Glycine* consists of three subgenera: (i) *Leptocyamus* (Benth.) F. J. Herm. which included six primarily Australian species; (ii) *Glycine* which included the *Glycine* complex from Africa and Southeastern Asia, and (iii) *Soja* (Moench) F. J. Herm., composed of the soybean and its wild annual counterpart described as *G. ussuriensis* by Regel and Maack (1861). In addition, Hermann found that name changes had to be made because of earlier homonyms. Thus, *G. sericea* became *G. canescens, G. tomentosa* became *G. tomentella,* and variety *latifolia* of *G. tabacina* was no longer considered distinct.

Table 2–2. The genus *Glycine* L. according to Bentham (1864, 1865) and subsequent classification.

Section *Leptocyamus*		
Glycine falcata		
Glycine clandestina		
Glycine clandestina var. *sericea*		
Glycine latrobeana		
Glycine tabacina		
Glycine tabacina var. *latifolia*		
Glycine tabacina var. *uncinata*		
Glycine sericea		
Glycine tomentosa		
Section *Johnia*		
Glycine javanica		
Section *Soja*		
Glycine soja (cultivated)		
Glycine hedysaroides	=	*Ophrestia*
Glycine pentaphylla	=	*Ophrestia*
Glycine lyalli	=	*Ophrestia*

Table 2–3. The genus *Glycine* L. according to Hermann (1962).

Subgenus *Leptocyamus*
Glycine clandestina
Glycine clandestina var. *sericea*
Glycine falcata
Glycine latrobeana
Glycine canescens (previously *G. sericea)*
Glycine tabacina
Glycine tomentella (previously *G. tomentosa)*
Subgenus *Glycine*
Glycine petitiana
Glycine javanica
Subgenus *Soja*
Glycine ussuriensis
Glycine max (cultigen)

Further revision became necessary when Verdcourt (1966) chanced to examine Linnaeus' specimen of *G. javanica* during the preparation of the *Flora of Tropical East Africa.* He discovered that the type specimen was not *G. javanica* but rather a *Pueraria* with an abnormal inflorescence. In order to avoid major alterations in nomenclature of economically important legume genera, Verdcourt proposed that the name *Glycine* be conserved from a later author, Willdenow (1802) and that *G. clandestina* should become the type for the genus (Table 2–4). Thus, the original type specimen became a synonym of *Pueraria montana* (Lour.) Merr. However, all those plants previously regarded as *G. javanica* L. were thus without a name, and for these Verdcourt adopted the name *G. wightii* (R. Grah. ex Wight and Arn.) Verdc. Verdcourt also altered the subgeneric names to reflect the change in type. In addition, Verdcourt apparently overlooked the possibility that *Soja* Moench (1794) ("Soia" p. 153, "Soja"

Table 2-4. The genus *Glycine* Willd. as revised by Verdcourt (1966, 1970) and subsequent classification.

Subgenus *Glycine*
Glycine clandestina
Glycine falcata
Glycine latrobeana
Glycine canescens
Glycine tabacina
Glycine tomentella
Subgenus *Bracteata*
Glycine wightii (previously *G. javanica)* = *Neonotonia*
Subgenus *Soja*
Glycine soja (previously *G. ussuriensis)*
Glycine max

index) had priority over Willdenow (1802). Therefore, Lackey (1978) proposed to conserve the generic name *Glycine* Willdenow over *Soja* Moench.

In 1970, Verdcourt proposed that *G. soja* is the valid designation of the wild annual relative of the soybean since it was described in 1846 by Siebold and Zuccarini as a new species and not based on *Dolichos soja* L. Therefore, *G. soja* predates *G. ussuriensis* Regal and Maack of 1861 (Verdcourt, 1970).

In 1977, Lackey proposed the removal of *G. wightii* from the genus and suggested a new designation *Neonotonia wightii* (R. Grah. ex Wight and Arn.) Lackey (Lackey, 1977a, 1977c). Thus the last Linnaean *Glycine* was removed from the genus. The evidence he presented for his actions was as follows. Cytological studies by several investigators (see Hadley and Hymowitz, 1973) showed that the *G. wightii* complex had chromosome numbers of $2n = 22$ or $2n = 44$, while the remaining members of the genus were $2n = 40$ or $2n = 80$. Pritchard and Wutoh (1964) reported that *G. wightii* had larger chromosomes than *G. max* and the wild perennial species from Australia. In addition, morphological characteristics, geographic distribution, seed protein electrophoretic banding patterns, presence of the free amino acid canavanine in the seeds, and production of isoflavone derivatives after fungal inoculation rather than pterocarpan derivatives all serve to separate *G. wightii* from the other two subgenera (Ingham et al., 1977; Lackey, 1977a, 1977b; Mies and Hymowitz, 1973).

In Verdcourt's (Gillett et al., 1971) treatment of *Glycine* there is a reference to "*Glycine* sp. A", a name assigned to two partially identifiable specimens from East Africa which seemed to belong to *Glycine.* Seeds of one specimen were sent to Iowa State University and germinated. The plants produced were described as *Neonotonia verdcourtii* Isly (Iseley et al., 1980).

In 1980, Newell and Hymowitz proposed a seventh species, *G. latifolia* (Benth.) Newell and Hymowitz, for the subgenus *Glycine.* Under Bentham's description of *G. tabacina* in *Flora Australiensis* of 1864 he

Table 2–5. The genus *Glycine* Willd., subgenus, and distribution (Hymowitz and Newell, 1981).

Species	2*n*	Distribution
	Subgenus *Glycine*	
Glycine clandestina Wendl.	40	Australia
Glycine clandestina var. *sericea* Benth.	—	Australia
Glycine falcata Benth.	40	Australia
Glycine latifolia (Benth.) Newell and Hymowitz	40	Australia
Glycine latrobeana (Meissn.) Benth.	40	Australia
Glycine canescens F. J. Herm.	40	Australia
Glycine tabacina (Labill.) Benth.	40, 80	Australia; south China; Taiwan; Mariana Island; Ryukyu Island; South Pacific Islands
Glycine tomentella Hayata	38, 40, 78, 80	Australia; south China; Taiwan; Philippines; Papua New Guinea
	Subgenus *Soja* (Moench) F. J. Herm.	
Glycine soja Sieb. & Zucc.	40	China; Taiwan; Japan; Korea; USSR
Glycine max (L.) Merr.	40	Cultigen

listed the variety *latifolia* (Table 2–2), originally described by him as a separate species but then relegated to varietal level under *G. tabacina*. Morphological, cytological, and seed protein electrophoretic data suggested that this former variety deserved specific rank, as it was quite distinct from both *G. tabacina* and *G. tomentella*, the two species with which it was usually confused (Newell and Hymowitz, 1980).

The genus *Glycine* Willd., as currently delimited, is divided into the two subgenera *Glycine* and *Soja* (Moench) F. J. Herm. (Table 2–5). The subgenus *Glycine* comprises seven wild perennial species (Hymowitz and Newell, 1981): *G. clandestina* Wendl.; *G. falcata* Benth.; *G. latifolia* (Benth.) Newell and Hymowitz; *G. latrobeana* (Meissn.) Benth.; *G. canescens* F. J. Herm.; *G. tabacina* (Labill.) Benth.; and *G. tomentella* Hayata. The subgenus *Soja* includes the cultigen *G. max* (L.) Merr. and its annual wild counterpart, *G. soja* Sieb. and Zucc.

2–2 *GLYCINE* SPECIES

2–2.1 *Glycine clandestina* Wendl.

Leaves of *G. clandestina* are digitate; the leaflets range from ovate-lanceolate or oblong to linear, and exhibit reticulate venation (Fig. 2–1 to 2–3). Flowers vary in color from pale pink to rose-purple, the wings often darker than the standard. Pods may be short, oblong with few seeds, or linear, curved, with many seeds. Seeds are oblong or suborbicular and grey-brown or black. The accesions studied cytologically are diploid (2*n* = 40) (Newell and Hymowitz, 1983).

Glycine clandestina is a slender twiner restricted to Australia (Fig. 2–1). The species has been reported from the Pescadores, Ryukyu, and

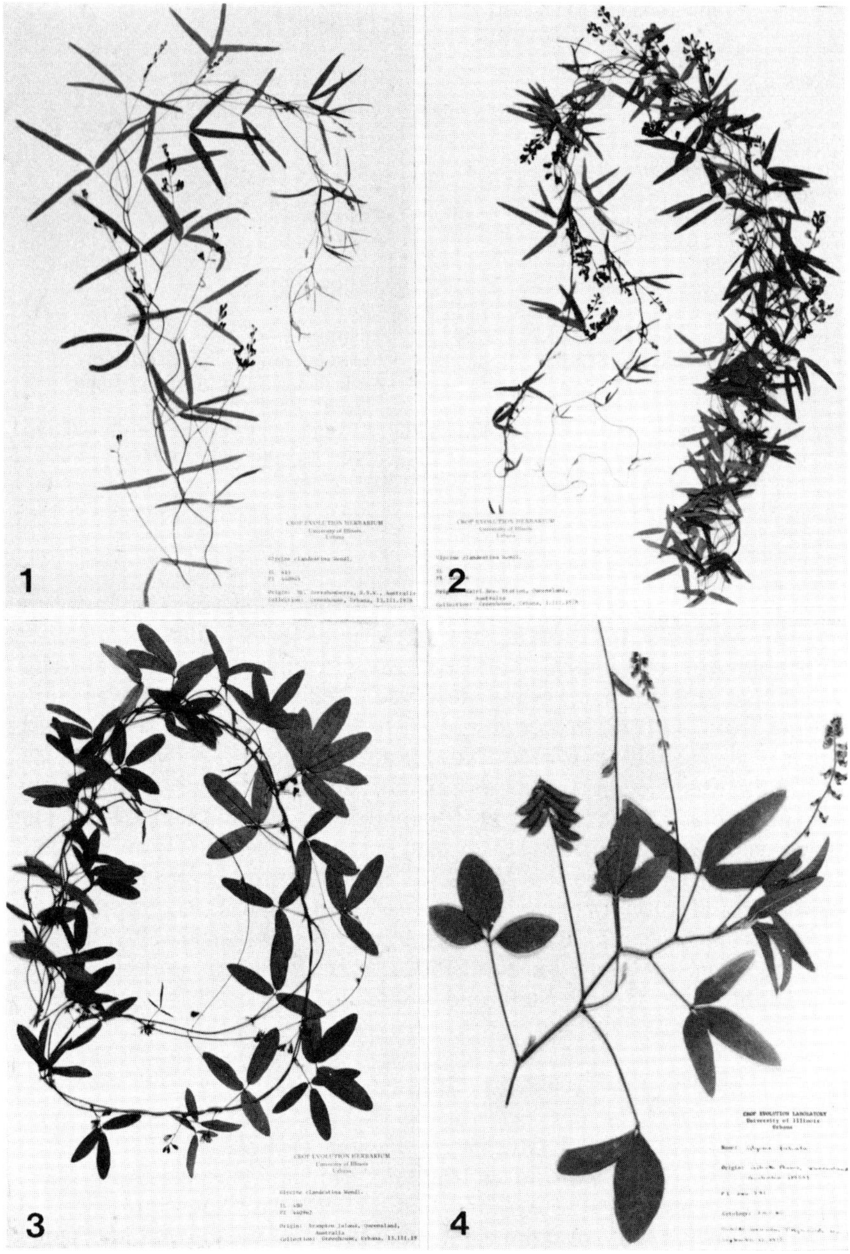

Fig. 2–1 to 2–4. Herbarium specimens of *Glycine*. *Fig. 2–1. G. clandestina* (long pod). *Fig. 2–2. G. clandestina* (short pod). *Fig. 2–3. G. clandestina* (curved pod). *Fig. 2–4. G. falcata.*

Mariana Islands in the West Central Pacific (Fosberg et al., 1979; Hayata, 1919; Hosokawa, 1935; Ohwi, 1943; Walker, 1976). However, morphological examination of herbarium material and living collections suggested that those specimens from the Pescadores and Ryukyus were not *G. clandestina* but rather narrow-leaved forms of *G. tabacina.* Living collections from the Marianas were found to be broad-leaved forms of *G. tabacina* found in the Ryukyus (Newell, 1981; Newell and Hymowitz, 1978a, 1978b). *Glycine clandestina* var. *sericea* Benth. was recognized by Bentham (1864) as a densely silky or villous pubescent form from Western and South Australia. The group is morphologically variable and grades into *G. tabacina* in many characteristics.

2–2.2 *Glycine falcata* Benth.

Glycine falcata is found in Australia. It has a decumbent or erect growth habit and strigose pubescence. The leaflets are broad, oblong-lanceolate, and digitately arranged. Flower racemes are long and stout with white to pale lilac flowers (Fig. 2–4). Pods are broadly falcate and hirsute-strigose with oblong or ovoid seeds. *Glycine falcata* differs from the other species in growth habit, seed protein banding pattern (Mies and Hymowitz, 1973), the presence of pods on underground rhizomes (Hymowitz and Newell, 1975), and absence of a persistent membraneous deposit on the seed coat (Newell and Hymowitz, 1978b). The few accessions studied cytologically are diploid ($2n = 40$).

2–2.3 *Glycine latifolia* (Benth.) Newell and Hymowitz

Glycine latifolia is restricted to Australia. It is a trailing or occasionally twining plant with elongated, robust stems. The leaves are pinnately trifoliolate; the leaflets range from broadly elliptic to rhombic or ovate to obovate. Flower racemes are long and slender with lavender or purple flowers (Fig. 2–5). Pods are short and densely villous, hirsute, or occasionally strigose. Plants are diploid ($2n = 40$) (Newell and Hymowitz, 1980).

2–2.4 *Glycine latrobeana* (Meissn.) Beth.

Glycine latrobeana is distributed in Australia. It is a small herb with a compact, decumbent, or somewhat twining growth habit (Fig. 2–6). Leaves are digitately trifoliolate with characteristically obovate or suborbicular leaflets. Flowers are larger than those of *G. clandestina* but the pods are similar (Bentham, 1864).

2–2.5 *Glycine canescens* F. J. Herm.

Glycine canescens is restricted to Australia. It is a twining herb with pinnately trifoliolate leaves. The leaflets are elliptic-linear to oblong-lan-

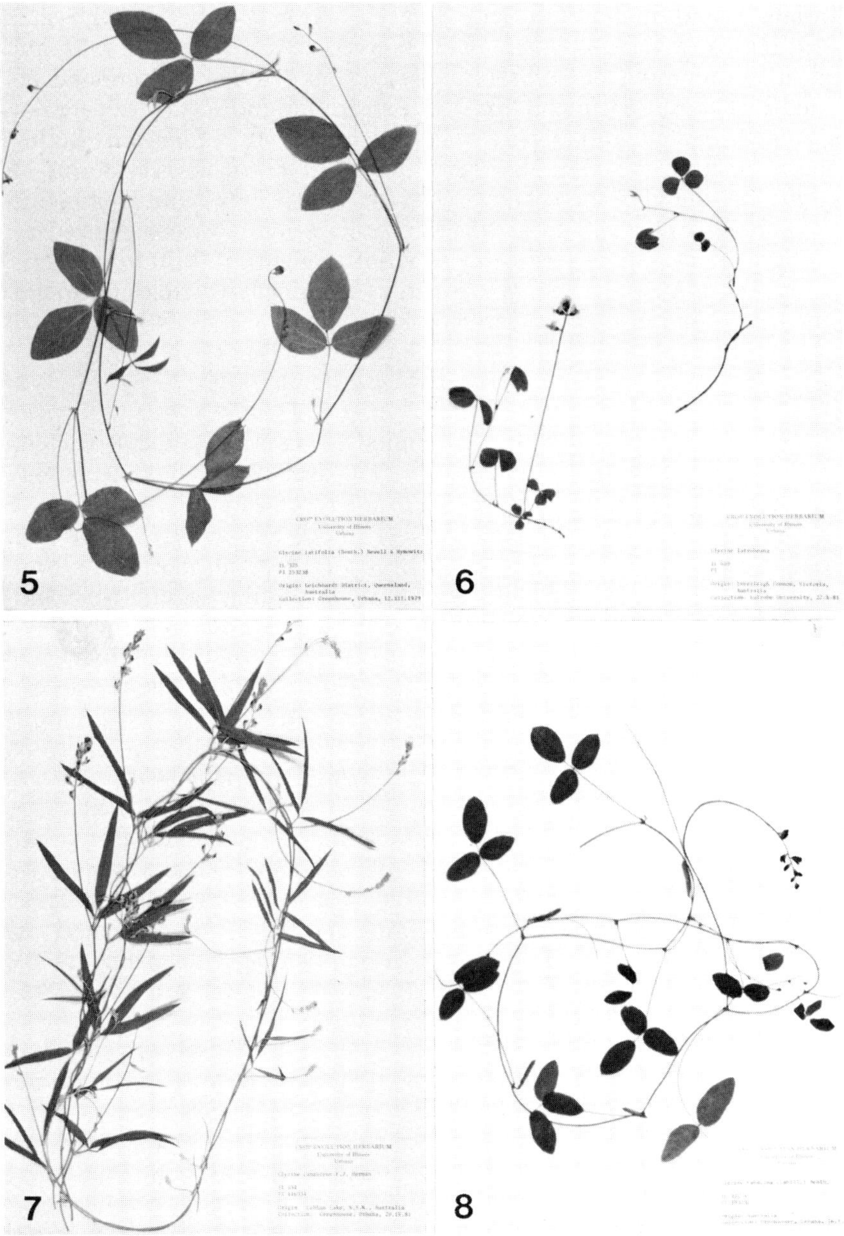

Fig. 2–5 to 2–8. Herbarium specimens of *Glycine. Fig. 2–5. G. latifolia. Fig. 2–6. G. latrobeana. Fig. 2–7. G. canescens. Fig. 2–8. G. tabacina* ($2n=80$).

ceolate and pinnately veined (Fig. 2–7). The whole plant possesses a hoary, silky-strigose pubescence which produces a silvery appearance in extreme cases. Flowers are pink, can be fragrant, and pods are linear.

Seeds are rectangular and somewhat flattened. Plants are diploid ($2n = 40$) (Newell and Hymowitz, 1975).

2–2.6 *Glycine tabacina* (Labill.) Benth.

Glycine tabacina is found in Australia, China, and the South Pacific Islands of New Caledonia, Vanuatu, Fiji, Tonga, and Niue (Dubois, 1971; Hermann, 1962; Hymowitz, 1982; Li et al., 1983; Sykes, 1970). In the West Central Pacific area it is distributed in Taiwan, Ryukyu, and Mariana Islands (Newell, 1981). In addition, the species has been reported to be present on Norfolk Island (Maiden, 1903).

The stems of *G. tabacina* trail or twine and bear pinnately trifoliolate leaves (Fig. 2–8 and 2–9). The leaflets have reticulate venation and range from ovate lanceolate or elliptic linear on the upper nodes to obovate on the lower nodes. Flower racemes are usually longer than those in *G. clandestina*, and the deep rose-purple flowers often are fragrant. Pods are stout and linear with oblong or ovoid black, sometimes brown, seeds. Diploids ($2n = 40$) and tetraploids ($2n = 80$) have been reported (Newell and Hymowitz, 1978a).

2–2.7 *Glycine tomentella* Hayata

Glycine tomentella is distributed in Australia, China, Papua New Guinea, Philippines, and Taiwan (Li et al., 1983; Newell and Hymowitz, 1978a, 1983; Verdcourt, 1979). It is a trailing or twining herb with a pronounced tomentous-villous pubescence (Fig. 2–10 and 2–11). The leaves are pinnately trifoliolate. The leaflets are usually oblong or ovate with strongly pinnate venation. Flowering peduncles are short, and flowers are more crowded towards the apex in comparison with other taxa. Flowers range in color from dark rose-purple to pale purple, the wings darker than the standard. Pods are straight and stout with constrictions between the seeds, which may be brownish or black. Chromosome counts of $2n = 38$, 40, 78, and 80 have been reported (Newell and Hymowitz, 1978a, 1983).

2–2.8 *Glycine soja* Sieb. and Zucc.

Glycine soja is distributed throughout China, the adjacent areas of the USSR, Korea, Japan, and Taiwan. It grows in fields, hedgerows, along roadsides and riverbanks. The plant is an annual procumbent or slender twiner having pinnately trifoliolate leaves and often tawny, strigose, or hirsute pubescence (Fig. 2–12). The leaflets may be narrowly lanceolate, ovate, or oblong-elliptic. The purple or rare white flowers (Li et al., 1983) are inserted on short, slender racemes. The pods are short with a strigose to hirsute pubescence and oval-oblong seeds (Hermann, 1962). Plants are diploid ($2n = 40$). Evidence accumulated from cytogenetic, and morphological studies karyotype, seed protein, and restriction endonuclease

Fig. 2–9 to 2–12. Herbarium specimens of *Glycine*. *Fig. 2–9. G. tabacina* ($2n=80$) with long narrow leaf. *Fig. 2–10. G. tomentella* ($2n=40$). *Fig. 2–11. G. tomentella* ($2n=80$). *Fig. 2–12. G. soja.*

fragment analysis of mitochondrial DNA studies supports the hypothesis that *G. soja* is the wild ancestor of the soybean (Ahmad et al., 1976; Hadley and Hymowitz, 1973; Hymowitz and Newell, 1980; Mies and Hymowitz, 1973; Pueppke et al., 1982; Sisson et al., 1978).

2–2.9 *Glycine max* (L.) Merr.

Glycine max (L.) Merr., the cultivated soybean has never been found in the wild. It is an annual generally exhibiting an erect, sparsely branched, bush-type growth habit with pinnately trifoliolate leaves. The leaflets are broadly ovate, oval to elliptic-lanceolate. The purple or white flowers are borne on short axillary racemes on reduced peduncles. The pods are either straight or slightly curved, usually hirsute. The one to three seeds per pod are usually ovoid to subspherical. The seed coats range in color from light yellow, olive green, and brown to reddish black. The seeds weigh from about 10 to 20 g per 100 seeds.

The domesticate is morphologically extremely variable. This is due primarily to the development of soybean "land races" in East Asia. Individual farm families have grown soybean containing specific traits generation after generation for use as food and feed, or for their religious, ceremonial, or medicinal value. Today, these land races provide the major sources of genetic diversity in soybean germplasm collections.

The subgenus *Soja* contains, in addition to *G. max* and *G. soja*, a form known as *G. gracilis*. This semicultivated or weedy form known only from Northeast China is somewhat intermediate in morphology between *G. max* and *G. soja* and was proposed as a new species of *Glycine* by Skvortzow (1927). Hermann (1962) considered *G. gracilis* a variant of *G. max* and was supported by Wang (1976). Based upon numerical taxonomic analyses, Broich and Palmer (1980a) recommended that since *G. gracilis* probably represents weedy races associated with the soybean and because *gracilis* phenotypes can be distinguished from *G. max*, the designation *G. max* forma *gracilis* be utilized. In addition, Broich and Palmer (1981) suggested since *G. max* and its semi-wild relative are in a sense the products of humans; they should be regarded taxonomically distinct from *G. soja*, a wild species. Since *G. max*, *G. soja*, and *G. gracilis*—the cultivated, wild and weedy forms—are "generally interfertile," however, (Ahmad et al., 1976, 1977, 1979; Broich and Palmer, 1980b; Erickson and Beversdorf, 1982; Erickson et al., 1981, 1982; Gai et al., 1982; Hadley and Hymowitz, 1973; Kaizuma et al., 1980a, 1980b; Leschenko and Sichkar, 1978, and make up the primary gene pool of the soybean (Harlan and de Wet, 1971) it would seem appropriate to rank them as subspecies (Smartt, 1984).

2–3 ORIGIN OF SOYBEAN

The soybean is a domesticate of China. Current evidence for the antiquity of the soybean lies in the pictographic analysis of the archaic Chinese word for the soybean "shu," the *Book of Odes*, and bronze inscriptions. All of the above lines of evidence point to the emergence of the soybean as a domesticate during the Chou Dynasty, that is, from the 11th to the 7th century B.C. Domestication, however, is a trial and error process and not an event. This process for soybeans, probably took place during the Shang Dynasty (ca. 1700–1100 B.C.) or earlier. For additional information of the historical, geographic, and archeological evidence for the origin of the soybean please refer to the following references (Hadley and Hymowitz, 1973; Ho, 1975; Hymowitz and Newell, 1980, 1981).

As mentioned in the previous section, the evidence is overwhelming that the wild annual soybean, *G. soja* is the ancestor of the cultivated soybean, *G. max*. The next logical question then is "Where is the probable ancestor of *G. soja*?" Newell and Hymowitz (1982) reported that they successfully hybridized soybean cv. Altona ($2n = 40$) with a wild perennial tetraploid relative, *G. tomentella* ($2n = 78$ and 80). The successful hybridization of the soybean with *G. tomentella* suggests that the two subgenera *Glycine* and *Soja* are closely related and justifies the taxonomic arrangement of the current nine species in one genus. Secondly, if indeed within *Glycine* perennialism is considered to be a primitive condition and annualism is a derived or evolutionarily advanced condition (Stebbins, 1950) then those perennial *Glycine* spp. with an overlapping geographical distribution with *G. soja* are perhaps the sources from which the annual *G. soja* evolved. At present, *G. tabacina* and/or *G. tomentella* overlap in their distribution with *G. soja* in Fujien Province, China; Ryukyu Islands, Japan; and Taiwan. Thus far, chromosome counts of all of a relatively few accessions of *G. tabacina* and *G. tomentella* from those areas have been $2n = 80$ (unpublished data). Thus, either a diploid *G. tabacina* or *G. tomentella* still exists in the region but has not been collected or such plants are now extinct. Only by extensive plant collecting expeditions to search for diploid *G. tabacina* and *G. tomentella* in China, Japan, and Taiwan will perhaps the answer to the origin of the soybean be elucidated.

Glycine is the only genus in the Phaseoleae where species have diploid chromosome number of 40 and 80 but not 20. According to Lackey (1980), "Considering the other counts in the tribe and subtribe, the unique chromosome number of *Glycine* is probably derived from diploid ancestors with base number 11, which have undergone aneuploid loss to base number 10 and subsequent polyploidy to give somatic counts of 40 or sometimes 80". Perhaps the soybean should be regarded as a stable tetraploid with diploidized genomes (Buttery and Buzzell, 1976; Gurley et al., 1979). For additional literature information please refer to Hadley and Hymowitz (1973).

2–4 CYTOTAXONOMIC STUDIES

Hermann's (1962) revision of *Glycine* clarified a previously confused taxonomic picture of the genus. Unfortunately, members of the subgenus *Glycine* are rare in nature and at the time of Hermann's revision living material was not available for intensive experimental studies. In recent years, however, collecting efforts by U.S. and Australian researchers have extended the range of germplasm available for experimentation and programs of cytogenetic and biochemical studies are currently in progress (Marshall and Broué, 1981; Newell and Hymowitz, 1983).

2–4.1 Intraspecific Crosses Within the Subgenus *Glycine*

1.*Glycine clandestina*—Newell and Hymowitz (1983) reported that *G. clandestina* encompasses at least three morphological groups which can be readily distinguished. The first group consists of plants with curved pods (curved pod group), trifoliolate first seedling leaves, and short racemes bearing few flowers (Fig. 2–3). The second group consists of plants with short pods (short pod group), simple first seedling leaves, long racemes with many flowers, a trailing growth habit and somewhat pinnate leaves (Fig. 2–2). The third group consists of twining plants with long straight pods (long pod grop), simple first seedling leaves, long racemes with many flowers, and digitate leaves (Fig. 2–1). Hybrids within each group were fertile. Attempts to cross the curve pod form to the other two pod forms were not successful. Hybrid F_1 plants between short × long pod groups were completely sterile and did not set seed. Cytological analysis indicated limited chromosome pairing (Fig. 2–13) at diakinesis with an average 24.9 I and 7.5 II per cell.

2. *Glycine falcata*—No intraspecific crosses have been reported.

3. *Glycine latifolia*—Various combinations of *G. latifolia* accessions and reciprocals were fully fertile with no apparent abnormalities (Newel and Hymowitz, 1983).

4. *Glycine latrobeana*—No intraspecific crosses have been reported.

5. *Glycine canescens*—Various combinations of hybrids within *G. canescens* accessions and reciprocals were fertile and produced seed. Hybrid plants studied cytologically showed almost normal meiosis with 20 bivalents (Broué et al., 1979; Newell and Hymowitz, 1983; Putievsky and Broué, 1979).

6. *Glycine tabacina*—Newell and Hymowitz (1983) reported that various combinations of hybrids between *G. tabacina* accessions at the tetraploid level produced fertile seed, seed which germinated abnormally and died, or plants which were vegetatively vigorous, flowered profusely but were sterile. Cytogenetic analysis of sterile plants revealed a low incidence of chromosome pairing at meiosis. One hybrid possessed an average of 53.5 I and 13 II and another hybrid possessed an average of 52.1 I and 13.8 II at diakinesis, instead of the expected 40 II. Hybrids with abnormal seedlings had a common parent PI 378706. Seedling death may have been due to a genetic imbalance when this particular accession was combined with cerain others.

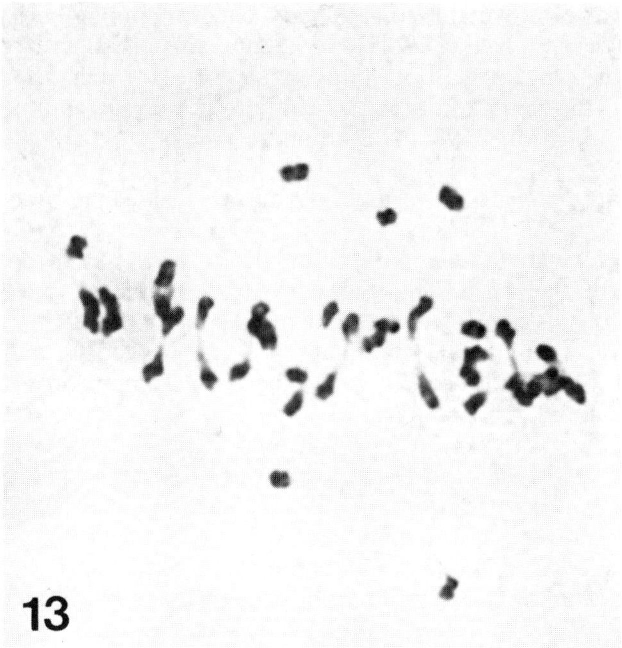

Fig. 2–13. Metaphase I in *G. clandestina* PI 440948 ($2n=40$) × PI 440956 ($2n=40$) showing 15II + 10I. ca. × 1600.

Putievsky and Broué (1979) reported that at the diploid level they obtained one *G. tabacina* × *G. tabacina* hybrid. The F_1 plant had normal meiosis with 20 II. The pollen stainability was 100%.

7. ***Glycine tomentella***—Within *G. tomentella* hybrids between parents sharing a similar Australian geographic location and having 78 or 80 chromosome number were fertile as expected. Hybrids between 78 chromosome aneuploids from New South Wales and tetraploids from Queensland possessed 79 chromosomes (Fig. 2–14). The plants were vegetatively vigorous and flowered profusely but the hybrids were completely sterile. Pollen stainability was generally less than 20%. Average chromosome configruations per hybrid ranged from 36.7 to 44.6 I, 15.8 to 21.0 II, 0 to 0.2 III and 0 to 0.3 IV at diakinesis (Broué et al., 1979; Newell and Hymowitz, 1983; Putievsky and Broué, 1979).

Hybrids resulting from crosses between tetraploid accessions from Taiwan and the Philippines and 78 chromosome aneuploids from New South Wales, Australia possessed 79 chromosomes. The plants grew and flowered profusely but were sterile. Average chromosome configurations ranged from 49.6 to 51.7 I, 13.4 to 14.6 II, 0 to 0.2 III, and 0 to 0.04 IV at diakinesis (Newell and Hymowitz, 1983).

Hybrids resulting from crosses between tetraploids from Taiwan and the Philippines and tetraploids from Queensland, Australia resulted in

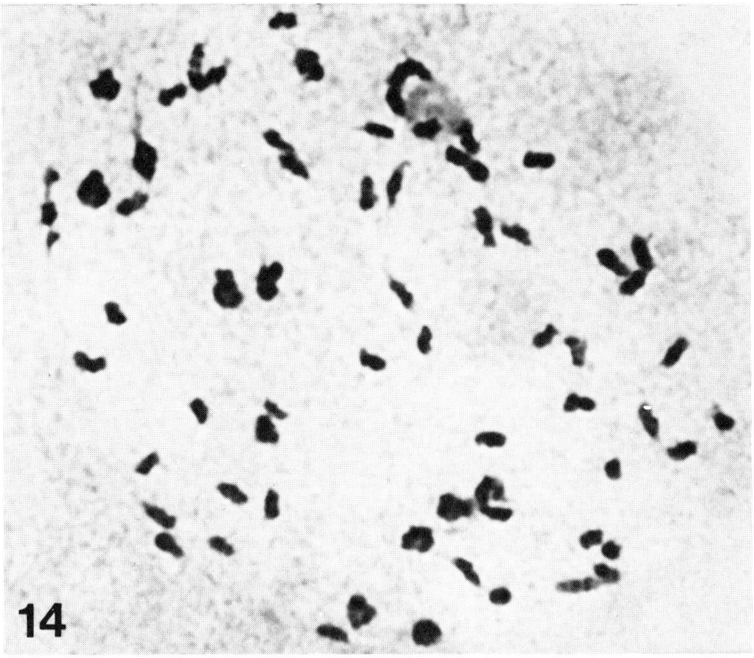

Fig. 2–14. Diakinesis in *G. tomentella* PI 441002 ($2n=80$) × PI 373980 ($2n=78$) showing $2n=79$ chromosomes with 15II + 49I. ca × 1600.

seedling lethality. The seeds germinated readily but development ceased once the radicle and cotyledons emerged (Newell and Hymowitz, 1983).

2–4.2 Interspecific Crosses Within the Subgenus *Glycine*

1.***Glycine falcata* and *G. clandestina***—Putievsky and Broué (1979) reported that F_1 plants of *G. falcata* × *G. clandestina* had 36I + 2 II, and only an occasional trivalent per cell. Most anthers examined lacked stainable pollen. The hybrid plants were sterile.

2. ***Glycine canescens* and *G. clandestina*.**— Successful *G. canescens* × *G. clandestina* hybrids and reciprocals were obtained by several investigators (Newell and Hymowitz, 1983; Putievsky and Broué, 1979; Sedova, 1982). Apparently, in all cases an accession from the *G. clandestina* long pod group was used as one of the parents. Cytogenetic analysis indicated a high degree of chromosome pairing (Fig. 2–15) at diakinesis with 0 to 8 I and 16 to 20 II per cell. The pollen stainability was from 90 to 98%. All hybrids set F_2 seed although seed fertility was reduced by 40 to 67%. The above data suggest that there is a close affinity between *G. canescens* and the long pod group of *G. clandestina*.

3. ***Glycine falcata* and *G. canescens***—Combinations between *G. falcata* and *G. canescens* and a reciprocal produced inviable hybrids. The

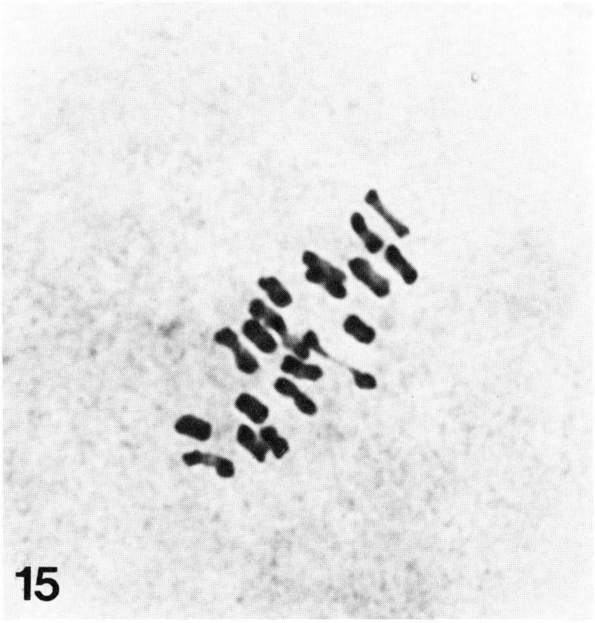

Fig. 2–15. Metaphase I in *G. canescens* PI 440932 × *G. clandestina* PI 440948 showing $2n=40$ chromosomes with 20II. ca × 1600.

seedlings developed into stunted plants that reached a maximum of 10 cm in height. All plants died within 6 months (Newell and Hymowitz, 1983). Inviable hybrids also were reported by Putievsky and Broué (1979).

4. *Glycine falcata* and *G. tomentella* (2*n* = 40)—Two hybrids between (*G. falcata* and diploid *G. tomentella* resulted in inviable plants. Small inflorescence buds were produced but these shriveled before they reached flowering stage. The hybrids ceased growth and died after 6 and 11 months (Newell and Hymowitz, 1983).

5. *Glycine latifolia* and *G. tabacina* (2*n* = 80)—Palmer and Hadley (1968) were able to obtain reciprocal hybrids using a colchicine-doubled *G. latifolia.* All F_1 hybrids were morphologically intermediate between the parents and were highly male and female sterile as estimated by pollen stainability and seed set. Trivalent and quadrivalent associations at meiosis suggested that a certain amount of pairing betwen parental genomes was taking place. They hypothesized the existence of three quite closely related genomes. *Glycine tabacina* was designated $2(A_0\,A_2)$ and *G. latifolia* $2(A_1)$, whereby the two genomes of *G. tabacina* were more distantly related to one another than either was to *G. latifolia.*

6. *Glycine latifolia* and *G. tomentella* (2*n* = 40)—Reciprocal crosses between *G. latifolia* and *G. tomentella* (2*n* = 40) yielded F_1 seeds that germinated but the seedlings died soon after the expansion of the cotyledons (Newell and Hymowitz, 1983).

7. ***Glycine latrobeana* and *G. tabacina* (2*n* = 40)**—A combination betwen *G. latrobeana* and *G. tabacina* (2*n* = 40) produced an inviable hybrid that did not flower (Putievsky and Broué, 1979).

8. ***Glycine tomentella* (2*n* = 80) *and G. canescens***—Putievsky and Broué (1979) obtained hybrid seed from tetraploid *G. tomentella* and *G. canescens* crosses. The F_1 plants were sterile triploids. Average chromosome configurations at meiosis (M1) ranged from 21.9 to 26.0 I, 17 to 18.8 II, 0.13 III, and 0.01 IV. The data suggest that almost all the chromosomes from *G. canescens* were involved in the formation of bivalents that perhaps one genome of tetraploid *G. tomentella* is derived from *G. canescens.*

9. ***Glycine tomentella* (2*n* = 80) and *G. tabacina* (2*n* = 80)**—Hybrids between *G. tomentella* and *G. tabacina* were vegetatively vigorous and flowered profusely but did not set seed. Cytological analysis revealed a low incidence of chromosome pairing at diakinesis (Fig. 2- 16) with a range of 73.2 to 74 I, 2.8 to 3.1 II, 0.05 to 0.19 III, and 0.01 IV per cell. The majority of the chromosome associations were end-to-end with a few ring bivalents commonly found in pollen mother cells of either species. No viable pollen was shed from the anthers. Apparently, the tetraploid forms of *G. tomentella* and *G. tabacina* are not closely related as evidenced by the high number of univalents at meiosis, no viable pollen and lack of seed set (Newell and Hymowitz, 1983; Putievsky and Broué, 1979).

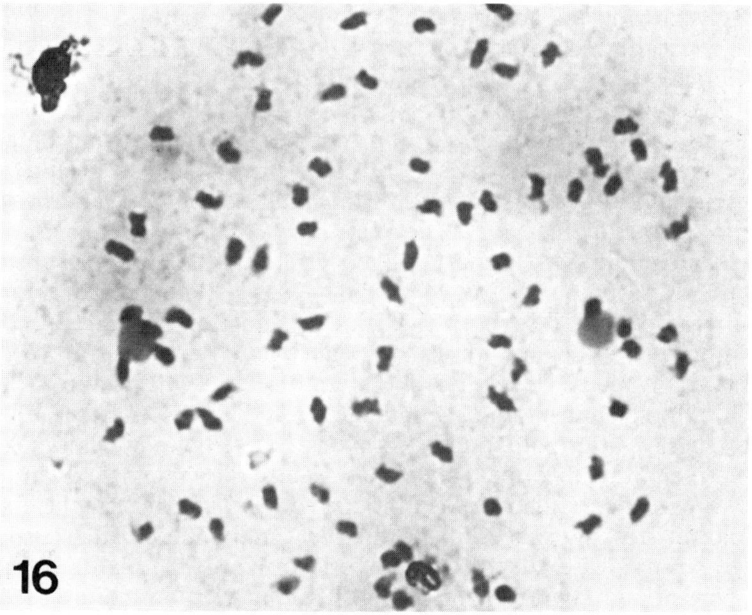

Fig. 2–16. Diakinesis in *G. tomentella* PI 441005 (2*n*=80) × *G. tabacina* PI 440991 (2*n*=80) showing 80I. ca × 1600.

10. ***Additional Hybrids***—In Table 2-1 of their publication, Broué et al. (1982) list the parents used in crossing attempts with members of the subgenus *Soja*. Included in the list were the following hybrids: *G. latrobeana* × *G. canescens*; *G. tabacina* × *G. canescens*; and *G. clandestina* × *G. tomentella*. Unfortunately no information was provided as to the accessions used, chromosome number of accessions or cytogenetic analysis of hybrids.

2-4.3 Crosses Between the Subgenera *Glycine* and *Soja*

Wild relatives of crop plants, as sources of genetic diversity, have been effectively employed in various breeding programs (Stalker, 1980). From a taxonomic standpoint the seven perennial members of the subgenus *Glycine* are candidates for gene exchange with the soybean, and therefore potentially useful for broadening the germplasm base of the crop. For example, investigations have shown that the wild perennial *Glycine* species carry resistance to diseases such as soybean rust (*Phakopsora pachyrhizi* Sydow), yellow mosaic virus and powdery mildew (*Microsphaera diffusa* Cke. & Pk.) (Burdon and Marshall, 1981a, 1981b; Mignucci and Chamberlain, 1978; Singh et al., 1974).

Palmer (1965) attempted to hybridize the soybean with wild perennial *Glycine* spp. but was unsuccessful. He obtained small pods that aborted at an early developmental stage. Hood and Allen (1980) and Ladizinsky et al. (1979) found that pollen germination and subsequent fertilization triggered pod initiation, but pods developed for 10 to 21 days before yellowing and abscising. Therefore, for any hybridization program to be successful an embryo or immature seed culture system had to be developed.

The culture of immature soybean embryos to maturity has been achieved with limited success (Chan and Lin, 1967; Cutter and Bingham, 1975; Vagera and Hanackova, 1979). However, it has not been possible to rescue hybrid *Glycine* embroys by this method (Chan, 1969; Hood and Allen, 1980; Ladizinsky et al., 1979).

Broué et al. (1982) reported five sterile hybrids ($2n$ = 59) between a synthetic amphiploid of *G. tomentella* ($2n$ = 38) and *G. canescens* ($2n$ = 40) as female and soybean cvs. Lincoln and Hark as males produced by embryo or ovule culture using a transplanted endosperm procedure and nutrient media developed by De Latour et al. (1978) and Williams (1978). Nurse endosperm was obtained from selfed seed of cv. Lincoln 10 to 15 days after pollination. All hybrids were twining perennials and vegetatively resembled the female parent more than the male. Plant traits used as evidence of their hybrid origin included intermediate leaf shape and flower size and shape. In addition, leaf indophenol oxidase isozyme patterns provided evidence of the hybrid origin of the plants.

Newell and Hymowitz (1982), utilizing in vitro immature seed culture, reported successful crosses resulting in seven viable plants between the soybean cv. Altona and the wild perennial *G. tomentella* ($2n$ = 78,

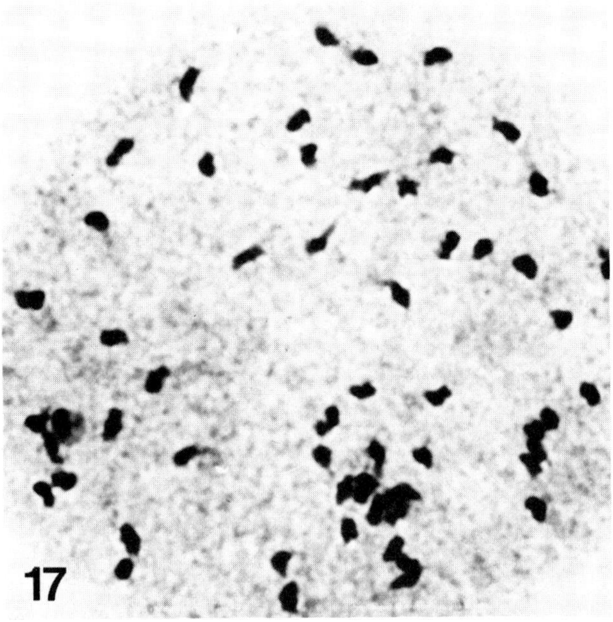

Fig. 2–17. Diakinesis in *G. max* cv. Altona ($2n$=40) × *G. tomentella* PI 483218 ($2n$=78) showing $2n$=3x=59 chromosomes with 2II + 55I. ca. 1600.

80). Pods resulting from crosses were left to develop on the plant from 2 to 5 weeks. When pod elongation stopped and slight yellowing of the pods was visible, the immature seeds were extracted under sterile conditions and cultured. The immature seeds were supported on filter paper bridges in test tubes containing 15 mL of liquid culture medium (Vagera and Hanackova, 1979). The modified medium used was based upon the B-5 medium developed by Gamborg et al. (1968). The F_1 plants were vegetatively vigorous and resembled the wild male parent morphologically. Root tips showed chromosome numbers of $2n$ = 59 or $2n$ = 60, consistent with a hybrid origin. Hybrids showed reduced chromosome pairing (Fig. 2–17) with an average 43.3 I, 7.3 II, 0.3 III, and 0.1 IV per cell chromosome configurations. The plants flowered profusely but were sterile. Doubling of the chromosome number of the above plant by colchicine treatment resulted in an amphiploid plant with $2n = 6x = 118$ chromosomes (Fig. 2–18). The amphiploid plant was essentially sterile however, occasionally produced seed.

2–5 MORPHOLOGICAL STUDIES

A cluster anslysis and principal components analysis were applied by Newell and Hymowitz (1978a) to morphological data on 31 vegetative and 27 inflorescence characters obtained from 58 accessions representing

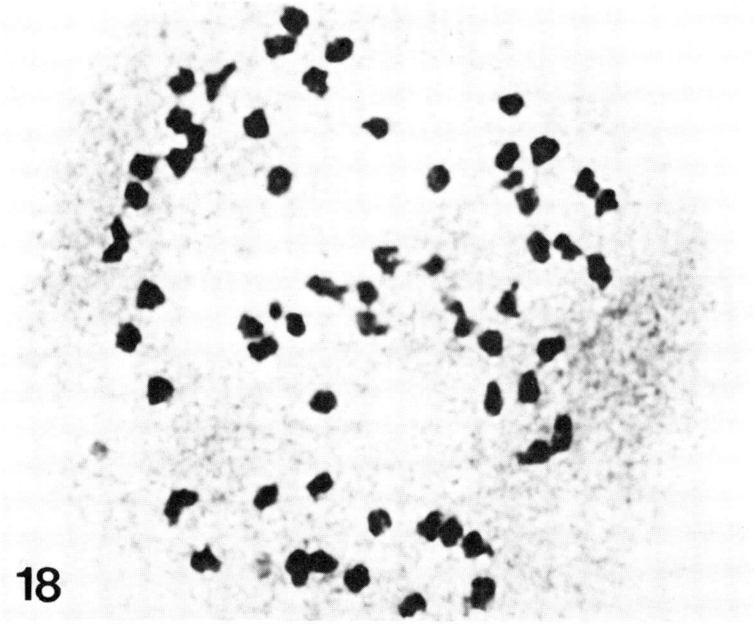

Fig. 2–18. Diakinesis in *G. max* cv. Altona $2n=40 \times$ *G. tomentella* PI 483218 ($2n=78$) amphiploid showing $2n=6x=118$ chromosomes with 59II. ca $\times$ 1600.

populations of *G. canescens* ($2n = 40$), *G. clandestina* ($2n = 40$), *G. falcata* ($2n = 40$), *G. tabacina* ($2n = 40, 80$), and *G. tomentella* ($2n = 38, 40, 78, 80$). Numerical analysis grouped the accessions essentially according to current species delimitations with some exceptions. *Glycine tabacina* specimens from Taiwan approached *G. clandestina* in several characteristics. The diploid *G. tomentella* specimens formed a separate cluster and appeared morphologically distinct from the remaining taxa.

Seed of the genus *Glycine* typically exhibits a muriculate appearance resulting from adherence to the true seed coat of the endocarp (Wolf et al., 1981) or inner pod wall layer. Thickened cell walls of the endocarp superimpose a reticulate network on the seed coat, the type of network ranging from alveolate to stellate depending on the shape of the endocarp cells. Tubercles distributed at intervals give the seed its roughened appearance. Seed lacking an attached endocarp appears smoother and shiny.

Seed morphology of 64 accessions representing six species of the subgenus *Glycine* were examined in detail by Newell and Hymowitz (1978b) to determine whether seed characteristics could be of taxonomic significance at the species level. Seed-coat surface patterning was observed by scanning electron microscopy (SEM). *Glycine canescens* and *G. clandestina* seeds possess a reticulate network and tubercles of irregular shape, the endocarp appearing granular in *G. clandestina*. Seeds of *G. latrobeana* and *G. tabacina* lack a distinct network and have stellate tubercles; the endocarp is granular in *G. latrobeana* and some plants of *G. tabacina*. A few collections of *G. clandestina* approach *G. tabacina* in seed appearance. *Glycine tomentella* seeds exhibit a regularly alveolate arrangement, while those of *G. falcata* lack an endocarp layer altogether. Diploid collections of *G. tomentella* ($2n = 40$) exhibit recognizable differences in seed morphology compared to tetraploids ($2n = 80$). An incompletely attached endocarp is accompanied by aneuploidy in several *G. tomentella* accessions, while other 38 and 78 chromosome aneuploids produce normal seeds.

Apparently each species of the subgenus *Glycine* possesses a characteristic seed morphology. *Glycine canescens*, *G. falcata* and *G. latrobeana* appear to be distinct. *Glycine clandestina*, as a group, exhibited the greatest variability, with some specimens approaching *G. tabacina*. *Glycine tabacina* was relatively uniform despite a wide geographical distribution. Diploid *G. tomentella* and tetraploids of the species exhibited similar seed morphology which did not vary with geographical location. Thus, in the majority of cases, seed surface pattern could enable identification as to species.

2–6 CHEMOSYSTEMATIC STUDIES

Isoenzymatic phenotypes of horizontal starch gel zymograms were used by Broué et al. (1977) to investigate phylogenetic relationships of populations of *G. clandestina*, *G. canescens*, *G. tabacina*, and *G. tomentella*. The following isozyme assays were utilized: alcohol dehydrogenase; anodal arylesterase; NADH-diaphorase; acid phosphatase; endopeptidase; urease; malate dehydrogenase; shikimate dehydrogenase; and glutamate dehydrogenase. Numerical analysis of the data showed that *G. clandestina*, *G. tabacina*, and *G. tomentella* reacted as well-defined taxa. *Glycine canescens* populations did not behave as a homogeneous group. Some populations tended to group with *G. clandestina* whereas others grouped less readily with *G. tomentella*. Other isozymes of possible chemosystematic interest include: glucose-6-phosphate dehydrogenase; glutamate oxaloacetic transaminase; NADP-active isocitrate dehydrogenase; leucine amino peptidase; and mannose-6-phosphate isomerase (Kiang and Gorman, 1983).

Vaughan and Hymowitz (1984) surveyed the leaf flavonoids and isoflavonoids of six species in the subgenus *Glycine* to determine if they could be utilized for interpreting species relationships and to determine hybridity in sterile plants. In *G. falcata* and *G. tomentella* only C-glycosylflavones were detected, whereas, in *G. canescens* only kaempferol was detected. *Glycine tabacina* had O-glycosylflavonols of kaempferol, and in two accessions quercetin also was detected. *Glycine tabacina* was the only species in which both flavonols and C-glycosylflavones were found. The isoflavonoids coumestrol, genistin, and daidzein were present in all species though not detected in all accessions.

C-glycosylation has been considered of more ancient origin than O-glycosylation (Swain, 1975). The distribution of the genus *Glycine* from

Tasmania to northern China suggests that *Glycine* may be an ancient genus. Since *G. tomentella* contains only C-glycosylflavones, is a widely distributed species, and thus far the most successful species in hybridization attempts with the soybean, perhaps ancestors of this species are ancestral to the genus.

Flavonoid patterns in two interspecific hybrids gave mixed results. In a cross between *G. tomentella* ($2n = 80$) × *G. tabacina* ($2n = 80$) only the flavonoids of the maternal parent were detected in the leaves of the F_1 hybrid. In a cross between *G. max* ($2n = 40$) × *G. tomentella* ($2n = 78$) the flavonoid pattern of the leaves of the F_1 hybrid plant was partially additive.

Using sodium dodecyl sulfate-gel electrophoresis, Staswick et al. (1983) compared the 7S and 11S seed storage protein banding patterns of *G. canescens*, *G. clandestina*, *G. tabacina*, and *G. tomentella* to those of the soybean. Each of the five species produced unique electrophoretic patterns that varied in the total number of bands and their relative mobilities. Three groups of proteins in the 11S fraction of *G. tomentella* were quite similar to those in the glycinin fraction in soybeans. In addition, the investigators suggest that comparison of the polypeptide composition of the glycinin complex in perennial species with those in the soybean may prove useful in predicting the origin of soybean.

Using polyacrylamide gel electrophoresis, Mies and Hymowitz (1973) compared trypsin inhibitor banding patterns in seed protein extracts from *Glycine* spp. Generally, there was greater similarily in trypsin inhibitor banding patterns within a subgenus than between subgenera. The trypsin inhibitor banding patterns of *G. max*, *G. soja*, and *G. gracilis* were indistinguishable. In the subgenus *Glycine*, the trypsin inhibitor banding patterns of *G. clandestina*, *G. tabacina*, and *G. tomentella* were similar, while the one accession of *G. falcata* was completely different.

2–7 CONCLUSION

This chapter covers the literature up to January 1984. An analysis of the biosystematic research reported in this chapter reveals the following: (i) the wild perennial *Glycine* spp. display considerable amounts of morphological, cytological, biochemical, and geographical diversity; (ii) investigations conducted since 1975 have considerably enhanced our knowledge of intra- and interspecific relationships within the subgenus *Glycine*; (iii) there are many gaps to be filled in our understanding of speciation among the wild perennial relatives of the soybean; and (iv) additional wide hybrids between the subgenera *Glycine* and *Soja* are needed to elucidate species affinities between the two subgenera and is the major step towards exploiting the genetic diversity represented in the wild perennial *Glycine* spp. for use in soybean breeding programs.

REFERENCES

Ahmad, Q.N., E.J. Britten, and D.E. Byth. 1976. Chromosome morphology in the soybean and a related species. Australian Oilseeds and Protein Grains Conference, Toowoomba. 3(b):23–24.

----, ----, and ----. 1977. Inversion bridges and meiotic behavior in species hybrids of soybeans. J. Hered. 68:360–364.

----, ----, and ----. 1979. Inversion heterozygosity in the hybrid soybean × *Glycine soja*. J. Hered. 70:358–364.

Bentham, G. 1864. Flora Australiensis, Vol 2. L. Reeve, London.

----. 1865. On the genera *Sweetia*, Sprengel, and *Glycine*, Linn., simultaneously published under the name of *Leptolobium*. J. Linn. Soc. Bot. 8:259–267.

Broich, S.L., and R.G. Palmer. 1980a A cluster analysis of wild and domesticated soybean phenotypes. Euphytica 29:23–32.

----, and ----. 1980b Segregation patterns of some simply inherited traits in *Glycine max* × *G. soja* crosses. Soybean Gent. Newsl. 7:62–64.

----, and ----. 1981. Evolutionary studies of the soybean: the frequency and distribution of alleles among collections of *Glycine max* and *G. soja* of various origin. Euphytica 30:55–64.

Broué, P., J. Douglass, J.P. Grace, and D.R. Marshall. 1982. Interspecific hybridisation of soybeans and perennial *Glycine* species indigenous to Australia via embryo culture. Euphytica 31:715–724.

----, D.R. Marshall, and J.P. Grace. 1979. Hybridization among the Australian wild relatives of the soybean. J. Aust. Inst. Agric. Sci. 45:256–257.

----, ----, and W.J.M. Müller. 1977. Biosystematics of subgenus *Glycine* (Verdc.). Isoenzymatic data. Aust. J. Bot. 25:555–566.

Burdon, J.J., and D.R. Marshall. 1981a. Evaluation of Australian native species of *Glycine* for resistance to soybean rust. Plant Dis. 65:44–45.

----, and ----. 1981b. Inter- and intra-specific diversity in the disease response of *Glycine* species to leaf rust fungus, *Phakopsora pachyrhizi* J. Ecol. 69:381–390.

Buttery, B.R., and R.I. Buzzell. 1976. Flavonol glycoside genes and photosynthesis in soybeans. Crop Sci. 16:547–550.

Chan, K.L. 1969. Methods of overcoming cross incompatibility and hybrid sterility in genus *Glycine*. J. Agric. Assoc. China N.S. 66:16–24.

----, and F.M. Lin. 1967. Studies on the method of embryo culture in the cultivated and wild form soybeans. J. Taiwan Agric. Res. 16:8–14.

Cutter, G.L., and E.T. Bingham. 1975. Soybean embryo culture studies. Soybean Genet. Newsl. 2:52–53.

De Latour, G., W.T. Jones, and M.D. Ross. 1978. Production of interspecific hybrids of *Lotus* aided by endosperm transplants. N.Z.J. Bot. 16:61–68.

Dubois, M.J. 1971. Ethnobotanique de Maré, Iles Loyaute (Nouvelle Caledonie). J. Agric. Trop. Bot. Appl. 18:310–369.

Erickson, L.R., and W.D. Beversdorf. 1982. Effect of selection for protein on lengths of growth stages in *Glcyine max* × *Glycine soja* crosses. Can. J. Plant Sci. 62:293–298.

----, ----, and S.T. Ball. 1982. Genotype × environmental interactions for protein in *Glycine max* × *Glycine soja* crosses. Crop Sci. 22:1099–1101.

----, H.D. Voldeng, and W.D. Beversdorf. 1981. Early generation selection for protein in *Glycine max* × *G. soja* crosses. Can. J. Plant Sci. 61:901–908.

Fosberg, F.R., M.-H. Sachet, and R. Oliver. 1979. A geographical checklist of the Micronesian Dicotyledonae. Micronesica 15:41–295.

Gai, J., W.R. Fehr, and R.G. Palmer. 1982. Genetic performance of some agronomic characters in four generations of a backcrossing program involving *Glycine max* × *Glycine soja*. (In Chinese.) Acta Genet. Sin. 9:44–56.

Gamborg, O.L., R.A. Miller, and K. Ojima. 1968. Nutrient requirements of suspension cultures of soybean root cells. Exp. Cell Res. 50:151–158.

Gillett, J.B., R.M. Polhill, and B. Verdcourt. 1971. Leguminosae (part 4) subfamily Papilionoideae (2). p. 528–533. *In* E. Milne-Redhead and R.M. Polhill (ed.) Flora of tropical East Africa. [Verdcourt. B. *Glycine*]. Crown Agents, London.

Gurley, W.B., A.G. Hepburn, and J.L. Key. 1979. Sequence organization of the soybean genome. Biochim. Biophys. Acta 561:167–183.

Hadley, H.H., and T. Hymowitz. 1973. Speciation and cytogenetics. *In* B.E. Caldwell (ed.) Soybeans: Improvement, production, and uses. Agronomy 16:97–116.

Harlan, J.R., and J.M.J. de Wet. 1971. Toward a rational classification of cultivated plants. Taxon 20:509–517.

Hayata, B. 1919. Incones Plantarum Formosanarum 9:26. Bureau of Productive Industry, Formosa.

Henderson, P. 1881. Henderson's handbook of plants. Henderson and Co., New York.

Hermann, F.J. 1962. A revision of the genus *Glycine* and its immediate allies. U.S.D.A. Tech. Bull. 1268:1–79.

Hitchcock, A.S., and M.L. Green. 1947. Species lectotypicae generum Linnaei. Brittonia 6:114–118.

Ho, P.T. 1975. The cradle of the East. Univ of Hong Kong, Hong Kong.

Hood, M.J., and F.L. Allen. 1980. Interspecific hybridization studies between cultivated soybean, *Glycine max* and a perennial wild relative, *G. falcata*. Agron. Abstr. American Society of Agronomy, Madison, WI, p. 58.

Hosokawa, T. 1935. Materials of the botanical research towards the flora of Micronesia III. Trans. Nat. Hist. Soc. Formosa 25:17–39.

Hymowitz, T. 1982. Exploration for a wild relative of the soybean on Vanuatu. Naika 7 (September):1–4.

----, and C.A. Newell. 1975. A wild relative of the soybean. Ill. Res. 17(4):18–19.

---- and ----. 1980 Taxonomy, speciation, domestication, dissemination, germplasm resources and variation in the genus *Glycine*. p. 251–264. *In* R.J. Summerfield and A.H. Bunting (ed.) Advances in legume science. Royal Botanic Gardens, Kew, UK.

----, and ----. 1981. Taxonomy of the genus *Glycine*, domestication and uses of soybeans. Econ. Bot. 35:272–288.

Ingham, J.L., N.T. Keen, and T. Hymowitz. 1977. A new isoflavone phytoalexin from fungus-inoculated stems of *Glycine wightii*. Phytochemistry 16:1943–1946.

Isely, D., R.W. Pohl, and R.G. Palmer. 1980. *Neonotonia verdcourtii* (Leguminosea): A new *Glycine*-like species from Africa. Iowa State J. Res. 55:157–162.

Kaizuma, N., Y. Kiuck, and F. Ono. 1980a. Breeding of high protein soybeans from a species cross between soybeans (*Glycine max* (L.) merrill) and wild soybeans (*G. soja* Sieb. and Zucc.)—necessity of repeated backcrossing method. (In Japanese) J. Fac. Agric. Iwate Univ. 15:11–28.

----, F. Ono, J. Fukui, and K. Sakurai. 1980b. Inheritance of some qualitative characters in F_2 generation of a species cross between soybeans (*Glycine max* (L.) Merrill) and wild soybeans (*G. soja* Sieb. and Zucc.) (In Japanese.) J. Fac. Agric. Iwate Univ. 15:1–9.

Kiang, Y.T., and M.B. Gorman. 1983. Soybean. p. 295–328. *In* S.D. Tanksley and T.J. Orton (ed.) Isozymes in plant genetics and breeding, Part B. Elsevier Science Publishing Co., New York.

Lackey, J.A. 1977a. A synopsis of Phaseoleae (Leguminosae, Papilionoideae). Ph.D. Diss. Iowa State Univ., Ames (Diss. Abstr. 77-16963).

----. 1977b. A revised classification of the tribe Phaseoleae (Leguminosae: Papilionoideae), and its relation to canavanine distribution. J. Linn. Soc. Bot. 74:163–178.

----. 1977c. Neonotonia, a new generic name to include *Glycine wightii* (Arnott) Verdcourt (Leguminosae, Papilionoideae). Phytologia 37:209–212.

----. 1978. Proposal to conserve the generic name 3864 *Glycine* Willdenow over *Soia* Moench. Taxon 27:560.

----. 1980. Chromosome numbers in the Phaseoleae (Fabaceae: Faboideae) and their relation to taxonomy. Am. J. Bot. 67:595–602.

----. 1981. Phaseoleae DC. p. 301–327. *In* R.M. Polhill and R.H. Raven (ed.) Advances in legume systematics, Part 1. Royal Botanic Gardens, Kew, UK.

Ladizinsky, G., C.A. Newell, and T. Hymowitz. 1979. Wide crosses in soybeans: prospects and limitations. Euphytica 28:421–423.

Lawrence, G.H.M. 1949. Name of the soybean. Science 110:566–567.

Leschenko, A.K., and V.I. Sichkar. 1978. Distant hybrids of soybean. Soybean Genet. Newsl. 5:81–85.

Li, F., R. Chang, and S. Shu. 1983. The plants of the genus *Glycine* in China. (In Chinese.) Soybean Sci. 2:109–116.

Linnaeus, C. 1737. Genera Plantarum. 1st ed. Conrad Wishoff, Leiden, The Netherlands.

----. 1753. Species plantarum, Vol. 2. Lars Salvius, Stockholm.

----. 1754. Genera plantarum. 5th ed. Lars Salvius, Stockholm.

----. 1767. Mantissa plantarum. *In* J. Cramer and H.K. Swann (ed.) Historiae naturalis classica, Vol. 7. Lars Salvius, Stockholm. Reprint 1961. Wheldon and Wesley, and Hafner Publishing Co., New York.

Maiden, J.H. 1903. Flora of Norfolk Island. Proc. Linn. Soc. NSW 28:692–725.

Marshall, D.R., and P. Broué. 1981. The wild relatives of crop plants indigenous to Australia and their use in plant breeding. J. Aust. Inst. Agric. Sci. 47:149–154.

Merrill, E.D. 1917. An interpretation of Rumphius's Herbarium Amboinense. Bureau of Printing, Manila, Philippines.

Mies, D.W., and T. Hymowitz. 1973. Comparative electrophoretic studies of trypsin inhibitors in seed of the genus *Glycine*. Bot. Gaz. 134:121–125.

Mignucci, J.S., and D.W. Chamberlain. 1978. Interactions of *Microsphaera diffusa* with soybeans and other legumes. Phytopathology 68:169–173.

Moench, C. 1794. Methodus plantas horti botanici et agri marburgensis, a staminum situ describendi. Academy of Marburg, Marburg, West Germany.

Newell, C.A. 1981. Distribution of *Glycine tabacina* (Labill.) Benth. in the West-Central Pacific. Micronesica 17:59–65.

----, and T. Hymowitz. 1975. *Glycine canescens* F.J. Herm., a wild relative of the soybean. Crop Sci. 15:879–881.

----, and ----. 1978a. A reappraisal of the subgenus *Glycine*. Am. J. Bot. 65:168–179.

----, and ----. 1978b. Seedcoat variation in *Glycine* Willd. subgenus *Glycine* (Leguminosae) by SEM. Brittonia 30:76–88.

----, and ----. 1980. A taxonomic revision in the genus *Glycine* subgenus *Glycine* (Leguminosae). Brittonia 32:63–69.

----, and ----. 1982. Successful wide hybridization between the soybean and a wild perennial relative, *G. tomentella* Hayata. Crop Sci. 22:1062–1065.

----, and ----. 1983. Hybridization in the genus *Glycine* subgenus *Glycine* Willd. (Leguminosae, Papilionideae) Am. J. Bot. 70:334–348.

Ohwi, J. 1943. Symbolae ad Floram Asiae Orientalis X1X. Acta Phytotax. Geobot. 12:107–113.

Paclt, J. 1949. Nomenclature of the soybean. Science 109:339.

Palmer, R.G. 1965. Interspecific hybridization in the genus *Glycine*. M.S. thesis. Univ. of Illinois, Urbana.

----, and H.H. Hadley. 1968. Interspecific hybridization in *Glycine*, subgenus Leptocyamus. Crop Sci. 6:557–563.

Piper, C.V. 1914. The name of the soybean: a chapter in its botanical history. J. Am. Soc. Agron. 6:75–84.

----, and W.J. Morse. 1923. The soybean. McGraw-Hill Book Co., New York.

Pritchard, A.J., and J.G. Wutoh. 1964. Chromosome numbers in the genus *Glycine* L. Nature (London) 202:322.

Pueppke, S.G., U.K. Benny, and T. Hymowitz. 1982. Soybean lectin from seeds of the wild soybean, *Glycine soja* Sieb. and Zucc. Plant Sci. Lett. 26:191–197.

Putievsky, E., and P. Broué. 1979. Cytogenetics of hybrids among perennial species of *Glycine* subgenus *Glycine*. Aust. J. Bot. 27:713–723.

Regel, E.A. von, and R. Maack. 1861. *In* E.A. Regel (ed.) Tentamen florae ussuriensis, Ser. 7. Vol. 4 no. 4 Mem. Acad. Imp. Sci. St.-Pétersbourg, Leningrad.

Ricker, P.L., and W.J. Morse. 1948. The correct botanical name for the soybean. J. Am. Soc. Agron. 40:190–191.

Sedova, T.S. 1982. Interspecific hybridization of cultivated and wild soybean species of the subgenera *Glycine* and *Soja*. (In Russian.) Genetika 18:1532–1536.

Siebold, P.F. de, and J.G. Zuccarini. 1846. Florae japonicae. Abh. Bayer. Akad. Wiss. Math.-Phys. 4:109–204.

Singh, B.B., C.C. Gupta, and B.D. Singh. 1974. Sources of field resistance to rust and yellow mosaic diseases of soybean. Indian J. Genet. Plant Breed. 34:400–404.

Sisson, V.A., C.A. Brim, and C.S. Levings, III. 1978. Characterization of cytoplasmic diversity in soybeans by restriction endonuclease analysis. Crop Sci. 18:991–996.

Skvortzow, B.V. 1927. The soybean-wild and cultivated in Eastern Asia. Proc. Manchurian Res. Soc. Pub. Ser. A. Nat. History Sec. 22:1–8.

Smartt, J. 1984. Gene pools in grain legumes. Econ. Bot. 38:24–35.

Stalker, H.T. 1980. Utilization of wild species for crop improvement. Adv. Agron. 33:112–147.

Staswick, P.E., P. Broué, and N.C. Nielsen. 1983. Glycinin composition of several perennial species related to soybean. Plant Physiol. 72:1114–1118.

Stebbins, Jr., G.L. 1950. Variation and evolution in plants. Columbia University Press, New York.

Swain, T. 1975. Evolution of flavonid compounds. p. 1096–1129. *In* J.B. Harborne et al. (ed.) The flavonoids, Part 2. Academic Press, New York.

Sykes, W.R. 1970. Contributions to the flora of Niue. N.Z. Dep. Sci. Ind. Res. Bull. 200:1–321.

Vagera, J., and H. Hanackova. 1979. Embryo culture of soybean-culture technique and possibilities of technical application. (In Czechoslovakian). Rostl. Vyroba 25:349–360.

Vaughan, D.A., and T. Hymowitz. 1984. leaf flavonoids of *Glycine* subgenus *Glycine* in relation to systematics. Biochem. Syst. Ecol. 12:189–192.

Verdcourt, B. 1966. A proposal concerning *Glycine* L. Taxon 15:34–36.

----. 1970. Studies in the Leguminosae-Papilionoideae for the Flora of Tropical East Africa: II. Kew Bull. 24:235–307.

----. 1979. A manual of New Guinea legumes. Office of Forests, Div. of Botany. Bot. Bull. 11:1–645.

Walker, E.H. 1976. Flora of Okinawa and the Southern Ryukyu Islands. Smithsonian Institute Press, Washington, DC.

Wang, C.L. 1976. Review on the classification of soybeans. (In Chinese.) Acta Phytotaxon. Sin. 14(1):22–30.

Willdenow, K.L. 1802. Species plantarum. 3rd ed. Berlin.

Williams, E.J. 1978. A hybrid between *Trifolium repens* and *T. ambiguum* obtained with the aid of embryo culture. N.Z. J. Bot. 16:499–506.

Wolf, W.J., F.L. Baker, and R.L. Bernard. 1981. Soybean seed-coat structural features: pits, deposits and cracks. Scanning Electron Microsc. 3:531–544.

3 Vegetative Morphology

Nels R. Lersten
Iowa State University
Ames, Iowa

John B. Carlson
University of Minnesota
Duluth, Minnesota

The soybean [*Glycine max* (L.) Merr.] is an annual plant 75 to 125 cm in height (Shibles et al., 1974). It may be sparsely or densely branched, depending on cultivar and growing conditions. First-order branching of the main stem is most common; second-order branching is rare (Dzikowski, 1936). Intrinsic genetic factors and environmental effects (e.g., daylength, spacing, and soil fertility) affect branching. The possible branching variations are shown in Fig. 3-1.

The root system is best described as diffuse. It consists of a taproot, which usually cannot be distinguished from other roots of similar diameter, and a large number of secondary roots which in turn support several orders of smaller roots. In addition, multibranched adventitious roots emerge from the lower portion of the hypocotyl. The first bacterial root nodules are visible about 10 days after planting and at maturity the root system is extensively nodulated (Fig. 3-2).

The horizontal and vertical extent of the root system varies depending on cultural conditions. The taproot may reach a depth of 200 cm, and the side roots a length of 250 cm, in plants grown singly in an open field (Dzikowski, 1936), but the root system is less extensive under typical competitive conditions in field plantings. Mitchell and Russell (1971) and Raper and Barber (1970) showed that soybean grown under normal field conditions lacks a distinct taproot. Most lateral roots emerge from the upper 10 to 15 cm of the taproot and remain more-or-less horizontal but some extend obliquely to a depth of 40 to 75 cm, then turn steeply downward, sometimes reaching 180 cm. In the confined area of a rhizotron compartment, varieties differed in rates of growth, but all produced roots capable of reaching the bottom (217 cm) of each compartment (Kaspar, et al., 1978). Under almost all conditions, however, most of the roots remained in the upper 15 cm of soil.

Soybeans: Improvement, Production, and Uses, 2nd ed.—Agronomy Monograph no. 16.

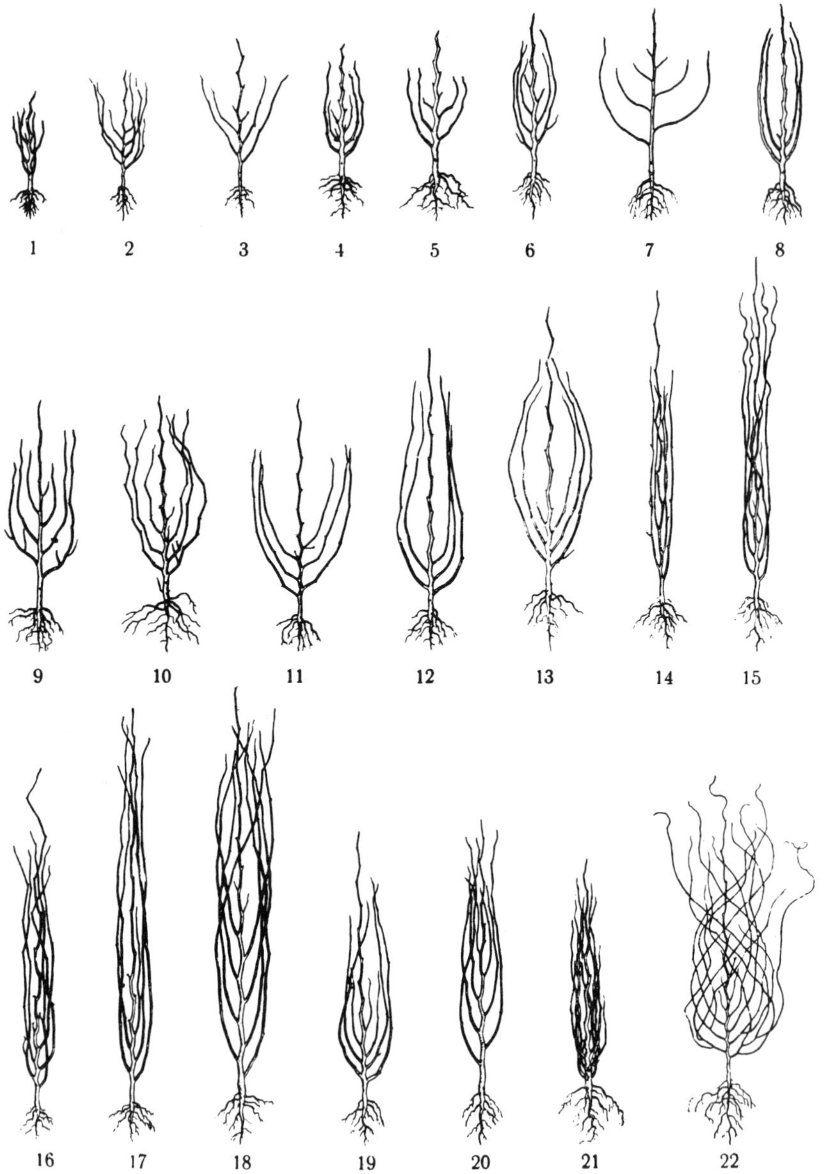

Fig. 3–1. Patterns of branching of the soybean. From Dzikowski (1936).

There are disadvantages to all extant methods of estimating the total root system (Bohm et al., 1975). An accurate three-dimensional reconstruction of the root system as it exists in the soil is perhaps impossible to attain.

Fig. 3–2. Portion of mature soybean root system bearing bacterial nodules.

3–1 LEAF

3–1.1 Gross Morphology

There are four different types of soybean leaves: the first pair of simple cotyledons or seed leaves, the second pair of simple primary leaves, trifoliolate foliage leaves, and the prophylls. Each of the oppositely arranged pair of simple primary leaves is ovate in outline form, with petioles 1 to 2 cm in length and a pair of stipules at its point of attachment to the stem. These leaves occur at the first node above the cotyledons. All leaves produced subsequently are trifoliolate and arranged alternately in two opposite rows (i.e., distichously) (Sun, 1957a). Although individual leaves do not show any preferred planes of orientation, leaves of some cultivars tend collectively to be somewhat more vertically oriented than those of other cultivars (Blad and Baker, 1972).

Leaflets of trifoliolate foliage leaves have entire margins and vary in outline from oblong to ovate to lanceolate (Dzikowski, 1936) (Fig. 3–3). Occasionally, four to seven leaflets may occur or lateral leaflets may fuse with the terminal leaflet (Fig. 3–3g and 3h) (Williams, 1950). Leaflets vary from 4 to 20 cm in length and 3 to 10 cm in width. Petiolules of lateral leaflets are 1 cm or less in length, considerably shorter than those of the terminal leaflet. The terminal leaflet has two small subtending stipels, each lateral leaflet a single one. At the base of the petiole there is a pair of stipules. These small, leaflike structures have parallel venation

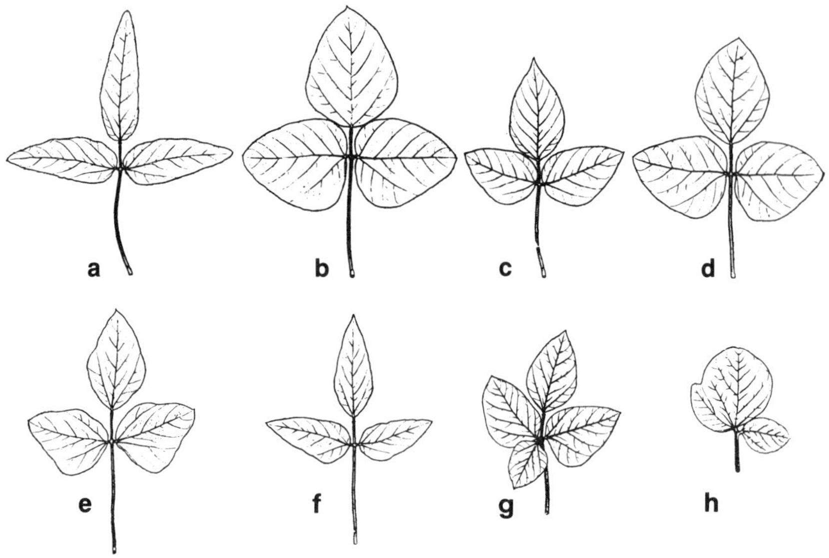

Fig. 3–3. Leaf variation in soybean, including extra leaflet (g) and fused leaflets (h). Stipules and stipels have been omitted. From Dzikowski (1936).

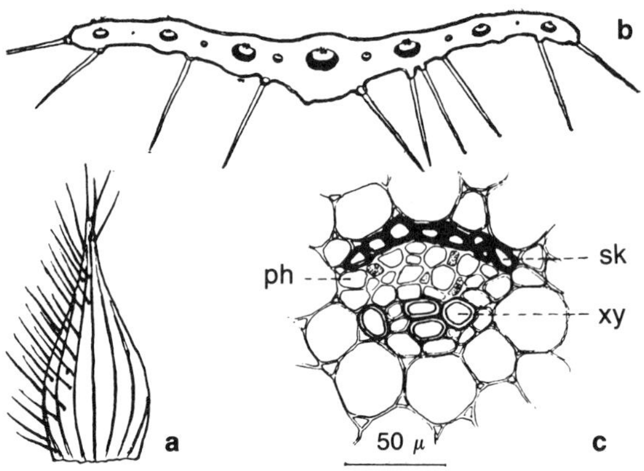

Fig. 3–4. Stipule structure. (a) Surface view showing parallel veins and long trichomes. (b) Transection. (c) Enlarged transection of one stipule bundle. Abbreviations: ph—phloem, sk—sclerenchyma, and xy—xylem. From Dzikowski (1937).

consisting usually of about seven main veins alternating with smaller veins (Fig. 3–4).

A pulvinus occurs at the base of the petiole. Another, but smaller, pulvinus is located at the base of the petiolule of each leaflet. The pulvini act as hinges, allowing an entire leaf and individual leaflets to move. Such movements are diurnal, occurring in response to changes in pulvinar osmotic pressure (Dzikowski, 1936, 1937).

The fourth type of leaf, the prophyll, occurs as a first tiny (rarely over 1 mm long) pair of simple leaves at the base of each lateral branch. Prophylls lack petioles and pulvini.

3–1.2 Petiole and Petiolules

Three vascular traces extend from the stem into the petiole at each node (Watari, 1934; Crafts, 1967). Dzikowski (1937) showed that they immediately merge into one concentric vascular bundle upon entering the basal pulvinus (Fig. 3–5a), which is an enlarged, cushion-like zone with an extensive cortex composed of parenchyma. The vascular tissue in the pulvinus consists of an outer cylinder of phloem and an inner dissected ring of xylem. Several parenchyma rays divide the xylem into wedge-shaped segments, as seen in a transectional view. The pith of the pulvinus appears similar to collenchyma tissue.

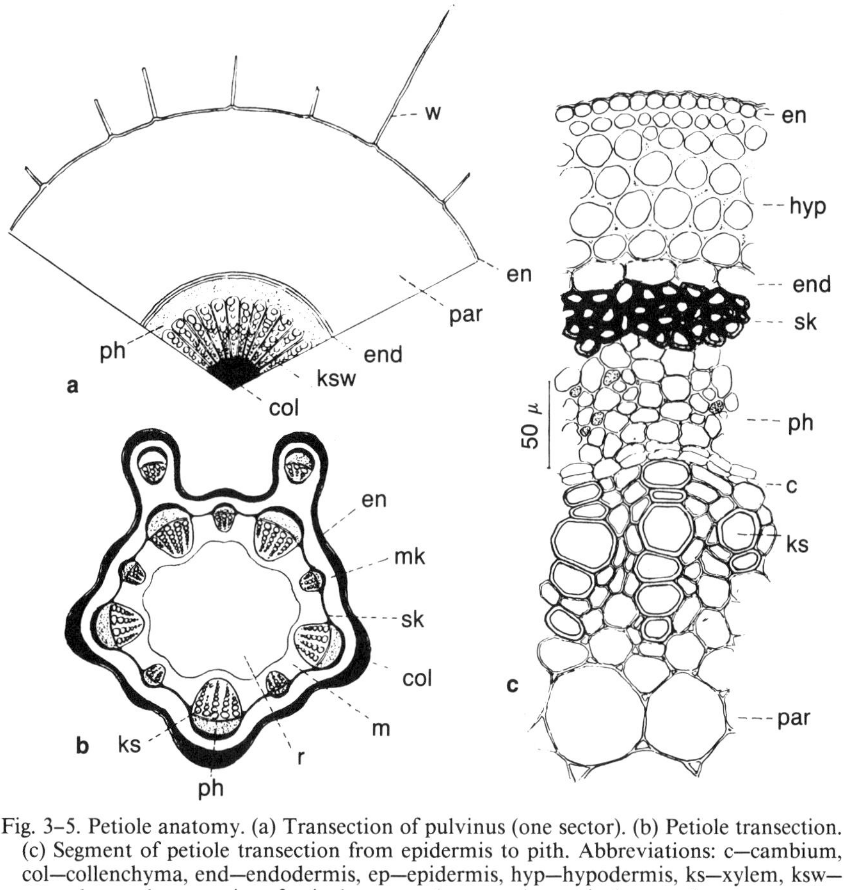

Fig. 3–5. Petiole anatomy. (a) Transection of pulvinus (one sector). (b) Petiole transection. (c) Segment of petiole transection from epidermis to pith. Abbreviations: c—cambium, col—collenchyma, end—endodermis, ep—epidermis, hyp—hypodermis, ks—xylem, ksw—secondary xylem, m—interfascicular parenchyma, mk—cortical parenchyma, par—parenchyma, ph—phloem, r—pith, sk—sclerenchyma, and w—trichomes. From Dzikowski (1937).

Just beyond the pulvinus, vascular tissue expands rather abruptly (Fig. 3–5b) to an eustele consisting of five large vascular bundles alternating with five small bundles (Dzikowski, 1937; Fisher, 1975). Two prominent vasculated ridges extend along each side of the adaxial surface of the petiole. The single vascular bundle of each ridge becomes part of the vascular tissue of the stipels at the base of each leaflet. Watari (1934) described complex vascular interconnections at the juncture of the three petiolules, beyond which about three bundles extended into each leaflet. According to Bostrack and Struckmeyer (1964), however, seven bundles enter the terminal leaflet: two supply the pair of stipels, the others merge in the leaflet pulvinus and are reduced to one strand.

3–1.3 Epidermis

The leaf epidermis consists mostly of flat, tabular cells with slightly thickened radial walls. A thin cuticle covers all epidermal surfaces (Williams, 1950). Epicuticular wax is a sparse layer of irregular rod-like particles except over guard cells and some of the immediately adjacent cells (Fig. 3–6) (Flores and Espinoza, 1977). Dzikowski (1937) noted that upper epidermal cells are somewhat larger than lower epidermal cells, and the latter also have fewer convolutions or infoldings of the radial walls.

Stomata occur on both epidermal surfaces (Fig. 3–14), but they are more abundant abaxially (Fig. 3–6). Unpublished data by Carlson indi-

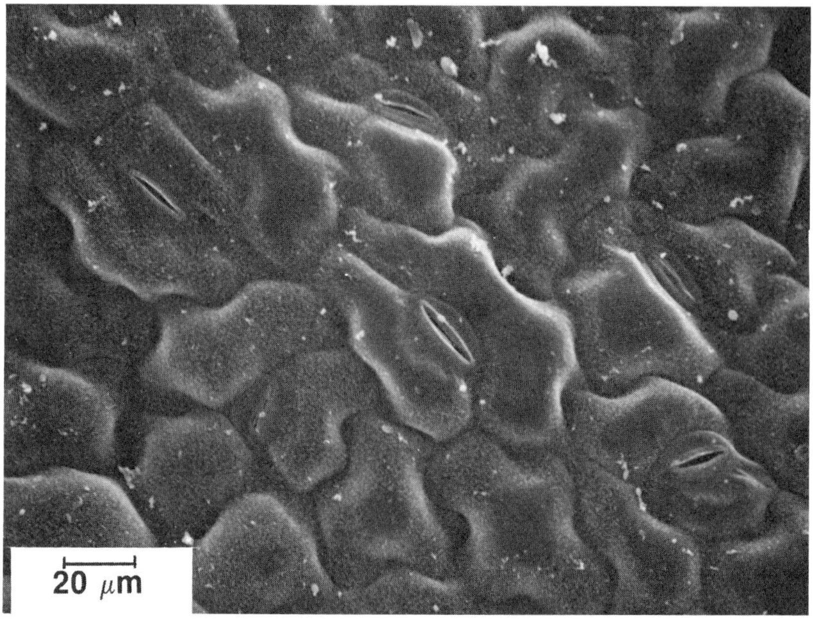

Fig. 3–6. Abaxial leaf epidermis showing stomata. Scanning electron micrograph from Flores and Espinoza (1977).

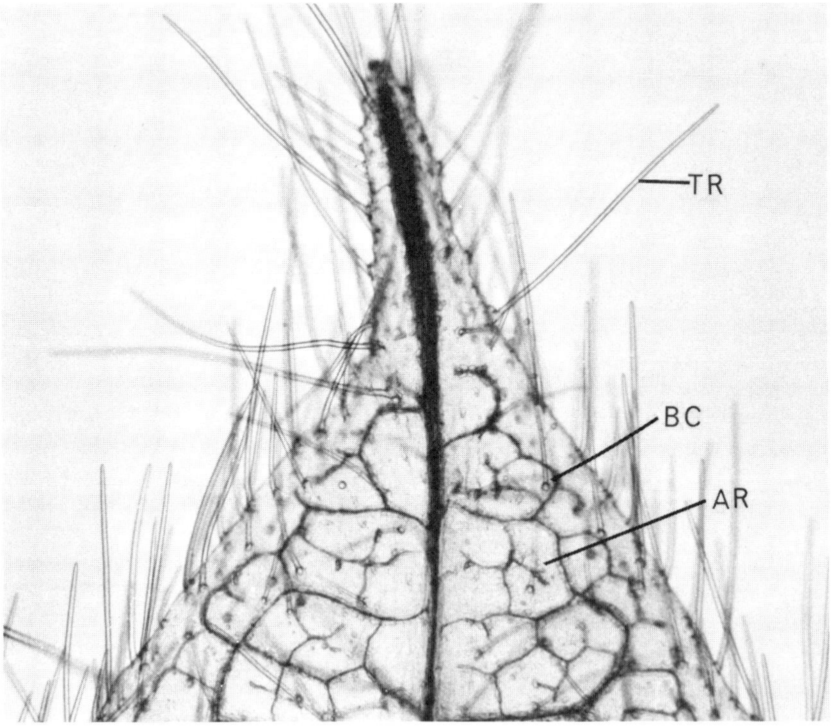

Fig. 3–7. Cleared, unstained tip of terminal leaflet of Ottawa-Mandarin cultivar showing areoles (AR) and trichomes (TR) with basal cells (BC). × 300.

cate that terminal leaflets of trifoliolate leaves of 'Ottawa-Mandarin' soybean have about three times as many stomata (17 000 stomata cm^{-2}) on the lower epidermis as on the upper (5400 stomata cm^{-2}). This agrees with data of Ciha and Brun (1975) from field-grown plants of 43 genotypes. They found variation among genotypes, but overall there was a mean number of 130 stomata mm^{-2} adaxially and 316 stomata mm^{-2} abaxially.

When closed, the two guard cells are about 12 μm wide and 24 μm long. When fully opened, the total width, including pore and guard cells, is about 16 μm. The stomatal pore is, therefore, about 4 μm, similar to measurements from photomicrographs in Flores and Espinoza (1977). There are no obvious subsidiary cells, although cells immediately adjacent to the guard cells may be more swollen and somewhat different in shape than ordinary epidermal cells (Fig. 3–6).

Trichomes vary in size, density, and color on aerial parts among soybean cultivars commonly grown in the USA. Glabrous strains occur in Japan, but none are grown commercially in North America (Bernard and Singh, 1969). Pubescent cultivars have closely packed uniseriate trichomes on young leaves and on the outer surface of stipules (Fig. 3–4, 3–7, 3–10). Woolley (1964) found that such hairs on the upper surface

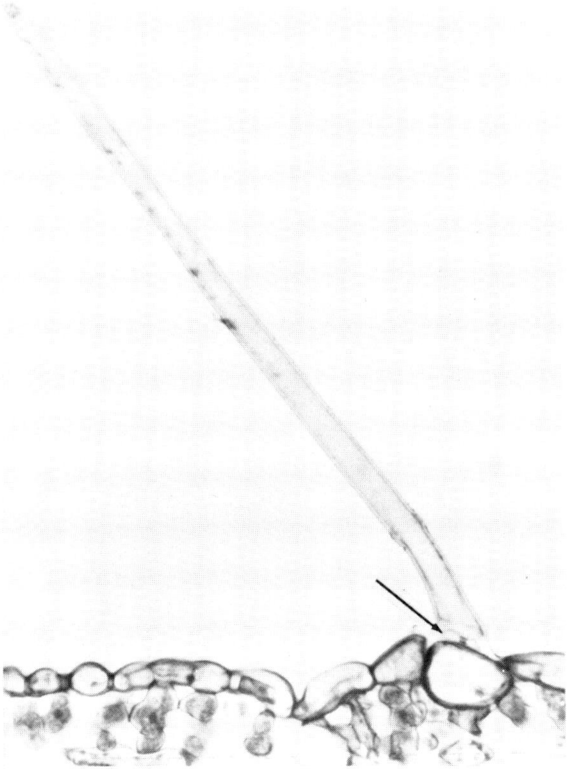

Fig. 3–8. Longisection of trichome showing long terminal cell and a short basal cell (*arrow*) attached to a somewhat enlarged epidermal cell. Compare this with Fig. 3–9.

of mature 'Hawkeye' soybean leaves were about 1 mm in length and spaced about 1 mm apart, accounting for perhaps 10% of the total leaf surface. Each hair consists of a long distal cell, 0.5 to 1.5 mm in length, and a short basal cell (Fig. 3–8) arising from a modified epidermal cell which is surrounded by a cushion of epidermal cells (Fig. 3–9) . The hairs are slanted slightly toward the tip and edges of the leaflet. Mature hairs dry out and become air-filled or flattened (Dzikowski, 1937).

In addition to these elongate uniseriate trichomes, small five-celled, club-shaped trichomes (Fig. 3–10 and 3–11) are abundant on all young organs (see Fig. 3 of Ali and Fletcher, 1970). These trichomes persist but gradually senesce on the mature leaf. Franceschi and Giaquinta (1983a) reported that the cuticle over the distal two cells becomes distended, indicating that a secretory product accumulates beneath it. They speculated from ultrastructural evidence that a volatile terpenoid compound is secreted which helps to protect developing leaflets against foraging insects.

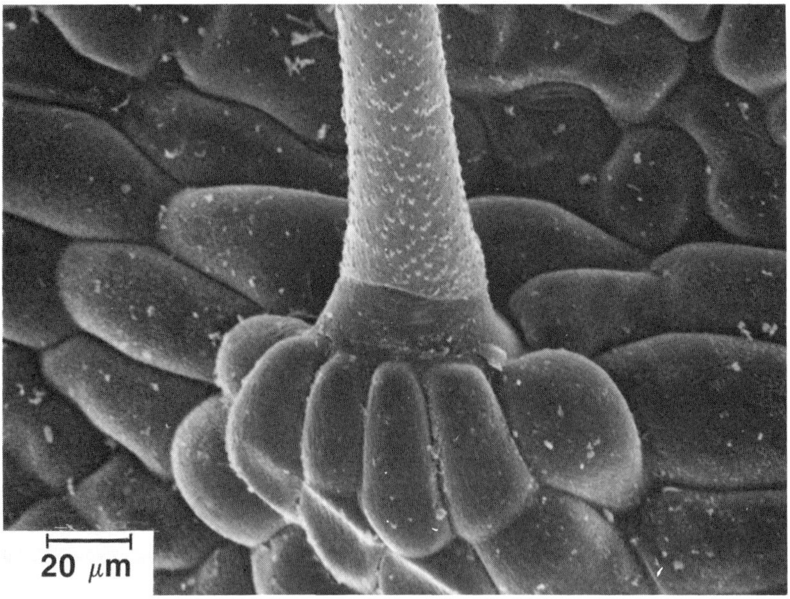

Fig. 3–9. Scanning electron micrograph of lower part of setaceous trichome. Short basal cell has thinner cuticle than distal cell. Pillow-like epidermal cells encircle trichome base. From Flores and Espinoza (1977).

3–1.4 Mesophyll

The mesophyll of a soybean leaf consists of five to six interior cell layers, except the vascular bundles (Fig. 3–12 and 3–14). The two layers just below the adaxial epidermis are palisade mesophyll, consisting of columnar cells each containing 15 to 30 chloroplasts (Duane Ford, 1983, personal communication). More insolated upper leaves may develop a third palisade layer (Lugg and Sinclair, 1980). Where stomata occur, one to three palisade cells are lacking, thus forming a substomatal chamber (Fig. 3–14). Spongy mesophyll consists of two or three cell layers interior to the lower epidermis. The cells are irregularly lobed, with conspicuous intercellular spaces and large substomatal chambers (Fig. 3–12 and 3–14), and have fewer chloroplasts than palisade cells.

Lying between palisade and spongy mesophyll is a single layer of flat, horizontally lobed cells at the level of the phloem (Fig. 3–12 and 3–13) This paraveinal mesophyll layer collects and conducts photosynthates laterally to the phloem (Fisher, 1967). It is not restricted to soybean; a similarly modified mesophyll layer occurs in other Phaseoleae and certain other legume tribes, and in a few other families (Lackey, 1978).

Paraveinal mesophyll in soybean is the first leaf tissue to differentiate; mature cells are six to eight times larger than either palisade or spongy mesophyll cells (Franceschi and Giaquinta, 1983b, 1983c). During vegetative growth, paraveinal mesophyll cells have only a few starch-free

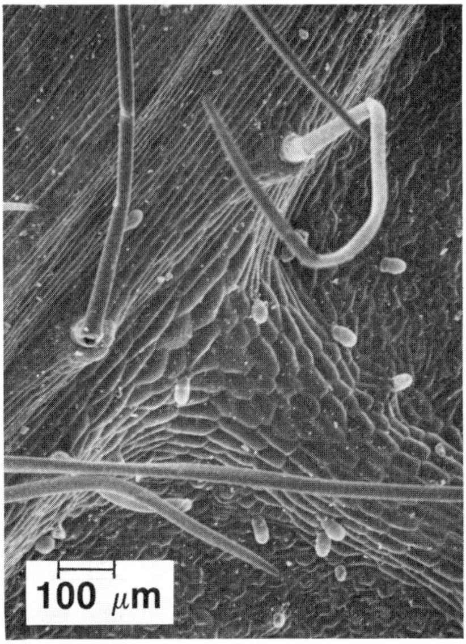

Fig. 3–10. Scanning electron micrograph of abaxial leaf surface at junction of secondary vein (*lower right*) and midvein. Long setaceous hairs and numerous small clavate hairs are present. From Flores and Espinoza (1977).

Fig. 3–11. Scanning electron micrograph of one clavate hair. Outlines of five cells can be seen. From Flores and Espinoza (1977).

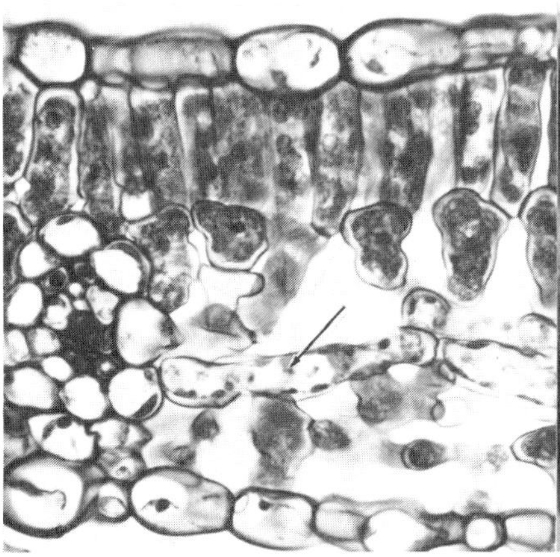

Fig. 3–12. Transection of soybean leaflet. Centrally located minor vein flanked by paraveinal mesophyll (*arrow*). Palisade mesophyll is above PVM, spongy mesophyll is below.

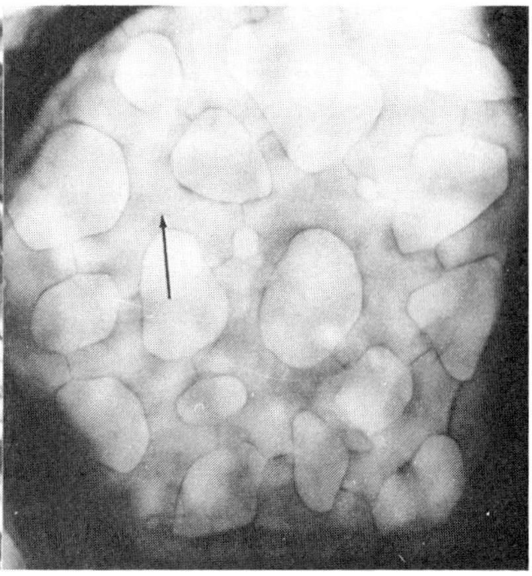

Fig. 3–13. Cleared leaflet with focus on paraveinal mesophyll, which consists of flat, lobed cells (*arrow*) with large intercellular spaces.

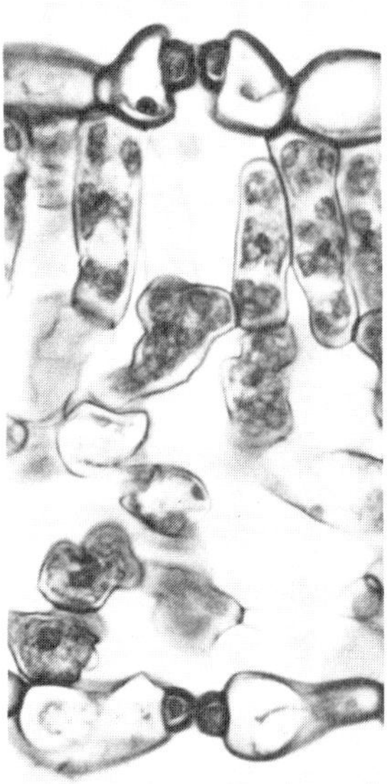

Fig. 3-14. Portion of leaf transection to show stomata on both leaf surfaces. Note substomatal chamber beneath each pair of guard cells.

chloroplasts. The large vacuole in each cell stores a glycoprotein that gradually diminishes during the first 14 days of seed filling.

3-1.5 Vascular Architecture

Each leaflet is richly vasculated (Fig. 3-15). As many as six orders of veins have been recognized (Bán et al., 1981). The midvein in the midrib develops a slight amount of secondary vascular tissue, and there is usually a small accessory bundle adaxial to it (Fig. 3-16).

Second order bundles depart alternately to each side from the midvein and extend as conspicuous lateral veins to the margin, where they form loops that usually connect to the next distal secondary bundle (Fig. 3-15). The progressively smaller orders of vascular bundles form an extensive minor vein reticulum and vein endings between second order veins (Fig. 3-17). All vascular bundles in a leaf are collateral, with adaxial xylem and abaxial phloem. The accessory bundle in the midrib, however, is inverted.

The larger veins are flanked by sclerenchyma; parenchymatous bundle sheath extensions connect these bundles above and below with each epidermis. Twin prismatic crystals occur in some of these parenchyma cells (Fig. 3-18 and 3-19), which occur in linear patches of various lengths along the third and fourth vein orders. In larger vein sheaths, crystals may occur on both sides, and considerable variation has been found in crystal distribution among cultivars (Bán et al., 1980).

In the smallest two vein orders, sclerenchyma and bundle sheath extensions are replaced by a completely encircling parenchymatous bun-

Fig. 3–15. Central leaflet with all veins included. From Bán et al. (1981).

dle sheath (Fig. 3–12). These ultimate veins and vein endings typically have only one to two xylem vessels and two to three sieve tubes. The latter are greatly reduced in diameter but their associated companion cells and phloem parenchyma cells are greatly enlarged (Fig. 3–12). This size reversal is related to the energy-intensive task of accumulating photosynthates from the surrounding mesophyll and forcing it into the sieve tubes. In many species these enlarged phloem cells also have internal convoluted cell wall ridges or pegs, which greatly increases cell membrane area and enhances their efficiency in this task. Such wall features, however, are said to be lacking in soybean and other members of the tribe Phaseoleae (Watson, et al., 1977).

3–2 STEM

3–2.1 Mature Primary Structure

Because the stem is either elongating, or increasing in diameter following internodal elongation, at any particular level it is only briefly at

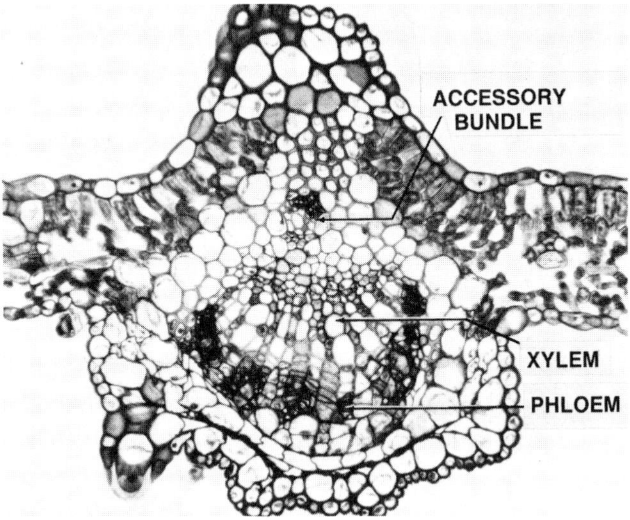

Fig. 3–16. Transection of leaflet midrib.

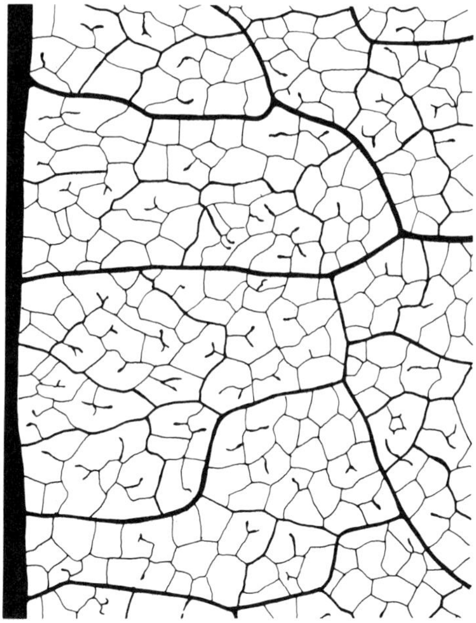

Fig. 3–17. Enlarged view of a small portion of Fig. 3–15 showing minor venation and vein endings. From Bán et al. (1981).

the stage of primary tissue maturation comparable to that of the leaf and much of the root system. For convenience, however, the stem will first be described in its mature primary condition, followed by the stages leading up to it and the subsequent events of secondary growth.

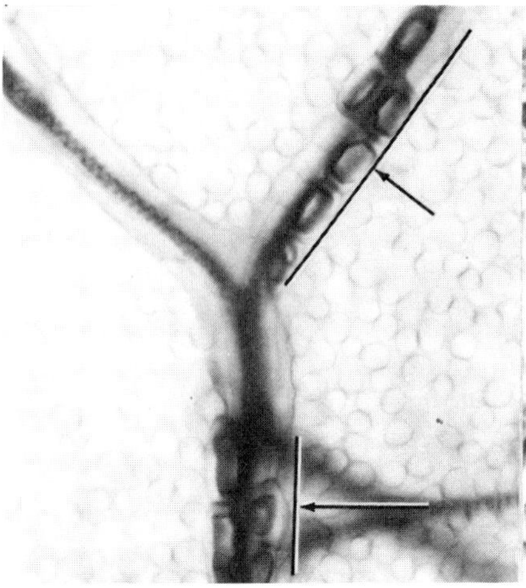

Fig. 3–18. Portion of a cleared leaflet with focus on adaxial bundle sheath cells. Twin prismatic crystals occur in some cells (*arrows*) but not others.

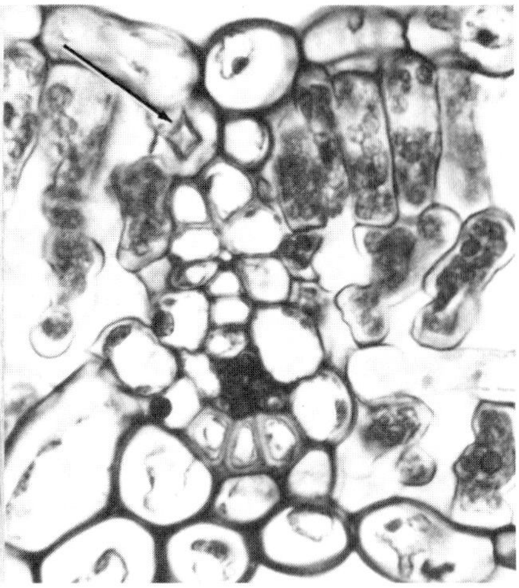

Fig. 3–19. Leaflet transection showing vascular bundle with bundle sheath extension above and below. Arrow indicates crystal in upper bundle sheath cell. Compare with Fig. 3–18.

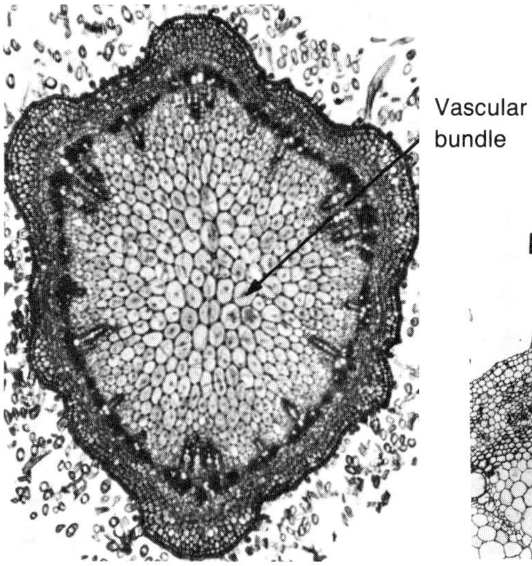

Fig. 3–20. Stem internode transection showing about 16 vascular bundles in a eustelic arrangement, with large pith and narrow cortex. From Curry (1982). × 15.

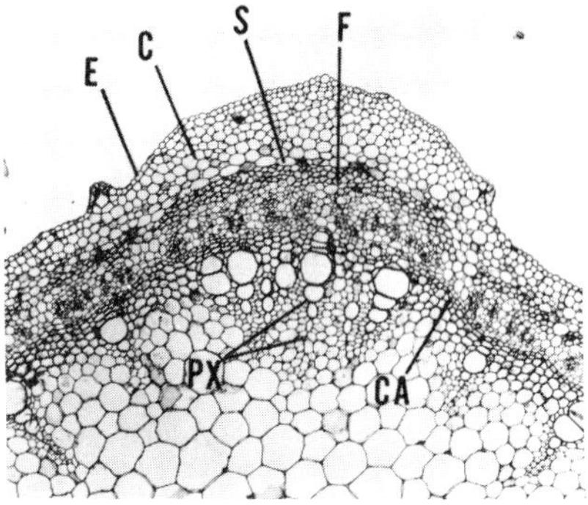

Fig. 3–21. Enlarged view of a segment of internode transection. Abbreviations: E—epidermis, C—collenchyma rib, CA—cambium, F—phloem fibers, PX—protoxylem, and S—starch sheath. From Miksche (1961).

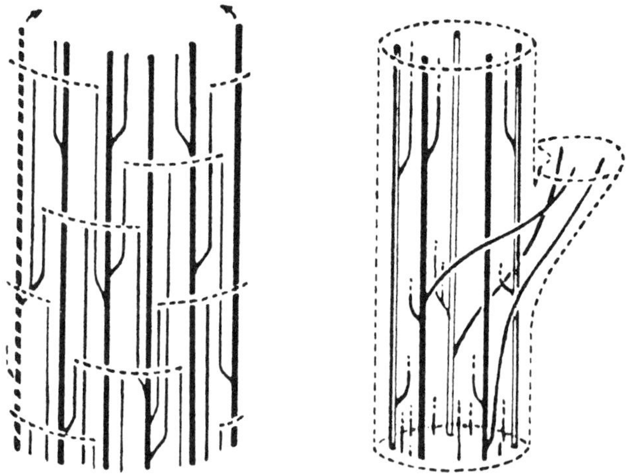

Fig. 3-22. Diagrammatic view of open type of primary stem vasculature found in soybean. Each of the three leaf traces entering the petiole (*right*) arises from a different cauline bundle in the stem (*left*).

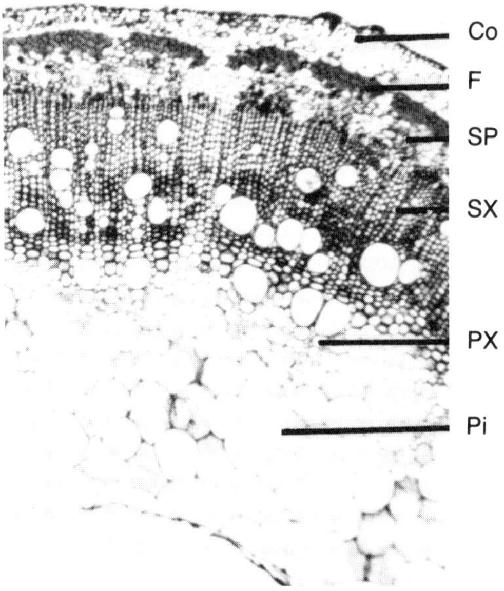

Fig. 3-23. Lower internode in which considerable secondary growth has occurred. Abbreviations: Co—cortex, F—primary phloem fibers, Pi—pith, Px—primary xylem, SP—secondary phloem, SX—secondary xylem.

Figure 3-20 illustrates a representative stem internode in which the major regions can be seen: central pith composed of large, thin-walled storage parenchyma cells; zone of vascular bundles arranged in a cylindrical pattern (eustele); and cortex, between eustele and epidermis. The stem appears somewhat irregular in outline because bulging strands of collenchyma occur to the outside of the larger vascular bundles.

The stem epidermis has the same cell types and trichomes as described earlier for the leaf. Epidermal cells are elongated and lack radial wall infoldings, and stomata are restricted to narrow strips of epidermis between collenchyma strands (N.R. Lersten, 1984, unpublished observations). The cortex includes these collenchyma strands, and below them a cylinder of chlorenchyma three to four cells thick; between collenchyma strands the chlorenchyma extends to the epidermis. Just interior to the chlorenchyma is a single-layered endodermoid layer or starch sheath, lacking chloroplasts but with abundant starch grains (Fig. 3–21). According to Cumbie (1960), these cells also lack casparian strips. The zone of pericyclic fibers (Bell, 1934) lying interior to the starch sheath was interpreted by Miksche (1961) as consisting of primary phloem fibers. The parenchyma tissue separating adjacent bundles constitutes the medullary rays, where cortex merges with pith. The central pith region is composed of large, chloroplast-free parenchyma cells.

In the internodes some vascular bundles are larger than others (Fig. 3–20). According to Curry (1982), the larger bundles are sympodia (bundles which will split off leaf traces at higher levels) whereas the smaller ones are incipient leaf traces. At each node three of these traces depart from the eustele, one from each of three sympodia. This fits the open type of stem vasculature said by Dormer (1945) to be representative of the Phaseoleae and certain other legume tribes (Fig. 3–22).

Stem vascular bundles are of the common collateral type, with a strand of xylem toward the pith flanked by a strand of phloem to the exterior, and a strip of potential cambial cells in between. Primary xylem consists of protoxylem, the ephemeral portion functioning only during internodal elongation, and metaxylem, the later-maturing permanent water-conducting tissue. Phloem is similarly divided into protophloem and metaphloem, although they are difficult to distinguish at bundle maturity because the thin-walled protophloem cells are largely obliterated by elongation and lateral expansion of surrounding tissue. Also considered to be part of the phloem (Miksche, 1961) is the bundle cap, a strand

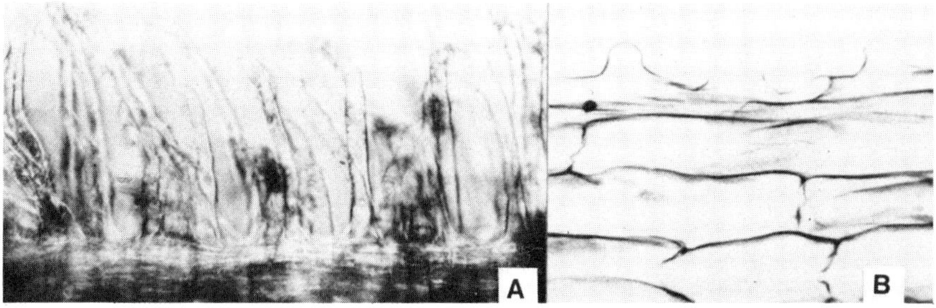

Fig. 3–24. (A) Root hairs extending from epidermis. × 120. (B) Surface view of root epidermis. From Anderson (1961).

of sclerenchyma fibers just to the exterior of the protophloem (Fig. 3–21).

The ultrastructure of soybean phloem has been described (Fisher, 1975; Wergin and Newcomb, 1970). The most notable feature is the large fusiform P-protein crystal that develops in each sieve tube member.

Soybean vascular bundles have xylem transfer cells with extensive wall ingrowths in the form of ridges or papillae (Kuo et al., 1980). Xylem transfer cells occur at nodes, where they are thought to aid the transfer of water and nutrients to axillary buds. Equally well-developed xylem transfer cells also occur in all vascular bundles in the mid-internode, where they may possibly be involved in the transfer of nitrogenous solutes from xylem to phloem within a bundle.

3–2.2 Secondary Growth

As more stem tissue, leaves and, later, reproductive structures continue to be added as the plant grows, lower portions of the stem must accommodate them by adding more vascular tissue and supporting tissue through the activity of the vascular cambium. This cylindrical meristem is initiated by fascicular cambium located between primary xylem and phloem within each bundle. These meristematic strips become connected later by interfascicular cambium, meristematic strips that form between each pair of bundles. During early cambial activity, secondary xylem and phloem are added only within existing bundles (Fig. 3–20). In nodes and internodes near the ground, where it continues for a long time, a complete cylinder of secondary xylem and phloem is produced (Cumbie, 1960) (Fig. 3–23).

In the lower internodes of stems that have undergone considerable secondary growth, pith cells collapse and the stem becomes hollow. Primary xylem strands are still present but considerable secondary xylem has been added. It consists of large pitted vessels and numerous smaller vessels and tracheids, diffuse-apotracheal axial parenchyma, sclerenchyma fibers scattered in patches, and uniseriate to multiseriate xylem rays (Datta and Saha, 1971). The primary phloem fibers are conspicuous even in these older woody stems. No periderm has been reported (Cumbie, 1960).

A sequence of cross sections from the lowest portion of the stem upward would show progressively less secondary growth, especially following the onset of flowering, and the upper internodes would have little or no secondary tissue. Struckmeyer (1941) noted that, prior to flowering, considerable cambial activity is evident in the stem. During flowering, however, cambial activity decreases and cells that had previously been produced by cambial activity develop thickened walls.

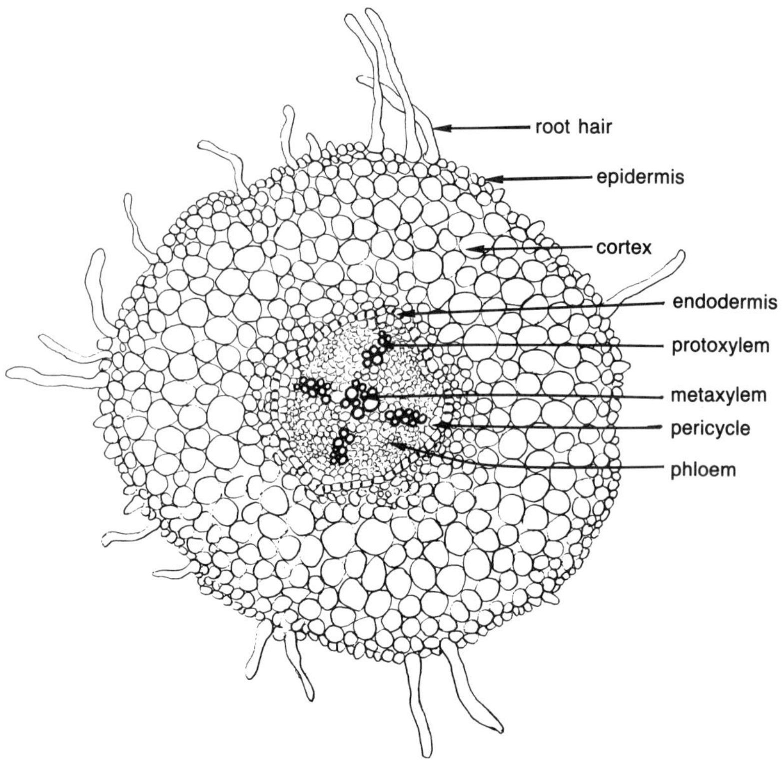

Fig. 3–25. Transection of primary root, 2 cm from apex. Camera lucida drawing of free-hand section. × 8.

3–3 ROOT SYSTEM

3–3.1 Primary Structure

The uniseriate epidermis consists of tabular cells elongated along the root axis (Fig. 3–24B). Cell walls are usually thin but outer and radial walls may be somewhat thickened. Intercellular spaces are lacking. Any epidermal cell may form a root hair; these first appear 4 days after germination about 1 cm behind the primary root tip (Anderson, 1961). Branch roots also form root hairs (Fig. 3–24A). Dittmer (1940), in a study of mature field-grown 'Illini' soybean, found root hairs on all roots except the tap root, where secondary growth had removed the epidermis.

Little is known about numbers of root hairs and the total root hair surface of field-grown soybean. Table 3–1 presents data from core samples of field-grown soybean. It is not possible from these data to estimate total root hair number or total area of absorbing surface for the entire root system.

The cortex is the broad zone between epidermis and stele. In the primary root it consists of 8 to 11 layers of slightly elongated cells with much intercellular space (Fig. 3–25). In branch roots the narrower cortex

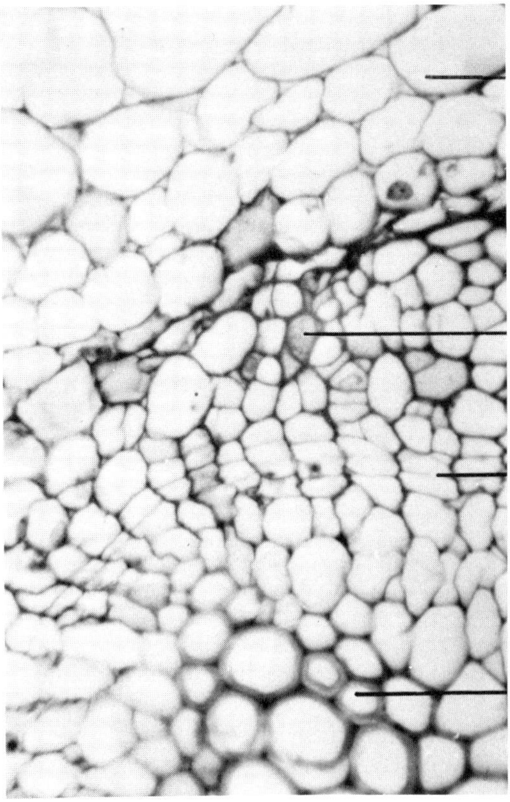

Fig. 3–26. Portion of primary root with newly formed vascular cambium. Abbreviations: camb—vascular cambium, co—cortex, px—protoxylem, and svpl—sieve plate. × 390.

Table 3–1. Data on roots and root hairs of mature, field-grown (Illini) soybean. From Dittmer (1940).

	Avg root diam	Avg root hair length × diam	No. root hairs mm^{-1} root length
	—mm—	—μ—	
Taproot†	2.50	—	—
Secondary roots	0.65	110×17	606
Tertiary roots	0.31	90×14	210
Quaternary roots	0.23	90×14	170

†Secondary growth resulted in loss of most of epidermis and root hairs.

has four to nine layers. Cortical cells close to the epidermis or near the endodermis are considerably smaller than intervening cortical cells. Little or no starch is stored in the cortex, because annual plants store their reserves mostly in seeds. The innermost cortical cell layer is the endodermis, with a continuous suberized casparian strip encircling its radial walls. The casparian strip is recognizable about 2.3 cm from the root

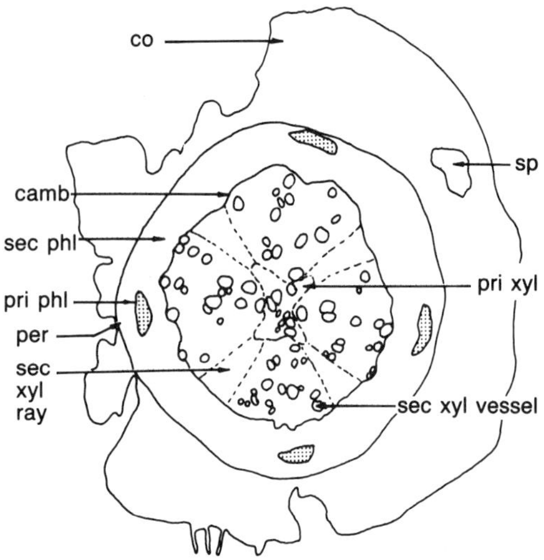

Fig. 3–27. Diagram of root transection with considerable secondary growth. Only largest vessels are indicated. Abbreviations: camb—cambium, co—cortex sloughing off, per—periderm, pri phl—primary phloem, pri xyl—primary xylem, sec phl—secondary phloem, sec xyl—secondary xylem. × 38.

apex (Sun, 1955). No other thickenings appear in the endodermis as the root matures.

The stele consists of phloem, xylem, associated stelar parenchyma, and surrounding pericycle (Byrne et al., 1977a). Xylem parenchyma cells have transfer cell-like wall ingrowths (Lauchli et al., 1974) which probably allow them to actively secrete ions into the xylem sap or to re-absorb ions from the sap. The pericycle delimits the stele; it has been reported to be two- to three-layers thick in places but Byrne et al., (1977a) described it as only one-layer thick. Pericycle cells have thin walls and they retain their meristematic potential during the maturation of xylem and phloem.

The xylem is almost always tetrarch (Bell, 1934; Byrne et al., 1977a, 1977b). Occasional lateral roots may be triarch, a condition also induced by pathogens (Byrne et al., 1977a, 1977b). Phloem occurs as a single strand midway between each pair of protoxylem ridges (Fig. 3–25).

Lateral roots are protuberances from the stele. They traverse the cortex but have no physiological connection with it. Lateral roots are diarch at their base because they are connected to one protoxylem ridge and adjacent metaxylem. The diarch xylem is surrounded by a cylinder of phloem arising from the two adjacent phloem strands. This stelar arrangement extends about 500 μm into the lateral root, where reorganization into a tetrarch pattern occurs and the root enlarges (Fig. 3–46). The lateral root distally is similar anatomically to comparable levels in the parent root (Byrne et al., 1977a).

The anatomy of individual roots has been shown to change in compacted soils. Baligar et al. (1975) reported these qualitative changes: wavy epidermis with mostly ruptured cells, more spherical cortical parenchyma with increased number of intercellular spaces, change in stelar shape from cylindical to somewhat flattened, and increased wall thickening in xylem cells.

3–3.2 Secondary Growth of Root

Secondary growth in soybean roots follows a common pattern among dicots. Vascular cambium is first visible in 4-day seedlings as bands of tangentially dividing stelar parenchyma cells between phloem and xylem 3 to 5 cm behind the root apex (Anderson, 1961; Sun, 1955). These cambial bands expand laterally by divisions in the pericycle opposite the xylem ridges; this activity results eventually in a continuous cambial layer surrounding the xylem (Fig. 3–26). Acropetal differentiation of the cambium continues as the primary root elongates.

Continued secondary growth eventually produces a cylinder of secondary xylem around the primary xylem, which is in turn surrounded by secondary phloem (Fig. 3–27 and 3–28). The characteristic tetrarch primary xylem pattern is somewhat obscured as secondary xylem is added, and the only evidence of primary phloem is the presence of groups of four strands of primary phloem fibers (Fig. 3–27 and 3–28). Secondary xylem consists of pitted vessels, parenchyma cells that may become sclerified, fibers, and parenchymatous rays that are most prominent opposite the original protoxylem ridges. Secondary phloem has parenchyma, sieve tubes, companion cells, and fibers. The cortex develops large intercellular spaces because parenchyma cells are disrupted as the stele increases in diameter.

Where extensive secondary growth occurs, such as in the upper portion of the taproot and the older secondary roots, the epidermis, cortex, and endodermis are disrupted and sloughed off. In these roots, the peri-

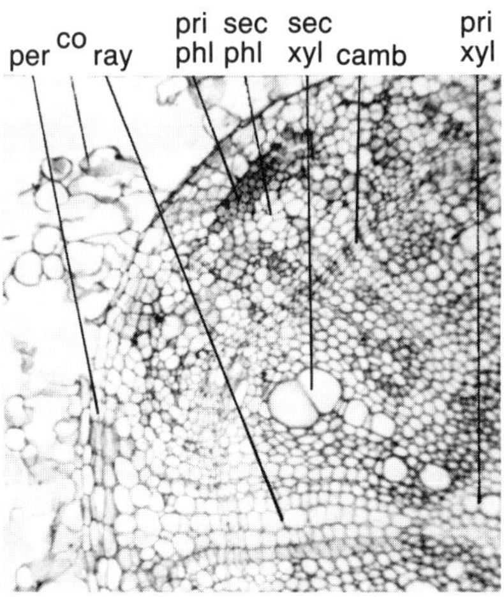

Fig. 3–28. Photomicrograph of Fig. 3–27. Labels same as Fig. 3–27.

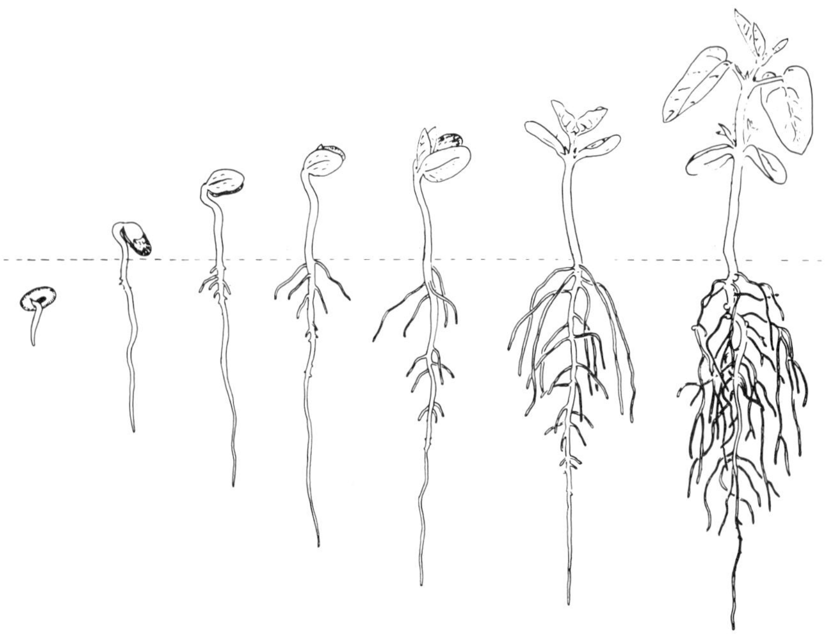

Fig. 3–29. Stages in germination and early seedling growth. Dotted line indicates soil level. Modified from Dzikowski (1936).

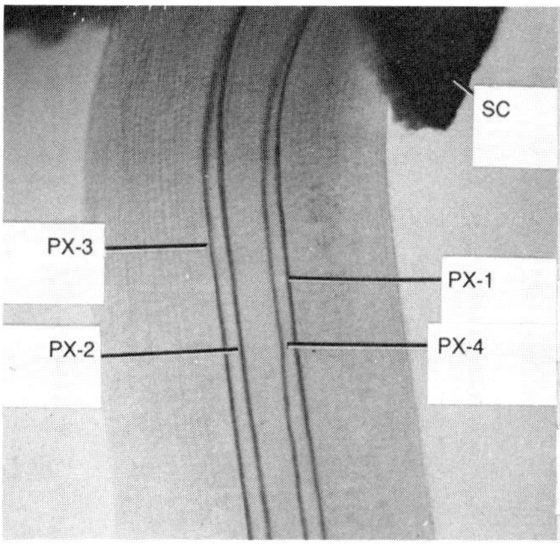

Fig. 3–30. Cleared, unstained upper hypocotyl 36 h after planting. PX indicates protoxylem ridges 1–4, SC—seed coat. × 14.

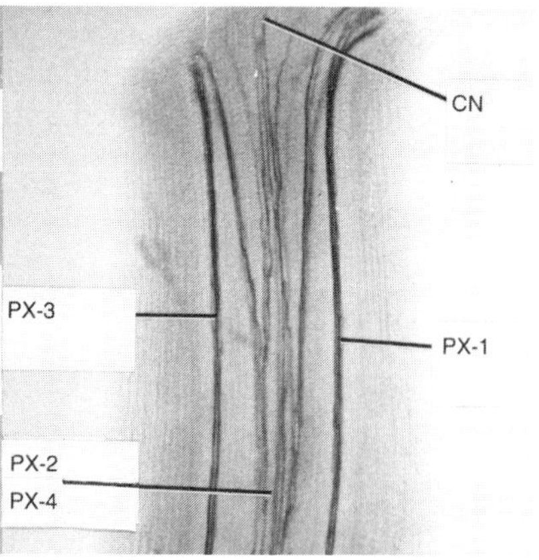

Fig. 3–31. Cleared, unstained upper hypocotyl 48 h after planting. Protoxylem strands PX-1 and PX-3 are seen from the side and appear as single strands each. PX-2 and PX-4 are in face view and appear superimposed. They now have additional primary xylem elements that pass into the cotyledonary node, CN. × 14.

cycle forms a cork cambium that divides periclinally and produces a protective periderm (Fig. 3–27 and 3–28) around the vascular tissue.

3–4 GERMINATION AND SEEDLING GROWTH

3–4.1 Establishment

Seeds of most soybean cultivars imbibe water rapidly following planting. Some genotypes, however, especially wild types, have a large proportion of hard seeds that are slow to take up water. The collective results of many studies on legumes indicate that water enters the otherwise impenetrable seed through the lens, a local area of modified palisade epidermal cells that occurs on the opposite side of the hilum from the micropyle (Gunn, 1981).

Under favorable conditions, the radicle emerges from the seed in 1 or 2 days (Dzikowski, 1936; Williams, 1950). Downward growth of the radicle (now called the *primary root*) is rapid and by the 4th or 5th day the first secondary roots emerge 4 to 5 cm behind the primary root apex (Anderson, 1961; Sun, 1955). The cotyledons emerge in 3 to 4 days, and are pulled out of the soil by rapid growth in the doubled-over crook in the upper region of the hypocotyl. Further growth and straightening of the hypocotyl elevates the pale cotyledons, which soon become green and photosynthetic in addition to supplying stored reserves to the seedling. Later, the cotyledons turn yellow and fall off.

Liu et al. (1971) found that microbodies, which they assumed to be glyoxysomes, appear in cotyledon cells about 48 h after planting. As cotyledons emerge and chloroplasts become functional, other microbodies, assumed to be peroxisomes, are associated with the chloroplasts. They speculated that the glyoxysomes help to convert stored lipids to hexose sugar, whereas the peroxisomes indicate that peroxisomal photorespiration occurs in cotyledons. As cotyledons senesce, both chloroplasts and peroxisomes show symptoms of degeneraton.

After cotyledons emerge, the two primary leaves expand from the apical bud and mature within a few days. Further growth of the seedling involves the formation of trifoliolate leaves. These early growth stages are shown in Fig. 3–29.

The vascular pattern of the hypocotyl, in the dormant embryo and up to about 36 h after germination, consists principally of a tetrarch arrangement of xylem. Each protoxylem ridge is prominent, with exarch maturation. At the distal end of the hypocotyl the four xylem ridges merge into two—one for each cotyledon (Fig. 3–30).

The characteristic crook in the upper hypocotyl develops because of increased mitotic activity (Bell, 1934). As new cells are added here, more vascular tissue develops. Figure 3–31, of the upper portion of the hypocotyl of Ottawa-Mandarin soybean 48 h after germination, shows additional vascular elements which have arisen from the procambium of the arch.

Fig. 3–32. Representation of vascular transition in soybean root. (Measurements in parentheses show lengths from each level to next above.)
1. Tetrach, exarch radial arrangement of root (centripetal differentiation). (8 mm.)
2. Vascular cambium internal to primary phloem strands. Xylem strands triangular in cross section. Phloem strands larger in diameter. (5.5 mm.)
3. Xylem strands show doubleness resulting from change in direction of differentiation from centripetal to tangential. Locations of the four points of first primary xylem differentiation have persisted and still occupy positions near the periphery of the axis. (11 mm.)
4. Phloem strands expanded tangentially and reduced radially. There are now eight points of first xylem (all primary) differentiation, that is, a dichotomy of primary xylem strands has occurred so that each tangential wing of primary xylem is now a separate strand. Vascular cambium does not occur at this level or above. (23 mm.) (This pattern was about 30-mm long, over half the distance from level 1 to 11.)
5. Strands differentiated in a concave pattern, with the latest formed vessels indicating a change from exarch to endarch differentiation. The points of initial primary xylem differentiation are still exterior. The convex strands of phloem have been extended tangentially. (2 mm.)
6. Primary xylem has established contact with phloem, the points of initial xylem formation have shifted inward and phloem strands have fused into two broad convex strands, leaving medullary rays in the cotyledonary plane only. Two small endarch xylem strands have differentiated against the phloem strands occupying positions in the intercotyledonary plane. These bundles of xylem are the traces to the plumule. (0.85 mm.)
7. The two endarch plumular traces are distinct from the four primary xylem strands with which they were associated at their origin. In addition to becoming entirely separate the *plumular* traces have increased in magnitude and appear double. This does not result in two strands. (2 mm.)
8. The plumular traces are clearly single collateral bundles. Formation of four new medullary rays at this level has transformed the two phloem strands into six, two of which are associated with plumular xylem strands. Each of the other four is associated with two of the primary xylem strands which have persisted from level 4. These eight xylem strands are almost completely endarch at level 8. (6 mm.)
9. Anastomosis of the phloem strands of the nonplumular bundles results in formation of two opposite collateral bundles, continuous with the cotyledonary bundles. Each exhibits four distinct xylem strands, which are entirely endarch at this level. The xylem strands of the two plumular traces have trifurcated, forming lateral strands of lesser magnitude than the central ones. (2 mm.)
10. Further development of the cotyledonary bundles. The xylem strands have fused in pairs, transforming the bundles into monophloic, bixylary units. Trifurcation of the phloem of the plumular traces has resulted in the establishment of six collateral plumular traces. Central traces are somewhat more massive than the lateral ones. (0.28 mm.)
11. Immediately below the cotyledonary node. Fusion of the xylem strands of the cotyledonary bundles has produced two massive bundles. The plumular traces are of about equal size and form groups of three in the intercotyledonary plane.

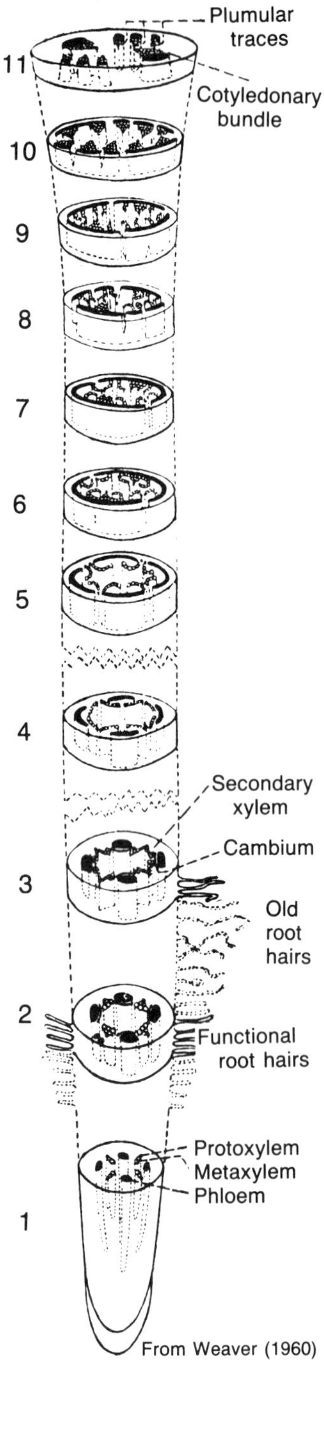

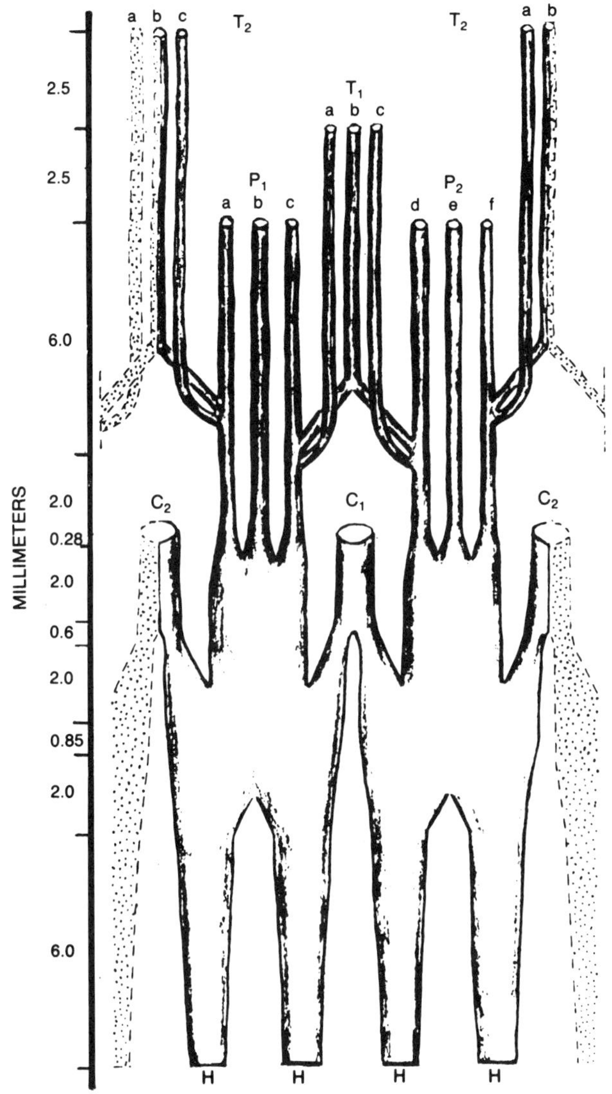

Fig. 3–33. Diagram of phloem distribution in a seedling from hypocotyl to second pair of trifoliolate leaves. From Crafts (1967).

Figure 3–32 is a diagrammatic representation of the vascular transition from the root apex through the hypocotyl and up to the cotyledonary node; the figure legend describes the salient features of the changing vascular pattern. The most significant fact is that all of the primary xylem of the primary root and hypocotyl passes into the two cotyledonary vascular bundles. There is also a gradual transition from the exarch pattern of xylem maturation characteristic of the root to the endarch maturation pattern characteristic of leaves and stem (Level 10, Fig. 3–32).

Weaver (1960) and Compton (1912) showed that xylem passing from the hypocotyl into the epicotyl and the primary leaves arises as a new tissue superimposed upon the preexisting xylem tissue of the root-hypocotyl-cotyledonary axis. These new vascular bundles are endarch in their pattern of maturation. Weaver noted that epicotylar phloem, unlike the xylem, is not superimposed upon the hypocotylar phloem. Some of the primary phloem present in the hypocotyl axis of the embryo extends up into the epicotyl during seedling development. Crafts (1967) reconstructed phloem distribution from hypocotyl to second trifoliolate leaf (Fig. 3–33), showing that primary phloem is continuous from the root up through the epicotyl. The upper portion of the embryo axis in the dormant embryo terminates in a short epicotyl consisting of two simple leaves, the primordium of the first trifoliolate leaf, and the stem apex.

Miksche (1961) reported that in the dormant seed there are two unifoliolate leaves with conduplicate vernation and two stipules at the base of each leaf. The first mitotic activity in the embryo occurs about 36 h after planting, in procambial elements within the corpus. After 40 h, divisions were noted in the tunica. By 48 h, mitotic figures were visible in the mesophyll of the unifoliolate leaves, the epidermal hairs, and in the primordium of the first trifoliolate leaf.

Soybean seedlings have red pigment bodies restricted to the subepidermal layer of the hypocotyl and young stem (Nozzolillo, 1973). The pigment is malvadin, known so far only from members of the tribe Phaseoleae.

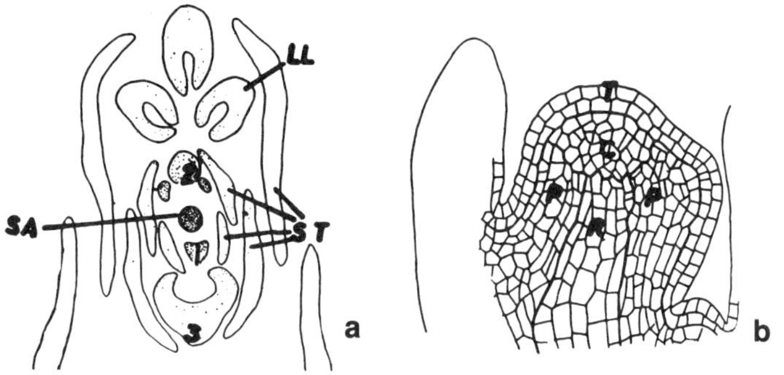

Fig. 3–34. (a) Transection of stem tip showing distichous leaf arrangement in alternate phyllotaxy. 1–3 indicates progressively older leaf primordia. Abbreviations: LL—leaflets of fourth leaf, SA—shoot apex, and ST—stipules. (b) Longisection through primordia 1 and 2 of Fig. 3–34a. Shoot apex shows tunica (T) and three zones within corpus: C—central initiation zone, P—peripheral meristem, R—rib meristem. From Sun (1957a).

3–5 PRIMARY MERISTEMS AND ORGAN DIFFERENTIATION

3–5.1 Stem Apical Meristem and Primary Stem

Sun (1957a) reported that the shoot apex of cv. Gibson consists of a two-layered tunica and a rather massive corpus subdivided into three areas: central initiation zone of rather large cells, peripheral zone of somewhat smaller cells, and a rib meristem immediately below the central initiation zone (Fig. 3–34). Grandet (1955) described two tunica layers and a single massive corpus in cv. Rouest 250, as did Bostrack and Struckmeyer (1964) in Hawkeye soybean. All of these investigators studied only seedlings.

Miksche (1961) found only a single-layered tunica and a massive corpus in the dormant embryo. Figure 3–35 shows two periclinal divisions in the layer of cells below the outer tunica layer. These divisions are either associated with the initiation of the first trifoliolate leaf, or else they indicate a change from a one-layered to a two-layered tunica as the embryo germinates.

Below the shoot apex, cell divisions virtually cease in the ground meristem as pith and cortical cells enlarge greatly. In the procambium, however, new cells produced by periclinal divisions differentiate into protophloem and protoxylem. These ephemeral conducting strands are obliterated by internodal elongation. As an internode attains maximum length, most of the remaining procambium differentiates into metaphloem and metaxylem. Figure 3–36 illustrates early stages of internodal elongation and concomitant stem differentation.

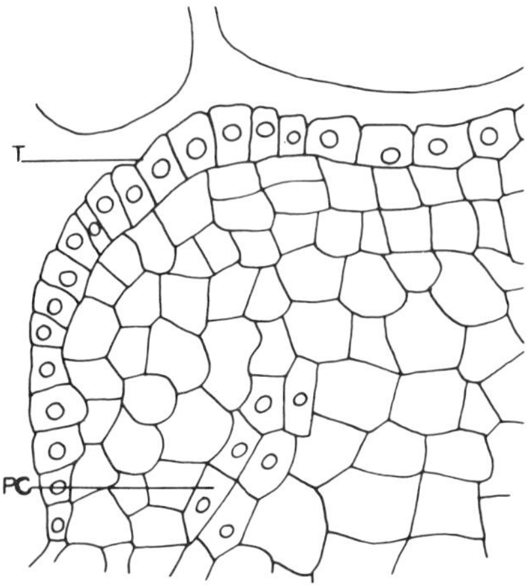

Fig. 3–35. Portion of stem apex region of dormant embryo showing periclinal walls in the layer below the tunica (T), and a procambial strand (PC) extending toward the first trifoliolate leaf primordium. From Miksche (1961).

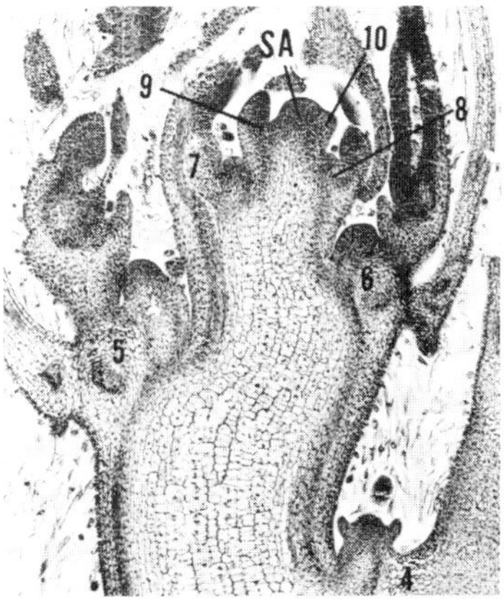

Fig. 3–36. Median longitudinal section of stem 21 days after planting. First three trifoliolate leaves now shown, others numbered consecutively from oldest (4) to youngest (10). Abbreviation: SA—shoot apex. From Miksche (1961).

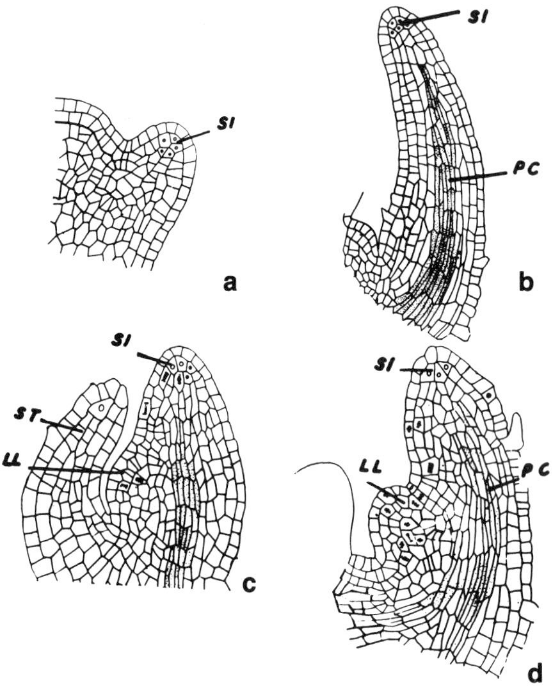

Fig. 3–37. Early stages of trifoliolate leaf development. (a) Very young stage, before vasculation. (b) Later stage, with procambium in future midrib. (c) Lateral leaflet initiation. (d) Young lateral leaflet. Abbreviations: LL—lateral leaflet primordium, PC—procambium, SI—subapical initials (nucleated cells), and ST—stipule. From Sun (1957a).

Axillary bud primordia are already present in the axil of the second trifoliolate leaf primordium below the stem apex (Sun, 1957a). Each axillary bud primordium is initiated by periclinal divisions of several cells in the outer layer of the corpus, and later involves divisions of the tunica as well.

3–5.2 Apical Meristems and Leaf Development

The initiation and development of trifoliolate leaves and axillary buds were described by Sun (1957a) and Decker and Postlethwait (1960). Figure 3–37 shows stages in the ontogeny of a trifoliolate leaf. The first evidence of leaf initiation occurs 30 to 50 μm below the summit of the shoot apex (Voroshilova, 1964), from periclinal divisions in the second tunica layer and divisions in the corpus just below this second tunica layer. The outer tunica layer divides only anticlinally and thereby maintains itself until it matures as the epidermis.

Shortly after a leaf primordium is initiated, it is penetrated by a procambial strand developing upward from below the shoot apex. When the primordium reaches 60 to 80 μm in length, stipules are initiated at both sides of its base by cell divisions in the outer tunica layer. The

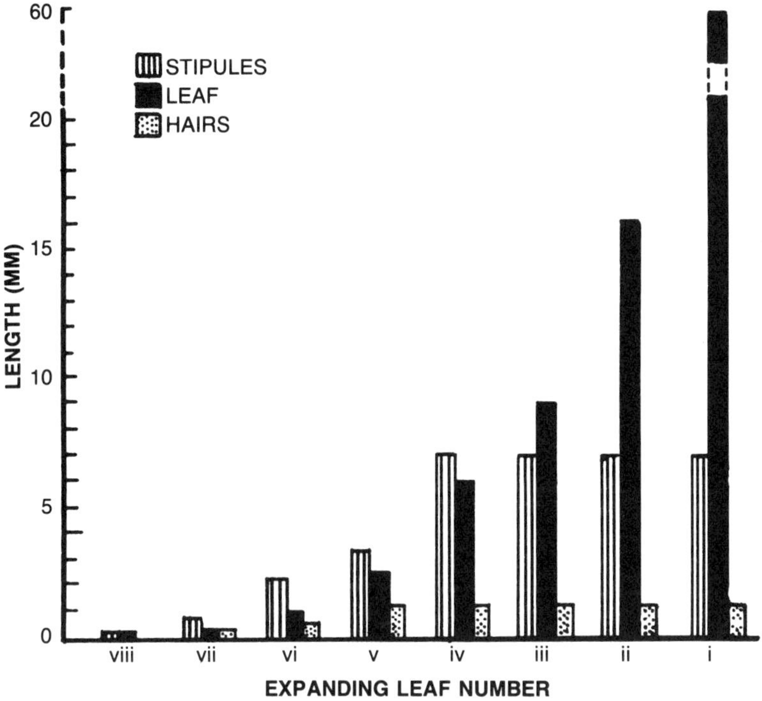

Fig. 3–38. Length of stipules, leaf, and hairs of successive developing soybean leaves. From Thrower (1962).

Table 3–2. Young, expanding soybean leaves. From Thrower (1962).

Expanding leaf no. (see Fig. 3–38)	Leaf length	Leaf form	Length of stipules	Length of leaf hairs	Area of leaf
	mm		——mm——		% of adult area
i	59.0	Trifoliate	7.0	1.3	31
ii	16.0	Trifoliate	7.0	1.3	<2
iii	9.0	Trifoliate	7.0	1.3	<1
iv	6.0	Trifoliate	7.0	1.3	<1
v	2.3	Transitional	3.4	1.1	<1
vi	0.9	3-lobed	2.2	0.6	<1
vii	0.2	3-lobed	0.7	—†	<1
viii	0.1	Single lobe	0.1	—‡	<1

†Hairs present only on stipule: hairs 0.2-mm long. ‡No hairs present.

adjacent outer part of the corpus soon contributes additional cells. The stipules grow faster than the leaf primordium until the lateral leaflets are initiated (Fig. 3–37c). When the leaf primordium is 140 to 200 μm in length, two new meristems form on opposite sides toward its base, thus initiating the two lateral leaflets. Table 3–2 and Fig. 3–38 show the relative development of the leaf proper, stipules, and hairs at different stages

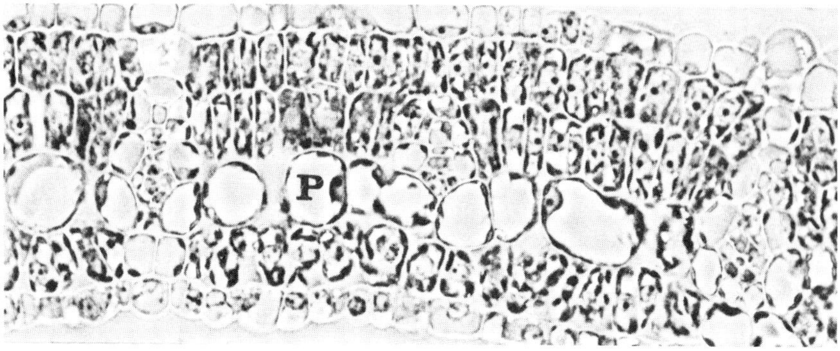

Fig. 3–39. Transection of immature soybean leaflet of 28-mm long leaf. Leaflet is irregularly six to seven cell layers thick. Paraveinal mesophyll layer (P) has differentiated precociously. From Franceschi and Giaquinta (1983b). × 385.

(Thrower, 1962). Stipules and hairs mature extremely early, before the leaf proper has attained even 1% of its mature size, thereby providing some protection and insulation.

When a leaf primordium is about 400 μm long, the lamina is initiated by increased mitotic activity in cells along the margins. Periclinal and anticlinal divisions in the marginal regions of the blade meristem produce six layers of cells (Decker and Postlethwait, 1960; Sun, 1957a). The top and bottom layers divide only anticlinally and eventually become the epidermis. The four central layers originate from a group of submarginal initial cells.

Sun (1957a) reported that the six organized layers later become seven by periclinal divisions of cells in the third abaxial layer. Decker and Postlethwait (1960) concluded that six layers are present in both immature and mature leaflets of Hawkeye soybean. Figures in Franceschi and Giaquinta (1983b) seem to demonstrate that either six or seven layers may occur (Fig. 3–39).

The adaxial layer becomes the upper epidermis and the second and third layers develop into palisade tissue. A portion of the fourth (middle) layer differentiates into veins but most of it becomes paraveinal mesophyll (Franceschi and Giaquinta, 1983b) (Fig. 3–39). The fifth and sixth layers become spongy mesophyll, and the abaxial layer becomes the lower epidermis (Sun, 1957a). Figure 3–12, of a transection of a mature leaflet, shows these layers.

In the midrib of an immature leaflet there is a cambium that forms a limited amount of secondary xylem and phloem. Other veins have only primary vascular tissue. In the mature midrib there is no evidence of a cambium (Decker and Postlethwait, 1960).

The time elapsed between the initiation of one leaf and another leaf on the opposite side of the stem apex is termed a *plastochron*. Miksche (1961) studied plastochronation of the first 10 trifoliolate leaves. Following germination, it took 3 to 3.5 days for the second trifoliolate leaf to be initiated. Subsequent plastochrons took approximately 2 days each.

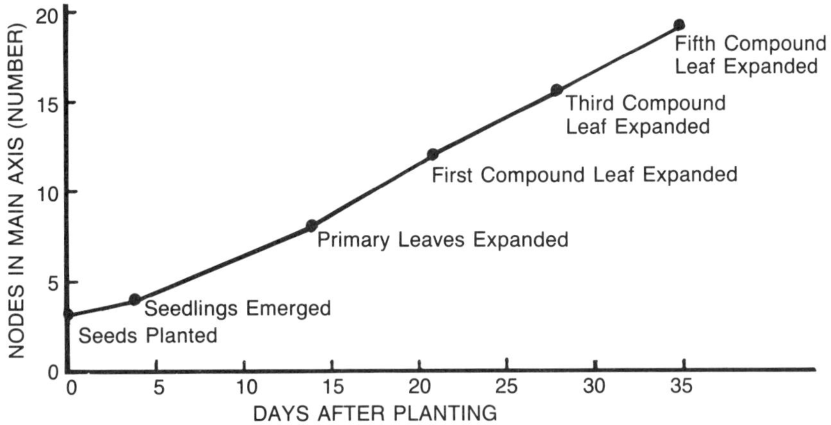

Fig. 3–40. Mean number of nodes of main stem of soybean plants at various stages of development and points in time under simulated normal growing conditions. From Johnson et al. (1960).

Johnson et al. (1960) found that, by 35 days after planting, when the fifth compound leaf was fully expanded, there were about 19 nodes on the main axis. Since three of these nodes were present in the mature embryo, the average of about 2 days per plastochron reported by Miksche seems to be verified, Johnson et al. also concluded that all nodes of the mature plant are already present by the 35th day. The change in the number of nodes on the main axis with reference to days after planting is indicated in Fig. 3–40.

3–5.3 Root Subapical Meristem and Primary Roots

For up to 4 days after germination the primary root tip has distinct stelar initials producing procambium, and another group of initials forming the ground meristem, protoderm, and columella (Miksche, 1961; Sun, 1955) (Fig. 3–4). Sun (1957b) and Anderson and Postlethwait (1960) observed that the apical meristem changes with increasing root age. The apical meristem of the primary root of an 8-day-old seedling has only one group of common initials from which all primary meristems originate (Fig. 3–42 and 3–43). Common initials were also reported by Patel et al. (1975).

Along with the histogens, the size of the quiescent center (an area of almost inert, infrequently dividing cells in the CI zone of Fig. 3–41 and 3–42) also changed in the primary root (Miksche and Greenwood, 1966). The center remained extremely small up to 24 h postgermination, thereafter increasing about 2.5 times to 40 h. After that it decreased dramatically to 60 h and shrunk to only about one third its size at 24 h. It stayed at this reduced size until 120 h, when the study ended. The significance of such volume changes is unknown.

The root cap continually wears away, but lost cells are replaced from within by divisions of the columella and protoderm. The life span of

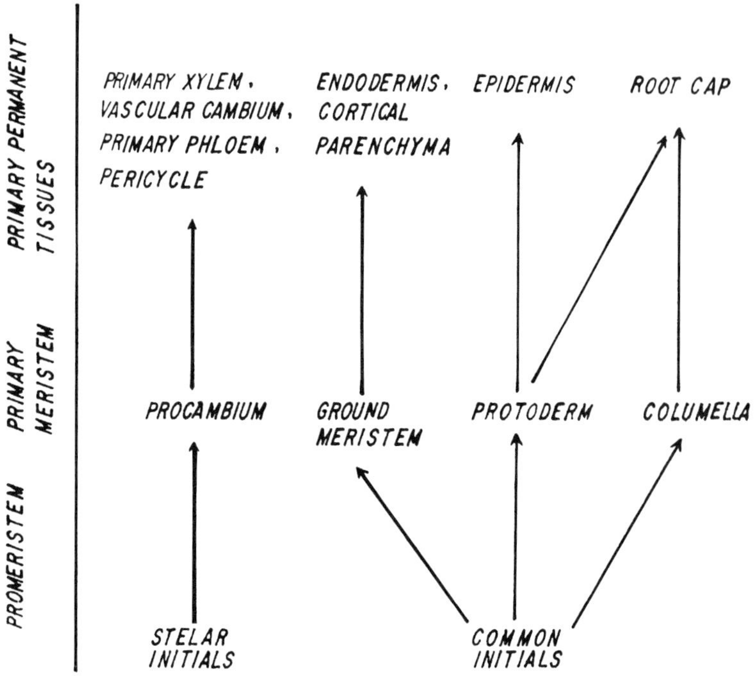

Fig. 3–41. Schematic representation of the pattern of primary tissue differentiation in soybean roots. From Sun (1957b).

individual soybean root cap cells is unknown but in other species they have been shown to exist for 7 to 21 days. Although regular growth patterns occur within the root apex, the shape of the root cap varies greatly (Anderson, 1961).

All root tip anatomical studies of soybean have used only the radicle and primary root. The assumption that all roots are identical seems warranted, except that Ransom and Moore (1983) have shown that there are some quantitative differences in subcellular components between the primary and lateral roots of *Phaseolus*. They tried unsuccessfully to find the anatomical basis for the vertically downward growth of the primary root as compared to the typically horizontal or obliquely downward growth of lateral roots. Their work is probably also applicable to soybean roots.

The level at which the various permanent tissues appear and mature is shown in Fig. 3–44, a composite of data from Sun (1955) and Anderson (1961). The pericycle originates about 160 μm from the stelar initials. Very slightly above this, at about 200 μm, some vacuolation of the metaxylem elements occurs in the center of the stele, and at the same level the four primary phloem strands become delimited.

Phloem matures first. The first mature sieve elements are present at about 410 μm from the apex. Slightly behind this, at least one mature sieve element is present for each phloem strand. At least three mature sieve elements are present in each phloem strand at about 900 μm from

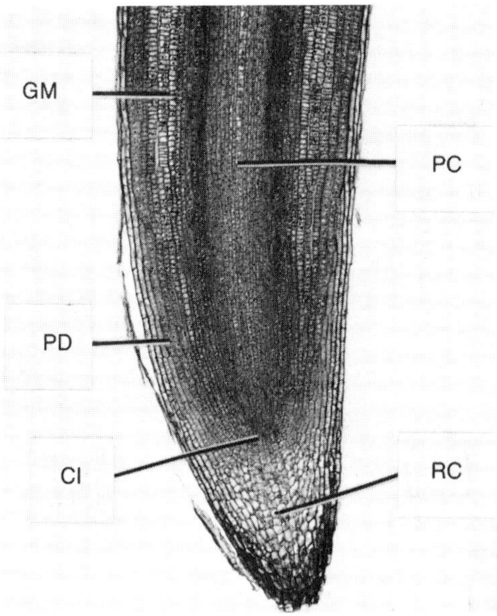

Fig. 3–42. Median longitudinal section of the primary root tip 8 days after planting. Labels as in Fig. 3–43.

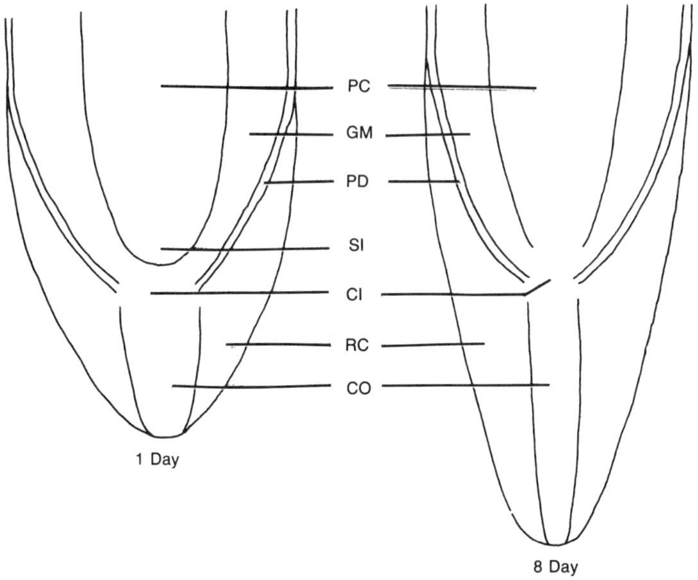

Fig. 3–43. Diagrammatic organization of 1- and 8-day primary root tips. Abbreviations: CI–common initials, CO–columella, GM–ground meristem, PC–procambium, RC–root cap, and SI–stelar initials. After Bell (1934) and Sun (1955).

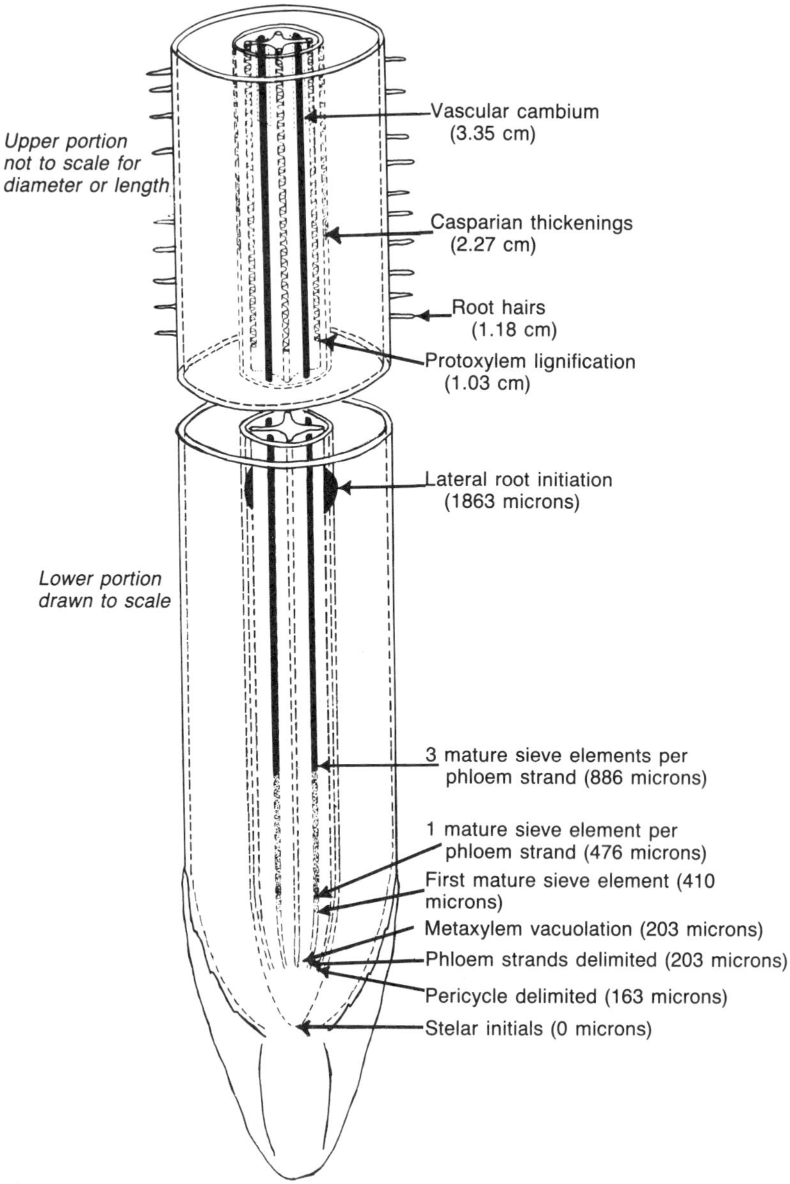

Fig. 3–44. Diagrammatic scheme of primary tissue differentiation in primary root of soybean. Based on data from Anderson (1961) and Sun (1955).

the apex. Lateral roots are initiated in the pericycle about 2 mm behind the stelar initials.

Maturation of the protoxylem elements occurs approximately 1 cm from the apex. In a slightly more mature area, root hairs can be seen. Approximately 2 cm from the apex, casparian strips are present in en-

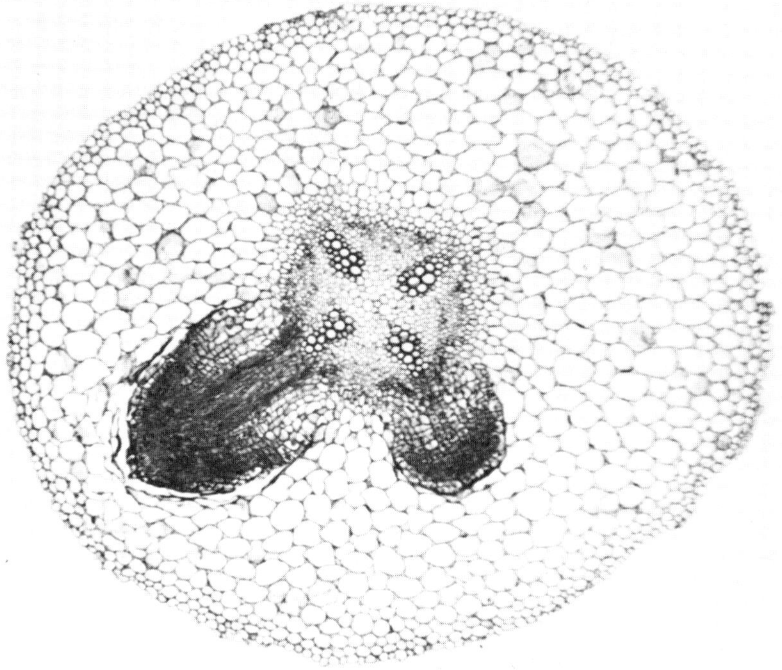

Fig. 3–45. Young branch roots, each opposite a protoxylem ridge, in process of growing through the parent tetrarch root. × 69.

dodermal cell walls, and at about 3 cm from the apex the vascular cambium becomes evident. At this level the characteristic tetrarch xylem pattern is visible, with mature protoxylem at the priphery of the xylem ridges and new elements maturing toward the center. The metaxylem is evident and some secondary wall thickening is occurring.

The four phloem strands, with mature sieve tube elements and companion cells, alternate with the protoxylem ridges. Vascular cambium is starting to form between the phloem strands and the metaxylem. Exarch maturation of the primary xylem continues until all metaxylem vessels have secondary walls.

The first vascular tissue to differentiate is the phloem. Mature sieve tube members are present 300 to 400 μm from stelar initials in the 3-day-old primary root. Four protophloem ridges are present in primary roots and normal branch roots (Byrne et al., 1977a) but triarch and diarch patterns have been reported. The first sieve tube element matures adjacent to the pericycle, and maturation proceeds centripetally until about six sieve tube members occur in each protophloem strand. These elements are vacuolate and similar in size, so that companion cells are not distinguishable (Anderson, 1961). The protophloem eventually collapses against the pericycle because of cambial activity. The cells immediately below crushed protophloem differentiate into primary phloem fibers, and cells located inward from the fibers mature into metaphloem elements. Metaphloem consists of scattered parenchyma cells, sieve tube members, and companion cells (Bell, 1934).

Primary xylem is first distinguishable at the same level as phloem, developing acropetally by enlargement and vacuolation of cells and lignification of walls. As in all roots, transverse maturation within the xylem is exarch, but vacuolation begins first in metaxylem elements at the center of the stele and proceeds centrifugally toward the protoxylem. In contrast, formation of secondary wall and lignification begins first in the protoxylem and then proceeds centripetally until the largest metaxylem elements differentiate in the center. Protoxylem cells usually have annular or spiral secondary wall thickenings, and they are longer and narrower than metaxylem elements. The first metaxylem elements to mature have a single row of pits on each side, later elements are shorter, larger in diameter, and have densely pitted walls. Some metaxylem parenchyma differentiates into thick-walled connective tissue (Bell, 1934). Anderson (1961) found that some xylem cells attain a large diameter but do not become vessel elements. Instead, they have large perfortions and bordered pits all over.

3–5.4 Branch Roots

Branch roots arise from cell proliferation in local areas of the pericycle just outside of protoxylem ridges. The xylem pattern, therefore, determines the number of rows of branch roots (Bell, 1934; Byrne et al., 1977a), so tetrarch roots have the potential for four longitudinal rows of branch roots and triarch roots three rows (Fig. 3–51). The first indication of branch root initiation is a radial elongation of pericycle cells followed by tangential and radial divisions (Sun, 1955). This results in a root apex with the same structural organization as the primary root.

Each branch root forces and/or digests its way (Fig. 3–45) through the endodermis, cortex, and epidermis. Flaps of epidermis are pushed aside where roots emerge, and there is a noticeable increase in diameter of the emergent part of the branch root (Fig. 3–46). This may result from the release of pressure by the parent root. This is also where the diarch pattern formed by the branch root's connection to its parent stele ordinarily changes to triarch (Byrne et al., 1977a). The branch root is usually somewhat smaller in diameter than the primary root, and ordinarily it has just one large metaxylem vessel instead of several in the center of the stele. Although the primary root is tetrarch, secondary branch roots may be triarch. Tertiary, quaternary, and successive orders of smaller branch roots may be either triarch or diarch.

No visible root primorida are ever found closer than 3 cm from the apex of the primary root (Sun, 1955). Branching from the secondary and higher-order roots, however, does occur closer to the apex than in the primary root (Anderson, 1961).

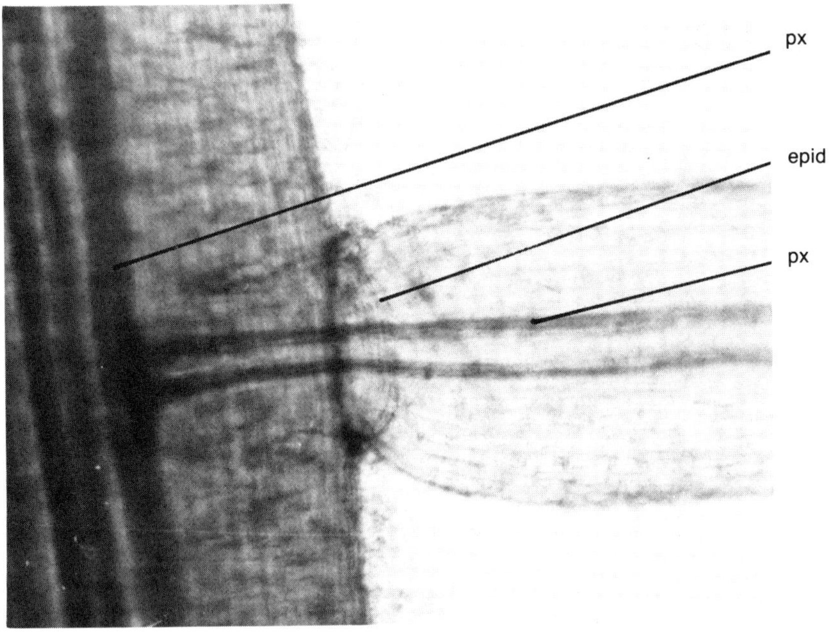

Fig. 3–46. Cleared unstained preparation showing branch root within parent root at left, and extending out to right. Beyond the point of emergence, branch root increases in diameter. Abbreviations: px—protoxylem strands, and epid—flaps of ruptured epidermis. × 55.

3–6 BACTERIAL ROOT NODULES

Root nodules are conspicuous spheroidal swellings of the root cortex inhabited by *Rhizobium japonicum*, a gram-negative rod-shaped soil bacterium capable of penetrating roots and establishing a N_2-fixing symbiotic relationship. There may be several hundred nodules on a mature plant, distributed at all levels to almost a meter below the surface (Grubinger et al., 1982).

Nodulation begins when rhizobia attach themselves to epidermal cells in a narrow annular zone just behind the root cap. Each bacterium becomes attached only at its tip, a swift recognition reaction occurring in less than 1 min after inoculation in the laboratory (Fig. 3–47). The bacteria remain attached even after repeated washing of the roots (Turgeon and Bauer, 1982).

Epidermal cells with immature or as yet unformed root hairs are the usual sites for penetration (Bhuvaneswari et al., 1980; Turgeon and Bauer, 1982). This suggests that a certain minimum period of contact between bacteria and host cell is needed, and that only actively growing root hairs can be penetrated. Infected hairs are always shorter than mature intact hairs, and they are markedly curled (Fig. 3–48) (Rao and Keister, 1978; Turgeon and Bauer, 1982; Pueppke, 1983).

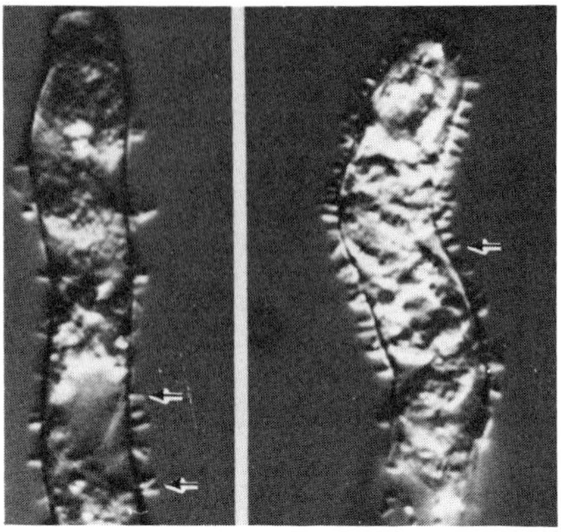

Fig. 3–47. Soybean root hairs after in vitro inoculation by rhizobia, as viewed by differential-interference contrast micrography. *Rhizobia* are attached at one end (*arrows*). (A) root hair with few rhizobia, and (B) root hair with dense rhizobial population. From Turgeon and Bauer (1982). × 1200.

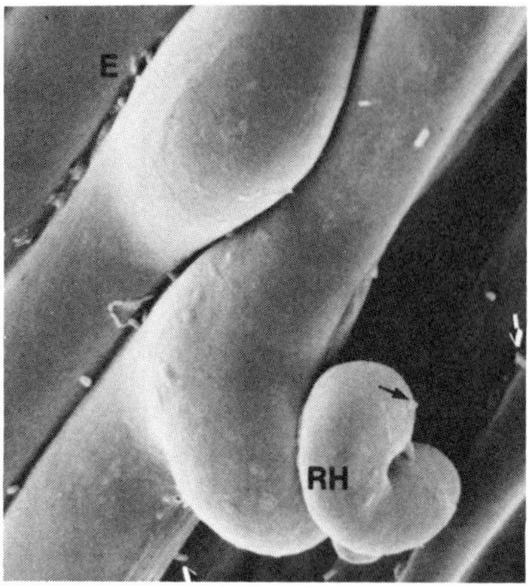

Fig. 3–48. Scanning electron micrograph of small area of root 12 h after inoculation. Curled root hair is the site of bacterial penetration. Arrows indicate bacteria, E—epidermis, and RH—root hair. From Turgeon and Bauer (1982). × 1400.

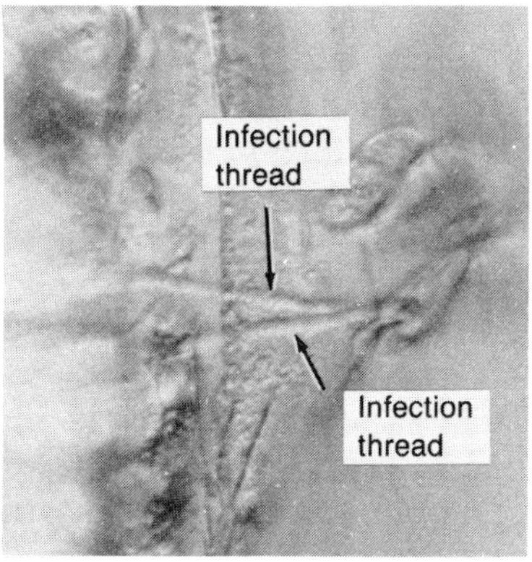

Fig. 3–49. Root hair with two infection threads (*arrows*) viewed by interference-contrast microscopy. From Pueppke (1983).

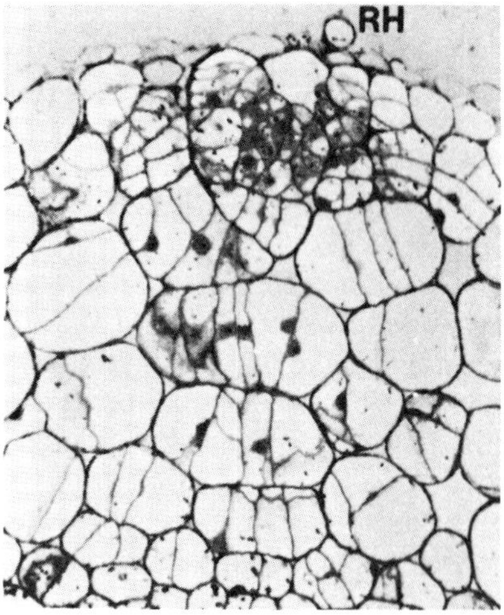

Fig. 3–50. Transection of soybean root at site of rhizobial penetration. Recent cell division is evident in the cortex adjacent to root hair (RH) even before penetration by infection threads. × 750.

At the point of infection, the root hair wall forms a depression that invaginates deeply, forming a tube (infection thread) lined by a continuation of the root hair cell wall and membrane (Goodchild and Bergersen, 1966). Although some root hairs have only one infection thread, two is more common and three or more have been seen (Fig. 3–49) (Rao and

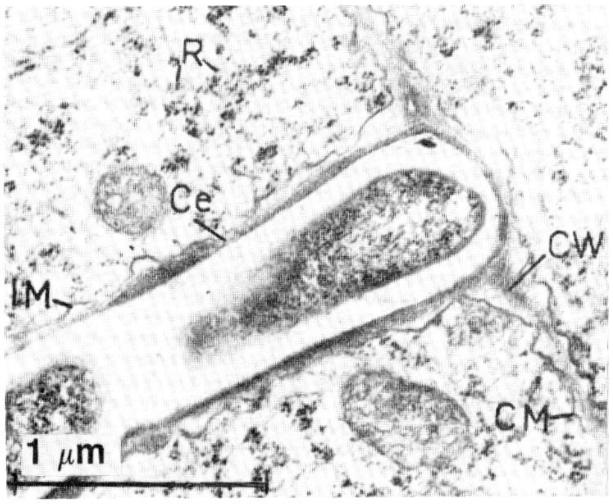

Fig. 3–51. Tip of infection thread penetrating cortical cell wall, viewed by transmission electron microscopy. Abbreviations: Ce—infection thread cellulose, CM—host cell membrane, CW—host cell wall, IM—infection thread membrane, R—ribosome. From Goodchild and Bergersen (1966).

Keister, 1978). Infection threads may branch within a root hair (Turgeon and Bauer, 1982; Pueppke, 1983).

It takes about 2 days for the infection thread, with its included dividing bacteria, to grow 60 to 70 μm to the base of the root hair cell (Bieberdorf, 1938; Turgeon and Bauer, 1982). The cortex adjacent to infected root hairs becomes meristematic and produces a wedge-shaped area of dividing cells even before any infection threads enter (Fig. 3–50) (Newcomb et al., 1979; Turgeon and Bauer, 1982). As a thread enters a cortical cell, an invagination of the wall and plasma membrane occurs in the same fashion as in the root hair (Fig. 3–51).

A second wave of mitoses increases cell number in the outer cortical layer, which then becomes the main area of infected cells (Newcomb et al., 1979). The combination of multiple threads and branching of threads in the cortex results in penetration of many, but not all, of these cells (Turgeon and Bauer, 1982). The peripheral uninfected area becomes the nodule cortex, which includes a sclereid layer and several vascular bundles. Numerous cells with calcium oxalate crystals form an irregular reticulum in the outer cortex (Sutherland and Sprent, 1984), a feature in common with other legumes having determinate nodules.

Whether or not the ploidy level of cortical cells increases before or after infection threads penetrate them is a controversial topic (Newcomb, 1981). The only published information about this in soybean is by Kodama (1970), who found only diploid cells in soybean and other Phaseoleae nodules, whereas polyploid cells were the rule in other legumes studied.

At some time during or following mitotic activity, rhizobia are released into cortical cells through small thin areas on the surface of the infection threads. These local areas bulge and release small droplets of thread matrix, with one bacterium per droplet. Each droplet is encased

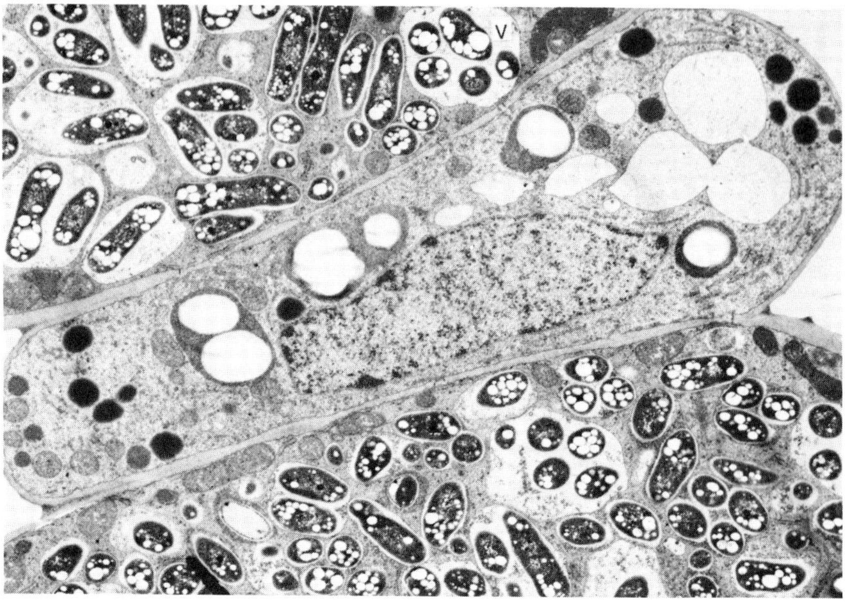

Fig. 3–52. Soybean nodule cells 25 days after inoculation. Middle cell is uninfected but upper and lower cells contain infection vacuoles (V), packets of bacteroids surrounded by a peribacteroid membrane. Transmission electron micrograph from Werner and Mörschel (1978).

in a peribacteroid membrane contributed by the host cell membrane (Verma et al., 1978; Newcomb, 1981). Because of this membrane and the infection thread membrane, rhizobia are never in direct contact with host cell cytoplasm until the nodule senesces.

Rhizobia are called *bacteroids* after their release into the host cell. They continue to divide, and gradually the host cell becomes filled with packets of bacteroids (Werner and Mörschel, 1978). Figure 3–52 illustrates this and also shows that some adjacent cells remain uninfected.

Mitosis in infected cortical cells ceases about 14 days after infection. Subsequent increases in the volume of infected tissue are due entirely to cell enlargement, which often causes infected cells to elongate. As the nodule matures, host cell nuclei become smaller and more regular in shape, and finally disintegrate (Bieberdorf, 1938). During this period oxygen- rich leghemoglobin develops gradually in the host tissue and the nodule becomes pink, remaining so until it begins to senesce. As leghemoglobin forms and bacteria cease dividing, N_2 fixation occurs (Bergersen, 1963).

The increasing volume of the developing nodule causes splitting and sloughing off of the epidermis, thus exposing the outer cortical layer. The second cortical layer becomes meristematic and forms the cortex of the nodule (Fig. 3–53). Division in this layer, which Ikeda (1955) termed the *inner meristem*, permits developing bacteroidal tissue deeper in the nod-

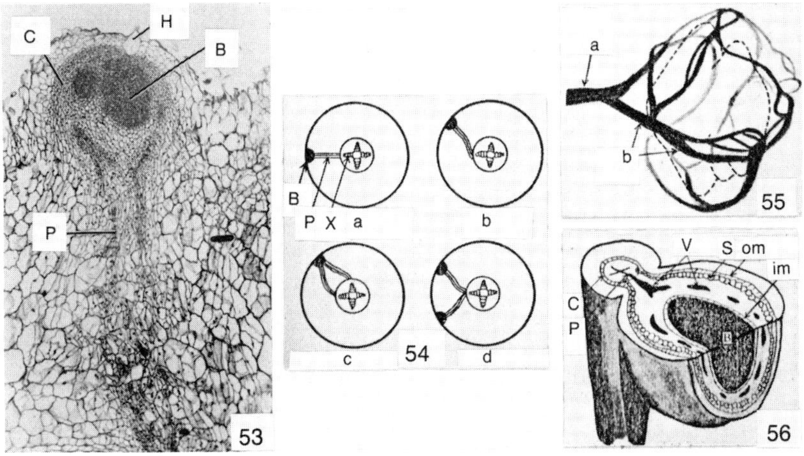

Fig. 3–53 to 3–56. *Fig. 3–53.* A portion of 12-day soybean nodule. Abbreviations: B—bacterial-infected cells, C—meristematic region of nodule cortex, H—hypertrophied root hair, and P—procambium. From Bieberdorf (1938). *Fig. 3–54.* Patterns of procambial strand development connecting nodule to parent stele. Abbreviations: B—developing nodule, P—procambium, and X—protoxylem. From Ikeda (1955). *Fig. 3–55.* Somewhat three-dimensional depiction of vascular network around nodule, a—vascular bundle connecting nodular network with parent root stele, b—one bundle of nodular vascular network. From Ikeda (1955). *Fig. 3–56.* Diagrammatic representation of mature nodule. Abbreviations: B—bacterioidal tissue, C—cork cambium of root, S—sclerenchyma, V—vascular tissue, im—inner meristem, and om—cork cambium of nodule. From Ikeda (1955).

ule to expand. About 8 days after infection, cork cambium forms in the outer layers of the nodule cortex and in nonnodulated areas of the root (Bieberdorf, 1938). The cork cambium keeps pace with nodule enlargement and continues to add more corky cells. Lenticels form on the nodule surface and are said to function in gas exchange (Pankhurst and Sprent, 1975).

After infection threads have penetrated two or three cortical layers (Bieberdorf, 1938; Ikeda, 1955), parenchyma cells between the nodule primordium and host root stele become meristematic and form a procambial strand that extends toward the protoxylem (Fig. 3–53 and 3–54). This procambium differentiates into pitted tracheids and scalariform vessels surrounded by starch-filled parenchyma cells capable of dividing and adding more xylem cells (Bieberdorf, 1938). Where the nodule vascular bundle meets the vascular system of the parent root, the new xylem from the nodule bundle becomes continuous with the existing secondary xylem. Within the cortex of the nodule, more procambium strands form and differentiate acropetally toward the nodule apex, branching and rebranching, and finally anastomosing into a continuous vascular network (Ikeda, 1955) (Fig. 3–55). Some cells near the xylem differentiate into phloem strands consisting of three to five sieve tubes. As each bundle matures, a well-defined endodermis with a casparian strip encircles it (Fraser, 1942). A mature vascular bundle has a central strand of xylem surrounded by a cylinder of phloem. A sclerenchyma sheath of isodiametric sclereids with simple pits differentiates just outside of the vascular network, maturing about 28 days after infection. It prevents any further increase in the size of the nodule (Bieberdorf, 1938).

A change occurs in vascular tissue of the parent root as rhizobia are being released from infection threads into cortical cells. Wall ingrowths characteristic of transfer cells appear in xylem parenchyma cells adjacent to metaxylem vessels (Newcomb and Peterson, 1979).

Mature nodules at the end of the 4th week after infection are spheroidal and 3 to 6 mm in diameter (Fig. 3–56). Nodules sometimes become irregular or lobed when two or more infected areas develop close together and merge during growth. Soybean nodules are determinate, lacking the apical meristem and extended terminal growth found in the indeterminate nodules of certain other legumes such as *Medicago*, *Melilotus*, and *Trifolium* (Sprent, 1980). The bacteroidal tissue is therefore homogeneous in age (Bergersen and Briggs, 1958).

Nodules retain their mature structure until the 6th or 7th week, then they begin to senesce (Bergersen, 1958). Since reinfection of younger portions of the root system may occur during the growing season, a mature soybean plant may have nodules of several age classes. A chronology of nodulation is presented in Table 3–3.

3–7 MYCORRHIZAL RELATIONS

In addition to bacterial nodules, soybean roots have been shown to form mycorrhizal associations (Bethlenfalvay et al., 1982). In plants in-

Table 3–3. Chronology of nodulation.† Compiled from Bergersen (1958), Bieberdorf (1938), and Ikeda (1955).

Age of nodule	Stage of nodulation
days	
0	Initial invasion of root hair or ordinary epidermal cell by *Rhizobium*.
1–2	Infection thread reaches base of epidermal cell and enters cortex.
3–4	Small mass of infected cells in nodule primordium; procambium strand extends from nodule to stele of root.
5	Very rapid bacterial and host cell division continues about 2 weeks.
7–9	Nodule visible; procambium of nodule vascular system arises at base of nodule and develops toward nodule apex.
12–18	Continued growth of all nodule tissues; periderm present; some mature cells in sclerenchyma layer; vascular system forms anastomosing network within nodule cortex; bacteroidal tissue is pink at the end of this period and N_2 fixation commences.
23	Most division has ceased in bacterial and host cells; nodule continues growth by cell enlargement for up to 2 more weeks; period of active N_2 fixation.
28–37	Nodule reaches maximum size; vascular tissue and sclerenchyma tissue mature; N_2 fixation continues until nodule degeneration begins.
50–60	Nodule degeneration.

†Absolute times might vary with cultural conditions, time of infection, soybean cultivar, and other factors.

oculated with the vesicular-arbuscular mycorrhizal fungus *Glomus fasciculatus*, hyphae penetrated host cells and formed highly branched endings. These intraradical hyphae, which do not kill the host cell, extended to new roots throughout the life span of the association. Fungal hyphae also extended into the soil (extraradical hyphae) to enhance nutrient absorption. The latter reached maximum development when the pods began rapid maturation and declined thereafter. Knowledge of the anatomical aspects of this association are almost lacking.

REFERENCES

Ali, A., and R.A. Fletcher. 1970. Xylem differentiation in inhibited cotyledonary buds of soybeans. Can. J. Bot. 48:1139–1140.

Anderson, C.E. 1961. The morphogenesis of the root of *Glycine max*. M.S. thesis. Purdue University, Lafayette, IN.

----, and S.N. Postlethwait. 1960. The organization of the root apex of *Glycine max*. Proc. Indiana Acad. Sci. 70:61–65.

Baligar, V.C., V.E. Nash, M.L. Hare, and J.A. Price, Jr. 1975. Soybean root anatomy as influenced by soil bulk density. Agron. J. 67:842–844.

Bán, A.D., L. Muller, B.H. de Souza, T. Strehl, and C.S.A. Martins. 1981. Soybean leaf architecture. (In Portuguese.) Agron. Sulriograndense 17:25–31.

----, T. Strehl, B.H., de Souza, C.S.A. Martins, and L. Muller. 1980. Anatomical study of crystals in soybean leaflets. (In Portuguese.) Agron. Sulriograndense 16:169–179.

Bell, W.H. 1934. Ontogeny of the primary axis of *Soja max*. Bot. Gaz. 95:622–635.

Bergersen, F.J. 1958. The bacterial component of soybean root nodules; changes in respiratory activity, dry cell weight and nucleic acid content with increasing nodule age. J. Gen. Microbiol. 19:312–323.

----. 1963. Iron in the developing soybean nodule. Aust. J. Biol. Sci. 16:916–919.

----. and M.J. Briggs. 1958. Studies on the bacterial component of soybean root nodules: Cytology and organization in the host tissue. J. Gen. Microbiol. 19:482–490.

Bernard, R.L., and B.B. Singh. 1969. Inheritance of pubescence type in soybeans: glabrous, curly, dense, sparse, and puberulent. Crop Sci. 9:192–197.

Bethlenfalvay, G.J., M.S. Brown, and R.S. Pacovsky. 1982. Relationships between host and endophyte development in mycorrhizal soybeans. New Phytol. 90:537–543.

Bhuvaneswari, T.V., B.G. Turgeon, and W.D. Bauer. 1980. Early events in the infection of soybean (*Glycine max* L. Merr.) by *Rhizobium japonicum*. I. Localization of infectible root cells. Plant Physiol. 66:1027–1031.

Bieberdorf, F.W. 1938. The cytology and histology of the root nodules of some Leguminosae. J. Am. Soc. Agron. 30:375–389.

Blad, B.L., and D.G. Baker. 1972. Orientation and distribution of leaves within soybean canopies. Agron. J. 64:26–29.

Bohm, W., H. Maduakor, and H.M. Taylor. 1975. Comparison of five methods for characterizing soybean rooting density and development. Agron. J. 69:415- 419.

Bostrack, J.M., and B.E. Struckmeyer. 1964. Effects of gibberllic acid on the anatomy of soybeans. Am. J. Bot. 51:611–617.

Byrne, J.M., T.C. Pesacreta, and J.A. Fox. 1977a. Development and structure of the vascular connection between the primary and secondary root of *Glycine max* (L.) Merr. Am. J. Bot. 64:946–959.

----, ----, and ----. 1977b. Vascular pattern change caused by a nematode, *Meloidogyne incognita*, in the lateral roots of *Glycine max* (L.) Merr. Am. J. Bot. 64:960–965.

Ciha, A.J., and W.A. Brun. 1975. Stomatal size and frequency in soybeans. Crop Sci. 15:309–313.

Compton, R.H. 1912. An investigation of seedling structure in the Leguminosae. J. Linn. Soc. Bot. 41:1–119.

Crafts, A.S. 1967. Bidirectional movement of labelled tracers in soybean seedlings. Hilgardia 37:625–638.

Cumbie, B.G. 1960. Anatomical studies in the Leguminosae. Trop. Woods 113:1–47.

Curry, T.M. 1982. Morphological and anatomical comparisons between fasciated and non-fasciated soybeans (*Glycine max* (L.) Merr.). M.S. thesis. Iowa State University, Ames.

Datta, P.C., and N. Saha. 1971. Secondary xylem of Phaseoleae. Acta Bot. Acad. Sci. Hung. 17:347–359.

Decker, R.D., and S.N. Postlethwait. 1960. The maturation of the trifoliate leaf of *Glycine max*. Indiana Acad. Sci. Proc. 70:66–73.

Dittmer, H.J. 1940. A quantitative study of the subterranean members of soybean. Soil Conserv. 6(2):33–34.

Dormer, K.J. 1945. An investigation of the taxonomic value of shoot structure in angiosperms with especial reference to Leguminosae. Ann. Bot. 9:141–153.

Dzikowski, B. 1936. Studia nad soja *Glycine hispida* (Moench) Maxim. Cz. 1. Morfologia. Pamietnik Panstwowego Instytutu Naukowego Gospodarstwa Wiejskiego w Pulawach. Tom XVI. zeszyt 2. Rosprawa Nr. 253:Oh 69–100.

----. 1937. Studia nad soja *Glycine hispida* (Moench) Maxim. Cz. 11. Anatomia. Mem. Inst. Natl. Pol. Econ. Rurale 258:229–265.

Fisher, D.B. 1967. An unusual layer of cells in the mesophyll of the soybean leaf. Bot. Gaz. 128:215–218.

----. 1975. Structure of functional soybean sieve elements. Plant Physiol. 56:555–569.

Flores, E.M., and A.M. Espinoza. 1977. Epidermis foliar de *Glycine soja* Sieb. y Zucc. Rev. Biol. Trop. 25:263–273.

Franceschi, V.R., and R.T. Giaquinta. 1983a. Glandular trichomes of soybean leaves: cytological differentiation from initiation through senescence. Bot. Gaz. 144:175–184.

----, and ----. 1983b. The paraveinal mesophyll of soybean leaves in relation to assimlate transfer and compartmentation. I. Ultrastructure and histochemistry during vegetative development. Planta 157:411–421.

----, and ----. 1983c. The paraveinal mesophyll of soybean leaves in relation to assimilate transfer and compartmentation. II. Structural, metabolic and compartmental changes during reproductive growth. Planta 157:422–431.

Fraser, H.L. 1942. The occurrence of endodermis on leguminous root nodules and its effect upon nodule function. Proc.-R. Soc. Edinburgh, sect. B:Biol. Sci. 61:328-343.

Goodchild, D.J., and F.J. Bergersen. 1966. Electron microscopy of the infection and subsequent development of soybean nodule cells. J. Bacteriol. 92:204–213.

Grandet, J. 1955. Sur le point vegetafif du *Soja hispida* Moench. C.R. Acad. Sci. 240:1003–1005.

Grubinger, V., R. Zobel, J. Vendeland, and P. Cortes. 1982. Nodule distribution on roots of field-grown soybeans in subsurface soil horizons. Crop Sci. 22:153–155.

Gunn, C.R. 1981. Seeds of Leguminosae. p. 913–925. *In* R.M. Polhill and P.H. Raven (ed.) Advances in legume systematics. Ministry of Agriculture, Fisheries and Food. Royal Botanic Garden, Kew, U.K.

Ikeda, H. 1955. Histological studies on the root nodules of soybean. (In Japanese.) Kagoshima U. Fac. Agron., Ser. B. 1955(4):54–64.

Johnson, H.W., H.A. Borthwick, and R.C. Leffel. 1960. Effects of photoperiod and time of planting on rates of development of the soybean in various states of the life cycle. Bot. Gaz. 122:77–95.

Kaspar, R.C., C.D. Stanley, and H.M. Taylor. 1978. Soybean root growth during the reproductive stages of development. Agron. J. 70:1105–1107.

Kodama, A. 1970. Cytological and morphological studies on the plant tumors. I. Root nodules of some Leguminosae. J. Sci. Hiroshima Univ., Ser. B., Div. 2 13:223–260.

Kuo, J., J.S. Pate, M. Rainbird, and C.A. Atkins. 1980. Internodes of grain legumes—new location for xylem parenchyma transfer cells. Protoplasma 104:181–185.

Lackey, J.A. 1978. Leaflet anatomy of Phaseoleae (Leguminosae: Papilionoideae) and its relation to taxonomy. Bot. Gaz. 139:436–446.

Lauchli, A., D. Kramer, and R. Stelzer. 1974. Ultrastructure and ion localization in xylem parenchyma cells of roots. p.363–371. *In* U. Zimmermann and J. Dainty (ed.) Membrane transport in plants. Springer-Verlag New York, New York.

Liu, K-C., A.J. Pappelis, and H.M. Kaplan. 1971. Microbodies of soybean cotyledon mesophyll. Trans. Ill. State Acad. Sci. 64:136–141.

Lugg, D.G., and T.R. Sinclair. 1980. Seasonal changes in morphology and anatomy of field-grown soybean leaves. Crop Sci. 20:191–196.

Miksche, J.P. 1961. Developmental vegetative morphology of *Glycine max*. Agron. J. 53:121–128.

----, and M. Greenwood. 1966. Quiescent centre of the primary root of *Glycine max*. New Phytol. 65:1–4.

Mitchell, R.L., and W.J. Russell. 1971. Root development and rooting patterns of soybean [*Glycine max* (L.) Merrill] evaluated under field conditions. Agron. J. 63:312–316.

Newcomb, W. 1981. Nodule morphogenesis and differentiation. Int. Rev. Cytol. Suppl. 13:247–298.

----, and R.L. Peterson. 1979. The occurrence and ontogeny of transfer cells assocated with root nodules and lateral roots in Leguminosae. Can. J. Bot. 57:2583- 2602.

----, D. Sippell, and R.L. Peterson. 1979. The early morphogenesis of *Glycine max* and *Pisum sativum* root nodules. Can. J. Bot. 57:2603–2616.

Nozzolillo, C. 1973. A survey of anthocyanin pigments in seedling legumes. Can. J. Bot. 51:911–915.

Pankhurst, C.E., and J.I. Sprent. 1975. Surface features of soybean root nodules. Protoplasma 85:85–89.

Patel, J.D., J.J. Shah, and K.V. Subbayamma. 1975. Root apical organization in some Indian pulses. Phytomorphology 25:261–270.

Pueppke, S.G. 1983. *Rhizobium* infection threads in root hairs of *Glycine max* (L.) Merr., *Glycine soja* Sieb. & Zucc., and *Vigna unguiculata* (L.) Walp. Can. J. Microbiol. 29:69–76.

Ransom, J.S., and R. Moore. 1983. Geoperception in primary and lateral roots of *Phaseolus vulgaris* (Fabaceae). I. Structure of columella cells. Am. J. Bot. 70:1048–1056.

Rao, R.V., and D.L. Keister. 1978. Infection threads in the root hairs of soybean (*Glycine max*) plants inoculated with *Rhizobium japonicum*. Protoplasma 97:311–316.

Raper, C.D., Jr., and S.A. Barber. 1970. Rooting systems of soybeans. I. Differences in root morphology among varieties. Agron. J. 62:581–584.

Shibles, R., I.C. Anderson, and A.H. Gibson. 1974. Soybean. *In* L.T. Evans (ed.) Crop physiology, some case histories. Cambridge University Press, Cambridge, UK.

Sprent, J.I. 1980. Root nodule anatomy, type of export product and evolutionary origin of some Leguminosae. Plant Cell Environ. 3:35–43.

Struckmeyer, B.E. 1941. Structure of stems in relation to differentiation and abortion of blossom buds. Bot. Gaz. 103:182–191.

Sun, C.N. 1955. Growth and development of primary tissues in aerated and non-aerated roots of soybean. Bull. Torrey Bot. Club 82:491–502.

----. 1957a. Histogenesis of the leaf and structure of the shoot apex in *Glycine max* (L.) Merrill. Bull. Torrey Bot. Club 84:163–174.

----. 1957b. Zonation and organization of root apical meristem of *Glycine max*. Bull. Torrey Bot. Club 84:69–78.

Sutherland, J.M., and J.I. Sprent. 1984. Calcium-oxalate crystals and crystal cells in determinate root nodules of legumes. Planta 161:193–200.

Thrower, S.L. 1962. Translocation of labelled assimilates in the soybean. II. The pattern of translocation in intact and defoliated plants. Aust. J. Biol. Sci. 15:629–649.

Turgeon, B.G., and W.D. Bauer. 1982. Early events in the infection of soybean by *Rhizobium japonicum*. Time course and cytology of the initial infection process. Can. J. Bot. 60:152–161.

Verma, D.P.S., V. Kazazian, V. Zogbi, and A.K. Bal. 1978. Isolation and chacterization of the membrane enclosing the bacteroids in soybean root nodules. J. Cell Biol. 78:919–936.

Voroshilova, G.I. 1964. The structure of the vegetative cone and the leaf development in soybeans (In Russian.) Bot. Zh. (Leningrad) 49(9):1329–1335.

Watari, S. 1934. Anatomical studies on some leguminous leaves with special reference to the vascular system in petioles and rachises. J. Fac. Sci. Imp. Univ. Tokyo, Sect. 3, 4:225–365.

Watson, L., J.S. Pate, and B.E.S. Gunning. 1977. Vascular transfer cells in leaves of Leguminosae–Papilionoideae. Bot. J. Linn. Soc. 74:123–130.

Weaver, H.L. 1960. Vascularization of the root-hypocotyl-cotyledon axis of *Glycine max* (L.) Merrill. Phytomorphology 10:82–86.

Wergin, W.P., and E.H. Newcomb. 1970. Formation and dispersal of crystalline P-protein in sieve elements of soybean (*Glycine max* L.). Protoplasma 71:365–388.

Werner, D., and E. Mörschel. 1978. Differentiation of nodules of *Glycine max*. Planta 141:169–177.

Williams, L.F. 1950. Structure and genetic characteristics of the soybean. p. 111–156. *In* K.S. Markley (ed.) Soybeans and soybean products. Interscience Publishing, New York.

Woolley, J.T. 1964. Water relations of soybean leaf hairs. Agron. J. 56:569–571.

4 Reproductive Morphology

John B. Carlson
University of Minnesota
Duluth, Minnesota

Nels R. Lersten
Iowa State University
Ames, Iowa

Following the period of vegetative growth, which varies depending upon cultivar and environmental conditions such as daylength and temperature, the plant enters the reproductive stage, during which axillary buds develop into flower clusters of 2 to 35 flowers each. There are two types of stem growth habit and floral initiation in soybean (Dzikowski, 1936; Guard, 1931; Williams, 1950).

One type is the indeterminate stem, in which the terminal bud continues vegetative activity during most of the growing season. In this type, the inflorescences are axillary racemes (Fig. 4–1) and the plant at maturity has a sparse and rather even distribution of pods on all branches with a diminishing frequency toward the tip of the stems. The stem may sometimes appear to have a terminal inflorescence but this apparently is a series of small one- or two-flowered axillary inflorescences crowded together by the short internodes at the stem tip.

The second type is the determinate stem, in which the vegetative activity of the terminal bud ceases when it becomes an inflorescence. This type has both axillary racemes and a terminal raceme (Fig. 4–2), and at maturity has pods distributed along the stem as well as a rather dense terminal cluster of pods.

The node of the first flower is related to the development stage of the plant. Since nodes of the cotyledons, the primary leaves, and the first two or three trifoliolate leaves are usually vegetative, the first flowers appear at nodes five or six and sometimes higher. Flowers form progressively toward the tip of the main stem and also toward the tips of the branches. The period of bloom is influenced by the time of planting and may extend from 3 to more than 5 weeks (Borthwick and Parker, 1938; Hardman, 1970).

Several investigators have reported that a soybean plant produces many more flowers than can develop into pods. From 20 to 80% of the

Soybeans: Improvement, Production, and Uses, 2nd ed.—Agronomy Monograph no. 16.

Fig. 4–1. Reduced inflorescence in axil of 'Ottawa-Mandarin' soybean.

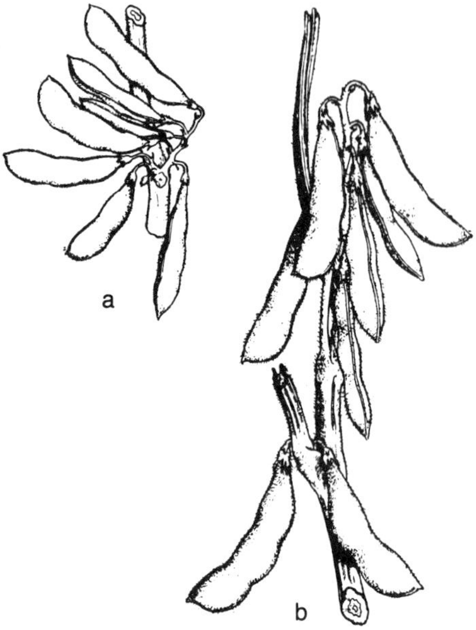

Fig. 4–2. Inflorescence types: (a) axillary inflorescence and (b) terminal inflorescence. From Dzikowski (1936).

flowers are reported to abscise for various cultivars (Hansen and Shibles, 1978; Hardman, 1970; Van Schaik and Probst, 1958; Wiebold et al., 1981). Most cultivars with many flowers per node have a higher percentage of flower abscission than those with few flowers per node. Abscission can occur at the time of bud initiation, during the development of floral organs, at the time of fertilization, during the early proembryo stage, or at any stage of cotyledon development. Flower or pod abscission occurs most often from 1 to 7 days after flowering (Kato and Sakaguchi, 1954; Kato et al., 1955; Pamplin, 1963; Williams, 1950).

Abernathy et al. (1977) reported that failure of fertilization is insignificant as a cause of floral abscission in soybean. Abscising flowers were mostly all fertilized and usually contained proembryos that had undergone two or three cell divisions. Hansen and Shibles (1978) found that, in two indeterminate cultivars, abscission was greatest on the lower stems, whereas pods were retained most often in the middle portions of the plant. In 11 determinate cultivars, in contrast, most harvestable pods were in the top third of the canopy and abscission increased in the lower portions (Weibold et al., 1981). In general, the earliest and latest flowers produced tend to abscise most often.

Individual ovules or entire ovaries may abort. Kato and Sakaguchi (1954) noted that the basal ovule, which is the last one to be fertilized, would frequently abort. Also, the terminal ovule would often abort because of its poorer ability to compete for available water. Thus, normal pods develop with some mature seeds although occasional ovules abort.

When inflorescences are initiated, there are marked changes from the normal vegetative development of the axillary buds. The two opposite prophyll primordia are differentiated as usual, but instead of the typical two-rowed (distichous) arrangement of the trifoliolate leaf primordia, the pattern of initiation changes to a spiral two-fifths phyllotaxy. These primordia do not develop into typical trifoliolate leaves with paired stipules, but instead become simple bracts that lack stipules. Growth of both prophylls and bracts is somewhat slower than their vegetative counterparts, but their axillary buds develop much more rapidly, each becoming the primordium of a flower.

In the development of each flower, a similar pair of prophylls is again produced. These prophylls, however, remain small and attached to the lower part of the pedicel of the flower. They become the bracteoles, which are inserted on each side of the calyx of the flower (Borthwick and Parker, 1938; Dzikowski, 1936, 1937; Murneek and Gomez, 1936) (Fig. 4–3).

When a bud in the axil of a trifoliolate leaf develops into an inflorescence, the stalk of that inflorescence remains stem-like, with typical stem anatomy, including epidermis, cortex, endodermis, vascular tissue, and considerable secondary growth from a vascular cambium (Dzikowski, 1937). In the development of an inflorescence, the bract of each flower is homologous to a trifoliolate leaf, and the two bracteoles (Fig. 4–3) are homologous to the prophylls that normally develop at the base of every branch. After forming the primordia of the bracteoles, the apical meristem of the flower gives rise directly to the floral organs.

4–1 FLOWER DEVELOPMENT

Soybean has a typical papilionaceous flower with a tubular calyx of five unequal sepal lobes, and a five-parted corolla consisting of posterior banner petal, two lateral wing petals, and two anterior keel petals in contact with each other but not fused (Fig. 4–3 and 4–4). The 10 stamens, collectively called the *androecium*, occur in two groups (diadelphous pattern) in which the filaments of nine of the stamens are fused and elevated as a single structure whereas the posterior stamen remains separate (Fig. 4–5). The single pistil is unicarpellate and has one to four campylotropous ovules alternating along the posterior suture (Fig. 4–21). The style is about half the length of the ovary and curves backward toward the free posterior stamen, terminating in a capitate stigma (Fig. 4–5). Trichomes occur on the pistil and also cover the outer surfaces of the calyx tube, the bract, and bracteoles. No trichomes are present on the petals or stamens (Dzikowski, 1937; Guard, 1931; Pamplin, 1963).

Guard (1931) gave a detailed description of floral organogeny. The future flower is at first merely a knob-like primordium in the axil of the bract (Fig. 4–6 to 4–8). The sepals are the first whorl of floral organs to

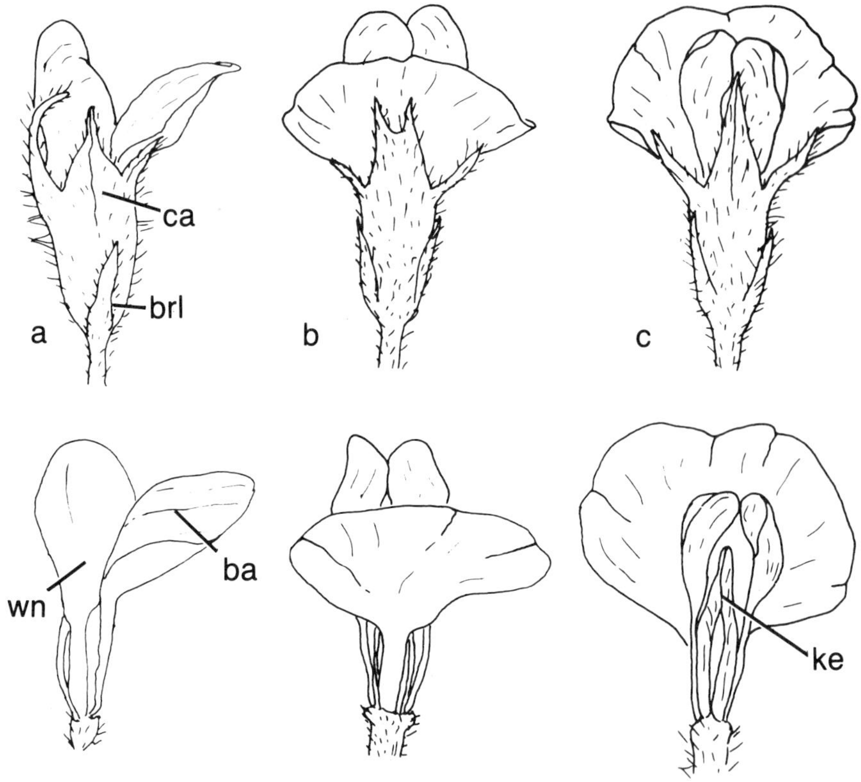

Fig. 4-3. Mature soybean flower. Right-hand column shows flower with calyx removed. (a) Side view showing calyx with bracteoles, banner, and wing petals. (b) Adaxial or top view showing four calyx lobes, both bracteoles, banner, and wing petals. (c) Abaxial or bottom view. Three calyx lobes visible. Abbreviations: ba—banner petal, brl—bracteole, ca—calyx, ke—keel petals, and wn—wing petal. × 6.

be initiated. The anterior, abaxial sepal lobe arises first on the abaxial side of the flower primordium and is followed in rapid succession by the two lateral lobes, and finally, by the two posterior, adaxial lobes. Very early, the bases of these lobes broaden and fuse, and later this becomes the calyx tube (Fig. 4-9 to 4-13).

The petals form next; their primordia alternate with the lobes of the calyx. The keel petals appear first, on the abaxial side, next the two wing petals form laterally and, finally, the banner petal forms on the adbaxial side of the flower (Fig. 4-14 and 4-15). Petal primordia develop slowly and are soon surpassed in growth by the stamens. An outer whorl of five stamens appears first, on the anterior side of the flower, just to the inside of the whorl of petal primordia and alternate with them. Initiation of stamen primordia proceeds toward the posterior side of the receptacle. Before the last stamen of the first whorl is visible, a second, inner, whorl of stamens appears, alternating with those of the first, and again starting

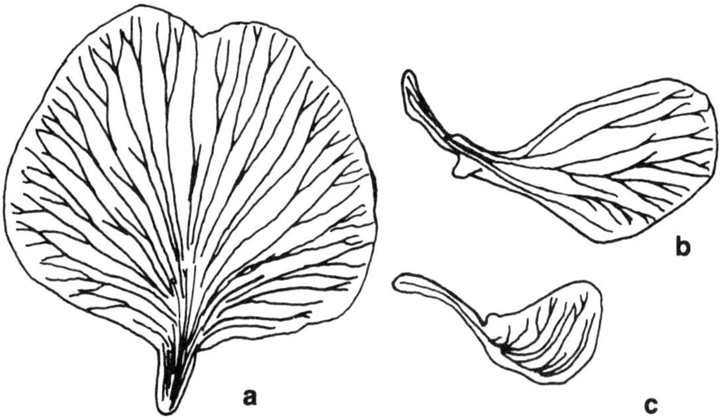

Fig. 4–4. Cleared petals showing venation. (a) Banner petal, (b) Wing petal, and (c) Keel petal. × 7.

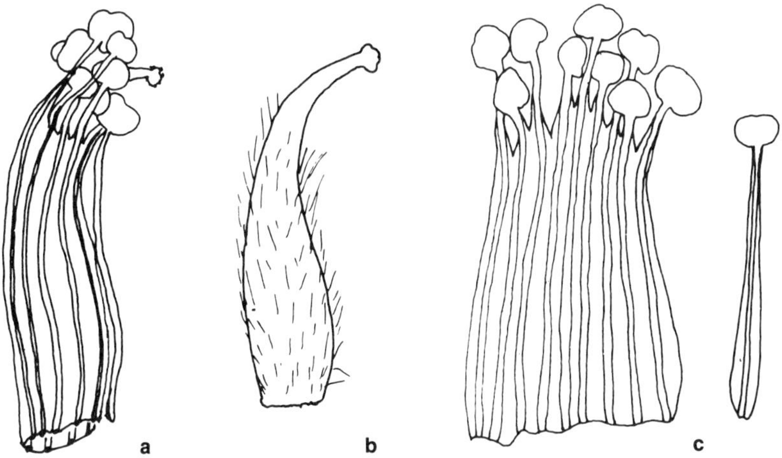

Fig. 4–5. Sexual structures of soybean flower. (a) Diadelphous stamens arranged around pistil, (b) Pistil, and (c) Nine fused stamens and one separate stamen. × 15.

on the anterior side of the receptacle and progressing toward the posterior, adaxial side. Because of tissue growth below them, these two whorls of stamens quickly align themselves into a single whorl on a staminal tube bearing nine stamens, with the larger and older stamens alternating with the smaller and younger stamens in sequence around the developing pistil. The single free stamen, located between the banner petal and the ventral suture of the pistil, is the last stamen to appear. Although remaining separate, it is a member of the inner whorl of stamens (Fig. 4–16 to 4–19).

The primordium of the pistil appears first as a U-shaped ridge about the same time as the initiation of the last whorl of stamens. The open part of the U, the ventral suture, is on the adaxial side of the flower (Fig.

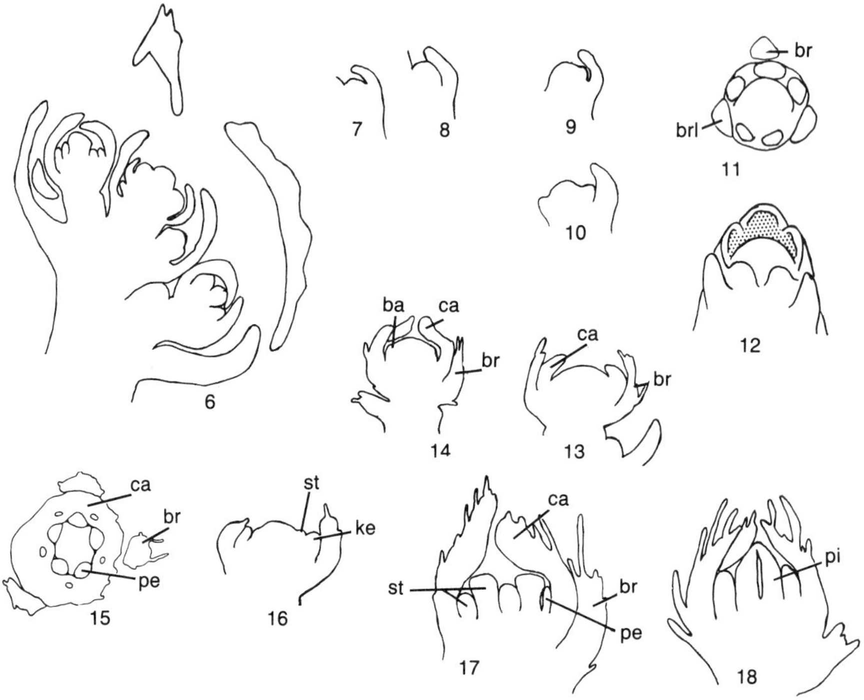

Fig. 4–6 to 4–18. Gross floral development in soybean. Abbreviations: ba—banner, br—bract, brl—bracteole, ca—calyx, ke—keel, ms—microsporangium, ne—nectary, ov—ovule, pe—petal, pi—pistil, and st—stamen. Fig. 4–7 to 4–18. × 60. *Fig. 4–6.* An L-section of raceme with several developing flowers, each in an axil of a bract. × 32. *Fig. 4–7 to 4–8.* Floral primordium in the axil of a bract. *Fig. 4–9.* First sepal primordium on the abaxial side of a flower meristem adjacent to a bract. *Fig. 4–10.* An L-section of a floral primordium showing two sepal primordia. *Fig. 4–11.* Top view of a flower with five sepal primordia. Three abaxial sepal primordia being elevated by growth of basal meristem. Two abaxial sepal primordia not yet being elevated. Abaxial bract and two lateral bracteoles are present. *Fig. 4–12.* Perspective view of same flower shown in Fig. 4–11. *Fig. 4–13.* An L-section through a flower showing bract and elongating calyx tube. Younger flower primordium visible below the upper flower. *Fig. 4–14.* An L-section of older flower through the center of banner petal primordium on adaxial side of flower. All other petal primordia are also present in this stage of floral development. *Fig. 4–15.* Cross section of flower in same stage of development as Fig. 4–14. The section passes through calyx tube, showing five principal veins, alternating with five petal primordia. Stamens and pistil not yet present. *Fig. 4–16.* First stamen primordium appears on the abaxial side between the keel petal primordia. *Fig. 4–17.* Lateral L-section showing outer floral organs. Both whorls of stamens are present, alternating with one another. The older stamens are larger and the younger, inner stamens are smaller. One small petal primordium is present. *Fig. 4–18.* Same flower showing a median L-section passing through the unicarpellate pistil.

4–18 and 4–19). All organs of the flower develop rapidly except the petals, which do not elongate much until the anthers have well-developed microsporangia. The staminal tube, the free stamen, and the style elongate at the same pace. Thus, the anthers at maturity are clustered around the stigma (Johns and Palmer, 1982) (Fig. 4–28). At this time the petals grow

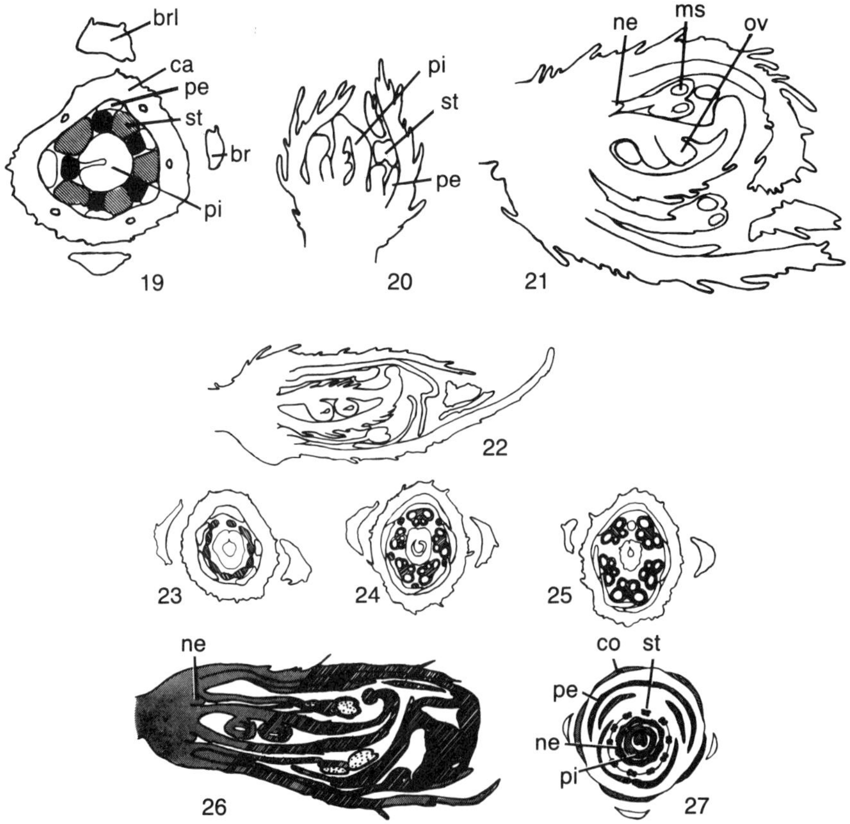

Fig. 4–19 to 4–27. Gross floral development (*continued*). Abbreviations: co—corolla, ne—nectary, pe—petal, pi—pistil, st—stamen. *Fig. 4–19.* Cross section of young flower with all floral organs established. Each whorl of organs is initiated abaxially and alternates with the previously established whorl. The margins of the carpel on adaxial side have not yet fused. × 33. *Fig. 4–20.* Older flower with first indication of ovule primordia in the pistil. The ventral sutures have now fused. × 33. *Fig. 4–21.* An L-section of a flower with ovules at the time of integument initiation. Megasporocytes present in ovule. Anthers have microsporangia with microsporocytes. Nectary is visible encircling the base of the pistil. × 33. *Fig. 4–22.* Older flower in which diadelphous arrangement of the stamens has been established. Ovules have four- or eight-nucleate embryo sacs about 2 days before opening of the flower. ×11. *Fig. 4–23, 4–24, and 4–25.* Cross sections of flower in Fig. 4–22 at different levels. *Fig. 4–23.* Cross section near the base of the flower showing stamen tube and typical diadelphous 9 + 1 arrangement of the stamens. × 11. *Fig. 4–24.* Section through the lower younger anthers. Also visible are the filaments of the upper anthers. × 11. *Fig. 4–25.* Cross section through the upper anthers. × 11. *Fig. 4–26.* Mature flower at the time of anthesis. Calyx, petals, stamen tube, nectary, and pistil with ovules present. × 11. *Fig. 4–27.* Floral diagram showing typical papilionaceous floral organ arrangement. Nectary is about 0.2 mm high at the base of the pistil. × 11.

very rapidly, soon surpassing the calyx, stamens, and pistil to become visible as the flower is in bloom (Fig. 4–20 to 4–27).

Before the margins of the leaf-like pistil fuse, two to four ovule primordia are produced alternately, and develop simultaneously, on the

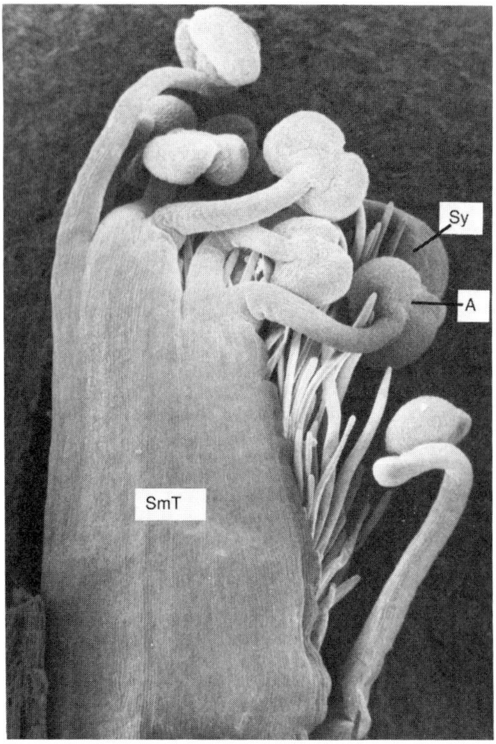

Fig. 4–28. Staminal tube and style growth are synchronized so that the anthers are lifted to the stigma. Stigma is located behind the labelled anther. Abbreviations: A—anther; SmT—staminal tube, and Sy—style. From Johns and Palmer (1982) × 27.

inner surface of the margins, on the placenta (Guard, 1931; Pamplin, 1963). Each ovule becomes campylotropous, with its micropylar end directed upward toward the stigma (Fig. 4–59).

The nectary is visible, about 10 days before anthesis, as a rim of tissue between the base of the pistil and the stamens (Fig. 4–21). At the time of anthesis, the discoidal nectary is a fully formed cup about 0.2 to 0.4 mm in height encircling the base of the staminal sheath (Carlson, 1973; Erickson and Garment, 1979).

The slightly oval nectary stomata are concentrated on each side of the adaxial indentation of the nectary where it contacts the filament of the free ventral stamen. Most of the stomata occur over the rim and ventral interior surface of the nectar cup, occasionally in groups of two or three. On the abaxial side of the cup there are only a few stomata (Erickson and Garment, 1979). Waddle and Lersten (1973) noted that nectaries are vasculated largely by phloem branching from the staminal base. Nectariferous tissue is, therefore, most closely associated with the stamens.

4-2 STAMEN DEVELOPMENT, MICROSPOROGENESIS, AND POLLEN MATURATION

The first whorl of five stamen primordia arises shortly after the initiation of the petal primordia, and is quickly followed by the second stamen whorl. The sequence of development is the same for both whorls of stamens, except that it occurs later in the inner whorl. Each stamen primordium contains a more or less homogenous mass of cells surrounded by a protoderm layer (Fig. 4–29 and 4–30). As the stamen develops, its apical portion forms a four-lobed anther and a short filament. Each anther lobe consists of a central region of archesporial (primary sporogenous) cells bounded peripherally by four to six layers of cells derived by periclinal divisions of the protoderm. These outer layers mature later into epidermis, endothecium, parietal layers, and tapetum. Toward the center of the anther the archesporium is bound by the centrally located connective tissue, in which the single stamen bundle occurs. The archesporial cells give rise to 25 to 50 microscope mother cells (MMC) per microsporangium, arranged in two to three columns (Palmer et al., 1978) (Fig. 4–31 to 4–38).

Each MMC secretes a callose sheath around itself, between plasmalemma and cell wall. Callose is a nonfibrillar carbohydrate composed of glucose units. The tapetal cells are reported to be uninucleate (Albertson & Palmer, 1979; Buss and Lersten, 1975) or uninucleate and binucleate (Kato et al., 1954; Prakash and Chan, 1976). As meiosis begins in the MMCs, the tapetal cells begin to enlarge and stain more intensely. Their inner tangential walls, bordering the anther locule, begin to disorganize. By the end of microsporogenesis, only the outer tangential walls remain. The cytoplasm is then limited only by the plasmalemma.

Osmiophilic orbicules and, later, mucilaginous substances are secreted into the locule among the young pollen grains (Madjd and Roland-Heydacker, 1978). The composition of the particles has not been determined. The tapetum becomes more vacuolate, with diffuse chromatin, and its nuclei are often lobed. The plasmalemma remains intact up to anther dehiscence even though the tapetal cells have mostly degenerated (Albertsen and Palmer, 1979). Cells of the two innermost parietal layers also become somewhat flattened and disorganized. The layer of cells immediately below the epidermis begins to elongate radially and will develop into the endothecium (Fig. 4–38 to 4–46).

During meiosis, the slender threads forming the reticulum within each MMC nucleus become shortened and thickened and, at first metaphase, the haploid number of $n=20$ chromosomes may be counted. Anaphase and telophase of first meiosis quickly follow. There is then a complete nuclear reorganization resulting in a two-nucleate dyad condition with no intervening cell walls. After a short interphase, a second nuclear division follows, resulting in four microspore nuclei sharing common cytoplasm bounded by the original MMC plasmalemma, the callose layer, and the MMC wall.

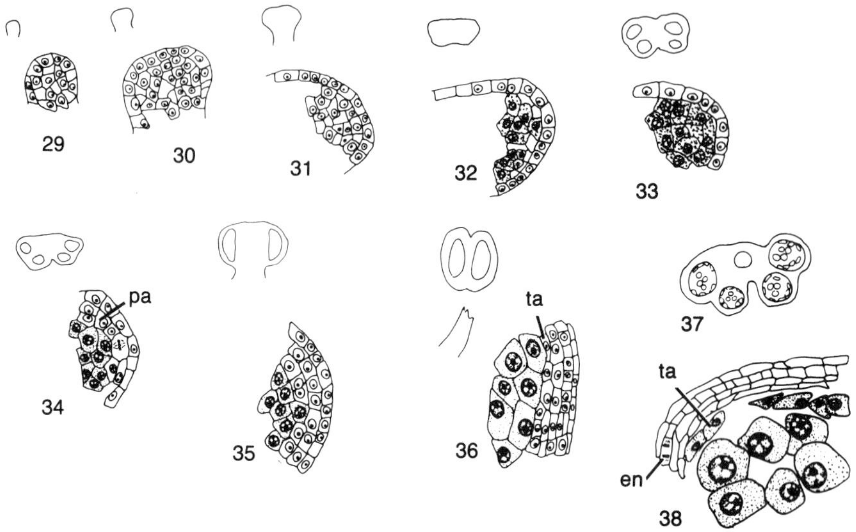

Fig. 4–29 to 4–48. Stamen development and microsporogenesis. Outlines, × 47. Details, × 232. Abbreviations: en—endothecium, pa—parietal layer, and ta—tapetum. *Fig. 4–29, 4–30, and 4–31.* Young stamens in outline and detail. Protoderm present but no sporogenous tissue evident. *Fig. 4–32.* Cross section of a young anther. Another becoming four-lobed. Archesporial cells present. *Fig. 4–33.* Anther with four microsporangia. Archesporial cells present in contact with epidermis. *Fig. 4–34.* Periclinal divisions establishing first and second parietal layers. *Fig. 4–35.* Same stage as Fig. 4–34 except in L-section. *Fig. 4–36.* Sporogenous cells are much larger than adjacent parietal and epidermal cells. Two to three parietal layers and future tapetal layer are present. *Fig. 4–37 and 4–38.* Cross section and details of anther. Microsporocytes present. Tapetal cells large. Two parietal layers becoming flattened. Hypodermal cells beginning to elongate radially, initiating the endothecium. *Fig. 4–39 to 4–48 continued on facing page.*

Cytokinesis is by simultaneous furrowing, thereby separating the four microspore nuclei into four microspores. The major exine regions are deposited in rudimentary form at this time. The callose disappears later by enzymatic dissolution, the MMC wall breaks, and the individual microspores separate from the tetrad. Each young microspore contains numerous vacuoles, large plastids lacking starch, and a large nucleus appressed to the microspore wall. The wall at this time has the tectum, columellae, and endexine. Three colpi are present (Albertsen and Palmer, 1979).

Mitosis and cytokinesis within the microspore produce the pollen grain, with a large vegetative cell (tube cell) and a small generative cell (Fig 3–47 and 3–48). The generative cell is at first in contact with the endexine but later is displaced inward by the formation of the intine. Many starch grains accumulate in each plastid of the vegetative cell during intine formation (Fig. 3–49). The mature pollen grain is subtriangular in polar view and spherical to oblate in equatorial view. The exine has both sculptured and smooth areas (Albertsen and Palmer, 1979).

At the time of maturation of the stamens, the anthers are yellow, as is the pollen. The average diameter of the pollen grains varies from 21 to 30 μm (Dzikowski, 1936; Murneek and Gomez, 1936).

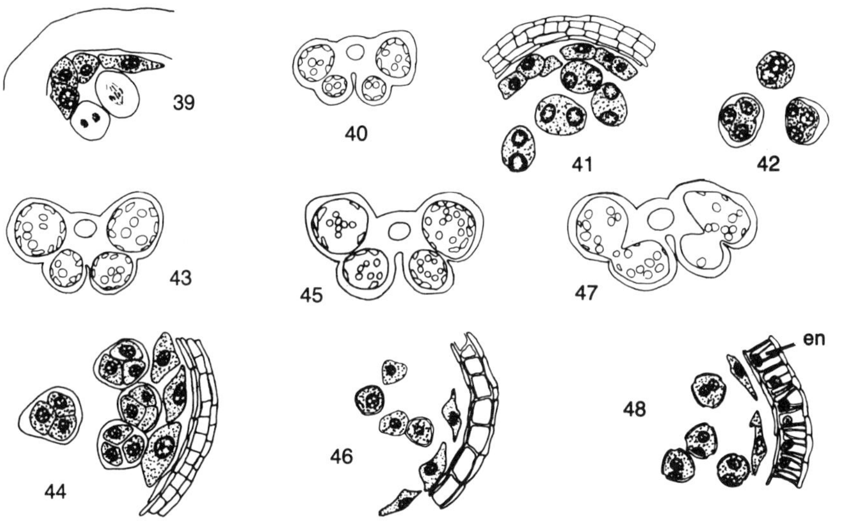

Fig. 4–39 to 4–48. Continued. Fig. 4–39. Meiosis in microsporocyte. *Fig. 4–40 and 4–41.* Cross section and details of older anther in diad stage. No cell plate forms. *Fig. 4–42.* Second division has formed tetrad stage. Furrowing of cytoplasm will form four microspores for each microsporocyte. *Fig. 4–43 and 4–44.* Later stage. Tetrads of microspores. Endothecium continuing to develop. Tapetal cells reach maximum development. *Fig. 4–45.* Septum between adjacent microsporangia becoming thinner. *Fig. 4–46.* Microspores have become separated and form thicker walls. Tapetal cells disorganized. Parietal cells are very flattened. Endothecium continuing to differentiate. *Fig. 4–47.* Almost mature anther. Septum between adjacent microsporangia has ruptured, forming two prominent pollen sacs. *Fig. 4–48.* Mature pollen grains with three germ pores. Tapetal cells disintegrating. Endothecium has strongly thickened portions along radial and inner tangential walls.

During pollen development, the two parietal layers of the microsporangium are crushed. The walls of the endothecium develop U-shaped thickenings on the radial and inner tangential walls (Fig. 4–48 to 4–50). The endothecium is well developed only on the outer walls of the anther and not along the line of dehiscence between the adjacent microsporangia of the anther. As the pollen matures, the septum separating the two microsporangia on either side of the anther ruptures, so that the mature anther has two pollen sacs. The mechanism of dehiscence probably consists of a turning outward of the endothecium as a result of a change in turgor of its cells. The line of dehiscence is along a thin layer of parenchyma tissue, which is easily ruptured under the tension developed by the endothecium.

Microsporogenesis in male sterile (ms_1) and partially male sterile (*msp*) soybean has been described in detail by Albertsen and Palmer (1979) and Stelly and Palmer (1982). In the ms_1 mutants, microsporogenesis is similar to that of fertile plants from anther ontogeny through telophase II of meiosis, but thereafter cytokinesis does not occur; instead, a four-nucleate coenocyte of microspore nuclei forms within the common MMC wall. This coenocyte later develops a pollen wall and accumulates starch and oil reserves as in normal pollen. The coenocyte is round to

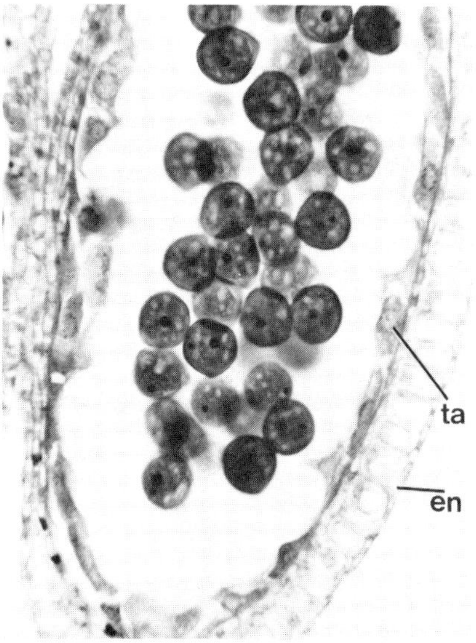

Fig. 4–49. Anther with mature pollen. Pollen grains contain many starch grains. Tapetal cells disintegrating. Endothecium with thickenings in the walls. Abbreviations: en—endothecium, and ta—tapetum. × 350.

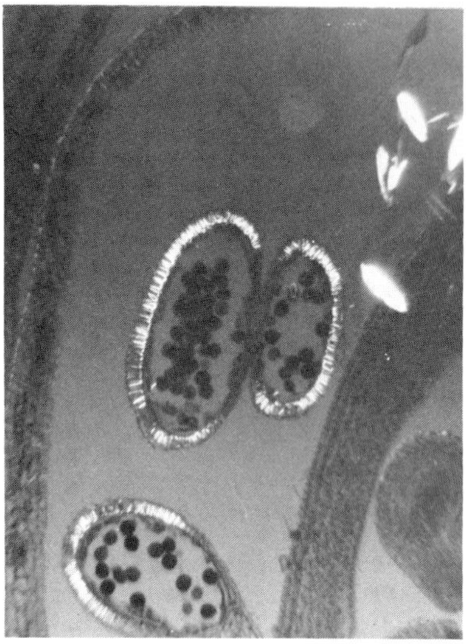

Fig. 4–50. Mature anther under polarized light showing crystalline thickenings in endothecial walls and trichome walls. × 94.

oval in shape, larger than a fertile pollen grain, and has a variable number of colpi. It ultimately collapses but some may first undergo mitosis-like events (Albertsen and Palmer, 1979).

In *msp* mutants, arrest of development and subsequent degeneration of sporogenous tissue may occur at any stage of anther development from sporogenous tissue to almost mature pollen. The onset of abnormalities is most commonly seen near pachytene and at the tetrad and free microspore stages. Cytomixis (movement of chromatin from one cell to another) occurs in sporangia in which development of sporogenous tissue is arrested at pre-meiosis or prophase I. In later stages, fusion of meiocytes results in multinucleate syncytes, which then degenerate. Degenerating sporogenous tissue is usually associated with degenerating tapetal tissue. Although abortion is almost 100% in *msp* anthers, some flowers do contain a few normal pollen grains, thus some successful seed set is possible (Stelly and Palmer, 1982).

4–3 OVULE DEVELOPMENT

Pamplin (1963) studied ovule development of pubescent and glabrous cultivars under both field and greenhouse conditions. He noted that the developmental stages were the same under all conditions for the four cultivars he used. George et al. (1979) also concluded that megagametophyte development under greenhouse or field conditions was remarkably similar.

The ovule of soybean has two integuments (bitegmic), and both ovule and embryo sac are bent back on themselves (campylotropous). Megaspores form deep in the nucellus (crassinucellate) (Prakash and Chan, 1976). As many as four ovules first appear as small masses of tissue on the placenta at alternate sides of the posterior suture of the unicarpellate pistil (Fig. 4–51 to 4–53). The cells of an ovule primordium are all about the same size and covered by a single-layered protoderm. Within 1 or 2 days after ovule initiation, several hypodermal archesporial cells are distinguishable (Fig. 4–60). These cells are larger than the neighboring cells and have more densely staining cytoplasm. Soon one of the archesporial cells surpasses the others in size and becomes the functional megasporocyte (Fig. 4–54 and 4–55). The neighboring cells of the archesporium become less prominent and soon resemble the rest of the cells of the young ovule. Periclinal divisions in the hypodermal region produce two parietal layers of nucellus between the elongate megasporocyte and the epidermis of the ovule (Fig. 4–61).

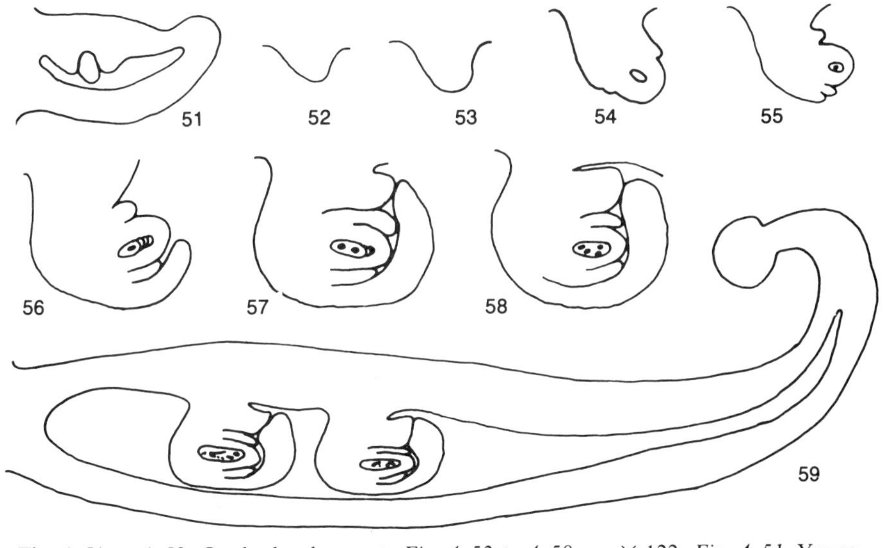

Fig. 4–51 to 4–59. Ovule development. *Fig. 4–52 to 4–58* are × 122. *Fig. 4–51.* Young pistil with three ovule primordia. × 67. *Fig. 4–52 to 4–53.* Ovules appearing as masses of cells alternating on placental margins. *Fig. 4–54.* Integuments being initiated. Ovule bending toward stylar end of pistil. *Fig. 4–55.* Further development of integuments. Megasporocyte present. *Fig. 4–56.* Outer integument has overtaken inner integument. Ovule continues to bend. Four megaspores present. The functional megaspore is larger and at the chalazal end. *Fig. 4–57.* Embryo sac in two-nucleate condition. Ovule now has typical campylotropous form. Outer integument has grown past inner integument and is almost in contact with the placenta. *Fig. 4–58.* Four nucleate embryo sac. Nucellus in contact with outer integument. *Fig. 4–59.* Pistil at time of fertilization. Stylar canal extends almost to capitate stigma. × 53.

Embryo sac (megagametophyte) development is of the normal or *Polygonum* type, which occurs in over 75% of angiosperms. Meiosis in the functional megasporocyte results in a linear tetrad of haploid megaspores (Fig. 4–56 and 4–67). Occasionally, a T-shaped tetrad forms (George et al., 1979). The chalazal (i.e., furthest from the micropyle) megaspore continues to enlarge while the three micropylar megaspores become disorganized and soon disintegrate (Fig. 4–57). The first mitotic division of the functional megaspore produces a two-nucleate megagametophyte (Fig. 4–57 and 4–71). These nuclei are displaced to opposite ends by the formation of a large central vacuole. A second mitosis produces the four-nucleate condition (Fig. 4–58 and 4–72). Two successive mitotic divisions result in an eight-nucleate megagametophyte with four nuclei located at the chalazal end and four at the micropylar end of the embryo sac (Fig. 4–62 to 4–73).

Following migration of one nucleus from each end toward the center, and subsequent cell wall formation, the mature megagametophyte consists of seven cells. It is commonly called the *embryo sac*, even though there is no embryo before fertilization. Three chalazal antipodals and an egg and two synergids at the micropylar end are all contained within the large central cell with its two polar nuclei. The egg nucleus is displaced toward the chalazal end of the cell by a large vacuole. Each of the two synergids also has a vacuole, typically at the chalazal end of the cell, which displaces the nucleus toward the micropyle. Tilton et al. (1984a) recently confirmed the presence of a filiform apparatus in each of the synergids. Further maturation of the embryo sac results in the gradual disintegration of the three antipodal cells (Fig. 4–74 and 4–75).

The ovule is supplied by a single vascular bundle, which extends from the posterior bundle of the carpel through the short funiculus. It

---

Fig. 4–60 to 4–75. *Glycine max.* L. Merr. From George et al. (1979). *Fig. 4–60.* Several hypodermal archesporial cells divided to form several primary sporogenous and primary parietal cells. *Fig. 4–61.* Megasporocyte just before Meiosis I with the nucleus located in the micropylar third of the cell. *Fig. 4–62.* Abnormally wide megasporocyte: note nucellar degeneration around the megasporocyte. *Fig. 4–63.* Dyad of unequal cells with the chalazal dyad member larger than the micropylar member. *Fig. 4–64.* Meiosis II; nonsynchronous division with the chalazal nucleus in metaphase and the micropylar nucleus in prophase. *Fig. 4–65.* A T-shaped tetrad; no megaspore degeneration. *Fig. 4–66.* Linear tetrad; functional *d* megaspore and degenerating *a, b,* and *c* megaspores. *Fig. 4–67.* Linear tetrad; note vacuoles in the large functional *d* megaspore, the degenerating megaspores and nucellar degeneration. *Fig. 4–68.* One of two linear tetrads in one ovule. *Fig. 4–69.* Isolated nucellus; two-nucleate megagametophyte; note the scattered vacuoles and degenerating megaspores. *Fig. 4–70.* Isolated nucellus; two-nucleate megagametophyte with the nuclei oriented in a line oblique to the long axis of the megagametophyte. *Fig. 4–71.* Isolated nucellus; two-nucleate megagametophyte with nuclei separated by a large central vacuole. *Fig. 4–72.* Isolated nucellus: four-nucleate megagametophyte with the chalazal and micropylar nuclei oriented in a plane almost parallel to the long axis of the megagametophyte. *Fig. 4–73.* Isolated nucellus: eight-nucleate megagametophyte. *Fig. 4–74.* Mature megagametophyte: two polar nuclei and two synergids in front of the egg: note the starch grains in the megagametophyte. *Fig. 4–75.* Mature megagametophyte; degenerating antipodals, two large polar nuclei, a large egg flanked by a synergid; note starch grains in the megagametophyte and the nucellar degeneration. ➡

enters the ovule at the extreme chalazal end just below the hypodermis, where it terminates at the exact center of the chalaza (Pamplin, 1963).

The two polar nuclei fuse before fertilization to form a single large diploid secondary nucleus within the large central cell and in close proximity to the egg apparatus. Starch begins to accumulate in the cytoplasm of the central cell, which frequently becomes tightly packed with starch grains. The starch begins to diminish following fertilization and usually has disappeared entirely 1 or 2 days later (Pamplin, 1963). At ovule maturity, antipodals have degenerated and disappeared, the central cell is engorged with starch, and a filiform apparatus (area of conspicuous wall ingrowths) is present in each synergid (Tilton et al., 1984b).

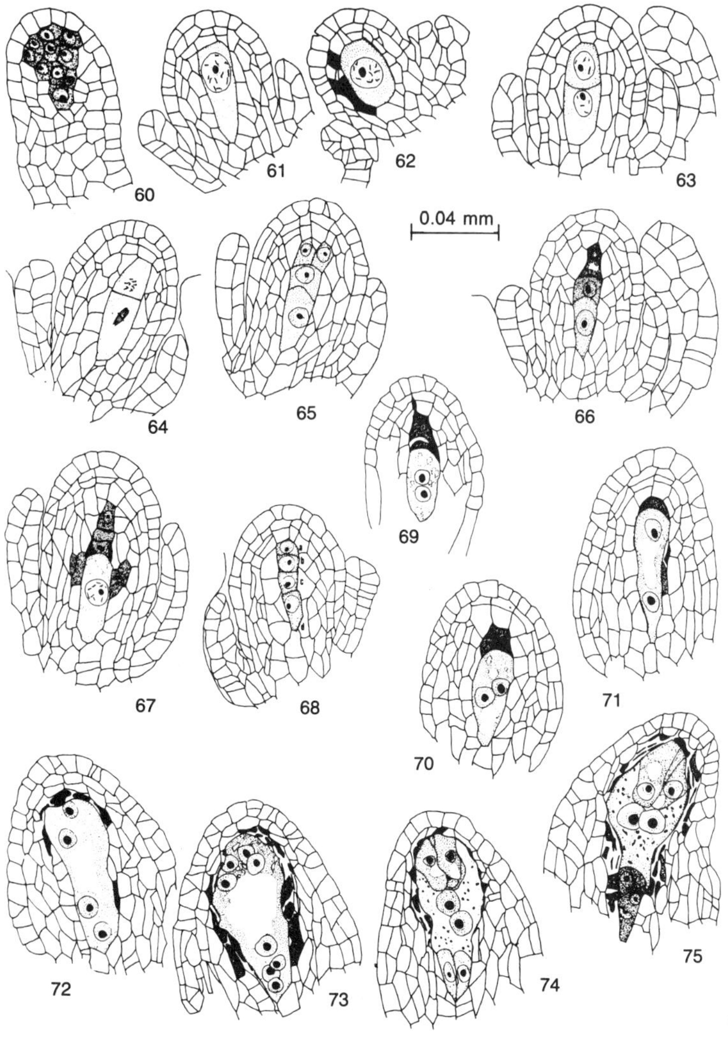

The integuments are initiated from the epidermis of the ovule about the time of the appearance of the megasporocyte. The inner integument arises first, but is quickly followed by the outer integument. Each integument is two cells thick at about the time of division of the functional megaspore (Fig. 4–66), then rapid periclinal divisions and elongation of cells occurs in the outer integument so that it surpasses the inner integument and grows over the apex of the nucellus. The micropyle formed by the outer integument is almost in contact with the placenta of the ventral suture, and has a form described as an inverted Y (Rembert, 1977). The rapid growth of the outer integument results in the apex of the nucellus being in direct contact with the epidermis of the outer integument. The inner integument never forms any part of the micropyle (Fig. 4–54 to 4–59.).

During ovule and embryo sac development, the nucellus increases in thickness by cell enlargement and some periclinal divisions. The nucellar cells in contact with the embryo sac become flattened and obliterated at a rate about equal to the production of more nucellar cells. The degeneration of the inner layers of the nucellus is most marked at the micropylar end of the ovule and is first visible when the functional megaspore is elongating (George et al., 1979). (Fig. 4–62).

At the time of fertilization the nucellus still surrounds the embryo sac, but only the epidermis remains intact at the micropylar end, in direct contact with the outer integument (Pamplin, 1963) (Fig. 4–75).

As the seed develops following fertilization, the nucellus ruptures at the micropylar end, exposing the embryo sac so that the suspensor of the embryo is now in direct contact with the epidermis of the outer integument (Fig. 4–84). The chalazal end of the nucellus persists for several days (Fig. 4–87), but continued development of the endosperm finally results in its complete obliteration by 14 days after fertilization (Pamplin, 1963).

4–4 POLLINATION AND DOUBLE FERTILIZATION

By the time of pollination the diadelphous stamens have been elevated so that the anthers form a ring around the stigma. The pollen thus is shed directly on the stigma, resulting in a high percentage of self-fertilization (Williams, 1950) (Fig. 4–28).

Natural crossing varies from $<0.5\%$ to about 1%. It has been noted that pollination may occur the day before full opening of the flower; i.e., pollination may occur within the bud (Dzikowski, 1936).

We have drawn freely from three recent comprehensive studies by Tilton et al., 1984a, 1984b, and 1984c in the remainder of this section. The wet stigma is overtopped by a proteinaceous film that originates from the cuticle. The film probably prevents dessication of the abundant quantities of lipoidal exudate present at the distal end of the stigma and confines the exudate to the stigmatic surface. It may also contain recognition factors.

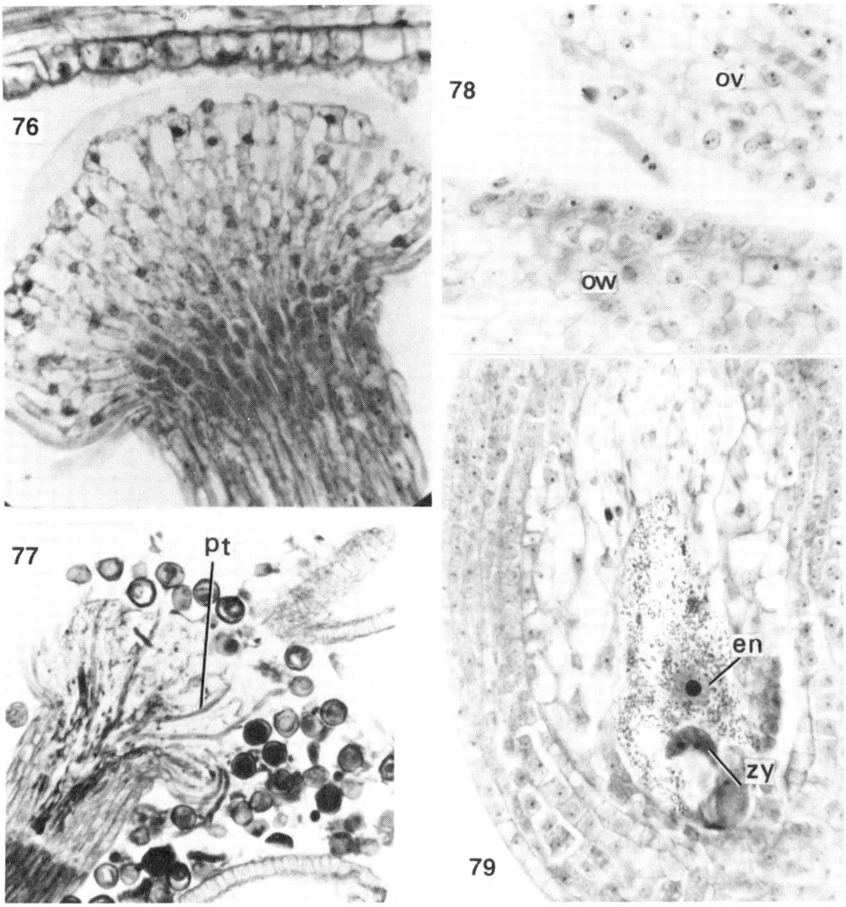

Fig. 4–76 to 4–79. *Fig. 4–76.* Stigma just prior to pollination. Thick layer of stigmatic substance on surface of stigma. × 520. *Fig. 4–77.* Stigma with many pollen tubes on day of flowering. Note broken pollen sacs in vicinity of stigma. × 236. *Fig. 4–78.* Pollen tube with two male gametes. Ovule to right and above; ovary wall below. × 380. *Fig. 4–79.* Zygote and primary endosperm nucleus. Abbreviations: en—primary endosperm nucleus, ov—ovule, ow—ovary wall, pt—pollen tube, zy—zygote. × 375.

The transmitting tissue of the stigma is made up of papillae with lateral protrusions that anastomose with each other. Papillae of this type occupy the distal end of the stigma and secrete most of the stigmal exudate (Fig. 4–76). Proximal to them are one to three whorls of free papillae lacking protrusions. These are also secretory (Fig. 4–80).

There are numerous exudate-filled channels in the stigma and style. Pollen tubes grow in these channels, which provide nutrition and mechanical guidance. At the base of the stigma, in the transition zone between stigma and style, there is a gradual increase in the amount of exudate between cells except in the center of the style. These cells comprise the stylar-transmitting tissue; they secrete an exudate similar in

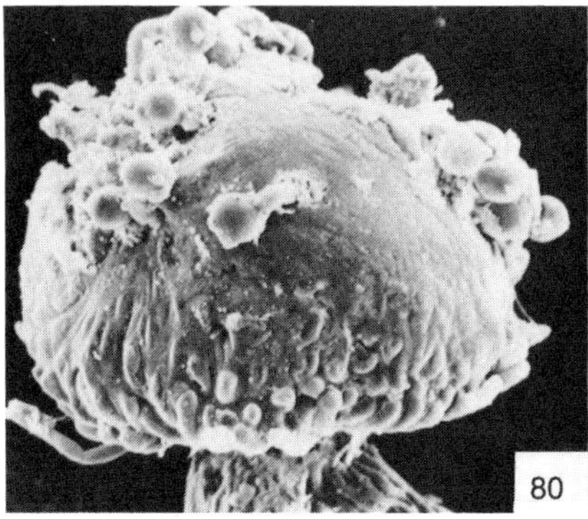

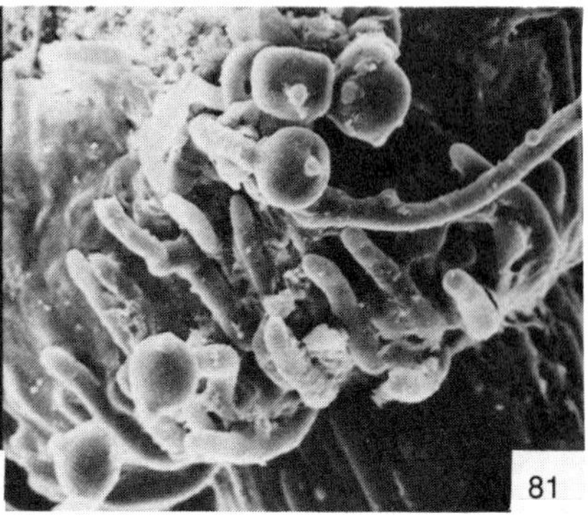

Fig. 4–80. Scanning electron micrograph view of pollen germinating on stigma. From Tilton et al. (1984c) × 230.

Fig. 4–81. Pollen grains lodged between lower-whorl papillae germinate but the pollen tubes grow into the main body of the stigma. From Tilton et al. (1984c). × 335.

appearance to that of the stigma. Stylar-transmitting tissue cells are mostly free from one another along their axial walls, thus forming a conduit through which pollen tubes grow.

Pollen usually germinates on the surface of the film overlying the stigmal exudate. Germination can also occur among the lower whorls of papillae but these tubes then grow into the stigma before entering the style (Fig. 4–77 and 4–81). Although many pollen grains are deposited on the stigma, and most of them germinate and grow into the stigma and upper style, perhaps as many as 90% of the tubes atrophy and die before reaching the distal end of the ovary. Only a few pollen tubes reach the locule and compete for ovules to fertilize.

Pollen tubes grow between the cells of the stylar-transmitting tissue. The ovarian transmitting tissue forms a secretory obturator on top of which the pollen tubes grow toward the ovules. Its exudate is pectinaceous, which perhaps controls the direction of pollen tube growth chemotactically.

During growth of the pollen tube toward the ovule, the generative cell divides and forms two male gametes, the sperm cells (Fig. 4–78). Finally, the pollen tube grows through the micropyle of the ovule, between nucellar epidermal cells, and enters the filiform apparatus of the degenerate synergid. Here the pollen tube tip bursts and releases the two sperm cells. One sperm fuses with the egg and forms the diploid zygote, the first cell of the embryo, while the other sperm fuses with the secondary nucleus, forming the primary endosperm nucleus (Fig. 4–79). Rustamova (1964) noted that the time from pollination to fertilization varies from about 8 to 10 h. Thus, the day of full opening of the flower is likely the day of fertilization or perhaps is 1 day after fertilization.

Following fertilization, a zone of extensive wall ingrowths called a Wandlabrinthe develops around the micropylar end of the central cell. The Wandlabrinthe encircles the synergids by forming a ridge of cell wall material that projects into the central cell toward the zygote and the synergids at the level of the filiform apparatus (Fig. 4–82). Degenerating nucellar cells are adjacent to, and contiguous with, the Wandlabrinthe, and adjacent cells of both integuments are rich in starch reserves. The Wandlabrinthe may increase the movement of nutrients into the central cell for endosperm and embryo nutrition.

4-5 EMBRYO DEVELOPMENT

The vacuole in the zygote becomes smaller and finally disappears entirely about the time of the first cell division, which occurs about 32 h after pollination (Pamplin, 1963; Rustamova, 1964). Soueges (1949) described soybean embryogeny from the first division of the zygote through the early cotyledon stages (Fig. 4–83). The first division of the zygote is transverse. The apical cell, facing the central cell, will become the embryo. The basal cell, facing the micropyle, forms the suspensor, an ephemeral structure that may aid early embryo growth (Fig. 4–83-1 and 4–84). Continued divisions of the derivatives of the apical cell produce the spherical proembryo at about 3 days (Fig. 4–83-14). The proembryo is approximately the same size as the somewhat conical suspensor (Fig. 4–85). A well-defined protoderm is present in the proembryo by 5 days after fertilization (Fig. 4–83–21 and Fig. 4–90). The suspensor in cv. Hawkeye

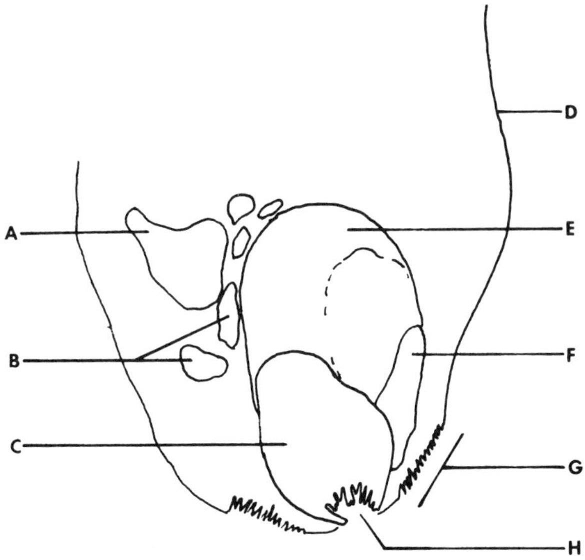

Fig. 4–82. Egg apparatus end of embryo sac, reconstructed in outline from transmission electron micrographs in Tilton et al. (1984) Abbreviations: A—endosperm initial, B—amyloplasts, C—persisting synergid, D—central cell wall, E—zygote, F—disintegrating synergid, G—Wandlabrinthe, and H—filiform apparatus.

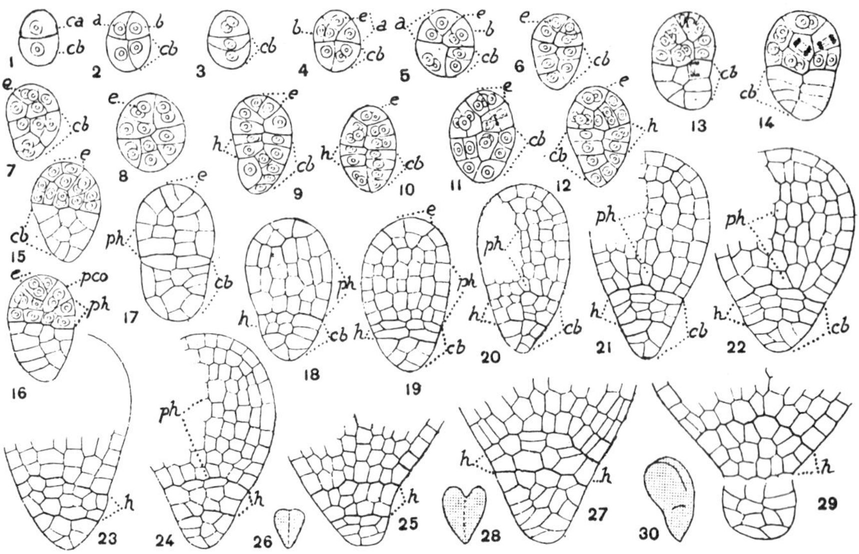

Fig. 4–83. *Glycine soja* Sieb. and Zucc. The principal stages of embryo development. From Soueges (1949). Abbreviations: *a* and *b*—daughter cells of the apical cell *ca*, and *cb*—basal cell, *e*—epiphysis, *h*—hypophysis, *ph*—hypocotyl proper, and *pco*—cotyledon primordium region.

at 6 days is several cells wide (Fig. 4–90), although suspensors of other cultivars are reported as tiny or rudimentary to small (Lersten, 1983).

About 6 to 7 days after fertilization, localized divisions at opposite sides of the proembryo just below the protoderm initiate the cotyledons. Pamplin (1963) observed that the cotyledon at the chalazal side of the embryo seems to be initiated first but is quickly followed by initiation of the second cotyledon, which grows rapidly and soon is the same size as the first cotyledon. In Fig. 4–91 and 4–92, the chalazal cotyledon appears to be slighty larger than the one toward the anterior surface of the ovule. This, however, may be a consequence of the plane of sectioning rather than an actual difference in size. The cotyledons in this so-called *heart stage* are initiated and developed in a plane that is approximately 90° displaced from their final position in the mature seed. As the cotyledons continue to develop, there is a gradual rotation such that the embryo, with its cotyledons, moves 90° and the cotyledons assume the position they will have in the mature seed, with their inner surfaces forming a plane to the sides of the ovule (Fig. 4–93 to 4–95).

At this stage the cotyledons appear circular in outline, but rapid growth along the margins, especially toward the chalazal end of the ovule, results in a pronounced elongation of the cotyledons, giving them their typical reniform shape. Ten to 12 days after fertilization, the tissue systems of the hypocotyl are well blocked out and consist of protoderm, the ground meristem of the cortex, and procambium. The derivatives of the hypophysis have formed the initials of the root which, until the time of

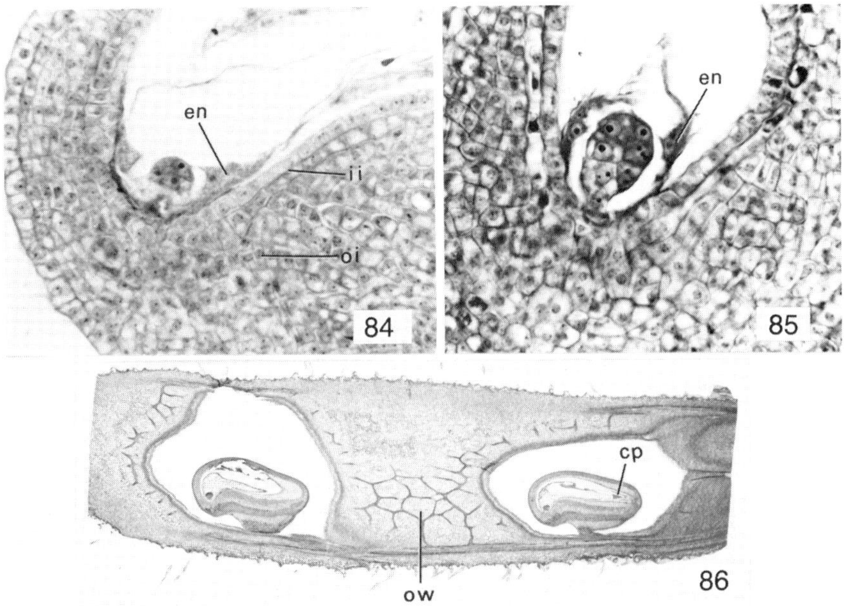

Fig. 4–84 to 4–86. *Fig. 4–84.* Four-celled embryo, 2 days after fertilization. Endosperm nuclei near embryo. × 320. *Fig. 4–85.* Club-shaped embryo, 3 days after fertilization. Endosperm surrounding embryo is acellular. × 320. *Fig. 4–86.* Pod with two ovules, 5 days after fertilization. Abundant endosperm in embryo sac is cellular only near embryo. Both ovules in the same development stage. Abbreviations: cp—chalazal process of endosperm, en—endosperm, ii—inner integument, oi—outer integument, and ow—ovary wall. × 10.

germination, remain limited to a small area at the end of the hypocotyl just above the point of attachment of the suspensor (Fig. 4–94).

The cotyledons, at about the time of the beginning of rotation of the embryo, already have procambium continuous with the procambium of the hypocotyl. The procambium of the cotyledons continues to develop and forms the finely divided vascular system present in the mature seed (Fig. 4–106).

The epicotyl is initiated simultaneously with the origin of the two cotyledons, as a residual meristem between them. Pamplin (1963) stated that it first appears as an elongated mound of deeply staining cells between the bases of the cotyledons. The outermost cell layer becomes the tunica. About 14 days after fertilization, the epicotyl forms the primordia of the two primary leaves at right angles to the point of attachment of the two cotyledons (Fig. 4–97 and 4–98). The primary leaves continue to enlarge, and by 30 days have reached their maximum dormant embryo size and have assumed the conduplicate vernation characteristic of the plumule of the mature seed (Kato et al., 1954) (Fig. 4–109).

The first trifoliolate leaf primordium, differentiated about 30 days after fertilization near the base of the two simple leaves, remains reduced in size and does not resume development until the time of germination.

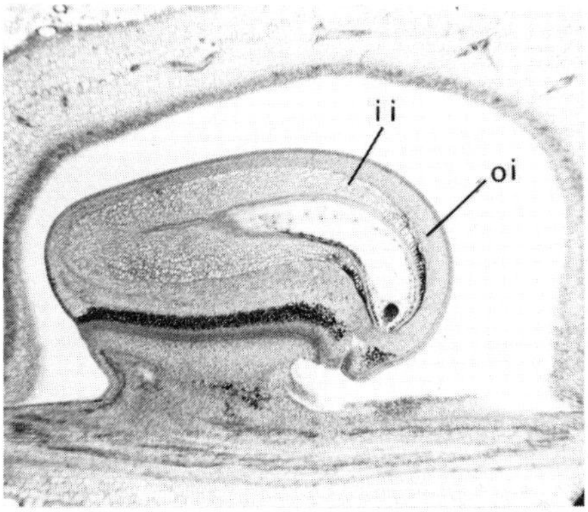

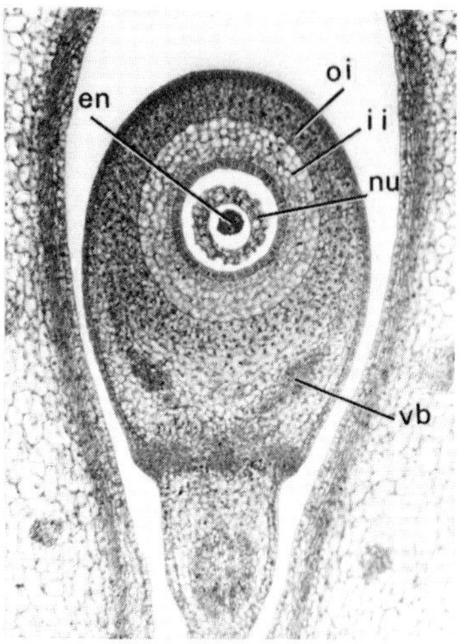

Fig. 4–87. Entire ovule with spherical embryo, 6 days after fertilization. Micropyle visible. Endosperm nuclei line outer surface of embryo sac. Large vacuole present in center of endosperm. × 28.

Fig. 4–88. Cross section of ovule at chalazal end, 6 days after fertilization. Chalazal process of endosperm and nucellus present. Abbreviations: en—chalazal process of endosperm, ii—inner integument, nu—nucellus, oi—outer integument, and vb—vascular bundles. × 73.

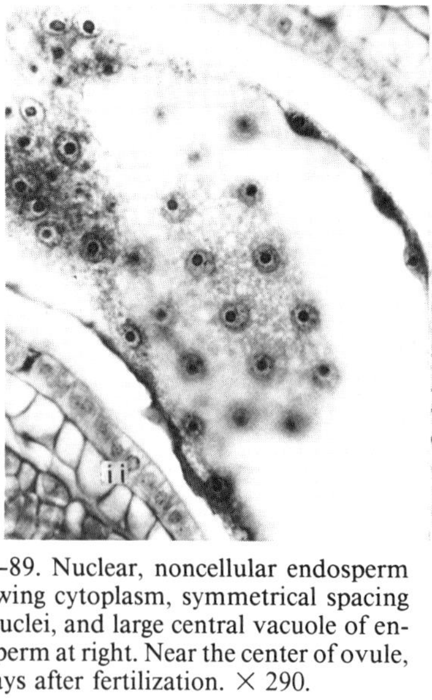

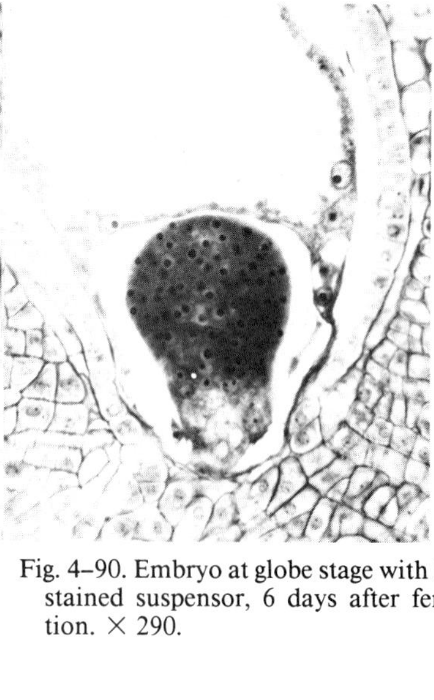

Fig. 4–89. Nuclear, noncellular endosperm showing cytoplasm, symmetrical spacing of nuclei, and large central vacuole of endosperm at right. Near the center of ovule, 6 days after fertilization. × 290.

Fig. 4–90. Embryo at globe stage with lightly stained suspensor, 6 days after fertilization. × 290.

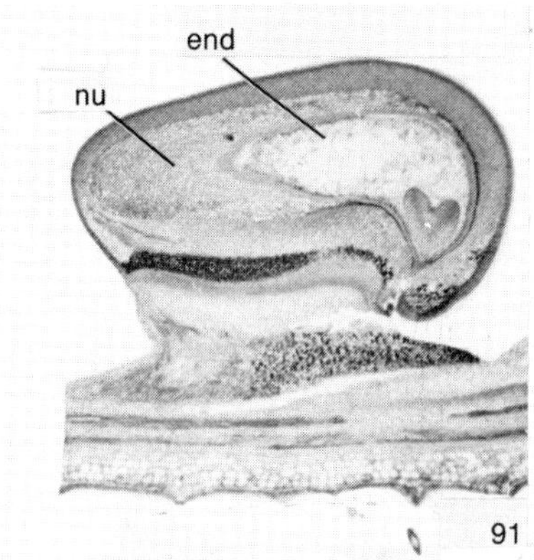

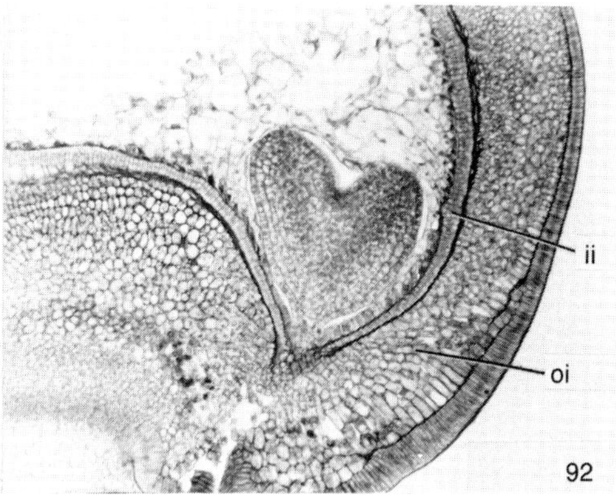

Fig. 4–91. Entire ovule, 8 days after fertilization. Cotyledons starting to develop. All endosperm is cellular. Abbreviations: end—endosperm and nu—nucellus. × 20.

Fig. 4–92. Enlarged portion of Fig. 4–91. Note small suspensor. Palisade epidermal cells of seed coat becoming elongated. Integumentary tapetum of inner integument prominent. Abbreviations: ii—inner integument and oi—outer integument. × 86.

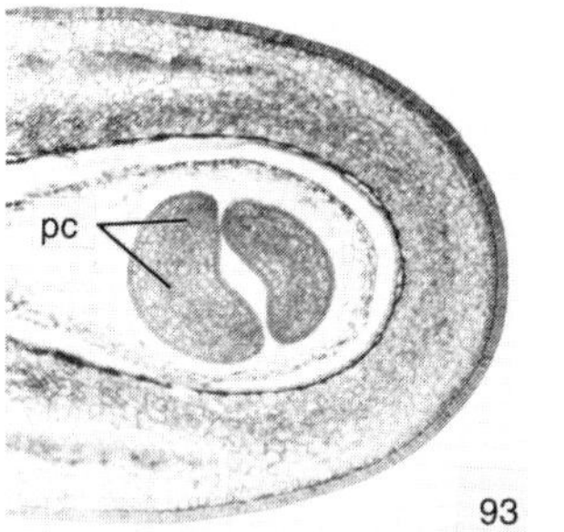

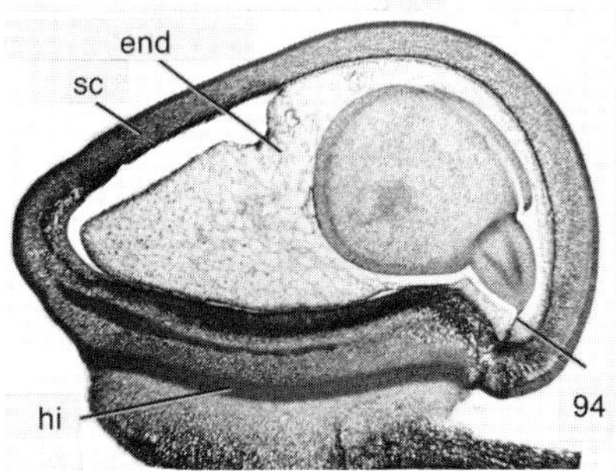

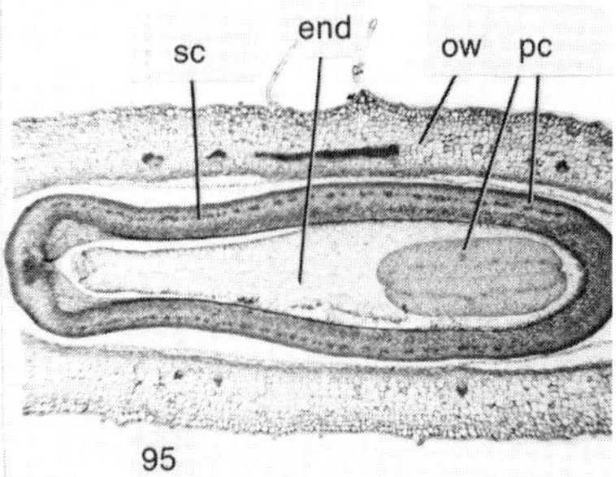

Fig. 4–93 to 4–95. *Fig. 4–93.* Cross section of cotyledons 12 days after fertilization. Cotyledons starting to rotate. × 50. *Fig. 4–94.* Late cotyledon stage, 14 days after fertilization. Abundant cellular endosperm. Suspensor still intact. Double layer of palisade cells visible at hilum. × 19. *Fig. 4–95.* An L-section of 14-day-old seed. Cotyledons have rotated 90° and are now in normal position. Endosperm fills most of the remainder of the seed. Extensive procambium network in seed coat. Abbreviations: end—endosperm, ow—ovary wall, pc—procambium, sc—seed coat, su—suspensor, and hi—hilum. × 19.

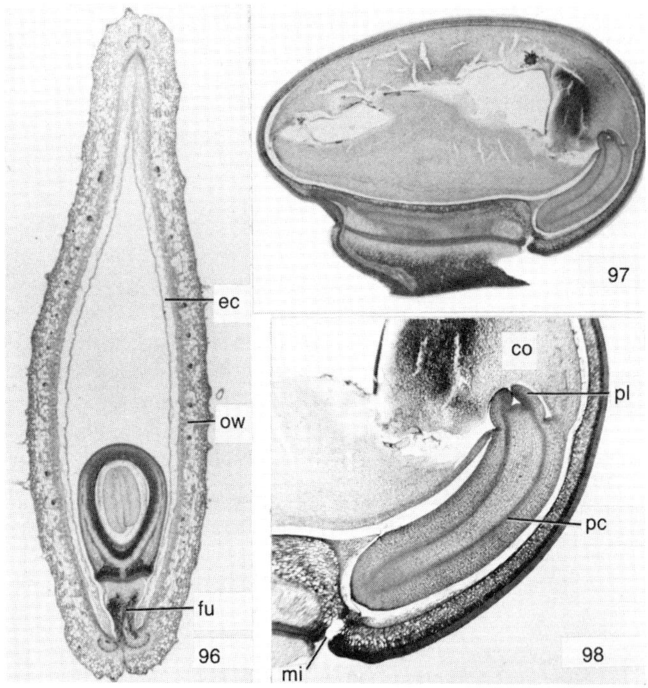

Fig. 4–96 to 4–98. *Fig. 4–96.* Cross section of pod and one seed 14 days after fertilization. Embryo in late cotyledon stage. Vascular bundle of funiculus shown connected to lateral bundle of ventral suture of carpel. × 11. *Fig. 4–97.* Median L-section of seed 18 days after fertilization. The section was parallel to inner epidermis of both cotyledons and resulted in some tissue breakage. × 9. *Fig. 4–98.* Same seed. Primary leaves are present and oriented at right angles to the point of attachment of the cotyledons. Abbreviations: co—cotyledon, ec—membranous endocarp, fu—funiculus, mi—micropyle, ow—ovary wall, pc—procambium, and pl—primary leaves. × 24.

Bils and Howell (1963) described biochemical and cytological changes during the development of soybean cotyledons. They noted that, about 15 to 18 days after flowering, plastids, mitochondria, and some lipid and protein globules were beginning to form. By 26 days, when the cotyledons had reached their maximum size, the cells contained many mitochondria, some lipid granules, and a few protein globules. As fresh weight began dropping during the final stages of maturation, the starch also started to decrease and, by the time of embryo dormancy, starch grains were gone. Lipids comprised 22% of the dry weight of the cotyledons, and proteins about 50%, at 60 days.

4–6 ENDOSPERM DEVELOPMENT

The primary endosperm nucleus divides almost immediately following fertilization. By the time of zygote division, the endosperm already has several free nuclei (Pamplin, 1963; Prakash and Chan, 1976). Divisions of the endosperm nuclei occur as simultaneous cycles for several days following fertilization. The nuclei and common cytoplasm of the endosperm are displaced toward the periphery of the embryo sac by the development of a large vacuole in the center of the mass of endosperm (Fig. 4–87). The free nuclei of the sac-like endosperm are spaced uniformly within the cytoplasm (Fig. 4–89). By 5 days after fertilization, the endosperm begins to become cellular around the embryo at the micropylar end of the embryo sac and, by 8 days, the heart-shaped embryo is completely embedded in cellular endosperm (Fig. 4–92) (Meng-Yuan, 1963; Prakash and Chan, 1976; Takao, 1962). Endosperm cell walls develop gradually toward the chalazal end of the embryo sac; by 14 days they extend almost to the chalazal end of the ovule (Fig. 4–94).

The chalazal end of the endosperm never becomes cellular but instead forms a rather darkly staining acellular mass, termed the *chalazal process* by Pamplin (1963). This can be seen in the basal ovule in Figures 4–86 and 4–88. The chalazal process is connected with the degenerated nucellus to form a so-called *chalazal haustorium.* This haustorium adheres to the nucellus, which is connected to the vascular bundles. Takao (1962) concluded that nutrients move from the vascular bundles through the chalazal end of the outer integument, the inner integment, the nucellus, and finally to the fluid cytoplasm in the embryo sac and the endosperm. The acellular chalazal process finally is crushed by the continuous growth of the cellular endosperm and, by 12 or 14 days after fertilization, it has been completely obliterated.

As the ovule continues to enlarge, both embryo and endosperm grow at approximately the same rate so that the relative proportion of endosperm to embryo tissue remains the same until about 14 days after fertilization (Pamplin, 1963). The rapidly growing cotyledons, after they have rotated completely, accumulate food reserves derived from the endosperm. Eighteen or 20 days after fertilization, only remnants of the endosperm remain. It should be emphasized that no evidence exists that cotyledons absorb endosperm directly. In the mature seed, the only evidence of endosperm is a thin aleurone layer and a few crushed endosperm cells (Fig. 4–103). In some cultivars, an aleurone layer is lacking (Prakash and Chan, 1976).

4–7 SEED-COAT DEVELOPMENT

The inner integument consists of two to three cell layers at the time of fertilization. After fertilization, periclinal divisions, especially in the chalazal end of the ovule, result in an increase in thickness of the inner integument to about 10 cell layers (Fig. 4–87 and 4–88). About 10 to 14 days after flowering, the inner layer of the inner integument becomes more densely staining and differentiates as an endothelium or integumentary tapetum, which presumably serves a nutritive function (Fig. 4–

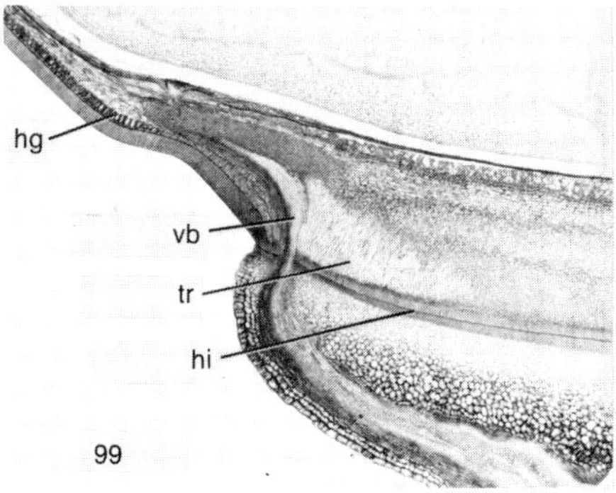

Fig. 4–99. Same seed as in Fig. 4–97 chalazal end of hilum. Single vascular bundle enters at the posterior edge of funiculus. Tracheid bar present adjacent to vascular bundle in the upper part of hilum. × 32.

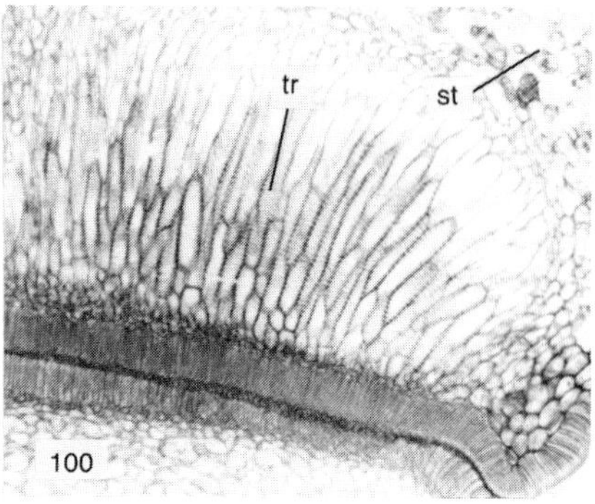

Fig. 4–100. Same seed. Details of tracheid bar at micropylar end of hilum. Note the reticulate thickenings of tracheid wall. Abbreviations: st—stellate parenchyma, tr—tracheids, and vb—vascular bundle. × 117.

89). Thorne (1981) noted the presence of a layer of multicellular tubules, each 50 by 200 μm long, between the seed coat and the cotyledons in soybean seeds nearing maturity. He suggested that these tubules help to transfer nutrients from the endothelium to the embryo. During development of the embryo, there is a gradual crushing of the inner integument, starting at the micropylar end and proceeding toward the chalazal end of the ovule. By 12 to 14 days after fertilization, most of the inner integument has completely disappeared (Kamata, 1952; Pamplin, 1963).

The outer integument at fertilization is two to four cell layers thick except in the region of the micropyle and the hilum, where it is considerably thicker (Pamplin, 1963). After fertilization, periclinal divisions occur and the outer integument becomes approximately 12 to 15 cell layers thick (Prakash and Chan, 1976). The epidermis of the outer integument consists of isodiametric cells at the time of fertilization. During growth and maturation of the seed, these cells elongate radially, especially near the hilum. The epidermal cells of the funiculus in the hilum region also elongate radially so that in the hilum there is a double layer of elongate thick-walled epidermal cells (Fig. 4–94, 4–99, and 4–100).

Extending the length of the hilum is a narrow, median strip of epidermal cells that never become thick-walled or elongate, and which separate when dry, leaving a narrow cleft (hilar groove) in the mature seed (Fig. 4–101). Just below this cleft, some cells of the outer integument differentiate into cells (tracheoids) that resemble tracheids, forming the so-called *tracheid bar* of the mature seed. The tracheoids are lignified, pitted, and oriented with their long axis perpendicular to the hilar groove. Lersten (1982) noted that tracheoids rarely had an intact pit membrane and, in the tribe Phaseoleae, subtribe Glycininae, the pits were warty to vestured. He speculated that the structure of the tracheoid pits enhances the efficiency of the tracheid bar and hilum in gas exchange. The tracheid bar may extend the entire length of the hilum or may be separated into

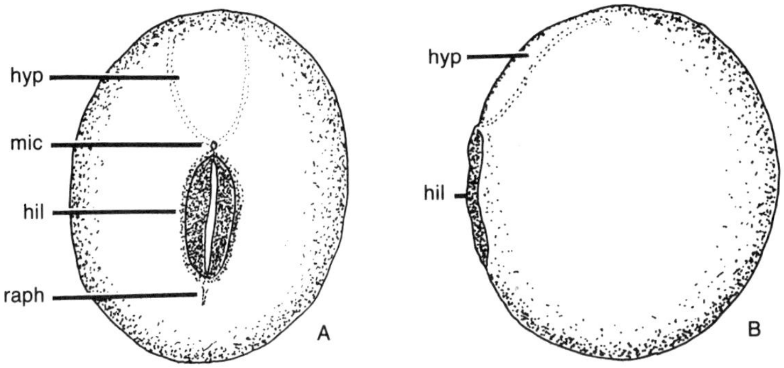

Fig. 4–101. Drawing of 'Chippewa' soybean seed. (A) Top view. (B) Side view. Abbreviations: hyp—hypocotyl-radicle axis, mic—micropyle, hil—hilum with central fissure, and raph—raphe. × 6.

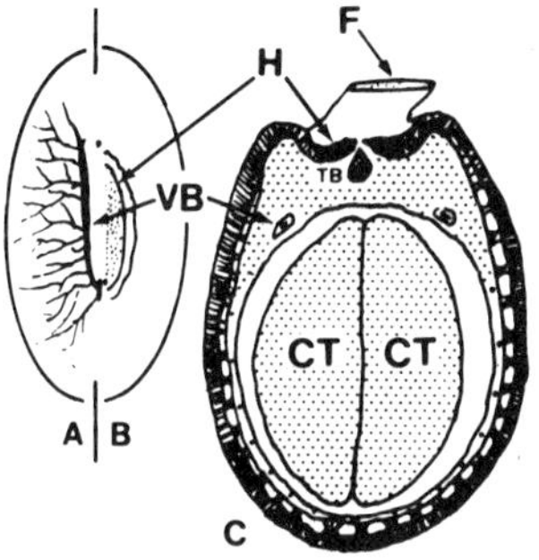

Fig. 4–102. Vascularization of the soybean seed coat. (A) Sketch of a lateral bundle, illustrating the approximate relationship to the hilum, illustrated in (B). The initial branching of the reticulate venation is also illustrated in (A). (C) Sketch of a typical transverse section of an entire seed attached to the funiculus, illustrating the approximate location of the lateral vascular bundles in relation to the tracheid bar, hilum, and cotyledons. Abbreviations: CT—cotyledon, F—funiculus, HL—hilum, P—parenchyma, TB—tracheid bar, and VB—vascular bundle. From Thorne (1981).

two groups of tracheids, one near the micropyle and the other near the chalazal end of the seed. Although the tracheoids have scalariform pitted walls and resemble true conducting tissue, they seem to have no conducting function (Dzikowski, 1937).

The hypodermal cells of most of the outer integument, by 28 days after fertilization, have differentiated into sclerified cells described in the section on the mature seed (Fig. 4–98 and 4–104).

The entire vascular system of the ovule at the time of fertilization consists of a single median vascular bundle that passes from the vascular bundle of the ventral suture of the pod through the funiculus and enters the ovule at the chalazal end (Fig. 4–96). By 4 or 5 days after fertilization, two lateral procambium strands appear in the outer integument above the hilum and extend most of the length of the ovule near the inner surface of the seed coat (Fig. 4–88). Maturation of these bundles proceeds from the chalazal end, where they are attached to the single median

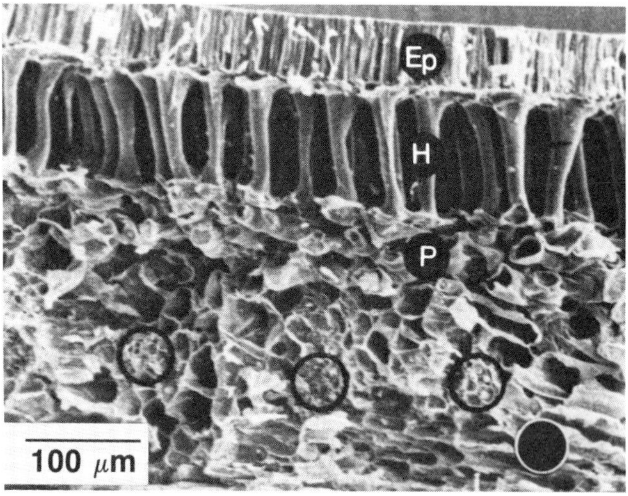

Fig. 4–103. Scanning electron micrograph of a seed-coat portion illustrating the reticulate venation (*circled*) embedded within the parenchyma layer. The distinctive spindle-shaped cells of the epidermis and hourglass-shaped cells of the hypodermis provide structural support to the seed. Abbreviations: Ep—epidermal cells (macrosclereids), H—hypodermis (osteosclereids). From Thorne (1981).

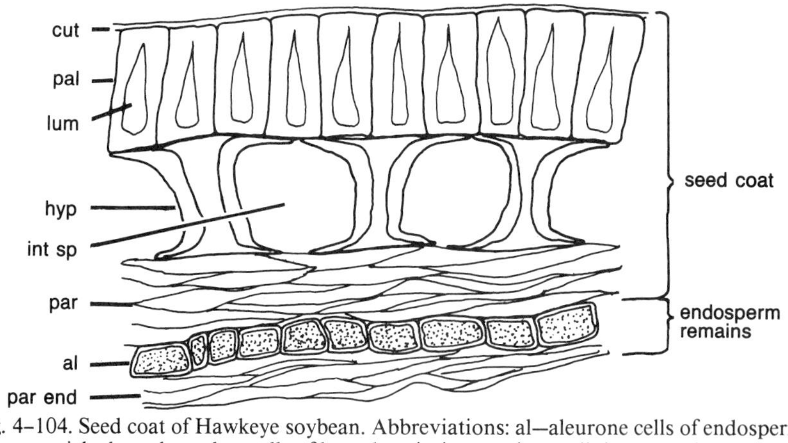

Fig. 4–104. Seed coat of Hawkeye soybean. Abbreviations: al—aleurone cells of endosperm, cut—cuticle, hyp—hourglass cells of hypodermis, int sp.—intercellular space, lum—lumen, pal—palisade, par—compressed parenchyma cells, and par end—remains of parenchyma cells of endosperm. × 535.

bundle, and progresses forward toward the micropylar end. A number of procambium branches develop from these two lateral veins and, by 12 days, a rather extensive anastomosing reticulate system of veins has formed throughout the entire outer integument (Pamplin, 1963; Thorne, 1981) (Fig. 4–95 to 4–102).

The veins are composed of small, thick-walled sieve tubes surrounded by a bundle sheath of small vascular parenchyma cells. Xylem

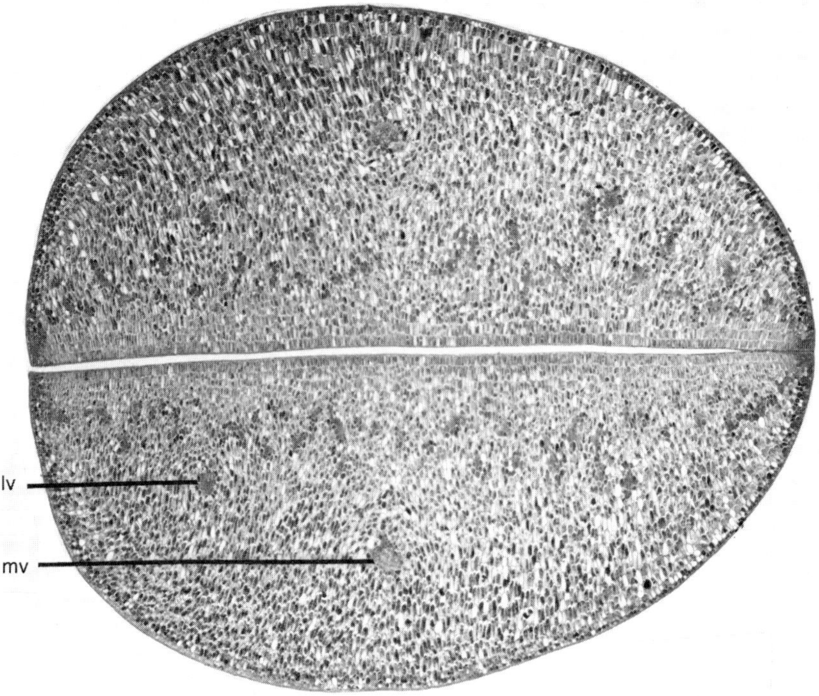

Fig. 4–105. Cross section, cotyledons of dormant Hawkeye seed. Abbreviations: mv—midvein, lv—lateral vein. × 16.

is absent (Thorne, 1981). As the seed coat continues to mature, the inner parenchyma cells become crushed and flattened, and eventually there is little evidence of a functional vascular system. Presumably the mature seed coat has no functional vascular tissue.

4–8 MATURE SEED

The mature soybean seed, like that of many other legumes, is essentially devoid of endosperm and consists of a seed coat surrounding a large embryo. Seed shape varies among cultivars from almost spherical to strongly flattened and elongate, but the seeds of most cultivars are oval in outline. The seed coat is marked with a hilum (seed scar) that varies in shape from linear to oval. At one end of the hilum is the micropyle, a tiny hole formed by the integuments during seed development (Fig. 4–97 and 4–98), but covered by a cuticle at maturity. The tip of the hypocotyl-radicle axis, often visible through the seed coat, is located just below the micropyle (Fig. 4–101). At the other end of the hilum is the raphe, a small groove extending to the chalaza, where the integuments were attached to the ovule proper. In most cultivars, the complete separation of the funiculus from the seed forms a hilum with a smooth surface except for a narrow fissure running lengthwise down its center. In a few varieties the funiculus remains attached to the hilum by a core of parenchyma (Dzikowski, 1936). When the funiculus finally separates in these cultivars, the hilum is rough and has a wide, white central scar formed by the parenchyma tissue.

The seed coat proper has three distinct layers: (i) epidermis, (ii) hypodermis, and (iii) inner parenchyma layer (Fig. 4–104). The epidermal layer consists of closely packed, thick-walled palisade cells (macrosclereids). These cells, 35 to 70 μm long, are elongated perpendicular to the surface of the seed and have thickened, pitted walls in the outer part of the cell. A cuticle is present on the outer wall of the macrosclereids. As is common in legumes, there is a particularly compact zone present in the walls of the upper part of the macrosclereids that refracts light more strongly than the rest of the wall (Esau, 1965). This characteristic *light line* is visible in seeds of many wild forms of the soybean, but is less prominent in the cultivated forms (Alexandrova and Alexandrova, 1935).

The hypodermis consists of a single layer of sclerified cells variously elongated and separated from each other. These cells range from 30 to 100 μm in length (Patel, 1976). The unevenly thickened cell walls are thin at the ends of the cell and very thick in the central, constricted portion of the cell. These cells thus form a strong supporting layer with considerable intercellular space (Fig. 4–103 and 4–104).

The inner parenchyma tissue consists of six to eight layers of thin-walled, flattened cells that lack contents. This parenchyma tissue is essentially uniform throughout the entire seed coat except at the hilum, where it forms three distinct layers: (i) an outer layer, formed of stellate parenchyma tissue with much intercellular space, in contact with the sclerified hypodermal cells; (ii) a middle parenchyma layer consisting of tiny, flattened cells and containing small bundles of spiral vessels that branch out around the hilum; (iii) an inner layer consisting of more or less typical parenchyma tissue (Dzikowski, 1936).

It is probably of physiological significance that the micropyle and the fissure in the center of the hilum are in direct contact with the loosely packed stellate parenchyma cells and that these are in contact with the extensive intercellular space formed by the sclereid layer. Since the cutinized palisade cells are essentially impermeable to gases, the principal pathway for gas exchange between the embryo and the external environment is through the hilum. Therefore, the structure of the hilum probably has an effect on the metabolism and moisture content of the embryo (Dzikowski, 1936).

The remnants of the endosperm are tightly appressed to the seed coat proper (Fig. 4–104). The outer endosperm layer, called the *aleurone layer*, is composed of small cuboidal cells filled with dense protein. Just to the interior to the aleurone layer are several layers of endosperm cells that have been flattened by growth of the embryo (Dzikowski, 1936; Williams, 1950). In some cultivars there is no distinct aleurone layer (Prakash and Chan, 1976).

Recently, Wolf et al. (1981) and Hill and West (1982) surveyed surface features of soybean seed coats. Many cultivars had abundant pits, up to 277 pits mm^{-2}, while others had few or none. Some "hard" seeds had pits; other seeds lacked them. Pits vary in shape from circular, 15 to 25 μm in diameter, to elongate, 3 $\times$ 40 μm. The pits penetrated about 20 to 35% of the thickness of the palisade layer. Oval-shaped cavities were often present below the pit, extending to the sclereid layer. Pits have been shown to provide an entry for fungal hyphae and they may also contribute to water uptake during germination.

The seed coats of some cultivars have a superficial reticulate or honeycomb pattern, which is formed by the residue of the epidermal cell walls of the inner layer of the ovary wall (endocarp), which adheres to the seed coat. Occasionally, small crystals are present as deposits on seed coats (Wolf et al., 1981).

The mature, dormant embryo consists of two large fleshy cotyledons, a plumule with two well-developed primary leaves enclosing one trifoliolate leaf primordium, and a hypocotyl-radicle axis that rests in a shallow depression formed by the cotyledons (Fig. 4–105, 4–108, and 4–109). The tip of the radicle is surrounded by an envelope of tissue formed by the seed coat (Miksche, 1961).

In cross section, each cotyledon is semicircular in outline, bounded by an epidermis of cuboidal cells containing aleurone grains (Fig. 4–105). Stomates are present on both surfaces. The mesophyll of the flat, adaxial side of the cotyledon is made up of one to three layers of palisade tissue that merge with a more spongy type of parenchyma in the central portion of the cotyledon (Fig. 4–106). The abaxial region of the cotyledon consists mainly of elongate parenchyma cells that do not form distinct layers. All cells of the mesophyll are filled with closely packed aleurone grains and oil droplets. Small crystals of calcium oxalate are scattered throughout the cotyledon.

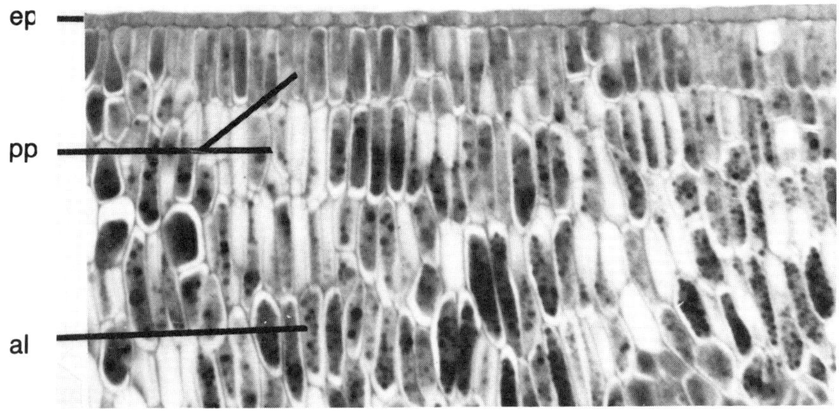

Fig. 4–106. Adaxial side of dormant cotyledon, Hawkeye soybean. Abbreviations: ep—epidermis, pp—palisade parenchyma, and al—aleurone grains. $\times$ 130.

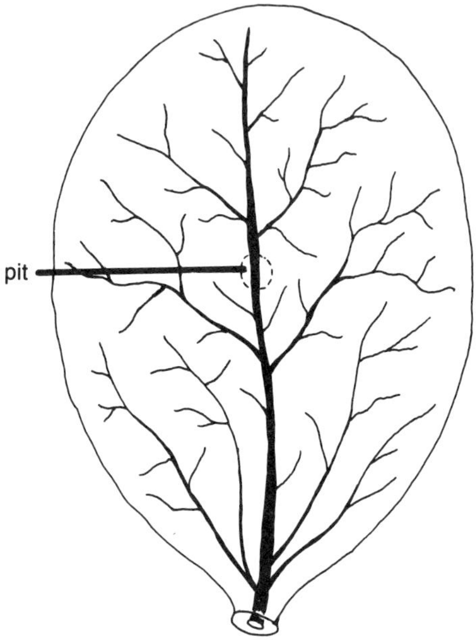

Fig. 4–107. Cleared dormant cotyledon of Hawkeye soybean, showing netted venation and pit. From Miksche (1959). × 7.

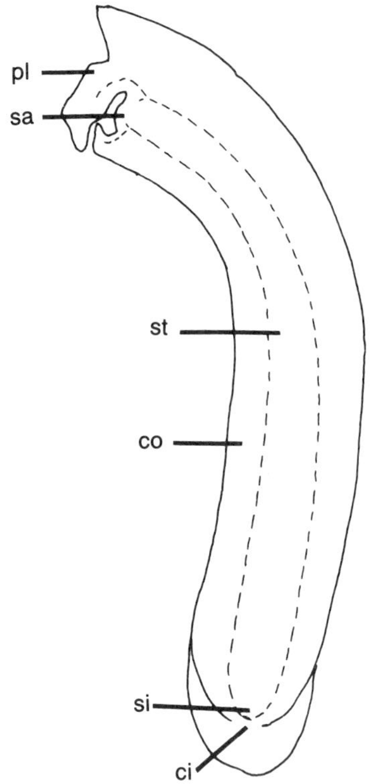

Fig. 4–108. Longitudinal section of embryo axis. Most of the primary leaves are removed. From Miksche (1959). Abbreviations: ci—common initials; co—cortex, pl—primary leaf base, sa—stem apex, si—stelar initials, and st—stele. × 23.

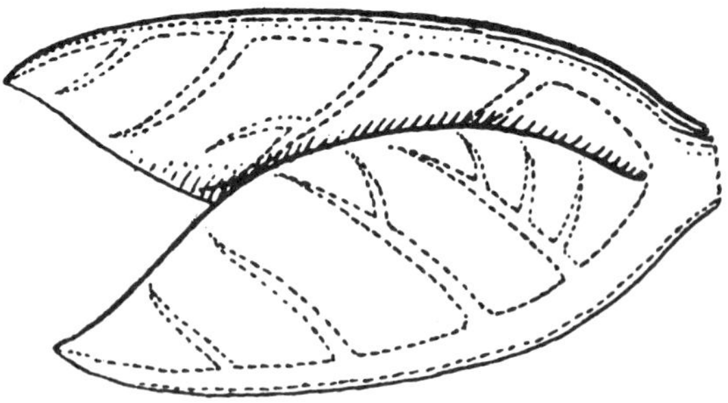

Fig. 4–109. Plumule with two simple leaves showing pinnate venation. From Miksche (1959).

The plasma membrane of parenchyma cells in cotyledons of dry seeds is disorganized, or at least not discernible, in many areas, while elsewhere it has a typical unit membrane structure. In these cells no endoplasmic reticulum can be discerned, and mitochondria are of irregular shape with little evidence of cristae. Nuclei appear round or lobed, bounded by a membrane in which nuclear pores are visible. A 20-min period of imbibition brings about an extensive reorganization of the membranes and organelles (Webster and Leopold, 1977).

A small pit is present in the center of the abaxial surface of the cotyledon above the midvein (Fig. 4–107). It is prominent in some cultivars but barely perceptible in others (Dzikowski, 1937; Miksche, 1961).

The plumule is about 2-mm long and has two opposite simple leaves, each with a pair of stipules at the base. The vascular system of the primary leaves is pinnate and consists of protoxylem and metaxylem initials, and some mature protophloem elements (Miksche, 1961). The stem apex has a uniseriate tunica and a massive corpus. Prior to seed maturation and dormancy, cell divisions in the corpus form the initials of the first trifoliolate leaf.

The hypocotyl-radicle axis is about 5-mm long and somewhat flattened both on the outer surface, which is in contact with the seed coat, and on the inner surface, which is tightly appressed to the cotyledons. The radicle, located at the tip of the embryo axis, consists of the stelar initials that produce the stele and a group of common initials that give rise to the root cap, epidermis, and cortex (Miksche, 1961). The transition from root to hypocotyl is not marked by any clear anatomical change in the dormant embryo. The tissue systems present in the hypocotyl are the epidermis, cortex, and stele (Fig. 4–108).

Soybean seeds vary in color from yellow, green, brown, to black, and they may be of one color, bicolored, or variegated. Seed-coat pigments, located mainly in the palisade layer, consist of anthocyanin in the vacuole, chlorophyll in the plastids, and various combinations of breakdown products of these pigments. Both the palisade layer and the stellate parenchyma are often pigmented in the hilum, thus giving a more intense coloration to that region (Alexandrova and Alexandrova, 1935).

The cotyledons of the mature embryo may be green, yellow, or chalky yellow, but in most genotypes they are yellow (Williams, 1950). The

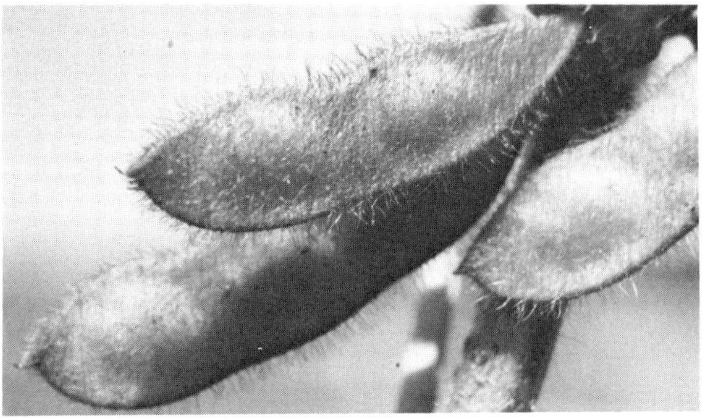

Fig. 4–110. Seventeen-day pods, Ottawa-Mandarin soybean.

various combinations of pigments in the seed coat and cotyledons are responsible for the wide range of colors of the seeds in wild and cultivated varieties of soybean.

4-9 POD DEVELOPMENT

The number of pods varies from 2 to more than 20 in a single inflorescence and up to 400 on a plant. A pod may contain from one to five seeds, but in most common cultivars it usually has two or three seeds per pod (Kato et al., 1954; Williams, 1950). Soybean pods are straight or slightly curved and vary in length from < 2 cm to up to 7 or more centimeters in some cultivars (Fig. 4–110). Mature pods vary from light yellow to yellow-gray, brown, or black. Pod coloration depends on the presence of carotene and xanthophyll pigments, the color of the trichomes, and the presence or absence of anthocyanin pigments (Dzikowski, 1936).

From the moment of fertilization the ovary starts developing into the fruit, but the style and stigma dry out. The calyx persists during fruit development; remnants of the corolla may also be present when the fruit is mature.

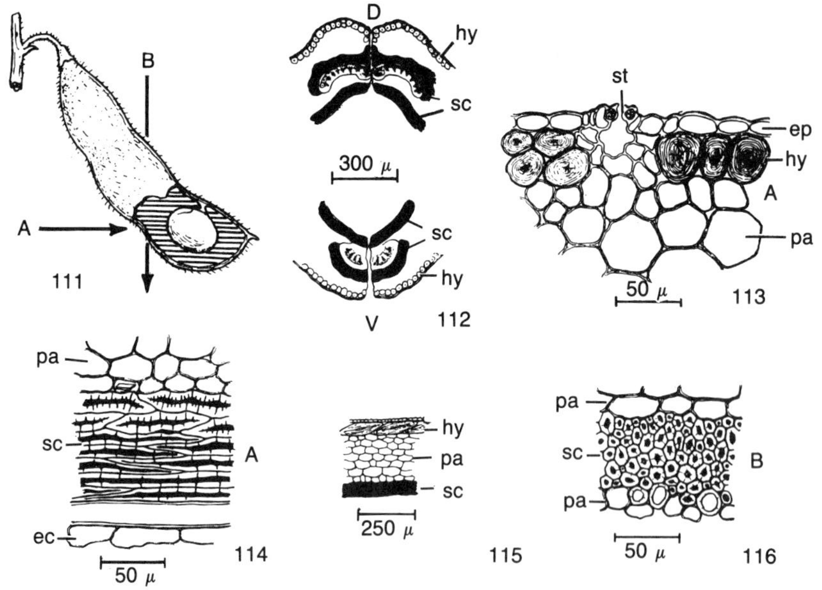

Fig. 4–111 to 4–116. *Fig. 4–111.* Soybean pod showing the direction of the fibers in the fibrous layers. *Fig. 4–112.* Diagram of ventral (V) and dorsal (D) sutures of a pod. *Fig. 4–113.* Portion of a transverse section of the wall of a pod, along the length of fibers (section A, Fig. 4–110). *Fig. 4–114.* The same part of the pod in section A. *Fig. 4–115.* Section of the pod wall, perpendicular to the fibers in the fibrous layer (section B, Fig. 4–110). *Fig. 4–116.* Inner layer of the pod wall. Portion of section B, Fig. 4–110. Abbreviations: ec—endocarp, ep—epidermis, hy—short fibers under the epidermis, pa—parenchyma, sc—sclerenchyma layer, and st—stoma. From Dzikowski (1937).

The soybean pod is similar to that of other legumes, consisting of two halves of the single carpel joined by dorsal and ventral sutures (Fig. 4–96 and 4–112). The dorsal suture constitutes the main vein of the former carpel, while the ventral suture consists of two principal bundles that correspond to marginal veins of the carpel. On both the dorsal and ventral sutures above the veins, the epidermis of the pod bends inward, forming deep grooves (Dzikowski, 1937) (Fig. 4–112). Extending below the grooves is a vertical layer of parenchyma that separates the conducting tissues into two regions. These layers of parenchyma later help the pods to dehisce.

The wall of a young pod consists of a variously hairy epidermis, a rather wide zone of parenchyma tissue in which the extensive vascular system is embedded, and an inner, very thin layer of parenchyma tissue destined to become the membranous endocarp (Fig. 4–86, 4–95, and 4–96).

As the pod matures, the epidermal cells develop strongly thickened walls covered by a well-developed cuticle. The surface of the epidermis has numerous elevated stomates, each connected by a substomatal chamber to the parenchyma of the inner portion of the pod. The clavate trichomes disappear but the setaceous trichomes develop thick walls and are persistent at maturity (Fig. 4–110) (Dzikowski, 1936, 1937).

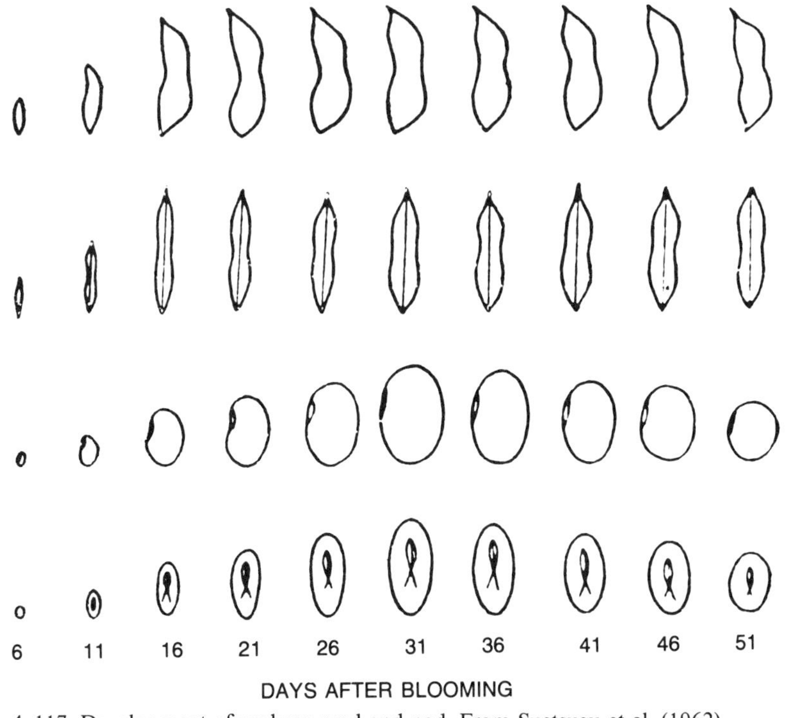

Fig. 4–117. Development of soybean seed and pod. From Suetsugu et al. (1962).

Table 4–1. Chronology of development of flower and ovule of soybean.†

Days before flowering	Morphological and anatomical features
25	Initiation of floral primordium in axil of bract.
25	Sepal differentiation.
20–14	Petal, stamen, and carpel initiation.
14–10	Ovule initiation; maturation of megasporocyte; meiosis; four megaspores present.
10–7	Anther initiation; male archesporial cells differentiate; meiosis; microsporogenesis.
7–6	Functional megaspore undergoes first mitotic division.
6–2	Second mitotic division results in four-nucleate embryo sac.
	Third mitotic division results in eight-nucleate embryo sac.
	Cell walls develop around antipodals and egg apparatus forming a seven-celled and eight-nucleate embryo sac.
	Polar nuclei fuse. Antipodal cells begin to degenerate. Nucellus begins to disintegrate at micropylar end and on sides of embryo sac.
	Single vascular bundle in ovule extends from chalaza through funiculus and joins with the carpellary bundle.
1	Embryo sac continues growth; antipodals disorganized and difficult to identify. Synergids with filiform apparatus; one synergid degenerating.
	Tapetum in anthers almost gone. Pollen grains mature; some are germinating.
	Nectary surrounding ovary reaches maximum height.
0	Flower opens; usually day of fertilization; resting zygote; primary endosperm nucleus begins dividing.
	Nectary starts collapsing.

†The times are a compilation of data from several soybean cultivars studied by Carlson (1973), Kato et al. (1954), Murneek and Gomez (1936), Pamplin (1961), Prakash and Chan (1976). The sequence of development is essentially the same regardless of cultivar but the absolute times vary with environmental conditions and with cultivars.

Just below the epidermis there is a hypodermal layer consisting of a single layer of short fusiform fibers with very thick walls and numerous small pits (Fig. 4–113 and 4–115). The parenchyma below the hypodermal layer consists of many large, thin-walled isodiametric cells. Within this parenchyma tissue, an extensive vascular system of anastomosing veins interconnects the principle bundles of the dorsal and ventral sutures. Below the parenchyma layer is a rather thick layer of elongate sclerenchyma fibers, which are thick-walled and of small diameter (Fig. 4–116). The orientation of the cells of the hypodermis is almost perpendicular to the orientation of the cells of the inner sclerenchyma layer (Dzikowski, 1936, 1937; Monsi, 1942) (Fig. 4–111). The innermost layer of the pod is a very thin endocarp composed of rather flattened parenchyma cells (Fig. 4–114).

When the pod dehisces, only the inner sclerenchyma layer, termed *motion tissue* by Monsi (1942), seems to be directly involved. A section cut parallel to the long axis of the sclerenchyma cells of the inner layer reveals that there are two distinct layers of cells present. The sclerenchyma cells closest to the central parenchyma tissue are considerably shorter, with blunt ends, and have abundant pits. Monsi (1942) and Dzikowski

Table 4-2. Chronology of development of seed and pod of soybean.†

Days after flowering	Morphological and anatomical features
0	Resting zygote. Several divisions of primary endosperm nucleus.
1	Two-celled proembryo. Endosperm with about 20 free nuclei.
2	Four- to eight-celled proembryo.
3	Differentiation into proembryo proper and suspensor. Endosperm in peripheral layer with large central vacuole.
4–5	Spherical embryo with protoderm and large suspensor. Endosperm surrounding embryo is cellular but elsewhere it is mostly acellular and vacuolate.
6–7	Initiation of cotyledons. Endosperm mostly cellular.
8–10	Rotation of cotyledons begins. Procambium appears in cotyledons and embryo axis. All tissue systems of hypocotyl present. Root cap present over root initials. Endosperm all cellular.
10–14	Cotyledons have finished rotation and are in normal position with inner surfaces of cotyledons parallel with sides of ovules. Cotyledons elongate toward chalazal end of ovule. Primary leaf primordia present. Endosperm occupies about half of seed cavity. Extensive vascularization of seed coat.
14–20	Continued growth of embryo and seed. Reduction of endosperm tissue by assimilation into cotyledons.
20–30	Primary leaves reach full size. Primordium of first trifolioate leaf present. Cotyledons reach maximum size. Endosperm almost gone.
30–50	Continued accumulation of dry matter, and loss in fresh weight of seeds and pod. Maturation of pod.
50–80	Various maturity times depending on variety and environmental factors.

†The times are a compilation of data from several soybean cultivars studied by Bils and Howell (1963), Carlson (1973), Fukui and Gotoh (1962), Meng-yuan (1963), Kamata (1952), Kato et al. (1956), Ozaki et al. (1956), Pamplin (1963), Suetsugu et al. (1962). The sequence of development is essentially the same regardless of cultivar but the absolute times vary with environmental conditions and with cultivars.

(1937) determined that these inner cells have an essentially transverse orientation of fibrils in their secondary wall. The sclerenchyma cells of the innermost layers are longer, narrower with more pointed ends, and have fibrils with an essentially longitudinal orientation (Fig. 4–114).

Separation of the two halves of the pod is preceded by the appearance of clefts through the parenchyma of the dorsal and ventral sutures. After separation, the halves twist spirally around the axis, that is, parallel to the direction of the fibers of the inner sclerenchyma layer (Dzikowski, 1937). Dehiscence of the pod must be directly due to differences in tension developed in the cells of the inner sclerenchyma layer as a result of loss of moisture. The innermost cells of the sclerenchyma layer, with a parallel orientation of fibrils along the longitudinal axis of the cells, shorten more during the drying process than the upper sclerenchyma cells, which have a transverse orientation of fibrils. The pod thus bends because of differences in changes in cell length in the two layers of the sclerenchyma tissue. Because of the slanted position of these fibers in relation to the axis of the pod, the two halves of the pod will twist. The parenchyma tissue of the pod, the hypodermis, and the epidermal layers, do not seem to have any direct connection with pod dehiscence.

Figure 4–117 is a diagrammatic representation of changes in pod and ovule length, width, and thickness with reference to days after blooming. Although times may vary among individual cultivars and under various environmental conditions, this table illustrates the sequence of changes occurring during pod and seed development. The maximum length of the pod is reached rather early in development, about 20 to 25 days after bloom (Andrews, 1966; Kamata, 1952). At this stage, the seeds have attained an average of 4% of their maximum dry weight (Fraser et al., 1982). The maximum width and thickness of the pod is reached about 30 days after bloom. This corresponds with the time that the seed reaches its maximum size in all dimensions. Maximum fresh seed weight and maximum seed size are reached 5 to 15 days later. As the maturing seed loses moisture, it changes from an elongate reniform shape to the more oval or spherical shape characteristic of the mature seed (Fig. 4–101).

4–10 CHRONOLOGY

A chronology of flower and ovule development is presented in Table 4–1, and of seed and pod development in Table 4–2. These tables represent a compilation of data from several of the investigations cited in this chapter. The sequence of events remains the same regardless of cultivar or environmental conditions but the absolute times between events may vary by several days as a function of environment and cultivar.

REFERENCES

Abernathy, R.H., R.G. Palmer, R. Shibles, and J.C. Anderson. 1977. Histological observations on abscising and retained soybean flowers. Can. J. Plant Sci. 57:713–716.

Albertsen, M.C., and R.G. Palmer. 1979. A comparative light and electron-microscopic study of microsporogenesis in male sterile (ms_1) and male fertile soybeans (*Glycine max* (L.) Merr.). Am. J. Bot. 66:253–265.

Alexandrova, V.G., and O.G. Alexandrova. 1935. The distribution of pigments in the testa of some varieties of soybeans, *Glycine hispida* Maxim. Bull. Appl. Bot. Genet. Plant Breed. 3(4):3–47.

Andrews, C.H. 1966. Some aspects of pod and seed development in Lee soybeans. Diss. Abstr. B 27(5):1347B.

Bils, R.F., and R.W. Howell. 1963. Biochemical and cytological changes in developing soybean cotyledons. Crop Sci. 3:304–308.

Borthwick, H.A., and W.M. Parker. 1938. Influence of photoperiods upon the differentiation of meristems and the blossoming of Biloxi soybeans. Bot. Gaz. 99:825–839.

Buss, P.A., and N.R. Lersten. 1975. A survey of tapetal number as a taxonomic character in Leguminosae. Bot. Gaz. 136:388–395.

Carlson, J.B. 1973. Morphology. *In* B.E. Caldwell (ed.) Soybeans: Improvement, production, and uses. Agronomy 16:17–95.

Dzikowski, B. 1936. Studia nad soja *Glycine hispida* (Moench) Maxim. Cz. 1. Morfologia. Mem. Inst. Natl. Pol. Econ. Rurale 254:69–100.

----. 1937. Studia nad soja *Glycine hispida* (Moench) Maxim. Cz. 11. Anatomia. Mem. Inst. Natl. Pol. Econ. Rurale 258:229–265.

Erickson, E.H., and M.B. Garment. 1979. Soya-bean flowers: Nectary ultrastructure, nectar guides and orientation on the flower by foraging honeybees. J. Agric. Res. 18:3–11.

Esau, K. 1965. Plant anatomy. 2nd ed. John Wiley and Sons, New York.

Fraser, J., D.B. Egli, and J.E. Leggett. 1982. Pod and seed development in soybean cultivars with differences in seed size. Agron. J. 74:81–85.

Fukui, J., and T. Gotoh. 1962. Varietal difference of the effects of day length and temperature on the development of floral organs in the soybean. I. Developmental stages of floral organs of the soybean. Jpn. J. Breed. 12:17–27.

George, G.P., A. George, and J.M. Herr, Jr. 1979. A comparative study of ovule and megagametophyte development in field-grown and greenhouse-grown plants of *Glycine max* and *Phaseolus aureus* (Papilionaceae). Am. J. Bot. 66:1033–1043.

Guard, A.T. 1931. Development of floral organs of the soybean. Bot. Gaz. 91:97–102.

Hansen, W., and R. Shibles. 1978. Seasonal log of the flowering and podding activity of field-grown soybeans. Agron. J. 70:47–50.

Hardman, L.L. 1970. The effects of some environmental conditions on flower production and pod set in soybean *Glycine max* (L.) Merrill var. Hark. Diss. Abstr. 31(5):2401-B.

Hill, H.J., and S.H. West. 1982. Fungal penetration of soybean seed through pores. Crop Sci. 22:602–605.

Johns, C.W., and R.G. Palmer. 1982. Floral development of a flower-structure mutant in soybeans, *Glycine max* (L.) Merr. (Leguminosae). Am. J. Bot. 69:829–842.

Kamata, E. 1952. Studies on the development of fruit in soybean 1–2. (In Japanese.) Crop Sci. Soc. Jpn. Proc. 20:296–298.

Kato, I., and S. Sakaguchi. 1954. Studies on the mechanism of occurrence of abortive grains and their prevention on soybeans, *Glycine max*. M. Bull. Div. Plant Breed. Cultiv. Tokai-Kinki Natl. Agric. Exp. Stn. 1:115–132.

----, ----, and Y. Naito. 1954. Development of flower parts and seed in soybean plant, *Glycine max* . M. Bull. Div. Plant Breed. Cultiv. Tokai-Kinki Natl. Agric. Exp. Stn. Bull. 1:96–114.

----, ----, and ----. 1955. Anatomical observations on fallen buds, flowers, and pods of soybean, *Glycine max*. M. Bull. Div. Plant Breed. Cultiv. Tokai-Kinki Natl. Agric. Exp. Stn. 2:159–168.

Lersten, N.R. 1982. Tracheid bar and vestured pits in legume seeds (Leguminosae: Papilionoideae). Am. J. Bot. 69:98–107.

----. 1983. Suspensors in Leguminosae. Bot. Rev. 49:233–257.

Linskens, H.F., P.L. Pfahler, and E.L. Knuiman-Stevens. 1977. Identification of soybean cultivars by the surface relief of the seed coat. Theor. Appl. Genet. 50:147–149.

Madjd, A., and F. Roland-Heydacker. 1978. Secretions and senescence of tapetal cells in the anther of *Soja hispida* Moench, Papilionaceae. (In French.) Grana 17:167–174.

Meng-Yuan, H. 1963. Studies on the embryology of soybeans. 1. The development of embryo and endosperm. (In Chinese.) Acta Bot. Sinica. 11:318–328.

Miksche, J.P. 1961. Developmental vegetative morphology of *Glycine max*. Agron. J. 53:121–128.

Monsi, M. 1942. Untersuchungen uber den Mechanismus der schleuderbewegung der Sojabohnen-Hulse. Jpn. J. Bot. 12:437–474.

Murneek, A.E., and E.T. Gomez. 1936. Influence of length of day (photoperiod) on development of the soybean plant, *Glycine max* var. Biloxi. Mo. Agric. Exp. Stn. Res. Bull. 242.

Ozaki, K., M. Saito, and K. Nitta. 1956. Studies on the seed development and germination of soybean plants at various ripening stages. (In Japanese.) Res. Bull. Hokkaido Natl. Agric. Exp. Stn. 70:6–14.

Palmer, R.G., M.C. Albertsen, and H. Heer. 1978. Pollen production in soybean with respect to genotype, environment, and stamen position. Euphytica 27:427–434.

Pamplin, R.A. 1963. The anatomical development of the ovule and seed in the soybean. Ph. D. diss. Univ. of Illinois, Urbana. (Diss. Abstr. 63–5128).

Patel, J.D. 1976. Comparative seed coat anatomy of some Indian edible pulses. Phyton 17:287–299.

Prakash, N., and Y.Y. Chan. 1976. Embryology of *Glycine max*. Phytomorphology 26:302–309.

Rembert, D.H., Jr. 1977. Contribution to ovule ontogeny in *Glycine max*. Phytomorphology 27:368–370.

Rustamova, D.M. 1964. Some data on the biology of flowering and embryology of the soybean under conditions prevailing around Tashkent. (In Russian.) Uzb. Biol. Zh. 8(6):49–53.

Soueges, R. 1949. Embryogénie des Papilionacées. Développement de l'embryon chez le *Glycine soja* Sieb. et Zucc. (*Soya hispida* Moench). C.R. Acad. Sci. 229:1183–1185.

Stelly, D.M., and R.G. Palmer. 1982. Variable development in anthers of partially male sterile soybeans. J. Hered. 73:101–108.

Suetsugu, I., I. Anaguchi, K. Saito, and S. Kumano. 1962. Developmental processes of the root and top organs in the soybean varieties. (In Japanese.) p. 89–96. *In* Bull. Hokuriki Agric. Exp. Stn. Takada 3.

Takao, A. 1962. Histochemical studies on the formation of some leguminous seeds. Jpn. J. Bot. 18:55–72.

Thorne, J.H. 1981. Morphology and ultrastructure of maternal seed tissues of soybean in relation to the import of photosynthate. Plant Physiol. 67:1016–1025.

Tilton, V.R., R.G. Palmer, and L.W. Wilcox. 1984a. The female reproductive system in soybeans, *Glycine max* (L.) Merr. (Leguminosae). p. 33–36. *In* Erdelskaan (ed.) Fertilization and embryogenesis in ovulated plants. VEDA, Bratislava, Czechoslovakia.

----, L.W. Wilcox, and R.G. Palmer. 1984b. Post-fertilization Wandlabrinthe formation and function in the central cell of soybean, *Glycine max* (L.) Merr. (Leguminosae). Bot. Gaz. 145:334–339.

----, ----, ----, and M.C. Albertsen. 1984c. Stigma, style, and obturator of soybean, *Glycine max* (L.) Merr. (Leguminosae) and their function in the reproductive process. Am. J. Bot. 71:676–686.

Van Schaik, P.H., and A.H. Probst. 1958. Effects of some environmental factors on flower productive efficiency in soybeans. Agron. J. 50:192–197.

Waddle, R., and N. Lersten. 1973. Morphology of discoid nectaries in Leguminosae, especially tribe Phaseoleae (Papilionoideae). Phytomorphology 23:152–161.

Webster, B.D., and A.C. Leopold. 1977. The ultrastructure of dry and imbibed cotyledons of soybean. Am. J. Bot. 64:1286–1293.

Wiebold, W.J., D.A. Ashley, and H.R. Boerma. 1981. Reproductive abscission levels and patterns for eleven determinate soybean cultivars. Agron. J. 73:43–46.

Williams, L.F. 1950. Structure and genetic characteristics of the soybean. p. 111–134. *In* K.S. Markley (ed.) Soybeans and soybean products. Interscience Publishers, New York.

Wolf, W.J., F.L. Baker, and R.L. Bernard. 1981. Soybean seedcoat structural features: Pits, deposits and cracks. Scanning Electron Microsc. 1981(3):531–544.

5 Qualitative Genetics and Cytogenetics

Reid G. Palmer
USDA-ARS
Iowa State University
Ames, Iowa

Thomas C. Kilen
USDA-ARS
Soybean Production Research
Stoneville, Mississippi

In the first edition of the Soybean monograph, Bernard and Weiss (1973) reported that seven linkage groups and 79 gene loci had been identified. During the next 10 yrs, the number of identified loci more than doubled. Two of the fastest growing aspects of qualitative genetics are the study of sterility mutants and electrophoretic analysis of isoenzymes. Hadley and Hymowitz (1973) stated that research in soybean cytogenetics was sporadic and secondary to that in other areas of soybean research. During the last 13 yrs, chromosome interchanges, inversions, and aneuploids have been described, and maps have been established for 13 linkage groups.

The close relationship among qualitative genetics, chromosome structure, and plant behavior was the impetus for considering qualitative genetics and cytogenetics within the same chapter of this monograph.

5-1 SOYBEAN GENETICS COMMITTEE

The Soybean Genetics Committee, established in 1955, carries out the following functions: (i) maintains a Genetic Type Collection including strains, preferably the author's original strain, carrying each qualitatively inherited gene that has been published; (ii) establishes guidelines and rules for assigning gene symbols; and (iii) acts as a review committee for manuscripts concerning qualitative genetic interpretation and gene symbols in the genus *Glycine*. The committee encourages scientists to submit

Soybeans: Improvement, Production, and Uses, 2nd ed.—Agronomy Monograph no. 16.

manuscripts to them and to send seeds of the genetic line under consideration so that orderly identification and nomenclature are followed and the line is preserved. Six elected members, the curator of the Soybean Genetics Collection, and the editor of the *Soybean Genetics Newsletter* constitute the Soybean Genetics Committee. The report of the committee and the rules for genetic symbols are published annually in the *Soybean Genetics Newsletter.*

5-2 *SOYBEAN GENETICS NEWSLETTER*

In 1974, the *Soybean Genetics Newsletter* was established as a means of communication at the international level on topics related to genetics and breeding of the soybean and immediate relatives. Information in the *Soybean Genetics Newsletter* is informal to stimulate thought and to exchange ideas. Newsletter articles can be preliminary and speculative. The Soybean Genetics Committee reviews articles concerning qualitative genetic interpretation and gene symbols. Genetic symbols reported in the newsletter have the same status as those published in refereed journals.

Current editor of the newsletter is R. G. Palmer, G301, Dep. of Agronomy, Iowa State University, Ames, IA 50011.

5-3 GENETIC TYPE COLLECTION

The Genetic Type Collection was started with genetic types, identified by T-numbers, that C. M. Woodworth and L. F. Williams had accumulated at the University of Illinois since 1921. In 1955, the Collection was revised and expanded when its maintenance was taken over by a USDA scientist. A list of the lines currently in this collection is given in Table 5-1. This table presents a summary of the Genetic Type Collection, from T16 through the current T279, listing the strain, genes, phenotype, prior designation, and the source of origin. All known and available qualitatively and cytoplasmically inherited genes are represented, except those carried in named cultivars and in plant introductions (PIs). Also included are a few combination types, and certain old T-strains are retained because they have been used in published research.

Table 5-1. Soybean genetic type collection.

Strain†	Gene‡	Phenotype	Prior designation	Source§
T16	–	Brown hilum on black seed		Found in Ebony before 1930
T31	*p2*	Puberulent		Found in Soysota × Ogemaw, by F.W. Wentz, Ames, IA, in 1926

(continued on next page)

Table 5-1. Continued.

Strain†	Gene‡	Phenotype	Prior designation	Source§
T41	*ln(d1 d2)*	Narrow leaflet		Unknown source, Urbana, IL
T43	*P1(cyt-G1)*	Glabrous	Progeny 435B	From Medium Green × "glabrous" before 1927
T48	-	Spread hilum		From Manchu × Ebony before 1930
T54	*dt1*	Determinate stem		Found in Manchu before 1927
T93	*v1 (D1 d2* or *d1 D2)*	Variegated leaves		Found in a hybrid population before 1931
T93A	*v1 (d1 d2)*	Variegated leaves		From T93
T102	*y4*	Greenish yellow leaves, weak plant		Found in Wilson-Five before 1932
T104	*d1 d2 G cyt-G1*	Green seed embryo; green seed coat; green seed embryo		From T42 (green cotyledon from H. Terao) × "Chromium Green" before 1932
T116	*y5*	Greenish yellow leaves, very weak plant		Found in radium-treated PI 65388 before 1934
T117	*Dt2 lw1 Lw2*	Semideterminate stem; nonwavy leaf	L34-602	From AK114 × PI 65394
T122	*lo (d1 d2)*	Oval leaflet, few-seeded pods		Unknown, before 1934, Urbana, IL
T134	*y5*	Greenish yellow leaves		Found in Illini × Peking in 1937
T135	*y9*	Bright greenish yellow leaves		Found in Illini in 1938
T136	*y6 (ln dt1)*	Pale green leaves		From PI 88351 × Rokusun in 1937
T138	*y7 y8*	Yellow growth in cool weather	L35-1156	Unknown source, Urbana, IL
T139	*g y3*	Yellow seed coat; leaves turn yellow prematurely		Found in Illini by Brunson in Kansas about 1936
T143	*Lf1 y3 y7 y8*	5-foliolate; leaves turn yellow prematurely; yellow growth in cool weather		From T138 × T137. (T137 is *y3* from a cross in PI 81029)
T144	*d1 d2 v1 y7 y8*	Green seed embryo; variegated leaves; yellow growth in cool weather		From LX431: T93A × T138
T145	*P1*	Glabrous		Unknown source, Urbana, IL
T146	*r-m*	Brown seed with black stripes		From LX286: PI 82235 × PI 91073
T152	*i*	Self dark seed coat		Mutant in Lincoln
T153	*k1*	Dark saddle on seed coat		Mutant in Lincoln

(continued on next page)

Table 5-1. Continued.

Strain†	Gene‡	Phenotype	Prior designation	Source§
T157	*i*	Self dark seed coat		Mutant in Richland
T160	—	Pale green leaves		Found in Hahto (Michigan)
T161	*y10*	Greenish yellow seedling		Found in L36-5 from Mandarin × Mansoy in 1940
T162	*y17*	Light yellowish green leaves		Found in Mandarin in 1940
T164	—	Slightly variegated leaves		Found in Morse in 1941
T171	—	Long peduncle		Unknown
T173	*f (ln)*	Fasciated stem		From Keitomame (f) × PI 88351 *(ln)*
T175	*E1 t*	Late maturity; gray pubescence		Unknown source, Urbana, IL
T176	*lw1 lw2 (Dt2)*	Wavy leaf		Unknown source, Urbana, IL
T180	*Rj1*	Nodulating	L46-1741-2	From same F_3 plant as T181
T181	*rj1*	Nonnodulating	L46-1743-2	Found in Lincoln[2] × Richland
T201	*rj1*	Nonnodulating		From L46-1743 × L46-1741
T202	*Rj1*	Nodulating		Sib of T201
T204	*ln lo*	Narrow leaflet, 4-seeded pods; oval leaflet, few-seeded pods	L48-101	From T136 × T122
T205	*lw1 lw2*	Wavy leaf	L48-163	From Dunfield × Manchuria 13177
T208	*Se*	Pedunculate inflorescence, small seeds		PI 196176
T209	—	Dwarf	L50-155	From Lincoln × "wild dwarf"
T210	*df2*	Dwarf	L49-738	Found in colchicine-treated Lincoln
T211H	*pm*	Dwarf, crinkled leaves, sterile	CX3941-844-2-5	From Kingwa × T161 at Lafayette, IN
T216	—	Reddish black seeds	L46-266	From PI 86038 × PI 88351
T218M	—	Chlorophyll chimera (resembles T225M)		Found in Illini in 1952
T219H	*y11*	Lethal yellow (heterozygote has greenish yellow leaves)	A691-1	Found in Richland × Linman 533 in 1941 at Ames, IA
T220	—	Greenish yellow leaves	L46-431	Found in Lincoln
T221	—	Yellowish green leaves	L46-426	Found in Peking
T223	—	Yellowish green leaves	L46-429	Found in Richland
T224	—	Greenish yellow leaves	L46-428	Found in Richland
T225M	*Y18-m*	Unstable allele resulting in chlorophyll chimera		Found in Lincoln in Iowa before 1955

(continued on next page)

Table 5-1. Continued.

Strain†	Gene‡	Phenotype	Prior designation	Source§
T225H	*y18*	Near-lethal, yellow leaves		From T225
T226	—	Greenish yellow leaves		Found in Lincoln in 1943 at Ames, IA
T227	—	Slightly greenish yellow leaves, becoming green		Found in Illini in 1943 at Kanawha, IA
T229	*y14*	Light green leaves		Found in F_4 Richland × Linman 533 in 1943 at Ames, IA
T230	*y13*	Whitish green seedling, greenish yellow leaves	A43K-643-1	Found in Mandell × Mandarin (Ottawa) in 1944 at Kanawha, IA
T231	—	Greenish yellow leaves, weak plant	A49-8414 (AX3015-55)	Found in Richland × Linman 533 in 1943 at Ames, IA
T232	—	Yellowish green leaves		Found in Hawkeye in 1950 at Ames, IA
T233	*y12*	Whitish primary leaves, yellowish green leaves		Found in Hawkeye in 1950 at Ames, IA in N2100
T234	*y15*	Pale yellowish green leaves		Found in L46-2132 (Clark progenitor) in 1952 at Ames, IA
T235	*wm*	Magenta flower	L58-274	Found in Harosoy in 1957
T236	*(Lf1 ln y6)*	Red-buff seed coat	L46-232	From T143 × "*y6 ln pc dt1 w1*"
T238	*k3*	Dark saddle on seed coat	S57-3416	Found in x-rayed Clark in 1956 at Columbia, MO
T239	*k2*	Tan saddle on seed coat	L63-365	Found in Harosoy in 1961
T241H	*st2*	Asynaptic sterile		Found in S54-1714 about 1956 at Columbia, MO
T242H	*st3*	Asynaptic sterile		Found in AX54-118-2-8 (Blackhawk × Harosoy) at Lafayette, IN
T243	*df2*	Dwarf		Found in colchicine-treated Lincoln at Ames, IA
T244	*df3*	Dwarf		Found in neutron-irradiated Adams at Ames, IA
T249H	*(P1)*	Whitish yellow seedling, lethal	L67-4408A	Found in F_3 (Clark^6 × PI 84987) × (Clark^6 × T145) in 1964
T250H	—	Lethal seedling	L67-4439	Found in F_2 Harosoy^5 × (Clark^6 × Chief) in 1964

(continued on next page)

Table 5-1. Continued.

Strain†	Gene‡	Phenotype	Prior designation	Source§
T251H	*mn*	Miniature plant	L67-4440A	Found in F_2 Harosoy[5] × T139 in 1961
T252	–	Pale green leaves	L64-2612	Found in F_3 Harosoy[6] × T139 in 1963
T253	*y20(k2)*	Yellowish green leaves, weak plant	L67-4415A	Found in T239 in 1963
T254	–	Greenish yellow leaves	L67-4412A	Found in F_2 Clark[6] × T176 in 1964
T255	*lf2*	7-foliolate		Found in Hawkeye in 1966 at Ames, IA
T256	*df4*	Dwarf		Found in Hark in 1966 at Ames, IA
T257H	*y16*	Near-lethal white		Found in C1128[8] × Mukden at Lafayette, IN
T258H	*st4*	Desynaptic sterile	A72-1103-6	Found in Hark in 1968 at Ames, IA
T259H	*ms2*	Male sterile	L71L-06-4	Found in F_3 of SL11 (Wayne-*r Rpm Rps1*) × L66L-177 [Wayne × (Hawkeye × Lee)] in 1971 at Eldorado, IL
T260H	*ms1* - North Carolina	Male sterile	N69-2774	Found in a farmer's field in 1966 in North Carolina
T261	*k2*	Tan saddle on seed coat	S56-26	Mutant found in Mandarin (Ottawa) before 1956 at Columbia, MO
T262	–	"Double pod"		Found in SRF200 (Hark-*ln*) about 1971 at Soybean Research Foundation, Mason City, IL
T263	*df5*	Dwarf	A76-2	Found in Harosoy 63 × PI 257435 in 1968 at the Iowa State University nursery in Hawaii
T264	*Pd2*	Dense pubescence	L58-2749	Found in neutron-irradiated Blackhawk in the N_2 generation in 1956
T265H	*y19*	Delayed albino	L75-0324	Found in Williams[6] × T259 in the F_2 generation in 1974-75 in the greenhouse
T266H	*ms1*-Urbana	Male sterile, (higher female fertility than T260, T267, and T268)		Found in an F_3 row of L67-533 (Clark[6] × Higan) × SRF300 in 1971

(continued on next page)

Table 5-1. Continued.

Strain†	Gene‡	Phenotype	Prior designation	Source§
T267H	*ms1*-Tonica	Male sterile	L56-292	Found in a field of Harosoy by F. M. Burgess, Tonica, IL, in 1955
T268H	*ms1*-Ames	Male sterile	A73g-21	Found in T258 at Ames, IA, in 1970
T269H	*fs1 fs2*	Structural sterile (T269H is from *Fs1 fs1 fs2 fs2* plants)	L70-8654	Found segregating in a plant-progeny row from PI 339868 (Yuwoltae) in 1970
T270H	—	Greenish yellow leaves, very weak plant	A76-518-3	Found segregating in an F_2-plant-progeny row from an outcross in T271H at Ames, IA
T271H	*msp*	Partial male sterile		Found in a 40-parent bulk population (AP6 [S1]C1) at Ames, IA in 1975
T272H	*st5*	Desynaptic sterile	A71-44-13	Found in Uniform Test entry W66-4108 (from Merit × W49-1982-32) in 1970 at Ames, IA. W49-1982-32 from Hawkeye × Manchu 3
T273H	*ms3*	Male sterile	A72-1711	Found in an F_3-plant-progeny row from Calland × Cutler in 1971 at Washington, IA
T274H	*ms4*	Male sterile	A74-4646	Semisterile plant found in Rampage in 1973 at Ames, IA
T275	*cyt-Y2*	Yellow leaves, becoming greenish yellow	A77-K150	Cytoplasmic yellow mutant from chimeric F_2 plant A75-1165-117 from T268H × (PI 101404B × Clark[6]) at Ames, IA
T276	*nr1*	Constitutive nitrate reductase absent		Selected in 1979 in the M_2 generation from Williams treated with EMS, nitrosoguanidine, and x-rays
T277H	*ms5*	Male sterile		Selected at Blacksburg, VA, in 1976 in the M_3 generation of Essex soybeans that had been neutron irradiated

(continued on next page)

Table 5-1. Continued.

Strain†	Gene‡	Phenotype	Prior designation	Source§
T278M	*cyt-Y3*	Yellow leaves, very weak plant (mutable plants are chlorophyll chimeras)		From chimeric plant (origin unknown) found at Ames, IA in 1972
T279	*lps*	Short petiole	D76-1609	Found at Stoneville, MS, in 1976 in F_3 of (Forrest[2] × PI 229358) × D71-6234. (D71-6234 is a selection from a high protein Lee type × PI 95960)

†For T-strains with an H suffix (e.g., T211H) the allele is carried as the heterozygote because the homozygote is lethal, sterile, or very weak. For T-strains with an M suffix (e.g., T225M) the trait is maintained by selecting the mutable genotype.

‡Cytoplasmically inherited traits are prefixed by *cyt-*. Genes for secondary traits are listed in parentheses.

§Selections were made at Urbana unless otherwise listed. Numerical superscripts are used to indicate backcrosses; e.g., Lincoln[2] × Richland means Lincoln × (Lincoln × Richland).

The Soybean Genetic Collection is divided into four categories. They are:

1. The Type Collection includes strains that collectively contain all published genes of soybean, preferably in the original strains. The U.S. and Canadian named cultivars, and PI strains are maintained in the U.S. Soybean Germplasm Collection. The Type Collection also includes mutants or strains considered by the Soybean Genetics Committee to have potential genetic interest.

2. The Isoline Collection includes near-isolines of single genes or combinations of genes backcrossed into the adapted cvs. Clark, Harosoy, and Lee. Also included are certain genes or gene combinations with other cultivars used as the recurrent parent.

3. The Linkage Collection includes linkage combinations and various genetic recombinations.

4. The Cytological Collection includes interchanges, inversions, deficiencies, trisomics, and tetraploids.

The Type, Isoline, and Linkage Collections are maintained by R.L. Bernard, Curator, Univ. of Illinois, Dep. of Agronomy, Turner Hall, 1102 S. Goodwin Ave., Urbana, IL 61801. The Cytological Collection is maintained by R. G. Palmer, Curator, G301, Dep. of Agronomy, Iowa State University, Ames, IA 50011.

5-4 SOYBEAN GERMPLASM COLLECTIONS

Collections of cultivated soybean [*G. max* (L.) Merr.] and wild soybean (*G. soja* Sieb. and Zucc.) are maintained by the USDA-ARS for use

Table 5-2. Collection of cultivated soybean germplasm.

Maturity group	Number of accessions			
	Old U.S. and Canadian cultivars	FC† strains	PI strains	Total
000‡	3	1	87	91
00	5	4	323	332
0	7	6	800	813
I	23	3	1059	1085
II	26	6	1115	1147
III	38	13	1052	1103
IV	38	18	2294	2350
V§	10	10	1511	1531
VI	17	10	443	470
VII	21	17	296	334
VIII	20	2	263	285
IX	0	0	124	124
X	0	0	151	151
Total	208	90	9518	9816

†FC = Forage crop and PI = Plant introduction.
‡Maturity Groups 000 to IV; Curator — R. L. Bernard; Location — Univ. of Illinois, Dep. of Agronomy, Turner Hall, 1102 S. Goodwin Ave., Urbana, IL 61801.
§Maturity Groups V to X; Curator — E. E. Hartwig; Location — Delta States Res. Ctr., P.O. Box 196, Stoneville, MS 38776.

by research scientists. Strains of cultivated soybean adapted to more northern latitudes (Maturity Groups 000 through IV) are maintained at the Univ. of Illinois, Urbana. Strains adapted to shorter-day conditions (Maturity Groups V through X) are maintained at Stoneville, MS. Accessions of wild soybean and perennial *Glycine* spp. are stored at Urbana, although entries of Maturity Group V and later of wild soybean are evaluated and increased at Stoneville.

The FC strains in the collection consist of foreign and domestic entries assigned identifying numbers by the former Forage and Range Research Branch, USDA, from 1911 to 1956. Plant introduction strains consist of foreign strains assigned identifying numbers by the Germplasm Resources Lab., Plant Genetics and Germplasm Institute, Agricultural Research Center-West, Beltsville, MD. New PI numbers for all species and information on sources, cultivar names, and other pertinent data are published annually by the Laboratory in *Plant Inventory*.

Germplasm accessions in the cultivated soybean, wild soybean, and perennial *Glycine* collections that are available for research purposes are listed in Tables 5–2, 5–3, and 5–4, respectively.

5–5 QUALITATIVE GENETICS

5–5.1 Diseases

Several loci controlling reaction to soybean diseases have been identified. The development of commercial cultivars with multiple-disease resistance has had a major impact in reducing economic losses.

Table 5-3. Collection of wild soybean germplasm.†

Country of origin	No. of accessions	Maturity group 00	0	I	II	III	IV	V	VI	VII	VIII	IX	X
China	30			1	18			5	3	1			2
Japan	183					1	7	44	85	46			
South Korea	313				2		40	243	27	1			
USSR	34	17	7	5	5								
Total	560	17	7	6	25	1	47	292	115	48	0	0	2

†Curator – R. L. Bernard; Location – Univ. of Illinois, Dep. of Agronomy, Turner Hall, 1102 S. Goodwin Ave., Urbana, IL 61801.

Table 5-4. Collection of perennial species of *Glycine*.†

Glycine species	No. of accessions	Country of origin
canescens	1	Australia
clandestina	13	Australia
falcata	2	Australia
latifolia	6	Australia
tabacina	27	Australia, Japan, Taiwan
tomentella	17	Australia, Philippines, Taiwan
Total	66	

†Curator — R. L. Bernard; Location — Univ. of Illinois, Dep. of Agronomy, Turner Hall, 1102 S. Goodwin Ave., Urbana, IL 61801

5-5.1.1 Bacterial Pustule

Bernard and Weiss (1973) summarized studies showing that resistance to bacterial pustule, caused by *Xanthomonas phaseoli* var. *sojensis* (Hedges) Starr and Burkh. found in the cv. CNS, was controlled by a single recessive allele (*rxp*). This allele also controls resistance to the disease wildfire, caused by *Pseudomonas tabaci* (Wolf & Foster) Stevens.

5-5.1.2 Bacterial Blight

A dominant allele (*Rpg1*) was reported by Mukherjee et al. (1966) to control resistance to race 1 of *P. glycinea* Coerper, the causal agent of bacterial blight. No additional loci for resistance have been reported.

5-5.1.3 Frogeye Leaf Spot

Bernard and Weiss (1973) summarized studies showing that one dominant allele (*Rcs1*) controlled resistance to race 1 of *Cercospora sojina* Hara, and that *Rcs2* controlled resistance to race 2. Additional loci for resistance have not been reported.

5-5.1.4 Downy Mildew

Bernard and Cremeens (1972) reported that all races of *Peronospora manshurica* (Naum.) Syd. were controlled by a single dominant allele (*Rpm*) found in the cv. Kanrich. A new race of *P. manshurica,* virulent on the cv. Union, which carries the allele *Rpm* for resistance to all previously known races of downy mildew, was observed in Illinois in 1981 (Lim et al., 1984).

5-5.1.5 Powdery Mildew

Powdery mildew, caused by *Microsphaera diffusa* Cke. & Pk., occurs frequently on soybean in greenhouses and occasionally in the field. However, because the disease occurs only sporadically on field-grown plants, little effort has been devoted to a search for resistance. Grau and Lawrence (1975) reported that resistance was controlled by a single dominant allele found in Chippewa 64, but did not propose a gene symbol. This may be the same allele, designated *Rmd* by Buzzell and Haas (1978), for adult-plant resistance in the cv. Blackhawk. No allelism tests have been done. The recessive allele (*rmd*) results in susceptibility at all stages of plant development.

5-5.1.6 Phytophthora Rot

Phytophthora rot, caused by *Phytophthora megasperma* Drechs. f. sp. *glycinea* Kuan & Erwin (*Pmg*), occurs in most of the soybean-producing areas of the United States and Canada. The disease can devastate a soybean crop when conditions favor the pathogen. Twenty-four physiologic races of *Pmg* have been identified (Keeling, 1984), based on pathogenicity of the isolates to differential cultivars. Bernard and Weiss (1973) reported on three alleles (*rps*, *rps-2*, and *Rps*) giving differential response to the three races of *Pmg* recognized at that time. Other alleles at the original locus and several additional loci have been identified since. Kilen et al. (1974) reported a second locus (*Rps2*) derived from CNS. Mueller et al. (1978) published data on a third locus (*Rps3*) for resistance from PI 86972-1. They suggested that *rps-2* become *Rps1-b* and that the Arksoy allele become *Rps1-c*. The allele *Rps1-k* was identified by Bernard and Cremeens (1981) while studying the inheritance of resistance to *Pmg* in the cv. Kingwa. *Rps4* was identified in PI 86050 by Athow et al. (1980), and *Rps5* was found in breeding line L62-904 by Buzzell and Anderson (1981). Athow and Laviolette (1982) identified *Rps6* in the cv. Altona. The identified resistance alleles at six loci are being utilized either singly or in combination in numerous breeding programs to develop cultivars having resistance to various races of *Pmg*.

5-5.1.7 Soybean Rust

Soybean rust, caused by *Phakopsora pachyrhizi* Syd., is an economically important fungal disease in many areas of the world, but there are no definitive reports of soybean rust in North America or Europe. McLean and Byth (1980) reported that Komata (PI 200492) has a single dominant allele (*Rpp1*) giving resistance to an Australian rust isolate. Singh and Thapliyal (1977) reported that Ankur (PI 462312) also has a single dominant allele for resistance. Hartwig and Bromfield (1983) studied the inheritance of resistance in PI 200492, PI 230970, and PI 462312 and concluded that each carried a single dominant allele at a different locus conferring resistance to soybean rust. The genotypes for rust reaction assigned to the three parental types were: PI 200492, *Rpp1 Rpp1 rpp2 rpp2 rpp3 rpp3*; PI 230970, *rpp1 rpp1 Rpp2 Rpp2 rpp3 rpp3*; and PI 462312, *rpp1 rpp1 rpp2 rpp2 Rpp3 Rpp3*.

5-5.1.8 Soybean Mosaic Virus

Soybean mosaic virus (SMV) is distributed worldwide and is an important soybean disease in many areas (Sinclair and Shurtleff, 1975).

The SMV has been recognized as a soybean disease almost as long as soybean has been grown in the USA. Koshimizu and Iizuka (1963) conducted inheritance studies using resistant and susceptible parents. In two crosses, the F_1 was resistant, and the F_2 population segregated in a ratio of three resistant to one susceptible. In another cross, the F_1 was susceptible, and the F_2 segregated in a ratio of seven resistant to nine susceptible. Kiihl and Hartwig (1979) identified two types of resistance in crosses between cultivars resistant and susceptible to a strain of SMV designated SMV-1. The highest level of resistance gave complete protection against SMV-1 and a variant of the strain, SMV-1-B. The lesser level of resistance gave protection against SMV-1 in the homozygous condition, but some of the heterozygous plants were necrotic. All homozygous recessive plants became necrotic after inoculation with SMV-1-B. They proposed the gene symbols *Rsv* for the highest level of resistance in PI 96983, *rsv-t* for the lesser level of resistance in the cv. Tokyo, and *rsv* for the susceptible reaction. They concluded that *Rsv, rsv-t* and *rsv* form an allelomorphic series with dominance in that order. Roane et al. (1983) found that resistance to SMV-VA (isolated from naturally infected plants growing in eastern Virginia) in York behaved as though controlled by a single dominant allele. They discussed the possibility that the dominant allele in York may be the same as the one (*rsv-t*) Kiihl and Hartwig (1979) assigned to the Ogden resistance because both cultivars have Tokyo as an ancestor. Buzzell and Tu (1984) reported that the cv. Raiden (PI 360844) has a single dominant allele for resistance at a separate locus from the allele in PI 96983. Their study showed that this newly characterized allele (*Rsv2*) conferred resistance to two strains of SMV, to which *Rsv* in a Williams background was susceptible.

5-5.1.9 Peanut Mottle Virus

Boerma and Kuhn (1976) studied soybean reaction to peanut mottle virus (PMV) and determined that resistance in the cvs. Dorman and CNS to a mild strain (M2) of the virus was controlled by a single dominant allele (*Rpv*). Shipe et al. (1979) identified a recessive allele (*rpv2*) in the cv. Peking that controls resistance to isolate PMV-S/V745.

5-5.1.10 Cowpea Chlorotic Mottle Virus

Boerma et al. (1975) determined that resistance to a distinct strain (S) of cowpea chlorotic mottle virus was controlled by a single dominant allele (*Rcv*) in the cvs. Lee, Bragg, and Hill.

5-5.2 Nematodes

Caviness and Riggs (1976) reported that as many as 50 species of plant-parasitic nematodes feed on soybean. The inheritance of resistance has been reported for the following three species: *Heterodera glycines*

Ichinohe (soybean cyst nematode), *Meloidogyne* spp. (root-knot nematode), and *Rotylenchulus reniformis* (reniform nematode).

No additional gene symbols have been designated for resistance to cyst nematode since they were summarized by Bernard and Weiss (1973), although additional inheritance studies have been conducted. Hartwig and Epps (1970) reported that a recessive allele from PI 90763 controlled resistance to race 2. Thomas et al. (1975) reported that the inheritance of resistance to race 4 of the soybean cyst nematode seems to be controlled by one dominant and two recessive alleles.

Although sources of resistance to several species of root-knot nematode have been reported, no loci for resistance have been designated. Williams et al. (1973) reported that a cultivar resistant to one species of root-knot nematode may not be resistant to another species or race and that the highest levels of resistance in enhanced breeding lines seem to result from a combination of alleles from several sources. Boquet et al. (1975) reported that the resistance to a specific race of root-knot nematode was controlled by one major allele with at least one modifying allele. Their data further suggested that susceptibility was partially dominant.

Williams et al. (1981) studied the inheritance of reaction to reniform nematodes in the cross Forrest (resistant) × Ransom (susceptible). They concluded that resistance is controlled by a single major recessive allele, designated *rrn* but that one or more minor alleles may contribute to the reaction, giving intermediate infection classes not shared by parental or F_1 plants.

5–5.3 Foliar-Feeding Insects

Although breeding for resistance to several insect species is being conducted at several locations, the genetic basis for resistance has not been clearly established, nor have gene symbols been assigned. Sisson et al. (1976) reported that the resistance of Miyako White (PI 227687) and Sodendaizu (PI 229358) was quantitatively inherited but suggested that it resulted primarily from additive gene action of two or three major alleles. Studies by Kilen et al. (1977) suggested partial dominance for susceptibility with the action of only a few major alleles controlling resistance.

Genes controlling reaction to bacteria, fungi, viruses, and nematodes are summarized in Table 5–5.

Table 5-5. Genes affecting pest reaction in soybean.

Gene	Phenotype	Strain†	Reference
	Bacterial pustule		
Rxp	Susceptible	Lincoln, Ralsoy	Hartwig and Lehman (1951),
rxp	Resistant	CNS	Feaster (1951), Bernard and Weiss (1973)

(continued on next page)

Table 5-5. Continued.

Gene	Phenotype	Strain†	Reference
	Bacterial blight		
Rpg1	Resistant, race 1	Norchief, Harosoy	Mukheıjee et al. (1966)
rpg1	Susceptible, race 1	Flambeau	
	Frogeye leaf spot		
Rcs1	Resistant, race 1	Lincoln, Wabash	Athow and Probst (1952) (as *Cs*); symbol by Probst et al. (1965)
rcs1	Susceptible, race 1	Gibson, Patoka, Hawkeye	
Rcs2	Resistant, race 2	Kent	Probst et al. (1965)
rcs2	Susceptible, race 2	C1043 (PI 70237 × Lincoln); C1270 [Mandarin (Ottawa) × Clark]	
	Downy mildew		
Rpm	Resistant	Kanrich	Bernard and Cremeens (1972)
rpm	Susceptible	Clark, Chippewa	
	Powdery mildew		
Rmd	Resistant	Blackhawk	Buzzell and Haas (1978)
rmd	Susceptible	Harosoy 63	
	Phytophthora rot		
Rps1	Resistant, races 1, 2, 10, 13, 16	Mukden	Bernard et al., 1957 (as *Ps*); Lam-Sanchez et al. (1968); Moots et al. (1983)
rps1	Susceptible	Lincoln, Harosoy	
Rps1-b	Resistant, races 1, 3–9, 13–15, 17, 18, 21, 22	FC 31745, Sanga, PI 84637	Hartwig et al. (1968) (as *rps-2*); Mueller et al. (1978), Laviolette and Athow (1983)
Rps1-c	Resistant, races 1–3, 6–11, 13, 15, 17, 21	Mack, PI 54615-1	Mueller et al. (1978), Laviolette and Athow (1983)
Rps1-k	Resistant, races 1–10, 13–15, 17–18, 21, 22	Kingwa	Bernard and Cremeens (1981), Laviolette and Athow (1983)
Rps2	Resistant, races 1–2	CNS	Kilen et al. (1974)
rps2	Susceptible	c	
Rps3	Resistant, races 1–5, 8, 9, 11, 13, 14, 16, 18	PI 171442, PI 86972-1	Mueller et al. (1978), Laviolette and Athow (1983)
rps3	Susceptible	Harosoy	
Rps4	Resistant, races 1–4, 10, 12–16	PI 86050	Athow et al. (1980)
rps4	Susceptible	Harosoy	
Rps5	Resistant, races 1–5, 8, 9, 11, 13, 14, 16	PI 91160	Buzzell and Anderson (1981)
rps5	Susceptible	Harosoy	
Rps6	Resistant, races 1-4, 10, 12, 14–16, 18-21	Altona	Athow and Laviolette (1982), Laviolette and Athow (1983)
rps6	Susceptible	Harosoy	

(continued on next page)

Table 5-5. Continued.

Gene	Phenotype	Strain†	Reference
	Soybean rust		
Rpp1	Resistant	Komata (PI 200492)	McLean and Byth (1980)
rpp1	Susceptible	Will, Davis	
Rpp2	Resistant	PI 230970	Hartwig and Bromfield (1983)
rpp2	Susceptible	c	
Rpp3	Resistant	Ankur (PI 462312)	Hartwig and Bromfield (1983)
rpp3	Susceptible	c	
	Soybean mosaic virus		
Rsv1	Resistant, SMV-1, SMV-1-B	PI 96983	Kiihl and Hartwig (1979)
rsv1-t	Resistant, SMV-1, Susceptible, SMV-1-B	Tokyo, Ogden	
rsv1	Susceptible, SMV-1, SMV-1-B	Hill	
Rsv2	Resistant, G7, G7A	Raiden	Buzzell and Tu (1984)
rsv2	Susceptible	Williams	
	Peanut mottle virus		
Rpv1	Resistant, M-2	Dorman, CNS	Boerma and Kuhn, (1976)
rpv1	Susceptible, M-2	Bragg, Pickett, Ransom	
rpv2	Resistant, PMV-S/V745	Peking	Shipe et al. (1979)
Rpv2	Susceptible, PMV-S/V745	PI 229315	
	Cowpea chlorotic mottle virus		
Rcv	Resistant	Bragg, Hill, Lee	Boerma et al. (1975)
rcv	Susceptible	Davis, Hood, Jackson	
	Cyst nematode		
rhg1 rhg2 rhg3	Resistant	Peking	Caldwell et al. (1960)
Rhg1, Rhg2, or *Rhg3*	Susceptible	Lee, Hill	
Rhg4 with *rhg1 rhg2 rhg3*	Resistant	Peking	Matson and Williams (1965)
rhg4	Susceptible	Scott	
	Reniform nematode		
rrn	Resistant	Forrest	Williams et al. (1981)
Rrn	Susceptible	Ransom	

†c = Indicates that the gene occurs in many cultivars.

5–5.4 Herbicide Reaction

Soybean cultivars frequently show differences in the degree of injury caused by herbicides, but the occurrence of genotypes giving highly sensitive herbicide reactions has been rare. Bernard and Wax (1975) found that the bentazon-sensitive reaction in Nookishirohana (PI 229342) was controlled by a single recessive allele, (*hb*). In that study, Clark 63 had the allele (*Hb*) for the tolerant reaction. Edwards et al. (1976) determined that sensitivity to metribuzin in the cv. Semmes was controlled by a single recessive allele (*hm*) and that the cv. Hood had the allele (*Hm*) for tolerance to metribuzin. In a later study, Kilen and Barrentine (1983) showed that Tracy and Semmes carry the same allele for sensitivity to the herbicide.

Genes affecting herbicide reaction in soybean are listed in Table 5–6.

5–5.5 Rhizobium Response

Four gene pairs affecting soybean response to *Rhizobium japonicum* (Kirchner) Buchanan are known. The recessive allele *rj1* prevents nodulation with most strains of *Rhizobium* under most environmental conditions. Certain strains of *R. japonicum* can nodulate *rj1* soybean plants (Clark, 1957). Such strains also show a propensity to produce rhizobitoxine-induced chlorosis symptoms on soybean (Devine and Weber, 1977) and have the ability to induce nodule-like swellings on peanut (*Arachis hypogaea* L.) roots (Devine et al., 1983a).

Three dominant alleles (*Rj2*, *Rj3*, and *Rj4*) cause ineffective nodule response with certain slow-growing strains of *Rhizobium*. Devine and Breithaupt (1981) found the *Rj2* allele in 2% and the *Rj4* allele in about 30% of 847 plant introductions from Asia. The geographical pattern of the allele frequencies among these Asian lines suggested that ecotypes of the host and microsymbiont co-evolved under pressure for mutual symbiotic compatibility (Devine and Breithaupt, 1980). In an examination of lines in the 1979 U.S. and Canadian regional preliminary and uniform soybean tests, Devine and Breithaupt (1981) found that frequencies of both dominant alleles declined progressively with advancing levels of breeding for agronomic performance.

Fast-growing strains of bacteria that nodulate and fix N in soybeans have been identified (Keyser et al., 1982). A dominant allele elicits ineffective nodulation with these fast-growing strains (Devine, 1984).

A summary of genes affecting *Rhizobium* response in soybean is presented in Table 5–7.

Table 5-6. Genes affecting herbicide reaction in soybean.

Gene	Phenotype	Strain	Reference
	Herbicide reaction		
Hb	Tolerant to bentazon	Clark 63	Bernard and Wax (1975)
hb	Sensitive to bentazon	Nookishirohana (PI 229342)	
Hm	Tolerant to metribuzin	Hood	Edwards et al. (1976)
hm	Sensitive to metribuzin	Semmes	

Table 5-7. Genes affecting *Rhizobium* response in soybean.

Gene	Phenotype	Strain†	Reference
Rj1	Nodulating	T180,T202,c	Williams and Lynch (1954)
rj1	Nonnodulating	T181,T201	(as *no*); symbol by Caldwell (1966)
Rj2	Ineffective by strains b7, b14, and b122	Hardee, CNS	Caldwell (1966)
rj2	Effective	c	
Rj3	Ineffective by strain 33	Hardee	Vest (1970)
rj3	Effective	Clark	
Rj4	Ineffective by strain 61	Hill, Dare, Dunfield	Vest and Caldwell (1972)
rj4	Effective	Lee, Semmes	

†c=Indicates that the gene occurs in many cultivars.

5–5.6 Growth and Morphology

5–5.6.1 Flowering and Maturity

Bernard (1971) reported two major gene pairs (*E1* and *E2*) that affected the time of flowering and maturity. Buzzell (1971) and Kilen and Hartwig (1971) described *E3*; the recessive allele (*e3*) did not respond to fluorescent daylength treatment. *E2* and *E3* are not linked. *E2* and *E3* do not have an equal effect in delaying maturity, and, when combined, they have less than an additive effect (Buzzell and Bernard, 1975).

Polson (1972) observed PI 297550 to be daylength neutral. A fourth allele affecting time of flowering (*E4*) was described by Buzzell and Voldeng (1980) in PI 297550. The recessive allele (*e4*) is insensitive to long daylength.

Shanmugasundaram (1977) reported that photoperiod insensitivity was inherited as a single-gene recessive, but no gene symbol was given. Data on days to flowering of parents, F_1, and F_2 from four crosses grown under 10- and 16-h photoperiods were given by Shanmugasundaram (1978a). Early flowering under a 10-h photoperiod was dominant to late flowering; however, partial dominance for late flowering was observed under a 16-h photoperiod.

Cultivars from Maturity Groups 00 to VIII have been screened for photoperiodic response to flowering (Shanmugasundaram, 1978b, 1981). The 3445 cultivars were placed into 10 different photoperiod-sensitivity classes. Photoperiod-insensitive cultivars generally were found in the early maturity groups, but some photoperiod-insensitive cultivars were identified in the late maturity groups (Shanmugasundaram, 1978b, 1981).

Nissly et al. (1981a) examined 515 strains of Maturity Group III for variation in photoperiod sensitivity and time of flowering. Thirty-two strains were identified as having low photoperiod sensitivity, and Kitamishiro (PI 317334B) exhibited almost no photoperiod sensitivity at either Urbana, IL, or Isabela, PR. No genetic studies were reported.

A summary of genes affecting time of flowering and maturity in soybean is presented in Table 5–8.

Table 5-8. Genes affecting growth and morphology in soybean.

Gene	Phenotype	Strain†	Reference
	1. Time of flowering and maturity		
E1	Late	T175	Owen (1927b), Bernard (1971)
e1	Early	Clark	
E2	Late	Clark	Bernard (1971)
e2	Early	PI 86024	
E3	Late and sensitive to fluorescent light	Harosoy 63	Buzzell (1971), Kilen and Hartwig (1971)
e3	Early and insensitive to fluorescent light	Blackhawk	
E4	Sensitive to long daylength	Harcor	Buzzell and Voldeng (1980)
e4	Insensitive to long daylength	Urozsajnaja (PI 297550)	
	2. Growth of stem, petiole, and inflorescence		
Dt1	Indeterminate stem	Manchu, Clark	Woodworth (1932, 1933), Bernard (1972)
dt1	Determinate stem	Ebony, PI 86024	
Dt2	Semideterminate stem	T117	Bernard (1972)
dt2	Indeterminate stem	Clark	
F	Normal stem	c	Nagai (1926), Takagi (1929), symbol by Woodworth (1932, 1933), Matsuura (1933), Albertsen et al. (1983)
f	Fasciated stem	T173, PI 83945-4, Shakujo (PI 243541)	
Lps	Normal petiole	Lee 68	Kilen (1983)
lps	Short petiole	T279	
S	Short, internode length decreased	Higan	Bernard (1975a)
s	Normal	Harosoy	
s-t	Tall, internode length increased	Chief	
Se	Pedunculate inflorescence	T208	VanSchaik and Probst (1958)
se	Subsessile inflorescence	PI 84631	
	3. Dwarfness		
Df2	Normal	c	Porter and Weiss (1948), symbol by Byth and Weber (1969)
df2	Dwarf	T210,T243	
Df3	Normal	c	Byth and Weber (1969)
df3	Dwarf	T244	
Df4	Normal	c	Fehr (1972a)
df4	Dwarf	T256	
Df5	Normal	c	Palmer (1984a)
df5	Dwarf	T263	
Mn	Normal	c	Delannay and Palmer (1984)
mn	Miniature plant	T251	
Pm	Normal	c	Probst (1950)
pm	Dwarf, crinkled leaves, sterile	T211	
Sb1 or *Sb2*	Normal	Davis	Kilen and Hartwig (1975), Kilen (1977), Boerma and Jones (1978)
sb1 sb2	Brachytic stem	Ya Hagi (PI 227224)	
	4. Leaf form		
Ab	Abscission at maturity	T161, c	Probst (1950)
ab	Delayed abscission	Kingwa	

(continued on next page)

Table 5-8. Continued.

Gene	Phenotype	Strain†	Reference
Lf1	5-foliolate	PI 86024	Takahashi and Fukuyama
lf1	Trifoliolate	c	(1919), symbol by Fehr (1972b)
Lf2	Trifoliolate	c	Fehr (1972b)
lf2	7-foliolate	T255	
Ln	Ovate leaflet	c	Takahashi and Fukuyama
ln	Narrow leaflet, 4-seeded pods	T41, PI 84631	(1919), Woodworth (1932, 1933), Takahashi (1934), Domingo (1945), symbol by Bernard and Weiss (1973)
Lo	Ovate leaflet	c	Domingo (1945)
lo	Oval leaflet, few-seeded pods	T122	
Lw1 Lw2	Nonwavy leaf	–	Rode and Bernard (1975b)
Lw1 lw2	Nonwavy leaf	Harosoy, Clark	
lw1 Lw2	Nonwavy leaf	T117	
lw1 lw2	Wavy leaf	T176, T205	
Lb1 Lb2	Nonbullate leaf	Harosoy	Rode and Bernard (1975c)
Lb1 lb2	Nonbullate leaf	Clark	
lb1 Lb2	Nonbullate leaf	T217	
lb1 lb2	Bullate leaf	L65-701 (Clark[6] × PI 196166)	
	5. Pubescence type		
Pa1 Pa2	Erect	Harosoy, Clark	Karasawa (1936), Ting (1946), symbol by Bernard (1975d)
Pa1 pa2	Erect	L70-4119 (Harosoy[6] × Higan)	
pa1 Pa2	Semi-appressed	Scott, Custer, Oksoy	
pa1 pa2	Appressed	Higan	
P1	Glabrous	T145	Nagai and Saito (1923)
p1	Pubescent	c	
P2	Normal	c	Stewart and Wentz (1926)
p2	Puberulent	T31	
Pb	Sharp hair tip	PI 163453 (wild soybean), Kingwa	Ting (1946)
pb	Blunt hair tip	Clark	
Pc	Normal	Clark, c	Bernard and Singh (1969)
pc	Curly (deciduous)	PI 84987	
Pd1	Dense	PI 80837	Bernard and Singh (1969)
pd1	Normal	Clark, c	
Pd2	Dense	T264	Bernard (unpublished)
pd2	Normal		
Ps	Sparse	PI 91160	Bernard and Singh (1969),
Ps-s	Semi-sparse	Higan	Bernard (1975c)
ps	Normal	c	
	6. Seed-coat structure		
B1 B2 B3	Bloom on seed coat	Sooty	Woodworth (1932, 1933),
b1, b2, or *b3*	No bloom	c	Tang and Tai (1962)

(continued on next page)

Table 5-8. Continued.

Gene	Phenotype	Strain†	Reference
N	Normal hilum abscission	c	Owen (1928b)
n	Lack of abscission layer	Soysota	

†c=Indicates that the gene occurs in many cultivars.

5-5.6.2 Growth of Stem, Petiole, and Inflorescence

Soybean plants show considerable variation in the abruptness of stem termination, but most are classified as either indeterminate (*Dt1*) or determinate (*dt1*). Differences observed between indeterminate and determinate types likely are due to the timing of termination of stem growth rather than the manner of termination (Bernard, 1972). Semideterminate (*Dt2*) soybean plants have a moderately abrupt stem termination. Both *Dt1* and *dt1* are represented in many cultivars; *Dt2* is represented by publicly and several privately developed cultivars.

Fasciated stems (*f*) are known in at least three strains, Shakujo (PI 83945-4), PI 243541, and Keitomame. Albertsen et al. (1983) found that these strains have the same allele controlling fasciation. They showed by grafting experiments that the genotype of the scion determined the phenotype of the scion; i.e., there was no effect of the rootstock. Curry (1982) found that the increased amount of total tissue mass in a fasciated plant consisted of an increased number of cells and was not the result of an increase in cell size.

Kilen (1983) identified a short-petiole plant in a segregating population. The mutant is true-breeding and inherited as a single-gene recessive (*lps*). Plants with the *Lps* allele have normal petiole lengths. Kilen (1983) suggested that the mutant may have value in narrow-row production.

A multiple allelic series affecting stem length has been identified (Bernard, 1975a). Plants with the dominant allele (*S*) have decreased internode length compared with *s s* plants. Tall plants (*s-t s-t*) have an increase in main-stem length, mainly due to a lengthening of the internodes.

A summary of genes affecting growth of stem, petiole, and inflorescence in soybean is given in Table 5–8.

5-5.6.3 Dwarfness

Four recessive dwarf mutants (*df*) are known. All mutants are markedly stunted and have low seed yield. The *pm* mutant, in addition to being of short stature, has crinkled leaves and produces no seed. Thus, the allele is maintained as the heterozygote T211H (Probst, 1950).

A short-internode mutant reported by Kilen and Hartwig (1975) was later designated brachytic and found to be controlled by a single recessive allele, *sb1* (Kilen, 1977). A second allele (*sb2*) is known to control the brachytic trait (Boerma and Jones, 1978). The *sb1 sb1 sb2 sb2* genotype is characterized by precocious expansion of the axillary buds that results in lateral displacement of the main shoot apex, giving a zigzag configuration, and by development of shortened internodes. Angularity of internodes is responsible for 10%, and shortening of the internodes is responsible for 35%, of the reduction in height (Walker and Boerma, 1978).

Genetic type T251 is characterized by fewer nodes, shorter internodes, and smaller leaves than normal plants, but internode length is greater than that of most dwarf mutants. Its growth is reduced, resulting in the phenotype of a miniature plant. The mutant is inherited as a single-gene recessive (*mn*) and is maintained as the heterozygote T251H (Delannay and Palmer, 1984).

Kilen (1975) reported a single recessive allele that, in a heterozygote, produces plants with curved leaves. The homozygous recessive genotype results in a miniature plant. These plants have small unifoliolate leaves, very short internodes, and minute, curved, rugose trifoliolate leaves. Mature homozygous plants are 2- to 3-cm tall, have cleistogamous flowers, and produce an occasional pod with seed. The miniature plant is temperature sensitive. In growth-chamber studies, with 38°C day and 15, 20, or 24°C night temperatures, recessive plants were 5, 26, and 99% as tall as the control cultivar (Kilen, 1979). No gene symbol has been assigned to this mutant.

A summary of genes involved in dwarfness in soybean is presented in Table 5–8.

5-5.6.4 Leaf Form

Information on genes that affect leaf abscission, number of leaflets per leaf, and the pleiotropic effect of *ln* on leaflet shape and number of seed per pod has been summarized by Bernard and Weiss (1973). Rode and Bernard (1975b, 1975c) described two leaf mutants, wavy and bullate.

Leaf waviness is observed mainly in the margin of leaflets. Wavy leaf is controlled by two recessive alleles (*lw1* and *lw2*); however, in the presence of tawny pubescence (*T*), wavy leaf is not expressed (Rode and Bernard, 1975b). They noticed different levels of waviness in certain hybrid populations; the inheritance of this variation is not known.

Bullate leaves have characteristic circular bumps or a blister-like upper surface. Two recessive alleles, *lb1* and *lb2*, are necessary for expression. Their expression and segregation are independent of *T* (Rode and Bernard, 1975c). Heterozygotes in some populations may appear as slightly bullate. Harosoy has the genotype *Lb1 Lb1 Lb2 Lb2*, whereas Clark has the genotype *Lb1 Lb1 lb2 lb2*.

Ten normal plants and two plants with severe crinkling and puckering of leaves were observed in an M_3 progeny row of irradiated Lee. Progeny rows of the 10 normal plants segregated about 3 normal: 1 abnormal leaves, whereas the two plants with crinkled and puckered leaves bred true. Inheritance of this trait seems to be a single recessive gene, but no gene symbols have been assigned (Singh et al., 1974).

A summary of genes involved in leaf form in soybean is presented in Table 5–8.

5-5.6.5 Pubescence Type

Many alleles affecting the presence, form, and number of trichomes are known. Appressed pubescence is more easily discernible on the upper leaf surface and is prevalent among accessions from Japan and Korea. It is common but not universal among *G. soja* accessions. Bernard (1975d) suggested that two alleles (*pa1* and *pa2*) are necessary for the appressed phenotype. In eastern and central North America, lines with appressed pubescence are often heavily fed upon and stunted by the potato leafhopper (*Empoasca fabae*). Semi-appressed phenotypes (*pa1* and *Pa2*) are found in a few cultivars.

Alleles controlling the density of pubescence are known. Dense pubescence phenotypes (*Pd1* and *Pd2*) show an increase in density, whereas sparse (*Ps*) and semi-sparse pubescence phenotypes (*Ps-s*) have fewer trichomes per unit area (Bernard, 1975c).

Sharp pubescence tip (*Pb*) was found in > 95% of the *G. soja* accessions, but was present in < 10% of the *G. max* cultivars and accessions examined by Broich and Palmer (1981). The puberulent phenotype occurs on *p2 p2* plants, which also express higher than normal rates of outcrossing (Singh, 1972). Outcrossing rates ranged from 11.1% to 58.3% with an average, across 31 entries, of 39.3%. Poor dehiscence of anthers was thought to be primarily responsible for the low level of self-pollination.

A summary of genes affecting pubescence type in soybean is presented in Table 5-8.

5-5.7 Sterility

Sterility systems have been classified as synaptic, structural, partial male sterile, or male sterile, female fertile. Synaptic mutants are those influencing chromosome pairing and include *st1* (Owen, 1928a), which has been lost, *st2* and *st3* (Hadley and Starnes, 1964), *st4* (Palmer, 1974b), and *st5* (Palmer and Kaul, 1983). An independent mutation at the *st2* locus has been described (Winger et al., 1977), and Brigham (1978) has reported a possible synaptic mutant. These mutants are characterized as being highly male and female sterile. The few seed that are found on the sterile plants, a result of self- or cross-pollination, often produce aneuploid or polyploid plants (Palmer, 1974a; Palmer and Heer, 1976; Palmer and Kaul, 1983).

Structural sterility includes mutants in which sterility is due to structural abnormalities in the flowers or in the reproductive organs. The flower-transformed mutant, *ft* (Singh and Jha, 1978), and the flower-structure mutant, *fs1 fs2* (Johns and Palmer, 1982) are structural steriles. In the *ft* mutant, fertile pollen is produced, but the plants are male sterile because of poor anther dehiscence. In the *fs1 fs2* mutant, anther filaments fail to elongate normally. Although fertile pollen is produced, self-pollination is precluded by a spatial separation between the stigma and anthers. Partial female sterility is due to abnormal ovule structure and position in the flower-structure mutant.

Partial male sterility refers to incomplete expression of male sterility. Partially male-sterile mutants include *p2* (Stewart and Wentz, 1926; Bernard and Jaycox, 1969; Singh, 1972), which is puberulent, the Arkansas male sterile (Caviness et al., 1970), and *msp* (Stelly and Palmer, 1980a, 1980b).

Growth-chamber experiments with the Arkansas mutant indicated that high temperature (35 °C) during illumination rendered partially male-sterile plants completely male sterile (Caviness and Fagala, 1973). Day temperatures of 29°C resulted in varied amounts of pod set on partially male-sterile plants. Stelly and Palmer (1980b) observed that high temperatures promoted male fertility in *msp* homozygotes. Development of the *msp* anthers was highly variable and included numerous cytological abnormalities (Stelly and Palmer, 1982). Additional single-gene recessive partially male-sterile mutants are known (Jha and Singh, 1978).

Several male-sterile, female-fertile mutants are known in soybeans. Brim and Young (1971) described *ms1*. Four additional independent occurrences of this or a similar mutant have been reported (Boerma and Cooper, 1978; Palmer et al., 1978b; Yee and Jian, 1983). Data thus far reported, however, do not discriminate between possible identical mutations and a possible multiple allelic series at the *ms1* locus. Male sterility is due to the failure of cytokinesis following telophase II (Patil and Singh, 1976; Rubaihayo and Gumisiriza, 1978; Skorupska and Nawracala, 1980), which results in a four-nucleate, pollen-like structure termed a *coenocytic microspore* (Albertsen and Palmer, 1979). Because of the failure of cytokinesis, only one-fourth the number of coenocytic microspores are expected in *ms1 ms1* plants as compared with the number of pollen grains per anther in fertile plants. Palmer et al. (1978a) found significant departures from this expected ratio, depending on the genetic background and environment where the plants were grown.

Although the *ms1* locus does not inhibit female function to the extreme extent that it does male function, various abnormalities have been noted. Kenworthy et al. (1973), Beversdorf and Bingham (1977), and Chen et al. (1985) reported polyembryony among progeny of *ms1* homozygotes. In addition, monoembryonic seedlings might be haploid or polyploid. Details regarding haploids and polyploids and cytological studies of *ms1* are presented in the haploid-polyploid section of this chapter.

The *ms2* mutant seems to be completely male sterile but has good female fertility, evidenced by cross-pollination success that is almost equal to the seed set on fertile sibling plants (Bernard and Cremeens, 1975). Cytologically, the first sign of abnormality is the appearance of large cytoplasmic vacuoles in tapetal cells, evident during prophase I. The reproductive cells abort at the tetrad stage of microsporogenesis. In external morphology, anthers initially appear normal. By the time of anthesis, the anthers are shrunken, distorted, and nondehiscent (Graybosch, 1984; Graybosch et al., 1984). The *ms2* allele may have a slight effect on female reproduction. Buss and Autio (1980) and Sadanaga and Grindeland (1981) noted a low frequency of polyploids and aneuploids, respectively, among progeny of *ms2* homozygotes.

A third male-sterile mutant, *ms3*, has good female fertility. Abortion of microspores occurs in anthers shortly after the initiation of the microspore wall (Palmer et al., 1980). Ultrastructurally, differences are detected between the tapeta of sterile plants and those of fertile plants as early as prophase I (Buntman and Horner, 1983). Tapetal cells of sterile plants also may accumulate fluorescent material of unknown composition. This property serves as an early marker of male sterility in *ms3 ms3* plants (Nakashima et al., 1984). Graybosch (1984) has determined that the 'Wabash' male sterile reported by Chaudhari and Davis (1977) is identical genetically and cytologically to the *ms3* male-sterile mutant described by Palmer et al. (1980).

In the *ms4* mutant, female fertility is good, and cytological development in anthers of sterile plants is normal until telophase II. Postmeiotic cytokinesis is absent, incomplete, or disoriented, resulting in cells with differing numbers of nuclei (Delannay and Palmer, 1982a; Graybosch, 1984). Some functional pollen may be produced; therefore, this mutant eventually may be classified as a partial male sterile (Graybosch and Palmer, 1984b).

Genetic data indicate that *ms5* is a single-gene recessive. Anthers are shriveled, and the pollen grains are small and collapsed (Buss, 1983). No cytological studies have been reported.

Several additional male-sterile mutants have been identified as single-gene recessives, but allelism tests with *ms1* through *ms5* have not been made (Patil and Singh, 1976). A review of the genetics and cytology of male-sterile, female-fertile soybean mutants has been prepared by Graybosch and Palmer (1984a).

Application of soybean male-sterile mutants to recurrent selection schemes in soybean breeding programs has been described (Brim and Stuber, 1973). Studies of seed set of male steriles have indicated a range from low levels of cross-pollination to seed yields only 8% below that of fertile sibling plants (Boerma and Moradshahi, 1975; Nelson and Bernard, 1979; Koelling et al., 1981; Carter et al., 1983; Palmer et al., 1983). Hybrid progeny of male-sterile plants can be identified readily if the female parent also is homozygous recessive for green cotyledon (*d1 d1 d2 d2*) and the male parent has yellow cotyledon (Sadanaga and Grindeland, 1981; Burton and Carter, 1983).

A summary of genes affecting sterility in soybean is presented in Table 5–9.

5–5.8 Physiology

5–5.8.1 Reaction to Nutritional Factors

Efficiency of Fe utilization has received attention because certain cultivars, when grown on calcareous soil, develop Fe-deficiency chlorosis

Table 5-9. Genes causing sterility in soybean.

Gene	Phenotype	Strain†	Reference
St2	Fertile	c	Hadley and Starnes (1964)
st2	Asynaptic sterile	T241	
St3	Fertile	c	Hadley and Starnes (1964)
st3	Asynaptic sterile	T242	
St4	Fertile	c	Palmer (1974b)
st4	Desynaptic sterile	T258	
St5	Fertile	c	Palmer and Kaul (1983)
st5	Desynaptic sterile	T272	
Fs1 or *Fs2*	Fertile	c	Johns and Palmer (1982)
fs1 fs2	Structural sterile	T269	
Ft	Fertile	c	Singh and Jha (1978)
ft	Structural sterile	Gamma ray-induced mutant	
Ms1	Fertile	c	Brim and Young (1971), Boerma and Cooper (1978), Palmer et al. (1978b)
ms1	Male sterile	T260, T266, T267, T268	
Ms2	Fertile	c	Bernard and Cremeens (1975), Graybosch et al. (1984)
ms2	Male sterile	T259	
Ms3	Fertile	c	Palmer et al. (1980)
ms3	Male sterile	T273	
Ms4	Fertile	c	Delannay and Palmer (1982a)
ms4	Male sterile	T274	
Ms5	Fertile	c	Buss (1983)
ms5	Male sterile	T277	
Msp	Fertile	c	Stelly and Palmer (1980a, 1980b)
msp	Partial male sterile	T271	

†c=Indicates that the gene occurs in many cultivars.

(Fehr, 1982). Weiss (1943) reported that, on the basis of plants tested in nutrient solution, the allele for Fe-utilization efficiency (*Fe*) was dominant to the allele for inefficiency (*fe*) found in PI 54619. Cianzio and Fehr (1980) indicated that the segregation they observed among field-grown plants from a cross between a cultivar resistant to Fe-deficiency chlorosis and a susceptible one could be explained by a major gene and modifying genes with minor effects. In a different study, Cianzio and Fehr (1982) reported that inheritance of Fe-deficiency chlorosis was quantitative and was controlled by additive gene action. They concluded that inheritance of resistance to Fe-deficiency chlorosis can vary, depending upon the parents and the test conditions used to evaluate the character.

Variability for tolerance and sensitivity to Mn is known. In grafting experiments, Heenan and Carter (1976) observed that the genotype of the rootstock had little or no control over tolerance or susceptibility to excess Mn. Ohki et al. (1980) found that sensitivity to Mn deficiency was partially dependent on the level of deficiency stress. Genetic control of tolerance or resistance to excess Mn seemed to be quantitative rather than qualitative, and differences between reciprocal crosses suggested a cytoplasmic influence (Brown and Devine, 1980).

Heenan et al. (1981) evaluated progeny of two crosses for inheritance of tolerance to levels of manganese. In the Lee (tolerant) × Bragg (susceptible) cross, F_1 plants had only moderate Mn toxicity symptoms, and the F_2 displayed a continuous distribution skewed towards tolerance. The F_6 progeny of Amredo (tolerant) × Bragg had a bimodal inheritance, suggesting that a single major gene and modifying genes with minor effects may influence tolerance to high Mn levels.

Variation for tolerance to Zn occurs among soybean cultivars. White et al. (1979a) grouped cultivars into sensitive and uptake classes. Within the sensitive class, subgroups tolerant, normal, and sensitive were described. Within the uptake class, subgroups accumulator, normal, and excluder were defined. The uptake and translocation of Zn, Mn, Fe, and P were studied under a range of Zn concentrations (White et al., 1979b). No genetic studies were done.

A nitrate-reductase mutant was found in the cv. Williams after mutagen treatment (Nelson et al., 1983). Nitrate reductase activity in mutant plants grown in a nitrate-containing nutrient solution seemed to be a summation of an inducible and constitutive activity. The *nr* mutant lacked the constitutive component (Ryan et al., 1983b).

A summary of genes affecting nutrition in soybean is given in Table 5-10.

5-5.8.2 Roots

Fluorescence in soybean roots first was reported by Chmelar (1934). Grabe (1957) evaluated 80 cultivars of *G. max* and found that Minsoy (PI 27890) was the only nonfluorescent cultivar. Fehr and Giese (1971) studied the inheritance of the lack of fluorescence in crosses with Minsoy and found that the trait was governed by a single recessive allele, *fr1*.

Delannay and Palmer (1982b) evaluated 572 *G. max* accessions for the presence or absence of root fluorescence. Lack of fluorescence was more common in accessions from Europe (19%) than from Asia (3.4%). Genetic studies of the nonfluorescent genotypes indicated that four independent alleles controlled the absence of fluorescence. Three were recessive alleles (*fr1*, *fr2*, and *fr4*), and one was dominant (*Fr3*). Geographical distribution of the four alleles was unequal; *fr1* was found in accessions from all geographical areas represented, and *fr2* was restricted to accessions from Europe. *Fr3* and *fr4* were found only in accessions from Asia.

Among 370 accessions of *G. soja*, Broich (1978) found that 79 lacked root fluorescence. In 39 accessions tested, nonfluorescence (*Fr3*) of the roots was dominant to fluorescence (*fr3*).

Root fluorescence is a useful marker gene in tissue-culture studies. Using herbicide isopropyl-*N*-*m*-chlorophenyl-carbamate (CIPC) to induce chromosome loss, Roth et al. (1982) tested for the presence or absence of the linkage groups containing *fr1* or *fr2*.

A summary of genes affecting root fluorescence in soybean is given in Table 5-10.

Table 5-10. Genes affecting physiology in soybean.

Gene	Phenotype	Strain†	Reference
	A. Reaction to nutritional factors		
Fe	Efficient Fe utilization	c	Weiss (1943)
fe	Inefficient	PI 54619	
Np	Phosphorus tolerant	Chief	Bernard and Howell (1964)
np	Sensitive to high P level	Lincoln	
Ncl	Chloride excluding	Lee	Abel (1969)
ncl	Chloride accumulating	Jackson	
Nr	Constitutive nitrate reductase present	Williams	Ryan et al. (1983a)
nr	Constitutive nitrate reductase absent	T276	
	B. Root		
Fr1	Fluorescent in UV light	c	Fehr and Giese (1971)
fr1	Nonfluorescent	Minsoy (PI 27890)	
Fr2	Fluorescent in UV light	c	Delannay and Palmer (1982b)
fr2	Nonfluorescent	Noir-l (PI 290136)	
Fr3	Nonfluorescent	PI 424078	Delannay and Palmer (1982b)
fr3	Fluorescent in UV light	c	
Fr4	Fluorescent in UV light	c	Delannay and Palmer (1982b)
fr4	Nonfluorescent	Dun-cuan (PI 404165)	
	C. Flavonol glycosides of leaves		
T	Quercetin and kaempferol present	c	Buttery and Buzzell (1973)
t	Quercetin absent, kaempferol present	c	
Wm	Glycosides present	c	Buzzell et al. (1977)
wm	Glycosides absent	T235	
Fg1	β(1-6)-glucoside present	T31, c	Buttery and Buzzell (1975)
fg1	β(1-6)-glucoside absent	Chippewa 64, c	
Fg2	α(1-6)-rhamnoside present	T31, c	Buttery and Buzzell (1975)
fg2	α(1-6)-rhamnoside absent	Chippewa 64, c	
Fg3	β(1-2)-glucoside present	T31, c	Buttery and Buzzell (1975)
fg3	β(1-2)-glucoside absent	Chippewa 64, c	
Fg4	α(1-2)-rhamnoside present	T31, c	Buttery and Buzzell (1975)
fg4	α(1-2)-rhamnoside absent	AK(FC 30761)	

†c = Indicates that the gene occurs in many cultivars.

5-5.8.3 Flavonols of Leaves

Flavonol glycosides in leaf tissue have been characterized genetically (Buzzell and Buttery, 1973, 1974) and biochemically (Buttery and Buzzell, 1975). Four flavonol glycoside alleles, *Fg1*, *Fg2*, *Fg3*, and *Fg4*, are known for the 16 classes of leaf flavonols. When *t* is homozygous, kaempferol glycosides occur in the leaves; when *T* is present, quercetin and analogous kaempferol glycosides are present. Thus, 32 phenotypes are possible. With the addition of glucose, *Fg1* codes for a β(1-6) carbon linkage and *Fg3* codes for a β(1-2) linkage. Similarly, with addition of rhamnose, *Fg2* codes for an α(1-6) linkage and *Fg4* codes for an α(1-2) linkage (Buttery and Buzzell, 1975). Their proposed scheme for the genetic control of flavonol glycoside biosynthesis in soybean leaves is given in Fig. 5–1.

The magenta flower allele (*wm*) is associated with low levels of flavonol glycosides in the leaves, which give a diminished rate of photosynthesis (Buttery and Buzzell, 1976). Leaf chlorophyll concentration or specific leaf weight are not affected. Plants with the *Fg1 fg2 Fg3 t* alleles have a lower rate of photosynthesis, lower leaf chlorophyll concentration, lower specific leaf weight, and lower seed yield (Buttery and Buzzell, 1976).

In *G.max* cultivars and *G. soja* accessions, *Fg1* and *Fg3* did not occur together in any accessions examined, either with *T* or with *t* (Buttery and Buzzell, 1976; Broich and Palmer, 1981). Distribution of *Fg1* and *Fg3* alleles is not random among populations from different geographic areas. In *G. max* accessions from India, *Fg1* occurred in high frequency, and *Fg3* in a low frequency, compared with accessions from China and Japan (Buttery and Buzzell, 1973, 1976). *Fg1* was almost nonexistent in accessions of *G. soja* from Japan, but occurred at higher frequencies in accessions from China, USSR, and Korea. *Fg3* was not present among 28 *G. soja* accessions from USSR, but was present among most accessions from China, Korea, and Japan (Broich and Palmer, 1981).

In past breeding work there seems to have been selection against *Fg1*, and in some cases against both *Fg1* and *Fg3* (Buzzell et al., 1980). Judicious selection of parents could eliminate some crosses that would produce a high proportion of low-yielding lines by avoiding parents that would bring together *Fg1* and *Fg3*. The deleterious effect of two complementary dominant alleles (*Fg1* and *Fg3*) suggested to Buttery and Buzzell (1976) that bringing together of alleles of similar function may be the result of allopolyploidy in the origin of soybeans.

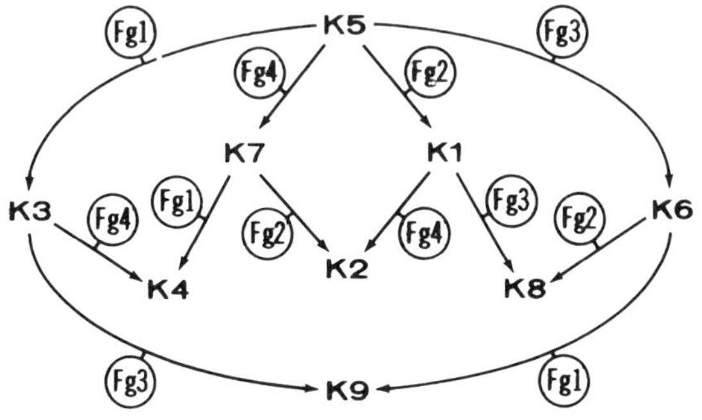

Fig. 5–1. Proposed scheme of the genetic control of the biosynthesis of flavonol glycosides K1-K9 in soybean leaves. The monoglucoside K5 is found in all soybean leaves except in the presence of *wm wm*. Glucose or rhamnose units are added by 1-2 or 1-6 linkages to the monoglucoside under the control of the dominant alleles *Fg1, Fg2, Fg3,* and *Fg4* to form diglycosides or triglycosides. The same scheme applies to the control of the formation of quercetin glycosides. From B. R. Buttery and R. I. Buzzell (1975).

A summary of genes affecting flavonols of leaves in soybean is given in Table 5-10.

5-5.8.4 Chlorophyll Deficiency—Nuclear

In a number of cultivars, chlorophyll is retained in the seed coat at maturity, giving it a distinct green color. Green seed coat is controlled by a single dominant allele, *G*. This allele is epistatic to *y3*, which controls a chlorophyll deficiency in which newly formed leaves or pods are nearly normal green but rapidly lose chlorophyll and become yellow.

Many single-gene recessive nuclear mutants with various levels of chlorophyll deficiency are known collectively as chlorophyll-deficient types. Only a few mutants have been added to the genetic collection in recent years, in spite of the frequency of appearance of such mutants, and few genetic studies have been reported.

Vig (1973, 1975), working with the *Y11 y11* genotype, noticed that dark green, yellow, and twin or double (dark green-yellow) spots often appeared on the unifoliolate or the first trifoliolate. The occurrence of the twin spots was attributed to somatic crossing over. Occurrence of light-green spots on yellow leaves could be a back mutation of *y11* to *Y11* (Vig, 1973, 1975), or, as suggested by Ashley (1978), this phenotype might be an expression of *y11* as a hemizygous ineffective recessive allele. The *Y11 y11* genotype has been used as a test system to determine the effect of environmental mutagens (Vig, 1982). Hatfield (1982) noticed that spots or chimeric sectors on tetraploid *Y11 Y11 y11 y11* plants and among their progeny were much more frequent than on diploid *Y11 y11* plants grown in the same environment.

On the basis of the appearance of spots on *y9 y9* plants and *Y11 y11* plants without treatment, or with mitomycin C treatment, and *y9 y9* plants treated with caffeine, Vig (1974) suggested that *y9* and *y11* were alleles. Bernard et al. (1983) made reciprocal crosses of *y9 y9* and *Y11 y11* and examined plants in the F_1, F_2, and F_3 generations. They concluded that *y9* and *y11* were nonallelic.

Nissly et al. (1981b) determined the inheritance of chlorophyll-deficient types T134 and T162 and their allelic relationships with other chlorophyll-deficient mutants. Chlorophyll-deficient phenotypes of T134 and T162 each were controlled by a single recessive nuclear gene. T134 carried the recessive *y5* allele (T116). T134 was found in a cross of Illini by Peking, and T116 was found in radium-treated PI 65388. T162 was nonallelic to the eight known chlorophyll-deficient mutants tested and was assigned gene symbol *y17*.

There is considerable scientific interest in mutable genes. The mutable chlorophyll allele *Y18-m* can mutate either to stable conditions, *Y18* or *y18*, or to different states of instability (Peterson and Weber, 1969). When *Y18-m Y18-m* plants were grown at 19 or 29 °C, more total mutant sectors and greater total mutant area were found at 19 °C than at 29 °C (Sheridan and Palmer, 1977). However, more total yellow tissue and more yellow tissue per sector were found at 29 °C than at 19 °C, indicating that the timing of mutation from *Y18-m* to *y18* was affected by temperature. Earlier mutation to *y18* at 29 °C resulted in fewer, larger sectors of yellow tissue.

Electron microscopic observations have been made of green, light-green, and yellow leaf tissues of *Y18-m* plants grown at two different temperatures and two different photosynthetic photon flux densities (Palmer et al., 1979). The ultrastructure of chloroplasts from green tissue was not influenced markedly by temperature or illuminance. Light-green leaf sectors were mixtures of tissue layers with either normal chloroplasts or aberrant chloroplasts. Major changes in chloroplast ultrastructure were evident from yellow tissue sampled from both temperature and illuminance regimes.

T265, the delayed albino mutant, has a phenotype that is distinct from all other chlorophyll-deficient mutants in soybean, except perhaps T270. T265 plants are green as seedlings, become green-yellow, and eventually lose pigmentation and die.

T253 has both a tan saddle (*k2*) and chlorophyll deficiency (*y20*). Palmer (1984b) examined about 23 000 F_2 and F_3 plants and did not detect cross-over genotypes between *y20* and *k2*. He suggested that tight linkage, perhaps due to a deficiency rather than pleiotropy, was the explanation for failure to recover recombinant genotypes.

A summary of nuclear genes affecting chlorophyll deficiency or retention in soybean is presented in Table 5–11.

Table 5-11. Nuclear genes affecting chlorophyll deficiency or retention in soybean.

Gene	Phenotype	Strain†	Reference
	A. Chlorophyll deficiency		
V1	Normal	c	Woodworth (1932, 1933)
v1	Variegated leaves	T93	
Y3	Normal	c	Nagai (1926), Takagi (1929, 1930), Terao and Nakatomi (1929), symbol by Morse and Cartter (1937)
y3	Green seedling, becoming yellow	Kura, T139	
Y4	Normal	c	Symbol by Morse and Cartter (1937), Woodworth and Williams (1938) (as *y5* by error)
y4	Greenish yellow leaves, weak plant	T102	
Y5	Normal	c	Symbol by Morse and Cartter (1937), Woodworth and Williams (1938) (as *y4* by error)
y5	Greenish yellow leaves	T116, T134	
Y6	Normal	c	Symbol by Morse and Cartter (1937), Woodworth and Williams (1938)
y6	Pale green leaves	T136	
Y7 or *Y8*	Normal	c	Morse and Cartter (1937), Probst (1950) (as *y8*), Williams (1950)
y7 y8	Yellow growth in cool weather	T138	
Y9	Normal	c	Probst (1950)
y9	Bright greenish yellow	T135	

(continued on next page)

Table 5-11. Continued.

Gene	Phenotype	Strain†	Reference
Y10	Normal	c	Probst (1950)
y10	Greenish yellow seedling	T161	
Y11	Normal	c	Weber and Weiss (1959)
y11	Lethal yellow	T219	
Y12	Normal	c	Weiss (1970a)
y12	Whitish primary leaves, yellowish green leaves	T233	
Y13	Normal	c	Weiss (1970e)
y13	Whitish green seedling, greenish yellow leaves	T230	
Y14	Normal	c	Nissly et al. (1976)
y14	Light green leaves	T229	
Y15	Normal	c	Nissly et al. (1976)
y15	Pale yellowish green leaves	T234	
Y16	Normal	c	Wilcox and Probst (1969)
y16	Nearly white lethal	T257	
Y17	Normal	c	Nissly et al. (1981b)
y17	Light yellowish green leaves	T162	
Y18	Normal	c	Peterson and Weber (1969)
Y18-m	Unstable allele resulting in chlorophyll chimera	T225M	
y18	Near-lethal yellow	T225H	Sheridan and Palmer (1975)
Y19	Normal	c	Soybean Genetics Committee (1978)
y19	Delayed albino	T265	
Y20 K2	Normal	c	Palmer (1984b)
y20 k2	Yellowish green leaves, weak plant; tan saddle on yellow seed coat	T253	
	B. Chlorophyll retention		
D1 or *D2*	Yellow seed embryo	c	Woodworth (1921), Owen (1927a), Veatch and Woodworth (1930)
d1 d2	Green seed embryo	Columbia, T104	
G	Green seed coat	Kura	Terao (1918), Takahashi and Fukuyama (1919), Nagai (1921), Woodworth (1921)
g	Yellow seed coat	c	

†c = Indicates that the gene occurs in many cultivars.

5-5.8.5 Chlorophyll Deficiency—Cytoplasmic

Uniparentally inherited mutants are known that cause either chlorophyll deficiency in the leaves or chlorophyll retention in the cotyledons, embryo axis, and seed coat. The cytoplasmic foliage mutant *cyt-Y2* was found among the progeny of a chimeric plant (Palmer and Mascia, 1980). Yellow plants became progressively green, were viable, and produced seed in field, greenhouse, and growth-chamber environments. Field-grown *cyt-Y2* plants had 38%, and growth-chamber-grown plants had 77%, as much total chlorophyll as green sibling *cyt-G2* plants. Chloroplast ultrastructure of *cyt-Y2* plants was similar to that of *cyt-G2* plants both from dark and light growth-chamber environments (Palmer and Mascia, 1980).

A nuclear-cytoplasmic interaction was evident between *y20 k2* and *cyt-Y2*. In crosses between *cyt-Y2* as female parent and *y20 k2* as male parent, Palmer and Cianzio (1985) failed to find the *k2* phenotype among F_2 or F_3 plants. In the reciprocal cross, *k2* was expressed. They reported that the genotype *cyt-Y2 y20 k2 y20 k2* was lethal under field conditions, but could survive under reduced light in growth-chamber or greenhouse environments. Furthermore, the interaction was unique to *y20 k2* because *Y20 k2* was viable and expressed the genotype *cyt-Y2 Y20 k2 Y20 k2*.

The cytoplasmic foliage mutant *cyt-Y3* was found among the progeny of a chimeric plant (Shoemaker et al., 1985). These plants are inviable in the field, but survive and flower under reduced light in the greenhouse. At low photosynthetic photon flux densities (PPFD), *cyt-Y3* plants exhibited a normal-appearing plastid ultrastructure and normal carotenoid levels, but had only about 33% of the chlorophyll content of normal *cyt-G3* plants. At medium and high PPFD, *cyt-Y3* plastids lacked a structural thylakoid. At these two PPFD, the total chlorophyll content of *cyt-Y3* was only 28 and 1% of normal, respectively, and the carotenoid levels of *cyt-Y3* also dropped to 33 and 2% of the levels of *cyt-G3* (Shoemaker et al., 1985).

In some lines (*cyt-G1*), chlorophyll is retained in the cotyledons, embryo axis, and seed coat, and the leaves and pods do not turn yellow during ripening (Terao, 1918). Leaf tissue killed by disease or other factors does not turn yellow as in other lines. Plants with genotype *cyt-G1 g g y3 y3* have a yellow seed embryo and chlorophyll-deficient foliage, rather than green seed embryo and chlorophyll-deficient foliage (Terao and Nakatomi, 1929). Plants that are *cyt-G1 G g y3 y3* or *cyt-G1 g g Y3 y3* bear 3 green: 1 yellow seeds upon self-pollination.

A summary of cytoplasmic factors affecting chlorophyll deficiency or retention in soybean is presented in Table 5–12.

5–5.9 Pigmentation

5–5.9.1 Flower

Most plant introductions and cultivars have either purple (*W1*) or white (*w1*) flowers. There is a pleiotropic effect of the *W1* allele, in that

Table 5-12. Cytoplasmic factors affecting chlorophyll deficiency or retention in soybean.

Gene	Phenotype	Strain†	Reference
	A. Chlorophyll deficiency		
cyt-G2	Normal	c	Palmer and Mascia (1980)
cyt-Y2	Yellow leaves, becoming yellowish green	T275	
cyt-G3	Normal	c	Shoemaker et al. (1985)
cyt-Y3	Yellow leaves, very weak plant, (mutable plants are chlorophyll chimeras)	T278M	
	B. Chlorophyll retention		
cyt-G1	Green seed embryo	T104, Medium Green	Terao (1918), Veatch and Woodworth (1930)
cyt-Y1	Yellow seed embryo	c	

†c = Indicates that the gene occurs in many cultivars.

purple pigmentation is evident on the hypocotyl of seedlings. The hypocotyl color of seedlings with the *w1* allele is green. The purple pigmentation is intensified by exposure to sunlight. Hartwig and Hinson (1962) reported that certain cultivars had dilute-purple flowers (*W1 W3 W4*) and others were near-white (*W1 w3 w4*).

Magenta flower color is controlled by mutant allele *wm* in the presence of *W1* (Buzzell et al., 1977). They reported that the *wm* allele reduces flavonol glycoside content of flowers and leaves and is a deleterious mutant in terms of lower photosynthetic rate, earlier leaf senescence, and lower seed yield.

In a cross of a Japanese cultivar with an American cultivar, Olivieri et al. (1980) reported that two complementary alleles were involved in the control of hypocotyl and flower pigmentation. Payne and Sundermeyer (1977) observed either presence or absence of bronze pigmentation on the hypocotyl of soybean cultivars with white flowers when grown in continuous light. Palmer and Payne (1979) reported that tawny (*T* __ *Td* __) pubescent genotypes and light-tawny (*T* __ *td td*) pubescent genotypes, when grown in continuous light, had bronze pigmentation on the hypocotyl shortly after emergence. Gray (*t t Td* __ or *t t td td*) pubescent genotypes had no detectable bronze pigmentation.

Several researchers have examined the chemistry of pigmentation of hypocotyl and flower color in soybean. Nozzolillo (1973) found that the anthocyanin malvidin was the predominant pigment responsible for the purple coloration in hypocotyls. Peters et al. (1984) observed that malvidin was present at a 40- to 60-fold higher concentration than delphinidin, whereas petunidin was present at a fourfold higher concentration than delphinidin in purple hypocotyl (purple flower, tawny, light-tawny, or gray pubescence) cultivars. The same three anthocyanins were present in bronze hypocotyl (white flower, tawny, or light tawny pubescence) cultivars, but in lesser amounts. No compounds with anthocyanin-like absorption spectra were detected in green hypocotyl (white flower, gray pubescence) cultivars.

A summary of the genes controlling pigmentation in soybean flowers is presented in Table 5–13.

5–5.9.2 Pubescence

Pubescence color among most plant introductions and cultivars is controlled by a single gene pair, with tawny or brown (*T*) dominant over gray (*t*). In tawny pubescent genotypes, the trichomes on the young plants are colorless, but after several weeks' growth, many of the trichomes on the stems, pods, and leaves have brown pigment. This pigment is retained in the mature plant and facilitates classification of tawny and gray pubescent genotypes. Among gray pubescent genotypes, most trichomes are without brown pigment, giving a distinct gray phenotype to the plants.

The *T* allele has a major effect on the production or regulation of an enzyme necessary for the formation of quercetin from kaempferol.

Table 5-13. Genes affecting pigmentation in soybean.

Gene	Phenotype	Strain†	Reference
	1. Flower		
W1	Purple	c	Takahashi and Fukuyama (1919), Woodworth (1923)
w1	White	c	
W3 w4	Dilute purple	Laredo	Hartwig and Hinson (1962)
w3 W4	Purple	c	
W3 W4	Dark purple	L70-4422 [L6^6 × (Laredo × Harosoy)]	
w3 w4	Near white	L68-1774 [L6^6 × (Laredo × Harosoy)]	
Wm	Purple	c	Buzzell et al. (1977)
wm	Magenta	T235	
	2. Pubescence		
T	Tawny (brown); quercetin and kaempferol present	c	Piper and Morse (1910), Nagai (1921) (as *C c*), Woodworth (1921), Williams (1950), Buttery and Buzzell (1973)
t	Gray; quercetin absent, kaempferol present	c	
Td	Tawny (brown); flavonol present	Clark	Buttery and Buzzell (1973), Bernard (1975b)
td	Light tawny (near-gray); flavonol absent	Grant, Sooty	
	3. Seed		
I	Light hilum	Mandarin, c	Nagai (1921), Nagai and Saito (1923), Owen (1928b), Woodworth (1932, 1933), Mahmud and Probst (1953)
i-i	Dark hilum	Manchu, c	
i-k	Saddle pattern	Black Eyebrow	
i	Self dark seed coat	Soysota, T157	
Im	Nonmottled seed	Merit	Cooper (1966)
im	Dark mottled seed	Harosoy	
K1	Nonsaddle	c	Takagi (1929, 1930), Williams (1958)
k1	Dark saddle on seed coat	Kura, T153, Agate	
K2	Yellow seed coat	c	Rode and Bernard (1975a), Palmer (1984b)
k2	Tan saddle on seed coat	T239, T253, T261	
K3	Nonsaddle	c	Bernard and Weiss (1973)
k3	Dark saddle on seed coat	T238	
O	Brown seed coat	Soysota, c	Nagai (1921), Weiss (1970b)
o	Reddish brown seed coat	Ogemaw	
R	Black seed coat	c	Nagai (1921), Woodworth (1921), Stewart (1930), Williams (1952)
r-m	Black stripes on brown seed	PI 91073, T146	Nagai and Saito (1923), Weiss (1970b)
r	Brown seed	c	
	4. Pod		
L1 L2	Black pod	Seneca	Bernard (1967)
L1 l2	Black pod	PI 85505	
l1 L2	Brown pod	Clark, c	
l1 l2	Tan pod	Dunfield, c	

†c = Indicates that the gene occurs in many cultivars.

Cultivars with the *T* allele have free quercetin (the aglycone) in the pubescence, and those with *t* have free kaempferol (Buttery and Buzzell, 1973). The *T* and *t* alleles are discussed in sections on flavonol glycosides, hypocotyl pigmentation, and seed-coat color.

Bernard (1975b) described another major gene pair affecting pubescence color. *Td* controlled dark-tawny and *td* controlled light-tawny (also called near-gray) in the presence of *T*. In contrast to the *T* allele, which affects hilum or seed coat pigments and flavonol glycosides, *Td* affects only pubescence color. In the presence of *td td* there is no, or markedly less, flavonol in the pubescence (Buttery and Buzzell, 1973).

A summary of genes affecting pubescence color in soybean is given in Table 5–13.

5–5.9.3 Seed

Loci affecting pigmentation of the seed have been discussed in detail by Bernard and Weiss (1973), and only the salient data presented by them will be given. Woodworth (1921) showed that there was a pleiotropic effect of the *T-t* gene for pubescence color. Seeds from tawny pubescent genotypes have black or brown pigment in the seed, whereas gray pubescent genotypes have imperfect black or buff pigment. The *Td td* gene pair does not affect seed pigmentation.

Hilum color and seed-coat color serve as useful marker genes for distinguishing between hybrid and self-pollinated progeny when making cross-pollinations. Williams (1952), Specht and Williams (1978), and Palmer and Stelly (1979) summarized the phenotypes as follows:

	Seed coat and hilum self-color	Saddle and hilum color	Hilum color	Hilum color
Genes	*i*	*i-k*	*i-i*	*I*
T R	black	black	black	gray
T r O	brown	brown	brown	yellow
T r o	red brown	red brown	red brown	yellow
t R W1	imperf. black	imperf. black	imperf. black	gray
t R w1	buff	buff	buff	yellow
t r	buff	buff	buff	yellow

The *k* loci affect seed-coat pigment distribution and produce a saddle pattern on the seed. The color of the saddle is controlled by the same alleles that control hilum color, *I*, *R*, *T*, and *W1*. A tan-saddle pattern is known, but is independent of the effects of the hilum color alleles (Rode and Bernard, 1975a). Tan saddle is nonallelic to *i-k* and *k1* and was designated *k2*. Five independent tan-saddle mutants are known that are *k2*. One of these, T253, is both chlorophyll deficient and tan saddle and was discussed in the section dealing with the nuclear chlorophyll-deficient mutants. Another mutant, nonallelic to *i-k*, *k1*, and *k2*, has been assigned the gene symbol *k3* (Bernard and Weiss, 1973), and it has black or brown saddle.

Yoshikura and Hamaguchi (1969) have identified delphinidin-3-monoglucoside and cyanidin-3-monoglucoside as anthocyanins of black-seeded soybean. Pelargonidin-3-glucoside has been identified in the red buff seed coats of T236 (Taylor, 1976). Buzzell and Buttery (1982) have related the chemistry to the genetics of seed-coat pigments. The *T* allele probably controls hydroxylation from pelargonidin to cyanidin-3-glucoside. The *W1* allele controls trihydroxylation to delphinidin-3-glucoside. *T* is not necessary for the formation of delphinidin, but when *T* is present, the production of delphinidin by *W1* is enhanced.

A summary of the genes controlling pigmentation in soybean seed is presented in Table 5–13.

5–5.10 Isoenzymes and Proteins

Since publication of the 1973 edition of *Soybeans: Improvement, Production, and Uses,* there has been a proliferation of reports describing isoenzymes and proteins in the genus *Glycine.* Guidelines for gene symbols have been revised recently by the Soybean Genetics Committee, and Table 5–14 reflects those changes. Genetic interpretations presented in this section are those of the authors cited. Two recent reviews of isoenzymes and protein variation in soybean have been given by Hymowitz (1983) and Kiang and Gorman (1983). The isoenzymes and proteins discussed in this section are seed or germinating seed components.

Table 5-14. Genes controlling inheritance of isoenzyme and protein variants in soybean.

Gene	Phenotype	Strain†	Reference
Ap-a	Acid phosphatase mobility variant	Ebony	Gorman and Kiang (1977), Hildebrand et al. (1980)
Ap-b	Acid phosphatase mobility variant	Amsoy 71, c	
Ap-c	Acid phosphatase mobility variant	Earlyana, Manchu	
Adh1	Alcohol dehydrogenase present	Altona, Wilson	Gorman and Kiang (1978), Kiang and Gorman (1983)
adh1	Alcohol dehydrogenase absent	A-100, Lindarin	
Adh2	Alcohol dehydrogenase present	Amsoy, Beeson	
adh2	Alcohol dehydrogenase absent	Cayuga, Grant	
Amy1	α-amylase band 1 present	Harosoy, Clark	Gorman and Kiang (1977, 1978), Kiang (1981)
amy1	α-amylase band 1 absent	Altona‡, PI 132201	
Amy2	α-amylase band 2 present	Harosoy, Clark	
amy2	α-amylase band 2 absent	Altona, PI 132201	

(continued on next page)

Table 5-14. Continued.

Gene	Phenotype	Strain†	Reference
Sp1-a	β-amylase mobility variant	Amsoy, Evans	Larsen (1967), Larsen and Caldwell (1968), Orf and Hymowitz (1976), Gorman and Kiang (1977, 1978), Hymowitz et al. (1979), Hildebrand and Hymowitz (1980a, 1980b), Kiang (1981)
Sp1-b	β-amylase mobility variant	Williams, Century, c	
Sp1-an	Seed protein band present, β-amylase activity weak or absent	Chestnut	
sp1	Seed protein band absent, β-amylase activity absent	Altona, PI 132201	
Cgy1	β-conglycinin subunit α′ present	c	Kitamura et al. (1984)
cgy1	β-conglycinin subunit α′ absent	Keburi	
Dia1-a	Diaphorase mobility variant	Evans, Elton	Gorman et al. (1983), Kiang and Gorman (1983)
Dia1-b	Diaphorase mobility variant (some bands weak)	Cayuga, Kingston	
Dia2-a	Diaphorase mobility variant	Amsoy, Elton	
Dia2-b	Diaphorase mobility variant	Wilson, Kingston	
Dia3	Diaphorase present	Kingston	
dia3	Diaphorase absent	Elton	
Ep	High peroxidase activity	Harosoy 63	Buzzell and Buttery (1969)
ep	Low peroxidase activity	Blackhawk	
Eu	Urease fast band	Blackhawk, Chippewa 64	Buttery and Buzzell (1971)
eu	Urease slow band	Corsoy, Midwest	
Gpd	Glucose-6-phosphate dehydrogenase present	Amsoy, Evans	Gorman et al. (1983), Kiang and Gorman (1983)
gpd	Glucose-6-phosphate dehydrogenase (weak)	Chestnut, Cayuga	
Gy4	Glycinin subunit $A_5A_4B_3$ present	c	Kitamura et al. (1984)
gy4	Glycinin subunit $A_5A_4B_3$ absent	Raiden	
Idh1-a	Isocitrate dehydrogenase mobility variant	Amsoy, Cayuga	Yong et al. (1981, 1982), Gorman et al. (1983), Kiang and Gorman (1983)
Idh1-b	Isocitrate dehydrogenase mobility variant	Wilson, Evans	
Idh2-a	Isocitrate dehydrogenase mobility variant	Amsoy, Cayuga	
Idh2-b	Isocitrate dehydrogenase mobility variant	Wilson, Evans	
Idh3-a	Isocitrate dehydrogenase mobility variant	Elton, Amsoy	
Idh3-b	Isocitrate dehydrogenase mobility variant	Agate, Wilson	
Lap1-a	Leucine aminopeptidase mobility variant	Norredo, Wilson	Gorman et al. (1982a, 1982b, 1983)

(continued on next page)

Table 5-14. Continued.

Gene	Phenotype	Strain†	Reference
Lap1-b	Leucine aminopeptidase mobility variant	Lindarin	
Lap2	Leucine aminopeptidase present	Amsoy	Kiang et al. (1984)
lap2	Leucine aminopeptidase absent	Jefferson	
Le	Seed lectin present	Harosoy	Pull et al. (1978), Orf et al. (1978), Stahlhut and Hymowitz (1980)
le	Seed lectin absent	T102	
Lx1	Lipoxygenase-1 present	Harosoy, Clark	Hildebrand and Hymowitz (1981, 1982)
lx1	Lipoxygenase-1 absent	Kedelee No. 367 (PI 133226), PI 408251	
Lx2	Lipoxygenase-2 present	Suzuyutaka	Davies and Nielsen (1984)
lx2	Lipoxygenase-2 absent	(PI 86023)	
Lx3	Lipoxygenase-3 present	Raiden, Century	Kitamura et al. (1983)
lx3	Lipoxygenase-3 absent	Wase Natsu (PI 417458), I-Higo-Wase, (PI 205085)	
Mpi-a	Mannose-6-phosphate isomerase mobility variant	Wilson, PI 65549 (wild soybean)	Gorman et al. (1983), Kiang and Gorman (1983)
Mpi-b	Mannose-6-phosphate isomerase mobility variant	Amsoy, Kingston	
Mpi-c	Mannose-6-phosphate isomerase mobility variant	Elton, Hark	
Pgd-a	Phosphogluconate dehydrogenase mobility variant	Agate, Kingston	Gorman et al. (1983), Kiang and Gorman (1983)
Pgd-b	Phosphogluconate dehydrogenase mobility variant	Elton, Hill	
pgd	Phosphogluconate dehydrogenase absent	Hidaka-1 (PI 406684), PI 65549 (wild soybean)	
Pgi-a	Phosphoglucose isomerase mobility variant	PI 135624 (wild soybean), PI 65549 (wild soybean)	Gorman et al. (1983), Kiang and Gorman (1983)
Pgi-b	Phosphoglucose isomerase mobility variant	Beeson, Hark	
Pgm1-a	Phosphoglucomutase mobility variant	Chestnut, Wells	Gorman et al. (1983), Kiang and Gorman (1983)
Pgm1-b	Phosphoglucomutase mobility variant	Amsoy, Hark	
Pgm2-a	Phosphoglucomutase mobility variant	PI 423990 (wild soybean), Shirosaya 1 (PI 423955)	
Pgm2-b	Phosphoglucomutase mobility variant	Amsoy, Wells	

(continued on next page)

Table 5-14. Continued.

Gene	Phenotype	Strain†	Reference
Sod	Superoxide dismutase bands 4 and 5 present	c	Gorman and Kiang (1978), Gorman et al. (1982b, 1984), Griffin and Palmer (1984)
sod	Superoxide dismutase bands 4 and 5 absent	Evans	
Sp1§			
Ti-a	Kunitz trypsin inhibitor mobility variant	Harosoy, Clark	Singh et al. (1969), Hymowitz and Hadley (1972), Orf and Hymowitz (1977, 1979)
Ti-b	Kunitz trypsin inhibitor mobility variant	Aoda	
Ti-c	Kunitz trypsin inhibitor mobility variant	PI 86084	
ti	Kunitz trypsin inhibitor absent	Kin-du (PI 157440), Baik Tae (PI 196168)	

†c = Indicates that gene occurs in many cultivars.
‡Altona is a mixture of several genotypes.
§See β-amylase which follows *amy2*.

Isoenzymes can be used for cultivar identification, in genetic linkage studies, as genetic markers in plant breeding or tissue culture, and in evolutionary studies of the genus *Glycine*. The latter subject is discussed in chapter 3 in this book. In addition, null alleles offer potential for the genetic improvement of animal feed and human food, because certain of the isoenzymes and proteins are antinutritional or undesirable components of soybean seed. This section describes the qualitative genetics of isoenzyme and protein variants. Additional polymorphisms have been described, for which genetic data are not currently available (Blogg and Imrie, 1982; Kiang and Gorman, 1983).

Several electrophoretic systems have been used to study isoenzyme and protein polymorphisms in the genus *Glycine*. These include starch gel electrophoresis (Cardy and Beversdorf, 1984) and vertical and horizontal polyacrylamide gel electrophoresis (PAGE). Detailed discussions of these techniques can be found in the publications cited.

Where appropriate, the isoenzymes described in this section are accompanied by Enzyme Commission (EC) numbers, as suggested by the International Union of Biochemists. The numbers serve to identify more precisely the enzyme activity under discussion.

5-5.10.1 Acid Phosphatase (EC 3.1.3.2)

Gorman and Kiang (1977) reported three acid phosphatase bands; two of the bands were invariant, but the third band exhibited three mobility variants. These three forms (*Ap-a*, *Ap-b*, and *Ap-c*) are inherited as codominant alleles at a single locus (Hildebrand et al., 1980). Other acid phosphatase polymorphisms have been reported in cultivated and wild soybean and perennial *Glycine* spp. but genetic studies are not yet completed (Broué et al., 1977; Kiang and Gorman, 1983).

5-5.10.2 Alcohol Dehydrogenase (EC 1.1.1.1)

Three alcohol dehydrogenase zymogram patterns from seed were observed by use of horizontal PAGE (Gorman and Kiang, 1977). The first type had seven alcohol dehydrogenase bands, the second type lacked the slowest band and fourth band, while the third type lacked the slowest, fourth, and fifth bands. Two dominant interacting loci (*Adh1* and *Adh2*) control the three types, and seem to be tightly linked. Each locus has dominant functional alleles (*Adh1* and *Adh2*) with recessive null alleles (*adh1* and *adh2*) (Gorman and Kiang, 1978). A fourth zymogram type, lacking the fourth and fifth bands but having the slowest band, was observed among segregating F_2 plants as a recombinant type. It has not been observed among *G. max* or *G. soja* accessions (Kiang and Gorman, 1983). Alcohol dehydrogenase polymorphisms have been observed among the perennial *Glycine* spp. (Broué et al., 1977; Kiang and Gorman, 1983).

5-5.10.3 Amylase (EC 3.2.1.1 and 3.2.1.2)

Mature soybean seed contains both α- and β-amylase. Three different loci have been suggested for the control of amylase in soybean seed (Gorman and Kiang, 1977; 1978). *Amy1* and *Amy2* control α-amylase (Kiang, 1981) and *Sp1* controls β-amylase (Hildebrand and Hymowitz, 1980a; Kiang, 1981). The *Sp1* locus is the same as that reported by Larsen (1967) and Larsen and Caldwell (1968), who designated their bands as an "A protein" and a "B protein" (Orf and Hymowitz, 1976).

Null variants were found at the *Amy1* and *Amy2* loci. Inheritance data showed that these nulls (*amy1* and *amy2*) are recessive to *Amy1* and *Amy2*, respectively (Kiang, 1981).

Four electrophoretic forms of β-amylase have been described (Hymowitz et al., 1979; Kiang, 1981). Two of the forms, *Sp1-a* and *Sp1-b*, are electrophoretically distinguishable from one another by their different mobilities (Orf and Hymowitz, 1976; Kiang, 1981). The third form, *Sp1-an*, has the *Sp1-a* seed protein band, but β-amylase activity is weak or absent. The fourth form, *sp1*, lacks the *Sp1* seed protein band and β-amylase activity (Hildebrand and Hymowitz, 1980b; Kiang, 1981).

Sp1-a and *Sp1-b* are codominant alleles for seed protein bands and β-amylase activity. Allele *sp1* is recessive to *Sp1-a* and to *Sp1-b* for seed protein bands and β-amylase activity. *Sp1-an* is codominant with *Sp1-b* for seed protein band and recessive to *Sp1-a* and *Sp1-b* for β-amylase activity. *Sp1-an* and *sp1* do not complement each other, because no seed with normal levels of β-amylase activity were recovered when the two were crossed. *Sp1-a*, *Sp1-b*, *Sp1-an*, and *sp1* form a multiple allelic series at a single locus (Hildebrand and Hymowitz, 1980b; Kiang, 1981). The significance of β-amylase in soybean seed has not been established; however, it seems not to be necessary for normal starch metabolism (Adams et al., 1981).

5-5.10.4 Diaphorase (EC 1.6.4.3)

Four diaphorase zymogram variants have been visualized by use of horizontal PAGE and starch gel electrophoresis among the cultivated and wild soybean and perennial *Glycine* spp. (Broué et al., 1977; Gorman et al., 1982a, 1982b, 1983; Kiang and Gorman, 1983). The first is controlled by a single locus, *Dia1* (Gorman et al., 1983).

The second diaphorase locus has two codominant alleles (*Dia2-a* and *Dia2-b*) which affect the mobility of two bands. The third diaphorase locus has dominant allele *Dia3*, and a recessive null allele (*dia3*) conditioning presence or absence of that band (Gorman et al., 1983). No genetic studies have been reported for the fourth diaphorase zymogram variant.

5-5.10.5 Glucose-6-phosphate Dehydrogenase (EC 1.1.1.49)

Two forms differing in band intensity were observed by horizontal PAGE among accessions of *G. max* and *G. soja*. Inheritance was controlled by a single nuclear gene (*Gpd*), with the high intensity type dominant to the weak intensity type (*gpd*) (Gorman et al., 1982b, 1983; Kiang and Gorman, 1983).

5-5.10.6 Isocitrate Dehydrogenase (EC 1.1.1.42)

Isocitrate dehydrogenase zymograms exhibit two zones of activity. The most mobile zone, which includes up to five bands in heterozygotes, is controlled by two loci (Yong et al., 1981, 1982; Gorman et al., 1982b, 1983). Yong et al. (1981, 1982) proposed that these represented duplicate loci. Gorman et al. (1983) and Kiang and Gorman (1985) proved that the two loci are independent. Each locus has one allele with a common electrophoretic mobility, as well as a variant allele with altered mobility. Various combinations of alleles at the two loci result in the observed one-, three-, or five-band phenotypes (Yong et al., 1981, 1982; Gorman et al., 1983; Kiang and Gorman, 1985).

There are two loci (*Idh1* and *Idh2*) that control the inheritance of the most mobile group of bands. Each locus has two alleles, *Idh1-a*, *Idh1-b* and *Idh2-a*, *Idh2-b*. The two alleles, *Idh1-b* at the first locus and *Idh2-a* at the second locus (intermediate protein mobility on electrophoretic gels), have the same mobility and may be the original form. The alleles *Idh1-a* (least mobile protein) and *Idh2-b* (most mobile protein) appear to be variant alleles derived from the original form.

Three mobility variants for one of the two bands in the least mobile zone have been described (Gorman et al., 1982b, 1983). Inheritance studies with only two of these variants indicate that they are under the control of a third locus (*Idh3*) with two codominant alleles (*Idh3-a* and *Idh3-b*) (Gorman et al., 1983).

5-5.10.7 Kunitz Trypsin Inhibitor

Accessions were screened by vertical PAGE for the presence or absence of the Kunitz trypsin inhibitor, an antinutritional component in the seed of most soybean lines. Four mobility variants were described. Three are distinguishable from one another by their different mobilities (Hymowitz, 1973; Orf and Hymowitz, 1979; Singh et al., 1969). These three forms are controlled by a codominant multiple allelic system (*Ti-a, Ti-b*, and *Ti-c*) at a single locus (Hymowitz and Hadley, 1972; Orf and Hymowitz, 1977). The fourth Kunitz trypsin inhibitor variant is a null that does not exhibit a protein band in the gels and is inherited as the recessive allele *ti* (Orf and Hymowitz, 1979).

5-5.10.8 Leucine Aminopeptidase (EC 3.4.1.1)

Gorman et al. (1982a, 1982b) and Kiang and Gorman (1983) reported three leucine aminopeptidase mobility variants in both cultivated and wild soybean. Two codominant alleles at a single locus (*Lap1-a* and *Lap1-b*) condition the difference for two of these variants. Gorman et al. (1983) reported a variant for a second, more mobile leucine aminopeptidase band. These variants are under the control of a second locus with *Lap2* dominant to the recessive null *lap2* (Kiang et al., 1984).

5-5.10.9 Lectin

Seeds were screened by polyacrylamide gel electrophoresis for the presence or absence of seed lectin. Two electrophoretic variants were found. One has a seed lectin band, and the second does not (Pull et al., 1978; Stahlhut and Hymowitz, 1980). Orf et al. (1978) reported that the presence of seed lectin is controlled by a single dominant allele *Le*, whereas the homozygous recessive genotype (*le le*) results in the lack of seed lectin. The *le* allele has been shown to contain an insertion sequence (Goldberg et al., 1983; Rhodes and Vodkin, 1985). Forty-nine percent (272 of 559) of *G. soja* accessions lacked seed lectin (Stahlhut et al., 1981). None of the 56 accessions examined from six perennial *Glycine* spp. contained seed lectin (Pueppke and Hymowitz, 1982). Accessions without seed lectin germinated normally. The significance of soybean seed lectins has not been established.

5-5.10.10 Lipoxygenase (EC 1.13.11.12)

Lipoxygenase has been implicated as a cause of undesirable flavors in soybean oil products. Soybean seeds contain at least three lipoxygenase isoenzymes. Seeds were screened for the presence or absence of lipoxygenase-1 by use of a spectrophotometric assay and results were confirmed by PAGE and by the Ouchterlony immunological double diffusion technique (Hildebrand and Hymowitz, 1981). Hildebrand and Hymowitz (1982) reported that the presence of lipoxygenase-1 is controlled by a single dominant allele (*Lx1*). The homozygous recessive genotype (*lx1 lx1*) results in the lack of lipoxygenase-1. For other traits, plants without lipoxygenase-1 were similar to plants that have lipoxygenase-1 activity.

I-Higo-Wase (PI 205085) and Wase Natsu (PI 417458) were found to lack lipoxygenase-3 by both immunological and PAGE tests. Genetic studies indicated that presence of lipoxygenase-3 is controlled by a single dominant allele (*Lx3*) and that the homozygous recessive genotype (*lx3 lx3*) results in the lack of lipoxygenase-3 (Kitamura et al., 1983).

Davies and Nielsen (1984) have identified a homozygous recessive genotype (*lx2 lx2*) that lacks lipoxygenase-2.

5-5.10.11 Mannose-6-phosphate Isomerase (EC 5.3.1.8)

Five homozygous mannose-6-phosphate isomerase zymogram patterns have been observed among *G. max* and *G. soja* accessions by horizontal PAGE (Gorman et al., 1982b, 1983). Inheritance studies with three of these types showed that a single locus with three codominant alleles (*Mpi-a, Mpi-b*, and *Mpi-c*) affected band mobility. Weak or null bands for mannose-6-phosphate isomerase were seen, but genetic data are not available for these variants (Kiang and Gorman, 1983).

5-5.10.12 Phosphogluconate Dehydrogenase (EC 1.1.1.44)

Gorman et al. (1983) reported two homozygous zymograms among accessions of *G. max* and two additional zymograms among accessions of *G. soja*. In two of the variants, mobility of two of the three bands differed. Inheritance studies indicated a single nuclear locus with two codominant alleles (*Pgd-a* and *Pgd-b*). In the third variant, the same two bands were absent and their absence was conditioned by the recessive null allele, *pgd*. A fourth variant, with altered mobility of all three bands, was observed; however, inheritance studies are not available (Kiang and Gorman, 1983).

5-5.10.13 Phosphoglucose Isomerase (EC 5.3.1.9)

Four homozygous zymograms for phosphoglucose isomerase have been found among cultivated and wild soybean (Gorman et al., 1983). Genetic data supported the hypothesis that the difference between two of the mobility variants was controlled by a single locus with two codominant alleles (*Pgi-a* and *Pgi-b*). A third mobility variant and a null variant have not been analyzed genetically (Kiang and Gorman, 1983).

5-5.10.14 Phosphoglucomutase (EC 2.7.5.1)

Gorman et al. (1982b, 1983) have observed two homozygous phosphoglucomutase zymograms in *G. max* and four in *G. soja*. They hypothesized that these types are inherited via two polymorphic phosphoglucomutase loci. The first locus has two codominant alleles (*Pgm1-a* and *Pgm1-b*) that affect the mobility (Gorman et al., 1983). The second variable locus has two alleles (*Pgm2-a* and *Pgm2-b*) that probably affect the mobility of the first phosphoglucomutase band (Kiang and Gorman, 1983).

5-5.10.15 Seed Storage Proteins

About 70% of soybean seed protein, on a dry weight basis, is accounted for by two components in the globulin fraction, glycinin, and β-conglycinin. The cv. Raiden lacks glycinin subunit $A_5A_4B_3$ (*gy4*), whereas the dominant allele (*Gy4*) is present in most soybeans (Kitamura et al., 1984). The cv. Keburi lacks the α'-subunit of β-conglycinin (*cgy1*), whereas most soybean germplasm contains the dominant allele *Cgy1* (Kitamura et al., 1984). The two loci are inherited independently from each other.

5-5.10.16 Superoxide Dismutase (EC 1.15.1.1)

Using vertical PAGE, Griffin and Palmer (1984) described nine superoxide dismutase bands. A null variant for bands 4 and 5 and a mobility variant for bands 8 and 9 were reported. They further concluded that superoxide dismutase isoenzymes were the source of the tetrazolium oxidase activity described by Gorman and Kiang (1977) and the INT-oxidase activity reported by Larsen and Benson (1970).

The presence or absence of bands 4 and 5 was shown by Gorman and Kiang (1978) to be due to a single locus (*Sod*) with a recessive null allele (*sod*). The inheritance of the mobility variant for bands 8 and 9 has not been reported.

5-5.10.17 Urease (EC 3.5.1.5)

Buttery and Buzzell (1971) reported that two urease isoenzymes were controlled by a single locus, with the fast form (*Eu*) dominant over the slow form (*eu*). In a survey of soybean germplasm, two *G. soja* accessions were seen to have slow migrating urease forms with low activity (Buzzell et al., 1974). They suggested that another locus besides *Eu-eu* may be involved.

A urease null mutant, which is phenotypically characterized by the lack of urease activity and antigen in the mature seed, was reported by Polacco et al. (1982). The genetics of the null is complex and seems to be a case of cis-acting, recessive epistasis with the *Eu* allele dominant and hypostatic to the null allele (Kloth et al., 1984).

Kloth and Hymowitz (1985) reported that crosses between an electrophoretically slow variant and an electrophoretically fast variant gave codominant inheritance.

5-6 CYTOGENETICS

5-6.1 Cytology

Cytological studies of members of the genus *Glycine* have only recently begun. The large number, small size, and similar morphology of the chromosomes, in addition to the lack of suitable techniques for mitotic and meiotic chromosome examinations, have discouraged scientists in the past.

Fukuda (1933), using meiotic chromosomes, and Sen and Vidyabhusan (1960), using mitotic chromosomes, karyotyped the soybean chromosomes as two large, 14 intermediate and four small. Biswas (1977) grouped somatic chromosomes of three *G. max* cultivars into four or six classes, depending upon cultivars. The somatic chromosomes of the Indian cultivar UPI were arranged into eight classes on the basis of the size of the chromosomes and the position of primary and secondary constrictions (Biswas and Bhattacharyya, 1972). Pillai (1976) organized the 20 somatic chromosomes of a "diploid" Indian native cultivar into 10 pairs. This report has not been verified. Ahmad et al. (1977a) found that the range of lengths of the somatic chromosomes of *G. max* and *G. soja* were similar for the two species, each exhibiting a continuous gradation. Ahmad et al. (1983) identified nine of the 20 chromosomes of the haploid complement of the cv. Daintree as two metacentric, six submetacentric, and one subtelocentric (satellite). The remaining 11 chromosomes were characterized into classes on the basis of chromosome length and arm ratio.

Oinuma (1952) and Palmer and Heer (1973) observed one pair of chromosomes with a satellite. Pillai (1976) has observed four satellite chromosomes in a 40-chromosome *G. max* Indian native cultivar. Zheng et al. (1984) observed a 40-chromosome *G. soja* line with four satellite chromosomes. Two satellites were on a pair of long chromosomes and the other two were located on a pair of short chromosomes.

Four groups of soybean chromosomes were distinguished after Giemsa staining (Ladizinsky et al., 1979). Three groups, composed of six pairs of chromosomes each, were characterized by a single band per chromosome. The fourth group, composed of two pairs, possessed two bands per chromosome. The relationship between the bands and the centromere in any particular chromosome could not be defined. This staining technique seems to have limited value for identification of individual soybean chromosomes (Ladizinsky et al., 1979).

The DNA content per haploid nucleus in soybean has been estimated at 1.97 pg, as determined through reassociation kinetics (Goldberg, 1978), and between 1.84 and 2.61 pg, as ascertained by cytophotometry of Feulgen-stained root-tip cell nuclei (Doerschug et al., 1978). Yamamoto and Nagato (1984) examined the DNA content of six *Glycine* spp. They concluded that cellular DNA content has been reduced during the course of soybean domestication.

Techniques for chromosome study of root-tip cells (Palmer and Heer, 1973) and from very young leaves (Sapra and Stewart, 1981) are suitable for members of the genus *Glycine*. Greater clarity of constrictions with root-tip cells usually is seen when 8-hydroxyquinoline, rather than paradichlorobenzene, is used as the pretreatment chemical (Palmer and Heer, 1973; Zheng et al., 1984).

Squash preparations of microspore mother cells are more difficult. Methods for preparation of meiocytes that routinely give good results are not known. However, fixation with Carnoy's (Newell and Hymowitz, 1983) or Pienaar's (Palmer and Kaul, 1983) seem to give the best preparations. Glasshouse-grown plants seem to yield a higher percentage of acceptable preparations, whereas plants grown under hot, dry conditions give very poor preparations.

Paraffin embedding of anthers for light microscopy has been successful with use of the method of Palmer et al. (1980). A stain-clearing technique for ovules (Stelly et al., 1984), with modifications for soybean (Kennell, 1984) was used in a light-microscope study of the *ms1* female reproductive system. Resin embedding of samples for light- and transmission-electron microscopy employ techniques developed by Albertsen and Palmer (1979), Buntman (1983), and Tilton et al. (1984). Preparation of samples for scanning electron microscopy follows the method of Buntman and Horner (1983).

Cytophotometric measurements of DNA content per root-tip cell nucleus are routine (Doerschug et al., 1978). Fluorescence microscopy has been used to study "pollen-tube formation" in *ms1* plants (Skorupska and Nawracala, 1980) and in the *ms3* mutant (Nakashima et al., 1984).

In summary, techniques have been developed that are useful in understanding soybean cytology and reproductive biology. No reliable techniques are known that consistently produce a high degree of success with squash preparations of soybean meiocytes.

5–6.2 Haploids and Polyploids

The discussion of haploids and polyploids will use the terminology $2n=2x=40$ chromosomes (diploid) and $2n=4x=80$ chromosomes (tetraploid). Haploids and polyploids are known to occur either from monoembryonic or polyembryonic seedlings.

Owen (1928c) reported that an accession from China gave twin seedlings at a frequency of 0.44%. A similar frequency of twins also was obtained among F_2 and F_3 seeds of crosses involving this accession. No chromosome counts were reported.

Shorter and Byth (1975) noticed that twinning occurred at a frequency of from 0 to 28.2% in the same genotype, depending upon environment. Among 36 000 seeds examined from the same cultivars used by Shorter and Byth (1975), Ahmad et al. (1977b) found 38 completely separated multiple seedlings.

Kenworthy et al. (1973) observed from 2.2 to 5.5% twin seedlings from 3485 seeds from *ms1 ms1* plants. Beversdorf and Bingham (1977) reported 2.3% polyembryony from 7206 seeds and Sorrells and Bingham (1979) reported 2.2% polyembryony from 15 500 seeds, from *ms1 ms1* plants. Kenworthy et al. (1973), Beversdorf and Bingham (1977), and Sorrells and Bingham (1979) used T260 as their *ms1* source. Chen et al. (1985) observed frequencies of polyembryony from 0.1 to 3.6% with four *ms1* types (T260, T266, T267, T268) from six different source populations. Buss and Autio (1980) noted two pairs of twins among 253 seeds examined from *ms2 ms2* plants.

Haploids (20 chromosomes) were reported to occur naturally (Pillai, 1976); however, there is no verification of this report. Most reported haploids have occurred among progeny of *ms1 ms1* plants. Kenworthy et al. (1973) observed one haploid among 47 seedlings from polyembryonic seed, Beversdorf and Bingham (1977) found five haploids among 167 polyembryonic seeds, and Sorrells and Bingham (1979) found 10 haploids among 341 polyembryonic seeds. Among 159 polyembryonic seedlings, Chen et al. (1985) found five haploids.

Haploids of parthenogenetic origin were found among progeny from *ms1 ms1* plants and were examined cytologically (Sorrells and Bingham, 1979; Crane et al., 1982). At pachytene, the paired regions generally were less than half the length of the chromosome and varied from 0 to 6 regions of pairing per cell. At diakinesis, the mean number of chromosome pairs per cell among the nine plants varied from 0.79 to 1.46. The amount of chromosome pairing in haploid soybean is compatible with the supposed tetraploid constitution of soybean, although the genetic control of pairing and genomic differences in the reduction of pairing remain unknown (Crane et al., 1982).

Ahmad et al. (1975) noted that one member of a twin pair from a non-*ms1* source was haploid. Treatment of branches with colchicine produced diploid sectors and the resulting seeds had 40 chromosomes.

Polyploids in soybean may be produced by application of colchicine, or may be found among progeny of *ms1 ms1* plants, *ms2 ms2* plants, or from synaptic mutants. The frequency of polyploids from synaptic mutants T241, T242, T258, and T272 is discussed in the section on aneuploids. Only two polyploids and one mixoploid have been reported among progeny of *ms2 ms2* plants (Buss and Autio, 1980; Sadanaga and Grindeland, 1981).

Biswas and Bhattacharyya (1972) treated seed and seedlings with various colchicine concentrations and for different time intervals. They found that seed treatment was ineffective for inducing polyploidy. A 0.5% colchicine in lanolin paste applied to the meristematic region of seedlings before and during the opening of the cotyledons was effective for inducing polyploidy (Sadanaga and Grindeland, 1981). Treatment of unifoliolate leaves and the apical bud with 0.1% aqueous colchicine resulted in polyploidy for many *Glycine* spp. (Cheng and Hadley, 1983).

In tetraploids, the frequency of stomata on the upper and lower leaf surfaces was reduced when compared with the diploid *G. max* cv. UPI (Biswas and Bhattacharyya, 1972). The mean size of guard cells was increased in the tetraploid. Beversdorf (1979) studied 1*x*, 2*x*, 3*x*, 4*x*, and 5*x* plants from *ms1 ms1* parent plants of *G. max*. He found that ploidy level was positively correlated with guard-cell length, number of chloroplasts per guard cell, leaflet width-to-length ratio, and flower size. He reported that ploidy level was negatively correlated with stomatal frequency and number of internodes.

Biswas and Bhattacharyya (1972) studied meiosis in three tetraploid plants of *G. max* and observed 40-40 chromosome distribution. Laggards and unequal separation of chromosomes were noticed only in a few cells. The range of chromosome associations and types per nucleus were 1.05 to 1.62 I's, 29.5 to 34.0 II's, 0.40 to 1.43 III's and 1.4 to 1.9 IV's. Forty pairs of chromosomes were observed in a few pollen mother cells. Sen and Vidyabhusan (1960) reported 5.2 I's, 14.9 II's, 1.8 III's, and 9.9 IV's. Tang and Lin (1963) saw no I's or III's and found IV's more numerous than II's. The discrepancy among these reports indicates that more research is necessary to understand chromosome behavior in tetraploids.

The first report of triploids and polyploids higher than tetraploids was by Gobs-Sonnenschein (1943). Sadanaga and Grindeland (1981) attempted to obtain triploids from natural cross-pollinations between tetraploid and diploid plants of *G. max*. They developed a tetraploid line with male sterility and green cotyledons (*ms2 d1 d2*). All yellow-cotyledon seed produced on these male-sterile plants would be hybrid. Across seven treatments, 816 hybrid seeds were obtained among 65 770 seeds harvested, but no triploids were found. The failure to find triploid seed from reciprocal tetraploid × diploid soybean crosses is well documented (Gobs-Sonnenschein, 1943; Porter and Weiss, 1948; Hu, 1968).

Polyploids can arise from either polyembryonic or monoembryonic seed. Kenworthy et al. (1973) reported three triploids among 47 polyembryonic *ms1 ms1* seedlings of T260. Beversdorf and Bingham (1977) observed 18 polyploids (3*x* to 4*x*) among 167 polyembryonic seed in progeny of T260. They found 27 polyploids (3*x* to 6*x*) among 217 monoembryonic seed that they examined for chromosome number. A total of 67 F_1 seedlings from seed of 10 haploids from T260 were analyzed cytologically by Sorrells and Bingham (1979). In addition to 52 diploids and four trisomics, they observed 11 polyploids, including a 70-chromosome plant. Chen et al. (1985) used T260, T266, T267, T268, and derived populations from T266 and T268, as sources and found 40 polyploids in 159 polyembryonic seed that ranged in ploidy level from 3*x* to 5*x*. Among the 376 monoembryonic seedlings they examined, they found 112 polyploids ranging from 3*x* to 10*x*.

Considerable variation in female fertility among the four *ms1* types (T260, T266, T267, and T268) is known, with T266 having the highest level of female fertility (Boerma and Cooper, 1978; Palmer et al., 1978b; Kennell and Horner, 1985). Female gametophytes of *Ms1 Ms1, Ms1 ms1,* and *ms1 ms1* plants were examined histologically to identify the mechanism(s) responsible for polyembryony, haploidy, and polyploidy (Cutter and Bingham, 1977; Kennell, 1984; Kennell and Horner, 1985).

Cutter and Bingham (1977) observed normal embryo sacs at anthesis in only 28% of the 225 male-sterile (T260) ovules examined. The most common abnormality was supernumerary nuclei in the regions of the egg apparatus and secondary nucleus. These nuclei could arise by proliferation of cells during embryo sac development. These supernumerary nuclei and their restitution nuclei provide a possible explanation for the occurrence of polyembryony, haploidy, and polyploidy among progeny of *ms1 ms1* plants.

Kennell and Horner (1985) noted that, in T266, T267, and T268 types, the *Ms1 Ms1* and *Ms1 ms1* plants were consistent in the appearance of the embryo sac and only a few aborted ovules were seen. Homozygous recessive *ms1 ms1* plants from the three types, however, varied significantly in their percentages of aborted ovules and the aberrant embryo sac structures they contained. The T266 type had 10.9% aborted, 8.6% abnormal, and 80.5% normal ovules. This is contrasted with the T267 type, which had 22.2, 38.2, and 39.6%, and T268 type, which had 24.8, 45.8, and 29.4% aborted, abnormal, and normal ovules, respectively. Up to four egg cells were present in some ovules. In other embryo sacs there were no megagametophyte cells, or the embryo sacs had the normal number of cells but in abnormal orientation.

The occurrence on *ms1 ms1* plants of seed that produce diploid-diploid twins that are heterogeneous for flower color and/or fertility indicates that twins arise from separate fertilizations or that one member develops parthenogenetically. Polyembryony with different ploidy levels is possible because endosperm resulting from a normal pollination could provide nutrition and the environment necessary for survival of a non-diploid embryo. Triploids (*ms1 ms1 Ms1*) and tetraploids (*ms1 ms1 ms1 Ms1*) are explained by fusion of extra nuclei in embryo sacs before or during fertilization by *Ms1* sperm as a result of cross-pollination (Chen et al., 1985).

The *ms1 ms1* genotype is a valuable source of haploids and polyploids for future soybean cytogenetic research. The mechanism(s) responsible for polyembryony, haploidy, and polyploidy are not completely understood; additional studies are necessary to elucidate these mechanisms.

5–6.3 Aneuploids

Abnormal plants were found among progeny of tetraploid soybean and were suspected to be aneuploids (Sen and Vidyabhusan, 1960; Tang and Lin, 1963). Hu (1968) found nine abnormal plants among 1059 progeny from tetraploid plants, but was unable to obtain definitive chromosome counts. Hu (1968) found aneuploids and polyploids among progeny of asynaptic mutants T241 and T242.

Palmer (1974a) determined the chromosome numbers of 79 seed from asynaptic T241 plants. Three plants had approximately 40 chromosomes; 76 plants had approximately 80 chromosomes and were sterile. From 138 seed of asynaptic T242 plants, eight plants were near 40 chromosomes; the remaining plants were near 80 chromosomes and were sterile. From 92 seed of desynaptic mutant T258, seven plants had 40 chromosomes, 72 plants had between 41 and 46 chromosomes, and only 13 had approximately 80 chromosomes and were sterile (Palmer and Heer, 1976). Palmer and Kaul (1983) obtained only three seed from almost 1000 desynaptic T272 plants. These seed gave rise to 40-, 42-, and 80-chromosome plants. The 40- and 80-chromosome plants were sterile, but the 42-chromosome plant produced five seed. All five progeny had 40 chromosomes. Four were fully fertile and one plant had about 50% aborted pollen and ovules and is discussed in the section on chromosome interchanges.

Six aneuploids (41-chromosome plants), out of 78 plants examined, have been found among progeny of haploid *ms1* plants crossed with diploid plants (Sorrells and Bingham, 1979; Crane et al., 1982).

Aneuploids and polyploids have been found among progeny of male-sterile, female-fertile mutant *ms2 ms2*. Sadanaga and Grindeland (1981) reported five trisomics, one monotelotrisomic, one tetraploid, and one mixoploid among 175 seed. Aneuploids have not been reported among progeny of *ms3 ms3, ms4 ms4,* or *ms5 ms5* plants.

Sadanaga and Grindeland (1979) harvested seeds from one- and two-seeded pods of neutron-irradiated plants. The M_3 progeny from M_2 plants with more than 20% aborted pollen grains were checked for chromosome number. Among the progeny, they found one 39-chromosome plant, 47 40-chromosome plants, 20 41-chromosome plants, and two 42-chromosome plants. This was the first report of a 39-chromosome plant. Some of the 40-chromosome plants were homozygous for a chromosome interchange and are discussed in the section on interchanges.

Triploid plants usually are a good source of aneuploids, but all attempts to produce them through crossing tetraploid by diploid plants, or the reciprocal, have failed. Triploids occur among the progeny of *ms1 ms1* plants and may be of two types, *ms1 ms1 ms1* or *Ms1 ms1 ms1.* Spontaneously occurring triploid plants have not been reported in soybean, but Stelly et al. (1979) noted aneuploids from naturally occurring sterile plants in a commercial field. Progeny testing of five plants failed to detect a genetic determinant for the sterility. These aneuploids could have arisen from parent plants that were either euploid or aneuploid.

Three primary trisomics, Tri A, Tri B, and Tri C, have been described and are currently being used in linkage studies (Palmer, 1976b). Tri A and Tri B came from T241, and Tri C from T258. Trisomics A and B cannot be distinguished morphologically from each other or from their respective diploid sibs, but Tri C shows slight morphological deviation from its diploid sib. There was no relationship between seed weight and somatic chromosome number among the progeny of the three trisomics

(Palmer, 1976b). The rate of ovule transmission of the extra chromosome was 34% for Tri A, 45% for Tri B, and 39% for Tri C. The rate of pollen transmission of the extra chromosome was 27% for Tri A, 22% for Tri B, and 43% for Tri C.

Gwyn (1984) has proposed that primary trisomics can be identified one from the other on the basis of the morphology of the 42-chromosome plants (tetrasomics) obtained by self-pollinating the parent 41-chromosome plants.

Data from trisomic inheritance studies have been summarized (Palmer, 1984c). Trisomic inheritance with use of primary trisomic A has been shown for a gene for variegated leaf and is the first example in soybean of the use of trisomics for locating genes on specific chromosomes (Newhouse et al., 1983).

Soybean can tolerate addition aneuploids, such as primary trisomics, double trisomics, and tetrasomics. With the high levels of fertility of trisomic plants and high levels of transmission of the extra chromosome through both male and female gametes, chromosome mapping studies are feasible. Because the cultivated and wild soybean are suspected to be polyploid, it is surprising that more deficiency aneuploids have not been found.

5–6.4 Interchanges

Chromosome aberrations such as reciprocal interchanges have been suspected, but have only recently been confirmed cytologically or genetically in soybean. Palmer and Heer (1984) determined that the pollen and ovule semisterility reported in the *G. max* × *G. soja* hybrid by Williams (1948) was the result of a heterozygous chromosome interchange. They also showed that, in plants heterozygous for a chromosome interchange, an ovule abortion or a normal seed occurred at random with respect to position within a pod. That is, in two-ovule pods or in three-ovule pods, the four or eight possible sequences of an ovule abortion or a normal seed occurred in equal frequencies. In plants heterozygous for a chromosome interchange, seed number per plant was reduced, but seed weight was increased. However, individual plant seed yields were similar whether from plants heterozygous for a chromosome interchange, plants homozygous for a chromosome interchange, or plants homozygous for noninterchange chromosomes.

Delannay et al. (1982) reported 41 suspected chromosome interchanges among F_1 plants of 142 different *G. max* × *G. soja* crosses. Of these suspected interchanges, 22 of 26 *G. soja* accessions from the USSR, 16 of 19 accessions from China, 1 of 59 accessions from South Korea, 1 of 37 accessions from Japan, and one accession of unknown origin (PI 212239) had a chromosome interchange. Palmer and Newhouse (1984) have shown that the interchanges from the USSR and China are identical to the interchange reported by Williams (1948) from *G. soja*. The interchanges from South Korea and Japan have not yet been tested.

These results with *G. soja* germplasm contrast with those of earlier reports in which, except for Williams (1948), complete fertility of F_1 hybrids between *G. max* and *G. soja* was observed. Previous studies used one or several accessions of *G. soja*, whereas recent research has included 142 accessions. Sadanaga and Newhouse (1982) have listed six homozygous interchanges that are used in linkage studies. The identification of the interchanges is based on chromosome association of the interchange chromosomes. One common chromosome (the satellite chromosome) is involved in an interchange in Clark T/T, KS172-11-3, and KS175-7-3. One common chromosome is involved in an interchange in L75-0283-4 and KS171-31-2.

Origin of six interchanges in soybean

Interchange	Origin
Clark T/T	Near-isogenic Clark with interchange from PI 101404B (*G. soja*). Interchange suspected by Williams (1948) and confirmed by Palmer and Heer (1984). Near-isogenic Clark developed by R. L. Bernard.
L75-2083-4	Spontaneous interchange suspected in an F_4 progeny row of a Beeson × Amsoy 71 cross in 1975 by R. L. Bernard. Confirmed by R. G. Palmer.
PI 189866	Semiwild accession from northeast China. Interchange confirmed by Delannay et al. (1982).
KS171-31-2	Interchange from an irradiated population of Hodgson. Confirmed by K. Sadanaga.
KS172-11-3	Interchange from an irradiated population of Hodgson. Confirmed by K. Sadanaga.
KS175-7-3	Interchange from an irradiation population of Steele. Confirmed by K. Sadanaga.

Forrai and Palmer (1984) used the six interchanges described by Sadanaga and Newhouse (1982) as a known tester set of interchanges to evaluate six unknown interchanges. Five unknown interchanges were from accessions identified from *G. max* × *G. max* crosses involving 626 different combinations between introduced cultivars of both northern and southern maturity (Palmer, 1985). The sixth interchange was found among progeny of a 42-chromosome plant that originally came from an *st5 st5* plant (Palmer and Kaul, 1983). Four of the five unknown interchanges from germplasm crosses were identical to the interchange from *G. soja* reported by Williams (1948). PI 323551, from India, was found not to have any chromosomes in common with the six known interchanges described by Sadanaga and Newhouse (1982). The interchange from an *st5 st5* plant was believed to be the result of a natural outcross with standard interchange line Clark T/T, which has the interchange chromosomes from *G. soja* PI 101404B (Forrai and Palmer, 1984).

Interchanges have been used in linkage studies and these results are given in the section on linkage groups.

5–6.5 Inversions

Several examples of inversions have been reported in soybean (Ahmad et al., 1977c, 1979; Delannay et al., 1982). Ahmad et al. (1977c,

1984) reported one or more paracentric inversions in a cross between *G. max* and a primitive *G. max* (or a *G. soja*; the correct species designation is uncertain). The frequency of meiotic irregularities and micronuclei formation was consistent between F_1 plants grown in the field in two different summers, but five to six times greater than glasshouse-grown plants. The difference in empty pollen-grain frequency between F_1 plants from the two field environments was considerably less than the seasonal differences for meiotic irregularities and micronuclei formation.

Delannay et al. (1982) reported seven suspected inversions from accessions identified from 361 *G. max* × *G. max* crosses, and 20 suspected inversions from 142 *G. max* × *G. soja* crosses. All the inversions found in *G. soja* were from accessions from South Korea and Japan.

5–6.6 Linkage Groups

Of 20 possible linkage groups in soybean, 13 have been identified. Linkage group 1 has six genetic loci, but placement of the loci with respect to one another is not yet complete. Linkage groups 2 through 7 remain as described in the 1973 edition of *Soybeans: Improvement, Production, and Uses.* Linkage group 8, with four loci, has the probable gene order *st5 w1 wm ms1.* Linkage groups 10 through 13 have only two loci per linkage group, and two have loci showing loose linkage. Roane et al. (1983) reported that reactions to peanut mottle and soybean mosaic viruses were conditioned by linked genes with 3.7 ± 0.8% recombination, but no linkage group was assigned. In this chapter, we have assigned these two loci to linkage group 13. Reactions to both viruses were inherited independently from *T*, pubescence color.

Devine et al. (1984) used six genetic traits and analyzed 15 distinct gene-pair combinations for linkage. All combinations gave evidence of independent assortment. A compilation of locus-to-locus linkage data is available (Yee and Palmer, 1984).

A true-breeding chimera mutant has been assigned to primary trisomic A (Newhouse et al., 1983). A summary of linkage data with trisomics A, B, and C has been completed (Palmer, 1984c).

Inversions have not been used in linkage studies in soybean to date. Chromosome interchanges have been used to place gene order for mutants of linkage group 8. Sadanaga and Grindeland (1984), and Sacks and Sadanaga (1984), with *G. max* interchange KS172-11-3, reported the order as *w1, wm,* interchange breakpoint, and *ms1.* Palmer and Kaul (1983), with *G. soja* PI 101404B interchange, reported the order as flower color mutants, *ms1*, breakpoint. With additional data, Palmer (1985) showed the order as *st5, w1, wm, ms1*, breakpoint. The same *G. soja* interchange was used with linkage group 6, and the order is *y11, df2*, breakpoint (Palmer, 1985).

Two or more linkage groups may in fact be the same linkage group. Additional studies are needed with primary trisomics to complete and substantiate the reported linkage groups.

Table 5-15. Genetic linkage groups in soybean.

Linkage group		Linked genes	Linkage intensity map†	Reference
1	*y12*	Chlorophyll deficient	*y12* — 20.2 ± 1.1 — *E1* — 3.9 ± 0.4 — *t*	Weiss, 1970a
	E1	Late flowering and maturity	*y12* — 21.6 ± 0.7 — *t*	
	t	Gray pubescence		Buzzell, 1974, 1977, 1979; Buzzell and Palmer, 1985
	fg4	Flavonol glycoside absent	*fg4* — 3.9 ±0.9 — *t*	
	fg3	Flavonol glycoside absent	*fg3* — 12.0 ± 1.8 — *fg4*	
	df5	Dwarf	*fg3* — 13.5 ± 5.9 — *t*	Palmer, 1977, 1984a
			df5 — 15.4 ± 1.0 — *t*	
2	*P1*	Glabrous	*P1* — 20.9 ± 2.4 — *r*	Weiss, 1970b
	r	Brown seed		
3	*G*	Green seed coat	*G* — 4.2 ± 0.6 — *d1*	
	d1	Green seed embryo		Weiss, 1970b
4	*v1*	Variegated leaves	*v1* — 35.6 ± 0.9 — *ln* — 26.4 ± 1.4 — *p2*	Weiss, 1970c
	ln	Narrow-leaflet, four-seeded pods	*v1* — independent — *p2*	
	p2	Puberulent		

(continued on next page)

Table 5-15. Continued.

Linkage group		Linked genes	Linkage intensity map†	Reference
5	*dt1*	Determinate stem	*dt1* — 39.4 ± 1.8 — *L1*	Weiss, 1970d
	L1	Black pod		
6	*df2*	Dwarf	*df2* — 12.1 ± 0.7 — *y11*	Weiss, 1970d
	y11	Chlorophyll deficient		
7	*y13*	Chlorophyll deficient	*y13* — 31.3 ± 1.9 — *o* — 17.8 ± 0.7 — *i*	Weiss, 1970e
	o	Reddish brown seed	*y13* — 41.1 ± 0.9 — *i*	
	i	Self dark seed		
8	*wm*	Magenta flower	*wm* — 2.2 ± 0.7 — *w1*	Buzzell et al., 1977; Sadanaga, 1983
	w1	White flower	*ms1* — 29.1 ± 0.9 — *wm*	Palmer, 1985
	ms1	Male sterile	*ms1* — 30.3 ± 0.9 — *w1*	Palmer, 1976a
	st5	Desynaptic sterile	*w1* — 29.9 ± 2.0 — *st5*	Palmer and Kaul, 1983

(continued on next page)

Table 5-15. Continued.

9	*Ap*	Acid phosphatase	*Ap* — 16.2 ± 1.5 — *Ti*	Hildebrand et al., 1980
	Ti	Kunitz trypsin inhibitor		Kiang et al., 1985
	Lap1	Leucine aminopeptidase	*Lap1* — 19.9 ± 0.1 — *Ap*	
10	*Rps1*	Resistant to phytophthora root rot	*Rps1* — 7.0 ± 1.2 — *hm*	Kilen and Barrentine, 1983
	hm	Metribuzin sensitive		
11	*rj1*	Nonnodulating	*rj1* — 40.0 ± 2.2 — *f*	Devine et al., 1983b
	f	Fasciated stem		
12	*fr1*	Nonfluorescent root	*fr1* — 42.6 ± 1.8 — *ep*	Palmer et al., 1984
	ep	Low seed coat peroxidase activity		
13	*rsv-t*	Resistant to soybean mosaic virus	*rsv-t* — 3.7 ± 0.8 — *Rpv1*	Roane et al., 1983
	Rpv1	Resistant to peanut mottle virus		

†Linkage intensity map given as percentage recombination with standard error.

A summary of linkage groups in soybean is presented in Table 5-15.

5-7 GENE SYMBOL INDEX

All known published gene symbols are listed alphabetically in Table 5-16. Each gene symbol is referred to the table in which it is identified. For other genes, explanation for their disuse or the currently used synonyms are given. Some gene symbol changes were made to eliminate duplication (two symbols for the same gene or two genes with the same symbol), to make the symbol conform to the other symbols in a series affecting the same trait, or to correct errors of interpretation. An example of gene symbol evolution can be given for the series controlling reaction to phytophthora rot. The original symbol *Ps* was changed to *Rps*, and appropriate suffixes were added as additional alleles at the locus and additional loci were identified.

Table 5-16. All gene symbols used and published for soybean.

Gene symbol	Reference and synonymy
A,a	Takagi (1929). = *G,g* or *Y3,y3* (Table 5-11).
Ab,ab	Table 5-8.
Adh1,adh1	Table 5-14.
Adh2,adh2	Table 5-14.
Adh1-+,adh1-n	Kiang and Gorman (1983). = *Adh1,adh1* (Table 5-14).
Adh4-+,adh4-n	Kiang and Gorman (1983). = *Adh2,adh2* (Table 5-14).
Am1-+,am1-n	Kiang (1981), Kiang and Gorman (1983). = *Amy1,amy1* (Table 5-14).
Am2-+,am2-n	Kiang (1981), Kiang and Gorman (1983). = *Amy2,amy2* (Table 5-14).
Am3-s,Am3-f,Am3-sw,am3-n1	Kiang (1981), Kiang and Gorman (1983). = *Sp1-a, Sp1-b,Sp1-an,sp1* (Table 5-14).
Amy1,amy1	Table 5-14.
Amy2,amy2	Table 5-14.
Ap2,ap2	Williams (1950). = *Pc,pc* (Table 5-8).
Ap-a,Ap-b,Ap-c	Table 5-14.
B,b	Woodworth (1921). = *R,r* (Table 5-13).
B,b	Takagi (1929). = *G,g* or *Y3,y3* (Table 5-11).
B1,B2,B3,b1,b2,b3	Table 5-8.
B2,b2	Ting (1946). = *Pb,pb* (Table 5-8).
Bl,bl	Ting (1946). = *Pb,pb* (Table 5-8).
Br,br	Matsuura (1933) based on Takahashi and Fukuyama (1919). = *Ln,ln* (Table 5-8).
C,c	Nagai (1921). = *T,t* (Table 5-13).
C,c	Terao and Nakatomi (1929). = *Y3,y3* (Table 5-11).
C1,c1; C2,c2	Matsuura (1933) based on Nagai (1926). (defective seed) Tentative.
Cgy1,cgy1	Table 5-14.
Cs,cs	Athow and Probst (1952). = *Rcs1,rcs1* (Table 5-5).
cyt-G1...3	Table 5-12.
cyt-Y1...3	Table 5-12.
D1,d1; D2,d2	Table 5-11.

(continued on next page)

Table 5-16. Continued.

Gene symbol	Reference and synonymy
De,de	Matsuura (1933) based on Nagai (1926). = *Sh,sh* of Morse and Cartter (1937). (pod dehiscence) Tentative.
De1,de1	Stewart and Wentz (1930). = *T,t* (Table 5-13).
De2,de2	Morse and Cartter (1937). = *P2,p2* (Table 5-8).
De2,de2	Liu (1949). = ? Uncertain
De3,de3; *De4,de4*	Liu (1949). Uncertain.
Df,df	Woodworth (1932) based on Stewart (1927). *df* is lost.
Df2...5,df2...5	Table 5-8.
Di1,di1	Gorman et al. (1983). = *Dia1-a,Dia1-b* (Table 5-14).
Di2-s,Di2-f	Gorman et al. (1983). = *Dia2-a,Dia2-b* (Table 5-14).
Di3,di3	Gorman et al. (1983). = *Dia3,dia3* (Table 5-14).
Dia1-+,dia1-n	Kiang and Gorman (1983). = *Dia1-a,Dia1-b* (Table 5-14).
Dia1, dia1	Palmer et al. (1985). = *Dia1-a, Dia1-b* (Table 5–14).
Dia2-s,Dia2-f	Kiang and Gorman (1983). = *Dia2-a,Dia2-b* (Table 5-14).
Dia3-+,dia3-n	Kiang and Gorman (1983). = *Dia3,dia3* (Table 5-14).
Dt1,dt1	Table 5-8.
Dt2,dt2	Table 5-8.
E,e	Owen (1927c). = *E1,e1* (Table 5-8).
E1...4,e1...4	Table 5-8.
Ep,ep	Table 5-14.
Eu,eu	Table 5-14.
Eu1-a,Eu1-b	Kloth and Hymowitz (1985). = *Eu,eu* (Table 5-14).
F,f	Table 5-8.
F,f	Takahashi (1934). = *Ln,ln* (Table 5-8).
Fe,fe	Table 5-10.
Fg1...4,fg1...4	Table 5-10.
Fl,fl	Morse and Cartter (1937). (flecked seed coat) No data presented.
Fr1...4,fr1...4	Table 5-10.
Fs1,Fs2; *fs1,fs2*	Table 5-9.
Ft,ft	Table 5-9.
G,g	Table 5-11.
Gpd,gpd	Table 5-14.
Gy4,gy4	Table 5-14.
H,h	Terao (1918), Terao and Nakatomi (1929). = *G,g* (Table 5-11).
H,h	Woodworth (1921). = *T,t* (Table 5-13).
H,h	Nagai and Saito (1923). = *I,i* (Table 5-13).
H,h	Ting (1946). (hard seed) Insufficient data.
Hb,hb	Table 5-6.
Hm,hm	Table 5-6.
Hy,hy	Sengupta (1975). = *W1,w1* (Table 5-13).
I,i-i,i-k,i	Table 5-13.
I,i	Nagai (1921). = *I* or *i-i,i* (Table 5-13).
I,i	Nagai and Saito (1923). = *i-i,i* (Table 5-13).
I,i	Woodworth (1921). = *D2,d2* (Table 5-11).
I-h,I-i,I-k,i	Owen (1928b). = *I,i-i,i-k,i* (Table 5-13).
i-h	Ting (1946). = *i-i* (Table 5-13).
I-de,i-de	Liu (1949). = *I* or *i-i,i* (Table 5-13).
Idh1-a,Idh1-b	Table 5-14.
Idh2-a,Idh2-b	Table 5-14.
Idh3-a,Idh3-b	Table 5-14.

(continued on next page)

Table 5-16. Continued.

Gene symbol	Reference and synonymy
Idh1-s,Idh1-f	Kiang and Gorman (1983). = *Idh1-a,Idh1-b* (Table 5-14).
Idh2-s,Idh2-f	Kiang and Gorman (1983). = *Idh2-a,Idh2-b* (Table 5-14).
Idh3-s,Idh3-m	Kiang and Gorman (1983). = *Idh3-a,Idh3-b* (Table 5-14).
Im,im	Table 5-13.
K1...3,k1...3	Table 5-13.
K2,k2	Table 5-11.
K,k	Nagai and Saito (1923). = *k1* (Table 5-13).
L,l	Woodworth (1923). = *L1.l1* or *L2,l2* (Table 5-13).
L1,l1; L2,l2	Table 5-13.
Lap1-a,Lap1-b	Table 5-14.
Lap2,lap2	Table 5-14.
Lap1-s,Lap1-f	Gorman et al. (1982a), Gorman et al. (1983), Kiang and Gorman (1983). = *Lap1-a,Lap1-b*, (Table 5-14).
Lb1,lb1; Lb2,lb2	Table 5-8.
Le,le	Table 5-14.
Lf1,lf1; Lf2,lf2	Table 5-8.
Ln,ln	Table 5-8.
Lo,lo	Table 5-8.
Lps,lps	Table 5-8.
Lt,lt	Matsuura (1933) based on Takahashi and Fukuyama (1919) and Nagai (1926). = *Lf1,lf1* (Table 5-8).
Lw1,lw1; Lw2,lw2	Table 5-8.
Lx1 . . . 3, lx1 . . . 3	Table 5-14.
M,m	Nagai and Saito (1923). = *r-m,r* (Table 5-13).
Mi1,mi1;Mi2,mi2;MiR,miR	Geeseman (1950). Downy mildew races lost.
Mn,mn	Table 5-8.
Mpi-a,Mpi-b,Mpi-c	Table 5-14.
Mpi-s,Mpi-m,Mpi-f	Gorman et al. (1983), Kiang and Gorman (1983). = *Mpi-a,Mpi-b,Mpi-c* (Table 5-14).
Ms1...5,ms1...5	Table 5-9.
Msp,msp	Table 5-9.
N,n	Table 5-8.
Na,na	Woodworth (1932). = *Ln,ln* (Table 5-8).
Ncl,ncl	Table 5-10.
No,no	Williams and Lynch (1954). = *Rj1,rj1* (Table 5-7).
Np,np	Table 5-10.
Nr,nr	Table 5-10.
O,o	Table 5-13.
O,o	Domingo (1945). = *Lo,lo* (Table 5-8).
P1,p1;P2,p2	Table 5-8.
Pa1,pa1;Pa2,pa2	Table 5-8.
Pb,pb	Table 5-8.
Pc,pc	Table 5-8.
Pd1,pd1;Pd2,pd2	Table 5-8.
Pgd-a,Pgd-b,pgd	Table 5-14.
Pgd-s,Pgd-f,pgd-n	Kiang and Gorman (1983). = *Pgd-a,Pgd-b,pgd* (Table 5-14).
Pgi-a,Pgi-b	Table 5-14.
Pgi-s,Pgi-f	Kiang and Gorman (1983). = *Pgi-a,Pgi-b* (Table 5-14).
Pgm1-a,Pgm1-b	Table 5-14.
Pgm1-s,Pgm1-f	Kiang and Gorman (1983). = *Pgm1-a,Pgm1-b* (Table 5-14).

(continued on next page)

Table 5-16. Continued.

Gene symbol	Reference and synonymy
Pgm2-p,Pgm2-n	Kiang and Gorman (1983). = *Pgm2-a,Pgm2-b* (Table 5-14).
Pgm2-a,Pgm2-b	Table 5-14.
Ph1,ph1; Ph2,ph2	Matsuura (1933). Probably *ph1,ph2* = *dt1,e1* (Table 5-8).
Pm,pm	Table 5-8.
Ps,Ps-s,ps	Table 5-8.
Ps,ps	Bernard et al. (1957). = *Rps1,rps1* (Table 5-5).
R,r-m,r	Table 5-13.
R,r	Takahashi (1934). = *Ln,ln* (Table 5-8).
R1,r1	Owen (1928b). = *R,r* (Table 5-13)
R2,r2,r′2	Owen (1928b). (seed color) Based on misinterpretation of data.
R2,r2	Probst (1950). (seed color) Tentative.
r1-o	Stewart (1930). = *o* (Table 5-13).
Rcs1,rcs1; Rcs2,rcs2	Table 5-5.
Rcv,rcv	Table 5-5.
Rhg1...4,rhg1...4	Table 5-5.
Ri,ri	Wang (1948). = *R,r-m* (Table 5-13).
Rj1...4,rj1...4	Table 5-7.
Rmd,rmd	Table 5-5.
Rpg1,rpg1	Table 5-5.
Rpm,rpm	Table 5-5.
Rpp1...3,rpp1...3	Table 5-5.
Rps,rps	Bernard and Weiss (1973). = *Rps1,rps1* (Table 5-5).
Rps1,rps1	Table 5-5.
Rps1	Lam-Sanchez et al. (1968). = *Rps1-c* (Table 5-5).
Rps-a	Mueller et al. (1978). = *Rps1* (Table 5-5).
Rps1-a	Mueller et al. (1978). = *Rps1* (Table 5-5).
Rps1-b	Table 5-5.
Rps1-c	Table 5-5.
Rps1-k	Table 5-5.
Rps-b	Mueller et al. (1978). = *Rps1-b* (Table 5-5).
Rps-c	Mueller et al. (1978). = *Rps1-c* (Table 5-5).
rps-2	Hartwig et al. (1968). = *Rps1-b* (Table 5-5).
Rps2...6,rps2...6	Table 5-5.
Rpv,rpv	Boerma and Kuhn (1976). = *Rpv1,rpv1* (Table 5-5).
Rpv1,rpv1	Table 5-5.
Rpv2,rpv2	Table 5-5.
Rrn,rrn	Table 5-5.
Rsv,rsv-t,rsv	Kiihl and Hartwig (1979). = *Rsv1,rsv1-t,rsv1* (Table 5-5).
Rsv1,rsv1-t,rsv1	Table 5-5.
Rsv2,rsv2	Table 5-5.
Rxp,rxp	Table 5-5.
S,s	Woodworth (1923). (stem height) Lost, possibly = *E1,e1* (Table 5-8).
S,s,s-t	Table 5-8.
Sb1,sb1;Sb2,sb2	Table 5-8.
Se,se	Table 5-8.
Sh,sh	Morse and Cartter (1937). (shattering) Tentative.
Sh2,sh2	Morse and Cartter (1937) based on Nagai (1926). (shattering) Tentative.
Sod,sod	Table 5-14.
Sp,sp	Matsuura (1933) based on Nagai (1926). (branching) Tentative.

(continued on next page)

Table 5-16. Continued.

Gene symbol	Reference and synonymy
Sp1-a,Sp1-b,Sp1-an,sp1	Table 5-14.
St,st	Woodworth (1932) and Matsuura (1933) based on Owen (1928a). *st* is lost.
St2...5,st2...5	Table 5-9.
T,t	Table 5-10, Table 5-13.
T1,t1	Owen (1928b), Probst (1950). = *T,t* (Table 5-13).
T2,t2	Owen (1928b). Based on misinterpretation of data.
T2,t2	Probst (1950). (pubescence color) Tentative.
T2,t2	Williams (1952). (pubescence color) Tentative.
Td,td	Table 5-13.
Ti-1,Ti-2,Ti-3	Singh et al. (1969); Hymowitz and Hadley (1972). = *Ti-a,Ti-b,Ti-c* (Table 5-14).
Ti-a,Ti-b,Ti-c,ti	Table 5-14.
To-3,to-3	Gorman and Kiang (1978). = *Sod,sod* (Table 5-14).
To4,to4-n	Kiang and Gorman (1983). = *Sod,sod* (Table 5-14).
V,v	Woodworth (1921). = *G,g* (Table 5-11).
V1,v1	Table 5-11.
W1,w1	Table 5-13.
W2w2	Matsuura (1933) based on Takahashi and Fukuyama (1919). (flower color) Tentative.
W3,w3	Table 5-13.
W4,w4	Table 5-13.
Wm,wm	Table 5-10, Table 5-13.
X,x	Woodworth (1932) based on Takahashi and Fukuyama (1919). = *Lf1,lf1* (Table 5-8).
Y1,y1	Morse and Cartter (1937) based on Nagai (1926). *y1* is lost.
Y2,y2	Morse and Cartter (1937), based on Nagai (1926). = *G,g* (Table 5-11).
Y3,y3	Table 5-11.
Y4,y4	Table 5-11.
Y4,y4	Woodworth and Williams (1938). = *Y5,y5* (Table 5-11).
Y5,y5	Table 5-11.
Y5,y5	Woodworth and Williams (1938). = *Y4,y4* (Table 5-11).
Y6,y6	Table 5-11.
Y7 or *Y8,y7,y8*	Table 5-11.
Y7,y7	Morse and Cartter (1937), Woodworth and Williams (1938). = *Y3,y3* (Table 5-11).
Y9...18,y9...18	Table 5-11.
Y18-m	Table 5-11.
Y20,y20	Table 5-11.

The Soybean Genetics Committee has established guidelines for selecting gene symbols, and these are published each year in the *Soybean Genetics Newsletter*. They include the following main points: (i) a gene symbol consists of a base of one to three letters (e.g., *t, pc, rps*), to which may be appended subscripts and/or superscripts as described in item iii. Gene symbols may, however, be written on one line; (ii) two or more loci that control the same or similar traits are differentiated by appending different numerical suffixes to the same letter base (e.g., *e1, e2, e3*); and (iii) when more than two alleles exist for a locus, the additional alleles or those symbolized subsequently to the pair first published are differentiated by adding one or two uncapitalized letters as a superscript, or preceded by a hyphen if written on one line (e.g., *Rps1-b, Rps1-c, Rps1-k*).

Groups of loci that control somewhat less closely related traits are assigned bases with the same initial letter. Thus, a series has been established for disease resistance (*rcs, rpg, rps*), for leaf morphology (*lf, ln, lo*), for pubescence type (*p, pc, pd*), etc.

Specific guidelines for assigning gene symbols have been established by the Soybean Genetics Committee and were published in the 1987 *Soybean Genetics Newsletter.*

REFERENCES

Abel, G.H. 1969. Inheritance of the capacity for chloride inclusion and chloride exclusion by soybeans. Crop Sci. 9:697–698.

Adams, C.A., T.H. Broman, and R.W. Rinne. 1981. Starch metabolism in developing and germinating soya bean seeds is independent of α-amylase activity. Ann. Bot. 48:433–440.

Ahmad, Q.N., E.J. Britten, and D.E. Byth. 1975. A colchicine-induced diploid from a haploid soybean twin. J. Hered. 66:327–330.

----, ----, and ----. 1977a. Cytogenetic relationships between soybean and a related species, *Glycine ussuriensis*. p. 9.18–9.22. *In* S. Ramanujam (ed.) Proc. Third SABRAO Congr., Canberra, Australia. Society for Advancement of Breeding Researches in Asia and Oceania. Bangi, Selangor, Malaysia.

----, ----, and ----. 1977b. Haploid soybeans, a rare occurrence in twin seedlings. J. Hered. 68:67.

----, ----, and ----. 1977c. Inversion bridges and meiotic behavior in species hybrids of soybeans. J. Hered. 68:360–364.

----, ----, and ----. 1979. Inversion heterozygosity in the hybrid soybean × *Glycine soja*. J. Hered. 70:358–364.

----, ----, and ----. 1983. A quantitative method of karyotypic analysis applied to the soybean, *Glycine max*. Cytologia 48:879–892.

----, ----, and ----. 1984. Effects of interacting genetic factors and temperature on meiosis and fertility in soybean × *Glycine soja* hybrids. Can. J. Genet. Cytol. 26:50–56.

Albertsen, M.C., T.M. Curry, R.G. Palmer, and C.E. LaMotte. 1983. Genetics and comparative growth morphology of fasciation in soybeans (*Glycine max* (L.) Merr.) Bot. Gaz. 144:263–275.

----, and R.G. Palmer. 1979. A comparative light- and electron-microscope study of microsporogenesis in male-sterile (*ms1*) and male-fertile soybeans (*Glycine max* (L.) Merr). Am. J. Bot. 66:253–265.

Ashley, T. 1978. A possible alternate explanation for light green spots on yellow leaves in the soybean *y11 y11* test system. Soybean Genet. Newsl. 5:15–16.

Athow, K.L., and F.A. Laviolette. 1982. *Rps6*, a major gene for resistance to *Phytophthora megasperma* f. sp. *glycinea* in soybean. Phytopathology 72:1564–1567.

----, ----, E.H. Mueller, and J.R. Wilcox. 1980. A new major gene for resistance to *Phytophthora megasperma* var. *sojae* in soybean. Phytopathology 70:977–980.

----, and A. H. Probst. 1952. The inheritance of resistance to frogeye leaf spot of soybeans. Phytopathology 42:660–662.

Bernard, R.L. 1967. The inheritance of pod color in soybeans. J. Hered. 58:165–168.

----. 1971. Two major genes for time of flowering and maturity in soybeans. Crop Sci. 11:242–244.

----. 1972. Two genes affecting stem termination in soybeans. Crop Sci. 12:235–239.

----. 1975a. An allelic series affecting stem length. Soybean Genet. Newsl. 2:28–30.

----. 1975b. The inheritance of near-gray pubescence color. Soybean Genet. Newsl. 2:31–33.

----. 1975c. The inheritance of semi-sparse pubescence. Soybean Genet. Newsl. 2:33–34.

----. 1975d. The inheritance of appressed pubescence. Soybean Genet. Newsl. 2:34–36.

----, and C.R. Cremeens. 1972. A gene for general resistance to downy mildew of soybeans. J. Hered. 62:359–362.

----, and ----. 1975. Inheritance of the Eldorado male-sterile trait. Soybean Genet. Newsl. 2:37–39.

----, and ----. 1981. An allele at the *rps* locus from the variety 'Kingwa'. Soybean Genet. Newsl. 8:40–42.

----, and R.W. Howell. 1964. Inheritance of phosphorus sensitivity in soybeans. Crop Sci. 4:298–299.

----, and E.R. Jaycox. 1969. A gene for increased natural crossing in soybeans. Agron. Abstr. American Society of Agronomy, Madison, WI, p. 3.

----, R.G. Palmer, and B.P. Giles. 1983. Genes *y9* and *y11* for similar chlorophyll deficiencies prove to be nonallelic. Soybean Genet. Newsl. 10:35–38.

----, and B.B. Singh. 1969. Inheritance of pubescence type in soybeans: glabrous, curly, dense, sparse, and puberulent. Crop Sci. 9:192–197.

----, P.E. Smith, M.J. Kaufmann, and A. F. Schmitthenner. 1957. Inheritance of resistance to phytophthora root and stem rot in the soybean. Agron. J. 49:391.

----, and L.M. Wax. 1975. Inheritance of a sensitive reaction to bentazon herbicide. Soybean Genet. Newsl. 2:46–47.

----, and M.G. Weiss. 1973. Qualitative genetics. *In* B.E. Caldwell (ed.) Soybeans: Improvement, production, and uses. Agronomy 16:117–154.

Beversdorf, W.D. 1979. Influences of ploidy level on several plant characteristics in soybeans. Can. J. Plant Sci. 59:945–948.

----, and E.T. Bingham. 1977. Male-sterility as a source of haploids and polyploids of *Glycine max*. Can. J. Genet. Cytol. 19:283–287.

Biswas, A.K. 1977. Karyotype in three cultivated varieties of *Glycine max* (L.) Merr. Curr. Sci. 46:195–196.

----, and H.K. Bhattacharyya. 1972. Induced polyploidy in legumes II. *Glycine max* (L.) Merr. Cytologia 37:605–617.

Blogg, D., and B.C. Imrie. 1982. Starch gel electrophoresis for soybean cultivar identification. Seed Sci. Technol. 10:19–24.

Boerma, H.R., and R.L. Cooper. 1978. Increased female fertility associated with the *ms1* locus in soybeans. Crop Sci. 18:344–346.

----, and B.G. Jones. 1978. Inheritance of a second gene for brachytic stem in soybeans. Crop Sci. 18:559–560.

----, and C.W. Kuhn. 1976. Inheritance of resistance to peanut mottle virus in soybeans. Crop Sci. 16:533–534.

----, C.W. Kuhn, and H.B. Harris. 1975. Inheritance of resistance to cowpea chlorotic mottle virus (soybean strain) in soybeans. Crop Sci. 15:849–850.

----, and A. Moradshahi. 1975. Pollen movement within and between rows to male-sterile soybeans. Crop Sci. 15:858–861.

Boquet, D., C. Williams, and W. Birchfield. 1975. Inheritance of resistance to the Wartelle race of root-knot nematode in soybeans. Crop Sci. 16:783–785.

Brigham, R.D. 1978. A sterile mutant in progeny of a Forrest × Kent cross (*Glycine max* L.). Soybean Genet. Newsl. 5:95.

Brim, C.A., and C.W. Stuber. 1973. Application of genetic male sterility to recurrent selection schemes in soybeans. Crop Sci. 13:528–530.

----, and M.F. Young. 1971. Inheritance of a male-sterile character in soybeans. Crop Sci. 11:564–566.

Broich, S.L. 1978. The systematic relationships within the genus *Glycine* Willd. subgenus *Soja* (Moench) F.J. Hermann. M.S. thesis. Iowa State University, Ames.

----, and R.G. Palmer. 1981. Evolutionary studies of the soybean: The frequency and distribution of alleles among collections of *Glycine max* and *G. soja* of various origin. Euphytica 30:55–64.

Broué, P., D.R. Marshall, and W.J. Muller. 1977. Biosystematics of subgenus *Glycine* (Verdc.): Isoenzymatic data. Aust. J. Bot. 25:555–566.

Brown, J.C., and T.E. Devine. 1980. Inheritance of tolerance or resistance to manganese toxicity in soybeans. Agron. J. 72:898–903.

Buntman, D.J. 1983. A comparative analysis of microsporogenesis between normal and *ms3* male sterile soybean (*Glycine max* L.) M.S. thesis. Iowa State University, Ames.

----, and H.T. Horner. 1983. Microsporogenesis of normal and *ms3* mutant soybean (*Glycine max*). Scanning Electron Microsc. 2:913–922.

Burton, J.W., and T.E. Carter, Jr. 1983. A method for production of experimental quantities of hybrid soybean seed. Crop Sci. 23:388–390.

Buss, G.R. 1983. Inheritance of a male-sterile mutant from irradiated Essex soybeans. Soybean Genet. Newsl. 10:104–108.

----, and W.R. Autio. 1980. Observations of polyembryony and polyploidy in *ms1* and *ms2* male-sterile soybean populations. Soybean Genet. Newsl. 7:94–97.

Buttery, B.R., and R.I. Buzzell. 1971. Properties and inheritance of urease isoenzymes in soybean seeds. Can. J. Bot. 49:1101–1105.

----, and ----. 1973. Varietal differences in leaf flavonoids of soybeans. Crop Sci. 13:103–106.

----, and ----. 1975. Soybean flavonol glycosides: Identification and biochemical genetics. Can. J. Bot. 53:219–224.

----, and ----. 1976. Flavonol glycoside genes and photosynthesis in soybeans. Crop Sci. 16:547–550.

Buzzell, R.I. 1971. Inheritance of a soybean flowering response to fluorescent-daylength conditions. Can. J. Genet. Cytol. 13:703–707

----. 1974. Soybean linkage tests. Soybean Genet. Newsl. 1:11–14.

----. 1977. Soybean linkage tests. Soybean Genet. Newsl. 4:12–13.

----. 1979. Soybean linkage tests. Soybean Genet. Newsl. 6:15–16.

----, and T.R. Anderson. 1981. Another major gene for resistance to *Phytophthora megasperma* var. *sojae* in soybeans. Soybean Genet. Newsl. 8:30–33.

----, and R.L. Bernard. 1975. *E2* and *E3* maturity gene tests. Soybean Genet. Newsl. 2:47–49.

----, and B.R. Buttery. 1969. Inheritance of peroxidase activity in soybean seed coats. Crop Sci. 9:387–388.

----, and ----. 1973. Inheritance of flavonol glycosides in soybeans. Can. J. Genet. Cytol. 15:865–867.

----, and ----. 1974. Flavonol glycoside genes in soybeans. Can. J. Genet. Cytol. 16:897–899.

----, and ----. 1982. Genetics of black pigmentation of soybean seed coats/hila. Soybean Genet. Newsl. 9:26–29.

----, ----, and R.L. Bernard. 1977. Inheritance and linkage of a magenta flower gene in soybeans. Can. J. Genet. Cytol. 19:749–751.

----, ----, and J.H. Haas. 1974. Soybean genetics studies at Harrow. Soybean Genet. Newsl. 1:9–11.

----, ----, and R.M. Shibles. 1980. Flavonol classes of cultivars in maturity groups 00–IV. Soybean Genet. Newsl. 7:22–26.

----, and J.H. Haas. 1978. Inheritance of adult plant resistance to powdery mildew in soybeans. Can. J. Genet. Cytol. 20:151–153.

----, and R.G. Palmer. 1985. Soybean linkage group 1 tests. Soybean Genet. Newsl. 12:32–33.

----, and J.C. Tu. 1984. Inheritance of soybean resistance to soybean mosaic virus. J. Hered. 75:82.

----, and H.D. Voldeng. 1980. Inheritance of insensitivity to long daylength. Soybean Genet. Newsl. 7:26–29.

Byth, D.E., and C.R. Weber. 1969. Two mutant genes causing dwarfness in soybeans. J. Hered. 60:278–280.

Caldwell, B.E. 1966. Inheritance of a strain-specific ineffective nodulation in soybeans. Crop Sci. 6:427–428.

----, C.A. Brim, and J.P. Ross. 1960. Inheritance of resistance of soybeans to cyst nematode, *Heterodera glycines*. Agron. J. 52:635–636.

Cardy, B.J., and W.D. Beversdorf. 1984. A procedure for the starch gel electrophoretic detection of isozymes of soybean. Univ. of Guelph Tech. Bull. 119/8401.

Carter, T.E., Jr., J.W. Burton, and E.B. Huie, Jr. 1983. Implications of seed set on *ms2 ms2* male-sterile plants in Raleigh. Soybean Genet. Newsl. 10:85–87.

Caviness, C.E., and B.L. Fagala. 1973. Influence of temperature on a partially male-sterile soybean strain. Crop Sci. 13:503–504.

----, and R.D. Riggs. 1976. Breeding for nematode resistance. p. 594–601. *In* L.D. Hill (ed.) World soybean research: Proceedings. The Interstate Printers and Publishers, Danville, IL.

----, H.J. Walters, and D.L. Johnson. 1970. A partially male sterile strain of soybean. Crop Sci. 10:107–108.

Chaudhari, H.K., and W.H. Davis. 1977. A new male-sterile strain in Wabash soybeans. J. Hered. 68:266–267.

Chen, L.-F.O., H.E. Heer, and R.G. Palmer. 1985. The frequency of polyembryonic seedlings and polyploids from *ms1* soybean. Theor. Appl. Genet. 69:271–277.

Cheng, S.H., and H.H. Hadley. 1983. Studies in polyploidy in soybeans: A simple and effective colchicine technique of chromosome doubling for soybean (*Glycine max* (L.) Merr.) and its wild relatives. Soybean Genet. Newsl. 10:23–24.

Chmelar, F. 1934. The possibilities of accelerating seed analysis and the determination of variety by employing luminescence tests in ultraviolet light. Proc. Int. Seed Test. Assoc. 6:435–445.

Cianzio, S. Rodriguez de, and W.R. Fehr. 1980. Genetic control of iron-deficiency chlorosis in soybeans. Iowa State J. Res. 54:367–375.

----, and ----. 1982. Variation in the inheritance of resistance to iron-deficiency chlorosis in soybeans. Crop Sci. 22:433–434.

Clark, F.E. 1957. Nodulation of two near-isogenic lines of the soybeans. Can. J. Microbiol. 3:113–123.

Cooper, R.L. 1966. A major gene for resistance to seed coat mottling in soybean. Crop Sci. 6:290–292.

Crane, C.F., W.D. Beversdorf, and E.T. Bingham. 1982. Chromosome pairing and associations at meiosis in haploid soybean (*Glycine max*). Can J. Genet. Cytol. 24:293–300.

Curry, T.M. 1982. Morphological and anatomical comparisons between fasciated and non-fasciated soybeans (*Glycine max* (L.) Merr.) M.S. thesis. Iowa State University, Ames.

Cutter, G.L., and E.T. Bingham. 1977. Effect of soybean male-sterile gene *ms1* on organization and function of the female gametophyte. Crop Sci. 17:760–764.

Davies, C.S., and N.C. Nielsen. 1984. Development of soybean germplasm with low lipoxygenase content. Agron. Abstr. American Society of Agronomy, Madison, WI, p. 63.

Delannay, X., T.C. Kilen, and R.G. Palmer. 1982. Screening of soybean (*Glycine max*) accessions and *G. soja* accessions for chromosome interchanges and inversions. Agron. Abstr. American Society of Agronomy, Madison, WI, p. 63.

----, and R.G. Palmer. 1982a. Genetics and cytology of the *ms4* male-sterile soybean. J. Hered. 73:219–223.

----, and ----. 1982b. Four genes controlling root fluorescence in soybean. Crop Sci. 22:278–281.

----, and ----. 1984. Inheritance of a miniature mutant in soybean. Soybean Genet. Newsl. 11:92–93.

Devine, T.E., 1984. Inheritance of soybean nodulation response with a fast-growing strain of *Rhizobium*. J. Hered. 75:359–361.

----, and B.H. Breithaupt. 1980. Significance of incompatibility reactions of *Rhizobium japonicum* strains with soybean host genotypes. Crop Sci. 20:269–271.

----, and ----. 1981. Frequencies of nodulation response alleles, *Rj2* and *Rj4*, in soybean plant introduction and breeding lines. USDA Tech. Bull. 1628. U.S. Government Printing Office, Washington, DC.

----, Y.T. Kiang, and M.B. Gorman. 1984. Simultaneous genetic mapping of morphological and biochemical traits in soybean. J. Hered. 75:311–312.

----, L.D. Kuykendall, and B.H. Breithaupt. 1983a. Nodule-like structures induced on peanut by chlorosis producing strains of *Rhizobium* classified as *R. japonicum*. Crop Sci. 23:394–397.

----, R.G. Palmer, and R.I. Buzzell. 1983b. Analysis of genetic linkage in soybean. J. Hered. 74:457–460.

----, and D.F. Weber. 1977. Genetic specificity of nodulation. Euphytica 26:527–535.

Doerschug, E.B., J.P. Miksche, and R.G. Palmer. 1978. DNA content, ribosomal-RNA gene number, and protein content in soybeans. Can. J. Genet. Cytol. 20:531–538.

Domingo, W.E. 1945. Inheritance of number of seeds per pod and leaflet shape in the soybean. J. Agric. Res. 70:251–268.

Edwards, C.J., Jr., W.L. Barrentine, and T.C. Kilen. 1976. Inheritance of sensitivity to metribuzin in soybeans. Crop Sci. 16:119–120.

Feaster, C.V. 1951. Bacterial pustule disease in soybeans: Artificial inoculation, varietal resistance, and inheritance of resistance. Mo. Agric. Exp. Stn. Res. Bull. 487.

Fehr, W.R. 1972a. Inheritance of a mutation for dwarfness in soybeans. Crop Sci. 12:212–213.

----. 1972b. Genetic control of leaflet number in soybeans. Crop Sci. 12:221–224.

----. 1982. Control of iron-deficiency chlorosis. J. Plant Nutr. 5:611–621.

----, and J.H. Giese. 1971. Genetic control of root fluorescence. Crop Sci. 11:771.

Forrai, L.G., and R.G. Palmer. 1984. Genetic and cytological studies of chromosome interchanges in soybean (*Glycine max* (L.) Merr.) and related species. p. 64. *In* R. Shibles (ed.) World soybean research conference III. Abstr. 296. Westview Press, Boulder, CO.

Fukuda, U. 1933. Cytogenetical studies on the wild and cultivated Manchurian soybeans (*Glycine* L.) Jpn. J. Bot. 6:489–506.

Geeseman, G.E. 1950. Inheritance of resistance of soybeans to *Peronospora manshurica.* Agron. J. 42:608–613.

Gobs-Sonnenschein, C. 1943. Die experimentelle Erzeugung polyploider Sojabohnen mit Alkaloidgemischen in Verbindung mit Kreuzungen polyploider Rassen. Zuechter 15:62–68.

Goldberg, R.B. 1978. DNA sequence organization in the soybean plant. Biochem. Genet. 16:45–68.

----, G. Hoschek, and L.O. Vodkin. 1983. An insertion sequence blocks the expression of a soybean lectin gene. Cell 33: 465–475.

Gorman, M.B., and Y.T. Kiang. 1977. Variety-specific electrophoretic variants of four soybean enzymes. Crop Sci. 17:963–965.

----, and ----. 1978. Models for the inheritance of several variant soybean electrophoretic zymograms. J. Hered. 69:255–258.

----, ----, Y.C. Chiang, and R.G. Palmer. 1982a. Preliminary electrophoretic observations from several soybean enzymes. Soybean Genet. Newsl. 9:140–143.

----, ----, ----, and ----. 1982b. Electrophoretic classification of the early maturity groups of named soybean cultivars. Soybean Genet. Newsl. 9:143–156.

----, ----, and Y.C. Chiang. 1984. Electrophoretic classification of selected *G. max* plant introductions and named cultivars in the late maturity groups. Soybean Genet. Newsl. 11:135–140.

----, ----, R.G. Palmer, and Y.C. Chiang. 1983. Inheritance of soybean electrophoretic variants. Soybean Genet. Newsl. 10:67–84.

Grabe, D.F. 1957. Identification of soybean varieties by laboratory techniques. Proc. Assoc. Off. Seed Anal. 47:105–119.

Grau, C.R., and J.A. Lawrence. 1975. Observations on resistance and heritability of resistance to powdery mildew of soybean. Plant Dis. Rep. 59:458–460.

Graybosch, R.A. 1984. Studies on the reproductive biology of male-sterile mutants of soybean (*Glycine max* (L.) Merr.). Ph. D. diss. Iowa State Univ., Ames (Diss. Abstr. 85-05820).

----, R.L. Bernard, C.R. Cremeens, and R.G. Palmer. 1984. Genetic and cytological studies of a male-sterile, female-fertile soybean mutant. J. Hered. 75:383–388.

----, and R.G. Palmer. 1984a. Male sterility in soybean, *Glycine max* (L.) Merr. p. 232–253. *In* S. Wang et al. (ed.) Proc. of the Second U.S.-China Soybean Symp. Office of Int. Coop. and Develop. U.S. Department of Agriculture, Washington, DC.

----, and ----. 1984b. Is the *ms4* male-sterile mutant partially fertile? Soybean Genet. Newsl. 11:102–104.

Griffin, J.D., and R.G. Palmer. 1984. Superoxide dismutase (SOD) isoenzymes in soybean. Soybean Genet. Newsl. 11:91–92.

Gwyn, J.J. 1984. Morphological discrimination among some aneuploids in soybean. M.S. thesis. Iowa State University, Ames.

Hadley, H.H., and T. Hymowitz. 1973. Speciation and cytogenetics. *In* B. E. Caldwell (ed.) Soybeans: Improvement, production, and uses. Agronomy 16:97–116.

----, and W.J. Starnes. 1964. Sterility in soybeans caused by asynapsis. Crop Sci. 4:421–424.

Hartwig, E.E., and K.R. Bromfield. 1983. Relationships among three genes conferring specific resistance to rust in soybeans. Crop Sci. 23:237–239.

----, and J.M. Epps. 1970. An additional gene for resistance to the soybean cyst nematode, *Heterodera glycines*. Phytopathology 60:584.

----, and K. Hinson. 1962. Inheritance of flower color of soybeans. Crop Sci. 2:152–153.

----, B.L. Keeling, and C.J. Edwards, Jr. 1968. Inheritance of reaction to phytophthora rot in the soybean. Crop Sci. 8:634–635.

----, and S.G. Lehman. 1951. Inheritance of resistance to the bacterial pustule disease in soybeans. Agron. J. 43:226–229.

Hatfield, P.M. 1982. Comparative physiology of a soybean chlorophyll mutant at two ploidy levels. M.S. thesis. Iowa State University, Ames.

Heenan, D.P., L.C. Campbell, and O.G. Carter. 1981. Inheritance of tolerance to high manganese supply in soybeans. Crop Sci. 21:625–627.

----, and O.G. Carter. 1976. Tolerance of soybean cultivars to manganese toxicity. Crop Sci. 16:389–391.

Hildebrand, D.F., and T. Hymowitz. 1980a. The *Sp1* locus in soybean codes for β-amylase. Crop Sci. 20:165–168.

----, and ----. 1980b. Inheritance of β-amylase nulls in soybean seeds. Crop Sci. 20:727–730.

----, and ----. 1981. Two soybean genotypes lacking lipoxygenase-1. J. Am. Oil Chem. Soc. 58:583–586.

----, and ----. 1982. Inheritance of lipoxygenase-1 activity in soybean seeds. Crop Sci. 22:851–853.

----, J.H. Orf. and T. Hymowitz. 1980. Inheritance of an acid phosphatase and its linkage with the Kunitz trypsin inhibitor in seed protein of soybeans. Crop Sci. 20:83–85.

Hu, S. 1968. Production, identification and characterization of heteroploids in *Glycine max* (L.) Merrill. M.S. thesis. University of Illinois, Urbana.

Hymowitz, T. 1973. Electrophoretic analysis of SBTI-A-2 in the USDA soybean germplasm collection. Crop Sci. 13:420–421.

----. 1983. Variation in and genetics of certain antinutritional and biologically active components of soybean seed. p. 49–60. *In* Better crops for food. CIBA Found. Symp. 97. Pitman Books, London.

----, and H.H. Hadley. 1972. Inheritance of a trypsin inhibitor variant in seed protein of soybeans. Crop Sci. 12:197–198.

----, N. Kaizuma, J.H. Orf, and H. Skorupska. 1979. Screening the USDA soybean germplasm collection for *Sp1* variants. Soybean Genet. Newsl. 6:30–32.

Jha, A.N., and B.B. Singh. 1978. Additional sterile and male-sterile mutants in soybean. Soybean Genet. Newsl. 5:30–35.

Johns, C.W., and R.G. Palmer. 1982. Floral development of a flower-structure mutant in soybeans, *Glycine max* (L.) Merr. (Leguminosae). Am. J. Bot. 69:829–842.

Karasawa, K. 1936. Crossing experiments with *Glycine soja* and *G. ussuriensis*. Jpn. J. Bot. 8:113–118.

Keeling, B.L. 1984. A new physiologic race of *Phytophthora megasperma* f. sp. *glycinea*. Plant Dis. 68:626–627.

Kennell, J.C. 1984. Microscopic investigations on the influence of the soybean male-sterile gene *ms1* on the development of the female gametophyte. M.S. thesis. Iowa State University, Ames.

----, and H.T. Horner. 1985. Influence of the soybean male-sterile gene (*ms1*) on the development of the female gametophyte. Can. J. Genet. Cytol. 27:200–209.

Kenworthy, W.J., C.A. Brim, and E.A. Wernsman. 1973. Polyembryony in soybeans. Crop Sci. 13:637–639.

Keyser, H.H., T.S. Hu, and D.F. Weber. 1982. Fast-growing rhizobia isolated from root nodules of soybean. Science 215:1631–1632.

Kiang, Y.T. 1981. Inheritance and variation of amylase in cultivated and wild soybeans and their wild relatives. J. Hered. 72:382–386.

----, Y.C. Chiang, and M.B. Gorman. 1984. Inheritance of a second leucine aminopeptidase locus and its linkage with other loci. Soybean Genet. Newsl. 11:143–145.

----, and M.B. Gorman. 1983. Soybean. p. 295–328. *In* S.D. Tanksley and T.J. Orton (ed.) Isozymes in plant genetics and breeding, Part B. Elsevier Science Publishing Co., New York.

----, and ----. 1985. Inheritance of NADP-active isocitrate dehydrogenase isozymes in soybeans. J. Hered. 76:279–284.

----, ----, and Y.C. Chiang. 1985. Genetic and linkage analysis of a leucine aminopeptidase in wild and cultivated soybeans. Crop Sci. 25:319–321.

Kiihl, R.A.S., and E.E. Hartwig. 1979. Inheritance of reaction to soybean mosaic virus in soybeans. Crop Sci. 19:372–375.

Kilen, T.C. 1975. An unusual miniature soybean. Crop Sci. 15:871–872.

----. 1977. Inheritance of a brachytic character in soybeans. Crop Sci. 17:853–854.

----. 1979. A temperature sensitive miniature soybean. Crop Sci. 19:405–406.

----. 1983. Inheritance of a short petiole trait in soybean. Crop Sci. 23:1208–1210.

----, and W.L. Barrentine. 1983. Linkage relationships in soybean between genes controlling reactions to phytophthora rot and metribuzin. Crop Sci. 23:894–896.

----, and E.E. Hartwig. 1971. Inheritance of a light-quality sensitive character in soybeans. Crop Sci. 11:559–561.

----, and ----. 1975. Short internode character in soybeans and its inheritance. Crop Sci. 15:878.

----, ----, and B.L. Keeling. 1974. Inheritance of a second major gene for resistance to phytophthora rot in soybeans. Crop Sci. 14:260–262.

----, J.H. Hatchett, and E.E. Hartwig. 1977. Evaluation of early generation soybeans for resistance to soybean looper. Crop Sci. 17:397–398.

Kitamura, K., C.S. Davies, N. Kaizuma, and N.C. Nielsen. 1983. Genetic analysis of a null-allele for lipoxygenase-3 in soybean seeds. Crop Sci. 23:924–927.

----, ----, and N.C. Nielsen. 1984. Inheritance of alleles for *Cgy1* and *Gy4* storage protein genes in soybean. Theor. Appl. Genet. 68:253–257.

Kloth, R.H., and T. Hymowitz. 1985. Re-evaluation of the inheritance of urease in soybean seed. Crop Sci. 25:352–354.

----, J.C. Polacco, and T. Hymowitz. 1984. The inheritance of a soybean seed urease null mutation. Agron. Abstr. American Society of Agronomy, Madison, WI, p. 75.

Koelling, P.D., W.J. Kenworthy, and D.M. Caron. 1981. Pollination of male-sterile soybeans in caged plots. Crop Sci. 21:559–561.

Koshimizu, S., and T. Iizuka. 1963. Studies on soybean virus diseases in Japan. Tohoku Natl. Agric. Exp. Stn., Rep. 27:1–103.

Ladizinsky, G., C.A. Newell, and T. Hymowitz. 1979. Giemsa staining of soybean chromosomes. J. Hered. 70:415–416.

Lam-Sanchez, A., A.H. Probst, F.A. Laviolette, J.F. Schafer, and K.L. Athow. 1968. Sources and inheritance of resistance to *Phytophthora megasperma* var. *sojae* in soybeans. Crop Sci. 8:329–330.

Larsen, A.L. 1967. Electrophoretic differences in seed proteins among varieties of soybean. Crop Sci. 7:311–313.

----, and W.C. Benson. 1970. Variety specific variants of oxidative enzymes from soybean seed. Crop Sci. 10:493–495.

----, and B.E. Caldwell. 1968. Inheritance of certain proteins in soybean seed. Crop Sci. 8:474–476.

Laviolette, F.A., and K.L. Athow. 1983. Two new physiologic races of *Phytophthora megasperma* f. sp. *glycinea*. Plant Dis. 67:497–498.

Lim, S.M., R.L. Bernard, C.D. Nickell, and L.E. Gray. 1984. New physiological race of *Peronospora manshurica* virulent to the gene *Rpm* in soybeans. Plant Dis. 68:71–72.

Liu, H.L. 1949. Inheritance of defective seed coat in soybeans. J. Hered. 40:317–322.

Mahmud, I., and A.H. Probst. 1953. Inheritance of gray hilum color in soybeans. Agron. J. 45:59–61.

Matson, A.L., and L.F. Williams. 1965. Evidence of a fourth gene for resistance to the soybean cyst nematode. Crop Sci. 5:477.

Matsuura, H. 1933. *Glycine soja*. p. 100–110. *In* A bibliographical monograph on plant genetics. 2nd ed. Hokkaido Imperial University, Tokyo.

McLean, R.J., and D.E. Byth. 1980. Inheritance of resistance to rust *Phakopsora pachyrhizi* in soybeans. Aust. J. Agric. Res. 31:951–956.

Moots, C.K., C.D. Nickell, L.E. Gray, and S.M. Lim. 1983. Reaction of soybean cultivars to 14 races of *Phytophthora megasperma* f. sp. *glycinea*. Plant Dis. 67:764–767.

Morse, W.J., and J.L. Cartter. 1937. Improvement in soybeans. p. 1154–1189. *In* Yearbook agriculture, USDA. U.S. Government Printing Office, Washington, DC.

Mukherjee, D., J.W. Lambert, R.L. Cooper, and B.W. Kennedy. 1966. Inheritance of resistance to bacterial blight in soybeans. Crop Sci. 6:324–326.

Mueller, E.H., K.L. Athow, and F.A. Laviolette. 1978. Inheritance to four physiologic races of *Phytophthora megasperma* var. *sojae*. Phytopathology 68:1318–1322.

Nagai, I. 1921. A genetico-physiological study on the formation of anthocyanin and brown pigments in plants. Tokyo Univ. Coll. Agric. J. 8:1–92.

----. 1926. Inheritance in the soybean. (In Japanese.) Nogyo oyobi Engei 1:14, 107–108.

----, and S. Saito. 1923. Linked factors in soybeans. Jpn. J. Bot. 1:121–136.

Nakashima, H., H.T. Horner, and R.G. Palmer. 1984. Histological features of anthers from normal and *ms3* mutant soybean. Crop Sci. 24:735–739.

Nelson, R.L., and R.L. Bernard. 1979. Pollen movement to male-sterile soybeans in southern Illinois. Soybean Genet. Newsl. 6:100–103.

Nelson, R.S., S.A. Ryan, and J.E. Harper. 1983. Soybean mutants lacking constitutive nitrate reductase activity. I. Selection and initial plant characterization. Plant Physiol. 72:503–509.

Newell, C.A., and T. Hymowitz. 1983. Hybridization in the genus *Glycine* subgenus *Glycine* Willd. (Leguminosae, Papilionoideae). Am. J. Bot. 70:334–348.

Newhouse, K.E., L. Hawkins, and R.G. Palmer. 1983. Trisomic inheritance of a chimera in soybean. Soybean Genet. Newsl. 10:44–49.

Nissly, C.R., R.L. Bernard, and C.N. Hittle. 1976. Inheritance in chlorophyll-deficient mutants. Soybean Genet. Newsl. 3:31–34.

----, ----, and ----. 1981a. Variation in photoperiod sensitivity for time of flowering and maturity among soybean strains of maturity group III. Crop Sci. 21:833–836.

----, ----, and ----. 1981b. Inheritance of two chlorophyll-deficient mutants in soybeans. J. Hered. 72:141–142.

Nozzolillo, C. 1973. A survey of anthocyanin pigments in seedling legumes. Can. J. Bot. 51:911–915.

Ohki, K., D.O. Wilson, and O.E. Anderson, 1980. Manganese deficiency and toxicity sensitivities of soybean cultivars. Agron. J. 72:713–716.

Oinuma, T. 1952. An artificial induced tetraploid soybean as green manure. (In Japanese, English summary.) Jpn. J. Breed. 2:7–13.

Olivieri, A.M., M. Lucchin, and P. Parrini. 1980. The monogenic and digenic control of hypocotyl and flower color in soybeans. Soybean Genet. Newsl. 7:64–66.

Orf, J.H., and T. Hymowitz. 1976. The gene symbols *Sp1-a* and *Sp1-b* assigned to Larsen and Caldwell's seed protein bands A and B. Soybean Genet. Newsl. 3:27–28.

----, and ----. 1977. Inheritance of a second trypsin inhibitor variant in seed protein of soybeans. Crop Sci. 17:811–813.

----, and ----. 1979. Inheritance of the absence of the Kunitz trypsin inhibitor in seed protein of soybeans. Crop Sci. 19:107–109.

----, ----, S.P. Pull, and S.G. Pueppke. 1978. Inheritance of a soybean seed lectin. Crop Sci. 18:899–900.

Owen, F.V. 1927a. Inheritance studies in soybeans. I. Cotyledon color. Genetics 12:441–448.

----. 1927b. Inheritance studies in soybeans. II. Glabrousness, color of pubescence, time of maturity, and linkage relations. Genetics 12:519–529.

----. 1927c. Hereditary and environmental factors that produce mottling in soybeans. J. Agric. Res. 34:559–587.

----. 1928a. A sterile character in soybeans. Plant Physiol. 3:223–226.

----. 1928b. Inheritance studies in soybeans. III. Seed coat color and summary of all other mendelian characters thus far reported. Genetics 13:50–79.

----. 1928c. Soybean seeds with two embryos. J. Hered. 19:373–374.

Palmer, R.G. 1974a. Aneuploids in the soybean, *Glycine max*. Can. J. Genet. Cytol. 16:441–447.

----. 1974b. A desynaptic mutant in the soybean. J. Hered. 65:280–286.

----. 1976a. Cytogenetics in soybean improvement. p. 56–66. *In* H.D. Louden and D. Wilkenson (ed.) Proc. Sixth Soybean Seed Res. Conf. American Seed Trade Association, Washington, DC.

----. 1976b. Chromosome transmission and morphology of three primary trisomics in soybeans (*Glycine max*). Can. J. Genet. Cytol. 18:131–140.

----. 1977. Soybean linkage tests. Soybean Genet. Newsl. 4:40–42.

----. 1984a. Genetic studies with T263. Soybean Genet. Newsl. 11:94–97.

----. 1984b. Pleiotropy or close linkage of two mutants in soybean. J. Hered. 75:457–462.

----. 1984c. Summary of trisomic linkage data in soybean. Soybean Genet. Newsl. 11:127–131.

----. 1985. Soybean cytogenetics. p. 337–344. *In* R. Shibles (ed.) World soybean research conference III: Proc. Westview Press, Boulder, CO.

----, M.C. Albertsen, and H. Heer. 1978a. Pollen production in soybeans with respect to genotype, environment, and stamen position. Euphytica 27:427–433.

----, M.C. Albertsen, and C.W. Johns. 1983. Pollen movement to two male-sterile soybean mutants grown in two locations. J. Hered. 74:55–57.

----, S.L. Broich, and X. Delannay. 1984. Linkage group 12. Soybean Genet. Newsl. 11:97–99.

----, J.D. Griffin, and B.R. Hedges. 1985. Current and obsolete gene symbols for isozymes and protein variants in soybean. Soybean Genet. Newl. 12:16–28.

----, and H. Heer. 1973. A root tip squash technique for soybean chromosomes. Crop Sci. 13:389–391.

----, and ----. 1976. Aneuploids from a desynaptic mutant in soybeans (*Glycine max* (L.) Merr.). Cytologia 41:417–427.

----, and ----. 1984. Agronomic characteristics and genetics of a chromosome interchange in soybean. Euphytica 33:651–663.

----, and S. Rodriguez de Cianzio. 1985. Conditional lethality involving nuclear and cytoplasmic chlorophyll mutants in soybean. Theor. Appl. Genet. 70:349–354.

----, C.W. Johns, and P.S. Muir. 1980. Genetics and cytology of the *ms3* male-sterile soybean. J. Hered. 71:343–348.

----, and M.L.H. Kaul. 1983. Genetics, cytology, and linkage studies of a desynaptic soybean mutant. J. Hered. 74:260–264.

----, and P.N. Mascia. 1980. Genetics and ultrastructure of a cytoplasmically inherited yellow mutant in soybeans. Genetics 95:985–1000.

----, and K.E. Newhouse. 1984. Evaluation of *Glycine soja* from The People's Republic of China and the USSR. Soybean Genet. Newsl. 11:105–111.

----, and R.C. Payne. 1979. Genetic control of hypocotyl pigmentation among white-flowered soybeans grown in continuous light. Crop Sci. 19:124–126.

----, M.A. Sheridan, and M.A. Tabatabai. 1979. Effects of genotype, temperature, and illuminance on chloroplast ultrastructure of a chlorophyll mutant in soybeans. Cytologia 44:881–891.

----, and D.M. Stelly. 1979. Reference diagrams of seed coat colors and patterns for use as genetic markers in crosses. Soybean Genet. Newsl. 6:55–57.

----, C.L. Winger, and M.C. Albertsen. 1978b. Four independent mutations at the *ms1* locus in soybeans. Crop Sci. 18:727–729.

Patil, A.B., and B.B. Singh. 1976. Male sterility in soybeans. Indian J. Genet. Breed. 36:238–243.

Payne, R.C., and E.W. Sundermeyer. 1977. Pigmentation differences of soybean cultivars with green hypocotyl color when grown in continuous light. Crop Sci. 17:479–480.

Peters, D.W., J.R. Wilcox, J.J. Vorst, and N.C. Nielsen. 1984. Hypocotyl pigments in soybeans. Crop Sci. 24:237–239.

Peterson, P.A., and C.R. Weber. 1969. An unstable locus in soybeans. Theor. Appl. Genet. 39:156–162.

Pillai, R.V.R. 1976. Diploids among the cultivars of soybean (*Glycine max* Linn.) in Manipur. Sci. Cult. 42:519–521.

Piper, C.V., and W.J. Morse. 1910. The soybean: History, varieties, and field studies. USDA Bureau of Plant Industry Bull. 197. U.S. Government Printing Office, Washington, DC.

Polacco, J.C., A.L. Thomas, and P.J. Bledsoe. 1982. A soybean urease null produces urease in cell culture. Plant Physiol. 69:1233–1240.

Polson, D.E. 1972. Day-neutrality in soybeans. Crop Sci. 12:773–776.

Porter, K.B., and M.G. Weiss. 1948. The effect of polyploidy on soybeans. J. Am. Soc. Agron. 40:710–724.

Probst, A.H. 1950. The inheritance of leaf abscission and other characters in soybeans. Agron. J. 42:35–45.

----, K.L. Athow, and F.A. Laviolette. 1965. Inheritance of resistance to race 2 of *Cercospora sojina* in soybeans. Crop Sci. 5:332.

Pueppke, S.G., and T. Hymowitz. 1982. Screening the genus *Glycine* subgenus *Glycine* for the 120,000 dalton seed lectin. Crop Sci. 22:558–560.

Pull, S.P., S.G. Pueppke, T. Hymowitz, and J.H. Orf. 1978. Soybean lines lacking the 120,000 dalton seed lectin. Science 200:1277–1279.

Rhodes, P.R., and L.O. Vodkin. 1985. Highly structured sequence homology between an insertion element and the gene in which it resides. Proc. Natl. Acad. Sci. USA 82:493–497.

Roane, C.W., S.A. Tolin, and G.R. Buss. 1983. Inheritance of reaction to two viruses in the soybean cross 'York' × 'Lee 68'. J. Hered. 74:289–291.

Rode, M.W., and R.L. Bernard. 1975a. Inheritance of a tan saddle mutant. Soybean Genet. Newsl. 2:39–42.

----, and ----. 1975b. Inheritance of wavy leaf. Soybean Genet. Newsl. 2:42–44.

----, and ----. 1975c. Inheritance of bullate leaf. Soybean Genet. Newsl. 2:44–46.

Roth, E.J., G. Weber, and K.G. Lark. 1982. Use of isopropyl-N (3-chlorophenyl) carbamate (CIPC) to produce partial haploid cells from suspension cultures of soybean (*Glycine max*). Plant Cell Rep. 1:205–208.

Rubaihayo, P.R., and G. Gumisiriza. 1978. The causes of genetic male sterility in three soybean lines. Theor. Appl. Genet. 53:257–260.

Ryan, S.A., R.S. Nelson, and J.E. Harper. 1983a. Selection and inheritance of nitrate reductase mutants in soybeans. Soybean Genet. Newsl. 10:33–35.

----, ----, and ----. 1983b. Soybean mutants lacking constitutive nitrate reductase activity. II. Nitrogen assimilation, chlorate resistance, and inheritance. Plant Physiol. 72:510–514.

Sacks, J.M., and K. Sadanaga. 1984. Linkage between the male sterility gene (*ms1*) and a translocation breakpoint in soybean, *Glycine max*. Can. J. Genet. Cytol. 26:401–404.

Sadanaga, K. 1983. Locating *wm* on linkage group 8. Soybean Genet. Newsl. 10:39–41.

----, and R. Grindeland. 1979. Aneuploids and chromosome aberrations from irradiated soybeans. Soybean Genet. Newsl. 6:43–45.

----, and ----. 1981. Natural cross-pollination in diploid and autotetraploid soybeans. Crop Sci. 21:503–506.

----, and ----. 1984. Locating the *w1* locus on the satellite chromosome in soybean. Crop Sci. 24:147–151.

----, and K. Newhouse. 1982. Identifying translocations in soybeans. Soybean Genet. Newsl. 9:129–130.

Sapra, V.T., and M.D. Stewart. 1981. Leaves as a source of somatic chromosomes for cytogenetic studies of soybean. Soybean Genet. Newsl. 8:22–23.

Sen, N.K., and R.V. Vidyabhusan. 1960. Tetraploid soybeans. Euphytica 9:317–322.

Sengupta, K. 1975. A new gene in soybean (*Glycine max* (L.) Merr.) conditioning hypocotyl pigmentation. Curr. Sci. 44:200–201.

Shanmugasundaram, S. 1977. Inheritance of photoperiod insensitivity to flowering in *Glycine max* (L.) Merr. Soybean Genet. Newsl. 4:15–18.

----. 1978a. Inheritance of time of flowering under short-day conditions. Soybean Genet. Newsl. 5:86–91.

----. 1978b. Variation in the photoperiodic response to flowering in soybean. Soybean Genet. Newsl. 5:91–94.

----. 1981. Varietal differences and genetic behavior for the photoperiodic responses in soybeans. p. 1–61. *In* Bull. Inst. Trop. Agric. Kyushu Univ. 4 (March).

Sheridan, M.A., and R.G. Palmer. 1975. Inheritance and derivation of T225H, *Y18 y18*. Soybean Genet. Newsl. 2:18–19.

----, and ----. 1977. The effect of temperature on an unstable gene in soybeans. J. Hered. 68:17–22.

Shipe, E.R., G.R. Buss, and S.A. Tolin. 1979. A second gene for resistance to peanut mottle virus in soybeans. Crop. Sci. 19:656–658.

Shoemaker, R.C., A. Cody, and R.G. Palmer. 1985. Characterization of a cytoplasmically inherited yellow foliar mutant (*cyt-Y3*) in soybean. Theor. Appl. Genet. 69:279–284.

Shorter, R., and D.E. Byth. 1975. Multiple seedling development in soybeans. J. Hered. 66:323–326.

Sinclair, J.B., and M.C. Shurtleff (ed.) 1975. Compendium of soybean diseases. American Phytopathological Society, St. Paul, MN.

Singh, B.B. 1972. High frequency of natural cross pollination in a mutant strain of soybean. Curr. Sci. 41:832–833.

----, S.C. Gupta, and B.D. Singh. 1974. An induced crinkled leaf mutant in soybean. Soybean Genet. Newsl. 1:16–17.

----, and A.N. Jha. 1978. Abnormal differentiation of floral parts in a mutant strain of soybean. J. Hered. 69:143–144.

----, and P.N. Thapliyal. 1977. Breeding for resistance to soybean rust in India. p. 62–65. *In* R.E. Ford and J.B. Sinclair (ed.) Rust of soybeans, the problem and research needs. Int. Agric. Publ. INTSOY Ser. 12. University of Illinois, Urbana.

Singh, L., C.M. Wilson, and H.H. Hadley. 1969. Genetic differences in soybean trypsin inhibitors separated by disc electrophoresis. Crop Sci. 9:489–491.

Sisson, V.A., P.A. Miller. W.V. Campbell, and J.W. Van Duyn. 1976. Evidence of inheritance of resistance to the Mexican bean beetle in soybeans. Crop Sci. 16:835–837.

Skorupska, H., and J. Nawracala. 1980. Observations of pollen grains of soybean plants in the male-sterile line Urbana *ms1*. Genet. Pol. 21:63–68.

Sorrells, M.E., and E.T. Bingham. 1979. Reproductive behavior of soybean haploids carrying the *ms1* allele. Can. J. Genet. Cytol. 21:449–455.

Soybean Genetics Committee. 1978. New gene symbols. Soybean Genet. Newsl. 5:13.

Specht, J.E., and J.H. Williams. 1978. Hilum color as a genetic marker in soybean crosses. Soybean Genet. Newsl. 5:70–73.

Stahlhut, R.W., and T. Hymowitz. 1980. Screening the USDA soybean germplasm collection for lines lacking the 120,000 dalton seed lectin. Soybean Genet. Newsl. 7:41–43.

----, ----, and J.H. Orf. 1981. Screening the USDA *Glycine soja* collection for presence or absence of a seed lectin. Crop Sci. 21:110–112.

Stelly, D.M., P.S. Muir, and R.G. Palmer. 1979. Spontaneously occurring sterile plants. Soybean Genet. Newsl. 6:45–47.

----, and R.G. Palmer. 1980a. A partially male-sterile mutant line of soybeans, *Glycine max* (L.) Merr.: Inheritance. Euphytica 29:295–303.

----, and ----. 1980b. A partially male-sterile mutant line of soybeans, *Glycine max* (L.) Merr.: Characterization of the *msp* phenotypic variation. Euphytica 29:539–546.

----, and ----. 1982. Variable development in anthers of partially male-sterile *msp* soybeans. J. Hered. 73:101–108.

----, S.J. Peloquin, R.G. Palmer, and C.F. Crane. 1984. Mayer's hemalum-methyl salicylate: A stain-clearing technique for observation within whole ovules. Stain Technol. 59:155–161.

Stewart, R.T. 1927. Dwarfs in soybeans. J. Hered. 18:281–284.

----. 1930. Inheritance of certain seed-coat colors in soybeans. J. Agric. Res. 40:829–854.

----, and J.B. Wentz. 1926. A recessive glabrous character in soybeans. J. Am. Soc. Agron. 18:997–1009.

----, and ----. 1930. A defective seed-coat character in soybean. J. Am. Soc. Agron. 22:658–662.

Takagi, F. 1929. On the inheritance of some characters in *Glycine soja*, Bentham (soybean). Sci. Rep. Tohoku Univ., Ser. 4 4:577–589.

----. 1930. On the inheritance of some characters in *Glycine soja*, Bentham (soybean). (In Japanese.) Jpn. J. Genet. 5:177–189.

Takahashi, N. 1934. Linkage relation between the genes for the forms of leaves and the number of seeds per pod of soybeans. (In Japanese, English summary.) Jpn. J. Genet. 9:208–225.

Takahashi, Y., and J. Fukuyama. 1919. Morphological and genetic studies on the soybean. (In Japanese.) Hokkaido Agric. Exp. Stn. Rep. 10.

Tang, W.T., and C.C. Lin. 1963. Artificial induction and practical value of tetraploid soybeans, Bot. Bull. Acad. Sin. 4:103–110.

----, and G. Tai. 1962. Studies on the qualitative and quantitative inheritance of an interspecific cross of soybean, *Glycine max* × *G. formosana*. Bot. Bull. Acad. Sin. 3:39–60.

Taylor, B.H. 1976. Environmental and chemical evaluation of variations in hilum and seedcoat colors in soybeans. M.S. thesis. University of Arkansas, Fayetteville.

Terao, H., 1918. Maternal inheritance in the soybean. Am. Nat. 52:51–56.

----, and S. Nakatomi. 1929. On the inheritance of chlorophyll colorations of cotyledons and seed-coats in the soybean. (In Japanese, English summary.) Jpn. J. Genet. 4:64–80.

Thomas, J.D., C.E. Caviness, R.D. Riggs, and E.E. Hartwig. 1975. Inheritance of reaction to race 4 of soybean-cyst nematode. Crop Sci:15:208–210.

Tilton, V.R., L.W. Wilcox, R.G. Palmer, and M.C. Albertsen. 1984. Stigma, style, and obturator of soybean, *Glycine max* (L.) Merr. (Leguminosae) and their function in the reproductive process. Am. J. Bot. 71:676–686.

Ting, C.L. 1946. Genetic studies on the wild and cultivated soybeans. J. Am. Soc. Agron. 38:381–393.

VanSchaik, P.H., and A.H. Probst. 1958. The inheritance of inflorescence type, peduncle length, flowers per node, and percent flower shedding in soybeans. Agron. J. 50:98–102.

Veatch, C., and C.M. Woodworth. 1930. Genetic relations of cotyledon color types of soybeans. J. Am. Soc. Agron. 22:700–702.

Vest, G. 1970. *Rj3*—a gene conditioning ineffective nodulation in soybean. Crop Sci. 10:34–35.

----, and B.E. Caldwell. 1972. *Rj4*—a gene conditioning ineffective nodulation in soybean. Crop Sci. 12:692–693.

Vig, B.K. 1973. Somatic crossing over in *Glycine max* (L.) Merrill: Mutagenicity of sodium azide and lack of synergistic effect with caffeine and mitomycin C. Genetics 75:265–277.

----. 1974. Exploitation of leaf mosaicism for determination of allelic relationship in *Glycine max*. Soybean Genet. Newsl. 1:42–44.

----. 1975. Soybean *Glycine max*: A new test system for study of genetic parameters as affected by environmental mutagens. Mutat. Res. 31:49–56.

----. 1982. Soybean (*Glycine max* (L.) Merr.) as a short-term assay for study of environmental mutagens. A report of the U.S. Environmental Protection Agency Gene-Tox Program. Mutat. Res. 99:339–347.

Walker, D.B., and H.R. Boerma. 1978. Morphological study of the mechanism causing brachytic stem in soybeans. Can. J. Plant Sci. 58:993–998.

Wang, S. 1948. A recessive mottling gene of soybeans. J. Agric. Assoc. China 186:35–38.

Weber, C.R., and M.G. Weiss. 1959. Chlorophyll mutant in soybeans provides teaching aid. J. Hered. 50:53–54.

Weiss, M.G. 1943. Inheritance and physiology of efficiency in iron utilization in soybeans. Genetics 28:253–268.

----. 1970a. Genetic linkage in soybeans. Linkage group I. Crop Sci. 10:69–72.

----. 1970b. Genetic linkage in soybeans. Linkage groups II and III. Crop Sci. 10:300–303.

----. 1970c. Genetic linkage in soybeans: Linkage group IV. Crop Sci. 10:368–370.

----. 1970d. Genetic linkage in soybeans: Linkage groups V and VI. Crop Sci. 10:469–470.

----. 1970e. Genetic linkage in soybeans: Linkage group VII. Crop Sci. 10:627–629.

White, M.C., R.L. Chaney, and A.M. Decker. 1979b. Differential cultivar tolerance in soybean to phytotoxic levels of soil Zn. II. Range of Zn additions and the uptake and translocation of Zn, Mn, Fe, and P. Agron. J. 72:126–131.

----, A.M. Decker, and R.L. Chaney. 1979a. Differential cultivar tolerance in soybean to phytotoxic levels of soil Zn. I. Range of cultivar response. Agron. J. 71:121–126.

Wilcox, J.R., and A.H. Probst. 1969. Inheritance of a chlorophyll-deficient character in soybeans. J. Hered. 60:115–116.

Williams, C., W. Birchfield, and E.E. Hartwig. 1973. Resistance in soybeans to a new race of root-knot nematode. Crop Sci. 13:299–301.

----, D.F. Gilman, D.S. Fontenot, and W. Birchfield. 1981. Inheritance of reaction to the reniform nematode in soybean. Crop Sci. 21:93–94.

Williams, L.F. 1948. Inheritance in a species cross in soybeans. Genetics 33:131–132 (Abstr.)

----. 1950. Structure and genetic characteristics of the soybean. p. 111–134. *In* K.S. Markley (ed.) Soybean and soybean products, Vol. 1. Interscience Publishers, New York.

----. 1952. The inheritance of certain black and brown pigments in the soybean. Genetics 37:208–215.

----. 1958. Alteration of dominance and apparent change in direction of gene action by a mutation at another locus affecting the pigmentation of the seedcoat of the soybean. Proc. Int. Congr. Genet., 10th 2:315–316 (Abstr.)

----, and D.L. Lynch. 1954. Inheritance of a non-nodulating character in the soybean. Agron. J. 46:28–29.

Winger, C.L., R.G. Palmer, and D.E. Green. 1977. A spontaneous mutant at the *st2* locus. Soybean Genet. Newsl. 4:36–40.

Woodworth, C.M. 1921. Inheritance of cotyledon, seed-coat, hilum, and pubescence colors in soy-beans. Genetics 6:487–553.

----. 1923. Inheritance of growth habit, pod color, and flower color in soybeans. J. Am. Soc. Agron. 15:481–495.

----. 1932. Genetics and breeding in the improvement of the soybean. Bull. Agric. Exp. Stn. (Ill.) 384:297–404.

----. 1933. Genetics of the soybean. J. Am. Soc. Agron. 25:36–51.

----, and L.F. Williams. 1938. Recent studies on the genetics of the soybean. J. Am. Soc. Agron. 30:125–129.

Yamamoto, K. and Y. Nagato. 1984. Variation of DNA content in the genus *Glycine*. Jpn. J. Breed. 34:163–170.

Yee, C.C., and L. Jian. 1983. Allelism tests of Shennong male-sterile soybean L-78-387. p. 241–242. (Abstr.) (In Chinese.) Second Assembly Symp. Genet. Soc. China. Sec. 4. no. 053.

----, and R.G. Palmer. 1984. Summary of locus-to-locus linkage data in soybean. Soybean Genet. Newsl. 11:105–126.

Yong, H.S., K.L. Chan, C. Mak, and S.S. Dhaliwal. 1981. Isocitrate dehydrogenase gene duplication and fixed heterophenotype in the cultivated soybean *Glycine max*. Experientia 37:130–131.

----, C. Mak, K.L. Chan, and S.S. Dhaliwal. 1982. Inheritance of isocitrate dehydrogenase in the cultivated soybean. Malay, Nat. J. 35:225–228.

Yoshikura, K., and Y. Hamaguchi. 1969. Anthocyanins of black soybeans. Jpn. Soc. Food Nutr. J. 22:15–18.

Zheng, H.Y., R.Y. Chen, and H. Sun. 1984. A diploid strain of wild soybean (*G. soja*) with four-satellited chromosomes. p. 96. *In* R. Shibles (ed.) World soybean research conference III. Abstr. 438. Westview Press, Boulder, CO.

6 Quantitative Genetics: Results Relevant to Soybean Breeding

J.W. Burton
North Carolina State University
Raleigh, North Carolina

Quantitative genetic and statistical research have provided the theoretical framework for modern plant and animal breeding. Most of the currently used breeding methods for manipulating metric traits have their origin or rationale in quantitative genetic science. For any given breeding objective, plant breeders are faced with questions which relate to appropriate genetic materials and methods. Dudley and Moll (1969) enumerated these questions as follows: (i) is there sufficient genetic variability in the germplasm pool, (ii) how much testing is necessary, (iii) which genetic population should be used, (iv) what breeding procedure will be the most rapid, efficient and at the same time, successful, (v) what type of variety will be the final goal, and (vi) will the manipulation of different traits require different procedures? An overall objective of quantitative genetic research is to help breeders find answers to these questions.

Johnson and Bernard (1963) and Brim (1973) defined quantitative genetic issues that are pertinent to soybean [*Glycine max* (L.) Merr.] breeding and summarized the results of such research in soybean. The purpose of this article is to review quantitative genetic research relevant to soybean breeding that has been reported since 1973 and also to summarize some of the important results from research prior to 1973.

6-1 PARTITION OF HEREDITARY VARIANCE

In soybean plants, hereditary variance has been partitioned through experiments using materials generated by two mating designs: nested (hierarchial) designs of self-fertilization and complete or partial diallel designs. With the former, various progeny sets in different inbred generations are generated through selfing and then tested in replicated experiments. Relationships among these progenies are equated with com-

Soybeans: Improvement, Production, and Uses, 2nd ed.—Agronomy Monograph no. 16.

ponents of variance and covariance among generations. This permits least squares estimation of genotypic variance components due to additivity, dominance, epistasis, or linkage depending upon the reference population and the model specified. With one exception (Hanson et al., 1967), the reference population for reported experiments have all been F_2 or F_7 lines derived from a cross between two inbred lines (Horner and Weber, 1956; Gates et al., 1960; Brim and Cockerham, 1961; Hanson and Weber, 1961; Croissant and Torrie, 1971). Details of this methodology have been provided by Horner and Weber (1956) and Cockerham (1963) for reference populations which have gene frequencies of 0.5. Hanson and Weber (1961) and Cockerham (1963) showed that all gene frequencies could be accommodated with genetic models which include only additive and additive × additive types of epistasis. Cockerham (1983) has recently modified the procedures for interpreting the covariances of self-fertilization relatives by using several identity by descent measures in addition to the inbreeding coefficient (F). This permitted the development of genetic models with additive and dominance effects (no epistasis) that are general for all gene frequencies.

6–1.1 Self-fertilization Designs

In general, results from self-fertilization design experiments have shown most of the genotypic variability in the reference population to be due to additive gene effects. Taking as their reference population a set of F_2 lines from a cross between 'Adams' and 'Hawkeye', Horner and Weber (1956) found that 96% of the genetic variability for the trait, maturity, was explained by the completely additive model. Using the same experimental material, Gates et al. (1960) fit a model with additive, dominance, and linkage effects. They found additive effects to be significant for all traits measured. Dominance effects were significant only for time of flowering and linkage was important for yield, time of flowering, and plant height. Hanson and Weber (1961) used the same data but considered the reference population to be a set of homozygous lines derived from the cross. Thus, a model was fit to the data which included only additive and additive × additive effects. Significant additive variation was found for maturity, height, seed weight, percent oil, and lodging. Additive × additive variation was significant only for percent oil.

Brim and Cockerham (1961) derived materials from two inbred line crosses, N48-4860 × Lee and Roanoke × Lee, through bulk self-fertilization and by paired matings of F_3 plants. Their reference populations were the two sets of F_2 plants. Using a regression analysis, they fit six models to the data, which included additive, dominance, and additive × additive effects taken singly, doubly, and all together. The all additive model accounted for between 97.5 and 99.6% of the regression sums of squares, depending on the trait. Inclusion of dominance and additive × additive epistasis accounted for only small additional percentages of the variation. In both populations, estimates of additive variance obtained with the complete model were significant for all traits measured. With this same model, dominance variance was significant for only fruiting period and unthreshed weight in the second population. Inbreeding depression existed in yield, however, which is good evidence that dominance was involved. Estimates of additive × additive epistasis were significant for maturity, height, yield, percent protein, and percent oil.

Using the models developed by Gates et al. (1960), Croissant and Torrie (1971) estimated additive, dominance, and linkage components of genotypic variance in two F_2 populations derived from the crosses Norchief × Clark and Norchief × Harosoy. They found significant additive effects for all traits measured and significant dominance effects for seed weight, plant height, and lodging. Linkage effects were significant for flowering time, plant height, seed size, and lodging.

Hanson et al. (1967) used as their reference population random homozygous lines from an eight-parent diallel which had been intermated for two generations. Using a model that included additive and additive × additive variance, they found highly significant additive × additive variance for yield and maturity and moderate levels for lodging and plant height. For yield and maturity, additive × additive epistasis accounted for 61 and 55%, respectively, of the variability partition.

6–1.2 Diallel Designs

The results of diallel experiments are more difficult to interpret genetically. Baker (1978) and Sokol and Baker (1977) pointed out that three of the underlying assumptions of diallel analysis are unrealistic for the nonrandom sets of inbred lines used in many diallel experiments. These assumptions are no epistasis, gene frequencies of one-half, and independent distribution of genes among the parents. With failure of all three assumptions, estimates of general combining ability (GCA) include dominance and epistatic effects as well as additive effects. Also, specific combining ability (SCA) estimates will include epistatic effects as well as dominance effects.

In soybean, diallel experiments have involved nonrandom sets of inbred lines. Therefore, while estimates of GCA and SCA are valid, a good genetic interpretation of those estimates is not possible. Even so, presence or absence of significant variation due to SCA should give some indication of the importance of nonadditive effects. Also, diallel crosses can be useful in evaluating the suitability of parent lines and cross combinations for given purposes.

Ratios of the GCA variance to the SCA variance from five experiments are shown in Table 6–1. Ratios are presented for only those traits where the SCA was statistically significant. Three of the studies (Weber et al., 1970; Singh et al., 1974; Kaw and Menon, 1980) reported significant SCA for yield which is evidence that nonadditive genetic effects were important. Significant SCA was also associated with seed size, height, and maturity in some of the experiments.

Table 6-1. Ratio of average general combining ability variance to average specific combining ability variance (GCA/SCA) from diallel analyses of traits for which SCA was significant.

Character	Leffel and Weiss (1958)	Weber et al. (1970)	Singh et al. (1974)	Paschal and Wilcox (1975)	Kaw and Menon (1980)
Yield	—	1.6	1.6	—	1.9
Seed size	4.0	—	—	21.1	37.5
Height	2.2	2.7	—	—	—
Maturity	15.8	6.2	—	8.8	—

Additive and additive × additive types of genetic effects are fixed in homozygous lines. Therefore, as Brim (1973) stated, when homozygous lines are the final product of a varietal breeding program, the presence of significant additive × additive epistasis should not greatly change breeding procedures. The major effect of epistasis is to make progeny evaluation in early generations less effective. Also, linkages become important with epistasis because epistatic components of variance are increased by gene linkage (Cockerham, 1963). Therefore, effects of epistasis might be more prevalent in populations with linkage disequilibrium, such as inbred generations from crosses between two homozygous lines. With the discovery of genetic male sterility in soybean (Brim and Young, 1971), it became practical to develop random mating populations. Such populations are now available and because they have been intermated for several generations, linkage disequilibrium should be minimal. These populations should be excellent sources of random lines for use in experiments designed to estimate components of genetic variance since epistatic effects would not be confounded with linkage effects. Methods developed by Stuber (1970) for estimating genetic variances with inbred relatives may be useful in this regard.

6–1.3 Estimate Bias Due to Intergenotypic Competition

When genetically heterogeneous lines (e.g., F_2, F_3, and F_4 lines) are used to estimate genetic parameters, intergenotypic competition may bias the estimates (Hanson and Weber, 1961; Hamblin and Rosielle, 1978). Intergenotypic competition is a recognized phenomenon ın soybean. Significant competition effects between rows and between hills of different genotypes have been found in several studies (Hanson et al., 1961b; Schutz and Brim, 1967; Lin and Torrie, 1970; Gedge et al., 1977). Schutz and Brim (1967) demonstrated that the response of one genotype to competition by another genotype was a linear function of the frequency of the competitor. Because of results like these, it is generally believed that the best field-plot evaluation of a genotype's yielding ability occurs when it is bordered by itself. In plots of early generation lines, where every plant is likely to be genetically different, intergenotypic competition undoubtedly occurs. Yet, the net effect of each competitive interaction between plants in the plot may be small compared to the overall direct effect of the line. This appears to be true with respect to mean performance. Gedge et al. (1978) compared the yield means of 40 random F_3 lines from two crosses with the means of four random F_6 lines from each F_3 line. They found that in both crosses F_3 line means were generally not significantly different from the means of their respective F_6 lines.

The effect of competition on estimates of genetic variance may be greater than the effect on means. Hamblin and Rosielle (1978) showed large biases in estimates of additive and dominance variance in two F_2 barley (*Hordeum vulgare* L.) populations which were segregating for a single locus. Based on these results, they concluded that estimates of genetic parameters obtained with heterogeneous lines may have little relevance for populations of pure lines. However, since their results were obtained in populations that were essentially mixtures of three genotypes, it may not be possible to generalize this result to more heterogeous lines. The self-fertilization designs suggested by Hanson and Weber (1961) would use homozygous lines to estimate additive and additive × additive variance which would eliminate competition as a factor and be directly applicable to pure line selection.

6–2 HETEROSIS

The evidence from the nested and diallel experiments indicates that while additive variance is the primary component of hereditary variance, nonadditive types of genetic effects can contribute significantly to the variation in some traits of some soybean populations. One expected manifestation of this nonadditivity is heterosis. A summary of experiments where heterosis has been investigated is shown in Table 6–2. The F_1 soybean seed are difficult to generate due to the time-consuming nature of hand pollinations along with the resulting low seed set. Consequently, most experiments have compared space planted F_1's with the space planted parents and often in only one environment. For seed yield, average high parent heterosis ranged from 20.2 to 3.3% and average midparent heterosis ranged from 35.5 to 7.9% (Table 6–2). In studies where enough seed was generated to test the F_1's in row plots and multiple environments (Brim and Cockerham, 1961; Nelson and Bernard, 1984; Hillsman and Carter, 1981), the range of average high parent heterosis for 35 hybrids was 20.4 to -6.9%. Over all studies, 85% of the F_1 crosses showed midparent heterosis and 62% showed high parent heterosis.

Significant dominance effects are usually needed to justify a large emphasis in a breeding program on F_1 hybrid development. Heterosis, which can result from all types of nonadditive gene action, may not indicate dominance. For this reason, Compton (1977) has suggested that inbreeding depression is better evidence of dominance than heterosis. In soybean and other self-pollinated species, inbreeding depression is not as easily observed as it is in cross pollinators such as corn (*Zea mays* L.) and estimates of dominance variance are usually lower in self-pollinated

Table 6-2. Average yield heterosis expressed as a percent of the midparent and as a percent of the high parent.

	Veatch (1930)†	Wiess et al. (1947)‡	Leffel and Weiss (1958)§	Brim and Cockerham (1961)¶	Weber et al. (1970)#	Paschal and Wilcox (1975)††	Kaw and Menon (1979)§§	Hillsman and Carter (1981)¶¶	Nelson and Bernard (1984)‡‡
Mean midparent percentage heterosis	35.5	20.8	13.7	27.5	25.1	—	17.4	12.9	7.9
Mean high parent percentage heterosis	19.6	14.5	—	20.2	13.4	8.0	4.3	6.2	3.3
Percentage F_1's > midparent	81.3	94.1	57.8	100.0	90.6	40.0	82.2	68.8	96.0
Percentage F_1's > high parent	56.3	52.9	31.1	100.0	76.5	7.0	55.5	62.5	68.0

†16 F_1's, spaced plants, 1 yr, between one and four plants.
‡17 F_1's, spaced plants, 1 yr, between 28 and 65 plants.
§45 F_1's, spaced plants 1 yr, two replications, nine plants/replication.
¶2 F_1's, rows, 2 yrs, two locations, two replications, 60 plants/replication (plot).
#85 F_1's, spaced plants, 1 yr, 1 location, between 3 and 47 plants.
††30 F_1's, spaced plants, 2 yrs, five replications/yr, five plants/replication (plot).
‡‡37 F_1's, 3-m rows, 2 yrs, and/or two locations, three replications, 78 plants/replication (plot).
§§45 F_1's, spaced plants, 1 yr, three replications, 15 plants/replication (plot).
¶¶ 8 F_1's, 3-m rows, 2 yrs, one location, three replications, 60 plants/replication (plot).

species. This is probably a result of natural selection against deleterious genes during the repeated inbreeding of its evolutionary history.

In soybean, the aforementioned experimental evidence clearly shows that given the proper genetic combinations, high parent heterosis occurs. It is not known how much of the heterotic effect is due to dominance and dominance types of epistasis and how much could be due to additive $\times$ additive epistasis. Presumably, a homozygous line with performance equal to the F_1 could be selected from segregating generations. The probability of finding such a line, however, would be influenced by the number of genes involved and the relative importance of dominance effects, which cannot be fixed, and additive $\times$ additive effects, which can. A homozygous line equal in performance to a heterotic F_1 might be quite difficult and expensive to identify if a large number of genes influence the trait and if dominance and dominance types of epistasis were significantly involved.

Currently, an economical way to produce F_1 hybrid seed in soybean for farm use does not exist. Additional research with various sources of male sterility may lead to the development of an economical method. In that event, it would be useful to know which hybrid combinations would be the most productive. Bailey et al. (1980) have described a design, using the F_2 and backcross generations to predict heterosis, that might be an economically useful way to test parental combinations. Burton and Carter (1983) have described a method for producing experimental quantities of F_1 hybrid soybean seed. This method involves the development of genetic male-sterile maintainer lines with the green cotyledon trait ($d_1d_1d_2d_2$) for use as female parents. Any yellow-seeded cultivar which flowers synchronously with the female parent can serve as the male parent. Additional research is needed to produce estimates of heterosis for a wider array of genotypes under commercial cultural conditions and to develop more information on the environmental stability of F_1 hybrids relative to pure lines. Such research will provide a better assessment than now exists as to the economic advantages or disadvantages of F_1 hybrids relative to pure lines.

6–3 HERITABILITY

Heritability is usually defined as the proportion of total or phenotypic variance for a given trait, that is strictly due to genetic variation. Hanson (1963) stated that heritability estimates serve two purposes in plant breeding. First, they show the relative ease with which different traits are selected under a given testing regime. Thus, it is often noted that percent protein is more highly heritable than seed yield. The other purpose for heritability estimates is prediction of selection progress. The change in population mean (ΔG) due to selection is a function of heritability (h^2) and the selection differential S

Table 6-3. Heritability estimates in percent.

Character	Johnson and Bernard (1963)	Anand and Torrie (1963)			Kwon and Torrie (1964)		Fehr and Weber (1968)†		Smith and Weber (1968)		Byth et al. (1969b)‡	
		Cross 1§	Cross 2	Cross 3	Cross 4	Cross 5	Cross 6	Cross 7	Cross 7	Cross 8	Cross 7	Cross 8
Yield	38¶	23	33	50	10	3	39	28	52	57	58	50
		(12)	(14)	(22)								
Seed weight	68	53	65	84	44	79	92	94	92	93	88	91
		(37)	(44)	(67)								
Height	75	82	84	73	70	67	66	85	82	80	90	86
		(94)	(62)	(55)								
Lodging	54	59	43	61	51	74	60	75	63	68	70	68
		(55)	(24)	(49)								
Days to flowering	84	91	65	87	75	76	—	—	—	—	—	—
		(162)	(51)	(98)								
Fruiting	65	67	46	48	66	81	—	—	—	—	—	—
		(95)	(17)	(27)								
Maturity	78	84	81	86	79	82	75	75	90	91	92	94
		(148)	(51)	(81)								
Protein	63	—	—	—	57	—	76	77	86	90	81	88
Oil	67	—	—	—	51	—	74	72	88	89	82	86

†Based on one environment only.
‡Averaged over maternal and daughter lines.
§Same numeral on cross designation indicates progenies were from crosses of the same parentage.
¶All estimates obtained from variance components except for those in parentheses which were obtained from regression of F_4 on F_3 and for those of Johnson and Bernard who presented expected heritabilities based on available data and observation of several soybean breeders and geneticists.

$$\Delta G = Sh^2$$

or alternatively

$$\Delta G = (S/\sigma_p)\sigma_p h^2 = k\ \sigma_p\ h^2$$

where k = the standardized selection differential and σ_p is the phenotypic standard deviation.

Dudley and Moll (1969) pointed out that each estimate of heritability in plant breeding relates specifically to the reference population of genotypes for which it was estimated. Also, each estimate relates only to the particular set of environments and experimental conditions in which the genotypes were tested. Thus, it is difficult to generalize heritability estimates from one population of genotypes to another, or from one set of experimental conditions (e.g., plot size) to another. When interpreting or comparing heritability estimates, it is of course necessary to note the unit on which they are based (e.g., plot mean, family mean, etc.).

In soybean breeding, most heritability estimates are made by evaluating a set of lines in one or more environments and then from an analysis of variance, estimating genotypic and phenotypic variances (Johnson et al., 1955a). Two other heritability estimation methods involve single plant evaluation: (i) estimation of genetic variance in a single environment by subtraction of nonsegregating generations (parents or F_1's) from segregating generations (F_2, F_3, etc.) (Powers, 1955) and (ii) parent-offspring regression (Falconer, 1960) or the "standard unit" modification suggested by Frey and Horner (1957). Cahaner and Hillel (1980) have discussed these two methods and their biases due to nonadditive variance. Bias of parent-offspring regression estimates due to genotype × environment interaction has been discussed by Casler (1982). Estimates of this type are of limited value for traits which are affected by intergenotypic competition (e.g., seed yield), because single plant performance is often not indicative of performance under the high plant density of economic cultural practice (Hanson, 1963). A fourth type of estimate is realized heritability, which is a narrow-sense estimate based on the ratio of selection response to selection differential, or the regression of selection response on cumulative selection differential (Falconer, 1960 p. 171–172).

Brim (1973) presented a representative sample of heritability estimates from eight populations, for nine quantitative traits that are commonly measured in soybean breeding populations (Table 6–3). Heritability was lowest (between 3 and 58%) for seed yield. With the exception of one population, all other traits had heritabilities > 50%.

In more recent experiments, researchers have investigated heritabilities of the percent and yield of oil and protein. Shannon et al. (1972) estimated heritability of yield, percent protein, and protein yield (yield × percent protein/100) in six populations of F_3 lines from crosses between two high and two low protein lines (Table 6–4). As in previous studies, the heritabilities for percent protein were higher than those for yield. Protein yield and seed yield heritabilities were similar. Predicted progress as a percentage of the population mean from selecting the highest 10%

Table 6-4. Heritability estimates in percent for seed protein, oil and yield.

Character	Shannon et al. (1972)†						Shorter et al. (1976)‡		Openshaw and Hadley (1984)§	
	$P_1 \times P_2$	$P_1 \times Y_1$	$P_1 \times Y_2$	$P_2 \times Y_1$	$P_2 \times Y_2$	$Y_1 \times Y_2$	Cross 9	Cross 10	Cross 11	Cross 12
Percentage protein	88	51	96	90	89	92	70(46)	86(56)	90	75
Percentage oil	—	—	—	—	—	—	84(66)	83(50)	93	71
(Protein + oil) percentage	—	—	—	—	—	—	—	74(39)	78	67
Yield	73	0	57	30	21	16	—	—	—	—
Protein yield kg ha^{-1}	74	3	58	1	17	23	55(39)	72(36)	—	—
(Protein + oil) kg ha^{-1}	—	—	—	—	—	—	60(42)	73(36)	—	—

†F_3 lines in the F_4 generation, two replications at two locations, P_1 and P_2 were homozygous high protein lines (43.3 and 44.1%, respectively), Y_1 and Y_2 were homozygous high yielding lines (>3200 kg ha^{-1}).

‡F_3 lines in the F_4 generation, two replications at one location, Cross 9 had parents with 41.6 and 40.7% protein, Cross 10 had parents with 45.9 and 40.7% protein. Standard unit heritability in parenthesis ($F_3 - F_4$ correlations).

§F_3 lines in the F_3 or F_4 generation, two replications at one location, Cross 11 had parents with 49.6 and 46.2% protein, Cross 12 had parents with 45.9 and 45.2% protein.

of each population (k=1.76) ranged from 4.7 to 3.3% for percent protein, 0 to 10.7% for yield and 0 to 10.7% for protein yield. In populations derived from a cross between two low protein lines and a cross between a high and low line, Shorter et al. (1976) found heritabilities for percent protein to be 70 and 86%, respectively. Heritabilities for protein yield were lower, 55 and 72% (Table 6–4). They also estimated heritabilities for percent oil and the sums of oil and protein percent and oil and protein yield. Predicted genetic advance as a percentage of the population mean was 3 and 5.8% for percent protein and 18.9 and 17.2% for protein yield. In two recurrent selection experiments, Brim and Burton (1979) calculated realized heritability estimates for percent protein of 29 and 34% over six cycles of selection. Response per cycle to selection was 0.7 and 1.6% of the base population mean. In a recurrent mass selection experiment, Burton and Brim (1981) estimated realized heritability for percent oil to be 21%.

Heritability has been determined for a variety of other traits. Several of these are presented in Table 6–5. The reference population, and the size of the experiment are also presented. Where tested, most of these heritability estimates accurately predicted selection response. Selection response in this context was usually the difference between performance of the population and performance of the selected lines in the selfing generation following the generation in which heritability was estimated.

6–4 CORRELATION AMONG TRAITS

Phenotypic correlation between two traits measured in a population, like phenotypic variance for a single trait, has genetic and environmental components. These components can be determined statistically from an analysis of covariance. Phenotypic and genotypic correlation coefficients can be calculated from the phenotypic and genotypic variances and covariances (Johnson et al., 1955b). In plant breeding, it is useful to know how traits are genetically correlated. When selection is being practiced on a single trait, it is important to know how other traits are being affected. For instance, Brim and Burton (1979) increased percent seed protein from 42.8 to 46.1% in five cycles of recurrent selection for high protein, but another result was a correlated decrease in percent oil from 19.5 to 17.5%. Genotypic correlations are also useful for simultaneous selection of more than one trait, either by index or by indirect selection. If the genetic correlation (r_A) between two traits (X and Y), and the heritabilities of the two traits (h^2_x and h^2_y) are known, then the correlated response of trait Y to selection on trait X ($\Delta_{GY \cdot X}$) is predicted by the following equation:

$$\Delta_{GY \cdot X} = k\sigma_{py}h_y h_x r_A$$

where k is the standardized selection differential and σ_{py} is the phenotypic standard deviation for trait Y (Falconer, 1960, p. 171–172). From this, it is evident that if $h_x r_A$ is larger than h_y, selection trait for X will result

Table 6-5. Heritability estimates and reference populations for various characters.

Character	Population	Heritability†	Reference population	Reference
Height of the first pod	1	52	F_3 lines in the F_3 from indeterminate by determinate two-line crosses	Martin and Wilcox (1973)§
	2	63		
	3	29		
Resistance to purple seed stain	4	51	F_4 lines in the F_4	Wilcox et al. (1975)§
Harvest index	5	82	41 cultivars of Maturity Groups 00 to IV	Buzzell and Buttery (1977)¶
Yield	5	68		
Percent seed protein	6	(29)	S_1 progenies from an intermating recurrent selection population	Brim and Burton (1979)§
	7	(34)		
Tolerance to A1	8	60	Random F_7 lines from a broad genetic based intermating population	Hanson and Kamprath (1979)§
	8	(50)		
Seed weight	9	71	F_3 lines from six two-way and three-way crosses	Bravo et al. (1980)¶
Pod width	9	92		
Seed weight	9	(32)		
Pod width	9	(69)		

(continued on next page)

Table 6-5. Continued.

Character	Population	Heritability†	Reference population	Reference
Yield in rows			40 F_5 and F_6 Maturity Group II lines	Weaver and Wilcox (1982)§
30 cm	10	53		
76 cm	10	47		
30 cm	11	54	40 F_5 and F_6 Maturity Group III lines	
76 cm	11	34		
Percent seed oil	12	(28)	Individual plants from an intermating recurrent mass selection population	Burton and Brim (1981)§
CAP‡	13	41	F_4 lines in the F_5 generation	Harrison et al. (1981)§
Yield	13	28		
CAP	14	65		
Yield	14	14		
Percent sugar in seeds	15	72	F_3 lines in the F_3 or F_4 generation	Openshaw and Hadley (1981)¶
	16	67		
Percent oleic acid in seed oil	17	(21)	Individual plants from an intermating recurrent mass selection population	Burton et al. (1983)¶

†All heritabilities were estimated from variance components except those in parentheses which are realized estimates.
‡Canopy apparent photosynthesis.
¶Based on data from one environment.
§Based on data from two environments.

Table 6-6. Estimates of genotypic and phenotypic (in parentheses) correlations of soybean yield with other characters.

Character	Johnson and Bernard (1963)†	Anand and Torrie (1963)‡			Kwon and Torrie (1964)§	
		Cross 1	Cross 2	Cross 3	Cross 4	Cross 5
Seed wt.	0.20	−0.27	0.02	−0.16	−0.59	0.22
	—	(0.03)	(−0.03)	(−0.07)	(−0.46)**	(0.20)
Height	0.30	0.65	0.57	0.43	0.82	0.54
	—	(0.41)**	(0.44)**	(0.32)**	(0.69)**	(0.44)**
Lodging	0.00	0.47	0.36	0.72	0.97	0.44
	—	(0.36)**	(0.07)	(0.36)**	(0.76)**	(0.27)*
Days to flower	0.00	0.76	0.26	0.45	0.87	0.69
	—	(0.37)**	(0.11)	(0.31)**	(0.68)**	(0.47)**
Fruiting period	0.20	0.71	−0.27	0.43	0.89	0.15
	—	(0.38)**	(−0.05)	(0.13)	(0.71)**	(0.13)
Maturity	0.40	1.05	0.01	0.47	0.95	0.52
	—	(0.48)**	(0.04)	(0.37)**	(0.75)**	(0.37)**
Protein	−0.20	—	—	—	—	−0.58
	—	—	—	—	—	(−0.42)**
Oil	0.10	—	—	—	—	0.07
	—	—	—	—	—	(0.05)

(continued on next page)

Table 6-6. continued.

	Byth et al. (1969b)¶				Byth et al. (1969a)††	Simpson and Wilcox (1983)‡‡			
	Cross 7		Cross 8						
	ML	DL	ML	DL		Cross 13	Cross 14	Cross 15	Cross 16
Seed wt.	0.07	0.15	−0.07	0.27	0.26	—	—	—	—
	(0.10)	(0.16)*	(−0.01)	(0.21)**	(0.21)	(0.00)	(0.21)*	(0.02)	(0.04)
Height	−0.28	−0.52	−0.15	−0.08	0.32	—	—	—	—
	(−0.13)	(−0.36)**	(−0.04)	(0.02)	(0.26)	(0.43)**	(0.37)**	(0.40)**	(0.35)**
Lodging	−0.14	−0.48	0.15	−0.17	−0.11	—	—	—	—
	(−0.21)*	(−0.41)**	(0.03)	(−0.20)**	(−0.26)	(0.45)**	(0.22)**	(0.36)**	(0.30)**
Days to flower	—	—	—	—	—	—	—	—	—
	—	—	—	—	—	—	—	—	—
Fruiting period	—	—	—	—	—	—	—	—	—
	—	—	—	—	—	—	—	—	—
Maturity	0.14	0.10	0.06	0.31	0.59	—	—	—	—
	(0.13)	(0.09)	(0.08)	(0.22)**	(0.37)	(0.54)**	(0.48)**	(0.51)**	(0.60)**
Protein	0.35	0.20	−0.55	−0.25	−0.23	0.54	−0.74	−0.40	−0.20
	(0.22)*	(0.13)	(−0.34)**	(−0.17)*	(−0.14)	—	—	—	—
Oil	−0.03	0.10	0.45	0.07	0.11	−0.22	0.20	0.25	−0.27
	(−0.01)	(0.09)	(0.26)*	(0.08)	(0.07)	—	—	—	—

*, **Exceeds the 5 and 1% levels of probability, respectively.
†Expected genotypic correlations based on data available to 1963.
‡Data based on two replications in two environments.
§Data based on two replications in five environments.
¶Data based on two replications in three environments. ML = maternal lines, DL = daughter lines.
††Data based on eight $F_3 - F_5$ populations grown in Iowa or North Carolina. Levels of significance not given.
‡‡Data based on two replications, one location, in 2 yrs.

in a greater change in trait Y then direct selection for Y. There have been two recent experimental examples of this. Bravo et al. (1980) showed that on an individual plant basis, selection for pod width was more effective for increasing seed size than direct selection for seed size. Harrison et al. (1981) found that indirect selection in one population for yield by selection for the correlated trait (r_A=0.72) canopy apparent photosynthesis (CAP), was 0.79 times less efficient than direct selection for yield, assuming no differences in economic costs of measuring the two. In a second population, however, which had a high genetic correlation (r_A=0.96) between yield and CAP, indirect selection was 1.46 times more efficient.

Examples like the above are rare. Usually direct selection is more efficient than indirect selection. Byth et al. (1969b) and Johnson et al. (1955b) found that indirect selection for yield using correlated agronomic traits was not as effective as direct selection for yield. Even so, indirect selection could be more efficient if the time required to complete one selection cycle were less than the time required to complete a cycle of direct selection. For instance, if one cycle of indirect selection required 1 yr while direct selection required two, then indirect selection response could be as little as one-half that of direct selection and still equal the yearly gain in yield.

Brim (1973) summarized phenotypic and genotypic correlations between yield and eight commonly measured traits (Table 6–6). The correlations were taken from five different studies of populations derived from two-line crosses and are representative of those found in soybean breeding literature. A sixth study (Simpson and Wilcox, 1983) which examined correlations in four high protein populations has been included in the table. It is evident from differences among the correlation coefficients from any particular pair of traits, that significance as well as direction of the correlations depended upon the population in which the traits were measured. With the exception of the populations investigated by Byth et al. (1969b), yield was positively correlated with plant height, lodging, days to maturity, and length of fruiting period. Correlations between yield and the seed traits, size and percent oil were generally low. Protein was negatively associated with yield in all but two of the populations investigated.

There has also been interest in correlations between yield and yield components and between yield and physiological traits. This interest has derived from the hope that a trait will be found with higher heritability than yield that can be used to indirectly select for yield. In two populations, Johnson et al (1955b) found that the genetic correlations between yield and pod number were 0.28 and 0.14. Correlations between yield and seed size were 0.66 and 0.43. Ecochard and Ravelomanantsoa (1982) in a segregating population of spaced plants found a genetic correlation of 0.95 between pod number and total seed yield and a correlation of 0.25 between seed size and total seed yield. This latter result is similar to other reports (Table 6–6). In a group of seven cultivars, Pandey and

Torrie (1973) found average correlations of 0.50 between yield and pods per unit area, 0.35 between yield and the number of seeds pod^{-1}, and 0.04 between yield and seed size.

Buzzell and Buttery (1977) found correlations of -0.44 and -0.19 between harvest index and yield in two populations grown in hill plots. As previously noted, Harrison et al. (1981) found a significant positive correlations between CAP and yield. The correlation, however, between photosynthesis and yield is apparently related to how and when photosynthesis is measured. Ford et al. (1983) found no significant correlation between yield and photosynthesis (rate of CO_2 uptake per unit leaf area), and Wiebold et al. (1981) were unsuccessful in selecting for increased carbon exchange rate (CER) in a population previously identified as being variable for CER. Buttery et al. (1981) found a positive relationship between photosynthetic rate per unit leaf area (P_A) and yield when measured 40 to 50 days after planting.

When the objective of selection is to alter seed composition, it is important to know how other seed components are likely to be affected. The negative relationship between percent seed protein and percent seed oil is well established (Hanson et al., 1961a; Shorter et al., 1976; Brim and Burton, 1979; Burton and Brim, 1981). Krober (1956) and Burton et al. (1982) have shown that percent protein in seeds and methionine content of the protein are not correlated. Openshaw and Hadley (1981) found that the correlation between percent oil and percent sugar in seeds was positive and significant, and that the correlation between percent protein and percent sugar was significant and negative. Correlations between sugar content and yield were nonsignificant. In a selection experiment, increases in the oleic acid fraction of soybean oil led to correlated decreases in the linoleic and linolenic acid fractions (Burton et al., 1983).

6–5 SELECTION

6–5.1 Pure Line Selection

Soybean cultivars are typically developed by hybridization of two or more lines followed by self-fertilization to the F_4 or later generation. Homozygous lines are isolated and tested to determine those with superior performance and cultivar potential. With this method, the major issue has been how material in the F_2, F_3, and F_4 segregating generations should be handled. Pedigree selection, which has been used in many breeding programs, involves visual selection of the best appearing families in each generation followed by within family selection of one or more plants for advancing the next generation. Brim (1966) suggested modified pedigree selection, i.e., single seed descent (SSD), as an economical way to develop homozygous lines and simultaneously preserve the genetic variation in the population. Bulk breeding has also been proposed in which the F_2 population is advanced in bulk by selfing with no selection until the F_5 to F_7 generations before lines are isolated for testing.

An important point to note in the discussion of these methods is that selection response or progress is obtained in homozygous lines derived by self-fertilization, without intermating, from the reference population. In any given population, expected response to selection depends, to a large degree, upon the level of inbreeding of the single plants from which lines are derived, and the generation in which they are tested. This may differ from the response in the outbred generation which results when the selected lines are intermated. Hanson et al. (1967) determined the expected response formula for selection among homozygous lines which had been derived from a random mating of homozygous lines. Using a more general genetic model that includes additive, dominance, and epistasis and any number of loci and alleles, Cockerham and Matzinger (1985) have developed formulations for the expected response to single seed descent selection both with and without the intermating of selected lines.

A major factor in the success of any of the aforementioned breeding methods is the choice of parents. If two homozygous parents of an F_2 population differ by an equal number of favorable independent loci controlling a multigenic trait (e.g., yield), then there is a high probability that a superior transgressive segregate can be found (Bailey and Comstock, 1976). If the two parents differ widely in the number of favorable alleles each carries, then it will be difficult to isolate a homozygous line with more favorable alleles than the better parent. Also, with no selection in early generations, there is a low probability that a line can be isolated that is homozygous for a large number of the favorable alleles (Sneep, 1977; Bailey, 1977). Assuming that 20 loci are segregating in an F_2 population, Bailey (1977) determined that < 2 in 10 000 inbred lines developed without selection could be expected to have as many as 18 loci homozygous for the favorable allele. Early generation selection, if successful, would increase the probability of isolating a superior genotype.

Recent simulation studies have helped to define this problem and put it into a useful perspective. Bailey and Comstock (1976) simulated selection in the selfing generations following a two-line cross with selection among F_2 plants followed by intraline selection in the F_3 and F_4 generations. Single seed descent was used to advance the lines to the F_8 generation. Their results showed that the probability of fixation of favorable alleles was increased by coupling linkages and decreased by repulsion linkages. Coupling linkages would be more prevalent in crosses between parents which differ widely in the number of favorable alleles they carry. Probabilities were also increased by selection and higher intraline heritabilities.

Curnow (1978) investigated the probability of allele fixation in a finite self-fertilizing population derived from a double heterozygous F_1. He assumed a two locus model with no linkage, equal genetic effects of the two loci, and only additive gene action. If a fixed number, 10, of high scores were selected each generation, the probability of fixing the most favorable alleles was 0.30 in generation 1. That value increased to 0.88

by generation 5 with little change thereafter. The probabilities were less if the genetic effects of the loci were less or if there was repulsion linkage between the two loci.

Casali and Tigchelaar (1975) simulated selection in an F_2 population of 400 individuals assuming 20 segregating loci and heritabilities of 75, 50, 25, or 10%. They compared SSD, bulk breeding, and pedigree breeding methods. At high heritabilities, 50 or greater, pedigree selection produced F_6 lines with the highest value. At low heritabilities (25 and 10), SSD was the most effective in producing superior F_6 lines.

Yonezawa and Yamagata (1981), by investigating the probabilities of obtaining desirable homozygous genotypes in later selfing generations, concluded that early generation selection of highly heritable traits followed by bulk breeding or SSD was more efficient than selection delayed until the F_5 or F_6. Brim (1973) warned that early generation selection for highly heritable traits, such as maturity, plant height, lodging, and disease resistance, could decrease the variability for yield in the lines that remained. Yonezawa and Yamagata (1981) recognized this problem and as a solution, suggested that extremely large F_2 populations be tested initially.

Snape and Riggs (1975) compared the F_2 distribution with the distribution of F_6 lines derived by SSD. They considered a trait controlled by 21 segregating loci with combinations of additive, complete dominance, and complementary and duplicate gene interactions. They demonstrated that transgressive segregates were produced in the F_6 generation in all cases. Even though the F_2 distributions differed greatly among the various genetic models, the F_6 distributions were similar. They concluded that given the expense and difficulty of early generation testing, SSD and selection among homozygous lines was the most efficient method.

With the exception of resistance to diseases and other traits which are qualitatively inherited, genotype × environment interaction make it difficult to identify useful selection criteria for single F_2 plants or F_3 rows in soybean breeding, particularly when the objective is higher yield. This was demonstrated in a comparison of early generation testing (EGT), pedigree selection (PS), and SSD made by Boerma and Cooper (1975). They simultaneously used the methods in four populations derived from two-way crosses. With the EGT method, yield tests were made in the F_3 and F_4 generations with limited plot size and replications. Visual selection was used in early generations of the PS method with a single two replicate test at the F_5 generation. The visual selection criteria were pod number and fullness, plant height, maturity, and overall vigor. There was no yield testing with the SSD method. They found no consistent differences among the populations between the three methods either in mean of selected lines or in mean yield of the five highest-yielding lines from each method. Based on these results, they recommended the SSD method since it was the least costly of the three.

Another related issue is the usefulness of intermating the F_2 and/or F_3 generation prior to selection of homozygous lines. In a simulation study, Pederson (1974) found that when a character is controlled by several loci that are distributed over three or more chromosomes, the genotypic variance and the frequency of desirable homozygous genotypes increased only slightly with intermating. He concluded that directional selection, even at low heritabilities, would be more effective. Likewise, Stam (1977) found that up through the F_4 or F_5 generations, selection response with selfing was equal to that with intermating, regardless of the number or linkage of loci.

6–5.2 Population Improvement—Single Trait Selection

Recurrent selection methods have had limited use with soybean and other autogamous species. This has been partially due to difficulties involved in random mating selected lines in each selection cycle by hand pollinations. Also, the breeding objective in autogamous species is usually a pure line cultivar, and recurrent selection is often perceived by breeders as a rather circuitous route toward such an objective. As previously discussed however, with multigenic traits there is a low probability of obtaining homozygous lines that have a large increase over the parents in the number of favorable alleles using pure line breeding methods. In this light, recurrent selection should be a useful way to increase the frequency of favorable alleles in a population and thereby increase the probability of isolating superior pure lines. Bailey and Comstock (1976) in advocating a cyclic program with intermating used the following probability argument. If the probability of fixation of favorable alleles for a pair of first cycle lines is $P=0.7$, then the probability that both lines will be homozygous for a favorable allele is 0.49, but the probability that both lines will be lacking is $(1\text{-}P)^2=0.09$. Generally, if n lines are chosen for intermating then the probability that none of the n lines is carrying the favorable allele is small, $(1\text{-}P)^n$.

Several recurrent selection schemes have been proposed for or used with soybean or other self-pollinators (Hanson et al., 1967; Compton, 1968; Matzinger and Wernsman, 1968; Kenworthy and Brim, 1979; Sumarno and Fehr, 1982). With the discovery of genetic male sterility in soybean (Brim and Young, 1971), it became possible to develop populations that were essentially allogamous. This allowed random mating by insect pollination and eliminated the need for hand pollination. Brim and Stuber (1973) described ways in which male sterility could be applied to recurrent selection schemes in soybean. Recurrent selection using male sterility for intermating has been used successfully in soybean to select for increased percent seed oil (Burton and Brim, 1981), higher percent oleic acid in seed oil (Burton et al., 1983) and yield (Koinange et al., 1981). When S_1 progeny selection is used in a population with genetic male sterility, male sterile plants segregate in the test population. This segregation obviously affects plot yields. Yield, however, is still heritable under these circumstances (Koinange et al., 1981). The heritability of yield in these populations can probably be increased by considering the number of sterile plants per plot as a covariate and adjusting plot means with an analysis of covariance (J. W. Burton, unpublished data).

In addition to developing a population with the necessary genetic variation and choosing a suitable selection method, another important concern is the size of the population to be tested and the proportion to be selected in each cycle. As Rawlings (1980) states, the problem lies in choosing a population size which is large enough to minimize the loss of favorable alleles due to random drift and yet small enough to maximize genetic gain within the constraints of available resources. Rawlings (1980) investigated this using the probability of fixation of an allele, $\mu(q)$, with initial gene frequency, q, which has the following form for an additive model:

$$\mu(q) = \frac{1 - e^{-2Nsq}}{1 - e^{-2N}}$$

where N is the effective population size and s is the selective advantage of the allele. The effective population size is approximately, but not necessarily equal to, the number of individuals selected in a cycle to serve as parents for the next generation of intermating. The selective advantage, s, is the product of the standardized selection differential and the genetic effect of the locus, a*. The a* is related to heritability, h^2, number of loci, m, and initial gene frequency as follows:

$$a^* = \frac{[2(h^2 m^{-1})]^{1/2}}{q(1 - q)}$$

Assuming a ratio of $h^2\ m^{-1}$ of 1/1000, the minimum effective population size needed for $\mu(q) \geqq 0.95$ decreased as q increased and decreased as the proportion selected decreased. If 10% of a population was selected, N was 19 when q=0.5 and 33 when q=0.25. Based on these results, he concluded that short-term selection progress could be made with effective populations of 20 to 30 without seriously limiting long-term selection gain.

Using similar methods, Bailey and Comstock (1976) determined how the probability of fixation [$\mu(q)$] is affected by effective population size in an initial F_2 population where q=0.5 at segregating loci and $h^2\ m^{-1}=1/250$. They found that for N=1, $\mu(q)=0.58$, for N=8, $\mu(q)=0.92$, and for N=16, $\mu(q)=0.99$. Thus, the probability of fixation would not be expected to increase much with an effective population size > 16.

Baker and Curnow (1969) used a genetic model in which a trait with heritability of 0.2 is affected by 150 loci where initial gene frequencies are 0.1, 0.2, and 0.3 at 50 each of the loci. Effects of genes at all loci were assumed to be equal and additive. Under these assumptions they concluded that with an effective population size of 16, the half-life of the selection process would be reached after 22 generations at which point 80% of the expected selection limit would have been realized. As effective population size increased over 16, many more generations are required to reach an equal percentage of the selection limit. With effective population size of 32, the half-life is reached after 44 generations, but only 68% of the selection limit would have been realized. For the genetic model, they estimated relative selections limits for effective populations of 16 and 32, to be 114 and 177. They showed that for the first 10 generations, expected selection progress for effective population sizes of 16 and 32 were similar. Brim and Burton (1979) obtained empirical results similar to these. In selection for higher percent seed protein, they found that the rate of progress when 12 lines were saved each generation was similar to the rate of progress when 45 lines were saved.

These results demonstrate that recurrent selection programs can be designed that are within the resources of most breeding projects. Populations need not be large to realize genetic progress. Baker and Curnow (1969) recommended short-term selection within several small populations, followed by selection and recombination of those most productive.

The foregoing arguments assume that adequate levels of genetic variability exist in the initial breeding population. In soybean, some questions about this point arise when one considers that most varietal development populations originated from a relatively small number of plant introductions. Looking at Maturity Groups 00 to IV, Specht and Williams (1984) identified only 41 different plant introductions in the pedigrees of 136 cultivars released between 1939 and 1981. Of the 57 cultivars released since 1970, 55% of the germplasm was contributed by four parents and an additional 35% was contributed by eight other parents. Delannay et al. (1983) has shown that only seven plant introductions contributed 82% of the germplasm for cultivars released between 1971 and 1981 in Maturity Groups V to IX. It is not known how genetically diverse the initial parents were, but considering the number of them alone the genetic base for current cultivar development appears to be somewhat narrow.

Soybean cultivar development in this century has been likened to a recurrent selection program which is currently in its third or fourth cycle (Luedders, 1977). Using coancestry measures, St. Martin (1982) found that 27 cultivars released between 1976 and 1980 (the fourth cycle selections) had an average inbreeding coefficient of 0.25. Using the formula, $F_t = 1/N + [(N\text{-}1)/N]F_{t\text{-}1}$ (Hanson et al., 1967), where N is the population size and t is the cycle of selection, he determined the effective population size was 15 assuming $t = 4$ cycles of selection and 11 assuming $t = 3$. These results suggest that some genetic variability has been lost from the original breeding populations, but there is no way of knowing how much or how much of that loss was due to favorable genes. However, it is reasonable to conclude that long-term selection progress in soybean cultivars will be limited by the small effective population size.

6–5.3 Index Selection

An agronomically acceptable soybean cultivar must meet a number of standards in addition to having good-yielding ability. These include resistance to shattering, lodging, diseases and stress, suitable height for mechanical harvest, and industrially acceptable seed composition. This multiplicity of breeding goals poses a problem because as Brim (1973) noted, advantage obtained from superiority in one trait must often be balanced against inferiority in other traits. Gardner (1977) and Comstock (1977) have discussed the need for simultaneous selection of several traits in order to improve the total phenotypic value and to prevent undesirable correlated responses to selection.

Three methods of simultaneous selection currently being used are independent culling, tandem selection, and index selection. Independent culling is probably the most widely used. In theory however, index selection is usually the most efficient. Young (1961) has shown that the superiority of index selection increases as the number of traits under selection increases but decreases as heritability increases. Index superiority also decreases as differences in relative importance of the traits increase. Pesek and Baker (1969) compared index selection of two negatively correlated traits in the F_6 and F_7 generations to tandem selection in those generations. Different heritabilities, environmental variances, and economic values of two traits were tested. Index selection was more efficient for all parameters, particularly at low heritabilities.

Even though theoretically superior, index selection has had limited use in plant breeding. One reason for this is the difficulty of assigning relative weights to the traits under selection. For instance, because of market fluctuations from season to season, it is not possible to know the relative economic values of soybean oil and protein. Another impediment is the difficulty of obtaining reliable estimates of genotypic and phenotypic variances and covariances. A comprehensive summary of index selection theory and procedures has been provided by Lin (1978) along with discussions of some of their limitations.

In soybean, the problem of assigning economic weights has often been solved by giving yield a weight of 1 and all other traits a weight of zero. This utilizes direct and indirect selection together. Essentially yield estimation is improved by adusting for variation in the other traits. Caldwell and Weber (1965) and Byth et al. (1969a) used this type of index, with yield as the primary trait and combinations of maturity, lodging, height, seed size, protein, and oil as secondary traits. Both applied the indexes to segregating generations of two-line crosses. In all cases, selection for yield with the index was either no better or only slightly better than selection for yield alone. Caldwell and Weber (1965) concluded that only highly heritable traits closely associated with yield should be used as covariates. Byth et al. (1969a) concluded that as long as precise estimates of yield could be made, selection for yield alone was equal to selection with the index. Pritchard et al. (1973) used the same kind of indexes involving combinations of yield (weight=1) and yield components (weights=0). They applied the index to the F_4 and F_5 generations of two line crosses. They found all indexes resulted in greater progress in both populations than selection for yield alone. However, index selection in the F_4 generation was only 5% more efficient. In the F_5 generation, there was a large increase in the efficiency of index selection due to a lower heritability for yield. They concluded, like Byth et al. (1969a), that inclusion of other traits in the index is helpful when heritability for yield is low due to large genotype × environment interactions.

Brim et al. (1959) used an index involving oil yield, protein yield, lodging resistance, seed weight, and fruiting period in two populations of random F_3 lines in the F_4 generation. Oil and protein yields were given relative economic weights of 1:1 1:0.6, and 1:0.2, all other traits were given economic weights of zero. In one population, differences in economic weight ratios did not change the expected genetic progress, while they had a marked effect in the other. In one population the index which inclued oil yield gave the greatest expected response, while in the other population the indexes which included seed weight were best. Population parameters are not used to derive this kind of index, which avoids index unreliability due to parameter estimation errors. However, the authors noted the difficulties in trying to pick realistic economic weights.

The problem of economic weighting can be circumvented by using indexes which arbitrarily restrict the response of one or more trait. Kempthorne and Nordskog (1959) proposed a restricted index which changed one trait while holding another constant. Miller and Fehr (1979) successfully used this type of index to increase percent protein and prevent a change in maturity. Such an index was also used by Matzinger and Wernsman (1980) with tobacco (*Nicotiana tabacum* L.) to increase yield while keeping alkaloids constant. Pesek and Baker (1970) proposed that restrictions, "desired genetic gains," be applied to all traits in the index. They applied two such indexes to a set of F_8 lines from a cross between two wheat (*Triticum aestivum* L.) cultivars. Genetic variance and covariances were determined for the traits days to head, days to ripening, height, and yield. Desired gains were based on percentage increases or decreases in the four traits. In the results, there was not a close agreement between expected gain and realized gain. This disparity was thought to be due to random error rather than a poor index. A drawback to the use of these indexes is that the gain in all traits can be lessened if a specified desired gain is unreasonable. In this regard, industrial requirements for crop quality from planting through harvest and processing should be useful guidelines in determining reasonable desired gains. Other restricted indexes have been developed by Cunningham et al. (1970) and James (1968).

If applied in early segregating generations of two-line crosses, indexes might help to make early generation selection more effective. Weber (1982) investigated the use of family indexes which incorporate information from relatives. For example, to select among F_4 families an index would be developed which would include information from the F_2 and/or F_3 generations. Weights for the family data are determined by maximizing the correlation between the index and the expected genotypic value of homozygous lines sampled from the selected families. Weber demonstrated with simulation that the family indexes maximized the gain from selection. Use of an index should also be quite effective in recurrent selection for or against more than one trait. Matzinger et al. (1977) successfully used an index in four cycles of recurrent mass selection with tobacco to decrease plant height and increase the number of leaves per plant.

6–5.4 Conclusion

Soybean breeding in the USA, from a historical perspective, can be viewed as a process of cyclical or recurrent selection in which superior cultivars are selected and released, then recombined and reselected. Luedders (1977) identified four cycles in the selection of Groups I, II, III, and IV cultivars between 1933 and 1971. He simultaneously tested sets of cultivars from each cycle and found a total increase of 50% or about 1% yr^{-1}. In a similar experiment, Wilcox et al. (1979) found the rate of progress for Maturity Group II and III cultivars released between 1923 and 1974 to be 0.6% yr^{-1}. Specht and Williams (1984) tested 240 cultivars in Maturity Groups 00 to IV which had been released between 1902 and 1977. They found that the average annual genetic gain in yield considering all releases was approximately 18.8 kg ha^{-1}. Considering only cultivars released since 1943 and excluding those which were forage, vegetable, or backcross derived, they found yield increases ranging from 8% in Maturity Group III to 34% in Maturity Group II with an overall average gain of 14%. Similarily, Boerma (1979) investigated the genetic improvement of Maturity Groups VI, VII, and VIII in the southeastern states between 1914 and 1973. He showed the rate of progress to be 0.7% yr^{-1}. It is evident from these results that genetic improvement in soybean, though substantial, has proceeded at a slow pace.

Probability arguments and evidence from simulation studies suggest that cyclical procedures of early generation selection followed by recombination should increase the frequency of favorable alleles in populations and result in more rapid rates of progress. This is supported by results from recurrent selection experiments. Kenworthy and Brim (1979) and Koinange et al. (1981) using recurrent selection with S_1 progeny testing showed increases in yield of 2.7 and 2.3% yr^{-1}, respectively. Koinange (1981) found that yield of the population derived in the third cycle of selection was 7.6% greater than the midparent and 4.5% greater than the highest parent.

Sumarno and Fehr (1982) subdivided a population derived from high-yielding lines of Maturity Groups 0 to IV into early, midseason, and late subpopulations. Three cycles of recurrent selection with F_5 or F_6 line testing resulted in yield increases of 2.1, 0, and 0.4% yr^{-1} in the early, midseason, and late populations, respectively.

Choice of the initial parents for any selection program is critical to its success. Methods need to be developed for better identification of superior parental genotypes. In the absence of other criteria, genetically unrelated parents of nearly equal productivity would probably contribute a large number of favorable alleles to a base population. Some type of index selection should be useful if multiple traits are to be changed.

6–6 GENOTYPE × ENVIRONMENT INTERACTION

In any plant population, variation in the expression of a quantitative trait is due to genetic and environmental variability and an interaction between the two. Variation due to genotype by environment interaction (G×E) is troublesome to plant breeders if it stems from differences in the ranking of genotypes among environments. Large interactions of this type decrease the heritability of a trait and make genotypic evaluation difficult. Given that such interactions occur, the breeder is faced with two basic questions, which environments should be used for testing and how many are necessary for adequate genotypic evaluation. The two questions are linked because often the number of necessary environments is dependent upon the kinds of environments chosen.

In defining issues related to G×E in plant breeding, it is helpful to consider environmental variation in a continuum from predictable to unpredictable (Allard and Bradshaw, 1964). Predictable variation is due to those conditions which are controlled in some way (greenhouses, growth chambers, irrigated) or those which have permanent characteristics (photoperiod, soil type, and endemic pathogen). Weather-related conditions generally contribute the most to unpredictable variation. If particular environments can be defined, then it may be possible to develop cultivars which are adapted specifically for that environment. Carter and Boerma (1979) and Boerma et al. (1982) have proposed that cultivars be developed specifically for late planting in the southeastern states. Cooper (1981) has developed cultivars specifically for definable high yield environments. Actually, disease or pest resistant cultivars are developed for use in particular pest or disease environments although such environments are sometimes unpredictable. To deal with unpredictable environments, emphasis is placed on the development of cultivars which are stable, i.e., perform well in a wide range of environments. Most soybean breeding programs have regional testing programs for evaluation of genotypes across environments and analyses have been developed to determine the relative stability of cultivars (Finlay and Wilkinson, 1963; Eberhart and Russell, 1966; Perkins and Jinks, 1968; Hanson, 1970; Shukla, 1972). In this sense, disease and pest resistance can be viewed as contributing to the stability of a cultivar. Genetically heterogeneous cultivars (blends, F_4 lines, etc.) have also been used to achieve stability.

Decisions concerning the choice of an appropriate breeding method for a particular objective require a knowledge of how the genotypic and environmental populations interact. In soybean G×E interaction has been investigated with two types of studies. In one type, direct estimates of G×E have been made with variance components from variance analyses of genotypic performances in several environments. In the other type of study, interactions between specific genotypes and environments have been analyzed using stability regression procedures. A summary of results and conclusions from some of these studies follows.

6–6.1 Estimates of Genotype × Environment Interaction

Two populations of F_3 lines in the F_4 and F_5 generations were evaluated by Johnson et al. (1955a or 1955b) at two or three locations in 2 yrs. For yield in population 1, the sum of the interaction components of variance was 2.1 times larger than the genotypic component, while in population 2, the genotypic component was 8.3 times larger. This difference in G×E interaction of the two populations could have been due either to genetic differences between the two or to the different sets of environment in which the two were grown. With three other traits measured, height, seed weight, and percent oil, both populations had genotypic variance components which were larger than the sum of the interaction components. The authors showed that genetic variability indicated by one location in 1 yr was reduced 71% when average performance over locations and years was considered, thus illustrating the need for tests in multiple environments.

Schutz and Bernard (1967) estimated the interaction variance for the Uniform Soybean Tests, Maturity Groups 0 to IV, VI and VII, in the years 1954 to 1956. Their results showed that the sum of interaction components for yield was as large or larger than the variance among lines. For protein, oil, lodging, maturity, height, and seed weight, the interaction components were much less. This is consistent with the lower heritabilities of yield relative to the other traits. The line × year variance components were as large as the line × location components and one-fifth as large as the three-way interaction (line × location × year). The three-way interaction was larger than either of the two-way interactions in all traits and tests. This was interpreted to mean that each year-location combination could be considered a separate environment and that locations could be substituted for years in the regional tests. In addition, they found that testing yield in more than 20 environments did not result in much least significant difference (LSD) reduction and that 10 environments were adequate for measuring the other traits.

Kwon and Torrie (1964) evaluated F_3, F_4, and F_5 generations of lines from two crosses. In both populations, the variance component for lines was greater than the interaction components for all traits except yield. Estimates of the line × year variance component were larger than either the line × location or line × location × year components for yield, seed weight, lodging, days to flowering, and percent oil. The three-way interaction was more similar in size to the line × year interaction than the line × location for all traits except flowering, fruiting period, and percent protein. The large line × year effect was thought to be a result of seasonal differences in rainfall and temperature.

Garland and Fehr (1981) investigated G×E interactions in hill and row plots using 50 lines chosen from a population with a broader genetic base than in the previous studies. The lines were tested at five locations in 2 yrs, and G×E variance components were estimated for yield, maturity, height, lodging, and phenotypic score. When the lines were evaluated in row plots, the line × year × location variance component was larger than either of the two-way interactions for all five traits except yield. With yield, the line × year and line × year × location components were equal. The line × location components for maturity, height, and lodging were zero. For all traits, the sum of the G×E components was less than the genotypic variance components. This was also true when the same lines were tested in hill plots.

Erikson et al. (1982) determined percent protein for 115 F_3 lines from crosses between *G. max* and *G. soja* in 2 yrs and two locations. As in the previous studies the line × location × year component was found to be significant and larger than either of the two-way interactions, but less than the variance component due to lines.

Hanson (1964) showed that considering *l* for locations and *y* for years as *yl* random environments would give biased estimates of genotypic and G×E variances unless the intraclass correlations

$$\rho_y = \sigma^2_{gy}/(\sigma^2_{gy} + \sigma^2_{gyl} + \sigma^2_{gy} + \sigma^2_{gl}) \text{ and } \rho_l = \sigma^2_{gl}/(\sigma^2_{gyl} + \sigma^2_{gy} + \sigma^2_{gl})$$

were zero. ρ_y and ρ_l were defined as the correlation of G×E effects within and among genotype-year classes and genotype-location classes, respectively. His analysis indicated that the bias was small as long as the sum of ρ_y and ρ_l was < 0.4. It is apparent from the above that intraclass correlations will be small if σ^2_{gy} and σ^2_{gl} are small relative to σ^2_{gyl}. Therefore, considering the published estimates of G×E variance components previously discussed, it is reasonable in most cases to consider *y* years and *l* locations as *yl* random environments.

In order to determine the extent to which G×E interaction is associated with performance in preliminary yield tests, Baihaki et al. (1976) evaluated 44 Maturity Group 0 and I lines and four cultivars at three locations in each of 2 yrs. Based on mean performance, the lines were placed in high-, medium-, and low-yielding groups. They found that the low-yielding group contributed 50.6% to the total line × location × year interaction. The high and medium groups contributed 23.7 and 25.1%, respectively. They concluded that testing in one environment would be adequate if the lower yielders always contribute more to the G×E interaction. Ten of the 16 highest-yielding lines were chosen by selection of the top one-third in one environment.

Genetics heterogeneous materials have generally been found to reduce variances due to G×E interactions. Schutz and Brim (1971) tested the yield of two and three line mixtures of four cultivars along with the cultivars in four locations and 2 yrs. They found that the mixtures had smaller interaction variance components than the cultivars. The line × location × year variance component was again larger than either the line

× location or line × year interaction components. Walker and Fehr (1978) obtained similar results in tests of pure lines and line mixtures.

Byth and Weber (1968) developed two populations of F_3 lines advanced in bulk to the F_7 generation (termed *maternal*). Within each population, single plants were sampled in the F_5 generation to get F_6 lines (termed *daughter*) from particular maternal lines. These lines were tested in three environments. There was no difference between the mean performance of the maternal and daughter lines; but the G×E components of variance were greater among the less heterogeneous daughter lines for all traits except percent oil. Also, the G×E variance within F_6 daughter pairs was greater than that among daughter pairs indicating that residual heterozygosity in F_5 lines may result in increased stability. Byth and Weber pointed out that environmentally stable genetic systems reduce bias in selection due to specific environmental influence and may require less testing. They cautioned that lower genetic variation within an environment due to heterogeneity might result in reduced genetic advance.

6–6.2 Analysis of Stability

Analyses developed by Finlay and Wilkinson (1963), Eberhart and Russell (1966) and Perkins and Jinks (1968) define G×E interaction as a linear function of environmental effects. Linear regression of individual genotype performance in each environment on the mean performance of all genotypes in each environment is used to measure stability. Eberhart and Russell (1966) suggested the use of deviations from linear regression as an additional stability measure. They defined a stable genotype as one with a linear regression coefficient of one and zero deviations from the regression. Perkins and Jinks (1968), using a different model, developed similar stability parameters. Several studies have been conducted with soybean using these stability analyses.

Smith et al. (1967) evaluated yield stability of 19 pure lines in the 1962 and 1963 Uniform I and II tests. Lines with highest mean yields generally had regression coefficients greater than 1.0 and the largest deviations from regression. The stability of random sets of heterogeneous homozygous F_3 lines was compared with the stability of homozygous F_6 lines from the same populations. The heterogeneous lines showed more stability by having lower deviations from regression. There was a significant positive correlation between linear regression coefficients of daughter and maternal lines indicating that the stability parameter is heritable. However, there was an extremely low correlation between deviations from regression for the two.

In stability analyses of cultivar mixtures and individual cultivars, Schutz and Brim (1971) found no differences in regression coefficients. None were significantly different from 1.0. However, lack of stability was indicated by significant deviations from regression for all cultivars and 8 of 10 mixtures. The deviations for the individual cultivars were greater than those for mixtures indicating the mixtures might have been more stable.

Walker and Fehr (1978) tested the stability of 28 cultivars and experimental lines individually and as 2, 10, 12, and 14 component mixtures in 12 environments. Regression coefficients were not heterogeneous. They ranged from 0.82 to 1.16 among all entries and were not significantly different from 1.0. There was a decrease in the deviation mean squares with an increase in the number of components in a mixture up to eight components. Thereafter, deviation mean squares were not different. Only 12 of the 80 entries in the experiment had significant deviation mean squares. Differences between the mixture yields and the multiple pure stands of the mixture components were nonsignificant. Deviation mean squares for the multiple pure stands were greater than those for mixtures indicating that they were less stable.

Wilcox et al. (1979) evaluated in 12 environments Maturity Groups II and III cultivars that had been released between 1923 and 1974. The newer lines in each group were descendants of the older lines. They found that yield potential had increased 25% with selection and that stability of the lines had not changed.

Funnah and Mak (1980) evaluated the yield potential of 20 genotypes in 12 Malaysian environments. Under the Perkins and Jinks (1968) regression model, 4 of the 20 had significant regression coefficients. They also estimated the stability variance (Shukla, 1972) portion of the G×E variance contributed by each genotype. With this parameter, 11 of 20 were responsible for a significant contribution to the G×E variance and were, therefore, classified as unstable. With another method proposed by Francis and Kannenburg (1978), the genotypes were grouped by plotting entry means against their CVs. Seven were stable by this criterion. Three were stable by all three criteria.

Beaver and Johnson (1981) tested 19 indeterminate and determinate cultivars in Maturity Groups II, III, and IV at eight Illinois environments. Stability analyses were also performed on yield data from the 1977, 1978, and 1979 USDA Uniform Maturity Group III tests. In all analyses, regressions of cultivar performance on environmental means showed that most of the variation attributable to G×E interaction was due to differences among the fitted lines rather than deviations from regression. Repeatability estimates across analyses were 0.48 for the regression coefficients and 0.61 for the deviations. This result is an indication that the response of a cultivar to environmental variation is a heritable characteristic.

Mungomery et al. (1974) pointed out that in regression approaches to the analysis of genotypic stability, there is an a priori assumption that the response to environmental variation is either linear or curvilinear. Since genotypic response is often not linear, they proposed using a pattern analysis to group cultivars according to their response over environments. To test the procedure, 58 soybean lines and cultivars were evaluated for yield at four locations in 2 yrs. An analysis of variance showed significant genetic variation among lines and also highly significant line × location and line × location × year interactions. A pattern analysis procedure was used to divide the lines into 10 groups. Lines were clustered by minimizing the Euclidean distance between pairs of lines and groups. The procedure successfully minimized the variance within groups and maximized among group variance. It was noted that the groups showed differences in productivity and also differences in origin of the lines. Mungomery et al. (1974) speculated that group membership might be a uselful criterion for choosing parents of a population, and thought it should be possible to predict average environmental response of progeny based on parental response patterns. Also, once a commercially acceptable response pattern is identified it might be possible to select genotypes using that pattern as a selection criterion. In cases where there is a large number of test environments, Byth et al. (1976) showed that the pattern analysis could be extended to produce groupings of environments as well as genotypes.

6–6.3 SUMMARY

The question of which environments to use in a testing program must be answered by every plant breeder. The answer will depend upon available economic and environmental resources, breeding objectives, and the nature of expected G×E interaction. Generally, significant line × year interactions suggest that unpredictable environments contribute most to the variability. Faced with this, the breeder usually tests in as many different environments as possible. Another strategy is to reduce variability with environmental management practices such as irrigation and fertilization. Cowley et al. (1981) showed that heritability estimates for yield in F_3 bulk soybean populations were higher in irrigated than in nonirrigated environments. They argued that genetic potential was more fully expressed in nonstress environments and would, therefore, be the most suitable for genotypic evaluation. Falconer (1952) stated the problem as follows: Will best results be achieved when selection is carried out under the conditions in which the organism will be living or under those which allow the greatest expression of a character? Like Cowley et al. (1981), he argued that expectation of greater heritability would favor selection in an environment other than the one in which the organism will live. He suggested that performance in two different environments be treated as two genetically correlated traits. Then index selection could be used to optimize genotypic performance in both, provided appropriate weights could be determined.

As previously described, several stability analyses have been proposed and tested; however, the information has generally not been put to practical use. Stability analyses are not routinely performed in most breeding programs. Three factors probably contribute to this. One is that a large enough data base may not be available on any given set of lines. Another is the uncertainty about which stability analysis to use and then how to interpret the analysis results. Finally, high mean yield over a wide range of environments is generally perceived to be an adequate indicator of line stability. However, stability analyses should be helpful in identifying high-yielding genotypes that respond differently to stressful environments. Hanson (1970) has suggested that stability parameters be calculated annually for the elite lines in regional yield tests. Such a practice is well within the computing capabilities of most soybean breeding programs and would give geneticists another criterion for genotypic evaluation. An analysis could be agreed upon by the plant breeders involved and experience with the analysis results should eventually lead to a useful interpretation of the parameters.

6–7 CONCLUSION

Soybean breeding over the past 40 yrs has produced cultivars with greater genetic yield potential that are adapted to modern cultural practices. Breeding has also protected the crop by incorporating disease and pest resistance into cultivars. Seed quality has also been improved. However, as Brim (1973) noted, "past success does not necessarily provide conclusive proof of the efficiency of present breeding procedures." Evaluating breeding methods within a quantitative genetic context provides a way to compare the efficiencies of different procedures and determine the likelihood that a new procedure will be successful. Future research should continue to focus on ways to improve method efficiency and ways to increase the rate of improvement. These include the following: (i) the development of breeding populations, taking into account the genetic origin of the parents, their overall phenotype, and their performance in diverse environments; (ii) the development of single and multiple trait selection schemes which incorporate a more rapid cycling of elite line identification, selection and recombination; (iii) the investigation of the relative importance of dominance and epistasis, particularly as they affect heterosis; (iv) the development of ways to manage genotype × environment interactions so that heritabilities are increased; and (v) the allocation of resources with respect to preliminary vs. advance testing.

REFERENCES

Allard, R.W., and A.D. Bradshaw. 1964. Implications of genotype-environmental interactions in applied plant breeding. Crop Sci. 4:503–508.

Anand, S.C., and J.H. Torrie. 1963. Heritability of yield and other traits and interrelationships among traits in the F_3 and F_4 generations of three soybean crosses. Crop Sci. 3:508–511.

Baihaki A., R.E. Stucker, and J.W. Lambert. 1976. Associations of genotype × environment interactions with performance level of soybean lines in preliminary yield tests. Crop Sci. 16:718–721.

Bailey, T.B., Jr. 1977. Section limits in self-fertilizing populations following the cross of homozygous lines. p. 399–412. *In* E. Pollack et al. (ed.) Proceedings of the international conference on quantitative genetics. Iowa State University Press, Ames.

----, and R.E. Comstock. 1976. Linkage and the synthesis of better genotypes in self-fertilizing species. Crop Sci. 16:363–370.

----, Jr., C.O. Qualset, and D.F. Cox. 1980. Predicting heterosis in wheat. Crop Sci. 20:339–342.

Baker, L.H., and R.N. Curnow. 1969. Choice of population size and use of variation between replicate populations in plant breeding selection programs. Crop Sci. 9:555–560.

Baker, R.J. 1978. Issues in diallel analysis. Crop Sci. 18:533–536.

Beaver, J.S., and R.R. Johnson. 1981. Yield stability of determinate and indeterminate soybeans adapted to the northern United States. Crop Sci. 21:449–454.

Boerma, H.R. 1979. Comparison of past and recently developed soybean cultivars in maturity groups VI, VII, and VIII. Crop Sci. 19:611–613.

----, and R.L. Cooper. 1975. Comparison of three selection procedures for yield in soybeans. Crop Sci. 15:225–229.

----, E.D. Wood, and G.B. Barrett. 1982. Registration of duocrop soybean. Crop Sci. 22:448.

Bravo, J.A., W.R. Fehr, and S.R. de Cianzio. 1980. Use of pod width for indirect selection of seed weight in soybeans. Crop Sci. 20:507–510.

Brim, C.A. 1966. A modified pedigree method of selection in soybeans. Crop Sci. 6:220.

----. 1973. Quantitative genetics and breeding. *In* B.E. Caldwell (ed.) Soybeans: Improvement, production, and uses. Agronomy 16:155–186.

----, and J.W. Burton. 1979. Recurrent selection in soybeans. II. Selection for increased percent protein in seeds. Crop Sci. 19:494–498.

----, and C.C. Cockerham. 1961. Inheritance of quantitative characters in soybeans. Crop Sci. 1:187–190.

----, H.W. Johnson, and C.C. Cockerham. 1959. Multiple selection criteria in soybeans. Agron. J. 51:42–46.

----, and C.W. Stuber. 1973. Application of genetic male sterility to recurrent selection schemes in soybeans. Crop Sci. 13:528–530.

----, and M.F. Young. 1971. Inheritance of a male-sterile character in soybeans. Crop Sci. 11:564–566.

Burton, J.W., and C.A. Brim. 1981. Recurrent selection in soybeans. III. Selection for increased percent oil in seeds. Crop Sci. 21:31–34.

----, and T.E. Carter, Jr. 1983. A method for production of experimental quantities of hybrid soybean seed. Crop Sci. 23:388–390.

----, A.E. Purcell, and W.M. Walter, Jr. 1982. Methionine concentration in soybean protein from populations selected for increased percent protein. Crop Sci. 22:430–432.

----, R.F. Wilson, and C.A. Brim. 1983. Recurrent selection in soybeans. IV. Selection for increased oleic acid percentage in seed oil. Crop Sci. 23:744–747.

Buttery, B.R., R.I. Buzzell, and W.I. Findlay. 1981. Relationships among photosynthetic rate, bean yield and other characters in field-grown cultivars of soybean. Can. J. Plant Sci. 61:191–198.

Buzzell, R.I., and B.R. Buttery. 1977. Soybean harvest index in hill-plots. Crop Sci. 17:968–970.

Byth, D.E., B.E. Caldwell, and C.R. Weber. 1969a. Specific and non-specific index selection in soybeans, *Glycine max* L. (Merrill). Crop Sci. 9:702–705.

----, R.L. Eisemann, and I.H. de Lacy. 1976. Two-way pattern analysis of a large data set to evaluate genotypic adaptation. Heredity 37:215–230.

----, and C.R. Weber. 1968. Effects of genetic heterogenity within two soybean populations. I. Variability within environments and stability across environments. Crop Sci. 8:44–47.

----, ----, and B.E. Caldwell. 1969b. Correlated truncation selection for yield in soybeans. Crop Sci. 9:699–702.

Cahaner, A., and J. Hillel. 1980. Estimating heritability and genetic correlation between traits from generations F_2 and F_3 of self-fertilizing species: a comparison of three methods. Theor. Appl. Genet. 58:33–38.

Caldwell, B.E., and C.R. Weber. 1965. General, average, and specific selection indices for yield in F_4 and F_5 soybean populations. Crop Sci. 5:223–226.

Carter, T.E., Jr., and H.R. Boerma. 1979. Implications of genotype × planting date and row spacing interactions in double-cropped soybean cultivar development. Crop Sci. 19:607–610.

Casali, V.W.D., and E.C. Tigchelaar. 1975. Computer simulation studies comparing pedigree, bulk, and single seed descent selection in self-pollinated populations. J. Am. Soc. Hortic. Sci. 100:364–367.

Casler, M.D. 1982. Genotype × environment interaction bias to parent-offspring regression heritability estimates. Crop Sci. 22:540–542.

Cockerham, C.C. 1963. Estimation of genetic variances. p. 53–94. *In* W.D. Hanson and H.F. Robinson (ed.) Statistical genetics and plant breeding. Pub. 982. National Academy of Sciences-National Researches Council, Washington, DC.

----. 1983. Covariances of relatives from self-fertilization. Crop Sci. 23:1177–1180.

----, and D.F. Matzinger. 1985. Selection response based on selfed progenies. Crop Sci. 25:483–488.

Compton, W.A. 1968. Recurrent selection in self-pollinated crops without extensive crossing. Crop Sci. 8:773.

----. 1977. Heterosis and additive × additive epistasis. Soybean Genet. Newsl. 4:60–62.

Comstock, R.E. 1977. Quantitative genetics and the design of breeding programs. p. 705–718. *In* E. Pollak et al. (ed.) Proceedings of the international conference on quantitative genetics. Iowa State University Press, Ames.

Cooper, R.L. 1981. Development of short-statured soybean cultivars. Crop Sci. 21:127–131.

Cowley, C.R., C.D. Nickell, and A.D. Dayton. 1981. Heritability and interrelationships of chemical and agronomic traits of soybeans (*Glycine max* (L.) Merr.) in diverse environments. Trans. Kansas Acad. Sci. 84:1–14.

Croissant, G.L., and J.H. Torrie. 1971. Evidence of nonadditive effects and linkage in two hybrid populations of soybeans. Crop Sci. 11:675–677.

Cunningham, E.P., R.A. Moen, and T. Gjedrem. 1970. Restriction of selection indexes. Biometrics 26:67–74.

Curnow, R.N. 1978. Selection within self-fertilizing populations. Biometrics 34:603–610.

Delannay, X., D.M. Rodgers, and R.G. Palmer. 1983. Relative genetic contributions among ancestral lines to North American soybean cultivars. Crop Sci. 23:944–949.

Dudley, J.W., and R.H. Moll. 1969. Interpretation and use of estimates of heritability and genetic variances in plant breeding. Crop Sci. 9:257–262.

Eberhart, S.A., and W.A. Russel. 1966. Stability parameters for comparing varieties. Crop Sci. 6:36–40.

Ecochard, R., and Y. Ravelomanantsoa. 1982. Genetic correlations derived from full-sib relationships in soybean (*Glycine max* L. Merr.). Theor. Appl. Genet. 63:9–15.

Erickson, L.R., W.D. Beversdorf, and S.T. Ball. 1982. Genotype × environment interactions for protein in *Glycine max* × *Glycine soja* crosses. Crop Sci. 22:1099–1101.

Falconer, D.S. 1952. The problem of environment and selection. Am. Nat. 86:293–298.

----. 1960. Introduction to quantitative genetics. The Ronald Press Co., New York.

Fehr, W.R., and C.R. Weber. 1968. Mass selection by seed size and specific gravity in soybean populations. Crop Sci. 8:551–554.

Finlay, K.W., and G.N. Wilkinson. 1963. The analysis of adaptation in a plant-breeding programme. Aust. J. Agric. Res. 14:742–754.

Ford, D.M., R. Shibles, and D.E. Green. 1983. Growth and yield of soybean lines selected for divergent leaf photosynthetic ability. Crop Sci. 23:517–520.

Francis, T.R., and L.W. Kannenberg. 1978. Yield stability studies in short-season maize. I. A descriptive method for grouping genotypes. Can. J. Plant Sci. 58:1029–1034.

Frey, K.J., and T. Horner. 1957. Heritability in standard units. Agron. J. 49:59–62.

Funnah, S.M., and C. Mak. 1980. Yield stability studies in soyabeans (*Glycine max*). Exp. Agric. 16:387–392.

Gardner, C.O. 1977. Quantitative genetic research in plants: Past accomplishments and research needs. p. 29–37. *In* E. Pollak et al. (ed.) Proceedings of the international conference on quantitative genetics. Iowa State University Press, Ames, IA.

Garland, M.L., and W.R. Fehr. 1981. Selection for agronomic characters in hill and row plots of soybeans. Crop Sci. 21:591–595.

Gates, C.E., C.R. Weber, and T.W. Horner. 1960. A linkage study of quantitative characters in a soybean cross. Agron. J. 52:45–49.

Gedge, D.L., W.R. Fehr, and D.F. Cox. 1978. Influence of intergenotypic competition on seed yield of heterogeneous soybean lines. Crop Sci. 18:233–236.

----, ----, and A.K. Walker. 1977. Intergenotypic competition between rows and within blends of soybeans. Crop Sci. 17:787–790.

Hamblin, J., and A.A. Rosielle. 1978. Effect of intergenotypic competition on genetic parameter estimation. Crop Sci. 18:51–54.

Hanson, W.D. 1963. Heritability. p. 125–139. *In* W.D. Hanson and H.F. Robinson (ed.) Statistical genetics and plant breeding. Pub. 982. National Academy of Sciences-National Research Council, Washington, DC.

----. 1964. Genotype-environment interaction concepts for field experimentation. Biometrics 20:540–552.

----. 1970. Genotypic stability. Theor. Appl. Genet. 40:226–231.

----, C.A. Brim, and K. Hinson. 1961b. Design and analysis of competition studies with an application to field plot competition in the soybean. Crop Sci. 1:255–258.

----, and E.J. Kamprath. 1979. Selection for aluminum tolerance in soybeans based on seedling-root growth. Agron. J. 71:581–586.

----, R.C. Leffel, and R.W. Howell. 1961a. Genetic analysis of energy production in the soybean. Crop Sci. 1:121–126.

----, A.H. Probst, and B.E. Caldwell. 1967. Evaluation of a population of soybean genotypes with implications for improving self-pollinated crops. Crop Sci. 7:99–103.

----, and C.R. Weber. 1961. Resolution of genetic variability in self-pollinated species with an application to the soybean. Genetics 46:1425-1434.

Harrison, S.A., H.R. Boerma, and D.A. Ashley. 1981. Heritability of canopy-apparent photosynthesis and its relationship to seed yield in soybeans. Crop Sci. 21:222–226.

Hillsman, K.J., and H.W. Carter. 1981. Performance of F_1 hybrid soybeans in replicated row trials. Agron. Abstr. American Society of Agronomy, Madison WI, p. 63.

Horner, T.W., and C.R. Weber. 1956. Theoretical and experimental study of self fertilized populations. Biometrics 12:404–414.

James, J.W. 1968. Index selection with restrictions. Biometrics 24:1015–1018.

Johnson, H.W., and R.L. Bernard. 1963. Soybean genetics and breeding. p. 1–73. *In* A.G. Norman (ed.) The soybean. Academic Press, New York.

----, H.F. Robinson, and R.E. Comstock. 1955a. Estimates of genetic and environmental variability in soybeans. Agron. J. 47:314–318.

----, ----, and ----. 1955b. Genotypic and phenotypic correlations in soybeans and their implications in selection. Agron. J. 47:477–483.

Kaw, R.N., and P.M. Menon. 1979. Heterosis in a ten-parent diallel cross in soybean. Indian J. Agric. Sci. 49:322-324.

----, and ----. 1980. Combining ability in soybean. Indian J. Genet. Plant Breed. 40:305–309.

Kempthorne, O., and A.W. Nordskog. 1959. Restricted selection indices. Biometrics 15:10–19.

Kenworthy, W.J., and C.A. Brim. 1979. Recurrent selection in soybeans. I. Seed yield. Crop Sci. 19:315–318.

Koinange, E.M.K. 1981. Recurrent selection for increased seed yield in soybean (*Glycine max* (L.) Merr.) using genetic male sterility. Unpublished Master's thesis. North Carolina State University, Raleigh.

Koinange, E.M.K., J.W. Burton, and C.A. Brim. 1981. Recurrent selection for yield in soybeans using genetic male-sterility. Agron. Abstr. American Society of Agronomy, Madison, WI. p. 64.

Krober, O.A. 1956. Methionine content of soybeans as influenced by location and season. J. Agric. Food Chem. 4:254–257.

Kwon, S.H., and J.H. Torrie. 1964. Heritability of and interrelationships among traits of two soybean populations. Crop Sci. 4:196–198.

Leffel, R.C., and M.G. Weiss. 1958. Analysis of diallel crosses among ten varieties of soybeans. Agron. J. 50:528–534.

Lin, C.C., and J.H. Torrie. 1970. Inter-row competitive effects of soybean genotypes. Crop Sci. 10:599–601.

Lin, C.Y. 1978. Index selection for genetic improvement of quantitative characters. Theor. Appl. Genet. 52:49–56.

Luedders, V.D. 1977. Genetic improvement of yield in soybeans. Crop Sci. 17:971–972.

Martin, R.J., and J.R. Wilcox. 1973. Heritability of lowest pod height in soybeans. Crop Sci. 13:201–203.

Matzinger, D.F., C.C. Cockerham, and E.A. Wernsman. 1977. Single character and index mass selection with random mating in a naturally self-fertilizing species, p. 503–518. *In* E. Pollak et al. (ed.) Proceedings of the international conference of quantitative genetics. Iowa State University Press, Ames.

----, and E.A. Wernsman. 1968. Four cycles of mass selection in a synthetic variety of an autogamous species, *Nicotiana tabacum* L. Crop Sci. 8:239–243.

----, and ----. 1980. Population improvement in self-pollinated crops. p. 191–199. *In* F.T. Corbin (ed.) World soybean research conference II: Proceedings. Westview Press, Boulder, CO.

Miller, J.E., and W.R. Fehr. 1979. Direct and indirect recurrent selection for protein in soybeans. Crop Sci 19:101–106.

Mungomery, V.E., R. Shorter, and D.E. Byth. 1974. Genotype × environment interactions and environmental adaptation. I. Pattern analysis-application to soya bean populations. Aust. J. Agric. Res. 25:59–72.

Nelson, R.L., and R.L. Bernard. 1984. Production and performance of hybrid soybeans. Crop Sci. 24:549–553.

Openshaw, S.J., and H.H. Hadley. 1981. Selection to modify sugar content of soybean seeds. Crop Sci. 21:805–808.

----, and ----. 1984. Selection indexes to modify protein concentration of soybean seeds. Crop Sci. 24:1–4.

Pandy, J.P., and J.H. Torrie. 1973. Path coefficient analysis of seed yield components in soybeans (*Glycine max* (L.) Merr.). Crop Sci. 13:505–507.

Paschal, E.H., II, and J.R. Wilcox. 1975. Heterosis and combining ability in exotic soybean germplasm. Crop Sci. 15:344–349.

Pederson, D.G. 1974. Arguments against intermating before selection in self-fertilising species. Theor. Appl. Genet. 45:157–162.

Perkins, J.M., and J.L. Jinks. 1968. Environmental and genotype-environmental components of variability. III. Multiple lines and crosses. Heredity 23:339–356.

Pesek, J. and R.J. Baker. 1969. Comparison of tandem and index selection in the modified pedigree method of breeding self-pollinated species. Can. J. Plant Sci. 49:773–781.

----, and ----. 1970. An application of index selection to the improvement of self-pollinated species. Can J. Plant Sci. 50:267–276.

Powers, L. 1955. Components of variance method and partitioning method of genetic analysis applied to weight per fruit of tomato hybrid and parental populations. USDA Tech. Bull. 1131. U.S. Government Printing Office, Washington, DC.

Pritchard, A.J., D.E. Byth, and R.A. Bray. 1973. Genetic variability and the application of selection indices for yield improvement in two soya bean populations. Aust. J. Agric. Res. 24:81–89.

Rawlings, J.O. 1980. Long- and short-term recurrent selection in finite populations—choice of population size. p. 201–215. *In* F.T. Corbin (ed.) World soybean research conference II: Proceedings. Westview Press, Boulder, CO.

Schutz, W.M., and R.L. Bernard. 1967. Genotype × environment interactions in the regional testing of soybean strains. Crop Sci. 7:125–130.

----, and C.A. Brim. 1967. Inter-genotypic competition in soybeans. I. Evaluation of effects and proposed field plot design. Crop Sci. 7:371–376.

----, and ----. 1971. Inter-genotypic competition in soybeans. III. An evaluation of stability in multiline mixtures. Crop Sci. 11:684–689.

Shannon, J.G., J.R. Wilcox, and A.H. Probst. 1972. Estimated gains from selection for protein and yield in the F_4 generation of six soybean populations. Crop Sci. 12:824–826.

Shorter, R., D.E. Blyth, and V.E. Mungomery. 1976. Estimates of selection parameters associated with protein and oil content of soybean seeds. (*Glycine max* (L.)Merr.) Aust. J. Agric. Res. 28:211–222.

Shukla, G.K. 1972. Some statistical aspects of partitioning genotype-environmental components of variability. Heredity 29:237–245.

Simpson, A.M., Jr., and J.R. Wilcox. 1983. Genetic and phenotypic associations of agronomic characteristics in four high protein soybean populations. Crop Sci. 23:1077–1081.

Singh, T.P., K.B. Singh, and J.S. Brar. 1974. Diallel analysis in soybean. Indian J. Genet. Plant Breed. 34:427–432.

Smith, R.R., D.E. Byth, B.E. Caldwell, and C.R. Weber. 1967. Phenotypic stability in soybean populations. Crop Sci. 7:590–592.

----, and C.R. Weber. 1968. Mass selection by specific gravity for protein and oil in soybean populations. Crop Sci. 8:373–377.

Snape, J.W., and T.J. Riggs. 1975. Genetical consequences of single seed descent in the breeding of self-pollinating crops. Heredity 35:211–219.

Sneep, J. 1977. Selection for yield in early generations of self-fertilizing crops. Euphytica 26:27–30.

Sokol, M.J., and R.J. Baker. 1977. Evaluation of the assumptions required for the genetic interpretation of diallel experiments in self-pollinating crops. Can. J. Plant Sci. 57:1185–1191.

Specht, J.E., and J.H. Williams. 1984. Contribution of genetic technology to soybean productivity-retrospect and prospect. p. 49–74. *In* W.R. Fehr (ed.) Genetic contributions to yield gains of five major crop plants. Spec. Pub. 7. Crop Science Society of America and American Society of Agronomy, Madison, WI.

Stam, P. 1977. Selection response under random mating and under selfing in the progeny of a cross of homozygous parents. Euphytica 26:169–184.

St. Martin, S.K. 1982. Effective population size for the soybean improvement program in maturity groups 00 to IV. Crop Sci. 22:151–152.

Stuber, C.W. 1970. Estimation of genetic variances using inbred relatives. Crop Sci. 10:129–135.

Sumarno, and W.R. Fehr. 1982. Response to recurrent selection for yield in soybeans. Crop Sci. 22:295–299.

Veatch, C. 1930. Vigor in soybeans as affected by hybridity. J. Am. Soc. Agron. 22:289–310.

Walker, A.K., and W.R. Fehr. 1978. Yield stability of soybean mixtures and multiple pure stands. Crop Sci. 18:719–723.

Weaver, D.B., and J.R. Wilcox. 1982. Heritabilities, gains from selection, and genetic correlations for characteristics of soybeans grown in two row spacings. Crop Sci. 22:625–629.

Weber, C.R., L.T. Empig, and J.C. Thorne. 1970. Heterotic performance and combining ability of two-way F_1 soybean hybrids. Crop Sci. 10:159–160.

Weber, W.E. 1982. Selection in segregating generations of autogamous species. I. Selection response for combined selection. Euphytica 31:493–502.

Weiss, M.G., C.R. Weber, and R.R. Kalton. 1947. Early generation testing in soybeans. J. Am. Soc. Agron. 39:791–811.

Wiebold, W.J., R. Shibles, and D.E. Green. 1981. Selection for apparent photosynthesis and related leaf traits in early generations of soybeans. Crop Sci. 21:969–973.

Wilcox, J.R., F.A. Laviolette, and R.J. Martin. 1975. Heritability of purple seed stain resistance in soybeans. Crop Sci. 15:525–526.

----, W.T. Schapaugh, Jr., R.L. Bernard, R.L. Cooper, W.R. Fehr, and M.H. Niehaus. 1979. Genetic improvement of soybeans in the midwest. Crop Sci. 19:803–805.

Yonezawa, K., and H. Yamagata. 1981. Selection strategy in breeding of self-fertilizing crops. I. Theoretical considerations on the efficiency of single plant selection in early segregating generations. Jpn. J. Breed. 31:35–48.

Young, S.S.Y. 1961. The use of sire's and dam's records in animal selection. Heredity 16:91–102.

7 Breeding Methods for Cultivar Development

Walter R. Fehr
Iowa State University
Ames, Iowa

Genetic improvement of cultivars has played a key role in establishing the soybean [*Glycine max* (L.) Merr.] as a major agricultural commodity. Soybean production in the USA began with cultivars introduced from other countries (Hartwig, 1973). Plant introductions were evaluated for their agronomic characteristics and the most desirable ones were released for commercial use. Pure-line cultivars also were obtained by selection within heterogeneous plant introductions.

The first cultivars selected from planned crosses between plant introductions were released in the 1940s. Cultivars selected from the first populations formed by hybridization were used as parents to form populations for the second cycle of selection. The process of utilizing superior progeny from one cycle of selection as parents to form populations for the next cycle continues up to the present time.

Cultivar development and related genetic research is conducted by the USDA, state agricultural experiment stations, and private companies. The involvement of the three agencies has evolved over the years as the importance of the soybean increased and as legislation was enacted to provide rights to the developer of a cultivar.

The USDA has the longest record of involvement in cultivar development. It was responsible for coordination of the introduction of soybean germplasm into the USA (Hartwig, 1973). The USDA conducted the evaluation and release of plant introductions or selections from them as cultivars. The first cultivars obtained by hybridization were selected by soybean breeders of the USDA.

The USDA coordinates the regional evaluation of advanced breeding lines, referred to as the Uniform Tests. The Uniform Tests have been instrumental in identifying cultivars from public institutions that contribute to improved soybean production in one or more states. The USDA is responsible for maintenance of plant introductions in the U.S. Soybean Germplasm Collections. More than 10 000 accessions of *G. max, G. soja*, and related species are maintained at Urbana, IL and Stoneville, MS.

Soybeans: Improvement, Production, and Uses, 2nd ed.—Agronomy Monograph no. 16.

The USDA maintains genotypes with unique characteristics in the Genetic Type Collection (Bernard and Weiss, 1973). The collection includes all known and available qualitative genes that are not found in cultivars.

Soybean breeding research by state agricultural experiment stations has expanded since the early 1960s. Their activities have been directed toward cultivar development and basic research on genetics and breeding. State breeders have also been involved in obtaining information for farmers on the performance of private and public cultivars.

Soybean breeding by private companies was rapidly expanded after passage of the Plant Variety Protection Act in 1970. The act gave the developer of a cultivar the right to control its distribution and marketing. Private companies have made a major investment in cultivar development throughout the major soybean-producing areas. The number of breeders in the private sector is similar to that of the USDA and state agricultural experiment stations combined. Private cultivars are being rapidly adopted for commercial production, particularly in the northern USA.

Cooperation between federal, state, and private agencies has included the evaluation of plant introductions to identify parents that may contribute to genetic diversity of cultivars in the future. A Soybean Breeders Workshop is held annually to discuss research of mutual interest to the three agencies.

7–1 OBJECTIVES OF CULTIVAR DEVELOPMENT

The first step in cultivar development is to identify the characters that are important for commercial soybean production. In assessing the current objectives of cultivar development, a review will be made of the progress that already has been achieved for improvement of the character.

7–1.1 Seed Yield

Yield is the character of paramount importance to the farmer. The amount of genetic improvement in yield that has been realized by hybridization and selection has been substantial. Luedders (1977) compared the performance of cultivars of Maturity Groups I to IV that represented the original plant introductions, cultivars developed from plant introductions in the first cycle of hybridization and selection, and cultivars from the second cycle. He observed a 26% increase in yield from the first cycle and 16% from the second cycle of selection. Wilcox et al. (1979) observed a 25% difference in yield between plant introductions of Maturity Groups II and III released before 1940 and cultivars released after 1970. Boerma (1979) reported a yield increase of 0.7% yr^{-1} for cultivars of Maturity Groups VI to VIII released from 1942 to 1973. Specht and Williams (1984) observed an average rate of yield improvement of 18.8 kg ha^{-1} yr^{-1} from 1902 to 1977 for Maturity Groups 00 to IV.

The four studies leave no doubt that major improvements in seed yield have been achieved. The average rate of yield increase that they reported was 15.1 kg ha^{-1} yr^{-1}. When the rate of yield increase is expressed as a percentage of average yield for the four experiments, the average increase has been 0.6% yr^{-1}.

The potential for future yield improvement may be influenced by the amount of resources expended, the importance that must be placed on characters other than yield, the high level of adaptability of the crop, and the genetic diversity of parents. The amount of resources devoted to cultivar development expanded markedly during the 1970s with the investment of the private sector. During the same period, mechanization and computerization permitted breeders to evaluate a large number of lines and the increased use of greenhouses and winter nurseries reduced the length of time for cultivar development (Fehr, 1976). These investments provide optimism for continued yield improvement in the future.

The amount of emphasis that can be placed on yield improvement depends on the number of other characters that must be considered. For example, pests are now creating economic loss that previously were not important or that were controlled by the genetic resistance of available cultivars. The soybean cyst nematode, caused by *Heterodera glycines* Ichinohe, recently was identified in the northern USA, and a new race of the nematode emerged in the South that was able to attack cultivars resistant to previous races. Boerma (1979) found that in the absence of important disease infestations, none of the cultivars of Maturity Groups VI to VIII released since 1970 yielded significantly more than 'Lee', which was released in 1954. Major emphasis has been placed on disease resistance for cultivars of those maturity groups.

The rate of genetic improvement may have declined since the time that cultivar development by hybridization began. Luedders (1977) and Specht and Williams (1984) noted that the first cycle of cultivar development by hybridization provided a larger percentage yield increase than subsequent cycles. The lower rate may reflect the difficulty in obtaining genetic improvement beyond that of cultivars that are already well adapted. It also may be associated with the limited amount of genetic diversity among the high-yielding cultivars used as parents for hybridization.

The ancestry of potential parents is considered in population development for yield improvement. Parents with diverse origin are considered more likely to produce superior progeny than those of similar ancestry. It has become increasingly difficult to identify high-yielding genotypes that do not have common parentage. St. Martin (1982) examined the parentage of 27 public cultivars of Maturity Groups 00 to IV that were released between 1976 and 1980. Only 20 plant introductions were in the pedigrees of the cultivars. Any pair of the 27 cultivars had at least one common parent in their ancestry. The coefficient of parentage among the 27 cultivars ranged from 0.06 to 0.76, with an average of 0.25. He concluded that soybean breeding has increased yield substantially over the past 50 yrs at the expense of the loss of a substantial portion of the genetic variability originally available. The conclusion reached by St. Martin would apply equally well to cultivars of all maturity groups, regardless if they were of public or private origin.

Plant introductions are being used as parents for population development in an attempt to expand the genetic diversity of cultivars. It is not likely that they will provide short-term yield improvement that exceeds that obtained by matings between high-yielding cultivars. Schoener and Fehr (1979) compared the performance of lines from populations with 0, 25, 50, 75, and 100% of plant introduction parentage. The populations were develped from four plant introductions with good yield potential and four high-yielding cultivars or experimental lines. The population with no plant introductions as parents had a significantly higher mean yield than any of the other populations and a higher frequency of high-yielding lines. Vello et al. (1984) compared intermated populations developed from 40 plant introductions and 40 high-yielding cultivars. Populations with 25, 50, 75, and 100% of plant introduction parentage had a lower mean yield and lower frequency of high-yielding segregates than the population developed from only the high-yielding cultivars.

7–1.2 Pest Resistance

The yield potential of a cultivar cannot be realized when it is injured by diseases, nematodes, and insects. Breeding for resistance to some pests is an objective of almost every soybean breeding program. The amount of emphasis placed on pest resistance is dependent on the prevalence and regularity with which injury of economic importance occurs on susceptible cultivars. Resistance may not be a prerequisite for cultivar release when the pest occurs sporadically in limited areas, but may be essential when major economic loss is observed frequently.

Specific resistance, general resistance, and tolerance are alternative means of protecting cultivars from major economic loss. Specific resistance controlled by major genes has been the primary source of pest control. It affords the advantage of simplicity in transferring resistance among genotypes. 'Union' was developed by transferring to 'Williams' the *Rpm* allele for resistance to downy mildew, caused by *Peronospora manshurica* (Naoum.) Syd. ex Gaum., and the Rps_1 allele for resistance to races 1 and 2 of phytophthora rot, caused by *Phytophthora megasperma* (Drechs.) var. *sojae* Hildebrand (Bernard and Cremeens, 1982). The disadvantage of specific resistance is the lack of protection it may provide to new races of a pest.

General or field resistance does not confer the immunity associated with specific resistance, but is responsible for a reduced level of infection. It is a desirable type of resistance because it provides protection against multiple races of a pest. Field resistance generally is a quantitative character controlled by multiple genes, which makes it more difficult to transfer among genotypes than specific resistance. A high degree of resistance to phytophthora rot was reported to be present in 'Davis' (Caviness and

Walters, 1966), and moderate resistance to brown stem rot, caused by *Phialophora gregata* (Allington and Chamberlain) W. Gams, was reported for 'BSR 201' (Tachibana et al., 1983).

A tolerant cultivar has less loss in productivity than a nontolerant one, even though they both have similar levels of infection. Screening for tolerance involves comparisons of yield under infested and uninfested conditions. When defined in this strict sense, tolerance is not commonly used in breeding for pest resistance.

7-1.3 Maturity

Soybean cultivars are classified into maturity groups ranging from 000 in the high latitudes to X at low latitudes (Hartwig, 1973). Soybean is a short-day species, and the variation among maturity groups is associated with differences in the photoperiod requirement of the cultivars. Soybean cultivars of Maturity Group 00 adapted to the long days at high latitudes have a long critical photoperiod or are photoperiod insensitive (Criswell and Hume, 1972). The critical photoperiod of cultivars progressively decreases from the high to the low latitudes. The photoperiod requirement limits the latitudes to which a cultivar is best adapted.

Maturity is an important consideration in a breeding program for a particular geographical area. Plants generally are considered mature when 95% of the pods have reached their mature color. Maturity is a quantitative character, although major genes for the character have been identified (Bernard and Weiss, 1973).

7-1.4 Lodging Resistance

The lodging of soybean plants is rated on a scale from 1 to 5, with 1 for erect plants, 3 for plants that are leaning at a 45° angle, and 5 for prostrate plants. Adequate lodging resistance is considered an important character in cultivar development. Luedders (1977) indicated that lodging resistance in Maturity Groups I to IV was increased by 17% in cultivars obtained from the first cycle of hybridization and 20% in the second cycle. Specht and Williams (1984) reported that lodging had decreased by a total of 1.0 unit during the past 75 yrs for Maturity Groups 00 to IV.

The amount of lodging that can be tolerated without a negative physiological impact on yield has been evaluated by artificially holding plants in a prostrate or upright position and by altering plant population. Severe lodging, created artifically or naturally, has caused significant yield loss (Leffel, 1961; Cooper, 1971a, 1971b). Moderate natural lodging of a 2.6 score resulted in 13% lower yield than plants artificially held upright (Weber and Fehr, 1966). With alteration in lodging by variation in plant population, an increase in lodging score from 1.3 to 2.0 in the cv. Essex did not cause a significant yield reduction but an increase in lodging scores from 2.1 to 2.3 in 'Forrest' and 2.3 to 2.7 in 'Mack' were associated with a significant yield loss (Hoggard et al., 1978). Although no absolute limits for lodging have been established, it is not generally considered necessary to have cultivars that are completely free of lodging.

Lodging resistance in the southern USA for Maturity Groups V and later is enhanced by selection of cultivars with the determinate growth habit. When cultivars with the indeterminate growth habit are planted in May or early June, they grow tall and exhibit excessive lodging (Boerma et al., 1982).

For Maturity Groups IV and earlier in the northern USA, adequate lodging resistance for most environments can be obtained by selection among indeterminate genotypes. Short-statured determinate cultivars of Maturity Groups II to III have been developed to provide excellent lodging resistance in high-yielding environments (Cooper, 1981). The semideterminate growth habit has been considered for obtaining improved lodging resistance in high-yielding environments. Although semideterminate genotypes are about 15% shorter than indeterminates, they exhibit only about a 10% reduction in lodging score (Bernard, 1972; Green et al., 1977; Wilcox, 1980; Hartung et al., 1981).

7–1.5 Plant Height

The majority of soybean cultivars have a plant height of 1 m, measured as the length of the main stem from the soil surface to the terminal node. Plant height is not generally a major factor in soybean breeding, except when cultivars are developed for particular environments. Short-statured determinate cultivars of Maturity Groups II to III have been developed for high-yielding environments (Cooper, 1981). Their shortness is a positive attribute for obtaining improved lodging resistance.

Increased plant height is considered a desirable attribute for cultivars used for late planting in the South. The indeterminate cultivar of Maturity Group VII, 'Duocrop', was described as being specifically adapted to planting after 20 June (Boerma et al., 1982). With late planting, its increased vegetative growth after the onset of flowering is considered an advantage over determinate cultivars for more efficient mechanical harvest and higher seed yields. Tallness also may be desirable for low-yielding environments. 'Amcor' is about 10- to 15-cm taller than other indeterminate cultivars of Maturity Group II (Cooper et al., 1981). Its taller growth has been associated with higher seed yields where stress conditions limit plant growth (Walker and Cooper, 1982).

7–1.6 Seed Size

Seed size is not usually a selection criterion, unless cultivars are developed to supply a special demand for unusually small or large seeds. The seed size of widely grown cultivars ranges from about 12 to 18 g 100 seeds^{-1} (Hartwig, 1973). A seed size of less than 10 g 100 seeds^{-1} is preferred for natto, a product consisting of fermented whole beans (Cowan, 1973). Seed size in excess of 22 g 100 seeds^{-1} is preferred when soybean is used for certain food products. 'Verde', a cultivar with green seeds that weigh 32 g 100 $seeds^{-1}$, was developed for processing as a canned or frozen vegetable (Crittenden, 1971). The yellow seeds of 'Prize' that weigh 27 g 100 $seeds^{-1}$ have been used for home gardens, for roasting as a confectionary product and for production of miso, a fermented product (Weber, 1967b).

7-1.7 Seed Quality

Two of the factors considered in seed quality are appearance and germinability. Shrunken, discolored, and broken seeds can cause a reduction in the market value of the crop. A high germination level is important for seed to be used for planting.

Undesirable seed quality can be caused by unfavorable weather conditions and disease. It is more often a problem for cultivars grown in the low latitudes than those grown in higher latitudes. Effective breeding for seed quality includes evaluation of visual appearance, laboratory and field emergence, and disease resistance (Green and Pinnell, 1968a, 1968b; Athow, 1973; Wilcox et al., 1975).

7-1.8 Protein and Oil Quantity and Quality

Soybean cultivars generally have about 40% protein and 20% oil. Although the protein and oil percentages could be altered readily by breeding, the characters are not emphasized in cultivar development. The price paid for commercial soybean is not based on seed composition, therefore the character is not considered by the farmer in the selection of a cultivar for planting.

There is a limited market for high protein seeds with yellow seed coats and hila for use in tofu production. Higher seed protein results in a higher protein percentage in the tofu. 'Vinton', a specialty cultivar, has 44.9% protein compared with about 40% for widely grown cultivars (Bahrenfus and Fehr, 1980c).

A number of quality factors are under investigation that may become selection criteria in the future. They include increased methionine and decreased antinutritional factors in the protein (Howell et al., 1972; Orf and Hymowitz, 1979). Oil quality would be improved by a reduction in linolenic acid and elimination of lipoxygenase activity (Hammond and Fehr, 1975; Wilson et al., 1981; Hildebrand and Hymowitz, 1982).

7-1.9 Shattering Resistance

The ability of a cultivar to retain its seed after harvest is an important characteristic. Adequate shattering resistance is available in most high-yielding parents currently used for population development. 'Nathan' is an example of a recently released cultivar that has good shattering resistance (Hartwig and Epps, 1982).

7–1.10 Resistance to Mineral Deficiencies and Toxicities

The performance of a cultivar can be influenced by its ability to overcome mineral deficiencies and toxicities. Resistance to Fe-deficiency chlorosis on soils with a high pH has received the most attention for cultivar development in the USA (Fehr, 1982). Selection is practiced for cultivars such as 'Swift' and 'Weber' that are able to effectively utilize soil Fe when grown on certain calcareous soils (Lambert and Kennedy, 1973; Bahrenfus and Fehr, 1980b).

7–1.11 Resistance to Herbicide Injury

Genetic differences have been observed among cultivars for resistance to injury by certain herbicides used for soybean production. Selection for resistance to metribuzin injury has been practiced by some breeders. 'Tracy-M' is a cultivar obtained by selection of plants with tolerance to metribuzin within the susceptible cv. Tracy (Hartwig et al., 1980).

7–2 POPULATION DEVELOPMENT

Cultivar development is accomplished by selection among individuals that are genetically different. In the early years of soybean breeding, selection was practiced among plant introductions that were genetically different. Genetic variability within heterogeneous plant introductions also permitted selection to be practiced. After selection among and within plant introductions had been exploited, artificial hybridization became the mechanism for obtaining genetic variability. Artificial mutagenesis has not been a productive method of obtaining genetic variability for cultivar development in the USA.

Populations can be developed with different numbers of parents, varying percentages of each parent, and different amounts of recombination before inbreeding is initiated. Several types of populations have been used successfully for cultivar development.

7–2.1 Types of Populations

7–2.1.1 Two-Parent Population

The majority of soybean cultivars have been selected from crosses involving two inbred parents. The first cultivars of hybrid origin were selected from populations developed by crossing two plant introductions as parents. Two-parent crosses are still the most common type of population in soybean breeding. An example of a recent cultivar obtained from a two-parent cross is 'Sparks', a selection from the cross Williams × 'Calland' (Nickell et al., 1983).

7-2.1.2 Multiple-Parent Population

Populations developed from more than two parents have been used for cultivar development. 'Ransom' was selected from the three-parent cross of (N55-5931 × N55-3818) × D56-1185 (Brim and Elledge, 1973). Cultivars have been selected from two types of four-parent populations. A double-cross population of ('Mandarin (Ottawa)' × 'Kanro') × ('Richland' × 'Jogun') was used to develop the large-seeded cv. Disoy (Weber, 1967a). Mandarin (Ottawa) and Richland were small-seeded, high-yielding parents and Kanro and Jogun were large-seeded parents with lower yield potential. Vinton, a specialty cultivar with larger seed and higher protein than widely grown cultivars, was selected from the population 'Hark' × ['Provar' × (Disoy × 'Magna')] (Bahrenfus and Fehr, 1980c; Fehr and Bahrenfus, 1984). The large-seeded parents Disoy and Magna were mated, the F_1 was crossed to the high-protein cv. Provar, and the three-parent F_1 was crossed to the small-seeded, high-yielding cv. Hark. 'Lakota' and 'Elgin' were selected from a population developed by intermating 40 high-yielding cultivars for use in recurrent selection to increase seed yield (Bahrenfus and Fehr, 1984; Fehr and Bahrenfus, 1984).

Multiple parents have been used to transfer genes for pest resistance from agronomically unacceptable parents into more desirable populations. The procedure, referred to as modified backcrossing, will be discussed in the next section.

7-2.1.3 Backcross Population

Useful segregating populations have been developed by the use of a limited number of backcrosses. One purpose for backcrossing is to increase the frequency of segregates beyond that available in a two-parent population. For yield improvement of large-seeded cultivars, it is desirable to obtain the higher yield potential available in commercial cultivars with smaller seed. A cross between a large- and a small-seeded parent does not provide an adequate frequency of large-seeded progeny (Bravo et al., 1981). A backcross to the large-seeded parent results in the necessary large-seeded segregates. This strategy was used in the selection of the large-seeded cvs. Kim and Kanrich (Weber, 1966). Kim was selected from the cross 'Sac' × (Sac × Richland) and Kanrich was a progeny of the cross Kanro × (Kanro × Richland). Sac and Kanro were the large-seeded parents and Richland was the small-seeded parent with high yield.

A limited number of backcrosses has been used to transfer genes from a plant introduction to a high-yielding genetic background. The technique is especially useful when the character being transferred is controlled by multiple genes, which necessitates the screening of a large number of segregates each generation to recover those with the desired genes for use in crossing. When different recurrent parents are used during backcrossing, the procedure is sometimes referred to as modified backcrossing. There are two reasons for changing recurrent parents during backcrossing. First, the original recurrent parent may be replaced by one with more desirable characteristics. Second, different parents may be used to accumulate different favorable characteristics from each.

Limited backcrossing has been used successfully to transfer resistance to the soybean cyst nematode into populations from which desirable progeny can be selected as cultivars. 'Bedford' was selected from the cross Forrest × [(Forrest × (D68-18 × PI 88788)] (Hartwig and Epps, 1978). D68-18, an agronomically acceptable parent, was mated to PI 88788, a parent with black seed and poor agronomic characteristics that has resistance to race 4 of the soybean cyst nematode. Resistant F_2 plants identified in the two-parent population were crossed to Forrest, a cultivar with agronomic characteristics superior to D68-18. Resistant F_2 plants from the backcross population were mated to Forrest. Bedford was selected from among 135 cyst-resistant lines obtained from the final backcross population.

The cv. Kirby represents an example of the use of modified backcrossing to increase the frequency of segregates for pest resistance when it is not practical to select for the character in a two-parent population (Hinson, 1983, personal communication). The objective was to develop a cultivar of Maturity Group VIII with resistance to race 3 of the soybean cyst nematode without extensive screening for the character. The backcross population used was 'Centennial' × [Forrest × ('Cobb' × D68-216)]. D68-216, Forrest, and Centennial of Maturity Groups V and VI were the sources of race 3 resistance and Cobb was the source of late maturity. The only selection practiced during backcrossing was for late maturity. Kirby was an F_5 plant selection from the backcross population.

When a desired character is controlled by a major gene that can be readily monitored during backcrossing, it is common to conduct enough backcrosses to avoid testing individual progeny for yield in extensive replicated tests. Many cultivars with specific resistance to phytopthora rot have been developed in this manner, including 'Corsoy 79', 'Beeson 80,' and 'Williams 82'.

Four or more backcrosses generally are used when there is no yield evaluation of progeny. Wilcox et al. (1971) indicated that a specific gene for resistance to phytophthora rot could be added to a cultivar by backcrossing for seven generations, followed by visual selection of lines from the BC_7 generation that have the maturity, height, lodging, and seed size of the recurrent parent. Lines with the phenotype of the recurrent parent could be composited without replicated tests for yield and protein and oil percentage. Cultivars with the yield potential of the recurrent parent have been developed with less than seven backcrosses, even though no selection for agronomic characteristics was practiced during backcrossing and no replicated tests of individual lines were conducted before compositing those with the phenotype of the recurrent parent. 'Vickery' was obtained after four backcrosses and performed as well as the recurrent parent, Corsoy (Fehr et al., 1981).

7–2.2 Hybridization

Segregating populations for cultivar development are initiated from hybrid seeds produced by manual crosses between male-fertile parents. The use of natural hybridization in association with genetic male sterility is restricted to recurrent selection programs.

The environmental conditions required for successful artificial hybridization and the techniques involved were described by Fehr (1980). The daylength required to obtain flowers suitable for hybridization is longer than that used to initiate rapid flowering. For example, a daylength of about 14.5 h produces suitable flowers on genotypes at Maturity Groups I to III, but flowers obtained at a 12-h daylength self-pollinate when they are still too small for artificial manipulation. Fluorescent light is not effective in controlling the flowering response of some genotypes (Buzzell, 1971). Night interruption does not provide the same control over floral initiation as does continuous light (Lawrence and Fehr, 1981). Successful artificial hybridization is enhanced when plants are free of moisture or nutrient stress.

Flowers of the female parent are pollinated artificially the day before their anthers have matured enough to shed pollen. A floral bud at the appropriate stage is swollen and the corolla is visible through the calyx or has just begun to emerge (Fig. 7–1A). The five sepals of the calyx are removed with a forceps to expose the corolla (Fig. 7–1B). The ring of 10 stamens and the pistil are visible after the corolla has been removed (Fig. 7-1C). Emasculation of the flower is not required (Walker et al., 1979).

Pollen is collected from open flowers on the day the females are prepared (Fig. 7–1D). Under humid conditions in the southern USA, flowers are collected in the morning, dried in a desiccator, and their pollen used in the afternoon. In northern USA, flowers are collected and used without drying during that part of the day when they shed pollen readily. The pollen is placed on the stigma by brushing the anthers against it (Fig. 7–1E). The pollinated flower is identified with a tag. If the pollination is successful, a pod will be visible in about 7 days (Fig. 7–1F). The scar made by removal of the sepals aids in differentiating pods produced by artificial hybridization from those resulting from natural self-pollination. A person with experience can achieve at least 50% success in obtaining a pod with one to three hybrid seeds.

7–3 INBRED LINE DEVELOPMENT

Two of the factors considered for inbreeding a population are: (i) the method to be used and (ii) the number of generations of self-pollination to conduct before lines are derived for evaluation as potential cultivars. The methods of inbred line development include pedigree, bulk, mass selection, single-seed descent, and early generation testing. Lines can be derived from a population in the F_2 or in any of the more advanced generations of inbreeding.

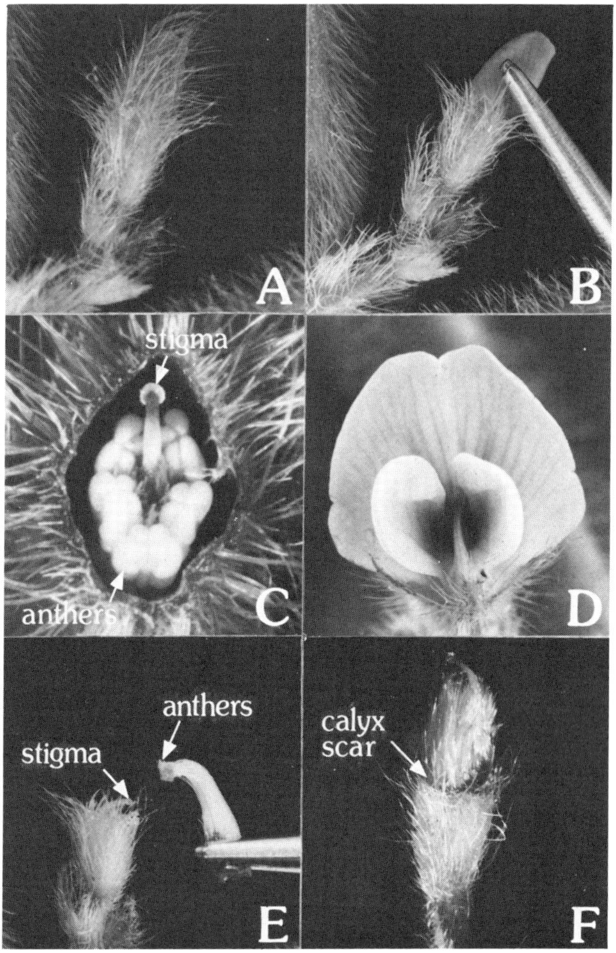

Fig. 7–1. Preparation and pollination of a soybean flower. (A) Flower at the stage for preparation and pollination. (B) Removal of the corolla from the female flower after the calyx has been removed. (C) Ring of 10 anthers and the stigma. (D) Flower with pollen available. (E) Pollination of the stigma. (F) A pod 7 days after pollination with the calyx scar differentiating it from self-pollinated pods. (Photographs A, B, D, E, and F by C. J. Deutsch and P. A. Krumhardt, Iowa State Univ., Ames; photograph C by G. L. Berkey, Ohio Agricultural Research and Development Center, Wooster, OH).

7–3.1 Methods

7–3.1.1 Pedigree

The pedigree method consists of selection among individual plants and their progeny during inbreeding (Fehr, 1978). Selection generally begins with the identification of desirable F_2 plants. The F_3 progenies of the selected plant are grown in rows, the most desirable progeny rows are chosen, and desirable F_3 plants are harvested individually from within selected progeny. The procedure of selection among and within progeny continues until the desired level of homozygosity is achieved.

Selection by the pedigree method most often involves visual evaluation for plant and seed characteristics. It could also be used, however, for characters that require laboratory analysis, such as protein and oil composition.

The usefulness of the pedigree method depends on the effectiveness of selection among individual plants and unreplicated progeny rows. Selection among plants and lines is considered to be effective for maturity and height (Weiss et al., 1947; Kalton, 1948). For lodging resistance, selection among plants has been found to be of limited value, but selection among lines is effective (Weiss et al., 1947).

Estimates of the effectiveness of selection among plants for yield have varied. Weiss et al. (1947) considered selection based on the yield of spaced F_2 plants to be moderately effective, while Kalton (1948) and Hinson and Hanson (1962) concluded that the yields of spaced plants were of little value. With respect to visual selection for yield, Raeber and Weber (1953) found that progeny from plants judged to be phenotypically superior had a significantly higher yield than random selections. Wilcox and Schapaugh (1980) did not observe any improvement in the mean yield of populations following visual selection for yield among plants from the F_2 to F_4 generations.

Visual selection among lines for yield has been more effective for eliminating lines that are inferior than for identifying the best lines (Hanson et al., 1962; Kwon and Torrie, 1964a). Byth et al. (1969) reported that selection for phenotypic desirability among homogeneous lines was more effective than among heterogeneous lines.

Effective selection by the pedigree method depends on the availability of environments in which genetic differences among segregates are expressed. The cv. Woodworth resulted from pedigree selection from the F_2 to F_5 generations in southern Illinois under conditions of severe stress on seed quality and pod-set (Bernard and Lindahl, 1977). The characters considered during selection each year were good seed quality, pod-set, plant vigor, and lodging resistance.

The pedigree method is not suited for inbreeding in greenhouses and tropical nurseries where the expression of agronomic characters is not typical of that observed under field conditions in the area to which the genotypes are adapted. The length of time required for cultivar development when only one generation is grown each year for pedigree selection generally exceeds that of methods that can take advantage of multiple generations each year (Fehr, 1978). This limits the current use of the pedigree method for cultivar development in the USA.

7-3.1.2 Bulk

For each generation of inbreeding by the bulk method, plants of a segregating population are harvested together and a sample of the seed is used to plant the next generation. When the desired level of inbreeding has been achieved, individual plants are harvested to initiate line evaluation.

Natural selection can cause changes in the characteristics of a bulk population during multiple generations of inbreeding. Tall height, late maturity, lodging susceptibility, and branching favor competitiveness and productivity. Plants with those characters tend to increase in frequency in a bulk population during inbreeding (Mumaw and Weber, 1957).

Procedures for minimizing an undesirable shift toward late maturity have been investigated. Empig and Fehr (1971) evaluated a procedure referred to as a maturity group bulk that consisted of subdividing populations into early, midseason, and late maturity. Plants of each maturity were threshed in bulk and a sample of the seed was used to plant each subpopulation separately. They also evaluated a restricted cross bulk procedure that involved sampling seed from a 10-cm section of the plants in a population, rather than bulking all the seed. The two procedures were compared with the traditional bulk method and single-seed descent. The mean percentage of early lines retained by the maturity group bulk (17.0%) was similar to single-seed descent (18.5%) and higher than the bulk (8.9%) or restricted cross bulk (4.4%). The percentage of late lines retained by the four methods was similar. Luedders (1978) attempted to decrease the difference in productivity between early and late-maturing plants in a bulk by delaying the planting date. Delayed planting did not change the relative seed number obtained from early and late plants. He suggested, however, that the improvement in seed quality and shattering resistance obtained when early genotypes are planted at later dates may favor their retention in a bulk.

The frequency of genotypes with disease resistance can be enhanced through natural selection by exposing bulk populations to a pathogen. Buzzell and Haas (1972) reported that genotypes susceptible to phytophthora rot were only 35% as productive as resistant genotypes in 1 yr and 75% as productive in another year. If the susceptible plants were 75% as productive as resistant ones from the F_2 to the F_5 generation, their frequency in the F_6 population would be 29%. In the absence of natural selection from F_2 to F_5, the percentage of susceptible F_6 plants would be 48%. Hartwig (1972) and Hartwig et al. (1982) indicated that the degree of field resistance to phytophthora rot, a quantitative character, influenced natural selection for individuals with major genes for specific resistance to the disease. Natural selection increased the percentage of plants with a major gene for resistance in populations derived from parents that did not have a high degree of field resistance. Natural selection was not effective when the population contained plants with field resistance that competed effectively with plants containing a major gene for resistance.

Luedders and Duclos (1978) observed that plants resistant to the soybean cyst nematode had a reproductive advantage over susceptible ones on infested soil. They postulated that the frequency of resistant plants should increase over generations in segregating populations grown on infested land. Hartwig et al. (1982) obtained results to support this hypothesis.

The bulk method has been used effectively to inbreed populations grown under field conditions in the area to which the segregates are adapted. 'GaSoy 17' was selected from a bulk population inbred in Georgia and 'Douglas' was obtained from a bulk population inbred in Kansas (Baker and Harris, 1979; Nickell et al., 1982). The population from which 'Dowling' was obtained had been advanced in bulk on low-lying, slowly drained clay soil in Texas from the F_3 to the F_5 generations (Craigmiles et al., 1978). The cultivar is considered best adapted to the prairie soils of the Texas Gulf Coast.

Use of the bulk method is restricted by the current emphasis on rapid cultivar development. It is not well suited for inbreeding populations in greenhouses or tropical nurseries where natural selection may be undesirable because the productivity of genotypes may be considerably different than under field conditions in their area of adaptation. Artificial selection among individuals or lines is more direct and rapid in breeding for pest resistance than relying on natural selection in a bulk population.

7-3.1.3 Mass Selection

The frequency of desirable segregates for some characters in a bulk population can be increased during inbreeding by selection of plants or seeds with an appropriate phenotype. Selection among plants or seeds can be practiced in association with either the bulk or single-seed descent methods.

The published reports on mass selection have involved direct and indirect selection for characters in association with the bulk method. Direct selection for seed size has been effective for altering the frequency of segregates with large or small seed (Fehr and Weber, 1968). Seeds harvested in bulk were passed over sieves to obtain the desired size. Direct mass selection for maturity was practiced as a means of reducing undesirable shifts for late maturity in a bulk population (Empig and Fehr, 1971). Plants of early, midseason, and late maturity were selected from a population, threshed as separate bulks, and grown as three subpopulations the next generation.

Indirect mass selection for specific gravity was used to change the frequency of segregates for protein and oil composition (Hartwig and Collins, 1962; Smith and Weber, 1968; Fehr and Weber, 1968). Seeds from a population were placed in a glycerol-water solution in which seeds with a high density sank and those with a low density floated. Selection of the high-density fraction increased the frequency of genotypes with high protein and selection of the low-density fraction favored segregates with high oil. Buzzell and Haas (1972) suggested the use of indirect mass selection for large seed in a population segregating for resistance to phytophthora rot when grown on a soil infested with the causal organism. The disease would cause susceptible segregates to produce smaller seeds than those obtained from resistant ones. They proposed that by selecting for large seed in a population harvested in bulk, the frequency of resistant segregates would increase. Hartwig et al. (1982) cautioned that the suggestion of Buzzell and Haas (1972) would not be appropriate if there was genetic segregation for seed size and small-seeded segregates were desired.

Direct or indirect mass selection can be used in association with single-seed descent. When populations are grown in environments where genetic differences are expressed for a character, seeds can be harvested from only those plants with the desired phenotype. By harvesting one or a few seeds from selected plants, genetic variability may be maintained for unselected characters more effectively than when selected plants are threshed together in bulk. Single-seed decent is used with direct mass selection for resistance to Fe-deficiency chlorosis at Iowa State University by growing populations on calcareous soil and harvesting seed from plants without excessive yellowing. Indirect mass selection for seed size can be carried out in tropical winter nurseries by harvesting one or a few seeds from plants with appropriate pod width (Bravo et al., 1980; Frank and Fehr, 1981; Cianzio et al., 1982).

7-3.1.4 Single-seed Descent

The concept of single-seed descent was described by Brim (1966), who referred to it as a modified pedigree method. Three procedures have been used to implement the method.

7-3.1.4.1 Single-Seed Procedure—Single-seed descent in the strict sense refers to planting a segregating population, harvesting a sample of one seed per plant, and using the one-seed sample to plant the next generation. When the population has been advanced from the F_2 to the desired level of inbreeding, the plants from which lines are derived will each trace to different F_2 individuals. The number of plants in a population declines each generation due to failure of some seeds to germinate or some plants to produce at least one seed. As a result, not all of the F_2 plants originally sampled in the population will be represented by a progeny when generation advance is completed. When the single-seed procedure is used, most breeders prefer to harvest and retain a reserve sample of seed each generation in case the one planted is destroyed.

7-3.1.4.2 Multiple-Seed Procedure—Soybean breeders commonly harvest one or more pods from each plant in a population and thresh them together to form a bulk. Part of the bulk is used to plant the next generation and part is put in reserve. The procedure has been referred to as modified single-seed descent or the pod-bulk technique.

The multiple-seed procedure has been used to save labor at harvest. It is considerably faster to thresh pods with a machine than to remove one seed from each by hand for the single-seed procedure. The multiple-seed procedure also makes it possible to plant the same number of seeds of a population each generation of inbreeding. Enough seeds are harvested to make up for those plants that did not germinate or produce seed.

Genetic variability in a population inbred by the multiple-seed procedure may be less than for the single-seed procedure. There is no way to control the number of seeds from each of the original F_2 individuals that are included in the sample planted each generation. Some inbred lines derived from the population may trace to the same plant in a previous generation.

7–3.1.4.3 Single-Hill Procedure—It can be desirable for certain types of breeding research to have each F_2 plant represented by a progeny when inbred lines are derived from a population (Fehr, 1978). This can be accomplished by growing progeny from each F_2 plant in a separate hill or short row each generation. Enough seeds are planted from an F_2 family each generation to assure that at least one will develop into a plant with mature seeds.

The single-hill procedure requires considerably more labor and space than either the single-seed or multiple-seed procedures. For these reasons, it is not generally used to inbreed populations for cultivar development.

7–3.1.4.4 General Considerations—Single-seed descent is currently the most widely used method of inbreeding soybean populations. 'Century' was an F_5 plant selection from a population advanced by single-seed descent in Indiana (Wilcox et al., 1980). 'Pella' was selected from a population that was inbred by single-seed descent in Puerto Rico and Iowa (Bahrenfus and Fehr, 1980a).

The popularity of single-seed descent is due to the widespread use of greenhouses and tropical nurseries to grow up to two generations during the winter months. In these conditions, the growth and production of plants frequently is not typical of that observed in their area of adaptation. Productivity of plants does not influence the genetic makeup of a population advanced by single-seed descent as long as plants produce at least one seed.

The conditions that favor a reduced generation time are short daylength and relatively high temperatures (Fehr, 1980). A 12-h daylength provides rapid flowering and seed production. Shorter daylengths do not reduce generation time for Maturity Groups VII or earlier, but 8- or 10-h daylengths may be beneficial for Maturity Groups VIII or later (Garner and Allard, 1920; Byth, 1968; Lawrence and Fehr, 1982). Temperatures of 26 to 32°C favor rapid flowering with a 12-h daylength (van Schaik and Probst, 1958).

Generation advance by single-seed descent in the greenhouse does not cause consistent changes in segregating populations for agronomically important characters (Martin et al., 1978; Sediyama and Wilcox, 1980). Plant height in the greenhouse can be reduced to 20 to 30 cm by spraying the antigibberellin Amo-1618 at a concentration of 0.25 g kg^{-1} (w/w) under a 12-h daylength or a concentration of 0.5 g kg^{-1} under a 14-h daylength (Wilcox, 1974). The chemical is applied when plants have one fully-developed trifoliolate leaf and again when there are two fully-developed trifoliolate leaves.

The locations used by U.S. soybean breeders for winter nurseries include southern Florida, Puerto Rico, Hawaii, Belize, and countries in the southern hemisphere. Locations in the northern hemisphere have daylengths of about 12 h or less from November to May, which facilitates the production of two crops during winter. One generation is obtained with the longer daylengths that occur in Chile and other locations used in the southern hemisphere during the same period.

The successful use of a tropical nursery for generation advance requires attention to temperature, pest control, and promptness of seed harvest. Temperatures of 21°C or less can delay the onset of flowering (van Schaik and Probst, 1958). Pod set can be prevented on some genotypes that are exposed to 18°C for 2 weeks after flower initiation (Hume and Jackson, 1981). Cool temperatures also delay maturation of the crop (Major et al., 1975). Diseases and insects can reduce the quality and quantity of seed in tropical climates. Systematic application of fungicides and insecticides may be necessary for successful production of the crop. High temperatures and relative humidity in tropical climates can lead to rapid loss of germination when seed is not harvested promptly.

7-3.1.5 Early Generation Testing

The goal of early generation testing is to identify desirable heterogeneous populations or lines from which superior homozygous individuals can be selected. Several different procedures for implementing early generation testing have been described and evaluated. Selection can be practiced among populations or among individuals within populations.

7-3.1.5.1 Selection Among Populations—

7-3.1.5.1 Heterosis of F_1 Plants

The degree of heterosis expressed by spaced F_1 plants from two-parent crosses was evaluated by Weiss et al. (1947) and Kalton (1948) as a means of identifying superior populations. Their results indicated that the degree of heterosis for seed yield was not a reliable indicator of those populations that would contain high yielding progeny.

7-3.1.5.2 Space-planted Populations

Weiss et al. (1947) evaluated the performance of populations in replicated tests based on the mean of measurements from individual F_2 plants spaced 20-cm apart. They compared the yield, maturity, and lodging of the F_2 populations with the average of F_4-derived lines in F_5 chosen from them by pedigree selection. They stated that the average performance of individual spaced F_2 plants provided significant information on yield, maturity, and lodging resistance. They indicated, however, that measurement of spaced F_2 plants did not provide an adequately accurate evaluation to permit rigorous selection among crosses, except to eliminate those with inferior lodging resistance or unsuitable variation for maturity.

7-3.1.5.3 Bulk Populations

The performance of populations can be evaluated in plots that are planted at normal densities and harvested in bulk. Populations with superior performance as a bulk are considered a potential source of superior inbred lines.

Selection among populations based on the performance of bulk populations in early generations was evaluated by Weiss et al. (1947) and Kalton (1948). They concluded that selection among bulk populations for yield was not effective. Populations with the highest yield in the F_2 generation did not consistently provide the highest yielding progeny. The relative performance among populations was not consistent when they were tested in different generations, as reflected by a significant population × generation interaction. Useful information from an early generation test of bulk populations was obtained for the variation of maturity, height, and lodging among and within populations.

7-3.1.5.2 Selection within Populations—The progeny of individual F_2 or F_3 plants have been evaluated in replicated tests to identify those with the potential to provide superior inbred progeny. Several procedures have been described for the evaluation of progeny from an individual plant, referred to as a family.

7-3.1.5.2.1 Bulk Family

The progeny of a plant can be maintained in bulk during 1 or more years of evaluation in replicated tests. A superior family selected from the test is advanced in bulk until an adequate level of homozygosity is attained. Progeny of inbred plants selected from the bulk family are subsequently evaluated as potential cultivars.

High- and low-yielding F_3-derived lines were selected by Raeber and Weber (1953) in a replicated test and F_5 plants were obtained from each. The F_5-derived lines from the high-yielding families were significantly superior to those from the low-yielding families.

The relationship observed between yield of F_2-derived lines evaluated by the bulk-family procedure and that of homogeneous lines selected from them seems to depend on the extent to which genotype × environment interaction can influence the comparison. Weiss et al. (1947) compared in separate environments the yield of F_2-derived lines with that of an F_3-derived line obtained from each of them. There was a significant negative correlation of −0.15 between the two generations due to markedly different relationships between yield and maturity in the two environments used for testing.

More positive results have been reported when the heterogeneous lines and their progeny are tested in the same environment. Byth and Caldwell (1970) compared the yield of F_2-derived lines with the average of two F_5-derived lines selected from them. There was a significant positive correlation of 0.60 between the yield of the F_2-derived lines and their progeny. Gedge et al. (1978) evaluated the yield of F_2-derived lines and four F_5-derived lines selected from each. The regression coefficients averaged 0.94 when the F_2-derived line performance was regressed on the mean of the F_5-derived lines.

7–3.1.5.2.2 Replicates of Individual Progeny

Each replicate in a test of a heterogeneous family can be planted with the progeny of a different plant from within that family. Voigt and Weber (1960) selected three agronomically desirable F_3 plants from F_2-derived lines. The progeny of each F_3 plant was grown in a separate replicate of a test to determine the yield potential of the F_2 family. The average of the three replicates was used to select the best F_2 families, then an F_4 plant was chosen from the highest yielding of the three replicates of a family.

7–3.1.5.2.3 Step-wise Selection Among and Within Families

The evaluation of a family in replicated tests over 2 or more yrs can involve step-wise selection among and within families. Weiss et al. (1947) evaluated F_2-derived lines in F_3 for yield, maturity, lodging, and height in three replications planted at the normal density. In separate plots, F_3 plants of each line were spaced 20-cm apart and 12 individuals were selected and threshed individually. The F_2-derived lines with the desired maturity, lodging resistance, height, and yield were selected, then one of the 12 F_3 plants from each selected line was chosen for evaluation the following season as an F_3-derived line in F_4. The F_3-derived lines were grown in a replicated test, the best lines were selected, 12 F_4 spaced plants were harvested from separate plots, and one of 12 plants was evaluated as an F_4-derived line in F_5 the following season.

7–3.1.5.2.4 Pure-line Family

The genetic potential of a heterogeneous line can be evaluated by the performance of one of its inbred progeny. Ivers and Fehr (1978) harvested individual F_2 plants from a population and advanced the bulk progeny of each to the F_5. The progeny of one random F_5 plant from each F_2 family was evaluated in a replicated test. Performance of the F_5-derived lines was used to identify F_2 families from which additional selections could be made.

7–3.1.5.3 General Considerations—The pure-line family and bulk-family procedures were compared by Ivers and Fehr (1978). The two procedures were similar for the number of high-yielding lines identified. They indicated that the choice between the procedures would depend on the breeding objective and the facilities available.

Selection among populations by early generation testing in replicated tests generally is not used by breeders in the USA. A visual assessment of the merits of populations may be made for characteristics such as maturity and lodging, but not in replicated tests. The average performance of a cross for yield and other characteristics can be estimated from the mean of the parents per se (Leffel and Hanson, 1961; Thorne and Fehr, 1970b).

Early generation testing for yield within populations is used to a limited extent. The bulk family procedure is the simplest and most widely adopted of the alternatives available. Cultivars developed by early generation testing of heterogeneous lines include 'Elf' and related determinate cultivars of Maturity Groups II and III (Cooper, 1981).

The effectiveness of selection among heterogeneous lines for yield is influenced by a number of factors. One of the important factors is the effect of genotype × environment interaction when lines are evaluated in a limited number of locations and years. Selection among heterogeneous lines is similar in effectiveness to selection among homogeneous lines when an equivalent amount of testing is utilized. Byth et al. (1969) measured actual yield advance from selection of the top 10% of F_2- and F_5-derived lines tested in the same environments. The percentage of yield increase over the population means averaged 10.2% for F_2-derived lines and 13.2% for F_5-derived lines when selection was based on performance in three environments. The ineffectiveness of selection for yield among F_2-derived lines reported in several papers may be due to insufficient testing (Weiss et al., 1947; Kalton, 1948; Kwon and Torrie, 1964b). Breeders do not expect to identify superior homogeneous lines with tests conducted in two replications at one environment; therefore, it is not surprising that limited tests of heterogeneous lines are relatively unsuccessful.

Most breeders are unwilling to devote the same amount of resources to evaluation of heterogeneous lines as they do for identifying superior homogeneous lines that can be released as cultivars. Part of the reason is that tests of heterogeneous lines in 2 or more yrs increase the length of time for cultivar development compared with single-seed descent when facilities are available to grow one or two generations during the winter (Fehr, 1976; Fehr, 1978). Another consideration is that tests of heterogeneous lines reduce the resources available for evaluation of homogeneous lines that are potential cultivars.

Interplot competition may bias selection among heterogeneous lines when unbordered plots are used for yield evaluation. Boerma and Cooper (1975b) reported that inbred lines obtained by early generation testing of F_2-derived lines were consistently later than those obtained by pedigree or single-seed descent. They suggested that late heterogeneous lines were favored in the unbordered plots used for the early generation test. Kenworthy and Brim (1979) used nine-hill plots to avoid interplot competition when testing the seed yield of S_1 lines in a recurrent selection program.

Proper classification of heterogeneous lines for maturity can be a problem. It is difficult to determine which lines have equivalent maturity when there are differences in segregation for the character. The range in maturity among plants may be 2 days for one line, 5 days for a second line, and 10 days for a third line; yet all may have the same average maturity. Determination of average maturity in a line that has broad segregation for the character is complicated by the tendency to be biased toward lateness by a few late-maturing plants (Byth and Weber, 1968). If maturity is based on 95% of the pods mature in a plot, a segregating line with some late plants is likely to be considered equivalent in maturity to one that is uniformly late.

Intergenotypic competition within heterogeneous lines was suggested as a possible source of bias in an early generation test for yield (Leffel and Hanson, 1961). An evaluation of intergenotypic competition by Gedge et al. (1978) indicated, however, that it was not an important factor in the performance of F_2-derived lines.

Selection among lines derived in the F_2 or later generations with limited testing can be effective for characters other than seed yield. Weiss et al. (1947) demonstrated that evaluation of F_2- or F_3-derived lines in a single environment was effective for maturity, height, and lodging. F_2-derived lines have been used successfully in population improvement by recurrent selection for protein composition (Brim and Burton, 1979; Miller and Fehr, 1979) and resistance to Fe-deficiency chlorosis (Prohaska and Fehr, 1981). Special precautions may be necessary to avoid undesirable shifts toward late maturity when early generation testing is used for recurrent selection (Miller and Fehr, 1979).

7–3.2 Comparisons of Inbreeding Methods

Alternative methods of inbred line development have been compared for their effectiveness in identifying superior genotypes for yield. The studies will be discussed in the order they were reported.

A study by Raeber and Weber (1953) was described as a comparison of the bulk and pedigree methods. With our current terminology, the pedigree method they used would be referred to as early generation testing of F_3-derived lines. They evaluated F_5-derived lines from the five highest and the five lowest-yielding F_3-derived lines. For the bulk method, they selected phenotypically superior plants and a random sample of plants in the F_5 generation. They found no significant difference between the mean yield of selections from the five highest-yielding F_3- derived lines and that of progeny from phenotypically superior plants from the bulk. Both methods were superior to random lines from the bulk, which was superior to selections from the five lowest-yielding F_3-derived lines.

The pedigree and bulk methods were compared by Torrie (1958). Mean seed yields of lines obtained by the two methods were similar for five out of six populations. For one population, the mean yield for the bulk method was superior to the pedigree method.

Voigt and Weber (1960) compared the performance of F_4-derived lines obtained by the pedigree, bulk and early generation testing of F_2-derived lines. Early generation testing was superior to the other two methods for the mean yield of the lines evaluated and the frequency of lines that exceeded a check cultivar in yield. The bulk and pedigree methods had similar results for yield selection. They indicated that superior high-yielding lines were obtained by all of the methods.

Empig and Fehr (1971) evaluated the single-seed descent, cross bulk, restricted cross bulk, and maturity group bulk methods. The mean yield of the lines tested did not differ significantly among methods, but single-seed descent and the maturity group bulk were more effective for retaining early maturing segregates. Single-seed descent was about twice as effective as the other methods for maintaining large-seeded segregates.

Luedders et al. (1973) compared lines obtained by the pedigree, bulk, and early generation testing methods. Their early generation testing procedure involved step-wise evaluation of F_2- and F_3-derived lines. They did not observe significant differences among methods for mean yield of the selected lines.

Pedigree, single-seed descent, and early generation testing of F_2-derived lines were compared by Boerma and Cooper (1975a). No consistent differences among methods were observed for the mean yield of all lines tested, mean of the top five lines, or the highest-yielding line in four populations.

Ivers and Fehr (1978) evaluated lines by four methods; pedigree, single-seed descent, early generation testing by the pure-line family procedure, and early generation testing of F_2-derived lines. The pure-line family and early generation testing methods produced the highest frequency of superior F_5-derived lines, but the yield of the best line obtained by the four methods was not different.

The conclusion that seems evident from the comparisons cited above is that no method of inbreeding is clearly superior in effectiveness for identifying high-yielding lines. Consequently, the efficiency of selection is a primary consideration in the choice of an inbreeding method. Factors included in efficiency are the length of time required to obtain inbred lines and the amount of labor that must be devoted to inbreeding (Fehr, 1976; Fehr, 1978). The length of time required to obtain inbred lines for replicated testing influences the rapidity of cultivar development and the amount of genetic improvement per year in a breeding program (Fehr, 1976). Soybean breeders in the USA routinely use greenhouses or winter nurseries in tropical locations to reduce the number of years for inbred line development. The yield, maturity, height, and lodging of genotypes in a greenhouse or a tropical winter nursery is not representative of their performance in locations where they would be grown commercially. This prevents effective visual selection of characters by the pedigree method, eliminates replicated yield evaluation by early generation testing, and may result in undesirable natural selection with the bulk method. The method best suited and most widely used for inbreeding in greenhouses and tropical winter nurseries is single-seed descent. It permits the inbreeding of plants that produce at least one seed, regardless of their other characteristics in an environment.

The labor devoted to inbreeding can reduce the effort that can be expended for the evaluation of inbred lines in replicated tests. Pedigree selection requires extensive labor for planting and maintaining plots, selection and threshing of desirable lines and plants, and record keeping. With early generation testing, replicated evaluation of lines that are not homogeneous reduces the number of plots available for testing inbred lines. The bulk and single-seed descent methods are substantially less time-consuming than either pedigree or early generation testing, therefore more homogeneous lines can be included in replicated tests (Ivers and Fehr, 1978).

7–3.3 Number of Inbreeding Generations

A cultivar can be derived from a plant in the F_2 or later generations of inbreeding. The positive effects of increasing the number of generations of inbreeding are the larger percentage of homozygous plants from which lines with uniform appearance can be obtained and greater genetic variability among lines. A potential negative effect is the additional years that may be required before line evaluation can be initiated (Fehr, 1978). The decision concerning the number of generations of inbreeding often represents a compromise between these positive and negative effects.

No record was found of a cultivar that was an F_2 plant selection. 'Scott' was described as an F_3 plant selection from a two-parent cross (Williams and Luedders, 1967). Selections in the F_4 are most common for cultivar development, examples of which are 'Simpson' and 'Mead' (Lambert et al., 1982; Williams et al., 1982). Davis was obtained from a plant selected in the F_5, the second most common generation for the derivation of cultivars (Caviness and Walters, 1966). Cultivars derived in later generations include Verde, an F_6 plant selection, Provar, an F_7 plant selection, and 'Corsoy', an F_8 plant selection (Crittenden, 1971; Weber and Fehr, 1970a; Weber and Fehr, 1970b).

7–4 INBRED LINE EVALUATION

The evaluation of a line as a potential cultivar begins when individual plants are obtained from an inbred population and ends when the line is discarded or released for commercial use. The most commonly used strategy is to select individual plants in one season, grow the progeny of each in an unreplicated plot and visually select desirable lines in the second season, and conduct replicated yield tests beginning in the third season. An alternative procedure used to a more limited extent is to select individual plants in one season and begin replicated yield tests of their progeny the next season. Visual selection may be practiced in the replicated yield test to discard lines before harvest.

7-4.1 Selection Before or During Replicated Yield Tests

The amount of selection that can be practiced among plants depends on the environment in which the population is grown and the heritability of the character. When grown in the area of adaptation, selection may be practiced for maturity, height, and overall phenotypic desirability. The seed yield of individual plants is not measured. Plants may be evaluated for pest resistance through the use of artificial or natural infection. No selection may be possible if plants are harvested from greenhouses or tropical nurseries.

Visual selection among lines in unreplicated or replicated plots is effective for maturity, height, lodging, and most other characters. The usefulness of visual selection for yield has been evaluated in several studies. Hanson et al. (1962) used three experienced soybean breeders to independently evaluated plots of random F_2-derived lines before the seed yield of each was determined. The breeders agreed on 22% of the plots that were in the top one-fifth based on actual seed yield and 60% of the plots in the bottom three-fifths. They indicated that when extreme types for yield are not present, visual selection should be used primarily for elimination of low-yielding lines, rather than identifying lines with the highest yield. Kwon and Torrie (1964a) also concluded that low-yielding plots were more readily identified than high-yielding ones.

Byth et al. (1969) rated heterogeneous F_2-derived lines and homogeneous F_5-derived lines at maturity with a score ranging from 1 for the least desirable to 5 for the most desirable phenotype. They observed that visual selection for agronomic desirability was effective for identifying high-yielding genotypes under certain conditions. The effectiveness of their visual selection decreased as mean yield of an environment was reduced due to stress. Visual selection was more effective among their homogeneous lines than among their heterogeneous ones. Garland and Fehr (1981) used phenotypic score for visual evaluation of random lines in hill and row plots. They selected the best 10 of 50 lines based on seed yield in replicated plots and the best 25 based on a visual rating. An average of 8 of the top 10 lines were included in the group that was visually selected in hill or row plots. Two of the high-yielding lines frequently were not selected visually due to their susceptibility to lodging and variability in maturity.

There has been extensive research to identify characters that can be used for indirect selection of yield. The goal has been to find a character highly correlated with yield that can be measured with less cost and greater accuracy on more genotypes than by replicated yield tests. At present, no physiological traits have been found that permit indirect selection to be used effectively for inbred line evaluation. Total photosynthesis measured on a single light-saturated leaflet and harvest index were not closely associated with seed yield (Ford et al., 1983; Buzzell and Buttery, 1977; Kenworthy and Brim, 1979; Schapaugh and Wilcox, 1980). Canopy-apparent photosynthesis was significantly correlated with seed yield, but the cost of measuring the character currently is more expensive and time consuming than evaluation of yield per se (Harrison et al., 1981). The potential usefulness of length of the seed-filling period is one of the characters currently under investigation (Dunphy et al., 1979; Reicosky et al., 1982).

7–4.2 Replicated Tests

Genetic potential for seed yield is the most difficult character to determine for lines selected from a segregating population. Weber and Horner (1957) indicated that yield required 33 times more replication than oil percentage and 50 times more than protein percentage to obtain comparable precision. The evaluation process must include as many lines as possible to increase the chance of obtaining a superior segregate. It must be precise enough to separate genetic differences from experimental error and genotype × environment interaction. With a fixed amount of resources available for yield evaluation, it is necessary to find an appropriate balance between number of lines evaluated and the precision of the tests.

7–4.2.1 Plot Type

Plot types used for yield evaluation range from unbordered hills to bordered rows over 10 m in length. Factors that influence the type of plot include the level of precision required, the amount of seed available, and the type of equipment used for planting and harvest.

7–4.2.1.1 Hill Plots—The smallest plot type currently in use for yield evaluation is a single unbordered hill. Hills are used by a limited number of breeders in the northern USA for the first yield evaluation of lines from a segregating population. One advantage of hill plots compared with larger row plots is that sufficient seed can be obtained from a plant to grow a replicated test without a generation of seed increase (Garland and Fehr, 1981). This can reduce by 1 yr the length of time required for cultivar development. A second advantage is the lower cost per plot, which becomes a consideration when a large number of lines are evaluated.

The optimum spacing between unbordered hill plots for genotypes of Maturity Groups II and IV was evaluated by Shannon et al. (1971a) with indeterminate, semideterminate, and determinate isolines of 'Harosoy' and 'Clark.' Intergenotypic competition between hills decreased as the spacing between hills increased until no competitive effects were evident for the Harosoy genotypes at 65 cm and for the Clark genotypes at 89 cm. Maximum yield differences among the three isolines of a cultivar occurred at spacings in excess of 65 cm for Harosoy and 76 cm for Clark.

Shannon et al. (1971b) studied the effect of plant density at different hill spacings for the cvs. Harosoy 63 and Clark 63. Yields were not significantly different among 3-, 6-, 9-, and 12-plant hills through spacings from 30 to 65 cm for Harosoy 63 and from 30 to 104 cm for Clark 63. At wider spacings than these, yields of 6-, 9-, and 12-plant hills were similar and significantly greater than 3-plant hills. Their data indicated that there would be no advantage to thinning hills to a constant stand when the number of plants per hill was between 6 and 12.

The relative performance of genotypes in hill and row plots has been compared in several studies. Torrie (1962) evaluated cultivars of Maturity Groups I and II in hill plots spaced 1-m apart with 10 or 15 plants per hill and in single-row plots 5.5-m long with the center 4.9-m harvested for yield. The cultivar × plot type interaction was significant for seed yield in 2 of 11 comparisons across 4 yrs. The coefficients of variability averaged 9.6% in rows, 14.2% in 15-plant hills, and 13.2% in 10-plant hills. He indicated that more replication would be required with hill than with row plots to obtain comparable precision for measuring differences among cultivars for seed yield, but not for height, maturity, lodging, protein percentage, or oil percentage.

Green et al. (1974) evaluated nine cultivars of Maturity Groups O to II in hill plots spaced 51- or 102-cm apart and in bordered plots with rows 18-, 51-, and 102-cm apart. The correlations between hill and row plots for yield ranged from 0.84 to 0.91 for the 102-cm rows, 0.29 to 0.84 for the 51-cm rows, and 0.44 to 0.74 for the 18-cm rows. Relative performance of lines in hills was consistent with that of the three types of row plots for height and maturity, but not for lodging and seed size.

Garland and Fehr (1981) compared 50 random lines of Maturity Group II in hill plots spaced 1-m apart and in two-row plots. The phenotypic correlation between hills and rows averaged across eight environments was 0.84 for yield. Hill plots were as effective as row plots for identifying the two highest-yielding lines. Phenotypic correlations between hill and row plots for maturity, height, and lodging were 0.86 or larger. They indicated that the hill plots are most useful for the first yield evaluation of lines, and are not considered a substitute for rows in more advanced phases of testing.

Hill plots have been considered for yield evaluation in the southern USA for genotypes in Maturity Groups V or later. To reduce the effects of intergenotypic competition that are realized in the South, Schutz and Brim (1967) suggested the use of a nine-hill plot for screening large groups of lines for yield in a limited space. The potential bias from intergenotypic competition for the nine-hill plots would be about 30% of that of a single unbordered hill. They compared four cultivars in nine-hill plots and in row plots in one environment and observed the same ranking for yield. Kenworthy and Brim (1979) successfully used the nine-hill plot for recurrent selection for yield in North Carolina. However, the nine-hill plot has not been widely adopted for yield evaluation up to the present time.

7–4.2.1.2 Single-Row Plots—The yield of a genotype in a single row can be biased by intergenotypic competition with adjacent plots. Hartwig et al. (1951) evaluated in North Carolina and Mississippi the yield of two cultivars bordered by genotypes with differing maturity, height, lodging, and plant type at row spacings of 91 to 112 cm. Seed yield of the cultivars was significantly increased or decreased by border genotypes. The border genotypes did not influence the two cultivars similarly, the response varied among locations, and grouping cultivars by maturity and plant type did not eliminate the border effect. They concluded that single-row plots do not give accurate estimates of yield performance in areas where soybean strains make rank growth. A study of intergenotypic competition for cultivars of Maturity Groups IV and V grown in 97- or 102-cm rows were conducted in Maryland by Hanson et al. (1961). They reported that competition between single-row plots could create major biases in yield.

The amount of intergenotypic competition between single rows for cultivars of Maturity Groups I and II was evaluated in two studies. Lin and Torrie (1970) evaluated competition between cultivars in rows spaced 46- and 91-cm apart in Wisconsin. Significant effects of competition were observed for 5 out of 12 comparisions at 46 cm and 1 out of 12 at 91 cm. Competition effects were considered important at the narrow spacing, but not in the wide spacing. Gedge et al. (1977) reported that the average change in yield associated with interplot competition between five cultivars in Iowa was 2.6% for rows 100-cm apart, 5.3% for 75 cm, 8.0% for 50 cm, and 17.6% for 25 cm. Significant changes in yield in 100-cm rows of up to 7.2% were associated with competition. The authors of both studies concluded that single-row plots were not suitable for narrow row spacings. Their use at a spacing of about 100 cm could be subject to bias due to interplot competition between some cultivars.

Single-row plots are not used for yield evaluation of genotypes of Maturity Groups V or later in the southern USA. For genotypes of Maturity Groups IV or earlier in the northern USA, single-row plots spaced about 1-m apart are sometimes used for the first replicated yield test. There is some experimentation with plots of about 1 m in length, but the majority of one-row plots for the first yield test are 3- to 5-m long.

7–4.2.1.3 Multiple-Row Unbordered Plots—The yield bias caused by intergenotypic competition between adjacent rows can be reduced by planting and harvesting a plot with two or more rows. If the bias caused by intergenotypic competition is expressed as 100%, the bias will be 50% in a two-row plot, 33% in a three-row plot, 25% in a four-row plot, and 20% in a five-row plot. Hanson et al. (1961) suggested the use of two rows 2.5-m long instead of one row 5-m long when seed from individual plants was limited. Gedge et al. (1977) indicated that an unequal row spacing within and between plots could minimize the land area required for multiple-row plots without increasing intergenotypic competition. For example, two-row plots at a spacing of 75 cm within and between plots would be subject to more bias from intergenotypic competition than two-row plots with a 75-cm space between rows within a plot and a 1-m space between plots.

Two-row plots are used by some breeders for the first replicated yield test in the southern USA for Maturity Groups V and later. They are used extensively in the northern USA for the 1st and 2nd yr of replicated tests.

7–4.2.1.4 Plots with a Common Border—Use of a single cultivar as a common border between all unbordered plots has been considered for yield evaluation. Thorne and Fehr (1970a) compared the variability among lines when bordered by a common cultivar and when grown in bordered plots. The differences among lines for yield were biased by the competition with the common border. They indicated that common borders seemed to be of little value in eliminating the effects of intergenotypic competition. Breeders in the USA have not adopted common borders for replicated yield tests.

7–4.2.1.5 Bordered Plots—Bordered plots permit the evaluation of yield without bias from intergenotypic competition (Hartwig et al., 1951). They are used for the first and subsequent replicated tests of lines in the southern USA. Bordered plots are used for all maturity groups in the final stages of yield evaluation before a line is considered for release.

7–4.2.1.6 Plot Shape and Size—The relative width and length of row plots can influence the precision of comparisons among genotypes. Weber and Horner (1957) evaluated the precision of different plot sizes and shapes composed of basic units 2.5-m long by 0.6-m wide (8 by 2 feet) from a uniformity trial. They concluded that the optimum plot size was 3.2 times the basic unit. Long, narrow plots gave greater precision than short, wide plots when incomplete block designs were utilized.

Plot shape and size generally are based on practical considerations. Small plots are necessary if seed from a single plant is used for a replicated test. Even when adequate seed is available, the length of plots is generally < 6 m. The primary exception would be plots 10 m or longer that are harvested with a commercial combine. Plot length generally is determined by the number of rows included. A length of about 5 m is common for the harvest of one row in a single unbordered plot or in a three-row bordered plot. A length of about 3 m is frequently utilized for the harvest of two or more rows in unbordered or bordered plots. Four-row bordered plots are preferred to three-row plots because a smaller percentage of the land area is devoted to borders. Five-row bordered plots are common for row spacings of 40 cm or less.

7–4.3 Resource Allocation for Yield Evaluation

The resources available for yield evaluation are divided among lines, replications, locations, and years. The breeder must decide on the number of lines to test, the number of replications and locations to use at each stage of the evaluation process, and the number of years of testing to conduct before a line is considered for release.

7.4.3.1 First Yield Evaluation

The primary emphasis in the first yield evaluation is to discard inferior lines. A total of two or three replications are used to permit the evaluation of as many lines as possible. The replications may be at one or more locations in the area to which the lines are adapted. One replication at each location has two primary advantages over multiple replications at a single location. First, strengths and weaknesses of a line in different environments can be examined. When it is important to test lines for response to factors that cannot be controlled, such as moisture, temperature, and pest infestations, the probability of obtaining favorable conditions for evaluation increases as the number of environments is increased. In statistical terms, the heritability for yield will be greater with one replication at several environments than several replications at one location when genotype × environment interaction is important. The heritability on an entry-mean basis can be expressed as:

$$h^2 = \frac{\sigma_g^2}{(\sigma_e^2/r\mathrm{E} + \sigma_{gE}^2/\mathrm{E} + \sigma_g^2)}$$

where h^2 = heritability, σ_g^2 = genotypic variance among lines, σ_e^2 = experimental error, σ_{gE}^2 = genotype × environment interaction, r = number of replications, and E = number of environments. When one replication is used at three locations, the denominator becomes $\sigma_e^2/3 + \sigma_{gE}^2/3 + \sigma_g^2$. For three replications at one location, the denominator is $\sigma_e^2/3 + \sigma_{gE}^2/1 + \sigma_g^2$. The potentially higher heritability with one replication at each environment increases the effectiveness of selection among lines.

The second advantage of more than one environment in the first yield test is the reduced risk of losing the lines or not being able to obtain any useful data during the season. A severe drought, hail, or other natural disaster can disrupt a breeding program if it occurs at the site where lines are being evaluated.

The primary disadvantage of growing one replication at two or more locations is the higher cost. Less time and labor are required when all data are obtained at one central location.

The percentage of lines to select from the first yield test has been estimated. Luedders et al. (1973) evaluated F_4-derived lines in four replications at one location during 2 yrs. Selection of the top 25% of the lines after the 1st yr retained the lines with the highest mean yield averaged over the 2 yrs. Baihaki et al. (1976) evaluated breeding lines of Maturity Groups O to I in three replications at each of three Minnesota locations during 2 yrs. Selection of the best 33% of the lines in any single-environment test identified 11 of the top 16 lines and three to five of the highest-yielding lines. Tests at three locations in 1 yr usually identified only one or two more of the superior lines. Selection of 50% of the lines instead of 33% identified only two or three more of the top 16 lines.

Garland and Fehr (1981) determined the percentage of lines that it was necessary to save in the first yield test to retain the highest-yielding lines. Fifty random lines were evaluated in three replications of random and nonrandom hill plots and two-row plots at four Iowa locations during 2 yrs. When selection was based on a single environment, the best line was retained when an average of 8% of the highest-yielding lines were selected in random and nonrandom hill plots and 9% were selected in rows. To retain the best five lines, it was necessary to select 56% of the lines in random hills, 42% of the nonrandom hills, and 35% of the rows.

7-4.3.2 Second Yield Evaluation

The emphasis in the 2nd yr of yield evaluation is on identification of lines that warant extensive testing as potential cultivars. Hanson and Brim (1963) suggested the use of replicated tests at five locations for the 2nd yr of yield evaluation. In addition to more locations, breeders may use larger plots with more control of intergenotypic competition than in the first yield test.

7-4.3.3 Third and Subsequent Yield Evaluations

Evaluation of yield in the third and subsequent yield evaluations involves tests at a relatively large number of locations in areas where a line could be grown commercially. Schutz and Bernard (1967) indicated that data from 10 to 15 locations in a single year should be adequate to eliminate low-yielding lines in a regional test. They considered 2 or 3 yrs of regional testing to be adequate for the final evaluation of lines before release.

Public soybean breeders cooperate in regional Uniform Tests that have been established for each maturity group. Lines of a maturity group from different public breeding programs are evaluated together by states that are interested in the possible release of superior genotypes as cultivars. This permits the collection of data from many different environments in a single year. For the 1st yr of regional evaluation, referred to as the preliminary test, two replications of bordered plots generally are grown at each location. In subsequent years, lines are grown in three or four replications at more locations than were used for the preliminary evaluation.

Private companies conduct their own tests over a large number of locations in the final years of evaluation before release. In addition to conventional yield test plots, some companies evaluate a limited number of lines over a broad geographical area with large plots that are planted and harvested by farmers. One replication of plots six to eight rows wide and over 15-m long are used at each farm. Although yield comparisons of entries at individual locations are not practical with a single replication, the genotype × location interaction can be used to test for differences in the mean performance of lines across locations. The tests provide useful information on lodging, harvestability, and other agronomic characters that may influence acceptance of a new cultivar by the farmer.

7-4.3.4 Tests for Unique Environments

The importance placed on the development of cultivars for unique environments can influence the allocation of resources for yield evaluation. In the South, soybeans are planted as a full-season crop and as a second crop after small grains are harvested. Carter and Boerma (1979) suggested that a separate breeding program was needed for the development of full-season and late-planted cultivars. Lines to be used as full-season cultivars would be planted in May or early June in a row spacing of about 1 m. A late-planted, narrow-row environment was suggested for development of double-cropped cultivars. Separate breeding programs have the potential liability of reducing by half the amount of available resources for the genetic improvement of cultivars for each planting date.

In the northern USA, soybean is planted as a full-season crop in rows 15-cm to 1-m apart. The importance of developing different cultivars for wide and narrow rows has been considered. When yield in narrow rows is expressed as a percentage of the yield in wide rows, cultivars with early maturity generally have a higher percentage of yield increase than those of late maturity. Within a particular maturity, however, there is no definitive evidence that cultivars with unique characteristics for narrow rows can be identified (Shibles and Green, 1979). Breeders generally evaluate lines in one row spacing until a limited number of high-yielding ones have been identified. The final stages of testing often includes an evaluation of performance in several row spacings.

The production of soybean in combination with other crop species is an alternative to monoculture in some parts of the world. In the USA, monoculture is the principal system of production, however, various types of intercropping are under investigation. Relay intercropping consists of planting soybeans in a field of small grains about 1 to 2 months before the crop is harvested. The planting system provides a longer growing season for the soybeans than can be obtained when planting is delayed until after the small grain is harvested. Research by McBroom et al. (1981a, 1981b) indicated that separate breeding programs are not required to develop cultivars for monoculture or intercropping. Soybeans that were intercropped with winter wheat (*Triticum aestivum* L.) or spring oats (*Avena sativa* L.) were shorter, lodged less, and produced lower yields than soybeans in monoculture, but no significant genotype × cropping system interaction was observed.

7-4.4 Techniques for Plot Management

The three primary aspects of plot management for yield evaluation are planting, end-trimming, and harvest. Major changes have occurred over time in the efficiency with which some of the operations are carried out. Increased efficiency has permitted a larger number of plots to be grown without an increase in labor.

7-4.4.1 *Planting*

The preparation of plots for planting includes the assignment of entry and plot designations, printing of field books, labeling of envelopes, and packaging of seed. Before computer facilities were available, all of these operations were done by hand. Breeders now use computers to carry out the preparation of field plans, the printing of field books, and the labeling of planting envelopes. The result has been a large increase in the number of plots that can be prepared and a reduction in human error. Packaging of seed is done by hand or with an electronic seed counter. The number of packets required for multiple-row plots has been reduced by the use of seed dividers on planting equipment (Clark and Fehr, 1973).

Plots are sown with tractor-powered planters of various designs. They permit plots to be planted rapidly at a controlled depth of seed placement. Systems are available that eliminate the need for marking alleys manually before plots are planted (Clark et al., 1978).

7-4.4.2 *End-trimming*

Plants at the end of a plot have more space available in the alley and are more productive than those in the center of a plot. The yield of plots will be inflated if the end plants are not removed immediately before harvest. Probst (1943) evaluated the yield of plots of the cvs. Illini, Dunfield, Mandell, and Mukden in Indiana when end-trimmed at emergence or at maturity. He reported that the yields of plots end-trimmed at emergence had an average of 16% more yield than those end-trimmed at maturity. The yield inflation could be removed from a plot by removing 30 cm from each end of the rows at maturity. He found that cultivars had different amounts of yield inflation when end-trimmed at emergence, but the differences were not enough to change their relative yields. Wilcox (1970) evaluated the yield of cultivars from Maturity Groups I to IV that were end-trimmed at different growth stages. The yield inflation at the end of a plot was least for Maturity Group I cultivars and greatest for those of Maturity Group IV. He observed that border effects could be adequately controlled for cultivars of Maturity Groups I to III by trimming 0.6 m from the end of each row at or beyond the stage when the top pods on plants contained seed approaching full size. For cultivars of Maturity Group IV, removal of the 0.6-m section would have to be delayed until leaves on plants were 30 to 50% yellow with a few falling. Boerma et al. (1976) studied the effect of end-trimming on the seed yield of cultivars from Maturity Groups V through VIII. Their results indicated that 0.76 m should be removed from both ends of yield plots at physiological maturity or a later stage to obtain reliable data for comparing genotypes in the southern USA.

In the early stages of yield evaluation, breeders in the northern USA commonly end-trim plots during vegetative or early reproductive stages for two reasons. First, the labor requirement is much less than when delayed until maturity. Second, trimming plots at maturity conflicts with the recording of data, selection activities, and harvest operations. The primary disadvantage of early end-trimming is its effect on the yield of lines differing in maturity. Three procedures are used in making line comparisons when plots are end-trimmed early. (i) Lines are not compared that differ appreciably in maturity. (ii) Yield is adjusted for maturity by regression analysis before comparisons are made among lines of different maturity. (iii) Yields are adjusted by multiplying the harvested weight by a factor that accounts for the yield inflation in plots end-trimmed early. Wilcox (1970) suggested the factors 0.874 for Maturity Group I cultivars, 0.840 for Maturity Group II, 0.796 for Maturity Group III, and 0.755 for Maturity Group IV.

The practice of early season end-trimming was not recommended by Boerma et al. (1976) for genotypes of Maturity Groups V to VIII. They observed significant changes in the rank and in the relative differences in yield among genotypes when early and late season end-trimming were compared. Plant height of genotypes was more closely associated with the extent of inflation in plot yields due to early season end-trimming than was maturity date.

Breeders throughout the USA usually end-trim plots at maturity for the evaluation of lines in the advanced stages of testing. To assure that the same length of row is harvested from each plot, the sections to be harvested may be staked early in the season or a rod of the desired length may be placed next to the row at the time of end-trimming. Plants beyond the stakes or the end of the rod are cut or pulled.

7-4.4.3 Harvest

Self-propelled combines are used extensively for the harvest of yield test plots. The combines are designed to provide rapid clean-out between plots and a minimum of seed mixture (Clark and Fehr, 1976). It is not possible to competely avoid seed mixtures, therefore stationary equipment is still used when pure seed is required.

The seed from a plot may be weighed and the moisture percentage determined on the combine. The data may be recorded manually or may be entered into an electronic data collector. Information can be transferred from the electronic data collector to the computer for analysis.

When the precision obtained by recording data on the combine is not considered adequate or seed must be saved for planting, the seed of plots is bagged and dried artifically or naturally before weighing. To avoid manually recording weight of the plots and entering it into the computer, plot and entry information can be printed on the harvest tag in a code that is read and recorded by the data collector. After the tag information has been recorded, the weight from an electronic scale is transferred to the data collector.

7-5 BREEDER SEED PRODUCTION

The production of pure seed for multiplication is the final step in the development of a cultivar. The method of purification and the stage in the testing program when seed purification is initiated varies among breeders.

7–5.1 Methods of Purification

Three methods of purification are used to obtain breeder seed, also referred to as basic seed. They differ in the purity of seed that is likely to be produced and in the time and labor required. The methods usually begin with a sample of seed or plants from replicated yield tests.

Purification of a cultivar can be based on selection of uniform seeds and plants, a method referred to as mass selection. Off-type seeds are removed from a sample of the cultivar, the selected sample is planted, off-type plants are removed, and the selected individuals are harvested in bulk to obtain breeder seed.

A second method is based on one generation of progeny testing. Uniform plants are selected, the seed characteristics of each are examined, off-type plants discarded, and the progeny of selected individuals are grown and inspected for plant characteristics. The homogeneous progeny rows with the same characteristics may be harvested in bulk. Alternatively, the progeny rows may be harvested separately, examined for uniformity of seed characteristics and other qualitative characters such as pest resistance, and then bulked as breeder seed.

The third method of purification involves two generations of progeny testing. The first generation of progeny testing is carried out in the same manner as described for the second method, except that selected progeny rows are not bulked. Each progeny row is harvested separately, the seeds are examined for uniformity, and the seed from each is grown in a separate seed increase the following season. Progeny with acceptable characteristics are harvested in bulk as breeder seed. Mixing of the harvested seed may be required as a separate step to assure that each bag of breeder seed contains a similar quantity from the individual progeny increases.

The number of seasons and the amount of labor required for breeder seed production is least for mass selection, intermediate for one generation of progeny testing, and greatest for two generations of progeny testing. The seed with the highest level of genetic purity is expected from two generations of progeny testing and the least purity from mass selection.

When either one or two generations of progeny testing are used to produce breeder seed, a decision must be made concerning the number of progeny to include in the breeder seed. The advantages of utilizing a large number of progeny include the greater chance of retaining heterogeneity within the cultivar for nonvisual characters that may favor reduced genotype × environment interaction and the greater quantity of breeder seed produced. There are no data on the relationship between number of progeny rows that are included in a cultivar and its stability across environments. Indirect information was obtained from a study of yield stability by Walker and Fehr (1978). They compared the stability of seed mixtures composed of 2 to 14 cultivars. Stability was measured by regression of entry yields on an environmental index (Eberhart and Russell, 1966). The number of components did not influence the differences among regression coefficients, but average mean squares for deviations from regression tended to decline with each additional component until eight were present. Their data indicated that a small number of progeny rows would be needed to maximize the effect of heterogeneity on stability. Breeders generally determine the number of progeny rows based on the quantity of breeder seed required.

7–5.2 Timing of Breeder Seed Production

The coordination of breeder seed production and line evaluation determines when a suitable quantity of seed will be available to farmers after the decision is made to release a new cultivar. The early initiation of purification permits seed to be available for commercial use sooner than when purification is delayed until a line is certain to be released. Conversely, early purification of a large number of lines is more expensive than waiting until the number has been substantially reduced. The method of purification also influences the time when breeder seed production is initiated. Purification can be delayed longer with mass selection than with one or two generatons of progeny testing.

7–6 COMMERCIAL USE OF SEED MIXTURES

A seed mixture of two or more cultivars is an alternative to growing each of the components individually in pure stand. The use of seed mixtures affects the breeder, the seed producer and merchandiser, and the farmer. Four primary reasons for considering mixtures will be discussed.

7–6.1 Marketing of Seed

One of the reasons mixtures have been considered is that they provide a means for seed producers to merchandise a product with a brand name of their choice. A mixture of cultivars, also referred to as a blend or multiline, can be sold under a brand designation without revealing the name of the components or their percentages, except as required by the developer or state seed laws. A seed producer without a breeding program can mix seed of cultivars, assign a brand name to the blend, and merchandise it as a unique product.

A blend can be used to encourage farmers to purchase seed each year. The frequency of components in the seed planted is not necessarily the same as in the seed harvested due to the effects of intergenotypic competition. As the frequency of components change, the performance of the blend may be altered. A seed merchandiser can indicate that the only effective way to be certain that a blend is the same from 1 yr to the next is to annually purchase seed with known percentages of the components.

7–6.2 Seed Yield

Blends have been evaluated as a means of increasing seed yield over that of cultivars grown in pure stand. The yields of a blend have been compared with their highest-yielding component and with the weighted mean of the components in pure stand when adjusted for their frequency in the blend. There is no evidence that a blend containing components with different yield potentials in pure stand will perform better than the highest-yielding component. Yields of blends have been reported that exceeded the highest-yielding component, but not by an amount that was statistically significant. For example, Brim and Schutz (1968) observed two- and three-component mixtures with yields 2 to 4% above the highest-yielding component, but the deviations in yield were less than the least significant difference among treatment means at the 5% probability level.

The frequency of blends that do not yield significantly more than the weighted mean of their components is much greater than the frequency of those with significant deviations. On the other hand, blends are more likely to exhibit overcompensation by exceeding the weighted mean yield of the components than they are to exhibit undercompensation by yielding less than the mean of the components.

Soybean breeders who are considering the development of blends must decide if there are breeding methods that will facilitate the selection of lines that will contribute to overcompensation in a blend. They also must determine if there are procedures that can be used to identify combinations of cultivars that exhibit overcompensation and the optimum frequency of each component. Finally, they must decide if yield increases can be achieved more effectively by expending resources to identify blends with overcompensation or by evaluation of pure lines with higher yield potential.

No special methods have been identified to permit the selection of lines specifically for use in blends rather than in pure stand. Gedge et al. (1978) evaluated inbred lines from heterogeneous F_2-derived lines for their tendency to promote overcompensation in a mixture. Pure lines from high-yielding heterogeneous lines did not result in more overcompensation when blended than pure lines from low-yielding heterogeneous lines.

Two general procedures have been proposed to identify combinations of genotypes that exhibit overcompensation when grown together in a mixture. One procedure is based on the measurement of intergenotypic competition between pairs of genotypes. Brim and Schutz (1968) indicated that intergenotypic competition between cultivars in a nine-hill plot was similar to the competition that occurred within a mixture. The four nine-hill plots required to evaluate intergenotypic competition between a pair of genotypes would be a pure stand of each, genotype 1 in maximum competition with 2, and genotype 2 in maximum competition with 1. Data on intergenotypic competition would be used to predict the performance of a mixture at different component frequencies. They observed a close relationship between predicted and observed yields of two- and three-component mixtures involving four cultivars adapted to the southern USA.

The intergenotypic competition between genotypes of Maturity Groups I and II in five- and nine-hill plots was evaluated by Cianzio (1970). The five-hill plots excluded the four hills of a nine-hill plot that were most distant from the center. The intergenotypic competition in the hill plots was not the same as observed within blends. As a result of their study, Fehr (1973) investigated the use of pairs of rows spaced 8-cm apart as an alternative to hill plots for measurement of intergenotypic competition. The paired-rows more effectively estimated the intergenotypic competition observed within a blend than did the hill plots studied by Cianzio (1970), but the competition in paired-rows did not agree entirely with that observed within a mixture.

A second procedure for identifying overcompensation between genotypes is to evaluate the yield of mixtures in replicated trials. This approach requires some restriction on the different frequencies of components that are evaluated. Fehr and Rodriguez (1974) inidicated that when yield is the primary consideration and one cultivar is consistently higher yielding in pure stand than another, it may be possible to limit the evaluation of mixtures to those in which the highest-yielding cultivar has a frequency of at least 70%. When two cultivars have similar yield potential, they suggested that the initial test be limited to the frequencies 0:4, 1:3, 1:1, 3:1, and 4:0.

A major difficulty in the evaluation of mixtures is distinguishing between experimental error and true overcompensation or undercompensation. Fehr and Rodriquez (1974) observed overcompensation in a blend of Corsoy and Amsoy at component frequencies of 1:3 and 3:1 and undercompensation at a 1:1 ratio. They further evaluated the apparent relationship between component frequency and compensatory response between the two cultivars by altering the frequencies in 2.5% increments (Fehr and Cianzio, 1980). Fluctuations in compensatory response at different component frequencies were not attributable to a true biological response, but instead were caused by random error. They suggested that the evaluation of blends for the purpose of obtaining significant overcompensation probably could not be justified. The extensive resources that must be expended to evaluate blends would more appropriately be spent on the identification of pure lines with greater yield potential.

7–6.3 Overcoming Deficiencies of High-Yielding Cultivars

A blend can be used to reduce risk in soybean production when the highest-yielding cultivar available is susceptible to a production problem. The use of a blend for this purpose is an appropriate consideration when the problem occurs sporadically and resistant plants compensate effectively for yield loss of the susceptible component. When the problem did not occur, the blend would yield less than a pure stand of the susceptible cultivar, but more than a pure stand of the resistant cultivar. When the problem was expressed, the resistant component in the blend would assure that some production was realized.

The concept has been applied to the problem of Fe-deficiency chlorosis of soybean (Fehr and Trimble, 1982; Trimble and Fehr, 1983). Cultivars with superior yield when grown on noncalcareous soils may suffer substantial yield reduction on calcareous soils due to their inability to utilize the soil Fe. Fields in affected areas generally have spots of calcareous soil that constitute varying percentages of the total production area. A farmer with fields containing calcareous soil may choose to plant a high-yielding susceptible cultivar in pure stand, a lower-yielding resistant cultivar, or a blend of the two. The most productive alternative depends on the percentage of calcareous soil in the field, the difference in yield between the susceptible and resistant cultivars on noncalcareous soil, and the degree of susceptibility in the high-yielding cultivar.

When a blend is used to overcome a deficiency in a high-yielding cultivar, the optimum frequency of the components will be determined by the maximum percentage of the high-yielding susceptible component that can be tolerated when the problem occurs (Fehr and Rodriquez, 1974). By using the highest frequency possible for the high-yielding component, the yield of the blend in the absence of the problem will be maximized. Trimble and Fehr (1983) determined the frequency of components that would be suitable to minimize yield loss from Fe-deficiency chlorosis on calcareous soils. The optimum percentage of the high-yielding susceptible component depended on its degree of susceptibility to Fe chlorosis; 80% for a moderately susceptible cultivar, 67% for a susceptible, and 61% for a highly susceptible cultivar.

The yield compensation of resistant plants for susceptible ones is not unique to Fe chlorosis. Ross (1983) evaluated the yield of mixtures of resistant and susceptible lines for soybean mosiac virus. He concluded that yield compensation by uninfected soybean plants probably plays a major role in moderating yield loss in fields with the disease.

7–6.4 Stability of Performance

Stability of performance across a range of environmental conditions is a desirable attribute for a soybean cultivar or blend. The concept of stability with respect to genotype × environment interaction is reviewed in chapter 6 in this book. From the viewpoint of identifying superior cultivars and blends, information obtained from the studies of Schutz and Brim (1971) and Walker and Fehr (1978) seem particularly relevant. The studies indicated that mixtures of cultivars are more stable on the average than individual cultivars in pure stand, but that some cultivars are as stable as mixtures. Walker and Fehr (1978) did not observe a consistent relationship between the stability of a pure line and the stability of mixtures in which it was a component. This may relate to the observation of Schutz and Brim (1971) that the stability of a mixture is associated with the types of intergenotypic competition that occur among the components. The implication of the two studies is that stability of performance for a blend cannot be assumed to be superior to cultivars in pure stand. If stability is an important consideration in identifying blends for commercial use, they must be evaluated over a broad range of environmental conditions.

7–6.5 Other Considerations

Identification of cultivars that perform well together in a blend involves consideration of traits in addition to seed yield. No appreciable change should be expected in the maturity, height, or lodging of a cultivar in a mixture compared with its performance in pure stand (Sumarno and Fehr, 1980). The difference in time of maturity between blend components must be limited to facilitate harvest. Large differences in seed size among components may create problems for planting and harvesting.

REFERENCES

Athow, K.L. 1973. Fungal diseases. *In* B.E. Caldwell (ed.) Soybeans: Improvement, production, and uses. Agronomy 16:459–489.

Bahrenfus, J.B., and W.R. Fehr. 1980a. Registration of Pella soybean. Crop Sci. 20:415.

----, and ----. 1980b. Registration of Weber soybean. Crop Sci. 20:415–416.

----, and ----. 1980c. Registration of Vinton soybean. Crop Sci. 20:673–674.

----, and ----. 1984. Registration of Lakota soybean. Crop Sci. 24:384.

Baihaki, A., R.E. Stucker, and J.W. Lambert. 1976. Association of genotype × environment interactions with performance level of soybean lines in preliminary yield tests. Crop Sci. 16:718–721.

Baker, S.H., and H.B. Harris. 1979. Registraton of GaSoy 17 soybeans. Crop Sci. 19:130.

Bernard, R.L. 1972. Two genes affecting stem termination in soybeans. Crop Sci. 12:235–239.

----, and C.R. Cremeens. 1982. Registration of Union soybean. Crop Sci. 22:688.

----, and D.A. Lindahl. 1977. Registration of Woodworth soybean. Crop Sci. 17:979.

----, and M.G. Weiss. 1973. Qualitative genetics. p. 117–154. *In* B.E. Caldwell (ed.) Soybeans: Improvement, production, and uses. Agronomy 16:117–154.

Boerma, H.R. 1979. Comparison of past and recently developed soybean cultivars in maturity groups VI, VII, and VIII. Crop Sci. 19:611–613.

----, and R.L. Cooper. 1975a. Comparison of three selection procedures for yield in soybeans. Crop Sci. 15:225–229.

----, and ----. 1975b. Effectiveness of early-generation yield selection of heterogeneous lines in soybeans. Crop Sci. 15:313–315.

----, W.H. Marchant, and M.B. Parker. 1976. Response of soybeans in Maturity Groups V, VI, VII, and VIII to end-trimming. Agron. J. 68:723–725.

----, E.D. Wood, and G.B. Barrett. 1982. Registration of Duocrop soybean. Crop Sci. 22:448–449.

Bravo, J.A., W.R. Fehr, and S.R. Cianzio. 1980. Use of pod width for indirect selection of seed weight in soybeans. Crop Sci. 20:507–510.

----, ----, and ----. 1981. Use of small-seeded parents for the improvement of large-seed cultivars. Crop Sci. 21:430–432.

Brim, C.A. 1966. A modified pedigree method of selection in soybeans. Crop Sci. 6:220.

----, and J.W. Burton. 1979. Recurrent selection in soybeans. II. Selection for increased percent protein in seeds. Crop Sci. 19:494–498.

----, and C. Elledge. 1973. Registration of Ransom soybean. Crop Sci. 13:130.

----, and W.M. Schutz. 1968. Inter-genotypic competition in soybeans. II. Predicted and observed performance of multiline mixtures. Crop Sci. 8:735–739.

Buzzell, R.I. 1971. Inheritance of a soybean flowering response to fluorescent-daylength conditions. Can. J. Genet. Cytol. 13:703–707.

----, and B.R. Buttery. 1977. Soybean harvest index in hill plots. Crop Sci. 17:968–970.

----, and J.H. Haas. 1972. Natural and mass selection estimates of relative fitness for the soybean *rps* gene. Crop Sci. 12:75–76.

Byth, D.E. 1968. Comparative photoperiodic responses for several soybean varieties of tropical and temperate origin. Aust. J. Agric. Res. 19:879–880.

----, and B.E. Caldwell. 1970. Effects of genetic heterogeneity within two soybean populations. II. Competitive responses and variance due to genetic heterogeneity for nine agronomic and chemical characters. Crop Sci. 10:216–220.

----, and C.R. Weber. 1968. Effects of genetic heterogeneity within two soybean populations. I. Variability within environments and stability across environments. Crop Sci. 8:44–47.

----, ----, and B.E. Caldwell. 1969. Correlated truncation selection for yield in soybeans. Crop Sci. 9:699–702.

Carter, T.E., Jr., and H.R. Boerma. 1979. Implications of genotype × planting date and row spacing interactions in double-cropped soybean cultivar development. Crop Sci. 19:607–610.

Caviness, C.E., and H.J. Walters. 1966. Registration of Davis soybeans. Crop Sci. 6:502.

Cianzio, S.R. 1970. Blend performance of soybean, *Glycine max.* L. Merrill, varieties estimated by hill and row plots. Unpublished M.S. thesis. Iowa State University, Ames.

----, S.J. Frank, and W.R. Fehr. 1982. Seed width to pod width ratio for identification of green soybean pods that have attained maximum length and width. Crop Sci. 22:463–466.

Clark, R.C., and W.R. Fehr. 1973. Seed divider for plot planters. Crop Sci. 13:763–764.

----, and ----. 1976. Seed cleaning system for plot combines. Crop Sci. 16:880–881.

----, ----, and J.C. Freed. 1978. Check-wire system for plot planters. Agron. J. 70:357–359.

Cooper, R.L. 1971a. Influence of early lodging on yield of soybean [*Glycine max* (L.) Merr.]. Agron. J. 63:449–450.

----. 1971b. Influence of soybean production practices on lodging and seed yield in highly productive environments. Agron. J. 63:490–493.

----. 1981. Development of short-statured soybean cultivars. Crop Sci. 21:127–131.

----, R.J. Martin, A.K. Walker, and A.F. Schmitthenner. 1981. Registration of Amcor soybeans. Crop Sci. 21:633.

Cowan, J.C. 1973. Processing and products. *In* B.E. Caldwell (ed.) Soybeans: Improvement, production, and uses. Agronomy 16:619–664.

Craigmiles, J.P., E.E. Hartwig, and J.W. Sij. 1978. Registration of Dowling soybean. Crop Sci. 18:1094.

Criswell, J.G., and D.J. Hume. 1972. Variation in sensitivity to photoperiod among early maturing soybean strains. Crop Sci. 12:657–660.

Crittenden, H.W. 1971. Registration of Verde soybeans. Crop Sci. 11:312.

Dunphy, E.J., J.J. Hanway, and D.E. Green. 1979. Soybean yields in relation to days between specific developmental stages. Agron. J. 71:917–920.

Eberhart, S.A., and W.A. Russell. 1966. Stability parameters for comparing varieties. Crop Sci. 6:36–40.

Empig, L.T., and W.R. Fehr. 1971. Evaluation of methods for generation advance in bulk hybrid soybean populations. Crop Sci. 11:51–54.

Fehr, W.R. 1973. Evaluation of intergenotypic competition with a paired-row technique. Crop Sci. 13:572–575.

----. 1976. Description and evaluation of possible new breeding methods for soybeans. p. 268–275. *In* L.D. Hill (ed.) World soybean research. The Interstate Printers and Publishers, Danville, IL.

----. 1978. Breeding. *In* A.G. Norman (ed.) Soybean physiology, agronomy, and utilization. Academic Press, New York.

----. 1980. Soybean. p. 589–599. *In* W.R. Fehr and H.H. Hadley (ed.) Hybridization of crop plants. American Society of Agronomy, Madison, WI.

----. 1982. Control of iron-deficiency chlorosis in soybeans by plant breeding. J. Plant Nutr. 5:611–621.

----, and J.B. Bahrenfus. 1984. Registration of Elgin soybean. Crop Sci. 24:385–386.

----, and S.R. Cianzio. 1980. Relationship of component frequency to compenstory response in soybean blends. Crop Sci. 20:392–393.

----, and S.R. Rodriguez. 1974. Effect of row spacing and genotypic frequency on the yield of soybean blends. Crop Sci. 14:521–525.

----, A.F. Schmitthenner, J.B. Bahrenfus, C.S. Schoener, A.K. Walker, and H. Tachibana. 1981. Registration of Vickery soybean. Crop Sci. 21:475.

----, and M.W. Trimble. 1982. Minimizing soybean yield loss from iron deficiency chlorosis. Iowa State Univ. Coop. Ext. Serv. Pm-1059.

----, and C.R. Weber. 1968. Mass selection for seed size and specific gravity in soybean populations. Crop Sci. 8:551–554.

Ford, D.M., R. Shibles, and D.E. Green. 1983. Growth and yield of soybean lines selected for divergent leaf photosynthetic ability. Crop Sci. 23:517–520.

Frank, S.J., and W.R. Fehr. 1981. Associations among pod dimensions and seed weight in soybeans. Crop Sci. 21:547–550.

Garland, M.L., and W.R. Fehr. 1981. Selection for agronomic characters in hill and row plots of soybeans. Crop Sci. 21:591–595.

Garner, W.W., and H.A. Allared. 1920. Effect of the relative length of day and night and other factors of the environment on growth and reproduction in plants. J. Agric. Res. 18:553–606.

Gedge, D.L., W.R. Fehr, and D.F. Cox. 1978. Influence of intergenotypic competition on seed yield of heterogeneous soybean lines. Crop Sci. 18:233–236.

----, ----, and A.K. Walker. 1977. Intergenotypic competition between rows and within blends of soybeans. Crop Sci. 17:787–790.

Green, D.E., P.F. Burlamaqui, and R. Shibles. 1977. Performance of randomly selected soybean lines with semideterminate and indeterminate growth habits. Crop Sci. 17:335–339.

----, and E.L. Pinnell. 1968a. Inheritance of soybean seed quality. I. Heritability of laboratory germination and field emergence. Crop Sci. 8:5–11.

----, and ----. 1968b. Inheritance of soybean seed quality. II. Heritability of visual ratings of soybean seed quality. Crop Sci. 8:11–15.

----, R.M. Shibles, and B.J. Moraghan. 1974. Use of hill plots and short rows to predict soybean performance under wide- and narrow-row management. Iowa State J. Res. 49:39–46.

Hammond, E.G., and W.R. Fehr. 1975. Oil quality improvement in soybeans, *Glycine max* (L.) Merr. Fette, Seifen, Anstrichm. 77:97–101.

Hanson, W.D., and C.A. Brim. 1963. Optimum allocation of test material for two-stage testing with an application to evaluation of soybean lines. Crop Sci. 3:43–49.

----, ----, and K. Hinson. 1961. Design and analysis of competition studies with an application to field plot competition in the soybean. Crop Sci. 1:255–258.

----, R.C. Leffel, and H.W. Johnson. 1962. Visual descrimination for yield among soybean phenotypes. Crop Sci. 2:93–96.

Harrison, S.A., H.R. Boerma, and D.A. Ashley. 1981. Heritability of canopy-apparent photosynthesis and the relationship to seed yield in soybeans. Crop Sci. 21:222–226.

Hartung, R.C., J.E. Specht, and J.H. Williams. 1981. Modification of soybean plant architecture by genes for stem growth habit and maturity. Crop Sci. 21:51–56.

Hartwig, E.E. 1972. Utilization of soybean germplasm strains in a soybean improvement program. Crop Sci. 12:856–859.

----. 1973. Varietal development. p. 187–210. *In* B.E. Caldwell (ed.) Soybean: Improvement, production, and uses. Agronomy 16:187–210.

----, W.L. Barrentine, and C.J. Edwards, Jr. 1980. Registration of Tracy-M soybeans. Crop Sci. 20:825.

----, and F.I. Collins. 1962. Evaluation of density classification as a selection technique in breeding soybeans for protein or oil. Crop Sci. 2:159–162.

----, and J.M. Epps. 1978. Registration of Bedford soybeans. Crop Sci. 18:915.

----, and ----. 1982. Registration of Nathan soybeans. Crop Sci. 22:1264.

----, H.W. Johnson, and R.B. Carr. 1951. Border effects in soybean test plots. Agron. J. 43:443–445.

----, T.C. Kilen, L.D. Young, and C.J. Edwards, Jr. 1982. Effects of natural selection in segregating soybean populations exposed to phytophthora rot or soybean cyst nematodes. Crop Sci. 22:588–590.

Hildebrand, D.F., and T. Hymowitz. 1982. Inheritance of lipoxygenase-1 activity in soybean seeds. Crop Sci. 22:851–853.

Hinson, K., and W.D. Hanson. 1962. Competition studies in soybeans. Crop Sci. 2:117–123.

Hoggard, A.L., J.G. Shannon, and D.R. Johnson. 1978. Effect of plant population on yield and height characteristics in determinate soybeans. Agron. J. 70:1070–1072.

Howell, R.W., C.A. Brim, and R.W. Rinne. 1972. The plant geneticists contribution toward changing lipid and amino acid composition of soybeans. J. Am. Oil. Chem. Soc. 49:30–32.

Hume, D.J., and A.K.H. Jackson. 1981. Pod formation in soybeans at low temperatures. Crop Sci. 21:933–937.

Ivers, D.R., and W.R. Fehr. 1978. Evaluation of the pure-line family method for cultivar development. Crop Sci. 18:541–544.

Kalton, R.R. 1948. Breeding behavior at successive generations following hybridization in soybeans. p. 671–732. *In* Iowa Agric. Exp. Stn. Res. Bull. 358.

Kenworthy, W.J., and C.A. Brim. 1979. Recurrent selection in soybeans. I. Seed yield. Crop Sci. 18:315–318.

Kwon, S.H., and J.H. Torrie. 1964a. Visual discrimination for yield in two soybean populations. Crop. Sci. 4:287–290.

----, and ----. 1964b. Heritability of and interrelationships among traits of two soybean populations. Crop Sci. 4:196–198.

Lambert, J.W., B.S. Kennedy, and J.H. Orf. 1982. Registration of Simpson soybean. Crop Sci. 22:1264.

----, and B.W. Kennedy. 1973. Registration of Swift soybeans. Crop Sci. 13:583.

Lawrence, B.K., and W.R. Fehr. 1981. Reproductive response of soybeans to night interruption. Crop Sci. 21:755–757.

----, and ----. 1982. Reproductive response of soybeans to short day lengths. Iowa State J. Res. 56:361–365.

Leffel, R.C. 1961. Plant lodging as a selection criterion in soybean breeding. Crop Sci. 1:346–349.

----, and W.D. Hanson. 1961. Early generation testing of diallel crosses of soybeans. Crop Sci. 1:169–174.

Lin, C., and J.H. Torrie. 1970. Inter-row competitive effects of soybean genotypes. Crop Sci. 10:599–601.

Luedders, V.D. 1977. Genetic improvement in yield of soybeans. Crop Sci. 17:971–972.

----. 1978. Effect of planting date on natural selection in soybean populations. Crop Sci. 18:943–944.

----, and L.A. Duclos. 1978. Reproductive advantage associated with resistance to soybean-cyst nematode. Crop Sci. 18:821–823.

----, ----, and A.L. Matson. 1973. Bulk, pedigree, and early generation testing breeding methods compared in soybeans. Crop Sci. 13:363–364.

Major, D.J., D.R. Johnson, J.W. Tanner, and I.C. Anderson. 1975. Effects of daylength and temperature on soybean development. Crop Sci. 15:174–179.

Martin, R.J., J.R. Wilcox, and F.A. Laviolette. 1978. Variability in soybean progenies developed by single seed descent at two plant populations. Crop Sci. 18:359–362.

McBroom, R.L., H.H. Hadley, and C.M. Brown. 1981a. Performance of isogenic soybean lines in monoculture and relay intercropping environments. Crop Sci. 21:669–672.

----, ----, ----, and R.R. Johnson. 1981b. Evaluation of soybean cultivars in monoculture and relay intercropping systems. Crop Sci. 21:673–676.

Miller, J.E., and W.R. Fehr. 1979. Direct and indirect recurrent selection for protein in soybeans. Crop Sci. 19:101–106.

Mumaw, C.R., and C.R. Weber. 1957. Competition and natural selection in soybean varietal composites. Agron. J. 49:154–160.

Nickell, C.D., F.W. Schwenk, and W.T. Schapaugh, Jr. 1982. Registration of Douglas soybean. Crop Sci. 22:160.

----, ----, and ----. 1983. Registration of Sparks soybean. Crop Sci. 23:598.

Orf, J.H., and T. Hymowitz. 1979. Inheritance of the absence of the Kunitz trypsin inhibitor in seed protein of soybeans. Crop Sci. 19:107–109.

Probst, A.H. 1943. Border effect in soybean nursery plots. J. Am. Soc. Agron. 35:662–666.

Prohaska, K.R., and W.R. Fehr. 1981. Recurrent selection for resistance to iron deficiency chlorosis in soybeans. Crop Sci. 21:524–526.

Raeber, J.G., and C.R. Weber. 1953. Effectiveness of selection for yield in soybean crosses by bulk and pedigree systems of inbreeding. Agron. J. 45:362–366.

Reicosky, D.A., J.H. Orf, and C. Poneleit. 1982. Soybean germplasm evaluation for length of the seed filling period. Crop Sci. 22:319–322.

Ross, J.P. 1983. Effect of soybean mosiac on component yields from blends of mosiac resistant and susceptible soybeans. Crop Sci. 23:343–346.

Schapaugh, W.T., Jr., and J.R. Wilcox. 1980. Relationships between harvest indices and other plant characteristics in soybeans. Crop Sci. 20:529–533.

Schoener, C.S., and W.R. Fehr. 1979. Utilization of plant introductions in soybean breeding populations. Crop Sci. 19:185–188.

Schutz, W.M., and R.L. Bernard. 1967. Genotype × environment interactions in the regional testing of soybean strains. Crop Sci. 7:125–130.

----, and C.A. Brim. 1967. Inter-genotypic competition in soybeans. I. Evaluation of effects and proposed field plot design. Crop Sci. 7:371–376.

----, and ----. 1971. Inter-genotypic competition in soybeans. III. An evaluation of stability in multiline mixtures. Crop Sci. 11:684–689.

Sediyama, T., and J.R. Wilcox. 1980. Response of populations to inbreeding under two photoperiods. Crop Sci. 20:499–501.

Shannon, J.G., J.R. Wilcox, and A.H. Probst. 1971a. Response of soybean genotypes to spacing in hill plots. Crop Sci. 11:38–40.

----, ----, and ----. 1971b. Population response of soybeans in hill plots. Crop Sci. 11:477–479.

Shibles, R., and D.E. Green. 1979. Plant types and management for narrow row soybeans in the North. *In* Proc. 9th Soybean Seed Res. Conf., American Seed Trade Association, Washington, DC.

Smith, R.R., and C.R. Weber. 1968. Mass selection by specific gravity for protein and oil in soybean populations. Crop Sci. 8:373–377.

Specht, J.E., and J.H. Williams. 1984. Contribution of genetic technology to soybean productivity—retrospect and prospect. p. 49–74. *In* W.R. Fehr (ed.) Genetic contributions to yield gains of five major crop plants. Spec. Pub. 7. Crop Science Society of America and American Society of Agronomy, Madison, WI.

St. Martin, S.K. 1982. Effective population size for the soybean improvement program in maturity groups 00 to IV. Crop Sci. 22:151–152.

Sumarno, and W.R. Fehr. 1980. Intergenotypic competition between determinate and indeterminate soybean cultivars in blends and alternate rows. Crop Sci. 20:251– 254.

Tachibana, H., J.B. Bahrenfus, and W.R. Fehr. 1983. Registration of BSR 201 soybean. Crop Sci. 23:186.

Thorne, J.C., and W.R. Fehr. 1970a. Effects of border row competition on strain performance and genetic variance in soybeans. Crop Sci. 10:605–606.

----, and ----. 1970b. Exotic germplasm for yield improvement in 2-way and 3-way soybean crosses. Crop Sci. 10:677–678.

Torrie, J.H. 1958. A comparison of the pedigree and bulk methods of breeding soybeans. Agron. J. 50:198–200.

----. 1962. Comparison of hills and rows for evaluating soybean strains. Crop Sci. 2:47–49.

Trimble, M.W., and W.R. Fehr. 1983. Mixtures of soybean cultivars to minimize yield loss caused by iron-deficiency chlorosis. Crop Sci. 23:691–694.

Van Schaik, P.H., and A.H. Probst. 1958. Effects of some environmental factors on flower production and reproductive efficiency in soybeans. Agron. J 50:192–197.

Vello, N.A., W.R. Fehr, and J.B. Bahrenfus. 1984. Genetic variability and agronomic performance of soybean populations developed from plant introductions. Crop Sci. 24:511–514.

Voigt, R.L., and C.R. Weber. 1960. Effectiveness of selection methods for yield in soybean crosses. Agron. J. 52:527–530.

Walker, A.K., S.R. Cianzio, J.A. Bravo, and W.R. Fehr. 1979. Comparison of emasculation and nonemasculation for artificial hybridization of soybeans. Crop Sci. 19:285–286.

----, and R.L. Cooper. 1982. Adaptation of soybean cultivars to low-yield environments. Crop Sci. 22:678–680.

----, and W.R. Fehr. 1978. Yield stability of soybean mixtures and multiple pure stands. Crop Sci. 18:719–723.

Weber, C.R. 1966. Registration of Kim and Kanrich soybeans. Crop Sci. 6:391.

----. 1967b. Registration of Disoy soybeans. Crop Sci. 7:403.

----. 1967b. Registration of Prize soybeans. Crop Sci. 7:404.

----, and W.R. Fehr. 1966. Seed yield losses from lodging and combine harvesting in soybeans. Agron. J. 58:287–289.

----, and ----. 1970a. Registration of Provar soybeans. Crop Sci. 10:728.

----, and ----. 1970b. Registration of Corsoy soybeans. Crop Sci. 10:729.

----, and T.W. Horner. 1957. Estimates of cost and optimum plot size and shape for measuring yield and chemical characters in soybeans. Agron. J. 49:444–449.

Weiss, M.G., C.R. Weber, and R.R. Kalton. 1947. Early generation testing in soybeans. J. Am. Soc. Agron. 39:791–811.

Wilcox, J.R. 1970. Response of soybeans to end-trimming at various growth stages. Crop Sci. 10:555–557.

----. 1974. Response of soybeans to Amo-1618 and photoperiod. Crop Sci. 14:700–702.

----. 1980. Comparative performance of semideterminate and indeterminate soybean lines. Crop Sci. 20:277–280.

----, K.L. Athow, T.S. Abney, F.A. Laviolette, and T.L. Richards. 1980. Registration of Century soybean. Crop Sci. 20:415.

----, F.A. Laviolette, and R.J. Martin. 1975. Heritability of purple seed stain resistance in soybeans. Crop Sci. 15:525–526.

----, A.H. Probst, K.L. Athow, and F.A. Laviolette. 1971. Recovery of the recurrent parent phenotype during backcrossing in soybeans. Crop Sci. 11:502–507.

----, and W.T. Schapaugh, Jr. 1980. Effectiveness of single plant selection during successive generations of inbreeding in soybeans. Crop Sci. 20:809–811.

----, ----, R.L. Bernard, R.L. Cooper, W.R. Fehr, and M.H. Neihaus. 1979. Genetic improvement of soybeans in the Midwest. Crop Sci. 19:803–805.

Williams, J.H., J.E. Specht, A.F. Dreier, and R.S. Moomaw. 1982. Registration of Mead soybean. Crop Sci. 22:449.

Williams, L.F., and V.D. Luedders. 1967. Registration of Scott soybeans. Crop Sci. 7:81.

Wilson, R.F., J.W. Burton, and C.A. Brim. 1981. Progress in the selection for altered fatty acid composition in soybeans. Crop Sci. 21:788–791.

8 Seed Production and Technology

Dennis M. TeKrony
University of Kentucky
Lexington, Kentucky

Dennis B. Egli
University of Kentucky
Lexington, Kentucky

Gerald M. White
University of Kentucky
Lexington, Kentucky

The efficient production of maximum soybean [*Glycine max* (L.) Merr.] yields requires adequate supplies of high-quality seed of improved cultivars. This large volume of planting seed is supplied primarily by a seed industry consisting of professional seed growers and seedsmen. Due to the uncertain quality and short longevity of soybean seed, however, nearly all of the seed planted must be produced and marketed on an annual basis. Traditionally, much of this seed was of publicly developed cultivars and was produced primarily through state seed certification agencies. With the passage of the Plant Variety Protection Act (PVPA) in 1970, however, a much greater proportion of the planting seed and the cultivars used have originated from private seed companies. This has resulted in a rapidly expanding seed industry which at times has had problems maintaining the quality of the seed produced.

Soybean seed ontogeny is a complicated biological process that begins with a fertilized ovule and continues until seed maturity. In the short period of 30 to 60 days that a seed is attached to the mother plant it must develop an embryonic axis and complete a complex series of biochemical and physiological events necessary for synthesis and storage of food reserves. At physiological maturity (maximum accumulation of seed dry weight) the seed reaches its maximum potential for germination and vigor. Unfortunately, this potential is short-lived compared to other grain crops and is often reduced prior to planting. This can result in poor field emergence and inadequate stands, especially under adverse soil conditions. Since the majority of the planting seed must be produced each year,

Soybeans: Improvement, Production, and Uses, 2nd ed.—Agronomy Monograph no. 16.

if seed quality problems occur the farmer's demands can exceed the supply of high-quality seed of the most popular cultivars.

8-1 ATTRIBUTES OF SEED QUALITY

Seed quality is a multiple criterion that encompasses several important attributes. A seed scientist may be concerned with the quality characteristics of an individual seed while the seed trade usually considers the quality components of a seed lot. Each individual soybean seed possesses certain measurable quality characteristics which include genetic and chemical composition, physical condition, physiological viability and vigor, size, appearance, and presence of seedborne microorganisms. When seeds are combined into a seed lot, these characteristics are averaged across the population and may be altered by contamination by other crops, cultivars, weed seeds, or inert material. The quality components of a seed lot commonly include crop purity, cultivar purity, weed and crop contaminants, germination, vigor, uniformity, moisture content, and seedborne pathogens. The most chronic quality problems in soybean seeds relate to germination and vigor; however, cultivar purity and weed seed contamination are also serious problems in some production areas.

8-1.1 Individual Seed Quality

The ontogeny of a soybean seed begins with double fertilization of the egg and polar nuclei (chapter 4 in this book) and continues until death of the embryo. Seed growth and development follows a sequence of cytological and metabolic events (chapter 16 in this book) which ends when the seed reaches its maximum dry weight at physiological maturity (PM) (Delouche, 1974; Crookston and Hill, 1978; TeKrony et al., 1979). At PM the seed is completely yellow and is no longer connected to the vascular system of the plant (TeKrony et al., 1979). Seed moisture at PM, however, is approximately 550 g of water kg^{-1} of fresh seed weight (Crookston and Hill, 1978; TeKrony et al., 1979) and a period of desiccation is required for the seed to dry to a harvestable moisture content (TeKrony et al., 1980b). Harvest maturity has been defined as the first time that the seed reaches a moisture content of 140 g kg^{-1} (TeKrony et al., 1980a). A soybean seed at harvest is almost spherical in shape with a large, well-developed embryo surrounded by a thin testa (seed coat).

The soybean seed is capable of germination when about 30% of the maximum dry weight has been accumulated and reaches maximum germination potential about midway between anthesis and PM (Adams and Rinne, 1981; Ackerson, 1984). The maximum vigor potential of soybean seed, as measured by accelerated-aging, speed of germination and conductivity, does not occur, however, until much later when the seed has accumulated nearly 90% of its maximum dry weight (Miles et al., 1983). Thus, the maximum germination and vigor potential of a soybean seed is not reached until just prior to PM. The seed cannot be harvested commercially at PM, however, because of high seed moisture (ca. 550 g kg^{-1}) and must remain on the plant for approximately 2 or more weeks until the seed reaches harvest maturity.

8-1.2 Seed Lot Quality

The quality of a population of soybean seeds (seed lot) represents their collective planting performance potential. High quality seeds usually meet or exceed minimum quality standards for a number of important characteristics. These characteristics are not equal in relative importance, but in combination provide a measure of seed lot quality. The standard of performance for each quality characteristic is established by the seed grower, seed company, seed certification agency, or regulatory inspector. The minimum recommended standards for soybean seed followed by seed certification agencies in North America (AOSCA, 1983) are shown in Table 8-1. The procedures followed for measuring these components are outlined in the Rules for Testing Seeds published by the Association of Official Seed Analysts and the International Seed Testing Association (ISTA, 1976; AOSA, 1981). All official (state and federal) seed testing laboratories and the registered seed technologists in commercial seed laboratories follow these rules.

A laboratory test which measures any attribute of seed lot quality can only be as representative as the seed sample submitted. Regardless of the seed lot size or storage container (bag, bulk), the sample tested must represent the entire seed lot. This requires precise procedures for sampling following guidelines established by official seed testing associations (ISTA, 1976; AOSA, 1981).

Table 8-1. Soybean seed standards recommended by the Association of Official Seed Certifying Agencies (AOSCA, 1983).

Factor	Standards for each class		
	Foundation	Registered	Certified
		%	
Pure seed (minimum)	NS†	98.00	98.00
Inert matter (maximum)	NS	2.00	2.00
Weed seeds (maximum)‡	0.05	0.05	0.05
Objectionable weed seeds (maximum)§	None	None	None
Total other crop seeds (maximum)	0.20	0.30	0.60
Other cultivars	0.10	0.20	0.50
Other kinds¶	0.10	0.10	0.10
Germination and hard seed (minimum)	NS	80.00	80.00

†NS = No standard.
‡Total weed seed shall not exceed 10 per 454 g.
§Designated by each state certifying agency.
¶Not to exceed 3 per 454 g in any class except corn and sunflower seed where maximum is; foundation—NS, registered—None, and certified—1 per 454 g.

8-1.2.1 Crop Purity

The crop purity indicates how much material in a seed lot is intact soybean seed, other crop seed, and weed seed. It is determined by conducting a purity test on a small laboratory sample. The sample size evaluated for soybean seed purity is 500 g, which is approximately 2500 seeds. In the purity analysis, a physical hand separation of components is made and the results are reported as a percentage by weight for (i) pure seed, (ii) other crop seed, (iii) weed seed, and (iv) inert matter (AOSA, 1981).

Pure seed is the percentage of soybean seed for the cultivar stated that occurs in the seed lot being tested. Unless excessive physical seed breakage has occurred or the seed lot is severely contaminated with crop or weed seed, the purity for soybean seed should exceed 98.0%. Many seed producers and companies strive for a purity of 99.0% or higher for all seed sold. If seed of another crop or soybean cultivar exceeds 5.0% by weight, the seed lot would be designated as a mixture by state seed laws and the Federal Seed Act (USDA, 1975).

Other crop seed is the percentage of crop seeds other than soybeans present in the seed lot tested. The most common crop contaminant in soybean seed is corn (*Zea mays* L.); however, sunflower (*Helianthus annuus* L.), cowpea [*Vigna unguiculata* (L.) Walp.], field bean (*Phaseolus vulgaris* L.), and other crop seed can also occur. No other crop seed should occur in high quality soybean seed sold for planting purposes.

Weed seed are those seeds commonly recognized by laws, regulations, customs, or general usage as weeds in the state or region. Since weed seeds vary greatly in size, they should be expressed as the number found in the purity analysis (500 g) in addition to the percentage by weight (as required by seed laws). Each state has established lists of noxious weeds which are determined to be troublesome and objectionable. Such lists are usually defined in two categories, primary (prohibited) and secondary (restricted) noxious weed seed. Soybean seed cannot usually be sold if primary noxious weed seed are present, while sales are restricted if secondary noxious weed seeds exceed established levels.

Weed seeds that are classified as troublesome in soybean seed will vary from one region to another due to adaptation of plant species. Those weed seeds that cause the greatest problems for soybean seed producers, however, are those that are either difficult to control in the field or to separate from soybean seeds during routine seed conditioning. Examples include: common cocklebur (*Xanthium pensylvanicum* Wallr.), giant ragweed (*Ambrosia trifida* L.), purple moonflower (*Ipomoea turbinata* Lag.), common morningglory [*Ipomoea purpurea* (L.) Roth], ballonvine (*Cardiospernum halicacabum* L.), and Johnsongrass (*Sorghum halepense* Pers.). Even though weed contamination can cause problems in some soybean production fields, soybean seed should not be sold with noxious or other objectionable weed seed present. The present availability of herbicides and seed cleaning equipment make this a realistic goal.

Inert matter denotes the percentage of material in the seed lot tested that is not seed. It includes pods, stems, small stones, soil particles, and pieces of broken seeds that are one-half or less than the original size. Splits (soybean seeds that are split directly in half at the juncture of the two cotyledons with the embryonic axis) are also classified as inert matter. Thus, if soybean seed are harvested, handled, or conditioned at low seed moistures, physical damage will occur and the percent inert matter could increase to high levels. A low percentage of inert matter ($< 2.00\%$) is desirable and this objective is often met or exceeded by good seed producers.

8-1.2.2 Germination

Even though crop purity is important, it means nothing if the seeds are incapable of germination. Thus, the single most recognized and accepted index of seed quality is germination. The germination capacity of a soybean seed lot is the percentage of pure seed that will produce a normal seedling (pure live seed) under optimum laboratory testing conditions. The procedures (ISTA, 1976; AOSA, 1981) followed when conducting a germination test have been carefully evaluated and standardized for many years. Thus, a test conducted by an official or registered seed analyst is commonly referred to as the standard germination test. The definition of germination is "the emergence and development from the seed embryo of those essential structures which are indicative of the ability to produce a normal plant under favorable conditions" (AOSA, 1981).

The Rules for Testing Seeds specify the optimum temperature and substratum for the germination test as well as the sample size. The time recommended for a soybean germination test is 8 days, however, a preliminary count can be made at 3 to 5 days, especially if seedborne fungi which cause moldy, diseased seedlings are present. One of the most critical evaluations made by a seed analyst during the standard germination test is the determination of normal and abnormal seedlings. A normal soybean seedling must have at least one intact cotyledon, a healthy epicotyl, and vigorous primary radicle or secondary root system. Malformed seedlings that do not meet these criteria or are severely diseased or have hypocotyl breaks extending into the conducting tissue are classified as abnormal. This rather subjective evaluation of seedling development is the primary difference between a layman's interpretation of germination and the seed analyst's classification of normal seedlings.

The official seed certifying agencies of North America (AOSCA, 1983) recommend a minimum germination of 80% for certified soybean seed (Table 8–1). This germination level tends to be the accepted standard for the industry, however, some seed company quality control programs have raised the minimum standard to 90%. Many state and federal seed laws prohibit the sale of soybean seed unless it has an acceptable germination and has been tested recently (within the past 26–52 weeks).

8-1.2.3 Cultivar Purity

Testing for soybean cultivar purity in seed laboratories received little attention until the passage of the PVPA in 1970 (PVPA, 1973). Since that time, the number of cultivars in use has expanded rapidly (Batcha, 1983; Perrin et al., 1983) which has placed more emphasis on cultivar purity as a measure of seed quality. However, contrary to crop purity and germination testing; uniform, standardized procedures are not available to seed analysts for determining cultivar purity. As a general rule, a soybean cultivar cannot be identified by examining only the seed's morphological characteristics in the laboratory. It is possible to conclude that the seed belongs to a certain group of cultivars, but it is seldom possible to identify the exact cultivar. Thus, the methods for evaluating seed for cultivar purity are changing from visual observations of seed and seedling morphology to detailed grow-out tests or the use of biochemical or cytological methods.

The prominent morphological seed identification character for soybean cultivars is hilum color which can range from clear, buff, brown, imperfect black to black. Keys have been developed which classify the hilum color of soybean cultivars (Dorchester, 1945; Payne, 1979); however, there are many cultivars within each color classification. Thus, precise identification of soybean cultivars on this basis is difficult if not impossible. Factors which have been shown to affect the expression of hilum or seed color include: fungal infection (Nittler et al., 1974), the production environment (Taylor and Caviness, 1982), and seed handling (Payne, 1979).

Hypocotyl color is another morphological characteristic that has been used in conjunction with hilum color to classify soybean cultivars. This trait is closely related to flower color with purple hypocotyls occurring in cultivars with purple flowers and green hypocotyls occurring in white-flowered cultivars (Bernard and Weiss, 1973). Payne and Morris (1976) evaluated over 60 soybean cultivars for hypocotyl color and classified them into six pigmentation groups. Payne (1979) cautioned, however, that hypocotyl color could be influenced by the length of the photoperiod and light intensity as well as the nutrient content of the growing medium which could lead to interpretation problems when using this procedure.

Because morphological characteristics of soybean seeds and seedlings are subjective and variable, seed scientists have examined chemical or biochemical parameters to evaluate seeds for cultivar purity. A rather simple chemical taxonomic technique was developed by Buttery and Buzzell (1968) to separate cultivars based on the presence or absence of the perioxidase enzyme in the seed coat. They were able to separate soybean cultivars into two groups, those having high (dark red color) or low (no color) perioxidase activity. Due to its simplicity and speed ($<$ 30 min), this procedure has been adapted by many seed testing laboaratories. Similar to hilum or hypocotyl color, perioxidase activity is limited to placing cultivars in two groups and does not positively identify each cultivar.

A more sophisticated and potentially more valuable laboratory method of cultivar verification is the electrophoretic analysis for proteins and isoenzymes. Early research by Larsen (1967) used this procedure and was able to divide most (but not all) soybean cultivars into two groups based on the presence of either the A or B protein band. More recently, electrophoresis of urease (Buttery and Buzzell, 1971), esterase (Payne and Koszykowski, 1978), and other enzymes (Gorman and Kiang, 1977) have been used in separating soybean cultivars.

A recurring problem, when evaluating the purity of soybean cultivars, is that no single test can accurately distinguish and classify all cultivars. Thus, a combination of morphological and chemical techniques must be used. Wagner and McDonald (1981) used five procedures (hilum color, hypocotyl color, peroxidase activity of seed coat, and two electrophoresis procedures) and were able to separate and identify 15 of the 36 soybean cultivars commonly grown in Ohio. Similarly many large seed companies have now finger-printed all of their own cultivars and others as a part of in-house quality control programs. Even when the most sophisticated laboratory procedures are used, however, not all cultivars can be positively identified. This forces seed companies, seed certification agencies, or regulatory officials into extensive greenhouse and field testing for final verification. Grow-out tests in the greenhouse or field are time consuming, however, and are usually conducted after the seed has been planted and used by many farmers.

The quality standard usually used as a guide for cultivar purity is that required for certified seed production for the foundation, registered, and certified classes (Table 8–1). Many argue, however, that current cultivars are released at an earlier generation than in the past and are not as homozygous as older cultivars. This has caused seed certification agencies to classify those plants that differ morphologically, but are acceptable within a cultivar, as variants and those plants that are unacceptable as off-types (AOSCA, 1983). If the trends for earlier release and greater variability within a cultivar continue, these agencies may also have to increase the minimum levels of contamination allowed in the classes of certified seed.

8–1.2.4 Vigor

During the 1970s, no term related to soybean seed quality has received more attention than *seed vigor*. Coordinated efforts, among seed analysts, seed scientists, and seedsmen have been made to establish certain criteria for identifying and classifying seed vigor. This culminated in 1983 when a comprehensive review entitled *Seed Vigor Testing* was published by the Seed Vigor Testing Committee of the Association of Official Seed Analysts (AOSA, 1983). This publication provided a usable definition of seed vigor; "Seed vigor comprises those properties which determine the potential for rapid, uniform emergence and development of normal seedlings under a wide range of field conditions." Thus, two seed lots having nearly identical standard germination levels may perform quite differently under poor field conditions due to differences in their vigor potential. Seed vigor evaluations have been classified as (i) seedling growth and evaluation tests, (ii) stress tests, and (iii) biochemical tests

(AOSA, 1983). Tests falling into the first category include seedling vigor classification, seedling growth rate, and speed of germination. Stress tests include accelerated aging germination and the cold test, while biochemical tests include tetrazolium chloride, electrical conductivity, respiraton, and other tests of metabolic potential. A survey of all seed testing laboratories in North America indicated that 74% of those laboratories responding were evaluating seed for vigor (TeKrony, 1983). Forty-four commercial and official laboratories were testing soybean seeds for vigor and nearly 40% of these laboratories were conducting over 200 tests yr^{-1}. The most popular vigor tests for soybean seeds were the accelerated aging and cold test, while other tests used included conductivity, tetrazolium, and seedling vigor classification.

The accelerated aging test was originally developed to estimate longevity of seed in storage (Delouche and Baskin, 1973), however, it has also been shown to relate well to stand establishment of soybeans (Byrd and Delouche, 1971; TeKrony and Egli, 1977). This vigor test stresses unimbibed soybean seeds with high temperature (41 °C) and relative humidity (100% RH) for short periods (3 to 4 days) prior to testing them for germination under optium conditions specified for the standard germination test (AOSA, 1981). The test is commonly conducted in either a large accelerated aging chamber (for multiple seed samples) or a small single sample aging chamber (AOSA, 1983). The single sample procedure involves placing the seeds (40 to 45 g) in a single layer on a wire mesh screen above water inside a small plastic germination box (McDonald and Phaneendranath, 1978). Many of these boxes can then be placed into an incubator set at the desired temperature and high relative humidity.

The accelerated aging test offers the advantages of being inexpensive, simple, and requiring little additional training of seed analysts. However, much variation in test results has been reported between laboratories (McDonald, 1977; Tao, 1978a, 1980a). Thus, precautions must be taken during aging to reduce variability. Tomes et al. (1981) reported that the interaction between aging temperature, seed moisture, and time during aging had the greatest effect on seed germination. They concluded that variability could be reduced by (i) evaluating seed on a weight (40 to 45 g) rather than number (200 seed) basis, (ii) precisely controlling aging temperatures at 41 °C, and (iii) measuring the initial and final moisture of the seed. Recent national vigor referees have incorporated some of these recommendations and have produced excellent test repeatability (Spain, 1982).

The cold test was originally developed to measure corn seed vigor (Clark, 1953; Svien and Isely, 1955), however, in recent years it has been used to evaluate seed vigor in other crops (McDonald, 1975) including soybeans (Johnson and Wax, 1978). The cold test simulates early spring field conditions by providing a seed environment of high soil moisture, low soil temperature, and microbial activity. Seeds are placed in soil or on kimpak or paper towels lined with soil and incubated at 10°C for a specified period (5 to 7 days). At the end of this stress period, the seeds are transferred (in the same planting medium) to the favorable temperatures prescribed for the standard germination test (AOSA, 1981) and the normal seedlings that develop are counted.

The greatest difficulty with the cold test is the inability to standardize the soil source from one testing location to another (Delouche, 1976; Burris and Navratil, 1979). This was supported by seed vigor referees which showed significant variability between laboratories when using different soil sources (McDonald, 1977; Tao, 1980a). The use of peat moss or vermiculite inoculated with *Phythium* spp. instead of soil in the cold test has not been successful. Burris and Navratil (1979) compared many cold testing procedures for corn inbreds and reported that the temperature of the cold stress environment was more important than the soil medium. They reported that successful seed vigor evaluations may be possible using a sterile cellulose medium (without soil). Even with the inherent problems of the planting medium, the cold test still is used more than any other vigor test (TeKrony, 1983) with consistent results often occurring within a seed testing laboratory. This provides a convenient inhouse test for quality control purposes.

The conductivity test is a measurement of electrolytes leaking from plant tissue. Poor membrane structure is usually associated with deteriorating, low vigor seeds. When these seeds are soaked in water, a greater loss in electrolytes (amino and organic acids) occurs and the conductivity of the soak water increases. The higher the conductivity of the soak water, the lower the seed vigor. The use of the conductivity test to measure the vigor of garden pea (*Pisum sativum* L.) has been established in Europe (Matthews and Bradnock, 1967, 1968). This test has been shown to correlate well with the vigor of soybean seed (Yaklich et al., 1979; McDonald and Wilson, 1979; Tao, 1980a; Loeffler, 1981).

Conductivity is measured by placing 25 to 50 uninjured soybean seed in a beaker containing 75 mL of distilled water. The seed are soaked for 24 h at 20 °C after which the conductivity of the soak water is measured using a dip cell (AOSA, 1983). The conductivity test provides a rapid, nonsubjective and inexpensive measure of seed vigor. It has been shown, however, that initial seed moisture (Pollock et al., 1969), seed size (Tao, 1978b; Loeffler, 1981) chemical seed treatment (Tao, 1980b) and seedborne disease (Loeffler, 1981) can influence conductivity results. Another concern is that, since the conductivity test measures the average conductivity of 25 to 50 seeds, a few low quality seeds may bias test results. A commercial instrument is now available which monitors the electrolyte leakage of individual seed (McDonald and Wilson, 1979, 1980). Conductivity estimated with this instrument was closely related to the conductivity of a composite sample (McDonald and Wilson, 1979; Loeffler, 1981).

Seedling vigor tests are usually conducted under the same environmental conditions as the standard germinaton test; however, seedling growth is measured or evaluated in two different ways; (i) seedling growth rate and (ii) seedling vigor classification. Both procedures offer certain advantages to seed testing laboratories in that they are inexpensive, require no specialized training or equipment and are relatively rapid. Distinct disadvantages of seedling vigor tests, however, are (i) moisture and temperature of the testing medium must be precisely controlled, (ii) the timing of evaluation is critical and (iii) additional evaluation of seedlings into weak or strong categories is too subjective. For these reasons seedling vigor tests have been difficult to standardize among laboratories (McDonald, 1977; Tao, 1978a).

A seedling vigor classification test has been described (AOSA, 1983) and is used in some laboratories for soybean, field bean, cotton (*Gossypium hirsutum* L.), and peanut (*Arachis hypogaea* L.). This test is an expansion of the standard germination test with the requirement that normal seedlings are further classified as *strong* and *weak*. The test is conducted at a constant temperature of 25 °C and a preliminary count and seedling classification are made 5 days after planting. The strong and weak classification of seedlings separates normal soybean seedlings free of deficiencies from those which have deficiencies. Seedlings which would be classified as weak would have: a primary root missing, one cotyledon missing, partial decay, one primary leaf missing, or are short, spindly and poorly developed.

Another type of seedling vigor test is the first count of a standard germination test which has been classified as a speed of germination test. The number of normal seedlings counted at the first count represents the faster germinating seeds and is a measure of seedling vigor. The first count vigor evaluation for soybean has been conducted at either 4 (Burris et al., 1969) or 3 days (TeKrony and Egli, 1977). TeKrony and Egli (1977) counted only those seedlings that were strong and at least 3.75-cm long at 3 days.

Seedling growth rate is a vigor evaluation based on the growth of seedlings under the same conditions of a standard germination test except that the moisture content of the paper towels is precisely controlled (AOSA, 1983). At the end of the germination period, growth of the normal seedlings is measured either by length or dry weight (excluding the cotyledons). Limitations of this test are (i) it is time consuming to remove and/or measure seedlings, (ii) it is influenced by slight variation in moisture and temperature, and (iii) differential cultivar responses can make accurate test interpretations difficult.

It has been shown that soybean seed lots which have nearly identical and acceptable (>80%) standard germination, may have quite different seed vigor levels. Even though seed vigor tests are not presently standardized among seed-testing laboratories, progress is being made toward reducing variability and some vigor tests are widely used (TeKrony, 1983). Thus, many seed growers and seedsmen routinely test soybean seed for seed vigor and use this information as a valuable in-house tool when monitoring the quality of seed lots. As seed vigor testing procedures become accepted and standardized soybean seed may eventually be labeled for vigor and the information related directly to farmers at the time of purchase. Caution must be exercised, however, as misconceptions or misunderstandings of seed vigor could seriously delay final acceptance of this important concept.

8- 1.2.5 Seed-borne Pathogens

A large number of fungi, bacteria, and viruses may attack soybean seed prior to harvest and reduce the quality of the seed (Neergaard, 1977; Sinclair, 1982). Even though seed infection can cause serious reductions in seed germination and vigor, little attempt has been made to identify or report the presence of seed-borne pathogens in seed testing laboratories in the USA. Much greater progress has been made toward developing quantitative laboratory testing procedures for seed-borne diseases in seed testing laboratories in Europe. This has resulted in the publication of procedures for seed health testing by the International Seed Testing Association in the official Rules for Seed Testing (ISTA, 1976) and in a handbook for Seed Health Testing (ISTA, 1959). A comprehensive review of seed-borne pathogens and seed health testing procedures has been published by Neergaard (1977).

Even though some soybean seed-borne pathogens can be detected by visual examination of dry seed, the most commonly used procedure is incubation of seeds on agar plates, ordinary germination blotters or cellulose pads. The symptoms of both the purple seed stain disease caused by the fungus *Cercospora kikuchii* (Mats and Tomoy) and soybean mosaic virus can be detected by direct inspection of dry soybean seed. Seeds that are severely infected with *Phomopsis* spp. can be detected visually by their chalky, shriveled appearance. Positive identification of these and other fungi is usually made, however, after several days of incubation on agar or blotter media (Kmetz et al., 1974; McGee et al., 1980; Shortt et al., 1981; TeKrony et al., 1984). Using the agar method, seeds are usually surface sterilized in sodium hypochlorite and plated on acidified (pH 4.5) potato dextrose agar. The plates are held under fluorescent light at room temperature (22 °C) for 10 to 14 days before fungal identification can be made based on colony morphology. An experienced analyst, familiar with the colony characteristics of *Phomopsis* spp. and other fungi, can identify and count colonies macroscopically by examining both sides of the plates.

The blotter method combines the identification procedures commonly used by plant pathologists with the procedures used by seed analysts for the germination test. The seeds are placed on moistened blotters (Neergaard, 1977) or cellulose pads (Shortt et al., 1981) in petri dishes, plastic boxes, or other suitable containers and incubated in fluorescent light for 7 days. The seeds or germinated seedlings are examined either macroscopically or with a stereomicroscope for the presence of the pathogen and the extent of disease. The blotter method offers an advantage to routine seed testing laboratories since most of the equipment needed for conducting the test is readily available. The agar method requires that the seed analyst train for a period of time with a pathologist experienced in the use of this method. Similar training is necessary for the detection and identification of fungi for both methods.

8-2 RELATIONSHIP OF SEED QUALITY TO PERFORMANCE

The quality of planting seed is determined to provide information on its potential performance in the field (seedling emergence and/or yield) or its ability to maintain quality during storage. The ability of seed quality parameters to accurately predict performance is hampered by a number of factors, including the many causes of variation in seed quality and the wide range in environmental conditions that may be encountered during seed storage or when the seed is planted.

8-2.1 Storability

The production of a soybean crop requires that the planting seed must be stored (at a minimum) from the time of harvest in the fall until the spring planting season. During storage, quality can remain at the initial level or decline to a level that may make the seed unacceptable for planting purposes. It is well known that seed moisture and temperature are primary determinants of quality changes during storage (Toole and Toole, 1946; Holman and Carter, 1952; McNeal, 1966). However, the deterioration of seed during storage is also related to the quality (germination and/or vigor) of the seeds placed into storage (Egli et al., 1979; Burris, 1980; Ellis et al., 1982).

Byrd and Delouche (1971) reported that, as seeds deteriorate during storage, their performance potential and vigor decline before there is any loss in viability (standard germination). This suggests that the standard germination of a seed lot placed in storage may not be a good indicator of its storage potential because it may not accurately indicate the degree of seed deterioration. Egli et al. (1979) found little relationship ($r=0.23$) between the initial standard germination of 12 seed lots and germination after 9 months of storage at 135 g kg^{-1} of moisture and ambient temperatures. Similar results were reported by Baskin and Vieira (1980).

Byrd and Delouche (1971) reported that several stress tests (accelerated aging, cold test, and germination after immersion in 75 °C water for 70 s) that measure seed vigor were more closely associated with the longevity of soybean seed in storage than the standard germination test. This association has been supported by a number of studies (Delouche and Baskin, 1973; Egli et al., 1979; Baskin and Viera, 1980; Burris, 1980). Wien and Kueneman (1981) found that the accelerated aging test (40 °C, 100% RH, 72 h), did not accurately predict seed emergence after 39 weeks in storage; however, a modified accelerated aging test (35 to 40 °C and 75% RH for 6 weeks) was a better predictor of deterioration during storage. Egli et al. (1979) reported that the correlation between standard germination of 12 lots of soybean seed after 39 weeks storage at 135 g

kg^{-1} of moisture and ambient temperatures and the initial accelerated aging germination (40 °C, 100% RH, 72 h) was $r = 0.69$. The initial accelerated aging germination was also closely associated with the accelerated aging germination after 26 weeks in storage. They concluded that the accelerated aging test was an excellent predictor of storability.

In some cases, it has been reported that the germination of soybean seed may increase during storage (TeKrony et al., 1982). This has been shown to occur in seed lots that initially have relatively high levels of infection by the pod and stem blight fungi (*Phomopsis* spp.). Germination is closely related to the level of pod and stem blight fungal infection (TeKrony et al., 1984) and, as the fungus dies during storage, the standard germination of the seed increases (TeKrony et al., 1982).

Mechanical damage is another factor that may influence seed deterioration during storage. White et al. (1976) reported that the physical damage caused by drying at high air temperatures resulted in a more rapid decline in germination during storage. Paulsen et al. (1981b) reported that mechanical damage that occurred during harvesting did not have a significant effect on the rate of deterioration during storage. This difference in results may be related to the different levels or types of physical injury imposed in the two studies. It is not yet clear whether mechanical damage can affect the storability of soybean seeds without reducing the initial standard germination or vigor level of the seed.

Several reports (Burris, 1980; Wien and Kueneman, 1981; Minor and Paschal, 1982; Ellis et al., 1982) have suggested that there are genotypic differences in the storability of soybean seed. Wien and Kueneman (1981) screened lines from Indonesia, the International Institute of Tropical Agriculture, and the USA for storability and found that several small seeded lines (80–100 mg $seed^{-1}$) from Indonesia maintained higher levels of germination in storage than other lines. Minor and Paschal (1982) screened 235 genotypes of potential tropical and subtropical adaptation and found a number of genotypes with potentially superior storability. They indicated that longer storage half-lives tended to be associated with higher initial germination, higher levels of hard seed, small seed, and earlier maturity. Egli et al. (1979) compared the storability of three seed lots from each of four cultivars adapted in Kentucky and concluded that the initial quality of the seed (viability and vigor) was the main determinate of seed storability and that there was no direct cultivar effect on storability per se. Starzing et al. (1982) evaluated black-and-yellow seed from a bulk F_4 population derived from the cross of black-and-yellow seeded parents and found that the germination of the yellow seed declined faster in storage than the black seed. They attributed this difference to lack of fungal growth on the black seeds, although the amount and kind of fungal growth was not determined. Since initial quality levels are frequently confounded with genotypes, it is difficult to determine if there are true genotypic differences in the ability of seed to resist deterioration and loss of quality during storage.

8–2.2 Field Emergence

Many attempts have been made to relate standard germination to field emergence (seedling emergence under field conditions) with widely varying results. Some workers have reported a close association between standard germination and field emergence (Sherf, 1953; Athow and Caldwell, 1956) while other studies have shown that standard germination consistently overestimates field emergence (Edje and Burris, 1971; TeKrony and Egli, 1977; Johnson and Wax, 1978; Yaklich and Kulik, 1979). This diversity of results has been partially explained by variation in field conditions, with standard germination providing accurate predictions of field emergence only under near ideal field conditions. Unfavorable seedbed conditions reduce field emergence and reduce the association between standard germination and field emergence (TeKrony and Egli, 1977; Johnson and Wax, 1978). Delouche (1974) concluded that the standard germination test is an insensitive and misleading measure of seed quality because it focuses primarily on the final consequences of deterioration and does not adequately take into account the very substantial loss in performance potential that can and does occur before germination capacity is lost.

The definition of seed vigor (AOSA, 1983) suggests that measures of seed vigor should provide a better relationship to field emergence than standard germination. This concept has been evaluated in a number of experiments with the general conclusion that estimates of seed vigor relate better to field emergence than standard germination (Edje and Burris, 1971; TeKrony and Egli, 1977; Johnson and Wax, 1978; Yaklich and Kulik, 1979; Clark et al., 1980). However, it was not possible to identify a single vigor test that consistently predicted field performance in all conditions. TeKrony and Egli (1977) reported that the 4-day germination was better than accelerated aging in predicting field emergence in years with adverse field conditions. Yaklich and Kulik (1979) and Kulik and Yaklich (1982) found that the accelerated aging germination, tetrazolium-viable seeds, and the cold test were the most consistent of the many tests they evaluated. Johnson and Wax (1978) obtained consistent results with the cold test but not with the accelerated aging test.

Johnson and Wax (1978) suggested that it may not be possible to develop a single test or array of tests that will predict field emergence for all field conditions to which the seed might be exposed. Several workers have suggested that a combination of vigor tests may relate better to a wide range in field conditions. TeKrony and Egli (1977) converted the results of laboratory tests to a vigor index and related combinations of these vigor indices to field emergence. The combined indices were better than a single index, but the relationship to field emergence was still variable across environments. Luedders and Burris (1979), Yaklich and Kulik (1979) and Loeffler (1981) used the results of various vigor tests to develop multiple regression equations to predict field emergence; however, the predictive ability of the equations was no better than the individual tests.

High-quality seed (high standard germination and high vigor) can be expected to produce better field emergence in a wide range of seed bed conditions than seed of marginal or low quality. However, it has not been possible to accurately predict field emergence under all seed bed conditions from the results of individual measures of quality or from combinations of tests. Soybean yields are relatively insensitive to plant population over a wide range of populations (Tanner and Hume, 1978), thus precise stands are not needed to maximize yield. Consequently, some reduction in field emergence below that predicted by laboratory tests can be tolerated without a significant effect on yield. Most recommended planting rates are higher than the populations required for maximum yield to compensate for possible reductions in emergence.

8–2.3 Yield

Seed quality can influence yield in two ways; indirectly by influencing emergence and final stand or directly through its influence on plant vigor. If inadequate plant populations are obtained as a result of the use of low-quality planting seed, yields will be reduced (Edje and Burris, 1971; Johnson and Wax, 1978). However, when seed lots that varied in quality were compared at populations adequate for maximum yield, there was no relationship between seed quality and yield (Edje and Burris, 1971; Egli and TeKrony, 1979; Kulik and Yaklich, 1982). These results suggest that the primary advantage for the use of high-quality (high germination and vigor) planting seed is to increase the probability of obtaining satisfactory plant populations under a wide range of field conditions.

There are, however, a number of indirect indications that seed quality may have a direct effect on yield, beyond the attaining of adequate plant populations. Some investigators have reported that seed size is positively correlated with yield (Fontes and Ohlrogge, 1972; Burris et al., 1973; Smith and Camper, 1975), while others show no relationship between the two variables (Singh et al., 1972; Johnson and Leudders, 1974). The seed quality of the various seed size classes used in these experiments was usually not reported, however, Burris et al. (1971, 1973) reported that large seed produced more vigorous seedlings than small seed. Fehr and Probst (1971) reported that the source (area of production) of planting seed significantly influenced yield although the differences were not large. Torrie (1958) found that yield from seed that had been stored for several years was lower than the yield from seed that had been stored for only 1 yr. Plants from seed that had suffered imbibitional injury at low temperatures (Hobbs and Obendorf, 1972) or plants that were the last to emerge (Pinthus and Kimel, 1979) showed reduced vigor and yield; however, the plants were grown at lower than normal populations which may give an advantage to more vigorous plants.

A consideration of the available data suggests that in normal production systems, there is little direct yield advantage to be expected from the use of high-quality planting seed. However, considering the cost of replanting or potential yield losses because of stand failure, the use of high-quality planting seeds is clearly justified.

8-3 FACTORS INFLUENCING SEED QUALITY

8-3.1 Environmental

Soybean seed quality is highly variable across locations and years indicating that environmental conditions during seed production have a significant effect on seed quality. Environmental conditions can influence seed quality during seed development, during the desiccation period (physiological maturity to harvest maturity) or after harvest maturity when the seed is essentially in storage in the pod in the field.

The effects of the environment during seed development and maturation have been demonstrated by a number of workers. Green et al. (1965) reported that seed produced from later dates of planting which reached maturity after hot, dry weather had ended generally exhibited higher germination and field emergence than seed which matured during hot, dry weather. Harris et al. (1965) reported similar results. TeKrony et al. (1980b) attributed lower initial germination and vigor at harvest maturity of one cultivar to high temperatures during the period from physiological maturity to harvest maturity.

Seed quality of earlier maturing cultivars at a given location is generally lower than that of later-maturing cultivars (Smith et al., 1961; Green et al., 1965; Mondragon and Potts, 1974; Ross, 1975; Grau and Oplinger, 1981; TeKrony et al., 1984). Delayed planting, especially of early maturing cultivars, has been shown to result in improved seed quality (Green et al., 1965; Nicholson and Sinclair, 1973; TeKrony et al., 1984). TeKrony et al. (1984) evaluated six cultivars of varying maturity in three planting dates for 4 yrs and found a positive linear relationship between the date of harvest maturity and standard germination and seed vigor (accelerated aging germination and speed of germination). They also reported a linear decline in seed infection by *Phomopsis* spp. as harvest maturity was delayed and the variation in standard germination was closely associated with levels of seed infection by *Phomopsis* spp. They concluded that the lower levels of standard germination associated with the early dates of harvest maturity were primarily a result of the high levels of *Phomopsis* spp. seed infection. Seed vigor was not as closely associated with *Phomopsis* spp. seed infection, suggesting that the environment was acting directly on the seed in terms of influencing seed vigor.

A number of workers have shown that seed quality deteriorates when the seed remains in the field after harvest maturity (Mondragon and Potts,

1974; Wilcox et al., 1974; Ellis and Sinclair, 1976; Alexander et al., 1978; TeKrony et al., 1980b). Early maturing cultivars have been shown to be affected more by delayed harvest than late-maturing cultivars (Wilcox et al., 1974). High temperatures, RHs, and precipitaton have been shown to enhance field deterioration (Wilcox et al., 1974; Alexander et al., 1978; TeKrony et al., 1980b). Moore (1971) attributed much of the decline in quality to physical damage to the seed as a result of alternate wetting and drying, although this relationship was not supported by TeKrony et al. (1980b). Mondragon and Potts (1974) concluded that deterioration was related to the rate and range in fluctuations in temperature and RH in the plant canopy rather than the absolute levels. Potts et al. (1978) investigated field deterioration using a soybean strain exhibiting high levels of hardseededness. They found that germination remained high longer during field exposure after maturity and this was associated with less fluctuation in seed moisture.

The declines in germination when seeds remain in the field after harvest maturity have also been associated with increases in the levels of seed infection by *Phomopsis* spp. and other fungi (Wilcox et al., 1974; Ross, 1975; Ellis and Sinclair, 1976; Alexander et al., 1978). The increase in *Phomopsis* spp. seed infection during field weathering was reduced by the use of the foliar fungicide benomyl (Ross, 1975; Ellis and Sinclair, 1976).

TeKrony et al. (1980b) found that after harvest maturity declines in seed vigor (accelerated aging germination) occurred before declines in standard germination. They suggested that seed vigor was more sensitive to field deterioration than seed viability. The loss in seed vigor was accelerated by warm, moist conditions leading to the suggestion that the deterioration was similar to that experienced during storage (TeKrony et al., 1980b).

Environmental conditions during seed development, the desiccation period, and after harvest maturity can influence the quality of harvested seed. The widely varying levels of seed quality encountered as a result of environmental effects suggest that the quality level of a seed lot should be measured as soon as possible after harvest to determine its potential for use as planting seed.

8–3.2 Genetic

Environmental conditions during seed development, maturation, and exposure of the seed on the plant in the field before harvest are an important determinant of seed quality. Thus, any evaluation of genetic differences in seed quality must consider environmental effects. Although there are a number of reports in the literature of cultivar differences in seed quality (Ross, 1975; Paschal and Ellis, 1978) it is not always clear whether the differences are due to differences in specific plant characteristics or a result of variation in environmental conditions at some time in the seed development process.

Cultivars that differ in maturity may show consistent differences in seed quality across years; but, the differences may be due to variations in environmental conditions during seed development and/or seed maturation. Altering planting dates so that the cultivars of different maturities mature at the same time has shown that environmental conditions are, in many cases, more important than other plant characteristics in determining seed quality (TeKrony et al., 1984).

Green and co-workers (Green and Pinnell, 1968a, 1968b; Green et al., 1971) evaluated progeny from crosses of three genotypes from Japan that exhibited high levels of seed quality with two adapted cultivars and reported narrow-sense heritabilities of 3 to 29% for field emergence and 2 to 60% for standard germination. They concluded that it should be possible to improve seed quality through plant breeding and that the most efficient method for evaluating segregating populations was by using a general visual rating of seed quality and a laboratory germination test in which normal seedlings were counted early in the test (Green et al., 1971). Singh et al. (1978) evaluated field emergence in the F_3 and F_4 generations of a diallel cross of six genotypes and also concluded that seed quality could be improved by hybridization and selection. TeKrony et al. (1984) found consistent differences in quality between two genotypes (OX-303 and 'Beeson') of similar maturity and concluded that the two genotypes differed in plant characteristics that were important in determining seed quality.

A number of seed characters that have been related to seed quality or performance have also been shown to be under genetic control. These characters may be useful to plant breeders attempting to improve soybean seed quality. Wide variations in seed size exist in the soybean germplasm (Hartwig, 1973). Paschal and Ellis (1978) and Singh (1976), when evaluating lines for potential tropical adaptation, reported that small seed were associated with higher seed quality. TeKrony et al. (1984) reported that a small seeded genotype (OX-303) was consistently of higher quality than Beeson even though they both matured at approximately the same time. However, no physiological basis for the relationship between seed size and quality has been suggested.

Potts et al. (1978) demonstrated that seeds from a line showing a high level of hardseededness were more resistant to field weathering. The small-seeded genotype that TeKrony et al. (1984) showed to exhibit higher-quality levels also showed higher than normal levels of hardseededness which may have contributed to its high-quality levels. Kilen and Hartwig (1978) suggested that the permeable-impermeable response of soybean seeds may be controlled by as few as three major genes.

Starzing et al. (1982) reported that black seeds from segregating plants of a cross of black-and-yellow seeded lines showed slower deterioration during storage at 100% RH than yellow seeds because of lower levels of fungal infection. The usefulness of this character may be limited, however, by the requirement of the soybean industry for yellow seed.

Caviness and Simpson (1974) reported genotypic differences in seed-coat thickness, but no relationship between seed-coat thickness and visual ratings of seed quality. Seed-coat thickness was not associated with seed size in their studies.

Reductions in seed quality have been related to fluctuations in seed moisture content in the pod (Moore, 1971). Yaklich and Cregan (1981) measured the movement of moisture into seed in mature pods of a number of cultivars from Maturity Groups II through VI and reported significant genotype differences. However, the relationship of this character to weathering of seed in the field has not been investigated.

8–3.3 Mechanical

The soybean seed is poorly designed to resist mechanical damage. The embryo is surrounded by a thin seed coat and the radicle-hypocotyl axis lies against the basal margins of the cotyledons. The position of the radicle-hypocotyl axis combined with the thin seed coat make the seed especially vulnerable to injury from mechanical abuse (Delouche, 1974; TeKrony et al., 1980a). Mechanical injury to the seed can occur at any time during harvesting, drying, and conditioning of the seed (Delouche, 1974). Mechanical damage to an individual seed can include the formation of cracks or breaks in the seed coat, cracks in the cotyledons, injury or breakage of the hypocotyl-radicle axis, and complete breakage of the seed to the point where it would no longer be classified as part of the pure seed fraction (Delouche, 1974; Rojanasaroj et al., 1976).

The amount of mechanical damage to the seed is inversely related to the seed moisture level (Green et al., 1966; Newberg et al., 1980; Paulsen et al., 1981a; Singh and Singh, 1981; Prakobboon, 1982). The optimum moisture level for harvesting or handling seed is between approximately 120 and 140 g kg^{-1}. Physical damage increases significantly as the moisture decreases below 120 g kg^{-1}. Although visible physical damage tends to decrease as the moisture level increases above 140 g kg^{-1}, the seed at the higher moisture level may be damaged internally and the germination reduced (Green et al., 1966). Large seeds tend to be more susceptible to mechanical damage than small seeds (Paulsen, 1978; Paulsen et al., 1981a) and seeds that have been exposed to weathering in the field or that have been dried at high temperatures are more susceptible to mechanical damage (Green et al., 1966; Rojanasaroj et al., 1976).

The effect of mechanical damage on seed viability and potential seed performance will depend upon both the amount and type of damage. In general, as the amount of mechanical damage increases, the standard germination decreases, usually as a result of an increase in the proportion of the seeds producing abnormal seedlings (Green et al., 1966; Stanway, 1974, 1978; Luedders and Burris, 1979; Paulsen et al., 1981a; Prakobboon, 1982). Mechanical damage also reduces field emergence (Green et al., 1966; Stanway, 1974, 1978; Luedders and Burris, 1979; Wall et al.,

1983) although Luedders and Burris (1979) concluded that the amount or severity of mechanical damage was not reliably related to field emergence. Wall et al. (1983) found that treatment of mechanically damaged seeds with fungicides did not improve field emergence and Paulsen et al. (1981a) reported that the use of damaged planting seeds did not influence yield if adequate stands were obtained.

Several tests have been developed to measure mechanical damage. The indoxyl acetate test (Paulsen and Nave, 1979) has been used to detect seed-coat cracks, scratches, abrasions, or other small imperfections in the seed coat. The sodium hypochlorite soak test is a rapid test that detects breaks in the seed coat that allow rapid imbibition by the cotyledons (Luedders and Burris, 1979; Paulsen, 1980). The tetrazolium test is also useful to detect mechanical damage to the cotyledons and the hypocotyl-radicle axis (Moore, 1972).

8–3.4 Seed-borne Diseases

Soybean seeds may be infected by a large number of fungi, bacteria, and viruses (Sinclair, 1975, 1982). Infection of the seed by pathogens may reduce seed quality by altering the appearance of the seed, reducing germination or the ability of the seed to produce a healthy vigorous seedling, or transmitting the pathogen to the next generation of plants. Thus, it is obvious that high-quality seed should be free of pathogens.

More than 30 fungi are listed as being seed borne in soybean (Sinclair, 1975). McGee et al. (1980), using a large number of seed lots produced in Iowa, identified nine genera of fungi as seed borne. Ellis et al. (1979) isolated 35 genera of fungi from seed of cultivars from Maturity Groups VIII, IX, and X grown in Puerto Rico. McGee et al. (1980) reported that only *Fusarium* and *Phomopsis* spp. reduced laboratory germination and only *Phomopsis* spp. reduced field emergence. Ellis et al. (1979), however, found that 19 of the 35 genera they isolated significantly reduced germination.

Phomopsis seed decay caused by *Phomopsis sojae* (Lehman) and *Diaporthe phaseolorum* (Cke and Ell) Scc. var. *sojae* (Lehman) Wehn is generally recognized as a major cause of low seed quality in the USA (Sinclair, 1975, 1982). Although *P. sojae, D. phaseolorum* var. *sojae* and *D. phaseolorum* var. *caulivora*, Athow and Caldwell are all associated with seed decay, *Phomopsis* spp. are the most common (Kmetz et al., 1978). All three organisms were isolated from symptomless young plants (Kmetz et al., 1978); however seed infection takes place near maturation (Hepperly and Sinclair, 1980). Germination of seed is reduced in direct proportion to the level of infection (Kmetz et al., 1978; McGee et al., 1980; Kulik and Schoen, 1981; TeKrony et al., 1984). Infected seeds may be shriveled, elongated, cracked, and appear white and chalky or they may show no visual symptoms (Sinclair, 1982). Estimation of infection levels from visual symptoms does not give an accurate estimate of actual seed infection levels (Jeffers et al., 1982b).

Levels of seed infection are increased when soybean residues are present from a previous crop to provide a source of inoculum (Kmetz et al., 1979) and when warm, wet conditions prevail during seed development and maturation, although moisture appears to be more important than temperature (Spilker et al., 1981; TeKrony et al., 1983). Infection can also increase rapidly if the seeds are allowed to remain in the field after harvest maturity (Wilcox et al., 1974; Ellis and Sinclair, 1976). Although germination and field emergence of infected seeds are severely reduced, there is no evidence of higher levels of seed infection on plants produced from infected seed (McGee et al., 1980). Foliar fungicides have been shown to be effective in controlling *Phomopsis* seed decay (Ellis et al., 1974; Jeffers et al., 1982b) and Wall et al. (1983) reported increased field emergence following fungicidal seed treatment of *Phomopsis*-infected seed.

Purple seed stain, caused by *Cercospora kikuchii* (T. Matsu and Tomoyasu) Gardener, occurs in all areas of soybean production (Sinclair, 1982). The purple discoloration of the seed results in a seed lot that is not visually appealing; however, it is not clear whether the fungus reduces germination or field emergence. Murakishi (1951) and Wilcox and Abney (1973) reported reductions in germination and field emergence of purple-stained seed. Hepperly and Sinclair (1981) reported a significant correlation ($r = 0.19$) between the level of purple seed stain and germination for a number of seed lots from Illinois but no significant relationship ($r = 0.12$) in seed lots produced in Puerto Rico. Lehman (1950) and Sherwin and Kreithow (1952) reported no effect of purple seed stain on germination. Several workers have reported an inverse relationship between levels of seed infection by *C. kikuchii* and *Phomopsis* spp. (Roy and Abney, 1977; McGee et al., 1980; Hepperly and Sinclair, 1981) which resulted in an increase in germination as the infection levels by *C. kikuchii* increased because of the decline in *Phomopsis* spp. (Roy and Abney, 1977). Perhaps this antagonistic relationship obscured the relationship between purple seed stain and germination in the past.

There are many other fungi that infect soybean seed, for example *Peronospora manshurica* (Naum.) Syd. or *Colletotrichum dematium* (Pers. ex Fr.) Grove var. *truncatum* (Schw.) Arx, and may reduce germination (Sinclair, 1975, 1982); however their occurrence may be sporadic and they are not usually considered major determinants of seed quality.

A number of bacteria and viruses have been reported to be seed borne in soybean (Sinclair, 1982). Seeds from plants infected with the soybean mosaic virus may be mottled and many exhibit reduced germination (Quiniones et al., 1971; Sinclair, 1982). The soybean mosaic virus is seed transmitted which suggests that virus-free seed would be more desirable for planting purposes (Dunleavy, 1973).

A more complete discussion of seed- borne diseases affecting soybeans can be obtained in the *Compendium of Soybean Diseases* (Sinclair, 1982).

8–3.5 Insects

There are many insects that attack soybean and may cause reductions in yield (Turnipseed and Kogan, 1976); however, only the stink bug complex has a direct significant effect on seed quality. The three most common members of the stink bug complex are the green stink bug [*Acrosternum hilae* (Say)], the southern green stink bug [*Nezara viridula* (L.)], and the brown stink bug [*Euschistus servus* (Say)] (Todd, 1976). Both nymphs and adults feed on soybean by puncturing plant tissues and extracting the juices and, although they may attack all parts of the plant, they prefer young tender growth and fruiting structures (Todd, 1976). Seeds damaged at an immature stage may be shriveled and greatly reduced in size, whereas seed damaged later in development may show only a puncture mark surrounded by a discolored area (Todd, 1976).

Germination and field emergence of injured seeds are reduced in direct proportion to the degree of injury (Daugherty et al., 1964; Todd and Turnipseed, 1974; Yeargan, 1977). Jensen and Newsom (1972) reported that the effect of the injury depended on the location of the puncture. If the puncture was on the hypocotyl-radicle axis, the seed would probably not germinate; however, if the puncture was located in the cotyledons, the seed would probably germinate but may show reduced vigor. Thomas et al. (1974) reported the largest amount of injury when plants were exposed to stink bugs beginning with pod set. There was, however, no effect on germination when exposure was started when the pods were turning yellow. Jenson and Newson (1972) pointed out that most seeds exhibiting moderate to heavy stink bug damage, and low germination, would probably be removed during seed conditioning.

Kilpatrick and Hartwig (1955) reported similar levels of fungal infection in seeds with and without stink bug injury suggesting that the puncture wounds did not facilitate fungal invasion. Stink bugs are able to transmit the causal organism of the yeast spot disease (*Nematospora coryli* Pegl.) (Daugherty, 1967) which may affect seed quality (Sinclair, 1975, 1982).

8–4 PRODUCING AND MAINTAINING HIGH-QUALITY SEED

8–4.1 Cultural Practices

The cultural practices used by seed producers are generally similar to those used in the production of commercial soybean. A major goal of the seed producer is to maximize yield; however, because of the specialized use of the seed crop, there are a number of cultural practices that have a direct influence on seed quality.

Rotating soybean with other crops is generally recommended to prevent build up of certain diseases and to aid in weed control (chapter 9 in this book; Pendleton and Hartwig, 1973; Tanner and Hume, 1978).

In seed production, rotations are frequently used to help maintain genetic purity by eliminating the possibility of cultivar contamination by volunteer soybean plants from the previous years crop. Levels of inoculum are an important determinant of the level of seed infection by *Phomopsis* spp. (Kmetz et al., 1978). Thus, the chances of having high levels of seed infection are greater if soybean residue is present from a previous crop compared with growing soybean in a rotation (Kmetz et al., 1979). Tillage practices that result in incorporation of the soybean residue have resulted in lower levels of seed infection by *Phomopsis* spp. than minimum tillage systems that leave large amounts of residue on the surface (Grau and Oplinger, 1981).

Cultural practices can also be manipulated to create less favorable environmental conditions for infection of seed by *Phomopsis* spp. Seed infection by *Phomopsis* spp. is enhanced by warm, wet conditions during seed development (Spikler et al., 1981; TeKrony et al., 1983). Delaying planting of early maturing cultivars or growing cultivars on the northern edge of their zone of adaptation (causing them to mature relatively late), results in generally cooler and drier and conditions during seed development and maturation and can significantly reduce seed infection by *Phomopsis* spp. (Grau and Oplinger, 1981; TeKrony et al., 1984). Trends for improved germination and vigor were reported by TeKrony et al. (1984) when the planting of early maturing cultivars was delayed. In some areas, planting early maturing cultivars in a double-cropped system after wheat (*Triticum aestivum* L.) has become a common practice to improve seed quality; however, the seed producer must be willing to accept the yield reduction usually associated with this practice (Egli, 1976).

The occurrence of gray moldy seed caused by infection of the seed by *Phomopsis* spp. and *D. phaseolorum* (Cke and Ell.) Sacc. var. *sojae* (Lehman) Wehn has been decreased by K fertilization (Crittenden and Svec, 1974). Jeffers et al. (1982a) reported that K fertilization decreased the incidence of moldy seed, increased germination in some cases, but had essentially no effect on the level of seed infection by *Phomopsis* spp. or *D. phaseolorum*. They also reported that fertilizer rates in excess of those required for maximum yield had little influence on germination or moldy seed.

A number of fungicides are available to control fungal leaf diseases on soybean and these fungicides have been shown to reduce levels of *Phomopsis* spp. seed infection and improve seed quality (Ellis et al., 1974; Jeffers et al., 1982b) in addition to increasing yields in some environments (Ross, 1975; Backman et al., 1979). To reduce the level of *Phomopsis* spp. seed infection, the fungicide must be applied before there is visual evidence of the presence of the disease. Thus, several point systems have been developed, including field history, cultivar maturity, planting date, and environmental conditions to provide a guide on when to use foliar fungicides to improve seed quality (Stuckey et al., 1981; Sinclair, 1982).

Weed-free soybean fields are the goal of all soybean producers; however, freedom from weeds or other crop species is particularly important for seed production. If these contaminants are present in the seed when harvested, it will create additional problems during seed conditioning and, in extreme cases, may make it impossible to condition the seed to the point where it can be marketed. Thus, field selection is one of the most important decisions a seed producer makes to reduce contamination from weeds, other crops, and potentially detrimental seed-borne diseases.

8–4.2 Harvesting

Threshing and conveying operations during harvest consist of dynamic events which often involve large momentum exchanges during collisions of seeds with machine components and other seeds (Bartsch et al., 1979). Paulsen et al. (1981a) stated that the common cause of damage in all grain-handling studies is the particle velocity immediately before impact and the rigidity of the surface against which the impact occurs. Several impact devices have been designed to evaluate the effect of impact velocity on soybean seed quality (Cain and Holmes, 1977; Bartsch et al., 1979; Paulson et al., 1981a). Paulsen et al. (1981a) reported that the percentage of splits and fine material increased as the impact velocity increased and as the seed moisture decreased from 170 to 80 g kg^{-1}. The seed that had low percentages of splits after impact also had high standard germination. Bartsch et al. (1979) reported that detectable levels of mechanical damage were observed at impact velocities as low as 5 m s^{-1} although significant increases in damage levels did not occur until impact velocities increased from 10 to 15 m s^{-1}. Thus, it was concluded that a reduction in the cylinder speed of a combine from 15 to 10 m s^{-1} peripheral velocity would result in a significant reduction in harvest damage. They also concluded that soybean seed harvesting and conditioning should be completed at the highest practical seed moisture content. Significant reductions in impact damage occurred as the moisture content increased from 80 to 180 g kg^{-1}. Analysis of temperature effects indicates that cold weather handling can also be expected to reduce seed quality (Burris, 1979b).

Bartsch et al. (1979) found that seed impacted near the radicle had the largest reduction in tetrazolium vigor index and that seed orientation was highly significant in influencing impact damage. However, seed orientation would appear to be impossible to control or even predict during harvesting or conditioning operations.

Cain and Holmes (1977) evaluated the impact damage to soybean seed as the result of a single high speed collision with a steel plate and concluded that impact damage is dependent on both seed moisture content and the velocity of impact. A single impact produced extensive external injury in relatively dry (107 g kg^{-1}) seed. Seeds at approximately 190 g kg^{-1} of moisture sustained the least impact damage. Seeds impacted at higher moisture levels (253, 302, and 353 g kg^{-1}) did not exhibit extensive external injury, but showed increased respiration rates and decreased germination after impact in comparison to control samples at the same moisture levels.

It is apparent that one method to improve the quality of soybean seed is to reduce the number and level of physical impacts imposed on the seed during harvesting, handling, and conditioning operations. The moisture content at which these operations are performed will also have a significant effect on the resulting impact damage. While it is not possible to totally eliminate impact damage to soybean seed, it is possible to reduce such damage by following management practices which reduce the level of impacts and their overall effect on seed quality.

Equipment used for combining soybean seed fields must achieve a high-harvesting efficiency and meet the same operational requirements as equipment used for commercial grain production. However, in addition, equipment selection, operational procedures, and management decisions must also consider final seed quality. Nave (1977, 1979) has reviewed the present status of soybean-harvesting equipment and those innovations which have been introduced to reduce header losses. A seed producer must be concerned with maintaining threshing and separating efficiency while avoiding undue impact damage to the seed. Efforts to reduce threshing damage while increasing capacity have resulted in the development of rotary threshing equipment (Nave, 1979). Rotary combines have one or more longitudinal rotors to replace the conventional cylinder and straw walkers for threshing and separating grain from crop material. Because the material is apparently subjected to less impact with the rotor, the harvested seed sustain much less breakage than with the conventional rasp-bar cylinder (DePauw et al., 1977).

Newberg et al. (1980) evaluated the damage to soybean caused by rotary and conventional threshing mechanisms. Three different combines (a single-rotor machine, a double-rotor machine, and a conventional rasp-bar cylinder machine) were tested under field conditions at four peripheral velocities. For the cultivar tested ('Amsoy 71'), the percentage of splits were significantly higher for the conventional cylinder than for either the single or double-rotor threshing mechanisms at similar peripheral speeds. For all three threshing mechanisms, the percentage of splits increased as peripheral threshing speed increased; however, the increase in splits was less with the rotary threshing mechanisms than the conventional cylinder. Threshing and separation losses with the rotary combines were significantly higher at the slowest rotor speeds relative to those at the higher speeds. Increasing concave clearance generally decreased the percentage of splits for all three combines; however, the effect was less than that caused by changes in cylinder or rotor speed. Newberg et al. (1980) also found a significant increase in the percentage of splits caused by the elevating mechanism used to move the soybean from the clean-grain auger to the grain tank in all three combines. This indicates that improvements in the design of the augers and elevators used to convey soybean into combine grain tanks are needed; especially for harvesting soybean for seed.

The viability and vigor of soybean seeds tend to decline in the field after reaching physiological maturity especially under adverse weather conditions. Thus, soybean seed should be harvested as soon as possible after they reach a practical harvest moisture content (usually 150 g kg^{-1} or less). If the harvest moisture content of the soybean is too low (below 120 g kg^{-1}), unacceptable levels of physical damage can be expected. Rotary combines tend to produce less physical damage to soybean during harvest; however, conventional combines can do a satisfactory job if properly adjusted. Adjustment of cylinder speed and concave clearances does not appear to be as critical for rotary-type combines. In either type, the cylinder speed should be high enough to achieve complete threshing and separation but not so high as to increase seed impact damage. Combine settings often need to be readjusted as harvest conditions change with time of day or with varying field environmental conditions.

8–4.3 Drying

As discussed above maximum seed quality is obtained when soybean seed is harvested as soon as possible after field drying to a suitable harvest moisture content (150 g kg^{-1}). Often, some drying will be required after harvest to maintain seed viability during storage. Drying temperatures, air flow rates, and drying times all need to be controlled within certain limits to maintain maximum seed quality. Improper drying conditions can reduce seed germination and physically damage the seed and decrease its quality. Prior to 1970, little research on soybean drying was reported in the literature. In 1945 and 1946, Holman and Carter (1952) conducted a limited numer of drying studies as a part of their work on soybean storage. They found that drying was best accomplished using forced-natural air in mild weather or forced-heated air when ambient temperatures were low and/or RHs high. For natural air drying they recommended that air temperatures should be above 16 °C and the RH below 75%. Matthes et al. (1974) found a definite correlation between drying time and seed germination in the upper levels of their batch dryer. They recommended minimum air flow rates of 9.9 to 13.2 m^3 min-t^{-1} to maintain seed germination for the moisture contents (227 to 280 g kg^{-1}) and drying conditions (of 42 to 55% RH) tested. Rodda (1974) stated that natural air drying at air flow rates of 2.2 to 3.3 m^3 min-t^{-1} was adequate for drying soybean seed harvested at moisture contents up to 160 g kg^{-1}.

Walker and Barre (1972) observed considerable cracking of the seed coat in soybean dried at high temperatures and/or low RH. There were significant differences between cultivars when the RH of the drying air was 40% or more. There was little effect of temperature on germination up to and including 54 °C; above that, however, there was a drastic reduction in germination. Similar results were reported by White et al. (1980) although they found that the incidence of seed-coat cracks did not approach zero until the drying air RH was 50% or higher. Data from these experiments was used to develop a thin-layer drying model (White et al., 1981) to describe the effect of initial moisture content, drying air temperature, and RH on the drying rate of fully exposed soybean seeds. Ting et al. (1980) studied the development of seed-coat cracks as a function of depth when soybeans were dried in a batch-type dryer and found that distance from the air inlet was the most significant factor affecting the development of seed-coat cracks. The further any location was from the air inlet, the lower the drying damage. Drying conditions were found to affect the magnitude of the seed-coat damage gradient existing in the soybeans after drying had been completed.

Pfost (1975) reported that crackage increased with an increase in drying air temperature, initial moisture content, and drying rate and decreased with an increase in final seed moisture and drying air RH. Drying air temperatures of 54 °C and lower had little or no effect on germination; however, germination was sharply reduced by 66°C air temperatures. Soybean seed at high initial moisture contents suffered greater losses of germination than those at low moisture contents from equal exposure to 66°C drying air.

In an effort to improve the quality of soybean seeds dried with heated air, Sabbah et al. (1977) investigated the use of a reversed-direction airflow procedure with a laboratory batch dryer. This procedure involved periodically changing the direction of the air flow through the drying bed according to a given set of drying conditions. This approach resulted in considerable improvement in soybean seed quality, however practical application would require some modification in the design of conventional batch-in-bin drying systems. Villa et al. (1978) used simulation techniques and developed mathematical models for predicting the drying process and the loss of soybean seed germination. Good agreement between experimental and simulated results was obtained for a limited range of drying conditions. Results showed that relatively high air flow rates are required for drying in hot and humid areas because of the adverse effect of temperature on germination. Air flow requirements for drying soybean seed in bins were found to be about one and a half times higher than those for drying soybean for the commercial grain market.

Ghaly and Sutherland (1983) reported that soybeans of 140 to 180 g kg^{-1} initial moisture can be dried for 4 h, using temperatures of 40 to 55 °C, without significantly reducing germination. An air temperature of 80 °C killed all the seed at the three moisture levels tested (140, 160, and 180 g kg^{-1}) as did 70 °C at 160 and 180 g kg^{-1} of moisture. It was clear that the soybean seeds became more sensitive to heat damage as the initial moisture increased from 140 to 180 g kg^{-1}. Maximum safe drying temperatures reported for soybean seed at 140, 160, and 180 g kg^{-1} of moisture were 65, 60, and 55°C, respectively. Heating soybean seeds to 60°C at a fixed seed moisture was found to increase the susceptibility to heat damage.

In drying soybean seeds, the objective is to reduce the moisture content of the seed to the desired level without undue loss of seed viability and vigor and without inflicting physical damage on the seed which could significantly reduce their quality and storability. To limit seed-coat cracks in the dried seed, the RH of the drying air should be above 40% (Walker and Barrer, 1972; White et al., 1980). The maximum RH must be sufficiently low (65% or lower) to dry the seed to the desired moisture.

When drying soybean seeds and most cereal grains, a maximum drying air temperature of 43 °C is generally recommended (Hall, 1980). Nellist (1980), however, pointed out that a range of recommended drying temperatures are being used in various countries and some recommendations are based on somewhat sparse experimental evidence. Several investigations (Walker and Barre, 1972; Pfost, 1975; Ghaly and Sutherland, 1983) have shown that under certain conditions drying air temperatures can exceed 43 °C without any obvious damage to soybean seed germination. Apparently, safe drying temperatures are affected by both soybean seed moisture and the time of exposure to the drying air. If one wishes to limit physical damage to soybean seed by controlling the RH of the drying air, then the question of a safe maximum drying air temperature is usually of little consequence. Limitations on temperature increases for humidity control will keep the drying air temperatures well below the generally recommended 43 °C maximum.

Natural air drying can be used to dry soybean seeds with moisture contents of 160 g kg^{-1} or less if the air temperature is above 10°C and the RH below 70%. Air flow rates of 2 to 3 m^3 min-t^{-1} are required. If the RH is higher, a few degrees of supplemental heat will be required. For seed between 160 and 190 g kg^{-1} moisture, air flow rates of 5 to 6 m^3 min-t^{-1} should be used with supplemental heat added as necessary to keep the drying air RH in the 55 to 65% range. For soybean seed moistures above 200 g kg^{-1}, Matthes et al. (1974) suggested supplemental heat to control the RH at 40 to 50% with air flow rates of 10 to 12 m^3 min-t^{-1}. At this air flow rate, soybean depths should be limited to 1.2 m or less.

Soybean seeds can be dried with any type of commerical grain dryer provided RH and temperature restrictions are observed. However, drying equipment which utilizes recirculators or stirring devices are not recommended because of potential seed damge. Because of the temperature and humidity restrictions on drying air, most soybean seeds will be dried with batch-in-bin or in-storage type drying systems (for a description of various drying systems see Hall [1980] and Justice and Bass [1978]). When the drying air RH is below 60%, the seeds can potentially dry below 100 g kg^{-1} of moisture which can increase damage in subsequent handling operations. Overdrying can be compensated for in batch-in-bin systems by blending the overdried seed with the underdried portion of the batch when emptying the bin. This is not possible with in-storage types of drying systems so the operator must limit overdrying in the bin by keeping the RH of the drying air above 55%.

8–4.4 Storage

Soybean seed must be properly stored in order to maintain an acceptable level of germination and vigor until needed for planting. Storage periods may vary from as little as 6 months, if the seeds are to be planted the next season, up to 20 months or longer if the seeds are to be carried over for one or more seasons. Longevity of seed in storage is influenced by the quality of the seed going into storage, seed moisture content, and storage temperature (Delouche, 1974; Justice and Bass, 1978; Egli et al., 1979; Burris, 1980).

The quality of soybean seed entering storage can be reduced by: adverse weather conditions prior to harvest; damage from pathogens, insects, and other pests; mechanical damage caused by harvesting and handling operations; and seed injury resulting from necessary drying operations. Seed deterioration from these causes should be minimized insofar as possible to increase potential seed longevity in storage.

Irrespective of initial seed quality, temperature and seed moisture are the two most important factors affecting seed deterioration in storage. Because seeds are hygroscopic, they will exchange moisture with the surrounding air until the vapor pressure of the seed and that of the air reach a state of equilibrium. If the seed comes to equilibrium with air maintained at a relatively constant moisture level, then its moisture content is referred to as the equilibrium moisture content (EMC) of that seed corresponding to the existing air conditions. On the other hand, if the seed are surrounded by a limited amount of air (such as occurs in the interstitial spaces among seeds in a storage bin) then the air will come to moisture equilibrium with the seed without any significant change in the seed moisture. The RH of the air in this situation is referred to as the equilibrium relative humidity (ERH) corresponding to the existing seed moisture at the prevailing temperature. All equilibrium moisture properties are a function of seed temperature.

Equilibrium moisture properties are specific for each type of seed and are important in developing storage recommendations and in the management of seed drying systems. The EMC values for soybean seed as measured by Alam and Shove (1973) and tabulated in the *ASAE Yearbook* (ASAE, 1983) are presented in Table 8–2. The values presented are for desorption (drying) conditions where the seed moisture decreases to reach equilibrium with the stated air conditions. For adsorption conditions, where the seed gains moisture to achieve equilibrium, equilibrium

Table 8–2. Equilibrium moisture content of soybeans under desorption conditions.†

Temperature	Relative humidity, %								
	10	20	30	40	50	60	70	80	90
°C	g kg^{-1}								
5	52	63	69	77	86	104	129	169	224
15	43	57	65	72	81	101	124	161	219
25	38	53	61	69	78	97	121	158	213
35	35	48	57	64	76	93	117	154	206
45	29	40	50	60	71	87	111	149	—

†All moisture contents are presented on a wet basis as grams of water per kilogram of seed weight.

moisture contents would be slightly lower than those presented. This means that the values from this table should predict the lowest moisture to which soybean seeds can be dried when using air of a specified temperature and RH; however, under adsorption conditions the expected seed moisture levels would be lower than those shown in Table 8–2.

Equilibrium moisture properties are useful in analyzing drying and storage systems. For example, soybean seed can be dried to a moisture content of 97 g kg^{-1} when using 25°C drying air having a RH of 60%. Seeds at a moisture content of 158 g kg^{-1} stored at 25°C will produce a RH of 80% in the air contained within the seed mass. This type of information can be used to predict the growth of microorganisms in the stored seed and the potential for seed deterioration. Most storage fungi cannot grow and reproduce on seeds in equilibrium with a RH $<$ 65% (Christensen and Kaufmann, 1969). As indicated in Table 8–2, this corresponds to a 109 g kg^{-1} seed moisture at 25 °C. Activity of storage insects can also be expected to drop at RHs below 50% (Delouche, 1974).

Temperature and moisture conditions are known to affect physiological, biochemical, and genetic changes in seeds during storage (Roos, 1980). Microbial activity is also closely related to these parameters (Christensen and Kaufmann, 1969). Most studies related to seed storage do not separate the influence of the above processes on seed deterioration; instead, they relate reductions in seed quality to the storage environment and storage time.

Ramstad and Geddes (1942) reported a close relationship between grade, chemical changes, germination changes, and the moisture of stored soybean seed. When soybean with seed moistures varying from 138 to 169 g kg^{-1} were stored for 15 weeks at room temperature and at 37.8 °C, none of the seed were viable. However, seed samples stored at the same moisture for the same length of time at approximately 2 °C retained a high degree of viability. After 18 months storage at 2 °C, germination levels of 85, 84, 78, and 42% were reported for samples at 138, 149, 158, and 189 g kg^{-1} moisture, respectively. They concluded that maintenance of high-germination soybean seeds required storage at a low moisture and a low temperature.

Toole and Toole (1946) studied the effect of temperature and seed moisture on soybean seed viability of two cultivars stored for periods of 10 yrs. Seeds with approximately 180 g kg^{-1} of moisture were dead in 1 to 3 months at 30 °C, in 22 to 39 weeks at 20 °C, and in 2 yrs at 10 °C. The seeds maintained high viability for 2 to 3 yrs at 2 °C, but were dead in 6 yrs. Nearly complete germination was obtained after 6 yrs at -10 °C. At a more typical storage moisture content of approximately 135 g kg^{-1} the seeds were dead after 22 weeks storage at 30 °C and after 2 yrs at 20 °C. High viability was maintained for 3 yrs at 10 °C and 10 yrs at 2 °C while little change in germination occurred after 10 yrs at -10 °C. At 80 to 90 g kg^{-1} moisture, the rate of seed deterioration was less at all storage temperatures with no change in germination over a 10-yr period at 10, 2, and -10 °C. Hukill (1963) used the data from Toole and Toole

(1946) to develop an "age index" to show the relationship between moisture, temperature, time, and germination for soybeans. His results, however, could not account for variation in the viability of the seed when placed in storage.

Burris (1980) stored seeds of six soybeans cultivars for 3 yrs at seed moisture levels of 80, 100, 120, and 140 g kg^{-1} and storage temperatures of −1, 10, 15, and 27 °C. The rate of seed deterioration in storage increased with increasing temperature and seed moisture for all cultivars. Burris (1980) used the combined data from all cultivars and developed constants for seed storage prediction equations initially proposed by Roberts (1960). These equations were based on the assumption that the frequency of individual deaths with time in a seed population stored under constant conditions could be described by a normal distribution. With the appropriate constants, these equations can be used to predict the percentage viability of a seed lot after any given period under any combination of temperature and seed moisture normally encountered. Roberts (1972) showed that such equations could be applied to a wide range of seed species as well as to a particular seed lot. A major disadvantage, however, was that they did not take into account variations in potential longevity between seed lots resulting from differences in genotype or differences in seed quality caused by prestorage treatments or conditions.

Ellis and Roberts (1980) improved the viability prediction equations (Roberts, 1960) to take into account variations in initial seed quality within a given species and to more accurately consider the influence of a wider range of storage environments. Constants for the improved equations were shown to be essentially the same for soybean seed lots of both high and low vigor levels. Three equations were employed by Ellis and Roberts (1980) to predict seed viability. They first described the seed survival curve in terms of the viability, ν (probit percentage viability) to be expected after a given storage period, p.

$$\nu = K_i - p/\sigma \qquad [1]$$

where K_i is a constant for the seed lot in question and σ is the standard deviation of the cumulative frequency distribution of seed deaths for the specified storage conditions. Differences between seed lots should not affect the value of σ and are accounted for by differences in the value of K_i.

The storage environment has no effect on K_i, but it affects σ according to the following equation:

$$\log \sigma = K_E - C_W \log m - C_H t - C_Q t^2 \qquad [2]$$

where m = percent seed moisture content on a wet basis, t = storage temperature in °C, and K_E, C_W, C_H, and C_Q are constants whose values are common for all seed lots of a given species.

Equations [1] and [2] can be combined to give

$$\nu = K_i - p/10^{K_E - C_W \log m - C_H t - C_Q t^2}. \qquad [3]$$

This equation describes the probit percentage viability to be expected for any seed lot after any time when stored at various temperatures and seed moisture. Ellis et al. (1982) summarized four essential features of seed physiology characterized by the above viability equations.

1. Although the survival of different seed lots, or cultivars within a seed species, may differ when stored under identical conditions, the seed survival curves are symmetrical sigmoids which can be described by negative cumulative normal distributions which, in a given species, have the same standard deviation in any given combination of temperature and seed moisture.
2. The relative difference between seed lots is maintained in all storage environments because the relative effect on longevity from altering either temperature or seed moisture is the same for all lots.
3. There is a negative logarithmic relationship between seed longevity and seed moisture.
4. Seed longevity increases slightly less than exponentially with a decrease in temperature so that the rate of loss in viability per 10 °C rise in temperature increases with temperature.

The K_i in the above viability equations is specific for each seed lot and is a measure of initial seed quality (Ellis and Roberts, 1980). Its value is dependent on genotype, the prestorage environment and their interaction. The K_i must be estimated before the viability equations can be applied. This may be accomplished by conducting a germination test at the start of the storage period, or more accurately by carrying out an initial rapid-aging test, in which a sample of seed is rapidly deteriorated under a constant adverse environment (Ellis and Roberts, 1980).

The survival of soybean seed in sealed storage has been investigated by Ellis et al. (1982). Various combinations of storage temperature (from −20 to 70 °C) and seed moisture (ranging from 50 to 250 g kg^{-1}), were studied for four different soybean cultivars. Viability constants for the above equations (with p expressed in days) were derived from the data as follows:

$K_E = 7.748$,
$C_W = 3.979$,
$C_H = 0.053$, and
$C_Q = 0.000228$.

This work confirmed the applicability of the viability equations in predicting the storage life of soybean seed under known environmental conditions. All soybean cultivars responded in the same fashion to storage temperature and moisture. The relative initial difference between seed lots in absolute longevity was maintained in all storage environments because the relative effect on longevity from altering either temperature or moisture was the same for all lots (Ellis et al., 1982). The value of K_i for a given seed lot is indicative of its quality and potential storage life. For known storage seed moistures and temperatures, seed with a high value of K_i will have higher viability and vigor after a predetermined storage period. This agrees with the recommendation of Burris (1980) that seeds of high moisture and average vigor be marketed first and that seeds of low moisture and high vigor be selected if necessary for carrying over to the next planting season.

The viability of soybean seed after storage for specified periods of time at given levels of moisture and temperature can be predicted using the viability equations of Ellis and Roberts (1980) along with the viability constants presented by Ellis et al. (1982) provided an accurate estimate of K_i, the initial viability constant, has been established. Without such an estimate, only generalized storage recommendations can be made.

Misra (1981) indicates that soybean seed at 120 g kg^{-1} of moisture should be stored for no longer than 9 months. This is generally accepted for the storage of seed until the next planting season although the actual level of seed deterioration will depend on initial seed quality and storage temperature. Delouche (1974) recommends that soybean seed be rapidly and properly conditioned to 100 to 120 g kg^{-1} of moisture after harvest for storage until the following spring. He recommended 100 g kg^{-1} of moisture content or less for carryover seed storage and air conditioning to reduce a summer storage temperatures. For longer-term storage soybean seed moistures of 80 g kg^{-1} would be advantageous (Burris, 1980); however, significant mechanical damage can be expected when handling and conditioning soybean seed at this moisture. In tropical areas, high temperatures and humidities make storage more difficult. Rodda and Ravalo (1978) stored soybean seed at 25 °C and found that only low moisture seed (65 g kg^{-1}) stored in sealed containers maintained adequate germination levels for 9 months.

Seeds stored in bulk should be preconditioned prior to storage, if possible, and aerated as necessary to maintain seed quality. Aeration reduces temperature gradients in storage and, thereby, reduces convective air currents which can cause moisture migration. Aeration systems need to provide at least 0.11 m^3 of air min^{-1} t^{-1} of seed. In the fall, the seed needs to be cooled as necessary to keep the average seed temperature within approximately 3 °C of the average monthly temperature until the seed temperature reaches 2 to 4 °C. It is not a good practice to freeze seed if it can be prevented since it will require a longer period to warm up and condensation can be a problem if frozen seed is moved or aerated during periods of high humidity. In the spring, the aeration system should be used to warm the seed to 10 to 12 °C but no higher in order to avoid unnecessary seed deterioration. A detailed description and analysis of aeration systems can be found in the literature (Burrell, 1974; Midwest Plan Service, 1980; Loewer et al., 1979).

8–4.5 Seed Conditioning

Seed conditioning is the final step that converts soybean seed into the finished product, high-quality seed for planting purposes. Depending upon the crop maturity and seed moisture at harvest, the final product may be overthreshed resulting in split seeds and seed fragments or under-threshed resulting in pods, stem portions, and other materials in the grain. In either case, the grain as it leaves the combine is not fit for planting and seed conditioning is necessary before the seed is sold to the farmer. Thus, there are several reasons for conditioning soybean seed to upgrade the quality. These include the following:

1. Remove other crop and weed seed.
2. Remove damaged, immature, and diseased soybean seed.
3. Remove foreign material (pods, stems, soil peds, and insects).
4. Apply seed protectants.
5. Improve seed lot appearance.
6. Maintain or improve seed germinability.

Several excellent reviews have been published which discuss the conditioning machines available for soybean seeds (Harmond et al., 1968; Vaughan et al., 1968; Greg et al., 1970; ISTA, 1977). The operator of a seed-conditioning plant must be able to exploit the differences between the physical characteristics of the soybean seed and the other components of the seed lot. A knowledge of the capabilities and the limitations of all equipment is important for successful seed conditioning. It is the intent of this section to briefly review the flow of soybean seed through a typical seed-conditioning plant and discuss the following steps:

(i) preconditioning, (ii) basic seed cleaning, (iii) seed separation and grading, and (iv) seed treatment.

The basic diagram that is commonly used for flow of seed through a seed-conditioning plant is shown in Fig. 8–1. Nearly all soybean seed is received at a conditioning plant in bulk and may be preconditioned over a scalper or aspirator prior to storage or basic seed cleaning. Seeds are most commonly conveyed from bulk storage bins to a receiving pit where they are elevated to a distribution point before being passed by gravity to the basic seed cleaner or other seed-cleaning machines (Fig. 8–2). Depending upon the design of the seed-conditioning plant (vertical or horizontal), the seeds may be elevated from one to several times and dropped from various heights into holding bins prior to cleaning, treating, and final bagging operations (Fig. 8–1).

8–4.5.1 Preconditioning

Prior to basic seed cleaning, a precleaning examintion of the rough seed is essential and preconditioning of the seed is often beneficial. As rough seed is received from the combine, but prior to storage, a seed sample should be taken and tested immediately to determine seed moisture. Seed that is too low (< 100 g kg^{-1}) in moisture or of extremely low quality (high percentage of splits or contaminants) may be unfit for seed purposes and should be rejected. If the seed moisture is too high (> 150 g kg^{-1}) it may have to be dried before bulk storage. Seeds that are high in moisture will often contain considerable trash and green material (pods, stems, etc.) which, if removed during preconditioning, may lower the seed moisture enough to allow storage without additional drying.

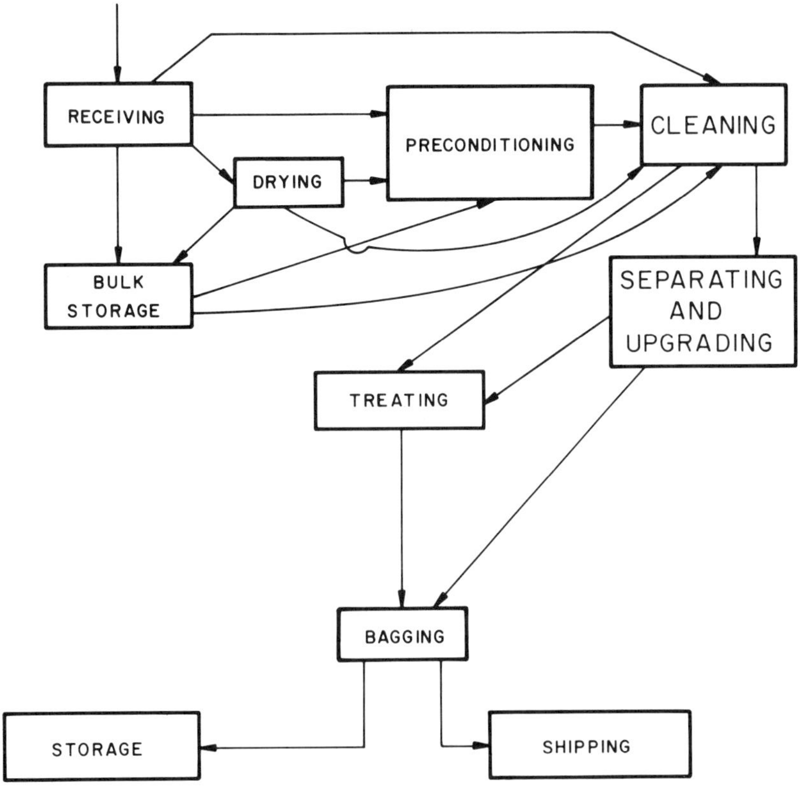

Fig. 8–1. Basic flow diagram of operations in seed conditioning.

A scalper is the most commonly used preconditioning machine. Many types of scalpers are available; however, the simplest types are the single-flat vibrating screen or a rotating reel screen, which allow the soybean seed to pass through and the rough, larger material to be scalped off. The maximum benefits from scalping rough seed are achieved during receiving (Fig. 8–1) before conveying the seed into drying or aeration bins. By removing trash from the seed lot at this time, it reduces the resistance of the seed to air flow, increases the rate of drying, and aids in the control of storage molds and insects. It also increases the efficiency and effectiveness of basic seed cleaning equipment and improves the chances for separation of crop and weed contaminants.

Prior to basic seed cleaning, a seed sample should be taken and examined for purity, by a visual examination or a complete purity test. This precleaning examination is conducted to determine the kinds and quantities of contaminants that need to be removed to achieve the desired purity. The use of hand screens will provide information on seed size and assist the operator in selecting screens to use in basic cleaning. Failure to conduct a precleaning examination (or to conduct it accurately) is often a primary reason for substandard seed quality and costly recleaning.

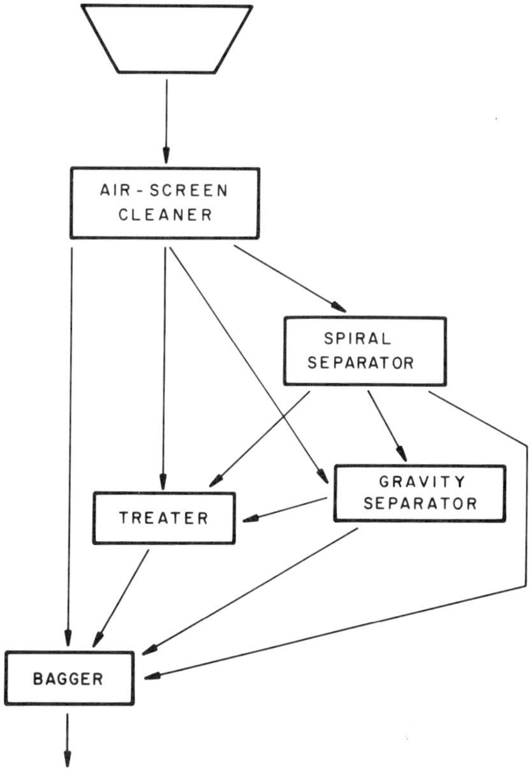

Fig. 8–2. Common sequence of machines used to condition soybean seed.

8-4.5.2 Basic Seed Cleaning

The basic machine used in seed-conditioning plants for soybean and most other crop seed is the air screen cleaner. Most modern soybean seed-conditioning plants will have air screen cleaners that range from two air separations and four screens (Fig. 8–3A) to multifan machines with up to eight screens. The air screen machine exploits the differences in seed size, shape, and density of the seed lot. The machine uses three cleaning elements: (i) aspiration, in which light material is removed from the seed mass, (ii) scalping, in which the good soybean seeds are dropped through screen openings, but larger material is carried over the screen and removed, and (iii) grading, in which the good soybean seeds ride over the screen openings, while smaller particles fall through. A flow diagram of seed through an air screen machine is shown in Fig. 8–3A with two air separations (at the seed entry and discharge) and two scalping screens (1 and 3) and two grading screens (2 and 4). The size of seed in each soybean seed lot will vary depending upon the environmental conditions during production and the cultivar being conditioned. A typical selection of screen sizes that may be used for each of the four screens when cleaning soybean has been recommended by the manufacturers of

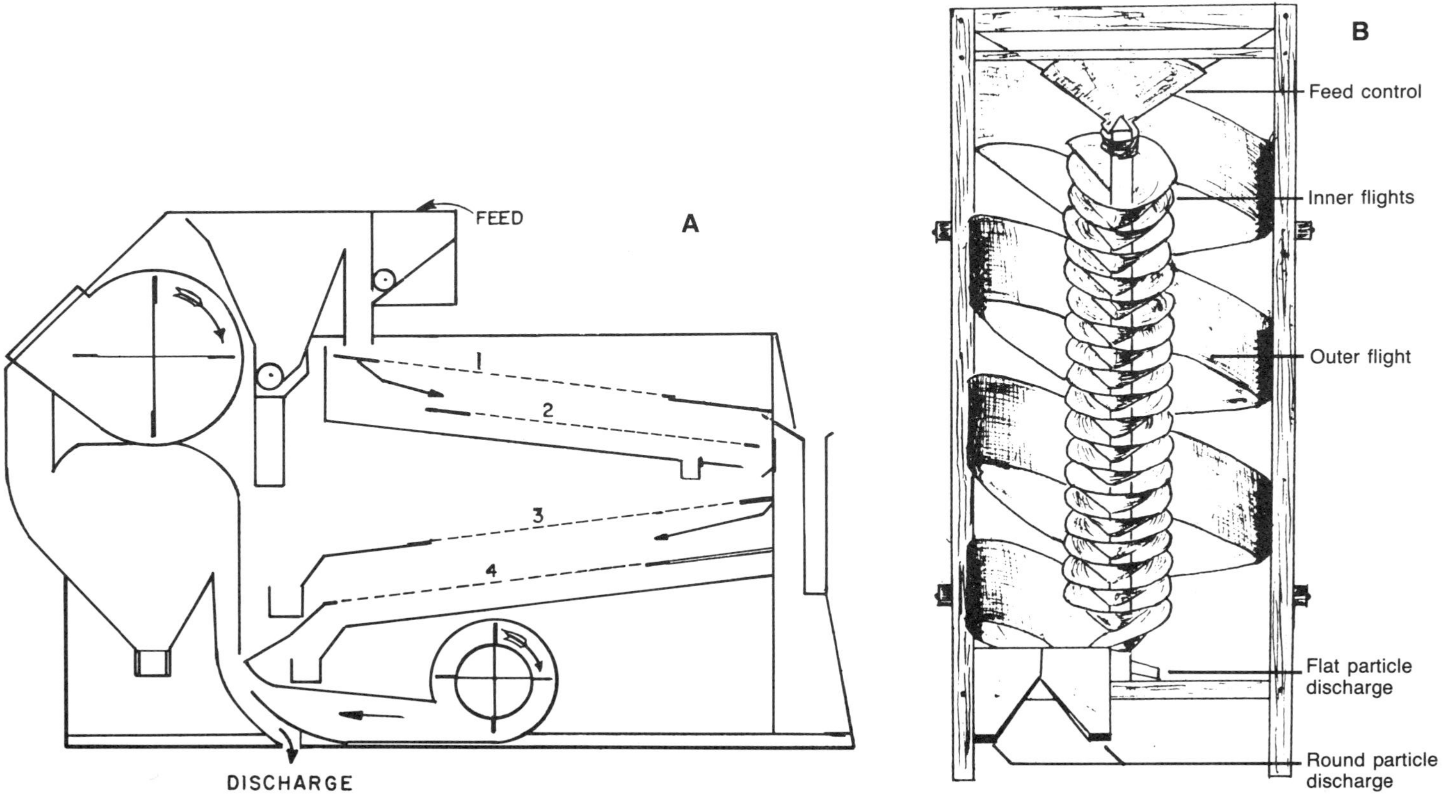

Fig. 8–3. Soybean seed conditioning machines; (A) cross section of a four-screen air-screen cleaner, (B) cutaway view of a spiral separator, and (C) diagram of a specific gravity separator. Drawings are the courtesy of Seed Technology Lab., Mississippi State University.

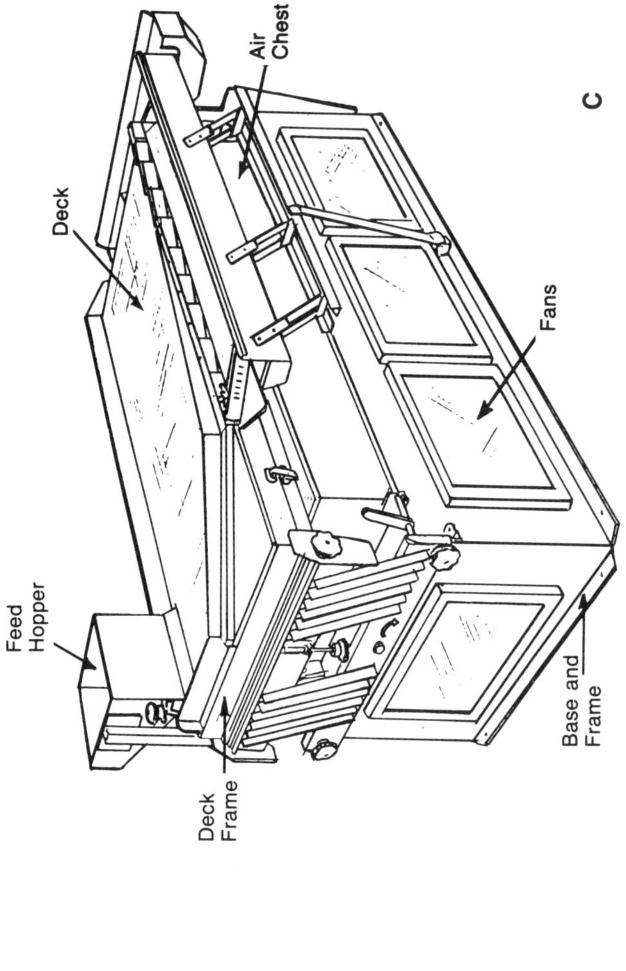

Fig. 8–3. Continued.

air screen machines and has been reviewed by Potts and Vaughan (1977) and Henderson and Vaughan (1980).

The precision of an air screen cleaner is determined by the purity level desired and the quality of rough seed received following harvest. In some cases, the operator of a seed-conditioning plant cannot afford the time and seed loss involved in cleaning all soybean seed lots to the highest purity level. Unfortunately, the most frequent cause of poor soybean seed cleaning on an air screen machine is the tendency to operate the machine at an excessive rate. Thus, many seed lots remain average in quality and do not reach the purity levels desired. The air screen cleaner is an excellent machine, however, which (if properly operated) can clean many seed lots to the desired purity level. When additional uniformity is needed it may be necessary to use more specialized seed separation equipment.

8-4.5.3 Seed Separation and Grading

In recent years, an increasing number of seed conditioning plants are cleaning soybean seeds over specialized equipment to improve both the quality and appearance of the seed. The two machines that are comonly used are: (i) spiral separators, and (ii) specific gravity separators (Fig. 8-3B, and 8-3C). Both offer much more precision in separation, however, all seed must initially be cleaned over the air screen machine.

A spiral separator consists of from one to several sheet metal flights spirally wound around a central vertical tube or axis (Fig. 8-3B). The seed lot is fed into the top and flows down the spiral so that the rounder and heavier soybean seeds tend to flow faster and in a wider arc of travel than the smaller immature or flattened soybean seeds or other contaminants. The seed separation is made according to the shape, density, and degree of roundness of the seed components. Weed seeds which can be separated from soybean seeds using a spiral separator include purple moonflower, common morningglory, common cocklebur, and giant ragweed (Potts and Vaughan, 1977). The spiral can also be used to remove those remaining broken soybean seeds (splits), corn seeds, misshapen or immature soybean seeds, and soil peds in soybean cyst nematode infected areas. Spirals are not effective, however, in separating soybean seeds from cowpea and balloonvine seed, or other soybean seeds having cracked seed coats.

Since a spiral separator has no moving parts, it is relatively inexpensive and easy to operate. Rate of seed flow into the separator is the only adjustment that is necessary; however, recently the addition of flight dams (small wooden or rubber strips) attached to the inner flight of the spiral has improved the separation of weed seed from soybean seed. Equipment companies manufacture spirals which are enclosed to reduce the noise level and have adjustable flight dams on each spiral unit. These companies also offer spirals in multiple units to increase capacity and allow the seed to flow continuously from various air screen machines.

The specific gravity separator is one of the most sophisticated machines in a seed-conditioning plant. The principle of separation for this machine is primarily air stratification on a flat deck (table) and vibrational conveying (Fig. 8–3C). Seed are separated by differences in seed density, size, and surface texture. For many years, little success was achieved using specific gravity separators on soybean seeds; however, with additional research and operator experience the machine has gained acceptance as a finishing machine. It has been particularly successful in removing soil peds from soybean seeds (Lasqueves et al., 1979), which is important in production areas which have the soybean cyst nematode. For this reason, they recommended the addition of a gravity table to the air screen, spiral separator line when conditioning soybean seed. It has also been used to separate de-spined common cocklebur seed from soybean seed; however, less success has been realized in separating machanically damaged and diseased soybean seeds from sound soybean seeds.

In summary, the primary purpose of seed conditioning is to upgrade the quality of soybean seed; however, precautionary measures must be taken during conditioning to prevent the quality from being lowered. Previous sections have discussed the susceptibility of soybean seeds to mechanical injury during seed-handling operations. The seed moisture, seed temperature, rigidity of the surface at which impact occurs, and the type of conveying and seed cleaning equipment can all influence the amount of physical breakage that occurs during conditioning (Bartsch et al., 1979; Burris, 1979a; Hoffman and McDonald, 1981; Paulson et al., 1981a). Delouche (1974) reported that seed cracking and splitting increased sharply as moisture decreased below 125 g kg^{-1}, while seed bruising injury may occur at seed moistures above 140 g kg^{-1}. Increases in mechanical damage and reductions in seed quality have also been reported when seeds are dropped or impacted at low temperatures (Bartsch et al., 1979; Burris, 1979b). Thus, seed conditioning and handling during freezing temperatures is not recommended.

Investigations have been conducted to determine where seed injury occurred as the seed flowed through a seed conditioning plant (Hoffman and McDonald, 1981; Misra, 1982). Mechanical injuries incurred at each step were dependent upon seed moisture but tended to be cumulative. The initial seed elevation from the receiving pit to a bin over the air screen machine created the greatest mechanical injury and reduction in seed quality. For most seed lots, however, the quality improved as the seed was cleaned and mechanically damaged seed were removed. It is important to remember that seed conditioning is only one step in the production of high-quality soybean seed. The seed producer can do many things during production, harvesting, drying, and bulk storage to improve the cleanliness and condition of combine-run seed. Close cooperation between the producer and conditioner will reduce seed conditioning costs and assure higher quality of all seed lots.

8-4.5.4 Seed Treatment

The quality of soybean seeds prior to conditioning is dependent on many environmental and management factors that may occur during production, harvesting, drying, and storage. It has been shown that treating diseased seeds with a fungicide can improve germination and emergence (Athow and Caldwell, 1956; Ellis et al., 1975). A fungicide can also protect the seed from some pathogenic soil microorganisms. This is especially important when seeds are planted in cold, wet soils or other field conditions that are not favorable for seed germination and growth.

A recent survey estimated that 48% of the soybean seeds planted in the USA in 1981 were treated with a fungicide (MacFarlane, 1980). Only 11% was commercially treated, however, while the remainder (37%) was treated in the hopper or planter box by farmers just before planting. Commercial seed treatment was most prevalent in the midwestern states (Indiana, Illinois, and Ohio) while planter box treatment occured most often in southern states (Arkansas, Mississippi, and Louisiana).

Fungicides for seed treatment have been formulated into four basic types; liquids, flowables (concentrated or ready-to-use), wettable powders, and dusts. These fungicides can all be applied commercially by seed conditioners using seed treaters designed to apply accurately measured quantities of fungicide to a given weight or volume of soybean seeds. Seed treatment chemicals are usually formulated for a specific type of application (i.e., dust, liquid). Attempts to apply a slurry formulation as a dust or planter box treatment is not recommended and could be hazardous. To perform accurately, a commercial seed treater must be adjusted correctly and given continuous maintenance.

Seed treatment machinery is classified in many ways, however, McFarlane and Hairston (1984, personal communication) have provided the following simplified outlines.

1. Wet-type treaters utilize the slurry, mist, and spray principles and employ the weight of the seed to operate a seed dump and chemical measuring system.

Slurry treaters utilize fluid formulations that are usually kept in a uniform suspension by continuous agitation to be applied as a slurry. Slurry formulations may be purchased in a ready-to-use liquid or flowable forms or may be prepared by mixing wettable powders or emulsifiable concentrates (flowable) with water. After the chemical is applied, the seeds and chemical are conveyed through a coating chamber (auger or revolving drum), spreading an even coat of chemical on each seed and allowing the moisture to evaporate. Slurry treaters provide accurate and thorough seed coverage, but have the disadvantage of requiring continuous agitation (especially on older models) and may produce considerable dust if wettable powders are used.

Mist or spray-type treaters do not require agitation and utilize true liquids or flowable materials of low viscosity. The chemical is fed directly into the treater and is applied directly to the seed as an atomized mist or with nonplugging nozzles as a spray after which the treated seeds move through a coating chamber. These treaters are recommended when small amounts of chemical must be applied to a relatively large quantity of seed. They require less space and provide excellent seed coverage with no dust problem.

2. Dry dust-type treaters are used for dry, powder formulations and are usually used for seeds that are fragile such as soybeans. Measured amounts of powdered chemical are continuously applied to the seeds using a vibrating feeder and the seeds and chemical are blended together in a coating chamber (auger-type or revolving drum). If the auger-type mixer is used, the element inside the chamber is a nylon brush for gentle movement of seeds through the chamber. Dust treaters are easy to clean and operate as no moisture is added to the seed during treatment. Dust-type treaters do not distribute the chemical as uniformly as a wet-type treater, however, and require controlled ventilation because of excessive dust in the working area.

Regardless which treater is used, the seed must move through a coating chamber of some sort before it is conveyed to a bin for final bagging. Because soybean seeds are rather fragile, a drum-type coater is better to use than the auger-type or film coater with rods. If an auger-type chamber is used, it should be kept three-fourths full of treated seeds to prevent the seeds from "banging" against each other and the chamber walls which can cause mechanical damage.

While most mechanized seed treatment is done by seed conditioners, smaller scale, less expensive equipment is available for on-the-farm treating. This equipment ranges from simple augers into which a metered supply of fungicide is pumped to small units that are similar to commercial treaters. The primary concern with on-the-farm systems is adequate supervision and adjustment to insure complete seed coverage. No matter what formulation or method of fungicide application is used, to be effective, thorough coverage of the seed is essential. Regardless of the method of seed treatment, it is important to follow good health and sanitary precautions and to apply fungicides only at labelled concentrations.

For planter box seed treatment, a measured amount of fungicide is mixed together with a predetermined weight or volume of seed in a planter box or outside container. It is absolutely essential to thoroughly mix the fungicide (usually a dust formulation) with the seed immediately before planting. Even though planter box seed treatment is inexpensive and widely used for soybean seeds (MacFarlane, 1980) it is the least effective and most nonuniform method of seed treatment.

MacFarlane (1980) reported that 95% of the plant pathology specialists at universities in major soybean producing states recommend soybean seed treatment at least in some situations. Yet only an estimated 11% of the soybean seeds planted in 1981 were commercially treated. The primary reason for not treating soybean seed is that treated seed not sold for seed purposes cannot be sold as grain. Secondly, in most production areas soybean seeds (treated or untreated) cannot be carried over into the next planting season. Thus, if the seed is treated with a fungicide, it must be used for planting purposes or destroyed.

8-5 SEED MULTIPLICATION

The development of superior soybean cultivars by plant breeders has been a major factor in the continued use and expansion of this crop. It takes many years to develop a cultivar and when it is released only a small amount of breeders (stock) seed is available. Thus, the seed multiplication program provides the critical seed increase link betwen the plant breeders who develop new soybean cultivars and the farmers who use them. It is essential that seed multiplication programs; (i) insure high levels of cultivar purity, (ii) increase seed supplies rapidly, and (iii) maintain high levels of seed quality.

For publicly developed soybean cultivars released by state agricultural experiment stations and the USDA, seed certification programs have provided an unbiased system of seed increase for many years. Seed certification agencies are available in each state and internationally to provide uniform procedures for field and laboratory inspection to insure genetic identity and purity for each cultivar produced. Seed companies have developed similar in-house multiplication programs for privately developed cultivars which are often increased with assistance from state seed certification programs. The seed multiplication program for both publicly and privately developed cultivars must be organized in an efficient, yet accurate manner to allow for rapid increase and distribution of seed. This usually means that several agencies or departments must be coordinated to work as an intermediate between the plant breeder and farmer.

8-5.1 Cultivar Release

Regardless of the developing agency, a procedure must be available for evaluating potential cultivars and recommending their release. Policy statements have been developed by state agriculture experiment stations and the USDA governing the development, release, and multiplication of publicly developed cultivars (ESCOP, 1972). Similar statements are available in seed companies. When a plant breeder has an experimental breeding line for which release is recommended, appropriate information on identifying characteristics, descriptive information, area of adaptation, agronomic performance, and other specific use information must be submitted to a review committee or board. The Experiment Station Committee on Organization and Policy (ESCOP, 1972) stated that a cultivar should not be released unless it is distinctly superior to existing cultivars in one or more characteristics or it is superior in overall performance in areas where adapted and is at least satisfactory in other major requirements. Due to university and company release policies as well as intense competition, most private and public soybean cultivars meet this criteria before release. A single major production hazard which a new cultivar can overcome, e.g., resistance to phytophthora root rot, may become an overriding consideration in releasing a cultivar.

The cultivar is given a permanent name, which is acceptable to all participating agencies and preferably one short word, although seed companies may use brand names and numerical designations. In most cases, a newly released cultivar is registered with the Crop Science Society of America and a seed sample submitted to the National Seed Storage Laboratory in Fort Collins, CO. A procedure is usually outlined at this time for the increase, maintenance, and distribution of breeder, foundation, and certified seed.

In some countries, tests and trials of new cultivars are organized on a national level by an authoritative body. Such testing may be voluntary or compulsory, but is usually conducted over a period of 2 to 3 yrs. After extensive testing for agronomic performance and crop quality, the cultivar may be recommended and/or registered for farmer use. Usually the decision to release the cultivar in that country is closely aligned to its performance in these tests and its recommendation (or registration) by the authoritative testing agency.

8–5.2 Plant Variety Protection

The PVPA became law in the USA in 1970 with the following purpose—"To encourage the development of novel cultivars of sexually reproduced plants and to make them available to the public, providing protection to those who breed, develop, or discover them, and thereby promoting progress in agriculture in the public interest" (PVPA, 1973). The Act provided protection to new cultivars of sexually reproduced crops such as soybeans that are novel; that is, they are distinct, uniform, and stable compared to existing cultivars. Performance testing was not a requirement for acceptance and all participation and application for protection was voluntary. If protection was granted the protection period was for 18 yrs and it was the owner's responsibility to protect the cultivar. The owner could specify, however, that the protected cultivar be sold by cultivar name only as a class of certified seed under Title V of the Federal Seed Act (USDA, 1975).

Before 1970, soybean cultivar development was done primarily by state agricultural experiment stations and the USDA. A recent chronicle of plant variety protection (Batcha, 1983) indicated that only six seed companies had soybean cultivar development programs in 1970 with six plant breeders employed by these companies. By 1983, 28 seed companies had cultivar development programs that employed 60 plant breeders. During the period from 1971 through 1982, plant variety protection certificates were issued for 247 soybean cultivars which was more than for any other crop. The impact of these privately developed cultivars on the American farmer was emphasized in a recent survey of 15 soybean producing states (which accounted for 88% of the acreage harvested) which showed that over 20% of the acreage was planted to privately developed cultivars in 1983 (Anonymous, 1983) compared to approximately 4% 5 yrs earlier. For the first time, all cultivars developed by one company ranked second nationally to the leading single cv. 'Williams,' and accounted for nearly 9% of the total acreage. Thus, plant variety protection has provided the incentive for private seed companies to invest in plant breeding programs and cultivar development. As these plant-breeding programs continue to exand, the trend for increased use of privately developed cultivars will continue.

8–5.3 Eligibility of Soybean Cultivars

To be eligible for seed certification, Plant Variety Protection, or marketing under state and federal seed laws a cultivar must be properly named and described. Guidelines have been developed for classifying plant populations, which include a definition of a cultivar that has been accepted by several public and private agencies.[1] The term *cultivar* is considered an exact equivalent of variety and means a subdivision of a kind which is distinct, uniform, and stable; distinct in the sense that the cultivar can be differentiated by one or more identifiable, morphological, physiological, or other characteristics from all other cultivars of public knowledge; uniform in the sense that variations in essential and distinctive characteristics are describable; and stable in the sense that the cultivar will remain unchanged to a reasonable degree of reliability in its essential and distinctive characteristics and its uniformity when reproduced.

Seed certification agencies are aided in determining the eligibility of newly released soybean cultivars by local and national cultivar (variety) review boards. The National Soybean Variety Review Board was established by the Association of Official Seed Certifying Agencies (AOSCA) of North America in 1973 and consists of six members representing the American Seed Trade Association, AOSCA, Crop Science Society of America, National Council of Commercial Plant Breeders, USDA and USDA-ARS. The function of this board is to review and evaluate information provided by plant breeders of public and privately developed cultivars on the acceptability of these cultivars for seed certification. The board carefully evaluates the descriptive information and performance data provided to determine that the cultivar will remain distinct, uniform, and stable when increased through seed certification programs. If a soybean cultivar is accepted by the national review board most state certification agencies will accept it in that state.

8–5.4 Seed Increase Programs

The primary responsibility of a seed increase program is to multiply seed of a cultivar in such a way as to maintain high levels of genetic

[1] Committee to develop guidelines for classifying cultivated plant populations, 1977.

purity. To accomplish this, minimum standards must be set for field inspection during production and seed purity following harvest by an official seed certifying agency. This agency may be a governmental department or agency, an association of seed growers, or a seed company.

Seed certification is a quality control program whereby seed and propagating materials of improved cultivars is maintained at a high level of genetic purity and made available to the public. Certified seed in the USA is produced by outstanding farmer-growers following the procedures outlined by the certification agencies in each state and AOSCA (AOSCA, 1983). In Canada, the seed certification program is administered by the Canadian Seed Growers Association. These procedures insure positive identification of stock seed planted, field inspection during the growing season and seed inspections following harvest to assure the genetic identification and purity of each cultivar. In some states and Canada certified seed must also meet minimum quality standards for germination, crop purity, and freedom from certain weeds and diseases as well as genetic purity.

8-5.4.1 Limited Generation System

Inherent in the certification concept is a generation system whereby the pedigree of soybean cultivars is maintained through subsequent seed production. A four-generation scheme has evolved for soybeans and seed of each generation is produced under different quality criteria and identified by a specially colored tag.

1. *Breeder seed* (white tag) is that limited amount of seed produced under the direct supervision of the originating plant breeder or a designated agency. It supplies a source for foundation seed and is not available to the general public.
2. *Foundation seed* (white tag) is the first generation progeny of breeder seed produced under the direct supervision of the foundation seed organization. It is sold directly to certified seed growers and is usually available in limited quantities.
3. *Registered seed* (purple tag) is the progeny of foundation seed and is produced by certified seed growers as another seed increase generation before the production of certified seed. It is not intended as a comercial class of seed. A few states do not recognize the registered class for soybeans and produce certified seed directly from foundation seed.
4. *Certified seed* (blue tag) is the progeny of foundation or registered seed and is produced by certified seed growers. It represents the final seed class in the certification program and is usually available in large quantities for use by commercial farmers.

The Canadian generation system is the same as the USA's except there is a select seed class between the breeder and foundation seed generation.

8-5.4.2 Foundation Seed Production

The success of the limited generation scheme of seed multiplication is largely due to the role of foundation seed organizations. These organizations insure a continuous supply of seed stock from which registered and/or certified seed are produced. The foundation seed organization may consist of a separate project within the agriculture experiment station, a private association of seed growers, or a private seed business. Regardless of the organization, close-working relationships are usually maintained between foundation seed organizations and the originating plant breeder or institution.

Foundation seed organizations receive breeder seed of newly released soybean cultivars and increase them to foundation seed. In succeeding years, breeder seed of the cultivar must be maintained and made available. This is usually done by the foundation seed organization in cooperation with the originating plant breeder or institution. Usually breeder seed is produced in a small portion of a foundation seed field which is carefully inspected and rogued for off-type plants. Less frequently, the releasing institution may grow small lots of breeder seed under the direct supervision of the plant breeder for annual release to the foundation seed organization.

Foundation seed organizations must plan production carefully to anticipate the demand of all soybean cultivars and avoid over production. Foundation seed organizations do not usually have adequate land or facilities to produce the necessary foundation seed. Thus, contract seed production is commonly done with careful selection of seed growers.

8–5.4.3 Certified Seed Production

Seed certification programs provide an unbiased, service-oriented method of maintaining genetic identity of seed on the open market. Certified seed production for the registered and certified classes is conducted by seed growers under procedures outlined by the seed certification agency (AOSCA, 1983). All state agencies have published minimum requirements for land history, field inspection, and seed standards which must be met for each seed lot. Seed certification has become important for publicly developed cultivars of soybeans and many other crops. It is of less importance for privately developed cultivars. Some larger seed companies have established their own seed multiplication programs and completely avoid seed certification while other seed companies utilize seed certification agenices as an unbiased third party to aid in their own quality control programs. Many smaller seed companies and seed cooperatives multiply most of their privately developed soybean cultivars through seed-certification channels.

8–5.4.3.1 Eligibility and Application for Certification—Farmers should be familiar with the soybean seed certification requirements and have adequate equipment and experience with soybean production before attempting certified seed production. An application for certification must be submitted to the seed-certification agency requesting field inspection and certification for all soybean fields. The grower must keep a tag and/or invoice of the class planted (foundation or registered) to document the seed source. Prior to planting the grower should carefully check the previous cropping history of the seed field to be certain it qualifies for the production of the certified class and cultivar intended.

8–5.4.3.2 Field Inspection and Harvesting—Since contamination from off-type soybean plants (AOSCA, 1983) cannot always be detected in the harvested seed, the field inspection is the most critical step in monitoring the genetic purity of each cultivar. Inspection of soybean seed fields is commonly made at leaf-fall when genetic differences in pubescence color and maturity are most obvious, however, many certification agencies also make an earlier inspection at full bloom. The maximum percentage of off-type soybean plants allowed in the foundation, registered, and certified classes at the time of field inspection is 0.1, 0.2, and 0.5%, respectively (AOSCA, 1983). Important criteria that the field inspector examines for each seed field are: (i) vertification of previous land history, (ii) identification and percentage of off-type soybean plants, (iii) sufficient border between adjoining soybean fields to prevent mechanical mixing at harvest, and (iv) contamination by other crops, weeds, and diseases that may influence the quality of the seed at harvest. Following field inspection each seed field is either accepted, rejected, or rejected subject to reinspection (providing contaminants could be rogued from the field by the grower). Seed growers must take the extra time and patience necessary at harvest to carefully clean all combines, trucks, and storage equipment to prevent mechanical mixtures of certified seed with soybean or other crop seed. Seed lot and cultivar identity is critical during harvesting and storage.

8–5.4.3.3 Conditioning, Sampling and Testing—Most seed certification agencies have an approved list of seed conditioners who have the necessary equipment and experience for cleaning certified seed. Established certified growers and seedsmen have their own facilities for seed conditioning and storage and are on this list, while other seed growers must have their seed conditioned at an approved plant. Extreme care must be taken during conditioning to clean all equipment before each seed lot and cultivar is conditioned. After the seed lot has completed all seed conditioning, a sample must be taken and submitted for seed analysis. Certified seed may be sampled by automatic samplers, but is commonly sampled from bagged or bulk seed by officials designated by the certifying agency.

8–5.4.3.4 Seed Testing and Tagging—Certified seed is either tested in the laboratory of the official agency or in official state laboratories or commercial laboratories that have a registered seed analyst. All certified seed must exceed the minimum genetic requirements for the class inspected before certification is completed (Table 8–1). Some certifying agencies also require that certified seed meet certain requirements for crop purity, germination, and freedom from crop or weed seed.

Certified seed is identified by official tags or labels which state the class of seed (foundation, registered, or certified) and other pertinent information including the name of the certification agency. Some agencies have a *one-tag system* in which the certification tag serves as a complete labeling tag with all information for certification and analysis (germination, purity, etc.) included. Other agencies have a *two-tag system*, where the analysis tag and certification tag are different and attached separately. Some certification agencies allow approved seed conditioners to *pretag* the certification and analysis tag (one-tag system) on the bag during conditioning, provided the seed is not marketed until testing is completed.

8–5.5 Quality Control Programs

Many seed producers and companies have relied entirely on the state seed certification agency for quality control and the certified blue tag was their assurance that the seeds met the quality standard. Other seed companies and seedsmen have developed their own quality control programs with standards that usually exceed the minimum standards for certified seed (Berkey, 1981). Such programs are concerned with quality at all phases of the seed business from planting through production, harvesting, drying, storage, and conditioning until final seed marketing. They require a commitment from management, an understanding of seed quality by all employees, and commonly designate one employee as the quality control coordinator for the entire program.

A good quality control program is established on the premise that poor soybean seed quality can be prevented. Such a program will usually include the following components:

1. Establishment of minimum acceptable standards for soybean seed quality.
2. Development of an organized system of sampling and evaluating seed quality to be certain these standards are met.
3. Isolation and prevention of seed quality problems.
4. Total commitment from management and all employees.

A seed quality control program should not be limited to a simple germination and purity standard, but should include all the major quality characteristics important for soybean seed. The first step should be the establishment of minimum standards acceptable for all quality characteristics.

Any good seed producer knows that quality is most often gained (or lost)in the production field long before the seed arrives at the seed conditioning plant. Thus, the quality control coordinator will select only the best farmers as contract growers and will monitor fields throughout the production cycle regarding recommended practices for weed, insect, and disease control to insure high seed quality. Seed fields will also be inspected several times throughout the growing season for stage of seed development and contamination from other crops, weeds, or off-type soybean plants.

Systematic sampling and testing of each seed lot is one of the most important factors in a quality control program. The first sample should be taken at (or before) harvest with additional samples taken at various stages before and after conditioning. The last sample will usually provide the information needed for final labeling purposes. The number of samples taken is determined by the control limits that the quality control coordinator establishes for each seed lot and cultivar. The control limits may be narrow and specify sampling before and after harvesting, conveying, drying and storage, and seed conditioning. Most established quality control programs take fewer samples including a rough seed sample at harvest, one or two samples during storage and/or conditioning and a final sample after conditioning.

The tests conducted on each sample are determined by the time of sampling and previous experience of the seed grower and quality control coordinator. Due to the importance of seed moisture on the mechanical integrity of soybean seed each sample taken should be evaluated for moisture. Quality control charts are estabished for each seed grower, cultivar and seed lot and the results are checked against standards. Quality control can be a powerful management technique which can be used to the advantage of a seed grower or company. It can result in an improvement in seed quality, a reduction in operating costs, and as a competitive tool by the progressive seed grower or company.

8–5.6 Changing Concepts

Farmer demands and attitudes regarding seed needs have changed substantially over the last 20 yrs. They are no longer content to merely buy and plant soybean seed, but instead insist upon improved, named cultivars and high-quality seeds. Such seeds must not only be genetically pure but also of high germination, vigor, and emergence potential when planted under a range of field conditions.

Soybean cultivar development, production and use is dependent upon the coordinated efforts of agricultural experiment stations and the private seed industry. Plant variety protection has not only resulted in greater numbers of cultivars, but increased competition for farmer acceptance and sales. Even though cultivars are no longer recommended by most agricultural experiment stations, the farmers still demand unbiased evaluations of performance before seed purchase. Thus, seed sales are still highly dependent upon cultivar performance, even though release and acceptance in plant variety protection is based on novelty.

For many years, certified seed of primarily publicly released cultivars provided farmers with the assurance that the seed purchased was true-to-type for the cultivar as labeled. Certified seed also provided farmers with a minimum quality standard for crop purity and germination needed to produce an adequate field stand with a minimum of additional weed and disease problems. With the passage of the PVPA in 1970, plant breeders of private seed companies were given the option of protecting their cultivars through seed certification. This has benefited many smaller seed companies who cannot afford the costly procedures of civil court action, but still want protection for their investment. Certified seed agencies have adjusted their procedures to accommodate the certification of these private varieties while maintaining the confidentiality of the closed pedigrees. Some larger seed companies have argued that the minimum-quality standards for all certified seed prevent them from gaining a competitive edge over other seedsmen by marketing high-quality seed. This has led to the dropping of the quality standards for certified soybean seed in many states and certification for genetic (cultivar) purity only. Only time and farmer satisfaction will determine if this change is to the advantage of the seedsmen or the farmer.

Cultivar blends of soybean seed have been developed by private seed companies in many states and are marketed aggressively in competition with named cultivars. The development and merits of such blends are discussed in chapter 7 in this book. In most states, these blends can be sold by company brand name without disclosing the components of the blend, while in other states and Canada the seed laws require labeling as to kind and cultivar. Thus, all blend components (in excess of 5% of the total crop purity) must be listed on the label. This has been a controversial issue in interstate sales of soybean blends and is hotly contested by seedsmen in some states. With the continued development of privately developed soybean cultivars, the future of soybean blends may depend entirely on performance and farmer acceptance.

8–6 SUMMARY

The production of high-quality seed requires a high level of management which must begin before planting of the seed crop and does not end until the seed is sold to the producer. In many respects, the technical knowledge and management information needed to produce high-quality seed is available as documented in this chapter. However, this does not mean that high-quality seed is always available to the producer. Unfavorable environmental conditions during maturation and/or harvest may lower seed quality to unacceptable levels prior to harvest. Under favorable environmental conditions, seed producers do not always utilize the technology and information available to them. In both cases, the efficiency of the production system is reduced and the seed produced may not be of marketable quality.

Considerable progress has been made in understanding the effect of environmental factors during production and storage on seed quality. Less information is available on the relationships between seed developmental processes and the ultimate germinability and vigor level of the seed and their interaction with the environment. The development of improved cultivars by plant breeders has had a significant impact on the soybean industry; however, there has been less attention given to developing cultivars with improved seed quality. The development of cultivars whose seeds are less susceptible to environmental stress or show slower rates of deterioration in storage would have a significant impact on the soybean industry.

The techniques used to measure seed quality have shown steady improvement and new techniques have been developed. Sufficient evidence has been published to confirm that soybean seed vigor is a separate, measurable entity of seed quality and progress is being made toward standardizing vigor-testing techniques among seed-testing laboratories. However, it is still difficult to relate the results of laboratory tests to the actual performance of the seed, either in terms of field performance (emergence and stand establishment) or storability. Progress on this problem will help insure the consistent availability of high-quality seeds to soybean producers.

REFERENCES

Ackerson, R.C. 1984. Abscisic acid and precocious germination in soybeans. J. Exp. Bot. 35:414–421.

Adams, C.A., and R.W. Rinne. 1981. Seed maturation in soybeans (*Glycine max* L. Merr.) is independent of seed mass and of the parent plants yet is necessary for production of viable seeds. J. Exp. Bot. 32:615–620.

Alam, A., and G.C. Shove. 1973. Hygroscopicity and thermal properties of soybeans. Trans. Am. Soc. Agric. Eng. 16:707–709.

Alexander, L.J., P. Decker, and K. Hinson. 1978. The effect of maturity, time of harvest, weather and storage conditions on the quality and deterioriation of soybean seed by fungi in Florida. Fla. Agric. Exp. Stn. Tech. Bull. 804.

Anonymous. 1983. Crop production. November. Crop Reporting Board, ECSA, USDA, Washington, DC.

Association of Official Seed Analysts. 1981. Rules for testing seeds. J. Seed Technol. 6:1–126.

----. 1983. Seed vigor testing handbook, no. 32. Association of Official Seed Analysts, Boise, ID.

Association of Official Seed Certifying Agencies. 1983. Certification standards. Raleigh, NC.

American Society of Agricultural Engineers. 1983. ASAE Yearbook. American Society of Agricultural Engineers, St. Joseph, MI.

Athow, K.L., and R.M. Caldwell. 1956. The influence of seed treatment and planting rate on the emergence and yield of soybeans. Phytopathology 46:91–95.

Backman, P.A., R. Rodriguez-Kabana, J.M. Hammond, and D.L. Thurlow. 1979. Cultivar, environment and fungicide effects on foliar disease losses in soybeans. Phytopathology. 69:562–564.

Bartsch, J.A., C.G. Haugh, K.L. Athow, and R.M. Peart. 1979. Impact damage to soybean seed. ASAE Paper 79–3037. American Society of Agricultural Engineers, St. Joseph, MI.

Baskin, C.C., and E.H.N. Vierira. 1980. Predicting the storability of soybean seed lots. J. Seed Technol. 5:1–6.

Batcha, J.A. 1983. A chronicle of Plant Variety Protection. Asgrow Seed Co., Kalamazoo, MI.

Berkey, D.A. 1981. Comprehensive quality control. Proc. Short Course Seedsmen 23:29–41.

Bernard, R.L., and M.G. Weiss. 1973. Qualitative genetics. *In* B.E. Caldwell (ed.) Soybeans: Improvement, production, and uses. Agronomy 16:117–153.

Burrell, N.J. 1974. Aeration. p. 454–480. *In* C.M. Christensen (ed.) Storage of cereal grains and their products. American Association of Cereal Chemists, St. Paul.

Burris, J.S. 1979a. Bulk handling of soybeans. p. 55–65. *In* Proc. 2nd Seed Technol. Conf., Ames, IA. 23–24 August. Iowa State University, Ames.

----. 1979b. Effect of conditioning environment on seed quality and field performance of soybeans. p. 79–85. *In* Proc. Soybean Seed Research Conf., 9th, Chicago, IL. 13–14 December. American Seed Trade Association, Washington, DC.

----. 1980. Maintenance of soybean seed quality in storage as influenced by moisture temperature and genotype. Iowa State J. Res. 54:337–389.

----, O.T. Edje, and A.H. Wahab. 1969. Evaluation of various indices of seed and seedling vigor in soybeans, *Glycine max* (L.) Merr. Proc. Assoc. Off. Seed Anal. 59:73–81.

----, ----, and ----. 1973. Effect of seed size on seedling performance in soybeans. II. Seedling growth and photosynthesis and field performance. Crop Sci. 13:207–210.

----, and R.J. Navratil. 1979. Relationship between laboratory cold-test methods and field emergence in maize inbreds. Agron.J. 71:985–988.

----, A.H. Wahab, and O.T. Edje. 1971. Effects of seed size on seedling performance in soybeans. I. Seedling growth and respiration in the dark. Crop Sci. 11:492–498.

Buttery, B.R., and R.I. Buzzell. 1968. Peroxidase activity in seeds of soybean varieties. Crop Sci. 8:722–725.

----, and ----. 1971. Properties and inheritance of urease isoenzymes in soybean seeds. Can. J. Bot. 49:1101–1105.

Byrd, H.W., and J.C. Delouche. 1971. Deterioration of soybean seed in storage. Proc. Assoc. Off. Seed Anal. 61:41–57.

Cain, D.F., and R.G.Holmes. 1977. Evaluation of soybean seed impact damage. ASAE Paper 71-1552. American Society of Agricultural Engineers, St. Joseph, MI.

Caviness, C.E., and A.M. Simpson. 1974. Influence of variety and location on seedcoat thickness of mature soybean seed. Proc. Assoc. Off. Seed Anal. 64:102–108.

Christensen, C.M., and H.H. Kaufmann. 1969. Grain storage—the role of fungi in quality loss. University of Minnesota Press, Minneapolis.

Clark, B.E. 1953. Relationship between certain laboratory tests and the field germination of sweet corn. Proc. Assoc. Off. Seed Anal. 1953:42–44.

----, G.E. Harman, T.J. Kenny, and E.C. Waters, Jr. 1980. Relationship between the results of certain laboratory tests and the field germination of soybean seeds. Newsl. AOSA 54:36–43.

Crittenden, H.W., and L.V. Svec. 1974. Effect of potassium on the incidence of *Diaporthe sojae* in soybean. Agron. J. 66:696–697.

Crookston, R.K., and D.S. Hill. 1978. A visual indicator of the physiological maturity of soybean seed. Crop Sci. 18:867–870.

Daugherty, D.M. 1967. Pentatomidae as vectors of yeast-spot disease of soybeans. J. Econ. Entomol. 60:147–152.

----, M.H. Neustadt, C.W. Gehrke, L.E. Cavanah, L.F. Williams, and D.E. Green. 1964. An evaluation of damage to soybeans by brown and green stink bugs. J. Econ. Entomol. 57:719–722.

Delouche, J.C. 1974. Maintaining soybean seed quality. p. 46–62. In Soybean, production, marketing and use. TVA Bull. V-19.

----, 1976. Standardization of vigor tests. J. Seed Technol. 1:75–85.

----, and C. C. Baskin. 1973. Accelerated aging techniques for predicting the relative storability of seed lots. Seed Sci. Technol. 1:427–452.

DePauw, R.A., R.L. Francis, and H.C. Snyder. 1977. Engineering aspects of axial-flow combine design. ASAE Paper no. 77-1550. American Society of Agricultural Engineers, St. Joseph, MI.

Dorchester, C.S. 1945. Seed and seedling characteristics in certain varieties of soybeans. J. Am. Soc. Agron. 37:223–232.

Dunleavy, J.M. 1973. Virus diseases. *In* B.E. Caldwell (ed.) Soybeans: Improvement, production, and uses. Agronomy 16:505–526.

Edje, O.T., and J.S. Burris. 1971. Effects of soybean seed vigor on field performance. Agron. J. 63:536–538.

Egli, D.B. 1976. Planting date, row width, population, growth regulators. p. 58–62. *In* L.D. Hill (ed.) World soybean research conference proceedings. The Interstate Printer and Publishers, Danville, IL.

----, and D.M. TeKrony. 1979. Relationship between soybean seed vigor and yield. Agron. J. 71:755–759.

----, G.M. White, and D.M. TeKrony. 1979. Relationship between seed vigor and the storability of soybean seed. J. Seed Technol. 3:1–11.

Ellis, M.A., M.B. Ilyas, and J.B. Sinclair. 1975. Effect of three fungicides on internally seedborne fungi and germination of soybean seeds. Phytopathology 65:535–556.

----, ----, F.D. Tenne, J.B. Sinclair, and H.L. Palm. 1974. Effect of foliar applications of benomyl on internally seedborne fungi and pod and stem blight in soybean. Plant Dis. Rep. 58:760–763.

----, E.H. Paschal, P.E. Powell, and F.D. Tenne. 1979. Internally seedborne fungi of soya bean in Puerto Rico and their effect on seed germination and field emergence. Trop. Agric. 56:171–174.

----, and J.B. Sinclair. 1976. Effect of benomyl field sprays on internally borne fungi, germination, and emergence of late-harvested soybean seeds. Phytopathology 66:680–682.

Ellis, R.H., K. Osci-bonsu, and E.H. Roberts. 1982. The influence of genotype, temperature and moisture on seed longevity in chickpea, cowpea, and soya bean. Ann. Bot. 50:69–82.

----, and E.H. Roberts. 1980. The influence of temperature and moisture on seed viability period in Barley (*Hordeum distichum* L.). Ann. Bot. 45:31–37.

Experiment Station Committee on Organization and Policy. 1972. A statement of responsibilities and policies relating to development, release and multiplication of publicly developed varieties of seed-propagated crops. U.S. Government Printing Office, Washington, DC.

Fehr, W.R., and A.H. Probst. 1971. Effect of seed source on soybean strain performance for two successive generations. Crop Sci. 11:865–867.

Fontes, L.A.N., and A.J. Ohlrogge. 1972. Influence of seed size and population on yield and other characteristics of soybeans [*Glycine max* (L.) Merr.]. Agron. J. 64:833–836.

Ghaly, T.F., and J.W. Sutherland. 1983. Quality aspects of heated-air drying of soybeans. J. Stored Prod. Res. 19:31–41.

Gorman, M.B., and Y.T. Kiang. 1977. Variety-specific electrophoretic variants of four soybean enzymes. Crop Sci. 17:963–965.

Grau, C.R., and E.S. Oplinger. 1981. Influence of cultural practices and foliar fungicides on soybean yield, *Phomopsis* seed infection and seed germination. Proc. Soybean Seed Res. Conf. 11th 11:1–8.

Green, D.E., L.E. Cavanah, and E.L. Pinnell. 1966. Effects of seed moisture content, field weathering and combine cylinder speed on soybean seed quality. Crop Sci. 6:7–10.

----, V.D. Luedders, and B.J. Moraghan. 1971. Heritability and advance from selection for six soybean seed quality characters. Crop Sci. 11:531–533.

----, and L.E. Pinnell. 1968a. Inheritance of soybean seed quality. I. Heritability of laboratory germination and field emergence. Crop Sci. 8:5–11.

----, and ----. 1968b. Inheritance of soybean seed quality. II. Heritability of visual ratings of soybean seed quality. Crop Sci. 8:11–15.

----, ----, L.E. Cavanah, and L.F. Williams. 1965. Effect of planting date and maturity date on soybean seed quality. Agron. J. 57:165–168.

Greg, B.R., A.G. Law, S.S. Verdi, and J.S. Balis. 1970. Seed processing. Avion Printers, New Delhi.

Hall, C.W. 1980. Drying and storage of agricultural crops. AVI Publishing Co., Westpoint, CT.

Harmond, J.E., N.R. Brandenberg, and L.M. Klien. 1968. Mechanical seed cleaning and handing Agric. Handb. 354. U.S. Department of Agriculture—Agricultural Research Service, Washington, DC.

Harris, H.B., M.B. Parker, and B.J. Johnson. 1965. Influence of molybdenum content of soybean seed and other factors associated with seed source on progeny response to applied molybdenum. Agron. J. 57:397–399.

Hartwig, E.E. 1973. Varietal development. p. 187–210. *In* B.E. Caldwell (ed.) Soybeans: Improvement, production, and uses. Agronomy 16:187–210.

Henderson, J., and C.E. Vaughan. 1980. Screen selection. Proc. Short Course Seedsmen 22:39–47.

Hepperly, P.R., and J.B. Sinclair. 1980. Detached pods for studies of *Phomopsis sojae* pods and seed colonization. J. Agric. Univ. P.R. 64:330–337.

----, and ----. 1981. Relationships among *Cercospora kikuchii*, other seed mycoflora, and germination of soybeans in Puerto Rico and Illinois. Plant Dis. 65:130–132.

Hobbs, P.R., and R.L. Obendorf. 1972. Interaction of initial seed moisture and imbibitional temperature on germination and productivity of soybean. Crop Sci. 12:664–667.

Hoffman, A., and M.B. McDonald, Jr. 1981. Maintaining soybean seed quality during conditioning. Proc. Soybean Seed Res. Conf., 11th 11:73–91.

Holman, L.E., and D.G. Carter. 1952. Soybean storage in farm-type bins: A research report. Ill. Agric. Exp. Stn. Bull. 553.

Hukill, W.V. 1963. Storage of seeds. Proc. Int. Seed Test. Assoc. 28:871–883.

International Seed Testing Association. 1959. Handbook for seed health testing. ISTA, Zurick, Switzerland.

----. 1976. International rules for seed testing, 1976. Seed Sci. Technol. 4:1–177.

----. 1977. Seed cleaning and processing. Seed Sci. Technol. 5:1–335.

Jeffers, D.L., A.F. Schmitthenner, and M.E. Kroetz. 1982a. Potassium fertilization effects on *Phomopsis* seed infection, seed quality, and yield of soybeans. Agron. J. 74:886–890.

----, ----, and D.L. Reichard. 1982b. Seed-borne fungi, quality, and yield of soybeans treated with benomyl fungicide by various application methods. Agron. J. 74:589–592.

Jenson, R.L., and L.D. Newson. 1972. Effect of stink-bug-damaged soybean seeds on germination, emergence, and yield. J. Econ. Entomol. 65:261–264.

Johnson, D.R., and V.D. Leudders. 1974. Effects of planting seed size on emergence and yield in soybeans (*Glycine max* (L.) Merr.). Agron. J. 66:117–118.

Johnson, R.R., and L.M. Wax. 1978. Relationship of soybean germination and vigor tests to field performance. Agron. J. 70:273–278.

Justice, O.L., and L.H. Bass. 1978. Principles and practices of seed storage. USDA-SEA Agric. Handb. 506, U.S. Government Printing Office, Washington, DC.

Kilen, T.C., and E.E. Hartwig. 1978. An inheritance study of impermeable seed in soybeans. Field Crops Res. 1:65–70.

Kilpatrick, R.A., and E.E. Hartwig. 1955. Fungus infection of soybean seed as influenced by stink bug injury. Plant Dis. Rep. 39:177–180.

Kmetz, K., C.W. Ellett, and A.F. Schmitthenner. 1974. Isolation of seedborne *Diaporthae phaseolorum* and *Phomopsis* from immature soybean plants. Plant Dis. Rep. 58:978–982.

----, ----, and ----. 1979. Soybean seed decay: Sources of inoculum and nature of infection. Phytopathology 69:798–801.

----, A.F. Schmitthenner, and C.W. Ellet. 1978. Soybean seed decay: Prevalence of interaction and symptom expression caused by *Phomopsis* sp., *Diaporthe phaseolorum* var. *sojae*, and *D. phaseolorum* var. *caulivora*. Phytopathology 68:836–840.

Kulik, M.M., and J.F. Schoen. 1981. Effect of seedborne *Diaporthe phaseolorum* var. *sojae* on germination, emergence, and vigor of soybean seedlings. Phytopathology 75:544–547.

----, and R.W. Yaklich. 1982. Evaluation of vigor tests in soybean seeds: Relationship of accelerated aging, cold, sand bench, and speed of germination tests to field performance. Crop Sci. 22:766–770.

Larsen, A.L. 1967. Electrophoretic differences in seed proteins among varieties of soybean, *Glycine max* (L.) Merrill. Crop Sci. 7:311–313.

Lasqueves, E.E., A.H. Boyd, and G.B. Wetch. 1979. Removal of soil peds from soybean (*Glycine max*) seeds. Seed Sci. Technol. 8:309–318.

Lehman, S.J. 1950. Purple seed stain of soybean seed. N.C. State Agric. Exp. Stn. Bull. 369.

Loeffler, T.M. 1981. The bulk conductivity test as an indicator of soybean seed quality. M.S. thesis. University of Kentucky, Lexington.

Loewer, O.J., I.J. Ross, and G.M. White. 1979. Aeration, inspection and sampling of grain storage bins. Coop. Ext. Serv., Univ. of Kentucky Bull. AEN-45.

Luedders, V.D., and J.S. Burris. 1979. Effects of broken seed coats on field emergence of soybeans. Agron. J. 71:877–879.

MacFarlane, J.J. 1980. Soybean seed treatment—U.S. Status Report. Proc. Soybean Seed Res. Conf., 10th 10:88–99.

Matthes, K.R., A.H. Boyd, and G.B. Welch. 1974. Heated air drying of soybean seeds. ASAE Paper 74-3001. American Society of Agricultural Engineers, St. Joseph, MI.

Matthews, S., and W.T. Bradnock. 1967. The detection of seed samples of wrinkle-seeded peas (*Pisum sativum* L.) of potentially low planting value. Proc. Int. Seed Test. Assoc. 32:553–563.

----, and ----. 1968. Relationship between seed exudation and field emergence in peas and french beans. Hortic. Res. 8:89–93.

McDonald, M.B., Jr. 1975. A review and evaluation of seed vigor tests. Proc. Assoc. Off. Seed Anal. 65:109–139.

----, 1977. AOSA vigor subcommittee report: 1977 vigor test "referee" program. Assoc. Off. Seed Anal. Newsl. 51:14–41.

----, and B.R. Phaneendranath. 1978. A modified accelerated aging seed vigor test for soybean. J. Seed Technol. 3:27–37.

----, and D.O. Wilson. 1979. An assessment of the standardization and ability of the ASA-610 to rapidly predict soybean germination. J. Seed Technol. 4:1–12.

----, and ----. 1980. ASA-610 ability to detect changes in soybean seed quality. J. Seed Technol. 5:56–66.

McGee, D.C., C.L. Brandt, and J.S. Burris. 1980. Seed mycoflora of soybeans relative to fungal interactions, seedling emergence and carryover of pathogens for subsequent crops. Phytopathology 70:615–617.

McNeal, X. 1966. Conditioning and storage of soybean. Ark. Agric. Exp. Stn. Bull. 714.

Midwest Plan Service. 1980. Managing dry grain in storage. Iowa State Univ. Bull. AED-20.

Miles, D.F., D.M. TeKrony, and D.B. Egli. 1983. Effect of the desiccation environment and seed maturation on soybean seed quality. Agron. Abstr. American Society of Agronomy, Madison, WI, p.119.

Minor, H.C., and E.H. Paschal. 1982. Variation in storability of soybeans under simulated tropical conditions. Seed Sci. Technol. 10:131–139.

Misra, M.K. 1981. Soybean and storage. Coop. Ext. Serv. Iowa State Univ. Bull. PM-1004.

----. 1982. Soybean seed quality during conditioning. Proc. Short Course Seedsmen 24:49–53.

Mondragon, R.L, and H.C. Potts. 1974. Field deterioration of soybeans as affected by environment. Proc. Assoc. Off. Seed Anal. 64:63–71.

Moore, R.P. 1971. Mechanisms of water damage to mature soybean seed. Proc. Assoc. Off. Seed Anal. 61:112–118.

----. 1972. Effects of mechanical injuries on viability. p. 94–113. *In* E.H. Roberts (ed.) Viability of seeds. Syracuse University Press, Syracuse, NY.

Murakishi, H.H. 1951. Purple seed stain of soybean. Phytopathology 41:305–318.

Nave, W.R. 1977. What's new in soybean harvesting equipment. Soybean News 29:3.

----. 1979. Soybean harvesting equipment: Recent innovations and current status. p. 433–449. *In* F.T. Corbin (ed.) World soybean research conference II: Proceedings. Westview Press, Boulder, CO.

Neergaard, P. 1977. Seed pathology, Vol. 1. The MacMillan Press Ltd., New York.

Nellist, M.E. 1980. Safe drying temperatures for seed grain. p. 371–388. *In* P.D. Hebblethwaite (ed). Seed production. Butterworths, London.

Newberg, R.S., M.R. Paulsen, and W.R. Nave. 1980. Soybean quality with rotary and conventional threshing. Trans. Am. Soc. Agric. Eng. 23:303–308.

Nicholson, J.F., and J.B. Sinclair. 1973. Effect of planting date, storage conditions and seed-borne fungi on soybean seed quality. Plant Dis. Rep. 57:770–774.

Nittler, L.W., G.E. Harman, and Bette Nelson. 1974. Hila discoloration of "Tranverse" soybean seeds: A problem in cultivar purity, analysis and a possible indication of low quality seeds. Proc. Assoc. Off. Seed Anal. 64:115–119.

Paschal II, E.H., and M.A. Ellis. 1978. Variation in seed quality characteristics of tropically grown soybeans. Crop Sci. 18:837–840.

Paulsen, M.R. 1978. Fracture resistance of soybeans to comprehensive loading. Trans. Am. Soc. Agric. Eng. 21:1210–1216.

----. 1980. Soybean damage detection. p. 493–500. *In* F.C. Corbin (ed.) World soybean research conference II: Proceedings. Westview Press, Boulder, CO.

----, and W.R. Nave. 1979. Improved indoxyl acetate test for detecting seed cost damage. Trans. Am. Soc. Agric. Eng. 22:1475–1479.

----, ----, and L.E. Gray. 1981a. Soybean seed quality as affected by impact damage. Trans. Am. Soc. Agric. Eng. 24:1577–1582, 1589.

----, ----, T.L. Mounts, and L.E. Gray. 1981b. Storability of harvest-damaged soybeans. Trans. Am. Soc. Agric. Eng. 24:1583–1589.

Payne, R.C. 1979. Some new tests and procedures for determining variety (soybeans). J. Seed Technol. 3:61–76.

----, and T. J. Koszykowski. 1978. Esterase isoenzyme differences in seed extracts among soybean cultivars. Crop Sci. 18:557–559.

----, and L.F. Morris. 1976. Differentiation of soybean cultivars by seedling pigmentation patterns. J. Seed Technol. 1:1–9.

Pendleton, J.W., and E.E. Hartwig. 1973. Management. *In* B.E. Caldwell (ed.) Soybeans: Improvement, production, and Uses. Agronomy 16:211–237.

Perrin, R.K., K.A. Kunnings, and L.A. Ihnen. 1983. Some effects of the U.S. Plant Variety Protection Act of 1970. Econ. Res. Rep. 46, Dep. Economy and Business, North Carolina University, Raleigh.

Pfost, D. 1975. Environmental and varietal factors affecting damage to seed soybeans during drying. Ph.D. thesis. The Ohio State Univ., Columbus (Diss. Abstr. 36:4076B).

Pinthus, M.J., and U. Kimel. 1979. Speed of germination as a criterion of seed vigor in soybeans. Crop Sci. 19:291–292.

Plant Variety Protection Act. 1973. Regulations and rules of practice. USDA, Agricultural Marketing Service, Washington, DC.

Pollock, B.M., E.E. Roos, and J.R. Manalo. 1969. Vigor of garden bean seeds and seedlings influenced by initial seed moisture, substrate, oxygen, and imbibition temperature. J. Am. Soc. Hortic. Sci. 94:577–584.

Potts, H.C., J. Duangpatra, W.G. Hairston, and J.C. Delouche. 1978. Some influences of hardseededness on soybean seed quality. Crop Sci. 18:221–224.

----, and C.E. Vaughan. 1977. Soybean seed processing. Proc. Short Course Seedsmen 19:61–75.

Prakobboon, N. 1982. A study of abnormal seeding development in soybean as affected by threshing injury. Seed Sci. Technol. 10:495–500.

Quiniones, S.S., J.M. Dunleavy, and J.W. Fisher. 1971. Performance of three soybean varieties inoculated with soybean mosaic virus and bean pod mottle virus. Crop Sci. 11:662–664.

Ramstad, P.E., and W.F. Geddes. 1942. The respiration and storage behavior of soybeans. Univ. of Minnesota Agric. Exp. Stn. Tech. Bull. 156.

Roberts, E.H. 1960. The viability of cereal seed in relation to temperature and moisture. Ann. Bot. 24:12–31.

----. 1972. Storage environment and the control of viability. p. 14–58. *In* E.H. Roberts (ed.) Viability of seeds. Syracuse University Press, Syracuse, NY.

Rodda, E.D. 1974. Soybean drying—seed, food, feed. ASAE Paper 74–3540. American Society of Agricultural Engineers, St. Joseph, MI.

----, and E.J. Ravalo. 1978. Soybean seed storage under constant and ambient tropical conditions. ASAE Paper 78–6030. American Society of Agricultural Engineers, St. Joseph, MI.

Rojanasoroj, C., G.M. White, O.J. Loewer, and D.B. Egli. 1976. Influence of heated air drying on soybean impact damage. Trans. Am. Soc. Agric. Eng. 19:372–377.

Roos, E.E. 1980. Physiological, biochemical, and genetic changes in seed quality during storage. Hortic. Sci. 15:781–784.

Ross, J.P. 1975. Effect of overhead irrigation and benomyl sprays on late-season foliar diseases, seed infection, and yields of soybean. Plant Dis. Rep. 59:809–813.

Roy, K.W., and T.S. Abney. 1977. Antagonism between *Cercospora kikuchii* and other seedborne fungi of soybeans. Phytopathology 67:1062–1066.

Sabbah, M.A., G.E. Meyer, H.M. Keener, and W.L. Roller. 1977. Reversed-direction-airflow drying for soybean seed. Trans. Am. Soc. Agric. Eng. 20:562–566, 570.

Sherf, A.F. 1953. Correlation of germination data of corn and soybean lots under laboratory greenhouse and field conditions. Proc. Assoc. Off. Seed Anal. 43:127–130.

Sherwin, H.S., and K.K. Kreitlow. 1952. Discoloration of soybean seeds by the frogeye fungus, *Cercospora sojina*. Phytopathology 42:568–572.

Shortt, B.J., A.P. Grybauskas, F.D. Tenne, and J.B. Sinclair. 1981. Epidemiology of *Phomopsis* seed decay of soybean in Illinois. Plant Dis. 65:62–64.

Sinclair, J.B. 1975. Compendium of soybean diseases. The American Phytopathology Society, St. Paul.

----. 1982. Compendium of soybean diseases. The American Phytopathology Society, St. Paul.

Singh, B.B. 1976. Breeding soybean varieties for the tropics. p. 11–17. *In* R.M. Goodman (ed.) Expanding the use of soybeans. Proc. Conf. for Asia and Oceania, Chiang Mai, Thailand. February. INTSOY Series 10. International Agricultural Publications, University of Illinois.

Singh, C.B., M.A. Dalal, and S.P. Singh. 1978. Genetic analysis of field germination in soybean (*Glycine max* (L). Merrill). Theor. Appl. Genet. 52:165–169.

Singh, J.N., S.K. Tripathi, and P.S. Negi. 1972. Note of effect of seed size on germination, growth and yield of soybean. Indian J. Aric. Sci. 42:83–86.

Singh, K.N., and Bachchan Singh. 1981. Effect of crop and machine parameters on threshing effectiveness and seed quality of soybean. J. Agric. Eng. Res. 26:349–355.

Smith, T.J., and J.M. Camper, Jr. 1975. Effects of seed size on soybean performance. Agron. J. 67:681–684.

----, M.T. Carter, G.D. Jones, and M.W. Alexander. 1961. Soybean performance in Virginia as affected by variety and planting date. Virginia Agric. Exp. Stn. Bull. 526.

Spain, G. 1982. Standardization of seed vigor tests. Proc. Soybean Seed Res. Conf., 12th 12:107–109.

Spilker, D.A., A.F. Schmitthenner, and C.W. Ellett. 1981. Effects of humidity, temperature, fertility, and cultivar on the reduction of soybean seed quality by *Phomopsis* sp. Phytopathology 71:1027–1029.

Stanway, V.M. 1974. Germination response of soybean seeds with damaged seed coats. Proc. Assoc. Seed Anal. 64:97–106.

----. 1978. Evaluation of 'Forrest' soybeans with damaged seed coats and cotyledons. J. Seed Technol. 3:19–26.

Starzing, E.K., S.H. West, and K. Hinson. 1982. An observation on the relationship of soybean seed coat color to viability maintenance. Seed Sci. Technol. 10:301–305.

Stuckey, R.E., R.M. Jacques, D.M. TeKrony, and D.B. Egli. 1981. Foliar fungicides can improve soybean seed quality. Kentucky Seed Improvement Association, Lexington.

Svien, T.A., and D. Isely. 1955. Factors affecting the germination of corn in the cold test. Proc. Assoc. Off. Seed Anal. 55:80–86.

Tanner, J.W., and D.J. Hume. 1978. Management and production. p. 157–217. *In* A.G. Norman (ed.) Soybean physiology, agronomy, and utilization. Academic Press, New York.

Tao, K.J. 1978a. The 1978 referee test for soybean and corn. Assoc. Off. Seed Anal. Newsl. 52:43–66.

----. 1978b. Factors causing variations in the conductivity test for soybean seeds. J. Seed Technol. 3:10–18.

----. 1980a. The 1980 vigor "referee" test for soybean and corn seed. Assoc. Off. Seed Anal. Newsl. 54:53–68.

----. 1980b. Effect of seed treatment on the conductivity vigor test for corn. Plant Physiol. 65:S–141.

Taylor, B.H., and C.E. Caviness. 1982. Hilum color variation in soybean seed with imperfect black genotype. Crop Sci. 22:682–683.

TeKrony, D.M. 1983. Current status of seed vigor testing. Proc. Soybean Seed Res. Conf., 12th 12:96–101.

----, and D.B. Egli. 1977. Relationship between laboratory indices of soybean seed vigor and field emergence. Crop Sci. 17:573–577.

----, ----, and John Balles. 1980a. The effect of the field production environment on soya bean seed quality. p. 403–425. *In* P.D. Hebblethwaite (ed.) Seed production. Butterworths, London.

----, ----, ----, T. Pfeiffer, and R.J. Fellows. 1979. Physiological maturity in soybeans. Agron. J. 71:771–775.

----, ----, ----, L. Tomes, and R.E. Stuckey. 1984. Effect of date of harvest maturity on soybean seed quality and *Phomopsis* sp. seed infection. Crop Sci. 24:189–193.

----, ----, and A.D. Phillips. 1980b. Effect of field weathering on the viability and vigor of soybean seed. Agron. J. 72:749–753.

----, ----, L. Tomes, and G. Henson. 1982. Changes in the viability of pod and stem blight fungus in soybean seed during storage. Agron. Abstr. American Society of Agronomy, Madison, WI, p. 137.

----, ----, R.E. Stuckey, and J. Balles. 1983. Relationship between weather and soybean seed infection by *Phomopsis* sp. Phytopathology 73:914–918.

Thomas, G.D., C.M. Ignoffo, C.E. Morgan, and W.A. Dickerson. 1974. Southern green stink bug: influence on yield and quality of soybeans. J. Econ. Entomol. 67:501–503.

Ting, K.C., G.M. White, I.J. Ross, and O.J. Loewer. 1980. Seed coat damage in deep-bed drying of soybeans. Trans. Am. Soc. Agric. Eng. 23:1293–1296, 1300.

Todd, J.W. 1976. Effects of stink bug feeding on soybean seed quality. p. 611–618. *In* L.D. Hill (ed.) World soybean research conference proceedings. The Interstate Publishers and Printers, Danville, IL.

----, and S.G. Turnipseed. 1974. Effects of southern green stink bug damage on yield and quality of soybeans. J. Econ. Entomol. 67:421–426.

Tomes, L., D.M. TeKrony, and D.B. Egli. 1981. The relationship of time, temperature and moisture content during accelerated aging to soybean seed germination. Agron. Abstr. American Society of Agronomy, Madison, WI, p. 102.

Toole, E.H., and V.K. Toole. 1946. Relation of temperature and seed moisture to viability of stored soybean seed. USDA Cir. 753. U.S. Government Printing Office, Washington, DC.

Torrie, J.H. 1958. Comparison of different generations of soybean crosses grown in bulk. Agron. J. 50:265–267.

Turnipseed, S.G., and M. Kogan. 1976. Soybean entomology. Ann. Rev. Entomol. 21:247–282.

U.S. Department of Agriculture. 1975. Federal Seed Act. American.

Vaughan, C.E., B.R. Gregg, and J.C. Delouche. 1968. Seed processing and handling handbook 1. Mississippi State University, Mississippi State, MS.

Villa, L.G., G. Roa, and I.C. Macedo. 1978. Minimum airflow for drying soybean seeds in bins with ambient and solar heated air. ASAE Paper 78-3017. American Society of Agricultural Engineers, St. Joseph, MI.

Wagner, C.K., and M.B. McDonald, Jr. 1981. Identification of soybean (*Glycine max* (L.) Merrill) cultivars using rapid laboratory techniques. Ohio Agric. Res. and Development Ctr. Res. Bull. 1133.

Walker, R.J., and H.J. Barre. 1972. The effect of drying on soybean germination and crackage. ASAE Paper 72-817. American Society of Agricultural Engineers, St. Joseph, MI.

Wall, M.T., D.C. McGee, and J.S. Burris. 1983. Emergence and yield of fungicide treated soybean seed differing in quality. Agron. J. 75:969–973.

White, G.M., T.C. Bridges, O.J. Loewer, and L.J. Ross. 1980. Seed coat damage in thin-layer drying of soybeans. Trans. Am. Soc. Agric. Eng. 23:224–227.

----, ----, ----, and ----. 1981. Thin-layer drying model for soybeans. Trans. Am. Soc. Agric. Eng. 24:1643–1646.

----, O.J. Loewer, I.J. Ross, and D.B. Egli. 1976. Storage characteristics of soybeans dried with heated air. Trans. Am. Agric. Soc. Eng. 19:302–310.

Wien, H.C., and E.A. Kueneman. 1981. Soybean seed deterioration in the tropics. II. Varietal differences and techniques for screening. Field Crops Res. 4:123–132.

Wilcox, J.R., and T.S. Abney. 1973. Effects of *Cercospora kikuchii* on soybeans. Phytopathology 63:796–797.

----, F.A. Laviolette, and K.L. Athow. 1974. Deterioration of soybean seed quality associated with delayed harvest. Plant Dis. Rep. 58:130–133.

Yaklich, R.W., and P.B. Cregan. 1981. Moisture migration into soybean pods. Crop Sci. 21:791–792.

----, and M.M. Kulik. 1979. Evaluation of vigor tests in soybean seeds: Relationship of the standard germination test, seedling vigor classification, seedling length, and tetrazolium staining to field peformance. Crop Sci. 79:247–252.

----, ----, and J.D. Anderson. 1979. Evaluation of vigor tests in soybean seeds: Relationship of ATP, conductivity, and radioactive tracer multiple criteria laboratory tests to field performance. Crop Sci. 19:806–810.

Yeargan, K.V. 1977. Effects of green stink bug damage on yield and quality of soybeans. J. Econ. Entomol. 70:619–622.

9 Crop Management

Richard R. Johnson
Deere & Company Technical Center
Moline, Illinois

The major challenge in soybean [*Glycine max* (L.) Merr.] management is to integrate all production variables to meet the unique characteristics of an individual farm. This integration involves more than managing genotypes and environments. Constraints from inputs such as capital, time, and labor can greatly influence the management system adopted. Sometimes availabilities of needed inputs such as irrigation water, drainage rights, or local markets also place restrictions on optimum cropping systems for a given farm. In short, several factors can affect crop management decisions and no one management system can be considered to be the optimum.

The objective of this chapter is to discuss crop management research that has been reported since the review of Pendleton and Hartwig (1973). Emphasis will be on general concepts. Another recent comprehensive reference concerning soybean management has been prepared by Scott and Aldrich (1983).

9-1 SEED SELECTION

9-1.1 Wide vs. Specific Adaptability

Improved cultivars have made substantial contributions to past increases in soybean yield (Luedders, 1977; Boerma, 1979; Wilcox et al., 1979; Boyer et al., 1980). To capitalize on genetic improvements in yield, newly released cultivars must continually be evaluated in different production environments. Breeders have traditionally emphasized development of cultivars that perform well over a wide range of climatic and edaphic conditions (Schutz and Bernard, 1967). The current importance of wide adaptability is recognized by a survey of 15 states accounting for about 88% of the 1983 U.S. soybean production (Crop Reporting Board, 1983b). In this survey, 'Williams' was the leading cultivar for the 8th consecutive year and accounted for 12.3% of the 1983 harvested area.

Soybeans: Improvement, Production, and Uses, 2nd ed.–Agronomy Monograph no. 16.

To produce widely adapted cultivars, plant breeders have emphasized traits such as major pest resistance, lodging resistance, nonshattering pods, and high average yields. As soybean cultivars have gained in importance and have been increasingly exposed to a broader range of cropping conditions, there has been a tendency to release cultivars for special growing situations. For example, short-statured cultivars have been released for high-yielding environments in the northern USA (Cooper, 1981); 'Amcor' has been identified as a Maturity Group II cultivar adapted to low yield environments (Walker and Cooper, 1982); and 'Narow' is a Maturity Group V cultivar especially adapted for planting in narrow rows at conventional spring planting dates (Caviness et al., 1983). Although these specialty cultivars can perform quite well in prescribed environments, they may be adversely affected by other major production variables. The cultivar Narow is not only less suited for late planting, but is also highly susceptible to injury from the herbicide metribuzin [4-amino-6-(1,1-dimethylethyl)-3-(methylthio)-1,2,4-triazin-5(4*H*)-one] (Caviness et al., 1983). When selecting cultivars, it is thus important to identify strengths and weaknesses. Compared to more stable yielding cultivars, those developed for special environments can provide exciting opportunities but can also expose a grower to greater risk if environmental conditions develop that are not consistent with the specialty cultivar's requirements.

9–1.2 Maturity Classification

Planting cultivars of differing maturity can assist in spreading risk and can also spread harvest time. Stage of plant development is influenced by temperature and photoperiod (Major et al., 1975). For this reason, soybean cultivars have been placed in 13 maturity groups ranging from 000 (earliest) to X (latest). Designation of early, mid-, and full-season cultivars is used to describe relative maturity on a local basis, but for full-season production these designations seldom span more than three maturity groups at a given location. For full-season production at a given location, it is usually possible to identify cultivars that will yield well yet differ in maturity by as much as 20 to 30 days. There has been some tendency, especially in areas with shorter growing seasons, for mid- and full-season cultivars to yield more than early season cultivars.

9–1.3 Determinate and Indeterminate Growth Types

Soybean cultivars differ in growth habit as well as maturity. Bernard (1972) reports that determinate cultivars have predominated in Japan, Korea, and the southern USA, whereas indeterminate soybean cultivars have been grown in northeast China and the northern USA. Indeterminate cultivars continue main stem elongation several weeks after beginning flowering, while determinate plants terminate main stem elongation at, or soon after, the onset of flowering. Semideterminates are an intermediate stem type that terminate stem growth fairly abruptly after a flowering period almost as long as that of indeterminate types. Shading or lodging can cause a semideterminate stem to appear indeterminate, and insect or other injury to an indeterminate stem tip may simulate determinateness.

Several determinate and semideterminate cultivars adapted to the northern USA have recently been released. Within a maturity group, these determinate and semideterminate cultivars are generally shorter, more lodging resistant and have lower basal pod heights than indeterminate cultivars (Hartung et al., 1981; Cooper, 1981; Beaver and Johnson, 1981a). Cooper (1981) reported that these short-statured determinates are adapted to high yield environments, but are not adapted to environments with early season stress. Beaver and Johnson (1981a) found that determinate and semideterminate cultivars yielded as well as indeterminate cultivars in a wide range of productivity levels, but determinate cultivars had a less predictable yield response to varying levels of productivity. At least three factors may account for the less stable performance of short-statured determinate cultivars. First, Fehr et al. (1981) reported that 100% defoliation of determinate cultivars causes greater yield loss at all reproductive stages than that which occurs with indeterminate cultivars. Secondly, Septoria brown spot (caused by *Septoria glycines* Hemmi) has been shown to make faster vertical progress and cause greater yield loss in a short determinate than in an indeterminate cultivar (Pataky and Lim, 1981). Thirdly, because there is a high correlation between plant height and lowest pod height, environments with early season stress can cause sufficient height reduction in short-statured cultivars to increase the potential for harvest losses (Beaver and Johnson, 1981a; Hartung et al., 1981).

Since semideterminate cultivars are generally intermediate in height between determinate and indeterminate cultivars, semideterminates may possess more yield stability potential than determinates when grown in northern environments. Green et al. (1977) and Wilcox (1980) have compared semideterminate and indeterminate breeding lines in narrow and wide rows. Narrow rows increased yields of both plant types in about equal proportions and both research groups concluded that either plant type could be successfully used in a midwestern U.S. breeding program. Availability of future semideterminate cultivars in the northern USA may thus depend upon which plant type plant breeders choose to place their emphasis.

9–1.4 Number of Cultivars and/or Blends

Ryder and Beuerlein (1979), Boquet et al. (1982) and Beatty et al. (1982) have emphasized that several different cultivars are often capable of producing high and nearly equal yields when planted at optimum dates under a range of cultural conditions. Walker and Fehr (1978) concluded that stable production could best be achieved by growing several rather than one cultivar.

Since the mid-1970s, mixtures of two or more pure line cultivars have been sold as blends. One objective in producting a blend is to combine pure lines so that the blend yield is greater than the weighted mean yield of the component cultivars in a pure stand (Fehr and Rodriguez, 1974). This might be accomplished by combining pure lines that complement each other, i.e., a lodging resistant but pest susceptible cultivar might be combined with a lodging susceptible but pest resistant cultivar. In practice, there is little evidence that a blend will yield more than the weighted mean yield of its components (Walker and Fehr, 1978; Fehr and Cianzio, 1980). At least two factors limit the use of blends. First, replanting harvested seed from a blend is not recommended because the genetic makeup of the blend may change. Secondly, blends containing cultivars differing in maturity do not allow for a spread in harvest time. For these reasons, blends have not been used as widely as pure line cultivars. A producer with small land area might find a blend to be a desirable way to spread risk and achieve more stable yields.

9–1.5 Seed Germination and Size

The warm germination test measures viability under relatively ideal conditions and has historically been a standard measure of seed quality. Yet, a loss in seed viability is often preceded by a loss in vigor. One or more of the vigor tests have been useful in identifying seed that will have stand establishment advantages under less than ideal field conditions (TeKrony and Egli, 1977; Johnson and Wax, 1978; Tao, 1978; Kulik and Yaklich, 1982). In the absence of stand differences, there has seldom been a yield advantage in using high vigor seed (Johnson and Wax, 1978; Egli and TeKrony, 1979). Nevertheless, the uncertain nature of weather and soil conditions provide adequate justification for planting seed of high germination and vigor. Seed of high quality seldom costs much more than that of lower quality. In a recent survey, TeKrony (1982) reported that the accelerated aging test, cold test, and tetrazolium tests are the vigor tests most often used by seed-testing laboratories. The survey also showed that since 1976 there had been a substantial increase in the number of vigor tests conducted. Standard warm germination tests of at least 80% are often considered a minimum for acceptable field use. Using vigor tests in conjunction with warm germination tests will help measure acceptable quality. A more thorough discussion of seed quality can be found in chapter 8 of this book.

In general, seed size has not had much effect on crop yields. Where differences have been reported, the advantages have usually been in favor of larger rather than smaller sizes (Fontes and Ohlrogge, 1972; Smith and Camper, 1975). Size uniformity rather than absolute size may be the most important factor affecting yield since this leads to plant uniformity (Fontes and Ohlrogge, 1972).

9–2 TILLAGE

Recent changes in tillage practices have greatly influenced crop management. An extensive review of tillage is presented in chapter 10 in this book. The following will briefly cover tillage from the standpoint of management.

Before 1960, clean tillage was used because it assisted in economic control of weed, insect, and disease pests. After World War II, advancing technology in the chemical industry began to furnish pesticides that provided alternate methods of pest control. Chemical pesticides along with improved seeding equipment have allowed production agriculture to adopt what is now becoming known as conservation tillage. Conservation tillage systems are designed to provide a rough, residue-covered soil surface that is resistant to wind and water erosion. No-tillage represents the extreme in conservation tillage since seed is planted in a previously undisturbed soil, and the only tillage used is that necessary to place seed in the soil. Less extreme forms of conservation tillage are usually referred to as reduced tillage since the entire field is often tilled, but in such a way that crop residue is still present on the soil surface at planting time. In a review of soil erosion control with conservation tillage, Laflen et al. (1981) have emphasized that it is generally the percentage residue cover rather than the particular tillage system per se that reduces erosion.

In addition to soil conservation, a number of other reasons are advanced to promote conservation tillage. These include a conservation of labor, moisture, energy, and money. Compared to clean tillage, the reduced number of tillage operations in conservation tillage systems generally do conserve soil, labor, and moisture. However, cost and energy reductions associated with reduced machinery operations are sometimes offset by an increased need for pesticides and fertilizers (Siemens and Oschwald, 1978; Lockeretz, 1983; Jolly et al., 1983). Soybean yields obtained with different tillage systems have differed (Siemens and Oschwald, 1978; Bauder et al., 1979; Nave et al., 1980; Colvin and Erbach, 1982; Touchton and Johnson, 1982; Gebhardt and Minor, 1983). In general, drought-prone and well-drained soils yield more with conservation tillage because of increased moisture conservation. However, fine-textured and poorly drained soils often yield less with conservation than with clean tillage—largely due to wetter and cooler soils that delay planting and reduce early season crop growth. Rotating corn (*Zea mays* L.) and soybean has helped eliminate yield reductions experienced in continuous corn when conservation tillage has been used (Triplett and Van Doren, 1977; Erbach, 1982; Mulvaney, 1984).

Use of conservation tillage in soybean production has steadily grown. No-Till Farmer (1983) has surveyed state agronomists of the Soil Conservation Service (SCS) each year since 1972. In 1972, about 2, 12, and 86% of the soybean area was reported to be in the no-, reduced, and clean tillage categories, respectively. By 1982, about 7, 38, and 55% of the soybean area was in these same categories. Of the 1982 soybean area that was no-till planted, nearly 80% was double-cropped soybean. A 1982 Office of Technology Assessment (OTA) report suggests that USDA estimates of conservation tillage use are somewhat lower than those of No-Till Farmer but follow the same general trend. The OTA report projects that 75% of U.S. cropland will eventually be in some form of conservation tillage, but cites other estimates ranging from 50 to 84% adoption. Pest control, particularly weeds, was the major factor given as limiting the rate of adoption of conservation tillage.

An important concept of conservation tillage is the mulch of residue left on the soil surface. Table 9–1 is from a review by Colvin et al. (1981) and shows the amount of residue remaining on the surface after a single tillage pass in different cropping situations. The previous crop, time of tillage, and sequence of tillage events affect residue remaining. Speed, tillage depth, and ground engaging attachments used also affect amount of residue left by a single pass of a particular tillage machine, and account for the ranges given for each implement in Table 9–1. When prior tillage has been conducted, some implements can return buried residue to the surface—thus accounting for values above 100% in Table 9–1. At least one company currently markets over 30 different sweeps, shovels, or spikes for use on chisel plows (Johnson, 1982). Sweeps tend to incorporate small amounts of residue while spikes and twisted shovels tend to incorporate intermediate and large amounts of residue, respectively. Some of the ranges for residue left on the surface in Table 9–1 appear to be conservative and will likely change as tillage machines are modified. For instance, moldboard plows are currently available that can be manually

Table 9–1. Percent pretillage residue cover remaining after a single tillage pass (Colvin et al., 1981).

Tillage implement	Fall without previous tillage		Spring without previous tillage		Spring following previous tillage	
	Avg	Range	Avg	Range	Avg	Range
	% pretillage surface cover after corn					
Moldboard plow	4	0–10	7	5–10	–	–
Disk	84	–	50	42–73	80	46–100
Chisel	56	40–85	56	44–68	–	–
Field cult.	–	–	–	–	84	–
Till plant (sweep)	–	–	62	59–66	–	–
Plant (double-disk opener)	–	–	90	82–100	80	–
	% pretillage surface cover after soybean					
Moldboard plow	2	–	3	–	–	–
Disk	–	–	–	–	58	56–60
NH_3 knife on 762-mm centers	–	–	39	27–54	44	43–45
Chisel	14	–	28	25–31	115	106–130
Till plant (sweep)	–	–	–	–	74	73–76
Plant (double-disk opener)	–	–	81	70–94	100	76–113

or hydraulically adjusted to vary width of cut from 36 to 61 cm per bottom. When operated at narrow widths, these plows can leave up to 25% residue cover in corn stubble, but will incorporate most residue if adjusted to the widest cut (Johnson, 1982). Thus, in conservation tillage, the manner in which a machine is equipped and operated can be as important as selection of the particular implement.

Compared to corn, soybean produces less residue which is subject to more rapid decomposition. Soil erosion following soybean is greater than following corn (Siemens and Oschwald, 1978; Laflen and Moldenhauer, 1979). Thus, where soybean is grown in rotation, tillage systems used after the soybean crop may be more important for erosion control than those used to prepare for the soybean crop. As shown in Table 9–1, a given tillage tool often incorporates a higher percentage of soybean than corn residue. Following the soybean crop, delaying all tillage until spring provides the most effective erosion control. In the corn-soybean rotation, anhydrous ammonia is often applied as the N fertilizer for corn. Anhydrous knife applicators may leave less surface residue cover than some other tillage tools and should be considered as a tillage tool when managing residue (Table 9–1).

Some soybean herbicides must be soil incorporated and others tend to give more consistent weed control if incorporated. Thompson et al. (1981) have provided an extensive review of the incorporation capabilities of several different types of tillage tools. As in residue management, successful herbicide incorporation is often as dependent on how the machine is equipped and operated as on the general type of implement used.

In most areas of the USA, soybean yields have not been increased by deep tillage. However, certain soils in the southeastern USA compact easily. On these soils, root penetration ceases at a bulk density of about 1.75 g cm^{-3} which is often found in the compacted layer, and soybean crops have typically responded to subsoiling that is deep enough to break the hard pan (Musen, 1977; Smith et al., 1978; Martin et al., 1979). In the south central Midwest, there are about 5 million ha of claypan soils that are poorly drained and often experience periods of excessive rainfall and drought within the same growing season. These soils are not improved by deep tillage, but soybean yields have increased with a combination of irrigation and improved surface or internal drainage (Walker et al., 1982). Drainage alone or irrigation alone had only slight effects on soybean yield but did improve corn yields.

Hanthorn and Duffy (1983) surveyed the 1980 cropping season costs and returns of clean, reduced, and no-tillage soybean producers in midwestern, midsouthern, and southeastern USA. Herbicide use and cost differed by region, but additional herbicide applications were generally substituted for any reductions in tillage and mechanical cultivation. Insecticide use and costs were not significantly different among tillage strategies. The number of mechanical cultivations did not differ between reduced and clean tillage systems but ranged from 1.12 to 1.79 cultivations per season among the three regions. No-tillage soybean averaged 0, 0.19, and 0.47 cultivations in the Southeast, Midsouth, and Midwest, respectively. Midwestern producers received the highest returns, but in this region no-tillage soybean generated significantly lower returns than clean tillage soybean—largely due to lower yields. Returns in the Midsouth and Southeast did not differ with tillage system. The authors concluded that no one tillage strategy shows a clear economic advantage over others.

In summary, the relatively uniform set of tillage practices of the past have evolved into more complex management systems. Optimum tillage practices have become site specific much like fertilizer and pesticide recommendations. Tillage systems will need to differ not only from one region to another, but from field to field, and in some cases, practices within a field will change from 1 yr to the next. The concept of rotating tillage systems on a given field has several merits. More thorough tillage may be necessary after a high residue-producing crop such as corn, than after a low-residue, erosion-prone crop such as soybean. Occasional use of clean tillage can sometimes reduce problems encountered in reduced or no-tillage systems. For example, deeper moldboard plowing can greatly reduce problems from shallow germinating annual weeds or carryover herbicides. Using more thorough tillage in years when relatively immobile nutrients (i.e., lime, P, or K) are applied can help reduce nutrient availability problems. One of the challenges researchers will face in coming years is the integration of other crop management practices with the changing tillage systems. It seems obvious that reduced and no-tillage systems will require better management on the part of the grower than is required under clean tillage systems.

9–3 FERTILITY

Soybean fertility is covered in detail in chapter 12 in this book. This section will present only a brief review of recent management research.

9–3.1 Nitrogen

Soybean is a legume and when well nodulated is capable of fixing its own N. Harper (1974) found that both symbiotic N_2 fixation and nitrate (NO_3^-) utilization appear essential for maximum yield. However, he found that excessive NO_3^- appears detrimental to maximum yield because symbiotic fixation is completely inhibited. Apparently, most soils can meet NO_3^- needs of the plant because soil applications of N show no yield advantage regardless of source of N or time, method, or rate of application (Rogers et al., 1971; Chesney, 1973; Welch et al., 1973; Pal and Saxena, 1976; Deibert et al., 1979; Nelson and Weaver, 1980; Porter et al., 1981). An exception to this rule has occurred on soils that are somewhat poorly drained, low in organic matter and strongly acid below the plow layer (Bhangoo and Albritton, 1976). These soils have sometimes responded to N rates in the range of 50 to 110 kg ha^{-1}.

In Iowa, Garcia and Hanway (1976) combined N with P, K, and S to form a relatively low salt NPKS solution. Significant yield increases were obtained from two to four NPKS foliar sprayings between developmental stages R5 to R7. Results of similar studies since the Iowa work have been discouraging, and the practice has not generally been recommended (Welch et al., 1979; Keogh et al., 1979; Poole et al., 1983). Labeled N was applied in NPKS treatments by Vasilas et al. (1980). From 44 to 67% of the total N applied was recovered in the plants, and a high proportion of the recovered N was found in the seed. Yield was increased in 1 out of 2 yrs. In summary, several attempts have been made to increase soybean yields with N fertilizer, but positive results have been elusive and economic returns rarely occur.

9–3.2 Lime, Phosphorous, and Potassium

Liming acid soils to a pH of 6.0 to 6.5 is an important prerequisite for profitable soybean production. Limestone differs greatly in neutralizing value and fineness of grind. These factors along with the soil depth that is being neutralized are important considerations in determining lime application rates. No-tillage and forms of reduced tillage that provide little soil mixing will often have different lime requirements than systems using deeper, more thorough tillage. Alkaline soils with a pH $>$ 7.5 can also cause problems in soybean production, but it is seldom economical to attempt to reduce pH. Availability of Fe, Mn, Cu, B, Zn, and P all decrease in alkaline soils. Micronutrient applications may be required to correct deficiencies. Soil-applied triazine herbicides are also more prone to cause soybean injury on alkaline soils (Ladlie et al., 1976). This injury can occur from triazine carryover from a previous crop or from direct application to the soybean crop. In short, growers must often manage around a high pH problem rather than correcting the high pH itself. For example, management of alkaline soils may involve factors such as greater use of micronutrients, more careful selection of herbicides and cultivars, and greater use of banding or starter fertilizers.

High soybean yields require adequate levels of P and K, and rates of application should be based on soil tests and local recommendations. Where soil test levels are high, method of P and K application is of little importance. On soils testing low in P and K, however, application method is more important. In Iowa, deMooy et al. (1973) concluded that soybean is less responsive to P and K fertilizer than is corn and that soybean yields showed little difference between the effect of direct and residual fertilizer. When using fall plow down, they suggested that in a corn-soybean rotation application to corn in the cropping sequence will be more effective than to soybean. Both P and K are relatively immobile nutrients and under conditions of conservation tillage there may be merit in the application of these elements at that point in the cropping sequence where more thorough tillage is used. Starter or band applications have tended to be helpful on cool or low testing soils. In Minnesota, Ham et al. (1973) reported the largest yield response from combinations of starter and broadcast fertilizer. Placing fertilizer directly with the seed can cause injury and is generally not recommended.

9-3.3 Micronutrients

Molybdenum is an essential element for N metabolism and its use on soybean has recently been reviewed by Boswell (1980). Positive yield responses to supplemental Mo application have been reported in the Far East (Japan, China, Taiwan), Europe, and at least 12 states in the USA. In the USA, positive responses have more frequently occurred east of the Mississippi River where rainfall is moderate to heavy and soils tend to be acid. Boswell (1980) further notes that critical tissue levels of Mo have not been well established although most leaf tissue contents have been <0.20 mg kg^{-1} where yield has been increased by added Mo. Rates must be higher for soil applications than for seed or foliar spray treatment. Responses have been obtained with seed treated Mo at rates as low as 17 g ha^{-1}, while soil application rates may need to be >800 g ha^{-1}. Liming to maintain soil pH above 6.2 may effectively correct or prevent Mo deficiencies.

Manganese deficiency is common on alkaline, sandy soils during cool, wet spring weather. In Wisconsin, Randall et al. (1975) reported that row application was somewhat more effective than broadcast application. Combined row and foliar application resulted in higher yields than either row or foliar treatments alone. Georgia research has confirmed the inefficient utilization of broadcast Mn (Wilson et al., 1981). In Florida, Robertson et al. (1973) obtained yield increases from both Mn and Cu fertilization. Copper has also increased yields on some soils in Indiana (Oplinger and Ohlrogge, 1974). In North Carolina, Barnes and Cox (1973) compared copper sulfate with chelated and/or complexed Cu materials, and found all sources were equally effective at increasing double-crop soybean yields when broadcast and incorporated into the soil before wheat (*Triticum aestivum* L.) planting.

Iron chlorosis is a common problem on calcareous soils. Iowa researchers have developed a visual score ranging from 1, no yellowing, to 5, severe yellowing. Froehlich and Fehr (1981) reported that average yield loss increased by 20% for each unit increase in chlorosis score. Thus, Fe chelate sprays (Randall, 1977) or cultivars resistant to Fe chlorosis (Neibur and Fehr, 1981) should be used to prevent chlorosis expression.

In Georgia, Touchton and Boswell (1975) reported yield increases from applications of 0.28 to 1.2 kg ha^{-1} of B, but observed yield decreases from 2.24 kg ha^{-1}. Similar results were obtained from broadcast soil applications and from foliar sprays during early bloom.

In summary, micronutrient deficiencies are the exception rather than the rule. However, if soil or plant tissue tests indicate a deficiency, applications should be considered.

9-4 PLANTING PRACTICES

9-4.1 Planting Date

In warm climates, cool soil temperatures seldom control planting time, but in cooler areas, temperature can play an important role. The optimum temperature for hypocotyl elongation is about 30°C which also corresponds to the optimum temperature for germination (Hatfield and Egli, 1974). However, soybean germination and growth typically begins at temperatures of 8 to 10°C. In Missouri, Major et al. (1975a) planted soybean at dates ranging from late April to early July and found that the number of days to emergence decreased from about 18 at early dates to 5 days at later dates. For emergence, the number of growing degree days above 10°C remained constant at about 100 indicating that temperature was the primary variable influencing days to emergence. If plants are frosted after emergence, Hume and Jackson (1981) found that plants damaged at the cotyledon stage generally survive better than those frosted at later stages. In most cases, a plant would regrow with 50% of its tissue damaged, but with 70% tissue damage, only an occasional plant regrew.

Soybean tolerance to relatively wide ranges in planting dates has no doubt helped the widespread acceptance of this crop. Nevertheless, soybean does have an optimum planting date that can differ by both region and cultivar. Several recent studies in the northern USA have included factorial combinations of planting dates, row widths, and cultivars (Ryder and Beuerlein, 1979; Beaver and Johnson, 1981b; Helsel et al., 1981). In each of these studies, planting date was the variable having the greatest impact on yield. Highest yields were generally obtained with early to mid-May planting dates, and yields began to drop off quite rapidly with planting dates beyond late May.

In some areas of the southern USA, soil moisture conditions for planting are most favorable during April before temperatures get too high. Yet, planting during the short photoperiods of early to mid-April often results in shorter plants and lower yields than planting between late April and early June (Caviness and Thomas, 1979; Parker et al., 1981; Boquet et al., 1982; Thurlow and Pitts, 1983; Griffen et al., 1983). Planting after early June generally causes reduced yields as plants again become shorter. On shallow soils with a limited water-holding capacity, non-irrigated soybean typically shows erratic planting date responses that are largely dependent on the timing of summer rains.

Several of the above studies reported that cultivars differed in response to planting date. If adapted cultivars are planted during early to mid-May, many of the cultivar interactions with planting date are minimized. Cultivar choice for earlier or later plantings are less clear cut, but most sudies have shown an advantage for using mid- to full-season cultivars for extremely late planting dates such as those associated with double-cropping.

Plants are generally taller when planted between mid-May and early June and decrease in height with either very early or late plantings. In Arkansas, Caviness and Thomas (1979) observed decreased lodging on the shorter plants resulting from very early or late plantings. In Illinois, Beaver and Johnson (1981b) observed that short determinate cultivars exhibited a general increase in lodging when planted later while indeterminate cultivars decreased in lodging as planting dates were delayed past early June.

9–4.2 Plant Density and Row Width

In nonstress environments, light interception by the crop canopy can limit crop yields. Equidistant plant spacings represent the ideal. At plant densities resulting in maximum yields, equidistant spacings would occur at 15- to 25-cm row widths and would result in maximum seasonal light interception. Several studies in the northern USA have shown a yield advantage for planting in rows narrower than 75 to 100 cm (Green et al., 1977; Ryder and Beuerlein, 1979; Costa et al., 1980; Wilcox, 1980; Cooper, 1981). When several row widths have been used within an experiment, intermediate row widths of about 50 cm have provided much of the yield advantage gained in going to row widths of 25 cm or less (Helsel et al., 1981; Beaver and Johnson, 1981b).

Some row width studies in the southern USA continue to show little yield advantage for row widths <90 to 100 cm (Doss and Thurlow, 1974; Heatherly, 1981). However, several southern studies have shown an advantage of planting in row widths of 45 to 50 cm compared with 90 to 100 cm (Akhanda et al., 1976; Parker et al., 1981; Beatty et al., 1982; Boquet et al, 1982; Thurlow and Pitts, 1983). Many of these southern studies showing a narrow row advantage have included May as well as later planting dates indicating that full-season seedings also have the potential to gain from narrow rows.

In general, planting date and cultivar selection have not caused large interactions with row-width response, but there has been some tendency for later planting dates and early flowering cultivars to be somewhat more responsive to narrower row widths.

Under irrigated conditions, Reicosky et al. (1982) concluded that early in the season, 25-cm rows had slightly higher evapotranspiration than 100-cm rows, whereas later in the season, row spacing had no effect on evapotranspiration. During 2 yrs of lower seasonal water supplies in western Iowa, Taylor (1980) observed no differences in yield among row widths. In a 3rd yr when water supply was high, narrow rows yielded more than 100-cm rows. Under severe drought conditions in North Dakota, Alessi and Power (1982) found that enhanced early season water use by soybean in narrow rows leaves less water available for pod-fill resulting in reduced yields. Using these and other findings, they concluded that narrow rows may be beneficial when water is not restricting; may have no effect on yield when moderate stress is encountered; and may reduce yields under extreme full-season water stress situations.

Several systems exist to control weeds in narrow-row seedings grown without cultivation and in wider rows grown with cultivation (Wax et al., 1977). These systems generally involve combinations of herbicides and may involve both soil-applied and postemergence compounds. Soybean plants in the early vegetative stages are quite tolerant to physical injury, and use of ground equipment for early season postemergence applications has not reduced the yield potential of narrow-row seedings. In Illinois, Nave et al. (1980) used full-scale equipment and larger plots to grow soybean crops under clean and reduced tillage in row widths of 18, 38, 51, and 76 cm. They found that weed control and stand establishment were more difficult to achieve with narrow rows and reduced tillage. Preemergence herbicides used in reduced tillage treatments were associated with severe weed problems in 3 of 4 yrs. Where these problems existed, cultivation was generally more effective for controlling weeds in 51- and 76-cm rows than was application of postemergence herbicides in 18- and 38-cm rows. The 18-cm rows averaged 4 to 6% higher yields than other row widths. This study illustrates that the 7 to 20% yield advantage often reported in small plot, narrow-row research in the northern USA is not always as easily achieved in full-scale production situations.

Indiana and Iowa researchers have compared erosion rates from soybean planted with clean tillage in solid seeded, 51-, and 76-cm row widths. Row width had only a minor effect on soil erosion, but tended toward less erosion in narrow rows (Mannering and Johnson, 1969; Colvin and Laflen, 1981). In Tennessee, Shelton et al. (1983) found that crop rotation and tillage system had more influence on erosion from soybean fields than did soybean row width.

Soybean seed size differs greatly and planting rates should be based on the number of seeds per unit area rather than weight per unit area. Whether planted in rows or in equidistant spacings, soybean plants can produce similar yields across a wide range of seeding rates (Wilcox, 1974; Lueschen and Hicks, 1977; Hoggard et al., 1978; Costa et al., 1980). As seeding rates increase, plant height, height of the lowest pod, and lodging all tend to increase. When using high-quality seed, seeding rates of 350 000 to 500 000 seeds ha^{-1} are generally sufficient to maximize yield, reduce potential harvest losses due to low basal pod height, and insure adequate stands in unfavorable seedbeds. The lower seeding rates are better suited to wider rows and lodging susceptible cultivars, while the higher rates are more suited to narrow rows and lodging resistant cultivars.

Soybean plants can tolerate some variation in spacing within a row. Stivers and Swearingin (1980) induced skips in 76-cm rows before the V-3 growth stage. Several short skips reduced yield less than one long skip of the same total length. With alternate skips in each row, yield reductions varied from 1.1% with 0.30-m skips to 15.3% with 1.22-m skips when skips constitute 50% of the entire row. This study was conducted under weed-free conditions. Eliminating small skips may be more important for helping control early season weed growth than for yield per se.

The main reasons for using narrower rows and higher plant densities are to intercept sunlight sooner in the season and to provide early season competition with weeds. The fact that soybean plants often have similar yields under a range of plant populations and narrower row widths implies that complete light interception early in the season is often not necessary to maximize yield. As discussed in secion 9–6.1, plant growth responses before reproductive development often have only minor effects on yield, but optimum growing conditions are crucial during pod fill. Johnson et al. (1982) have used this concept and planting pattern research discussed above to draw the following summary points on planting patterns.

1. The objective of choosing a planting pattern should be to have full canopy closure by the time all plants are flowering.
2. In the USA, optimum row widths are narrower as planting progresses northward.
3. Late-planted and double-cropped soybean are often more responsive to narrower rows than are soybean planted at conventional spring planting dates.
4. Within a region, soybean cultivars that benefit most from narrow rows are those that flower earlier or do not "spread out" into row centers.
5. Within a region, fields that are consistently under stress from weeds, drought, fertility, disease, or insects will be less likely to respond to narrow rows.
6. Planting in 15- to 25-cm row widths approaches the ideal pattern of equidistant spacing and should result in maximum yields if stand is adequate and pests are controlled.
7. If postemergence pest control and improved stand establishment are needed, producers can realize the majority of the narrow-row advantage by using row widths of 40 to 50 cm and leaving out rows to provide clearance for tractor wheels.

Considering the above summary points, it is of interest to review row widths in the 10 leading soybean-producing states in the USA (Table 9–2). These states account for about three-fourths of the harvested U.S. soybean plants and data in Table 9–2 are from annual surveys of 90 to 140 fields in each state. During the past 5 yrs, average row widths have steadily narrowed in each state due to increased use of solid seeded and intermediate row widths. It would appear that average row widths in several states are still too wide for optimum yields. The availability of improved herbicides and improved narrow-row planting equipment should encourage continued adoption of narrow rows.

9–4.3 Inoculation

Adequate populations of *Rhizobium japonicum* must be present to produce a well-nodulated soybean crop that will not require N fertiliz-

Table 9–2. Soybean row widths for 1979 to 1983 in the leading 10 soybean producing states in the USA. Adapted from data of the Crop Reporting Board (1981, 1983b).

State	Hectares harvested (1979–1983 Avg)	Year	25 cm or less	26 to 72 cm	73 cm and greater	Avg width
	millions		———	% of fields	———	cm
Illinois	3.8	1979	5	7	88	81
		1980	5	10	85	78
		1981	18	10	72	71
		1982	18	8	74	71
		1983	17	13	70	68
Iowa	3.3	1979	1	5	94	87
		1980	3	6	91	84
		1981	3	9	88	82
		1982	3	7	91	83
		1983	4	11	84	80
Missouri	2.2	1979	5	5	90	83
		1980	13	7	81	76
		1981	18	7	75	73
		1982	22	10	68	70
		1983	27	13	60	66
Minnesota	1.9	1979	8	10	82	80
		1980	12	10	78	75
		1981	10	10	80	75
		1982	17	18	65	68
		1983	13	15	71	69
Arkansas	1.8	1979	13	1	86	92
		1980	21	4	75	88
		1981	12	12	76	86
		1982	12	14	75	86
		1983	13	15	73	85
Indiana	1.8	1979	3	7	90	83
		1980	9	12	79	77
		1981	12	6	82	76
		1982	14	14	72	74
		1983	18	9	73	71
Ohio	1.5	1979	28	16	56	60
		1980	32	13	55	59
		1981	37	18	45	54
		1982	35	15	50	56
		1983	38	9	53	55
Mississippi	1.5	1979	10	7	83	90
		1980	25	7	69	86
		1981	21	10	68	86
		1982	24	11	65	84
		1983	14	16	70	80
Louisiana	1.3	1979	24	3	73	94
		1980	27	3	70	91
		1981	33	4	62	90
		1982	36	10	55	85
		1983	26	9	65	88
Tennessee	0.9	1979	10	15	74	84
		1980	16	15	69	81
		1981	18	22	60	76
		1982	26	20	54	72
		1983	22	20	58	73
Ten state total	20.0					
U.S. total	27.2					

ation. In rhizobia-free tropical soils, Smith et al. (1981) determined that inoculum levels above 1 $\times$ 10^5 rhizobia per centimeter of row were necessary to establish effective nodulation. On rhizobia-free soils in North Carolina, Mahler and Wollum (1981) found that four different serogroups produced adequate nodulation, but some serogroups resulted in higher grain yields than others. In Wisconsin, Brill (1981) has also identified superior N_2-fixing strains when used in fields where legumes had never been grown. However, the superior strains were unable to compete with indigenous strains when introduced into fields with a history of soybean production.

To increase the probability of establishing nodules from the more efficient rhizobia strains, granular soil inoculants have been developed to be applied in the seed furrow using insecticide attachments on planters. These soil inoculants have the capability of supplying many times more rhizobia per seed than is attained with seed treatment inoculums. When applied to soils where soybean are grown with some regularity, however, both the seed and soil inoculants have failed to increase yields in Arkansas (Thompson and Pongsakul, 1976), Illinois (Johnson and Boone, 1976), Indiana (Nelson et al., 1978), and Louisiana (Dunigan et al., 1980).

Surveys conducted during 1979 in Arkansas and South Carolina found that 54 and 70%, respectively, of the farmers were inoculating soybean (Wolf and Nester, 1980). In Arkansas, those using inoculants treated 45% of their planted area; 96% used planter-box inoculants, 2% used pre-inoculated seed, and 2% used a granular-type inoculant. About one-fourth of the inoculants used in both surveys also contained Mo. Apparently, many producers use inoculants as a carrier of Mo or as a cheap insurance to insure adequate nodulation.

The quality of planter box inoculants being sold to farmers in South Carolina and Georgia was surveyed by Skipper et al. (1980). Viable rhizobial counts ranged from 6.9×10^9 to $< 1.0 \times 10^3$ per gram of inoculant. The low viable rhizobial counts were associated with poor nodulation under greenhouse conditions. Nonpeat-base products and combination products containing Mo and/or fungicides were inferior to peat-base inoculants. In Alabama, Hiltbold et al. (1980) also surveyed commercial inoculants available for sale and found a wide range of efficacy among the products. When used on fields with low indigenous supplies of rhizobia, yield increased with products providing more than 10^3 rhizobia per seed. Smith et al. (1983) studied the effects of shipping conditions on quality maintenance of granular soil inoculants shipped around the world. They found that prolonged shipping times, higher temperatures, and reduced moisture all decreased the final population of rhizobia; but final inoculant moisture content exerted the largest influence.

In summary, there is little need to inoculate succeeding crops if a well-nodulated soybean crop has been grown. Where used, fresh inoculants in sealed containers should be maintained under cool conditions until used in the field. Rhizobia strains differ in their capability to contribute to high yields, and one of the challenges of N research is to determine how to manage these differences under field conditions.

9–4.4 Fungicidal Seed Treatments

Seed treatments provide one means to control seed-borne and soil-borne diseases. The ability of soybean to compensate for wide differences in stand, the availability of high-quality seed, and the spotty nature of seed and soil-borne diseases all reduce the need for fungicide seed treatments. A survey by MacFarlane (1980) reported that of the 1981-planted soybean seed, about 47% was to be treated with a fungicide treatment—11% by a commercial seed treatment firm, local elevator, or farmer using a mechanical seed-treating device, and 36% by planter box treatments in the field. Most university plant pathology departments surveyed recommended seed treatment when seed germination was $<$ 80% and/or when cool, wet soil conditions were anticipated. Wall et al. (1983) have criticized such recommendations as being too vague for grower use. They found captan [*cis-N* ((trichloromethyl) thio)-4 cyclohexene-1, 2-dicarboximide] and carboxin-thiram (5,6-dihydro-2-methyl-*N*-phenyl-1, 4 oxathiin-3-carboxamide; tetramethylthiuram disulfide) seed treatments to be equally effective in increasing emergence of seedlots with more than 15% *Phomopsis* spp. No seed treatment, however, consistently improved field emergence of seeds with reduced quality caused by mechanical damage, age, or size. They also found no obvious differences in fungicide performance in relation to planting dates or soil types.

Rushing (1982) has emphasized that fungicide seed treatments must be matched to the potential problem. He lists *Pythium* spp., *Phytophthora* spp., and *Rhizoctonia* spp. as the three major soil-borne diseases which can reduce soybean yields. Captan and thiram provide nonsystemic activity against *Pythium* spp. with virtually no control against other pathogens listed. Carboxin-thiram controls all but *Phytophthora* spp. Use of highly tolerant cultivars have been effective in controlling *Phytophthora* spp. under mild and moderate disease pressure, but does not always offer control during seed germination and seedling stages of development (Schmitthenner and Kroetz, 1982). Metalaxyl [*N*-(2,6-dimethylphenyl)-*N*-(methoxyacetyl)-alanine methyl ester] is a systemic seed fungicide that will control *Phytophthora* and *Pythium* but because of its narrow spectrum of control will require inclusion of protectants such as captan, thiram, or carboxin-thiram to provide broad spectrum control.

In summary, when high-quality seed is used, adequate stands are generally achieved without the need for fungicide seed treatments. However, under stress conditions seed fungicides can aid stand establishment if the proper chemicals are used for the anticipated pathogen.

9–4.5 Planting Equipment

Properly designed planting equipment should provide population control, accurate seed spacing in the row, seed depth control, and adequate seed-soil contact (Agness and Luth, 1975). Factors such as operating speed, seedbed condition, and machine cost can all influence how well a given planting device meets these requirements. Since the early 1970s, at least three significant changes have occurred in soybean planting equipment. First, a full range in row width capabilities has become commercially available. Throckmorton (1980) reviewed machines available for narrow, intermediate, and wide row planting. He noted that cost restrictions have largely restricted narrow (15 to 25-cm widths) plantings to various types of drills. Intermediate row widths (30–70 cm) can be seeded with either drills or row crop planters, and if skips are left for tractor tires, cultivators are also available for use. Row widths > 70 cm are almost always seeded with row crop planters. Several different metering devices are used on soybean planters including feed cups, air drums, fluted rollers, horizontal plates, and air-disk meters. Nave and Paulsen (1979) tested these five meters and concluded that all provided about equal seed-spacing accuracy while causing only minimal seed damage.

A second significant change on planters has been improved depth gauging. During the early 1970s, most row crop planters gauged depth from the press wheel while drills used spring-loaded depth control rods (Baumheckel, 1976). Depth bands were available on both drills and row crop planters, but bands had to be changed to alter depth. In 1975, a row crop planter was introduced that gauged depth with a double-disk opener at the point of seed release. This concept provided excellent uniformity in depth control (Agness and Luth, 1975), and similar concepts have since been adopted on a number of other commercial row crop planters as well as at least one commercial grain drill.

A third change in soybean planting equipment has been the introduction of machines capable of planting in reduced and no-till seedbeds. Colvin and Erbach (1982) reported that any one of several different drills could be used to successfully plant in narrow row widths under reduced tillage conditions. Nave et al. (1977) have stressed that when a drill does not have coulters, it is helpful to use a tillage tool between the tractor and drill to remove wheel tracks. Use of depth bands and seed-firming wheels was also reported to be of value in obtaining target stands with drills.

No-till seedbeds often have an abundant quantity of surface residue and are firmer than prepared seedbeds. No-till double-crop seedbeds are generally the most adverse because the small grain crop has often depleted the surface soil moisture and the residue is fresh. Machines differ in their capacity to plant in the wide range of no-till seedbeds. In general, row crop planters designed to plant in row widths wider than 30 cm have the capability of no-till planting in a wider range of seedbeds than grain drills designed to plant in row widths of 15 to 25 cm. Most planters capable of operating under no-till conditions use some type of coulter to open the soil for the seeding device. Each coulter requires a great deal of weight, up to several hundred kilograms, to guarantee penetration in dry, firm seedbeds. Heavy-duty drills with close row spacings are available but require more planting units and weight—both requirements that increase machine cost. A second option to open a seed furrow is to use powered tillage blades, a factor that also increases machine cost. Thus heavy-duty drills capable of no-till planting under adverse conditions are expensive per unit of width.

A number of lighter-duty drills are also available with coulters. These machines have limited ability to no-till plant, but are effective in fields where some full-width tillage has been conducted. In looser soils or fields where irrigation can be used to moisten the topsoil, these machines can effectively serve as no-till planters. Several row crop planters capable of no-till planting in row widths of 36 to 50 cm have recently been introduced. Compared to no-till drills these machines offer the advantage of more successful operation in adverse seedbeds at a lower cost per unit width.

Selection of proper coulter type is important for no-till planting. Smooth coulters require the least down pressure for penetration, but prepare an extremely narrow furrow that must be followed by an aggressive furrow opener. Rippled coulters have a straight sharp edge, but ripples located beyond the coulter edge do some limited soil loosening. Fluted coulters have a curved edge that loosens soil in a band 2- to 5-cm wide. Smooth or rippled coulters generally work better in surface residue and cover a wider range of soil conditions than do fluted coulters. Compared to fluted coulters, smooth or ripple coulters:

1. Require less weight to penetrate hard, dry soil.
2. Incorporate less crop residue into the seed zone.
3. Operate at higher speeds and in wetter soils without removing soil from the weed zone.

The wider seed zone prepared by fluted coulters helps decrease misalignment problems with the seed opener and can be especially advantageous when planting contoured rows.

9–4.6 Seeding Depth

Cultivars differ in their emergence capability but a seeding depth of 2.5 to 4 cm is optimum for most cultivars and soils. Shallower depths may be justified on crust-prone soils, and deeper planting may be justified on loose sands. Some cultivars have an inhibition of hypocotyl elongation at 25 °C, but all are unaffected by higher or lower temperatures (Burris and Knittle, 1975). High-quality seed and rotary hoeing are of little help in aiding emergence if extensive hypocotyl swelling occurs under these conditions. If sensitive cultivars are planted in soils near 25 °C, planting depth should be < 2.5 cm.

Some planting guides recommend that soybean should not be planted in dry soil. However, if the seed zone is uniformly dry and if the planting date is moving past the optimum, it seems that a shallow planting in dry soil can often be justified. High-quality seed should remain viable for 10 to 14 days. If sufficient time has passed for the soil to dry, there are many regions that would have a high probability of rainfall within an additional 10 to 14 days. The penalty for late planting, especially with a prolonged wet period after an initial rain, may exceed the risk of rain not coming soon enough for emergence. A planting depth that places all seed in either dry or moist soil in the upper 4 cm of soil should be selected. This will avoid varied emergence times within a field and associated maturity differences at harvest time.

9–5 CROPPING SYSTEMS

9–5.1 Rotations

It is a common practice to rotate soybean with crops such as corn, wheat, or cotton (*Gossypium hirsutum* L.). There are a number of reasons to support growing soybean crops in rotation rather than in a continuous soybean sequence. These reasons include: (i) higher yields, (ii) a decreased need for N fertilizer on subsequent crops, (iii) breaking up of pest cycles, and (iv) spreading labor and machine requirements over a larger portion of the growing season.

In a previous review, Pendleton and Hartwig (1973) noted that soybean has been grown continuously in some areas without serious yield losses. Yet, most recent research has shown a yield advantage for rotation. In a 10-yr central Illinois study, Slife (1976) found that soybean yields averaged 14% higher when grown in a corn-corn-soybean or corn-soybean-wheat sequence as compared to a continuous soybean sequence. In Minnesota, Hicks and Peterson (1981) found that soybean yields in a corn-soybean rotation were 11% higher than in a continuous soybean cycle. In a 4-yr northern Illinois study, yields in a corn-soybean rotation were 21% higher than continuous soybean when grown under clean tillage and were 26% higher under reduced tillage (Mulvaney, 1984). In all three of the above studies, corn in rotation with soybeans yielded more than continuous corn. In Ohio, Jeffers et al. (1970) compared continuous cropping with 2- and 3-yr rotations of soybean, corn, and sugarbeet (*Beta vulgaris* L.). Compared to continuous cropping, soybean crops grown in alternate years yielded 3% more, while soybean grown every 3rd yr yielded 6% more. Sugarbeet and corn both responded more to rotation than did soybean. In Arkansas, Hinkle (1970) compared continuous soybean crops with various rotations involving cotton and soybean double-cropped after wheat. Soybean crops grown in rotation averaged 14% higher yields than continuous soybean and were more responsive to rotation than was cotton. In summary, these studies show that rotation soybean generally yield more than continuous soybean and several other crops also benefit from rotations that include soybean.

Soybean crops can cut fertilizer costs by reducing the need for N by subsequent crops (Higgs et al., 1976). The amount of N credit from soybean will depend on the cropping sequence and region of the country. Beuerman et al. (1982) emphasize that soybean plants cause a net removal of N from the soil, but Illinois recommendations call for an N reduction of 40 kg ha^{-1} for corn and 10 kg ha^{-1} for wheat when grown after soybean.

Rotations can be highly effective in breaking pest cycles. When herbicides were used, Slife (1976) observed fewer weed seeds after 10 yrs where soybean crops had been in the rotation than where continuous corn had been grown. Rotating herbicides within a continuous corn sequence caused a similar reduction in weed seeds. If soybean crops are rotated with corn, there is seldom a need to apply a corn rootworm (*Diabrotica* spp.) insecticide unless extensive infestations of volunteer corn were present in the soybean crop. In Arkansas, Hinkle (1970) reported that continuous soybean crops are subject to yield reductions from charcoal rot (*Sclerotium bataticola*), but rotations will reduce problems with this pest.

The soybean cyst nematode (*Heterodera glycines*) is often controlled by resistant cultivars, but more than one race is now present in some areas. Crop rotations can be effectively used to help prevent the excessive buildup of new races. In Arkansas, Price et al. (1976) have recommended a 3-yr rotation to help limit the spread of new races. In the 1st yr a nonsusceptible crop such as cotton, corn, or sorghum is grown. In the 2nd yr, a cultivar resistant to the prevalent race is grown. The 3rd yr, a cultivar susceptible to all races is grown.

Crop rotations can also effectively spread labor and machinery requirements. In the northern USA where the corn-soybean rotation is common, soybean planting can often be delayed until corn planting is complete without any subsequent reduction in yield potential. Yet, in this rotation early season soybean cultivars are often ready for harvest before corn. In areas with a longer growing season, double-cropping rotations use more of the total growing season in addition to spreading labor requirements.

9–5.2 Double-cropping

In double-cropping, soybean crops are planted after the harvest of a previous crop. The most common case in the USA is planting after a small grain such as wheat, but soybean plants are also double-cropped after vegetable crops, corn, and the initial harvest of a forage crop. During 1980 to 1983, 9 to 16% of the U.S. soybean area was double-crop planted (Crop Reporting Board, 1983a). Of the 1982 double-cropped soybean crops, No-Till Farmer (1983) reports that 39, 39, and 22% were planted with clean, reduced, and no-tillage, respectively. Since double-cropping after wheat is the most common case, this system will be emphasized in this discussion.

Planting and maintaining a vigorous weed-free wheat crop will maximize total yields and minimize weed problems in the soybean crop. Early wheat cultivars have been developed especially for double-cropping and can assist in timely soybean planting (Collins and Jones, 1975). Although timing of fertilizer application is not critical for soybean, most states recommend that P and K for both crops be applied at or before small grain planting (Herbek, 1982). If lime is needed, it is best applied in the fall to allow more time for neutralizing activity to occur. If drying and storage facilities are available, harvesting high moisture wheat may gain several days in the planting of soybean, a process that is more important in shorter-season areas.

Management of small grain straw differs greatly. Most agree that where a market exists, it is generally profitable to harvest and remove the straw. Otherwise, it should be chopped and uniformly spread behind the combine. In some areas of the southern USA, straw burning is common. Advantages of burning straw are destroying weeds and surface weed seeds, reducing potential for seedling disease, easier cultivation, and minimizing phytotoxic effects of wheat residue. These advantages must be weighed against the disadvantages of greater air pollution and possible long-term effects such as reduced soil organic matter and increased compaction (Collins, 1982). In Mississippi, Sanford (1982) obtained the highest and most consistent yields when soybean were no-till seeded into burned stubble, but these yields did not differ from systems where straw was burned followed by conventional tillage or where straw was shredded followed by no-till planting. Lowest yields were obtained when straw was incorporated into the soil or soybean was no-till planted into standing straw. In Arkansas, Collins (1982) has noted that wheat straw contains phytotoxins that can slow soybean growth and potentially reduce yields. The growth-retardant capacity dissipates after about 3 weeks of decomposition.

Having an adequate moisture supply for rapid soybean germination and emergence is crucial to successful double-cropping success. In many areas, maintaining wheat residue is important for moisture conservation and soil erosion control. Under no-tillage conditions in Virginia, Hovermale et al. (1979) concluded that a 20-cm wheat stubble resulted in optimum yields, but additional straw mulch showed no benefit.

Baldwin (1982) has reviewed weed control for double-cropping. Weed control programs should be based on past weed history, tillage program, and careful scouting after emergence. Often a program of preplant or preemergence (soil applied) and postemergence herbicides is necessary. Due to absorption by straw or ash, the activity of most soil-applied herbicides will be lower when applied following small grains. Therefore, the highest labeled rate for a specific soil texture and organic matter should usually be used. In a no-tillage system, all existing vegetation should be controlled before soybean emerges. No-tillage planting should not be considered if an emerged weed problem is so severe that complete control with burn-down herbicide is uncertain.

Compared to earlier planting dates, the later planting dates associated with soybean double-cropping cause large enough genotype × planting date interactions to justify evaluating double-crop soybean cultivars under the double-crop environments (Akhanda et al., 1976; Carter and Boerma, 1979). Traits which increase vegetative growth are desirable in double-cropped cultivars and may account for the reason that mid- to full-season cultivars often perform better than earlier season cultivars. Later planting dates also enhance the need for narrower rows and may justify slightly higher seeding rates. As with full-season production, double-cropped soybean has generally not responded to N fertilizer (Herbek, 1982).

Some areas in the southern USA have attempted double-cropping soybean after soybean. Cultivars are available which when planted shortly before the spring equinox, will mature near the summer solstice and permit a second crop (Boote, 1981). Timely planting, use of irrigation, and narrow rows are among the intensive management practices required for these systems (Woodruff, 1980; Boerma and Ashley, 1982). The higher risk nature of this system has caused Woodruff (1980) to conclude that this system may have more merit if crops other than soybean are considered for the second crop.

9–5.3 Intercropping

In the midwestern USA, researchers have used relay intercropping in an attempt to extend multiple cropping further northward or to obtain higher yields than obtained with double-cropping (Jeffers and Triplett, 1978; Chan et al., 1980; McBroom et al., 1981). In this system, soybean crops are planted into the growing small grains, so both crops occupy the same area until small grain harvest. Soybean crops finish out the season—much like a relay team in a track event. The risks in this system are greater than with double-cropping. Generally, the small grain rows are widened to accommodate soybean seeding equipment. Thus, small grain yields are often reduced in accordance with how much the rows are widened and how much the small grain plants are injured during soybean planting. Early season moisture is critical since the small grain crop is actively growing and can easily remove moisture from the soybean seedling root zone.

Soybean crops in relay systems sometimes grow tall enough to interfere with small grain harvest. Thus, soybean planting date should be selected with regard to growth stage of the small grain crop rather than calendar date. In central Illinois, soybean crops interplanted during late boot or early heading stages of small grain growth produced greater soybean yields than soybean planted during early jointing or early grain fill (Johnson and Brown, 1979). Choice of soybean cultivar was found to be important by Jeffers and Triplett (1978) and Johnson and Brown (1979). A period of vegetative growth after small grain harvest is essential to allow plants to fill in the canopy and support a reasonable seed load. Fuller-season indeterminate cultivars can have an advantage in this regard. On the other hand, McBroom et al. (1981) found no major differences in intercrop yields among a broad range of soybean cultivars.

As in any successful cropping system, early season weeds must be controlled in relay intercrop systems. Some preemergence soybean herbicides can be safely applied to small grains, or postemergence chemicals compatible with both crops can be used. In summary, relay intercropping requires a high level of management and adequate early season moisture. Further research will be required to determine if the system can be profitable in mechanized agriculture.

Another intercropping system that has received some attention is the planting of corn and soybean in alternate strips. In Minnesota, 1, 3, 6, and 12 rows each of corn and soybean have been planted in alternate strips using 76-cm row widths (Crookston and Hill, 1979). Although corn yields were impoved in some combinations, accompanying soybean yields were always reduced to the extent that none of the combinations made better use of the land area than did sole cropping.

9-6 POSTEMERGENCE CROP MANAGEMENT

9-6.1 Critical Stages

Establishing an adequate uniform stand is critical for high yields. Once this has been accomplished, a major management objective is to minimize crop stress throughout the remainder of the growing season. Final crop yield will be a product of the seasonal dry matter production and the partitioning of this dry matter into grain production. By minimizing stress the crop will be given the optimum opportunity to intercept sunlight and to convert this solar energy into dry matter. The proportion of the dry matter partitioned into grain yield will largely depend on the cultivar and planting date that were used.

Not all growth stages are as important as others in influencing final yield. In general, the early portion of pod filling (reproductive stages R4 and R5) is the period most responsive to an optimum growth environment while the vegetative period and later portions of pod filling are the least responsive. Optimum growth environment is of intermediate importance during flowering and early pod formation (reproductive stages R1-R3). The sensitive nature of early pod filling to stress has been shown by experiments that have induced defoliation (Caviness and Thomas, 1980), water stress (Sionit and Krammer, 1977), lodging (Woods and Swearingin, 1977), and shade (Schou et al., 1978). In all of these studies, stress during early pod fill reduced yield more than stress at any other growth stage. Other approaches have enhanced the dry matter supply by increasing photosynthesis beyond normal. Hardman and Brun (1971) found that CO_2 enrichment during vegetative and flowering stages did not increase yield whereas similar treatment during pod fill increased yield. Schou et al. (1978) found that light enrichment treatments increased yield most during late flowering to early pod fill.

Stresses during vegetative growth that do not affect stand will generally have only minimal effects on yield when compared to stresses occurring at pod-filling and pod-formation stages. Stresses during vegetative growth that will not soon be outgrown should be corrected early to assure optimum conditions by reproductive development. For example, several postemergence herbicides will control their target weeds only if applied early in the season. Or, if nodulation fails and N deficiency develops early, fertilizer N should be applied. Other early season stresses may be temporary and are often outgrown without affecting yield. Such stresses may include minor crop injury due to herbicides or insect feeding as well as short-term drought.

Other chapters will discuss in detail postemergence crop management for irrigation (chapter 10 in this book), weed control (chapter 11 in this book), diseases (chapters 17 and 18 in this book), and insects (chapter 20 in this book). The general management concept that should be emphasized is that the most critical time to minimize stress is during early pod fill. The remainder of this section will review research with growth regulators.

9–6.2 Chemical Growth Regulators

Growth regulators continue to receive considerable attention, but compounds causing economic soybean yield increases are not currently available. Response to TIBA (2,3,5-triiodobenzoic acid) has been variable (Clapp, 1973; Stutte and Rudolph, 1971). Tanner and Ahmed (1974) reported that TIBA increased seed yield under good growth conditions but did not affect yield when conditions were poor for growth. Johnson and Anderson (1974) reported yield increases by applying low rates of 2,4-D (2,4-dichlorophenoxyacetic acid) 1 to 2 weeks before TIBA application at beginning bloom. However, they concluded that the feasibility of using 2,4-D with TIBA to increase yield is limited since optimum rates vary with environment and cultivar.

Morphactin-containing compounds can delay senescence and have been studied for their growth regulating properties with soybean. Clapp (1975) applied methyl-2-chloro-9 hydroxylfluroene-(9)-carboxylate at flower initiation and increased seed yield during 1 of 2 yrs. Dybing and Lay (1981) also used various morphactins to delay senescence but obtained either no change or a reduction in yield. Dybing and Lay (1982) found that a morphactin could increase soybean oil concentration but also resulted in a decrease in protein concentration.

Several additional growth regulating compounds from the chemical industry have been evaluated for effects on crop yield and have exhibited varying responses (Stutte and Rudolph, 1971; Blomquist et al., 1973; Stutte et al., 1975; Oplinger et al., 1978; Fuehring and Finkner 1978). To date, none of these compounds have achieved commercial use on soybean plants. The sometimes critical nature of application rate and timing as well as the difficulty of screening for yield will provide ample challenges as efforts continue in the future to identify growth regulators that may have potential to increase yield.

9–7 HARVESTING

9–7.1 Late-Season Frost Injury

Physiological maturity occurs when the grain reaches its maximum dry matter accumulation. In Kentucky, TeKrony et al. (1981) found that physiological maturity occurred when one normal pod on the main stem had reached its mature pod color (stage R7). In Minnesota, Gbikpi and Crookston (1981) reported that loss of green color in all pods is the best indicator of physiological maturity and occurs slightly later than appearance of one normal pod at mature pod color. Seed moisture content at physiological maturity will usually be in the range of 40 to 60%.

Frost after physiological maturity generally does not damage plants if pods remain intact. However, some regions have a limited number of frost-free days, and frost before physiological maturity can lead to crop damage. Premature death from frost can leave a cross section of the seed green. To grade U.S. no. 1 or 2, the percentage of green bean (*Phaseolus vulgaris* L.) cannot exceed 1 and 2%, respectively. In addition, early frost can reduce seed yield. Saliba et al. (1982) exposed plants to freezing temperatures at various growth stages after stage R4 (full pod). Freezing injury was first observed at temperatures ranging from −2.8 to −3.9°C and was positively associated with the concentrations of epiphytic ice nucleation active bacteria that were present on plant leaves. The latest growth stage at which freezing caused significant yield reductions differed with cultivar and varied from R6.0 to R7.2. Frost-injured plants reached maturity earlier, but had seed moisture equivalent to nonfrosted plants. Protein concentration was not affected by frost, but seed oil concentration was reduced if frost occurred before R6 (full seed). Judd et al. (1982) found that temperatures required to cause reductions in seed germination and vigor decreased as seed maturation progressed. Seed in yellow pods (55% moisture) showed reductions in germination and vigor following an 8-h exposure at −7°C whereas germination of seed in brown pods at 35% moisture was reduced only by exposure to −12°C.

9–7.2 Chemical Dessication

Chemical dessication prior to physiological maturity can reduce soybean yield (Whigham and Stoller, 1979). If applied after physiological maturity, dessicants can speed the rate of field drying without affecting yield. In Ohio, Byg and Walker (1974) found that a dessicant could advance harvest 2 to 5 days with early cultivars and 1 day or less with later-maturing cultivars. Thus, use of dessicants to advance harvest time can seldom be justified unless excessive green weeds are present.

9–7.3 Harvesting Equipment

As recently as the early 1970s, soybean-harvesting losses were generally more than 8% of crop yield and the majority of the losses occurred at the combine header (Nave et al., 1973; Ayres, 1973). At this time, floating cutter bar headers which reduced harvest losses over a rigid platform by 25 to 30% were available and were beginning to be used. Nevertheless, it was apparent that significant reductions in harvest losses could be attained with further header improvements.

During the mid-1970s, several header improvements were provided that have led to the capability of reducing harvest losses by about 50%. Bichel and Hengen (1978) have reviewed these improvements and research contributing to their development. Several companies now offer flexible cutterbar headers that feature full-width skid shoes, improved cutterbar flexibility and long-floating dividers that assist in dividing the crop at the header edge. The flexible platforms can be locked straight for harvesting small grain, and have the advantage of being suited to any row width. In 1976, a flexible cutterbar platform was introduced that had guard and sickle spacings half the normal 7.6 cm, thus doubling the number of cutting edges. This feature lowers shatter losses and allows faster travel speeds. In 1975, a low-profile, row-crop header with individual row units that float independently to follow uneven ground was introduced. Corrugated meshed belts grip plants before they are cut by a rotary knife. Compared to flexible platforms, row-crop headers have the advantages of higher travel speeds, elimination of reel shatter and lower harvest losses, but are currently restricted to row widths of 76 cm or greater. Automatic height control is available for all header types. Nave et al. (1980) has reported that harvesting losses can be reduced to $< 4\%$ by the use of these improved combine headers.

Another development in harvesting has been the introduction of rotary combines. These machines have one or more rotors that replace the conventional cylinder and straw walkers for threshing and grain separation. In Illinois, Newbery et al. (1980) compared conventional-cylinder, single-rotor, and double-rotor machines at three seed moisture contents. Each threshing mechanism was operated at four peripheral velocities. Total threshing and separation losses were under 0.6% for all three machines. For all three mechanisms, percentages of split seeds increased as peripheral threshing speed was increased, but the increase was less with rotary threshing. When operated in the manufacturers' recommended cylinder or rotor-speed range, the percentage of splits was well below the allowable 10% limit for U.S. no. 1 soybean. Soybean susceptibility to breakage and seed-coat crack percentage did not differ as a result of the type of threshing mechanism or the cylinder or rotor speed.

Once harvest maturity is reached, it is not unusual for seed moisture to vary by several percentage points within a few hours. Drier grain is more susceptible to splitting. Newbery et al. (1980) have emphasized that a reduction in percentage of splits as a result of decreased cylinder or rotor speed may be offset by an increase in threshing and separating losses, especially for rotary threshing mechanisms. Grain loss monitors and on-the-go cylinder or rotor speed adjustments are features on modern combines that can help reduce grain loss and damage—especially at lower seed moistures.

With the increased interest in conservation tillage and soil erosion control, another important aspect of the combine is straw redistribution behind the machine. Straw choppers are more effective at this process than straw spreaders, and choppers with extended vanes are capable of spreading the straw over a full-header width.

9–7.4 Storage and Drying

Storage and drying needs have been outlined by Pepper et al. (1982). Soybean plants having 10% moisture or less remain in generally good condition up to 4 yrs. Market-grade soybean with about 12% moisture retain their grade for nearly 3 yrs, although germination and other qualities of the seed gradually decline over that period. Seed with 13% moisture can be safely stored from harvest to late spring but if moisture is 14%, the safe period is limited to the winter months. Usually, soybean above 15% moisture should not be stored without drying.

A few days in the harvest season may be gained if harvest begins at 18 to 20% moisture and the grain is later dried or aerated to a moisture level safe for storage. Humidity of drying air should be kept above 40% to avoid seed-coat cracking. Soybean for seed should not be dried at temperatures above 43°C. To avoid splits and other damage, maximum temperatures above 54 to 60°C are seldom recommended. To reduce respiration and the activity of mold or insects, grain should be cooled if the temperature of the grain mass is 6°C above outside air temperature, but grain temperature should not be reduced below 2 to 5°C.

9–8 CROP MANAGEMENT OUTSIDE THE USA

This chapter has emphasized management research from North America. Yet, the concepts discussed are applicable to other areas of the world. For example, extensive planting date and planting pattern studies have recently been conducted in two areas of Australia (Constable, 1977; Lawn et al. 1977). The results were quite similar to those obtained in the southern USA at similar latitudes.

A unique program that developed during the 1970s was the International Soybean program (INSOY). This program of the University of Illinois at Urbana-Champaign and the University of Puerto Rico, Mayaqüez Campus, cooperates with international and national organizations to expand the use of soybean. Since the inception of the INSOY cultivar testing program in 1973, more than 250 cultivars have been tested in over 100 countries by some 500 cooperators. Each year, INSOY publishes results of cultivar trials in its publication series. The INSOY Newsletter (INSOY, 1982) summarized a number of generalizations that can be drawn from tests to date:

1. In the tropics, yields tend to be larger from later-maturing than from earlier-maturing cultivars.
2. Yields are somewhat lower in tropical and subtropical than in temperate regions.
3. Plants are affected more by changes in altitude than by changes in latitude.
4. Shattering and lodging are seldom serious problems.
5. Size of harvested seed is not related to yield.
6. Yields from a newly introduced crop are usually good.
7. Poor nodulation is a major problem in popularizing soybean cultivation in the tropics.
8. Chemical composition of seed is comparable in all environmental zones.
9. Seed quality is a universal problem, but small seeded cultivars have better seed quality than large seeded cultivars.
10. Protein and oil concentration of a cultivar remains stable in different sites and environments.

9–9 SUMMARY

Since the early 1970s, several developments have greatly influenced soybean management. Plant breeders continue to release improved higher-yielding cultivars for all maturity zones. These cultivars must continually be evaluated under different cropping systems to determine the best cultivars for use in each cropping system. Development of improved herbicides as well as improved tillage and planting equipment, have allowed increased adoption of conservation tillage. These same developments have also allowed the increased use of narrower row widths. As further improvements occur in chemical pesticides and farm equipment, there will no doubt be an even greater adoption of these two important production practices.

Strong evidence has developed supporting the need to rotate soybean with other crops. Double-crop production of soybean has grown to become a significant crop rotation in itself. Several innovations in combine headers have resulted in reduced harvest losses. Nitrogen fertilization and growth regulators are two areas of management research that have received considerable attention, yet remain elusive in terms of providing any additional new tools to increase yields.

As soybean plantings have increased and as the management tools available for production have become more complex, a greater need has developed for educational efforts that carry these management alternatives to the producer. Future gains in soybean productivity will become even more dependent on imaginative, interdisciplinary research.

REFERENCES

Agness, J.B., and H.J. Luth. 1975. Planter evaluation techniques. ASAE Paper 75-1003. American Society of Agricultural Engineers, St. Joseph, MI.

Akhanda, A.M., G.M. Prine, and K. Hinson. 1976. Influence of genotype and row width on late-planted soybeans in Florida. Proc. Soil Crop Sci. Soc. Fla. 35:21–25.

Alessi, J., and J.F. Power. 1982. Effects of plant and row spacing on dryland soybean yield and water use efficiency. Agron. J. 74:851–854.

Ayres, G.E. 1973. Profitable soybean harvesting. Iowa State Univ. Ext. Serv. Pm-573.

Baldwin, F.L. 1982. Weed management systems for doublecropped soybeans. p. 8–11. *In* Proceedings soybean doublecrop conference. American Soybean Association, St. Louis.

Barnes, J.S., and F.R. Cox. 1973. Effects of copper sources on wheat and soybeans grown on organic soils. Agron. J. 65:705–708.

Bauder, J.W., G.W. Randall, J.B. Swan, J.A. True, and C.F. Halsey. 1979. Tillage practices in south-central Minnesota. Minn. Coop. Ext. Serv. Pamplet PM-864.

Baumheckel, R.E. 1976. Planting equipment and the importance of depth control. p. 190–196. *In* L.D. Hill (ed.) World soybean research conference proceedings. The Interstate Printers and Publishers, Danville, IL.

Beatty, K.D., I.L. Eldridge, and A.M. Simpson. 1982. Soybean response to different planting patterns and dates. Agron. J. 74:859–862.

Beaver, J.S., and R.R. Johnson. 1981a. Yield stability of determinate and indeterminate soybeans adapted to the northern United States. Crop Sci. 21:449–454.

----, and ----. 1981b. Response of determinate and indeterminate soybeans to varying cultural practices. Agron. J. 73:833–838.

Bernard, R.L. 1972. Two genes affecting stem termination in soybeans. Crop Sci. 12:235–239.

Beuerman, R.A. et al. 1982. Illinois agronomy handbook 1983–84. Univ. of Illinois Coop. Ext. Serv. Circ. 1208.

Bhangoo, M.S., and D.J. Albritoon. 1976. Nodulating and non-nodulating Lee soybean isolines response to applied nitrogen. Agron. J. 68:642–645.

Bichel, D.C., and E.J. Hengen. 1978. Development of soybean harvesting equipment in the USA. p. 200–208. *In* Grain and forage harvesting proceedings first international grain and forage harvesting conference. American Society of Agricultural Engineers, St. Joseph, MI.

Blomquist, R.V., C.A. Kust, and L.E. Schrader. 1973. Effect of ethrel on seasonal activity of three enzymes and lodging resistance in soybeans. Crop Sci. 13:4–7.

Boerma, H.R. 1979. Comparison of past and recently developed soybean cultivars in maturity groups VI, VII, and VIII. Crop Sci. 19:611–613.

----, and D.A. Ashley. 1982. Irrigation, row spacing, and genotypic effects on late and ultra-late planted soybeans. Agron. J. 74:995–999.

Boote, K.J. 1981. Response of soybeans in different maturity groups to March plantings in the southern USA. Agron. J. 73:854–859.

Boquet, D.J., K.L. Koonce, and D.M. Walker. 1982. Selected determinate soybean cultivar yield responses to row spacings and planting dates. Agron. J. 74:136–138.

Brill, W.J. 1981. Agricultural microbiology. Sci. Am. 245(3):199–215.

Boswell, F.C. 1980. Factors affecting the response of soybeans to molybdenum application. p. 417–432. *In* F.T. Corbin (ed.) World soybean research conference II: Proceedings. Westview Press, Boulder, CO.

Boyer, J.S., R.R. Johnson, and S.G. Saupe. 1980. Afternoon water deficits and grain yields in old and new soybean cultivars. Agron. J. 72:981–986.

Burris, J.S., and K.H. Knittle. 1975. Partial reversal of temperature-dependent inhibition of soybean hypocotyl elongation by cotyledon excision. Crop Sci. 15:461–462.

Byg, D.M., and R.J. Walker. 1974. Effect of paraquat on harvesting soybean seed. ASAE Paper 74-1560. American Society of Agricultural Engineers, St. Joseph, MI.

Carter, T.E., and H.R. Boerma. 1979. Implications of genotype × planting date and row spacing interactions in doublecropped soybean cultivar development. Crop Sci. 19:607–610.

Caviness, C.E., K.D. Beatty, H.J. Walters, R.D. Riggs, and R.P. Nester. 1983. 'Narow' a new soybean variety adapted for narrow rows. Arkansas Farm Res. 32(5):3.

----, and J.D. Thomas, 1979. Influence of planting date on three soybean varieties. Arkansas Farm Sci. 28(2):8.

----, and ----. 1980. Yield reduction from defoliation of irrigated and non-irrigated soybeans. Agron. J. 72:977–980.

Chan, L.M., R.R. Johnson, and C.M. Brown. 1980. Relay intercropping soybeans into winter wheat and spring oats. Agron. J. 72:35–39.

Chesney, H.A.D. 1973. Performance of soybeans in the wet tropics as affected by N, P, and K. Agron. J. 65:887–889.

Clapp, J.G. 1973. Response of Bragg soybean to TIBA (2,3,5-triiodobenzoic acid). Agron. J. 65:41–43.

----. 1975. Response of soybeans to morphactin. Crop Sci. 15:157–158.

Collins, F.C. 1982. Straw management in double crop systems. p. 7–8. *In* Proceedings soybean doublecrop conference. American Soybean Association, St. Louis.

----, and J.P. Jones. 1975. Doublecrop, a new early wheat. Arkansas Farm Res. 24(2):3.

Colvin, T.S., and D.C. Erbach. 1982. Soybean response to tillage and planting methods. Trans. ASAE 25:1533–1535.

----, ----, and J.M. Laflen. 1981. Effect of corn and soybean row spacing on plant canopy, erosion, and runoff. Trans. ASAE 24:1227–1229.

----, J.M. Laflen and D.C. Erbach. 1981. A review of residue reduction by individual tillage implements. *In* Crop production with conservation in the 80's. American Society of Agricultural Engineers, St. Joseph, MI.

Constable, G.A. 1977. Effect of planting date on soybeans in the Naomi Valley, New South Wales. Aust. J. Exp. Agric. Anim. Husb. 17:148–155.

Cooper, R.L. 1981. Development of short-statured soybean cultivars. Crop Sci. 21:127–131.

Costa, J.A., E.S. Oplinger, and J.W. Pendleton. 1980. Response of soybean cultivars to planting patterns. Agron. J. 72:153–156.

Crookston, R.K., and D.S. Hill. 1979. Grain yields and land equivalent ratios from intercropping corn and soybeans in Minnesota. Agron. J. 71:41–44.

Crop Reporting Board, USDA-SRS. 1981. Crop production. 10 November. U.S. Government printing Office, Washington, DC.

----. 1983a. Crop production. 10 June. U.S. Government Printing Office, Washington, DC.

----. 1983b. Crop production. 10 November. U.S. Government Printing Office, Washington, DC.

Deibert, E.J., M. Bijeriego, and R.A. Olson. 1979. Utilization of ^{15}N fertilizer by nodulating and non-nodulating soybean isolines. Agron. J. 71:717–723.

deMooy, C.J., J.L. Young, and J.D. Kaap. 1973. Comparative response of soybeans and corn to phosphorous and potassium. Agron. J. 65:851–855.

Doss, D.B., and D.L. Thurlow. 1974. Irrigation, row width, and plant population in relation to growth characteristics of two soybean varieties. Agron. J. 66:620–623.

Dunigan, E.P., O.B. Sober, J.L. Rabb, and D.J. Boquet. 1980. Effects of various inoculants on nitrogen fixation and yield of soybeans. La. State Univ. Bull. 726.

Dybing, C.D., and C. Lay. 1981. Yield and yield components of flax, soybean, wheat and oats treated with morphactins and other growth regulators for senescence delay. Crop Sci. 21:904–908.

----, and ----. 1982. Oil and protein in field crops treated with morphactens and other growth regulators for senescence delay. Crop Sci. 22:1054–1058.

Egli, D.B., and D.M. TeKrony. 1979. Relationship between soybean seed vigor and yield. Agron. J. 71:755–759.

Erbach, D.C. 1982. Tillage for continuous corn and corn-soybean rotation. Trans. ASAE 25:906–911.

Fehr, W.R., and S.R. de Cianzio. 1980. Relationship of component frequency to compensatory response in soybean blends. Crop Sci. 20:392–393.

----, B.K. Lawrence, and T.A. Thompson. 1981. Critical stages of development for defoliation of soybean. Crop Sci. 21:259–262.

----, and S.R. Rodriguez. 1974. Effect of row spacing and genotypic frequency on the yield of soybean blends. Crop Sci. 14:521–525.

Fontes, L.A.N., and A.J. Ohlrogge. 1972. Influence of seed size and population on yield and other characteristics of soybeans. Agron. J. 64:833–836.

Froehlich, D.M., and W.R. Fehr. 1981. Agronomic performance of soybeans with differing levels of iron deficiency chlorosis on calcarious soil. Crop Sci. 21:438–441.

Fuehring, H.D., and R.E. Finkner. 1978. Soybeans: variety, fertilizer, growth regulator, antitranspirant, and planting-after-wheat studies on the high plains of eastern New Mexico. N.M. State Univ. Agric. Exp. Stn. Rep. 372.

Garcia, R., and J.J. Hanway. 1976. Foliar fertilization of soybeans during the seed-filling period. Agron. J. 68:653–657.

Gebhardt, M.R., and H.C. Minor. 1983. Soybean production systems for claypan soils. Agron. J. 75:532–537.

Gbikpi, P.J., and R.K. Crookston. 1981. A whole-plant indicator of soybean physiology maturity. Crop Sci. 21:469–473.

Green, D.E., P.F. Burlamaqui, and R. Shibles. 1977. Performance of randomly selected soybean lines with semideterminate and indeterminate growth habits. Crop Sci. 17:335–339.

Griffen, J.L., R.M. Lawrence, R.J. Habetz, and D.K. Babcock. 1983. Response of soybeans to planting date in southwest Louisiana. La. Agric. Exp. Stn. Bull. 747.

Ham, G.E., W.W. Nelson, S.D. Evans, and R.D. Frazier. 1973. Influence of fertilizer placement on yield response of soybeans. Agron. J. 65:81–84.

Hanthorn, M., and M. Duffy. 1983. Corn and soybean pest management practices for alternative tillage strategies. p. 14–17. *In* Inputs outlook and situation report. 21 October. USDA-ERS ISO-2. U.S. Department of Agriculture, Washington, DC.

Hardman, L.L., and W.A. Brun. 1971. Effect of atmospheric carbon dioxide enrichment at different developmental stages on growth and yield components of soybeans. Crop Sci. 11:886–888.

Harper, J.E. 1974. Soil and symbiotic nitrogen requirements for optimum soybean production. Crop Sci. 14:255–260.

Hartung, R.C., J.E. Specht, and J.H. Williams. 1981. Modification of soybean plant architecture by genes for stem growth habit and maturity. Crop Sci. 21:51–56.

Hatfield, J.L., and D.B. Egli. 1974. Effect of temperature on the rate of soybean hypocotyl elongation and field emergence. Crop Sci. 14:423–426.

Heatherly, L.G. 1981. Soybean response to tillage of Sharkey clay soil. Miss. State Univ. Exp. Stn. Bull. 892.

Helsel, Z.R., T.J. Johnston, and L.P. Hart. 1981. Soybean production in Michigan. Mich. State Univ. Bull. E-1549.

Herbek, J.H. 1982. Fertilization decisions in doublecropping soybeans. p. 12–19. *In* Proceedings soybean doublecrop conference. American Soybean Association, St. Louis.

Hicks, D.R., and R.H. Peterson, 1981. Effect of corn variety and soybean rotation on corn yields. p. 89–94. *In* H.D. Loden and Dolores Wilkinson (ed.) Proc. 36th Corn and Sorghum Industry Res. Conf., Chicago IL. 9–11 December. American Seed Trade Association, Washington, DC.

Higgs, R.L., W.H. Paulson, J.W. Pendleton, A.F. Peterson, J.A. Jackobs, and W.D. Shrader. 1976. Crop rotations and nitrogen. Univ. of Wisconsin Bull. R2761.

Hiltbold, A.E., D.L. Thurlow, and H.D. Skipper. 1980. Evaluation of commercial soybean inoculants by various techniques. Agron. J. 72:675–681.

Hinkle, D.A. 1970. Effect of crop rotations on cotton and soybean yields. Arkansas Farm Res. 19(2):10.

Hoggard, A.L., J.G. Shannon, D.R. Johnson. 1978. Effect of plant population on yield and height characteristics in determinate soybeans. Agron. J. 70:1070–1072.

Hovermale, C.H., H.M. Camper, and M.W. Alexander. 1979. Effects of small grain stubble height and mulch on no-tillage soybean production. Agron. J. 71:644–647.

Hume, D.J., and A.K.H. Jackson. 1981. Frost tolerance in soybeans. Crop Sci. 21:689–692.

International Soybean Program. 1982. Improved varieties for developing countries. INSOY Newsl. 30 (August). Univ. of Illinois, Urbana.

Jeffers, D.L., H.J. Mederski, R.H. Miller, P.E. Smith, E.W. Stroube, and G.B. Triplett. 1970. Attacking soybean production problems. Ohio Rep. 55(1):3–10.

----, and G.B. Triplett. 1978. Management for relay intercropping wheat and soybeans. p. 63–70. *In* Proc. 8th Soybean Seed Res. Conf., Chicago, IL. 14–15 December. American Seed Trade Association, Washington, DC.

Johnson, R.R. 1982. The impact of changing soybean cultural practices on farm equipment. p. 43–50. *In* Proc. 12th Soybean Seed Res. Conf., Chicago, IL. 7–8 December. American Seed Trade Association, Washington, DC.

----, and I.C. Anderson. 1974. Interaction of 2,3,5-triiodobenzoic acid and 2,4-dichlorophenoxyacetic acid on growth and yield of soybeans. Crop Sci. 14:381–384.

----, and L.V. Boone. 1976. Soybean inoculation: is it necessary? Ill. Res. 18(4):3–4.

----, and C.M. Brown. 1979. Systems for relay intercropping soybeans into winter wheat and spring oats. Agron. Abstr. American Society of Agronomy, Madison, WI, p. 102.

----, D.E. Green, and C.W. Jordan. 1982. What is the best soybean row width? A U.S. perspective. Crops Soils 34(4):10–13.

----, and L.M. Wax. 1978. Relationship of soybean germination and vigor tests to field performance. Agron. J. 70:273–278.

Jolly, R.W., W.M. Edwards, and D.C. Erbach. 1983. Economics of conservation tillage in Iowa. J. Soil Water Conserv. 38:291–294.

Judd, R., D.M. TeKrony, D.B. Egli, and G.M. White. 1982. Effect of freezing temperatures during soybean seed maturation on seed quality. Agron. J. 74:645–650.

Keogh, J.L., R. Maples, and K.D. Beatty. 1979. Foliar feeding of soybean. Arkansas Farm Res. 28(5):9.

Kulik, M.M., and R.W. Yaklich. 1982. Evaluation of vigor tests in soybean seeds: relationship of accelerated aging, cold, sand bench, and speed of germination tests to field performance. Crop Sci. 22:766–770.

Ladlie, J.S., W.F. Meggitt, and D. Penner. 1976. Effect of pH on metribuzin activity in the soil. Weed Sci. 24:505–507.

Laflen, J.M., and W.C. Moldenhauer. 1979. Soil and water losses from corn-soybean rotations. Soil Sci. Soc. Am. J. 43:1213–1215.

----, ----, and T.S. Colvin. 1981. Conservation tillage and soil erosion on continuously row-cropped land. *In* Crop production with conservation in the 80's. American Society of Agricultural Engineers, St. Joseph, MI.

Lawn, R.J., D.E. Byth, and V.E. Mungomery. 1977. Response of soybeans to planting date in south-eastern Queensland. III. Agronomic and physiological response of cultivars to planting arrangements. Aust. J. Agric. Res. 28:63–79.

Lockeretz, W. 1983. Energy implications of conservation tillage. J. Soil Water Conser. 38:207–211.

Luedders, V.D. 1977. Genetic improvement in yield of soybeans. Crop Sci. 17:971–972.

Lueschen, W.E., and D.R. Hicks. 1977. Influence of plant population on field performance of three soybean cultivars. Agron. J. 69:390–393.

MacFarlane, J.J. 1980. Soybean seed treatment—U.S. status report. p. 88–99. *In* Proc. 10th Soybean Seed Res. Conf., Chicago, IL. 11–12 December. American Seed Trade Association, Washington, DC.

Mahler, R.L., and A.G. Wollum, II. 1981. The influence of irrigation and *Rhizobium japonicum* strains on yields of soybeans grown in a Lakeland sand. Agron. J. 73:647–651.

Major, D.J., D.R. Johnson, and V.D. Luedders. 1975a. Evaluation of thermal unit methods for predicting soybean development. Crop Sci. 15:172–174.

----, ----, J.W. Tanner, and I.C. Anderson. 1975b. Effects of daylength and temperature on soybean development. Crop Sci. 15:174–179.

Mannering, J.V., and C.B. Johnson. 1969. Effect of crop row spacing on erosion and infiltration. Agron. J. 61:902–905.

Martin, C.K., D.K. Cassel, and E.J. Kamprath. 1979. Irrigation and tillage effects on soybean yield in a coastal plain soil. Agron. J. 71:592–594.

McBroom, R.L., H.H. Hadley, C.M. Brown, and R.R. Johnson. 1981. Evaluation of soybean cultivars in monoculture and relay intercropping systems. Crop Sci. 21:673–676.

Mulvaney, D.L. 1984. Conservation tillage in northern Illinois. p. 57–59. *In* Thirty-sixth Illinois custom spray operator training school manual. Coop. Ext. Serv., Univ. of Illinois, Urbana.

Musen, H.L. 1977. Soybean rooting patterns. p.44–49. *In* Proc. 7th Soybean Seed Res. Conf., Chicago, IL. 8–9 December. American Seed Trade Association, Washington, DC.

Nave, W.R., R.L. Cooper, and L.M. Wax. 1977. Tillage-planter interaction in narrow-row soybeans. Trans. ASAE 20:9–12.

----, and M.R. Paulsen. 1979. Soybean seed quality as affected by planter meters. Trans. ASAE 22:739–745.

----, D.E. Tate, J.L. Butler, and R.R. Yoerger. 1973. Soybean harvesting. USDA-ARS ARS-NC-7. U.S. Department of Agriculture, Washington, DC.

----, L.M. Wax, and J.W. Hummel. 1980. Tillage for corn and soybeans. ASAE Paper 80-1013. American Society of Agricultural Engineers, St. Joseph, MI.

Nelson, A.N., and R.W. Weaver. 1980. Seasonal nitrogen accumulation and fixation by soybeans grown at different densities. Agron. J. 72:613–616.

Nelson, D.W., M.L. Swearingin, and L.S. Beckman. 1978. Response of soybeans to commercial soil-applied inoculants. Agron. J. 70:517–518.

Newbery, R.S., M.R. Paulsen, and W.R. Nave. 1980. Soybean quality with rotary and conventional threshing. Trans. ASAE 23:303–308.

Niebur, W.S., and W.R. Fehr. 1981. Agronomic evaluation of soybean genotypes resistant to iron deficiency chlorosis. Crop Sci. 21:551–554.

No-Till Farmer. 1983. 1982–1983 tillage survey. February. No-Till Farmer, Brookfield, WI.

Office of Technology Assessment. 1982. Croplands. p. 91–133. *In* Impacts of technology on U.S. cropland and rangeland productibility. U.S. Government Printing Office, Washington, DC.

Oplinger, E.S., I.C. Anderson, and R.R. Johnson. 1978. Effect of seed and foliar applications of Ergostim on soybeans and corn. Plant Growth Regul. Proc. 2:124–128.

----, and A.J. Ohlrogge. 1974. Response of corn and soybeans to field applications of copper. Agron. J. 66:568–571.

Pal, U.R., and M.C. Saxena. 1976. Relationship between nitrogen analysis of soybean tissues and soybean yields. Agron. J. 68:927–932.

Parker, M.B., W.H. Marchant, and B.J. Mullinix. 1981. Date of planting and row spacing effects on four soybean cultivars. Agron. J. 73:759–762.

Pataky, J.K., and S.M. Lim. 1981. Effects of row width and plant growth habit on Septoria brown spot development and soybean yield. Phytopothology 71:1051–1056.

Pendelton, J.W., and E.E. Hartwig. 1973. Management. *In* B.E. Caldwell (ed.) Soybeans: Improvement, production, and uses. Agronomy 16:211–237.

Pepper, G.E. et al. 1982. Illinois growers guide to superior soybean production. Univ. of Illinois Coop. Ext. Serv. Circ. 1200.

Poole, W.D., G.W. Randall, and G.E. Ham. 1983. Foliar fertilization of soybeans. I. effect of fertilizer sources, rates, and frequency of application. Agron. J. 75:195–200.

Porter, O.A., M.S. Bhangoo, D.J. Albritton, and J.G. Burleigh. 1981. Influence of nitrogen fertilizer on nodulating and non-nodulating soybeans. Arkansas Farm Res. 30(1):8.

Price, M., R.D. Riggs, and C.E. Caviness. 1976. Races of soybean-cyst nematodes compete. Arkansas Farm Res. 25(2):16.

Randall, G.W. 1977. Iron chlorosis in soybeans. Univ. of Minn. Soils Ser. 95:124–127.

----, L.E. Schulte, and R.B. Corey. 1975. Effect of soil and foliar-applied manganese on the micronutrient content and yield of soybeans. Agron. J. 67:502–507.

Reicosky, D.C., H.R. Rowse, W.K. Mason, and H.M. Taylor. 1982. Effect of irrigation and row spacing on soybean water use. Agron. J. 74:958–964.

Robertson, W.K., L.G. Thompson, and F.G. Martin. 1973. Manganese and copper requirements for soybeans. Agron. J. 65:641–644.

Rogers, H.T., D.L. Thurlow, F. Adams, C.E. Evans, and J.I. Wear. 1971. Fertility requirements. p. 39–47. *In* Soybean production in Alabama. Auburn Univ. Agric. Exp. Stn. Bull. 413.

Rushing, K.W. 1982. The future potential of seed treatments and coatings. p. 62–66. *In* Proc. 12th Soybean Seed Res. Conf., Chicago, IL. 7–8 December. American Seed Trade Association, Washington, DC.

Ryder, G.J., and J.E. Beuerlein. 1979. A study of soybean production systems. Ohio Rep. 64(2):19–22.

Saliba, M.R., L.E. Schrader, S.S. Hirano, and C.D. Upper. 1982. Effects of freezing field-grown soybean plants at various stages of podfill on yield and seed quality. Crop Sci. 22:73–78.

Sanford, J.O., 1982. Straw and tillage management practices in soybean-wheat doublecropping. Agron. J. 74:1032–1035.

Schmitthenner, A.F., and M.E. Kroetz. 1982. Fungicide cleared for disease control in Ohio soybeans. Ohio Rep. 67 (2):21–24.

Schou, J.B., D.L. Jeffers, and J.G. Streeter. 1978. Effects of reflectors, black boards, or shades applied at different stages of plant development on yield of soybeans. Crop Sci. 18:29–34.

Schutz, W.M., and R.L. Bernard. 1967. Genotype × environment interactions in regional testing of soybean strains. Crop Sci. 7:125–130.

Scott, W.O., and S.R. Aldrich. 1983. Modern soybean production. S and A Publications, Champaign, IL.

Shelton, C.H., F.D. Tompkins, and D.D. Tyler. 1983. Soil erosion from five soybean tillage systems. J. Soil Water Conserv. 38:425–428.

Siemens, J.C., and W.R. Oschwald. 1978. Corn-soybean tillage systems, erosion control, effects on crop production, costs. Trans. ASAE 21:293–302.

Sionit, N., and P.J. Krammer. 1977. Effect of water stress during different stages of growth of soybean. Agron. J. 69:274–278.

Skipper, H.D., J.H. Palmer, J.E. Giddens, and J.M. Woodruff. 1980. Evaluation of commercial soybean inoculants from South Carolina and Georgia. Agron. J. 72:673–674.

Slife, F.W. 1976. Economics of herbicide use and cultivar tolerance to herbicides. p. 77–82. *In* Proc. 31st Corn and Sorghum Res. Conf. Chicago, IL. 7–9 December. American Seed Trade Association, Washington, DC.

Smith, E.S., D.H. Vanghan, and D.E. Brann. 1978. Under-row ripping with various tillage practices for corn and soybeans. ASAE Paper 78-1512. American Society of Agricultural Engineers, St. Joseph, MI.

Smith, R.S., M.A. Ellis, and R.E. Smith. 1981. Effect of *Rhizobium japonicum* inoculant on soybean nodulation in a tropical soil. Agron. J. 73:505–508.

----, W.H. Judy, and W.C. Stearn. 1983. International inoculant shipping evaluation. INSOY Series 23. International Soybean Program. University of Illinois, Urbana.

Smith, T.J., and H.M. Camper. 1975. Effects of seed size on soybean performance. Agron. J. 67:681–684.

Stivers, R.K., and M.L. Swearingin. 1980. Soybean yield compensation with different populations and missing plant patterns. Agron. J. 72:98–102.

Stutte, C.A., J.T. Cothren, and S.D. Bryant. 1975. Evaluation of soybean growth regulators 1974. Arkansas Farm Res. 24(4):13.

----, and R.D. Rudolph. 1971. Growth regulators increase soybean yields. Arkansas Farm Res. 20(2):16.

Tanner, J.W., and S. Ahmed. 1974. Growth analysis of soybeans treated with TIBA. Crop Sci. 14:371–374.

Tao, K.J. 1978. Effects of soil water holding capacity on the cold test for soybeans. Crop Sci. 18:979–982.

Taylor, H.M. 1980. Soybean growth and yield as affected by row spacing and by seasonal water supply. Agron. J. 72:543–547.

TeKrony, D.M. 1982. The present status of seed vigor testing. p. 96–101. *In* Proc. 12th Soybean Seed Res. Conf., Chicago, IL. 7–8 December, American Seed Trade Association, Washington, DC.

----, and D.B. Egli. 1977. Relationship between laboratory indices of seed vigor and field emergence. Crop Sci. 17:573–577.

----, ----, and G. Henson. 1981. A visual indicator of physiological maturity in soybean plants. Agron. J. 73:553–556.

Thompson, L., and P. Pongsakul. 1976. Soil inoculation for soybeans investigated. Arkansas Farm Res. 25(2):15.

----, W.A. Skroch, and E.O. Beasley. 1981. Pesticide incorporation: distribution of dye by tillage implements. N.C. Agric. Ext. Serv. Publ. AG-250.

Throckmorton, R.I. 1980. Equipment for narrow-row and solid plant soybeans. p. 485–491. *In* F.T. Corbin (ed.) World soybean research conference II: Proceedings. Westview Press, Boulder, CO.

Thurlow, D.L., and J.H. Pitts. 1983. Planting date and row spacing affects growth and yield of soybeans. Highlights Agric. Res. 30(3):12.

Touchton, J.T., and F.C. Boswell. 1975. Effects of B application on soybean yield, chemical composition, and related characteristics. Agron. J. 67:417–420.

----, and J.W. Johnson. 1982. Soybean tillage and planting method effects of double-cropped wheat and soybeans. Agron. J. 74:57–59.

Triplett, G.B., and D.M. Van Doren. 1977. Agriculture without tillage. Sci. Am. 236(1):28–33.

Vasilas, B.L., J.D. Legg, and D.C. Wolf. 1980. Foliar fertilization of soybeans: absorption and translation of ^{15}N labeled urea. Agron. J. 72:271–275.

Walker, A.K., and R.L. Cooper. 1982. Adaptation of soybean cultivars to low-yield environments. Crop Sci. 22:678–680.

----, and W.R. Fehr. 1978. Yield stability of soybean mixtures and multiple pure stands. Crop Sci. 18:719–723.

Walker, P.N., M.D. Thorne, E.C. Benham, and S.K. Sipp. 1982. Yield response of corn and soybeans to irrigation and drainage on claypan soil. Trans. ASAE 25:1617–1621.

Wall, M.T., D.C. McGee, and J.S. Burris. 1983. Emergence and yield of fungicide-treated soybean seed differing in quality. Agron. J. 75:969–973.

Wax, L.M., W.R. Nave, and R.L. Cooper. 1977. Weed control in narrow and wide-row soybeans. Weed Sci. 25:73–78.

Welch, L.F., L.V. Boone, C.G. Chambliss, A.T. Christiansen, D.L. Mulvaney, M.G. Oldham, and J.W. Pendleton. 1973. Soybean yields with direct and residual nitrogen fertilization. Agron. J. 65:547–550.

----, C.M. Brown, and R.R. Johnson. 1979. Foliar fertilization of wheat, oats and soybeans. Ill. Res. 21(3):5–6.

Whigham, D.K., and E.W. Stoller. 1979. Soybean desiccation by paraquat, glyphosate, and ametryn to accelerate harvest. Agron. J. 71:630–633.

Wilcox, J.R. 1974. Response of three soybean strains to equidistant spacings. Agron. J. 66:409–412.

----, 1980. Comparative performance of semideterminate and indeterminate soybean lines. Crop Sci. 20:277–280.

----, W.T. Schapaugh, R.L. Bernard, R.L. Cooper, W.R. Fehr, and M.H. Niehaus. 1979. Genetic improvement of soybeans in the midwest. Crop Sci. 19:803–805.

Wilson, D.O., F.C. Boswell, K. Ohki, M.B. Parker, and L.M. Shuman. 1981. Soil distribution and soybean plant accumulation of manganese in manganese-deficient and manganese-fertilized field plots. Soil Sci. Am. J. 45:549–552.

Wolf, D.C., and R.P. Nester. 1980. Soybean inoculation practices in Arkansas. Arkansas Farm Res. 29(5):9.

Woodruff, J.M. 1980. Double cropping soybeans behind soybeans: potential and problems in the southeast. p. 61–69. *In* Proc. 10th Soybean Seed Res. Conf., Chicago, IL. 11–12 December. American Seed Trade Association, Washington, DC.

Woods, S.J., and M.L. Swearingin. 1977. Influence of simulated early lodging upon soybean seed yield and its components. Agron. J. 69:239–242.

10 Tillage and Irrigation

D. M. Van Doren, Jr.
Ohio State University and the
Ohio Agricultural Research
and Development Center
Wooster, Ohio

D. C. Reicosky
USDA-ARS-NCR
North Central Soil Conservation
Research Laboratory
Morris, Minnesota

Tillage and irrigation have been treated in this chapter as separate topics, even though both deal with management of soil water and effects of water on soybean growth. Literature cited to support various discussions will be illustrative rather than exhaustive, with the majority published after 1970. In the process of condensing the many disparate studies into a manageable size, some tangetial information has been omitted, with a regrettable loss of overall content.

10-1 TILLAGE

In order to illustrate the effects of tillage over a wide range of practices, five tillage treatments will be specified. There are infinite variations of these five as well as other general systems not listed herein, but these systems produce a wide range of soil environments for soybean growth for comparative purposes. Wherever possible all five will be documented during subsequent discussions, but since no single study encompasses all five systems, there is some incompleteness in the discussion.

1. Moldboard plow overall to 20-cm depth (fall or spring) followed by 10-cm overall secondary tillage with disc, harrow, etc. preplant and possibly postemergence cultivation.
2. Chisel plow overall to 20-cm depth (fall or spring) possibly followed by the same secondary tillage as with moldboard plowing.

Soybeans: Improvement, Production, and Uses, 2nd ed.—Agronomy Monograph no. 16.

3. Disk overall preplant to 5- to 15-cm depth (fall or spring) and possibly postemergence cultivation.
4. Strip tillage with a chisel or subsoiler to 30- to 50-cm depth in or between the rows to be planted. May be used with any other system.
5. No tillage between harvest of one crop and planting of the succeeding crop.

10–1.1. Effect on Soil Properties

Basically, tillage alters the position of solid material on or in the soil. Depending on type of tillage tool and depth of action, such repositioning alters soil strength, aeration, water characteristics, and thermal properties, as well as changes the position of crop residues, fertilizers, and pesticides. Whether or not such repositioning is necessary depends upon pretillage soil conditions, crop needs, and weather.

Specific examples of how tillage alterations of soil properties might influence soybean [*Glycine max* (L.) Merr.] growth, economics of production, or soil erosion will be given during discussion of the specific topic. Readers desiring further information on the effects of tillage on soil properties are referred to Unger and Van Doren (1982) as a starting point.

10–1.2 Effect on Soybean Root Growth

The root system of any crop provides support for the shoot, absorbs nutrients and water for use by the entire plant, produces enzymes and hormones and otherwise controls some physiological processes of the shoot, and in the case of legumes, aids in the production of plant-available N. Growth of the shoot thus depends in part upon past root growth and current root health.

Soil can be too dense and strong (Mazurak and Pohlman, 1968) and have pores too small for soybean roots to penetrate at all (Aubertin and Kardos, 1965). Under less severe density restraints root growth may still be less than the maximum rate (Pierce et al., 1983). To the extent that traffic and/or tillage create such restrictive soil conditions, soybean root growth will be restricted, thereby potentially limiting the above-mentioned supportive functions of the root system. The closer to the soil surface such restrictions occur, the greater the effect on crop growth (Pierce et al., 1983). Nodulation and nodule activity apparently have not been as severely affected by compact soil as has root extension (Lindemann et al., 1982; Voorhees et al., 1976). To the extent that tillage loosens soil and creates lower strength and larger pores, root growth can be more extensive and thereby potentially more supportive of overall plant growth (Campbell et al., 1974; Kamprath et al., 1979; Trouse, 1979).

10–1.3 Effect on Soybean Seed Yields

This discussion will focus on those reported data representing conditions of more or less equal plant population and equal weed control, which were considered to be true if no comments were made by the authors to the contrary. Great advances in ability to plant soybean into residue-covered or rough seedbeds and in providing season-long weed control without tillage have been made during the last 20 yrs, and further improvements are expected. It was, therefore, decided not to confound the effects of tillage on soil properties and soybean growth with situations of differential plant population or weed control.

This approach is not meant to deny the necessity of establishing satisfactory stands and providing suitable weed control, nor to deny that such accomplishments can be influenced by tillage. For example, according to Goyal et al. (1981), the time for soybean seeds to reach maximum emergence force is greatly influenced by soil water content. A 1% increase in water content near the permanent wilting percentage decreased time by 7%, over twice as great an effect as caused by the same change in water content near field capacity. Soil temperature had little influence on this time. Conversely, a 1 °C increase in soil temperature near 12 °C increased the maximum emergence force by 26%, over six times as much as caused by the same temperature change near 20 °C. Soil water content had little influence on the magnitude of the maximum emergence force. Therefore, as tillage influences soil properties of temperature and water content, emergence capabilities of soybean seedlings can be considerably altered. Tillage also influences the strength of soil and ability to withstand changes in strength (e.g., resist crusting). One of the more difficult tillage-related research problems is to develop a comprehensive relationship between tillage, soil properties, climate, seed characteristics, planting equipment, and emergence of any crop, including soybean.

Whether competing vegetation is present by design, as with strip killing or induced short-term dormancy of sod (Elkins et al., 1982, 1983) or by weed presence due to incomplete control by tillage or herbicides (Burnside, 1978; Gebhardt and Minor, 1983), dry matter of such competing vegetation is produced at the expense of soybean dry matter production. Only Elkins et al. (1983) showed the best soybean grain yield at less than nearly complete suppression of competing vegetation.

Other variables interacting with tillage to influence soybean yields include soil properties of drainage, surface texture, and subsoil strength, plus cropping systems, residue management, duration of use of a tillage practice, and precipitation. Yield data from the surveyed literature are summarized in Table 10–1. A least squares analysis (Harvey, 1960) was performed on data from each drainage-texture-cropping combination using each crop year as a "replicate" and treatment means as data entries. If the sum of standard errors of least squares means of two treatments was less than the difference between the least squares means of those treatments, the two treatments were considered to be different, which is so noted in Table 10–1. The yield response pattern seems to be most closely associated with soil drainage, with soil texture and cropping variables sometimes also significant. The following discussion is arranged in order from poor to good soil drainage. All comparative values given

Table 10–1. Effects of soil properties, cropping practices, and tillage on soybean seed production.

Soil properties: Drainage†	Soil properties: Texture‡	Cropping§	MP (15–25 cm)¶ Fall	MP (15–25 cm)¶ Spring	Chisel (15–25 cm) Fall	Chisel (15–25 cm) Spring	Chisel (30–50 cm) Fall	Chisel (30–50 cm) Spring	D or FC¶ (8–12 cm)	NT¶ <6 yrs	NT¶ ≥6 yrs
Poor	Fine	Rotation	2.98(51)#	(=)−0.03(9)	(=)−0.07(20)	(<)−0.12(10)		(=)−0.03(2)	(<)−0.14(14)	(<)−0.23(29)	(<)−0.42(19)
		Continuous	3.44(14)	(=)−0.02(5)	(<)−0.11(11)		(=) 0(2)		(<)−0.17(6)	(<)−0.38(10)	(<)−0.29(4)
	Silt loam	All††	(=)−0.12(2)	2.87(5)	(=)+0.01(2)	(=)−0.19(2)	(=)+0.02(2)	(=)−0.10(2)	(=)−0.17(7)	(=)+0.03(5)	
	Fine	Double-crop		1.38(3)					(≥)+0.36(2)	(=)+0.21(4)	
Imperfect	All††	All††	2.81(11)		(=) 0(8)	(>)+0.20(2)		(≥)+0.16(2)	(>)+0.10(12)	(=)+0.06(2)	
	Fine	Double-crop		1.50(5)						(=)+0.08(5)	
Well-1	Silt loam	All††	(>)+0.32(2)	2.19(32)	(>)+0.26(3)	(=)+0.08(10)		(=)+0.07(5)	(=)+0.02(17)	(=)+0.04(24)	(>)+0.24(13)
Well-2	Coarse	All††		2.23(17)		(>)+0.32(6)		(>)+0.38(18)	(=)+0.02(3)		

†Data from poorly drained soils are from the Aquoll, Aqualf, and Aquept suborders; Data from imperfectly drained soils are from the Aquic or Aquentic Hapludult, Chromudert, Hapludoll, Argiudoll, and Fragiudalf subgroups; Well-1 drained are soils from the Fragiudalf, Argiudoll, Chromudert, Hapludult, and Paleudalf great groups; Well-2 drained soils are from the Paleudult great group.

‡Fine textures are silty clay loams, clay loams, and clays; coarse textures are sands, loamy sands, sandy loams, and loams.

§Sole crop full-season soybean grown in rotation with any other crop (rotation) or in monoculture (continuous). Double-crop soybean follows small grains.

¶MP (15–25 cm) is moldboard plowing to 15 to 25 cm depth; D = disk; FC = field cultivate; NT = no tillage for < 6 consecutive yrs on the same plot or 6 yrs or greater in the same plot.

#Underlined numbers are the least squares mean soybean yield of the base treatment for a given line in Mg ha^{-1}. Other numbers in the same line are the yield of that treatment minus yield of the base treatment in Mg ha^{-1}. Symbols in parenthesis preceding a number indicate if the treatment difference equals the base treatment (=), is less than (<) or greater than (>) the base treatment by the sums of the standard errors of the two least squares means in question. Numbers in parenthesis following a yield value or yield difference are the number of experiment years included in the least squares mean.

††The designation "All" indicates insufficient data for separating texture or full-season cropping into discrete categories. Some or all of the other categories may be included.

in the text are derived from Table 10–1 unless accompanied by a specific citation.

10–1.3.1 Poorly Drained Soils

The poorly drained soils surveyed were Aquolls (Bone et al., 1977, 1978; Crabtree and Rupp, 1980; Griffith, 1982; Jeffers et al., 1973; Nelson, 1973; Randall, 1974, 1978; Richey et al., 1977; Van Doren, unpublished data), Aqualfs (Bone et al., 1977; Jeffers et al., 1973; Kapusta, 1979; Tupper, 1978; Tyler and McCutchen, 1980; Van Doren et al., 1977; Van Doren, unpublished data), and Aquepts (Heatherly, 1981). These soils have relatively level topography. Many have high clay contents, and all have a slowly permeable layer in or below the rooting zone, all of which cause excessive wetness, especially during times when precipitation exceeds evapotranspiration. Tile or other drainage aids are necessary for good soybean production.

10–1.3.1.1 Poorly Drained, Fine-textured Soils—Use of moldboard plowing at 15 to 25 cm depths on fine-textured, poorly drained soils for soybean in rotation with other crops creates superior soil conditions for support of full-season soybean growth. No other primary tillage treatment produced higher yields, including deeper (30–50 cm) tillage at random or under the row. Reducing tillage depth to 12 cm or less or spring chisel plowing <25-cm deep reduced soybean yield, with the greatest reduction being associated with no-tillage (8%), especially with prolonged use of no-tillage (17.5%). Yield reductions for nonmoldboard plowing for soybean grown in monoculture tended to be greater than for soybean grown in rotation, especially for continuous use of no-tillage (5.7%).

Perhaps the greater yields associated with moldboard or similar tillage were due to lower bulk density and subsequent lower soil strength, or with greater amounts of freely drainable pores at near zero water potential (Klute, 1982). No such data were presented along with the soybean yield data. One study supported this reasoning indirectly in that yields for fall moldboard plowing, with and without tile, and no-tillage, with and without tile averaged 2.62, 2.49, 2.49, and 1.95 mg ha^{-1}, respectively, with an $LSD_{0.05}$ of 0.52 Mg ha^{-1} (Bone et al., 1977). No-tillage reduced yield, especially without tile.

Increased incidence of diseases (root rot organisms) which flourish under wet conditions could account for reduced yields, especially for continuous soybean, under the more compact, less aerated no-tillage environment. Again there is one, indirect, bit of evidence to support this possibility. In a study of 21-yrs duration, Van Doren (unpublished data) split plots which had been continuously fall moldboard plowed (MP) and others continuously not tilled (NT) in a 2-yr corn (*Zea mays* L.)-soybean rotation with root rot tolerant (NS) and susceptible (S) soybean cultivars. Yield averages for the 16th, 18th, and 21st yrs were 2.97 (MP-NS), 2.55 (NT-NS), 2.65 (MP-S), and 1.92 (NT-S) Mg ha^{-1} with an $LSD_{0.05}$ of 0.36

Mg ha^{-1}. The yield differential between tillage treatments was especially pronounced with the root rot susceptible cultivars.

There appears to be little effect of crop residues on soybean yield on these soils, at least for the latitudes where the studies were conducted (Sanford, 1982 from Mississippi; Van Doren, unpublished data from Ohio). The expected advantage of residues for improved water conservation might not pertain in level topography and for soils which tend to crack when dry, at least under moderate to high rainfall conditions. Adequate infiltration will probably occur with or without residues under such circumstances. In the only study surveyed in which soil water contents were measured (Heatherly, 1981), little differences were found in stored soil water during the growing season as a function of tillage depth.

There has been little evidence that preplanting secondary tillage is of any benefit to soybean yields if a reasonable stand can be established without such tillage (Bone et al., 1978). However, if spring primary tillage leaves the soil with large clods, or if the soil dries too much before planting or before planned secondary tillage, structural stability of these soils may be too great to achieve the desired seed-soil contact with available planting equipment without prior secondary tillage. Transmission of water via loosely packed aggregates (clods) to seeds is a slow process that often results in reduced germination and reduced stand. Soils receiving fall tillage or no-tillage benefit from natural weathering processes over winter and in the early spring, which tend to produce more finely granulated soil conditions, which in turn probably need less secondary tillage to produce the necessary seed-soil contact to obtain the desired stand.

All of the studies reviewed had a common timing for similar tillage for all treatments. For example, where deep spring tillage was performed, it was performed at the same time for all treatments receiving deep spring tillage. Similarly, all treatments in a given experiment were planted on the same day. This approach almost assured that those treatments which required a greater time for drying during wet springs dictated the planting date for al!. If the delay caused planting beyond the optimum date, those treatments which could have been planted earlier did not receive due credit in terms of their potential for soybean production. Spring tillage of mulch-covered land deeper than 10 to 15 cm tends to be the combination that causes such delays. The practical solution to such potential delay is to substitute shallow or no tillage for planned deeper spring tillage should undesirable delays occur.

In contrast to the effect of tillage for full-season soybean production, tillage from 15- to 25-cm deep caused reduced soybean yields when performed in June or July following a small grain. It seems that by June or July these soils were no longer wet enough to be as detrimental to soybean growth on reduced- or nontilled land as for full-season soybean planted in April or May. Large annual variability occurred among and within the three studies cited (Crabtree and Rupp, 1980; Jeffers et al., 1973; Sanford, 1982). For example, Crabtree and Rupp (1980) reported that in 1 yr moldboard plowing produced 0.34 Mg ha^{-1} greater soybean yield than did no-tillage, while in the 2nd yr no-tillage had 0.54 Mg ha^{-1} greater yield. The common feature in their study was the greater the soil water content shortly after planting, the greater the soybean yield. Why no-tillage had greater soil water content 1 yr and less the other was not explained. One would expect reduced tillage with the aid of surface residues to retain a greater amount of the soil water present at planting and that which was derived from subsequent precipitation than the rougher, bare surface associated with moldboard plowing.

No soybean data were found for the semiarid USA without irrigation. Because of the shortage of available soil water most years, perhaps full-season soybean yields might be affected by tillage on the more poorly drained soils there as have been double-cropped soybean in the semihumid areas of the USA. This should be considered as soybean production moves into drier climates.

10–1.3.1.2 Poorly Drained, Medium-textured Soils—The three studies included here were all from silt loam Aqualfs (Tyler and McCutchen, 1980; Van Doren et al., 1977; Van Doren, unpublished data). All tillage treatments appeared to have the same yields as fall moldboard plowing. There were too few studies available to develop any trends associated with timing, depth, and intensity of tillage, or interaction with crop sequence or use of residues. Tyler and McCutchen (1980) found no difference in end of season soil water contents as a function of tillage depth in an experiment with a bare soil surface for all treatments.

These soils are expected to have considerably lower structural stability than the finer-textured, poorly drained soils discussed previously. Perhaps natural weathering associated with freezing and precipitation, plus vehicular traffic effects created similar enough soil conditions for all tillage treatments that soybean growth was affected more or less the same.

10–1.3.2 Imperfectly Drained Soils

The imperfectly drained soils surveyed were Aquic and Aquentic Hapludults (Hovermale and Camper, 1979), Hapludolls (Colvin and Erbach, 1982; Erbach, 1982; Sanford et al., 1973), Chromuderts (Sanford, 1982), Argiudolls (Siemens and Oschwald, 1976), and Fragiudalfs (Tyler and McCutchen, 1980). In a few cases, these soils were the larger part of a mixture of better and/or more poorly drained soils included in the study. These soils may have a little more slope and have somewhat less duration and intensity of wetness in the spring than the poorly drained soils. Tile or other drainage aids are recommended for good crop production.

Full-season soybean yielded the same for all depths and timing of tillage, with the exception of possibly greater yields with spring chisel plowing. This equality extended to soybean planted on ridges prepared the previous season with little other tillage (Erbach, 1982). Soil water content as measured after harvest in one study was not influenced by tillage depth (Tyler and McCutchen, 1980). Residue cover did not significantly affect 3-yr average soybean yields in a double-cropping no-tillage situation (Hovermale and Camper, 1979).

These soils are intermediate in drainage and topography between the poorly and well-drained soils. Soybean perhaps suffers less from wetness than on poorly drained soils having shallow or no tillage, and perhaps benefits less from water conservation or enhanced rooting depth than will be discussed for well-drained soils. The net result seems to be a consistently uniform yield potential for a given site-year situation regardless of tillage on imperfectly drained soils.

10-1.3.3 Well-drained Soils (except Paleudults)

Well-drained soils will be discussed in two parts: (i) all except Paleudults, and (ii) Paleudults. Non-Paleudult soils include Fragiudalfs (McKibben, 1980; Tyler and McCutchen, 1980; Tyler and Overton, 1982; Van Doren, unpublished data); Argiudolls (Siemens and Oschwald, 1976); Chromuderts (Young et al., 1979); Hapludults (Camper and Lutz, 1977; Hovermale and Camper, 1979; Touchton and Johnson, 1982; Touchton et al., 1982); and Paleudalfs (Tyler et al., 1983), the majority having silt loam surface textures. These soils are well enough drained because of sloping topography and/or lack of naturally occurring slowly permeable subsoil horizons that drainage aids such as tile are not necessary for maximum productivity.

Fall tillage which leaves a bare surface is not recommended for well-drained soils where slope steepness and length present a soil erosion hazard. Consequently, moldboard plowing of these soils was done in the spring for all except Siemens and Oschwald (1978). The highest soybean yields on these soils (Paleudults excluded) are associated with long-term, continuous no-tillage (2.43 Mg ha^{-1}). Yields for all other spring tillage treatments surveyed were significantly lower than for long-term no-tillage and similar to spring moldboard plowing (2.19 Mg ha^{-1}).

The overall equality of soybean yields as a function of spring tillage (if any) masks substantial year to year variability. Tyler and Overton (1982) showed 0.31 Mg ha^{-1} higher yield for no-tillage than all tilled treatments in 1 yr and no difference in another. Van Doren (unpublished data) found in 4 of 6 yrs that disked treatments with residue cover of 5 and 60% and no-till treatments with residue cover of 9 and 72% yielded 2.53, 2.73, 2.14, and 2.67 Mg ha^{-1}, respectively, with an $LSD_{0.05}$ of 0.20 Mg ha^{-1}. Lack of residue cover significantly reduced yields an average of 0.36 Mg ha^{-1}, with the greatest effect on nontilled land. There was no effect of residue cover for the other 2 yrs. The general explanation is that residues help conserve water for use by the crop, and in those years with substantial soil water deficit such water conservation can be translated into increased yields. Soil water content data have been obtained often on these soils but only Tyler and Overton (1982) measured soil water content during the soybean-growing season. They found that no-tillage following a winter wheat [*Triticum aestivum* (L.) em. Thell.] cover crop averaged 2 to 4% (w/w) greater soil-water content in mid-August than all of the tilled treatments. The magnitude of effect varied with years.

10-1.3.4 Well-drained Paleudults

All Paleudults surveyed were well-drained soils of sand to sandy loam surface texture which can easily form a compacted layer or layers 10- to 15-cm thick associated with implement or vehicular traffic (Kashirad et al., 1967). Such layers may not be impermeable to water, but can be extremely restrictive to penetration by roots (Cassel et al., 1978). Literature reviewed includes Beale and Langdale, 1967; Gallaher, 1977; Kamprath et al., 1979; Martin et al., 1979; Parker et al., 1975; Rhoads, 1978; Suman and Peele, 1974; and Wolf et al., 1981.

Unlike any other soil class surveyed, tillage deeper than 25 cm on the well-drained Paleudults produced significantly greater yield (0.38 Mg ha^{-1}) than moldboard plowing. Shallow tillage tended to produce yields equal to moldboard plowing, while chisel plowing about 25-cm deep produced yields almost as great as the deeper, chisel-type tillage. The greatest success with deep tillage was with a subsoil shank or tine directly under the row to be planted (Rhoads, 1978), without subsequent traffic over the tillage zone. Deep tillage at some distance from the planted row was less effective (Rhoads, 1978).

Deep tillage can often break through dense tillage pans or other types of compacted layers produced by tillage tools and traffic. The particle-size distribution in the coarse-textured Paleudults is such that soil movement and compression evidently cause such layers to form on these structurally weak soils to a much greater degree than on other soils surveyed. Once the compacted layers had been ruptured, and not subsequently reformed by traffic, crop roots could penetrate to moist soil below the compaction zone. Greater amounts of accessible soil water was considered the major reason for the measured increased crop yields (Kamprath et al., 1979), though there may be other factors involved in certain circumstances, such as high subsoil P (Rhoads, 1982) or something not yet identified (Martin et al., 1979). For example, supplying 5 to 28 cm of water by irrigation during various portions of the vegetative stages of growth did not improve conditions sufficiently on 25-cm deep moldboard plowed plots to equal yields on non-irrigated plots moldboard plowed plus under-row chiseled to 45-cm depth (Martin et al., 1979). Introduction of pests such as nematodes into subsoil layers by deep tillage can complicate interpretation of tillage results (Parker et al., 1975).

Longevity of deep tillage effects has generally not been studied, as most studies report annual deep tillage. Rhoads (1978) found that soybean planted over 1-yr-old, 40-cm deep subsoil slots averaged the same yield as soybean in plots without prior year subsoiling. Under-row subsoiling increased yields 0.73 Mg ha^{-1} in the same year. Evidently, 1 yr of traffic and weather eliminated the effectiveness of the prior year's subsoiling.

Effect of crop residues on Paleudults has not been consistent. Beale and Langdale (1967) found no effect of burning or not burning small grain straw prior to tillage and planting double-cropped soybean, even with a lister treatment with some (unspecified) residues left on the surface over a 4-yr period. Gallaher (1977) showed an increase of 0.88 Mg ha^{-1} for soybean grown with residues vs. bare ground in a 1-yr study. Gallaher (1977) measured a 3-day delay in total depletion of available soil water in June and delayed senescence associated with residue cover.

10–1.4 Interactions with Other Cultural Practices

As tillage methods are altered to produce less soil mixing and/or a greater accumulation of previous crop residues at or near the soil surface, most other aspects of soybean production also will need to be altered. For example, planting equipment must be able to penetrate residues without plugging and must place seed in contact with soil—not residues. High organic matter accumulation at the surface of nontilled land (Dick, 1983) or interception of chemical sprays by previous crop residues (Banks and Robinson, 1982) may reduce the efficacy of surface-applied pesticides. Residues of previous crops or weeds may leach and/or decompose to form chemicals either harmful or beneficial to soybean (Bhowmik and Doll, 1982), which may dictate changes in rotations and weed control. Broadcast application of fertilizer P and K will accumulate these nutrients near the surface of nontilled (Dick, 1983) and disked systems (Tyler et al., 1983), causing positional unavailability in dry seasons or climates. As tillage alters growing conditions to reduce stress on the developing soybean, such characteristics as seed quality and germinability may also be affected (Tyler and Overton, 1982). These and other aspects of soybean production associated with tillage other than moldboard plowing will be discussed in greater detail in other chapters.

10–1.5 Tillage, Time, and Energy

Two primary advantages of reducing tillage are the general reductions in total time and fossil energy required to establish and protect the crop. With comparably sized equipment for all operations, the relative time/energy relationships among cultural operations per unit area are: moldboard plow 20 cm (100%/100%) > chisel plow 20 cm (55%/65%) > disk 10 cm (42%/26%) ⩾ planting (60%/21%) ⩾ cultivating (65%/12%) > spraying (35%/4%) (Richey et al., 1977; Vaughan et al., 1976). These approximations ignore subtleties such as a 14% increase in planting time on nontilled vs. plowed plus disked land (Vaughan et al., 1976), the range of energy values about the mean to account for differences in soil strength (±22% for moldboard plowing, ±19% for chisel plowing, ±9% for disking,and ±4% for no-tillage; Richey et al., 1977), and extra energy associated with the usually greater quantities of herbicides used as tillage is reduced.

10–1.6 Economics of Tillage

Combining soybean yields with labor, energy, cost of machinery, and other associated production costs generally results in about equal net profit (±$20 ha^{-1}) for the five tillage systems discussed herein (German et al., 1977; Siemens and Oschwald, 1978), *so long as soybean yields are equal!* The overriding considerations in relative profit or loss among tillage systems is ability to sustain good yields without developing problems requiring added expense to solve (e.g., a problem perennial weed).

10–1.7 Effect on Runoff and Soil Erosion

Many factors influence erodibility of soil by water. Factors not within our control for a specific site include precipitation characteristics, soil texture and mineralogy, and land slope steepness. Factors which can be controlled to some degree include crop residue placement, sequence and types of crops grown, soil surface roughness and organic matter content, land slope length, and direction of tillage relative to land slope direction. Of the controllable factors, the latter two can be applied equally to any combination of factors related to tillage and to soybean as previous crop. Also equally applicable would be alterations of crop canopy by reducing row width. Colvin and Laflen (1981) and Mannering and Johnson (1969) have shown small decreases in measured soil erosion during the middle of the vegetative growth period when row widths were reduced from 75 to 50 or 25 cm.

There is not unanimous agreement concerning the total effect of soybean on soil erosion and reasons for the effect. According to Wischmeier and Smith (1978), with equal surface residue cover, soybean as previous crop leave the soil 25% more susceptible to erosion than does corn as previous crop. McGregor (1978) indicates no difference. Whether or not soybean creates soil physical properties that cause soil to be more erosion susceptible than does corn (Fahad et al., 1982), soybean does produce significantly less plant tissue. This fact alone makes land following soybean more erosion susceptible than following corn for any tillage practice. A summary of the expected effects of tillage and previous crop on soil erosion is presented in Table 10–2.

Expected soil erosion is directly proportional to the C (cover and management) and K (soil erodibility) factors from the Universal Soil Loss Equation (USLE) of Wischmeier and Smith (1978), as well as proportional to their scaled product (Table 10–2). The greater residue following corn contributes to lower expected soil erosion, especially if tillage keeps those residues at or on the surface. A small additional benefit for maintaining residues at the surface is the buildup of organic matter, which results in a reduction in the K factor. Tillage which tends to bury residues has some redeeming quality if the soil is left in a roughened condition (large random roughness) and if the soil has sufficient structural stability

Table 10–2. Relative expected soil erosion as a function of tillage and of corn or soybean as previous crop.

Previous crop	Spring tillage†	Residue distribution after tillage‡		Fractional surface cover§	Surface random roughness¶	Organic matter (0–1 cm)#	Crop management subfactors††			K‡‡	Scaled product§§
		Surface	1–10 cm depth				$C_{surface}$	C_{sub}	C_{rough}		
		Mg ha^{-1}			cm	%					
Soybean	MP	0.08	0.11	0.05	2.3	2.1	0.84	0.98	0.63	0.37	1.00
	CP	0.51	1.19	0.24	1.6	2.9	0.44	0.85	0.78	0.34	0.52
	D	0.85	0.85	0.37	1.3	3.6	0.28	0.89	0.85	0.31	0.34
	NT	1.70	0	0.61	0.6	4.6	0.12	0.65	0.97	0.28	0.11
Corn	MP	0.56	0.56	0.17	2.3	2.1	0.56	0.93	0.68	0.37	0.68
	CP	2.24	3.36	0.52	1.6	2.9	0.28	0.64	0.87	0.34	0.28
	D	3.90	1.70	0.72	1.3	3.6	0.09	0.80	0.93	0.31	0.11
	NT	5.60	0	0.84	0.6	4.6	0.055	0.65	0.97	0.28	0.05

†MP = moldboard plow ~20 cm; CP = chisel plow ~20 cm; D = disk ~10 cm; NT = no-tillage.

‡Assumed residues from 7.50 Mg ha^{-1} corn crop and 2.70 Mg ha^{-1} soybean crop after 30% overwinter loss for each crop. Remainder of MP treatment residues are below 10-cm depth. Distribution of residues as suggested by Voorhees et al. (1981).

§Fraction of the soil surface covered by residues computed from mass of surface residues (Gregory, 1982).

¶Average roughness after tillage and before significant weathering, after Voorhees et al. (1981).

#Values for surface 1 cm for MP and NT in long-term tillage study after Dick (1983); CP and D treatments extrapolated based on surface residues.

††Subfactors computed as per Laflen et al. (1983); $C_{surface}$ = residue cover subfactor; C_{sub} = residue anchoring subfactor; C_{rough} = roughness subfactor.

‡‡Soil erodibility factor after Wischmeier and Smith (1978) computed for a silt loam soil having organic matter contents included in this table.

§§Product of $C_{surface} \cdot C_{sub} \cdot C_{rough} \cdot K$ scaled so that the largest value in this column equals unity.

to maintan the roughened condition at least until the crop canopy has fully formed.

Use of appropriate tillage and/or use of cover crops to supplement soybean residues following soybean harvest reduce the erosion susceptibility following soybean enough that other considerations are of much less practical concern. Note that use of no-tillage after soybean creates a lower erosion hazard than moldboard or chisel plowing after corn because of more favorable residue placement, greater surface soil organic matter content, and the greater degree of surface cover per unit weight of soybean tissue than for corn tissue. Research is needed to truly define the total effects of soybean on soil erosion potential to enable refinement of land management. Equal or greater need is for research to design tillage systems which manage crop residues for minimizing erosion potential (no-tillage or disk only) as reliable for crop establishment and protection as is now possible with more conventional tillage such as moldboard plowing.

Runoff of water from land in or previously in soybean is not influenced as dramatically or as predictably by tillage, rotations, or row width as is soil erosion (Table 10–3). In only one reported instance was runoff greater for the preferred narrow (<0.75 m) rows than for wider rows. Though increased runoff for a given set of soil and residue conditions would result in increased erosion, runoff does not necessarily dictate the magnitude of erosion among sets of soil and residue conditions.

The above discussion has dealt with water erosion. Similar discussion and results could be generated regarding wind erosion. In addition, cloddiness of the surface is an important factor in wind erosion control (Woodruff and Siddoway, 1965). Wind erosion is usually not a widespread problem for soybean production in the USA. Where the climate is dry enough that wind erosion is a widespread problem, it is too dry for soybean without irrigation. Irrigation would help produce necessary

Table 10–3. Ratios of measured erosion and runoff for selected variables.

Source†	Management variable	Ratio description	Ratio values: Erosion	Runoff
1	No-tillage	After soybean/After corn	2.8	2.0
1	Moldboard plow	After soybean/After corn	2.1	1.6
1	Soybean after corn (June)	0.75-m rows/0.25-m rows	1.1	1.0
2	Continuous row crop	After soybean/After corn	2.5	1.5
2	Corn-soybean rotation	After soybean/After corn	1.6	1.3
3	Corn-soybean rotation	After soybean/After corn	1.5	1.3
4	Soybean (June)	0.75-m rows/<0.75-m rows	1.5	0.8
5	Soybean (June-July)	1.02-m rows/0.51-m rows	1.1	1.1
		Avg	1.8	1.3
		SD/avg	0.4	0.3

†1 = Laflen and Colvin (1982) (two soils, simulated rain, 1 yr).
2 = Laflen and Colvin (1981) (two soils, simulated rain, 3 yrs).
3 = Laflen and Moldenhauer (1979) (natural rain, 6 yrs).
4 = Colvin and Laflen (1981) (two soils, simulated rain, 1 yr).
5 = Mannering and Johnson (1969) (simulated rain, 1 yr).

plant tissue to permit reasonable control of erosion. When soil texture is suitable for excessive wind erosion in humid climates, there is usually sufficient rainfall to produce residues for satisfactory erosion control.

10–1.8 Summary

In general, as soil becomes better drained and coarser in texture, the need or desirability of tillage for producing satisfactory soybean seed yields decreases in the absence of root growth-limiting layers in the normal rooting zone. This assumes that an adequate plant population and satisfactory weed control have been obtained. Tillage deeper than about 25 cm has had little effect on soybean yields except where tillage disrupts a root-growth limiting soil layer.

As the need for tillage is diminished, the desirability of surface residue cover increases for increased yield and decreased soil erosion. Soybean may leave land more susceptible to soil erosion than do other crops, but if the soil surface can be kept reasonably well covered with growing crops or crop residues, potential soil erosion problems will be appreciably reduced.

10–2 IRRIGATION

One of the most frequent and severe environmental limitations on soybean seed yields is water supply. As the value of soybean as a cash crop increases, the area of irrigated soybean is also likely to increase. Optimal water-use efficiency is crucial for irrigated soybean production to maintain economic viability and to insure that water resources are used effectively. To achieve optimal water-use efficiency requires a complete understanding of soybean water requirement and response to water deficits during each stage of growth.

10–2.1 Soybean Water Requirement as a Function of Plant Development

Total seasonal water use reported for soybean grown in the midwestern USA has typically ranged from 330 to 766 mm (Carter and Hartwig, 1962; Whitt and van Bavel, 1955; Herpich, 1963; Kanemasu et al., 1976; Somerhalder and Schleusener, 1960; Musick et al., 1976). Doorenbos and Pruitt (1977) have indicated that the seasonal water use for soybean can range from 450 of 825 mm of water where the growing season ranges from 100 days at low altitudes to up to 190 days in the higher altitudes. Total seasonal water use for soybean in New South Wales, Australia ranged from 451 to 748 mm per growing season (Mason et al., 1981). Such water use is strongly affected by the length of growing season available, the rate of crop development before reaching full ground cover, and the amount of available water. These water-use values for soybean indicate a fairly broad range that is highly dependent on the rainfall distribution, the total evaporative demand, and soil water-holding characteristics for each of the locations. The large range of reported water use could also reflect different water balance accounting methods among experiments.

Water use is not uniform throughout the growing season. A typical seasonal water-use pattern which consists of both soil evaporation and plant transpiration for soybean grown in the midwestern USA is shown schematically in Fig. 10–1. In conjunction with the water-use data is the soybean classification scheme of Fehr et al. (1971) showing the approximate time of the various vegetative and reproductive stages of growth. Water-use rate is generally low during the germination and seedling stages due to partial canopy cover and with a large portion of the water lost due to soil evaporation. However, as the plant moves into the rapid growth stage from V3 to V6, there is a rapid increase in water use. Maximum water use occurs during the R1 to R6 reproductive stages when full canopy is attained. Once the plant has started to mature and the pods are filled, there is a rapid decrease in water use associated with leaf senescence at the end of the season and a concurrent reduction in evaporation demand. The length of season represented in this figure is approximate and can range from 120 to 160 days after planting depending on the location. Water use expressed as the ratio of evapotranspiration to open pan evaporation shows a trend similar to that in Fig. 10–1 (Shaw and Laing, 1966; Mason et al., 1981; Thompson, 1977).

The foregoing water-use values are averages over a range of environmental conditions and have included evaporation from soil and plant tissue. Actual daily evaportion of water from plant tissue depends upon the amount of tissue exposed to radiation and wind, available water in the root zone, and evaporation potential. As available soil water content becomes insufficient to supply water for plant evaporation, plant water content may be reduced so that carbohydrate fixation, flowering and pod set, and leaf senescence may be unfavorably affected by the resulting plant water stress.

Soybean susceptibility to water deficits or stress differs among growth stages. Hiler et al. (1974) have developed a stress degree index which accounts for these differences. The crop's susceptibility values were determined experimentally as the fractional reduction in yield resulting from maximum water deficit during a given growth stage. For soybean, the corresponding crop susceptibility factors were 0.12, 0.24, 0.35, and 0.13 for the vegetative, early to peak flowering, late flowering to early pod development, and late pod development to maturity stages, respectively. Using these different crop susceptibility factors to avoid stress, irrigation is needed when the soil-water depletion reaches 80% in the vegetative stage, 45% in the early to peak flowering, 30% in the late flowering to early pod development, and 80% in the late pod to maturity period. The yield susceptibility factor for soybean developed from various literature sources given by Sudar et al. (1981) is similar and shows a

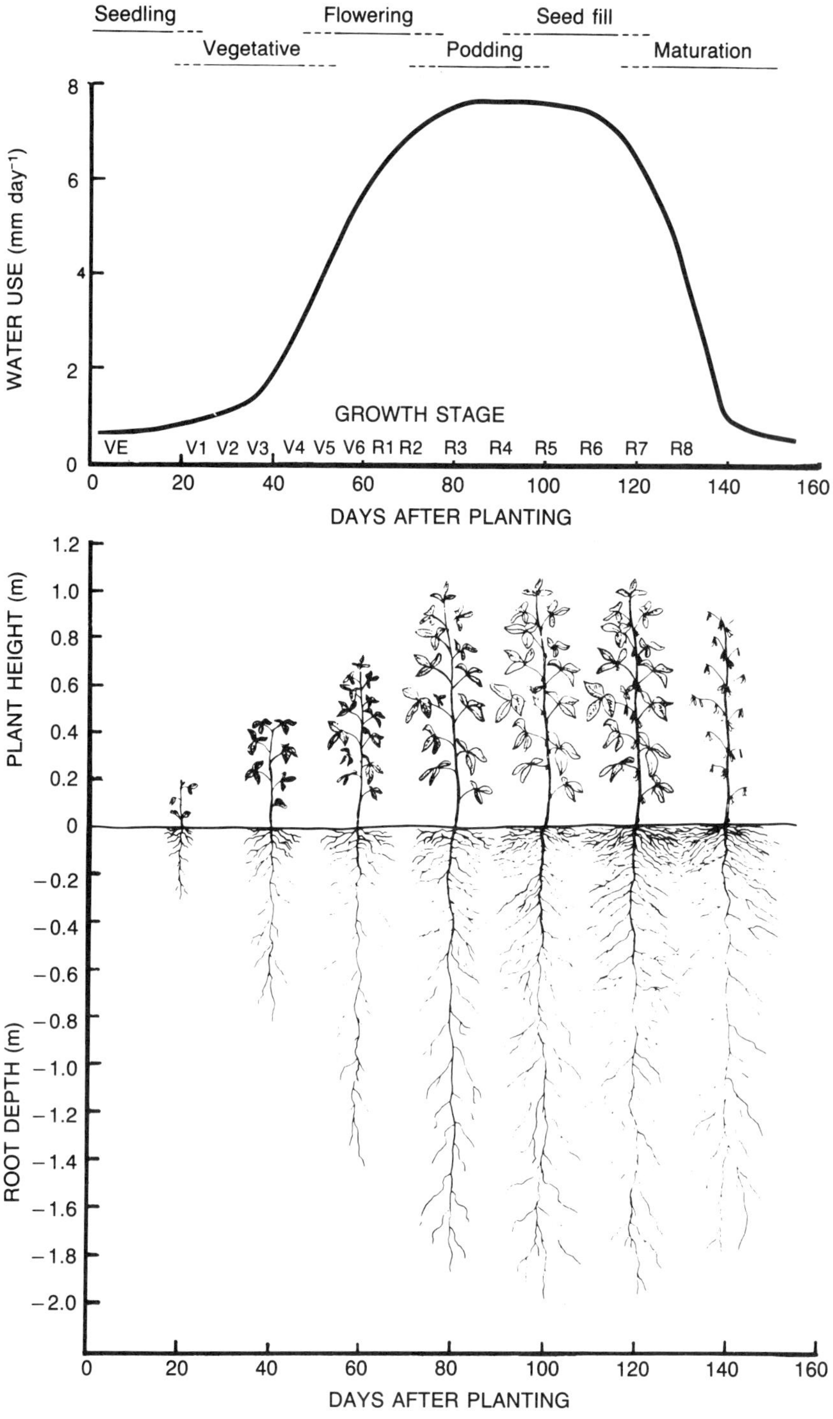

Fig. 10–1. Seasonal water use and growth patterns of soybean.

maximum during the pod-filling period. In other words, the soil water content should not be low during the late flowering and early pod development period and water should be applied to minimize the crop stress during this time. Corroborative data have been obtained by Doss et al. (1974) and by Korte et al. (1983b).

In trying to assess the importance of irrigation during the vegetative stages, it is important to understand that many soybean crops are planted into the soil when the water content is near field capacity. The large amount of available water may preclude the development of any severe water deficit during the early vegetative period. However, for soybean crops planted on a sandy soil with low water-holding capacity, additional water may be required to establish sufficient vegetative growth prior to the flowering period. Another important consideration during the vegetative period is that additional water enhances vegetative growth resulting in taller plants that lodge easily. Thus, there is an optimum where adequate vegetation is produced to generate the maximum seed yield but not excessive vegetative growth to result in severe lodging and subsequent harvest losses. If limited irrigation water is available, it is generally not recommended to irrigate during the vegetative stages.

10–2.2 Probability of the Need for Irrigation

For efficient and effective utilization of land and water resources, it is necessary to predict the amount of water available for plant growth and the likely occurrence of soil water deficits. This information is essential for long-term planning of agricultural activities, for making decisions regarding the purchase of irrigation equipment, and for assessing feasibility and usefulness of providing water by irrigation.

Under ideal soil water conditions, plant water extraction is generally related to root distribution. However, as surface layers dry, the pattern of water extraction shifts and a small portion of the root system in the subsoil is responsible for a major portion of the water uptake. Allmaras et al. (1975) found that the maximum soybean-rooting depths coincided with the maximum depth of water extraction but the corn-rooting depths ranged from 0.15- to 0.30-m deeper than the depth of maximum water extraction. Thus, the available water in the potential rooting zone and the time of maximum root penetration can influence the time soybean plants experience soil water deficits sufficient to reduce seed yield.

The occurrence of soil water deficits depends on a number of factors in addition to root proliferation. These include the amount and distribution of rainfall, meteorological conditions that determine the evaporative demand, and critical growth stages when soil water deficit affects yields. The rate that soil water is used by the crops, the capacity of the expanding root system to make soil water available to plants during periods of inadequate rainfall, surface runoff, and the type or nature of the hydraulic characteristics of the soil all contribute to the availability of soil water.

Rojiani et al. (1982) present a procedure to predict upper and lower bounds of the probability of occurrence of soil water deficits for any given period during the growing season for a given soil type and plant species. Basically, the model uses meteorological information for the site, a soil water balance model and data on plant available water to develop probability distribution functions of the plant available soil water on a given day in the growing season. The soil water distribution and the upper and lower bounds of the probability of having no soil water deficit are evaluated. They found that the correlation between the amount of plant available water in successive days needs to be considered. Their results indicate that the computed probability bounds are usually within 5 to 10% of the observed values and can be useful in making decisions relating to irrigation needs.

Sudar et al. (1981) developed a soil-plant-atmosphere-water model for estimating daily available soil water and evapotranspiration using readily available climate, crop, and soil data. This model was expanded to include estimates of crop water deficits and the effects of stress on the canopy development, plant phenology, and grain yield. They identified a crop yield susceptibility factor represented by a bell-shaped weighting factor with a peak at flowering for soybean. A crop yield water-stress index was developed based on the ratio of actual to potential transpiration and a variable yield susceptibility relationship during the growing season. A water-stress index was computed for the growing season as an accumulated daily water stress multiplied by the daily yield susceptibility factor. The calibration of this scheme for soybean research sites in Iowa and Missouri showed the method was a practical way for approaching water deficit effects on crop yields. While these results show promise for the method, the authors point out the need to evaluate the yield susceptibility factor and to derive accurate values for different parts of the country.

Should water deficits not develop during the growing season, however, there is no reason to expect a yield response to irrigation unless it is a negative response to excess water. Soils with high available water-holding capacities exhibit lower probability for developing water deficits than sandy soils with low available water-holding capacities. In many parts of the Midwest where the average annual precipitation ranges from 940 to 1170 mm, soybean response to irrigation has been erratic because of the generally sufficient water supply. Overall probabilities for success with irrigation in the USA appear best in the Southeast region which has high evaporation demand and a dominance of soils with low available water-holding capacities, and in semiarid to arid regions with low rainfall.

The probability of the need for irrigation also depends on economic considerations including soybean prices, the expected yield increase from irrigation, the amount of investment per acre, and the management ability of the individual farmer. It is difficult to determine whether an irrigation system used only for soybean will be economically feasible for a particular farm. However, if the farmer has already invested in an irrigation system for corn, then irrigating soybean may be profitable by spreading the investment over several crops. The other intangible factor is that resulting from corn and soybean rotations. Soybean in a corn and soybean rotation under irrigation can contribute to the overall yield increases and productivity with the response from the irrigated rotation system being larger than irrigation of corn and soybean in monoculture.

10–2.3 Irrigation Scheduling

Irrigation scheduling generally reduces to two questions; when to irrigate and how much water to apply? These are the primary irrigation management decisions that need to be made and can have the most effect on the crop yield and efficient water use. Many of the scheduling techniques range from simple soil water monitoring to using sophisticated computer programs to predict crop water use. Determining how much water to apply is directly related to the soil texture, the rooting depth obtained in a particular soil, and the crop's water-use rate. In a practical situation, the amount of water that can be applied is also directly related to the equipment capacity for supplying the water.

Regardless of the irrigation scheduling method used, best yields and most efficient water use are generally obtained when the available soil water in the root zone is not depleted more than 50 to 60% (Brady et al., 1974). During the early part of the soybean-growing season, when stress is less important than later in the season or when the rainfall frequency may be greater, it is often most efficient and economical to refill to only 80% of available water-holding capacity. This allows some available storage for rainfall that might fall shortly after the irrigation in those areas subject to high intensity rainfall.

Irrigation scheduling can be oriented towards maximizing either the yield per unit land area or yield per unit of applied irrigation. Generally, the farmer is interested in maximizing the yield per unit land area, which may be economically justified when water supplies are readily available and irrigation costs are low. However, these situations are rare at this time of increasing scarcity of water resources and increasing costs of energy for pumping. Thus, there is a need for improved irrigation scheduling to increase water-use efficiency defined here as the yield per unit of water applied.

Several methods are used to determine when the soil moisture depletion is at 50% or other predetermined levels. Three of the most common methods are (i) the hand or the feel method, (ii) the use of tensiometers or gypsum blocks (Cassel et al., 1978), and (iii) the checkbook balance sheet method (Brady et al., 1974; Gregory and Schottman, 1982; Woodruff et al., 1972). All three have been used with varying degrees of success and none has shown an advantage over others in terms of maximizing the yield increase from irrigation. In general, the results have been quite variable with all three methods of scheduling irrigation and it is generally recommended that several years of experience and intense management be utilized to determine the method that best suits the individual operator.

In order to study the need for irrigation as well as irrigation management strategies, simulation models are used to summarize and optimize the existing knowledge about soybean water use and yield, weather patterns, soil properties, and the economics into a conceptual framework compatible with irrigation objectives. Jones and Smajstrla (1982) have developed a modeling approach to the irrigation management of soybean. They focus on modeling at three different levels of sophistication. First is a simple soil-water balance method, second is a crop yield response model that includes a soil-water balance model, and the third level utilizes a dynamic crop growth model sensitive to plant water relations and other weather and management variables. Within each of these three levels of modeling, there is an increased level of complexity in data required. Irrigation scheduling using a soil-water balance model is based on maintaining optimum water levels in the root zone for maximum productivity. For practical application to irrigation scheduling, they conclude the state-of-the-art appears to be the combination of the water balance and yield response models.

An irrigation scheduling model for the prediction of soil water use by soybean was presented by Mason and Smith (1981). The model was developed and tested against the results from 3 yrs with good agreement between the measured and predicted water use and yield of two soybean cultivars. In all three seasons, the total water-use efficiency and the efficiency of supplemental water use was increased by increasing the time between irrigations. Better timing of the irrigations could, therefore, reduce the amount of water needed to produce good soybean yields, particularly in years with significant rainfall. The accuracy of the results and the general agreement between predicted and measured yields and the predicted and the actual water use show the potential advantages of a computer-based system for irrigation scheduling. While the ultimate use of dynamic growth models in irrigation management appears to have merit, their application to study irrigation rates awaits further development.

Under ideal conditions, the plants should signal when to irrigate and the soil should indicate how much water to apply. Hiler et al. (1974) have shown that irrigation water can be used more efficiently if plant water deficit criteria are available upon which to base the need for irrigation. Stegman et al. (1976) used the xylem potential and stomatal conductance as plant water deficit criteria for scheduling irrigation on soybean. Plant water deficit development was evaluated relative to variables indicative of the prevailing soil and atmospheric environments. Data collected from six crops grown on two different soil types indicated that the leaf xylem potentials were best correlated with air temperature and available soil water. They conducted simulation tests using water-balance methods and concluded that irrigation scheduling with plant stress criteria can reduce annual irrigation requirements. For soybeans, they found that the critical xylem potential was −1200 kPa and that, after sugarbeet (*Beta vulgaris* L.), soybean was the most sensitive crop to the available water. One disconcerting aspect in the multiple linear regression analysis was that soybean had the largest standard error of the estimate for leaf water potential when regressed on air temperature and available water. Leaf xylem potential may not be the most sensitive parameter as indicated by the scatter in the data.

It has long been recognized that plant temperatures could be used to assess the water status of plants and hence be applied to irrigation scheduling. With the advent of the infrared thermometer as a noncontact, nondestructive technique for measuring leaf and canopy temperatures, the potential benefits from utilizing this technique are just being evaluated. Jackson (1982) presents an excellent review of the development of canopy temperature measuring techniques for characterizing crop water stress. Idso et al. (1981) developed a plant water stress index that normalizes the stress degree parameters for environmental variability. The prime determinant is the ambient vapor pressure deficit. The utilization of this stress index depends upon knowing the crop's specific nonwater-stressed baseline defined as the relationship between the foliage temperature minus air temperature differential and the vapor pressure deficit under conditions of nonlimiting soil moisture. The data collected by Idso (1982) for soybean at several locations around the country follow the theoretical trend indicated by several investigators. While the infrared thermometer shows promise, much more research is necessary before operators can use this tool to schedule irrigation.

10–2.4 Overview of Soybean Response to Irrigation

The variation of soybean response to water is evident from the water-use and irrigation efficiencies at several locations within the USA summarized in Table 10–4. The data represent samples of irrigation studies where rainfall and irrigation have been quantified and cover a range of soil types and locations. The data selection was restricted to full-season varieties for the specific locations and to wide row spacings (0.76–1.00 m). No attempt was made to separate studies which included irrigation timing as a variable from those which did not. Both the water-use efficiency and irrigation water-use efficiency as defined in Table 10–4 were calculated based on the total amount of rain or irrigation applied during the growing season. From Table 10–4, the water-use efficiency ranged from a low of 0.83 to a high of 8.6 kg ha^{-1} mm^{-1}. The wide range in water-use efficiency values reflects seasonal, experimental, and varietal differences and the error for not including soil water extraction in the water-use term. The occasional negative values for the irrigation efficiency (i.e., where the irrigated yields were smaller than the nonirrigated yields) usually occurred only when the rainfall was not limiting. The range of the irrigation efficiencies of the limited data set was from −5.28 to 15.88 kg ha^{-1} mm^{-1}. The fact that these negative values exist suggest a lack of

Table 10–4. Summary of soybean water-use efficiency and yield increase per unit of irrigation from selected literature citations.

Location	Soil	Year	Cultivar	Rain	Irrigation	Non-irrigated yield	Irrigated yield range	Water-use† efficiency	Irrigation‡ efficiency	Reference
				mm		kg ha^{-1}		kg ha^{-1}mm^{-1}		
Florida	Lakeland sand	1978	Bragg	550	132–226	821	2245–2766	1.49	8.61–10.79	Jones et al. (1982)
		1979	Bragg	694	59–124	2822	2894–3164	4.07	0.81– 3.78	
		1980	Bragg	517	122–255	2007	3230–3938	3.88	5.08–13.73	
Florida	Kendrick fine sand	1977	Cobb	300	70–170	2500	2490–2640	8.33	−0.14– 0.82	Robertson et al. (1980)
South Carolina	Norfolk loamy sand	1975	Davis	587	262	1990	2341	3.39	1.34	Reicosky and Deaton (1979)
			McNair 800	587	262	2041	2405	3.48	1.39	
South Carolina	Norfolk loamy sand	1978	Bragg	403	202	1277	3089	3.17	8.97	Hunt et al. (1981)
South Carolina	Norfolk loamy sand	1978	Bragg	403	202	1550	2890	3.85	6.63	Karlen et al. (1982)
			Coker 388	403	202	1590	3070	3.95	7.33	
			Ransom	403	202	2020	3520	5.01	7.43	
South Carolina	Norfolk loamy sand	1979	Lee	722	244	1650	1790	2.29	0.57	Matheny and Hunt (1983)
		1980	Lee	415	671	1400	2800	3.37	2.09	
North Carolina	Lakeland sand	1979	Ransom	750	250	1198	1761	1.60	2.25	Mahler and Wollum (1981)
North Carolina	Wagram loamy sand	1977	Ransom	499	25–280	1720	1610–2110	3.45	−4.40– 2.69	Martin et al. (1979)
Texas	Pullman clay loam	1967	Clark	152	152–381	854	1278–2764	5.62	2.79– 6.26	Dusek et al. (1971)
			Hill	152	152–381	599	558–2327	3.94	−0.27– 7.16	
Arkansas	Crowley silt loam	1974	Lee 64	632	38	2919	3107	4.62	4.95	Caviness and Thomas (1980)
		1975	Lee 64	329	152	1957	2650	5.95	4.56	
		1976	Lee 64	418	228	1769	2630	4.23	3.78	
Arkansas	Crowley silt loam	1978	Forrest	455	270	907	3456	1.99	9.44	Jung and Scott (1980)

Iowa	Ida silt loam	1979	Wayne	297	250	2104	2112	7.08	0.03	Mason et al. (1982)
Michigan	Owasso Loam	1978	Corsoy	381	118–173	2287	2959–3161	6.00	3.88– 7.41	Vitosh et al. (1983, pers. comm.)
Nebraska	Sharpsburg silty clay loam	1974	Amsoy 71	400	200–350	3183	3758–3869	7.96	1.96– 2.88	Al-Ithawi et al. (1980)
		1976	Amsoy 71	325	125–250	2091	2784–3232	6.43	4.56– 5.54	
Georgia	Greenville sandy loam	1972	Hampton 266A	471	125–234	390	1942–2265	0.83	7.44–10.25	Ashley and Ethridge (1978)
		1973	Ransom	392	99–196	2043	3575–3649	5.21	8.19–15.88	
			Coker 102	392	99–196	1203	2150–2668	3.07	7.47– 9.98	
		1974	Hampton 266A	511	99–165	1931	1946–2926	3.78	0.15– 0.47	
			Ransom	511	99–165	2371	2926–2986	4.64	3.36– 6.21	
Alabama	Lucedale fine sandy loam	1968	Bragg	230	178–302	1980	2490–2680	8.61	2.32– 2.87	Doss and Thurlow (1974)
			Hampton	230	178–302	1760	2290–2510	7.65	1.76– 4.21	
		1969	Bragg	410	159–248	2090	3000–3210	5.10	4.52– 5.72	
		1970	Bragg	540	70–184	2540	2380–2530	4.70	−2.29– 0.05	
			Hampton	540	70–184	2070	2310–2360	3.83	2.58– 3.43	
Alabama	Lucedale fine sandy loam	1970	Bragg	543	108–377	2610	2140–2790	4.81	−2.96– 0.48	Doss et al. (1974)
		1971	Bragg	555	108–377	2060	2100–3320	3.71	0.37– 4.12	
		1972	Bragg	321	108–377	1800	1230–2940	5.61	−5.28– 3.38	
Mississippi	Sharkey clay	1979	Bedford	675	189	3423	3262	5.07	−0.85	Heatherly (1984)
			Tracy	675	193	4022	4035	5.96	0.07	
			Brady	675	151	3712	3907	5.50	1.29	
		1980	Bedford	406	531	989	2730	2.44	3.28	
			Tracy	406	627	1237	3053	3.05	2.90	
			Bragg	406	690	1332	3524	3.28	3.18	
		1981	Bedford	417	321	982	2777	2.35	5.59	
			Braxton	417	420	1029	3275	2.47	5.35	
		1982	Bedford	337	293	975	2246	2.89	4.34	
			Braxton	337	391	1009	2717	2.99	4.37	
	Dubbs silt loam	1979	Forrest	675	76	3510	3625	5.20	1.51	
		1980	Forrest	406	172–309	1123	2448–2979	2.77	5.94– 7.70	

(continued on next page)

Table 10–4. Summary of soybean water-use efficiency and yield increase per unit of irrigation from selected literature citations.

Location	Soil	Year	Cultivar	Rain	Irrigation	Non-irrigated yield	Irrigated yield range	Water-use† efficiency	Irrigation‡ efficiency	Reference
				mm		kg ha^{-1}		kg ha^{-1}mm^{-1}		
Minnesota	Hamerly clay loam	1978	Clay	321	287	3122	3448	9.73	1.14	Reicosky (unpub. data)
			Evans	321	287	3030	3562	9.44	1.86	
			Hodgson	321	287	3126	3285	9.74	0.55	
			Swift	321	287	3087	3385	9.62	1.04	
		1979	Clay	362	182	2987	3028	8.25	0.23	
			Evans	362	182	2997	3106	8.28	0.60	
			Hodgson	362	182	3263	3217	9.01	−0.25	
			Swift	362	182	3012	3050	8.32	0.21	
		1980	Clay	409	158	2889	2903	7.06	0.09	
			Evans	409	158	3094	3244	7.56	0.95	
			Hodgson	409	158	3054	3310	7.47	1.62	
			Swift	409	158	2985	3203	7.30	1.38	
		1981	Evans	384	156	2909	2740	7.58	−1.08	
		1982	Evans	420	316	2619	2678	6.24	0.19	
		1983	Evans	402	217	3420	3404	8.53	−0.07	
	Sioux sandy loam	1980	McCall	401	345	1742	2346	4.34	1.75	
			Evans	401	345	2243	3067	5.59	2.39	
			Hodgson	401	345	2328	2646	5.81	0.92	
		1981	McCall	396	149	1708	2381	4.31	4.52	
			Evans	396	149	1977	2777	4.99	5.37	
			Hodgson	396	149	2253	2596	5.69	2.30	
		1982	McCall	430	303	1673	2112	3.89	1.45	
			Evans	430	303	1836	1969	4.27	0.44	
			Hodgson	430	303	1706	1585	3.97	−0.40	
		1983	McCall	383	434	1456	2252	3.80	1.83	
			Evans	383	434	1671	2533	4.36	1.99	
			Hodgson	383	434	2065	2867	5.39	1.85	

†Water-use efficiency is defined as seed yield divided by the total rainfall for the growing season.
‡Irrigation water-use efficiency is defined as (irrigated seed yield − nonirrigated seed yield)/amount of irrigation applied.

understanding of soybean response to applied water, and further points out the need for high level management when irrigating soybean.

There are limited quantitative data on the various components of the water balance for soybean useful in evaluating water-use efficiency. Most studies report rainfall. However, many report seasonal rainfall only as normal, above normal, or below normal for the given location. Many studies report only the number of irrigations and give no indication of the total amount of water applied in each irrigation. More important, there are only a few studies where the amount of soil water extracted during the growing season is reported. Two studies that provide most of the detailed information necessary in the water balance are those of Cassel et al. (1978) and Mason et al. (1981) summarized in Table 10–5. Both rainfall and irrigation during growing seasons are presented as well as soil water extracted as measured with the neutron probe. Cassel et al. (1978) did not measure runoff but assumed it to be negligible. Mason et al. (1981) calculated runoff from the difference in the calculated ET and the other components of the water balance equation. Neither study measured the subsurface drainage component which probably can be assumed negligible. The water-use efficiencies, defined here as the mass of grain per hectare divided by total water used, ranged from 0.98 to 5.79 kg ha^{-1} mm^{-1} and are surprisingly similar considering the location and variety differences in the two studies. For the nonirrigated treatments, the range of values was wider on the sandy loam soil (Cassel et al., 1978) and the values generally higher on the clay soil (Mason et al., 1981) and indicated that the major component of this water-use efficiency term was the yield in the numerator.

One measure of irrigation water-use efficiency is the yield increase above that of the nonirrigated control per unit of irrigation water applied. If we apply this definition to the data in Table 10–5, the yield increase per unit of water applied ranges from 0.73 to 17.80 kg ha^{-1} mm^{-1}. In general, the largest grain yield increase per unit of irrigation applied occurred when irrigation was limited and applied only during the pod-fill period. The irrigation water-use efficiency generally decreased as the amount of irrigation increased primarily because of the associated decrease in soil water deficiency. The variation in the irrigation water-use efficiency within each of these studies illustrates our limited understanding how soybean respond to additional water. An alternate conclusion is that we cannot measure all components of the water balance equation with sufficient accuracy to relate it to subtle changes in yield.

Stressed plants do not always exhibit yield reductions commensurate with the observed stress. Reicosky and Deaton (1979) studied the effect of irrigation on yields of soybean grown on Norfolk fine sandy loam (fine-loamy, siliceous, Thermic Typic Paleudults) in South Carolina. Two determinate cultivars—McNair 800 (Maturity Group VIII) and Davis (Maturity Group VI) were grown with conventional fertilizer and herbicide treatments. Water was applied through a trickle irrigation system when the matric potential at the 0.15 m depth was equal to −20 kPa. While

Table 10–5. Summary of total and irrigation water-use efficiency in selected soybean studies where soil water extraction was measured.

Cultivar	Year	Treatment	Rainfall	Irrigation	Soil water extraction	Runoff	Total water use	Yield	Water-use efficiency Total§	Water-use efficiency Irrigation¶	Reference
			mm					kg ha^{-1}	kg ha^{-1}mm^{-1}		
Anoka and† SRF-100	1972	W1	323	0	185	–‡	508	1067	2.10§	–	Cassel et al. (1978)
		W2	323	305	124	–	752	1601	2.13	1.75	
		W3	323	432	132	–	887	1874	2.11	1.87	
		W4	323	546	94	–	963	1464	1.52	0.73	
	1973	W1	195	0	21	–	216	212	0.98	–	
		W2	195	102	53	–	350	2028	5.79	17.80	
		W3	195	324	34	–	553	3105	5.61	8.93	
		W4	195	318	4	–	517	2791	5.40	8.11	
	1974	W1	126	0	100	–	226	349	1.54	–	
		W2	126	133	113	–	372	1840	4.95	11.21	
		W3	126	318	78	–	522	2305	4.42	6.15	
		W4	126	406	71	–	603	2237	3.71	4.65	
Ruse	1975–1976	FF	412	260	20	130	562	2737	4.87	3.96	Mason et al. (1981)
		MM	412	170	20	72	530	2660	5.02	5.61	
		Nil	403	0	125	72	456	1707	3.74	–	
	1976–1977	FF	551	420	6	334	643	2612	4.06	2.12	
		MM	551	290	6	235	612	2359	3.86	2.20	
		Nil	482	0	85	74	493	1720	3.49	–	
	1977–1978	FF	335	401	20	34	722	3387	4.69	4.92	
		MM	335	250	11	18	578	2809	4.86	5.58	
		Nil	335	0	120	0	455	1413	3.11	–	
Bragg	1975–1976	FF	412	320	0	150	582	2943	5.06	3.23	
		MM	412	170	55	90	547	2750	5.03	4.94	
		Nil	412	0	142	90	464	1910	4.12	–	
	1976–1977	FF	551	400	40	352	639	3042	4.76	2.57	
		MM	551	252	35	242	596	2681	4.50	2.64	
		Nil	551	0	68	93	526	2015	3.83	–	
	1977–1978	FF	335	451	10	48	748	3597	4.81	4.44	
		MM	335	250	5	37	553	3070	5.55	5.89	
		Nil	335	0	135	19	451	1597	3.54	–	

†All data in this study represent the mean of two cultivars.

§Total water-use efficiency is defined as the mass of grain per hectare per unit of water where the total water is the sum of rainfall + irrigation + Δ storage.

‡Not measured and assumed to be zero.

¶Irrigation water-use efficiency is defined as (irrigated grain yield − nonirrigated grain yield)/amount of irrigation applied.

the total rainfall from planting to harvest was adequate on the seasonal basis (400 mm) a period of about 32 days without significant rainfall resulted in drought stress and a slight depression in yield. The addition of 262 mm of irrigation water resulted in only a 0.35 and a 0.36 Mg ha^{-1} increase in the seed yield of Davis and McNair 800, respectively. The slight reduction in yield as a result of intermittent drought stress is partially explained by the subsoil water extraction. Evapotranspiration (ET) measured with a portable chamber on the nonirrigated plots after 25 and 33 days without significant rainfall ranged from 40 to 60% of the ET on the irrigated plots. Even though the nonirrigated plants exhibited severe wilt symptoms during the drought, small differences between irrigated and nonirrigated leaf water potentials were measured during the period, indicating that the nonirrigated plants had partially adjusted to the drought. Air temperature within the canopy was as much as 6.7 °C higher on the nonirrigated plots as a result of dissipating the radiant energy as increased sensible heat. Part of the yield difference attributed to soil water deficit may be a result of increased canopy air temperature on the nonirrigated treatment.

Information is just becoming available on the effect of irrigation on the development of soybean vegetative and fruiting components. Ashley and Ethridge (1978) measured the effect of water application on vegetative components, fruiting patterns, and seed yields. The variables studied included four water regimes and three cultivars. The four water regimes included nonirrigated, full-season irrigation, irrigation begun at the bloom stage, and irrigation begun at the pod-fill stage. Full-season and bloom stage irrigation treatments produced higher seed yields than the nonirrigated check except for 'Hampton 266A' in 1974. Seed yields from plants receiving the pod stage irrigation were higher than the nonirrigated check in the drier seasons of 1972 and 1973, but not in 1974. Irrigation beginning at pod filling produced yields that were generally equal to the full-season and bloom stage treatments. The magnitude of the response was determined to a large extent by the rainfall distribution during the reproductive development. The relatively short-statured cultivars provided the highest yields and the greatest response to irrigation and had less lodging problems. Number of pods per plant during reproductive development, and dry weights of stem, branches, and leaves followed a pattern similar to seed yields.

In analyzing the effect of irrigation on yield components, Korte et al. (1983a, 1983b) found that a single irrigation during flowering increased the number of pods per plant and the seeds per plant, but that there was an offsetting decrease in the weight of 100 seeds resulting in little change in the final soybean seed yield. A single irrigation during the pod elongation period had no effect on the 100-seed weight but increased the number of pods per plant and the seeds per plant resulting in subsequent large increase in the seed yield. Irrigation during the seed enlargement stage resulted in only a slight increase in the number of pods and seeds per plant but greatly increased the 100-seed weight. These results suggest the need for care in interpreting the soybean yields because of the compensation within each of the yield components. After the plants experience water deficit, the yield components can readjust so that the final yield is not dramatically affected. Irrigation early in the reproductive ontogeny greatly reduced the flower and pod abortion, whereas, irrigation later in the ontogeny reduced seed abortion within the developing pods. The consistent yield reductions from two irrigations at flowering and podfill indicate physiological effects not yet understood. They concluded that irrigation timing differentially affected different components of the yield, and as a result, had significant effects on the final yield.

Sammons et al. (1981) evaluated the performance of 20 soybean cultivars under suboptimal soil water on a Bloomfield sand (coarse-loamy, mixed, mesic Psammentic Hapludalfs) in central Illinois. They used a solid set overhead irrigation system throughout the season to apply contrasting water treatments of optimal irrigation (approximately 6.1 mm day^{-1}) and suboptimal (18.3 mm week^{-1}) to detect genetic variation for the tolerance to water deficit imposed throughout the reproductive period. The mean yield of 20 cultivars irrigated under the optimum water regime was 2.99 Mg ha^{-1}, while the suboptimal regime yielded 2.48 Mg ha^{-1}. They concluded that the yield potential appeared to be more strongly associated with the overall vegetative performance than how the yield was distributed among the plants' reproductive organs. The yield and vegetative response to soil water deficit were not always consistent. They concluded that the response to soil water deficit was more accurately measured under field conditions and that laboratory or greenhouse data are at best only supportive of the efforts under field conditions.

Soybean response to irrigation has often been disappointing to many growers in the Midwest. Frequently, they have found that the increase in yield did not pay for the cost of irrigation and seldom has the soybean response been as great or profitable as responses obtained from other agronomic crops. One of the reasons for the limited soybean yield response to irrigation is that the timing of irrigation or natural rainfall is considerably more flexible for soybean than for corn or other agronomic crops. Soybean and corn have basic physiological differences which regulate the duration of the critical period for soil water deficits. The critical period in corn is relatively small and centered around anthesis. However, in contrast, soybean plants have a period of 3 to 5 weeks during which new flowers are formed and normal yields may be produced by pod formation from only one-fourth of the potential flowers on the plant if the water and nutrient supplies are optimum during the remainder of the seed-filling period. This flexibility of soybean gives the impression of limited or lack of response to irrigation and is often mistakenly construed as soybean plants being drought tolerant. Soybean has more of a drought escape mechanism because of the duration of the pod filling and flowering period. However, if the soil water deficit is severe at any time during this flowering and critical pod-filling period, a depression in yield can be expected.

It is generally recognized that most plants prefer adequate soil oxygen for maximum productivity. However, there is some speculation on how soybean responds to water logging or poor aeration associated with flood irrigation. Recent work in Australia has shown that soybean can produce higher yields when grown under wet soil culture (Hunter et al., 1980; Troedson et al., 1983; Nathanson et al., 1984) when compared to conventional methods. Briefly, the seed is planted in raised beds and furrow irrigated by continuous low volume water flow when the first trifoliolate appears. The water level is maintained near the soil surface throughout the season. After an initial acclimation period, when soybean exhibited chlorotic symptoms characteristic of N stress, the plants recovered and yielded an average of 22% more for 14 cultivars than conventionally grown soybean that yielded 3.44 Mg ha^{-1} in a field study (Troedson et al., 1983). The yield increase for one cultivar was as large as 68%, and is presently attributed to the complete absence from soil water deficit, root proliferation, and nodule dry matter accumulation in the saturated zone above the water table. While this method requires specific soil conditions, the soybean response is intriguing. The practical consequence of this work awaits further research where the water used is quantified. However, it does open up exciting prospects on some of the other factors that may limit soybean production.

10–2.5 Interaction with Other Cultural Practices

10–2.5.1 Row Spacing and Plant Population

Water availability and water-use rates as affected by row spacing have become major considerations in soybean production, particularly in the drier regions of the country. Results of numerous studies have shown that soybean yields are frequently increased by planting in narrow rows (Cooper, 1977; Donovan et al., 1963; Reiss and Shorewood, 1965; Safo-Kantanka and Lawson, 1980; Weber et al., 1966). Peters and Johnson (1960) reported a higher yield in 0.5-m rows than 1.0-m rows and suggested that the root system of the soybean does not fully make use of the available water stored in the soil between the wide rows. They concluded that evaporation from the soil surface alone was responsible for most of the total water loss from the soil profile under the soybean crop when the soil surface was kept moist. The yield increase is usually attributed to development of full canopy with 95% light interception before rapid seed development (Shibles and Weber, 1966). The full canopies intercept more radiation and have a greater photosynthetic rate than partial canopies associated with wider row spacing. The complete canopy cover earlier in the season results in some additional reduction in soil evaporation losses.

Increased yield from narrow rows can vary with seeding rate and variety. Basnet et al. (1974) studied the effect of two interrow and two intrarow spacings of five cultivars on an irrigated Muir silt loam (fine-silty, mixed, mesic Cumulic Haplustolls) in Kansas. The plants in the

0.46-m row width or the 3.8 cm within row spacing were taller, lodged more, and produced fewer nodes, branches and pods on the main stem. The highest yields were obtained from the lowest plant density in the narrow rows in 1969, and the highest plant density in wider rows in 1970. The row width × spacing and the row width × cultivar interactions were significant.

One would conclude that there ought to be a sizable interaction between row spacing, plant population, and water-use efficiency in response to irrigation. Those combinations which increase use of natural precipitation (narrow rows) should result in decreased benefits from irrigation. The available data do not universally support this hypothesis. Doss and Thurlow (1974) evaluated soybean response to irrigation on a Lucedale fine sandy loam (fine-loamy, siliceous, thermic Rhodic Paleudults) to determine the effect of three soil water regimes, two row widths and three plant population levels on water use, the rate of plant growth and the seed yield of two cultivars. Though daily water-use rates were lower for the 0.6-m than the 0.9-m row spacings, there was no interaction of irrigation and row width on plant height for any of the 3 yrs. There was an irrigation × row width interaction for seed yields in only 1 of 3 yrs. Yields were greatest for 0.6-m row spacing, especially at the intermediate level of irrigation. Matson (1964) also showed no consistent interactions among irrigation, row spacing, and varieties of soybean on seed yields.

Recently, Taylor (1980) has attempted to relate row spacing and population effects, soil water depletion, and evapotranspiration rates for soybean. In the 3-yr experiment with 'Wayne' soybean, an indeterminate cultivar, Taylor found no difference in yield among the various row spacings during 1976, a drier than normal growing season. During 1975, a growing season with normal rainfall, yields tended to increase as row spacings decreased, but the differences were not significant at the 5% level. During the 1977 year with greater than normal rainfall and preplant irrigation, soybean in the 0.25-m rows outyielded those in the 1-m rows by 17%. Heatherly (1984) found that in an extremely dry year, irrigated narrow row spacings had yields above the irrigated wide rows. In wet or moderately dry years, however, row spacing did not affect the yield of irrigated soybean.

Taylor et al. (1982) found that irrigation increased leaf area, total biomass, plant height, number of vegetative nodes and crop growth rates, but delayed reproductive development and did not affect the seed yield. The 0.25-m row spacings yielded an average of 2470 kg ha^{-1} while the 1.0-m row spacing yielded only 2108 kg ha^{-1}. They attributed this increase in yield resulting from row spacing to increased radiation interception by narrow rows. The difference in the radiation interception during the late seed development period probably caused the differences in yield between the two row spacings.

In the same study, Reicosky et al. (1982) used two methods for evaluating water use, (i) measuring evapotranspiration with a portable chamber and (ii) calculating water use from soil water extraction data. Irrigation resulted in the expected higher water-use rate for both row spacings. The row spacing effect early in the season was related to the leaf area index with the 0.25-m irrigated treatment having a slightly higher ET than the 1.0-m irrigated treatment. However, later in the season, row spacing effects were not evident. While some small differences due to row spacing were noted earlier before complete canopy closure on the wider row spacings, there was not sufficient soil water deficit to affect the final yields at the end of the season (Taylor et al., 1982). Thus, the small differences in response to soil water status and irrigation for this 1 yr's data suggest that soil water status played a minor role in determining the soybean yields. The main differences due to irrigation were attributed to seed size and duration of reproductive development.

In most cases where soybean yields have been increased by narrow row spacings, the results have been obtained on soils with a large water-holding capacity. In a drier region, Alessi and Power (1982) found no effect of row spacing on soybean yields in years of normal or above normal precipitation. They did observe reduction in yield of the narrow rows when there was severe soil water deficit encountered during years of below normal precipitation. The data suggested that narrow rows may be beneficial for soybean production when water is not restricting, i.e., with irrigation. Soil water depletion was generally confined to the upper 0.9 m of the soil profile. The average water use for the 4-yr study period was 236 mm and was not significantly affected by plant spacing. In 2 of the 4 yrs, total water use was greatest and average soybean yields were least from the 0.15-m row width. Within-row spacing affected yields only 1 of the 4 yrs. Water-use efficiency was least for the 0.15-m rows in 3 of the 4 yrs suggesting that planting 0.15-m rows enhances water use prior to flowering. In an extreme drought, this enhanced early season water use leaves less water available during the pod-fill period and seed yields may be reduced accordingly.

10-2.5.2 Double-Cropping

Double-cropping has proven successful under the longer growing seasons in the southeastern USA. Between 25 and 40% of the soybean-cropping area is double-cropped with soybean usually following a small grain. Guilante et al. (1974) found irrigation was vital for double-cropping on sandy soils and earliest possible planting of the second crop resulted in higher yields. Some soybean cultivars matured late when planted late and diseases prevented normal yields. They suggested seed yields of both the first and second crops could be improved with higher populations and narrower rows. Double-cropping has been successful in eastern Virginia (Camper et al., 1972) and in Mississippi (Sanford et al., 1973) with soybean usually being the second crop. In both studies, the authors indicated that the weather played a major role each year in determining both the yield and the quality of grain and soybean produced. Variable results from 1 yr to the next were directly related to the amount of rainfall between planting and emergence, suggesting the increased importance of water management in double-cropping systems. Camper et al. (1972) found that soybean appeared to be more dependable than either corn or sorghum [*Sorghum bicolor* (L.) Moench] if planted in July, suggesting that the soybean is more tolerant of environmental stresses. The consistent performance of soybean can result in stable production in double-cropping systems.

Boerma and Ashley (1982) evaluated the effects of late and ultra-late planting dates in two row widths (0.51 and 0.91 m) for irrigated and nonirrigated treatments over 3 yrs. The late and ultra-late planting dates ranged from 5 to 10 weeks later than the full-season crop. They found that delays of planting from early July to late July resulted in an average yield reduction on the irrigated plots of 53 kg ha^{-1} day^{-1} (1.8%) and 19 kg ha^{-1} day^{-1} (1.2%) for the nonirrigated treatments. In a dry season, irrigation increased yield 355% in the late planting and 115% in the ultra-late planting. In an intermediate rainfall season, irrigation increased yields 38% for the late and 15% for the ultra-late planting. The 0.51-m rows averaged 17% higher yields than the 0.91-m rows when averaged over all years. It should be noted that both irrigated and nonirrigated treatments received an initial irrigation to obtain a stand, pointing out the importance of timely water application. Without the initial irrigation, the yield decrease from the delay until adequate rainfall was received would be even larger. Under present management schemes, the late-planted soybean indicated a low-yield potential. They concluded that the magnitude of the yield reduction from delays in planting can sometimes be offset by irrigation during the dry season and indicated the importance of intensive management to obtain maximum yields from late-planted soybean.

10–2.5.3 Nutrient Placement

Little information is available on irrigation effects and the placement of nutrients on soybean yield. Lutz and Jones (1975) evaluated the effects of irrigation and the placement of lime, P, K, and micronutrients on the yield and composition of soybean on a Davidson clay loam (clayey, kaolinitic, thermic Rhodic Paleudults) in Virginia. They found that soybean yields were increased each year with irrigation with an average annual increase for 3 yrs of 0.51 Mg ha^{-1} or 22%. Yields were unaffected by P and K treatments during the first 2 yrs, but in the 3rd yr, yields were lower where the P and K had not been applied. Fertility treatments did not affect the oil or protein concentration in the seed, nor did they affect the yield. Irrigation appeared to have little influence on the oil and the protein concentration of the soybean seed. No interactions of irrigation and nutrient placement were reported for seed yields or oil and protein concentrations. Matson (1964) showed no interaction between irrigation and fertilizer rates on seed yields on a sandy soil.

10-3 SUMMARY

Soybean is gaining in economic importance and competing favorably with other major grain crops. Yield increases in soybean have not appeared to keep pace with other species, and their yield responses due to irrigation are perplexing and give the impression that soybean is a flexible species capable of adapting to limited soil water deficits. The seasonal water requirement can range from 330 to 766 mm with the period of maximum water use starting near canopy closure at the onset of flowering and continuing through podfill. Available evidence indicates the late flowering and pod-fill periods are the most sensitive to soil water deficits.

The probability of need for irrigation during this critical period depends on the interaction of several soil and climatic factors. The available soil water and rooting depth determine the supply capacity of the soil and the climatic factors that control the evaporative demand and water use all interact to affect plant performance. Satisfactory economic returns from irrigated soybean require a high level of management and precise irrigation scheduling. To date, irrigation scheduling of soybean has been moderately successful on sandy soils with limited natural rainfall in the arid areas. However, season-to-season variation in yield increases due to irrigation in the humid areas with fine-textured soils points out the need to improve our understanding of soybean water management. The limited amount of data available to accurately calculate the water-use efficiency indicates a narrow range of water-use efficiencies that can be best increased by increasing grain yields and prudent water management. The available data have led to a better understanding of soybean response to water and water deficits in the field. However, much detailed research is needed to fully understand soybean response to irrigation for optimizing yields and water-use efficiencies.

ACKNOWLEDGMENT

The authors gratefully acknowledge the reprints, reports, and data provided by several colleagues and the assistance of S. A. Gausman in preparing the figures.

REFERENCES

Alessi, J., and J.F. Power. 1982. Effects of plant and row spacing on dryland soybean yield and water-use efficiency. Agron. J. 74:851–854.

Al-Ithawi, B., E.J. Deibert, and R.A. Olson. 1980. Applied N and moisture level effects on yield, depth of root activity and nutrient uptake by soybeans. Agron. J. 72:827–832.

Allmaras, R.R., W.W. Nelson, and W.B. Voorhees. 1975. Soybean and corn rooting in southwestern Minnesota: II. Root distributions and related water inflow. Soil Sci. Soc. Am. J. 39:771–777.

Ashley, D.A., and W.J. Ethridge. 1978. Irrigation effects on vegetative and reproductive development of three soybean cultivars. Agron. J. 70:467–471.

Aubertin, G.M., and L.T. Kardos. 1965. Root growth through porous media under controlled conditions. I. Effect of pore size and rigidity. Soil Sci. Soc. Am. Proc. 29:290–293.

Banks, P.A., and E.L. Robinson. 1982. The influence of straw mulch on the soil reception and persistence of metribuzin. Weed Sci. 30:164–168.

Basnet, B., E.L. Mader, and C.D. Nickell. 1974. Influence of between and within row spacing on agronomic characteristics of irrigated soybeans. Agron. J. 66:657–659.

Beale, O.W., and G.W. Langdale. 1967. Tillage and residue management practices for soybean production in a soybean-small grain rotation. Agron. J. 59:31–33.

Bhowmik, P.C., and J.D. Doll. 1982. Corn and soybean response to alleopathic effects of weed and crop residues. Agron. J. 74:601–606.

Boerma, H.R., and D.A. Ashley. 1982. Irrigation, row-spacing and genotype effects of late and ultra-late planted soybeans. Agron. J. 74:995–999.

Bone, S.W., D.M. Van Doren, and G.B. Triplett, Jr. 1977. Tillage research in Ohio: A guide to the selection of profitable tillage systems. Ohio State Univ. Coop. Ext. Serv. Bull. 620.

----, ----, and ----. 1978. Corn and soybean production potential of selected tillage practices on a Typic Argiaquoll (Brookston) soil. Ohio Agric. Res. Dev. Ctr. Res. Bull. 1099.

Brady, R.A., L.R. Stone, C.D. Nickell, and W.L. Powers. 1974. Water conservation through proper timing of soybean irrigation. J. Soil Water Conserv. 29:266–268.

Burnside, O.C. 1978. Mechanical, cultural and chemical control of weeds in a sorghum-soybean rotation. Weed Sci. 26:362–369.

Campbell, R.B., D.C. Reicosky, and C.W. Coty. 1974. Physical properties and tillage of Paleudults in the Southeastern Coastal Plains. J. Soil Water Conserv. 29:220–224.

Camper, H.M., Jr., C.F. Genter, and K.E. Loope. 1972. Double cropping following winter barley harvest in eastern Virginia. Agron. J. 64:1–3.

----, and J.A. Lutz, Jr. 1977. Plowsole placement of fertilizer for soybeans and response to tillage of plowsole. Agron. J. 69:701–704.

Carter, J.L., and E.E. Hartwig. 1962. The management of soybeans. Adv. Agron. 14:359–412.

Cassel, D.K., A. Bauer, and D.A. Whited. 1978. Management of irrigated soybeans on a moderately coarse textured soil in the upper Midwest. Agron. J. 70:100–104.

----, H.D. Bowen, and L.A. Nelson. 1978. An evaluation of mechanical impedance for three tillage treatments on Norfolk sandy loam. Soil Sci. Soc. Am. J. 42:116–120.

Caviness, C.E., and J.D. Thomas. 1980. Yield reduction from defoliation of irrigated and non-irrigated soybeans. Agron. J. 72: 977–980.

Colvin, T.S., and D.C. Erbach. 1982. Soybean response to tillage and planting methods. ASAE Paper 82-1022, American Society of Agricultural Engineering, St. Joseph, MI.

----, and J.M. Laflen. 1981. Effect of corn and soybean row spacing on plant canopy, erosion, and runoff. Trans. ASAE 24:1227–1229.

Cooper, R.L. 1977. Response of soybean cultivars to narrow rows and planting rates under weed free conditions. Agron. J. 69:89–92.

Crabtree, R.J., and R.N. Rupp. 1980. Double and monocropped wheat and soybeans under different tillage and row spacings. Agron J. 72:445–448.

Dick, W.A. 1983. Organic carbon, nitrogen, and phosphorous concentrations and pH in soil profiles as affected by tillage intensity. Soil Sci. Soc. Am. J. 47:102–106.

Donovan, L.S., F. Dimmock, and R.B. Carson. 1963. Some effects of planting pattern on yield, percent oil and percent protein in Mandarin [Ottawa] soybeans. Can. J. Plant Sci. 43:131–140.

Doorenbos, J., and W.O. Pruit. 1977. Guidelines for predicting crop water requirements. FAO Irrigation and Drainage paper 24. Food and Agricultural Organization of the United Nations, Rome.

Doss, B.D., R.W. Pearson, and H.T. Rogers. 1974. Effect of soil water stress at various growth stages on soybean yield. Agron. J. 66:297–299.

---- and D.L. Thurlow. 1974. Irrigation, row width, and plant population in relation to growth characteristics of two soybean varieties. Agron. J. 66:620–623.

Dusek, D.A., J.T. Musick, and K.B. Porter. 1971. Irrigation of soybeans in the Texas High Plains, Tex. Agric. Exp. Stn. Rep. MP973.

Elkins, D., D. Frederking, R. Marashi, and B. McVay. 1983. Living mulch for no-till corn and soybeans. J. Soil Water Conserv. 38:431–433.

----, J.D. George, and G.E. Birchett. 1982. No-till soybeans in forage grass sod. Agron. J. 74:359–363.

Erbach, D.C. 1982. Tillage for continuous corn and corn-soybean rotation. Trans. Am. Soc. Agric. Eng. 25:906–911.

Fahad, A.A., L.N. Mielke, A.D. Flowerday, and D. Swartzendruber. 1982. Soil physical properties as affected by soybeans and other cropping sequences. Soil Sci. Soc. Am. J. 46:377–381.

Fehr, W.R., C.E. Caviness, D.T. Burmood, and J.S. Pennington. 1971. Stage of development descriptions of soybeans (*Glycine max* L. Merrill). Crop Sci. 11:929–931.

Gallaher, R.N. 1977. Soil moisture conservation and yield of crops no-till planted in rye. Soil Sci. Soc. Am. J. 41:145–147.

Gebhardt, M.R., and H.C. Minor. 1983. Soybean production systems for claypan soils. Agron. J. 75:532–537.

German, L., K. Schneeberger, H. Workman, and J. McKinsey. 1977. Economic and energy efficiency comparisons of soybean tillage systems. p. 277–287. *In* William Lockeretz (ed.) Agriculture and energy. Academic Press, New York.

Goyal, M.R., G.L. Nelson, L.O. Drew, T.G. Carpenter, and T.J. Logan. 1981. Moisture and soybean seedling emergence. Trans. ASAE 24:1432–1435.

Griffith, D.R. 1982. A guide to no-till planting after corn or soybeans. Purdue Univ. Coop. Ext. Serv. Circ. ID-154.

Gregory, J.M. 1982. Soil cover prediction with various amounts and types of crop residue. Trans. Am. Soc. Agric. Eng. 25:1333–1337.

----, and R.W. Schottman. 1982. Irrigation scheduling procedure for subhumid areas. Trans. Am. Soc. Agric. Eng. 25:1290–1294.

Guilante, T.C., R.E. Perez-Levy, and G.M. Prine. 1974. Some double cropping possibilities under irrigation during the warm season in North and West Florida. Soil Crop Sci. Soc. Fla Proc. 34:138–143.

Harvey, W.R. 1960. Least squares analysis of data with unequal subclass numbers. USDA-ARS 20-8 (July) U.S. Government Printing Office, Washington, DC.

Heatherly, L.G. 1981. Soybean response to tillage of Sharkey clay soil. Mississippi Agric. For. Exp. Stn. Bull. 892.

----. 1984. Soybean response to irrigation of Mississippi River Delta soils. USDA-ARS ARS-18. U.S. Government Printing Office, Washington, DC.

Herpich, R.L. 1963. Irrigating soybeans. Kansas State Univ. Agric. Ext. Engineering in balanced farming. Land Reclamation 12.

Hiler, E.A., T.A. Howell, R.B. Lewis, and R.T. Boos. 1974. Irrigation timing by the stress day index method. Trans. ASAE 17:393–398.

Hovermale, C.H., and H.M. Camper. 1979. Effects of small grain stubble height and mulch on no-tillage soybean production. Agron. J. 71:644–647.

Hunt, P.G., A.G. Wollum II, and T.A. Matheny. 1981. Effects of soil water on *Rhizobium japonicum* infection, nitrogen accumulation and yield in Bragg soybeans. Agron. J. 73:501–505.

Hunter, M.N., P.L.M. de Jabrun, and D.E. Byth. 1980. Response of nine soybean lines to soil moisture conditions close to saturation. Aust. J. Exp. Agric. Anim. Husb. 20:339–345.

Idso, S.B. 1982. Non-water-stressed baselines: A key to measuring and interpreting plant water stress. Agric. Meteorol. 27:59–70.

----, R.D. Jackson, P.J. Pinter, Jr., R.J. Reginato, and J.L. Hatfield. 1981. Normalizing the stress degree parameter for environmental variability. Agric. Meteorol. 24:44–55.

Jackson, R.D. 1982. Canopy temperature and crop water stress. p. 43–85. *In* Daniel Hillel (ed.) Advances in irrigation, Vol. 1. Academic Press, New York.

Jeffers, D.L., G.B. Triplett, and J.E. Beuerlein. 1973. Double-cropped soybeans. Ohio Rep. 58:67–69.

Jones, J.W., L.C. Hammond, and K.J. Boote. 1982. Predicting crop yield in response to irrigation practices. Fla. Agric. Res. 82:20–22.

----, and A.G. Smajstrla. 1982. Application of modeling to irrigation management of soybeans. p. 571–599. *In* F.T. Corbin (ed.) World soybean research conference II: Proceedings. Westview Press, Boulder, CO.

Jung, P.K., and H.D. Scott. 1980. Leaf water potential, stomatal resistance and temperature relations in field-grown soybeans. Agron. J. 72:986–990.

Kamprath, E.J., D.K. Cassel, H.D. Gross, and D.W. Dibb. 1979. Tillage effects on biomass production and moisture utilization by soybeans on Coastal Plain soils. Agron. J. 71:1001–1005.

Kanemasu, E.T., L.R. Stone, and W.L. Powers. 1976. Evapotranspiration model tested for soybean and sorghum. Agron. J. 68:569–572.

Kapusta, G. 1979. Seedbed tillage and herbicide influence on soybean weed control and yield. Weed Sci. 27:520–526.

Karlen, D.L., P.G. Hunt, and T.A. Matheny. 1982. Accumulation and distribution of P, Fe, Mn, and Zn by selected determinate soybean cultivars grown with and without irrigation. Agron. J. 74:297–303.

Kashirad, A., J.G.A. Fiskell, V.W. Carlisle, and C.E. Hutton. 1967. Tillage pan characterization of selected coastal plain soils. Soil Sci. Soc. Am. Proc. 31:534–541.

Klute, A. 1982. Tillage effects on the hydraulic properties of soil: A review. p. 29–44. *In* P.W. Unger, and D.M. Van Doren, Jr. (ed.) Predicting tillage effects on soil physical properties and processes. Spec. Pub. 44. American Society of Agronomy and Soil Science Society of America, Madison, WI.

Korte, L.L., J.E. Specht, J.H. Williams, and R.C. Sorensen. 1983a. Irrigation of soybean genotypes during reproductive ontogeny. II. Yield component responses. Crop Sci. 23:528–533.

----, ----, ----, and ----. 1983b. Irrigation of soybean genotypes during reproductive ontogeny. I. Agronomic responses. Crop Sci. 23:521–527.

Laflen, J.M., and T.S. Colvin. 1981. Effect of crop residue on soil loss from continuous row cropping. Trans. ASAE 24:605–609.

----, and ----. 1982. Soil and water loss from no-till narrow-row soybeans. ASAE Paper 82-2033. American Society of Agricultural Engineering, St. Joseph, MI.

----, and W.C. Moldenhauer. 1979. Soil and water losses from corn-soybean rotations. Soil Sci. Soc. Am. J. 43:1213–1215.

Lindemann, W.C., G.E. Hamm, and G.W. Randall. 1982. Soil compaction effects on soybean nodulation, $N_2(C_2H_4)$ fixation and seed yield. Agron. J. 74:307–310.

Lutz, Jr., J.A., and G.D. Jones. 1975. Effect of irrigation, lime and fertility treatments on the yield and chemical composition of soybeans. Agron. J. 67:523–526.

Mahler, R.L., and A.G. Wollum II. 1981. The influence of irrigation and *Rhizobium japonicum* strains on yields of soybeans grown in a Lakeland sand. Agron. J. 73:647–651.

Mannering, J.V., and C.B. Johnson. 1969. Effect of row crop spacing on erosion and infiltration. Agron. J. 61:902–905.

Martin, C.K., D.K. Cassel, and E.J. Kamprath. 1979. Irrigation and tillage effects on soybeans yields in a Coastal Plains soil. Agron. J. 71:592–594.

Mason, W.K., G.A. Constable, and R.C.G. Smith. 1981. Irrigation for crops in a subhumid environment. II. The water requirement of soybeans. Irrig. Sci. 2:13–22.

----, H.R. Rouse, A.T.P. Bennie, T.C. Kaspar, and H.M. Taylor. 1982. Responses of soybeans to two-row spacings and two soil water levels. II. Water use, root growth and plant water status. Field Crops Res. 5:15–29.

----, and R.C.G. Smith. 1981. Irrigation for crops in a subhumid environment. III. An irrigation scheduling model for predicting soybean water use and crop yield. Irrig. Sci. 2:89–101.

Matheny, T.A., and P.G. Hunt. 1983. Effects of irrigation on accumulation of soil and symbiotically fixed N by soybean grown on a Norfolk loamy sand. Agron. J. 75:719–722.

Matson, A.L. 1964. Some factors affecting the yield response of soybeans to irrigation. Agron. J. 56:552–555.

Mazurak, A.P., and K. Pohlman. 1968. Growth of corn and soybean seedlings as related to soil compaction and matric suction. Trans. Int. Congr. Soil Sci. 9th 1:813–822.

McGregor, K.C. 1978. C factors for no-till and conventional-till soybeans from plot data. Trans. ASAE 21:1119–1122.

McKibben, G.E. 1980. A three-year comparison of O-till, conventional and plow-plant corn and soybeans following eleven years of continuous corn. Ill. Agric. Exp. Stn. DSAC Rep. 8:46–48.

Musick, J.T., L.L. New, and D.A. Dusek. 1976. Soil water depletion-yield relationships of irrigated sorghum, wheat and soybeans. Trans. Am. Soc. Agric. Eng. 19:489–493.

Nathanson, K., R.J. Lawn, P.L.M. de Jabrun, and D.E. Byth. 1984. Growth, nodulation and nitrogen accumulation by soybean in saturated soil culture. Field Crops Res. 8:73–92.

Nelson, W.W. 1973. Tillage for soybeans—Lamberton. 1972. Univ. of Minn. Dep. Soil Sci. Soil Ser. 89:86.

Parker, M.B., N.A. Minton, O.L. Brooks, and C.E. Perry. 1975. Soybean yields and Lance Nematode populations as affected by subsoiling, fertility, and nematicide treatments. Agron. J. 67:663–666.

Peters, D.B., and L.C. Johnson. 1960. Soil moisture use by soybeans. Agron. J. 52:687–689.

Pierce, F.J., W.E. Larson, R.H. Dowdy, and W.A.P. Graham. 1983. Productivity of soils: Assessing long-term changes due to erosion. J. Soil Water Conserv. 38:39–44.

Randall, G.W. 1974. Corn-soybean tillage. Univ. Minn. Agric. Ext. Serv. Soil Ser. 91:160–166.

----. 1978. Corn-soybean tillage. Univ. Minn. Agric. Ext. Serv. Soil Ser. 103:127–133.

Reicosky, D.C., and D.E. Deaton. 1979. Soybean water extraction, leaf water potential and evapotranspiration during drought. Agron. J. 71:45–50.

----, H.R. Rouse, W.K. Mason, and H.M. Taylor. 1982. Effect of irrigation and row spacings on soybean water use. Agron. J. 74:954–964.

Reiss, W.D., and L.V. Shorewood. 1965. Effect of row spacing, seeding rate, and potassium and calcium hydroxide additions on soybean yields on soils of southern Illinois. Agron. J. 57:431–433.

Rhoads, F.M. 1978. Response of soybeans to subsoiling in northern Florida. Proc. Soil Crop Sci. Soc. Fla. 37:151–154.

Richey, C.B., D.R. Griffith, and S.D. Parsons. 1977. Yields and cultural energy requirements for corn and soybeans with various tillage-planting systems. Adv. Agron. 29:141–182.

Robertson, W.K., L.C. Hammond, J.T. Johnson, and K.J. Boote. 1980. Effects of plant-water stress on root distribution of corn, soybeans and peanuts in sandy soil. Agron. J. 72:548–550.

Rojiani, K.B., B.B. Ross, F.E. Woeste, and V.O. Shamholtz. 1982. A probabilistic model for the assessment of plant water availability. Trans. ASAE 25:1576–1588.

Safo-Kantanaka, O., and N.C. Lawson. 1980. The effect of different row spacings and plant arrangements on soybeans. Can. J. Plant Sci. 60:227–231.

Sammons, D.J., D.B. Peters, and T. Hymowitz. 1981. Screening soybeans tolerance to moisture stress: A field procedure. Field Crops Res. 3:321–335.

Sanford, J.O. 1982. Straw and tillage management practices in soybean-wheat double-cropping. Agron. J. 74:1032–1035.

----, D.L. Myhre, and N.C. Merwine. 1973. Double cropping systems involving no-tillage and conventional tillage. Agron. J. 65:978-982.

Shaw, R.H., and D.R. Laing. 1966. Moisture stress and plant response. p. 73–94. *In* W.H. Pierre et al. (ed.) Plant environment and efficient water use. American Society of Agronomy, Madison, WI.

Shibles, R.M., and C.R. Weber. 1966. Interception of solar radiation and dry matter production by various soybean planting patterns. Crop Sci. 6:55–59.

Siemens, J.C., and W.R. Oschwald. 1976. Tillage practices for soybean production. p. 63–73. *In* L.D. Hill (ed.) World soybean research conference proceedings. The Interstate Printers and Publishers, Danville, IL.

----, and ----. 1978. Corn-soybean tillage systems: Erosion control, effects on crop production, costs. Trans. Am. Soc. Agric. Eng. 21:293–302.

Somerhalder, B.P., and P.E. Schleusener. 1960. Irrigation can increase soybean production. Nebr. Exp. Stn. Q. 7:16–17.

Stegman, E.C., L.H. Schiele, and A. Bauer. 1976. Plant water stress criteria for irrigation scheduling. Trans. Am. Soc. Agric. Eng. 19:850–855.

Sudar, R.A., K.E. Saxton, and R.G. Spomer. 1981. A predictive model of water stress in corn and soybeans. Trans. Am. Soc. Agric. Eng. 24:97–102.

Suman, R.F., and T.C. Peele. 1974. Limiting agronomic factors in soybean production. S.C. Agric. Exp. Stn. Tech. Bull. 1051.

Taylor, H.M. 1980. Soybean growth and yield as affected by row spacing and by seasonal water supply. Agron. J. 72:543–547.

----, W.K. Mason, A.T.P. Bennie, and H.R. Rouse. 1982. Responses of soybeans to two-row spacings and two soil water levels. I. An analysis of biomass accumulation, canopy development and solar radiation, interception and components of seed yield. Field Crops Res. 5:1–14.

Thompson, J.A. 1977. Effect of irrigation termination on the yield of soybeans in southern New South Wales. Aust. J. Exp. Agric. Anim. Husb. 17:156–160.

Touchton, J.T., W.L. Hargrove, R.R. Sharpe, and F.C. Boswell. 1982. Time, rate and method of phosphorous application for continuously double-cropped wheat and soybeans. Soil Sci. Soc. Am. J. 46:861–864.

----, and J.W. Johnson. 1982. Soybean tillage and planting method effects on yield of double-cropped wheat and soybeans. Agron. J. 74:57–59.

Troedson, R.J., R.J. Lawn, D.E. Byth, and G.L. Wilson. 1985. Saturated soil culture—an innovative water management option for soybeans in the tropics and subtropics. p. 171–180. *In* S. Shanmugasundaram and E.W. Sulzberger (ed.) Soybean in tropical and subtropical cropping systems. Asian Vegetable Research and Development Center, Shanhua, Taiwan.

Trouse, A.C., Jr. 1979. Some advantages of under-the-row subsoiling. p. 321–325. *In* Proc. Int. Soil Tillage Res. Org. (8th Conf.) Stuttgart, Fed. Rep. of Germany. 10–15 September. Univ. of Hohenheim Press, Stuttgart.

Tupper, G.R. 1978. Soybean response to deep tillage method and date on a silty clay soil. Miss. Agric. For. Exp. Stn. Res. Rep. 4.

Tyler, D.D., and T.C. McCutchen. 1980. The effect of three tillage methods on soybeans grown on silt loam soils with fragipans. Tenn. Farm Home Sci. 114:23–26.

----, and J.R. Overton. 1982. No-tillage advantages for soybean seed quality during drought stress. Agron. J. 74:344–347.

----, ----, and A.Y. Chambers. 1983. Tillage effects on soil properties, diseases, cyst nematodes, and soybean yields. J. Soil Water Conserv. 38:374–376.

Unger, P.W., and D.M. Van Doren, Jr. (ed.) 1982. Predicting tillage effects on soil physical properties and processes. Spec. Pub. 44. American Society of Agronomy and Soil Science Society of America, Madison, WI.

Van Doren, D.M., Jr., G.B. Triplett, Jr., and J.E. Henry. 1977. Influence of long-term tillage and crop rotation combinations on crop yields and selected soil parameters for an Aeric Ochraqualf soil. Ohio Agric. Res. Dev. Ctr. Res. Bull. 1091.

Vaughan, D.H., E.S. Smith, and H.A. Hughes. 1976. Energy requirements of reduced tillage practices for corn and soybean production in Virginia. p. 245–254. *In* William Lockeretz (ed.) Agriculture and energy, Academic Press, New York.

Voorhess, W.B., R.R. Allmaras, and C.E. Johnson. 1981. Alleviating temperature stress. p. 217–266. *In* G.F. Arkin and H.M. Taylor (ed.) Modifying the root environment to reduce crop stress. ASAE Monograph 4. American Society of Agricultural Engineering, St. Joseph, MI.

----, V.A. Carlson, and C.G. Senst. 1976. Soybean nodulation as affected by wheel traffic. Agron. J. 68:976–979.

Weber, C.R., R.M. Schibles, and D.E. Byth. 1966. Effect of plant population in row spacing on soybean development and production. Agron. J. 58:99–102.

Whitt, D.M., and C.H.M. van Bavel. 1955. Irrigation of tobacco, peanuts and soybeans. p. 376–381. *In* Water. USDA Yearbook. U.S. Government Printing Office, Washington, DC.

Wischmeier, W.H., and D.D. Smith. 1978. Predicting rainfall erosion losses. USDA Agric. Handb. 537. U.S. Government Printing Office, Washington, DC.

Wolf, D., T.H. Garner, and J.W. Davis. 1981. Tillage mechanical energy input and soil-crop response. Trans. ASAE 24:1412–1419.

Woodruff, C.M., M.R. Peterson, D.H. Schnarre, and C.F. Cromwell. 1972. Irrigation scheduling with planned soil moisture depletion. ASAE Paper 72-772. American Society of Agricultural Engineers, St. Joseph, MI.

Woodruff, N.P., and F.H. Siddoway. 1965. A wind erosion equation. Soil Sci. Soc. Am. Proc. 29:602–608.

Young, J.K., F.D. Whisler, and H.F. Hodges. 1979. Soybean leaf N as influenced by seedbed preparation methods and stages of growth. Agron. J. 71:568–573.

11 Weed Control

T. N. Jordan
Purdue University
West Lafayette, Indiana

H. D. Coble
North Carolina State University
Raleigh, North Carolina

L. M. Wax
USDA-ARS
Urbana, Illinois

The intensity and distribution of weed species in the soybean crop are functions of a complex interaction among soil properties, rainfall patterns, temperature, and cultural practices. These factors vary according to region, and the soybean-growing areas of the USA may be divided into four regions: Mississippi Delta, Southeast, Mid-Atlantic, and North Central. The states comprising each of these areas and the weeds of importance in each region are included in Table 11–1. Within each region, some 15 to 20 species make up the predominant weed population; over 40 species are found competing with soybean throughout the USA (Elmore, 1983; Shaw, 1982).

Table 11–1. Common and scientific names of major weeds found in soybean and areas of importance.

Common name	Scientific name	Areas of importance†
Annual grass		
Barnyardgrass	*Echinochloa crusgalli* (L.) Beauv.	Ms De, No Ce
Broadleaf signalgrass	*Brachiaria platyphylla* (Griseb.) Nash	Ms De, So Ea, Mi At
Crabgrass	*Digitaria* spp.	Ms De, So Ea, MiAt, No Ce
Fall panicum	*Panicum dichotomiflorum* Michx.	Mi At, No Ce
Foxtails	*Setaria* spp.	Mi At, No Ce
Goosegrass	*Eleusine indica* (L.) Gaertn.	Ms De, So Ea, Mi At
Red rice	*Oryza sativa* L.	Ms De
Red sprangletop	*Leptochloa filiformis* (Lam.) Beauv.	Ms De, So Ea
Shattercane	*Sorghum bicolor* (L.) Moench	No Ce
Texas panicum	*Panicum texanum* Buckl.	So Ea

(continued on next page)

Soybeans: Improvement, Production, and Uses, 2nd ed.—Agronomy Monograph no. 16.

Table 11-1. Continued.

Common name	Scientific name	Areas of importance†
Field sandbur	*Cenchrus incertus* M.A. Curtis	So Ea
Annual broadleaf		
Eastern black nightshade	*Solanum ptycanthum* Dun.	Mi At, No Ce
Common cocklebur	*Xanthium strumarium* L.	Ms De, So Ea, Mi At, No Ce
Common lambsquarters	*Chenopodium album* L.	Ms De, So Ea, Mi At, No Ce
Common ragweed	*Ambrosia artemisiifolia* L.	Ms De, So Ea, Mi At, No Ce
Florida beggarweed	*Desmodium tortuosum* (Sw.) DC.	So Ea
Florida pusley	*Richardia scabra* L.	So Ea
Giant ragweed	*Ambrosia trifida* L.	No Ce
Hemp sesbania	*Sesbania exaltata* (Raf.) Cory	Ms De
Jimsonweed	*Datura stramonium* L.	Mi At, No Ce
Mexicanweed	*Caperonia palustris* (L.) Stihil	Ms De
Morningglories	*Ipomoea* spp.	Ms De, So Ea, Mi At, No Ce
Pigweeds	*Amaranthus* spp.	Ms De, So Ea, Mi At, No Ce
Prickly sida	*Sida spinosa* L.	Ms De, So Ea
Sicklepod	*Cassia obtusifolia* L.	Ms De, So Ea, Mi At
Smartweeds	*Polygonum* spp.	Mi At, No Ce
Nodding spurge	*Euphorbia nutans* Lag.	Ms De
Spurred anoda	*Anoda cristata* (L.) Schlect.	Ms De, So Ea
Velvetleaf	*Abutilon theophrasti* Medik.	Mi At, No Ce
Sunflower	*Helianthus annuus* L.	No Ce
Wild mustards	*Brassica* spp.	No Ce
Wild poinsettia	*Euphorbia heterophylla*	Ms De
Perennials		
Bermudagrass	*Cynodon dactylon* (L.) Pers.	Ms De, So Ea
Canada thistle	*Cirsium arvense* (L.) Scop.	No Ce
Common milkweed	*Asclepias syriaca* L.	No Ce
Field bindweed	*Convolvulus arvensis* L.	No Ce
Honeyvine milkweed	*Ampelamus albidus* (Nutt.) Britt.	Ms De, No Ce
Johnsongrass	*Sorghum halepense* (L.) Pers.	Ms De, So Ea, Mi At, No Ce
Purple nutsedge	*Cyperus rotundus* L.	Ms De, So Ea
Quackgrass	*Agropyron repens* (L.) Beauv.	Ms De, No Ce
Trumpet creeper	*Campsis radicans* (L.) Seem.	Ms De, So Ea, Mi At
Yellow nutsedge	*Cyperus esculentus* L.	Mi At, So Ea, Mi At, No Ce
Redvine	*Brunnichia ovata* (Walt.) Shinners.	Ms De
Special weed problems		
Balloonvine	*Cardiospermum halicacabum* L.	Ms De, So Ea
Crotalaria	*Crotalaria spectabilis* Roth	Ms De, So Ea
Itchgrass	*Rottboellia exaltata* L.f.	Ms De
Southern cowpea	*Vigna unguiculata* (L.) Walp.	So Ea
Wild proso millet	*Panicum miliaceum* L.	No Ce

†Ms De (Mississippi Delta) includes LA, MS, AR, TN, MO bootheel; So Ea (Southeast) includes FL, AL, GA, SC, southern NC; Mi At (Mid-Atlantic) includes NC, VA, DE, MD, NJ, PA; and No Ce (North Central) includes OH, KY, IN, MI, IL, WI, MN, IA, MO, KS, NE.

11-1 WEED DISTRIBUTION IN THE MISSISSIPPI DELTA REGION

The warm temperatures, relatively high rainfall, and nutrient-rich alluvial soils of the Mississippi Delta area provide an environment suitable for a wide array of weed species. Among the most common species infesting soybean in the area are johnsongrass [*Sorghum halepense* (L.) Pers.], bermudagrass [*Cynodon dactylon* (L.) Pers.], common cocklebur (*Xanthium strumarium* L.) and several species of annual morningglories (*Ipomoea* spp.), with pitted, entireleaf, and ivyleaf the most prevalent. Other troublesome broadleaf weeds in the Delta include hemp sesbania [*Sesbania exaltata* (Raf.) Cory], sicklepod (*Cassia obtusifolia* L.), nodding spurge (*Euphoria nutans* Lag.), spurred anoda [*Anoda cristata* (L.), pigweeds (*Amaranthus* spp.), prickly sida (*Sida spinosa* L.) common lambsquarters (*Chenopodium album* L.), wild poinsettia (*Euphorbia heterophylla*) mexicanweed [*Caperonia palustrus* (L.) and common ragweed (*Ambrosia artemisiifolia* L.). Common annual grass species include barnyardgrass, (*Echinochloa crusgalli* (L.) Beauv.), broadleaf signalgrass [*Brachiaria Platyphylla* (Griseb.) Nash], crabgrass (*Digitaria* spp.), and goosegrass [*Eleusine indica* (L.) Gaertn.] with localized infestations of red rice (*Oryza sativa* L.) and red sprangletop [*Leptochloa filiformis* (Lam.) Beauv.]. Perennial weeds found in soybean in the Delta in addition to johnsongrass and bermudagrass include yellow and purple nutsedge (*Cyperus esculentus* L. and *C. rotundus* L.) as well as honeyvine milkweed [*Ampelamus albidus* (Nutt.) Britt], trumpet creeper [*Campsis radicans* (L.) Seem] and redvine (*Brunnichia cirrhosa* Gaertn.) in scattered areas.

11-2 WEED DISTRIBUTION IN THE SOUTHEAST REGION

The Southeast is similar to the Mississippi Delta with respect to temperature and rainfall, but the soil characteristics are more varied. The soils in the region range from the coarse-textured sands and sandy loams of the coastal plain to the fine-textured clay soils of the piedmont plateau. In addition, small pockets of highly organic muck soils exist over the area near the coast. Major annual grass weeds in the region include crabgrass, goosegrass, and broadleaf signalgrass over the entire region with Texas panicum (*Panicum texanum* Buckl.) and red sprangletop causing more problems in the southern half of the region. In the coarse-textured sandy soils, field sandbur (*Cenchrus incertus* M.A. Curtis) is often a problem, and fall panicum (*Panicum dichotomiflorum* Michx.) is increasing in the northern part of the region.

The most troublesome broadleaf weeds in the Southeastern region are common cocklebur, sicklepod, Florida beggarweed [*Desmodium tortuosum* (SW.) DC], pigweeds, and the annual morningglories. Other common broadleaf weeds include common ragweed, common lambsquarters, Florida pusley (*Richardia scabra* L.), prickly sida, and spurred anoda.

Major perennial weed problems of the region include johnsongrass, bermudagrass, and the nutsedges, with scattered infestations of trumpet creeper.

11-3 MID-ATLANTIC REGION

The Mid-Atlantic region is characterized by relatively uniform sandy loam soils and adequate rainfall. Temperatures are somewhat lower than those in the Delta and Southeast, and the growing season is shorter. Major annual grass species in the area include fall panicum, crabgrass, and giant foxtail (*Setaria faberii*); broadleaf signalgrass and goosegrass are present in the southern part of the region.

Annual broadleaf weeds causing major losses in soybean include common cocklebur, smartweeds, pigweeds, jimsonweed [Datura stramonium (L.)], Eastern black nightshade (*Solanum ptycanthum* Dun.) and the annual morningglories. Common ragweed, common lambsquarters, and velvetleaf (*Abutilon theophrasti* Medik.) are common in parts of the region, and sicklepod is moving into the southern section of the area.

The major perennial weed in the area is johnsongrass, although scattered infestations of quackgrass [*Agropyron repens* (L.) Beauv.] yellow nutsedge, and trumpet creeper are also found.

11-4 WEED DISTRIBUTION IN THE NORTH CENTRAL REGION

The North Central region is typified by medium- to fine-textured soils with relatively high organic matter levels and high fertility levels. The region receives less rainfall than the other soybean-growing areas, but water availability from subsoil is much greater there. The major annual grass weed in the region is giant foxtail, while some of the other foxtails are predominant in scattered areas. Fall panicum has increased since the mid-1970s and is now considered a major weed pest. Crabgrass and barnyardgrass are also abundant in the region. Shattercane [*Sorghum bicolor* (L.) Moench.] is a serious problem in some of the bottom land near rivers in the southern and western part of the region.

Annual broadleaf weeds causing the greatest loss in the North Central region are common cocklebur, velvetleaf, smartweeds (*Polygonum* spp.) and pigweeds. Other common infestants include common lambsquarters, common ragweed, jimsonweed, Eastern black nightshade, and tall and ivyleaf morningglory. Scattered infestations of giant ragweed (*Ambrosia trifida* L.), wild mustard (*Brassica* spp.), and sunflower (*Helianthus annuus* L.) are also found and cause serious problems when they are present.

There are several perennial weeds in soybean in the North Central region. In the southern part johnsongrass causes problems, while quackgrass and Canada thistle [*Cirsium arvense* (L.) Scop.] are pests in the northern part of the region. Field bindweed (*Convolvulus arvensis* L.) is abundant in the western states of the region with yellow nutsedge, honeyvine milkweed, and common milkweed (*Asclepias syriaca* L.) found throughout the region.

11–5 SPECIAL WEED PROBLEMS

The unique characteristics of certain weeds cause them to receive special attention, even though they may not be widespread throughout a region of the soybean-growing area. One such weed is balloonvine (*Cardiospermum halicacabum* L.), an annual broadleaf weed with a vining growth habit found in the Mississippi Delta and Southeast region. Balloonvine does not cause major yield losses, but its seeds are essentially the same size and shape as soybean. Therefore, it is impossible to remove the weed seeds from soybean seeds by conventional methods, and soybean grown for seed will be contaminated if balloonvine is present in the field. A similar problem occurs in the Southeast with southern cowpea [*Vigna unguiculata* (L.) Walp.], a cultivated plant that has escaped to become a weed in localized areas. Cowpea is difficult to control in soybean, and the similarity of seed size and shape makes soybean seed contamination a problem.

Crotalaria (*Crotalaria spectabilis* Roth.) is an annual broadleaf weed found in the Southeast and Mississippi Delta areas. It is of special concern because of the high toxicity of its seed to animals, especially chickens (*Gallus gallus domesticus*). Soybean contaminated with crotalaria seed cannot be sold, because feed products made from those lots would be unusable. Crotalaria is also difficult to control and germinates over a long period of time, thus requiring special efforts for control.

Itchgrass (*Rottboellia exaltata* L.f.) is a robust, profusely tillering annual grass of tropical origin that has become established in isolated areas of the lower Mississippi Delta near the Gulf Coast. It can exert devastating effects on soybean and, although presently confined to isolated infestations, could spread over much of the soybean-growing area of the USA (Patterson et al., 1979).

In the North Central region wild proso millet (*Panicum miliaceum* L.) presents a problem similar to that posed by itchgrass in the South. Presently confined to areas of Wisconsin, Minnesota, Iowa, and Illinois, wild proso millet can cause devastating loss in soybean and has the ability to survive in most soybean-growing areas.

11–6 WEED POPULATION SHIFTS

In order for a plant to become a weed, it must be capable of invading and colonizing a disturbed site, such as a soybean field. The invasion of a disturbed site by a potential weed occurs when its reproductive propagules are introduced into the area either by natural dissemination or by man. The invasion process is relatively simple and usually occurs annually, at least to some degree. Colonization of a disturbed site is much more complex and is largely influenced by cultural practices, especially tillage and weed control (Baker, 1974).

In soybean production, tillage is used to prepare a good seedbed as well as provide weed control. Therefore, successful weeds in soybean are usually those with the same life cycle and growth habits as the soybean. In minimum or no-till situations, weed populations usually increase both in number and diversity unless adequate chemical control is substituted for the tillage (Worsham, 1970). There is usually a rapid increase in annual grass populations, followed in a relatively short time by perennial species (Triplett and Little, 1972).

The objective of weed control for the soybean producer is to grow his crop with as few weeds as economically feasible. To accomplish this objective, a variety of weed control practices are utilized. Herbicides and cultivation are the most successful tools for weed control early in the growing season with crop competition providing most of the control later in the year. If a particular group of weed species is able to survive the control practices imposed, it will quickly become the predominant weed complex in the field. A change in control practices will cause a change in the weed population: those species surviving the new practices becoming the major weed problem (Altieri, 1982). As long as successful colonizers with tolerance to changing weed control programs exist, continual weed population shifts will occur.

11-7 LOSSES DUE TO WEEDS

Soybean yield losses resulting from weed interference and the cost of weed control constitute some of the highest costs involved in the production of the crop (Anonymous, 1979). Nationwide, the monetary losses due to weeds in the past few years have averaged about 17% of the crop value, or approximately $1.9 billion (Chandler et al., 1984). By adding an estimated cost of control of $1.1 billion, the total cost of weeds in the soybean crop for the USA can be calculated at approximately $3.0 billion per year. The cost is greater than that for all other pests combined (Shaw, 1978).

Weeds compete directly with soybean for light, nutrients, and moisture, and may interfere indirectly through the production and release of allelopathic chemicals that inhibit crop growth (Lolas and Coble, 1982). Weeds often serve as alternate hosts for insects and plant pathogens that attack soybean, and the physical presence of weeds in the crop may interfere with the control of other pests. In addition, the efficiency of operation of harvesting equipment is reduced by the presence of significant numbers of weeds (Nave and Wax, 1971). The quality of the harvested crop is directly influenced by weeds. Increases in moisture content, foreign matter, and splits have been documented when high levels of weeds were present at harvest (Anderson and McWhorter, 1976).

The degree of interference between crops and weeds is directly proportional to the density and duration of the weed infestation in the crop. These factors have been investigated for a number of specific weeds in soybean (Thurlow and Buchanan, 1972; Oliver et al., 1976; Eaton et al., 1973; Coble and Ritter, 1978; Coble et al., 1981). In general, this research has shown that a period of 4 to 6 weeks without weed competition at the beginning of the growing season will allow production of maximum yields under most environmental conditions. Any weeds emerging in the crop after this initial weed-free period will not compete effectively with soybean and will not affect yield potential. Similarly, a period of 4 to 6 weeks of weed interference at the beginning of the season usually can be tolerated by soybean with no significant yield loss, provided that the crop is maintained weed-free for the remainder of the season. The onset of significant interference with crop yield potential is highly correlated with light interference (Coble et al., 1981).

Weed density relationships to crop yield reduction are weed species dependent but follow the same general pattern of decreased crop yield with increasing weed density. The relationship between crop yield and weed density has been schematically depicted as a sigmoid curve over a range of weed populations from 0 to maximum weed density (Zimdahl, 1980). The sigmoidal relationship most likely results from (i) researchers' inability to measure the true effects of low weed populations on crop yield, and (ii) intraspecific competition among weeds at high populations. If a single weed competes with soybean for some resource, then some of that resource will not be available for soybean growth, resulting in yield loss, however slight. Adding more weeds to the system will produce an incrementally greater yield loss in direct proportion to the number of weeds until the individual weed effects begin to overlap. Thus, the true relationship between soybean yield and weed density should be linear up to the point where weeds begin to interfere with each other.

The relationship between crop yield and weed density will cease to be linear when weeds begin to influence each other so that the overall effect of each weed is lessened. Soybean yield will not usually reach zero, even though maximum weed density is reached, because of intraspecific competition among weeds.

Natural weed populations in most fields are high enough to cause devastating yield losses if left uncontrolled. Loss figures of 50 to 90% are common for soybean grown in natural weed populations (Anderson and McWhorter, 1976; Barrentine, 1974; Coble and Ritter, 1978; Coble et al., 1981). The overall soybean yield loss estimate of 17% nationwide indicates that growers do a relatively good job of weed control, but that some residual weed population is left after control measures have been implemented.

It is in dealing with these residual weed populations that questions arise concerning the economics of additional control practices. The residual populations are usually low and the weeds are mostly scattered over the field and not intraspecifically competitive. Therefore, linear relationships between crop yield loss and weed density should be valid.

The relative competitive abilities of selected annual broadleaf weeds in soybean are presented in Table 11–2. In this table, data taken from the references cited were subjected to linear regression analysis and the soybean yield loss per weed per 10 m^2 calculated. These data illustrate the relative differences among weed species in their competitive effect on soybean. Common cocklebur is the most competitive species evaluated thus far with one weed every 10 m^2 resulting in over 66 kg ha^{-1} loss in yield. Velvetleaf, common ragweed, hemp sesbania, and Pennsylvania smartweed all caused similar yield reductions and were about one-half as competitive as common cocklebur with losses ranging from 31 to 33 kg ha^{-1} per weed per 10 m^2. Tall morningglory was slightly less than one-third as competitive as common cocklebur but caused almost twice the yield reduction of Venice mallow, prickly sida, or sicklepod. Relatively little work has been conducted with grass competition in soybean, but published results indicate grass weeds are far less competitive on a per plant basis than broadlead weeds (Slife, 1979; McCarty, 1983).

11–8 CONTROL PRACTICES

11–8.1 Non-chemical Control

Nonchemical control of weeds in soybean, or in any crop for that matter, is a primary function of the overall management input into the production practices for the crop. This input may include any number of managerial choices such as mechanical control (tillage or mowing), biological control of weeds, crop rotation with the associated herbicides

Table 11–2. Relative competitive abilities of selected broadleaf weeds in soybean.

Weed species	Yield loss† per weed per 10 m^2	Weed-free yield	Percentage yield reduction	Reference
	kg ha^{-1}		%	
Common cocklebur	66.4	1960	3.4	Barrentine (1974)
Velvetleaf	33.3	3110	1.1	Eaton et al. (1976)
Common ragweed	33.0	1661	2.0	Coble et al. (1981)
Hemp sesbania	31.8	1981	1.6	McWhorter and Anderson (1979)
Pennsylvania smartweed	31.0	2798	1.1	Coble and Ritter (1978)
Tall morningglory	18.5	2000	0.9	Oliver et al. (1976)
Venice mallow	9.5	2042	0.5	Eaton et al. (1973)
Prickly sida	9.5	3110	0.3	Eaton et al. (1976)
Sicklepod	9.2	2400	0.4	Thurlow and Buchanan (1972)

†Data derived from linear regression analysis of data in each reference.

compatible with the subsequent crop, or preventive control. Preventive control includes all measures taken to prevent the introduction and spread of weeds. Basic among these measures are prevention of weed seed production and planting of weed-free crop seeds.

Practices to help insure planting weed-free seed include: i) choosing certified seed when possible and examining the label for weed seed information, ii) obtaining information on the weed seed content of unidentified seed, and iii) cleaning all home-grown seed. All seed that contains weed seeds should be recleaned.

Reproductive parts of weed plants such as rhizomes, bulbs, and tubers, are often transported from infested to uninfested areas by farm machinery, from tillage implements to harvesting equipment. Care should be taken to clean all pieces of equipment thoroughly before moving from an infested area to an uninfested area. Weed seed harvested and removed from the crop seed should be destroyed so that it will not reinfest fields.

Crop rotation is an excellent way to reduce weed populations in soybean. The rotation should include a strongly competitive crop grown in each part of the rotation, including both i) summer row crops such as corn, or drill crops such as rice and ii) winter or early spring grain crops such as wheat, oat, or barley. The competition from the alternate crop as well as the associated herbicide usually provide effective weed control (Klingman and Ashton, 1982).

Biological control of weeds in agronomic crops is not well understood and its use is better suited for large areas such as rangeland or waterways which are infested with only one major weed problem. In cultivated crops such as soybean, a complex of eight or more different weed species is not uncommon, and the control of a single weed by biological means will have little if any impact on crop yields. The exception is the control of perennial weeds such as nutsedges, canada thistle, quackgrass, bermudagrass, or perennial morning glory species. Biological control of these weeds would be a major contribution to more effective crop production (Klingman and Ashton, 1982). The judicious use of tillage practices and crop competition is also an effective nonchemical control practice. These latter two areas are discussed in more detail in other sections of this chapter.

11–8.2 Chemical Control

Herbicides discussed in this section are primarily those that have been registered for use in soybean production since the first edition of this monograph in 1973. For a detailed discussion of a history and development of soybean herbicides, as well as a discussion of the timing of application of soybean herbicides, the reader should refer to the first edition (1973). The herbicides used most effectively on soybean in the USA are listed (by common name and chemical name) in Table 11–3. For a more detailed discussion of herbicides, the Weed Science Society of America's *Herbicide Handbook* (Weed Science Society of America, 1983) or one of a number of weed science textbooks dealing with prin-

Table 11–3. Herbicides used to control weed in soybean in the USA.

Common name	Chemical name
Acifluorfen	Sodium 5-{2-chloro-4-(trifluoromethyl)phenoxy}-2 nitrobenzoic acid
Alachlor	2-chloro-*N*-(2,6-diethylphenyl)-*N*-(methoxymethyl) acetimide
Bentazon	3-(1-methylethyl)-(1*H*)-2,1,3-benzothiadiazin-4(3*H*)-one 2, 2-dioxide
Chloramben	3-amino-2,5-dichlorobenzoic acid
Chlorimuron	Ethyl 2-[[[[(-chloro-6-methoxypryrimidin-2-y1)amino]carbonyl]amino]sulfonyl]benzoate
Dalapon	2,2-dichloropropionic acid
Diclofop-methyl	Methyl ± 2-{4-(2,4-dichlorophenoxy)phenoxy}propanoic acid
Fluaziflop-butyl	butyl 2-{4-(5-trifluoromethyl-2-pyridinyl oxy) phenoxy}propanoic acid
Glyphosate	*N*-(phosphonomethyl)glycine
Imazaquin	2-[4,5-dinydro-4-methyl-4-(1-methylethyl)-5-oxo-1H-imidazol-2-y1]-3-quinolinecarboxylic acid
Linuron	*N′*-(3,4-dichlorophenyl)-1-methoxy-1-methylurea
Mefluidide	*N*-[2,4-dimethyl-5[[trifluoromethyl)sulfonyl] amino] phenyl] acetamide
Metolachlor	2-chloro-*N*-(2-ethyl-6-methylphenyl)-*N*-(2-methoxy-1-methylethyl)acetamide
Metribuzin	4-Amino-6-(1,1-dimethylethyl)-3-(methylthio)-1,2,4-triazin-5(4*H*)-one
Naptalam	Sodium 2-[(-1naphthalenylamino) carbonyl] benzoic acid
Norflurazon	4-chloro-5-(methylamino)-2-(trifluoromethyl)phenyl)-3(2*H*)-pyridazinone
Oryzalin	4-(dipropylamino)-3,5-dinitrobenzenesulfonamide
Paraquat	1,1′-dimethyl-4,4′-bipyridinium ion
Pendimethalin	*N*-(1-ethylpropyl)-3,4-dimethyl-2,6-dinitrobenzenenamine
Propachlor	2-chloro-*N*-(1-methylethyl)-*N*-phenylacetamide
Sethoxydim	2{1-(ethoxyimino)butyl}-5-{2-(ethylthio)propyl}-3-hydroxy-2-cyclohexen-1-one
Trifluralin	2,6-dinitro-*N*,*N*,-dipropyl-4-(trifluromethyl) benzenamine
Vernolate	*S*-propyl dipropylcarbamothioate
2,4-DB	4-(2,4-Dichlorophenoxy)butyric acid

†Common and chemical names are those used by the Weed Science Society of America.

ciples and practices of weed control or mode and mechanism of action of herbicides are suggested.

11–8.2.1 Current Herbicides

There are currently over 30 herbicides used on soybean in the USA. Many of these are limited to specific situations. The herbicides that are most extensively used are listed in Table 11–3. Those herbicides and practices that have made or that have shown the potential to make a significant contribution to the solution of weed control problems in soybean since the first edition of this monograph will be discussed. This discussion is not intended to serve as a recommendation. The herbicides discussed herein are those that are currently used in soybean weed control. Some will probably be replaced in the future as safe and/or more effective treatments are developed. Information on recommendations is provided and distributed by local and state extension personnel who are familiar with the problems and solutions in specific geographical areas.

In the early 1970s, the primary weed control programs were centered around soil-applied incorporated herbicides such as trifluralin, followed by an early postemergence application of dinoseb plus naptalam, or centered around a preemergence application of chloramben. In the North Central states, alachlor was used for annual grass control where a corn-soybean rotational system were common. Metribuzin became available in the early 1970s as a preemergence herbicide for control of annual broadleaf weeds, and in combination with trifluralin or alachlor, a broad-spectrum herbicide program could be developed for controlling annual grasses and broadleaf weeds with a tank mix of two herbicides applied simultaneously.

Broadcast applications of postemergence herbicides were generally considered for use as a last resort or salvage treatment if the soil-applied herbicide program did not perform as expected. The use of postemergence herbicide as a primary treatment became popular with the development and sale of bentazon for selectively controlling many broadleaf weeds, including common cocklebur, prickly sida, lambsquarter, velvetleaf, jimsonweed, smartweeds, morning glories, common ragweed, and wild mustard. Application of bentazon at 0.56 to 1.12 kg ha^{-1} over the top of soybean from the unifoliolate to the second trifoliolate stage of growth provided control of many of the broadleaf weeds that were left by soil-applied herbicide combinations.

By 1978, the use of glyphosate in recirculating sprayers and rope wick applicators was popular for the tall-growing grass species such as johnsongrass, volunteer corn, and shattercane. Additional herbicides were developed in the mid- to late 1970s for both surface-applied preemergence and selective over-the-top broadcast postemergence weed control. Metolachlor was added to the list of soil-applied herbicides that controlled annual grasses and small seed broadleaf weeds. Acifluorfen was developed as a selective postemergence herbicide which controlled morning glories and pigweed better than bentazon. The ability to apply herbicides preemergence to the crop without incorporating the products and the use of postemergence broadleaf herbicides as primary rather than salvage treatments led to an increase in reduced tillage soybean production systems.

By the early 1980s, three and four herbicides were being tank-mixed to provide weed control for reduced and no-tillage soybean production. Tank-mixing postemergence herbicides for broadspectrum control of weeds became a standard practice by the early 1980s. In 1983, two postemergence grass herbicides were registered for use in soybean. Sethoxydim and fluaziflop-butyl are both selective herbicides that are specific in their herbicidal action for control of both annual and perennial grasses. With the development of the selective postemergence grass herbicides a total postemergence herbicide program could finally be developed for no-tillage and solid-seeded soybean.

11-8.2.2 Factors Influencing Soil-applied Herbicide Treatments

Many factors influence the performance of soil-applied herbicides, including methods of application. Traditionally, most herbicides have been applied with ground application equipment either broadcast or in narrow bands centered over the row. However, aerial application of herbicides in row crops has been increasing. With proper calibration, application techniques, and weather conditions, herbicides applied with aerial equipment have performed as well as those applied with ground equipment (Bovey and Burnside, 1965). The use of aerial application equipment allows treament of an area when excessive rainfall has precluded application by other means. However, aerial application has certain limitations. Banded or directional treatments cannot be applied, and the likelihood of uneven distribution, drift, and air pollution may be greater than with ground equipment.

Perhaps no other single factor enhances the effectiveness of soil-applied herbicides as much as rainfall. One reason for this appears to be a requirement for the leaching of the herbicides into the soil. However, the amount and timeliness of rainfall required for optimum results depends on the characteristics of the herbicide, the soil texture, moisture and physical conditions, plant species, temperature, and other factors. Some herbicides perform well when incorporated into the soil without rainfall, but most of these herbicides are more effective with rainfall than without rainfall when left on the surface (Greer and Santelmann, 1968).

Herbicides are usually applied either as aqueous sprays or in granular form. Under optimum irrigation or rainfall, the granular formulation and the aqueous spray of the same herbicide usually perform similarly (Cardenas and Santelmann, 1966). However, in some instances granular formulations do not perform as well under limited rainfall as aqueous sprays. With some highly volatile herbicides, activity is not lost as rapidly from granular formulations as from aqueous sprays.

Texture and organic matter of the soil affect most herbicides; they markedly influence the behavior of a few herbicides, such as linuron, metribuzin, and trifluralin. As clay and organic matter content of the soil increase, there is greater adsorption of the herbicides onto clay and organic matter colloids; thus, rates must be increased to provide effective weed control. On some low organic, coarse-textured soils certain herbicides leach so readily that they may injure crops or fail to control weeds. The various factors influencing adsorption of herbicides by soils have been reviewed in detail (Bailey and White, 1964; Harris and Warren, 1964).

For most of the common herbicides used in soybean, organic matter is relatively more important than clay content in influencing herbicide activity. Manufacturers and public agencies recommend herbicide rates based on organic matter content of the soil. A laboratory analysis of organic matter, however, is not always possible on each different soil type on the grower's farm, and sampling is time consuming. Alexander (1969) developed a color chart that allows the grower to estimate quickly the organic matter content of his soil. The color chart is reliable on many Midwestern soils, but may not apply to darker soil with low organic matter in the Southern soybean regions. Neither does it apply to sandy soils, peats, or mucks. Careful adjustment of rates on the basis of organic matter can provide optimum weed control and minimize crop injury and cost.

Another factor that affects tolerance of soybean plants to herbicide injury is soil pH. In the Midwestern states, soybean is largely grown in rotation with corn. Triazine herbicides are standardly used in both crops {atrazine [2-chloro-4-(ethylamino)-6-(isopropylamino)-*s*-triazine] in corn and metribuzin in soybean} for controlling annual weeds. The weed control obtained from this group of herbicides is dependent on the pH of the soil. Soils with pH of 5.8 or lower decrease the activity of triazine herbicides by increasing the adsorption of herbicide moleculars to the soil particles, rendering the chemical unavailable for weed control. When the soil pH is above 7.2, an excessive amount of chemical is released from the soil particle to the soil solution to be taken up by plant root systems, increasing both weed control and soybean plant injury. This high pH condition is especially critical when atrazine residues are present from the previous year's corn crop.

The addition of metribuzin to soils already having atrazine will cause the total concentration of triazine to be greater than that which can be tolerated by young soybean plants. Ladlie et al. (1977) showed that low concentrations of atrazine, such as those found in residual amounts after a year of corn production, increased the uptake of metribuzin in soybean plants by increasing the stomatal aperture, causing an increase in transpiration by the plants. They stated that the conditions which favored the synergistic herbicidal action of atrazine and metribuzin were: low atrizine residue levels which caused an increase in transpiration, high metribuzin rates, and high soil pH levels. As the atrazine carryover/metribuzin interaction became widespread by the late 1970s, atrazine rates were lowered in fields that were to be rotated to soybean the following year. This reduction of atrazine rates in corn has at times caused an increase in problem weeds such as cocklebur, velvetleaf, and jimsonweed which are best controlled in soybean with application of postemergence herbicides.

11-8.2.3 Factors Influencing Foliar Herbicide Treatments

Treatments applied to the foliage of plants may be influenced by temperature, relative humidity, rainfall, herbicide rate and uniformity of application, growth stage of crop and weed, soil moisture availability, and many other factors. These factors influence the adsorption translocation and eventual fate of herbicides within the plant and thus impact the effectiveness of the treatment. Timing postemergence herbicides for maximizing the optimum influence by the environment is essential.

The addition of surfactants or crop oils to the spray solution markedly increases the effectiveness of foliar-applied herbicides because they enhance herbicide retention on the leaf surface and subsequent penetration or adsorption into the leaf (Gentner, 1966; Hill et al., 1965; Jansen et al., 1961; Smith et al., 1967). Other plant-adjuvant-herbicide interactions may account for increased phytotoxicity beyond that attributed to improve surface wetting and penetration. Studies on the influence of surfactant structure on herbicidal activity demonstrated considerable differences in activity among various surfactant structures (Jansen, 1965; Smith et al., 1966). An inappropriate surfactant may be of little value or may actually decrease the effectiveness of the herbicide in question. The nonphytotoxic crop oil concentrate (which includes a surfactant as an emulsifier) increases wetting and penetration of herbicide sprays. These nonphytotoxic oils are either of petroleum or vegetable base and usually increase effectiveness of the herbicide as much or more than the traditional surfactants.

The volume of water in which postemergence herbicides are applied has been of interest to researchers and farmers alike in recent years. This interest has been largely the result of the development of new application equipment capable of applying low volumes in the range of 5 to 50 L ha^{-1}. The application equipment is discussed in a different section of this chapter. With most postemergence herbicides, both contact and translocatable, the volume of diluent does not influence the effectiveness of herbicides provided the application is accurate and uniform. With some translocated herbicides such as glyphosate, reducing the volume of diluent increases the herbicidal activity. Jordan (1981) found that bermudagrass control with 0.84 kg ha^{-1} glyphosate could be increased from approximately 20 to 90% by decreasing the diluent volume from 374 to 47 L ha^{-1}. Other such research has been reported with translocatable herbicides. It is generally felt that the increase in herbicide activity by reducing the volume of diluent is the result of increased concentration of herbicides as well as any formulation adjuvant in each spray droplet.

Another factor affecting the effectiveness of postemergence herbicides is the recent practice of tank mixing the postemergence grass herbicides, sethoxydim, or fluazifop-butyl, with the selective broadleaf herbicides, bentazon, or acifluorfen. The tank mixture of any of these combinations results in a decrease in activity of the postemergence grass herbicide. Several researchers have reported on this antagonistic interaction (Banks and Tripp, 1983, Rhodes and Coble, 1982; Hartzler and Foy, 1983). The antagonism appears to be related to a chemical incompatibility between the herbicides when mixed in the same solution. The antagonism can be overcome by applying the two chemicals separately at intervals of as little as 1.5 h.

The development of new application technology and selective postemergence herbicides has given farmers an alternative to conventional tillage and planting practices as well an alternative to a soil-applied herbicide weed control program. Since postemergence herbicides are not influenced as readily by soil characteristics, tillage practices, or crop residues, there is a move toward no-tillage, narrow row, drill soybean, or double-cropping soybean after small grains throughout the entire U.S. soybean production region.

11-9 TILLAGE AND CROPPING PRACTICES

Timely, effective tillage is a good means of weed control. In many of the soybean-producing areas, weeds emerge in the spring before soybean is planted. Up until the mid-1970s when safe postemergence herbicides were developed, good weed control took planning and timely operations of tillage and preemergence herbicides. Today it is still important to control weeds early in the development of the soybean plant, before the flowering stage. Since control methods are more effective and economical on small weeds than on large ones, early control is essential (Scott and Aldrich, 1970). In some seasons, a pound of soybean dry matter (weight of stem, leaves, and beans) may be lost for every pound of weed dry matter produced. In other seasons, the ratio is not as great.

After the crop shades the ground, weeds that germinate usually affect yield only by contributing to harvest loss. In the 1950s and 1960s, it was felt that several shallow cultivations prior to planting were needed to destroy as many annual weeds as possible and to allow planting of soybean in a weed-free, well-prepared seedbed. The results of preplanting tillage may vary somewhat with weed species and location. Studies in Mississippi showed that 6 to 10 preplanting cultivations control johnsongrass in soybean (McWhorter and Hartwig, 1965), while in Minnesota it was shown with several common species that repeated tillage at intervals prior to planting was not necessary for control of annual weeds in soybean (Robinson and Dunham, 1956). The number of weeds that emerge after the soybean were planted was about the same regardless of the number of preplanting tillage operations. The general conclusion of weed scientists was that enough tillage was needed to prepare a weed-free seedbed in order to obtain the maximum benefit from soil-applied herbicides. Much of the primary tillage has been accomplished with the mold-board plow or the disc, but the use of chisel plows and field cultivators for tillage has increased. In addition to providing weed-free seedbeds, disc and field cultivators have been the major tools for incorporating herbicides. The harrow, while not always essential, is often helpful behind a disc or field cultivator to improve uniformity of incorporation. The early 1980s brought a move to more shallow incorporation of grass herbicides such as trifluralin, alachlor, and metolachlor with one-pass incorporation by implements such as the field cultivator. Many different manufacturers are now selling multipurpose one-pass incorporation equipment for herbicide incorporation on reduced tillage acreage.

Cultivation after the soybean plants have emerged has also been a valuable part of the total weed control program. The rotary hoe is a good mechanical weed control tool in soybean. In addition, it performs shallow tillage to break soil crust to allow uniform crop emergence and to incorporate surface-applied herbicides when rains are not timely (Knake et al., 1965; Peters et al., 1959). The rotary hoe has been used in soybean in the USA for about 50 yrs, but its maximum effectiveness was not realized until power sources were developed to operate at speeds of 10 to 12 kmph. The objective of the rotary hoe is to throw the shallowly rooted weeds out of the ground or to disturb the soil enough to seriously interfere with with water uptake by the weeds. Soybean plants deeply rooted should lose enough early morning turgidity to prevent the tillage equipment from causing mechanical damage to the crop.

Although the rotary hoe is most effective when the soil is relatively dry, it is somewhat effective even if the soil is moderately wet. It can be operated so that the soybean stand is reduced about 10% without any appreciable yield reduction (Lovely et al., 1958). The rotary hoe is most effective when combined with other postemergence operations such as sweep cultivation and/or herbicides. Use of the rotary hoe after herbicide treatment may be necessary for weed control when the herbicide is not effective.

Although growers commonly use the row cultivator once or twice, or occasionally not at all, it will likely continue to compliment the use of herbicides, especially in conventional tillage, wide-row soybean practices (Knake, 1975). Sweep cultivation is helpful for controlling broadleaf annual weeds such as annual morning glories, common cocklebur, jimsonweed, velvetleaf, and several perennial weeds that many of the herbicides of the 1960s and early 1970s did not control. Two sweep cultivations were usually required when no other means of control was used. Generally, shallow sweep cultivation is best to minimize soybean root pruning and to avoid disturbing weed seed at deeper levels, although throwing soil into the row to cover smaller weeds may be desirable. It may also cause excessive ridging of soil in the row. This ridging may prove troublesome at harvest and in the application of direct postemergence herbicides. Even with more selective postemergence herbicides available today for both grass and broadleaf weed control, cultivation is still used to supplement herbicide treatments and has been shown to significantly improve weed control of species such as giant foxtail, velvetleaf, common ragweed, Pennsylvania smartweed, large crabgrass, common lambsquarter, as well as that of common cocklebur (Gebhart, 1981; Regehr, 1982; McWhorter and Barrentine, 1975). The improved control obtained by cultivation was reflected in increased soybean yields and in net profit.

11.9.1 Conservation Tillage

Much confusion exists over the terminology of tillage practices referred to as conventional, reduced, and no-tillage. Many writers use several terms interchangeably, including minimum-tillage, reduced-tillage, mulch-tillage, no-tillage, etc. The term most often used today is *conservation tillage* (Mannering and Fenster, 1983). They defined conservation tillage as any tillage system that reduces loss of soil or water relative to conventional tillage, which is defined as the combined primary and secondary tillage operations performed in preparing a seedbed for a given crop grown in a given geographical area. Several conservation tillage systems make use of both surface residue and roughness in reducing runoff and soil loss.

Conservation tillage practices that greatly reduce preplanting tillage operations, especially no-tillage, increase reliance on herbicides for weed control. Systems that leave a large amount of crop residue on the soil surface restrict the ability to use preplanting incorporated herbicides, and surface-applied preemergence herbicides in combination with selective postmergence herbicides are necessary to obtain satisfactory weed control. Without the ability to incorporate herbicides, timely rainfall and favorable environmental conditions to maximize the herbicide effects is essential.

Lack of weed control is often cited as the main reason for farmers' reluctance to adopt conservation tillage (Ritchie and Follett, 1983; Brock, 1982). The weed spectrum and infestation levels of annual weeds in many tillage systems, short of no-tillage, may not be too different from those found in conventional tillage if the herbicide program works effectively. This may necessitate increasing the use of herbicides in the reduced tillage program. Even in no-tillage, the annual weed pressure could decline in time, since the weed seeds remain on or near the soil surface. If, however, the herbicide program fails, then the resulting weed problems can be devastating. The biggest deterrent to conservation tillage with greatly reduced primary tillage, and especially no-tillage, is the increase of perennial weeds.

In the regions where corn is produced a corn-soybean rotation is ideally suited to no-tillage. In the Delta states region, soybean is often rotated with rice production. These rotations have advantages over monocultures of each crop. Rotation interrupts the life cycle of pest organisms and also has weed control advantages (King, 1983). If perennial broadleaf weeds such as common milkweed, hemp dogbane, Canada thistle, or one of the viney perennial weeds becomes a problem in soybean, rotating the land into corn production and using 2,4-D [2,4-dichlorophenoxy) acetic acid] or dicamba (3,6-dichloro-o-anisic acid) on these weeds can greatly reduce their competitive effects, and help eliminate them. Crop rotation and its associated weed control practices is also beneficial for the control of annual weed species, since it provides a broader spectrum of weed control than monoculture (Burnside, 1978).

With the expanded use of a herbicide program on a continuous crop, shifts occur in annual weed populations (Young and Evans, 1976). Examples of these shifts include the increase in grass species in corn and soybean with the continuous use of atrazine and the increase in large seeded broadleaf weeds in soybean with the continuous use of trifluralin. Good weed control each year in crop rotations is beneficial for effective weed control and higher yields in subsequently planted crops. Allowing weeds to produce seed any year of a crop rotation will reduce the crop production potential in the subsequent years (Burnside, 1978).

Crop rotation is not without problems. In the Great Plains and Southwestern regions of the USA, atrazine residues from the previously year's corn or sorghum crops can injure soybean plants in rotation. The degree of atrazine residue problems depends on the tillage system used (Burnside and Wicks, 1980; Burnside and Schiltz, 1978). Deep tillage with the moldboard plow mixes the herbicide with large volumes of soil and decreases the concentration, which can protect seedlings of susceptible crops from chemical injury. Changing a tillage practice can modify the soil environment enough to require a change in herbicide selection or rates in order to achieve satisfactory weed control or desirable residual activity. Tillage systems with crop residue left on the soil surface may call for increased herbicide rates, since the residue intercepts part of the applied chemical. The mulch left on the soil surface over a number of years will increase the organic matter level in the surface from 2 to 3 cm to as much as twice the level of the underlying soil (Triplett and Wiese, 1979).

Rates of most of the commonly used soybean herbicides are based on organic matter content. Thus, these herbicides must be applied at increased rates to achieve satisfactory weed control. The surface mulch likewise retains soil moisture for longer periods than does a residue-free soil surface. This increased moisture can increase the rate of herbicide breakdown, thus decreasing the length of residual activity. Tillage systems which greatly reduce deep mixing of the soil, especially no-tillage, restrict fertility programs to surface application. In corn-soybean rotations, surface N applications in corn can greatly increase the acidity of the soil surface layer, causing the pH in this top zone to drop to 5 or lower. At these acid levels the triazine herbicides, such as atrazine in corn or metribuzin in soybean, hydrolize rapidly and the normal use rates will not control weeds. Liming the soils can correct these acid conditions; however, overliming can cause the normal use rate of triazine herbicides to be more active than usual, causing injury to the soybean seedlings. If tillage practices are to change in a given field, the herbicide programs must be carefully planned to maximize the efficiency of the chemical for controlling weeds without causing injury to the crop.

11–9.2 Cropping Practices

With the onset of conservation tillage and expanded production of soybean in the late 1970s, interest in growing soybean in narrow rows or in solid seeded stands increased. Under high performance crop management systems where weed control and stand establishment is consistently achieved, solid-seeded drilled soybean or soybean planted in rows in 25 cm or less have increased yield in the northern sections of the soybean-producing states under full-season conditions, in small grain stubble systems where soybean followed small grain, and in double-cropping practices. In these areas the growing season is shorter but the summer days are longer, and the short-season cultivars are better positioned to intercept the maximum sunlight in June and July when the days are longest (Swearingin, 1980). As reduced row width soybean practices move south with longer season varieties, much of the yield advantage over conventional wide-row soybean production is lost. In the Southeastern states narrow-row soybean production has had maximum benefits when some limited production factors are dominant, such as late planting, a poor stand establishment, or a shortage of water.

Inadequate weed control is cited as the major factor for the failure of solid seeded drill or narrow row soybean practices. This could be directly due to failure to achieve weed control from the inability of the herbicide program to control weeds or from the failure to get a uniform stand which allows weeds to become established in the areas where stands are thin. The recent development of selective postemergence herbicides for both broadleaf species and for grasses has given new importance to the need for and advantages of reducing tillage and row spacings in soybean production. Herbicide combinations for broader spectrum control are used in the majority of the soybean fields in the USA. These advances in weed control technology provide the safest and most effective methods of weed control in soybean to date. Even with these increased combinations of preplant incorporated, preemergence and postemergence herbicides, weed populations continue to shift, necessitating readjustments in weed control programs to stay abreast of weed problems. Unpredictable weather at planting and for some time after planting often results in poor weed control. This is especially true of the effect of dry weather on treatments applied preemergence to the soil surface. Where tillage systems allow, preplant incorporation is increasing in popularity because it greatly reduces the need for rain to activate the herbicides.

Early canopy closure by narrow rows complements a good weed control program. However, the inability to control weeds with postemergence cultivation or herbicide applications still add a degree of uncertainty to wide-scale adaptation of narrow soybean production. Many of the other types of reduced tillage systems, such as double-cropping soybean into small grain stubble and ridge tillage, have weed control problems similar to those of narrow-row and other conservation tillage practices.

11–10 INTEGRATED WEED MANAGEMENT

Alterations of the natural environment to favor human interests is a basic goal of many of scientist's endeavors. Deriving and maintaining these alterations, such as growing soybean crops, requires energy inputs. These inputs are directed at many things, including managing weed populations. Attempts to make the most efficient use of energy inputs have fostered an integrated approach to weed management.

Shaw (1982) has suggested the following as a formal definition of integrated weed management: "a directed agroecosystem approach for the management and control of weed populations at threshold levels that prevent economic damage in the current and future years." Several important facets of integrated weed management are implied by this definition. First, the concept of the agroecosystem is introduced, indicating that both cropland and noncrop areas are important to the management of weeds in a given field. Of particular importance here is the influence of adjacent areas on weed population dynamics in a crop field, especially through movement of weed reproductive propagules within the system.

Shaw's definition also embraces the idea of weed population management, with weed control as a necessary part of a management strategy. The need for increased knowledge of weed biology and crop-weed interactions becomes apparent when one considers how these interactions influence the particular management strategy to be employed.

Another important area addressed is the concept of economic thresholds. Although there are no reports of beneficial weeds in soybean, there are certainly situations in which the weed population present would cost more to remove than the loss it would cause if left in the crop. These threshold levels should be identified for multispecies weed complexes for most efficient utilization by growers.

Finally, Shaw's definition implies that integrated weed management is a long-term process with strategies and tactics utilized in 1 yr influencing subsequent crop years.

Currently, weed management strategies can be classified as either preventive, in which growers attempt to prevent a population from becoming established in an area; or suppressive, in which control tactics are utilized on an existing population. In rare circumstances eradication of a species is attempted as a strategy, but eradication is seldom feasible because of the difficulty and high cost involved (Klingman and Ashton, 1975). Tactics available to growers for use in preventive strategies include planting weed-free crop seed, cleaning equipment to avoid transferring weed seeds or other reproductive parts from one field to another, controlling weeds in adjacent noncrop areas to avoid spread into crop fields, and preventing weed populations in a given field from reproducing. These tactics usually have little direct effect on weed populations during the crop year in which they are practiced but play an important role in population management in subsequent years.

Control tactics utilized in population suppression strategies include cultural, mechanical, biological, and chemical control. Cultural control tactics include the manipulation of the soybean crop in ways that have some influence on weed populations. An excellent review of this subject has been presented by Walker and Buchanan (1982). Cultural control tactics include crop rotations and crop competition enhancement. The main utility of crop rotations for weed control lies in the ability it affords the grower of utilizing different control practices on the same field in the different crops. For example, more effective or more economical control measures may be available for control of certain broadleaf weeds in corn than in soybean. By utilizing corn in the rotation, a grower can reduce populations of these broadleaf weeds, thereby requiring lower inputs for management of the same species in soybean. In addition, certain weed species tend to be associated more with one crop than another. Buchanan et al. (1975) showed common cocklebur to occur in higher populations in soybean than in corn, regardless of control method.

Crop competition enhancement involves several cultural practices, including row spacing, seeding rate, fertilization, and cultivar selection (Walker and Buchanan, 1982). Narrow rows obviously produce a full canopy sooner and provide maximum shading at an earlier date, thereby requiring less weed control input to achieve maximum potential yield (Walker et al., 1981; Wax and Pendleton, 1968; Wax et al., 1977). High soybean seeding rates have been shown to reduce weed growth (Staniforth and Weber, 1956). However, probably the most important practical aspect of seeding rates is to avoid skips in the stand and to have a uniform population within an acceptable range. High seeding rates often lead to excess lodging and result in more serious weed problems at harvest than normal seeding rates.

Fertilization, although perhaps not as effective as a direct weed control tactic, is important in that a properly fertilized soybean crop will become competitive with weeds at an earlier stage than a poorly fertilized crop. Also, evidence exists to support a detrimental effect on the crop-weed response to N fertilization in soybean, with fertilization increasing weed competitiveness (Staniforth, 1962).

Cultivar selection may influence the degree of weed-induced yield loss in some cases. McWhorter and Hartwig (1972) and Burnside (1972) found differential tolerance to weed competition among cultivars. In an integrated weed management system, the interaction of row spacing, seeding rate, fertilization, and cultivar can be manipulated to give the crop a competitive edge on weeds, thus requiring lower energy inputs from other sources for adequate weed management.

Mechanical control tactics for soybean include primary tillage for seedbed preparation and secondary tillage or cultivation after the crop is established. The primary purpose of seedbed preparation with respect to weed control is to assure that weeds do not become established before the crop, creating an unmanageable situation. Cultivation was the primary weed control tactic used by growers until the mid-1960s, when herbicide use became widespread. Since that time, cultivation has become a supplemental tool in most cases, used mainly to remove weeds that escape control tactics earlier in the growing season. The use of cultivation as a supplemental tool in addition to herbicides usually results in better weed control and higher yields than cultivation practiced alone (Slife, 1979).

Biological control tactics utilize natural enemies, usually phytophagous insects or plant pathogens, for weed population suppression. In general, biological control in soybean has received relatively little attention in the past because the organisms used were host specific, and weeds tend to occur in multispecies complexes (Andres, 1982). However, renewed interest has been shown recently in using biological agents, mainly plant pathogenic fungi, for control of weeds that are difficult to manage by more conventional methods (Quimby and Walker, 1982). Field evaluation of pathogens is now underway for control of Northern jointvetch [*Asechynomene virginica* (L.) B.S.P.] (Boyette et al., 1979), prickly sida (Templeton, 1974), sicklepod (Walker and Riley, 1982), and spurred anoda and velvetleaf (Walker, 1981).

Chemical control involves the use of herbicides. This subject will be treated in-depth in section 11–11 of this chapter and will only be covered generally here. Herbicides can be classified as either prophylactic or remedial with regard to their influence on weed populations within the crop. Prophylactic treatments are those designed to prevent a weed population from becoming established in the crop. These treatments are usually applied either preplant and incorporated into the soil or after planting but preemergence to both the crop and weeds. They are designed to kill weeds either during germination or in early growth stages so that the net effect is the prevention of weed establishment. Preplant and preemergence treatments are important parts of integrated weed management in fields where economically damaging populations of susceptible weeds are expected to occur from the soil seed reserve.

Remedial treatments are those made after the weed population has become established but before significant yield loss due to weed competition has occurred. Most studies show that weeds removed by 4 to 6 weeks after crop emergence will not adversely influence soybean yield. Most remedial treatments however, are made earlier in the growing season because the postemergence herbicides used are more effective in controlling weeds at the earlier stages of growth. The most precise use of postemergence herbicides as remedial treatments could be made in conjunction with economic threshold information for weeds. Currently however, data on useful economic thresholds for weeds in soybean are scarce. The economic thresholds of some single weed species can be calculated from research conducted on weed interference in soybean (Table 11–2), but most growers face situations in which a multispecies weed complex infests the crop. Threshold values for multispecies weed complexes are beginning to receive attention (Johnson and Coble, 1981). Once they are developed, sounder economic decisions can be made using remedial treatments for weed control.

Future efforts to improve integrated weed management should include integration with other pest management systems. Just as weeds usually occur in multispecies complexes, pests of more than one type usually occur at the same time or at least in the same crop. It is not unusual to find weeds, insects, nematodes, and fungal or bacterial plant pathogens interacting with a soybean crop at the same time. These pests, and the respective control tactics used for each, interact with the crop and with each other in ways that generally are poorly understood. However, some important pest interactions have been elucidated recently.

In the lower Mississippi Delta, the velvetbean caterpillar (*Anticarsia gemmatalis* Hübner) is a major insect pest on soybean. In the same area, hemp sesbania is a common weed problem. Situations in which infestations of velvetbean caterpillar occur in circular areas around large isolated hemp sesbania plants have been observed. (L.D. Newsom, 1983, personal communication). The weeds in these instances have been scattered and would not have caused economic loss by themselves. However, the weeds obviously serve as an attractant to the adult insects, which lay eggs on or near the weeds. The velvetbean caterpillar larvae then cause severe damage to the crop if not controlled. Thus, the real economic damage from the weed population is much more serious than its direct effects would suggest.

Evidence now exists to show that one pest population may exert an influence on a crop that allows the resurgence of another pest population so that the combined effect is greater than the sum of the individual effects. Studies comparing the effects of soybean looper [*Pseudoplusia includens* (Walker)] defoliaton and velvetleaf competition demonstrate this relationship (Table 11–4). These studies show that 30 to 60% soybean defoliation by the insects caused 10 and 16% yield reduction, respectively. With no defoliation, four velvetleaf plants per 3.25 m^2 caused a 28% reduction in yield and eight weeds per 3.25 m^2 led to a 37% reduction. The combined effects of four weeds per 3.25 m^2 with soybean looper defoliation at both 30 and 60% were additive. However, with eight weeds per 3.25 m^2, the influence on yield was far greater. The most likely reason for this effect is the increase in velvetleaf biomass in defoliated plots. After defoliation, more light was available to the weeds, enabling them to become more competitive compared to the soybean.

Table 11–4. Influence of soybean looper defoliation on weed resurgence and soybean yield.†

Treatment	Weed biomass	Soybean yield	Yield reduction
	g 3.25 m^{-2}	kg ha^{-1}	%
None	–	3407	–
30% defoliation	–	3071	10
60% defoliation	–	2876	16
4 weeds 3.25 m^{-2}	439	2466	28
8 weeds 3.25 m^{-2}	924	2164	37
30% defol. + 4 weeds	556	2164	37
60% defol. + 4 weeds	659	1875	45
30% defol. + 8 weeds	1185	1431	58
60% defol. + 8 weeds	1113	1176	66

†Helm, C.G., M.R. Jeffords, and M. Kogan. Unpublished data. Univ. of Illinois, Urbana.

Populations of one pest class have been shown to influence the tolerance of the soybean crop to control tactics utilized for another pest class. Recent studies have shown an interaction between soybean thrips [*Sericothrips variabilis* (Beach)] and the herbicide acifluorfen as expressed in soybean injury from the herbicide application (Table 11–5). In this study, acifluorfen at recommended application rates caused more than twice as much crop injury and defoliation when thrips were present than when the insect was controlled before the herbicide was applied. Insect feeding on the soybean leaves apparently stressed the crop plant so the tolerance to the herbicide was less than adequate.

The types of interactions presented here, as well as others that may exist, need to be understood if a fully integrated approach to pest managment is to be achieved.

11–11 CULTIVAR RESPONSE TO HERBICIDES

Cultivar response to crop plants, particularly soybean, has been documented over several decades (Shaw et al., 1948; Dunham, 1951; Bucholtz, 1953; Hayes and Wax, 1975; Barrentine et al., 1982). Many of these findings show that the response is genetically controlled. Andersen (1976) wrote an in-depth review of the differential tolerance of soybean cultivars to herbicides and discussed much of the work that has been done in this area up to the time of this writing. Andersen stated that the tolerance of soybean cultivars to herbicides had been studied to determine (i) if important cultivars can be treated with a herbicide of marginal safety (e.g., 2,4-DB or metribuzin), (ii) if some cultivars might be susceptible to a generally innocuous herbicide (e.g., bentazon), (iii) what cultivars, if any, might tolerate an herbicide that is generally toxic to soybean but that controls certain difficult weed problems (e.g., glyphosate), (iv) what cultivars, if any, might tolerate residual amounts of an herbicide used on preceding crops (e.g., atrazine), and (v) what cultivars, if any, might tolerate spray-drift from applications on other crops [e.g., propanil (3′,4′-dichloropropionanilide)].

The earliest evidence for the interspecific differential response of soybean to an herbicide was found in 1948 when 2,4-D was applied

Table 11–5. Influence of soybean thrips control with carbaryl on crop injury from acifluorfen.†

Treatment	Herbicide injury rating	Percentage soybean defoliation	Fresh wt.
	%		g plant^{-1}
None	0	0.9	5.2
Carbaryl	0	0.4	6.1
Acifluorfen	36	30.2	2.8
Carbaryl + acifluorfen	17	13.2	3.8

†Huckaba, R.M., and H.D. Coble. Unpublished data. North Carolina State Univ., Raleigh.

preemergence to five cultivars of soybean. In this study, Shaw et al. (1948) observed severe injury to the crop plants whenever the chemical gave good to excellent weed control. They noted that one cultivar, Hawkeye, was conspicuously less injured than the other cultivars tested. Work that followed in the mid-1950s showed that large differences in tolerance to foliar applications of 2,4-D and 2,4,5-T [(2,4,5-trichlorophenoxy)acetic acid] existed among cultivars of soybean (Fribourg and Johnson, 1955; Slife, 1956). Even though these reports indicated that selected cultivars could tolerate rates of 2,4-D and 2,4,5-T that are high enough to control certain weed species, these herbicides have never been registered for use in soybean. Although not registered for use of soybean, 2,4-D is still an important herbicide in soybean production because of the damage it causes from drifting off target due to poor application techniques. Along with dicamba, 2,4-D presents a major injury problem to soybean crops in the Corn Belt states each year. In addition to yield losses, Thompson and Egli (1973) showed that progeny of plants treated with these herbicides exhibited reductions in germination, emergence, and dry weights, as well as malformation of the unifoliolate and first trifoliolate leaves.

The only phenoxy herbicide that has been registered for use in soybean is 2,4-DB, which is used as a postemergence treatment primarily for control of common cocklebur and annual morning glories. As with the other phenoxy compounds, 2,4-D and 2,4,5-T, soybean tolerance has varied greatly with 2,4-DB (Walters and Caviness, 1968; Hartwig, 1974; Wax et al., 1974). One cultivar Tracy, was developed from work involving the selection of breeding lines which showed tolerance to 2,4-DB (Hartwig, 1974).

The most studied herbicide for interspecific differential responses for soybean is metribuzin, a triazine. In his review, Andersen (1976) gave several reasons for this interest: (i) it (metribuzin) is extremely effective as a preemergence or preplanting treatment on a number of weed species, (ii) the margin of safety to soybean is not great, and (iii) a high degree of differential response has been found among cultivars. Some cultivars show a high degree of susceptibility to metribuzin and several are listed on the product labeling as cultivars that should not be planted when metribuzin is used. It is interesting to note that Tracy, which was developed for its resistance to 2,4-DB, is a cultivar that is highly susceptible to metribuzin (Barrentine et al., 1976). Through the efforts of single plant selection a cultivar tolerant to metribuzin, 'Tracy M', was developed. In addition to genetic factors which control cultivaral responses to metribuzin, environmental factors and seed quality also play a role in cultivaral responses to this herbicide.

Bentazon is a postemergence herbicide used to control certain broadleaf weeds in soybean. Andersen (1976) pointed out that in studies involving more than 40 cultivars, it appears that soybean is, in general, tolerant to Bentazon. Wax et al. (1974) found one cultivar Hurrelbrink, to be slightly injured (10%) at Bentazon rates of 0.1 kg ha^{-1}. Hayes and Wax (1975) studied the differential interspecific responses of two strains

PI 229.342 (sensitive) and 'Clark 63' (tolerant). They found that the tolerant cultivar metabolized bentazon into two metabolites (I and II) at a rapid rate, while the sensitive cultivar metabolized bentazon at a much slower rate into only one metabolite (II). They concluded that the tolerant soybean had the genetic capability to produce the system responsible for the rapid metabolism of bentazon into metabolite I. The sensitive cultivar does not contain the system, and the rate at which it forms metabolite II is too slow to prevent phytotoxic levels of bentazon from reaching the active site(s) in the chloroplast. Thus, sensitivity is related to the lack of rapidly detoxifing bentazon.

Persistent soybean injury problems occur in a corn-soybean rotation system when atrazine is used for broad-spectrum weed control in corn production. Andersen (1976) wrote that work in Kansas and Canada showed that soybean cultivars could tolerate rates of atrazine up to 1.12 kg ha^{-1} without sustaining severe yield losses. These studies do not suggest that the amount of atrazine carried over in the soil from 1 yr to the next should cause serious problems. Andersen, however, continues to point out that atrazine residues do pose a serious problem in many of the upper midwestern states. To determine if the sources of atrazine resistance could be found in soybean germplasm, about 2700 strains of soybean were screened for response to atrazine. No outstanding sources of atrazine tolerance were found, but a strong association of atrazine tolerance and seed size (the larger the seed, the greater its tolerance to atrazine) was observed. It was concluded from these studies that atrazine injury to soybean might be minimized by planting large seeded cultivars or planting the largest seeds from a lot of a given cultivar. These conclusions were verified in field studies using various sizes of seed from given soybean cultivars (Andersen, 1970).

Soybean grown in the rice-producing areas in the southeastern USA is often injured by propanil drift or accidental spray by aircraft. A wide diversity in cultivaral tolerances to propanil has been demonstrated (Eastin, 1979). In addition to differential soybean cultivar tolerance, the rate of the herbicide and the growth stage of the soybean plant influenced the degree of injury. Smith and Carviness (1973) suggested that tolerant cultivars should be planted next to rice fields that are likely to be treated with propanil.

While genetic control of herbicide tolerance characteristic of soybean and other crops is well documented, the mechanism of control is quite variable. In some instances the herbicide response was simply inherited, while in others it was complexly inherited.

In recent years, another type of cultivaral response to herbicides has been reported. This is the resistance of biotypes of weeds to certain herbicides, especially the triazine herbicides. LeBaron (1983) noted that although triazine-resistant weeds were not discovered first, they have been of the greatest interest because of the importance of this herbicide group in agronomic crop production and the number and relatively wide distribution of weeds resistant to the triazines. Since first reported in 1970, triazine resistance has been confirmed in at least 24 genera and 37 species of weeds located in a least 25 states (mostly in the northern region of the USA), 4 provinces of Canada, and 10 additional countries.

It takes longer for a resistant weed biotype to predominate than it does for insects or plant diseases to demonstrate resistance because weeds take a full year to complete their life cycles and do not move about as freely as insects and diseases. Triazine resistance occurs in areas were a herbicide has been used alone for several consecutive years, especially where little or no-tillage is used. Adequate cultivation or tillage tends to delay or even prevent resistance from occurring. Likewise, crop rotations with associated herbicides delay or prevent the establishment of resistant weeds in a given area.

LeBaron (1983) pointed out that farmers in the USA are fortunate to have a large number of herbicides with various modes of action to control weeds. This fact will probably become increasingly important in the years to come, especially if the same weed biotypes develop resistance to more than one type of herbicide.

11-12 SPECIALIZED EQUIPMENT AND TECHNIQUES

The importance of application of herbicides has acquired new interest. Herbicides are now routinely applied through irrigation systems, a process known as *herbigation*, and applied simultanously with dry fertilizer by impregnating fertilizer with the chosen herbicide. Other techniques being actively pursued by researchers include the development of safening agents to protect the crop seeds from certain herbicides, as well as the development of formulations that control the release of soil applied herbicides, preventing rapid degradation of the chemical and maintaining a rate of herbicide that is biologically effective for a longer period of time. The most extensive work, however, has been done with application equipment that reduces the diluent volume and minimizes drift problems. The trend toward reduced volumes of carrier as well as alternative diluents to water have been evaluated and in most cases proven useful techniques for herbicides application.

Postemergence applications have always presented problems of drift and insufficient selectivity to the crop. Only in the last 5 yrs have truly selective and safe herbicides been developed for postemergence application to soybean. In the mid-1970s, the development of the recirculating sprayer (RCS) and a variety of "wipe on" applicators improved the problems of drift with nonselective but highly effective herbicides such as Glyphosate (Wills and McWhorter, 1981). This new generation of application equipment also ushered in the idea of conserving herbicides by applying then recapturing a majority of the total solution volume. The rope wick applicator was by far the most widely accepted piece of application equipment for the control of tall growing perennial species such as johnsongrass and common milkweed, as well as annuals such as volunteer corn and shattercane. Its popularity grew primarily due to the simplicity of the apparatus (which had no moving parts) and the low cost of construction. The two major disadvantages of the rope wick applicator over similar equipment such as the RCS were the slow wicking action of the ropes and the slow operating speed needed to allow the wicks to recharge with herbicide solutions. The ability to apply nonselective herbicides through recirculating sprayers and wipe-on applicators revolutionized the thinking of many researchers. A new thrust was begun to develop equipment that would conserve herbicides and reduce the carrier volumes.

The trend toward reduced volumes gave farmers and commercial applicators the ability to cover large areas with the equipment with less down-time for filling and mixing herbicide solutions. It also allowed the use of smaller and lighter equipment, which improved the compaction problems and/or allowed getting in the fields sooner after rains for more timely postemergence applications. With the use of nozzles with smaller orifices and lower pressure volumes, as little as 80 L ha^{-1} of herbicide solution can be applied with conventional nozzles (Taylor, 1981). The smaller nozzles, however, can produce smaller droplets, potentially causing an increased drift problem. An alternative technique has been the use of the rotary atomizer, popularly known as the controlled droplet applicator (CDA). The rotating unit of the CDA produces highly uniform droplets in a range large enough to prevent long distance drift problems yet small enough to provide excellent foliar coverage of weeds. These applicators have been used to apply both soil-applied and foliar-applied herbicides in volumes of 50 L ha^{-1} or less with results that are comparable to conventional application equipment delivering 190 L ha^{-1} or more.

The actual relative effectiveness of the low-volume application in comparison to the more conventional equipment is not well documented. Taylor (1981) reported that there is no obvious reason why the performance of soil-applied compounds should be sensitive to application techniques. With foliar applied herbicides he reported that, with the exception of glysophate, there has been no evidence for increased activity with systemic compounds through the CDA's, while contact herbicides may be slightly less effective through pieces of equipment.

Matthews (1981) reported that the drift problems associated with droplets smaller than 100 μm can be overcome and the small particle size turned to good use by imparting an electrical charge to the droplets as they are formed. these electrostatic sprayers apply specially formulated herbicides in volumes < 1 L ha^{-1}. The electrically changed droplets improve the coverage of the undersurface of leaves, especially in the upper regions of the plant. Since the droplets are magnetically attracted to plants, the penetration deep into the crop canopy is restricted, and a minimum amount of herbicide reaches the lower foliage or the ground. The use of these excellent coverages has proven more beneficial for insect and disease control than for weed control. The new application technology utilizing low volumes of carrier and precision droplet size will continue to advance into the 1980s. The concern about drift and the need to maximize the efficacy of herbicide applications will dictate that considerable research efforts be given to the application of herbicides.

REFERENCES

Alexander, J.D. 1969. A color chart for organic matter. Crops Soils 21(8):15–17.

Altieri, M.A. 1982. Ecology and management of weed populations. Eugene Memmler Publisher, Glendale, CA.

Andersen, R.N. 1970. Influence of soybean seed size and response to atrazine. Weed Sci. 18:162–164.

----. 1976. Differential soybean variety tolerance to herbicide. p. 444–452. *In* L.D. Hill (ed.) World soybean research conference proceedings. The Interstate Printers and Publishers, Danville, IL.

Anderson, J.M., and C.G. McWhorter. 1976. The economics of common cocklebur control in soybean production. Weed Sci. 24:397–400.

Andres, L.A. 1982. Integrating weed biological control agents into a pest management program. Weed Sci. (Suppl.) 30:25–30.

Anonymous. 1979. A look at world pesticide markets. Farm Chem. 142 (9):61–68.

Bailey, G.W., and J.L. White. 1964. Review of absorption and desorption of organic pesticides by soil colloids, with implications concerning pesticide bioactivity. J. Agric. Food Chem. 12:324–332.

Baker, H.G. 1974. The evolution of weeds. Ann. Rev. Ecol. Syst. 5:1–24.

Banks, P.A., and T.N. Tripp. 1983. Control of Johnsongrass (*Sorghum halapense*) in soybeans (*Glycine max*) with foliar-applied herbicides. Weed Sci. 31:628–633.

Barrentine, W.L. 1974. Common cocklebur competition in soybeans. Weed Sci. 22:600–603.

----, C.J. Edwards, Jr., and E.E. Hartwig. 1976. Screening soybeans for tolerance to metribuzin. Agron. J. 68:351–353.

----, E.E. Hartwig, C.J. Edwards, Jr., and T.C. Kilen. 1982. Tolerance of three soybean (*Glycine max*) cultivars to metribuzin. Weed Sci. 30:344–348.

Bovey, R.W., and O.C. Burnside. 1965. Aerial and ground applications of preemergence herbicides in corn, sorghum, and soybeans. Weeds 13:334–336.

Boyette, C.D., G.E. Templeton, and R.J. Smith. 1979. Control of winged waterprimrose and northern jointvetch with fungal pathogens. Weed Sci. 27:497–501.

Brock, B.G. 1982. Weed control versus soil erosion control. J. Soil Water Conserv. 37:73–76.

Buchanan, G.A., C.S. Hoveland, V.L. Brown, and R.H. Wade. 1975. Weed population shifts influenced by crop rotations and weed control programs. Proc. South. Weed Sci. Soc. 28:60–71.

Bucholtz, K.P. 1953. Varietal responses of corn to herbicide. Proc. North Cent. Weed Control Conf. 10:47–49.

Burnside, O.C. 1972. Tolerance of soybean cultivars to weed competition and herbicides. Weed Sci. 20:294–297.

----. 1978. Mechanical, cultural, and chemical control of weeds in a sorghum-soybean (*Sorghum bicolor*)-(*Glycine max*) rotation. Weed Sci. 26:362–369.

----, and M.E. Schiltz. 1978. Soil persistance of herbicides for corn, sorghum, and soybeans during the year of application. Weed Sci. 26:108–115.

----, and G.H. Wicks. 1980. Atrazine carryover in soil in a reduced tillage crop production system. Weed Sci. 28:661–666.

Cardenas, J., and P.W. Santelmann. 1966. Influence of irrigation and formulation on activity of NPA, Amiben, and DCPA. Weeds 14:309–312.

Chandler, J.M., A.S. Hamill, and A.G. Thomas. 1984. Crop losses due to weeds in Canada and the United States. Spec. Rep. (May). Weed Science Society of America, Champaign, IL.

Coble, H.D., and R.L. Ritter. 1978. Pennsylvania smartweed interference in soybeans. Weed Sci. 26:556–559.

----, F.M. Williams, and R.L. Ritter. 1981. Common ragweed interference in soybeans. Weed Sci. 29:339–342.

Dunham, R.S. 1951. Differential responses in crop plants. p. 195–206. *In* F. Skoag (ed.) Plant growth substances. University of Wisconsin Press, Madison.

Eastin, E.F. 1979. Soybean (*Glycine max*) cultivar response to propanil. Weed Sci. 27:4–6.

Eaton, B.J., K.C. Feltner, and O.G. Russ. 1973. Venice mallow competition in soybeans. Weed Sci. 21:89–94.

----, O.G. Russ, and K.C. Feltner. 1976. Competition of velvetleaf, prickly sida, and Venice mallow in soybeans. Weed Sci. 24:224–228.

Elmore, C.D. 1983. Weed survey—Southern states. Research report. South. Weed Sci. Soc. 36:148–184.

Fribourg, H.A., and I.J. Johnson. 1955. Response of soybean strains to 2,4-D and 2,4,5-T. Agron. J. 47:171–174.

Gebhardt, M.R. 1981. Preemergence herbicides and cultivation for soybeans (*Glycine max*). Weed Sci. 29:165–168.

Gentner, W.A. 1966. Influence of acetone of the herbicidal properties of chloroxuron. Weeds 14:95–96.

Greer, H.A., and P.W. Santelmann. 1968. Influence of environmental factors on herbicide activity in soybeans. Proc. South Weed Sci. Soc. 21:104.

Harris, C.I., and G.F. Warren. 1964. Absorption and desorption of herbicides by soil. Weeds 12:120–126.

Hartwig, E.E. 1974. Registration of Tracy soybeans. Crop Sci. 14:777.

Hartzler, R.G., and C.L. Foy. 1983. Compatibility of BAS 90520H with acifluorfen and bentazon. Weed Sci. 31:597–599.

Hayes, R.M., and L.M. Wax. 1975. Differential intraspecific response of soybean cultivars to bentazon. Weed Sci. 23:516–521.

Hill, G.D., Jr., I.J. Belasco, and H.L. Poloeg. 1965. Influence of surfactants on the activity of diuron, linuron, and bromacil as foliar sprays on weeds. Weeds 13:103–106.

Jansen, L.L. 1965. Effects of structural variation in ionic surfactants on phytotoxicity and physical-chemical properties of aqueous sprays of several herbicides. Weeds 13:117–123.

----, W.A. Gentner, and W.C. Shaw. 1961. Effects of surfactants on the herbicidal activity of several herbicides in aqueous spray systems. Weeds 9:381–405.

Johnson, W.C., III, and H.D. Coble. 1981. A new method to determine weed competition. Proc. South. Weed Sci. Soc. 34:102.

Jordan, T.N. 1981. Effects of diluent volumes and surfactants on the phytotoxicity of glyphosate to bermudagrass (*Cynodon dactylon*). Weed Sci. 29:79–83.

King, A.D. 1983. Progress in no-till. J. Soil Water Conserv. 38:160–161.

Klingman, G.C., and F.M. Ashton. 1975. Weed science: Principles and practices. John Wiley and Sons, New York.

----, and ----. 1982. Weed science: Principles and practices. John Wiley and Sons, New York.

Knake, E.L. 1975. Weed control for soybeans. *In* Soybean production in the central U.S.A. Proceeding of the conference on soybean production in central U.S.A. Deere and Co., Moline, IL.

----, F.W. Slife, R.D. Seif. 1965. The effect of rotary hoeing on performance of preemergence herbicides. Weeds 13:72–74.

Ladlie, J.S., W.F. Meggett, and D. Penner. 1977. Effect of atrazine on soybean tolerance to metribuzin. Weed Sci. 25:115–121.

LeBaron, H.M. 1983. Herbicide resistance in plants—An overview. Weeds Today 14(2):4–6.

Lolas, P.C., and H.D. Coble. 1982. Noncompetitive effects of johnsongrass (*Sorghum halepense*) on soybean (*Glycine max*). Weed Sci. 30:588–593.

Lovely, W.G., C.R. Weber, and D.W. Staniforth. 1958. Effectiveness of the rotary hoe for weed control in soybeans. Agron. J. 50:621–625.

Mannering J.V., and C.R. Fenster 1983. What is conservation tillage? J. Soil Water Conserv. 38:141–143.

Matthews, G.A. 1981. Developments in pesticide application for the small-scale farmer in the tropics. Outlook Agric. 10:345–349.

McCarty, M.T. 1983. Economic thresholds of annual grasses in agronomic crops. Ph.D. thesis. North Carolina State Univ., Raleigh (Diss. Abstr. DA8318959).

McWhorter, C.G., and J.M. Anderson. 1979. Hemp sesbania competition in soybeans. Weed Sci. 27:58–64.

----, and W.L. Barrentine. 1975. Cocklebur control in soybeans as affected by cultivars, seeding rates, and methods of weed control. Weed Sci. 23:386–390.

----, and E.E. Hartwig. 1965. Effectiveness of preplanting tillage in relation to herbicides in controlling johnsongrass for soybean production. Agron. J. 57:385–389.

----, and ----. 1972. Competition of johnsongrass and cocklebur with six soybean varieties. Weed Sci. 20:56–59.

Nave, W.R., and L.M. Wax. 1971. Effect of weeds on soybean yield and harvesting efficiency. Weed Sci. 19:533–535.

Oliver, L.R., R.E. Frans, and R.E. Talbert. 1976. Field competition between tall morningglory and soybean. I. Growth analysis. Weed Sci. 24:482–488.

Patterson, D.T., C.R. Meyer, E.P. Flint, and P.C. Quimby, Jr. 1979. Temperature responses and potential distribution of itchgrass (*Rottboellia exaltata*) in the United States. Weed Sci. 27:77–82.

Peters, E.J., D.L. Klingman, and R.E. Larson. 1959. Rotary hoeing in combination with herbicides and other cultivations for weed control in soybeans. Weeds 7:449–458.

Quimby, P.C., Jr., and H.L. Walker. 1982. Pathogens as mechanisms for integrated weed management. Weed Sci. (Suppl) 30:30–34.

Regehr, D.L. 1982. Analysis of weed control components for full-season, conventional-tillage soybeans in Delaware. Proc. North East. Weed Sci. Soc. 36:45–49.

Ritchie, J.C., and R.F. Follett. 1983. Conservation tillage: where to from here. J. Soil Water Conserv. 38:267–269.

Rhodes, G.N., Jr., and H.D. Coble. 1982. Interactions of sethoxydim (Poast) and bentazon (Basagran). Proc. South. Weed Sci. Soc. 35:346.

Robinson, R.G., and R.S. Dunham. 1956. Pre-planting tillage for weed control in soybeans. Agron. J. 48:493–495.

Scott, W.O., and S.A. Aldrich. 1970. Modern soybean production. S and A Publications, Champaign, IL.

Shaw, W.C. 1978. Herbicides: The cost/benefit ratio—The public view. Proc. South. Weed Sci. Soc. 31:28–43.

----. 1982. Integrated weed management systems technology for pest management. Weed Sci. (Suppl.) 30:2–12.

----, L.C. Sahoe, and C.J. Willard. 1948. Effects of preemergence 2,4-D on different varieties of soybeans. Res. Rep. North Cent. Weed Control Conf. 5: Sect. IV, Abstr. 42.

Slife, F.W. 1956. the effect of 2,4-D and several other herbicides on weeds and soybeans when applied as postemergence sprays. Weeds 4:61–68.

----. 1979. Weed control systems in the corn belt states. p. 393–398. *In* F.T. Corbin (ed.) World soybean research conference II: Proceedings. Westview Press, Boulder, CO.

Smith, L.W., C.L. Foy, and D.E. Bayer. 1966. Structure-activity relationships of alkyl-phenol ethylene oxide ether non-ionic surfactants and three water-soluble herbicides. Weed Res. 6:233–242.

----, ----, and ----. 1967. Herbicidal enhancement by certain new biodegradable surfactants. Weeds 15:87–89.

Smith, R.J., Jr., and C.E. Caviness. 1973. Differential response of soybean cultivars to propanil. Weed Sci. 21:279–281.

Staniforth, D.W. 1962. Response of soybean varieties to weed competition. Agron. J. 54:11–13.

----, and C.R. Weber. 1956. Effects of annual weeds on the growth of soybeans. Agron. J. 48:467–471.

Swearingin, M.L. 1980. Managing soybeans in solid stands. p. 17–21. *In* Solid seeded soybeans systems for success. Conf. Proc. Am. Soybean Assoc. 21–22 January. St. Louis, MO. American Soybean Association, St. Louis.

Taylor, W.A. 1981. Controlled drop application of herbicides. Outlook Agric. 10:333–336.

Templeton, G.E. 1974. Endemic fungus disease for control of prickly sida in cotton and soybeans. Agron. J. 48:467–471.

Thompson, L. Jr., and C.B. Egli. 1973. Evaluation of seedling progency of soybeans treated with 2,4-D, 2,4-DB, and dicamba. Weed Sci. 21:141–144.

Thurlow, D.L., and G.A. Buchanan. 1972. Competition of sicklepod with soybeans. Weed Sci. 20:379–384.

Triplett, G.B., Jr., and G.D. Little. 1972. Control and ecology of weeds in continuous corn grown without tillage. Weed Sci. 20:453–457.

----, and A.F. Wiese. 1979. Influencing the action of herbicides-tillage. Crops Soils 32(3):8–9.

Walker, H.L. 1981. Fusarium lateritium: A pathogen of spurred anoda, prickly sida, and velvetleaf. Weed Sci. 29:629–631.

----, and J.A. Riley. 1982. Evaluation of *Alternaria cassiae* for the biocontrol of sicklepod. Weed Sci. 30:651–654.

----, and G.A. Buchanan. 1982. Crop manipulation in integrated weed management systems. Weed Sci. (Suppl.) 30:17–24.

Walker, R.H., T. Whitwell, J.R. Harris, D.L. Thurlow, and J.A. McGuire. 1981. Sicklepod control in soybean with herbicides, row spacings, and planting dates. Highlights Agric. Res. 28(1):18.

Walters, H.J., and C.E. Caviness. 1968. Response of Phytophthora resistance and susceptible soybean varieties 2,4-DB. Plant Dis. Rep. 52:355–357.

Wax, L.M., R.L. Bernard, and R.M. Hayes. 1974. Response of soybean cultivars to bentazon, bromoxynil, chloroxuron, and 2,4-DB. Weed Sci. 22:35–41.

----, W.R. Nave, and R.L. Cooper. 1977. Weed control in narrow and wide-row soybeans. Weed Sci. 25:73–78.

----, and J.W. Pendleton. 1968. Effect of row spacing on weed control in soybeans. Weed Sci. 25:73–78.

Wills, G.D., and C.G. McWhorter. 1981. Developments in post-emergence herbicide applicators. Outlook Agric. 10:337–341.

Worsham, A.D. 1970. Herbicide systems in no tillage in the southeast. *In* No-tillage crop production Natl. Conf. Proc., Univ. of Kentucky, Lexington, KY.

Young, J.A., and R.A. Evans. 1976. Responses of weed populations to human manipulations of the natural environment. Weed Sci. 24:186–190.

Zimdahl, R.L. 1980. Weed-crop competition. A review. International Plant Protection Center, Oregon State University, Corvallis.

12 Soil Fertility and Liming

David B. Mengel
Purdue University
West Lafayette, Indiana

William Segars
University of Georgia
Athens, Georgia

George W. Rehm
University of Minnesota
Minneapolis, Minnesota

Soybean [*Glycine max* (L.) Merr.] production for grain has increased rapidly in the USA since 1950. Research in many areas has shown that soybean plants respond well to fertile soils, and in many cases, to direct fertilization. In many major soybean-growing areas of the USA, however, soybean plants are not directly fertilized, but rather rely on residue fertility from crops such as corn (*Zea mays* L.). This chapter will review the mineral nutrient requirements of soybean, together with the diagnostic techniques used to predict fertilizer and lime requirements. It also will summarize some of the information available on soybean fertilization and liming practices currently used in the USA.

12-1 SOYBEAN MINERAL NUTRITION

12-1.1 Mineral Nutrients Essential for Soybean

Using Arnon's strict definitions of nutrient essentiality, there are 16 elements currently agreed upon as being essential for plant growth. Of these elements, three (C, H, and O) are the principal components of plant dry matter and are obtained from, or absorbed as, carbon dioxide (CO_2), water, and free atmospheric O_2. The remaining 13 are commonly referred to as the essential mineral nutrients. The essential mineral nutrients are: N, P, K, Ca, Mg, S, Fe, Mn, Mo, Cu, B, Zn, and Cl. In addition, Co has been shown beneficial for N_2 fixation in many free-living bacteria and legumes, and is considered essential using less restrictive definitions.

Soybeans: Improvement, Production, and Uses, 2nd ed.—Agronomy Monograph no. 16.

Table 12–1 lists the approximate nutrient composition of a 4000 kg ha^{-1} soybean crop. These values, which Ohlrogge and Kamprath (1968) developed, are given as examples. These values however, can differ substantially with changes in cultivar, climate, and inherent soil fertility.

12–1.2 Nutrient Uptake by Soybean

The relative uptake of N, P, and K by indeterminate soybean in the field, together with the relative amounts accumulating in various plant

Table 12–1. Approximate nutrient composition of a 4000 kg ha^{-1} soybean crop. Source: Ohlrogge and Kamprath (1968).

	Grain	Straw	Stubble and roots	Total
	kg ha^{-1}			
Dry matter	3360	3920	1680	8960
N	250	80	40	370
P	25	10	5	40
K	65	42	23	130
Ca	–	–	–	90
Mg	–	–	–	40
S	–	–	–	28
Cl	–	–	–	11
Fe	–	–	–	1.9
Mn	–	–	–	0.71
Zn	–	–	–	0.2
Cu	–	–	–	0.1
Bo	–	–	–	0.1
Mo	–	–	–	0.01

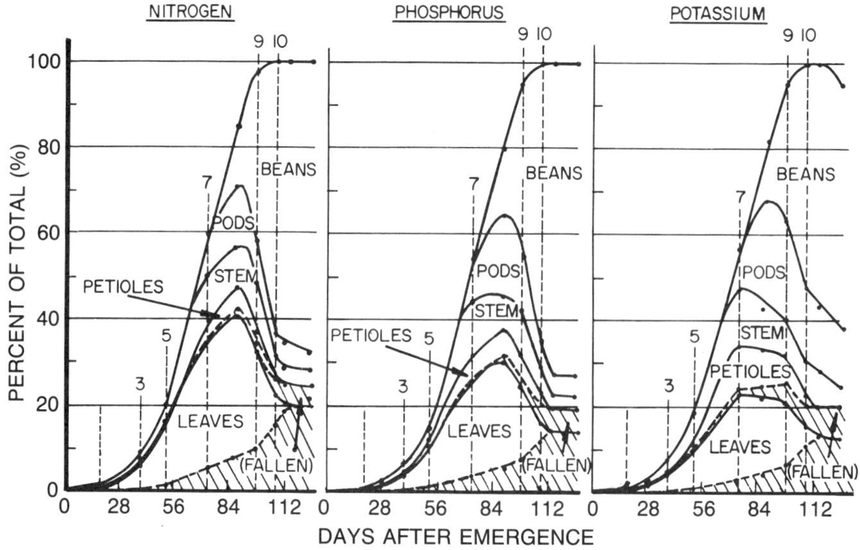

Fig. 12–1. Relative rate of N, P, and K accumulation in soybean plant parts during the growing season. From Hanway and Weber (1971b).

parts is illustrated in Fig. 12–1. This study, which Hanway and Weber (1971) conducted on a Nicollet silty clay loam (fine-loamy, mixed, mesic, Aquic Hapludolls), showed similar patterns of total N, P, and K accumulation in the plant, with maximum accumulations occurring near physiological maturity.

Henderson and Kamprath (1970) working with determinate soybean cultivars in North Carolina, noted an increasing rate of nutrient uptake for N, P, K, Ca, and Mg through early seed set. Maximum uptake rates observed were 7.7, 0.41, 4.6, 2.4, and 0.77 kg ha^{-1} day^{-1} for N, P, K, Ca, and Mg, respectively.

12–2 SOYBEAN FERTILIZATION PRACTICES CURRENTLY USED IN THE USA

While soybean has been shown to respond to fertilization and/or fertile soils, the practice of direct soybean fertilization is not universally accepted. Fertilizer use data collected by the National Fertilizer Development Center, Tennessee Valley Authority (Hargett and Berry, 1983) (Table 12–2), shows that while average fertilizer use rates for soybean has increased since 1965, rates applied to soybean are still quite low in relation to those applied to other agronomic crops such as corn. In 1982, U.S. soybean yields averaged 2117 kg ha^{-1}, and if one assumes a P content of 5.5 g of P kg^{-1}, and a K content of 17 g of K kg^{-1}, average P and K fertilizer application rates were substantially below the 11.6 kg of P ha^{-1} and 36 kg of K ha^{-1} removed in the soybean grain.

If the fertilizer use pattern by geographic regions is examined, however (Table 12–3), it is discovered that fertilizer use more closely ap-

Table 12–2. Fertilizer use for soybean in the USA: 1965–1982. From Hargett and Berry (1983).

Nutrient applied	1965	1970	1975	1980	1982
			kg ha^{-1}		
N	1.1	3.4	3.4	4.5	3.4
P	2.4	4.9	4.9	7.3	5.9
K	5.6	13.0	13.0	23.2	18.5

Table 12–3. Fertilizer applied to soybean in the USA by geographic region in 1982.† From Hargett and Berry (1983).

Nutrient applied	South Atlantic	East North Central	West North Central	East South Central	West South Central
			kg ha^{-1}		
N	9.0	3.4	2.2	5.6	2.2
P	9.8	6.8	2.9	10.8	5.4
K	37.2	30.7	7.4	27.0	13.0

†South Atlantic - VA, NC, SC, GA, FL; East North Central - WI, IL, IN OH, MI; West North Central - ND, SD, NE, KS, MN, IA, MO; East South Central - KY, TN, AL, MS; West South Central - TX, AK, OK, LA.

proximates removal of P and K in the eastern USA. Data collected by the USDA-Economic Research Service (ERS) (Table 12–4) also show that the percentage of soybean fields fertilized is much higher in the southeastern USA than the southwest or western Corn Belt, with the eastern Corn Belt, and the east North Central region being intermediate.

The differences in fertilizer use patterns reflect a number of factors, including soil and climatic differences that influence soybean response to fertilizer. They also reflect differences in cropping patterns and crop rotations which can result in substantial amounts of residual fertility being present for the use of the soybean crop. The soils of the southeastern USA, predominately Ultisols, are more highly weathered and inherently lower in P and K than the less leached Alfisols and Mollisols of the Corn Belt. In many cases, the cation exchange capacity (CEC) of many southern soils is also lower because of a relatively low clay and organic matter content. Thus, the ability of these soils to retain nutrients such as K is lower than those of the Midwest. The inherent need for fertilization for soybean production in the Southeast is much higher than in the Midwest.

Cropping patterns are also reflected in current fertilization practices. In the Midwest, soybean is commonly grown in rotation with corn. Traditionally, corn has been heavily fertilized and soybean has been grown on the residual fertility. While direct fertilization of soybean is increasing in the Midwest, the majority of the soybean crops are not fertilized at this time.

Time of fertilizer application for soybean in the USA does not vary widely across regions. Generally, those fields that receive fertilizer are fertilized at or before seeding. In the majority of cases, the fertilizer is broadcast and incorporated with tillage operations.

12–3 DIAGNOSING FERTILIZER AND LIMING NEEDS

The two most beneficial diagnostic techniques used in determining soybean fertilizer needs are soil testing and plant analysis. Agronomists

Table 12–4. Percent of soybean fertilized in USA by state, 1984, and rates of NPK applied to fertilized acres. From USDA-ERS Bull. 105-7, *Inputs — Outlook and Situation Report*, (Feb. 1985).

	Percentage of fields receiving				Rate applied			Percentage of fields fertilized at or before seeding
States	Any fertilizer	N	P	K	N	P	K	
	%				kg/ha			%
NC,GA,SC	65	46	61	64	16	18	66	97
IL,IN,OH	42	22	3	39	21	25	83	97
MN,IA,MO,NE	19	15	16	17	21	22	51	91
AL,MS,TN	60	30	58	59	20	23	58	97
AK,LA	30	11	29	26	19	22	53	89
15 State total	34	20	30	32	19	22	67	95

have repeatedly emphasized that the ultimate objective of a soil testing and plant analysis program is to develop an effective and efficient liming and fertilization program.

To be most effective, soil and plant analyses must be used in such a way that they support and supplement each other. Soil analyses are most often used prior to planting to evaluate the fertility level of the soil; whereas, plant analyses are used during the growing season to monitor the seasonal nutrient levels of plants, evaluate the effectiveness of fertilizer treatments and aid in the diagnosis of abnormal plant growth. In troubleshooting in-season growth disorders, both soil tests and plant analyses are used to determine if nutrient deficiencies, toxicities, or imbalances are the causative factor.

Within the scope of this publication, it is not possible to provide an in-depth discussion of soil testing and plant analysis and how the test results are correlated and calibrated. Rather, the practical application of these diagnostic methods are emphasized, as well as how they are utilized to achieve a lime and fertilizer recommendation for soybean producers.

Numerous excellent reviews and research papers have been published that discuss the principles and procedures utilized in soil testing and plant analysis. For a general examination of these subjects, refer to the following references: *Soil Testing: Correlating and Interpreting the Analytical Results* (Peck 1978) and *Soil Testing and Plant Analysis* (Hardy, 1973).

12–3.1 Soil Testing

12–3.1.1 Objectives of Soil Testing

There are two major objectives of soil testing. First, utilize a chemical extractant to remove a portion of the plant nutrients from the soil which is highly correlated with the amount the plant can absorb. This must be done with a test which is reproducible, and can be incorporated into a laboratory routine. However to be of practical benefit, a soil test must also meet a second criteria or objective, and must be capable of prescribing or predicting fertilizer or lime treatments. The importance of these objectives was best summarized by Colwell's (1967) definition of a soil test as *a measurement qualifies to be termed a soil test for a particular nutrient if, and only if, it provides information on the fertilizer requirements of a crop for that nutrient.*

With modern analytical techniques, the chemical analysis of soils can be extremely accurate. To make these results useful, however, growers must still strive to secure representative samples, and agronomists must calibrate these tests with crop responses.

12–3.1.2 Current Status of Soil Testing Programs

Soil testing programs are conducted by both public and private laboratories (Welch and Wiese, 1973). The magnitude of the soil testing effort in the USA is shown in the following table (Welch and Wiese, 1973; Owens, 1980; Jones, 1981).

Year	Public laboratories	Private laboratories	Total
	no. of soil samples analyzed (000)		
1960	1560	525	2085
1968	1295	2242	3537
1979	1485	1545	3030
1981	1792	1628	3420

Most states offer soil testing services by state-operated agencies. In addition, private laboratories serve almost all areas of the USA. The laboratory one selects to perform soil analysis is largely a matter of personal preference since most laboratories are well equipped and have qualified analytical staffs. However, one should determine that the selected laboratory develops fertilizer and lime recommendations based on valid research trials conducted on soils and climatic conditions representative of one's geographic area.

12-3.1.3 Soil Sampling

It has been documented that soils are extremely heterogeneous in both physical and chemical properties (Beckett and Webster, 1971). Because of soil variability, many researchers have suggested that the most important aspect of soil testing is the matter of obtaining a soil sample that is representative of the area sampled. Peck and Melsted (1973) illustrated soil P test variability in a 16.2-ha Illinois field. The soil P test ranged from a low of 8.9 kg ha^{-1} to a high of 147 kg ha^{-1}, with the mean value of 26.2 kg ha^{-1}. A similar range in variability may also be found in small-scale research plots.

There are many factors that contribute to soil heterogeneity. One must recognize that, in addition to chemical and physical differences, variation within a field may result from lime and fertilizer applications, method of fertilizer application (i.e., banding vs. broadcast), tillage practices, cropping sequences, etc. When collecting the soil sample, one must consider these sources of variation.

Peck and Melsted (1973) note that soil sampling technique has been essentially standardized according to these criteria: (i) random selection of coring sites and (ii) sampling to a specific depth. Generally, most soil scientists suggest compositing 12 to 15 cores to make a representative sample. There seems to be no standard area that constitutes the size of the sampling unit. Many laboratories suggest cores be collected from 4- to 8- ha sites if the soil is considered uniform; however, we must emphasize that the sampling area depends on soil uniformity rather than area per se. All laboratories provide soil sampling instructions. These instructions provide details on how to achieve a representative sample.

Olson, et al. (1984) note that traditionally soil testing has been done on samples acquired from the tillage layer of the soil. They suggest that the reasons for collecting samples from this layer are that (i) the layer contains the major portion of the crop's root system, (ii) nonmobile nutrients (such as P and K) accumulate there and (iii) samples are easier to collect from the tillage layer than from the subsoil. In addition, most soil test calibration studies have been based on tillage layer samples. While the subsoil nutrient content is important, there are few calibration studies available such that one can properly interpret the values. We suggest that one follow the guidelines for sample depth released by the chosen soil test laboratory since their interpretation is based on those guidelines.

Since soil sample results are often used to monitor fertility levels, it is desirable to collect samples at approximately the same time each year. Seasonal variations, especially in soil pH, have been observed. For soybean farmers, it may be most practical to soil sample immediately after harvest.

12-3.1.4 Soil Test Interpretation

Chemical values obtained by extraction have no absolute meaning concerning nutrient supply available to plants (Cope and Rouse, 1973). The test values reported are an index of nutrient availability. Therefore, the numbers can only be interpreted and have "real" meaning if they have been correlated and calibrated with crop response.

Ozus and Hanway (1966) note that correlation studies must be conducted to provide the basis for selecting the laboratory test that will provide the best index of nutrient availability to plants.

Cope and Rouse (1973) describe calibration of soil test values as the process of determining the crop-soil relationships. The calibrated soil test value for a nutrient will indicate the degree of deficiency of that nutrient and the amount of the nutrient to be applied as a fertilizer to correct the deficiency. A description of the extensive types of field studies required for soil test calibration is given in Cope (1983, 1984). Basically, calibration studies involve the application of various rates of nutrients on soils with varying soil test values and measuring crop response to the nutrient applied.

Since soils vary in their capacity to supply nutrients at specific soil test values, calibration of soil tests is complex and requires much field research and laboratory study (Johnson, 1984). Unfortunately, all soils in the soybean-producing states have not been adequately calibrated, through numerous calibration studies are continuously being conducted.

Since soils vary widely in their chemical, physical, and biological characteristics (Brady, 1974), one should select a soil testing laboratory that has properly correlated and calibrated the results for the type of soil sampled. Unfortunately, some soybean producers select laboratories that use analytical procedures that may be unacceptable for their particular soil.

In the USA, a technique commonly employed to interpret soil test values is to classify the values into fertility levels or categories (Johnson, 1984; Tisdale et al., 1985). Another commonly used method of classification is an index based on percentage sufficiency (Rouse, 1968).

An example of a rating scale, is given below (Johnson, 1984). We must emphasize that a number of different classification systems are currently in use; however, their objective remains the same—to relate the quantity of residual soil nutrients measured to the fertilizer requirements needed to achieve optimum soybean yield.

Interpretive category	Specific definition
Very low	Less than 50% of crop yield potential is expected without addition of the nutrient. Yield increase to added nutrient is always expected.
Low	50 to 75% of crop yield potential is expected without addition of the nutrient. Yield increase to added nutrient is expected.
Medium	75 to 100% of crop yield potential is expected without addition of the nutrient. Yield increase to added nutrient is expected if test value is in lower end of range.
High	Soil can supply sufficient quantities of the nutrient for the crop. Yield increase to added nutrient is not expected. Test again next year if the nutrient is not applied.
Very high	Soil can supply the nutrient in greater quantities than considered adequate. Yield increase to added nutrient is not expected. Addition of P or K will be wasteful, could induce nutrient imbalances, and could decrease yields in some situations.

The most extensive interpretative data for soybean soil tests are available for soil pH, lime requirement, P, K, and Mg. Land-grant university soil scientists have conducted thousands of correlation and calibration trials for these soil tests. Private laboratories (Castenson, 1973) generally depend on research by public soil scientists to provide them with interpretive data.

Soil tests are often utilized for the mobile nutrients N and S; however, numerous authors have emphasized that interpretations for these tests are frequently less reliable than those for P and K because of the difficulty in obtaining reliable correlation and calibration data.

Generally, there is poor correlation and calibration data for the micronutrient soil tests for soybean plants (Mortvedt, 1978). However, there are exceptions. For example, many Southeastern states now are interpreting soil test values for Mn. Extensive research trials are being conducted to correlate and calibrate micronutrient soil tests in many soybean-producing states.

12-3.1.5 Making Fertilizer Recommendations Based on Soil Tests

The soil test serves as the foundation for making lime and fertilizer recommendations. However, many other factors must be considered in achieving the final nutrient rates. An optimum recommendation may not be achieved unless consideration is given to information such as soil productivity potential, yield goals, cropping systems, management capability of the farmer, and method of fertilizer application (Melsted, 1967; Barber, 1973, Potash and Phosphate Institute, 1982). Brady (1974) emphasizes that test results must be correlated with crop response before reliable fertilizer recommendations can be made, and that recommendations must be made in light of a practical knowledge of the crop to be grown, the characteristics of the soil and other environmental conditions. Smith and Lamborn (1982) emphasized the importance of using the soil test as a guide. Additionally, Melsted (1967) observed that soil tests are a part of the factual data and must be reported in finite terms, while the resulting fertilizer recommendations will include interpretive judgements. Thus, in the final analysis, an agronomist may use a combination of scientific knowledge and practical experience to achieve a recommendation. Additionally, we must emphasize that lime and fertilizer recommendations must be expected to either result in an increase in soybean yields which is economically beneficial to the farmer, or the recommendation must maintain adequate lime and fertilizer levels so that yields do not decrease. It becomes obvious that persons charged with the responsibility of making recommendations must be sufficiently trained so that they understand the scientific principles of soil test interpretation and can also effectively evaluate the multitude of other factors associated with optimum plant growth and development.

It is of little value to list the various lime and fertilizer recommendations used in the soybean-producing states. It is important to emphasize that recommendations for soybean vary across the USA depending on the philosophy of the interpreter (Barber, 1973; Melsted, 1967; Rouse, 1968; McLean, 1978; Cope, 1984; Jones, 1973) and because the interpreter utilizes calibration studies conducted on soil types similar to his geographic region. De Mooy et al. (1973) have suggested that "soil properties and climatic conditions are correlated: a higher degree of leaching, coarser textures, lower soil reaction, and larger yield responses from fertilization are found in regions of higher rainfall . . . large responses are reported from the Atlantic coastal states of the USA . . . responses are variable and often small on the more productive soils of the North Central USA . . . ."

Fertilizer recommendations differ widely, not only as a result of calibration studies, but as a result of philosophical differences in soil fertility management. Olson et al. (1984) emphasize that differences exist among the many organizations and individuals that provide soil testing services as to proper fertilizer management and the interpretation of soil test results.

There are two broad philosophies for making fertilizer recommendations in the USA that can contribute to significant differences in the quantity of fertilizer recommended. These have been summarized by

Colliver (1982, p. 10–16) as fertilize-the-soil and fertilize-the-crop concepts.

The fertilize-the-soil concept has as its objectives: (i) build up low fertility soils to optimum soil test levels (based on soil test calibration data) and (ii) maintain optimum soil test levels once they are achieved. This is accomplished by applying fertilizer in quantities exceeding crop needs until the optimum soil test level is reached. Afterwards, fertilizers are supplied in amounts equal to crop removal. Using this philosophy, the recommendation can vary considerably, depending on how rapidly one desires to build up low testing soil to an optimum test level (i.e., accomplish build-up over a 1-, 2-, or 4-yr period).

The fertilize-the-crop philosophy has the objectives of (i) emphasis on meeting crop nutrient needs with minimum amount of fertilizer, with greater dependence on soil reserves; (ii) build-up only for low-testing soils and then only to medium test levels. The main difference between the two philosophies is that less fertilizer is recommended for increasing soil test levels in the fertilize-the-crop as compared to fertilize-the-soil. Certainly, both concepts have merit and both systems can be used to develop effective fertilizer recommendations for soybean.

We must emphasize that simply because a soil is chemically analyzed for a particular nutrient or factor, does not mean a valid interpretation can be made. The goal of the reliable interpreter should be to only recommend a nutrient to farmers when an economic crop response can be expected as predicted by soil test calibration studies.

12-3.1.6 Reliability of Soil Testing

Chemical soil testing is widely recognized by scientists as an effective means of determining rapidly and routinely the soil's nutrient content. Analytical procedures currently being utilized will provide values that can be reproduced, provided the same or an identical soil sample is subjected to analysis.

Most questions regarding the reliability of soil testing are raised by farmers who split samples and send them to two or more laboratories in order to compare test results. This has lead to confusion and misunderstanding when the various test values were not similar. It must be understood that laboratory values can only be accurately compared if the soil samples are truly identical, and the laboratories use the same analytical and reporting procedures. Various laboratories utilize different extractants, extraction processes, and reporting systems that will result in dissimilar values. This should not be construed as a weakness in soil testing; rather, it emphasizes that there are many analytical procedures that can be utilized to evaluate the nutrient levels in soils. The significant end result is not the numbers generated but the fertilizer and limestone recommendations made.

Most soil testing laboratories have quality control programs to ensure that their analytical results have a high degree of confidence. The differences in recommendations commonly seen between different laboratories are normally due to differences in interpretation. A reputable laboratory develops recommendations based on properly correlated and calibrated field research, but recommendations of various laboratories may vary due to philosophical differences.

12–3.2 Plant Analysis

12–3.2.1 Principles and Objectives

Munson and Nelson (1973) defined plant analysis as *the determination of the concentration of an element or extractable fraction of an element in plant tissue.* The concentration is usually expressed on a dry matter basis. Plant analysis is based on the principle that the quantity of a nutrient within the plant is an integral value of all the factors that have interacted to affect it. Thus, by analyzing plant tissue, one can compare the nutrient content with accepted norms to determine if the nutrient uptake is sufficient to produce optimum plant growth.

In the 1960s and 1970s, tissue testing was often used in the field to diagnose problems. These tests were generally used only for N, P, and K. The weakness of this system was that it gave only a semiquantitative value of the macronutrient content of the plant. Agronomists now prefer to collect plant samples in the field and return them to a laboratory in order that the plant nutrient content can be quantitatively determined using state-of-the-art analytical equipment. Also, all essential nutrients, in addition to the macronutrients, can be analyzed. Using modern analytical equipment and procedures, laboratory chemists can rapidly and efficiently analyze plants and return the results to the investigator. This reduces the need for in-field tissue testing and permits one to achieve a high degree of accuracy in the analysis.

As currently used, a plant analysis has two major applications (Plank, 1979): (i) confirm a nutrient deficiency when visual symptoms are present and (ii) monitor the plant nutrient status in order to determine if the nutrients are in a sufficient concentration for optimum yield. Plant analysis may also be used to indicate interactions or antagonisms among nutrients, identify incipient deficiencies, determine nutrient toxicities, or suggest that additional tests be performed to determine the cause of growth disorders.

12–3.2.2 Sampling Plants

As with soil sampling, proper sampling of plants is critical if an accurate interpretation of the analysis is to be achieved. The concentration of nutrients within the soybean plant can change rapidly with time and physiological maturity and vary depending on the plant part sampled. Thus, interpretation of the results of plant analysis requires an understanding of the growth stage at which the plants were sampled and the plant parts collected.

It is critical that the soybean farmer or investigator be familiar with the sampling instructions issued by the plant analysis laboratory. These instructions will show the particular plant part to be sampled, the number of plants to sample as well as the locations within the field. This will ensure that a sufficient quantity of plant tissue is submitted and that the sample is representative of the area (Plank, 1979). It is also important that the stage of growth at sampling be relayed to the lab so proper interpretation can be made. Many labs will specify the proper stages for sampling for monitoring purposes. For troubleshooting purposes, however, samples may be collected at any growth stage.

When plant analysis is used simply to monitor the nutrient status of plants during the growing season, plants are sampled at random throughout the field, generally at the growth stage or stages with the best interpretive data. When used as an aid in troubleshooting growth disorders, however, most agronomists use comparative samples. For example, samples collected from the affected plants and from normal plants in the immediate area. Comparative analyses are difficult to interpret if the plants are not at the same stage of growth, and have not received the same treatment.

Plant analysis laboratories will normally recommend that a soil sample be collected at the same time a plant sample is taken. This soil test information can be invaluable to the interpreter of the plant analysis since his objective is to determine if the nutrient content of plants is deficient, sufficient, or excessive.

12-3.2.3 Plant Analysis Interpretation

As with soil tests, plant analysis values must be calibrated by extensive research trials. It is imperative that the laboratory performing the plant analysis have skilled individuals who can provide interpretations based on documented calibration studies.

When plant samples are submitted, it is important that communication exist between the sampler and the interpreter. Normally, this is accomplished in the form of a history form or questionnaire. The questionnaire requests specific information regarding the conditions in the field at sampling time. Typical questions may be the fertilization practices used, soil test data, soil texture, drainage, rainfall received, appearance of plants, and cultivar. Additionally, the interpreter must establish that causative factors other than residual soil fertility and/or fertilization programs are not responsible for affecting nutrient absorption. It becomes obvious that interpretations of analyses made on plants infested with insects, diseases, or nematodes would be of little value. Interpretations are also meaningless if the root system development has been restricted due to soil compaction, poor drainage or other related factors.

The trained interpreter will normally use a critical or sufficiency concentration to evaluate plant nutrient levels. Ulrich and Hills (1967) defined a critical concentration as that concentration of a specific nutrient within a specified plant part at which growth begins to decline. Many agronomists use the sufficiency range approach (Plank, 1979). The concentration of each element analyzed is reported as less than, greater than, or within the sufficiency range. An example of sufficiency range values is given in Table 12–5. Normally, there is little difference between a critical value and a sufficient value that most interpreters use today. Both systems were developed by assigning a single concentration value where the plant nutrient status shifts from deficient to adequate. Dow and Roberts (1982) have proposed that a critical nutrient range be employed that may have more practical use than using a single value as the cut-off point (Fig. 12–2). In addition to defining deficient levels, research trials have also revealed the level where nutrient toxicity will occur. For obvious reasons, more attention has been devoted to the determination of deficient levels rather than toxic levels.

One weakness of the use of critical or sufficient levels to interpret nutrient concentrations is that it only reveals a single deficiency at a time and at a specified sampling period. Ulrich and Hills (1967) have emphasized that a second nutrient may be in short supply but, due to reduced

Table 12–5. Soybean sufficiency ranges for upper fully developed trifoliate leaves sampled prior to pod set. Adapted from Plank (1979) and Small and Ohlrogge (1973).

	Sufficiency range	
Element	Ohio	Georgia
	$g\ kg^{-1}$	
N	4.26–5.50	4.25–5.00
P	0.26–0.50	0.25–0.50
K	1.71–2.50	1.75–2.50
Ca	0.36–2.00	0.50–1.50
Mg	0.26–1.00	0.25–0.80
S	–	0.20–0.60
	$mg\ kg^{-1}$	
Mn	21–100	15–200
Fe	51–350	50–300
B	21–55	20–60
Cu	10–30	5–30
Zn	21–50	20–70
Mo	1.0–5.0	0.1–5.0

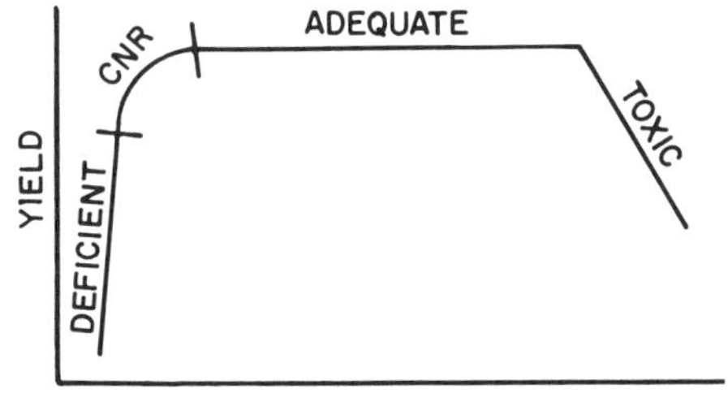

Fig. 12–2. Relation between nutrient concentration in plant tissue and crop yield showing the proposed critical nutrient range (CNR). From Dow and Roberts (1982).

growth as a result of a deficiency of the primary nutrient, may accumulate in plant tissue. When the primary deficiency is corrected, increased growth and a concomitant decrease in concentration of the second nutrient may quickly reveal its deficiency.

Many interpreters are now examining the use of the Diagnosis and Recommendation Integrated System (DRIS) concept that was developed by Beaufils (1971, 1973) and expanded by the work of Sumner (1977, 1978, 1979) and others. The DRIS technique, unlike any current method, effectively incorporates nutrient balance by diagnosing each nutrient in terms of all the others (Beverly, 1979). Research trials are continuing to develop norms for soybean for use in the DRIS approach to interpretation.

The trained interpreter of plant analysis recognizes the necessity of having calibrated research trials to document yield responses to corrective treatments. As Beverly (1979) has emphasized, the plant is the integrator of its environment, and plant analysis reflects the influence of past factors on the growth of the plant. The interpreter evaluates where the plant is and his recommendations must be based on a judgment of where the plant will go with and without treatment.

12-3.2.4 Fertilizer Recommendations Based on Plant Analysis

Severe nutritional disorders of soybean are almost always easily identified and diagnosed with the aid of plant analysis (Small and Ohlrogge, 1973). Under the conditions of severe deficiencies, visual symptoms of the nutrient disorders may well be present and a diagnosis based on symptoms, confirmed by plant analysis is a very reliable method of diagnosing nutrient disorders (Bould et al., 1984).

If corrective treatment for a nutritional disorder is to be recommended, the applied fertilizer must be cost effective, for example, increase yields in sufficient quantity to recover the cost of treatment. This will depend on the growth stage at which the deficiency is discovered, the severity of the deficiency, and the nutrient involved. The earlier in the growing season a disorder is identified the more likely corrective treatments can be effectively and economically utilized. However, in situations such as a soil pH problem, it may be impossible to recommend any effective treatments during the growing season. For some nutrients such as Mg or Fe, interpreters will recommend foliar application of the nutrient to correct the nutrient deficiency as rapidly as possible. However, for this fertilizer recommendation to be effective and reliable it must be based on calibration. The interpretation of plant analysis data and making corrective recommendations is a complex task that requires skill on the part of the interpreter and sufficient knowledge of the site conditions (Plank, 1979).

Ulrich and Hills (1967) stressed that one of the greatest values of plant analysis is the prevention of deficiencies rather than their correction after they occur. Plant analysis is useful in detecting "hidden-hunger" (i.e., no visible symptoms of growth disorder) (Tisdale et al., 1985). This use of plant analysis is often referred to as crop monitoring. Today's top producer is concerned with the prevention of nutrient deficiencies and is attracted to this concept. Using this technique, the interpreter may recommend in-season corrective treatments if a treatable deficiency is detected early in the growing season. More often, the farmer, and researcher uses crop monitoring to evaluate the effects of fertilizer treatments, or other practices, and make necessary adjustments for the next cropping sequence.

12-3.2.5 Plant Analysis as a Diagnostic Tool for Producers

Plant analysis can be effectively used as a diagnostic tool (Plank, 1979). Farmers are urged to confirm suspected nutrient deficiencies by a plant analysis before applying a corrective treatment. For one to effectively use a plant analysis in diagnosing growth disorders, however, specific sampling methods must be employed.

Plant samples from affected, as well as normal plants should be collected whenever possible. However, if plants have been under nutrient stress for a long period, a comparative analysis may be misleading if the abnormal and nearby normal plants are at markedly different stages of growth. Therefore, it is desirable that the plants be sampled as soon as possible following the discovery of the problem.

Plants selected for sampling and exhibiting symptoms of the suspected nutrient element deficiency should be similar in appearance and all at the same development stage. Dead or severely affected tissues should not be included in the sample. Confining the sampling area to plants in close proximity to each other is also desirable. Identical sampling procedures should be used for those plants selected as the normal counterparts. Soil samples should be collected from the affected and normal areas.

By comparing the analysis of both soil and plant tissue from the normal and affected areas, differences in concentration of specific nutrients may be detected. A comparison of test values may be far more useful in interpretation than using known interpretive standards.

The most frequent error made when a plant analysis is used as a diagnostic tool relates to the failure to use sufficient care when collecting the plant and soil samples. Therefore, those who collect these samples must follow the sampling procedures prescribed with great care. Failure to do so can significantly limit the effectiveness of the evaluation and may lead the interpreter into drawing incorrect conclusions.

12-4 LIMING

Controlling soil acidity is an important part of any soybean fertilization program in the USA. Liming, the neutralization of soil acidity (primarily by adding ground limestone to acid soils) has a number of potential benefits for soybean production. These include:

1. The reduction in the concentration of potentially toxic elements such as H, Al, and Mn.
2. The increased availability of plant nutrients such as Ca, Mg, and Mo.
3. Improved nodulation and N_2 fixation.

The exact benefit derived from liming soils for soybean production found at a given location will vary with the initial soil pH; soil properties such as the amount and type of clay present and the soil organic matter content; and the relative acid tolerance of the soybean genotype being grown and rhizobium strains present for nodulation. Therefore, the objective of this section is to outline some of the changes in soils which can occur with liming, and where these changes are likely to occur and result in increased soybean yields. For further information see the monograph *Soil Acidity and Liming* (Adams, 1984).

12–4.1 Changes in Soils That Occur upon Liming

Generally, the term *acid soil* refers to a soil with a low pH, or a high concentration of H in the soil solution. In addition, acid soils are characterized by large quantities of exchangeable H and/or Al on the cation exchange complex. Most acid inorganic soils have little exchangeable H, that which is replaceable by a neutral unbuffed salt solution. This is particularly true in the highly weathered soils of the southeastern USA. In many less-weathered soils of the Midwest, however, significant quantities of H may be found. The objective of liming, in addition to neutralizing the H in the soil solution, is to replace this exchangeable H and/or Al with Ca and Mg. This is normally done by adding ground limestone, either calcitic, calcium carbonate ($CaCO_3$), or dolomitic, a mixture of Ca and magnesium carbonate ($MgCO_3$). Other materials such as calcium hydroxide ($Ca(OH)_2$) or calcium oxide (CaO) could also be used.

Thomas and Hargrove (1984) have provided the following equations to illustrate the lime reactions occurring in an acid soil. In water, $CaCO_3$ dissolves and hydrolyzes as follows

$$CaCO_3 + H_2O \rightarrow Ca^{2+} + HCO_3^- + OH^-. \quad [1]$$

The OH^- ions formed can subsequently react with H^+ or Al^{3+} to remove these acidic cations from the soil system. The following generalized reaction of lime with an acid soil can be written (Thomas and Hargrove, 1984):

$$2\text{Al-soil} + 3CaCO_3 + 3H_2O \rightarrow 3\text{Ca-soil} + 2Al(OH)_3 + 3CO_2. \quad [2]$$

In this reaction, exchangable Al^{3+} reacts with OH^- to form an $Al(OH)_3$ precipitate and the Ca^{2+} from the lime replaces the Al^{3+} on the exchange complex. A similar equation can be written to illustrate the reaction of $CaCO_3$ with exchangeable H.

Thus by liming acid soils, exchangeable H and Al are removed and replaced by Ca or a mixture of Ca and Mg. If this reaction were taken to completion, and the soil were totally Ca saturated, the pH would be about 8.3. Since most soils are only limed to pH 6 to 6.5, however, the limed soil is only partially Ca (or Mg) saturated. The degree to which the soil is saturated with basic cations such as Ca, Mg, or K is referred to as the base saturation of the soil, and is commonly expressed as a percent of the CEC. The percent base saturation of a soil limed to pH 6.5 can vary widely, and is a function of properties such as clay and organic matter content. Mehlich (1942, 1943) found that North Carolina soils varied from 40 to 90% base saturated at pH 6.5.

12-4.2 Benefits of Liming for Soybean Production

12-4.2.1 Reduction in the Concentration of Toxic Elements in the Soil

One of the principal benefits of liming acid soils is the neutralization of Al and Mn and the reduction in H in the soil solution. While it is generally agreed that in mineral soils the principal acidic cation is Al, in organic soils, or mineral soils where an appreciable part of the CEC is from organic matter, large amounts of exchangeable H can be present. The addition of lime can reduce the concentration and toxic effects of both.

12-4.2.1.1 Hydrogen Toxicity—The concentration of H ions in the soil solution plays an important role in determining the rooting environment and, thus, influencing nutrient uptake and plant growth. In addition to having direct effects on roots, soil pH can also influence other components of the root environment such as the population and activity of beneficial and pathogenic soil microorganisms, the availability of essential inorganic nutrients and the presence of toxic substances (Jackson, 1967). The direct effect of soil pH on roots is difficult to measure in acid soils due to these interactions, and thus much of the research on H ion effects on roots has been conducted in solution culture (Moore, 1974).

High concentrations of H in the root environment can cause direct injury to plant roots and limit root growth. Work with a number of species (Arnon and Johnson, 1942) has shown that at pH 3, roots are severely damaged and collapse shortly after exposure to this high H concentration. At pH 4, however, no significant injury to existing roots occurs. Additional Ca in the solution at pH 4 and 5 enhanced growth suggesting that Ca may help offset the harmful effects of H. Sutton and Hallsworth (1958) reported a similar offset of H effects by Ca in nutrient solution. Lund (1970) working in a split root system showed a tremendous enhancement of soybean tap root elongation by Ca at low pH.

In addition to effects on root growth, Ekdahl (1957) reported that pH had a more pronounced effect on root hair development. Working at higher pH ranges with wheat (*Triticum aestivum* L.), Ekdahl found that root hair growth was increased by 40% as pH increased from 5.5 to 7.2, while root elongation increased only 10%.

In addition to effects on growth, pH has also been shown to affect the permeability of cell membranes, allowing leakage of previously absorbed nutrients. At pH 4 or less, sizeable losses of K (Jacobson et al., 1950, 1957, 1960), Mg (Moòre et al., 1961b), and Ca (Moore et al., 1961a) have been observed in short-term experiments. As with root growth, the addition of Ca or increase in concentration of Ca in solutions has been reported to reduce these losses of ions from roots at low pH (Marsehner et al., 1966).

High H ion concentrations have also been shown to adversely affect the absorption of cations by roots. Reductions in uptake of K (Fawzy et al., 1954; Jacobson et al. 1957, 1960), Mg (Moore et al., 1961b; Islam et al., 1980; Blamey et al., 1982), Ca (Maas, 1969), Mn (Maas et al., 1968), and Zn (Rashid et al., 1976) by excised roots have been observed at pH 5 or below. Kinetic studies by Rains et al. (1964) and Hagen and Hopkins suggest that this reduction in plant uptake of cations by H is due to competitive inhibition of the absorption process.

While low soil pH, or high H concentration in the root environment, has been shown to have a number of effects on plants, the potential for direct pH effects is fairly low in most mineral soils. On acid organic soils, however, where the toxic effect of metals such Al can be reduced by organic matter chelation, and Ca and Mg contents are naturally higher, direct H ion effects on soybean production are more likely to be observed.

12–4.2.1.2 Aluminum Toxicity—Aluminum toxicity is probably the most important yield-limiting factor in acid soils (Foy, 1974; McLean, 1976). Aluminum has been known to be toxic to plants since the early 1900s (Hartwell and Pember, 1918). In general, Al toxicity occurs on soils with pH < 5.5 (McCart and Kamprath, 1965), where the predominant exchangeable ion on the cation exchange complex in Al (Evans and Kamprath, 1970).

High concentrations of Al can have both direct and indirect effects on soybean growth. The principal direct effects of Al toxicity appear on the roots of plants growing in acid soils. Aluminum-injured roots are generally short, stubby, with thickened root tips and a reduction of lateral branching. Aluminum-injured roots are also less efficient in absorbing nutrients and water. Thus, foliar symptoms of Al injury can include wilting and deficiencies of nutrients such as P and Ca (Foy, 1984).

A number of approaches have been used to study the mechanisms of Al injury to plant roots. Staining techniques have shown that Al accumulates in the nucleus of plant root cells (McLean and Gilbert, 1927). Levan (1945) reported that the accumulation of Al in the nucleus resulted in severe cytological abnormalities. Clarkson (1965) showed that reduction of root elongation resulting from exposure to Al corresponded to a reduction in cell division. Later work (Clarkson, 1969; Clarkson and Sanderson, 1969) indicted that the reduction in cell division resulted from the interaction of Al with the DNA in the nucleus preventing complete replication of the genetic materials. More recent work (Matsumato et al.,

1979; Naidoo, 1977; Ulmer, 1979) suggests that the Al in the cell nuclei binds with the P in nucleic acids and this binding interferes with nucleic acid replicating and cell division.

In addition to interfering with cell division and root growth, Al has also been associated with the alteration of root membrane structure and function (Hecht-Bucholz and Foy, 1981). Aluminum has also been shown to interfere with water use by plants (Kaufman and Gardner, 1978).

Aluminum toxicity has also been associated with reduced nutrient uptake and translocation. The symptoms of Al toxicity in shoots are classically those of P deficiency (Foy and Brown, 1963, 1964). The interaction of Al with P; Ca and Mg; K; and metals such as Fe, Cu, and Zn have all been reported (Duncan et al., 1980; Clark et al., 1981; Furlani and Clark, 1981; Lance and Pearson, 1969; Johnson and Jackson, 1964; Hiatt et al., 1963).

Several factors have been shown to ameliorate this effect of Al on plants. Considerable variation between species and among varieties within species in Al tolerance has been reported. Several mechanisms appear to be involved in this differential tolerance (Foy, 1984). In soybean, Al-tolerant lines appear to accumulate less Al in roots (Foy, 1974) and are able to accumulate more Ca, Mg, and P than more sensitive cultivars (Sartain, 1974).

12–4.2.1.3 Manganese Toxicity—Manganese toxicity commonly occurs on strongly acid soils with large quantities of easily reducible Mn. Next to Al toxicity, Mn toxicity is probably the second most important growth-limiting factor on acid soils (Foy, 1984). Like Al toxicity, Mn toxicity occurs most commonly below pH 5.5, but can occur at higher pH on poorly drained or compacted soils. Many soils, however, do not contain adequate Mn to produce Mn toxicity regardless of soil pH (Adams, 1984).

Manganese toxicity symptoms of soybean primarily occur on the leaves. These include a terminal leaf chlorosis and leaf crinkling or cupping (Heenan and Carter, 1977). The leaf distortion is thought to be caused by lower growth rates of leaf margins as compared to the remainder of the leaf (Hewitt, 1963). Symptoms on roots are limited, though in solution culture a browning of root tissue has been observed at high Mn concentrations (Morris and Pierre, 1949).

Leaf Mn concentration appears to be a rather effective indicator of Mn toxicity. Toxicity symptoms appear in soybean with leaf Mn concentrations of 300 to 500 mg kg^{-1} (Adams, 1984); though reductions in growth due to Mn toxicity at lower levels in young plants have been reported (Heenan and Carter, 1976).

Manganese toxicity has been associated with a number of problems in plant growth. Manganese is a cofactor for the acitivity of many enzymes, and serves as an activator of several enzymes in the Krebs cycle, several oxidases, and nitrate reductase (Helyar, 1978). The concentration of a mineral activator such as Mn, in a cell or organelle can theoretically be controlled through active uptake and excretion to control enzyme activity (Atkinson, 1966). Managanese toxicity appears to send some of these enzyme systems out of control and normal regulation is no longer possible resulting in cell death (Helyar, 1978). The interaction of Mn with the uptake and transport of other nutrients has also been reported. Manganese toxicity is often associated with a decrease in tissue Ca and Fe contents, and increased Ca in the growth medium has been reported as reducing the incidence of Mn toxicity (Foy, 1984).

A number of factors will influence the availability of Mn to soybean. The amount of easily reducible Mn in the soil is critical, as is the soil pH. But in addition the soil organic matter content, CEC, and redox potential of the soil will also effect Mn concentrations in the soil solution.

12–4.2.2 Increased Availability of Required Nutrients

12–4.2.2.1 Calcium—Liming acid soils increases the amount of Ca adsorbed on the CEC and the Ca concentration in the soil solution. An increase in Ca availability may have little affect on soybean growth in soils, however, since Ca deficiencies of soybean are rather rare. In fact, under many acid soil conditions, the addition of soluble Ca, or Ca or Mg salt will increase the concentration of Al in the soil solution and, therefore, can reduce crop growth (Mengel, 1975; Adams and Wear, 1957; Ragland and Coleman, 1959).

The ability of a soil to supply the Ca required for adequate soybean growth is a function of CEC, the amounts of Ca adsorbed on the exchange complex and the rate at which that exchangeable Ca is released to the soil solution to replace Ca removed by plant uptake or leaching. Some acid, low CEC soils may have inadequate Ca for crop growth (Gammon, 1957), however, these are rather unique situations since the concentration of Ca required n the soil solution is relatively low. Adams and Moore (1983), working in dilute soil solutions observed Ca deficiency of cotton (*Gossypium hirsutum* L.) taproot when solution Ca was < 0.3 m*M*. Lund (1970) reported that a concentration of 0.05 mg Ca L^{-1} was adequate for soybean root growth in a nutrient solution at pH 5.6.

Thus while liming increases Ca availability, this may have only limited direct benefit to the soybean plant.

12–4.2.2.2. Magnesium—Magnesium deficiencies of soybean have been reported on acid soils in several states (Adams, 1975; Rogers et al., 1973; Key and Kurtz, 1960; Schlegal, 1983). Conditions where Mg deficiencies are commonly found in acid soils include: (i) low CEC, (ii) a history of liming with calcitic limestone, and (iii) high rates of NH_4^+ and/or K^+ fertilizers (Adams, 1984).

In Alabama, soybean yields have been increased with the additions of magnesium sulfate ($MgSO_4$) or dolomitic lime on soils with an extractable Mg level < 15 mg kg^{-1} (Adams, 1975; Rogers et al., 1973). In Indiana, liming an acid Nineveh loam, Typic Argiudoll, with pH 4.3 and 0.05 cmol of Mg kg^{-1} with calcitic lime increased soybean yields from

1060 to 1870 kg ha^{-1}. The use of magnesium potassium sulfate [$MgK_2(SO_4)_2$] increased yields on the unlimed soil to 2030 kg ha^{-1} and the combination of calcitic lime and $MgK_2(SO_4)_2$ increased yields to 3280 kg ha^{-1}. The use of dolomitic lime produced 3000 kg ha^{-1}.

Liming with dolomitic limestone, or adding a soluble Mg source such as $MgSO_4$ or $MgK_2(SO_4)_2$ or finely ground (> 100 mesh) of magnesium oxide (MgO) in conjuction with calcitic limestone appear to be adequate ways to correct Mg deficiencies on acid, low Mg soils.

12–4.2.2.3 Molybdenum—Molybdenum deficiencies of soybean on acid soils correctable by liming have been reported in a number of states (Mortvedt, 1981; Parker and Harris, 1978; Sedberry et al., 1973; Anthony, 1967; Smith and Cornell, 1973). Similar responses with forage legumes have been reported in a number of other states (Foy and Barber, 1959; Evans et al., 1951; Kliewer and Kennedy, 1960).

Molybdenum availability appears to be influenced by the amount of Mo present, low soil pH and high soil Fe content (Lucas and Knezek, 1971). Liming to soil pH 5.5 to 6 appears to be adequate in soils where Mo supplies are satisfactory. On some low Mo soils, however, response to additional Mo has been reported. Seed treatment is an effective means of applying Mo for soybean production.

12–4.2.3 Effects on Nodulation and Dinitrogen Fixation

Symbiotic N_2 fixation is a key to the N nutrition of legumes. For this process to function properly, adequate numbers of rhizobia need to be present in the soil, the root system of the host plant must be present and the infection of the root system by the rhizobium and subsequent nodule development must occur. Each of the components of the N_2 fixation process have been shown to be sensitive to soil acidity to some degree. Thus, soil acidity can be a major limitation to the N nutrition of soybean.

Soil acidity has been shown to have a marked effect on the growth and survival of *Rhizobium* spp. in soils. This is especially true of the species responsible for the nodulation of forage legumes. Loneragan and Dowling (1958) showed that *R. trifolii* failed to grow at pH < 4.5 in solution culture, while increasing the pH to 5.0 resulted in growth of the bacteria. A large proportion of the strains of *R. trifolii* present in acid soils were found ineffective for nodulation by Holding and King (1963). Liming, however, increased the proportion of effective strains and improved nodulation.

Similar findings have been made with *R. japonicum*, the bacteria responsible for nodulation of soybean, though the critical pH levels and Al and Mn concentrations appear to be somewhat different for each rhizobium species. Doolas (1930) reported depressed nodulation of soybeans below pH 4.6 while Spurway (1941) concluded that the optimum pH for soybean was 6.0. Mengel and Kamprath (1978) found that the growth of soybean grown on organic soils was increased significantly as soil pH increased from 4 to 5. Nodule numbers and weight and total N uptake increased markedly with critical pH for root and shoot growth falling in the range of 4.6 to 4.8. The lime response was attributed to reduced water soluble Al, increased Ca and more favorable pH for rhizobia.

Rhizobial strains within a specie have also been shown to differ in their ability to nodulate a host plant at a given pH (Munns, 1978; Keyser and Munns, 1979a, 1979b). Several factors can account for this reduction in nodulation including low pH, Al and Mn toxicity, and Ca or Mg deficiency.

The processes of root infection and nodulation appear to be more sensitive to soil acidity and require a higher pH than rhizobium survival (Munns, 1978). The Ca requirement of nodulation also appears linked to the pH of the growth medium (Albrecht and Davis, 1929; Loneragan and Dowling, 1958). Munns (1970) found that Ca concentrations below 0.2 m*M* and pHs below 4.8 inhibited nodulation of alfalfa (*Medicago sativa* L.) grown in solution.

Freire (1975, as cited by Foy, 1984) concluded that in the absence of toxic concentrations of Al and Mn soybean nodulation was not affected at pH 4.0 to 4.5. With toxic levels of Al and Mn and low levels of P and Ca, however, nodule formation and N_2 was reduced.

In recent work, Munns et al. (1981) suggest that soybean growth may not be limited by nodulation problems on acid soils but rather that direct effects on the host plant may be more of a problem.

12–4.3 Liming Practices in the USA

Liming acid soils is an important part of the overall fertility program for soybean production in the USA. The decision of whether lime is needed or not is generally based on the soil pH. Critical pH, or the pH below which yield responses to liming would be expected, will vary between regions and soils within regions. Critical pH levels currently in use for soybean vary from pH 5.2 to 6.0.

While soil pH will indicate the need for lime, it will not adequately measure the rate of lime needed to raise the pH of an acid soil above the critical pH. There are three principal methods currently in use in the USA to estimate the amount of exchangeable acidity which must be neutralized to raise the pH to the desired level. The first involves the estimation of lime requirement from soil properties such as soil pH, texture, type of clay, and organic matter content. This procedure is used in a few states (Adams, 1984; Jackson and Reisenauer, 1984). The second method, direct titration of soils with $Ca(OH)_2$ is an extremely accurate method, but is time consuming and not well suited to a large volume of samples. The third and most common procedure used involves the use of a buffer solution to estimate the amounts of titratable acidity in a soil. A number of buffer solutions are currently in use. These include the Woodruff (Woodruff, 1948); Mehlich (Mehlich, 1976); SMP (Shoemaker et al., 1961); Adams and Evans (Adams and Evans, 1962); and TEA (Greweling and Peech, 1965).

12-5 SOYBEAN RESPONSE TO FERTILIZER APPLICATION

12-5.1 Nitrogen

The soybean has been characterized as being rather nonresponsive to the application of fertilizer N. This characteristic provides the basis for a considerable amount of conflicting research reported in the literature.

Being a legume, soybean, if properly inoculated, is capable of fixing substantial amounts of the required N from the atmosphere. In addition, this crop is also capable of utilizing both soil and fertilizer N. This ability to utilize the various sources of N has served as a basis for several studies designed to evaluate the effect of the use of fertilizer N on growth, yield, and quality of the soybean crop.

There is ample evidence to suggest that the application of fertilizer N just prior to planting has an adverse effect on symbiotic N_2 fixation. Beard and Hoover (1971) reported that the number of nodules per plant was linearly and inversely related to the rate of fertilizer N applied. Nitrogen rates in excess of 56 kg ha^{-1} applied at planting produced fewer nodules, but nodule number was not affected by rates of up to 112 kg ha^{-1} if the fertilizer N was applied at flowering. The adverse effect of fertilizer N on nodulation is supported by Minnesota research where three readily available N sources and two slow-release N sources decreased N_2 fixation, plant nodule weight, nodule number, and weight per nodule (Ham et al., 1975).

The reader should not be left with the impression that the presence of any amount of N in the root zone has a negative effect on nodulation. In fact, greenhouse research utilizing sand or solution culture has demonstrated the importance of the presence of some soil or fertilizer N for the initial growth of soybean, even in the presence of adequate inoculation (Hatfield et al., 1974).

The use of small amounts of N in a band at planting has been suggested to stimulate early growth of soybean, which would in turn provide for earlier cultivation and better weed control. This concept of "starter N" has been researched, and data collected over three growing seasons showed that the use of 16.8 and 50.4 kg N ha^{-1} in a band prior to planting had no significant effect on leaf area, plant height, fresh weight, or soybean yield (Sij et al., 1979).

Results of studies dealing with the broadcast applications of fertilizer N for soybean have not been conclusive. Studies reported by Beard and Hoover (1971), Deibert et al. (1979) and Mulvaney et al. (1973) are typical of several reporting the lack of a soybean response to fertilizer N. Ham et al. (1975), however, reported that N fertilization increased seed yield, weight per seed, seed protein percentage, and the kilogram of protein produced per hectare. They report that the increase in seed yield and/or seed protein percentage suggests that N_2 fixation failed to supply amounts of N essential for maximum seed yield and/or protein percentage.

Soybean response to fertilizer N may be related to the amount of residual NO_3-N in the root zone as suggested by Al-Ithawi et al. (1980). They concluded that the use of N fertilizer significantly increased soybean seed yield at several soil moisture levels when the amount of residual NO_3-N in the root zone was low. While amounts of residual NO_3-N in the root zone and soil moisture levels have been identified as factors that might affect the response of soybean to N fertilization, these, alone, cannot completely explain the results reported. Working with 13 fields in eastern Nebraska, Sorensen and Penas (1978) measured increases in seed yield from the use of fertilizer N at nine sites. Through regression analysis they identified soil pH and organic matter content as factors which were related to a response to fertilizer N. Response to fertilizer was highly probable where the soil was acid and the organic matter content was < 29 mg kg^{-1}.

It should be recognized that the vast majority of the research with fertilizer N for soybean production has been conducted when the fertilizer N was placed in the upper 15 to 20 cm of the root zone. This is also the portion of the root zone where the largest number of nodules occur. The utilization of soil and/or fertilizer N from below the 15- to 20-cm zone has received little attention. In work with soil columns, Harper and Cooper (1971) reported that placement of N fertilizer at a concentration of 150 mg kg^{-1} below 30 cm had no inhibitory effect on nodulation.

Limited field research (Rehm and Sorensen, 1977) has also indicated that yields of inoculated soybean can be increased if there is an accumulation of NO_3-N below 20 cm. These limited observations need to be verified by additional field research.

12–5.2 Phosphorus

While effects of fertilizer P on soybean production have not been ignored, published results of the effect of this nutrient on soybean growth are somewhat limited. Some early research focused on the percentage of P in the plant derived from either the fertilizer or the soil. Use of radioisotope techniques led to the conclusion that the percentage of P in the plant derived from the fertilizer was inversely relatd to the level of soil P and directly related to the rate of P applied (Welch et al., 1949). Total uptake of P, however, was greater at higher soil P levels. This relationship was verified in field studies involving several sources of fertilizer P (Bureau et al., 1953). In addition to the soil test level for P, soybean response to fertilizer P has also been related to environment and variety (Ham et al., 1973).

Phosphorus has a major role in nodule development in soybean. De Mooy and Pesek (1966) working in pots, reported that maximum nodulation required P additions of 400 to 500 mg kg^{-1} with even higher levels required for maximum activity of the nodule. Use of high rates of fertilizer P, however, might also cause some problems. With establishment of high soil test P levels and high pH values, Adams et al. (1982) reported a P-induced Zn deficiency in soybean with yield reduction being directly related to the degree of Zn deficiency.

Phosphorus uptake by soybean and its response to P fertilization has also been related to moisture stress (Marais and Wiersma, 1975). Phosphorus uptake was limited under conditions of severe moisture stress. After watering, plants previously exposed to a high moisture stress resumed P uptake at a rate exceeding that of plants exposed to normal moisture regime. Moisture stress appeared to affect P uptake largely through its influence on diffusion rather than mass flow.

12–5.3 Potassium

The harvested soybean crop is considered to remove substantial amounts of K from the soil system. Therefore, it would be reasonable to expect this crop to respond to the application of fertilizer K on some soils. The nature of the response reported by Terman (1977) is typical of responses to K usage reported in the literature. He found that maximum dry matter yield and nutrient uptake occurred during the early pod-filling stage. Concentrations of K in leaves, topgrowth, and grain increased with rate of K applied.

Potassium, like P, is essential for maximum nodule development. De Mooy and Pesek (1966), from the results of an outdoor pot study, reported that maximum nodulation required K applications of 600 to 800 mg kg^{-1}.

More recently, researchers have conducted field studies to evaluate the effect of fertilizer K in combination with other nutrients on soybean growth and production. The studies reported by Jones et al. (1977) and Lutz and Jones (1974) are typical of several such studies. These authors reported yield increases from the use of both P and K individually, but highest yields were recorded when combinations of P and K were used.

Potassium uptake has been reduced by an increase in bulk density. Silberbush et al. (1983) reported that this reduction in uptake could be overcome by the addition of fertilizer P.

12–5.4 Sulfur

There is general agreement that legume crops, in general, require substantial amounts of S for growth and development. Therefore, it would be reasonable to expect soybean to respond to the application of S fertilizers. This is especially true for soils where positive responses to S have been reported for other crops.

Solution culture and greenhouse research has shown that the S nutrition of soybean can be influenced by the concentration of other essential nutrients in soils (Wooding et al., 1972). In addition, S is closely related to the N nutrition of the soybean plant (Wooding et al., 1970; Dev and Saggar, 1974). Yet, in field studies, there are no consistent reports of the beneficial effects of S fertilization on soybean growth and yield.

For example, S applied to the Coastal Plain sandy soils has increased the yield of several crops. Matheny and Hunt (1981), however, found no effect of S fertilizers applied for soybean grown on these sandy soils.

Crop responses to S have also been reported on fine-textured soils. Brown et al., (1981), however, found no effect of S applied to soybean grown on a Mexico silt loam.

Based on the results of studies reported to date, it would appear that the incidence of soybean response to S fertilization would be quite limited.

12–5.5 Micronutrients

Soybean response to the use of various micronutrients has been the focus of several studies. As would be expected, the importance of a specific micronutrient for soybean production is often related to soil characteristics.

Soil pH has a major influence on the need for specific micronutrients. Iron chlorosis is a serious concern for soybean producers on some calcareous soils. Attempts to overcome or correct this problem have been described by Frank and Fehr (1983) and Wallace and Mueller (1978). Foliar applications of various Fe-containing materials have the potential to correct the chlorosis problem. Kaap (1973) has summarized several studies which deal with foliar application of these materials.

Variety selection is another management tool that can be used to reduce the severity of the chlorosis problem. The physiology of Fe accumulation for contrasting varieties has been researched by Elmstrom and Howard (1970).

The importance of Mn for soybean production has been identified for some soils. Sources, rates, and method of Mn application were studied by Randall et al. (1975a). Band applications were more effective than broadcast rates of $MnSO_4$. The foliar applications were most effective when applied at early blossom or early pod set or in multiple applications at both stages. These researchers also reported that Mn deficiency was alleviated by the application of either monoammonium phosphate or diammonium phosphate (Randall et al., 1975b). The slight decrease in soil pH resulting from the use of these two products was apparently responsible for an increase in the availability of soil Mn.

In Georgia studies, additions of Mn increased yields, Mn concentration in the tissue and seed weight among several cultivars when soil pH was increased to 7.0 (Parker et al., 1981).

The influence of Mo on soybean production was investigated in Iowa studies. When soil pH was in the range of 5.8 to 6.7, seed treatment with Mo produced a highly significant yield response (De Mooy, 1970). This same seed treatment reduced yields when soybean crops were grown on a calcareous soil.

Critical levels of micronutrients in soybean leaf tissue have not been well defined and several studies involving research in the greenhouse have focused on the problem of establishing these critical levels. Studies with Mn have been reported by Ohki et al. (1977) and Mask and Wilson (1978). Greenhouse studies involving Zn have been summarized by Ohki (1977) and Demeterio et al. (1972). Woodruff (1979) and Touchton and Boswell (1975) evaluated the use of B for soybean production but found no effects on either growth or yield.

In any discussion of micronutrients, there is always concern for the potential depressive effects of high rates of one or more micronutrients. Studies summarized by Martins et al. (1974) show that soybean is fairly tolerant to relatively high applications of B, Cu, and Zn. Five annual applications of 3.3 kg of B ha^{-1} and 11.1 kg of Zn ha^{-1} had no depressive effect on soybean yields.

12–6 FERTILIZER APPLICATION METHOD FOR SOYBEAN

12–6.1 Row vs. Broadcast

The placement of fertilizer in relation to the seed has been extensively studied for most agronomic crops. There is, however, limited information concerning the placement of fertilizer for soybean production. In addition, placement studies have not been conducted with all nutrients.

Placement of fertilizer materials that supply immobile nutrients is of greatest concern. Therefore, most of the studies that have evaluated the effect of fertilizer placement on soybean production have focused on P with some attention given to the placement of micronutrients.

Some early research utilizing radioactive tracer techniques showed that early absorption of P was enhanced when the P fertilizer was applied in a band at planting rather than broadcast and incorporatd before planting (Welch et al., 1949). Effects of placement on seed yields in this study, however, were inconsistent.

Bullen et al. (1983) reported that placement of P 2.5 cm below the seed resulted in yields that were 64 and 50% higher (two sites) than those obtained from the control plots. Sidebanding P 2.5 cm below and 2.5 cm away from the seed increased yields by 40 and 39%. Seed placement and broadcast applications of P were not as effective.

Ham et al. (1973) reported that the response to P placement was affected by environmental factors. With low rainfall and a low soil test P level, the largest response to P was recorded from broadcast applications. With adequate rainfall and a low soil test P level, the largest yield increase resulted from the combination of row applied and broadcast fertilizer. When soil P levels were high or very high, yields were not increased by P usage.

Placement of fertilizer P also affected the yield of irrigated soybean crops in Nebraska (Rehm and Weise, 1980). At all sites where soybean crops responded to the application of fertilizer P, the response was greater when the P was broadcast and incorporated before planting rather than applied in a band near the row at planting.

It is also possible to place some fertilizer in direct contact with the seed for many crops. Studies with soybean, however, have shown that even small amounts of a mixed fertilizer can substantially reduce germination if placed in direct contact with the seed (Clapp and Small, 1970; Hoeft, et al., 1975).

12–6.2 Foliar Fertilization

In 1976, Garcia and Hanway (1976) reported an increase in soybean yields of up to 31% with a foliar application of a fertilizer containing N, P, K, and S. The authors postulated that this foliar-applied mixture apparently supplemented the transfer of N, P, K, and S from leaves during senescence. They further speculated that this foliar application allowed the leaves to continue to function photosynthetically and thus fix more C leading to the filling of normally unharvestable seeds. The optimum proportion of N:P:K:S was reported to be 10:1:3:0.5.

Following this positive report, several researchers initiated studies to measure the effect of foliar fertilization on soybean production. Results were mixed. For example, Vasilas et al. (1980) reported a response to the practice in 1 of 2 yrs. Syverud et al. (1980) reported a response to the use of N but little effect from the NPKS combination was noted.

The problem of leaf burn or damage from the foliar fertilizer application has been identified by several researchers. This damage was severe enough to reduce yields in studies reported by Parker and Boswell (1980). Boote et al. (1978) found that foliar fertilization had no effect on soybean yield or the number of harvestable seeds produced per acre. They also reported that the foliar fertilizer produced leaf burning.

Expanding on the foliar fertilization concept, Poole et al. (1983) evaluated the effects of the use of an NPKS material with micronutrients, a fungicide, and a growth regulator. They observed a positive relationship between leaf injury and yield depression. This was especially true when materials were applied during midday rather than in the early morning or late afternoon hours. The inclusion of micronutrients, a fungicide and a growth regulator did not produce significant increases in yield.

Considering the results of the majority of the research that has been reported at this time, it would appear that foliar fertilization of soybean should not be a recommended practice.

12–7 RESIDUAL EFFECTS OF FERTILIZER APPLICATIONS

The research referred to in the preceding sections was conducted in situations where yields were recorded in the year in which the fertilizer was applied. Yet, it is a common practice to give special attention to adequate fertilization of the crop which precedes soybean in a rotation and to allow the soybean to benefit from the residual effects of this practice.

Soybean response to residual effects of applied fertilizers has been verified in several studies.

Thompson and Robertson (1969) observed a quadratic response to fertilizer applied in previous years. The residual effect of the application of 1100 kg ha^{-1} of a 15–4.4–12 produced a significant increase in seed yield. Yields leveled off from residual effects of a 220 kg ha^{-1} application of this material and were decreased by a 4400 kg ha^{-1} application of this fertilizer in a previous year.

Fink et al. (1974) reported yield increases from a previous application of 336 kg of N ha^{-1} and 155 kg of P ha^{-1}. Boswell and Anderson (1976) suggested that residual effects of P and K applications are more consistent than yield responses occurring in the year of fertilizer application.

The study of residual effects of fertilizer application on soybean production, however, is not complete. Additional research is needed.

REFERENCES

Adams, F. 1975. Field experiments with magnesium in Alabama—cotton, corn, soybeans, peanuts. Ala. Agric. Exp. Stn. Bull. 472.

----. 1984. Crop response to lime in the Southern United States. *In* Fred Adams (ed.) Soil acidity and liming. 2nd ed. Agronomy 12:211- 265.

----. (ed.) 1984. Soil acidity and liming. 2nd ed. Agronomy 12. American Society of Agronomy, Crop Science Society of America, Soil Science Society of America, Madison, WI.

----, and C.E. Evans. 1962. A rapid method for measuring lime requirement of Red-Yellow Podzolic soils. Soil Sci. Soc. Am. Proc. 26:355–357.

----, and B.L. Moore. 1983. Chemical factors affecting root growth in subsoil horizons of coastal plain soils. Soil Sci. Soc. Am. J. 47:99–192.

----, and J.I. Wear. 1957. Manganese toxicity and soil acidity in relation to crinkle leaf of cotton. Soil Sci. Soc. Am. Proc. 21:305–308.

Adams, J.F., F. Adams, and J.W. Odom. 1982. Interaction of phosphorus rates and soil pH on soybean yield and soil solution composition of two phosphorus-sufficient ultisols. Soil Sci. Soc. Am. J. 46:323–328.

Albrecht, W.A., and F.L. Davis. 1929. Physiological importance of calcium in legume inoculation. Bot. Gaz. 88:310–321.

Al-Ithawi, B., E.J. Deibert, and R.A. Olson. 1980. Applied N, and moisture level effects on yield, depth of root activity, and nutrient uptake by soybeans. Agron. J. 72:827–832.

Anthony, J.L. 1967. Fertilizing soybeans in the hill section of Mississippi. Miss. Agric. Exp. Stn. Bull. 743.

Arnon, D.I., and C.M. Johnson. 1942. Influence of hydrogen ion concentration on the growth of higher plants under controlled conditions. Plant Physiol. 17:525–539.

Atkinson, D.E. 1966. Regulation of enzyme activity. Ann. Rev. Biochem. 35:85–124.

Barber, S.A. 1973. The changing philosophy of soil test interpretations. p. 201–211. *In* L.M. Walsh and J.D. Beaton (ed.) Soil testing and plant analysis. Soil Science Society of America, Madison, WI.

Beard, B.H., and R.M. Hoover. 1971. Effect of nitrogen on nodulation and yield of irrigated soybeans. Agron. J. 63:815–816.

Beaufils, E.R. 1971. Physiological diagnosis—a guide for improving maize production based on principles developed for rubber trees. J. Fert. Soc. S. Afr. 1:1–39.

----. 1973. Diagnosis and recommendation integrated system (DRIS). Univ. of Natal, South Africa Soil Sci. Bull. 1.

Beckett, P.H.T., and R. Webster. 1971. Soil variability: A review. Soils Fert. 34:1–15.

Beverly, R.B. 1979. Application of the diagnosis and recommendation integrated system to soybeans. M.S. thesis. Dep. of Agronomy, Univ. of Georgia, Athens.

Blamey, F.P.C., D.G. Edwards, C.J. Asher, and M.K. Kim. 1982. Response of sunflower to low pH. p. 66–71. *In* A. Scaife (ed.) Proc. 9th Int. Plant Nutrition Colloq., Vol. 1, Warwick, UK. 22–27 August. Commonwealth Agriculture Bureaux, Farnham House, Fornum Royal, Slough, UK.

Boote, K.J., R.N. Gallaher, W.K. Robertson, K. Hinson, and L.C. Hammond. 1978. Effect of foliar fertilization on photosynthesis, leaf nutrition, and yield of soybeans. Agron. J. 70:787–791.

Boswell, F.C., and O.E. Anderson. 1976. Long-term residual fertility and current N-P-K application effects on soybeans. Agron. J. 68:315–318.

Bould, C., E.J. Hewitt, and P. Needham. 1984. Diagnosis of mineral disorders in plants—Principles, Vol 1. Chemical Publishing Co., New York.

Brady, N.C. 1974. The nature and properties of soils. MacMillan Publishing Co., New York.

Bray, R.H. 1954. A nutrient mobility concept of soil-plant relationships. Soil Sci. 78:9–22.

Brown, J.R., W.O. Thom, and L.L. Wall, Sr. 1981. Effects of sulfur application on yield and composition of soybeans and soil sulfur. Commun. Soil Sci. Plant Anal. 12:247–261.

Bullen, C.W., R.J. Soper, and L.D. Bailey. 1983. Phosphorus nutrition of soybeans as affected by placement of fertilizer phosphorus. Can. J. Soil Sci. 63:119–210.

Bureau, M.F., H.J. Mederski, and C.E. Evans. 1953. The effect of phosphate fertilizer material and soil phosphorus level on the yield and phosphorus uptake of soybeans. Agron. J. 45:150–154.

Castenson, R.F. 1973. Operation and management of a commercial soil testing and plant analysis laboratory p. 473–488. *In* L.M. Walsh, and J.D. Beaton (ed.) Soil testing and plant analysis. Soil Science Society of America, Madison, WI.

Clapp, J.G., Jr., and H.G. Small, Jr. 1970. Influence of "Pop-up" fertilizers on soybean stands and yield. Agron. J. 62:802–803.

Clark, R.B., P.A. Pier, D. Knudsen, and J.W. Maranville. 1981. Effect of trace element deficiencies and excesses on mineral nutrients in sorghum. J. Plant Nutr. 3:357–374.

Clarkson, D.T. 1965. The effect of aluminum and some other trivalent metal cations on cell division in the root apices of *Allium cepa*. Ann. Bot. N.S. 29:311–315.

----. 1969. Metabolic aspects of aluminum toxicity and some possible mechanisms for resistance. p. 381–397. *In* I.H. Rorison (ed.) Ecological aspects of the mineral nutrition of plants. Blackwell Scientific Publications, Oxford, UK.

----, and John Sanderson. 1969. The uptake of a polyvalent cation and its distribution in the root apices of *Allium cepa*. Tracer and autoradiographic studies. Planta 89:136–154.

Colliver, G. 1982. Soil test recommendations—why they differ. TVA Bull. Y-174.

Colwell, J.D. 1967. The calibration of soil tests. J. Aust. Inst. Agric. Sci. 33:321–330.

Cope, J.T. 1983. Soil test evaluation experiments at 10 Alabama locations 1977–1982. Auburn Univ. Agric. Exp. Stn. Bull. 550.

----. 1984. Long-term fertility experiments on cotton, corn, soybeans, sorghum and peanuts 1929–1982. Auburn Univ. Agric. Exp. Stn. Bull. 561.

----, and R.D. Rouse. 1973. Interpretation of soil test results. p. 35–54. *In* L.M. Walsh and J.D. Beaton (ed.) Soil testing and plant analysis. Soil Science Soceity of America, Madison, WI.

Deibert, E.J., M. Bijeriego, and R.A. Olson. 1979. Utilization of 15 N fertilizer by nodulating and non-nodulation soybean isolines. Agron. J. 71:717–723.

Demeterio, J.L., R. Ellis, Jr., and G.M. Paulsen. 1972. Nodulation and nitrogen fixation by two soybean varieties as affected by phosphorus and zinc nutrition. Agron. J. 64:566–568.

De Mooy, C.J. 1970. Molybdenum response of soybeans (*Glycine max* (L.) Merrill) in Iowa. Agron. J. 62:195–197.

----, and J. Pesek. 1966. Nodulation responses of soybeans to added phosphorus, potassium, and calcium salts. Agron. J. 58:275–280.

----, ----, and E. Spaldon. 1973. Mineral nutrition. *In* B.E. Caldwell (ed.) Soybeans: Improvement, production, and uses. Agronomy 16. 267–352.

Dev, G., and S. Saggar. 1974. Effect of sulfur fertilization on the N:S ratio in soybean varieties. Agron. J. 66:454–456.

Doolas, G.G. 1930. Soil acidity and soybean inoculation. Soil Sci. 30:275–288.

Dow, A.I., and S. Roberts. 1982. Proposal: critical nutrient ranges for crop diagnosis. Agron. J. 74:401–403.

Duncan, R.R., J.W. Dobson, Jr., and C.D. Fisher. 1980. Leaf elemental concentrations and grain yield of sorghum grown on acid soil. Commun. Soil Sci. Plant Anal. 11:699–707.

Ekdahl, I. 1957. The growth of root hairs and roots in nutrient media and distilled water, and the effects of oxylate. Wantbrukshoegsk. Ann. 23:497–518.

Elmstrom, G.W., and F.D. Howard. 1970. Promotion and inhibition of iron accumulation in soybean plants. Plant Physiol. 45:327–329.

Evans, C.E., and E.J. Kamprath. 1970. Lime response as related to percent Al saturation, solution Al, and organic matter content. Soil Sci. Soc. Am. Proc. 34:893–896.

Evans, H.J., E.R. Purvis, and R.E. Bear. 1951. Effect of soil reaction on availability of molybdenum. Soil Sci. 71:117–124.

Fawzy, H., R. Overstreet, and L. Jacobson. 1954. Influence of hydrogen ion concentration on cation absorption by barley roots. Plant Physiol. 29:234–237.

Fink, R.J., G.L. Posler, and R.M. Thorup. 1974. Effect of fertilizer and plant population on yield of soybeans. Agron. J. 66:465–467.

Foy, C.D. 1974. Effects of aluminum on plant growth. p. 601–642. *In* E.W. Carson (ed.) The plant root and its environment. University Press of Virginia, Charlottesville.

----. 1984. Physiological effects of hydrogen, aluminum and manganese toxicities in acid soil. p. 57–97. *In* Fred Adams (ed.) Soil acidity and liming, 2nd ed. American Society of Agronomy, Madison, WI.

----, and S.A. Barber. 1959. Molybdenum response of alfalfa on Indiana soils in the greenhouse. Soil Sci. Soc. Am. Proc. 23:36–39.

----, and J.C. Brown. 1963. Toxic factors in acid soils: I. Characterization of aluminum toxicity in cotton. Soil Sci. Soc. Am. Proc. 27:403–407.

----, and ----. 1964. Toxic factors in acid soils: II. Differential aluminum tolerance of plant species. Soil Sci. Soc. Am. Proc. 28:27–32.

Frank, S.J., and W.R. Fehr. 1983. Band application of sulfuric acid or elemental sulfur for control of Fe-deficiency chlorosis of soybeans. Agron. J. 75:451–454.

Freire, J.R.J. 1975. Compartamento de soja e do seu rizobia ao Al e Mn nos solos do Rio Grande do Sul. Cienc. Cult. (Sao Paolo) 28:169–170.

Furlani, P.R., and R.B. Clark. 1981. Screening sorghum for aluminum tolerance in nutrient solutions. Agron. J. 73:587–594.

Gammon, J., Jr. 1957. Root growth responses to soil pH adjustments made with carbonates of calcium, sodium, or potassium. Proc. Soil Crop Sci. Soc. Fla. 17:249–254.

Garcia, R.L., and J.J. Hanway. 1976. Foliar fertilization of soybeans during the seed-filling period. Agron. J. 68:653–657.

Greweling, T., and M. Peech. 1965. Chemical soil tests. Revised ed. Cornell Univ. Agric. Exp. Stn. Bull. 960.

Ham, G.E., I.E. Liener, S.D. Evans, R.D. Frazier, and W.W. Nelson. 1975. Yield and composition of soybean seed as affected by N and S fertilization. Agron. J. 67:293–297.

----, W.W. Nelson, S.D. Evans, and R.D. Frazier. 1973. Influence of fertilizer placement on yield response of soybeans. Agron. J. 68:81–84.

Hanway, J.J., and C.R. Weber. 1971. N, P, and K percentages in soybeans (*Glycine max* (L.) Merrill) plant parts. Agron. J. 63:286–290.

----, and ----. 1971b. Accumulation of N, P, and K by soybean [*Glycine max* (L.)] plants. Agron. J. 63:406–408.

Hargett, N.L., and J.T. Berry. 1983. 1982 Fertilizer summary data. TVA Bull. Y-165.

Harper, J.E., and R.L. Cooper. 1971. Nodulation response of soybeans (*Glycine max* (L.) Merr.) to application rate and placement of combined nitrogen. Crop Sci. 11:438–440.

Hartwell, B.L., and F.R. Pember. 1918. The presence of aluminum as a reason for the difference in the effect of so-called acid soil on barley and rye. Soil Sci. 6:259–281.

Hatfield, J.L., D.B. Egli, J.E. Leggett, and D.E. Peaslee. 1974. Effect of applied nitrogen on the nodulation and early growth of soybeans (*Glycine max* (L.) Merr.). Agron. J. 66:112–114.

Hecht-Bucholz, C., and C.D. Foy. 1981. Effect of aluminum toxicity on root morphology of barley. p. 343–345. *In* R. Brouwer et al. (ed.) Structure and function of plant roots. Nijhoff/Junk, The Hague.

Heenan, D.P, and O.G. Carter. 1976. Tolerance of soybean cultivars to manganese toxicity. Crop Sci. 16:389–391.

----, and ----. 1977. Influence of temperature on the expression of manganese toxicity in two soybean cultivars. Plant Soil 47:219–227.

Helyar, K.R. 1978. Effects of aluminum and manganese toxicity on legume growth. p. 207–231. *In* C.S. Andrew and E.J. Kamprath (ed.) Mineral nutrition of legumes in tropical and subtropical soils. Commonwealth Scientific and Industrial Research Organization Commonwealth Scientific and Industrial Research Organization (CSIRO), Melbourne, Australia.

Henderson, J.B., and E.J. Kamprath. 1970. Nutrient and dry matter accumulation by soybeans. N.C. Agric. Exp. Stn. Tech. Bull. 197.

Hewitt, E.J. 1963. The essential nutrient elements: Requirements and interactions in plants. p. 137–360. *In* F.C. Steward (ed.) Plant physiology: A treatise, Vol. 3. Inorganic nutrition of plants. Academic Press, New York.

Hiatt, A.J., D.F. Amos, and H.F. Massey. 1963. Effect of aluminum on copper sorption by wheat. Agron. J. 55:284–287.

Hoeft, R.G., L.M. Walsh, and E.A. Liegel. 1975. Effect of seed-placed fertilizer on the emergence (germination) of soybeans (*Glycine max* L.) and snapbeans (*Phaseolus vulgaris* L.). Commun. Soil Sci. Plant Anal. 6:655–664.

Holding, A.J., and J. King. 1963. The effectiveness of indigenous populations of *Rhizobium trifolium* in relation to soil factors. Plant Soil 18:191–198.

Islam, A.K.M.S., D.G. Edwards, and C.J. Asher. 1980. pH optima for crop growth: Results of a flowing solution culture experiment with six species. Plant Soil 54:339–357.

Jackson, T.L., and H.M. Reisenauer. 1984. Crop response to lime in the Western United States. p. 333–347. *In* F. Adams (ed.) Soil acidity and liming. 2nd ed. American Society of Agronomy, Madison, WI.

Jackson, W.A. 1967. Physiological effects of soil acidity. p. 43–124. *In* R.W. Pearson and F. Adams (ed.) Soil acidity and liming. American Society of Agronomy, Madison, WI.

Jacobson, L., D.P. Moore, and R.J. Hannapel. 1960. Role of calcium in absorption of monovalent cations. Plant Physiol. 35:352–358.

----, R. Overstreet, H.M. King, and R. Mandley. 1950. A study of potassium absorption by barley roots. Plant Physiol. 25:639–647.

----, ----, R.M. Carlson, and J.A. Chastain. 1957. The effect of pH and temperature on the absorption of potassium and bromide by barley roots. Plant Physiol. 32:658–662.

Johnson, G.V. (chair) 1984. Procedures used by state soil testing laboratories in the Southern Region of the U.S. Southern Coop. Ser. Bull. 190.

Johnson, R.E., and W.A. Jackson. 1964. Calcium uptake and transport by wheat seedlings as affected by aluminum. Soil Sci. Soc. Am. Proc. 28:381–386.

Jones, G.D., J.A. Lutz, Jr., and T.J. Smith. 1977. Effects of phosphorus and potassium on soybean nodules and seed yield. Agron. J. 69:1003–1006.

Jones, J.B. 1973. Soil testing—changing role and increasing need. Commun. Soil Sci. Plant Anal. 4(4):241.

----. 1981. Soil testing and plant analysis survey. The Council on Soil Testing and Plant Analysis, Athens, GA.

Kapp, J. 1973. Iron studies on soybeans. p. 1–4. *In* Proc. Iowa Fertilizer and Ag Chemical Dealers Conf, Des Moines, IA.

Kaufman, M.D., and E.H. Gardner. 1978. Segmental liming of soil and its effect on the growth of wheat. Agron. J. 70:331–336.

Key, J.L., and L.T. Kurtz. 1960. Response of corn and soybeans to magnesium fertilizers. Agron. J. 52:300.

Keyser, H.H., and D.N. Munns. 1979a. Effects of calcium, manganese and aluminum on growth of rhizobia in acid media. Soil Sci. Soc. Am. J. 43:500–503.

----, and ----. 1979b. Tolerance of rhizobia to acidity, aluminum and phosphate. Soil Sci. Soc. Am. J. 43:519–523.

Kliewer, W.H., and W.K. Kennedy. 1960. Studies on response of legumes to molybdenum and lime fertilization on Mardin silt loam soil. Soil Sci. Soc. Am. Proc. 24:377–388.

Lance, J.C., and R.W. Pearson. 1969. Effects of low concentrations of aluminum on growth and water and nutrient uptake by cotton roots. Soil Sci. Soc. Am. Proc. 33:95–98.

Levan, A. 1945. Cytological reactions induced by inorganic solutions. Nature (London) 156:751–752.

Loneragan, J.F., and E.J. Dowling. 1958. The interaction of calcium and hydrogen ions in the nodulation of subterranean clover. Aust. J. Agric. Res. 9:464–472.

Lucas, R.E., and B.D. Knezek. 1971. Climatic and soil conditions promoting micronutrient deficiencies in plants. p. 265–288. *In* J.J. Mortvedt et al. (ed.) Micronutrients in agriculture. Soil Science Society of America, Madison, WI.

Lund, Z.F. 1970. The effect of calcium and its relation to some other cations on soybean root growth. Soil Sci. Soc. Am. Proc. 34:456–459.

Lutz, J.A., Jr., and G.D. Jones. 1974. Effects of deep placement of limestone and fertility treatments on the chemical composition of soybeans and on certain properties of a Tatum soil. Commun Soil Sci. Plant Anal. 5:399–411.

Maas, E.V. 1969. Influence of calcium and magnesium on manganese absorption. Plant Physiol. 44:796–800.

----, D.P. Moore, and B.J. Mason. 1968. Manganese absorption by excised barley roots. Plant Physiol. 43:527–530.

Marais, J.N., and D. Wiersma. 1975. Phosphorus uptake by soybeans as influenced by moisture stress in the fertilized zone. Agron. J. 67:777–781.

Marsehner, H., R. Handley, and R. Overstreet. 1966. Potassium loss and changes in fine structure of corn root tips induced by H^+-ion. Plant Physiol. 41:1725–1735.

Martins, D.C., M.T. Carter, and G.D. Jones. 1974. Response of soybeans following six annual applications of various levels of boron, copper, and zinc. Agron. J. 66:82–84.

Mask, P.L., and D.O. Wilson. 1978. Effect of Mn on growth, nodulation, and nitrogen fixation by soybeans grown in a greenhouse. Commun. Soil Sci. Plant Anal. 9:653–666.

Matheny, T.A., and P.G. Hunt. 1981. Effects of irrigation and sulphur application on soybeans grown on a Norfolk loamy sand. Commun. Soil Sci. Plant Anal. 12:147–159.

Matsumoto, H., S. Marimuro, and E. Hirasawa. 1979. Localization of absorbed aluminum in plant tissues and its toxicity: Studies in the inhibition of pea root elongation. p. 171–194. *In* K. Kudrev et al. (ed.) Mineral nutrition of plants, Vol. 1. Proc. 1st Int. Symp. on Plant Nutrition, Varna, Bulgaria. 24–29 September. Bulgaria Academy of Science, Institute of Plant Physiology, Sofia, Bulgaria.

McCart, G.D., and E.J. Kamprath. 1965. Supplying Ca and Mg for cotton on sandy, low cation exchange capacity soils. Agron. J. 57:404–406.

McLean, E.O. 1976. Chemistry of soil aluminum. Commun. Soil Sci. Plant Anal. 7:619–636.

----. 1978. Contrasting concepts in soil test interpretation: sufficiency levels of available nutrients versus basic cation saturation ratios. p. 39–54. *In* T.R. Peck (chair) Soil testing: Correlating and interpreting the analytical results. Spec. Pub. 29. American Society of Agronomy, Crop Science Society of America, and Soil Science Society of America, Madison, WI.

McLean, F.T., and B.E. Gilbert. 1927. The relative aluminum tolerance of crop plants. Soil Sci. 24:163–175.

Mehlich, A. 1942. Base saturation and pH in relation to soil type. Soil Sci. Soc. Am. Proc. (1941)6:150–156.

----. 1943. The significance of percentage base saturation and pH in relation to soil differences. Soil Sci. Soc. Am. Proc. 7:167–174.

----. 1976. New buffer pH method for rapid estimation of exchangeable acidity and lime requirement of soils. Commun. Soil Sci. Plant Anal. 7:637–652.

Melsted, S.W. 1967. The philosophy of soil testing. p. 13–23. *In* Soil testing and plant analysis, Part 1. Spec. Pub. 2. Soil Science Society of America, Madison, WI.

Mengel, D.B. 1975. Effect of soil acidity and liming on growth, nodulation and mineral nutrition of soybeans in Histosols. Ph.D. diss. North Carolina State Univ., Raleigh (Diss. Abstr. 378:22).

----, and E.J. Kamprath. 1978. Effect of soil pH and liming on growth and nodulation of soybeans in Histosols. Agron. J. 70:959–963.

Moore, D.P. 1974. Physiological affects of pH on roots. p. 135–152. *In* E.W. Carson (ed.) The plant root and its environment. University of Virginia Press, Charlottesville, VA.

----, L. Jacobson, and R. Overstreet. 1961a. Uptake of calcium by excised barley roots. Plant Physiol. 36:53–57.

----, R. Overstreet, and L. Jacobson. 1961b. Uptake of magnesium and its interaction with calcium in excised barley roots. Plant Physiol. 36:290–295.

Morris, H.D., and W.H. Pierre. 1949. Minimum concentrations of manganese necessary for injury to various legumes in culture solutions. Agron. J. 41:107–112.

Mortvedt, J.J. 1978. Micronutrient soil test correlation and interpretations. p. 99–117. *In* Soil testing: Correlating and interpreting the analytical results. Spec. Pub. 29. American Society of Agronomy, Crop Science Society of America, and Soil Science Soceity of America, Madison, WI.

----. 1981. Nitrogen and molybdenum uptake and dry matter relationships of soybeans and forage legumes in response to applied molybdenum on acid soil. J. Plant Nutr. 3:245–256.

Mulvaney, D.L., M.G. Oldham, and J.W. Pendleton. 1973. Soybean yields with direct and residual nitrogen fertilization. Agron. J. 65:547–550.

Munns, D.N. 1970. Nodulation of *Medicago sativa* in solution culture. V. Calcium and pH requirements during infection. Plant Soil 32:90–102.

----. 1978. Legume-Rhizobium relations. p. 247–263. *In* C.S. Andrew and E.J. Kamprath (ed.) Mineral nutrition of legumes in tropical and subtropical soils. Commonwealth Scientific and Industrial Research Organization (CSIRO), Melbourne, Australia.

----, J.S. Huhenberg, T.L. Righetti, and D.T. Lauter. 1981. Soil acidity tolerance of symbiotic and nitrogen fertilized soybeans. Agron. J. 73:407–410.

Munson, R.D., and W.L. Nelson. 1973. Principles and practices in plant analysis. p. 223–248. *In* L.M. Walsh and J.D. Beaton (ed.) Soil testing and plant analysis. Soil Science Society of America, Madison, WI.

Naidoo, G. 1977. Aluminum toxicity in two snapbean varieties. Ph.D. diss. Univ. of Tennessee, Knoxville (Diss. Abstr. 27B:5478).

Ohki, K. 1977. Critical zinc levels related to early growth and development of determinate soybeans. Agron. J. 69:969–974.

----, D.O. Wilson, F.C. Boswell, M.B. Parker, and L.M. Shuman. 1977. Mn concentration in soybean leaf related to bean yields. Agron. J. 69:497–600.

Ohlrogge, A.J., and E.J. Kamprath. 1968. Fertilizer use in soybeans. p. 273–295. *In* L.B. Nelson (ed.) Changing patterns in fertilizer use. Soil Science Society of America, Madison, WI.

Olson, R.A., R.D. Voss, R.C. Ward, and D.A. Whitney. 1984. The philosophy of soil testing. National Corn Handbook Fact Sheet. Purdue University, West Lafayette, IN.

Owens, H.I. 1980. Soil testing and plant analysis report. USDA-SEA—Extension, Washington, DC.

Ozus, T., and J.J. Hanway. 1966. Comparisons of laboratory and greenhouse tests for N and P availability in soils. Soil Sci. Soc. Am. Proc. 30:224.

Parker, M.B., and F.C. Boswell. 1980. Foliar injury, nutrient intake, and yield of soybeans as influenced by foliar fertilization. Agron. J. 72:110–113.

----, ----, K. Ohki, L.M. Shuman, and D.O. Wilson. 1981. Manganese effects on yield and nutrient concentration in leaves and seed of soybean cultivars. Agron. J. 73:643–646.

----, and H.B. Harris. 1978. Molybdenum studies on soybeans. G. Agric. Exp. Stn. Res. Bull. 215.

Peck, T.R. (chair) 1978. Soil testing: Correlating and interpreting the analytical results. Spec. Pub. 29. American Society of Agronomy, Crop Science Society of America, and Soil Science Society of America, Madison, WI.

----, and S.W. Melsted. 1973. Field sampling for soil testing. p. 67–75. *In* L.M. Walsh, and J.D. Beaton (ed.) Soil testing and plant analysis. Soil Science Society of America, Madison, WI.

Plank, O.C. 1979. Plant analysis handbook for Georgia. Univ. of Georgia Ext. Serv. Bull. 735.

Poole, W.D., G.W. Randall, and G.E. Ham. 1983. Foliar fertilization of soybeans. I. Effect of fertilizer sources, rates, and frequency of application. Agron. J. 75:195–200.

Potash and Phosphate Institute. 1982. Soil testing in high yield agriculture. Potash and Phosphate Institute, Atlanta, GA.

Ragland, J.L., and N.T. Coleman. 1959. The effect of soil solution aluminum and calcium on root growth. Soil Sci. Soc. Am. Proc. 23:355–357.

Rains, D.W., W.E. Schmidand, and E. Epstein. 1964. Absorption of cations by roots. Effects of hydrogen ions and essential role of calcium. Plant Physiol. 39:274–278.

Randall, G.W., E.E. Schulte, and R.B. Corey. 1975a. Effect of soil and foliar applied manganese on the micronutrient content and yield of soybeans. Agron. J. 67:502–507.

----, ----, and ----. 1975b. Soil Mn availability to soybeans as affected by mono and diammonium phosphate. Agron. J. 67:705–709.

Rashid, A., F.M. Chandry, and M. Sharif. 1976. Micronutrient availability to cereals from calcareous soils: III. Zinc absorption by rice and its inhibition by important ions of submerged soils. Plant Soil 45:613–623.

Rehm, G.W., and R.C. Sorensen. 1977. Soybean yields on a silt loam soil as influenced by prior fertilization, row spacings and plant populations for corn. Soil Sci. 124:235–240.

----, and R.A. Weise. 1980. Rate and placement of phosphate fertilizer for soybean production. p. 18. *In* Univ. of Nebraska, Lincoln Soil Sci. Res. Rep.

Rogers, H.T., F. Adams, and D.L. Thurlow. 1973. Lime needs of soybeans on Alabama soils. Ala. Agric. Exp. Stn. Bull. 452.

Rouse, R.D. 1968. Soil test theory and calibration for cotton, corn, soybeans and coastal bermudagrass. Auburn Univ. Agric. Exp. Stn. Bull. 375.

Sartain, J.B. 1974. Differential effects of aluminum on top and root growth, nutrient accumulation and nodulation of several soybean varieties. Ph.D. diss. North Carolina State Univ., Raleigh (Diss. Abstr. 358:641).

Schlegal, A. 1983. A study of magnesium deficiency of corn in Indiana. M.S. thesis. Purdue University, West Lafayette, IN.

Sedberry, J.E., T.S. Dharmaputra, R.H. Brupbacher, S.A. Phillips, J.G. Marshall, L.W. Sloane, D.R. Melville, J.L. Rabb, and J.H. Davis. 1973. Molybdenum Investigations with soybeans in Louisiana. Agric. Exp. Stn. Bull. 670.

Shoemaker, H.E., E.O. McLean, and P.F. Pratt. 1961. Buffer methods for determining lime requirement of soils with appreciable amounts of extractable aluminum. Soil Sci. Soc. Am. Proc. 25:274–277.

Sij, J.W., F.T. Turner, and J.P. Craigmiles. 1979. "Starter nitrogen" fertilization in soybean culture. Commun. Soil Sci. Plant Anal. 10:1451–1457.

Silberbush, M., W.B. Hallmark, and S.A. Barber. 1983. Simulation of effects of bulk density and P addition on K uptake by soybeans. Commun. Soil Sci. Plant Anal. 14:287–296.

Small, H.G., and A.J. Ohlrogge. 1973. Plant analysis as an aid in fertilizing soybeans and peanuts. p. 315–327. *In* L.M. Walsh, and J.D. Beaton (ed.) Soil testing and plant analysis. Soil Science Society of America, Madison, WI.

Smith, H.C., and J. Cornell. 1973. Lime and molybdenum in soybean production. Tenn. Farm Home Sci. Prog. Rep. 85:28–31.

Smith, C.M., and R.E. Lamborn. 1982. Diagnostic methods. p. 179–207. *In* The fertilizer handbook. The Fertilizer Institute, Washington, DC.

Sorensen, R.C., and E.J. Penas. 1978. Nitrogen fertilization of soybeans. Agron. J. 70:213–216.

Spurway, C.H. 1941. Soil reaction (pH) preference of plants. Mich. Agric. Exp. Stn. Bull. 306.

Sumner, M.E. 1977. Preliminary N, P and K foliar diagnostic norms for soybeans. Agron. J. 69:226–230.

----. 1978. Interpretation of nutrient ratios in plant tissue. Commun. Soil Sci. Plant Anal. 9(4):335.

----. 1979. Interpretation of foliar analyses for diagnostic purposes. Agron. J. 71:343–348.

Sutton, C.D., and E.G. Hallsworth. 1958. Studies on the nutrition of forage legumes. I. The toxicity of low pH and high manganese supply to lucerne as affected by climatic factors and calcium supply. Plant Soil 9:305–317.

Syverud, T.D., L.M. Walsh, E.S. Oplinger, and K.A. Kelling. 1980. Foliar fertilization of soybeans (*Glycine max* L.). Commun. Soil Sci. Plant Anal. 11:637–651.

Terman, G.L. 1977. Yields and nutrient accumulation by determinate soybeans as affected by applied nutrients. Agron. J. 69:234–238.

Thomas, G.W., and W.L. Hargrove. 1984. The chemistry of soil acidity. p. 3–56. *In* Fred Adams (ed.) Soil acidity and liming. 2nd ed. American Society of Agronomy, Madison, WI.

Thompson, L.G., Jr., and W.K. Robertson. 1969. Effects of residual fertilizer, row width, and spacing within the row on soybean (*Glycine max*) yields and soil fertility. Soil Crop Sci. Soc. Fl. 29:49–57.

Tisdale, S.L., W.L. Nelson, and J.D. Beaton. 1985. Soil fertility and fertilizers. 4th ed. MacMillan Publishing Co., NY.

Touchton, J.T., and F.C. Boswell. 1975. Effects of boron accumulation in soybean seed on several characteristics of germinated seedlings. Agron. J. 67:577–578.

Ulmer, S.E. 1979. Aluminum toxicity and root DNA synthesis in wheat. Ph.D. diss. Iowa State Univ., Ames. (Diss Abstr. 40:2933).

Ulrich, A., and F.J. Hills. 1967. Principles and practices of plant analysis. p. 11–24. *In* Soil testing and plant analysis, Part 2. Spec. Pub. 2. Soil Science Society of America, Madison, WI.

Vasilas, B.L., J.O. Legg, and D.C. Wolf. 1980. Foliar fertilization of soybeans: absorption and translocation of ^{15}N-labeled urea. Agron. J. 72:271–275.

Wallace, A., and R.T. Mueller. 1978. Complete neutralization of a portion of calcareous soil as a means of preventing iron chlorosis. Agron. J. 70:888–890.

Walsh, L.M., and J.D. Beaton (ed.) 1973. Soil testing and plant analysis. Soil Science Society of America, Madison, WI.

Welch, C.D., N.S. Hall, and W.L. Nelson. 1949. Utilization of fertilizer and soil phosphorus by soybeans. Soil Sci. Soc. Am. Proc. 14:231–235.

----, and R.A. Wiese. 1973. Opportunities to improve soil testing programs. p. 1–11. *In* L.M. Walsh, and J.D. Beaton (ed.) Soil testing and plant analysis. Soil Science Society of America, Madison, WI.

Wooding, F.J., G.M. Paulsen, and L.S. Murphy. 1970. Response of nodulated and nonnodulated soybean seedlings to sulfur nutrition. Agron. J. 62:277–280.

----, ----, and ----. 1972. Sulfur composition of soybeans as affected by macronutrient deficiencies. Commun. Soil Sci. Plant Anal. 3:151–158.

Woodruff, C.M. 1948. Testing soils for lime requirement by means of a buffered solution and the glass electrode. Soil Sci. 66:53–63.

Woodruff, J.R. 1979. Soil boron and soybean leaf boron in relation to soybean yield. Commun. Soil Sci. Plant Anal. 10:941–952.

13 Nitrogen Metabolism

J.E. Harper
USDA-ARS
University of Illinois
Urbana, Illinois

The interdependence and interaction of nitrate (NO_3^-) metabolism and symbiotic dinitrogen (N_2) fixation provides the basis for consideration of overall N metabolism as one topic. The symbiotic association of *Bradyrhizobium japonicum* (Kirchner) Jordan with roots of soybean [*Glycine max* (L.) Merr.] provides one N input system to the plant. Alternatively, nitrate is readily taken up from the soil solution by soybean roots and provides a second N input system. Although each N input system has independent pathways and control points, the soybean plant under most cultivated conditions assimilates N from both sources, and these inputs are interdependent. Photosynthetically derived energy is utilized to drive both input systems and may be one control point in determining the relative magnitude of N_2 fixation and NO_3^- uptake and metabolism. This chapter seeks to bring together literature dealing with nitrate uptake and metabolism, symbiotic N_2 fixation, and the interaction and limitations of these two N sources with specific emphasis on soybean. Processes involving N transformation and movement in the soil, which precede uptake by the root and nodule, will not be covered. Metabolism of N by the root, shoot, and nodule components of soybean plants, and the transport processes and compounds involved in N distribution from assimilation sites to centers of growth and/or storage, will be emphasized. More general reviews detailing differences in NO_3^- metabolism and symbiotic N_2 fixation among species should be consulted.

The first section of this chapter will deal with NO_3^- uptake, reduction, and incorporation by soybean plants. For additional information on NO_3^- and nitrite (NO_2^-) metabolism in higher plants and on the nitrate reductase (NR) enzyme in general, refer to reviews by Beevers and Hageman (1983), Duke and Duke (1984), Dunn-Coleman et al. (1984), Guerrero et al. (1981), and Hewitt and Notton (1980).

Soybeans: Improvement, Production, and Uses, 2nd ed.—Agronomy Monograph no. 16.

13-1 NITRATE METABOLISM

13-1.1 Uptake

The net rate of NO_3^- uptake by intact plants depends on (i) availability of NO_3^- to the uptake sites and (ii) on the relative rates of influx and efflux processes across the root cell plasma-lemma. The availability of NO_3^- to the uptake site is dependent on NO_3^- level in the soil solution, water availability and rate of flow to roots, mineralization rates, and root penetration into previously unexploited soil areas. Since NO_3^- is freely mobile, the moisture status of a soil is likely the predominant factor controlling NO_3^- availability to the root (Olson and Kurtz, 1982). Both influx and efflux of NO_3^- readily occur in soybean roots (Nicholas and Harper, 1985), similar to that shown for other crop species (MacKown et al., 1981, 1983). The relative rates of influx and efflux are dependent on previous NO_3^- nutrition of the root, and the external concentration to which the root is exposed. Under field conditions where soil NO_3^- is gradually depleted with plant growth, NO_3^- efflux will likely be minimal due to continued metabolism within, and translocation from, the plant root.

Nitrate uptake by soybean occurs in both light and dark conditions (Nicholas and Harper, 1985; Rufty et al., 1984). Uptake of NO_3^- appears to be saturated at relatively low (0.5 m*M*) solution NO_3^- levels during the seedling (7–21 days after planting [DAP]) (Gibson and Harper, 1985) and early vegetative (31–34 DAP) (Wych and Rains, 1978) growth stages. Peak uptake rates per plant occur at the time of early to mid-pod fill under conditions of adequate water supply and continuous NO_3^- availability (Harper, 1971). Increasing rates of NO_3^- uptake with plant age (Harper, 1971) are likely due in large part to increasing root mass rather than to an increase in specific uptake rate. Rufty et al. (1982a) have shown that soybean roots maintain a constant NO_3^- uptake rate per unit root mass at least during the early vegetative growth phase. Other work, however, (Wych and Rains, 1979) had indicated that the rate of NO_3^- uptake per unit root mass increased up to 33 DAP and then declined with further plant growth and time. Rabie et al. (1981) showed that uptake and metabolism of $^{15}N\text{-}NO_3^-$ was greater at the pod-initiation stage than during the initial pod-filling stage under controlled conditions. Under field conditions, and particularly during later growth stages, nitrate uptake is more likely to be limited by water and NO_3^- availability than by the kinetics of the uptake process. Soybean plants have been shown to remove as much N from the soil as does a corn (*Zea mays* L.) crop (Johnson et al., 1975), even though a portion of the overall N input in soybean is met via symbiotic N_2 fixation. The extensive literature on N utilization by soybean indicates that NO_3^- is readily and preferentially removed from the soil, and that only upon gradual depletion of plant-available soil N does N_2 fixation functionally contribute to the N demands of the soybean plant.

13–1.2 Translocation and Compartmentalization

Nitrate which is taken up may be either reduced in the roots (discussed in section 13–1.3), stored in the roots, or translocated to the shoots. Nitrate translocation has been primarily studied through monitoring of xylem sap composition. Nitrate does not appear to move in the phloem of soybean, based on lack of ^{15}N-NO_3^- movement from leaves following petiole girdling (Ohyama and Kawai, 1983). McClure and Israel (1979) reported that NO_3^- comprised 58% (average over entire growing season) of the total N in the xylem sap of non-inoculated soybean supplied with high (20 m*M*) NO_3^-. This verifies that a portion of the NO_3^- taken up is translocated to the shoots in an unreduced state, but also indicates that active NO_3^- reduction likely occurs in the roots before translocation. Xylem exudate analysis must be viewed with caution, however, since reduced-N compounds in the xylem may result from recycling from shoots to roots and back, as well as result from breakdown products of previously assimilated N in the root. Nitrate which remains in the root system and which is not directly reduced is likely stored in the vacuole (Oaks, 1979), although no specific report of vacuolar NO_3^- storage in soybean was found. The concentration of NO_3^- in soybean roots exceeds that found in either the leaf or the stem (including petioles and cotyledons) fractions (Crafts-Brandner and Harper, 1982). This indicates that roots may serve as temporary storage sites for NO_3^-. However, in view of the greater dry matter content of stems (including petioles and cotyledons) compared with roots, the absolute quantity of NO_3^- storage is greater in the stems. The mechanisms which control partitioning of NO_3^- in roots between storage, reduction, and translocation are not known. For a more extensive treatment of transport and partitioning of nitrogenous compounds, refer to a review by Pate (1980).

13–1.3 Reduction

The initial step of NO_3^- reduction is catalyzed by the NR enzyme, which is considered a cytoplasmic enzyme in higher plants (Beevers and Hageman, 1983). The NR enzyme has been found in essentially all plant parts of soybean with major activities in leaves and roots (Crafts-Brandner and Harper, 1982; Hunter et al., 1982; Hunter, 1983). Appreciable NR activity has also been found in stems and petioles (Andrews et al., 1984; Crafts-Brandner and Harper, 1982) and in cotyledons (Kakefuda et al., 1983; Orihuel-Iranzo and Campbell, 1980). Based on xylem exudate analysis, distribution of NR activity between roots and shoots varies among soyban cultivars (Hunter et al., 1982), but as pointed out earlier xylem exudate analysis must be viewed with caution. Root nodules also exhibit NR activity (Randall et al., 1978) which has been associated with both bacteroid and cytosol fractions (Streeter, 1982). The contribution of nodule NR activity to overall NO_3^- assimilation by soybean, however, has not been definitively established.

Although the magnitude of NO_3^- reduction in various soybean plant parts has not been definitively established, evidence is mounting that appreciable NO_3^- is reduced in soybean roots (Crafts-Brandner and Harper, 1982; Hunter et al., 1982; Nicholas and Harper, 1985). Previous to these reports, root NR was not detected which led to the conclusion by Randall et al. (1978) that nodule NR contributed significantly to NO_3^- reduction by the whole plant. Hunter (1983), however, recently concluded that it is improbable that the nodule is a major site of NO_3^- reduction. It still appears that the leaf fraction is the major site of NO_3^- reduction in soybean (Andrews et al., 1984; Crafts-Brandner and Harper, 1982; Harper and Hageman, 1972; Hunter et al., 1982).

Recent cytological evidence with soybean cotyledons indicated that NR was localized in small vesicles throughout the cytoplasm (Vaughn and Duke, 1981). Immunochemical localization, however, indicated only weak association of NR with organelles and greater association with the cytoplasm (Vaughn et al., 1984). No evidence of an association of NR with vesicles in leaves or roots of soybean has been reported.

Early studies with soybean indicated that NADH and NADPH were equally effective in supporting NO_3^- reduction in leaves (Evans and Nason, 1953). The NADH-NR enzyme was later reported to predominate in cysteine stabilized leaf extracts (Beevers et al., 1964; Wells and Hageman, 1974). Jolly et al. (1976) subsequently separated NR from soybean leaves into two fractions on DEAE-cellulose columns. One fraction exhibited predominant activity with NADPH while the other fraction was more active with NADH; in neither case was activity exclusively dependent on one electron donor suggesting incomplete separation and/or bispecific enzymes. The NADPH enzyme was reported to have a Sephadex molecular wt. of 220 000, pH optimum of 6.5, K_m (NO_3^-) of 4.5 m*M*, K_m (NADPH) of 1.5 μM, and K_m (NADH) of 3.9 μM. The NADH enzyme had a Sephadax molecular wt. of 330 000, pH optimum of 6.5, K_m (KNO_3) of 0.11 m*M*, K_m (NADH) of 8.1 μM, and K_m (NADPH) of 200 μM. The fact that the NADH-NR-dependent enzyme has a lower apparent K_m for NO_3^- (0.11 m*M*), compared with the NADPH enzyme (4.5 m*M*), indicates that the former enzyme is the primary enzyme involved in NO_3^- metabolism.

It has been known for some time that soybean leaves contain both a substrate (NO_3^-) inducible NR and a "constitutive" (present in the absence of NO_3^-) NR form (Aslam, 1982; Harper, 1974). That two distinct NR enzymes exist has been verified by selecting a soybean mutant which lacks constitutive- but retains inducible-NR activity (Nelson et al., 1983; Ryan et al., 1983a). Subsequent genetic characterization of the mutant plant, designated nr_1, has verified that the absence of constitutive NR activity was due to a single recessive nuclear gene (Ryan et al., 1983b). Biochemical characterization has revealed several differences in the two enzymes (Nelson et al., 1984). The pH optimum of the inducible and constitutive enzymes were 7.5 and 6.8, respectively. The apparent K_m (NO_3^-) was similar (0.15 m*M*) for the two enzymes under standard conditions, but the K_m (NO_3^-) of the constitutive enzyme increased to 0.51 m*M* in the presence of K_2CO_3, while the K_m (NO_3^-) of the inducible enzyme was unaffected (Nelson et al., 1984). Both enzymes also exhibit similar levels of diaphorase activity (Nelson et al., 1984). The inducible-NR enzyme was strongly inhibited by tungstate addition to the growing medium while the constitutive NR enzyme was only slightly inhibited (Aslam, 1982; Harper and Nicholas, 1978). The differential response of the constitutive- and inducible-NR enzymes to tungstate and potassium carbonate (K_2CO_3) indicate that the Mo cofactor portion is likely different between the two enzymes. Complementation studies have, however, indicated that the Mo cofactor is functional in both enzyme forms (Nelson et al., 1986).

Immunological studies (Robin et al., 1985) have recently indicated that the constitutive- and inducible-NR forms were equivalent, respectively, to the NADPH and NADH NR forms previously obtained by affinity chromatography isolation from NO_3^--grown soybean plants (Campbell, 1976). Subsequently, the differences in Michaelis constants for NO_3^- have also been resolved with the finding that soybean expresses two forms of constitutive NR activity (Streit et al., 1985); one with a high (5.0 m*M*) K_m (NO_3^-) and one with a low (0.19 m*M*) K_m (NO_3^-). Isolation of two additional NR mutants which lack one of the two constitutive NR forms further verifies that soybean contains three NR isozymes, namely, constitutive NADP-NR (c_1NR), constitutive NADH-NR (c_2NR), and NO_3^--inducible NR (iNR) (Streit and Harper, 1985).

Soybean cotyledons also exhibit inducible and constitutive forms of NR (Carelli and Magalhaes, 1981; Kakefuda et al., 1983) while soybean roots express only the inducible-NR form (Kakefuda et al., 1983; Nelson et al., 1983). The appearance of a NO_3^--inducible NR in soybean (leaves, cotyledons, and roots) which is similar in molecular weight, kinetic parameters, and pH optima to that reported for other higher plants (Beevers and Hageman, 1983), indicates that this enzyme form has been conserved among higher plant species. Smarrelli and Campbell (1981) have concluded, based on immunological studies, that assimilatory NR enzymes from a diverse range of eukaryotic species, including higher plants, algae, and fungi, have diverged from a common ancestor. The presence of a constitutive-NR enzyme in soybean is quite unique among higher plants and additional studies are needed to define the role of this enzyme in overall N metabolism of soybean.

In addition to soybean having inducible- and constitutive-NR enzymes, both of which are partially active with NADH and NADPH (Nelson et al., 1982), evidence has been presented that NR activity is further regulated by a specific inhibitor as well as a possible stimulator (Jolly and Tolbert, 1978). The inhibitor appeared to be a small-molecular-weight (31 000) heat-labile protein which existed in either an activated or inactivated form if plants were grown in dark or light conditions, respectively. Thus, the evaluation of NR activity in soybean is made difficult due to the existence of isoforms of the enzymes, capability of accepting electrons from either NADH or NADPH, and by possible presence of an inhibitor and a stimulator of activity.

The subsequent reduction of NO_2^- to ammonium (NH_4^+) is mediated by nitrite reductase. This enzyme is present in soybean when grown on NO_3^- but not when grown on zero N (Aslam, 1982), unlike the expression of NR activity in the presence and absence of NO_3^-. Measurement of low levels of nitrite reductase activity in leaves of soybean grown on urea (Nelson et al., 1984) may be due to slow interconversion of urea to NO_3^- with subsequent NO_3^- induction of nitrite reductase. The level of nitrite reductase activity in soybean was shown to be approximately five-fold greater than activity of the NR enzyme (Aslam, 1982), indicating that nitrite reductase is not likely a limitation in the metabolism of NO_3^- to NH_4^+; an observation which is consistent among plant species (Beevers, 1976).

13–1.4 Incorporation

The major product of ammonia (NH_3) assimilation is usually considered to be amino N. Early work had suggested that glutamate dehydrogenase was the major route of entry of NH_3 into amino-N compounds. Since 1970, however, many reports indicate that NH_3 assimilation proceeds through the glutamine synthetase/glutamate synthase pathway (see review by Miflin and Lea, 1980). It now appears that glutamate dehydrogenase has little role in assimilation of NH_3 in eukaryotic green plants. The localization of one of the forms of glutamine synthetase enzymes, as well as the glutamate synthase enzyme, in the chloroplasts further implicates involvement of the glutamine synthetase/glutamate synthase pathway. Furthermore, nitrite reduction to NH_3 has also been associated with the chloroplast (Beevers and Hageman, 1983), while glutamate dehydrogenase appears to be primarily a mitochondrial enzyme (Miflin and Lea, 1980).

Asparagine plays a key role in the metabolic pathway of amino N and is the major amino acid involved in xylem transport in soybean (McClure and Israel, 1979; Streeter, 1972a). The greatest amount of ^{15}N label from ^{15}N-NO_3^- was also shown to be incorporated into asparagine in soybean roots (Ohyama and Kumazawa, 1979). The reincorporation of NH_3 formed from asparagine breakdown also appeared to occur via glutamine synthetase/glutamate synthase reactions (Lea and Miflin, 1980). These authors made a strong point that initial NH_3 assimilation was followed by several release/reassimilation reactions before NH_3 is ultimately incorporated into seed protein. Ohyama (1983) concluded that the pathway of NO_3^- assimilation into seed protein involved intitial incorporation into leaf and root protein followed by breakdown and redistribution to reproductive plant parts. Ureides do not appear to play a major role in N transport and metabolism in NO_3^--fed plants. A discussion of the role of ureides in transport and assimilation of N derived from N_2 fixation is presented in section 13–2.4.5 of this chapter.

13–1.5 Foliar Nitrogen Loss

Recent reports (Stutte and Weiland, 1978; Stutte et al., 1979) have indicated that soybean plants lose appreciable quantities of nitrogenous compounds directly to the atmosphere. The form of N which is lost through foliar evolution remains unclear, although Weiland and Stutte (1979) reported that the majority was in a reduced-N form. An extrapolated estimate of 45 kg of N ha^{-1} loss in one crop season has been made (Stutte and Weiland, 1978). Additional verification of the magnitude of foliar N loss under other field environments is needed. Klepper (1979) observed that soybean leaves treated with herbicides accumulated NO_2^- and subsequently evolved gaseous N compounds which were considered to be a mixture of NO and NO_2, primarily NO. Subsequent work (Harper, 1981) indicated that soybean leaves had the capability of evolving gaseous N compounds during the in vivo NR assay if the system was purged with a gas phase; greater losses were observed with anaerobic (N_2) than with aerobic gas phases. With young leaf tissue of soybean, the amount of gaseous N compounds evolved and subsequently trapped as NO_2^- (following an oxidation step of the gaseous stream) was in some cases nearly equal to the amount of NO_2^- accumulating in the aqueous in vivo NR assay medium. Boiled controls indicated that this reaction was enzymatic. The nitrogenous gas compounds evolved from the in vivo NR assay system were initially thought to be NO and NO_2. Acetaldehyde oxime was proposed by Mulvaney and Hageman (1984) to be the compound evolved during the in vivo NR assay. Based on the solubility of this compound in water, and on GC/MS data, it is currently concluded that NO is the initial compound being evolved (Klepper, 1984, personal communication; Dean and Harper, 1986). Studies have linked the evolution of nitrogenous gasses during the in vivo NR assay with the presence of the constitutive NR enzyme; a soybean mutant which lacked constitutive NR also lacked the ability to evolve gaseous N compounds, even if NO_2^- was supplied (Nelson et al., 1983). The close linkage between these two activities has been suggested to be controlled by a single gene or closely linked genes (Ryan et al., 1983b).

The relationship, if any, between nitrogenous gas evolution from intact plants (Stutte and Weiland, 1978), from in vivo NR assay conditions (Harper, 1981), and from leaves treated with herbicides (Klepper, 1979) has not been established. Mulvaney and Hageman (1982) have confirmed evolution of nitrogenous N compounds from shoot portions of intact soybean seedlings, with greater evolution occurring during light than dark periods. The evidence does indicate that loss of nitrogenous compounds through gaseous evolution from soybean leaves can occur and this area needs further research.

13–1.6 Environmental Effects

The effect of environment on NO_3^- uptake, movement, and metabolism has not been extensively studied with soybean. Short-term (50 min)

NO_3^--uptake studies indicated that similiar rates of $^{15}N\text{-}NO_3^-$ uptake occurred in light and dark conditions (Rufty et al., 1982b). This was confirmed using longer-term uptake studies (Rufty et al., 1984). These workers also concluded that > 20% of the $^{15}N\text{-}NO_3^-$ taken up was reduced in the roots in both light and dark before transport through the xylem stream. Studies have shown near linear uptake of $^{15}N\text{-}NO_3^-$ during a 12-h dark period and continued uptake through a 48-h dark period (Nicholas and Harper, 1985). The results with intact plants (Nicholas and Harper, 1985; Rufty et al., 1984) and with leaf sections (Reed et al., 1983) showed that the $^{15}N\text{-}NO_3^-$ taken up was subsequently reduced to soluble compounds in the dark. Darkness does, however, result in decreased rates of NR activity which has been attributed to both a decline in reductant levels to drive the reaction and to a possible net loss of enzyme due to continued turnover (Nicholas et al., 1976b). Alternatively, Jolly and Tolbert (1978) reported that a dark-activated protein existed in soybean leaves which was inhibitory to the NADH-NR enzyme. Little doubt remains, however, that mechanisms do exist in soybean for nitrate reduction and subsequent N incorporation into reduced-N compounds during darkness. The question of what reductant is involved in nitrite reduction in the dark remains unresolved.

Light is known to stimulate NR induction in soybean roots and cotyledons (Duke et al., 1982; Kakefuda et al., 1983) and light appears to be an absolute requirement, a requirement which appears to be linked to red/far red reactions rather than being related to photosynthetic activity (Duke and Duke, 1984). Light also has been shown to increase NR activity in leaves following a dark period (Nicholas et al., 1976a), likely through increased availability of reductant energy as well as net enzyme synthesis. The recovery of NR activity during a light period, following a loss of NR activity during dark, was dependent on light flux. The increase in NR activity under high light could be reversed if plants were transferred to low light or to dark. Although the mechanisms of light stimulation of NR activity are not understood, it does seem apparent that light influences protein synthesis, the availability of electron donors, and possibly the movement of NO_3^- to the site of reduction.

High temperature (30 vs. 20 °C) has been shown to accelerate loss of NR activity during darkness but the subsequent recovery of NR activity in light appears independent of temperature (Nicholas et al., 1976a). The NR enzyme is known to be thermally labile, with a sharp increase in inactivation at about 35 °C. In addition to direct effects of temperature on the protein per se, NR activity is likely affected indirectly through temperature effects on reductant energy sources to drive the reaction. The optimum temperature during a light period for maximum NO_3^- uptake and metabolism appears to be about 30 °C (Magalhaes, 1977).

Water availability is an important environmental consideration in evaluating factors affecting NO_3^- metabolism. The movement of NO_3^- to the root system is largely a function of convective flow of soil water to plant roots in response to transpiration in the aboveground portion of the crops (Olson and Kurtz, 1982). The decrease in amount of tissue NO_3^- during later growth stages of soybean in field environments (Harper, 1974) is likely due to declining concentration of NO_3^- in the soil solution and concomitant declines in available soil water at that time of the growing season. That NO_3^- uptake is dependent on soil moisture has been repeatedly observed through a stimulation in NR activity of soybean on days following rainfall (J.E. Harper, 1972, 1974, 1981, 1986, unpublished data).

13–2 NODULATION AND DINITROGEN FIXATION

This section will be primarily confined to a review of literature since 1973. The focus will be on infection by the bacteria, on development of the nodules, and on biochemical and physiological aspects of nodulation and N_2 fixation. The more ecological aspects of soybean nodulation were covered previously (Vest et al., 1973). Several excellent reviews have recently expanded various aspects of legume nodulation in general (Bauer, 1981; Dart, 1977; Dazzo and Hubbell, 1982; Ljunggren, 1969; Meijer, 1982), and more specifically with soybean nodulation (Bergersen, 1982). Attempts have been made in this section to discuss nodulation as a symbiotic process, and to emphasize the requirement that the host plant must supply energy to the nodule and bacteroids in return for the reduced N supplied to the plant by the bacteroids.

13–2.1 Recognition and Infection

Recent observation that fast-growing strains of *Rhizobium japonicum* are capable of nodulating an unimproved line (cv. Peking) from China (Keyser et al., 1982) and two South African cultivars (Geduld and Usutu) (Van Rensburg et al., 1983) provides an opportunity for new approaches to soybean nodulation. The slow-growing bacterial strains which nodulate most cultivars of soybean has recently been proposed to be transferred to a newly established genus of slow-growing root nodule bacteria (*Bradyrhizobium* gen. nov.), with *Bradyrhizobium japonicum* as the newly designated species (Jordon, 1982).

Recognition is the initial step in a multistep sequence leading to formation of root nodules on legumes. The recognition process involves a specific or selective response between bacterial cells and the legume host root-hair cell. The involvement of plant lectins (carbohydrate binding proteins or glycoproteins) in this recognition process has been proposed by Bohlool and Schmidt (1974), based on binding of fluorescent-labeled soybean seed lectin (SBL) to several, but not all, *B. japonicum* strains. These observations were confirmed independently by Bhuvaneswari et al. (1977), who also used lectin derived from soybean seed but which was more highly purified. Similarities between seed and root lectins of soybean have been reported (Gade et al., 1981) further indicating that lectins may be involved in recognition. However, several other investigators (see review by Bauer, 1981) have failed to establish a close relationship betwen specific legume lectins and respective bacterial strains which nodulate that host plant. Brethauer and Paxton (1977) showed that SBL which had been partially purified by affinity chromatography and then labeled with ^{125}I did not consistently bind to all strains which nodulated soybean. In addition to using purified SBL, Brethauer and Paxton (1977) used a sugar hapten control to eliminate nonspecific binding of the SBL to the strains. They concluded that no strong correlation existed between binding ability of SBL and nodulation ability among the strains. The observation that several soybean lines lack the 120 kD seed lectin (Pull et al., 1978) and yet were similar to lines containing seed lectin with respect to recognition and infection by *Bradyrhizobium* (Pueppke, 1983) further questioned the validity of the lectin recognition hypothesis. The "lectinless lines", however, do have small quantities of a different type of lectin (Dombrink-Kurtzman et al., 1983), which may explain the capability of these lines to recognize *B. japonicum*. Thus, lectins remain a probable, yet unproven, factor in the recognition process between specific bacterial/host associations. The site of lectin accumulation has been localized in the root hair tip of *Glycine soja* (Stacey et al., 1980), and in other legume species (Dazzo and Hubbell, 1982). The bacterial receptor sites have been reported to occur in the capsule as well as in the cell wall polysaccharide in *B. japonicum* cells (Bal and Shantharam, 1981).

Infection in soybean generally occurs in portions of the primary root between the root tip and the smallest emergent root hair (Halverson and Stacey, 1984). These workers also observed that a mutant strain, which normally nodulated only root sections which developed after infection, could be induced to nodulate earlier if preincubated with root exudate. A protein factor in the exudate was implicated. These results indicated that in addition to the root hair being susceptible to infection only during specific growth stages, the *Bradyrhizobium* strain played a finite role in determining infection timing. A transient appearance and disappearance of the receptor on *B. japonicum* which specifically binds to soybean lectin has been observed (Bhuvaneswari et al., 1977). Thus, culture age of bacteria as well as stage of root hair development may be key factors in the recognition process leading ultimately to a successful infection. It does remain to be proven, however, that the lectin-receptor interaction occurs in situ at the sites of infection and is in fact responsible for the specific cell to cell recognition prerequisite for host specificity.

13–2.2 Nodule Initiation

The initial step in nodule initiation is successful penetration of the bacteria into the root hair and the formation of an infection thread. It is not clear whether the cell wall is actually penetrated following successful infection or whether an invagination process occurs where cell wall growth is inverted (see review of Dazzo and Hubbell, 1982). Nutman (1956) proposed the invagination process where no wall penetration is involved but rather the infection thread is an interior tubular structure with an open pore at the initiation site. The bacteria within the infection thread are still exterior to the root hair cytoplasm. Other work (Callaham, 1979) indicated that the root hair cell wall was hydrolyzed at the site of bacterial adhesion and was followed by synthesis of a new cell wall which precedes the advance of the infection thread. Although the mechanism is not as yet clear, an infection thread does form and ultimately grows to the base of the root hair. The nucleus of the root hair appears to direct the growth of the infection thread (Meijer, 1982).

13–2.3 Nodule Development

Following penetration of the cortical cell area of the root by the infection thread, extensive branching of the infection thread occurs, ultimately resulting in infection of many cells by the same infection thread (Dart, 1977). That nodules may result from multiple infection threads, or double infections from a single thread, is indicated by results showing nodules containing more than one strain of *B. japonicum* (Lindemann et al., 1974). The frequency of multistrain nodules can be as high as 20 to 30% (Kuykendall and Weber, 1978). Prior to bacterial release into the cytoplasm of a host cell, a swelling develops near the tip of the infection thread. The bacteria adhere to the inner membrane surface and then pass into the cytoplasm of the host cell by budding off, similar to the process of endocytosis (Goodchild and Bergersen, 1966). Each bacterium is enclosed within a membrane, termed *peribacteroid membrane*, to form a vesicle (Bergersen, 1982). The vesicle is important in the development of effective soybean-*Bradyrhizobium* N_2-fixing symbiosis. If bacteria are released into the cytoplasm of the host without being enclosed by peribacteroid membranes, then an ineffective nodule results (Newcomb et al., 1977). The bacteria then undergo rapid multiplication, accompanied by matched membrane synthesis, until the host cell becomes densely packed with bacteria-containing vesicles. The bacteria within a vesicle undergo multiplication until several bacteria, or bacteroids as they are usually termed at this stage, are present. The nature of differentiation into bacteroids is unknown but appears to result from interactions with the host cell (Meijer, 1982). Early work concluded that bacteroids were incapable of reproduction (Almno, 1933) and this was supported by experiments with soybean bacteroids (Bergersen, 1968). However, more recently it has been shown that bacteroids from several legume species, including soybean, can be induced to grow on culture media (Gresshoff et al., 1977; Sutton et al., 1977; Tsien et al., 1977). In soybean, the bacteroid may be only slightly larger than culture-grown rhizobia (Bergersen, 1982) and typically contains large amounts of poly-B-hydroxybutyric acid (Klucas and Evans, 1968). It is the bacteroid form that expresses nitrogenase activity in the nodule (Bergersen and Turner, 1967; Koch et al., 1967).

The soybean nodule is determinate in growth habit, oblate in shape, with a spherical meristem and a narrow nongirdling attachment to the root. The central volume of nodule tissue is comprised of large parenchyma cells densely packed with bacteroid-containing vesicles with 1 to 10 bacteroids per vesicle (Bergersen, 1982). The bacteroid-containing vesicles occupy 80% of the cell volume, with the remaining 20% being devoted to cytoplasm, nucleus, amyloplasts, and mitochondria, primarily oriented around the cell periphery (Bergersen and Goodchild, 1973). Within the central bacteroid zone, there remain uninfected (interstitial) cells and mitotically active cells which develop into vascular strands (Meijer, 1982). The interstitial cells provide the interface between the bacteroid-containing cells and the host cells. These interfaces appear to be the active sites of metabolic exchange between the host and the N_2-fixing bacteroids (Bergersen, 1982). Intercellular spaces between these cell types appear to be the primary route for gas exchange (Bergersen and Goodchild, 1973). A vascular system provides interchange between the nodule and the root. One or two main vascular traces originate from the root stele and then branch and radiate through the nodule cortex, terminating near the nodule apex (Newcomb et al., 1979; Sprent, 1980).

The functional life of a soybean nodule appears limited (Bergersen, 1982) although no reports were found which provide a definitive time span of an individual nodule. Bergersen (1982) does indicate that a soybean nodule can still increase in size up through 60 days after initiation and Klucas (1974) has shown loss of acetylene (C_2H_2) reduction activity of tap root nodules from 58 to 75 DAP, depending on cultivar. Nodule senescence begins at the center and progresses outward with loss of ultrastructural integrity followed by changes in the host nucleus and bacteroid decay.

13–2.4 Dinitrogen Fixation

13–2.4.1 Cellular Localization

The bacteroids of the legume root nodule have been established as the site of N_2 fixation (Bergersen and Turner, 1967; Koch et al., 1967). The central location of the densely packed bacteroid-containing cells requires a means of moving carbon substrates from the host cells to the bacteroids and subsequent return of reduced-N compounds out of the nodule to the host plant. This is accomplished by the network of vascular traces which pass through the central tissue and join at the base of the nodule with the root vascular system.

13–2.4.2 Nitrogenase Structure and Function

The structure of the nitrogenase enzyme is remarkably consistent across the many N_2-fixing systems which have been studied. Nitrogenase is localized in the bacteroids and in soybean it comprises 2 to 5% of the total nodular protein (Burns and Hardy, 1975). The nitrogenase enzyme is comprised of two components; a Mo-Fe protein (referred to as component 1 or dinitrogenase) and an Fe-protein (referred to as component 2 or dinitrogenase reductase) (Burris et al., 1980). The molecular weight of the Mo-Fe- and Fe-proteins are about 200 to 500 kD, respectively. Electrons pass from the Fe-protein to the Mo-Fe protein and then to N_2 for reduction to NH_3. The source of electrons to drive nitrogenase has not been definitely established, although Carter et al. (1980) have recently characterized an Fe-S-containing ferridoxin from soybean bacteroids, and have shown that this protein can donate electrons to bacteroid nitrogenase. The steps which may precede the reduction of ferridoxin have not been elucidated and may involve membrane-bound functions within the bacteroids.

With the observation that nitrogenase could be induced in *B. japonicum* cells which were separated from suspensions of cultured soybean cells by a 0.2 to 0.4 micron filter (Reporter and Hermina, 1975), it was soon shown that nitrogenase could be induced in rhizobia when cultured with cells of nonlegumes (Child, 1975; Scowcroft and Gibson, 1975). From this it was independently established in several laboratories that free-living cultures of rhizobia, including *B. japonicum*, could be induced to fix N_2 (see review by Gibson et al., 1977). This then confirmed that the genes necessary for nitrogenase expression reside in the bacteria rather than in the host.

13-2.4.3 Hydrogen Evolution and Uptake

It has been firmly established that nitrogenases from all known sources, including soybean, catalyze H_2 evolution during N_2 fixation (see review by Evans et al., 1981). Hydrogen evolution from soybean nodules was initially reported by Hoch et al. (1957) and has been the subject of intensive study since Schubert and Evans (1976) showed that ATP-dependent H_2 evolution was accompanied by significant energy loss during N_2 fixation. It had been shown earlier that evolution of H_2 from soybean nodules was stimulated by increased O_2 concentration and that N_2 was a competitive inhibitor of H_2 evolution (Bergersen, 1963). The effect of O_2 is likely associated with production of ATP via electron transport (Bergersen and Turner, 1967).

The extent of H_2 evolution from nodules is determined by the effectiveness of the H_2 recycling system (uptake hydrogenase) in the nodule. Evans et al. (1981) summarized results from several studies which reported losses of approximately 1 to 30% of the electron flux through nitrogenase to H_2 evolution, depending on whether the plants were inoculated with hydrogenase positive (Hup^+) or hydrogenase negative (Hup^-) *Bradyrhizobium* strains, respectively. Merberg and Maier (1983) recently reported selection of mutant strains of *B. japonicum* which express greater uptake hydrogenase activity than the wild-type parent strains. The bacterial strain was initially thought to carry the genetic information regulating H_2 uptake (Carter et al., 1978), however, more recent information has suggested that the host plant may influence the expression of hydrogenase activity (Keyser et al., 1982; Lopez et al., 1983). A survey of *B. japonicum* strains indicated that 17% of the 61 tested produced an effective uptake hydrogenase system with soybean (see review by Evans et al., 1981). Lim et al. (1980) and Uratsu et al. (1982) surveyed indigenous *B. japonicum* strains from soybean production areas in the USA and found $< 25\%$ of the strains isolated were Hup^+. The serogroups with relatively high frequencies of Hup^+ phenotypes (serogroups 122 and 110) comprised $< 10\%$ of the total *B. japonicum* field population (Keyser et al., 1984).

It now seems clear that the uptake hydrogenase provides a mechanism for energy conservation by transferring electrons from H_2 into the respiratory electron transport chain. Although it has been concluded that H_2 utilization by hydrogenase in heterocysts exposed to light may provide electrons to support the nitrogenase reaction (Eisbrenner and Bothe, 1979), evidence of this mechanism existing in non-photosynthetic N_2-fixing organisms, such as soybean nodules, was not found. The economic benefit of inoculating soybean with Hup^+ *B. japonicum* strains, as opposed to Hup^- strains, has been extensively studied by Evans and co-workers (see review by Evans et al., 1981). Both dry matter and total N input increases were reported for plants grown in controlled environments, while under field conditions significant increases were noted for only total N input when soybean was inoculated with Hup^+ vs. Hup^- strains. In view of the theoretical recovery by Hup^+ strains of approximately 9% of the energy involved in reduction of N_2 and protons (Evans et al., 1981), a beneficial effect on grain yield under practical field conditions will be difficult to demonstrate given the magnitude of background biological variability.

13-2.4.4 Role of Leghaemoglobin

There appears to be little doubt that the major function of leghaemoglobin is to serve as a carrier of O_2 in the nodule tissues (Bergersen, 1982). Leghaemoglobin has long been established as being confined to the bacteroid-containing cells (Smith, 1949). Bergersen and Appleby (1981) recently established that leghaemoglobin was localized in both the cytoplasm surrounding the bacteroids and in the bacteroid envelope, the concentrations being on average 3.0 and 0.34 m*M*, respectively. Leghaemoglobin belongs to a group of nodule-specific host polypeptides which have been termed *nodulins* (Legocki and Verma, 1980). For additional details on the in vivo role of leghaemoglobin in the maintenance of the critical O_2 concentrations and fluxes within the bacteroid, the reader is referred to a recent chapter on leghaemoglobin by Bergersen (1980).

13-2.4.5 Products of Dinitrogen Fixation

The primary product of N_2 fixation in nodules is NH_3 (Bergersen, 1965; Kennedy, 1966). Ammonia does not appear to be assimilated in bacteroids of root nodules including soybean (Brown and Dilworth, 1975). Free-living N_2-fixing *B. japonicum* excrete most (94%) of the N_2 fixed as NH_3 (O'Gara and Shanmugam, 1976). Fixed N as NH_3 is released from the bacteroid (Bergersen and Turner, 1967) where it is incorporated into glutamine and glutamate in the host cytosol (McParland et al., 1976; Ohyama and Kumazawa, 1980a). It has also been suggested that NH_3 may undergo oxidation to NO_3^- in the bacteroid (Ohyama and Kumazawa, 1978, 1980a). This was based on appearance of ^{15}N atom percent excess in NO_3^- when nodulated soybean roots were exposed to ^{15}N-N_2. Substantiation of this observation by other workers was not found in the literature, but these results may relate to observations of appearance of cytosolic NR activity in soybean nodules in the absence of NO_3^- in the growth medium (Streeter, 1982). At the present time, it cannot be ruled out that the nodular cytosolic NR enzyme is not similar to the constitutive NR enzyme present in young soybean leaves (Nelson et al., 1983) and cotyledons (Kakefuda et al., 1983). However, this seems unlikely since it is known that soybean roots do not contain the constitutive NR enzyme (Kakefuda et al., 1983; Nelson et al., 1983). It does not appear that oxidation of NH_3 to NO_3^- is of major importance as a N metabolic pathway, since ureides play a major role in movement of symbiotically fixed N_2 from soybean nodules.

Soybean, along with a number of tropical legumes, falls into that group of plants which are classed as ureide transporters when dependent on symbiotic N_2 fixation as a N source. Allantoin and allantoic acid are the principal form of N transported from the nodules to the shoots, although the xylem sap usually contains appreciable asparagine and other amino acids (Herridge, 1982a, 1982b; Matsumoto et al., 1977a; McClure and Israel, 1979; Minamisawa et al., 1983; Schubert, 1981; Streeter, 1979). It is noteworthy that Minamisawa et al. (1983) have presented data which indicates that asparagine accounted for 3 to 5% more of the total N transported in the xylem stream of nodulated plants infected with Hup^+ than with Hup^- *Bradyrhizobium* strains. These plants were totally dependent on N_2 fixation since there was no N added to the solution. The results indicate that an alteration in asparagine metabolism occurs with presence or absence of the uptake hydrogenase.

An intriguing proposal has been made by Shearer et al. (1982) that ureide transporters, such as soybean, may undergo reactions during the process of N_2 fixation which result in isotope discrimination and ^{15}N enrichment of the soybean nodule, relative to other plant parts. Other legume species, however, which are known to be amide transporters were recently shown to have ^{15}N enrichment in their nodules (Steele et al., 1983), which argues against the proposal that ureide synthesis is responsible for ^{15}N enrichment in nodules.

The pathway of ureide synthesis involves purine synthesis and subsequent oxidation (see review by Reynolds et al., 1982a, 1982b). The studies of Fujihara and Yamaguchi (1978) revealing that allopurinol, a specific inhibitor of xanthine-oxidizing enzymes, resulted in a decrease in amount of ureides in stems and nodules and an increase in nodule xanthine, provided the first evidence that purine oxidation was involved. Current models of ureide synthesis in soybean nodules (Boland et al., 1982) show purine synthesis as occurring in proplastids of the host cell. Subsequent oxidations of xanthine to allantoin and allantoic acid are proposed to occur in the peroxisome and endoplastic reticulum, respectively. These ureides (allantoin and allantoic acid) are then released from the nodule for translocation through the xylem to other plant parts. A portion of the ureides move directly to the pods for incorporation, rather than being metabolized in vegetative plant parts for subsequent retranslocation to the pods and seeds. The proportion of N as ureides in the xylem sap of soybean varied from a seasonal average of 78% in plants totally dependent on N_2 fixation to as little as 6% in noninoculated plants fed 20 m*M* nitrate (McClure and Israel, 1979). The amount of ureide N in the xylem sap among different soybean cultivars has been shown to be strongly correlated with amount of N_2 fixation (Herridge, 1982a, 1982b; McClure et al., 1980), and has been proposed as a method of assessing N_2-fixation inputs by soybean plants. It does appear that concentration of ureides in xylem sap is more indicative of short-term changes in N_2-fixation rates than is ureide concentration or content of various stem segments (Patterson and LaRue, 1983a). A general positive relationship between ureide concentration of a young stem segment (segment of stem above uppermost fully-expanded trifoliolate, including petioles) and the integrated seasonal C_2H_2 reduction activity has also been shown (Patterson and LaRue, 1983b). However, ureide concentration of the xylem sap (or tracheal sap as measured by Herridge, 1984) appears to be a better indicator of N_2 fixation in soybean than does ureide concentration or content of plant parts.

It is not as yet clear why soybean nodules, and nodules from several other legume species, undergo the rather complex sequence of reactions involving purine synthesis and oxidation to obtain N in the ureide form. The fact that another group of legumes (e.g., red clover, *Trifolium pratense* L.; alfalfa, *Medicago sativa* L.; fava bean, *Vicia faba* L.; peanut, *Arachis hypogaea* L.; and pea, *Pisum sativum* L.) do not synthesize large quantities of ureides indicates that differing selection pressures have been exerted among various legume species.

Metabolism of the ureides, allantoin, and allantoic acid, has recently been reviewed by Thomas and Schrader (1981a). The pathway of ureide metabolism was proposed to involve hydrolysis to urea and glyoxylate by allantoinase and allantoicase (Herridge et al., 1978; Thomas and Schrader, 1981a). Allantoinase has been shown to occur in stems and leaves of soybean (Zengbé and Salsac, 1983), while allantoicase occurs in roots, stems, petioles, leaves, pods, and seeds (Thomas and Schrader, 1981b, Zengbé and Salsac, 1983). A recent report (Shelp and Ireland, 1985) strongly indicates that the pathway of allantoate metabolism is via allantoicase. The ureidoglycolate formed is subsequently converted to glyoxylate and two molecules of urea by ureidoglycolase. Prior to pod initiation, leaves contain appreciable levels of allantoinase, but the level declines with pod formation (Serres, 1982). As pods form, ureides initially accumulate in the pod wall and then accumulate in the seed with further development (Matsumoto et al., 1977b). Urea, which appears to be formed from ureide metabolism is further metabolized by ureases (Kerr et al., 1983) for subsequent incorporation into amino acids. Ureases are likely present in all plant parts of soybean although the urease in leaf tissue or tissue culture appears to be an isoform of that in the seed (Kerr et al., 1983; Polacco and Havir, 1979). Soybean has been included with a group of plants with high urease activity in both nodulated and nonnodulated lines (Hogan et al., 1983). Urea toxicity symptoms can develop in soybean if urease is inhibited due to N deficiencies (Eskew et al., 1983) and Ni has been shown to stimulate urease activity (Klucas et al., 1983). Urea toxicity symptoms developed in the absence of Ni regardless of whether soybean plants were grown on N_2, NO_3^-, or NH_4^+ (Eskew et al., 1983), suggesting that urea is a N metabolic product under N_2 fixing and nonfixing growth conditions. Additional work in this area is needed to fully elucidate the metabolic pathways of ureide and urea synthesis and metabolism.

13–2.5 Limitations and Environmental Effects

The symbiotic relationship between the host plant and the bacterial strain in the root nodule predisposes the system to limitations which affect either partner. Any factor which limits plant vigor, such as mineral deficiency, disease, or adverse environmental conditions, can result in depressed N_2 fixation.

Dinitrogen fixation in soybean undergoes diurnal variation under field (Hardy et al., 1968; Sloger et al., 1975) and greenhouse (Bergersen, 1970) growth conditions. From this it was interpreted that current photosynthate was necessary to maintain active N_2 fixation (Vest et al., 1973). However, more recent data (Schweitzer and Harper, 1980) indicated that soybean could withstand extended dark periods without any change in C_2H_2 reduction activity of nodulated roots in situ, provided that temperature was maintained constant. Lower temperatures during the dark period, as frequently encountered in field conditions, did, however, result in decreased C_2H_2 reduction. This indicated that temperature plays a stronger role in diurnal variation in N_2-fixation capability of soybean than does light. Sloger et al. (1975) reported that the average specific activity of nodules from field-grown soybean sampled 10 times during the season was significantly correlated with average air temperature but not significantly correlated with average soil temperature. Continued exposure of soybean plants to low temperatures (13 °C) has been shown to totally inhibit nodule development (Duke et al., 1979). Likewise, transfer of nodulated plants from 20 to 13 °C resulted in an initial decline in C_2H_2 reduction, but activity recovered to initial levels following 2 days exposure to the lower temperature. The Arrhenius plot of loss of C_2H_2

reduction with declining temperature showed a sharp break at 15 °C. Symbiotic N_2 fixation by soybean appeared more sensitive to low (10 °C) (Matthews and Hayes, 1982) and high (40 °C) (Munevar and Wollum, 1981) temperatures than did comparably treated NO_3^--fed plants. This led Matthews and Hayes (1982) to the conclusion that soybean sown under cool environmental conditions would benefit from applied N at sowing.

Enhancement of net photosynthesis by CO_2 enrichment (Hardy and Havelka, 1975a) and by increased light (Lawn and Brun, 1974) has been shown to stimulate symbiotic N_2 fixation. These responses occurred over longer time intervals (weeks) and are considered to reflect overall growth responses of the plant and the nodules, rather than short-term enhancement of photosynthate movement to the nodules.

Molecular O_2 is required for ATP production in leguminous nodules, but can also result in inactivation of the nitrogenase enzyme (Bergersen, 1971). Limited O_2 availability to the nodulated root has been shown to decrease N_2-fixing activity of intact nodulated soybean roots (Pankhurst and Sprent, 1975). Decreasing O_2 concentrations, however, appear to have less affect on the initial step of N_2 fixation to NH_3 than on subsequent assimilation of NH_3 into other nitrogenous compounds (Ohyama and Kumazawa, 1980b). It has been shown that although nitrogenase activity is sensitive to short-term deviations in external O_2 concentrations, nodules of intact soybean plants can partially adapt to adverse pO_2 concentrations and resume near normal nitrogenase activities with time (Criswell et al., 1976).

Obaton et al. (1982) concluded that symbiotic N_2 fixation was more profoundly affected by soil water deficit than was NR activity. The stress period, however, imposed in that study was at a point in time when NR activity had reached a peak and prior to any appreciable increase in symbiotic N_2 fixation. Sprent (1972) concluded that water supply is probably the main environmental factor affecting N_2 fixation; not considering high plant-available soil N as an environmental stress. Loss of 20% of the water present in turgid nodules was sufficient to result in complete and irreversible inhibition of N_2 fixation (Sprent, 1971). It is known, however, that nodules can derive water from the root and therefore N_2 fixation can continue when nodules are in dry soil as long as water is available to the root system (Hume et al., 1976). Excess water can also have detrimental effects on nitrogenase activity (Mague and Burris, 1972), likely due to limitations in gas diffusion. Soybean does, however, have the ability to adapt to excess water by forming nodules on lateral roots at the soil surface, or actually growing in water with only a portion of the nodule being submerged (1982, personal observation).

Water stress may also inhibit N_2 fixation indirectly through decreasing photosynthetic rates (Huang et al., 1975a, 1975b). These workers demonstrated that carbon dioxide (CO_2) deprivation of the shoots or imposition of a water stress resulted in similar inhibitions of C_2H_2 reduction activity. A subsequent study (Finn and Brun, 1980), however, showed a severe inhibition of CO_2 exchange rate within 1 h following initiation of a water-stress treatment, while nodule activity (C_2H_2 reduction) was unaffected for a period of 6 h. The different responses noted in these two studies may involve the type of water stress imposed: water withheld from soil-grown plants (Huang et al., 1975a, 1975b) or changing nutrient solution medium to one containing polyethylene glycol (Finn and Brun, 1980). Thus, although it is clear that water deficit and excess are detrimental to N_2 fixation, the mechanism of inhibition remains unclear.

13-3 INTERACTIONS OF NITRATE METABOLISM AND SYMBIOTIC DINITROGEN FIXATION

13-3.1 Inhibitory Effects of Nitrate

The detrimental effect of nitrate on nodulation of soybean specifically, and legumes in general, has been the subject of numerous studies, and yet the mechanism(s) of NO_3^- inhibition to the symbiosis remains elusive. Nitrate has been shown to separately inhibit at least four aspects of legume nodulation including root hair infection, nodule development, nitrogenase activity, and bacteroid integrity (see reviews by Munns, 1977; Gibson and Jordan, 1983). The following section will emphasize data relating to NO_3^- effects on soybean nodulation, but will also draw on literature from other legume species where appropriate.

13-3.1.1 Root Hair Infection

Nitrate retards or inhibits root hair curling and infection by rhizobia in several legume species (Thornton, 1936; Munns, 1968), presumably including soybean although no direct evidence of this was found in the literature. Tanner and Anderson (1964) presented data indicating that NO_2^-, produced by bacterial reduction of NO_3^-, catalytically destroys indole-3-acetic acid (IAA). Libbenga and Torrey, (1973) have suggested that IAA is involved in the infection process. From these results, it has been proposed that NO_3^- affected the infection process via NO_2^- control over IAA levels. Munns (1968) demonstrated that addition of IAA could partially alleviate the effect of NO_3^- inhibition of infection in alfalfa. Observations, however, that NO_3^- inhibition of nodulation was not overcome by using NR-deficient mutants of rhizobia (Gibson and Pagan, 1977; Manhart and Wong, 1980; Streeter, 1982) questioned the involvement of bacterially produced NO_2^- in initial nodulation. This did not, however, eliminate plant produced NO_2^- as a possible inhibitor. A second mechanism of NO_3^- inhibition of infection was proposed by Dazzo and Brill (1978). They suggested that NO_3^- decreased the production of lectin or lectin-binding sites by the host plant, thus preventing attachment of the rhizobia. These studies need additional verification. Recent studies (Harper and Gibson, 1984) indicated that considerable variability exists among legume species with respect to NO_3^- tolerance of the infection and initial nodule development phases. Thus, a caution is urged in making generalizations based on observations with only one legume species. Soybean was one of the species most adversely affected by NO_3^-, with respect to nodule initiation (Harper and Gibson, 1984).

13-3.1.2 Nodule Development

The delay in nodule development on *Trifolium subterranium*, *Pisum arvense* L., and *P. sativum* L., upon exposure of the nodulated root to NO_3^- has been attributed to photosynthate deprivation of the nodules (Oghoghorie and Pate, 1971; Small and Leonard, 1969). Gibson's (1974) data supported this concept in that a decrease in ^{14}C movement to nodules of soybean occurred within 24 h of NO_3^- feeding. Likewise, supplying inorganic N from 3 to 10 days prior to sampling decreased the amount of ^{14}C movement to nodules of soybean (Brun, 1976; Latimore et al., 1977; Rabie et al., 1980), and inhibited C_2H_2 reduction activity (Streeter, 1981). Ursino et al. (1982) also recently concluded that NO_3^- assimilation in leaves of NO_3^--treated plants restricted nodulation and N_2 fixation by decreasing export of recent photosynthate to the nodules. Although the effect of NO_3^- on nodule development may be mediated through C deprivation to support nodular cell growth and function, the evidence supporting this is not definitive and awaits additional study.

Recently, a soybean mutant was selected in which nodulation development and nitrogenase activity (AR) was more tolerant to NO_3^- than in the parent 'Bragg' (Carroll et al., 1985a). Preliminary studies using mutagenized 'Williams' progeny have also resulted in selection of putative mutants which have increased nodulation tolerance to NO_3^- (Gremaud and Harper, unpublished). These studies provide evidence that nodulation tolerance to NO_3^- in soybean can be manipulated.

13-3.1.3 Nitrogenase Activity

Nitrite is known to be a potent inhibitor of nitrogenase activity from soybean bacteroids (Kennedy et al., 1975; Rigaud et al., 1973) and of purified enzyme preparations (Trinchant and Rigaud, 1980). The mechanism of NO_2^- inhibition of nitrogenase likely involves the binding of NO_2^- to the Mo-Fe component which is reversible (Trinchant and Rigaud, 1980). Nitrite may also oxidize the ferrous leghaemoglobin which would alter O_2 binding and transport functions of the leghaemoglobin (Rigaud and Puppo, 1977). The possibility that nitrogenase may be inhibited by nitric oxide, a reduction product of NO_2^-, has been proposed for *Clostridium pasteurianum* (Meyer, 1981). However, subsequent work with soybean bacteroids (Trinchant and Rigaud, 1982) ruled out involvement of nitric oxide (NO) based on the K_i values for NO_2^- and NO along with the low levels of NO generated by nitrite reduction.

Addition of NO_3^- to soybean with established nodules resulted in a decrease in nitrogenase activity within 24 to 48 h of supplying NO_3^- in the nutrient medium (Gibson, 1976; Neyra and Stephens, 1985). Further nodule development also ceased. Following removal of the NO_3^- and elution of the pots, complete recovery of specific nitrogenase activity was realized in 4 days. Whether this loss of nitrogenase activity upon NO_3^- addition was due to depriviation of photosynthetic energy to the nodule or to a more direct effect of NO_3^- mediated through reduction to NO_2^- is not known. Streeter (1982) has shown that NR activity is expressed by the cytosolic portion of the nodule, regardless of whether the *Bradyrhizobium* strain did or did not express NR activity. In addition, NR activity exists in the cytosol of soybean nodules of plants grown in the absence of NO_3^-. Whether this NR enzyme is induced by NO_3^- resulting from oxidation of NH_3 (Ohyama and Kumazawa, 1978) is not clear. The cytosolic localization of this NR activity would mean that the resulting NO_2^- would have to penetrate the bacteroidal membrane if it were to inhibit the nitrogenase enzyme directly. With intact detached nodules, Stephens and Neyra (1983), showed that nitrogenase activity (C_2H_2 reduction) was inhibited in nodules formed by NR-deficient mutants of *B. japonicum*, but that the isolated bacteroids were not inhibited by NO_3^-. This indicated that the cytosolic NR enzyme was involved in nitrate reduction and inhibition of nitrogenase. When the plant was infected with a NR-containing *Bradyrhizobium* strain, the inhibition of nitrogenase in response to NO_3^- addition was more severe than observed in the presence of the NR deficient strain (Stephens and Neyra, 1983). Thus, both the cytosolic and the bacteroid (when present) NR enzymes appear to contribute to inhibition of nitrogenase activity upon exposure of the nodule to NO_3^-. The lack of difference in nodulation response to NO_3^- between soybean (Stephens and Neyra, 1983; Streeter and DeVine, 1983) and other legume species (Antoun et al., 1980; Gibson and Pagan, 1977; Kiss et al., 1979; Manhart and Wong, 1980) when inoculated with NR-containing and -deficient bacterial strains may be due to expression of cytosolic NR activity, a possibility which remains to be verified. Streeter (1985) compared NR-containing and NR-deficient strains of *Bradyrhizobium* and observed similar inhibition of nitrate on nodule growth and C_2H_2 reduction activity in spite of markedly different accumulations of NO_2^-. From this it was concluded that NO_2^- did not play a role in inhibition of nodule growth and nitrogenase activity by NO_3^-. It does seem clear that nitrogenase within the intact bacteroids is not directly inhibited by NO_3^- per se (Stephens and Neyra, 1983), leaving one to conclude that NO_3^- metabolism is a prerequisite for inhibition of nodulation by NO_3^-.

Attempts to circumvent NO_3^- inhibition of nodulation by selection of more tolerant strains of *B. japonicum* have not been promising (Gibson and Harper, 1985; McNeil, 1982). These investigators independently concluded that manipulations of the plant, rather than the bacterial strain, may provide greater potential for overcoming NO_3^- inhibition of nodulation and N_2 fixation.

13-3.1.4 Bacteroid Integrity

Short-term studies with NO_3^- inhibition of nodule development indicate that nitrogenase may be adversely affected without alteration of the fine structure of nodules (Gibson, 1976). Longer-term exposure of nodulated plants to NO_3^- does, however, result in ultimate deterioration of the nodule structure followed by nodule senescence and disintegration.

As a general conclusion of the mechanism of NO_3^- inhibition of nodulation, the evidence is quite compelling that carbohydrate deprivation is involved in certain aspects of nodular inhibition. Equally compelling, however, is evidence that NO_3^- mediates detrimental effects on the recognition/infection stage of nodulation which are likely due to causes other than carbohydrate deprivation.

13-3.2 Energetics of Nitrogen Metabolism

Theoretical calculations have indicated that energy costs for NO_3^- metabolism and N_2 fixation are similar (Hardy and Havelka, 1975b), both requiring approximately 24 ATP equivalents to form two ammonium ions from two nitrate ions or from one N_2 molecule. However, Phillips (1980) and Minchin et al. (1981) have recently pointed out that the cost of N_2 fixation may range from 25.5 to 49 ATP equivalents, depending on the electrons proportioned to H_2 production, the status of the uptake hydrogenase system, and the type of organic carbon compounds exported. Studies with soybean have indicated that whole-plant energy costs for symbiotic N_2 fixation are greater than for NO_3^- metabolism (Finke et al., 1982; Ryle et al., 1979). Ryle et al. (1979) reported that 7.4 and 8.3 mol of C were respired for each mol of N assimilated from NO_3^- and N_2, respectively. Finke et al. (1982) also reported 8.3 mol of C respired per mol of N assimilated from N_2, but only 3.2 to 5.0 mol of C respired per mol of N assimilated from NO_3^-, depending on plant age. With the current data available, it would appear that symbiotic N_2 fixation may be more energy expensive than NO_3^- metabolism. Soybean plants grown under field conditions with supplemental fertilizer N, however, have rarely been shown to yield more than plants symbiotically fixing at least a portion of their N needs (Welch et al., 1973). These data strongly indicate that if symbiotic N_2 fixation is a more energy intensive process than is NO_3^- metabolism then the difference is insufficient to be realized in terms of soybean yield agains the background of biological variability, or that photosynthetic availability is not a primary limitation. Based on the positive responses of soybean growth to conditions conducive to enhanced photosynthesis (Hardy and Havelka, 1975a), it would appear that the difference in energy expenditure for NO_3^- and N_2 assimilation is insufficient to be realized in terms of affecting seed yield.

13-3.3 Seasonal Profiles of Nitrogen Assimilation

Seasonal profiles of NO_3^- assimilation by soybean have been reported by several workers (Harper, 1974; Harper and Hageman, 1972; Hatam,

1980; Obaton et al., 1982; Streeter, 1972b; Thibodeau and Jaworski, 1975). Nitrate reductase activity on a unit weight basis and within a specific leaf position, reaches a maximum at full-leaf expansion and declines thereafter (Harper and Hageman, 1972). Evaluation of profiles of NR activity of sequentially developing trifoliolate leaves results in a family of curves showing individual maxima at full-leaf expansion, but with decreasing levels of the absolute maximum attained with later developing trifoliolate leaves (Harper and Hageman, 1972). Although the maximum NR activity on a unit weight basis occurs at full-leaf expansion, the continued increase in leaf mass often results in higher total NR activities on a leaf basis at some point in time past full-leaf expansion (Harper and Hageman, 1972). The NR profiles do reflect levels of available NO_3^- and invariably NR activity is lower in soybean plants grown in field environments than in controlled environments with continuous NO_3^- supply (Harper, 1974; Streeter, 1972b; Thibodeau and Jaworski, 1975). Thus, NO_3^- and moisture availabilities are critical in determining the overall level of N assimilation in soybean. The general conclusion drawn by most workers is that NO_3^- availability rather than NR enzyme level is the primary limitation to overall NO_3^- assimilation.

Although highest specific NR activities (on unit leaf weight basis) occur in seedling plants, on a total plant canopy basis overall NO_3^- assimilation increases up to about the full-bloom growth stage (Harper and Hageman, 1972; Obaton et al., 1982), at least under conditions where plant-available soil N levels are appreciable such as in the Corn Belt area of the midwestern USA. The decline in total NO_3^- assimilation during pod-fill stages is most likely due to declining soil NO_3^- availability (Streeter, 1972a) and to increasing moisture stress conditions which also adversely affect NO_3^- availability to the plant root.

Symbiotic N_2-fixation profiles over the growing season largely reflect the lack of ability of the soybean plant to meet its N demands through NO_3^- metabolism (Harper, 1974). Under field conditions, such as in the Corn Belt, N_2 fixation does not contribute much to overall N input until 35 to 40 DAP (Harper, 1974; Obaton et al., 1982; Thibodeau and Jaworski, 1975). Under amended soil conditions which limited the NO_3^- availability (Lawn and Brun, 1974) and under low N-containing soils (Hinson, 1975), N_2 fixation profiles are initiated earlier than generally noted in the midwestern USA. Thus, the evidence is overwhelming that an inverse relationship exists between NO_3^- availability and symbiotic N_2 fixation and this ultimately controls the timing and magnitude of input profiles of NO_3^- assimilation and symbiotic N_2 fixation. In addition, involvement of hormonal control of N_2 fixation has been suggested by Peat et al. (1981). These workers showed that removal of floral buds at first appearance prevented the increase in N_2 fixation normally seen during the early reproductive growth stage.

It is noteworthy that small amounts of starter N are desirable and actually stimulate symbiotic N_2 fixation over that found in controls grown in the complete absence of any external N (Eaglesham et al., 1983; Harper,

1974). This may be of more importance under cool temperature conditions at sowing (Matthews and Hayes, 1982). Soybean plants totally deprived of any starter N source undergo a severe N stress period following depletion of cotyledonary N reserves and before functional nodule development occurs. This results in a less vigorous plant shoot, which is not capable of supplying sufficient photosynthate to support optimum N_2 fixation, and hence optimum plant growth is never attained (Harper, 1974; Wych and Rains, 1979). Thus, a small amount of starter N has a synergistic effect on subsequent plant growth and functional nodule development. In most field soils, there is sufficient residual plant-available N to meet this initial requirement for starter N, thus precluding the necessity of adding any supplemental N fertilizer to soybean to optimize symbiotic N_2 fixation. No evidence exists that reduced-N compounds derived from either NO_3^- metabolism or symbiotic N_2 fixation are preferentially available for synthesis of seed protein. Kato (1981) reported that the ability of soybean to transfer N from vegetative plant parts to seeds was similar regardless of whether the N was initially incorporated from NO_3^- or atmospheric N_2.

Dinitrogen fixation has been shown to reach a maximum during early to mid-pod-fill growth stages (Hardy et al., 1968; Harper, 1974; Obaton et al., 1982; Sloger et al., 1975). With amended soil conditions which limit availability of plant-available soil N, the onset of and maximum rates of N_2 fixation occur earlier in the growth cycle (Lawn and Brun, 1974). The decline in N_2 fixation during later pod-fill stages has been attributed to redirection of available photosynthate to pod development at the expense of the nodules. Malik (1983) has provided some support for this concept where he demonstrated a reversal in loss of N_2 fixation during pod fill by grafting young scions near the base of the plant during fruit development. That this procedure may also alter hormonal balance which affects maintenance of nodule functionality cannot, however, be ruled out. Studies which have involved male-sterile isolines of soybean (Imsande and Ralston, 1982; Riggle et al., 1984; Schweitzer and Harper, 1985; Wilson et al., 1978) or depodding of soybean (Crafts-Brandner et al., 1984; Lawn and Brun, 1974) have indicated that decreased pod load may slightly delay the decline in N_2-fixation rates, but it is obvious that factors other than carbohydrate availability for nodule function are also involved in senescence of nodules. Sheehy (1983) concluded, on the basis of calculated gross photosynthesis rate, that sufficient C was assimilated to support crop maintenence, seed growth, and N_2 fixation. He further suggested that loss of nodule effectiveness was due to a signal being transmitted to the nodule, rather than insufficient C availability.

Estimates of the amount of N symbiotically fixed range from nil under conditions of high N fertility (Johnson et al., 1975) to upwards of 400 kg of N ha^{-1} with CO_2 enrichment studies (Hardy and Havelka, 1975a). Under typical unamended field conditions N_2 fixation by soybean ranges from 80 to 100 kg of N ha^{-1} in Corn Belt soils (Johnson et al., 1975) to a high of 311 kg of N ha^{-1} on a sandy loam soil (Bezdicek et al., 1978). With midwestern USA soils which have appreciable residual NO_3^- levels, some 25 to 50% of total plant N has been attributed to N_2 fixation (Harper, 1976) while upwards of 80 to 94% of the N may be derived from symbiotic N_2 fixation under lower soil N conditions (Bezdicek et al., 1978; Schroder and Hinson, 1975).

Estimates of N input to soybean plants from soil and atmospheric N sources have been based on comparison of nodulated and nonnodulated isolines in several studies (Deibert et al., 1979; Domenach et al., 1979; Fried and Broeshart, 1975; Ham and Caldwell, 1978; Harper, 1974; Kohl et al., 1980; Legg and Sloger, 1975; Ruschel et al., 1979; Weber, 1966). These results must be viewed with caution in that the nonnodulating isoline may not be removing the same amount of soil N as removed by the nodulating isoline (Boddey et al., 1984), a critical assumption to estimating symbiotic N_2 fixation. Rennie (1982) and Wagner and Broeshart (1980) have concluded that the better nonfixing control plant may be a nodulating soybean line in a nonfixing mode, that is, infected with an ineffective *Bradyrhizobium* strain or not inoculated. This will, however, not be feasible on most soils due to indigenous *Bradyrhizobium* populations. Talbott et al. (1982) on the other hand, found reasonably good correlations ($r=0.89$ and 0.92 for 2 yrs) between ^{15}N methods and difference methods based on nodulating and nonnodulating soybean plants. This relationship was confirmed by Vasilas and Ham (1984). Thus, the use of the nonnodulating soybean isoline as the nonfixing control will likely continue, in spite of known limitations. Use of other crop species as the nonfixing control suffer the same limitations noted for the nonnodulating soybean line, that is, both require the assumption that similiar amounts of soil N are taken up relative to the nodulating soybean. The C_2H_2 reduction assay has been widely used to estimate symbiotic N_2 fixation in soybean since Hardy et al. (1968) popularized this approach. This method does have limitations, as used by most laboratories, of being point-in-time measurements. However, Denison et al. (1983) have reported on an in situ C_2H_2 reduction assay which enables multiple analyses over time of nodulated root systems. This approach may enable establishment of a better relationship between C_2H_2 reduction and N_2 fixation, however, it suffers from requiring extensive instrumentation which may preclude widespread adoption.

13–3.4 Prospects of Enhanced Nitrogen Assimilation

The possibility of increasing symbiotic N_2 fixation by selecting for more efficient mutant strains of *B. japonicum* was suggested by Maier and Brill (1978). Under field conditions, only one report was found of increased (about 12%) seed yield and seed N content when a strain selected for increased N_2-fixation efficiency was used (Williams and Phillips, 1983). Additional research efforts are underway to exploit *B. japonicum* mutants to increase symbiotic N_2 input with an ultimate goal to increase soybean productivity. Mutant strains of *B. japonicum* have been selected with enhanced hydrogenase activity which may have potential agronomic significance if nitrogenase activity can also be enhanced (Merberg and Maier, 1983). Thomas et al. (1983) tested *B. japonicum* mutants, previously selected for enhanced capacity for N_2 fixation at the seedling stage (Maier and Brill, 1978), for N_2-fixation capability up through 5 weeks of growth. Although the mutant strains resulted in higher rates of C_2H_2 reduction and ureide transport in the xylem at 3 weeks after planting and inoculation, this effect did not continue, leading to the conclusion that these strains would be beneficial only under early N stress where early nodulation might be beneficial. Efforts to select increased tolerance of the symbiotic association to NO_3^- would appear to be a feasible goal in attempts to enhance N_2 fixation. The success along this line with *Pisum sativum* (Jacobsen and Feenstra, 1983) and with soybean (Carroll et al., 1985a 1985b; Gremaud and Harper, unpublished) lends support to this approach in soybean.

13–4 CONCLUSION

The ability of grain legumes to symbiotically fix atmospheric N has provided a mechanism for these species to survive on N-deficient soils. However, much of the soybean production area in the USA involves soils which have appreciable levels of residual N and, thus, the symbiotic N_2-fixation system is functioning at less than maximum rates, due to inhibitory effects of NO_3^-. In spite of the fact that inhibition of nodulation by NO_3^- has been known for many years, the mechanism(s) of this inhibition remains elusive and thus efforts to circumvent this inhibition have not been successful to date. Continued efforts to enhance symbiotic N_2 fixation by soybean with the view that subsequent crops grown in rotation may benefit from greater residual plant-available soil N remains as a viable research area.

In regard to NO_3^- metabolism the observation that soybean expresses NR activity in both the presence and absence of NO_3^- is somewhat unique among higher plant species where usually only a NO_3^--inducible enzyme exists. The true in situ function of the constitutive or non-NO_3^--inducible enzymes remains to be evaluated. The association of the constitutive enzyme with the apparent enzymatic capability to evolve nitrogenous gases is also intriguing. Whether this reaction sequence relates to observations of foliar N evolution from intact plants is not known.

Reports that products of N_2 fixation are translocated from nodules to plant tops in the form of ureides have been one area where major advancement was realized since 1973. Many of the key biochemical steps in ureide synthesis and metabolism have been elucidated. Why soybean, and a number of other legume species, have evolved to synthesize and translocate ureide forms of N, when a number of other legume species do not, remains to be resolved. The most attractive proposal may be conservation of C involved in N translocation, a hypothesis yet to be proven.

Another development since 1970 was the renewed interest in use of *B. japonium* strains which express an active uptake hydrogenase, functioning to recycle H released during N_2 fixation. This provided hope that use of Hup^+ strains would be beneficial in terms of energy conservation and increased plant productivity. Although field studies have been somewhat disappointing with respect to a consistent positive benefit of Hup^+ strains on seed N and dry matter production, this attribute may prove more beneficial in future years as more physiological limitations are recognized and eliminated. It is obvious that N metabolism is but one of the many integrated processes involved in soybean productivity, and progress in this area of research will be realized only with continued progress in other areas and disciplines.

REFERENCES

Almon, L. 1933. Concerning the reproduction of bacteroids. Zentralbl. Bakteriol. Parasitenkd. Abt. 2. 87:289–297.

Andrews, M., J.M. Sutherland, R.J. Thomas, and J.I. Sprent. 1984. Distribution of nitrate reductase activity in six legumes: The importance of the stem. New Phytol. 98:301–310.

Antoun, H., L.M. Bordeleau, D. Prevost, and R.L. Lachance. 1980. Absence of a correlation between nitrate reductase and symbiotic nitrogen fixation efficiency in *Rhizobium meliloti*. Can. J. Plant Sci. 60:209–212.

Aslam, M. 1982. Differential effect of tungsten on the development of endogenous and nitrate-induced nitrate reductase activities in soybean leaves. Plant Physiol. 70:35–38.

Bal, A.K., and S. Shantharam. 1981. Lectin receptor sites in the cell wall of *Rhizobium japonicum*. Microbios Lett. 16:141–144.

Bauer, W.D. 1981. Infection of legumes by rhizobia. Annu. Rev. Plant Physiol. 32:407–449.

Beevers, L. 1976. Nitrogen metabolism in plants. E.J.W. Barrington and A.J. Willis (ed.) Elsevier Science Publishing, New York.

----, D. Flesher, and R.H. Hageman. 1964. Studies on pyridine nucleotide specificity of nitrate reductase in higher plants and its relationship to sulfhydryl level. Biochim. Biophys. Acta 89:453–464.

----, and R.H. Hageman. 1983. Uptake and reduction of nitrate: Bacteria and higher plants. p. 351–375. *In* A. Läuchli and R.L. Bieleski (ed.) Inorganic plant nutrition. Encyclopedia of plant physiology new series, Vol. 15B. Springer-Verlag New York, New York.

Bergersen, F.J. 1963. The relationship between hydrogen evolution, hydrogen exchange, nitrogen fixation, and applied oxygen tension in soybean root nodules. Aust. J. Biol. Sci. 16:669–680.

----. 1965. Ammonia—an early stable product of nitrogen fixation by soybean root nodules. Aust. J. Biol. Sci. 18:1–9.

----. 1968. The symbiotic state in legume root nodules: Studies with the soybean system. Int. Congr. Soil Sci. Trans., 9th. 2:49–63.

----. 1970. The quantitative relationship between nitrogen fixation and the acetylene-reduction assay. Aust. J. Biol. Sci. 23:1015–1025.

----. 1971. Biochemistry of symbiotic nitrogen fixation in legumes. Annu. Rev. Plant Physiol. 22:121–140.

----. 1980. Leghaemoglobin, oxygen supply and nitrogen fixation: Studies with soybean nodules. p. 139–160. *In* W.D.P. Stewart and J.R. Gallon (ed.) Nitrogen fixation. Academic Press, New York.

----. 1982. Root nodules of legumes: Structure and functions. Research Studies Press, New York.

----, and C.A. Appleby. 1981. Leghaemoglobin within bacteroid-enclosing membrane envelopes from soybean root nodules. Planta 152:534–543.

----, and D.J. Goodchild. 1973. Aeration pathways in soybean nodules. Aust. J. Biol. Sci. 26:729–740.

----, and G.L. Turner. 1967. Nitrogen fixation by the bacteroid fraction of breis of soybean root nodules. Biochim. Biophys. Acta. 141:507–515.

Bezdicek, D.F., D.W. Evans, B. Abede, and R.E. Witters. 1978. Evaluation of peat and granular inoculum for soybean yield and N fixation under irrigation. Agron. J. 70:865–868.

Bhuvaneswari, T.V., S.G. Pueppke, and W.D. Bauer. 1977. Role of lectins in plant-microorganism interactions. I. Binding of soybean lectin to rhizobia. Plant Physiol. 60:486–491.

Boddey, R.M., P.M. Chalk, R.L. Victoria, and E. Matsui. 1984. Nitrogen fixation by nodulated soybean under tropical field conditions estimated by the ^{15}N isotope dilution technique. Soil Biol. Biochem. 16:583–588.

Bohlool, B.B., and E.L. Schmidt. 1974. Lectins: A possible basis for specificity in the *Rhizobium*-legume root nodule symbiosis. Science 185:269–271.

Boland, M.J., J.F. Hanks, P.H.S. Reynolds, D.G. Blevins, N.E. Tolbert, and K.R. Schubert. 1982. Subcellular organization of ureide biogenesis from glycolytic intermediates and ammonium in nitrogen-fixing soybean nodules. Planta 155:45–51.

Brethauer, T.S., and J.D. Paxton. 1977. The role of lectin in soybean-*Rhizobium japonicum* interactions. p. 381–387. *In* B. Solheim and J. Raa (ed.) Cell wall biochemistry related to specificity in host-plant pathogen interactions. Columbia University Press, New York.

Brown, C.M., and M.J. Dilworth. 1975. Ammonia assimilation by *Rhizobium* cultures and bacteroids. J. Gen. Microbiol. 86:39–48.

Brun, W.A. 1976. The relation of N_2 fixation to photosynthesis. p. 135–143. *In* L.D. Hill (ed.) World soybean research conference proceedings. The Interstate Printers and Publishers, Danville, IL.

Burns, R.C., and R.W.F. Hardy. 1975. Nitrogen fixation in bacteria and higher plants. p. 1–89. *In* A. Kleinzeller et al. (ed.) Molecular biology biochemistry and biophysics. Springer-Verlag New York, New York.

Burris, R.H., D.J. Arp, D.R. Benson, D.W. Emerich, R.V. Hageman, T. Ljones, P.W. Ludden, and W.J. Sweet. 1980. The biochemistry of nitrogenase. p. 37–54, *In* W.D.P. Stewart and J.R. Gallon (ed.) Nitrogen fixation. Academic Press, New York.

Callaham, D.A. 1979. Structural basis for infection of root hairs of *Trifolium repens* by *Rhizobium trifolii.* M.Sc. thesis. Univ. of Massachusetts, Amherst.

Campbell, W.H. 1976. Separation of soybean leaf nitrate reductases by affinity chromatography. Plant Sci. Lett. 7:239–254.

Carelli, M.L.C., and A.C. Magalhaes. 1981. Development of nitrate reductase activity in green tissues of soybean seedlings (*Glycine max* L. Merr.). Z. Pflanzenphysiol. 104:17–24.

Carter, K.R., N.T. Jennings, J. Hanus, and H.J. Evans. 1978. Hydrogen evolution and uptake by nodules of soybeans inoculated with different strains of *Rhizobium japonicum.* Can. J. Microbiol. 24:307–311.

----, J. Rawlings, W.H. Becker, R.R. Becker, and H.J. Evans. 1980. Purification and characterization of a ferredoxin from *Rhizobium japonicum* bacteroids. J. Biol. Chem. 255:4213–4233.

Carroll, B.J., D.L. McNeil, and P.M. Gresshoff. 1985a. A supernodulation and nitrate-tolerant symbiotic (*nts*) soybean mutant. Plant Physiol. 78:34–40.

----, ----, and ----. 1985b. Isolation and properties of novel soybean (*Glycine max* (L.) Merr.) mutants that nodulate in the presence of high nitrate concentration. Proc. Natl. Acad. Sci. U.S.A. 82:4162–4166.

Child, J.J. 1975. Nitrogen fixation by a *Rhizobium* sp. in association with non-leguminous plant cell cultures. Nature (London) 253:350–351.

Crafts-Brandner, S.J., F.E. Below, J.E. Harper, and R.H. Hageman. 1984. Effects of pod removal on metabolism and senescence of nodulating and nonnodulating soybean isolines. II. Enzymes and chlorophyll. Plant Physiol. 75:318–322.

----, and J.E. Harper. 1982. Nitrate reduction by roots of soybean (*Glycine max* [L.] Merr.) seedlings. Plant Physiol. 69:1298–1303.

Criswell, J.G., U.D. Havelka, B. Quebedeaux, and R.W.F. Hardy. 1976. Adaptation of nitrogen fixation by intact soybean nodules to altered rhizosphere pO_2. Plant Physiol. 58:622–625.

Dart, P. 1977. Infection and development of leguminous nodules. p. 367–472. *In* R.W.F. Hardy and W.S. Silver (ed.) A treatise on dinitrogen fixation. Sect. III: Biology. John Wiley and Sons, New York.

Dazzo, F.B., and W.J. Brill. 1978. Regulation by fixed nitrogen of host-symbiont recognition in the *Rhizobium*-clover symbiosis. Plant Physiol. 62:18–21.

----, and D.H. Hubbell. 1982. Control of root hair infection. p. 274–310. *In* W.J. Broughton (ed.) Nitrogen fixation, *Rhizobium* Vol. 2. Oxford University Press, New York.

Dean, J.V., and J.E. Harper. 1986. Nitric oxide and nitrous oxide production by soybean and winged bean during the in vivo nitrate reductase assay. Plant Physiol. 82:718–723.

Deibert, E.J., M. Bijeriego, and R.A. Olson. 1979. Utilization of ^{15}N fertilizer by nodulating and non-nodulating soybean isolines. Agron. J. 71:717–723.

Denison, R.F., T.R. Sinclair, R.W. Zobel, M.N. Johnson, and G.M. Drake. 1983. A non-destructive field assay for soybean nitrogen fixation by acetylene reduction. Plant Soil 70:173–182.

Domenach, A.M., A. Chalamet, and C. Pachiandi. 1979. Estimation de la fixation d'azote par le soja á l'aide de deux methodes d'analyses isotopiques. C.R. Hebd. Seances Acad. Sci. 289:291–293.

Dombrink-Kurtzman, M.A., W.E. Dick, Jr., K.A. Burton, M.C. Cadmus, and M.E. Slodki. 1983. A soybean lectin having 4-0-Methyl-D-glucuronic acid specificity. Biochem. Biophys. Res. Commun. 111:798- 803.

Duke, S.H., and S.O. Duke. 1984. Light control of extractable nitrate reductase activity in higher plants. Physiol. Plant. 62:482–493.

----, L.E. Schrader, C.A. Henson, J.C. Servaites, R.D. Vogelzang, and J.W. Pendleton. 1979. Low root temperature effects on soybean nitrogen metabolism and photosynthesis. Plant Physiol. 63:956–962.

Duke, S.O., K.C. Vaughn, and S.H. Duke. 1982. Effects of norflurazon (San 9789) on light-increased extractable nitrate reductase activity in soybean [*Glycine max* (L.) Merr.] seedlings. Plant Cell Environ. 5:155–162.

Dunn-Coleman, N.S., J. Smarrelli, Jr., and R.H. Garrett. 1984. Nitrate assimilation in eucaryotic cells. Int. Rev. Cytol. 92:1–50.

Eaglesham, A.R.J., S. Hassouna, and R. Seegers. 1983. Fertilizer-N effects on N_2 fixation by cowpea and soybean. Agron. J. 75:61–66.

Eisbrenner, G., and H. Bothe. 1979. Modes of electron transfer from molecular hydrogen in *Anabaena cylindrica.* Arch. Microbiol. 123:37–45.

Eskew, D.L., R.M. Welch, and E.E. Cary. 1983. Nickel: An essential micronutrient for legumes and possibly all higher plants. Science 222:621–623.

Evans, H.J., and A. Nason. 1953. Pyridine nucleotide-nitrate reductase from extracts of higher plants. Plant Physiol. 28:233–254.

----, K. Purohit, M.A. Cantrell, G. Eisbrenner, S.A. Russell, F.J. Hanus, and J.E. Lepo. 1981. Hydrogen losses and hydrogenases in nitrogen-fixing organisms. p. 84–96. *In* A.H. Gibson and W.E. Newton (ed.) Current perspectives in nitrogen fixation. Elsevier/North Holland, New York.

Finke, R.L., J.E. Harper, and R.H. Hageman. 1982. Efficiency of nitrogen assimilation by N_2-fixing and nitrate-grown soybean plants (*Glycine max* [L.] Merr.). Plant Physiol. 70:1178–1184.

Finn, G.A., and W.A. Brun. 1980. Water stress effects on CO_2 assimilation, photosynthate partitioning, stomatal resistance, and nodule activity in soybean. Crop Sci. 20:431–434.

Fujihara, S., and M. Yamaguchi. 1978. Probable site of allantoin formation in nodulating soybean plants. Phytochemistry 17:1239–1243.

Fried, M., and H. Broeshart. 1975. An independent measurement of the amount of nitrogen fixed by a legume crop. Plant Soil 43:707–711.

Gade, W., M.A. Jack, J.B. Dahl, E.L. Schmidt, and F. Wold. 1981. The isolation and characterization of a root lectin from soybean [*Glycine max* (L.) cultivar Chippewa]. J. Biol. Chem. 256:12905–12910.

Gibson, A.H. 1974. Consideration of the growing legume as a symbiotic association. Proc. Indian Natl. Sci. Acad. 40B:741–767.

----. 1976. Recovery and compensation by nodulated legumes to environmental stress. p. 385–403. *In* P.S. Nutman (ed.) Symbiotic nitrogen fixation in plants. Cambridge University Press, London.

----, and J.E. Harper. 1985. Nitrate effect on nodulation of soybean by *Bradyrhizobium japonicum.* Crop Sci. 25:497–501.

----, and D.C. Jordon. 1983. Ecophysiology of nitrogen-fixing systems. p. 301–390. *In* O.L. Lange et al. (ed.) Encyclopedia of plant physiology new series, Vol. 12C. Physiological plant ecology III. Springer-Verlag New York, New York.

----, and J.D. Pagan. 1977. Nitrate effects on the nodulation of legumes inoculated with nitrate-reductase-deficient mutants of *Rhizobium*. Planta 134:17–22.

----, W.R. Scowcroft, and J.D. Pagan. 1977. Nitrogen fixation in plants: An expanding horizon? p. 387–417. *In* W. Newton, et al. (ed.) Recent developments in nitrogen fixation. Academic Press, New York.

Goodchild, D.J., and F.J. Bergersen. 1966. Electron microscopy of the infection and subsequent development of soybean nodule cells. J. Bacteriol. 92:204–213.

Gresshoff, P.M., M.L. Skotnicki, J.E. Eadie, and B.G. Rolfe. 1977. The viability of *Rhizobium trifolii* bacteroids from clover root nodules. Plant Sci. Lett. 10:299–304.

Guerrero, M.B., J.M. Vega, and M. Losada. 1981. The assimilatory nitrate-reducing system and its regulation. Annu. Rev. Plant Physiol. 32:169–204.

Halverson, L.J., and G. Stacey. 1984. Host recognition in the *Rhizobium*-soybean symbiosis. Detection of a protein factor in soybean root exudate which is involved in the nodulation process. Plant Physiol. 74:84–89.

Ham, G.E., and A.C. Caldwell. 1978. Fertilizer placement effects on soybean seed yield, N_2 fixation, and ^{33}P uptake. Agron. J. 70:779–783.

Hardy, R.W.F., and U.D. Havelka. 1975a. Photosynthate as a major factor limiting nitrogen fixation by field-grown legumes with emphasis on soybeans. p. 421–439. *In* P.S. Nutman (ed.) Symbiotic nitrogen fixation in plants. Cambridge University Press, London.

----, and ----. 1975b. Nitrogen fixation research: A key to world food? Science 188:633–643.

----, R.D. Holsten, E.K. Jackson, and R.C. Burns. 1968. The acetylene-ethylene assay for N_2 fixation: laboratory and field evaluations. Plant Physiol. 43:1185–1207.

Harper, J.E. 1971. Seasonal nutrient uptake and accumulation patterns in soybeans. Crop Sci. 11:347–350.

----. 1974. Soil and symbiotic nitrogen requirements for optimum soybean production. Crop Sci. 14:255–260.

----. 1976. Contribution of dinitrogen and soil or fertilizer nitrogen to soybean (*Glycine max* L. Merr) production. p. 101–107. *In* L.D. Hill (ed.) World soybean research conference proceedings. The Interstate Printers and Publishers, Danville, IL.

----. 1981. Evolution of nitrogen oxide(s) during *in vivo* nitrate reductase assay of soybean leaves. Plant Physiol. 68:1488–1493.

----, and A.H. Gibson. 1984. Differential nodulation tolerance to nitrate among legume species. Crop Sci. 24:797–801.

----, and R.H. Hageman. 1972. Canopy and seasonal profiles of nitrate reductase in soybeans (*Glycine max* L. Merr.). Plant Physiol. 49:146–154.

----, and J.C. Nicholas. 1978. Nitrogen metabolism of soybeans. I. Effect of tungstate on nitrate utilization, nodulation, and growth. Plant Physiol. 62:662–664.

Hatam, M. 1980. Seasonal and diurnal variations in nitrate reductase activity of soybean (*Glycine max* (L.) Merr.). Plant Soil 56:27–32.

Herridge, D.F. 1982a. Relative abundance of ureides and nitrate in plant tissues of soybean as a quantitative assay of nitrogen fixation. Plant Physiol 70:1–6.

----. 1982b. Use of the ureide technique to describe the nitrogen economy of field-grown soybeans. Plant Physiol. 70:7–11.

----. 1984. Effects of nitrate and plant development on the abundance of nitrogenous solutes in root-bleeding and vacuum-extracted exudates of soybean. Crop Sci. 24:173–179.

----, C.A. Atkins, J.S. Pate, and R.M. Rainbird. 1978. Allantoin and allantoic acid in the nitrogen economy of the cowpea (*Vigna unguiculata* [L.] Walp). Plant Physiol. 62:495–498.

Hewitt, E.J., and B.A. Notton. 1980. Nitrate reductase systems in eukaryotic and prokaryotic organisms. p. 275–325. *In* M. Coughlan (ed.) Molybdenum and molybdenum containing enzymes. Pergamon Press, Elmsford, NY.

Hinson, K. 1975. Nitrogen fertilization of soybeans (*Glycine max* (L.) Merr.) in peninsular Florida. Soil Crop Sci. Soc. Fla. Proc. 34:97–101.

Hoch, G.E., H.N. Little, and R.H. Burris. 1957. Hydrogen evolution from soybean root nodules. Nature (London) 179:430–431.

Hogan, M.E., I.E. Swift, and J. Done. 1983. Urease assay and ammonia release from leaf tissues. Phytochemistry 22:663–667.

Huang, C.-Y., J.S. Boyer, and L.N. Vanderhoef. 1975a. Acetylene reduction (nitrogen fixation) and metabolic activities of soybean having various leaf and nodule water potentials. Plant Physiol. 56:222–227.

----, ----, and ----. 1975b. Limitation of acetylene reduction (nitrogen fixation) by photosynthesis in soybean having low water potentials. Plant Physiol. 56:228–232.

Hume, D.J., J.G. Criswell, and K.R. Stevenson. 1976. Effects of soil moisture around nodules on nitrogen fixation by well watered soybeans. Can. J. Plant Sci. 56:811–815.

Hunter, W.J. 1983. Soybean root and nodule nitrate reductase. Physiol. Plant. 59:471–475.

----, C.J. Fahring, S.R. Olsen, and L.K. Porter. 1982. Location of nitrate reduction in different soybean cultivars. Crop Sci. 22:944–948.

Imsande, J., and E.J. Ralston. 1982. Dinitrogen fixation in male-sterile soybeans. Plant Physiol. 69:745–746.

Jacobsen, E., and W.J. Feenstra. 1983. A new pea mutant with efficient nodulation in the presence of nitrate. Plant Sci. Lett. 33:337–344.

Johnson, J.W., L.F. Welch, and L.T. Kurtz. 1975. Environmental implications of N fixation by soybeans. J. Environ. Qual. 4:303–306.

Jolly, S.O., W. Campbell, and N.E. Tolbert. 1976. NADPH- and NADH-nitrate reductases from soybean leaves. Arch. Biochem. Biophys. 174:431–439.

----, and N.E. Tolbert. 1978. NADH-nitrate reductase inhibitor from soybean leaves. Plant Physiol. 62:197–203.

Jordan, D.C. 1982. Transfer of *Rhizobium japonicum* Buchanan 1980 to *Bradyrhizobium* gen. nov., a genus of slow-growing, root nodule bacteria from leguminous plants. Int. J. Syst. Bacteriol. 32:136–139.

Kakefuda, G., S.H. Duke, and S.O. Duke. 1983. Differential light induction of nitrate reductases in greening and photobleached soybean seedlings. Plant Physiol. 73:56–60.

Kato, Y. 1981. Studies on nitrogen metabolism of soybean *Glycine max* cultivar Koganedaizu plants 6. Utilization and distribution of nitrogen derived from nitrate and symbiotic fixation. Jpn. J. Crop Sci. 50:282–288.

Kennedy, I.R. 1966. Primary products of symbiotic nitrogen fixation. I. Short-term exposures of serradella nodules to $^{15}N_2$. Biochim. Biophys. Acta 130:285–294.

----, J. Rigaud, and J.C. Trinchant. 1975. Nitrate reductase from bacteroids of *Rhizobium japonicum*: Enzyme characteristics and possible interaction with nitrogen fixation. Biochim. Biophys. Acta 130:295–303.

Kerr, P.S., D.G. Blevins, B.J. Rapp, and D.D. Randall. 1983. Soybean leaf urease: comparison with seed urease. Physiol. Plant. 57:339–345.

Keyser, H.H., B.B. Bohlool, T.S. Hu, and D.F. Weber. 1982. Fast-growing rhizobia isolates from root nodules of soybean. Science 215:1631–1632.

----, D.F. Weber, and S.L. Uratsu. 1984. *Rhizobium japonicum* serogroup with hydrogenase phenotype distribution in 12 states. Appl. Environ. Microbiol. 47:613–615.

Kiss, G.B., E. Vincze, Z. Kálmán, T. Forrai, and A. Kondorosi. 1979. Genetic and biochemical analysis of mutants affected in nitrate reduction in *Rhizobium meliloti*. J. Gen. Microbiol. 113:105–118.

Klepper, L. 1979. Nitric oxide (NO) and nitrogen dioxide (NO_2) emissions from herbicide-treated soybean plants. Atmos. Environ. 13:537–542.

Klucas, R.V. 1974. Studies on soybean nodule senescence. Plant Physiol. 54:612–616.

----, and H.J. Evans. 1968. An electron donor system for nitrogenase-dependent acetylene reduction by extracts of soybean nodules. Plant Physiol. 43:1458–1460.

----, F.J. Hanus, S.A. Russell, and H.J. Evans. 1983. Nickel: A micronutrient element for hydrogen-dependent growth of *Rhizobium japonicum* and for expression of urease activity in soybean leaves. Proc. Natl. Acad. Sci. 80:2253–2257.

Koch, B., H.J. Evans, and S. Russell. 1967. Properties of the nitrogenase system in cell-free extracts of bacteroids from soybean root nodules. Proc. Natl. Acad. Sci. U.S.A. 58:1343–1350.

Kohl, D.H., G. Shearer, and J.E. Harper. 1980. Estimates of N_2 fixation based on differences in the natural abundane of ^{15}N in nodulating and non-nodulating isolines of soybeans. Plant Physiol. 66:61–65.

Kuykendall, L.D., and D.F. Weber. 1978. Genetically marked *Rhizobium* identifiable as inoculum strain in nodules of soybean plants grown in fields populated with *Rhizobium japonicum*. Appl. Environ. Microbiol. 36:915–919.

Latimore, M., Jr., J. Giddens, and D.A. Ashley. 1977. Effect of ammonium and nitrate nitrogen upon photosynthate supply and nitrogen fixation by soybeans. Crop Sci. 17:399–404.

Lawn, R.J., and W.A. Brun. 1974. Symbiotic nitrogen fixation in soybeans. I. Effect of photosynthetic source-sink manipulations. Crop Sci. 14:11–16.

Lea, P.J., and B.J. Miflin. 1980. Transport and metabolism of asparagine and other nitrogen compounds within the plant. p. 569–607. *In* P.K. Stumpf and E.E. Conn (ed.) The biochemistry of plants. Academic Press, New York.

Legg, J.O., and C. Sloger. 1975. A tracer method for determining symbiotic nitrogen fixation in field studies. p. 661–666. *In* E.R. Klein and P.D. Klein (ed.) Proc. Int. Conf. Stable Isotopes. Academic Press, New York.

Legocki, R.P., and D.P.S. Verma. 1980. Identification of "nodule specific" host proteins (nodulins) involved in the development of *Rhizobium*-legume symbiosis. Cell 20:153–163.

Libbenga, K.R., and J.G. Torrey. 1973. Hormone-induced endoreduplication prior to mitosis in cultured pea root cortex cells. Am. J. Bot. 60:293–299.

Lim, S.T., K. Andersen, R. Tait, and R.C. Valentine. 1980. Genetic engineering in agriculture: hydrogen uptake (hup) genes. Trends Biochem. Sci. (Pers. Ed.) 5:167–170.

Lindemann, W.C., E.L. Schmidt, and G.E. Ham. 1974. Evidence for double infection within soybean nodules. Soil Sci. 118:274–279.

Ljunggren, H. 1969. Mechanism and pattern of *Rhizobium* invasion into leguminous root hairs. Physiol. Plant. Suppl. 5:1–82.

Lopez, M., V. Carbonero, E. Cabrera, and T. Ruiz-Argueso. 1983. Effects of host on the expression of the hydrogen uptake hydrogenase of *Rhizobium* in legume nodules. Plant Sci. Lett. 29:191–200.

MacKown, C.T., W.A. Jackson, and R.J. Volk. 1983. Partitioning of previously-accumulated nitrate to translocation, reduction, and efflux in corn roots. Planta 157:8–14.

----, R.J. Volk, and W.A. Jackson. 1981. Nitrate accumulation, assimilation, and transport by decapitated corn roots: effects of prior nitrate nutrition. Plant Physiol. 68:133–138.

Magalhaes, A.C. 1977. Effect of root temperature on nitrate reduction in the soybean plant. Cienc. Cult. (Sao Paulo) 29:63–65.

Mague, T.H., and R.H. Burris. 1972. Reduction of acetylene and nitrogen by field-grown soybeans. New Phytol. 71:275–286.

Maier, R.J., and W.J. Brill. 1978. Mutant strains of *Rhizobium japonicum* with increased ability to fix nitrogen for soybean. Science 201:448–450.

Malik, N.S.A. 1983. Grafting experiments on the nature of the decline in N_2 fixation during fruit development in soybean. Physiol. Plant. 57:561–564.

Manhart, J.R., and P.P. Wong. 1980. Nitrate effect on nitrogen fixation (acetylene reduction). Activities of legume root nodules induced by rhizobia with varied nitrate reductase activities. Plant Physiol. 65:502–505.

Matthews, D.J., and P. Hayes. 1982. Effect of root zone temperature on early growth, nodulation and nitrogen fixation in soya beans. J. Agric. Sci. 98:371–376.

Matsumoto, T., M. Yatazawa, and Y. Yamamoto. 1977a. Distribution and change in the contents of allantoin and allantoic acid in developing nodulating and non-nodulating soybean plants. Plant Cell Physiol. 18:353–359.

----, ----, and ----. 1977b. Effects of exogenous nitrogen-compounds on the concentrations of allantoin and various constituents in several organs of soybean plants. Plant Cell Physiol. 18:613–624.

McClure, P.R., and D.W. Israel. 1979. Transport of nitrogen in the xylem of soybean plants. Plant Physiol. 64:411–416.

----, ----, and R.J. Volk. 1980. Evaluation of the relative ureide content of xylem sap as an indicator of N_2 fixation in soybeans. Plant Physiol. 66:720–725.

McNeil, D.L. 1982. Variations in ability of *Rhizobium japonicum* strains to nodulate soybeans and maintain fixation in the presence of nitrate. Appl. Environ. Microbiol. 44:647–652.

McParland, R.H., J.G. Guevara, R.R. Becker, and H.J. Evans. 1976. The purification and properties of the glutamine synthetase from the cytosol of soya-bean root nodules. Biochem. J. 153:597–606.

Meijer, E.G.M. 1982. Development of leguminous root nodules. p. 311–331. *In* W.S. Broughton (ed.) Nitrogen fixation, *Rhizobium*, Vol. 2. Oxford University Press, New York.

Merberg, D., and R.J. Maier. 1983. Mutants of *Rhizobium japonicum* with increased hydrogenase activity. Science 220:1064–1065.

Meyer, J. 1981. Comparison of carbon monoxide, nitric oxide, and nitrite as inhibitors of the nitrogenase from *Clostridium pasteurianum*. Arch. Biochem. Biophys. 210:246–256.

Miflin, B.J., and P.J. Lea. 1980. Ammonia assimilation. p. 169–202. *In* P.K. Stumpf and E.E. Conn (ed.) The biochemistry of plants. Academic Press, New York.

Minamisawa, K., Y. Arima, and K. Kumazawa. 1983. Transport of fixed nitrogen from soybean nodules inoculated with H_2-uptake positive and negative *Rhizobium japonicum* strains. Soil Sci. Plant Nutr. 29:85–92.

Minchin, F.R., R.J. Summerfield, P. Hadley, E.H. Roberts, and S. Rawsthorne. 1981. Carbon and nitrogen nutrition of nodulated roots of grain legumes. Plant Cell Environ. 4:5–26.

Mulvaney, C.S., and R.H. Hageman. 1982. Evolution of N compounds from soybean seedlings under ambient conditions. Agron. Abstr. American Society of Agronomy, Madison, WI, p. 105.

----, and R.H. Hageman. 1984. Acetaldehyde oxime, a product formed during *in vivo* nitrate reductase assay of soybean leaves. Plant Physiol. 74:118–124.

Munevar, F., and A.G. Wollum. 1981. Effect of high root temperature and *Rhizobium* strain on noduation, nitrogen fixation, and growth of soybeans. Soil Sci. Soc. Am. J. 45:1113–1120.

Munns, D.N. 1968. Nodulation of *Medicago sativa* in solution culture. III. Effects of nitrate on root hairs and infection. Plant Soil 29:33–47.

----. 1977. Mineral nutrition and the legume symbosis. p. 353–391. *In* R.W.F. Hardy and A.H. Gibson (ed.) A treatise on dinitrogen fixation, Section IV: Agronomy and ecology. John Wiley and Sons, New York.

Nelson, R.S., S.A. Ryan, and J.E. Harper. 1982. Chlorate susceptibility and electron donor specificity of a soybean mutant with decreased leaf nitrate reductase activity. Plant Physiol. Suppl. 69:641.

----, ----, and ----. 1983. Soybean mutants lacking constitutive nitrate reductase activity. I. Selection and initial plant characterization. Plant Physiol. 72:503–509.

----, L. Streit, and J.E. Harper. 1984. Biochemical characterization and nitrate and nitrite reductase in the wild-type and a nitrate reductase mutant of soybean. Physiol. Plant. 61:384–390.

----, ----, and ----. 1986. Nitrate reductases from wild-type and nr_1-mutant soybean [*Glycine max* (L.) Merr.] leaves: II. Partial activity, inhibitor, and complementation analyses. Plant Physiol. 80:72–76.

Newcomb, W., Sippell, and R.L. Peterson. 1979. The early morphogenesis of *Glycine max* and *Pisum sativum* root nodules. Can. J. Bot. 57:2603–2616.

----, K. Syono, and J.G. Torrey. 1977. Development of an ineffective pea root nodule: Morphogenesis, fine structure, and cytokinin biosynthesis. Can. J. Bot. 55:1891–1907.

Neyra, C.A., and B.D. Stephens. 1985. Interactions between nitrate reduction and nitrogen fixation in grain legumes. p. 12–22. *In* J.E. Harper et al. (ed.) Exploitation of physiological and genetic variability to enhance crop productivity. American Society of Plant Physiologists, Rockville, MD.

Nicholas, J.C., and J.E. Harper. 1985. Nitrate uptake and reduction by soybean seedlings during dark periods. Plant Physiol. 77:365–369.

----, ----, and R.H. Hageman. 1976a. Nitrate reductase activity in soybeans (*Glycine max* [L.] Merr.). I. Effects of light and temperature. Plant Physiol. 58:731–735.

----, ----, and ----. 1976b. Nitrate reductase activity in soybeans (*Glycine max* [L.] Merr.). II. Energy limitations. Plant Physiol. 58:736–739.

Nutman, P.S. 1956. The influence of the legume in root-nodule symbiosis. A comparative study of host determinants and functions. Biol. Rev. 31:109–151.

Oaks, A. 1979. Nitrate reductase in roots and its regulation. p. 217–226. *In* E.J. Hewitt and C.V. Cutting (ed.) Nitrogen assimilation of plants. Academic Press, New York.

Obaton, M., M. Miquel, P. Robin, G. Conejero, A.-M. Domenach, and R. Bardin. 1982. Influence du déficit hydrique sur l'activité nitrate réductase et nitrogenase chez le soja (*Glycine max* L. Merr. cv. Hodgson). C.R. Acad. Sci. 294:1007–1012.

O'Gara, F., and K.T. Shanmugam. 1976. Regulation of nitrogen fixation by *Rhizobia*: Export of fixed N as NH_4^+. Biochim. Biophys. Acta 437:313–321.

Oghoghorie, C.G.O., and J.S. Pate. 1971. The nitrate stress syndrome of the nodulated field pea (*Pisum arvense* L.). p. 185–202. *In* T.A. Lie and E.G. Mulder (ed.) Biological nitrogen fixation in natural and agricultural habitats. Nijhoff, The Hague.

Ohyama, T. 1983. Comparative studies on the distribution of nitrogen in soybean plants supplied with N_2 and NO_3^- at the pod filling stage. Soil Sci. Plant Nutr. 29:133–145.

----, and S. Kawai. 1983. Nitrogen assimilation and transport in soybean leaves: Investigation by petiole girdling treatment. Soil Sci. Plant Nutr. 29:227–231.

----, and K. Kumazawa. 1978. Incorporation of ^{15}N into various nitrogenous compounds in intact soybean nodules after exposure to $^{15}N_2$ gas. Soil Sci. Plant Nutr. 24:525–533.

----, and ----. 1979. Assimilation and transport of nitrogenous compounds originated from $^{15}N_2$ fixation and $^{15}NO_3$ absorption. Soil Sci. Plant Nutr. 25:9–19.

----, and ----. 1980a. Nitrogen assimilation in soybean nodules. I. The role of GS/GOGAT system in the assimilation of ammonia produced by N_2-fixation. Soil Sci. Plant Nutr. Jpn. 26:109–115.

----, and ----. 1980b. Nitrogen assimilation in soybean nodules. III. Effects of rhizosphere pO_2 on the assimilation of $^{15}N_2$ in nodules attached to intact plants. Soil Sci. Plant Nutr. 26:321–324.

Olson, R.A., and L.T. Kurtz. 1982. Crop nitrogen requirements, utilization, and fertilization. *In* Nitrogen in agricultural soils. Agronomy 22:567–604.

Orihuel-Iranzo, B., and W.H. Campbell. 1980. Development of NAD(P)H: and NADH:nitrate reductase activities in soybean cotyledons. Plant Physiol. 65:595–599.

Pankhurst, C.E., and J.I. Sprent. 1975. Effects of water stress on the respiratory and nitrogen-fixing activity of soybean root nodules. J. Exp. Bot. 26:287–304.

Pate, J.S. 1980. Transport and partitioning of nitrogenous solutes. Annu. Rev. Plant Physiol. 31:313–340.

Patterson, T.G., and T.A. LaRue. 1983a. N_2 fixation (C_2H_2) and ureide content of soybeans: Environmental effects and source-sink manipulations. Crop Sci. 23:819–824.

----, and ----. 1983b. N_2 fixation (C_2H_2) and ureide content of soybeans: Ureides as an index of fixation. Crop Sci. 23:825–831.

Peat, J.R., F.R. Minchin, B. Jeffcoat, and R.J. Summerfield. 1981. Young reproductive structures promote nitrogen fixation in soya bean. Ann. Bot. 48:177–182.

Phillips, D.A. 1980. Efficiency of symbiotic nitrogen fixation in legumes. Annu. Rev. Plant Physiol. 31:29–49.

Polacco, J.C., and E.A. Havir. 1979. Comparisons of soybean urease isolated from seed and tissue culture. J. Biol. Chem. 254:1707–1715.

Pueppke, S.G. 1983. *Rhizobium* infection threads in root hairs of *Glycine max* (L.) Merr., *Glycine soja* Sieb. & Zucc., and *Vigna unguiculata* (L.) Walp. Can. J. Microbiol. 29:69–76.

Pull, S.P., S.G. Pueppke, T. Hymowitz, and J.H. Orf. 1978. Soybean lines lacking the 120,000-dalton seed lectin. Science 200:1277–1279.

Rabie, R.K., Y. Arima, and K. Kumazawa. 1980. Effect of combined nitrogen on the distribution pattern of photosynthetic assimilates in nodulated soybean plant as revealed by ^{14}C. Soil Sci. Plant Nutr. 26:79–86.

----, ----, and ----. 1981. Effect of application time of labeled combined nitrogen on its absorption and assimilation by nodulated soybeans. Soil Sci. Plant Nutr. 27:225–235.

Randall, D.D., W.J. Russell, and D.R. Johnson. 1978. Nodule nitrate reductase as a source of reduced nitrogen in soybean *Glycine max*. Physiol. Plant. 44:325–328.

Reed, A.J., D.T. Canvin, J.H. Sherrard, and R.H. Hageman. 1983. Assimilation of [^{15}N]nitrate and [^{15}N]nitrite in leaves of five plant species under light and dark conditions. Plant Physiol. 71:291–294.

Rennie, R.J. 1982. Quantifying dinitrogen (N_2) fixation in soybeans by ^{15}N isotope dilution: the question of the non-fixing control plant. Can. J. Bot. 60:856–861.

Reporter, M., and N. Hermina. 1975. Acetylene reduction by transfilter suspension cultures of *Rhizobium japonicum*. Biochem. Biophys. Res. Commun. 64:1126–1133.

Reynolds, P.H.S., D.G. Blevins, M.J. Boland, K.R. Schubert, and D.D. Randall. 1982a. Enzymes of ammonia assimilation in legume nodules: A comparison between ureide- and amide-transporting plants. Physiol. Plant. 55:255–260.

----, M.J. Boland, D.G. Blevins, D.D. Randall, and K.R. Schubert. 1982b. Ureide biogenesis in leguminous plants. Trends Biochem. Sci. 7:366–368.

Rigaud, J., F.J. Bergersen, G.L. Turner, and R.M. Daniel. 1973. Nitrate dependent anaerobic acetylene-reduction and nitrogen-fixation by soybean bacteroids. J. Gen. Microbiol. 77:137–144.

----, and A. Puppo. 1977. Effect of nitrite upon leghemoglobin and interaction with nitrogen fixation. Biochim. Biophys. Acta 497:702–706.

Riggle, B.D., W.J. Wiebold, and W.J. Kenworthy. 1984. Effect of photosynthate source-sink manipulation on dinitrogen fixation of male-fertile and male-sterile soybean isolines. Crop Sci. 24:5–8.

Robin, P., L. Streit, W.H. Campbell, and J.E. Harper. 1985. Immunochemical characterization of nitrate reductase forms from wild-type (cv. Williams) and nr, mutant soybean. Plant Physiol. 77:232–236.

Rufty T.W., Jr., D.W. Israel, and R.J. Volk. 1984. Assimilation of $^{15}NO_3^-$ taken up by plants in the light and in the dark. Plant Physiol. 76:769–775.

----, C.D. Raper Jr., and W.A. Jackson. 1982a. Nitrate uptake, root and shoot growth, and ion balance of soybean plants during acclimation to root-zone acidity. Bot. Gaz. 143:5–14.

----, R.J. Volk, P.R. McClure, D.W. Israel, and C.D. Raper, Jr. 1982b. Relative content of NO_3^- and reduced N in xylem exudate as an indicator of root reduction of concurrently absorbed $^{15}NO_3^-$. Plant Physiol. 69:166–170.

Ruschel, A.P., P.B. Vose, R.L. Victoria, and E. Salati. 1979. Comparison of isotope techniques and non-nodulating isolines to study the effect of ammonium fertilization on dinitrogen fixation in soybean, *Glycine max.* Plant Soil 53:513–525.

Ryan, S.A., R.S. Nelson, and J.E. Harper. 1983a. Soybean mutants lacking constitutive nitrate reductase activity. II. Nitrogen assimilation, chlorate resistance, and inheritance. Plant Physiol. 72:510–514.

----, ----, and ----. 1983b. Selection and inheritance of nitrate reductase mutants in soybeans. Soybean Genet. Newsl. 10:33–35.

Ryle, G.J.A., C.E. Powell, and A.J. Gordon. 1979. The respiratory costs of nitrogen fixation in soyabean, cowpea, and white clover. II. Comparisons of the cost of nitrogen fixation and the utilization of combined nitrogen. J. Exp. Bot. 30:145–153.

Schroder, V.N., and K. Hinson. 1975. Soil nitrogen from soybeans (*Glycine max* (L.) Merr.) Soil Crop Sci. Fla. Proc. 34:101–103.

Schubert, K.R. 1981. Enzymes of purine biosynthesis and catabolism in *Glycine max.* I. Comparison of activities with N_2 fixation and composition of xylem exudate during nodule development. Plant Physiol. 68:1115–1122.

----, and H.J. Evans. 1976. Hydrogen evolution: A major factor affecting the efficiency of nitrogen fixation in nodulated symbionts. Proc. Natl. Acad. Sci. U.S.A. 73:1207–1211.

Schweitzer, L.E., and J.E. Harper. 1980. Effect of light, dark, and temperature on root nodule activity (acetylene reduction) of soybeans. Plant Physiol. 65:51–56.

----, and ----. 1985. Leaf nitrate reductase, D-ribulose-1,5-bisphosphate carboxylase, and root nodule development of genetic male-sterile and fertile soybean isolines. Plant Physiol. 78:61–65.

Scowcroft, W.R., and A.H. Gibson. 1975. Nitrogen fixation by *Rhizobium* associated with tobacco and cowpea cell cultures. Nature (London) 253:351–352.

Serres, E. 1982. Variations des uréides glyoxyliques an cours d'un cycle de développment chez le soja (*Glycine max* [L] Merr.). C.R. Acad. Sci. Ser. 3 295:143–146.

Shearer, G., L. Feldman, B.A. Bryan, J.L. Skeeters, D.H. Kohl, N. Amarger, F. Mariotti, and A. Mariotti. 1982. ^{15}N abundance of nodules as an indicator of N metabolism in N_2-fixing plants. Plant Physiol. 70:465–468.

Sheehy, J.E. 1983. Relationships between senescence, photosynthesis, nitrogen fixation and seed filling in soya bean *Glycine max* (L.) Merr. Ann. Bot. 51:679–682.

Shelp, B.J., and R.J. Ireland. 1985. Ureide metabolism in leaves of nitrogen-fixing soybean plants. Plant Physiol. 77:779–783.

Sloger, C., D. Bezdicek, R. Milberg, and N. Boonkerd. 1975. Seasonal and diurnal variations in $N_2(C_2H_2)$-fixing activity in field soybeans. p. 271–284. *In* W.D.P. Stewart (ed.) Nitrogen fixation by free-living micro-organisms. Cambridge University Press, London.

Small, J.G.C., and O.A. Leonard. 1969. Translocation of C^{14}-labeled photosynthate in nodulated legumes as influenced by nitrate nitrogen. Am. J. Bot. 56:187–194.

Smarrelli, Jr. J., and W.H. Campbell. 1981. Immunological approach to structural comparisons of assimilatory nitrate reductases. Plant Physiol. 68:1226–1230.

Smith, J.D. 1949. The concentration and distribution of haemoglobin in the root nodules of leguminous plants. Biochem. J. 44:585–591.

Sprent, J.I. 1971. The effects of water stress on nitrogen-fixing root nodules. I. Effects on the physiology of detached soybean nodules. New Phytol. 70:9–17.

----. 1972. The effects of water stress on nitrogen-fixing root nodules. IV. Effects on whole plants of *Vicia faba* and *Glycine max.* New Phytol. 71:603–611.

----. 1980. Root nodule anatomy, type of export product, and evolutionary origin in some *Leguminosae.* Plant Cell Environ. 3:35–43.

Stacey, G., A.S. Paau, and W.J. Brill. 1980. Host recognition in the *Rhizobium*-soybean symbiosis. Plant Physiol. 66:609–614.

Steele, K.W., P.M. Bonish, R.M. Daniel, and G.W. O'Hara. 1983. Effect of rhizobial strain and host plant on nitrogen isotopic fractionation in legumes. Plant Physiol. 72:1001–1004.

Stephens, B.D., and C.A. Neyra. 1983. Nitrate and nitrite reduction in relation to nitrogenase activity in soybean nodules and *Rhizobium japonicum* bacteroids. Plant Physiol. 71:731–735.

Streeter, J.G. 1972a. Nitrogen nutrition of field-grown soybean plants: I. Seasonal variations in soil nitrogen and nitrogen composition of stem exudate. Agron. J. 64:311–314.

----. 1972b. Nitrogen nutrition of field-grown soybean plants: II. Seasonal variations in nitrate reductase, glutamate dehydrogenase, and nitrogen constituents of plant parts. Agron. J. 64:315–319.

----. 1979. Allantoin and allantoic acid in tissues and stem exudate from field-grown soybean plants. Plant Physiol. 63:478–480.

----. 1981. Effect of nitrate in the rooting medium on carbohydrate composition of soybean nodules. Plant Physiol. 68:840–844.

----. 1982. Synthesis and accumulation of nitrite in soybean nodules supplied with nitrate. Plant Physiol. 69:1429–1434.

----. 1985. Nitrate inhibition of legume nodule growth and activity. I. Long term studies with a continuous supply of nitrate. Plant Physiol. 77:321–324.

----, and P.J. DeVine. 1983. Evaluation of nitrate reductase activity in *Rhizobium japonicum.* Appl. Environ. Microbiol. 46:521–524.

Streit, L., and J.E. Harper. 1985. Biochemical characterization of soybean mutants lacking constitutive NADH-nitrate reductase. Plant Physiol. Suppl. 77:31.

----, R.S. Nelson, and J.E. Harper. 1985. Nitrate reductases from wild-type and nr_1-mutant soybean (*Glycine max* [L.] Merr.) leaves. I. Purification, kinetics, and physical properties. Plant Physiol. 78:80–84.

Stutte, C.A., and R.T. Weiland. 1978. Gaseous nitrogen loss and transpiration of several crop and weed species. Crop Sci. 18:887–889.

----, ----, and A.R. Blem. 1979. Gaseous nitrogen loss from soybean foliage. Agron. J. 71:95–97.

Sutton, W.D., N.M. Jepsen, and B.D. Shaw. 1977. Changes in the number, viability, and amino-acid-incorporating activity of *Rhizobium* bacteroids during lupin nodule development. Plant Physiol. 59:741–744.

Talbott, H.-J., W.J. Kenworthy, and J.O. Legg. 1982. Field comparison of the nitrogen-15 and difference methods of measuring nitrogen fixation. Agron. J. 74:799–804.

Tanner, J.W., and I.C. Anderson. 1964. External effect of combined nitrogen on nodulation. Plant Physiol. 39:1039–1043.

Thibodeau, P.S., and E.G. Jaworski. 1975. Patterns of nitrogen utilization in the soybean. Planta 127:133–147.

Thomas, R.J., and L.E. Schrader. 1981a. Ureide metabolism in higher plants. Phytochemistry 20:361–371.

----, and ----. 1981b. The assimilation of ureides in shoot tissues of soybeans. I. Changes in allantoinase activity and ureide contents of leaves and fruits. Plant Physiol. 67:973–976.

----, K. Jokinen, and L.E. Schrader. 1983. Effect of *Rhizobium japonicum* mutants with enhanced N_2 fixation activity on N transport and photosynthesis of soybeans during vegetative growth. Crop Sci. 23:453–456.

Thornton, H.G. 1936. The action of sodium nitrate upon the infection of lucerne root-hairs by nodule bacteria. Proc. R. Soc. London, B 119:474–492.

Trinchant, J.-C., and J. Rigaud. 1980. Nitrite inhibition of nitrogenase from soybean bacteroids. Arch. Microbiol. 124:49–54.

----, and ----. 1982. Nitrite and nitric oxide as inhibitors of nitrogenase from soybean bacteroids. Appl. Environ. Microbiol. 44:1385–1388.

Tsien, H.C., P.S. Cain, and E.L. Schmidt. 1977. Viability of *Rhizobium* bacteroids. Apl. Environ. Microbiol. 34:854–856.

Uratsu, S.L., H.H. Keyser, D.F. Weber, and S.T. Lim. 1982. Hydrogen uptake (Hup) activity of *Rhizobium japonicum* from major U.S. soybean production areas. Crop Sci. 22:600–602.

Ursino, D.J., D.M. Hunter, R.D. Laing, and J.L.S. Keighley. 1982. Nitrate modification of photosynthesis and photoassimilate export in young nodulated soybean plants. Can. J. Bot. 60:2665–2670.

Van Rensburg, H.J., B.J. Strijdom, and C.J. Otto. 1983. Effective nodulation of soybeans by fast-growing strains of *Rhizobium japonicum*. S. Afr. J. Sci. 79:251–252.

Vasilas, B.L., and G.E. Ham. 1984. Nitrogen fixation in soybeans: An evaluation of measurement techniques. Agron. J. 76:759-764.

Vaughn, K.C., and S.O. Duke. 1981. Histochemical localization of nitrate reductase. Histochemistry 72:191–198.

----, ----, and E.A. Funkhouser. 1984. Immunochemical characterization and localization of nitrate reductase in nonflurazon-treated soybean cotyledons. Physiol. Plant. 62:481–484.

Vest, G., D.F. Weber, and C. Sloger. 1973. Nodulation and nitrogen fixation. *In* B.E. Caldwell (ed.) Soybeans: Improvement, production, and uses. Agronomy 16:353–390.

Wagner, G.H., and H. Broeshart. 1980. Estimates of nitrogen fixation by legumes using various reference crops in the ^{15}N A-value technique. Agron. Abstr. American Society of Agronomy, Madison, WI, p. 162.

Weber, C.R. 1966. Nodulating and nonnodulating soybean isolines: II. Response to applied nitrogen and modified soil conditions. Agron. J. 58:46–49.

Weiland, R.T., and C.A. Stutte. 1979. Pyro-chemiluminescent differentation of oxidized and reduced N forms evolved from plant foliage. Crop Sci. 19:545–547.

Welch, L.F., L.V. Boone, C.G. Chambliss, A.T. Christiansen, D.L. Mulvaney, M.G. Oldham, and J.W. Pendleton. 1973. Soybean yields with direct and residual nitrogen ferilization. Agron. J. 65:547–550.

Wells, G.N., and R.H. Hageman. 1974. Specificity for nicotinamide adenine dinucleotide by nitrate reductase from leaves. Plant Physiol. 54:136–141.

Williams, L.E., and D.A. Phillips. 1983. Increased soybean productivity with a *Rhizobium japonicum* mutant. Crop Sci. 23:246–250.

Wilson, R.F., J.W. Burton, J.A. Buck, and C.A. Brim. 1978. Studies on genetic male-sterile soybeans I. Distribution of plant carbohydrate and nitrogen during development. Plant Physiol. 61:838–841.

Wych, R.D., and D.W. Rains. 1978. Simultaneous measurement of nitrogen fixation estimated by acetylene-ethylene assay and nitrate absorption by soybeans. Plant Physiol. 62:443–448.

----, and ----. 1979. Nitrate absorption and acetylene reduction by soybeans during reproductive development. Physiol. Plant. 47:200–204.

Zengbé, M., and L. Salsac. 1983. Variations des teneurs en uréides et enzymes du catabolisme purique chez *Glycine max*. Physiol. Veg. 21:67–76.

14 Carbon Assimilation and Metabolism

Richard Shibles
Iowa State University
Ames, Iowa

Jacob Secor
Dow Chemical USA
Walnut Creek, California

Duane Merlin Ford
Northeast Missouri State University
Kirksville, Missouri

In this chapter, an attempt is made to collate and interpret information on soybean [*Glycine max* (L.) Merr.] carbon metabolism from all levels of organization—subcellular to field-plant community. Since 1973 when the first edition of the Soybean monograph was published, many advances in knowledge occurred in all phases of this subject. Still, information is seriously lacking about certain aspects that are of singular importance to the understanding of soybean growth and productivity. Thus, whereas a good deal is known about carbon assimilation, the mechanism and regulation of assimilate partition have only recently received attention. The significance and mechanism of sink influences on source activity remain unresolved. And crop respiration is almost virgin territory. These are but three examples of areas currently under intense investigation. This review is intended to codify what is known, as well as point out problems that need research.

14-1 ASSIMILATION

14-1.1 Leaf Carbon Dioxide Assimilation

The apparent photosynthetic rate per unit leaf area (AP) expressed by a leaf is determined by the amount of photosynthetic apparatus per unit leaf area and prevailing environmental conditions. Under optimum temperature and moisture conditions, and with light-saturation, an ap-

Soybeans: Improvement, Production, and Uses, 2nd ed.—Agronomy Monograph no. 16.

parent photosynthetic rate, potential AP, can be attained that reflects the amount of photosynthetic apparatus. Environmental conditions during leaf development (persistent environmental effects), the phase of leaf ontogeny, the plant developmental stage during which the leaf was produced (i.e., nodal position), and genotype all influence the amount of photosynthetic apparatus per unit leaf area. The effects of these factors, as well as that of photorespiration, will be reported in later sections. Initially, the effects of transient environmental conditions are detailed. Transient fluctuations in irradiance, temperature, CO_2 concentration, or moisture do not delimit potential AP, unless extreme and debilitating conditions occur, but they do influence the proportion of the potential that is expressed. Water stress is not included because it is covered in chapter 15.

14-1.1.1 Transient Environmental Effects

At low irradiances, AP of individual leaves is linearly dependent on photosynthetic photon flux density (PPFD). Among nonstressed leaves, the initial slope, or quantum yield, varies little in response to growing conditions (Bowes et al., 1972; Gourdon and Planchon, 1982), nodal position (Kumura, 1969), or genotype (Dornhoff and Shibles, 1974). Variation may occur in response to water or heat stress, or perhaps, excessively high PPFD (Hirata, 1983a, 1983b). The AP saturates under high PPFD. The PPFD required for saturation is determined by the amount of photosynthetic apparatus per unit leaf area.

Leaf orientation influences the irradiance environment experienced by a leaf. Leaves orient primarily in response to the direction of the source. Terminal leaflets orient mostly by raising or lowering their tips; laterals adjust by rotation on an axis through the midvein. The range of angles for terminal leaflets may be from horizontal to 50°, whereas deviations from horizontal of up to 80° may be observed for laterals (Kawashima, 1969a, 1969b). There is genotypic variation for the degree of leaf inclination, and the response evidently varies with growth stage as well. Blad and Baker (1972) found that the leaves of 'Chippewa 64' and 'Hark' were more or less horizontal throughout the growing season. Wofford and Allen (1982) observed that, at growth stage V10 (Fehr and Caviness, 1977), all three cultivars that they evaluated adjusted their leaflet angles for maximum exposure to the sun, but at R3, orientation resulted in nearly minimum exposure. Kawashima (1969b) suggested that leaf orientation does not always operate on the principle of full exposure for individual leaves but, rather, for equal distribution of light on all leaves. On cloudy days, leaves are generally displayed at angles ranging form horizontal to 30°. Leaflet orientation is brought about by osmotic pressure changes within the pulvinus, and the major osmoticum seems to be K (Wofford and Allen, 1982).

The AP is rather insensitive to changes in air temperature between 15 and 30 °C (Fukui et al., 1965). Dornhoff and Shibles (1974) noted no decline until leaf temperature reached 35 °C.

Apparent photosynthesis is linearly dependent on atmospheric CO_2 concentrations up to 300 μL CO_2 L^{-1} air (Dornhoff and Shibles, 1974) or beyond (Brun and Cooper, 1967). The saturating CO_2 concentration depends on PPFD (Brun and Cooper, 1967) and the amount of photosynthetic apparatus per unit leaf area.

14-1.1.2 Photorespiration

All C_3 plants, those that fix CO_2 initially into three carbon sugars via the carboxylation of ribulose 1,5-bisphosphate (RuBP), suffer an O_2-dependent reduction in photosynthesis. For soybean, photosynthesis is inhibited in air (340 μL CO_2 L^{-1} air, 21% O_2) by about 30% relative to the rate in 1 or 2% O_2 (Forrester et al., 1966; Hitz and Stewart, 1980; Creach and Stewart, 1982).

There are two separate but related causes for this inhibition. First, the initial reaction of CO_2 fixation, which is catalyzed by RuBP carboxylase/oxygenase (rubisco) (EC 4.1.1.39), is competitively inhibited by O_2 (the Warburg effect). As its name implies, the enzyme can catalyze either the carboxylation of RuBP—yielding two molecules of 3-phosphoglycerate—or the oxygenation of RuBP—yielding a molecule each of 3-phosphoglycerate and phosphoglycolate (Bowes et al., 1971; Bowes and Ogren, 1972). When CO_2 and O_2 are present, both reactions proceed simultaneously at rates closely dependent on the relative concentrations of CO_2 and O_2. Second, photorespiration, the light-dependent evolution of CO_2, further reduces photosynthate production. Photorespiration is a direct consequence of the oxygenation of RuBP because that reaction produces phosphoglycolate, the subsequent metabolism of which results in the evolution of photorespiratory CO_2. Together these reactions form the photosynthetic carbon oxidation (PCO) pathway, the term we shall use hereinafter to refer to these reactions. (In a sense, photorespiration is a misnomer because it implies, contrary to fact, net conservation of energy in the metabolism of phosphoglycolate.)

Laing et al. (1974) calculated the approximate stoichiometry of the PCO pathway based on the kinetics of isolated soybean rubisco. They suggest that, in air at 25 °C, leaves fix 1 mol of O_2 for every 4 mol of CO_2. As a consequence, 0.5 mol of CO_2 are subsequently evolved. Accordingly, photosynthesis is inhibited by 30% (1.5/5.0) in air at 25 °C relative to the rate in the absence of O_2, which is consistent with the leaf data just mentioned. Of the total inhibition of photosynthesis, then, oxygenation of RuBP accounts for 67%; the evolution of CO_2 accounts for the remainder.

According to the Laing et al. (1974) model, CO_2 evolution in the PCO pathway amounts to about 14% (0.5/3.5) of net CO_2 uptake. Indeed, it has been estimated to average about 3.2 μmol m^{-2} s^{-1} or 15% of net CO_2 uptake by measurements of CO_2 evolution into CO_2-free air by leaves (Dornhoff and Shibles, 1970; Bhagsari et al., 1977) or by extrapolation of leaf CO_2-exchange rate (CER) to zero (Forrester et al., 1966). Such measurements, however, are generally assumed to be underestimates because they do not account for refixation of CO_2.

Evidently, PCO capacity varies with photosynthetic capacity. This follows from the facts that (i) the initial reactions of both processes are catalyzed by the same enzyme and (ii) the relative rates of RuBP carboxylation and oxygenation are only known to depend on relative O_2-CO_2 concentrations (Jordan and Ogren, 1981a), temperature (Laing et al., 1974), relative Mg^{2+}-Mn^{2+} concentrations (Jordan and Ogren, 1981a), and perhaps leaf age (Secor et al., 1982a). Therefore, fully expanded leaves with rapid or slow rates of photosynthesis should have rapid or slow PCO rates, respectively. Indeed, CER in normal atmospheres and in CO_2-free air has been found to be correlated among cultivars (Dornhoff and Shibles, 1970, 1976; Bhagsari et al., 1977).

The bifunctional nature of rubisco and the existence of the PCO pathway limits photosynthesis under normal conditions. Can this inhibition be reduced or eliminated? Some believe that chemical inhibition of the PCO cycle is futile (Servaites and Ogren, 1977; Créach and Stewart, 1982); any phosphoglycolate produced must be obligatorily metabolized through the PCO cycle.

The CO_2-compensation point is a manifestation of the PCO pathway, and Menz et al. (1969) proposed a screening method that would identify plants with low CO_2-compensation points. However, when Cannell et al. (1969) applied the method to 2458 soybean genotypes, representing the known gene pool at that time, no significant differences in CO_2-compensation point were found. Others (Curtis et al., 1969; Dornhoff and Shibles, 1970; Bowes et al., 1972) have found that the CO_2-compensation point, which typically is between 40 and 60 $\mu L\ CO_2\ L^{-1}$ air, does not vary among leaves within or among genotypes.

Because genotypic variation for PCO is associated with variation for photosynthesis, breeding programs designed to retard the former likely would retard photosynthesis as well. Stimulated by their finding that the specificity of rubisco for CO_2 relative to O_2 has increased during evolution (Jordon and Ogren, 1981b), Ogren's group has been using mutagens to improve specificity for CO_2, but without any success thus far (Ogren and Chollet, 1982).

14-1.1.3 Limitation of Photosynthetic Apparatus

The preceding two sections indicate the optimum environmental conditions for leaf CO_2 assimilation in soybean: saturating PPFD, moderate temperatures (20 to 30 °C), and ambient CO_2 and O_2 concentrations (340 $\mu L\ CO_2\ L^{-1}$ air; 21% O_2). Of course, faster rates would be expressed in high CO_2 and/or low O_2 environments, but a leaf is not normally subject to such conditions.

A recently developed model of photosynthesis suggests that, under optimum conditions, leaf apparent photosynthesis is colimited by (i) the amount and activity of rubisco and (ii) the ability of the leaf to regenerate

RuBP (Farquhar et al., 1980; von Caemmerer and Farquhar, 1981). The RuBP is regenerated via the PCO pathway and the photosynthetic carbon reduction (PCR) cycle. These pathways depend upon the supply of ATP and (NADPH + H^+) produced by the photosynthetic electron transport system. In other words, it is suggested that leaf potential AP depends on the capacity of the entire photosynthetic system. The Farquhar et al. (1980) model seems to explain the gas exchange characteristics of bean (*Phaseolus vulgaris* L.) leaves (von Caemmerer and Farquhar, 1981), but there is no direct evidence that it explains soybean leaf AP. Some of the predictions of the model, however, can be checked against available data.

For leaves in air with subambient CO_2, the model predicts that RuBP utilization is low. Therefore, rubisco is saturated with RuBP, and AP depends on the kinetics of the enzyme with respect to CO_2 and O_2. However, in air with supraambient CO_2, RuBP is used rapidly. Therefore, rubisco is not saturated with RuBP, and AP depends on RuBP regeneration. No research reported to date (Hitz and Stewart, 1980; Creach and Stewart, 1982; Vu et al., 1983) has shown that the steady-state pool of RuBP in soybean leaves varies in response to atmospheric CO_2 concentration as predicted by the model.

Although it is not certain that, in low CO_2 environments, RuBP levels in soybean leaves are saturating with respect to rubisco, it has been found that AP depends on the kinetics of rubisco under such conditions. Laing et al. (1974) used the in vitro response of rubisco to temperature, the atmospheric O_2 and CO_2 concentrations, and the carboxylation efficiency of intact leaves in the absence of O_2 to predict the temperature response of AP in subambient CO_2 environments. In others words, the response of AP depended on the response of rubisco to temperature. Seemann and Berry (1982) grew soybean plants under differing irradiances and nutrient regimes to induce AP differences. By use of the kinetic constants of purified rubisco and the amount of rubisco per unit leaf area, they predicted AP at 100 μL CO_2 L^{-1} intercellular air and 2% O_2. Furthermore, they noted that the catalytic efficiency of purified soybean rubisco is only 60% of the efficiency of the spinach (*Spinacea oleracea* L.) enzyme. This difference was reflected in the relative photosynthetic rates per unit leaf protein expressed by the two species in low CO_2, low O_2 environments. These data support the suggestion that, when CO_2 and/or O_2 concentration is subambient, the amount and activity of rubisco determines AP.

Despite the lack of total consistency between the predictions of the Farquhar model and the soybean data with respect to RuBP, the model remains a compelling and useful conceptual approach to the information available about soybean photosynthesis. Other information supports our conclusion that potential AP is defined by the amount of photosynthetic apparatus. In the following sections, we discuss the effect of environmental conditions during leaf growth, the phase of leaf ontogeny, the developmental stage during which the leaf was produced (i.e., nodal position), and genotype on AP. These factors seem to influence AP by determining the amount of photosynthetic apparatus per unit leaf area.

14-1.1.4 Persistent Environmental Effects

Soybean leaves adapt to their growth regime primarily by adjusting leaf dimensions. High irradiance during growth results in thicker leaves (Lugg and Sinclair, 1980; Gourdon and Planchon, 1982) with more photosynthetic apparatus per unit leaf area. The AP increases, but photosynthesis per unit leaf volume does not change, with increasing growth irradiance. Also, light saturation occurs at PPFDs similar to those received during growth (Bowes et al., 1972). Soybean leaves seem to alter the amount of photosynthetic material beneath each unit of leaf area so as to utilize the maximum available energy. Therefore, experimental results from low irradiance environments (growth cabinets or glasshouses) cannot be extrapolated to field performance.

Gourdon and Planchon (1982) found that AP, as measured at 22 °C, decreased with increasing growth temperature (20 to 30 °C). Fukui et al, (1965), however, found no clear trend. Specific leaf mass (mass per unit leaf area) is relatively greater at very low (ca. 18 °C) and very high (ca. 36 °C) growth temperatures, but varies little at intermediate temperatures (Hofstra and Hesketh, 1975). Interestingly, Gourdon and Planchon confirmed this effect, but also found that mesophyll resistance varied most among leaves grown at the intermediate temperatures. (*Mesophyll resistance*, as used herein, encompasses intracellular transfer resistances plus limitations imposed by carboxylation and other biochemical reactions. It is the residual obtained after subtraction of stomatal and laminar resistances from total resistance to CO_2 flux and, thus, corresponds to the original meaning of mesophyll resistance as defined by Gaastra [1962]). A clear understanding of the effects of growth temperature on AP is lacking.

The response of soybean, and other species, to elevated CO_2 has received much attention. It is clear that overall growth is enhanced by exposure to supraambient CO_2. This response is related to increases in leaf area (Cooper and Brun, 1967; Mauney et al., 1978), specific leaf mass (Cooper and Brun, 1967; Hofstra and Hesketh, 1975), and net assimilation rate (Cooper and Brun, 1967; Imai and Murata, 1979; Clough and Peet, 1981; Sionit et al., 1982). Leaves grown in elevated CO_2 concentrations have greater AP than leaves grown under ambient conditions, *if* AP is measured using high and ambient CO_2 concentrations, respectively (Mauney et al., 1978; Clough et al., 1981). However, if the leaves from each environment are measured at a single CO_2 concentration, AP for those grown under ambient conditions is greater (Hofstra and Hesketh, 1975; Clough et al., 1981), perhaps because increased starch levels under high CO_2 (Hofstra and Hesketh, 1975; Clough and Peet, 1981; Clough et al., 1981) produce a feedback inhibition of photosynthesis. The relationship between starch content and AP is equivocal, however, and will be discussed later (see section 14-1.5). It suffices to state that leaves grown in supraambient CO_2 levels have more starch and usually, but not always (cf. Mauney et al., 1979), slower photosynthetic rates at a given CO_2 concentration.

Ojima et al. (1965) applied several different fertilizer treatments to a nonnodulating soybean line. Leaves on plants that received N fertilizer expressed greater AP than leaves on plants that received no N. There was no response to P or K fertilization. The AP was positively correlated with percentage (by mass) of leaf N. Interestingly, Sionit (1983) showed that elevated atmospheric CO_2 concentrations can partly overcome nutrient deficiencies inasmuch as more biomass was produced by plants in 550 μL CO_2 L^{-1} air with one-eighth strength Hoagland's solution than by plants in 350 μL L^{-1} with one-half-strength Hoagland's.

14-1.1.5 Leaf Ontogeny and Nodal Position

Each leaf on a soybean plant follows a similar ontogenetic pattern with respect to CO_2 assimilation (Fig. 14-1). During leaf expansion, AP increases. Because both AP and leaf area growth are linear functions of time, during rapid leaf expansion, the total CO_2-fixing ability of the leaf

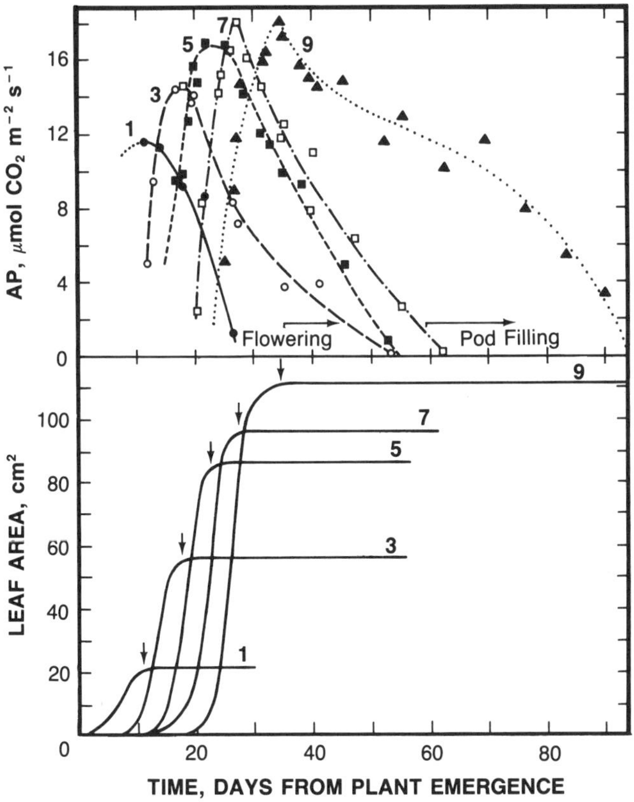

Fig. 14-1. Apparent photosynthetic rate (AP) and area of leaves 1, 3, 5, 7, and 9 on the main stem of 'Lee' grown in a controlled environment cabinet [30/25 °C day/night temperature, PPFD = 590 $\mu mol\ m^{-2}s^{-1}$]. Arrows indicate time when maximum AP was attained. Adapted from Woodward (1976).

increases exponentially. Greatest photosynthetic capacity is attained at about the same time that the leaf reaches full expansion. Maximum AP is maintained for a few to several days before beginning to decline gradually. This pattern—expansion, most active, senescence—is invariate among leaves, though differences occur in the duration of each phase, the maximum AP expressed, and the rate of decline. Figure 14–2 shows the life span of each leaf on plants of two cultivars.

14–1.1.5.1 Leaf Expansion Phase—Individual leaves normally expand rapidly (Fig. 14–1). Expansion rates of 15 to 30 cm^2 day^{-1} are not uncommon for leaves formed during the late vegetative or early reproductive stages. Earlier- and later-formed leaves expand more slowly (Kumura and Naniwa, 1965; Kumura, 1969; Woodward, 1976; Lugg and Sinclair, 1980). The rate of expansion is also closely dependent on moisture regime (Bunce, 1977). The rate at which AP increases during expansion is quite variable (Woodward, 1976).

With expansion, the amounts of chlorophyll and soluble protein per unit leaf area have been found to increase steadily (Silvius et al., 1978) or decrease slightly (Secor et al., 1984). Secor et al. (1982a, 1984) observed that the carboxylase activity of rubisco per unit leaf area increased during leaf expansion. And Woodward and Rawson (1976) have shown that both stomatal and mesophyll resistances decline as the leaf expands. The decrease in mesophyll resistance probably is related to the increases in rubisco amount and/or activity.

In sum, the photosynthetic capacity of a leaf increases during leaf expansion. The AP during this phase is dependent primarily on the rate at which photosynthetic material accumulates per unit leaf area, as determined by environment and genotype. Because this phase is so short and the leaves, for the most part, are small, the contribution of expanding leaves to canopy AP is small.

14–1.1.5.2 Most Active Phase—As previously mentioned, maximum AP is achieved at about the time of full expansion (Woodward, 1976; Lugg and Sinclair, 1981). Figure 14–3 shows the typical variation in leaf area and petiole length associated with nodal position. Among genotypes,

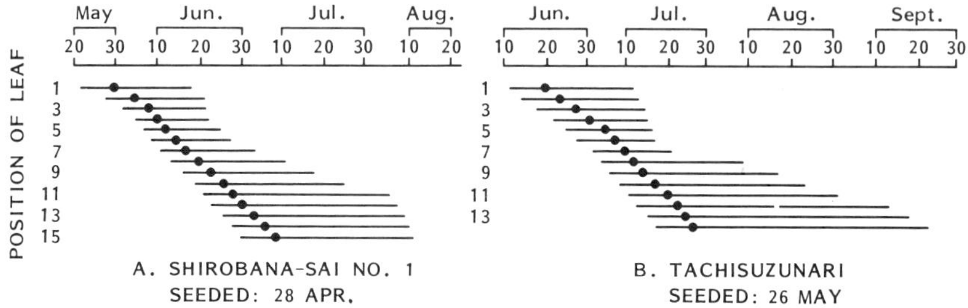

Fig. 14–2. Duration of each leaf on the main stem of two cvs. Shirobana-Sai no. 1 and Tachisuzunari seeded 28 April and 26 May, respectively. ● = date of full leaf expansion. From Kumura (1969).

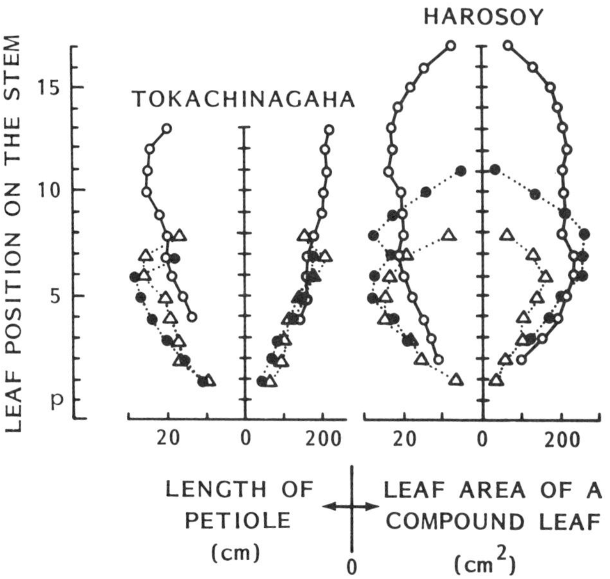

Fig. 14–3. Leaf areas and petiole lengths for a determinant 'Tokachinagaha' and an indeterminant 'Harosoy' type plant. Leaves on the main stem (○——○), on the branch from the primary node (●-----●), and on the branch from the first node (△-----△). P indicates the primary node. From Nakeseko et al. (1979).

a range of 120 to 400 cm^2 has been reported for area (one surface) of recently expanded upper leaves (Hesketh et al., 1981). Leaves reach full expansion and maximum AP in acropetal succession (Fig. 14–1).

Lower leaves generally have slower AP during their most active period than do middle or upper leaves (Fig. 14–1; Kumura, 1969; Gordon et al., 1982). For indeterminant types, leaves produced during vegetative growth have APs ranging from 19 to 25 μmol m^{-2} s^{-1}, but leaves produced during flowering and pod growth have rates ranging from 30 to 40 μmol m^{-2}s^{-1} (Dornhoff and Shibles, 1970; Gordon et al., 1982). Although Lugg and Sinclair (1981) found no greater AP for leaves developed during reproduction, they and Gordon et al. (1982) observed that such leaves retained their maximum activity for longer periods than lower leaves. Moreover, the most active phase was longer for 'Corsoy' upper leaves than for Chippewa 64 upper leaves.

The greater AP of later-formed leaves is associated with leaf thickening and increased specific leaf mass and, therefore, with more photosynthetic material per unit area. Leaf thickness and specific leaf mass of recently expanded leaves increase acropetally (Dornhoff and Shibles, 1970; Lugg and Sinclair, 1980, 1981). Lugg and Sinclair (1980) found that upper leaves of Corsoy and Chippewa 64 developed a third palisade mesophyll cell layer by periclinal divisions of the uppermost palisade cells. The per unit leaf area amounts of chlorophyll, soluble protein and rubisco, and the carboxylase activity of rubisco, peak during the most active phase of a leaf, though the timing of the peak with respect to attainment of maximum AP is somewhat variable (Wittenbach et al., 1980; Wittenbach, 1982, 1983a; Secor et al., 1983, 1984). The peak amounts/activities of these photosynthetic components increase acropetally in the same fashion as specific leaf mass, and leaf thickening.

14–1.1.5.3 Senescence Phase—Senescence is characterized by steadily declining AP. Lower leaves senesce in the same order in which they develop (Kumura and Naniwa, 1965), and this progressive senescence is a response to self-shading as well as leaf age (Johnston et al., 1969; Beuerlein and Pendleton, 1971). Upper leaves cease their functional lives at about the same time as a consequence of plant (monocarpic) senescence.

The rate at which AP declines during senescence is unpredictable. Gordon et al. (1982) found that, for eight genotypes, the rate of decline ranged from 0.6 to 4.1% per day and was proportional to the maximum AP expressed. Generally, upper leaves, which maintain their maximum capacity longer, senesce more rapidly than lower leaves (Secor et al., 1984). Cool nights (Sinclair, 1980) and other stresses hasten senescence.

Chlorophyll, soluble protein, rubisco activity, total N, specific leaf mass, and starch all decline more or less coincidently with AP decline during leaf senescence (Okatan et al., 1981; Mondal et al., 1978a; Wittenbach et al., 1980; Secor et al., 1983, 1984). Visible yellowing and eventually leaf abscission follow these declines. The onset of decline in AP seems more closely coupled with initial losses of rubisco and chlorophyll than with other leaf parameters (Secor et al., 1984). Nodal position or type of senescence (progressive vs. monocarpic) makes little difference in the pattern of declines (Secor et al., 1983, 1984).

Wittenbach et al. (1980) found that osmiophilic granules began to enlarge just before the onset of senescence and continued to enlarge during the remainder of the leaf's functional life. At first, there was an increase in the number of mitochondria and a later decrease. Chloroplasts declined in number, and the membranes of the envelope and stroma lamellae of the remaining chloroplasts appeared to degenerate. Vacuolation of the cells increased. Ford (1984), however, observed no decline in chloroplast number during early leaf senescence, the period over which $^{14}CO_2$ uptake declined by 50%. Decline in photosynthesis was associated with loss of chloroplast activity.

Late in the functional life of leaves on 'Fiskeby V', Peet and Kramer (1980) found that the percentage enhancement of photosynthesis by 2% O_2 increased. Also, the ratio of rubisco carboxylase to oxygenase activities (measured simultaneously) decreased 38% during the latter part of senescence (Secor et al., 1982a). These data suggest that part of the senescence decline in AP may be caused by an increase in PCO cycle function. Just before leaf yellowing, stomatal resistance increases markedly (Wittenbach et al., 1980; Secor et al., 1983, 1984), which may be related to an increase in abscisic acid (ABA) (Oritani and Yoshida, 1973; Samet and Sinclair, 1980). Lindoo and Noodén (1978) showed that cytokinin

Table 14–1. Genotypic variation in leaf photosynthesis. Adapted from Elmore (1980).

Source	Range min.	Range max.	Measure†	n‡	Leaf§	Environment¶
	$\mu mol\ m^{-2}\ s^{-1}$					
Ojima and Kawashima (1968)	13.6	19.5	AP	39	4,6	–
Curtis et al. (1969)	7.6	15.1	AP	39	1	GC
Dreger et al. (1969)	11.8	14.5	AP	9	1–5	GC
Dornhoff and Shibles (1970)	18.3	27.1	AP	20	REL	P
Watanabe (1973a)	15.8	20.8	AP	5	U	P
Watanabe (1973b)	12.6	19.6	AP	6	REL	P
Dornhoff and Shibles (1976)	21.2	27.5	AP	6	REL	P
Kaplan and Koller (1977)	16.4	23.3	AP	16	REL	GC
Buttery and Buzzell (1977)	18.1	23.6	TPs	48	REL	F
Bhagsari et al. (1977)	16.4	21.5	AP	16	REL	GH
Wiebold et al. (1981)	10.5	27.5	AP	117	REL	F
Hesketh et al. (1981)	22.7	46.1	AP	29	REL	F
Secor et al. (1982b)	18.3	28.2	TPs	110	REL	F

†AP = Apparent photosynthesis, TPs = Total photosynthesis.
‡Number of genotypes tested.
§Trifoliolate leaf assayed counting acropetally, U = Unifoliolate, REL = Recently expanded leaves.
¶GC = Growth cabinets, P = Pot culture outdoors, GH = Glasshouse, F = Field.

activity decreased and ABA activity increased in extracts of shoot tissue before yellowing. But Ciha et al. (1978) did not find similar increases in ABA.

Genotypes with delayed leaf senescence have been identified (Abu-Shakra et al., 1978). Although not measured, delayed AP decline was inferred because chlorophyll, rubisco, and N_2-fixing potential were retained longer than for normal types. Mature pods were observed on nodes still bearing green leaves. However, further work has shown that this response is unpredictable and strongly dependent upon environment.

14–1.1.6 Genotype

Besides the prevailing environmental conditions, the growth environment, leaf ontogeny, and plant developmental stage, genotype determines light-saturated photosynthetic rate (Table 14–1). There always remains some question as to whether such differences are dependent solely on genotype. In an experiment with many cultivars or experimental lines, it is difficult to hold the effects of leaf ontogeny, plant developmental stage, and environment constant for all lines. If cultivars from diverse maturity groups are used, the problem of different plant developmental stages becomes particularly trying. Where leaves of similar ontogeny on plants of similar seasonal duration have been compared, however, genotypic differences for AP have been found.

Many attempts have been made to associate leaf photosynthetic rates with various other leaf parameters among genotypes. The motivation for this work has been twofold. First, a selection criterion more easily measured than photosynthesis, per se, has been sought. However, no broadly useful trait yet has been identified. Second, a study of the interrelationships of various leaf traits may provide some insight about control of leaf photosynthesis.

A most interesting fact about the various leaf parameters is that they are all, more or less, interrelated. Watanabe (1973a, 1973b) showed that AP is positively correlated with amount of chlorophyll, number of chloroplasts, and rate of electron transport (where each is expressed per unit leaf area) of primary leaves. Furthermore, in vegetative stage trifoliolate leaves, he found a consistent balancing of photochemical activity with dark reaction activity among genotypes. Hesketh et al. (1981) found that AP was positively correlated with specific leaf mass and the per unit leaf area amounts of chlorophyll, soluble protein, and rubisco activity among 29 cultivars. It may be inferred that the latter leaf parameters are all interrelated.

In a population of 110 experimental lines, significant correlations of soluble protein or chlorophyll per unit leaf area with $^{14}CO_2$ uptake were reported by Secor (1979), though the correlation coefficients were low (ca. 0.30). The former pair of leaf traits were only inconsistently associated. Similarly, for the recently expanded vegetative leaves of 12 cultivars, percentage N was not, but chlorophyll was, significantly correlated with $^{14}CO_2$ uptake, and during reproductive growth, percentage leaf N, but not chlorophyll, was correlated with $^{14}CO_2$ uptake (Buttery et al., 1981). Earlier, Buttery and Buzzell (1977), working with 48 cultivars at flowering, found a highly significant correlation of $^{14}CO_2$ uptake with chlorophyll per unit leaf area.

Specific leaf mass has been shown to be correlated with chlorophyll, but not percentage leaf N (Buttery et al., 1981), and with soluble protein but not chlorophyll (Secor, 1979). Though specific leaf mass is widely noted for its association with photosynthesis (Dornhoff and Shibles, 1970, 1976; Hesketh et al., 1981; Wiebold et al., 1981; Buttery et al., 1981), in other experiments, the correlation has been nonexistent (Watanabe and Tabuchi, 1973), very poor (Secor et al., 1982b), or inconsistent (Ojima and Kawashima, 1968; Bhagsari et al., 1977; Kaplan and Koller, 1977). Wiebold et al. (1981) established that genotypic differences in specific leaf mass are associated with changes in leaf thickness and leaf density. This, plus the negative correlation of specific leaf mass with leaf area reported by Hesketh et al. (1981), indicates that leaves thicken at the expense of growth in area.

In short, other leaf parameters often are correlated with AP as well as with one another. Generally, genotypes with small, thick leaves tend to have more chlorophyll, protein, rubisco activity, and photochemical activity per unit leaf area than do those with large, thin leaves. The former also tend to express more rapid photosynthetic rates. As Hesketh et al. (1981) suggested, photosynthetic material may be diluted in large, compared with small, leaves.

Inconsistencies in correlations among traits may be ascribed to several factors. During leaf ontogeny, the various other leaf traits and quantities generally follow an ontogenetic pattern similar to that of AP, but there is some divergence. For example, AP may begin its senescence decline several weeks before soluble protein begins to decline. Also, maximum AP and maximum chlorophyll, protein, etc., per unit area do not always occur at the same time (see the section on leaf ontogeny). This would affect the correlation of AP with any of the latter.

Another reason for inconsistent correlations is that AP, as measured, may not reflect the total capacity of the system. Stomatal resistance generally explains some of the variation in AP among cultivars (Dornhoff and Shibles, 1970, 1976; Bhagsari et al., 1977; Kaplan and Koller, 1977).

In summary, genotypic differences in AP exist. This variation seems related to differences in leaf dimensions and the concentration of photosynthetic components, but correlations are inconsistent and often poor.

14–1.1.7 Diurnal Course of Leaf AP

As a summary of this section, it might be useful to reflect on the diurnal course of AP of a nonstressed soybean leaf. Early and late in the day, AP is directly related to incident PPFD. Over the midday hours on clear days, PPFDs are adequate to light-saturate leaves that are exposed to the sun. Under these conditions, leaf AP reaches a plateau during midday. The plateau rate varies among leaves, depending on leaf ontogeny, nodal position, environment during leaf expansion, and genotype. Small thick leaves with relatively more cholorophyll, soluble protein, rubisco, dry mass, etc., beneath each unit of area usually have more rapid AP during the midday hours than have large, thin leaves. Because of environmental and ontogenetic variations, maximum AP of a given leaf, as measured on one day, may (Buttery et al., 1981; Secor et al., 1982b) or may not (Dornhoff and Shibles, 1970; Watanabe and Tabuchi, 1973; Bhagsari et al., 1977; Ford et al., 1983) predict the maximum rate for other leaves of the same genotype on another day.

14–1.2 Pod and Stem Carbon Dioxide Assimilation

Though pods can fix atmospheric CO_2, as well as recycle CO_2 respired during seed growth, their CER is rarely, if ever, positive. Net CO_2 evolution is the rule (Quebedeaux and Chollet, 1975; Andrews and Svec, 1975; Sambo et al., 1977; Spaeth and Sinclair, 1983a). However, that 50 to 70% of the daily CO_2 respired by seeds may be refixed by the carpels (Sambo et al., 1977) is of great importance to plant carbon balance.

Pod CER responds to increasing irradiance much as leaf AP does, being linearly dependent on PPFD at low irradiances and saturating at PPFDs approaching those of full sun. But pod CER for field-grown plants is more closely associated with pod temperature than PPFD (Spaeth and Sinclair, 1983a, 1983b). This probably reflects the relative importance of respiration vs. photosynthesis to overall pod C balance.

Under constant conditions of temperature, PPFD, and CO_2 concentration, relative changes in photosynthetic potential during pod ontogeny can be observed. Quebedeaux and Chollet (1975) showed that pod respiration increased during early ontogeny, CER decreasing from -6.3 μmol of CO_2 h^{-1} g^{-1} fresh weight at anthesis to -14.1 μmol of CO_2 h^{-1} g^{-1} at the beginning of rapid seed growth. During this same period, gross photosynthesis increased from 5.8 to 11.0 μmol CO_2 h^{-1} g^{-1}. After the onset of rapid seed growth, gross photosynthesis and respiration declined at nearly the same rate, so net pod CER in the light remained constant at about -4.7 μmol h^{-1} g^{-1} until maturity. On the other hand, Sambo et al. (1977) observed that pod CER decreased continually with pod age. Stems contribute little if any to net CO_2 uptake in soybean (Kumura and Naniwa, 1965; Quebedeaux and Chollet, 1975).

14.1.3 Canopy Carbon Dioxide Assimilation

Canopy apparent photosynthesis per unit of land area (CAP) is primarily influenced by PPFD, leaf area index (LAI), prevailing environmental conditions, plant ontogeny, and genotype. The light compensation point for soybean canopies is about 200 μmol photons (400 to 700 nm) m^{-2} s^{-1}, though it increases for maturing canopies (Larson et al., 1981;

Baldocchi et al., 1981). At LAIs < 4.0, canopies light saturate (Jeffers and Shibles, 1969). At greater LAIs, light saturation does not occur in nonstressed canopies (Egli et al., 1970; Baldocchi et al., 1981, 1983; Larson et al., 1981; Christy and Porter, 1982), although others have observed that CAP does not change much over the midday hours (Sakamoto and Shaw, 1967; Kanemasu and Hiebsch, 1975; Schulze et al., 1976; Wells et al., 1982). Larson et al. (1981) have shown that late morning or early afternoon water deficits may cause apparent light saturation. These deficits, and the resultant decline in CAP, can occur even in irrigated plots, though to a lesser extent than in unirrigated plots, and may (Baldocchi et al., 1983) or may not (Larson et al., 1981) be associated with increased stomatal resistance.

Other indirect evidence indicates that fully developed, nonstressed canopies do not light saturate. Ninety-five percent of incident solar radiation is intercepted when the LAI is only 3 to 5 (Shibles and Weber, 1965, 1966; Heilman et al., 1977; Nakaseko and Gotoh, 1981). Yet LAIs of 6 to 7 are not uncommon. So most of the incident radiation is intercepted by the periphery of the canopy (Shaw and Weber, 1967; Hatfield and Carlson, 1978), though the normally shaded, lower leaves have considerable photosynthetic capacity when exposed to high light (Johnston et al., 1969). Furthermore, Kumura (1968) showed that, even under high irradiance, CAP varies with the proportion of diffuse light. Because canopies with LAIs > 4.0 would seem able to utilize additional light, saturation is not indicated.

The CAP seems insensitive to temperatures between 25 and 32 °C (Jeffers and Shibles, 1969; Hansen, 1972; Baldocchi et al., 1981). On hot, clear days, extreme temperatures can suppress CAP during the midday hours (another reason for apparent light saturation). The slow CAP rates observed on cool, overcast days may reflect both reduced PPFDs and suboptimum temperatures (Baldocchi et al., 1981).

Depending upon developmental stage, plant and row spacing, and PPFD, the critical LAI required for maximum CAP ranges between three and six (Hansen, 1972). When CAP is measured as a function of progressive canopy development (i.e., on successive dates during plant growth), it seems to be a linear function of LAI, even to LAIs of 6 to 8 (Jeffers and Shibles, 1969; Larson et al., 1981). However, this effect likely can be attributed to the coincidences in other seasonal trends; e.g., increasing PPFD and leaves at upper nodes having greater photosynthetic capacity, leading to higher canopy potential photosynthesis.

Figure 14–4 shows the ontogenetic trend in soybean CAP. During vegetative growth, daily maximum CAP increases rapidly as the canopy develops and PPFD increases. Denser populations develop leaf area more quickly and tend to have more rapid CAP and greater crop growth rates (CGR) before canopy closure than do sparser populations (Ojima and Fukui, 1966; Shibles and Weber, 1966; Buttery, 1969; Nakaseko and Gotoh, 1981; Christy and Porter, 1982). The effect of genotype during this period is not clear. However, both crop growth rate (CGR) (Hanway

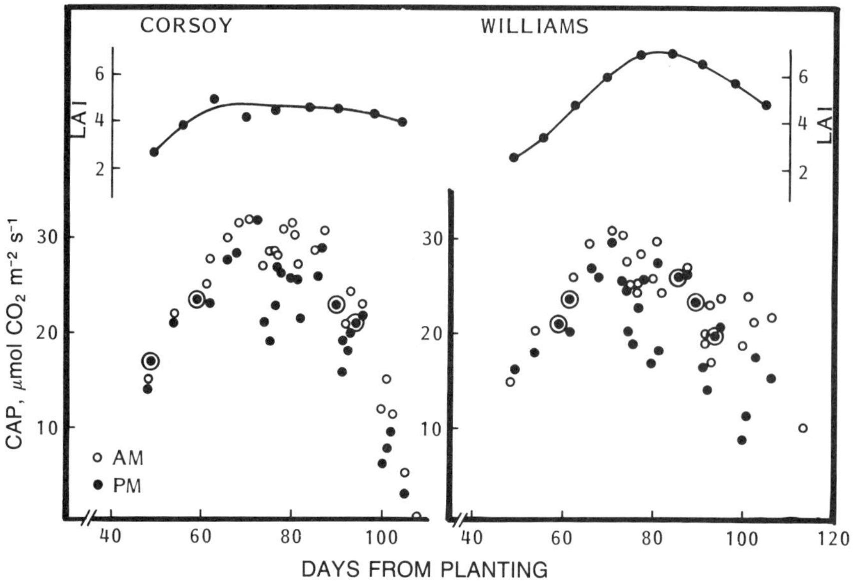

Fig. 14–4. Seasonal canopy apparent photosynthesis (CAP) for Corsoy and Williams soybean cultivars. Data are normalized for PPFD (1600 μmol $m^{-2}s^{-1}$. ○ and ● indicte AM and PM values, respectively. From Larson et al. (1981)

and Weber, 1971; Kaplan and Koller, 1977) and leaf expansion rate (Kaplan and Koller, 1977) seem to differ among cultivars, suggesting that there are also CAP differences.

Seasonal maximum CAP occurs during the late vegetative or early reproductive period. The timing likely depends on when the canopy closes, and delays will occur for sparse populations or slowly developing genotypes. The seasonal maximum CAP, or a slightly slower rate, will be maintained until the commencement of seed filling. The CAP rates during this period are commonly between 22 (Kanemasu and Hiebsch, 1975; Christy and Porter, 1982) and 38 μmol m^{-2} s^{-1} (Sakamoto and Shaw, 1967; Hansen, 1972). Except for extremely sparse populations, CAP differences among different plant densities disappear during this period if the critical LAI has been attained and canopies have closed (Christy and Porter, 1982).

When rapid seed filling commences (ca. R5.5), daily maximum CAP begins a steady decline, reaching zero at physiologic maturity. The decline seems independent of decreases in LAI and likely reflects reduced photosynthetic competence of the leaves due to senescence and the lower PPFD late in the season. The rate of decline depends principally on maturity time of the genotype. At the same location, earlier maturing cultivars begin senescing sooner and senesce faster than later maturing ones (Wells et al., 1982). If CAP is measured on the same day for several genotypes, each in a different stage of development, genotypic differences are confounded with plant ontogeny effects. Developmental variation may have been responsible for the CAP differences among genotypes observed by Egli et al. (1970) and Schulze et al. (1976). However, significant CAP differences among genotypes of similar maturity have been found during seed filling (Wells et al., 1982; Hole, 1983).

Others also have observed genotypic variation for mean CAP of fully developed canopies under full sunlight. Jeffers and Shibles (1969) and Hansen (1972) noted such differences from anthesis to late senescence. Seasonal mean CAP rates were shown to differ among 13 cultivars by Larson et al. (1981). However, the differences were not large inasmuch as the slowest rate was 87% of the most rapid.

In sum, CAP increases as PPFD increases until about the middle of the morning. Whether the diurnal course of CAP continues to follow the course of PPFD throughout the midday hours depends on the LAI and the prevailing environmental conditions. It seems that genotype mainly influences seasonal mean CAP early and late in the growing season. The major early effect is on leaf area development, whereas differences in rate of senescence account for most of the later effect.

14–1.4 Breeding for Enhanced Photosynthesis

Photosynthetic rate is a quantitatively inherited trait, and heritability of either leaf AP or CAP, calculated from variance components, is about 50% (Butter and Buzzell, 1972; Harrison et al., 1981; Wiebold et al., 1981). Ojima and Kawashima (1970) found leaf AP of F_3 lines to be correlated with that of F_2 plants (0.45 and 0.66) in two crosses. The mean leaf AP of an F_1 population has been reported to be lower than (Ojima et al., 1969) or the same as (Wiebold et al., 1981) the midparent rate, but dominance for rapid photosynthesis has never been reported for soybean. Lines with either rapid or slow leaf AP or CAP can be bred (Secor et al., 1982b; Harrison et al., 1981), but the task is arduous, and progress is slow, mainly because photosynthesis is strongly influenced by environment and ontogeny and is internally modulated by sink metabolism.

Calculations of heritability and predicted gain indicate that early generation selection for leaf AP should be successful, but results have been mixed. Ojima and Kawashima (1970) were able to select F_3 lines with rapid or slow AP from F_2 plants with rapid or slow AP. Wiebold et al. (1981), however, found actual gain from F_2 or F_3 selection to be nil: lines selected for rapid and slow AP failed to maintain their relative rates in the next generation. Perhaps this disagreement can be explained by the nature of the crosses. Ojima and Kawashima (1970) used wide crosses (Asian by North American cultivars), whereas Wiebold et al. (1981) crossed North American cultivars with differing leaf AP.

Using 110, F_6-derived progeny from one of the same crosses used by Wiebold et al. (1981), Secor et al. (1982b) selected two groups of lines on the basis of leaf $^{14}CO_2$ uptake rates. These rapid and slow groups maintained divergence in the F_8 generation and, for F_7-derived lines, in the F_8 and F_9 generations (Ford et al.,1983). Harrison et al. (1981) concluded that selection for rapid CAP is possible in later generations. So traditional breeding practices can lead to genotypes with enhanced photosynthetic rates.

Those interested in selection for photosynthesis need to be aware of flavonal glycosides. There are six simple inherited gene pairs (Fg_1/fg_1, $Fg_2/fg/_2$, Fg_3/fg_3, Fg_4/fg_4, *T/t*, and *Wm/wm*) that control flavonol glycoside production in soybean leaves. The *Fg* genes determine the type of flavonol glycoside produced. The presence of *T* (tawny pubescence) causes both quercetin and kaempferol flavonol glycosides to be present in leaves; *t* (grey pubescence) results in kaempferol glycosides. *Wm* allows normal expression of these phenotypes, but *wm* causes low levels of quercetin/kaempferol glycosides. Where Fg_1-Fg_3-*t*-*Wm* occurs, leaves will have high levels of kaempferol 2G-glucosylgentiobioside (K9), and AP, chlorophyll content, and specific leaf mass will be reduced relative to other genotypes. However, K9 evidently does not directly inhibit AP or chlorophyll production, because Cosio and McClure (1984) could detect no traces of K9 in mesophyll cells isolated from fully expanded leaves. They suggest that the compound and other kaempferol glycosides and their derivatives mainly occur in the epidermis. Therefore, the presence of K9 may indicate, rather than cause, reduced photosynthetic capacity. Substitution of *T* for *t* partly corrects the problem. Substitution of *wm* for *Wm* causes reduced AP, but chlorophyll content and specific leaf mass are unaffected (Buzzell et al., 1977), so high or low levels of K9 are indicative of a detrimental effect on photosynthesis. Breeders should avoid crosses that would produce a high proportion of progeny with both Fg_1 and Fg_3 (Buttery and Buzzell, 1973, 1976). Buzzell et al. (1980) have classified many lines and cultivars for flavonal glycoside genotype.

Measuring photosynthesis on the hundreds of lines usually developed in breeding programs is an arduous task, so a simply measured selection index would be useful. Nearly everyone who has found a correlation of photosynthesis with some leaf or canopy trait has suggested its use as a selection criterion. But few studies have taken the next step: assessing the gain to be made in photosynthesis by selection for the index. Wiebold et al. (1981) found that the gain in leaf AP from hypothetical selection for specific leaf mass in the F_2 and F_3 was about 5 to 11% and, for leaf thickness in the F_3, about 5%. Selection for leaf density would have been ineffective. However, the heritabilities of these traits proved lower than for leaf AP, and the phenotypic correlations with AP were poor to moderate (r = 0.12 to 0.65). They concluded that lines might be screened in early generations on the basis of specific leaf mass.

Selection for rapid and slow leaf $^{14}CO_2$ uptake has resulted in lines with rapid and slow CAP as well (Ford et al., 1983; Hole, 1983). But there are not enough data to suggest leaf AP as a selection index for CAP.

Leaf AP of a cultivar selected for greater yield is likely to exceed that of its parents. Ojima et al. (1968) reported that 70% of the cultivars they tested had leaf AP equal to or greater than the cultivar's parent with the most rapid AP. Also, Buttery and Buzzell (1972) found that, of 25 cultivars, 17 had more rapid net assimilation rates than either parent. The majority of the same cultivars had more rapid leaf $^{14}CO_2$ uptake than the midparent value (Buttery et al., 1981). Du et al. (1982) found AP of improved cultivars to be 60% better than that of the best parents, and Dornhoff and Shibles (1970) noted that newer, greater-yielding cultivars had more rapid leaf AP than older, lesser-yielding cultivars. Gay et al. (1980) report the lone exception: 'Lincoln' and 'Dorman' are older, lesser-yielding cultivars with leaf $^{14}CO_2$ uptake rates similar to those of 'Williams' and 'Essex', respectively, which are newer and greater yielding. In general, one cause of yield improvement in soybean seems to be enhanced leaf photosynthesis, but as we discuss later, breeding for faster photosynthesis has not always led to yield improvement. Although single-leaf and canopy photosynthesis can be enhanced by standard breeding techniques, the utility of such an endeavor remains to be proved.

14–1.5 Sink-Source Relationships

14–1.5.1 Evidence for a Relationship

The presence of a fruiting sink enhances the rate of photosynthesis expressed by leaves. Evidence for a sink-source relationship comes from two types of experiments: (i) where AP or CAP is measured before and after the beginning of rapid seed growth and (ii) where the photosynthetic rates of control plants with artificially altered sink-source ratios are compared.

The AP of leaves formed after R5 has been found 23 (Dornhoff and Shibles, 1970), 89 (Ghorashy et al., 1971), and 59% (Gordon et al., 1982) greater than that of leaves formed during vegetative growth. Because (i) leaf growth ceases or slows dramatically at fruiting and (ii) upper leaves of indeterminate types generally are smaller and thicker (Fig. 14–3), the greater AP of these later leaves might be a consequence of a sink-induced change in their developmental pattern rather than a direct effect of the reproductive sink on their photosynthetic apparatus. However, most evidence in soybean argues against this possibility. For example, Gordon et al. (1982) found AP of leaves that had reached full expansion during vegetative growth to be more rapid after the onset of seed growth than before.

By using four, isogenic determinant lines representing combinations of nodulating or nonnodulating ability with male sterility or male fertility, Huber et al. (1983b) were able to measure AP on leaves of similar ontogeny that experienced different sink demands. Maximum seasonal AP occurred at flowering and was similar for the four lines. Within nodulation groups, AP declined more rapidly for the male-sterile than for the male-fertile lines, and within fertility groups, AP declined more rapidly for the nonnodulating lines, especially for the male-fertile pair. So the loss of AP associated with senescence was slowed by the presence of a sink. This agrees with the data of Lugg and Sinclair (1981) who found that leaves of reproductive and vegetative plants had similar AP during their most active phases, but that duration of the most active phase was longer for reproductive leaves. Secor et al. (1984) indicate that onset of seed growth seemed to delay AP decline of the 8th and 13th trifoliolate leaves of Amsoy 71, compared with the 3rd leaf which senesced before flowering. Huber et al. (1983b) found that the presence or absence of the reproductive sink had a greater effect on AP than did N source.

An interesting experiment of the second type was performed by Enos et al. (1982). For field-grown 'Cobb', a cultivar grown in southern USA and hence adapted to short days, they altered the photoperiod experienced in Illinois by covering it each night from 1730 to 0730 until (i) midflowering or (ii) early pod set. The AP measured on 11 July, when all plants were vegetative, did not differ among treatments and averaged 25.8 μmol m^{-2} s^{-1}. But, in early August, the rates for vegetative plants (not covered) and plants averaging 47 and 80 pods per plant (treatments 1 and 2, respectively) were 18.3, 28.3, and 36.5, respectively. So AP was positively associated with pods per plant.

More intrusive manipulations designed to alter the ratio of sink to source capacities also have been used. For example, depodding reduces photosynthesis within a few days (Lawn and Brun, 1974; Mondal et al., 1978a; Setter et al., 1980a; Wittenbach, 1982, 1983a), and shading all but a few leaves increases the AP of the uncovered leaves markedly, also within days (Thorne and Koller, 1974; Peet and Kramer, 1980). In these experiments, the rapid response of the leaves, which had already reached full expansion, argues for a direct effect on the photosynthetic apparatus. However, the intrusive nature of such plant manipulations makes it difficult to speculate on the mechanism of the sink-source relationship or to judge whether the result truly represents a sink effect.

14-1.5.2 Plant Manipulations Alter Leaf Metabolism

Depodding a plant results a complete alteration of leaf physiology. Within 48 h, AP is depressed by up to 70% (Setter et al., 1980a). Girdling the petiole, so as to inhibit phloem translocation, has a similar but quicker and more drastic effect. The decrease in AP was associated with a 230% increase in stomatal resistance, whereas mesophyll resistance was unaffected. Both treatments resulted in an increased concentration of ABA in the leaf (two to five times), as a consequence of reduced ABA export (Setter et al., 1980b, 1981). The increase in stomatal resistance may be maintained for 50 days or more after depodding (Wittenbach, 1982, 1983a).

Depodding also causes a rapid loss of rubisco activity per unit leaf area, while starch and soluble carbohydrates increase. Soluble protein content either increases or remains constant, with the loss of rubisco being offset by an increase in the amount of a leaf storage glycoprotein (Wittenbach, 1983a, 1983b). The normal loss of chlorophyll associated with senescence is delayed (Lindoo and Noodén, 1976, 1977; Mondal et al., 1978a). Wittenbach suggests that depodding causes leaves to change their function from that of a source to that of a storage organ.

14-1.5.3 Mechanism of Sink-Source Relationships

The mechanism by which sinks stimulate photosynthesis has been at issue for some time, but no mechanism is yet generally accepted. It is often hypothesized that sinks reduce the concentration of soluble carbohydrates or starch in the leaves, thus mitigating end-product inhibition, or they alter the hormone content of leaves, which stimulates photosynthesis.

The idea that leaf soluble carbohydrates accumulate in the absence of a sink, and inhibit photosynthesis, is untenable in soybean. Huber et al. (1983b), using the aforementioned four isogenic lines differing in N source and male fertility, found no difference in sugar content of the leaves, though sink capacities and AP of these lines were radically different. Moreover, AP was not associated with soluble carbohydrate content in experiments in which the latter was altered either by depodding (Mondal et al., 1978a), exposure to 54 h of continuous light (Potter and Breen, 1980), or exposure to various CO_2 concentrations (Hofstra and Hesketh, 1975; Nafziger and Koller, 1976).

Evidence that an accumulation of starch can inhibit photosynthesis is more equivocal. Increasing the starch content of leaves by adjusting the sink load genetically (Huber et al., 1983b), by simply following the diurnal increase in starch content (Upmeyer and Koller, 1973), or by exposing treated plants to elevated CO_2 concentrations (Hofstra and Hesketh, 1975; Nafziger and Koller, 1976) has resulted in correlated declines in AP. On the other hand, after Potter and Breen (1980) exposed plants to 54 h of continuous light, starch content was increased 14-fold, but AP declined only 10%. Furthermore, Mondal et al. (1978a) found that, at mid-pod fill, continuous depodding since midbloom had resulted in 1.75 times more starch in the leaves than depodding only 32 h previously, but there was no difference in AP. And in Nafziger and Koller's experiment, the response of AP to starch content was curvilinear, such that AP was little affected below 15 g of starch m^{-2}. Leaves of field-grown plants generally have < 10 g of starch m^{-2} (Mondal et al., 1978a; Egli et al., 1980). So, if a sink could further reduce the starch content, one should expect little enhancement of AP.

How starch might inhibit photosynthesis is not clear. Generally, when starch content and AP are negatively correlated, mesophyll resistance also is negatively associated with AP (Upmeyer and Koller, 1973; Thorne and Koller, 1974; Nafziger and Koller, 1976; Hofstra and Hesketh, 1975). Whether this indicates that starch inhibits certain enzymatic reactions of photosynthesis, alters the CO_2 diffusion pathway, or operates through some other physical or biochemical mechanism is not known. Indeed, mesophyll resistance may be affected by some other mechanism of sink-source relationships, with starch accumulation being coincidental.

Evidence that sink-produced hormones enhance photosynthesis is nonexistent for soybean. In bean (*Phaseolus vulgaris* L.) application of indole-3-acetic acid (IAA) or cytokinin to leaf tissue stimulates photosynthesis (Turner and Bidwell, 1965; Borzenkova and Bortnikova, 1978; Wareing et al., 1968). This effect seems to result from the enhanced activity and/or synthesis of photosynthetic enzymes. In soybean, cytokinin-like activity has been noted in extracts from seeds and leaves (Lindoo and Noodén, 1978), and leaf levels of IAA are greatest during rapid seed growth when AP is also at a maximum (Mondal et al., 1978b). However, because the source of these hormones in vivo is not known, their role in sink-source relationships cannot be defined.

Sinks are reported to stimulate photosynthesis indirectly by relieving leaves of excess abscisic acid (ABA). Setter et al. (1980a, 1980b, 1981) reported that any treatment that deprived leaves of a way to get rid of ABA—for example, depodding, petiole girdling—caused its accumulation, a rise in stomatal resistance, and a decline in AP. Further evidence implicating ABA in sink-source relations has been reported by Schussler et al. (1984), who found in vivo seed growth rates, endogenous seed ABA levels, and capacity for in vitro sucrose uptake by soybean embryos all to be positively related (see section 14.2.6.7). In field-grown plants, they found that seed ABA and growth rate rose in parallel, peaked 25 and 28 days after anthesis, then declined in parallel. Additionally, Samet and Sinclair (1980) found that the ABA content of leaves rises late in senescence—when seed growth slows—but this was not corroborated by Ciha et al. (1978). Whether ABA movement from leaves to seeds in the phloem (Setter et al., 1981) is one causative factor in sink-source relations or merely a correlative event remains to be proved. As yet, there are no reports correlating both sink and source activities with ABA movement.

Another potential regulator of source activity in soybean is the putative senescence signal proposed by Noodén (Lindoo and Noodén, 1976; 1977). In bean (*P. vulgaris* L.) Seth and Wareing (1967) and Tamas et al. (1981) found that IAA, applied in lanolin paste in place of the seeds, hastened the decline in leaf chlorophyll and protein contents associated with senescence. Lindoo and Noodén (1978) applied a similar treatment to soybean plants and observed a slight stimulation of senescence. Also, Noodén et al. (1979) found that spraying the foliage with IAA prevented abscission but not leaf yellowing, spraying with cytokinin prevented yellowing but not abscission, and spraying with both delayed senescence. Whether or not IAA is the senescence signal proposed by Lindoo and Noodén remains equivocal.

In summary, plant hormones have been shown to influence photosynthetic rates, but as yet, no specific mechanism by which hormones may mediate sink-source relationships can be suggested.

14–2 PARTITION AND TRANSPORT OF ASSIMILATES

The leaf is a dynamic system with various separate but integrated processes responsible for carbon assimilation, utilization, storage, and export. Each process and the interrelationships among processes are subject to regulatory stimuli from within the leaf, from other parts of the plant, and from the external environment. For example, changes in length of photosynthetic period can affect plants such that metabolic pathways are reprogrammed after just one cycle of a new photosynthetic period (Carlson and Brun, 1984). The relative proportions of nonstructural compounds in the leaf change over the day and throughout the season, indicating that both short- and long-term regulation occurs. Nevertheless, there is a range of concentrations for particular compounds in the leaf. Soybean leaves, on a dry weight basis, contain 5 to 34% starch (Upmeyer and Koller, 1973; Silvius et al., 1978; Chatterton and Silvius, 1979; Giaquinta et al., 1981; Egli et al., 1980; Potter and Breen, 1980), 1 to 6% sucrose (Chatterton and Silvius, 1979; Egli et al., 1980; Potter and Breen, 1980) and 8 to 36% protein (Wittenbach et al., 1980; Sesay and Shibles, 1980; Secor et al., 1983), with the remainder of the dry matter mostly in structural compounds.

14–2.1 Compounds

In C_3 species like soybean, the primary products of photosynthesis are partitioned between two major pathways, one leading to chloroplastic starch synthesis, the other to cytoplasmic sucrose synthesis (Fig. 14–5). Starch and sucrose are the most heavily labeled products from either pulse-chase or steady-state $^{14}CO_2$ uptake, with starch often accounting for most of the radioactivity (Fisher, 1970a, 1970b; Vernon and Aronoff, 1952; Nelson et al., 1961). Monosaccharides such as glucose, fructose, and hexose phosphates are present, but usually in concentrations of $<$ 1% (Thorne and Koller, 1974; Fisher, 1970b; Streeter and Jeffers, 1979). Sterol glucosides and probably esterified sterol glucosides also are present (Fisher, 1970a, 1970b), as well as glycoproteins (Franceschi and Giaquinta, 1983a, 1983b), which may be important in N and carbon translocation during pod filling. In young, expanding leaves, more radioactive carbon can be in serine than in sucrose (Nelson et al., 1961), which might be the result of more photorespiration in very young leaves (Secor et al., 1982a).

Strong emphasis has been placed on investigating the control of starch and sucrose synthesis because they are the important storage and transport forms, respectively. Starch and sucrose levels can change rapidly in response to certain conditions, indicating that their respective pathways are sensitive to immediate regulation. Not only is there environmental regulation, but also genotypic. Under a similar environment, 'Wells' accumulates much less starch than does 'Amsoy 71', whereas both cultivars have similar levels of sucrose (Fader and Koller, 1983).

14–2.2 Variation in Starch and Sucrose Levels

There is a daily rhythm in leaf starch content, with accumulation occurring during daytime and hydrolysis and translocation resulting in

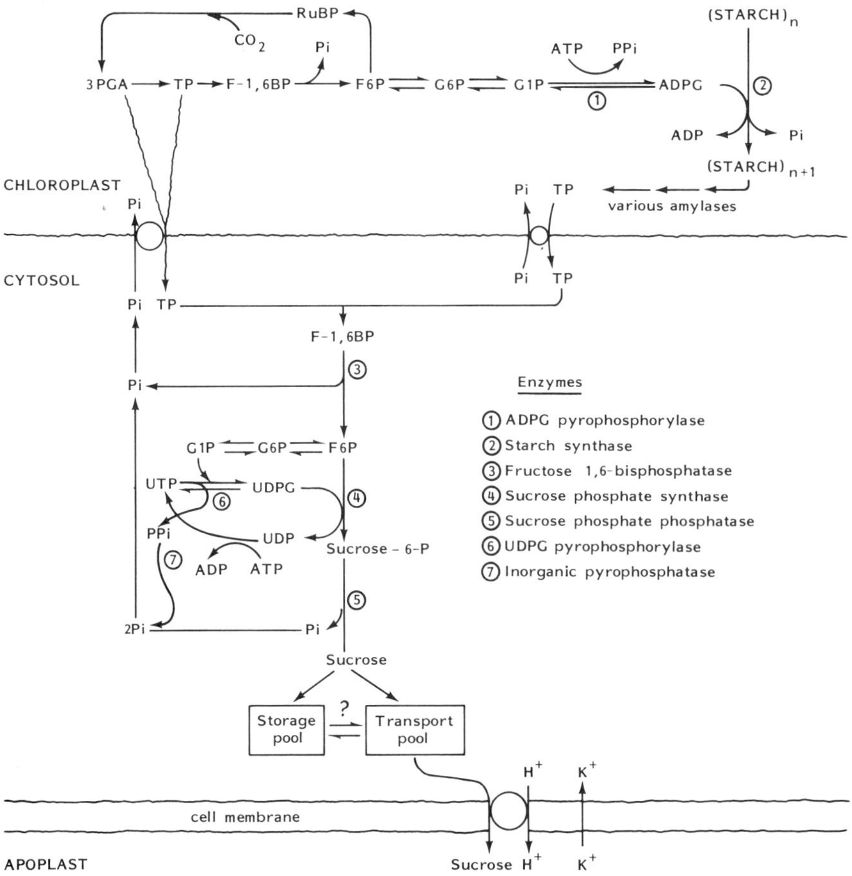

Fig. 14–5. Model of starch and sucrose synthesis and export in leaves. (PGA = 3 phosphoglyceric acid; TP = triose phosphate (either dihydroxyacetone phosphate or glyceraldehyde 3-P); F-1, 6BP = Fructose-1,6-bisphosphate; F6P = Fructose-6-phosphate; G6P = Glucose-6-phosphate; GlP = Glucose-1-phosphate; ATP = Adenosine-5′-triphosphate; P_i = orthophosphate; PP_i = pyrophosphate; ADPG = adenosine-5′-diphosphate glucose; RuBP = ribulose-1,5-bisphosphate; UDPG = uridine-5′-diphosphate glucose; and UDP (UTP) = uridine-5′-diphosphate (triphosphate).

a progressive reduction at night. Under normal photoperiods (longer than 12 h), the diurnal increase in leaf starch content occurs at a rate of approximately 3 to 3.5 mg dm^{-2} h^{-1} regardless of nodal position (Chatterton and Silvius, 1979, 1981; Upmeyer and Koller, 1973; Giaquinta et al., 1981; Rufty et al., 1983). Because rates are similar among unifoliolate leaves (Upmeyer and Koller, 1973), third and fourth trifoliolate leaves (Chatterton and Silvius, 1979, 1981), and older leaves from flowering plants (Giaquinta et al., 1981), it seems that starch accumulation rates are well regulated. Sucrose and other water soluble carbohydrates accumulate at a much slower rate (about 90% slower than starch) in leaves of plants grown at photoperiods longer than 12 h (Chatterton and Silvius,

1979; Upmeyer and Koller, 1973; Potter and Breen, 1980; Rufty and Huber, 1983). In general, though, sucrose levels in leaves remain low for the first 2 h of the photoperiod before rising (Upmeyer and Koller, 1973; Rufty and Huber, 1983; Dunphy and Hanway, 1976; Rufty et al., 1983). Both sucrose and starch levels decline at night (Rufty et al., 1983).

Other organs—petioles, stems, and roots—are reported to accumulate starch and sucrose in the daytime and mobilize them at night. Accumulation and utilization rates differ among organs; leaves accumulate the most and degrade the greatest percentage of their starch, whereas roots degrade the greatest percentage of their accumulated sucrose (Kerr et al., 1985b).

Starch accumulation in leaves seems to follow a specific tissue order. Generally, it initially accumulates in the second (abaxial) layer of palisade parenchyma, then in the spongy mesophyll cells, and lastly in the first (adaxial) palisade parenchyma and paraveinal mesophyll (Vendeland et al., 1982; Franceschi and Giaquinta, 1983b). However, there seem to be modifications in this pattern with plant ontogeny. Franceschi and Giaquinta (1983b) found that, during flowering, at the beginning of the photoperiod the second layer of palisade parenchyma, and, to a lesser extent, the spongy mesophyll cells contained some starch. By the end of the light period, however, there had been a large increase in the first palisade parenchyma. The second palisade and paraveinal mesophyll had increased in starch only slightly, and there was no starch in the paraveinal mesophyll. Thus, during this developmental stage, starch seems to be preferentially remobilized in the dark from the second palisade mesophyll, and it reaccumulates in the light. From flowering to mid-pod filling, starch accumulated in all cells, including the paraveinal mesophyll. During mid-pod filling, there was little diurnal variation in starch content. However, by late pod filling, at the end of the light period, almost no starch had accumulated in the first palisade parenchyma, whereas there was yet unmobilized starch in the second palisade parenchyma and spongy mesophyll. Altered patterns may reflect carbon translocation response to changes in sink demand.

As a leaf develops and ages, its carbohydrate balance changes. Whereas daily starch accumulation is greatest in expanding leaves and declines just before the leaf attains full expansion, daily sucrose accumulation increases continually as leaves age (Silvius et al., 1978; Nelson et al., 1961). Changes in leaf carbohydrate levels during plant ontogeny have received much attention. It seems that ^{14}C partitioning into carbohydrates, organic acids, and amino acids is similar in pre- and post-flowering plants (Housley et al., 1979), but no clear pattern has emerged for plants filling seed. Leaf carbohydrates are reported to increase continuously throughout seed filling (Dunphy and Hanway, 1976), or not to change (Ciha and Brun, 1978; Streeter and Jeffers, 1979), or to decrease from a maximum at or near midbloom (Giaquinta et al., 1981; Egli et al., 1980; Mondal et al., 1978a; Saito et al., 1970). Some of these discrepancies may be due to differences in temperature (Dunphy and Hanway, 1976), source to sink ratios (Egli et al., 1976; Egli and Leggett, 1976; Franceschi and Giaquinta, 1983b; Ciha and Brun, 1978), variation in irradiance or methods of carbohydrate analysis (Streeter and Jeffers, 1979).

14–2.3 Factors Regulating Carbon Partitioning

14–2.3.1 Light

Light regulates partitioning through changes both in duration and level. Chatterton, Silvius and co-workers (Chatterton and Silvius, 1979, 1981; Silvius et al., 1979) demonstrated that length of photosynthetic period, rather than photoperiod, was a key factor in determining the immediate fate of recently assimilated carbon. Leaves of plants grown under a 14-h photoperiod with a 7-h photosynthetic period (7 h at PPFD = 640, plus 7 h at PPFD = 10 μmol m^{-2} s^{-1}) accumulated more starch and at a faster rate than did leaves of plants grown under a 14-h photosynthetic photoperiod (PPFD = 640 μmol/m^{-2} s^{-1}). For the first 2 h of both treatments, water-soluble, nonstructural carbohydrates accumulated, but at different rates—approximately 1.5 and 0.5% of leaf dry w. h^{-1} for the 7- and 14-h photosynthetic periods, respectively. In addition to promoting more rapid starch and soluble carbohydrate accumulation, the shorter photosynthetic period resulted in less carbon being partitioned into leaf structural material, as indicated by a lower specific leaf mass (2.0 mg cm^{-2} vs. 3.4 mg cm^{-2}). This fact, coupled with a 33% increase in net photosynthetic rate on a leaf basis, means that leaves are more carbon-efficient when grown under short photosynthetic periods. Changes in carbon partitioning occurred within 4 days after the plants were transferred from long to short photosynthetic periods. Because the carbon is partitioned differently at the beginning of the photoperiod, Chatterton and Silvius (1979) concluded that starch accumulation is a programmed process. Huber and Israel (1982) also have demonstrated a similar effect of photosynthetic period on the change in starch accumulation, but additionally noticed that rates of starch and sucrose accumulation were inversely related.

There is also an effect of light level on carbon partitioning. Starch accumulation rates in leaves are similar for plants grown under 600 to 950 μmol m^{-2} s^{-1} PPFD, indicating that plants acclimated to a specific PPFD may have different amounts of sucrose produced and translocated (Silvius et al., 1979). However, when plants previously acclimated to 600 μmol $^{-2}$ s^{-1} are shifted to 950, the additional carbon assimilated as a result of the higher PPFD is partitioned into starch. The PPFD has a weaker regulatory effect than photosynthetic period, however. Regardless of whether plants are grown under low (320 μmol $^{-2}$ s^{-1}) or moderate (640 μmol m^{-2} s^{-1}) PPFD, more carbon is partitioned into starch under 7- than 14-h photosynthetic periods (Chatterton and Silvius, 1981). The programming of carbon partitioning seems not to be localized, but rather is a whole-plant phenomenon. When single leaves of plants grown under a 14-h photosynthetic period were subjected to 7-h photosynthetic periods, while the remaining leaves received 14 h of light for 4 days, their carbon-partitioning patterns were unchanged (Chatterton and Silvius, 1981).

The influence that photosynthetic period exerts on partitioning of carbon can be strongly modified by the developmental stage of plants. Whereas Chatterton and Silvius (1979, 1981) investigated this phenomenon using vegetative plants, Carlson and Brun (1984) studied plants during reproduction. As reproductive stage advances, it exerts a stronger influence over partitioning of carbon than does photosynthetic period. For example, at beginning of seed filling, short days promote preferential starch synthesis compared with soluble carbohydrates, a finding in agreement with Chatterton and Silvius. In contrast, however, by late seed filling (later R6) short photosynthetic periods decrease, not increase, the carbon flow into starch (Carlson and Brun, 1984). Thus, Carlson and Brun conclude that the increased sink demand, as a consequence of reproductive growth, may be of greater importance in regulation of leaf starch accumulation than is length of daily photosynthetic period.

14-2.3.2 Carbon Dioxide

Elevated CO_2 concentrations increase, whereas subnormal concentrations decrease starch accumulation in leaves. Soluble carbohydrate concentrations are less afflected. Nelson et al. (1961) reported that an increase in CO_2 concentration from 300 to 3000 $\mu L\ L^{-1}$ air resulted in a 10-fold increase in radiolabeled carbon in starch, but less than a threefold increase in sucrose. Likewise, others have reported a much smaller response in soluble-sugar concentration (Nafziger and Koller, 1976; Finn and Brun, 1982). Both long- and short-term exposures to elevated CO_2 concentrations result in increases in rate of starch accumulation and sucrose concentration in leaves (Huber et al., 1984b). Following cessation of elevated CO_2 treatment, the higher levels of starch in leaves return to control levels (Ackerson et al., 1984). However, because the extra carbon assimilated, as a result of elevated CO_2 concentration, is preferentially partitioned into starch rather than into sucrose suggests that some limitation exists for sucrose biosynthesis or transport.

There is evidence that ontogenetic patterns of carbohydrates in leaves respond differently to continuous CO_2 enrichment beginning shortly after emergence. For indeterminate types, a temporary, large increase in starch occurs shortly before flowering. Thereafter, starch levels are steady, but higher than for plants under atmospheric CO_2 levels (Havelka et al., 1984). In contrast, for determinate types, under enriched CO_2 starch begins to accumulate at about V6 and declines to similar levels as controls during the later phases of seed development (Ackerson et al., 1984).

14-2.3.3 Temperature

The work of Thomas et al. (1981) suggests that there is an inverse relationship between temperature and soluble-carbohydrate concentration in leaves. However, higher temperature hastened plant ontogenetic development in their study, thus making a simple temperature response difficult to interpret.

14–2.4 Metabolic Basis of Carbon Partitioning

Soybean leaves normally have a greater concentration of starch than of soluble sugar (mostly sucrose). Starch accumulation is commonly thought to represent a means of storing carbon when photosynthesis exceeds the rate of sucrose export or when photosynthetic periods are short (Chatterton and Silvius, 1979). Thus, there must be a system of metabolic regulators within the cell to channel carbon according to current demands and conditions.

A schematic diagram of starch and sucrose pathways is shown in Fig. 14–5. The first important branch point in carbon metabolism is the fate of the triose phosphates produced in the chloroplast during photosynthesis. Because triose phosphates (TP) and 3-phosphoglyceric acid (PGA) are the phosphorylated metabolites that most freely pass through the chloroplast membrane (Heber and Heldt, 1981), they play an essential role in the allocation of carbon between the chloroplast and the cytosol. The TP either can be transported from the chloroplast or remain inside to be converted to starch or other compounds.

The transport of TP and PGA from the chloroplast occurs with a strict exchange for cytoplasmic orthophosphate (Pi). Two important consequences are that the sum of Pi and phosphorylated compounds in the chloroplast is constant, and the exchange process is the major controlling process in carbon partitioning between the sites of synthesis and of export.

A balance between carbon export and Pi import is necessary for optimal photosynthesis and carbon partitioning. On one hand, high cytoplasmic concentrations of Pi will enhance carbon flow from the chloroplast, resulting in more sucrose and less starch synthesis. Too much carbon export from the chloroplast could result in insufficient carbon to regenerate RuBP, leading to a decline in photosynthesis. On the other hand, low cytoplasmic concentrations of Pi lead to increased starch accumulation (Nafziger and Koller, 1976; Thorne and Koller, 1974; Giaquinta et al., 1981). Giaquinta et al. (1981) reported that the starch-to-sucrose ratio in soybean leaves during pod filling increased as Pi concentration of nutrient solutions was lowered from 5 m*M* to 1 μM. There is also an inverse relationship between an increase in starch synthesis and decreasing cytosolic Pi throughout the day (Giaquinta et al., 1981). In addition to nutrient levels, there also should be metabolic regulators of Pi concentration. Likely candidates are those reactions that release or incorporate Pi. These would be important under conditions in which total phosphate levels are constant. Rate of sucrose synthesis may control starch formation indirectly by affecting cytosolic Pi concentrations (Huber and Israel, 1982; Huber, 1981b). For each mole of sucrose formed, 4 mol of Pi are released into the cytosol. Thus, the faster sucrose is synthesized, the more Pi is released, leading to more rapid TP export from and Pi import into the chloroplast, resulting in reduced starch synthesis.

In soybean leaves, much carbon is retained in leaves during the day, and the bulk of it occurs as starch in chloroplasts. The reason for such a large accumulation of starch in leaves in unclear, but may be the consequence of a highly active starch-synthesizing system or a limitation to transport of carbon through the chloroplast membrane. The regulatory enzyme in starch synthesis is adenosine diphosphate glucose (ADPG) pyrophosphorylase (EC 2.7.7.9), which is activated by a high PGA-to-Pi ratio (Preiss, 1982). Thus, during photosynthesis and under conditions of low carbon export from chloroplasts, the enzyme would be highly active, because PGA is being produced while the Pi pool is being depleted via photophosphorylation and is not being replenished because of slow Pi influx into the chloroplast.

Sucrose formation occurs in the cytosol principally by the sequential reactions of fructose-1,6,-bisphosphatase (FBPase; EC 3.1.3.11), sucrose phosphate synthase (SPS; EC 2.4.1.14), and sucrose phosphate phosphatase (SP, EC 3.1.3.00) (Huber and Israel, 1982; Preiss, 1982). Sucrose synthase (EC 2.4.2.14), which catalyzes the reversible reaction of uridine-5′-diphosphate glucose (UDPG) and fructose to yield uridine-5′-diphosphate glucose (UDP) and sucrose, is not suffieiently active in leaves to account for sucrose synthesis (Huber and Israel, 1982; Silvius et al., 1978; Preiss, 1982). Uridine-5′-diphosphatase, which catalyzes the conversion of UDP to uridine-5′-monophosphate and Pi, has been identified and characterized in fully expanded soybean leaves (Huber and Israel, 1982). Its activity seems too high for it to play an important regulatory role (Huber and Israel, 1982).

Silvius et al. (1978) and Huber and colleagues (Huber, 1981a, 1981b; Huber et al., 1983a) postulated that SPS might be important as a regulator of carbon partitioning into sucrose. Its in vitro activity is in the range equal to sucrose formation rates (Rufty et al., 1983), and it is sensitive to a variety of compounds, making it a good candidate for a regulatory role. Known inhibitors of the enzyme include Pi (Preiss, 1982; Amir and Preiss, 1982), sucrose-6-phosphate (Preiss, 1982), UDP (Huber and Israel, 1982), abscissic acid (Huber et al., 1984a), and sucrose (Huber, 1981b; Preiss, 1982; Huber, 1983), whereas activators include Mg^{2+} (Huber, 1981b; Preiss, 1982) and glucose-6-phosphate (Doehlert and Huber, 1983). Sucrose concentrations in vivo may inhibit SPS activity (Preiss, 1982), but the effect in vitro varies among species (Huber, 1981b).

Most of the evidence in support of the postulated regulatory role of SPS in carbon partitioning, however, is based on correlative evidence relating its activity in crude extracts to events associated with changes in assimilate demand and partitioning of carbon (Huber, 1981b; 1983; Huber and Israel, 1982; Rufty and Huber, 1983; Rufty et al., 1983; Kerr et al., 1985a). Among various species other than soybean, generally there is a strong negative relationship betweeen starch concentration and SPS activity (Huber, 1981b). In one study (Huber, 1983), soybean leaves did show a higher SPS activity and higher starch concentration than did leaves of wheat (*Triticum aestivum* L.), red beet (*Beta vulgaris* L.), or tobacco (*Nicotiana tobacum* L.). However, when the data were represented on a photosynthetic rate basis, the relationship in soybean tissue was similar to that of the other species. Among nodulated and nonnodulated soybean plants, SPS activity was negatively correlated with leaf starch accumulation. Removing all but one leaf on a plant caused that leaf to cease accumulating starch while increasing sucrose accumulation as well as SPS and FBPase activities. That sucrose levels and SPS activity are not always closely associated is evident from their respective daily profiles. One reason is that SPS activity in soybean leaves is controlled by an endogenous rhythm with a period of about 24 h. The SPS activity peaks late in the day or early in the dark period, whereas sucrose levels rise rapidly upon illumination and decline while SPS activity is rising. Furthermore, while SPS activity is declining during the night, starch degradation rate remains constant, implying that factors other than, or in conjunction with, SPS activity are important in carbon partitioning. The lack of consistent relationships among SPS activity, sucrose levels, and starch levels has led Huber and associates to propose that there may be two levels of control of SPS—a fine control that coordinates sucrose synthesis with photosynthesis and a coarse control that responds to changes in the demand for sucrose.

In support of Huber's idea of a fine (metabolic) control is the work of Stitt et al. (1983a, 1983b), whose data from wheat show that sucrose synthesis probably is controlled more by cytosolic FBPase, the enzyme responsible for the first irreversible step leading to sucrose formation (Fig. 14–5). The activity of FBPase is effectively retarded by fructose-2,6-bisphosphate, which acts as a competitive inhibitor with FBP as well as by increasing the sensitivity of FBPase to inhibition by Pi and adenosine-5′-monophosphate (Céske et al., 1982; Stitt et al., 1982). By regulating the rate at which TP is converted to sucrose, FBPase indirectly regulates the rate at which carbon leaves the chloroplast. Under conditions of low irradiance or low CO_2, the activities of both FBPase and SPS decline considerably, thus preventing a depletion of carbon from the chloroplast. However, when carbon substrates become available, FBPase activity increases, thus allowing sucrose synthesis to proceed. There is a strong parallel between increases in the potent modulator, fructose-2,6-bisphosphate, and starch accumulation. However, fructose-2,6-bisphosphate levels are not always negatively associated with sucrose levels, which led Stitt et al. (1983a) to suggest that interactions among adenosine-5′-monophosphate, Pi, and fructose-2,6-bisphosphate are important in regulating sucrose synthesis. A large increase in cytosolic TP would result in the synthesis of large amounts of FBP, which would overwhelm the effects of fructose-2,6-bisphosphate, resulting in sucrose accumulation. However, as more fructose-6-phosphate is produced, so is more fructose-2,6-bisphosphate, thus eventually leading to an inhibition of FBPase. Thus, there are several levels of control of sucrose synthesis, including supply of TP produced and exported from the chloroplast, the level of cytosolic modulators that may regulate enzyme activities, and cytosolic accumulation of sucrose resulting, perhaps, from a limitation of export. Knowledge of the effects of cytosolic modulators of sucrose synthesis in soybean leaves is lacking. At least one report has noted technical difficulties in extracting fructose-2,6-bisphosphate from soybean leaves (Huber and Bickett, 1984).

14–2.5 Nitrogen Partitioning in Leaves

The partitioning of N in soybean leaves may depend upon the form of N delivered to the leaf from the roots. It has been established that roots of soybean plants dependent upon symbiotically derived N export the ureides, allantoin, and allantoic acid, whereas those dependent upon mineral N export mostly amides and nitrate (chapter 13 in this book). Once inside the leaf cell, the N is incorporated into amino acids, and subsequently, proteins. Ureide metabolism in leaf cells is only beginning to become known, whereas amino acid metabolism is better understood (Schrader and Thomas, 1981).

Most leaf N is in protein, which constitutes about 8 to 36% of leaf dry weight, depending upon leaf age. Approximately half the soluble protein in soybean leaves is in rubisco (Wittenbach et al., 1980). Protein levels rapidly increase as the leaf expands and change little until leaf senescence is initiated, when protein is degraded (Burton et al., 1979; Wilson et al., 1978; Cure et al., 1982; Wittenbach et al., 1980; Secor et al., 1984). The rate of protein degradation is hastened by the onset of reproductive growth (Burton et al., 1979; Wilson et al., 1978; Secor et al., 1984). Evidently, there is genetic variation for ontogenetic patterns of leaf N concentration (Cure et al., 1982), for, when grown under short days, leaf N concentration of a line with a high seed N concentration (D72-8126) declined faster than it did in 'Ransom.'

There is little agreement regarding a definitive amino acid composition in leaves. This can be partly explained by the variety of $^{14}CO_2$ feeding times, leaf positions and ages, leaf-killing and extraction techniques, plant ages and developmental stages, and harvest times after the $^{14}CO_2$ feedings. Indeed, a recent report by Wallace et al. (1984) demonstrates that a small, rapid change in temperature, or slight physical injury to soybean leaves, results in a rapid change in leaf amino acid composition.

Generally, $< 20\%$ of the total ^{14}C incorporated into a leaf is found in amino acids. This percentage is similar among vegetative flowering, nodulating and nonnodulating plants (Housley et al., 1979). The predominant radiolabeled amino acids are alanine, serine, and glutamic acid (Vernon and Aronoff, 1952; Nelson et al., 1961; Clauss et al., 1964; Housley et al., 1979; Schrader et al., 1980). In some reports γ-aminobutyric acid is heavily radiolabeled (Housley et al., 1979), but this is most likely an artifact of the experimental system (Wallace et al., 1984), as explained in the next paragraph. Relative concentrations of amino acids vary, depending upon the time between $^{14}CO_2$ feeding and harvest. Serine levels, for example, were reported to decline from about 30% to < 10% of the total ^{14}C incorporated over a period of 2 h (Clauss et al., 1964; Nelson et al., 1961). One drawback of radiolabeling studies is that only amino acids synthesized from current photosynthesis are radiolabeled, this biasing results against unlabeled sources such as protein turnover. In an experiment aimed at determining amino acid composition in soybean leaves, Wallace et al. (1984) showed that glutamic acid was the most predominant amino acid, followed by glycine, aspartic acid, and, in lesser amounts, alanine, serine, and threonine, followed by the remaining amino acids.

Soybean leaves normally have a small quantity of γ-aminobutyric acid, yet the level can rise 20-fold in 2 min after an abrupt (< 5 min) lowering of temperature from 33 to 22 °C, or after a gentle rolling of leaflets (Wallace et al., 1984). Alanine levels can increase equally rapidly in response to a change from high to moderate PPFD (Wallace et al., 1984). Thus, to avoid artifacts, it is necessary to be exceptionally cognizant of experimental conditions when plant material is used for analytical and metabolic studies.

14–2.6 Transport from Leaves

14–2.6.1 Pathways

Giaquinta (1983) reviewed current literature regarding sucrose movement in leaves and believes that, although little definitive data exist on the subject, sucrose is transferred toward the phloem in the symplasm via plasmodesmata. It is likely that assimilates leave the symplasm from mesophyll cells, enter the apoplast, and then are loaded by companion cells into the phloem via a proton-mediated pump. Giaquinta cautiously points out that the evidence is based on a few species and that generalizations may not be warranted. This may be particularly true for soybean leaves, which contain a specialized layer of cells, known as the paraveinal mesophyll, that is important in assimilate transfer (for anatomical details, see chapter 3 in this book) (Fisher, 1967; Franceschi and Giaquinta, 1983a, 1983b; Franceschi et al., 1983). The cells are the first to differentiate and to form vacuoles, and they reach a final size two to four times greater than other cells. By maturity, these cells have a symplastic continuity with the bundle sheath cells of the phloem and the other mesophyll cells. This symplastic continuity, coupled with the fact that they contain only a few chloroplasts, which are not photosynthetically active, suggests that they are important in assimilate transfer. Franceschi and Giaquinta (1983b) presented evidence that the paraveinal mesophyll "functions in the synthesis, compartmentation, and remobilization of nitrogen reserves before and during pod filling." Of particular interest is that these cells contain glycoproteins that act as N-storage material, the levels of which are influenced by sink demand. These glycoproteins may be similar to the glycosides reported by Fisher (1970a, 1970b). A central role for paraveinal mesophyll cells in assimilate transfer in soybean levels is becoming increasingly evident.

14-2.6.2 Mesophyll Export and Phloem Loading

There is a dearth of information about the export of assimilates from soybean mesophyll cells. Until recently, protoplasts could be prepared only from young, expanding, nonexporting leaves. Information on export from cells isolated from mature leaves of reproductive plants indicates that total amino acid efflux can account for published rates of protein degradation and that efflux is a selective process (Secor and Schrader, 1984). Amino acids seem not to be exported at rates directly associated with internal pool sizes. Those exported fastest are alanine, lysine, leucine, and glycine, while γ-aminobutyric acid is exported at a slow rate. Those amino acids may not be the ones exported from intact leaf cells in situ, however. As previously mentioned, Wallace et al. (1984) showed that amino acid composition in leaves changes rapidly in response to rather mild handling of leaves and plants; so isolation methods may alter synthesis and export patterns.

In his review of phloem loading, Giaquinta (1983) pointed out that different molecules may have different pathways and loading mechanisms, and that differences may exist among species. Sucrose is the predominant sugar translocated in soybean plants, and Giaquinta (1983) suggests that sucrose is released via a facilitated transfer into the apoplast near the sieve element-companion cell complex. Sucrose is then loaded into the phloem by a sucrose-proton cotransport mechanism, with K^+ counter movement occurring in response to the change in membrane potential. The loading is very specific, with specificity recognition occurring at the sieve element-companion cell level. This is a general model, however, and variations may occur. No specific model has yet been proposed for soybean leaves.

Far less is known about the loading of nitrogenous compounds, particularly amino acids. Phloem transport of amino acids has been well established (Housley et al., 1977), and amino acids are the major nitrogenous compounds in the petiolar phloem sap of soybeans grown either on nitrate or symbiotically fixed N (Layzell and LaRue, 1982). There is evidence that uptake of amino acids into the phloem occurs from the apoplast and that the loading is energy dependent, distinct from sucrose, and nonspecific for amino acid (Housley et al., 1977; Servaites et al., 1979; Schrader et al., 1980). Various metabolic inhibitors that increase the efflux of amino acids from isolated cells (Secor and Schrader, 1984) will decrease the uptake of leucine applied to an abraded leaf area (Servaites et al., 1979). This suggests that the ability of the phloem to load amino acids is energy dependent. Further evidence of a metabolically dependent carrier for amino acid loading is that 100 m*M* KCl effectively inhibits leucine loading (Servaites et al., 1979) but has no effect on amino acid efflux from cells (Secor and Schrader, 1984), thus pinpointing the effect at the level of phloem loading. But phloem loading of amino acids is somewhat nonspecific because three exogenously applied amino acids with different functional groups (glutamate, γ-aminobutyric acid, and leucine) had similar transport velocities and mass transfer rates (Housley et al., 1977; Servaites et al., 1979). The transport velocities were similar to that for sucrose, but mass transfer rates were slower (Schrader et al., 1980). Because no barrier exists to prevent the uptake of various amino acids into the phloem, selectivity of uptake may depend simply upon what is available. Availability of a compound is, thus, a matter of what is exported from cells, with control resting in the transport properties of the plasmalemma. There can be preferential export from cells and preferential uptake into cells (which would reduce availability for phloem uptake). Compounds that are brought into a leaf via the xylem and are not transported into cells are likely to be exported from the leaf via the phloem.

14–2.6.3 Compounds in the Phloem

The most abundant (more than 90%) compound in the phloem is sucrose (Vernon and Aronoff, 1952; Layzell and LaRue, 1982), most of it derived from current photosynthesis (Fisher, 1970a). Up to 6% of the carbon and 50% of the N entering soybean pods may arise from xylem-to-phloem transfer (Layzell and LaRue, 1982). Compounds also may enter the phloem from cells and xylem along the vascular pathway. Once inside the phloem, compounds can be further metabolized or withdrawn (Layzell and LaRue, 1982). Thus, composition of phloem sap is ever-changing, but it differs from mesophyll cells (Housley et al., 1979) and among plant organs (Layzell and LaRue, 1982). Aside from sucrose, other compounds identified as phloem constituents include organic acids (malonic, succinic, fumaric, malic, tartaric, and citric), many amino acids, amides, sugar alcohols (pinitol, *chiro*-inositol), sugars (arabinose, fructose, glucose, and maltose) and ureides (Vernon and Arnoff, 1952; Housley et al., 1977, 1979; Nelson et al., 1961; Layzell and LaRue, 1982). Differences reported in phloem sap composition may be a consequence of differences in genotypes, environments, experimental conditions, and analytical techniques.

14–2.6.4 Translocation

The flux of carbon from a leaf and throughout the plant is a dynamic process that is strongly influenced by plant and leaf age as well as environmental conditions. Likewise, the ultimate destination of the carbon is influenced by similar factors.

The velocity of carbon flow in stem phloem is remarkably constant, at about 1 cm min^{-1}, considering the varied experimental systems used to estimate it (Fisher, 1970a; Silvius et al., 1978; Servaites et al., 1979; Housley et al., 1977; Blomquist and Kust, 1971; Vernon and Aronoff,

1952). Fisher, though, has estimated the velocity in petiole phloem to be faster, about 2.5 cm min^{-1}, because of its smaller cross-sectional area. Whereas velocity is rather constant, the quantity of carbon exported from leaves is highly variable, depending strongly on biological and environmental conditions. Reported values for percentage of total incorporated ^{14}C that is exported from leaves range from < 1% to > 50% (Thorne and Koller, 1974; Gallaher et al., 1976; Clauss et al., 1964; Nelson et al., 1961; Thrower, 1962; Blomquist and Kust, 1971). Because transport velocity is somewhat inflexible, variability in percentage export may indicate differences in partitioning of recently assimilated carbon into transport pools, as well as the possibility that transport capacity varies with experimental conditions. High PPFD increases the amount of ^{14}C exported (Thrower, 1962), and there are strong effects of leaf age and plant developmental stage on ^{14}C export from leaves (Blomquist and Kust, 1971; Hume and Criswell, 1973). Export rates from leaves of reproducing 'Bragg' soybean plants were approximately threefold faster than from leaves of vegetative plants (Huber et al., 1984b). Fisher (1970a) reported a sucrose efflux rate from leaves of about 3 mg h^{-1} dm^{-2} leaf area, whereas Chatterton and Silvius (1979) reported rates of 0.9 and 1.3 mg sucrose h^{-1} dm^{-2} for plants grown under 7- and 14-h photosynthetic periods within a 14-h photoperiod. With respect to photosynthetic rate, plants grown under 7- and 14-h photosynthetic periods exported 64 and 72% of recently assimilated carbon.

The translocation of a particular substance is dependent upon the integration of many events, including (i) photosynthetic rate, (ii) synthesis of the particular substance, (iii) mixing of various pools, (iv) intracellular metabolism, (v) translocation from site of synthesis to site of vein loading, (vi) radial diffusion from the phloem, (vii) accumulation, metabolism, and withdrawal from the phloem, and (viii) xylem-to-phloem transfer. It is widely accepted that export of sucrose is partly regulated by sucrose production (Chatterton and Silvius, 1979; Thorne and Koller, 1974; Silvius et al., 1978; Fader and Koller, 1983; Giaquinta, 1983; Huber et al., 1984b, 1984c), although increases in translocation can occur without measurable increases in sucrose pool size because of a high flux of sucrose through the pool (Silvius et al., 1979). When photosynthesis is slow and carbon export from the chloroplast is less, sucrose synthesis very likely also is minimal, resulting in low translocation flux. But increases in sucrose concentration, as occur for example under CO_2 enrichment, do not always lead to faster rates of export (Huber et al., 1984b, 1984c). Huber et al (1984c) reported that, below a sucrose content of about 12 mg dm^{-2}, export was positively related with sucrose content, whereas above that level, export rate was unaffected by sucrose concentration. None of the several photosynthetic parameters examined by Kerr et al. (1985a) was correlated with assimilate export, which led them to conclude that export may be controlled by interactions among several events. What limits transport at high sucrose concentrations is a matter for speculation.

Gallaher et al. (1976) compared C_4 and C_3 (including soybean) species and observed that translocation rates were more closely related to cross-sectional areas of phloem in leaf lamina than to photosynthetic rate. However, some of their data (Table 2 of Gallaher et al., 1976) indicate that, in soybean, the percentage of ^{14}C translocated from a leaf was not related to phloem cross-sectional area. Hanson and Kenny (1985) have shown that, although genotypic variation exists for phloem cross-sectional area in soybean petioles, there is sufficient flexibility in transport capacity to accommodate increased translocation. Supply of photosynthate from a leaf does not affect the cross-sectional area of the phloem (Sanders et al., 1977). Most likely, as suggested by Fisher (1970a), translocation flux from leaves is controlled by factors operating within the leaf mesophyll rather than in vascular tissue. A possible site for control of vascular traffic is the paraveinal mesophyll (Fisher, 1970a; Franceschi and Giaquinta, 1983a, 1983b).

Aside from the complicated sink/source balance, other factors are known to influence export from leaves. Low temperatures reduce efflux (Vernon and Aronoff, 1952). Plant growth regulators may play a role, but no consistent effects have been reported. Increases in translocation velocity and a changed distribution pattern have been attributed to the actions of IAA and gibberellic acid (Hew et al., 1967), whereas n-dimethylamic succinamic acid (ALAR) and 2-chloroethyl phosphoric acid (Ethrel) had little or no effect on translocation (Blomquist and Kust, 1971). Thrower (1962) has reported diurnal changes in translocation. Late in the day, less ^{14}C was translocated from a leaf than earlier. Plants grown under high PPFD translocate more carbon from leaves than do those grown under moderate levels, but transferring plants from moderate to high PPFD had no effect on translocation (Silvius et al., 1979). This implies that translocation is a programmed event that requires an acclimation period before a change is effected. Calcium-deficient plants had 56% less ^{14}C-labeled carbon transported from leaves than did plants with sufficient Ca (Gallaher et al., 1976).

14-2.6.5 Distribution Patterns

There is little lateral movement of recently assimilated carbon among mesophyll cells of a leaflet and no movement between leaflets of a leaf (Aronoff, 1955). Sugars are loaded into the phloem near the site of carbon fixation. Sugars can be forced to move acropetally within a leaflet, but the normal movement is basipetal (Thaine et al., 1959). There is evidence that some assimilates are diverted to darkened areas of a leaf blade and to localized areas of necrosis (Thaine et al., 1959).

Leaf age and plant developmental stage are important factors determining the destiny of translocated carbon. As leaves expand, they change from being importers to being exporters of carbon. According to Thrower (1962), when leaves are between 30 and 50% expanded there is simultaneous import and export of carbon, but once a leaf is 50% expanded, it normally no longer imports carbon (Thaine et al., 1959).

Usually, carbon is translocated from a mature leaf to the nearest actively growing region (Hume and Criswell, 1973; Nelson et al., 1961; Thaine et al., 1959; Stephenson and Wilson, 1977a, 1977b; Vernon and Aronoff, 1952). Upper, mature leaves of vegetative plants feed expanding leaves and the apical region; lower leaves feed the roots. Carbon from middle leaves goes in both directions. This pattern can be changed when leaves are removed or shaded (Thrower, 1962; Fellows et al., 1979). Sink requirements also dictate the direction and flux of carbon. Carbon can move long distances in response to increased sink demand, as demonstrated by Gent's (1982) experiments with Y-shaped plants.

Nodulating and nonnodulating isolines have similar translocation patterns when grown on NO_3^-, but more carbon will be transferred to roots when the nodulating isoline is dependent solely on N_2 (Russell and Johnson, 1975). Other source-sink manipulations have demonstrated flexibility of translocation patterns. Removing half the pods from a plant causes a decline in carbon export from leaves. Conversely, removing two-thirds of the leaflets causes an increase in the export of carbon from the remaining leaves; however, the distribution pattern of the carbon is unaffected by either treatment (Egli et al., 1976).

There seems to be a phyllotactic influence on the destination of translocated carbon. During vegetative development, an expanding leaf receives more carbon from the leaf two nodes below than from the one immediately below it (Thrower, 1962). During reproduction, carbon exported from a leaf is preferentially transported to pods at the same node and, secondarily, to pods two nodes above or below (Stephenson and Wilson, 1977a; Blomquist and Kust, 1971). When competition exists between vegetative and reproductive sinks, as commonly occurs for indeterminate genotypes, a rapidly growing reproductive sink has priority (Blomquist and Kust, 1971).

14-2.6.6 Remobilization of Stored Carbon to Seeds

Carbohydrate levels in leaves usually remain high until abscission, leading many researchers to believe that leaves do not serve as an important source of stored carbon for seed growth (Dunphy and Hanway, 1976; Wilson et al., 1978; Streeter and Jeffers, 1979; Ciha and Brun, 1978; Stephenson and Wilson, 1977a). The amount of carbon lost in abscised leaves is small, perhaps being equivalent to a loss in yield of about 2 to 3% (Egli et al., 1980). Stems, however, do act as a site for carbon storage, although the importance of remobilization of carbon from stems for final yield is equivocal (Streeter and Jeffers, 1979). That seed growth rate is stable and not related directly to short-term supply of photosynthates further supports the premise that storage carbohydrates are important for seed yield (Egli and Leggett, 1976). Other evidence is that pods from fertile isolines (low source-to-sink ratio) have rates of seed growth similar to those of pods from sterile plants (high source-to-sink ratio) (Burton et al., 1979). With determinate genotypes, leaves at middle nodes on the stem export ^{14}C to stems, where it is stored for later mobilization during seed growth (Stephenson and Wilson, 1977a). Further, remobilization of N from senescing leaves to developing seeds necessitates a considerable transfer of carbon with the N.

Evidence contrary to the significance of remobilized carbon comes from an extensive study utilizing several periodic feedings of $^{14}CO_2$ throughout the growing season and several subsequent harvests (Hume and Criswell, 1973). Because, at maturity, stems contained the ^{14}C that they had earlier accumulated, Hume and Criswell concluded that there was little evidence that stems acted as temporary storage sites for assimilates destined for seeds.

Carpels may be storage sites for carbohydrates and nitrogenous compounds, contributing up to 13% of the final seed weight for the cv. Norman but only 2% for Amsoy 71 (Thorne, 1979).

14-2.6.7 Phloem Unloading in Seeds

Considerable amounts of sucrose must be delivered and unloaded to the seeds during rapid growth. Recent research has advanced our understanding of the pathway and mechanism by which sucrose is transferred to developing embryos. Sucrose passes through the two major pod vascular bundles, along which seeds are alternately attached, and directly into the seed coat. Sucrose is rapidly and evenly distributed throughout the seed coat through a network of minor veins (Thorne, 1980), where it is unloaded from the symplasm (seed coat) into the apoplastic free space between the seed coat and the embryo (Thorne, 1981). The sucrose diffuses across the free space to the cotyledon, where it is taken up into the symplasm of the cotyledonary cells (Thorne, 1982b; Lichtner and Spanswick, 1981b).

There seem to be two distinct carrier-mediated steps. One is the export from the seed coat. Thorne (1982a) demonstrated that this step is reduced by low temperature or low pO_2, whereas accumulation by the embryo was far less sensitive. Further work has shown that this step is energy dependent, carrier mediated, and may involve ATPase (Thorne, 1982b; Thorne and Rainbird, 1983). The other step is the uptake of sucrose into developing embryos. On the basis of the work of Thorne (1982b) and Lichtner and Spanswick (1981a, 1981b), it seems that the uptake of sucrose involves an electrogenic, carrier-mediated, cotransport of sucrose and protons. However, in addition to the carrier-facilitated uptake, there also is a nonsaturated diffusion component to the overall process (Thorne, 1982b; Lichtner and Spanswick, 1981b). Sucrose constitutes the major source of carbon delivered to the embryo by the seed coat, with that delivered in nitrogenous compounds largely making up the balance.

Evidence suggests that ABA may stimulate sucrose unloading in soybean seeds. Schussler et al. (1984) determined that seeds from a large-seeded soybean line had more rapid growth rates, greater endogenous

ABA contents, and more rapid in vitro sucrose accumulation by their embryos than did seeds from medium-seeded or small-seeded lines. Because the greatest levels of endogenous ABA occurred in seed coats and cotyledons, Schussler et al. speculate that ABA may stimulate rapid unloading of sucrose into the seed coat apoplast and, perhaps also, enhance sucrose movement into cotyledons.

In addition to sucrose, large amounts of N must be delivered to developing soybean seeds. Although xylem sap of symbiotic soybean plants contains high concentrations of ureides, it seems that ureides are not utilized directly by seeds. On the basis of several lines of evidence, Rainbird et al. (1984) have shown that N nutrition for seed development in symbiotic soybean plants is provided primarily by glutamine and, to a lesser extent, by asparagine, followed by various other amino acids. According to Rainbird et al. (1984), ureides may not be utilized by embryos because ureides (i) are rapidly metabolized by maternal tissue, (ii) are inefficiently used for seed growth and storage protein biosynthesis and (iii) are slowly taken up by the embryo. Other researchers (Hsu et al., 1984) also have reported that seed coats have a low ureide content and that a high glutamine-to-glutamate conversion occurs in the seed coat via the enzyme glutamate synthase (EC 2.6.1.53). Further research is needed to fully elucidate the mechanisms of N unloading in pods and uptake into seeds.

14-3 RESPIRATION

Whereas carbon assimilation has received much attention by researchers, carbon dissimilation, with the exception of that associated with the PCO pathway, has been almost completely ignored. Dark (mitochondrial) respiration has been regarded as an obligatory carbon expenditure associated with metabolism essential for production of the complex compounds needed for growth and maintenance of tissues. On the other hand, crop scientists often have expressed concern that carbon may be "burned-off" under excess sugar levels.

Penning de Vries et al. (1974) believe that the respiratory efficiency for synthesis of specific compounds is similar for all species, carbon expenditure for growth processes varying only to the extent that species and cultivars differ in content of various compounds. Maintenance respiration would vary with environmental conditions and, though it has not been shown to do so, conceivably could differ among soybean genotypes. Maintenance is a relatively small component of total respiration and is a linear function of plant N content (McCree and Silsbury, 1978; McCree, 1983). The proportions and levels of growth and maintenance respiration have not been estimated for soybean, or most species.

Conventional opinion holds that the coincidence of high levels of soluble carbohydrates and high temperatures results in excessive respiratory carbon oxidation not coupled to energy production. In spinach and wheat, there seems to be a correlation between carbohydrate level and activity of the nonphosphorylating alternate oxidase electron pathway (Azcon-Bieto et al., 1983), but research in soybean is lacking.

14–3.1 Plant Respiration

In darkness, photosynthetically active leaves respire at about 3 μmol CO_2 m^{-2} s^{-1}, or about 10 to 20% of photosynthetic rate (Secor et al., 1983), although much slower rates of about 0.2 μmol CO_2 m^{-2} s^{-1} have been reported (Sambo et al., 1977). In field-grown plants, Kumura and Naniwa (1965) observed that respiration rate, on a dry-weight basis, was most rapid in leaf tissue, slowest in the stem, and intermediate in the root. Kishitani and Shibles (1986), however, reported that respiration rates of glasshouse-grown, nonnodulated plants were reasonably similar among vegetative organs at comparable stages of plant development, except that stems declined more rapidly than other tissues after rapid seed growth began.

Soybean plants show a diurnal pattern in dark CO_2 evolution, not associated with changes in temperature, that peaks between noon and late afternoon (Mkandawire, 1984) or early in the dark period (Warrington et al., 1977). The peak in respiration may be a consequence of increased carbohydrate substrate or a circadian rhythm. Respiration rate and carbohydrate content are well correlated in actively growing tissues, such as young leaves and meristems, but are poorly related in more mature tissues, such as older leaves and full-sized pods before rapid seed growth begins (Coggeshall and Hodges, 1980).

Cultivars that differ in yield and leaf photosynthetic rates have similar respiratory characteristics on a unit mass basis, but may differ on a plant basis. For example, glasshouse-grown, nonnodulated 'A3127' had less root but more stem respiration than 'Wayne', due entirely to less root and more stem tissue of A3127. Among these, and two other cultivars ('Hodgson' and 'Chippewa') studied by Kishitani and Shibles (1986), 36% of assimilated carbon was expended in respiration for 1 week during flowering.

14–3.2 Pod Respiration

Pods with rapidly growing seeds contribute more to respiration than do leaves (Kumura and Naniwa, 1965; Silvius et al., 1978), contributing 35 to 50% of whole-plant respiration (Kishitani and Shibles, 1986). The reported magnitude of carbon loss by detached pods varies widely, ranging from < 10 to > 50 μmol CO_2 h^{-1} g^{-1} fresh weight in the light, and 1 to 10 times faster in the dark at constant temperature (Andrews and Svec, 1975; Quebedeaux and Chollet, 1975) (see section 14–1.2). As seeds develop, changes occur in carbon exchange rates, but with no consistent pattern. On a weight basis, dark respiration has been observed to decline either continuously (Kumura and Naniwa, 1965; Andrews and Svec, 1975;

Egli and Wardlaw, 1980) or after reaching a maximum at mid-pod filling (Quebedeaux and Chollet, 1975).

The response of intact, attached pods to light and temperature has been investigated by Spaeth and Sinclair (1983a, 1983b). Under natural conditions, intact soybean pods respire more CO_2 during the day because the increase in pod temperature due to solar radiation hastens respiration more than the increase in solar radiation increases pod photosynthesis. Between 0 and 20 °C, the temperature response is linear, not exponential, as would be expected for most biological systems. Furthermore, the temperature coefficient changes with pod development, reaching a maximum of approximately 1 μmol of CO_2 h^{-1} pod^{-1} $°C^{-1}$ at mid-seed development.

In contrast, Egli and Wardlaw (1980) found that respiration rates on a dry weight basis were similar among seeds excised from plants grown at day/night temperatures ranging from 19/13 to 33/28 °C. The relative insensitivity of excised seed respiration to temperatures in this range correlates well with similar temperature insensitivities of seed growth rate and duration (Egli and Wardlaw, 1980).

According to Miller et al. (1983), the cytochrome oxidase pathway is the predominant electron pathway operating in various tissues from excised, developing soybean seeds. The cyanide-insensitive, alternative oxidase pathway constituted $< 6\%$ of total respiration regardless of plant growth conditions. Thus, seed respiration is energetically beneficial, leading to the production of ATP. However, there is preliminary evidence for a large and potentially significant active cyanide-insensitive respiratory capacity in leaf tissue (Sesay et al., 1985).

In pods with rapidly developing seeds, CO_2 efflux is related to seed growth rate, and both are strongly dependent upon assimilate supply. With growth chamber-grown plants (PPFD = 450 μmol m^{-2} s^{-1}), reduction in PPFD or CO_2 level diminished whole-fruit respiration within 24 h (Satterlee and Koller, 1984).

Several researchers (Sambo et al., 1977; Layzell and LaRue, 1982) have attempted to calculate a carbon budget for developing pods, which would lead to a better appreciation of the processes occurring within the pods. A summary of the reports is shown in Table 14–2, which expresses the data in terms of percentage of total carbon input. The ranges reflect

Table 14–2. Carbon balance sheet for soybean pods. Values are the percentage of total carbon input. Data derived from Sambo et al. (1977) and Layzell and LaRue (1982).

Inputs	——— % ———
Carbon imported from leaves	63–84
Gross photosynthesis	16–37
Total carbon input	100
Output	
Total respiration†	28–52
Net gain	48–72

†Total respiration is a summation of reported day and night respiration rates.

differences among reports in experimental methods and calculations. Regardless of the individual report, 50 to 70% of the carbon lost during respiration is refixed by photosynthesis. This carbon refixation represents the manner by which pods significantly contribute to their carbon economy; otherwise, they would lose considerably more than 50% of their carbon input.

14–4 CARBON METABOLISM, GROWTH RATE, AND YIELD

In soybean, lines of evidence suggest that plant growth and seed yield are directly dependent on photosynthesis. First, there is evidence from experiments in which variation in photosynthate production and yield was generated by altering growth environment. Shibles and Weber (1965, 1966) varied the amount of solar radiation intercepted by planting at different densities; most of the resultant variation in crop growth rate (CGR) was explained by variation in the interception of solar radiation. This work suggests that growth depends on CAP. Others have altered photosynthetic rate by exposing soybean to supplemental light (Johnston et al., 1969; Lawn and Brun, 1974) or CO_2-enrichment (Cooper and Brun, 1967; Hardman and Brun, 1971; Mauney et al., 1978); both treatments increase seed yield. Shading reduces seed yield (Lawn and Brun, 1974; Peet and Kramer, 1980; Christy and Porter, 1982).

Second, there is evidence that one cause of yield improvement in soybean has been enhanced photosynthesis. For example, 6 or 7 out of 10 newer, greater-yielding cultivars have a more rapid leaf AP than either of their parents (Ojima et al., 1968; Buttery and Buzzell, 1972; Du et al., 1982), the midparental rate (Buttery et al., 1981), or the older, lesser-yielding cultivars that they replaced (Dornhoff and Shibles, 1970), but see also Gay et al. (1980).

Table 14–3 presents a third line of evidence: the correlation between leaf $^{14}CO_2$ uptake, net assimilation rate, or CAP and seed yield. As would be expected, correlations are better for the integrated estimators, net assimilation rate and CAP. Not listed in Table 14–3 are correlations between photosynthesis and total plant growth, which generally are not encouraging. Among 16 cultivars grown in the glasshouse, leaf AP and crop growth rate were negatively correlated ($r = -0.44$; Kaplan and Koller, 1977). Crop growth rate was closely related to leaf area growth rate ($r = 0.85$), but leaves with large areas had smaller specific leaf mass and slower AP than small leaves; thus, the negative correlation of AP and crop growth rate. Ford et al. (1983) found that selection for rapid or slow leaf TPs produced no associated changes in crop growth rate. On the other hand, Shibles and Weber (1965, 1966) observed that crop growth rate depends on the percentage interception of solar radiation. And Heilman et al. (1977) were able to predict total canopy photosynthesis, respiration, and dry matter accumulation from measurements of PPFD and LAI. It is difficult to believe that photosynthesis could be related to seed yield (Table 14–3), but not to dry matter accumulation.

Table 14–3. Correlations of photosynthesis with soybean seed yield.

Source	r†	n	Period‡	Remarks
		Leaf TPs		
Buttery et al. (1981)	⩽0.37	12	July	Vegetative growth period
	⩾0.65*	12	August	Reproductive growth period
Ford et al. (1983)	0.53	9	R5+	Selected for rapid TPs
	0.18	9	R5+	Selected for slow TPs
		Net assimilation rate		
Chang (1979)	0.24	15	R1–R3	Means of best plots from
	0.57*	15	R3–R5	eight experiments
Zhang and Song (1982)	0.33	16	V3–R2	Spaced plants
	0.54*	16	R2–R5	
	0.68*	16	R5–R7	
		Canopy apparent photosynthesis		
Harrison et al. (1981)	0.55*	34	R4 & R5	F_3-derived lines Group V
	0.54*	34	R5 & R6	F_3-derived lines Group VI
Wells et al. (1982)§	–¶	16	R2.5–R4.5	0.13 in 1975; 0.45* in 1976
	0.45*	16	R4.5–R5.5	
	0.62*	16	R5.5 & over	
	0.64*	16	R4+	
Christy and Porter (1982)	0.99*	4,5	V3+	'Wells', means from different row spacings
	0.82	4	V3+	'Williams', means from shading treatments

*Significant at the 0.05 probability level.
†Correlations among means for genotypes unless otherwise indicated in Remarks.
‡Stages according to Fehr and Caviness (1977); '+' denotes an integrated rate over the period from the indicated stage to R7.
§Partial correlations holding the effects of maturity date constant.
¶No r calculated for combined years.

The closest relationships between seed yield and photosynthesis are found when photosynthesis is measured after flowering (Table 14–3). This agrees with the results of experiments using supplemental light, shade, or CO_2-enrichment to alter photosynthesis. These treatments affect seed yield when applied during pod filling; often have an effect if applied from flowering to the onset of pod filling, but rarely, if ever, influence seed yield if applied before flowering. On the other hand, Christy and Porter (1982) found that CAP varied among plant densities during vegetative but not reproductive growth; yet, the integral of CAP over the growing season was correlated with seed yield (Table 14–3).

Although, it seems that seed growth is largely supported by current photosynthesis, rather than by stored and remobilized assimilate, carbon gain during vegetative growth and flowering should have some role in determining seed yield. That role might be the maximization of floral initiation (Parker and Borthwick, 1940), pod set (Hardman and Brun, 1971), and seed set (Christy and Porter, 1982). Nevertheless, photosynthetic rates measured during reproduction are the only ones that correlate with seed yield (Table 14–3), and not always then (Ford et al., 1983; Hole, 1983).

There can be several reasons why photosynthesis might not correlate with seed yield or plant growth. Growth and yield are functions of (i) apparent photosynthesis per unit land area integrated over the growth period and (ii) the efficiency with which assimilate is converted to growth (carbon-use efficiency; Tanaka and Yamaguchi, 1968) or seed yield (photosynthetic conversion efficiency; Christy and Porter, 1982). Because growth seems to depend on current photosynthate, any failure to estimate accurately the integral of CAP over the growth period could result in a poor relationship. Instantaneous measurements of photosynthesis on single leaves, or even on whole canopies, may not reflect total photosynthate production over space or time. Another problem with measuring single-leaf AP is the negative correlation of leaf area and AP previously mentioned (see section 14–1.1). Thus, canopies with small, rapidly photosynthesizing leaves may or may not have an advantage over those with large, slowly photosynthesizing leaves, depending on the relative rates and total leafage. This point should be considered by those breeding for improved leaf AP; the problem is avoided by measuring CAP, which is a very cumbersome process, however, It is much easier to measure yield itself.

To date, variation in photosynthesis during reproductive growth has been found to explain, at most, 50% of the variation in seed yield among genotypes (Table 14–3). So the efficiency of photosynthate utilization seems equally important, yet it has not received equal attention. Kishitani and Shibles (1986) found no differences among cultivars for dark respiration rates of tissues. Hole (1983) found no differences in canopy respiration rates among six genotypes. However, these were exploratory studies, and we do not judge these data conclusive enough to state that variation for respiration among soybean cultivars does not exist. Christy and Porter (1982) found that CAP integrated over the growing season explained yield difference *within* cultivars, but suggested that the efficiency with which assimilate is converted to seed yield explained differences *among* cultivars.

In the absence of more definitive data, we can conclude only that photosynthesis is and must be an important determinant of soybean yield, but the efficiency of assimilate utilization may be equally important.

REFERENCES

Abu-Shakra, S.S., D.A. Phillips, and R.C. Huffaker. 1978. Nitrogen fixation and delayed leaf senescence in soybeans. Science 199:973–975.

Ackerson, R.C., U.D. Havelka, and M.G. Boyle. 1984. CO_2-enrichment effects on soybean physiology. II. Effects of stage-specific CO_2 exposure. Crop Sci. 24:1150–1154.

Amir, Jacob, and Jack Preiss. 1982. Kinetic characterization of spinach leaf sucrose-phosphate synthase. Plant Physiol. 69:1027–1030.

Andrews, A.K., and L.C. Svec. 1975. Photosynthetic activity of soybean pods at different growth stages compared to leaves. Can. J. Plant Sci. 55:501–505.

Aronoff, S. 1955. Translocation from soybean leaves. II. Plant Physiol. 30:184–185.

Azcon-Bieto, Joaquin, Hans Lambers, and D.A. Day. 1983. Effect of photosynthesis and carbohydrate status on respiratory rates and the involvement of the alternative pathway in leaf respiration. Plant Physiol. 72:598–603.

Baldocchi, D.D., S.B. Verma, and N.J. Rosenberg. 1981. Mass and energy exchanges of a soybean canopy under various environmental regimes. Agron. J. 73:706–710.

----, ----, ----, B.L. Blad, A. Garay, and J.E. Specht. 1983. Influence of water stress on the diurnal exchange of mass and energy between the atmosphere and a soybean canopy. Agron. J. 75:543–548.

Beuerlein, J.E., and J.W. Pendleton. 1971. Photosynthetic rates and light saturation curves of individual soybean leaves under field conditions. Crop Sci. 11:217–219.

Bhagsari, A.S., D.A. Ashley, R.H. Brown, and H.R. Boerma. 1977. Leaf photosynthetic characteristics of determinate soybean cultivars. Crop Sci. 17:929–932.

Blad, B.L., and D.G. Baker. 1972. Orientation and distribution of leaves within soybean canopies. Agron. J. 64:26–29.

Blomquist, R.V., and C.A. Kust. 1971. Translocation pattern of soybeans as affected by growth substances and maturity. Crop Sci. 11:390–393.

Borzenkova, R.A., and I.F. Bortnikova. 1978. Light dependence of kinetin action in the process of chloroplastogenesis. Sov. Plant Physiol. 25:201–206.

Bowes, George, and W.L. Ogren. 1972. Oxygen inhibition and other properties of soybean ribulose 1,5-diphosphate carboxylase. J. Biol. Chem. 247:2171–2176.

----, ----, and R.H. Hageman. 1971. Phosphoglycolate production catalysed by ribulose diphosphate carboxylase. Biochem. Biophys. Res. Commun. 45:716–722.

----, ----, ----. 1972. Light saturation, photosynthesis rate, RuDP carboxylase activity, and specific leaf weight in soybeans grown under different light intensities. Crop Sci. 12:77–79.

Brun, W.A., and R.L. Cooper. 1967. Effects of light intensity and carbon dioxide concentration on photosynthetic rates of soybean. Crop Sci. 7:451–454.

Bunce, J.A. 1977. Leaf elongation in relation to leaf water potential in soybean. J. Exp. Bot. 28:156–161.

Burton, J.W., R.F. Wilson, and C.A. Brim. 1979. Dry matter and nitrogen accumulation in male-sterile and male-fertile soybeans. Agron. J. 71:548–552.

Buttery, B.R. 1969. Analysis of the growth of soybeans as affected by plant population and fertilizer. Can. J. Plant Sci. 49:675–684.

----, and R.I. Buzzell. 1972. Some differences between soybean cultivars observed by growth analysis. Can. J. Plant Sci. 52:13–20.

----, and ----. 1973. Varietal differences in leaf flavonoids of soybeans. Crop Sci. 13:103–106.

----, and ----. 1976. Flavonol glycoside genes and photosynthesis in soybeans. Crop Sci. 16:547–550.

----, and ----. 1977. The relationship between chlorophyll content and rate of photosynthesis in soybeans. Can. J. Plant Sci. 57:1–5.

----, ----, and R.L. Bernard. 1977. Inheritance and linkage of a magenta flower gene in soybeans. Can. J. Genet. Cytol. 19:749–751.

----, ----, and W.I. Findlay. 1981. Relationships among photosynthetic rate, bean yield and other characters in field-grown cultivars of soybean. Can. J. Plant Sci. 61:191–198.

----, ----, and R.M. Shibles. 1980. Flavonol classes of cultivars in Maturity Groups 00–IV. Soybean Genet. Newsl. 7:22–26.

Cannell, R.Q., W.A. Brun, and D.N. Moss. 1969. A search for high net photosynthetic rate among soybean genotypes. Crop Sci. 9:840–841.

Carlson, D.R., and W.A. Brun. 1984. Effect of shortened photosynthetic period on ^{14}C-assimilate translocation and partitioning in reproductive soybeans. Plant Physiol. 75:881–886.

Chang, J. 1979. Correlation of leaf area and net photosynthesis rate to the yield of soybean. Chung-Kuo Nung Yeh K'o Hsueh May 1979:40–46.

Chatterton, N.J., and J.E. Silvius. 1979. Photosynthate partitioning into starch in soybean leaves. I. Effects of photoperiod *versus* photosynthetic period duration. Plant Physiol. 64:749–753.

----, and ----. 1981. Photosynthate partitioning into starch in soybean leaves. II. Irradiance level and daily photosynthetic period duration effects. Plant Physiol. 67:257–260.

Christy, A.L., and C.A. Porter. 1982. Canopy photosynthesis and yield in soybean. p. 499–511. *In* Govindjee (ed.) Photosynthesis: Development, carbon metabolism, and plant productivity, Vol. II. Academic Press, New York.

Ciha, A.J., M.L. Brenner, and W.A. Brun. 1978. Effect of pod removal on abscisic acid levels in soybean tissue. Crop Sci. 18:776–779.

----, and W.A. Brun. 1978. Effect of pod removal on nonstructural carbohydrate concentration in soybean tissue. Crop Sci. 18:773–776.

Clauss, Heinz, D.C. Mortimer, and P.R. Gorham. 1964. Time-course study of translocation of products of photosynthesis in soybean plants. Plant Physiol. 39:269–273.

Clough, J.M., and M.M. Peet. 1981. Effects of intermittent exposure to high atmospheric CO_2 on vegetative growth in soybean. Physiol. Plant. 53:565–569.

----, ----, and P.J. Kramer. 1981. Effects of high atmospheric CO_2 and sink size on rates of photosynthesis of a soybean cultivar. Plant Physiol. 67:1007–1010.

Coggeshall, B.M., and H.F. Hodges. 1980. The effect of carbohydrate concentration on the respiration rate of soybean. Crop Sci. 20:86–90.

Cooper, R.L., and W.A. Brun. 1967. Response of soybeans to a carbon dioxide enriched atmosphere. Crop Sci. 7:455–457.

Cosio, E.G., and J.W. McClure. 1984. Kaempferol glycosides and enzymes of flavonol biosynthesis in leaves of a soybean strain with low photosynthetic rates. Plant Physiol. 74:877–881.

Créach, Elisabeth, and C.R. Stewart. 1982. Effects of amino-acetonitrile on net photosynthesis, ribulose-1, 5-bisphosphate levels, and glycolate pathway intermediates. Plant Physiol. 70:1444–1448.

Cséke, Csaba, N.F. Weeden, B.B. Buchanan, and Kosaku Uyeda. 1982. A special fructose bisphosphate functions as a cytoplasmic regulatory metabolite in green leaves. Proc. Natl. Acad. Sci. USA 79:4322–4326.

Cure, J.D., R.P. Patterson, C.D. Raper, Jr., and W.A. Jackson. 1982. Assimilate distribution in soybeans as affected by photoperiod during seed development. Crop Sci. 22:1245–1250.

Curtis, C.E., W.L. Ogren, and R.H. Hageman. 1969. Varietal effects in soybean photosynthesis and photorespiration. Crop Sci. 9:323–327.

Doehlert, D.C., and S.C. Huber. 1983. Spinach leaf sucrose phosphate synthase: Activation by glucose-6-phosphate and interaction with inorganic phosphate. FEBS Lett. 153:293–297.

Dornoff, G.M., and R.M. Shibles. 1970. Varietal differences in net photosynthesis of soybean leaves. Crop Sci. 10:42–45.

----, and ----. 1974. Soybean leaf net CO_2-exchange as influenced by experimental environment and leaf age. Iowa State J. Res. 58:311–317.

----, and ----. 1976. Leaf morphology and anatomy in relation to CO_2-exchange rate of soybean leaves. Crop Sci. 16:377–381.

Dreger, R.H., W.A. Brun, and R.L. Cooper. 1969. Effect of genotype on the photosynthetic rate of soybean. Crop Sci. 9:429–431.

Du, Weiguang, Wang Yumin, and Tan Kehui. 1982. Varietal difference in photosynthetic activity of soybean and its relation to yield. Acta Agron. Sin. 8:131–135.

Dunphy, E.J., and J.J. Hanway. 1976. Water-soluble carbohydrate accumulation in soybean plants. Agron. J. 68:697–700.

Egli, D.B., D.R. Gossett, and J.E. Leggett. 1976. Effect of leaf and pod removal on the distribution of ^{14}C labeled assimilate in soybeans. Crop Sci. 16:791–794.

----, and J.E. Leggett. 1976. Rate of dry matter acumulation in soybean seeds with varying source-sink ratios. Agron. J. 68:371–374.

----, ----, and Audrey Cheniae. 1980. Carbohydrate levels in soybean leaves during reproductive growth. Crop Sci. 20:468–473.

----, J.W. Pendleton, and D.B. Peters. 1970. Photosynthetic rate of three soybean communities as related to carbon dioxide levels and solar radiation. Agron. J. 62:411–414.

----, and I.F. Wardlaw. 1980. Temperature response of seed growth characteristics of soybeans. Agron. J. 72:560–564.

Elmore, C.D. 1980. The paradox of no correlation between leaf photosynthetic rates and crop yields. p. 155–167. *In* J.D. Hesketh and J.W. Jones (ed.) Predicting photosynthesis for ecosystem models, Vol. II. CRC Press, Boca Raton, FL.

Enos, W.T., R.A. Alfich, J.D. Hesketh, and J.T. Woolley. 1982. Interactions among leaf photosynthetic rates, flowering and pod set in soybeans. Photosynth. Res. 3:273–278.

Fader, G.M., and H.R. Koller. 1983. Relationship between carbon assimilation, partitioning, and export in leaves of two soybean cultivars. Plant Physiol. 73:297–303.

Farquhar, G.D., S. von Caemmerer, and J.A. Berry. 1980. A biochemical model of photosynthetic CO; assimilation in leaves of C_3 species. Planta 149:78–90.

Fehr, W.R., and C.E. Caviness. 1977. Stages of soybean development. Iowa Coop. Ext. Serv. Spec. Rep. 80.

Fellows, R.J., D.B. Egli, an J.E. Leggett. 1979. Rapid changes in translocation patterns in soybeans following source-sink alterations. Plant Physiol. 64:652–655.

Finn, G.A., and W.A. Brun. Effect of atmospheric CO_2 enrichment on growth, nonstructural carbohydrate content, and root nodule activity in soybean. Plant Physiol. 69:327–331.

Fisher, D.B. 1967. An unusual layer of cells in the mesophyll of the soybean leaf. Bot. Gaz. 128:215–218.

----. 1970a. Kinetics of C-14 translocation in soybean. I. Kinetics in the stem. Plant Physiol. 45:107–113.

----. 1970b. Kinetics of C-14 translocation in soybean. II. Kinetics in the leaf. Plant Physiol. 45:114–118.

Ford, D.M. 1984. Leaf physiology and anatomy of soybean genotypes differing in leaf photosynthetic ability. Ph.D. diss. Iowa State Univ., Ames (Diss. Abstr. Int. 45:1965B).

----, Richard Shibles, and D.E. Green. 1983. Growth and yield of soybean lines selected for divergent leaf photosynthetic ability. Crop Sci. 23:517–520.

Forrester, M.L., G. Krotkov, and C.D. Nelson. 1966. Effect of oxygen on photosynthesis, photorespiration and respiration in detached leaves. I. Soybean. Plant Physiol. 41:422–427.

Franceschi, V.R., and R.T. Giaquinta. 1983a. The paraveinal mesophyll of soybean leaves in relation to assimilate transfer and compartmentation. I. Ultrastructure and histochemistry during vegetative development. Planta 157:411–421.

----, and ----. 1983b. The paraveinal mesophyll of soybean leaves in relation to assimilate transfer and compartmentation. II. Structure, metabolic and compartmental changes during reproductive growth. Planta 157:422–431.

----, V.A. Wittenbach, and R.T. Giaquinta. 1983. Paraveinal mesophyll of soybean leaves in relation to assimilate transfer and compartmentation. III. Immunohistochemical localization of specific glycoproteins in the vacuole after depodding. Plant Physiol. 72:586–589.

Fukui, Juro, Mutsuo Ojima, and Iwao Wantanabe. 1965. Studies on the seed production of soybean. I. Effect of temperature on photosynthesis of soybean. Proc. Crop Sci. Soc. Jpn. 33:432–436.

Gaastra, P. 1962. Photosynthesis of leaves and field crop. Neth. J. Agric. Sci. Spec. Issue 10(5):311–324.

Gallaher, R.N., R.H. Brown, D.A. Ashley, and J.B. Jones Jr. 1976. Photosynthesis of, and $^{14}CO_2$-photosynthate translocation from, calcium-deficient leaves of crops. Crop Sci. 16:116–119.

Gay, Scott, D.B. Egli, and D.A. Reicosky. 1980. Physiological aspects of yield improvement in soybeans. Agron. J. 72:387–391.

Gent, M.P.M. 1982. Effect of defoliation and depodding on long distance translocation and yield in Y-shaped soybean plants. Crop Sci. 22:245–250.

Ghorashy, S.R., J.W. Pendleton, D.B. Peters, J.S. Boyer, and J.E. Beuerlein. 1971. Internal water stress and apparent photosynthesis with soybeans differing in pubescence. Agron. J. 63:674–676.

Giaquinta, R.T. 1983. Phloem loading of sucrose. Annu. Rev. Plant Physiol. 34:347–387.

----, B.Quebedeaux, and V. Wittenbach. 1981. Alterations in photosynthesis and assimilate partitioning between starch and sucrose in soybean leaves during seed filling. p. 549–550. *In* G. Akoyunoglou (ed.) Photosynthesis 4, Regulation of carbon metabolism. Proc. 5th Int. Congr. Photosynthesis, Halkidiki, Greece. (7–13 Sept. 1980). Balaban International Science Services, Philadelphia.

Gordon, A.J., J.D. Hesketh, and D.B. Peters. 1982. Soybean leaf photosynthesis in relation to maturity classification and stage of growth. Photosynth. Res. 3:81–93.

Gourdon, F., and C. Planchon. 1982. Responses of photosynthesis to irradiance and temperature in soybean, *Glycine max* (L.) Merr. Photosynth. Res. 3:31–43.

Hansen, W.R. 1972. Net photosynthesis and evapotranspiration of field-grown soybean canopies. Ph.D. diss. Iowa State University Library, Ames (Diss. Abstr. Int. 33:4619B).

Hanson, W.D., and S.T. Kenny. 1985. Genotypic differences in soybean affecting the rates of assimilate transport from the leaf. Crop Sci. 25:229–234.

Hanway, J.J., and C.R. Weber. 1971. Dry matter accumulation in eight soybean [*Glycine max* (L.) Merrill] varieties. Agron. J. 63:227–230.

Hardman, L.L., and W.A. Brun. 1971. Effect of atmospheric carbon dioxide enrichment at different developmental stages on growth and yield components of soybeans. Crop Sci. 11:886–888.

Harrison, S.A., H.R. Boerma, and D.A. Ashley. 1981. Heritability of canopy-apparent photosynthesis and its relationship to seed yield in soybeans. Crop Sci. 21:222–226.

Hatfield, J.L., and R.E. Carlson. 1978. Photosynthetically active radiation, CO_2 uptake, and stomatal diffusive resistance profiles within soybean canopies. Agron. J. 70:592–596.

Havelka, U.D., R.C. Ackerson, and M.G. Boyle. 1984. CO_2-enrichment effects on soybean physiology. I. Effects of long-term CO_2 exposure. Crop Sci. 24:1146–1150.

Heber, Ulrich, and H.W. Heldt. 1981. The chloroplast envelope: Structure, function, and role in leaf metabolism. Annu. Rev. Plant Physiol. 32:139–168.

Heilman, J.L., E.T. Kanemasu, and G.M. Paulsen. 1977. Estimating dry-matter accumulation in soybean. Can. J. Bot. 55:2196–2201.

Hesketh, J.D., W.L. Ogren, M.E. Hageman, and D.B. Peters. 1981. Correlations among leaf CO_2-exchange rates, areas and enzyme activities among soybean cultivars. Photosynth. Res. 2:21–30.

Hew, C.S., C.D. Nelson, and G. Krotkov. 1967. Hormonal control of translocation of phtosynthetically assimilated ^{14}C in young soybean plants. Am. J. Bot. 54:252–256.

Hirata, Masahiko, Ryuichi Ishii, Atsuhiko Kumura, and Yoshio Murata. 1983a. Photoinhibition of photosynthesis in soybean leaves. I. Effects of different intensities and durations of light irradiation on light response curve of photosynthesis. Jpn. J. Crop Sci. 52:314–318.

----, ----, ----, and ----. 1983b. Photoinhibition of photosynthesis in soybean leaves. II. Leaf orientation-adjusting movement as a possible avoiding mechanism of photoinhibition. Jpn. J. Crop Sci. 52:319–322.

Hitz, W.D., and C.R. Stewart. 1980. Oxygen and carbon dioxide effects on the pool size of some photosynthetic and photorespiratory intermediates in soybean (*Glycine max* [L.] Merr.). Plant Physiol. 65:442–446.

Hoffstra, G., and J.D. Hesketh. 1975. The effects of temperature and CO_2 enrichment on photosythnesis in soybean. p. 71–80. *In* R. Marcelle (ed.) Environmental and biological control of photosynthesis. Junk, The Hague.

Hole, D.J. 1983. Canopy characteristics, carbon dioxide balance, and yield in soybeans. Unpublished M.S. thesis. Iowa State University Library, Ames.

Housley, T.L, D.M. Peterson, and L.E. Schrader. 1977. Long distance translocation of sucrose, serine, leucine, lysine, and CO_2 assimilates. I. Soybean. Plant Physiol. 59:217–220.

----, L.E. Schrader, Marna Miller, and T.L. Setter. 1979. Partitioning of ^{14}C-photosynthate, and long distance translocation of amino acids in preflowering and flowering, nodulated and nonnodulated soybeans. Plant Physiol. 64:94–98.

Huber, S.C. 1981a. Inter- and intra-specific variation in photosynthetic formation of starch and sucrose. Z. Pflanzenphysiol. 101:49–54.

----. 1981b. Interspecific variation in activity and regulation of leaf sucrose photsphate synthetase. Z. Pflanzenphysiol. 102:443–450.

----. 1983. Role of sucrose-phosphate synthase in partitioning of carbon in leaves. Plant Physiol. 71:818–821.

----, and D.M. Bickett. 1984. Evidence for control of carbon partitioning by fructose 2,6-bisphosphate in spinach leaves. Plant Physiol. 74:445–447.

----, D.C. Doehlert, T.W. Rufty, and P.S. Kerr. 1984b. Regulation of sucrose phosphate synthase activity in leaves. p. 605–608. *In* C.N. Sybesma (ed.) Advances in photosynthesis research. Proc. 6th Int. Congr. Photosynthesis, Brussels, Belgium. 1–6 Aug. 1983. Nijhoff, The Hague.

----, and D.W. Israel. 1982. Biochemical basis for partitioning of photosynthetically fixed carbon between starch and sucrose in soybean (*Glycine max* Merr.) leaves. Plant Physiol 69:691–696.

----, H. Rogers, and D.W. Israel. 1984b. Effects of CO_2 enrichment on photosynthesis and photosynthate partitioning in soybean (*Glycine max*) leaves. Physiol. Plant. 62:95–101.

----, and F.L. Mowry. 1984c. Effects of water stress on photosynthesis and carbon partitioning in soybean (*Glycine max* [L.] Merr.) plants grown in the field at different CO_2 levels. Plant Physiol. 76:244–249.

----, T.W. Rufty, Jr., P.S. Kerr, and D.C. Doehlert. 1983a. Different mechanisms for the regulation of sucrose phosphate synthase—a key enzyme in photosynthetic sucrose formation. p. 20–34. *In* D.D. Randall, et al. (ed.) Current topics in plant biochemistry and physiology 2. University of Missouri, Columbia.

----, R.F. Wilson, and J.W. Burton. 1983b. Studies on genetic male-sterile soybeans. II. Effect of nodulation on photosynthesis and carbon partitioning in leaves. Plant Physiol. 73:713–717.

Hume, D.J., and J.G. Criswell. 1973. Distribution and utilization of ^{14}C-labelled assimilates in soybeans. Crop Sci. 13:519–524.

Hsu, F.C., A.B. Bennett, and R.M. Spanswick. 1984. Concentrations of sucrose and nitrogenous compounds in the apoplast of developing soybean seed coats and embryos. Plant Physiol 75:181–186.

Imai, Katsu, and Yoshio Murata. 1979. Effect of carbon dioxide concentration on growth and dry matter production of crop plants. VII. Influence of light intensity and temperature on the effect of carbon dioxide-enrichment in some C3- and C4-species. Jpn. J. Crop Sci. 48:409–417.

Jeffers, D.L., and R.M. Shibles. 1969. Some effects of leaf area, solar radiation, air temperature, and variety on net photosynthesis in field-grown soybeans. Crop Sci. 9:762–764.

Johnston, T.J., J.W. Pendleton, D.B. Peters, and D.R. Hicks. 1969. Influence of supplemental light on apparent photosynthesis, yield, and yield components of soybeans (*Glycine max* L.). Crop Sci. 9:577–581.

Jordon, D.B., and W.L. Ogren. 1981a. A sensitive assay procedure for simultaneous determination of ribulose-1,5-bisphosphate carboxylase and oxygenase activities. Plant Physiol. 67:237–245.

----, and ----. 1981b. Species variation in the specificity of ribulose bisphosphate carboxylase/oxygenase. Nature (London) 291:513–515.

Kanemasu, E.T., and C.K. Hiebsch. 1975. Net carbon dioxide exchange of wheat, sorghum, and soybean. Can. J. Bot. 53:382–389.

Kaplan, S.L., and H.R. Koller. 1977. Leaf area and CO_2-exchange rate as determinants of the rate of vegetative growth in soybean plants. Crop Sci. 17:35–38.

Kawashima, Ryoichi. 1969a. Studies on the leaf orientation-adjusting movement in soybean plants. I. The leaf orientation-adjusting movement and light intensity on leaf surface. Proc. Crop Sci. Soc. Jpn. 38:718–729.

----. 1969b. Studies on the leaf orientation-adjusting movement in soybean plants. II. Fundamental pattern of the leaf orientation-adjusting movement and its significance for the dry matter production. Proc. Crop Sci. Soc. Jpn. 38:730–742.

Kerr, P.S., T.W. Rufty, Jr., and S.C. Huber. 1985a. Endogenous rhythms in photosynthesis, sucrose phosphate synthase activity, and stomatal resistance in leaves of soybean (*Glycine max* [L.] Merr.). Plant Physiol. 77:275–280.

----, ----, and ----. 1985b. Changes in nonstructural carbohydrates in different parts of soybean (*Glycine max* [L.] Merr.) plants during a light/dark cycle and extended darkness. Plant Physiol. 78:576–581.

Kishitani, Sachie, and Richard Shibles. 1986. Respiration rates of soybean cultivars. Crop Sci. 26:580–583.

Kumura, Atsuhiko. 1968. Studies on dry matter production of soybean plant. 3. Photosynthetic rate of soybean plant population as affected by proportion of diffuse light. Proc. Crop Sci. Soc. Jpn. 37:570–582.

----, and Isao Naniwa. 1965. Studies on dry matter production of soybean plant. I. Ontogenetic changes in photosynthetic and respiratory capacity of soybean plant and its parts. Proc. Crop Sci. Soc. Jpn. 33:467–472.

----. 1969. Studies on dry matter production in soybean plant. V. Photosynthetic system of soybean plant population. Proc. Crop Sci. Jpn. 38:74–90.

Laing, W.A., W.L. Ogren, and R.H. Hageman. 1974. Regulation of soybean net photosynthetic CO_2 fixation by the interaction of CO_2, O_2, and ribulose 1,5-diphosphate carboxylase. Plant Physiol. 54:678–685.

Larson, E.M., J.D. Hesketh, J.T. Woolley, and D.B. Peters. 1981. Seasonal variations in apparent photosynthesis among plant stands of different soybean cultivars. Photosynth. Res. 2:3–20.

Lawn, R.J., and W.A. Brun. 1974. Symbiotic nitrogen fixation in soybeans. I. Effect of photosynthetic source-sink manipulations. Crop Sci. 14:11–16.

Layzell, D.B., and T.A. LaRue. 1982. Modeling C and N transport to developing soybean fruits. Plant Physiol. 70:1290–1298.

Lichtner, F.T., and R.M. Spanswick. 1981a. Electrogenic sucrose transport in developing soybean cotyledons. Plant Physiol. 67:869–874.

----, and ----. 1981b. Sucrose uptake by developing soybean cotyledons. Plant Physiol. 68:693–698.

Lindoo, S.J., and L.D. Noodén. 1976. The interrelation of fruit development and leaf senescence in 'Anoka' soybeans. Bot. Gax. 137:218–223.

----, and ----. 1977. Studies on the behavior of the senescence signal in Anoka soybeans. Plant Physiol. 59:1136–1140.

----, and ----. 1978. Correlation of cytokinins and abscisic acid with monocarpic senescence in soybeans. Plant Cell Physiol. 19:997–1006.

Lugg, D.G., and T.R. Sinclair. 1980. Seasonal changes in morphology and anatomy of field-grown soybean leaves. Crop Sci. 20:191–196.

----, and ----. 1981. Seasonal changes in photosynthesis of field-grown soybean leaflets. 1. Relation to leaflet dimensions. Photosynthetica 15:129–137.

Mauney, J.R., K.E. Fry,. and G. Guinn. 1978. Relationship of photosynthetic rate to growth and fruiting of cotton soybean, sorghum, and sunflower. Crop Sci. 18:259–263.

----, Gene Guinn, K.E. Fry, and J.D. Hesketh. 1979. Correlation of photosynthetic carbon dioxide uptake and carbohydrate accumulation of cotton, soybean, sunflower, and sorghum. Photosynthetica 13:260–266.

McCree, K.J. 1983. Carbon balance as a function of plant size in sorghum plants. Crop Sci. 23:1173–1177.

----, and J.H. Silsbury. 1978. Growth and maintenance requirements of subterranean clover. Crop Sci. 18:13–18.

Menz, K.M., D.N. Moss, R.Q. Cannell, and W.A. Brun. 1969. Screening for photosynthetic efficiency. Crop Sci. 9:692–694.

Miller, M.G., A.C. Leopold, and R.L. Obendorf. 1983. Respiration during seed maturation. Physiol. Plant. 57:397–401.

Mkandawire, A.B.C. 1984. Respiration rates of soybean cultivars in the field. Unpublished M.S. thesis. Iowa State University Library, Ames.

Mondal, M.H., W.A. Brun, and M.L. Brenner. 1978a. Effects of sink removal on photosynthesis and senescence in leaves of soybean (*Glycine max* L.) plants. Plant Physiol. 61:394–397.

----, ----, and ----. 1978b. IAA levels and photosynthesis in leaves of control and depodded soybean plants. Plant Physiol. Suppl. 61:8.

Nafziger, E.D., and H.R. Koller. 1976. Influence of leaf starch concentration on CO_2 assimilation in soybean. Plant Physiol. 57:560–563.

Nakaseko, Kimio, and Kanji Gotoh. 1981. Comparative studies on dry matter production, plant type and productivity in soybean, azuki bean, and kidney bean. III. Dry matter production of soybean plant at various population densities. Jpn. J. Crop Sci. 50:38–46.

----, ----, and Koichiro Asanuma. 1979. Comparative studies on dry matter production, plant type and productivity in soybean, azuki bean, and kidney bean. II. Relationships between vertical distribution of leaf area and some morphological characteristics. Jpn. J. Crop Sci. 48:92–98.

Nelson, C.D., Heinz Clauss, D.C. Mortimer, and P.R. Gorham. 1961. Selective translocation of products of photosynthesis in soybean. Plant Physiol. 36:586–588.

Noodén, L.D., G.M. Kahanak, and Y. Okantan. 1979. Prevention of monocarpic senescence in soybeans with auxin and cytokinin: An antidote for self-destruction. Science 206:841–843.

Ogren, W.L., and Raymond Chollet. 1982. Photorespiration. p. 191–230. *In* Govindjee (ed.) Photosynthesis, Vol. 2. Development, carbon metabolism, and plant productivity. Academic Press, New York.

Ojima, Mutsuo, and Juro Fukui. 1966. Studies on the seed production of soybean. 3. An analytical study of dry matter production in the soybean plant community. Proc. Crop Sci. Soc. Jpn. 34:448–452.

----, ----, and Iwao Wantanabe. 1965. Studies on the seed production of soybean. II. Effect of three major nutrient elements supply and leaf age on the photosynthetic activity and diurnal changes in photosynthesis of soybean under constant temperature and light intensity. Proc. Crop Sci. Soc. Jpn. 33:437–442.

----, and Ryoichi Kawashima. 1968. Studies on the seed production of soybean. 5. Varietal differences in photosynthetic rate of soybean. Proc. Crop Sci. Soc. Jpn. 37:667–675.

----, and ----. 1970. Studies on the seed production of soybean. VIII. The ability of photosynthesis in F_3 lines having different photosynthesis in their F_2 generation. Proc. Crop Sci. Soc. Jpn. 39:440–445.

----, ----, and Shin-ichi Sakamoto. 1968. Studies on the seed production of soybean. 6. Relationship between the activity of photosynthesis of improved varieties and that of the parent ones. Proc. Crop Sci. Soc. Jpn. 37:676–679.

----, Juro Kawashima, and Kimito Mikoshiba. 1969. Studies on the seed production of soybean. VII. The ability of photosynthesis in F_1 and F_2 generations. Proc. Crop Sci. Soc. Jpn. 38:693–699.

Okatan, Yener, G.M. Kahanak, and L.D. Noodén. 1981. Characterization and kinetics of soybean maturation and monocarpic senescence. Physiol. Plant. 52:330–338.

Oritani, Takashi, and Ryuji Yoshida. 1973. Studies on nitrogen metabolism in crop plants. XII. Cytokinins and abscisic acid-like substance levels in rice and soybean leaves during their growth and senescence. Proc. Crop Sci. Soc. Jpn. 42:280–287.

Parker, M.W., and H.A. Borthwick. 1940. Floral initiation in Biloxi soybeans as influenced by photosynthetic activity during the induction period. Bot. Gaz. 102:256–268.

Peet, M.M., and P.J. Kramer. 1980. Effects of decreasing source/sink ratio in soybeans on photosynthesis, photorespiration, transpiration and yield. Plant Cell Environ. 3:201–206.

Penning de Vries, F.W.T., A.H.M. Brunsting, and H.H. van Laar. 1974. Products, requirements, and efficiency of biosynthesis: A quantitative approach. J. Theor. Biol. 45:339–377.

Potter, J.R., and P.J. Breen. 1980. Maintenance of high photosynthesis rates during the accumulation of high leaf starch levels in sunflowers and soybean. Plant Physiol. 66:528–531.

Preiss, Jack. 1982. Regulation of the biosynthesis and degradation of starch. Annu. Rev. Plant Physiol. 33:431–454.

Quebedeaux, Bruno, and Raymond Chollet. 1975. Growth and development of soybean (*Glycine max* [L.] Merr.) pods. CO_2 exchange and exzyme studies. Plant Physiol. 55:745–748.

Rainbird, R.M., J.H. Thorne, and R.W.F. Hardy. 1984. Role of amides, amino acids, and ureides in the nutrition of developing soybean seeds. Plant Physiol. 74:329–334.

Rufty, T.W. Jr., and S.C. Huber. 1983. Changes in starch formation and activities of sucrose phosphate synthase and cytoplasmic fructose-1,6-bisphosphatase in response to source-sink alterations. Plant Physiol. 72:474–480.

----, P.S. Kerr, and S.C. Huber. 1983. Characterization of diurnal changes in activities of enzymes involved in sucrose biosynthesis. Plant Physiol. 73:428–433.

Russel, W.J., and D.R. Johnson. 1975. Carbon-14 assimilate translocation in nodulated and nonodulate soybeans. Crop Sci. 15:159–161.

Saito, Masataka, Yadashi Yamaimoto, Kazuo Gato, and Koji Hashimoto. 1970. The influence of cool temperature before and after anthesis, on pod-setting and nutrients in soybean plants. Proc. Crop Sci. Soc. Jpn. 39:511–519.

Sakamoto, C.M., and R.H. Shaw. 1967. Apparent photosynthesis in field soybean communities. Agron. J. 59:73–75.

Sanders, T.H., D.A. Ashley, and R.H. Brown. 1977. Effects of partial defoliation on petiole area, photosynthesis, and ^{14}C translocation in developing soybean leaves. Crop Sci. 17:548–550.

Sambo, E.Y., J. Moorby, and F.L. Milthorpe. 1977. Photosynthesis and respiration of developing soybean pods. Aust. J. Plant Physiol. 4:713–721.

Samet, J.S., and T.R. Sinclair. 1980. Leaf senescence and abscisic acid in leaves of field-grown soybean. Plant Physiol 66:1164–1168.

Satterlee, L.D., and H.R. Koller. 1984. Response of soybean fruit respiration to changes in whole plant light and CO_2 environment. Crop Sci. 24:1007–1010.

Schulze, L.L., D.A. Ashley, H.R. Boerma, and R.H. Brown. 1976. Canopy photosynthesis in determinant soybeans. Agron. Abstr. American Society of Agronomy, Madison WI, p. 77.

Schrader, L.E., T.L. Housley, and J.C. Servaites. 1980. Amino acid loading and transport in phloem. p. 101–109. *In* F.T. Corbin (ed.) World soybean research conference II: Proc., North Carolina State Univ., Raleigh. 26–29. Mar. 1979. Westview Press, Boulder, CO.

----, and R.J. Thomas. 1981. Nitrate uptake reduction and transport in the whole plant. p. 49–93. *In* J.D. Bewley (ed.) Nitrogen and carbon metabolism. Nijhoff/Junk, The Hague.

Schussler, J.R., M.L. Brenner, and W.A. Brun. 1984. Abscisic acid and its relationship to seed filling in soybeans. Plant Physiol. 76:301–306.

Secor, Jacob. 1979. Photosynthesis and other leaf characteristics in selected soybean lines. Unpublished M.S. thesis. Iowa State University Library, Ames.

----. D.M. Ford, and Richard Shibles. 1982a. Ontogenetic changes in ribulose-1,5-bisphosphate carobxylase-oxygenase activity in soybean leaves. Plant Sci. Lett. 27:147–154.

----, D.R. McCarty, Richard Shibles, and D.E. Green. 1982b. Variability and selection for leaf photosynthesis in advanced generations of soybeans. Crop Sci. 22:255–259.

----, and L.E. Schrader. 1984. Characterization of amino acid efflux from isolated soybean cells. Plant Physiol. 74:26–31.

----, Richard Shibles, an C.R. Stewart. 1983. Metabolic changes in senescing soybean leaves of similar plant ontogeny. Crop Sci. 23:106–110.

----, ----, and ----. 1984. A metabolic comparison between progressive and monocarpic senescence of soybean. Can. J. Bot. 62:806–811.

Seemann, J.R., and J.A. Berry. 1982. Interspecific differences in the kinetic properties of RuBP carboxylase protein. Year Book—Carnegie Inst. Washington 81:78–83.

Servaites, J.C., and W.L. Ogren. 1977. Chemical inhibition of the glycolate pathway in soybean leaf cells. Plant Physiol. 60:461–466.

----, L.E. Schrader, and D.M. Jung. 1979. Energy-dependent loading of amino acids and sucrose into the phloem of soybean. Plant Physiol. 64:546–550.

Sesay, A., and R. Shibles. 1980. Mineral depletion and leaf senescence in soya bean as influenced by foliar nutrient application during seed filling. Ann. Bot. 45:47–55.

----, ----, and Cecil Stewart. 1985. Cyanide-insensitive respiration in soybean leaves. Plant Physiol. (Suppl.) 77:94.

Seth, A.K., and P.F. Wareing. 1967. Hormone-directed transport of metabolites and its possible role in plant senescence. J. Exp. Bot. 18:65–77.

Setter, T.L., W.A. Brun, and M.L. Brenner. 1980a. Stomatal closure and photosynthetic inhibition in soybean leaves induced by petiole girdling and pod removal. Plant Physiol. 65:884–887.

----, ----, and ----. 1980b. Effect of obstructed translocation on leaf abscisic acid, and associated stomatal closure and photosynthesis decline. Plant Physiol. 65:1111–1115.

----, ----, and ----. 1981. Abscisic acid translocation and metabolism in soybeans following depodding and petiole girdling treatments. Plant Physiol. 67:774–779.

Shaw, R.H., and C.R. Weber. 1967. Effects of canopy arrangements on light interception and yield of soybeans. Agron. J. 59:155–159.

Shibles, R.M., and C.R. Weber. 1965. Leaf area, solar radiation interception and dry matter production by soybeans. Crop Sci. 5:575–577.

----, and ----. 1966. Interception of solar radiation and dry matter production by various soybean planting patterns. Crop Sci. 6:55–59.

Silvius, J.E., N.J. Chatterton, and D.F. Kremer. 1979. Photosynthate partitioning in soybean leaves at two irradiance levels. Comparative responses of acclimated and unacclimated leaves. Plant Physiol. 64:872–875.

----, D.F. Kremer, and D.R. Lee. 1978. Carbon assimilation and translocation in soybean leaves at different stages of development. Plant Physiol. 62:54–58.

Sinclair, T.R. 1980. Leaf CER from post-flowering to senescence of field-grown soybean cultivars. Crop Sci. 20:196–200.

Soinit, Nasser. 1983. Response of soybean to two levels of mineral nutrition in CO_2-enriched atmosphere. Crop Sci. 23:329–333.

----, H. Hellmers, and B.R. Strain. 1982. Interaction of atmospheric CO_2 enrichment and irradiance on plant growth. Agron. J. 74:721–725.

Spaeth, S.C., and T.R.Sinclair. 1983a. Carbon exchange rate of intact individual soya bean pods. 1. Response to step changes in light and temperature. Ann. Bot. 51:331–338.

----, and ----. 1983b. Carbon exchange rate of intact individual soya bean pods. 2. Ontogeny of temperature sensitivity. Ann. Bot. 51:339–346.

Stephenson, R.A., and G.L. Wilson. 1977a. Patterns of assimilate distribution in soybeans at maturity. I. The influence of reproductive developmental stage and leaf position. Aust. J. Agric. Res. 28:203–209.

----, and ----. 1977b. Patterns of assimilate distribution in soybeans at maturity. II. The time course of changes in ^{14}C distribution in pods and stem sections. Aust. J. Agric. Res. 28:395–400.

Stitt, Mark, Richard Gerhardt, Birgit Kurzel, and H.W. Heldt. 1983a. A role for fructose 2,6-bisphosphate in the regulation of sucrose synthesis in spinach leaves. Plant Physiol. 72:1139–1141.

----, Gottfried Mieskes, Hans-Dieter Soling, and H.W. Heldt. 1982. On a possible role of fructose-2,6-bisphosphate in regulating photosynthetic metabolism in leaves. FEBS Lett. 145:217–222.

----, Wolfgang Wirtz, and H.W. Heldt. 1983b. Regulation of sucrose synthesis by cytoplasmic fructose bisphosphatase and sucrose phosphate synthase during photosynthesis in varying light and carbon dioxide. Plant Physiol. 72:767–774.

Streeter, J.G., and D.L. Jeffers. 1979. Distribution of total nonstructural carbohydrates in soybean plants having increased reproductive load. Crop Sci. 19:729–734.

Tanaka, A., and Y. Yamaguchi. 1968. The growth efficiency in relation to the growth of the rice plant. Soil Sci. Plant Nutr. 14:110–116.

Tamas, I.A., C.J. Engels, S.L. Kaplan, J.L. Ozbun, and D.H. Wallace. 1981. Role of indoleacetic acid and abscisic acid in the correlative control by fruits of axillary bud development and leaf senescence. Plant Physiol. 68:476–481.

Thaine, R., S.L. Ovenden, and J.S. Turner. 1959. Translocation of labelled assimilates in the soybean. Aust. J. Biol. Sci. 12:349–372.

Thomas, J.F., C.D. Raper, Jr., and W.W. Weeks. 1981. Day and night temperature effects on nitrogen and soluble carbohydrate allocation during early reproductive growth in soybeans. Agron. J. 73:577–582.

Thorne, J.H. 1979. Assimilate redistribution from soybean pod walls during seed development. Agron. J. 71:812–816.

----. 1980. Kinetics of ^{14}C-photosynthate uptake by developing soybean fruit. Plant Physiol. 65:975–979.

----. 1981. Morphology and ultrastructure of maternal seed tissues of soybean in relation to the import of photosynthate. Plant Physiol. 67:1016–1025.

----. 19082a. Temperature and oxygen effects on ^{14}C-photosynthate unloading and accumulation in developing soybean seeds. Plant Physiol. 69:48–53.

----. 1982b. Characterization of the active sucrose transport system of immature soybean embryos. Plant Physiol. 70:953–958.

----, and H.R. Koller. 1974. Influence of assimilate demand on photosynthesis, diffusive resistances, translocation, and carbohydrate levels of soybean leaves. Plant Physiol. 54:201–207.

----, and R.M. Rainbird. 1983. An *in vivo* technique for the study of phloem unloading in seed coats of developing soybean seeds. Plant Physiol. 72:268–271.

Thrower, S.L. 1962. Translocation of labelled assimilates in the soybean. II. The pattern of translocation in intact and defoliated plants. Aust. J. Biol. Sci. 15:629–649.

Turner W.B., and R.G.S. Bidwell. 1965. Rates of photosynthesis in attached and detached bean leaves, and the effect of spraying with indoleacetic acid solution. Plant Physiol. 40:446–451.

Upmeyer, D.J., and H.R. Koller. 1973. Diurnal trends in net photosynthetic rate and carbohydrate levels of soybean leaves. Plant Physiol. 51:871–874.

Vendeland, J.S., D.K. Bruck, and T.R. Sinclair. 1982. Differential starch accumulation in the leaf mesophyll layers of soybean. Crop Sci. 22:1251–1252.

Vernon, L.P., and S. Aronoff. 1952. Metabolism of soybean leaves. IV. Translocation from soybean leaves. Arch. Biochem. Biophys. 36:383–398.

von Caemmerer, S., and G.D. Farquhar. 1981. Some relationships between the biochemistry of photosynthesis and the gas exchange of leaves. Planta 153:376–387.

Vu, C.V., L.H. Allen, Jr., and George Bowes. 1983. Effects of light and elevated atmospheric CO_2 on the ribulose bisphosphate level of soybean leaves. Plant Physiol. 73:729–734.

Wallace, William, Jacob Secor, and L.E. Schrader. 1984. Rapid accumulation of γ-aminobutyric acid and alanine in soybean leaves in response to an abrupt transfer to lower temperature, darkness, or mechanical manipulation. Plant Physiol. 75:170–175.

Wareing, P.F., M.M. Khalifa, and K.J. Treharne. 1968. Rate-limiting processes in photosynthesis at saturating light intensities. Nature (London) 220:453–457.

Warrington, I.J., M. Peet, D.T. Patterson, J. Bunce, R.M. Haslemore, and H. Hellmers. 1977. Growth and physiological responses of soybean under various thermoperiods. Aust. J. Plant Physiol. 4:371–380.

Watanabe, Iwao. 1973a. Mechanism of varietal differences in photosynthetic rate of soybean leaves. I. Correlations between photosynthetic rates and some chloroplast characters. Proc. Crop Sci. Soc. Jpn. 42:377–386.

----. 1973b. Mechanism of varietal differences in photosynthetic rate of soybean leaves. II. Varietal differences in the balance between photochemical activities and dark reaction activities. Proc. Crop Sci. Soc. Jpn. 42:428–436.

----, and Kosei Tabuchi. 1973. Mechanism of varietal differences in photosynthetic rate of soybean leaves. III. Relationship between photosynthetic rate and some leaf-characters such as fresh weight, dry weight or mesophyll volume per unit leaf area. Proc. Crop Sci. Soc. Jpn. 42:437–441.

Wells, R., L.L. Schulze, D.A. Ashley, H.R. Boerma, and R.H. Brown. 1982. Cultivar differences in canopy apparent photosynthesis and their relationship to seed yield in soybeans. Crop Sci. 22:886–890.

Wiebold, W.J., Richard Shibles, and D.E. Green. 1981. Selection for apparent photosynthesis and related leaf traits in early generations of soybeans. Crop Sci. 21:969–973.

Wilson, R.F., J.W. Burton, J.A. Buck, and C.A. Brim. Studies on genetic male-sterile soybeans. 1. Distribution of plant carbohydrate and nitrogen during development. Plant Physiol. 61:838–841.

Wittenbach, V.A. 1982. Effect of pod removal on leaf senescence in soybeans. Plant Physiol. 70:1544–1548.

----. 1983a. Effect of pod removal on leaf photosythesis and soluble protein composition of field-grown soybeans. Plant Physiol. 73:121–124.

----. 1983b. Purification and characterization of a soybean leaf storage glycoprotein. Plant Physiol. 73:125–129.

----, R.C. Ackerson, R.T. Giaquinta, and R.R. Herbert. 1980. Changes in photosynthesis, ribulose bisphosphate carboxylase, proteolytic activity and ultrasturcture of soybean leaves during senescence. Crop Sci. 20:225–231.

Wofford, T.J., and F.L. Allen. 1982. Variation in leaflet oreintation among soybean cultivars. Crop Sci. 22:999–1004.

Woodward, R.G. 1976. Photosynthesis and expansion of leaves of soybean grown in two environments. Photosythentica 10:274–279.

----, and H.M. Rawson. 1976. Photosynthesis and transpiration in dictoyledonous plants. II. Expanding and senescing leaves of soybeans. Aust. J. Plant Physiol. 3:257–267.

Zhang, Yong-qui, and Yu Song. 1979. Correlation of leaf area and net photosynthesis rate to the yield of soybean. Chin. Agric. Sci. 2:40–46.

15 Stress Physiology

C. David Raper, Jr.
North Carolina State University
Raleigh, North Carolina

Paul J. Kramer
Duke University
Durham, North Carolina

It is difficult to define precisely plant stress because the term *stress* is used in reference to perturbations from normal conditions at various levels including the environment, the whole plant, and even the cellular and subcellular level. For example, drought as an environmental stress causes water stress, or water deficit, to develop in plants. In turn, a water deficit affects such physiological processes as translocation at the whole plant level; leaf expansion and gas exchange at the organ level; and photosynthesis at the subcellular level. Since the final result is reduction in growth and yield, perhaps stress can be defined simply as any condition that reduces yield below the maximum attainable level.

What constitutes a stress varies with the genotype and development stage of a plant. For example, a temperature low enough to injure the soybean [*Glycine max* (L.) Merr.] cv. Ransom at flowering might not injure a more chilling-tolerant cultivar such as Fiskeby V, and a night temperature injurious during flowering of either cultivar might be harmless or even beneficial during seed filling. Some environmental stresses, such as temperature and soil water potential, can be measured more accurately than others, such as soil aeration, mineral deficiencies, or biotic stresses. Likewise, the effects of a stress such as drought can be measured more accurately than the effects of certain other stresses such as chilling or deficiencies and toxicities of mineral elements.

Stresses operate at all levels from whole plants to the cellular and subcellular level, and in time scales ranging from weeks to hours, or even seconds in the case of sun flecks. For example, while upper leaves of a crop canopy are exposed to full sun, lower leaves are shaded and photosynthesis is reduced. At the same time, the water stress that often develops in the exposed upper part of the canopy reduces photosynthesis on a relatively immediate time scale by causing closure of stomata and

Soybeans: Improvement, Production, and Uses, 2nd ed.–Agronomy Monograph no. 16.

by affecting electron transport and carboxylating enzymes. While turgor loss causes closure of stomata in minutes, on a longer time scale of days it also decreases the leaf area available for photosynthesis by reducing leaf expansion. The major objective of this chapter is to identify some of the important stresses and describe how they alter physiological processes to affect growth and yield.

The stresses to which plants are subjected can be grouped into three categories of soil, atmospheric, and biotic stresses. Among the soil conditions that affect plant growth are available water content of the soil, soil aeration, soil structure, soil temperature, soil pH, mineral deficiencies and toxicities, and salinity. The major atmospheric factors include light, air temperature, humidity, wind, carbon dioxide (CO_2) concentration, and air pollutants. Biotic stress includes competition with weeds and other plants of the same crop, insects, nematodes, and various pathogenic organisms. All of these environmental factors reduce yield by creating physiological stresses within the plants. Lack of space limits discussion in this chapter to only a few among these important stresses. We have considered effects of stresses caused by temperature, water, light, CO_2 concentration, and metal toxicity. Effects of others, including biotic stresses, soil pH and mineral deficiencies, and soil structure are topics of other chapters.

It is difficult to estimate the relative importance of individual stresses because each is more or less random in occurrence and is influenced by interactions with the others. Droughts, unseasonably late or early frosts, and outbreaks of insects or diseases of epidemic proportions are largely unpredictable and the damage they cause depends on their duration and the stage of plant development at which they occur. In modern agricultural systems that incorporate the best cultivars and pest management practices on good soil, atmospheric conditions (or weather) continues to be the principal factor limiting the yield of soybean and other crops. As Watson (1963) wrote, climate determines what crops farmers can grow, but weather determines the yield that they obtain. The importance of weather with respect to crop yields is emphasized by Hudson (1977) and others in Landsberg and Cutting (1977). According to Boyer (1982), the average farm yield of soybean in the USA is only 27% of the record yield, and most of the difference between record and average yields is caused by water and temperature stresses.

15–1 TEMPERATURE STRESS

The well-defined, seasonal cycle of temperature in the Corn (*Zea mays* L.) and Soybean Belts determines the times of planting and harvesting of soybean. Cold soil; low air temperatures; and the danger of late spring frosts limits early planting, while late summer droughts and early autumn frosts are threats to late plantings. In addition, the large temperature perturbations above and below normal during the growing season (Fig. 15–1) constitute relatively unevaluated hazards to the growing crop. The potential for damage from these temperature variations depends on the stage of crop development at which they occur (Table 15–1). Both air and soil temperatures need to be considered. Low soil temperature early in the season is a limiting factor for germination and seedling development and early frosts truncate pod development, while high air temperatures in July and August affect physiological processes adversely. Important interactions occur among temperature, solar radiation, and water stress (Dale, 1983; Mederski, 1983; Shaw, 1983). High temperature often is associated with drought and with high vapor pressure deficit and rapid transpiration.

The length of the growing season for photoperiodically sensitive crops such as soybean is defined by a complex interaction between temperature and photoperiod (Jones and Laing, 1978; Thomas and Raper, 1978, 1983a, 1983b). Nevertheless, the range in genotypic diversity for photoperiodic sensitivity is sufficient for soybean in North America to be cultivated from the long growing season of the southern Cotton Belt to the short frost-free season of southern Canada and worldwide from the tropics of

Table 15–1. Temperature requirements during various stages of soybean development. From Holmberg (1973).

Stage of development	Temperature range		
	Minimum	Adequate	Optimum
	°C		
Germination	6– 7	12–14	20–22
Emergence	8–10	15–18	20–22
Formation of reproductive organs	16–17	18–19	21–23
Flowering	17–18	19–20	22–25
Seed formation	13–14	18–19	21–23
Ripening	8– 9	14–18	19–20

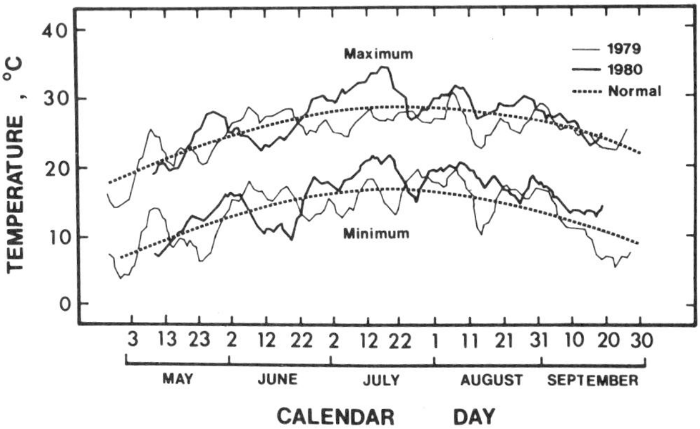

Fig. 15–1. Seven-day moving averages of daily maximum and minimum air temperatures for middle day of period for May through September 1979 and 1980, and normal maximum and minimum temperatures for 1953 to 1979, for West Lafayette, IN. From Dale (1983).

Brazil to subarctic Sweden and Siberia. Given the range of photoperiodic diversity within soybean for time of flowering and maturation, length of the frost-free growing season may be less of a determinant for acceptable production areas than other temperature characteristics of climate. For example, in Oregon the frost-free season is long enough but low night temperatures limit bean yields (Seddigh and Jolliff, 1984).

15–1.1 Seed Germination

Soybean seed generally will germinate between 10 and 40°C (Mederski, 1983) and seed of the more chill-tolerant cultivars such as Fiskeby V and 'Amurakaja 310' will germinate at temperatures of 6 to 8°C (Holmberg, 1973). There is at least one report of germination at 2 to 4°C (Inouye, 1953). Although there are interactions between temperature and both cultivar and depth of planting on rate of emergence (Grabe and Metzer, 1969), the most rapid emergence generally occurs at 25 to 30°C (Mederski, 1983). Final emergence is not greatly affected at temperatures between 16 and 32°C (Stuckey, 1976). The slower emergence at low temperatures, however, increases the probability of injury to seedlings from fungi and insects. Especially when seed are planted deeply, low temperatures decrease the ability of seedlings to overcome the mechanical resistance of soil (Tanner and Hume, 1978).

Little has been published on the physiological basis for slow germination of soybean seed at low temperatures. Perhaps it can be explained by the slow rate of enzyme-mediated processes involved in respiration and hydrolysis of food reserves and the slow rates of translocation of metabolites at low temperatures. However, Arrhenius plots for enzyme systems extracted from germinating seed are quite different from those for seed germination and seem to refute all the usual explanations for poor germination at low temperature except the possibility of protein denaturation (Simon, 1979). Seed germination involves activity of a wide range of enzymes, and it is not surprising that the composite result fails to fit the curve for simple enzyme systems. The extensive injury to membranes that often occurs when seeds imbibe water at low temperatures (Chabot and Leopold, 1982; Leopold, 1980) also may be a factor in poor germination and subsequent poor growth of soybean in cold soil (Hobbs and Obendorf, 1972).

15–1.2 Vegetative Growth

Late frosts can cause freezing injury to soybean seedlings. While seedlings were not killed at −4°C, seedlings of several soybean cultivars were killed with brief exposure to −6°C (Abel, 1970). Chilling injury, however, can be caused by less extreme temperatures. In reviews by McWilliam (1983) and Wilson (1983), many examples are cited of injury to plants of tropical and subtropical origin, including soybean, by exposure to temperatures between 10 and 15°C. The degree of injury depends on the severity and duration of exposure to low temperature and on the level of irradiation during exposure. Plants can withstand lower temperatures without injury to chloroplast ultrastructure during darkness at a high humidity than during exposure to light (Wise et al., 1983). Since the extent of chilling injury was aggravated in darkness as well as in light by a lowered humidity (Wise et al., 1983), the involvement of light in susceptibility to chilling injury may be at least partly associated with promotion of leaf water deficits.

There is scant doubt that water stress is involved in chilling injury. There is uncertainty, however, about the importance of the reduction in water absorption caused by chilling. Taylor and Rowley (1971) concluded that the direct effects of chilling on photosynthesis and the indirect effects from inhibited water absorption are about equally important. While low root temperatures certainly inhibit water absorption (Fig. 15–2), shoot chilling has more serious effects than root chilling on growth (Musser et al., 1983a). When root-zone temperature was maintained at 25°C, chilling shoots of 2-week-old soybean at 10°C for 1 week resulted in temporary reduction in leaf water potential, leaf elongation, rate of leaf emergence, stomatal conductance, and CO_2 uptake (Table 15–2). All of these characteristics returned to control levels after rewarming. Root chilling had smaller effects on the vegetative characteristics than shoot chilling, and simultaneous root and shoot chilling resulted in no more injury than

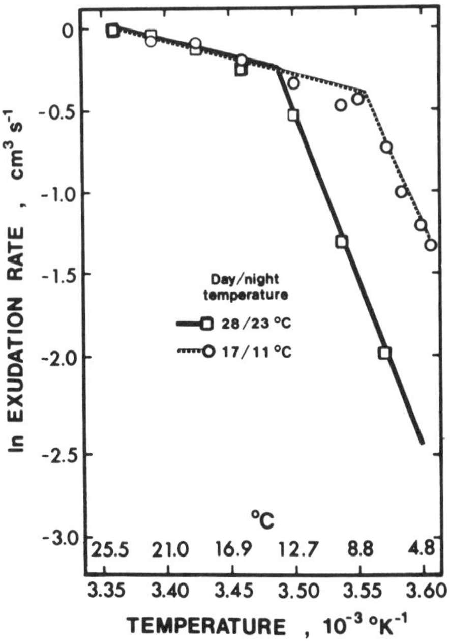

Fig. 15–2. An Arrihenius plot of water flow through excised soybean roots under pressure for root systems grown at day/night temperatures of 28/23 and 17/11°C. From Markhart et al. (1979).

Table 15–2. Physiological responses of Ransom soybeans to 7 days of root or shoot chilling during vegetative growth. From Musser et al. (1983a).

Physiological processes	Chilling treatment		
	Control, 25°C	Root chilled to 10°C	Shoot chilled to 10°C
Time between emergence of leaves, days	1.7	2.0	3.7
Rate of leaflet elongation, mm day^{-1}	18.7	10.1	3.7
Leaf water potential, MPa	−0.8	−0.9	−0.3
Leaf osmotic potential, MPa	−1.1	−1.0	−1.8
Leaf turgor potential, MPa	0.2	0.1	1.6
Abaxial stomatal conductance, cm s^{-1}	1.5	0.2	0.3
Carbon dioxide uptake rate, % of control	—	70	40–45

shoot chilling alone. It appears that vegetative growth can recover from short periods of chilling with a minimum of injury.

The opinion that a night temperature lower than the day temperature is favorable for growth of most plants (Went, 1953) is questionable as a generality. Vegetative growth of soybean at a mean daily temperature of 23°C was altered little whether the mean temperature was achieved by day/night temperatures of 26/20 or 29/17°C or kept constant at 23/23°C (Warrington et al., 1977). Apparently, if a constant day and night temperature is near the optimum for the plant, warm days and cool nights provide no additional benefits for vegetative growth. If, however, net photosynthesis during the day is limited by above optimum temperature or low radiation levels, a cooler night temperature can partially compensate for the reduced carbon fixation rate by the lowered respirational loss of carbon (Wann and Raper, 1979, 1984). But while these simple relationships with temperature may apply during vegetative growth, the interactions between temperature and photoperiod on reproductive morphology and physiology of soybean plants have a more complex relationship with growth (Thomas and Raper, 1978, 1983b; Thomas et al., 1981).

Growth of a plant involves numerous processes with different optimum temperatures, and the optimum temperature for growth of the whole plant can be quite different from that for individual processes. For example, photosynthesis of individual soybean leaves increases with increasing temperature to between 35 and 40°C and then begins to decline (Hofstra and Hesketh, 1969), while respiration usually continues to increase with temperature above the optimum for photosynthesis (Ormrod, 1964). Yet, dry matter accumulation in the whole plant begins to decline as air temperature increases above about 28 to 30°C (Raper et al., 1977; Wann and Raper, 1979). The lower temperature optimum for growth of the whole plant probably is a result of the complex behavior at different levels of plant organization and of imbalances among different physiological processes (Gold and Raper, 1983). For example, increasing temperature affects dry matter accumulation on the morphological level

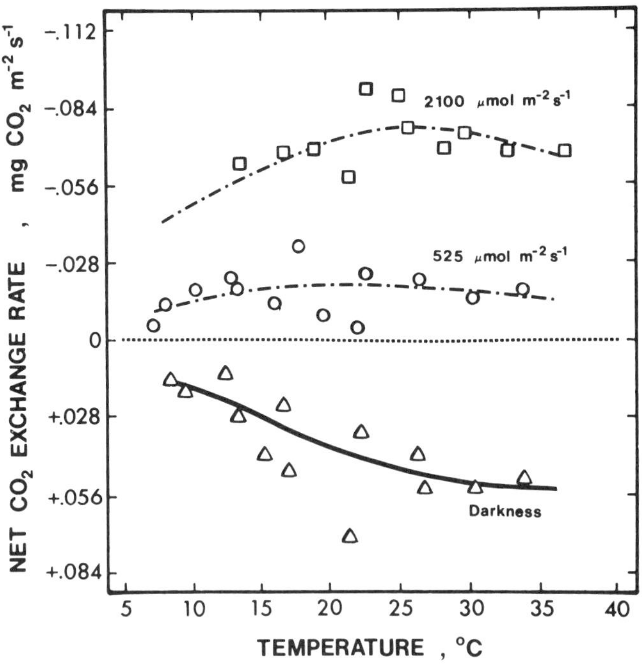

Fig. 15–3. Effects of temperature on CER of 30-day-old bean plants at 31.4 and 125 klux (about 525 and 2100 μmol m^{-2} s^{-1}) and in darkness. From Ormrod (1964).

through both direct effects on leaf initiation (Thomas and Raper, 1976) and indirect effects of increased leaf water deficits on leaf expansion (Hsiao, 1973) as well as on the metabolic level of photosynthesis and respiration. The effects of temperature at the morphological level of leaf initiation and expansion are irreversible and occur on a time scale of hours or days while on the level of physiological processes such as photosynthesis and respiration, responses often are reversible on a time scale of seconds or minutes. Moreover, within the same level of organization, physiological processes such as photosynthesis and respiration have different Q_{10} values (Wann and Raper, 1979). For example, dark respiration rate continues to increase with temperature beyond the temperature for maximum rate of photosynthesis (Fig. 15–3). It seems unlikely that the explanation of differences in high temperature responses of plant growth will be found by research only at the subcellular or molecular level.

Changes in temperature affect growth through altered partitioning of photosynthate between organs (Wann and Raper, 1984). Translocation rates of organic compounds in the phloem, however, are relatively insensitive to temperatures above 10°C (Wardlaw, 1974; Wardlaw and Bagnell, 1981). Localized chilling of stems of *Phaseolus vulgaris* (L.) at about 10°C does not directly inhibit the driving force of translocation, but results in damage to the cytoplasm lining the sieve tubes to cause blockage of sieve plates (Giaquinta and Geiger, 1973) or damage to the cytoplasm of companion cells (Geiger, 1976; Gunning and Robards, 1976). Translocation rates thus are not directly dependent on temperature within the growing range, but rather can be considered to be responsive to temperature indirectly through the effects of temperature on photosynthetic and respiration rates that affect concentration of soluble carbohydrates in the source pool and the size and metabolic activity of the receiving (sink) organ (Wann and Raper, 1984). For example, low temperature reduces translocation of photosynthate into soybean seed both by decreasing unloading into the seed coat and by reducing embryo growth (Thorne, 1982).

Most studies of the relationship between temperature and growth, at least after seedling emergence, are based on aerial temperatures. Since seasonal changes in soil temperature at depths of 10 cm or more usually lag somewhat behind air temperature, surface soil is usually cooler than the air in the spring and warmer in the autumn. Mederski (1983) reported several experiments indicating that a root temperature of about 25°C resulted in maximum nodule development and plant growth. Since average soil temperature at a 10 cm depth usually is below 25°C in northern Ohio until mid-July (Mederski, 1983), soil temperatures may be below the optimum during much of the growing season in a considerable part of the production areas for soybean. In most experiments in controlled environments, soil and air temperatures are varied simultaneously; however, the effects of root and shoot chilling on growth may be quite different (Table 15–2). Shoot dry weight of 'Bragg' soybean grown for 37 days at a soil temperature of 15°C was less than that at 30°C (Trang and Giddens, 1980), and the vegetative dry weight of 'Wells' soybean dependent on N_2 fixation was 10 times greater after 63 days at a root temperature of 20°C than at 13°C for a 28/21°C aerial temperature (Duke et al., 1979). This temperature effect probably involved both a delay in establishment of functional nodules and a reduction in nodule activity (Jones et al., 1981; Raper and Patterson, 1980). However, for hydroponically grown Ransom soybean supplied with adequate nitrate, vegetative dry weight after 26 days was reduced 20% for plants grown at a root temperature of 18°C relative to 24°C (Rufty et al., 1981). The plants demonstrated a capability for acclimation to the cool root temperature since relative growth rates during the latter part of the experimental period were similar at both 18 and 24°C.

Low root temperatures cause immediate reduction in water absorption, development of water stress (Musser et al., 1983a), and reduction in rate of leaf expansion (Table 15–2). At temperatures of 20 and 14.5°C, water flow through excised root systems under pressure was only 60 and 30%, respectively, of that measured at 25°C (Markhart et al., 1979). Thus, even moderate cooling of the soil early in the growing season might cause enough water stress to reduce leaf expansion. That there is some acclimation of root functioning at low temperatures is evident from the occurrence of a distinct break in Arrhenius plots (Fig. 15–2) of water flow under pressure at about 14°C for soybean grown at 28/23°C and at 8.7°C for soybean grown at 17/11°C (Markhart et al., 1979). The acclimation root systems kept at low temperatures for several days apparently is associated with an increased proportion of unsaturated fatty acids of new roots developed during exposure (Markhart et al., 1980; Osmond et al., 1982). Since acclimation may require several days, sudden cooling of the soil can produce severe water stress in shoots. The effects of root chilling are temporary, however, and no permanent damage results (Musser et al., 1983a).

The decreased flow of water through chilled root systems will decrease that fraction of ion uptake that depends on mass flow to reach the root surface. Both Clarkson (1976) and Markhart et al. (1979), however, found that activity of the ion pump responsible for transport of ions into the xylem increased with low temperatures in slowly transpiring plants. Acclimation of uptake rates of nutrients occurs with the production of new roots during exposure to cool root temperatures (Osmond and Raper, 1982; Osmond et al., 1982; Rufty et al., 1981). The reduction in rate of shoot growth that accompanies cool root temperatures, at least partly in response to water deficits, temporarily reduces the requirement for mineral nutrients before roots acclimate. A similar case may be made for N_2 fixation. While low root temperatures reduce nodule formation and activity (Lindeman and Ham, 1979), as well as shoot growth, nodules are developed and shoot growth recovers as roots are warmed (Duke et al., 1979). A greater potential problem is posed when stimulation of shoot growth by high aerial temperature results in a N requirement of seedlings that exceeds cotyledonary reserves or soil supply before active nodules are formed (Jones et al., 1981). The resulting N stress, with reduced leaf production and plant growth, may be exaggerated if soil temperatures remain cool and further delay nodule activity.

15–1.3 Reproductive Growth

The effects of temperature on flowering and pod and seed development have received less attention than the effects of photoperiod. Several studies, however, have shown important interactions between temperature and photoperiod with respect to both flowering and pod set. More flowers and pods occurred on Ransom soybean grown at day/night temperatures of 26/22 and 22/18°C than on those grown at 30/26 and 18/14°C (Thomas and Raper, 1977). In fact at 18/14 and 30/26°C very few pods were set although the plants flowered profusely, suggesting that both low and high temperatures resulted in pod abortion. Plants at the intermediate temperatures produced both more nodes and axillary branches and more flowers and pods per node. Similarly, a delay in photoinduction of flowering permits greater production of nodes and branches, leading to increased numbers of pods and seed yield (Patterson et al., 1977: Thomas and Raper, 1977).

Temperatures below 15°C prevent pod formation on many soybean cultivars, although some can set pods at temperatures as low as 10°C (Hume and Jackson, 1981). Based on 10 yrs of observations, Holmberg

(1973) reported that a mean temperature of about 17°C during and after flowering is required for good flowering and pod set of cultivars grown in Sweden. However, varieties from the east coast of Hokkaido and the islands of the Okhotsk Sea, as well as some of their descendents such as Fiskeby V, have a somewhat lower minimum temperature.

Temperature has separate effects on floral induction and pod development. 'Chippewa' soybean grown in cool locations flower profusely, but set few pods (Soldati and Keller, 1977). When day or night temperatures were 10 to 14°C following application of an inductive photoperiod, floral initiation and anthesis occurred for Ransom, but no pods were set (Thomas and Raper, 1977, 1978, 1981). A chilling temperature of 10°C applied during the 1st week of inductive photoperiod reduced the number of floral primordia 22% and delayed anthesis by a week (Musser et al., 1983b). Chilling during either the 2nd or 3rd week of induction also delayed anthesis by a week, but had no effect on the number of floral primordia produced. Regardless of which week chilling stress was applied, number of pods was less than for control plants. Reduced production of pods for plants chilled during the 1st week of reproductive development was associated with reduction in number of floral primordia, while the reduction in pods of plants chilled during the 3rd week was associated with increasing flower abortion. Chilling at anthesis during the 4th week of inductive photoperiod also caused flower abortion and reduced pod set. In rice (*Oryza sativa* L.), chilling injury at temperatures below 15°C results from male sterility caused by damage to developing pollen grains (Lin and Peterson, 1975; Nishiyama and Sataka, 1979). Chilling also causes male sterility in sorghum (*Sorghum bicolor* L. Moench) (Brooking, 1976). In soybean, however, pollen is formed between 2 and 3 weeks after initiation of the floral primordium (Kato et al., 1954). Since chilling during the 3rd and 4th week of induction resulted in flower abortion (Musser et al., 1983b), male sterility cannot be the only cause of flower abortion. Other potential causes of chilling injury need to be investigated.

Soybean yields in the field might be affected by both low and abnormally high temperatures. Temperature data (Fig. 15–1) from West Lafayette, IN, indicate several occasions when temperatures below 15°C might have injured reproductive development. Severe abscission of pods occurs in the absence of water stress at 40°C (Mann and Jaworski, 1970). While chilling injury may be more probable in the field than direct injury from such extremely high temperatures, especially in the cooler regions of major production areas (Fig. 15–1), even the relatively mild temperatures of 30/26°C can reduce pod set in Ransom (Thomas and Raper, 1977, 1981).

The effects of temperature during seed maturation can be carried over to the next growing season through altered seed quality. Optimum temperatures for seed ripening are reported (Holmberg, 1973; Sato and Ikeda, 1979) to be within a range of 25°C during the day and 15°C during the night. The higher temperatures that occur during seed ripening of early maturing soybean in midwestern (see Fig. 15–1) and southeastern regions of the USA can reduce subsequent germination (Green et al., 1965) and may be responsible for wide variations in germinability and seedling vigor from year to year. This view is supported by observations that tobacco (*Nicotiana tabacum* L.) seed matured at temperatures of 22/18 or 26/22°C produce more vigorous seedlings than seed matured at 30/26°C (Thomas and Raper, 1975). But perhaps a more serious problem is frost injury to ripening seed. Danger of injury from frost decreases rapidly as the water content decreases (Judd et al., 1982). Seed in green pods, which contain about 650 g kg^{-1} of water, are injured at −2°C while seed in brown pods, which contain 350 g kg^{-1} of water, are not injured at −12°C. Injury by both high temperatures and frost may involve interference with synthesis of specific proteins formed during slow drying in the late stages of seed development (Adams et al., 1982, 1983).

15–2 WATER STRESS

Water stress can be caused either by an excess of water that injures root systems by deficient aeration or by a deficit in available soil moisture that results in dehydration of the shoots. Water stress, whether caused by flooding or drought, affects every aspect of plant growth, including the anatomy, morphology, physiology, and biochemistry (Fig. 15–4). Severity of water stress varies from transient midday reductions in cell and leaf expansion and temporary reduction in stomatal conductance, to

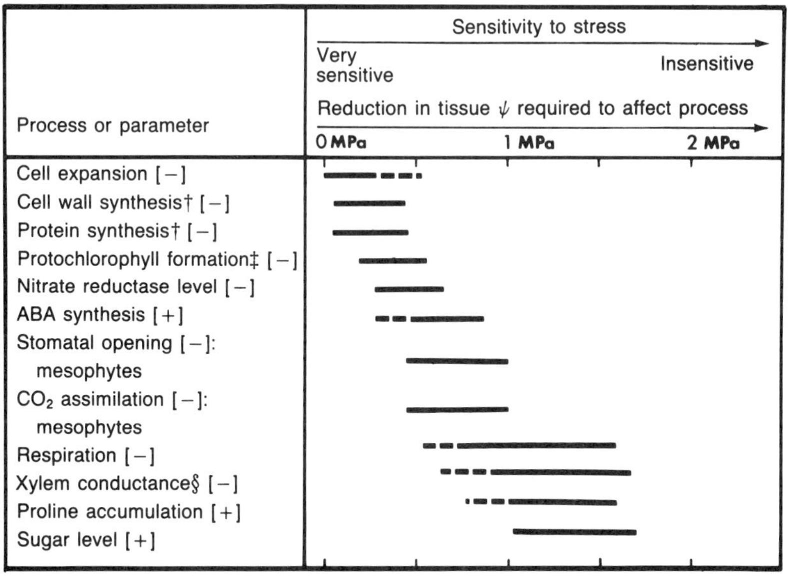

† Fast-growing tissue. ‡ Etiolated leaves. § Should depend on xylem dimension.

Fig. 15–4. Approximate sensitivity to water stress of a number of plant processes. From Hsiao (1973).

physical disruption of chloroplast structure accompanied by reduced photosynthesis and growth, and finally to death from tissue dehydration. Water stress at every stage of plant development can reduce yield directly or indirectly, but the extent of yield reduction from water stress varies with stage of development. This is particularly true of the brief and randomly occurring periods of drought or flooding that characterize the major production areas of midwestern and southeastern USA.

15–2.1 Seed Germination

The process of germination requires imbibition of water and is accompanied by a high rate of respiration. Excessive soil moisture limits the oxygen available for the respiratory processes of germination. A combination of cold soil and excess water may be especially damaging to germination and emergence. Cell membranes of seeds allowed to imbibe water at low temperature can be injured (Bramlage et al., 1978; Chabot and Leopold, 1982; Willing and Leopold, 1983), and immersion of soybean seed in cold water for periods as short as an hour results in reduced germination and seedling growth (Hobbs and Obendorf, 1972; Simon, 1979).

Germination and emergence are reduced more often by dry soil than by flooding. The water content of soybean seeds must reach about 500 g kg^{-1} of dry weight to insure germination, in contrast to corn seeds which require only about 300 g kg^{-1} (Hunter and Erickson, 1952). Seeds planted in soil that is too dry for germination often absorb enough water to be invaded and damaged by soil fungi. The data on the soil water potential limiting seed germination are unsatisfactory because of differences in terminology and methods. For example, Hanks and Thorp (1957) reported that when the soil water content was one-half of field capacity in a fine sandy loam emergence of soybean, wheat, and grain sorghum seedlings was reduced below emergence when the soil was at field capacity; however, they provided no data on soil water potential. Hunter and Erickson (1952) reported that soybean germinated at a soil moisture tension of -0.6 MPa, while corn germinated at -1.25MPa. In contrast, Heatherly and Russell (1979) reported that an acceptable rate of emergence was not obtained below -0.06 MPa in a silt loam nor below -0.07 MPa in a fine clay soil. Seed germination is influenced both by the soil water potential and by the area of contact between seed and soil, which increases with decreasing particle size (Collis-George and Hector, 1966). This may contribute to the inconsistency among reports. Also, emergence of germinated seedlings in drying soil is hindered by formation of a surface crust (Hanks and Thorp, 1957).

15–2.2 Vegetative Growth

Plant growth depends on the photosynthetic rate per unit leaf area, the total leaf area available for photosynthesis, and leaf area duration.

These and other processes are affected by water stress (Fig. 15–4). Thus total photosynthesis of water-stressed plants is decreased by reductions in rate of carbon fixation per unit leaf area resulting from premature stomatal closure and nonstomatal inhibition of the photosynthetic machinery, and by reductions in photosynthetic surface area caused by decreased leaf enlargement and hastened leaf senescence. Stomatal conductance and rates of photosynthesis and transpiration decrease simultaneously in water-stressed soybean (Fig 15–5). There is, however, discussion about the extent to which stomatal closure reduces photosynthesis vs. the possibility that nonstomatal reduction in photosynthesis causes closure of stomata in water-stressed plants (Farquhar and Sharkey, 1982). The large decrease in stomatal conductance reported (Bunce, 1978a) for soybeans at 40% relative humidity without any reduction in net photosynthetic rate indicates that stomatal closure can occur independently of photosynthetic rate.

Leaf water potentials lower than −0.5 MPa seem to affect chlorophyll formation (Alberte et al., 1975, 1977) and at about −1.0 MPa cause disorganization of chloroplast structure (Mohanty and Boyer, 1976; Vieira da Silva et al., 1974). This reduces electron transport and both cyclic and noncyclic photophosphorylation and decreases activity of ribulose bisphosphate carboxylase and other important enzymes involved in carbon assimilation. The effects of water stress on photosynthetic rates of soybean leaves are readily detectable at leaf water potentials of about −1.0 to −1.2 MPa (Boyer, 1970a; Boyer et al., 1980, Cure et al., 1983). Photosynthetic rate declines rapidly with further reductions in leaf water potential to about −1.8 MPa, and then continues to decline gradually with decreasing potential (Fig. 15–5 and 15–6). The effects of water stress on photosynthetic rate are reversible down to leaf water potentials of about −1.6 MPa or less in growth chamber conditions when other environmental stresses are absent (Cure et al., 1983; Mohanty and Boyer, 1976).

Respiration also decreases with leaf water potential, but in a different pattern than photosynthesis (Fig. 15–7). Dark respiration rate declines from about −0.6 to −1.6 MPa, then remains unchanged to the most severe stress imposed of −4.0 MPa. From these data it appears that photosynthesis is decreased more by water stress than is respiration, and

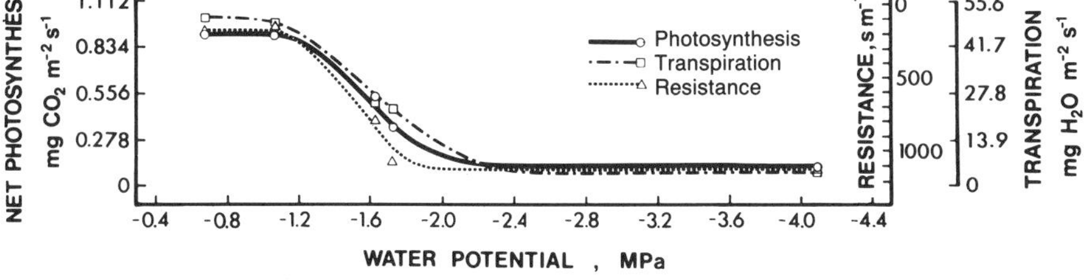

Fig. 15–5. The relationships among leaf water potential, photosynthesis, transpiration, and stomatal resistance in soybean. From Boyer (1970a).

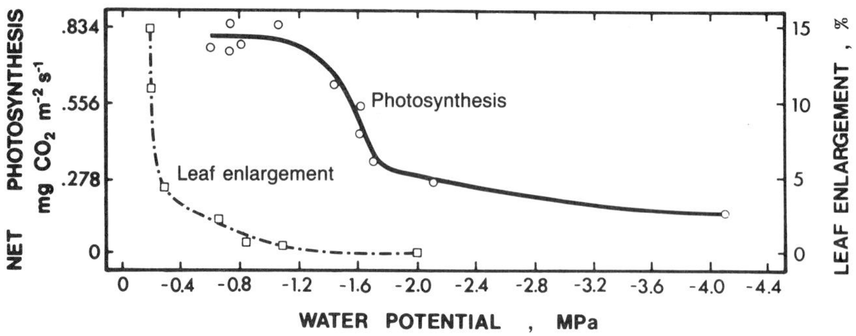

Fig. 15–6. The relationships among leaf water potential, leaf elongation, and photosynthesis of soybean. From Boyer (1970b).

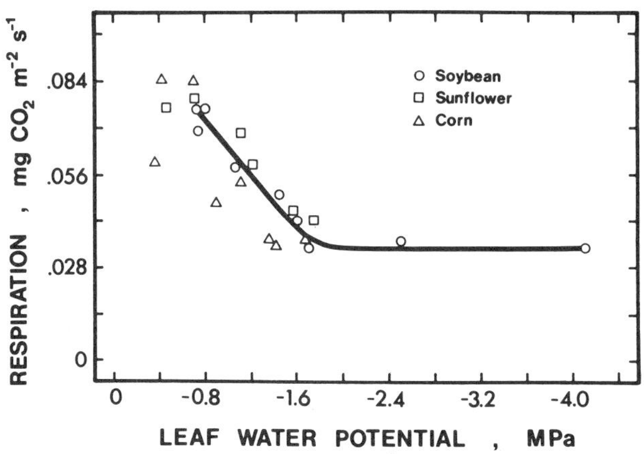

Fig. 15–7. Effect of decreasing leaf water potential on the rate of dark respiration of soybean, sunflower, and corn. From Boyer (1970b).

high temperatures have a greater impact than water stress on respiration (Fig. 15–3).

Translocation continues at water stresses lower than those inhibiting photosynthesis. Wardlaw (1974) concluded that the velocity of translocation is not greatly reduced until water potential drops as low as −2.0 to −3.0 MPa. This conclusion has been contested (see Ashley, 1983); however, the direct effects of water stress on translocation are difficult to determine. While velocity may be unaltered by leaf water potentials in the range of −1.0 to −2.0 MPa, the inhibition of photosynthesis in this range (Fig. 15–5 and 15–6) greatly reduces the amount of assimilate available for translocation (Silvius et al., 1977). Also, water stress directly shifts partitioning of assimilates between shoot and root by altering the relative size and activity of these sinks. Sharp and Davies (1979) observed with corn seedlings grown in pots that the proportion of total dry matter translocated to roots was increased during a single drying cycle, and also during the middle of the drying cycle the dry weight and length of roots were greater for stressed than for well-watered plants. Although decreases in stomatal conductance and leaf and root water potentials of the seedlings coincided with the decline in substrate water potential during the interval of enhanced root growth, root osmotic and total water potentials fell in unison; thus, root turgor remained nearly constant so that root expansion continued in the drying substrate. Other data (Eavis and Taylor, 1979; Hsiao and Acevedo, 1974; Osmond and Raper, 1982) also can be interpreted to indicate an absolute increase in translocation of assimilates to roots during a slight moisture stress in the shoot.

The importance of water stress in altering size of cells and tissues often is underestimated relative to the direct effects on metabolic activity. Cell and leaf expansion are more sensitive to water stress than is photosynthetic rate (Fig. 15–4 and 15–6), and in fact a decrease in rate of leaf enlargement usually is the first visible evidence of developing water deficit. Thus, a slight stress of −0.2 to −0.8 MPa that does not affect the current photosynthetic rate can have a long-term impact on plant growth, especially if the stress occurs early in development, by irreversibly reducing total leaf area available for photosynthesis. The interrelationships among leaf water potential, leaf expansion, and photosynthesis are discussed by Bunce (1977, 1978a).

Cell expansion generally is assumed to be closely correlated with turgor pressure, and it is true that some minimum turgor pressure is necessary for cell expansion (Cleland, 1971). Above the threshold value, however, cell expansion is not necessarily closely related with turgor, and the relationship between plant water status and cell expansion often seems complex. Cell expansion depends on biochemical factors affecting extensibility or modulus of elasticity of cell walls and on physical factors controlling diffusion of water into cells. The influx of water into cells is dependent on a gradient of decreasing water potential toward the growing region, which in turn depends on a supply of solutes to the expanding cells. It is possible for the turgor pressure to be similar in growing and nongrowing tissues if the osmotic and water potentials are lower in the growing tissue (Cavalieri and Boyer, 1982). Since water potential becomes zero in fully turgid cells, preventing further influx of water, cell enlargement should be increased by slight loss of turgor. In fact, Bunce (1977) observed larger soybean leaves with larger epidermal cells under conditions where mild water stress developed than under conditions which did not produce measurable water stress. More severe stress reduced the size of epidermal cells and leaves.

Soybean shoots grow more at night than during the day, but roots grow more during the day (Bunce, 1978b). This probably is because they are less subject to daytime water stress than the shoots (Sharp and Davies, 1979). Roots maintain a higher turgor than leaves during the day. Thus, during the day when leaf expansion is limited by the lowered leaf-water potential, photosynthates are more readily available for translocation to roots. At night stomatal closure relieves leaf water stress, resulting in increased turgor and more cell expansion, and the shoots become stronger sinks for carbohydrate than the roots, resulting in reduced root growth. In contrast to the low leaf expansion during the day for soybean grown in controlled environments (Bunce, 1978b), leaves of rice enlarged more during the day when turgor was low than at night when it was high (Cutler et al., 1980). The reduction in enlargement at night was related to lower temperatures at night, and when night temperatures were raised to approach day temperature, leaf enlargement at night was increased. Chilling temperatures can decrease leaf expansion of soybean even though the turgor of chilled leaves remains high (Table 15–2).

Water stress reduces N_2 fixation. Since a very close association exists between leaf-water potential, photosynthetic rate, and acetylene (C_2H_2) reduction activity of intact soybean plants subjected to a single cycle of water stress and recovery (Huang et al., 1975a, 1975b; Patterson et al., 1979), part of the reduction can be attributed to reduced availability of photosynthate for translocation to the nodules. Part of the reduction in N_2 fixation by water stress is a direct effect on water potential within the nodules. Dinitrogen fixation activity of nodules is reduced as water potential decreases (Huang et al., 1975a, 1975b) and ceases irreversibly in detached nodules when fresh weight drops below about 80% of the fully turgid weight (Sprent, 1971). Perhaps the physical alterations of nodules caused by water stress (Parkhurst and Sprent, 1975) are responsible for the delayed recovery of nodule activity upon rewatering relative to recovery of either photosynthetic activity of leaves (Huang et al., 1975b) or energy charge of the adenylate pool of the nodules (Patterson et al., 1979).

15–2.3 Reproductive Growth

The reproductive stage of plant growth is particularly sensitive to water stress (Begg and Turner, 1976), and soybean is no exception (Doss et al., 1974; Martin et al., 1979). A major portion of the variation in soybean yield can be attributed statistically to seasonal variations in rainfall during flowering and fruiting (Runge and Odell, 1960; Thompson, 1970). The extent of yield reduction from a single incident of water stress increases as the reproductive stage advances toward maturity (Table 15–3). The component of yield associated with the reduction also changes from number of seeds and pods for a stress during flowering and early pod-fill stages to size of seed for a stress during seed fill (Constable and Hearn, 1978; Martin et al., 1979; Sionit and Kramer, 1977).

Abscission of flowers, pods, and seeds of water-stressed plants, as well as the later reduction in seed size, may be at least partially a response to water stress through the effects of leaf water deficits on photosynthetic rates that decrease the concentration of assimilates in vegetative pools. Phloem transport to abortive flowers of European blue lupine (*Lupinus angustifolius* L.) is reduced several days before abscission occurs and these flowers accumulate less ^{14}C-labeled assimilate from leaves than flowers that progressed to pods (Pate and Farrington, 1981). Certainly translo-

Table 15–3. Effects on vegetative and seed weight at maturity of withholding water at various growth stages until plants were stressed to a leaf water potential of −2.3 MPa before rewatering. From Sionit and Kramer (1977).

Stage when stressed	Time after emergence	Vegetative dry weight	Seed air-dried weight
	days	g plant^{-1}	
At flower induction	30	65.6	46.9
During flowering	44	76.1	49.8
Start of pod formation	57	80.4	36.5
During seed filling	65	84.6	32.5
Unstressed control	—	89.9	58.5

cation to reproductive sinks should be responsive to the decrease in assimilate concentration of vegetative pools that occurs during water stress (Cure et al., 1985). If the stress is relieved prior to seed filling, photosynthesis can recover, growth can resume, and the remaining seeds can continue to develop to normal size (Patterson et al., 1979; Sionit and Kramer, 1977). If, however, photosynthesis is reduced by short periods of water stress in the range of −1.0 and −1.6 MPa after pod and seed set, size of the reproductive sink demand is not adjusted during the stress period (Cure et al., 1985). The associated reduction in rate of N_2 fixation, that even in the absence of stress is exceeded by the rate of N accumulation in seed, results in an increased rate of N remobilization from leaf pools (Cure et al., 1985). As a consequence, during a late-season water stress N in leaves can be reduced below the concentration necessary for maintenance of photosynthetic capacity, the period of seed filling abbreviated, and weight per seed reduced (Boote et al., 1978; Cure et al., 1985).

15–3 LIGHT

Light affects plant growth chiefly through photosynthesis and photomorphogenesis. While the flux of radiation within the photosynthetically active range of the spectrum is of primary importance in growth and yield of plants, the effect of light on plant form, and hence on distribution of photosynthetically fixed carbon, cannot be ignored. Included among pigment systems known to be involved in photomorphogenic responses are phytochrome, β-carotene, and riboflavins. The involvement of the phytochrome system in photoperiodic regulation of reproductive development, however, is the most studied of these photomorphogenic systems and is a major influence altering growth and soybean yield.

15–3.1 Photosynthetically Active Radiation

Total energy entering the plant system is dependent in part on the maximum photosynthetic rate per unit leaf area and in part on interception of photosynthetically active radiation (PAR) by the total leaf area of the plant. While the maximum photosynthetic rate depends on age and N level of the leaves, water status, temperature, and CO_2 concentration, interception of PAR is affected both by flux density of radiation above the canopy and distribution within the canopy. For field-grown soybean, most radiation is intercepted by leaves near the surface of the canopy (Sakamoto and Shaw, 1967a). While increased vertical orientation of leaves (Duncan, 1971) can enhance interception of radiation at leaf area indices (LAIs) (leaf area per unit ground area) greater than about 3.0, dry matter production by field-grown soybean reaches a maximum as LAI approaches 4.0 (Shibles and Weber, 1965). A LAI of 4.0 normally occurs late in vegetative growth. During much of the vegetative growth phase, photosynthesis can be limited by periods of reduced radiation as well as by factors such as temperature and water stress that both reduce the rate of leaf expansion and limit maximum photosynthetic rate. Although photosynthetic efficiency declines as leaves age even in an open canopy, photosynthetic efficiency of each successive newly expanded trifoliolate remains high (Fig. 15–8). Increases in leaf number after a LAI of 4.0 is attained thus may contribute to continued high levels of canopy photosynthesis until vegetative growth ceases.

Once the canopy of soybean is closed, photosynthetic rates measured for individual leaves (Sinclair, 1980) or the whole canopy (Sakamoto and Shaw, 1967b) of field-grown soybean do not respond to increases in diurnal radiation above about 50 to 60% of the maximum noon-time solar radiation during the summer months. This plateau probably is influenced by the midday deficits in leaf water potential that are associated with

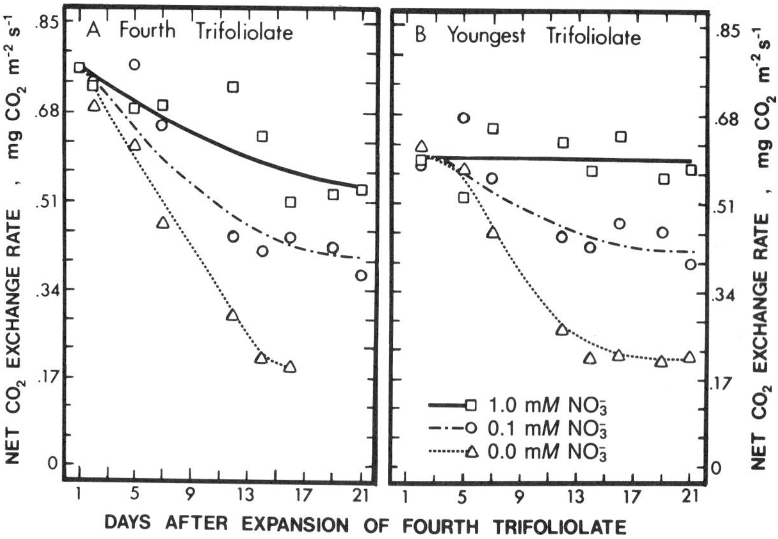

Fig. 15–8. Net CER for soybean at three levels of NO_3^- in hydroponic culture: (A) the fourth trifoliolate leaf measured over a 21-day period after expansion, and (B) the most recently expanded mainstem trifoliolate during the 21-day period. Unpublished data from the study of Rufty et al. (1984).

excessive transpiration and are common even when soil moisture is readily available (Boyer et al., 1980; Jung and Scott, 1980). Whether or not the plateau in photosynthetic response represents light saturation, photosynthetic photon flux density does not appear to be the limiting factor for photosynthetic performance of field-grown soybean for much of the day. In controlled environments, however, net CO_2 exchange rates (CER) of fully expanded upper leaves of soybean were reduced from 0.74 to 0.52 mg CO_2 $m^{-2}s^{-1}$ when photosynthetic photon flux density was reduced from 700 to 325 μmol $m^{-2}s^{-1}$ (Rufty et al., 1981). Over a 4-week period of vegetative growth beginning at the V2 developmental stage, partitioning of the reduced photosynthate under the low radiation level had a synergistic effect on reducing total plant growth and photosynthetic rates. The 30% reduction in net CER when radiation was lowered from 700 to 325 μmol $m^{-2}s^{-1}$ resulted in a 55% reduction in leaf area and a 60% reduction in total plant dry matter accumulation. Clearly, periods of reduced radiation during early vegetative development can reduce growth through combined effects of lowered photosynthetic rates per unit leaf area and consequentially reduced rates of leaf area expansion. While photosynthetic rate per unit leaf area following a period of low radiation may be increased as radiation returns to normal levels, delayed canopy closure may still reduce total light interception and vegetative growth.

Intense solar radiation also can be a stress. Increases in leaf temperature of field-grown soybean follow the diurnal pattern of solar radiation, with the differential between leaf and air temperatures increasing during the morning to a maximum at midday and then decreasing in the afternoon (Jung and Scott, 1980). Transpiration rates during the day follow the same diurnal pattern, and during the period of maximum midday solar radiation they often exceed flow rates of water through roots even when availability of soil water is not limiting (Boyer, 1971; Boyer et al., 1980; Ghorashy et al., 1971). The high midday radiation during summer months thus often reduces photosynthetic activity and seed yields by increasing thermal load and transpiration rate of leaves (Boyer et al., 1980).

15–3.2 Photomorphogenic Radiation

Regulation of development by photomorphogenic radiation, particularly by wavelengths in the 660 to 730 nm range that activate the phytochrome system, have a major impact on growth and yield of soybean. During vegetative growth, a low ratio of 660 to 730 nm wavelengths stimulates leaf enlargement and stem and petiole elongation of many species (Downs, 1955; Kasperbauer, 1971; Parker et al., 1949), including soybean (J.F. Thomas and C.D. Raper, unpublished data). Since leaves transmit more 730 than 660 nm radiation, Kasperbauer (1971) has demonstrated with tobacco that as the canopy closes the partially shaded plants within the canopy develop longer internodes and elongated leaves. While such photomorphogenic reactions of plants may enhance interception of PAR as the canopy closes, the photoperiodic reactions of soybean during reproductive development have a greater influence on yield.

Soybean generally is a short-day species, and when all aspects of reproductive development are considered, few cultivars are insensitive to photoperiod (Murfet, 1977). The course of reproductive development after evocation of the flowering response includes as discrete physiological events the initiation of floral primordia, development of floral structures leading to anthesis, embryo and seed development, and seed maturation (Kato et al., 1954). Sensitivity to photoperiod is subject to change throughout reproductive development, and variations in photoperiod have been observed to alter rates of reproductive development both before (Thomas and Raper, 1983a, 1984) and after (Cure et al., 1982; Johnson et al., 1960; Raper and Thomas, 1978; Thomas and Raper, 1976) anthesis.

Of all the reproductive events, floral initiation is perhaps the least sensitive to photoperiod length. The first flower primordia appears almost as soon under a long, 16-h photoperiod as under a short, 10-h photoperiod (Thomas and Raper, 1983a). Subsequent flower development, however, is much slower under long photoperiods, and the time to anthesis is more than doubled (Thomas and Raper, 1983a). Floral initiation always occurs first in a meristem in an axil of a leaf along the mainstem (Borthwick and Parker, 1938b; Nielsen, 1942; Thomas and Raper, 1983a) and then proceeds acropetally and basipetally on the mainstem and out along branches. Given an insufficient number of inductive short days, soybean initiates floral primordia at only a few mainstem nodes while other nodes and the shoot apex continue to initiate only vegetative structures (Borthwick and Parker, 1938b; Nielsen, 1942). In contrast, under continuous 10- and 12-h photoperiods floral initiation in the determinate cv. Ransom occurs quickly, leading within 7 to 10 days to transformation of the terminal shoot apex to its reproductive phase of development (Thomas and Raper, 1983a) and cessation of node production by all meristems (Fig. 15–9). The first discrete floral primordia occurs under long 14- to 16-h photoperiods almost as soon as under 10- to 12-h photoperiods, but floral initiation proceeds much more slowly under long photoperiods and plants retain the capacity for vegetative node production concomitantly with reproductive growth for an extended time (Fig. 15–9). Eventually, plants under the longer photoperiods switch from the dual mode of vegetative and reproductive growth to only reproductive growth, but with a greatly increased number of mainstem (Fig. 15–10) and total (Fig. 15–9) nodes. Along with the slow completion of the floral initiation process under the long photoperiods, time to anthesis of the first flower at a mainstem node of Ransom increased from 28 days at 10- and 12-h photoperiods to 68 days at the 16-h photoperiod, and time to anthesis at the terminal shoot apex increased from 35 to 78 days (Thomas and Raper, 1983a). These results confirm that, while reproductive development leading to anthesis is accelerated with decreasing daylength, the response is quantitative rather than absolute. It also appears from the data for Ransom that the concept of an absolute critical photoperiod, which generally

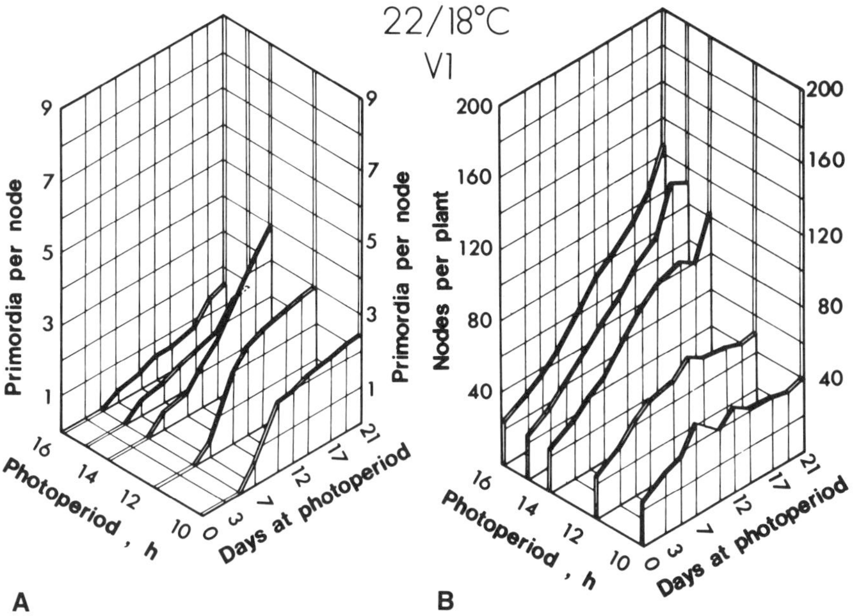

Fig. 15–9. Mean number of (A) floral primordia per node and (B) total nodes per plant produced by Ransom soybeans transferred to 10-, 12-, 14-, 15-, and 16-h photoperiods at the V1 developmental stage and grown for 21 days at 22/18°C day/night temperatures. From Thomas and Raper (1983a, 1984).

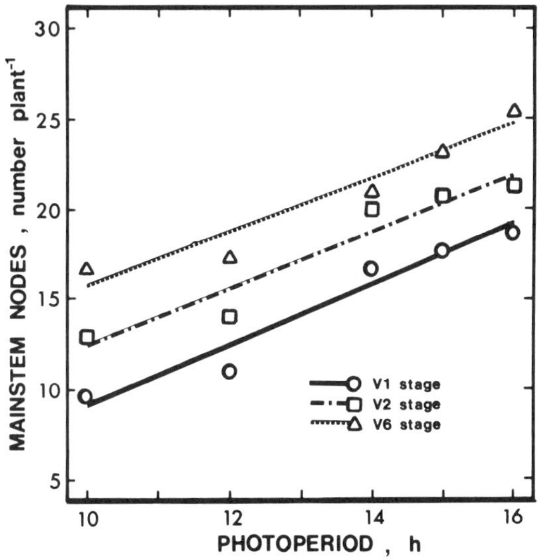

Fig. 15–10. Mean number of mainstem nodes per plant produced during a 21-day period after being transferred to 10, 12, 14, 15, and 16-h photoperiods at V1, V2, and V6 developmental stages. From Thomas and Raper (1984).

is considered as the daylength beyond which floral initiation fails to occur and less than which floral initiation abruptly terminates vegetative development, is misleading for soybean.

Changing daylengths should be considered as gradients of a stress. Although floral initiation may occur at the most undifferentiated axillary meristems (Borthwick and Parker, 1983a) as daylength after planting increases toward the summer solstice, the increasing photoperiods result in deceleration of the floral initiation process and promote continued vegetative development concomitantly with reproductive development. After the summer solstice, however, decreasing photoperiods accelerate completion of floral initiation and cessation of vegetative meristem development. The decreasing photoperiods also are associated with suppressed apical dominance as evidenced by increased internode elongation of branches relative to mainstem of Ransom when photoperiod was reduced from 16 to 10 h (Thomas and Raper, 1983b).

Photoperiod duration during vegetative growth and the transition from vegetative to reproductive growth phase has little direct effect on partitioning of dry matter among leaves, stems, and roots of Ransom soybean (Thomas and Raper, 1983b). Indirectly, however, the increased production of vegetative meristems under longer photoperiods (Fig. 15–9) establishes the potential for differences in dry matter accumulation and partitioning which occur as soybean that were moved from long to short photoperiods at varying ages approach maturity (Raper and Thomas, 1978; Thomas and Raper, 1977). This is emphasized when the effect of photoperiod on rate of continued reproductive development is considered. The successively shorter days that occur during reproductive development after the summer solstice accelerate the rate of seed growth and maturity (Cure et al., 1982; Johnson et al., 1960; Raper and Thomas, 1978; Shibles, 1980). In fact, for several indeterminate cultivars grown under natural photoperiods in field culture, the earlier developing seeds from pods at lower nodes had slower growth rates than seeds from pods set later at upper nodes as photoperiod became shorter (Gbikpi and Crookston, 1981).

Photoperiod has a relatively greater effect on accumulation rate of N than total carbon in seeds. The N concentration in seeds decreases as photoperiod increases (Cure et al., 1982, 1985; Gbikpi and Crookston, 1981). The lower rate of N accumulation by seeds under longer photoperiods is associated with retention of higher concentrations of N in leaves (Cure et al., 1982, 1985) and greatly reduced senescence and abscission of leaves at seed maturity (Cure et al., 1982, 1985; Raper and Thomas, 1978). Conversely, concentration of total nonstructural carbohydrates in leaves during reproductive growth is greater under the shorter photoperiod (Cure et al., 1985). This apparent effect of photoperiod on partitioning of carbohydrate possibly is regulated by activity of sucrose-phosphate synthase that alters the partitioning of photosynthate between nontranslocatable starch and readily translocatable sucrose (Chatterton and Silvius, 1979; Huber et al., 1984; Rufty et al., 1983).

The rapidly declining photoperiods during seed growth in field culture may be involved in the late-season declines in N_2-fixation activity frequently noted in field experiments (Hardy et al., 1968; Harper, 1974; Klucas, 1974; Lawn and Brun, 1974), but not necessarily in glasshouse (Israel, 1981) or growth chamber (Cure et al., 1985) studies. For nodulated Ransom soybean grown in controlled environments and entirely dependent on symbiotic fixation as the source of N, accumulation of N in the plant continued at near constant rates throughout and after completion of reproductive growth in well-watered plants under both long- and short-day photoperiods (Fig. 15–11). Nitrogen accumulation rates under long- and short-day photoperiods declined rapidly in response to a water stress during early seed filling. However, while the N_2-fixation rates remained low throughout the remainder of reproductive growth under short-day photoperiods, under long-day photoperiods N_2-fixation rates recovered after rewatering (Fig. 15–11). The lower rate of N accumulation by seeds under long-day photoperiods apparently averted the depletion of N reserves in vegetative tissues during the stress that occurred under short-day photoperiods (Cure et al., 1985). Conservation of N reserves in leaves

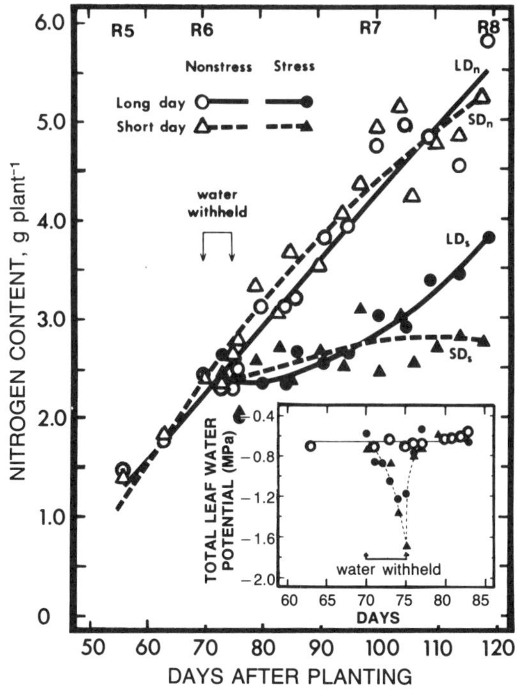

Fig. 15–11. Nitrogen accumulation in the combined vegetative and reproductive organs of the shoots of Ransom soybeans grown under long-day (LD) and short-day (SD) photoperiods after pod set. Half of the plants at each photoperiod were subjected to a single episode of water stress (SD_s and LD_s) at the R6 developmental stage and half were nonstressed controls (SD_n and LD_n). Leaf water potentials are shown in inset. From Cure et al. (1982, 1985).

under long-day photoperiods may explain the ability of plants to more fully recover photosynthetic and N_2-fixation capacity following a stress. When one considers the transient midday water stresses that occur even in well-watered plants in field culture (Boyer et al., 1980), but not necessarily in glasshouses or growth chambers, the interaction between stress-reduced photosynthetic capacity and the enhanced N partitioning to seeds by decreasing daylengths may explain late season declines in N_2 fixation in field culture.

It has not been established whether the photoperiodic signal alters partitioning during seed growth by affecting the export from the source leaves or import by reproductive tissues. The embryo itself is responsive to photoperiod under in vitro culture; however, in contrast to the response observed for seeds in attached pods, growth rates of in vitro cultured embryos increased as photoperiod increased (Raper et al., 1984). Several explanations are possible for this contrast. One could be that photoperiod directly alters source activities or growth regulator production in leaves (Cure et al., 1982). Another could be that regulation of substrate transport from pod to embryo through the seed coat is subject to photoperiodic regulation. From experiments in which attached flowers and young pods were shaded or given supplemental light, it appears that light perceived by these structures themselves had a role in regulating their abscission and accumulation of ^{14}C-photoassimilates (Heindl and Brun, 1983). Also, the in vitro growth of both excised seeds and seeds in excised pods requires light, but the growth rate of seeds in pod culture is slower than that of excised seeds (Hsu and Obendorf, 1982). Certainly, the import of sucrose by attached embryos from the vascular supply of the pod involves energy-dependent transport through the seed coat (Thorne, 1981, 1982). Since seed coats of excised seeds rupture and are shed within a few days of in vitro culture, more work is needed to elucidate the possible role of seed coat and pod wall tissues in photoperiodic regulation of seed growth rate. Finally, photoperiodic control of in vivo seed growth could involve regulation at both source and sink sites. Regardless of the tissues that perceive the light response, it appears that, as duration of photoperiod decreases during seed development, the ability of the embryo to utilize substrates decreases. Thus, the enhanced growth rate of seed that was observed in vivo as photoperiod decreased perhaps was regulated by the rate at which substrates from the source organs were made available to the embryos.

15–4 CARBON DIOXIDE

The CO_2 concentration in the global atmosphere has been increasing since 1958 (Keeling et al., 1976) and is expected to reach about 600 $\mu L\ L^{-1}$ by 2025 (Gribbin, 1981), or double the concentration existing early in this century. The first reaction to the projected increase in CO_2 concentration often has been to assume that crop yields will be increased by as much as 30% because of increased rates of photosynthesis (Kimball, 1983). However, inspection of data from research on effects of enriched-CO_2 concentrations on various species of plants indicates that the effects are far from simple (Kramer, 1981; Raper and Peedin, 1978; Thomas et al., 1975). Firstly, photosynthesis under field conditions is limited by environmental factors in addition to CO_2, including water, light and nutrients. Secondly, growth and yield depend on morphological and physiological factors in addition to the rate of photosynthesis (Evans, 1975, 1980; Gold and Raper, 1983). The absolute limit on enhancement of growth by CO_2 enrichment must reflect the finite limitations imposed by rates of leaf and flower initiation. The maximum rates for cell division and elongation establish the minimum intervals between initiation of successive leaves or floral primordia from a meristem and the development of axillary meristems. In the absence of environmental stresses, growth rate cannot be increased by increased CO_2 levels once the morphological limit is reached. The morphological limit at which CO_2 enrichment no longer accelerates growth can be readily approached under controlled-environment conditions (Rufty et al., 1981). Carbon dioxide enrichment under field conditions, however, may increase growth and yield by enhancing physiological processes that often are limited by environmental stresses.

15–4.1 Photosynthesis and Growth

Short-term experiments involving exposure of plants to enriched levels of CO_2 for a few hours or days have contributed to a frequent, but erroneous, impression that net CER per unit leaf area is always increased. Experiments lasting several weeks or months, however, indicate that while the net CER usually is increased, especially in the seedling or vegetative stages, it often declines after a few days or weeks (Clough et al., 1981; Kramer, 1981; Mauney et al., 1978). An important reaction of the plant to this initial enhancement of photosynthetic rate in response to CO_2 enrichment is a rapid increase in leaf area. The increased leaf area frequently persists through later growth so that even when net CER and net assimilation rates decline during later growth, dry matter accumulation per plant continues to increase. Such is the case for apically dominant sunflower (Mauney et al., 1978) and tobacco (Raper and Peedin, 1978). Net CER for sunflower after several weeks of CO_2 enrichment was similar to that at ambient concentration, and for tobacco was slightly less than that at ambient concentration. However, dry weights of plants of both species were increased because of the increase in leaf area. Conversely, soybean lacks strong apical dominance, and can maintain enhanced photosynthetic rates over prolonged periods of CO_2 enrichment (Hardman and Brun, 1971; Mauney et al., 1978; Rogers et al., 1983; Sionit, 1983; Sionit et al., 1984). Additionally, leaf area of soybean is increased under CO_2 enrichment (Jones et al., 1984; Rogers et al., 1984).

The response to CO_2 enrichment generally appears to be largest in seedling and juvenile stages of development and decreases as plants age (Thomas et al., 1975). Even for soybean, although net CER may be greater in both vegetative and reproductive stages when atmospheric CO_2 is enriched, the rates are higher during vegetative growth than during reproductive growth (Hardman and Brun, 1971; Mauney et al., 1978). Part of the decline with age in responsiveness to elevated CO_2 level may be attributed to the declines that normally occur in photosynthetic rates as leaves growing at ambient levels age (Fig. 15–8). Part, however, may be related to the existence of strong sinks for photosynthate. If the ratio of leaves to pods is varied, high rates of photosynthesis are prolonged when the reproductive sink is emphasized (Peet, 1984). There also is a strong sink in rapidly growing seedlings. Soybean leaves do not attain their maximum photosynthetic activity until after full expansion (Ojima et al., 1965; Woodward and Rawson, 1976), and young expanding leaves for a time are net importers of photosynthate rather than exporters (Thaine et al., 1959; Thrower, 1962). With its general lack of strong apical dominance and potential for profuse branching (Thomas and Raper, 1977, 1978, 1983b), the proportion of juvenile to mature leaf tissue can remain quite high in soybean, and combined with the associated expansion of stem and root tissues (Rufty et al., 1981), vegetative soybean plants have constantly expanding sink capacity for utilization of photosynthetically produced carbohydrates. The high sink capacity continues with the utilization of photosynthate to meet the high energy requirement of developing seed with their high protein and oil composition (Sinclair and de Wit, 1975).

When grown in CO_2-enriched atmosphere, the decline in photosynthetic rate of soybean was most rapid in plants with low sink demand (Clough et al., 1981; Peet, 1984). This decline has been related to the extent of build-up of nonstructural carbohydrates in leaves (Mauney et al., 1979). Part of the increase of nonstructural carbohydrates, especially starch, in leaves perhaps can be attributed to limits in the sucrose-phosphate synthase system because the normally low activity of sucrose-phosphate synthase is reduced further at high CO_2 levels (Huber et al., 1984). However, part may be attributable to a limited sink demand even during seed development (Peet, 1984). Under ambient CO_2 levels in controlled-environment studies when stresses were nearly absent, nonstructural carbohydrates in leaves and stems increased during seed filling (Cure et al., 1985), even though fruit loads for these plants represented more than 70% of shoot dry weight at physiological maturity (Cure et al., 1985). It thus seems that reproductive sink demand does not necessarily exceed photosynthetic capacity for soybean under ambient CO_2. Moreover, leaf mass increases under elevated CO_2 and, unless pod set is increased proportionately, CO_2 enrichment may actually increase the ratio of photosynthetic source to reproductive sink for soybean. Another sink for photosynthate in soybean, however, is the nodulated root system. In many plants, relative root mass increases under elevated CO_2 levels (Thomas et al., 1975), even though most experiments with elevated CO_2 have been conducted using plants grown in pots where restricted root growth may underestimate the importance of roots as sinks. For nodulated root systems, not only is the increased growth a sink for photosynthate (Finn and Brun, 1982), but also the high respiratory requirement for maintenance (Minchin and Pate, 1973) and functioning (Ryle et al., 1979a, 1979b, 1983; Williams et al., 1982) of nodules increases the sink activity of root systems and can alter the source-to-sink ratio of soybean during both reproductive and vegetative growth. Studies of long-term effects of CO_2 level on physiology and productivity of soybean must include determinations of relative changes in source and sink activities before cause-and-effect relationships can be established.

15–4.2 Interaction of Carbon Dioxide Concentration with Other Stresses

Many studies of effects of elevated concentrations of CO_2 are made under favorable growing conditions, but under field conditions effects are complicated by availability of water, nutrients, and light, and by competition with weeds. In general, drought tolerance appears to be increased by an increased atmospheric concentration of CO_2. For example, water-stressed wheat (*Triticum aestivum* L.) under high CO_2 in growth-chamber conditions yielded about 50% more than nonstressed wheat under normal CO_2 levels (Sionit et al., 1980, 1981). The growth of soybean also was reduced less by water stress under elevated than under normal concentration of atmospheric CO_2 (Huber et al., 1984; Rogers et al., 1984). Part of the increase in drought tolerance at increased atmospheric CO_2 levels is associated with decreased stomatal conductance (Fig. 15–12). Transpiration rate per unit leaf area is reduced by the decreased stomatal conductance caused by elevated levels of CO_2 (Egli et al., 1970, Rogers et al., 1984), but elevated CO_2 promotes increased photosynthetic rates (Sionit et al., 1984) despite the decreased stomatal conductance (Fig. 15–12). This increases water-use efficiency (Jones et al., 1984; Rogers et al., 1984). Although the decreased transpiration rate can enhance drought tolerance by decreasing water utilization from the drying soil, the increased leaf area resulting from elevated CO_2 levels can result in an increased rate of water utilization per plant. Part of the increased drought tolerance at elevated CO_2 levels, particularly at low plant populations, also must be attributed to increased root growth and proliferation that exploits a greater volume of soil for available water.

The requirements for N and other mineral nutrients to support long-term increases in growth are increased with increased levels of atmospheric CO_2. Additional assimilation of nutrients, particularly N, are required to support initiation of new tissues and continued activity of physiological processes, although to some extent concentration of N and other nutrients in expanded leaf tissues can be diluted by increased levels of nonstructural carbohydrates (Thomas et al., 1975; Williams et al., 1981;

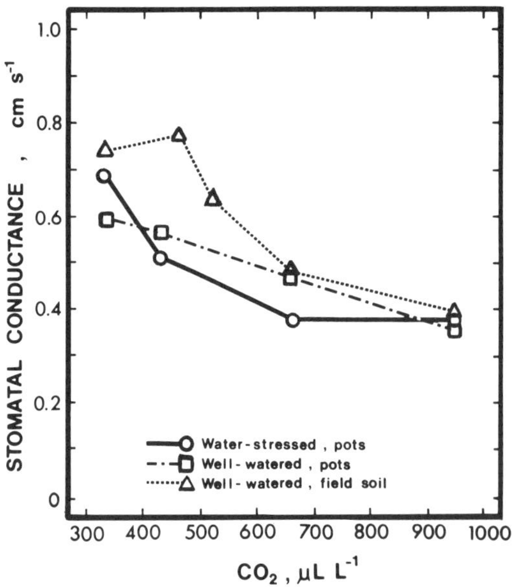

Fig. 15–12. Effect of CO_2 on stomatal conductance of water-stressed and well-watered soybean grown in pots and of well-watered soybean grown in field soil. From Sionit et al. (1984).

Wong, 1979). Obviously, growth and yield even under elevated CO_2 levels are subject to the availability of nutrients. An increased production of photosynthate, however, can somewhat increase accumulation of nutrients through increased partitioning of carbohydrates to roots (Sionit, 1983) and the resultant increase in N_2-fixation activity (Phillips et al., 1976) and volume of soil exploited by increased root length. While CO_2-enhanced photosynthesis might increase nutrient uptake from soils with low nutrient concentrations by increasing root mass and extension, this does not lessen the nutrient requirements for maximum growth and yield nor the necessity for adequate fertilization of nutrient-deficient soils to obtain a response to CO_2 enrichment.

The greatest benefits of an increase in global CO_2 probably will occur in sunny climates where irradiance is high rather than in regions with cloudy weather during the growing season. The response of plants to increased CO_2 concentration generally is greater at high irradiance levels, but there are differences among species. The dry weights of both soybean, a C_3 species, and corn, a C_4 species, were greater at 675 $\mu L\ L^{-1}$ than at 350 $\mu L\ L^{-1}$ CO_2 with both 600 and 1200 $\mu mol\ m^{-2}s^{-1}$ of PAR (Sionit et al., 1982). However, the response of soybean was greater than that of corn. The increases in dry weight for soybean between 350 and 675 $\mu L\ L^{-1}$ CO_2 were 72.7 and 76.4% at the low and high radiation levels, while the increases for corn were only 18.9 and 18.6%.

A depletion of CO_2 may occur within soybean canopies on sunny, windless days (Allen, 1975; Baldocchi et al., 1983). There have been only limited studies on the long-term effects of CO_2 depletion on crop plants. These studies indicate differences in response among species to CO_2 stress. For tobacco (Raper and Downs, 1973; Raper et al., 1973), leaf area and specific weight were not greatly altered when plants were allowed to reduce CO_2 to about 200 $\mu L\ L^{-1}$ in a growth chamber without supplementation to maintain the concentration, but stem elongation and dry weight, as well as carbohydrate content of leaves, were significantly reduced relative to plants grown at CO_2 levels maintained at 350 to 400 $\mu L\ L^{-1}$. For soybean (data of H.D. Gross as reported by Downs, 1983) and snap bean (*Phaseolus vulgaris* L.), leaf size and number were reduced when CO_2 in growth chambers was depleted to 200 $\mu L\ L^{-1}$ (Downs, 1980). While the 150 to 200 $\mu L\ L^{-1}$ differentials of these studies were far greater than the 10 to 30 $\mu L\ L^{-1}$ differentials that occur within soybean canopies in the field (Allen, 1975; Baldocchi et al., 1983), the results do establish the possibility that part of the response of plants to elevated CO_2 levels may be a result of reduced stress within the canopy.

Increases in the atmospheric CO_2 concentration likely will change the relative competitive capacities among crop plants and weeds (Patterson and Flint, 1980). As CO_2 concentration increases, C_3 weeds probably will become more competitive with C_4 crops, but C_4 weeds will become less competitive with C_3 crops such as soybean. Because the effects of CO_2 on growth usually are greatest during the seedling stage when crop and weed competition is most important, the probable reduction in competitive capacity of C_4 weeds relative to soybean with increased CO_2 is particularly important when considering the effects of biotic stress on yields. Also, the increased concentration of carbohydrates and decreased concentration of nitrogenous compounds in soybean leaves as atmospheric CO_2 concentration is increased may result in more extensive damage to leaves by insects (Lincoln et al., 1984).

15–5 METAL TOXICITY

Most of the metals can cause toxicities in plants growing in nutrient culture or in limited sites where soils have been contaminated by mine spoils or certain fungicides (Foy et al., 1978). Few metals, however, frequently cause phytotoxicity in soybeans under field conditions. The exceptions are Al and Mn which often can be important causes of stress in soybean on acid soils (Brown and Jones, 1977a).

15–5.1 Aluminum Toxicity

Aluminum toxicity is associated with increased susceptibility to drought stress and reduced accumulation of P, Ca, Mg, K, Fe, and N (Foy et al., 1969; Johnson and Jackson, 1964; Plant, 1956; Rorison, 1965; Sartain and Kamprath, 1975, 1978; Wright and Donahue, 1953). The primary processes affected by Al toxicity in crop plants are cell elongation in roots (Matsumoto et al., 1977; Wallace and Anderson, 1984) and cell division in the root apices, probably caused by formation of strong complexes with nucleic acids (Trim, 1959) during mitosis (Clarkson, 1965; Horst et al., 1983; Matsumoto and Morimura, 1980). This results in production of stubby lateral roots lacking the fine branching necessary for efficient absorption of nutrients and water from soil (Foy et al., 1978; Sartain and Kamprath, 1975). Nodule numbers also are reduced by high Al concentration (Sartain and Kamprath, 1975).

Aluminum toxicity occurs in acid soils with a high exchangeable Al saturation (Adams and Lund, 1966; Lund, 1970; Sartain and Kamprath, 1975). Such soils are prevalent in the southeastern USA (Cassel, 1983), as well as in extensive areas of Africa, South America, and Southeast Asia (Van Wambeke, 1976). While liming is effective in reducing the exchangeable Al in the plow layer (Armiger et al., 1968; Kamprath, 1970; Sartain and Kamprath, 1975), liming subsoils is difficult. Thus, penetration of roots into subsoil layers and efficient exploitation for water and nutrient reserves may be restricted by Al toxicity. As is evident from studies in which chisel plowing was used to disrupt a tillage-induced pan at the base of the Ap horizon of a Wagram loamy sand (loamy, siliceous, thermic *Arenic Paleudult*), there is a potential for enhanced yields of soybean with more extensive development of roots in subsoil horizons. Root growth and water extraction from the subsurface horizons and grain yield of soybean were increased when the mechanical impedance of the pan was reduced by tillage (Cassel, 1983; Kamprath et al., 1979; Martin et al., 1979). These studies, which were conducted at a site with a long history of cultivation and management, have the additional implication that continued applications of chemical fertilizers and lime to the Ap horizon of sandy soils in humid regions can result in sufficient leaching to increase the base saturation of subsurface horizons (Chaiwanakupt and Robertson, 1976; Juo and Ballaux, 1977; Terry and McCants, 1968, 1970; Volk and Bell, 1945).

A management-induced change in soil chemistry is only one possible explanation for increased proliferation and activity of roots in a subsoil with high exchangeable Al in its native state. Another possibility is the differential tolerance to Al that exists among soybean cultivars (Brown and Jones, 1977a; Hanson and Kamprath, 1979; Sartain and Kamprath, 1978). The difficulty and time involved in increasing base saturation and reducing exchangeable Al in subsoils makes selection for Al tolerance a more feasible goal for enhancing rooting activity in acid soils than changing the chemistry of the subsoil. There are several mechanisms for Al tolerance. One is cell membranes, possibly the plasmalemma, that are differentially permeable to Al and prevent entrance of Al into the root cells (Ali, 1973). There also is evidence that the gelatinous mucilage secreted by roots, particularly at the root tips, protects root meristems by adsorption of Al (Hecht-Buchholz and Foy, 1981; Horst et al., 1982) or chemically similar metal ions (Clarkson and Sanderson, 1969). Another mechanism is internal chelation of Al within the cytoplasm, possibly by organic acids (Grime and Hodgson, 1969; Jones, 1961; Lunt and Kofranek, 1970). Other studies show that plants vary in ability to increase the pH of the rhizosphere (Marschner and Römheld, 1983). Cultivars that increase the pH absorb less Al because of its lower solubility at high pH (Foy et al., 1967). In nodulated root systems of soybean, however, N_2 fixation decreases pH of the rhizosphere by uptake of cations in excess of anions (Israel and Jackson, 1982). None of the these mechanisms have been identified as involved in the differential tolerance to Al among soybean cultivars. Nevertheless, tolerance among cultivars, as established by rates of root growth of 5-day-old seedlings in nutrient solution containing Al, is heritable (Hanson and Kamprath, 1979). However, the relationship between responses of seedlings, which sometimes are more susceptible to Al toxicity (Thawornwong and Van Diest, 1974), and of plants grown in soil is not established (Hanson and Kamprath, 1979; Sartain and Kamprath, 1978). Increased tolerance of older roots may be a more pertinent criterion of ability of roots to penetrate into the unlimed subsurface horizons.

15–5.2 Manganese Toxicity

The primary symptoms of Mn toxicity, in contrast to those of Al, involve leaf abnormalities. Symptoms in soybean include crinkling, chlorosis, and necrotic lesions of leaves (Heenan and Campbell, 1980; Heenan and Carter, 1976, 1977). Reciprocal grafts between rootstocks and shoots of two soybean cultivars differing in tolerance to Mn have demonstrated that the tolerance is related to shoot factors rather than root activity (Heenan and Carter, 1976). The reduction in growth and yield associated with Mn toxicity apparently is caused by disruption of photosynthetic processes (Jackson, 1967) through biochemical disorders (Foy et al., 1978) or reduction in leaf area by decreased cell division or expansion (Terry et al., 1975).

While Mn toxicity is a problem primarily in acid soils, susceptibility of plants growing on acid soils high in extractable Mn is influenced by weather. Manganese toxicity of tobacco frequently occurs in the cooler mountain regions of North Carolina, but seldom occurs in the warmer piedmont regions despite similarities in soil pH and levels of extractable Mn (Rufty et al., 1979). More direct evidence of an interaction between temperature and susceptibility to Mn toxicity is available from glasshouse experiments. Löhnis (1951) observed that severity of Mn toxicity in bean growing in a glasshouse without temperature control was greater in the cooler areas of the glasshouse than in warmer areas. For both tobacco (Rufty et al., 1979) and soybean (Heenan and Carter, 1977), levels of Mn in nutrient solution that produced toxicity symptoms at temperatures of about 20°C did not produce toxicity symptoms at temperatures of about 28 to 31°C, and the concentration of Mn in leaves at the warmer temperature was similar to, or exceeded, that at the cooler temperatures.

Thus, the effect of temperature on susceptibility to Mn toxicity is not directly related to uptake and accumulation of Mn by the plant.

At high external supply, Mn concentration in leaves generally increases with leaf age, and within a sampling date younger leaves of a plant have lower concentrations of Mn than older leaves (Heenan and Campbell, 1980; Rufty et al., 1979). However, Mn toxicity symptoms in tobacco were most pronounced in younger leaves (Rufty et al., 1979). This suggests that the greatest sensitivity to Mn is during the period of cell division and elongation (Terry et al., 1975). Manganese tolerance related to both cultivar differences (Brown and Jones, 1977b; Heenan and Carter, 1975, 1977) and temperature during growth (Heenan and Carter, 1977; Rufty et al., 1979) likely is related to morphological development of leaves. Rufty et al. (1979) suggested that temperature-dependent tolerance may be associated with rate of vacuolar expansion within leaf cells. Manganese can accumulate in vacuoles (Munns et al., 1963), and increased availability of vacuoles for sequestering Mn (MacRobbie, 1971) away from physiologically active regions of cytoplasm might result in greater tolerance. Similarly, the genotypic tolerance may be associated with time or rate of development of vacuoles in emerging and expanding leaves. Certainly, the relationship between toxic concentrations of Mn and morphological development of leaves needs investigation.

15–6 STRESS TOLERANCE

Generally, identification of stress tolerance has depended on observation of the behavior of plants subjected to the particular stress under investigation (Burton, 1983; Castleberry, 1983). However, it seems probable that more rapid progress could be made in breeding for stress tolerance if plant breeders could identify and concentrate on specific morphological or physiological characteristics conferring tolerance of a specific stress. The chief difficulties with this approach are lack of sufficient information concerning the physiological bases of tolerance and lack of good tests for screening large populations for the desired physiological characters (Blum, 1983; Nelson, 1983).

15–6.1 Chilling Tolerance

Chilling injury provides an example of the difficulties in developing screening procedures to select for specific physiological mechanisms of tolerance. Soil temperatures below about 15°C cause injury to germinating seeds (Hobbs and Obendorf, 1972), and cool soil reduces water absorption (Markhart et al., 1979). Air temperatures below about 13 to 15°C reduce or prevent flowering and pod set (Hume and Jackson, 1981; Musser et al., 1983a; Thomas and Raper, 1978) and affect the photosynthetic apparatus (Musser et al., 1984). But how do low temperatures produce these effects? Chilling injury is generally attributed to damage to cell membranes (Bramlage et al., 1978; Lyons et al., 1979; McWilliam, 1983), but there is uncertainty concerning the nature of the damage. For several years emphasis was placed on phase changes in the lipids of the cell membranes (Lyons et al., 1979), but this is now questioned (O'Neill and Leopold, 1982; Wolfe, 1978). In any event, it seems likely that chilling injury occurs because cell membranes are unable to maintain their structural organization at low temperatures (Bramlage et al., 1978; Willing and Leopold, 1983). This is particularly plausible for explaining the damage to germinating seeds allowed to imbibe water at low temperatures. It also is important to note that some tissues such as developing pollen grains are more sensitive to chilling than other tissues, resulting in male sterility in chilled sorghum (Brooking, 1976) and rice (Lin and Peterson, 1975; Nishiyama and Sataka, 1979), and possibly soybean (Lawn and Hume, 1985). The reason for this is unknown. However, all of the various chilling injuries associated with reduced floral development probably result from damage to cell membranes (McWilliam, 1983). Until the site and nature of this injury is definitely established, it will be difficult to develop a breeding program based on physiological tolerance.

15–6.2 Drought Tolerance

The causes of injury by water stress are better identified. The behavior of plants with respect to drought can be classified as follows: drought avoidance where plants are not subjected to drought conditions; and drought tolerance either by dehydration postponement or by dehydration tolerance. Plants sometimes can be placed in more than one of these categories.

Complete avoidance of drought generally is impossible in the central and eastern USA where drought occurrence is largely random (Decker, 1983). Where late summer droughts are common, the best cultural approach to drought avoidance is early planting. This requires seed that germinate well and produce vigorous seedlings in cold soil. Apparently, little selection has been done for this character, although cultivars from Hokkaido and eastern Siberia and their progeny seem to possess this characteristic (Holmberg, 1973).

Selection for characteristics that postpone dehydration seems more promising. One of the most common is deep, much branched root systems that absorb water from a large volume of soil. Jordan and Miller (1980) and Taylor (1980) agreed on the desirability of deep rooting. But while Taylor (1980) doubted if cotton (*Gossypium hirsutum* L.) or soybean would benefit from increased root density, Boyer et al., (1980) found that high-yielding soybean cultivars had higher root densities and higher afternoon leaf water potentials than lower-yielding cultivars. Raper and Barber (1970), in an investigation of comparative root systems of 26 genotypes of soybean, found considerable genotypic differences. In single-plant plots, 'Harosoy 63' had nearly twice the root surface and 1.5 times the root length of 'Aoda', and therefore occupied the soil more thoroughly. Thus, there appears to be some genetic variability available to use in breeding programs to produce whatever type of root system seems most desirable for particular soil conditions and cultural regimes. There also are differences in axial and radial resistance to water flow into and through roots (Newman, 1976), and soybean root systems have a much higher resistance to entrance of water than corn or sunflower (*Helianthus annuus* L.) (Boyer, 1971). It is not known, however, whether there are sufficient differences among soybean cultivars to be of any significance.

The role of the physical properties of soil in determining the amount of available water and the development of roots was discussed by Cassel (1983), Kramer (1983), and Ritchie (1983). Cassel presented considerable data indicating the importance of impermeable soil layers, often caused by farm machinery, on root penetration. The effect on soybean root penetration of deep tillage to disrupt a traffic pan at a depth of about 25 cm is shown in Table 15–4. There is need for root systems that can penetrate soil layers with a high bulk density and that can proliferate in acid subsoils containing high concentrations of exchangeable Al (Cassel, 1983).

The loss of water from leaves depends primarily on stomatal opening and secondarily on leaf orientation and factors affecting transpiration. Stomata that close promptly as leaf water stress increases reduce both transpiration and photosynthesis, but reduce the former more than the latter. The writers are not aware of any systematic survey to determine if there are important differences among soybean cultivars with respect to stomatal response to stress. However, the environmental history of soybean may be more important than varietal differences, because there is evidence that stomata of plants previously stressed close at lower leaf water potentials than stomata of nonstressed plants (Van Volkenburgh and Davies, 1977). There may be sufficient differences in the amount of wax on leaves to affect water loss. Van Volkenburgh and Davies (1977) found wax on the lower surfaces of field-grown soybean leaves but not on leaves of chamber-grown plants unless the latter were grown with cool nights. The cuticular transpiration of detached chamber-grown soybean

Table 15–4. Effect of conventional tillage with a moldboard plow to a depth of 25 cm and in-row subsoiling to a depth of 45 cm on distribution of secondary roots of soybean in a Wagram loamy sand containing a tillage-induced pan between 20 and 30 cm below the surface. From Kamprath et al. (1979).

Depth	Dry weight of roots	
	Conventional tillage	Subsoiling
cm	$g\ m^{-3}$	
0–10	334	326
10–20	219	198
20–30	64	101
30–45	14	65
45–60	10	74
60–75	6	87

leaves also was higher than that of detached field-grown leaves. Heavy wax deposits decrease transpiration from sorghum leaves (Blum, 1979) more than photosynthesis (Chatterton et al., 1975), thereby increasing water-use efficiency. There also were differences among soybean cultivars in amount of pubescence, but since no important differences occurred in leaf water potential among three isolines of 'Clark' which vary from densely pubescent to glabrous (Ghorashy et al., 1971), it is doubtful that pubescence has much effect on water relations.

Osmotic adjustment refers to a decrease in osmotic potential greater than that caused by loss of water (Turner and Jones, 1980). Such a decrease permits cells to enlarge and stomata to remain open to a lower water potential than would be possible in its absence and postpones injury from dehydration. Osmotic adjustment has been observed in some soybean cultivars, but not in others (Sionit and Kramer, 1977; Turner and Jones, 1980; Wenkert et al., 1978). Apparently there has been no systematic survey for this adaptation in soybean, nor has there been any serious evaluation of its importance in decreasing injury from drought.

15–6.3 Tolerance and Recovery from Stress

No matter how good the plant characteristics that postpone dehydration, plants eventually suffer injury or death from dehydration if the drought is long enough. There is little information available concerning differences among soybean cultivars in ability to recover after severe dehydration. Blum (1979) regards ability to recover after stress as important in sorghum, and it should be important in other crops. The long-flowering period for many soybean cultivars should decrease the possibility of total crop loss from a single episode of drought or chilling during reproductive growth. Musser et al. (1983a) reported some recovery after chilling for Ransom soybean chilled to 10°C for 1 week during floral induction, and Schmid and Keller (1980) reported differences among cultivars in ability to recover after chilling. The chilled plants lost most of the flowers that opened during the chilling period, but after several weeks they resumed flowering and set pods. Water stress during flowering reduced the length of the flowering period of Bragg and Ransom soybean, but stress during early pod formation and seed filling reduced yield more than stress during flower induction and flowering (Table 15–4). It is possible from results of these and other experiments to predict the stage of development at which most injury will result from stress. However, because of inability to predict the time when stress will occur, it is doubtful if farmers can benefit much from this information unless irrigation is available. Perhaps more attention should be paid to selection for longer flowering periods and better recovery after stress.

15–6.4 Performance of Stress Tolerant Cultivars in the Absence of Stress

An important consideration in the breeding of plants for stress tolerance is whether genotypes that yield well when stressed will also yield well in the absence of stress. This characteristic, known as yield stability, obviously is important where stress occurs more or less randomly and one cannot predict at planting time whether or not a crop will be subjected to stress.

There is some difference of opinion concerning the possibility of the same genotype having a relatively high yield in both the presence and the absence of stress. Orians and Solbrig (1977) stated that "there is an inevitable correlation between ability to photosynthesize rapidly when soil moisture is readily available and inability to extract moisture when soils are drier, and vice versa." They base their argument on the questionable assumption that xeromorphic leaves cannot have a high rate of photosynthesis under favorable conditions, but that under water stress they have a higher rate than crop plants. Some woody plants with xeromorphic leaves, however, have high rates of transpiration (Caughey, 1945), indicating high stomatal conductance, and presumably they also have high rates of photosynthesis. Brigalow (*Acacia harpophylla*) has xeromorphic leaves, but its rate of photosynthesis is as high as that of trees with mesomorphic leaves (van den Driesche et al., 1971).

The data for yield stability of crop plants are conflicting. The thick leaves of soybean plants produced in response to water stress have higher rates of photosynthesis per unit of leaf surface after stress is removed than thin leaves of plants that have not undergone stress (Davies et al., 1977; Van Volkenburgh and Davies, 1977). Plants of soybean, sunflower, and buckwheat (*Fagopyrum esculentum* Moench) genotypes that had been selected for relatively good growth and high rates of photosynthesis when water-stressed produced less dry matter and had relatively lower rates of photosynthesis when grown in moist soil than plants lacking stress tolerance (Bunce, 1981). Similarly, Fischer (1981) observed that increase in yield was positively correlated with increase in susceptibility to drought injury among several wheat (*Triticum aestivum* L.) cultivars. On the other hand, Reitz (1974) stated that hard winter wheats fall into three groups: (i) those yielding relatively well only under stress; (ii) those yielding well only in the absence of stress; and (iii) those yielding well with and without stress, i.e., having a high degree of stability. Although absolute seed yields of eight soybean cultivars were reduced by water stress, the relative ranking for final yields among the cultivars were similar when grown to maturity at deficient soil moisture conditions and when grown at optimum soil moisture conditions (Mederski and Jeffers, 1973).

There also is evidence of yield stability for chilling-tolerant soybean. In one set of experiments, the chilling-tolerant cv. Fiskeby V outyielded four other cultivars when the mean temperature during the 30 days after anthesis were both cooler and warmer than average (Table 15–5). In another comparison, Fiskeby V outyielded four other cultivars in cool and average seasons and in the warmest season was outyielded by only one of the more chilling-sensitive cultivars (Holmberg, 1973). Experiments by Schmid and Keller (1980) also indicated a high degree of yield

Table 15–5. Yield stability for chilling-tolerant cv. Fiskeby V and four less tolerant soybean cultivars in relation to mean temperature during the 30 days immediately following anthesis at Fiskeby, Sweden. From Holmberg (1973).

Cultivar	Seed yield		
	1970	1971	1972
	kg ha^{-1}		
Toshi-dai 7910	710	1150	1475
Karafuto 1	790	1195	1270
Chishima	825	995	1260
Kamishunbetsu	880	990	1115
Fiskeby V	1390	1500	2150
	°C		
Mean temperature during first 30 days after anthesis†	15.7	17.0	18.8

†Between 1960 and 1972 the average temperature for the first 30 days after anthesis was 16.4°C.

stability over a wide temperature range for certain cultivars of Hungarian origin.

Perhaps the conflict among these reports is related to the mechanisms of stress tolerance involved. None of the specific mechanisms of tolerance were identified. Selection for a physiological mechanism such as osmotic adjustment of chloroplasts during stress, for example, may well provide increased tolerance during stress without reducing photosynthesis in the absence of stress. On the other hand, selection for increased partitioning of photosynthate to roots regardless of stress conditions may enhance yields if drought occurs, but in the uncertain drought conditions of humid regions may represent a wasteful diversion of photosynthates away from seed production in the absence of drought. In selecting for stress tolerance, recognition of the specific mechanism of tolerance and the climatic characteristics of the production area are important.

The performance of stress-tolerant plants in the absence of stress is important in connection with the troublesome problem of screening progeny for tolerance of stress. Blum (1979) suggested that there are three methods available. The one most frequently used is based on the assumption that genotypes which are relatively high yielding in the absence of stress also will be relatively high yielding in the presence of stress. This approach is attractive because differences among genotypes are much greater under favorable conditions and more easily observed. Frey (1964) and Johnson and Frey (1967) found this method satisfactory for oat (*Avena sativa* L.) and Mederski and Jeffers (1973) used it in selecting for drought tolerance of soybean. A second method is to select for high yield under stress and disregard the maximum possible yield. This seems a logical method when the occurrence of stress is certain, but unfortunately differences among genotypes under stress conditions may be small and a large amount of material is required to establish the existence of significant differences (Blum, 1979, 1983). A third, and perhaps better approach (Blum, 1979, 1983), is to assume that yield and stress tolerance are separate characteristics, just as yield and resistance to various diseases are inherited separately. Once morphological or physiological characteristics that can contribute to stress tolerance are determined, they can be introduced into high-yielding genotypes to combine yield and stress tolerance. For example, soybeans from the east coast of Hokkaido and adjacent islands have considerable chilling tolerance, but are small and low yielding. When crossed with high-yielding cultivars from Europe, however, the result was Fiskeby V and other cultivars with chilling tolerance and relatively stable yield (Holmberg, 1973).

The principal difficulty with a breeding strategy based on physiological characteristics is that we do not yet know enough about the morphological and physiological characteristics that confer stress tolerance, or how they confer it (Eastin et al., 1983). For example, it is not yet fully understood why flower initiation and pod development of soybean are so sensitive to chilling and water stress. Furthermore, good methods are lacking for rapidly screening large populations of plants for stress tolerance.

15–7 RESEARCH NEEDS

The ultimate objective of agricultural research is to find ways to increase the efficiency of crop production. Plant breeders already have provided farmers with soybean cultivars having the physiological and morphological characteristics for potentially high yields, but the full potential for yield seldom is attained because of the various environmental stresses (Boyer, 1982). Thus, the primary need of the farmer is for cultivars that yield well in favorable seasons and suffer the minimum reduction in yield when subjected to the common climatic and soil-related stresses. Research to improve productivity should be based, on one hand, on an understanding of the nature and timing of stresses and, on the other hand, on an understanding of how the stresses disrupt physiological and morphological processes to cause reduction in yield. This requires the cooperative efforts of agronomists, soil scientists, crop climatologists, and plant physiologists to provide the information needed by plant breeders.

15–7.1 Stress Tolerance vs. Avoidance

Once a specific stress is identified as a major factor in preventing attainment of the physiological potential of a cultivar, two approaches are possible for the agronomist and plant breeder. One is to increase tolerance of the stress, and the other is to find ways of avoiding it. In areas with well-defined wet and dry seasons, it sometimes is possible to time plantings so that crops mature before severe drought develops. In the eastern and central USA, however, drought can occur at almost any time during the growing season (Decker, 1983; Shaw, 1983) making avoidance difficult. Emphasis thus must be placed on finding plants with tolerance to drought either through postponement of dehydration or tolerance of dehydration. Tolerance of dehydration is important for survival in natural vegetation, but is less important in crop plants where economic yield rather than survival is of primary concern and severe dehydration usually severely limits crop yields. However, there are good possibilities of postponing dehydration by deeper, more extensively branched root systems and better control of water loss.

Sometimes the possibility of a stress can be eliminated. Water stress can be eliminated by irrigation where water is available, weeds can be eliminated by cultivation or the use of herbicides, and some insects and diseases can be controlled by chemicals and development of resistant cultivars. In many instances, consideration should be given to the ratio of costs to benefits in choosing methods of dealing with stresses. For example, it may be more profitable to sacrifice some yield in order to cut the cost of irrigation or of using pest controls. A low pH of topsoil can be corrected by liming, but it may be more practical to find root systems tolerant of high concentrations of exchangeable Al and Mn in acid subsoils than to attempt to reduce the concentration in the soil (Hanson and Kamprath, 1979; Heenan and Carter, 1976; Sartain and Kamprath, 1978).

15–7.2 Genetic Variability in Response to Stress

Fortunately, soybean has considerable genetic variability for such diverse characteristics as depth and density of rooting, determinate vs. indeterminate growth habit, length of life cycle (maturity group), sensitivity to photoperiod, and tolerance of lower temperature. Other desirable characters would include ability to penetrate soils of high bulk density, tolerance of Al and Mn and ability to tolerate dehydration and resume flowering after a severe drought. There seems to be enough genetic variability for most of these characteristics to justify the expectation that plant breeders will be able to produce cultivars with greater tolerance to most of the stresses that reduce yield. The problem is to determine which among these characters are most important in relation to stress tolerance and yield and to find ways to screen progeny for their presence.

The best approach to take in breeding for plants yielding well under stress depends on the options available. If considerable genetic variability in tolerance of a particular stress exists, then development of stress-tolerant cultivars usually is possible. However, it is no longer satisfactory to ask plant breeders for such broad characteristics as drought or chilling tolerance, because tolerance may depend on any one of several characters or on some combination of the characters. Drought tolerance, for example, might result singularly or in combination from a deep and profusely branched root system, good stomatal control of transpiration, early maturity, osmotic adjustment, tolerance of dehydration, or ability to resume growth after a drought. Agronomists and physiologists must identify for plant breeders the characters that are most important under various soil and climatic conditions.

In summary, we should not ask plant breeders for stress tolerance, but rather to incorporate the specific characteristics that provide the tolerance into high-yielding cultivars. This requires better information about the characteristics that provide tolerance and better methods of screening plant populations for the characteristics than is presently available. There are problems in this approach to providing stress tolerance (Boyer, 1983; Nelson, 1983; Ritchie, 1983; Vieira da Silva, 1983). Although much research has been done on effects of environmental stresses on primary physiological processes, these seldom have been positively correlated with yield, and research on physiological processes must be better related to critical stages in plant development (Eastin et al., 1983). Thus, it seems that there must be a better understanding of the interaction between environmental stresses, plant processes, and stage of plant development at which stress occurs.

During the late 1960s, genetic engineering has opened up new possibilities for enhancing stress tolerance by transfer of genes from one kind of organism to another (Csonka et al., 1983). Although gene splicing or transfer of genes is difficult in seed plants, it should eventually facilitate improvement in resistance to stresses. Protoplast fusion seems to have considerable potential (Shepherd et al., 1983), and the isolation of useful mutants from cell cultures already is being exploited (Chaleff, 1983). Realization of this developing potential, however, still depends on the ability of agronomists and physiologists to identify specific mechanisms of tolerance and to develop suitable screening techniques.

15–7.3 Screening Techniques

One of the difficult problems in developing stress-tolerant cultivars based on specific characteristics is to find methods of screening large numbers of progeny for the desired characteristics. Screening may require physical measurement of entire root systems under actual or simulated stress conditions, determination of physiological processes such as photosynthesis of plant canopies or of single leaves, measurement of the water status of specific organs such as leaves or stem apices, or the observation of changes in specific cells or organelles for injury to membranes during chilling or osomotic stress. Observation of root systems often requires laborious excavation or expensive measurements of soil water depletion at various distances and depths. Although occasionally gross measurements are made of the water status of plants with infrared thermometers to compare the water status of plots of a large number of genotypes (Blum, 1979, 1983), detailed measurement of stomatal behavior, rates of photosynthesis or transpiration, and leaf water potential in response to water stress are time-consuming and can be done on only a relatively small number of plants.

At one time, it was supposed that plants with high rates of photosynthesis per unit leaf area should be high yielding, but numerous experiments have shown that this is not necessarily correct (Curtis et al., 1969; Evans, 1975; Ford et al., 1983; Sinclair, 1980). The relationship between photosynthesis and biomass production depends on the total leaf area and the length of time during which leaves maintain a high rate of photosynthesis, as well as the rate of photosynthesis per unit leaf area. Biomass production of soybean usually is related to canopy photosynthesis (Egli et al., 1970; Jeffers and Shibles, 1969; Wells et al., 1982), and since this is related to LAI, rapid development of leaves and early closure of the canopy is desirable (Potter and Jones, 1977; Shibles and Weber, 1966). However, increased seed yields cannot result from increased photosynthesis unless the size of the reproductive sink is increased (Clough et al., 1981; Peet and Kramer, 1980). Thus, selection for high rates of photosynthesis measured on single leaves will not necessarily result in increased yields (Ford et al., 1983).

Even if yield were correlated to short-term measurements of photosynthesis, it is difficult to screen large populations from selected crosses for differences in rates. An alternative to the use of whole plants is to devise tests to use on tissue samples such as leaf disks. Sullivan and Ross (1979) tested sorghum cultivars for tolerance of heat and desiccation by subjecting discs of leaf tissue to heat or osmotic stress and then measuring the amount of leakage of electrolytes out of the tissue. Tests for cold and heat tolerance can be made on seeds (Duke et al., 1983). Extensive correlation studies are needed to evaluate the usefulness of these and other rapid-screening techniques. The use of cell and tissue cultures to screen for stress tolerance currently is receiving much attention because it is speedy and requires much less time and space than field tests (Csonka et al., 1983). Cells in culture often exhibit great variability and sometimes develop tolerance to stresses not found in the plants from which they came. They can be subjected to such stresses as heat, cold, salinity, herbicides, and toxins produced by plant pathogens, and then plants can be regenerated from the survivors. Some success has been attained in selection for salt, herbicide, and disease tolerance by this method and more will doubtless be obtained in the future. Use of this method, however, presently is limited by the difficulty in regenerating plants from cell cultures.

Methods involving selection in cell cultures are only effective for tolerance at the cellular level. While tolerance of dehydration operates at the cellular level, postponement of dehydration involves root systems, stomatal behavior, and leaf structure, which can be expressed only at the level of whole plants. Some stresses, such as salinity, operate at both the plant and the cell level. Some kinds of plants succeed in saline environments because they exclude salt, and others succeed because they can tolerate it (Osmond et al., 1980). The salt tolerance of barley (*Hordeum vulgare* L.) callus cultures is similar to that of intact plants, but callus cultures of a halophytic Salicornia are as sensitive to salt as those of nonhalophytes (Chaleff, 1983). Thus, although cell culture has considerable promise as a method of screening, its application is limited to stresses that operate at the cellular level. There will continue to exist a need for screening techniques that can select for stress tolerance at the more complex levels of the whole plant.

REFERENCES

Abel, G.H. 1970. Winter and summer soybean growth in southern California. Agron. J. 62:118–120.

Adams, C.A., M.C. Fjerstad, and R.W. Rinne. 1983. Characteristics of soybean seed maturation: Necessity for slow dehydration. Crop Sci. 23:265–267.

----, S.W. Norby, and R.W. Rinne. 1982. Protein modification and utilization of starch in soybean (*Glycine max* L. Merr.) seed maturation. J. Exp. Bot. 33:279–287.

Adams, F., and Z.F. Lund. 1966. Effect of chemical activity of soil solution aluminum on cotton root penetration of acid subsoils. Soil Sci. 101:195–198.

Alberte, R.S., E.L. Fiscus, and A.W. Naylor. 1975. The effects of water stress on the development of the photosynthetic apparatus in greening leaves. Plant Physiol. 55:317–321.

----, J.P. Thornber, and E.L. Fiscus. 1977. Water stress effects on the content and organization of chlorophyll and bundle sheath chloroplasts of maize. Plant Physiol. 59:351–353.

Ali, S.M.E. 1973. Influence of cations on aluminum toxicity in wheat (*Triticum aestivum* Vill., Host). Ph.D. thesis. Oregon State Univ., Corvallis (Diss. Abstr. 73-21304).

Allen, L. H., Jr. 1975. Shade-cloth microclimate of soybeans. Agron. J. 67:175–181.

Armiger, W.H., C.D. Foy, A.L. Fleming, and B.E. Caldwell. 1968. Differential tolerance of soybean varieties to an acid soil high in exchangeable aluminum. Agron. J. 60:67–70.

Ashley, D.A. 1983. Soybean. p. 389–422. *In* I.D. Teare and M.M. Peet (ed.) Crop-water relations. John Wiley and Sons, New York.

Baldocchi, D.D., S.B. Verma, and N.J. Rosenberg. 1983. Microclimate in the soybean canopy. Agric. Meteorol. 28:321–337.

Begg, J.E., and N.C. Turner. 1976. Crop water deficits. Adv. Agron. 28:161–217.

Blum, A. 1979. Genetic improvement of drought resistance in crop plants: A case for sorghum. p. 429–445. *In* H. Mussell and R.C. Staples (ed.) Stress physiology in crop plants. John Wiley and Sons, New York.

----. 1983. Breeding programs for improving crop resistance to water stress. p. 263–275. *In* C.D. Raper, Jr. and P.J. Kramer (ed.) Crop reactions to water and temperature stresses in humid, temperate climates. Westview Press, Boulder, CO.

Boote, K.J., R.N. Gallaher, W.K. Robertson, K. Hinson, and L.C. Hammond. 1978. Effect of foliar fertilization on photosynthesis, leaf nutrition, and yield of soybeans. Agron. J. 70:787–791.

Borthwick, H.A., and M.W. Parker. 1983a. Photoperiodic perception in Biloxi soybeans. Bot. Gaz. 100:374–387.

----, and ----. 1983b. Effectiveness of photoperiodic treatments of plants of different age. Bot. Gaz. 100:245–249.

Boyer, J.S. 1970a. Differing sensitivity of photosynthesis to low leaf water potentials in corn and soybean. Plant Physiol. 46:236–239.

----. 1970b. Leaf enlargement and metabolic rates in corn, soybean, and sunflower at various leaf water potentials. Plant Physiol. 46:233–235.

----. 1971. Resistance to water transport in soybean, bean, and sunflower. Crop Sci. 11:403–407.

----. 1982. Plant productivity and environment. Science 218:443–448.

----. 1983. Mechanisms controlling plant performance. p. 347–349. *In* C.D. Raper, Jr. and P.J. Kramer (ed.) Crop reactions to water and temperature stresses in humid, temperate climates. Westview Press, Boulder, CO.

----, R.R. Johnson, and S.G. Saupe. 1980. Afternoon water deficits and grain yields in old and new soybean cultivars. Agron. J. 72:981–986.

Bramlage, W.J., A.C. Leopold, and P.J. Parrish. 1978. Chilling stress to soybeans during inhibition. Plant Physiol. 61:525–529.

Brooking, I.R. 1976. Male sterility in *Sorghum bicolor* (L.) Moench induced by low night temperature. I. Timing of the stage of sensitivity. Aust. J. Plant Physiol. 3:586–589.

Brown, J.C., and W.E. Jones. 1977a. Fitting plants nutritionally to soils. I. Soybeans. Agron. J. 69:399–404.

----, and ----. 1977b. Manganese and iron toxicities dependent on soybean variety. Commun. Soil Sci. Plant Anal. 8:1–15.

Bunce, J.A. 1977. Leaf elongation in relation to leaf water potential in soybean. J. Exp. Bot. 28:156–161.

----. 1978a. Effects of water stress on leaf expansion, net photosynthesis and vegetative growth of soybeans and cotton. Can. J. Bot. 56:1492–1498.

----. 1978b. Interrelationships of diurnal expansion rates and carbohydrate accumulation and movement in soya bean. Ann. Bot. 42:1463–1466.

----. 1981. Relationship between maximum photosynthetic rates and photosynthetic tolerance of low leaf water potential. Can. J. Bot. 59:769–744.

Burton, G.W. 1983. Breeding programs for stress tolerance in forage and pasture crops. p. 289–296. *In* C.D. Raper, Jr. and P.J. Kramer (ed.) Crop reactions to water and temperature stresses in humid, temperate climates. Westview Press, Boulder, CO.

Cassel, D.K. 1983. Effects of soil characteristics and tillage practices on water storage and its availability to plant roots. p. 167–186. *In* C.D. Raper, Jr. and P.J. Kramer (ed.) Crop reactions to water and temperature stresses in humid, temperate climates. Westview Press, Boulder, CO.

Castleberry, R.M. 1983. Breeding programs for stress tolerance in corn. p. 277–287. *In* C.D. Raper, Jr. and P.J. Kramer (ed.) Crop reactions to water and temperature stresses in humid, temperate climates. Westview Press, Boulder, CO.

Caughey, M.G. 1945. Water relations of pocosin or bog shrubs. Plant Physiol. 20:671–689.

Cavalieri, A.J., and J.S. Boyer. 1982. Water potentials induced by growth in soybean hypocotyls. Plant Physiol. 69:492–496.

Chabot, J.F., and A.C. Leopold. 1982. Ultrastructural changes in membranes with hydration in soybean seeds. Am. J. Bot. 69:623–633.

Chaiwanakupt, P., and W.K. Robertson. 1976. Leaching of phosphate and selected cations from sandy soils as affected by lime. Agron. J. 68:507–511.

Chaleff, R.S. 1983. Isolation of agronomically useful mutants from plant cell cultures. Science 219:676–682.

Chatterton, N.J., W.W. Hanna, J.B. Powell, and D.R. Lee. 1975. Photosynthesis and transpiration of bloom and bloomless sorghum. Can. J. Plant Sci. 55:641–643.

----, and J.E. Silvius. 1979. Photosynthate partitioning into starch in soybean leaves. Plant Physiol. 64:749–753.

Clarkson, D.T. 1965. The effect of aluminum and some other trivalent metal cations on cell division in the root apices of *Allium cepa*. Ann. Bot. 29:311–315.

----. 1976. The influence of temperature on the exudation of xylem sap from detached root systems of rye (*Secale cereale*) and barley (*Hordeum vulgare*). Planta 132:297–304.

----, and J. Sanderson. 1969. The uptake of a polyvalent cation and its distribution in the root apices of *Allium cepa*: traces and autoradiographic studies. Planta 89:136–154.

Cleland, R. 1971. Cell wall extension. Annu. Rev. Plant Physiol. 22:197–223.

Clough, J.M., M.M. Peet, and P.J. Kramer. 1981. Effects of high atmospheric CO_2 and sink size on rates of photosynthesis of a soybean cultivar. Plant Physiol. 67:1007–1010.

Collis-George, N., and J.B. Hector. 1966. Germination of seeds as influenced by matric potential and by area of contact between the seed and soil water. Aust. J. Soil. Res. 4:145–164.

Constable, G.A., and A.B. Hearn. 1978. Agronomic and physiological responses of soybean and sorghum crops to water deficits. I. Growth, development, and yield. Aust. J. Plant Physiol. 5:159–167.

Csonka, L., D. Le Rudulier, S.S. Yang, A. Valentine, T. Croughan, S.J. Stavarek, D.W. Rains, and R.C. Valentine. 1983. Genetic engineering for osmotically tolerant microorganisms and plants. p. 245–261. *In* C.D. Raper, Jr. and P.J. Kramer (ed.) Crop reactions to water and temperature stresses in humid, temperate climates. Westview Press, Boulder, CO.

Cure, J.D., R.P. Patterson, C.D. Raper, Jr., and W.A. Jackson. 1982. Assimilate distribution in soybeans as affected by photoperiod during seed development. Crop Sci. 22:1245–1250.

----, C.D. Raper, Jr., R.P. Patterson, and W.A. Jackson. 1983. Water stress recovery in soybeans as affected by photoperiod during seed development. Crop Sci. 23:110–114.

----, ----, ----, and W.P. Robarge. 1985. Dinitrogen fixation in soybean in response to leaf water stress and seed growth rate. Crop Sci. 25:52–58.

Curtis, P.E., W.L. Ogren, and R.H. Hageman. 1969. Varietal effects in soybean photosynthesis and respiration. Crop Sci. 9:323–327.

Cutler, J.M., P.L. Steponkus, M.J. Wach, and K.W. Shahan. 1980. Dynamic aspects and enhancement of leaf elongation in rice. Plant Physiol. 66:147–152.

Dale, R.F. 1983. Temperature perturbations in the midwestern and southeastern United States important for corn production. p. 21–32. *In* C.D. Raper, Jr. and P.J. Kramer (ed.) Crop reactions to water and temperature stresses in humid, temperate climates. Westview press, Boulder, CO.

Davies, W.J. 1977. Stomatal responses to water stress and light in plants grown in controlled environments and in the field. Crop Sci. 17:735–740.

Decker, W.L. 1983. Probability of drought for humid and subhumid regions. p. 11–19. *In* C.D. Raper, Jr. and P.J. Kramer (ed.) Crop reactions to water and temperature stresses in humid, temperate climates. Westview Press, Boulder, CO.

Doss, B.D., R.W. Pearson, and H.T. Rogers. 1974. Effect of soil water stress at various growth stages on soybean yield. Agron. J. 66:297–299.

Downs, R.J. 1955. Photoreversibility of leaf and hypocotyl elongation of dark grown red kidney bean seedlings. Plant Physiol. 30:468–473.

----. 1980. Phytotrons. Bot. Rev. 46:447–489.

----. 1983. Climate simulations. p. 351–368. *In* W.J. Meudt (ed.) Strategies of plant reproduction. Allanheld, Osmun and Co. Publishers, Totowa, NJ.

Duke, S.H., G. Kaketuda, and T.M. Harvey. 1983. Differential leakage of intracellular substances from imbibing soybean seeds. Plant Physiol. 72:919–924.

----, L.E. Schrader, C.A. Henson, J.C. Servaites, R.D. Vogelzang, and J.W. Pendleton. 1979. Root temperature effects on soybean nitrogen metabolism and photosynthesis. Plant Physiol. 63:956–962.

Duncan, W.G. 1971. Leaf angles, leaf area, and canopy photosynthesis. Crop Sci. 11:482–485.

Eastin, J.D., R.M. Castleberry, T.J. Gerik, J.H. Hultquist, V. Mahalakshmi, V.B. Ogunlela, and J.R. Rice. 1983. Physiological aspects of high temperature and water stress. p. 91–112. *In* C.D. Raper, Jr. and P.J. Kramer (ed.) Crop reactions to water and temperature stresses in humid, temperate climates. Westview Press, Boulder, CO.

Eavis, B.W., and H.M. Taylor. 1979. Transpiration of soybeans as related to leaf area, root length, and soil water content. Agron. J. 71:441–445.

Egli, D.B., J.W. Pendleton, and D.B. Peters. 1970. Photosynthetic rates of three soybean communities as related to carbon dioxide levels and solar radiation. Agron. J. 62:411–414.

Evans, L.T. 1975. The physiological basis of crop yield. p. 327–355. *In* L.T. Evans (ed.) Crop physiology. Cambridge University Press, London.

----. 1980. The natural history of crop yield. Am. Sci. 68:388–397.

Farquhar, G.D., and T.D. Sharkey. 1982. Stomatal conductance and photosynthesis. Annu. Rev. Plant Physiol. 33:317–345.

Finn, G.A., and W.A. Brun. 1982. Effect of atmospheric CO_2 enrichment on growth, nonstructural carbohydrate content, and root nodule activity in soybean. Plant Physiol. 69:327–331.

Fischer, R.A. 1981. Optimizing the use of water and nitrogen through breeding of crops. Plant Soil 58:249–279.

Ford, D.M., R. Shibles, and D.E. Green. 1983. Growth and yield of soybean lines selected for divergent leaf photosynthetic ability. Crop Sci. 23:517–520.

Foy, C.D., R.L. Chaney, and M.C. White. 1978. The physiology of metal toxicity in plants. Annu. Rev. Plant Physiol. 29:511–566.

----, A.L. Fleming, G.R. Burns, and W.H. Armiger. 1967. Characterization of differential aluminum tolerance among varieties of wheat and barley. Soil Sci. Soc. Am. Proc. 31:513–521.

----, ----, and W.H. Armiger. 1969. Aluminum tolerance of soybean varieties in relation to calcium nutrition. Agron. J. 61:505–511.

Frey, K.J. 1964. Adaptation reaction of oat strains selected under stress and non-stress environmental conditions. Crop Sci. 4:55–58.

Gbikpi, P.J., and R.K. Crookston. 1981. Effect of flowering date on accumulation of dry matter and protein in soybean seeds. Crop Sci. 21:652–655.

Geiger, D.R. 1976. Phloem loading in source leaves. p. 167–183. *In* I.F. Wardlaw and J.B. Passioura (ed.) Transport and transfer processes in plants. Academic Press, New York.

Ghorashy, S.R., J.W. Pendleton, D.B. Peters, J.S. Boyer, and J.E. Beuerlein. 1971. Internal water stress and apparent photosynthesis with soybeans differing in pubescence. Agron. J. 63:674–676.

Giaquinta, R.T., and D.R. Geiger. 1973. Mechanism of inhibition of translocation by localized chilling. Plant Physiol. 51:372–377.

Gold, H.J., and C.D. Raper, Jr. 1983. Systems analysis and modeling in extrapolation of controlled environment studies to field conditions. p. 315–325. *In* C.D. Raper, Jr. and P.J. Kramer (ed.) Crop reactions to water and temperature stresses in humid, temperate climates. Westview Press, Boulder, CO.

Grabe, D.F., and R.B. Metzer. 1969. Temperature-induced inhibition of soybean hypocotyl elongation and seedling emergence. Crop Sci. 9:331–333.

Green, D.E., E.L. Pinnell, L.E. Cavanah, and L.F. Williams. 1965. Effect of planting date and maturity date on soybean seed quality. Agron. J. 57:165–168.

Gribbin, J. 1981. The politics of carbon dioxide. New Sci. 90:82–84.

Grime, J.P., and J.G. Hodgson. 1969. An investigation of the significance of lime chlorosis by means of large scale comparative experiments. p. 67–99. *In* I.H. Rorison (ed.) Ecological aspects of mineral nutrition of plants. Blackwell Publisher, Oxford, UK.

Gunning, B.E.S., and A.W. Robards. 1976. Plasmodesmata and symplastic transport. p. 15–41. *In* I.F. Wardlaw and J.B. Passioura (ed.) Transport and transfer processes in plants. Academic Press, New York.

Hanks, R.J., and F.C. Thorp. 1957. Seedling emergence of wheat, corn, grain sorghum and soybeans as influenced by soil crust strength and moisture content. Soil Sci. Soc. Am. Proc. 21:357–359.

Hanson, W.D., and E.J. Kamprath. 1979. Selection for aluminum tolerance in soybeans based on seedling-root growth. Agron. J. 41:581–586.

Hardman, L.L., and W.A. Brun. 1971. Effect of atmospheric carbon dioxide enrichment at different development stages on growth and yield components of soybeans. Crop Sci. 11:886–888.

Hardy, R.W.F., R.D. Holsten, E.K. Jackson, and R.C. Burns. 1968. The acetylene-ethylene assay for N_2 fixation: Laboratory and field evaluation. Plant Physiol. 43:1185–1207.

Harper, J.E. 1974. Soil and symbiotic nitrogen requirements for optimum soybean production. Crop Sci. 14:255–260.

Heatherly, L.G., and W.J. Russell. 1979. Effect of soil water potential of two soils on soybean emergence. Agron. J. 71:980–982.

Hecht-Buchholz, Ch., and C.D. Foy. 1981. Effect of aluminum toxicity on root morphology of barley. p. 343–345. *In* R. Brouwer et al. (ed.) Structure and function of plant roots. Nijhoff/Junk, The Hague.

Heenan, D.P., and L.C. Campbell. 1980. Transport and distribution of manganese in two cultivars of soybean (*Glycine max* (L.) Merr.). Aust. J. Agric. Res. 31:943–949.

----, and O.G. Carter. 1975. Response of two soybean varieties to manganese toxicity as affected by pH and calcium levels. Aust. J. Agric. Res. 26:967–974.

----, and ----. 1976. Tolerance of soybean cultivars to manganese toxicity. Crop Sci. 16:389–391.

----, and ----. 1977. Influence of temperature on the expression of manganese toxicity by two soybean varieties. Plant Soil 47:219–227.

Heindl, J.C., and W.A. Brun. 1983. Light and shade effects on abscission and ^{14}C-phosphate partitioning among reproductive structures in soybean. Plant Physiol. 73:434–439.

Hobbs, P.R., and R.L. Obendorf. 1972. Interaction of initial seed moisture and imbititional temperature on germination and productivity of soybean. Crop Sci. 12:664–667.

Hofstra, G., and J.D. Hesketh. 1969. Effect of temperature on the gas exchange of leaves in light and dark. Planta 85:228–237.

Holmberg, S.A. 1973. Soybeans for cool temperature climates. Agric. Hort. Genet. 31:1–20.

Horst, W.J., A. Wagner, and H. Marschner. 1982. Mucilage protects root meristems from aluminum injury. Z. Pflanzenphysiol. 105:435–444.

----, ----, ----. 1983. Effect of aluminum on root growth, cell-division rate and mineral element contents in roots of *Vigna unguiculata* genotypes. Z. Pflanzenphysiol. 109:95–103.

Hsiao, T.C. 1973. Plant responses to water stress. Annu. Rev. Plant Physiol. 24:519–570.

----, and E. Acevedo. 1974. Plant responses to water deficits, water-use efficiency, and drought resistance. Agric. Meteorol, 14:59–84.

Hsu, F.C., and R.L. Obendorf. 1982. Compositional analysis of *in vitro* matured soybean seeds. Plant Sci. Lett. 27:129–135.

Huang, C.-Y., J.S. Boyer, and L.N. Vanderhoeff. 1975a. Acetylene reduction (nitrogen fixation) and metabolic activities of soybean having various leaf and nodule water potentials. Plant Physiol. 56:222–227.

----, ----, and ----. 1975b. Limitation of acetylene reduction (nitrogen fixation) by photosynthesis in soybean having low water potentials. Plant Physiol. 56:228–232.

Huber, S.C., H.H. Rogers, and F.L. Mowry. 1984. Effects of water stress on photosynthesis and carbon partitioning in soybean (*Glycine max* [L.] Merr.) plants grown in the field at different CO_2 levels. Plant Physiol. 76:244–249.

----, T.W. Rufty, Jr., and P.S. Kerr. 1984. Effect of photoperiod on photosynthate partitioning and diurnal rhythms in sucrose phosphate synthase activity in leaves of soybean (*Glycine max* L. [Merr.]) and tobacco (*Nicotiana tabacum* L.). Plant Physiol. 75:1080–1084.

Hudson, J.P. 1977. Plants and weather. p. 1–20. *In* J.J. Landsberg and C.V. Cutting (ed.) Environmental effects on crop physiology. Acadmic Press, New York.

Hume, D.J., and A.K.H. Jackson. 1981. Pod formation in soybeans at low temperatures. Crop Sci. 21:933–937.

Hunter, J.R., and A.E. Erickson. 1952. Relation of seed germination to soil moisture tension. Agron. J. 44:107–109.

Inouye, C. 1953. Influence of temperature on the germination of seeds. 9. Soybean. Jpn. J. Crop Sci. 21:276–277.

Israel, D.W. 1981. Cultivar and *Rhizobium* strain effects on nitrogen fixation and remobilization by soybeans. Agron. J. 73:509–516.

----, and W.A. Jackson. 1982. Ion balance, uptake, and transport processes in N_2-fixing, and NO_3^- and urea dependent soybean plants. Plant Physiol. 69:171–178.

Jackson, W.A. 1967. Physiological effects of soil acidity. p. 43–124. *In* R.W. Pearson and F. Adams (ed.) Soil acidity and liming. American Society of Agronomy, Madison, WI.

Jeffers, D.L., and R.M. Shibles. 1969. Some effects of leaf area, solar radiation, air temperature, and variety on net photosynthesis in field grown soybeans. Crop Sci. 9:762–764.

Johnson, G.R., and K.J. Frey. 1967. Heritability of quantitative attributes of oats (*Avena* sp.) at varying levels of environmental stresses. Crop Sci. 7:43–46.

Johnson, H.W., H.A. Borthwick, and R.C. Leffel. 1960. Effects of photoperiod and time of planting on rates of development of the soybean in various stages of the life cycle. Bot. Gaz. 122:77–95.

Johnson, R.E., and W.A. Jackson. 1964. Calcium uptake and transport by wheat seedlings as affected by aluminum. Soil. Sci. Soc. Am. Proc. 28:381–386.

Jones, L.H. 1961. Aluminum uptake and toxicity in plants. Plant Soil 13:297–310.

Jones, P., L.H. Allen, Jr., J.W. Jones, K.J. Boote, and W.J. Campbell. 1984. Soybean canopy growth, photosynthesis, and transpiration responses to whole-season carbon dioxide enrichment. Agron. J. 76:633–636.

Jones, P.G., and D.R. Laing. 1978. The effects of phenological and meteorological factors in soybean yield. Agric. Meteorol. 19:485–496.

Jones, R.S., R.P. Patterson, and C.D. Raper, Jr. 1981. The influence of temperature and nitrate on vegetative growth and nitrogen accumulation by nodulated soybeans. Plant Soil 63:333–344.

Jordan, W.R., and F.R. Miller. 1980. Genetic variability in sorghum root systems: Implications for drought tolerance. p. 388–399. *In* N.C. Turner and P.J. Kramer (ed.) Adaptation of plants to water and high temperature stress. John Wiley and Sons, New York.

Judd, R., D.M. Tekrony, D.B. Egli, and G.M. White. 1982. Effect of freezing temperatures during soybean seed maturation on seed quality. Agron. J. 74:645–650.

Jung, P.K., and H.D. Scott. 1980. Leaf water potential, stomatal resistance, and temperature relations in field-grown soybeans. Agron. J. 72:986–990.

Juo, A.S.R., and J.C. Ballaux. 1977. Retention and leaching of nutrients in a limed Ultisol under cropping. Soil Sci. Soc. Am. J. 41:757–761.

Kamprath, E.J. 1970. Exchangeable aluminum as a criterion for liming leached mineral soils. Soil Sci. Soc. Am. Proc. 34:252–254.

----, D.K. Cassel, H.D. Gross, and D.W. Dibb. 1979. Tillage effects on biomass production and moisture utilization by soybeans on coastal plain soils. Agron. J. 71:1001–1005.

Kasperbauer, M.J. 1971. Spectral distribution of light in a tobacco canopy and effect of end-of-day light quality on growth and development. Plant Physiol. 47:775–778.

Kato, I., S. Sakaguchi, and Y. Naito. 1954. Development of flower parts and seed in soybean plant, *Glycine max* M. p. 96–114. *In* Takai-Kinki Natl. Agric. Exp. Stn. Bull. 1.

Keeling, C.B., R.B. Bacastow, A.E. Bainbridge, C.A. Ekdahl, Jr., P.R. Guenther, L.S. Waterman, and J.F.S. Chin. 1976. Atmospheric carbon dioxide variations at Mauna Loa Observatory, Hawaii. Tellus 28:538–551.

Kimball, B.A. 1983. Carbon dioxide and agricultural yield: An assemblage and analysis of 430 prior observations. Agron. J. 75:779–788.

Klucas, R.V. 1974. Studies on soybean nodule senescence. Plant Physiol. 54:612–616.

Kramer, P.J. 1981. Carbon dioxide concentration, photosynthesis, and dry matter production. BioScience 31:29–33.

----. 1983. Water relations of plants. Academic Press, New York.

Landsberg, J.J., and C.V. Cutting (ed.) 1977. Environmental effects of crop physiology. Academic Press, New York.

Lawn, R.J., and W.A. Brun. 1974. Symbiotic nitrogen fixation in soybeans. I. Effect of photosynthetic source-sink manipulations. Crop Sci. 14:11–16.

----, and D.J. Hume. 1985. Response of tropical and temperate soybean genotypes to temperature during early reproductive growth. Crop Sci. 25:137–142.

Leopold, A.C. 1980. Temperature effects on soybean imbibition and leakage. Plant Physiol. 65:1096–1098.

Lin, S.S., and M.L. Peterson. 1975. Low temperature-induced floret sterility in rice. Crop Sci. 15:657–660.

Lincoln, D.E., N. Sionit, and B.R. Strain. 1984. Growth and feeding responses of *Pseudoplasia includens* (Lepidoptera: Noctoidae) to host plants grown in controlled carbon dioxide atmospheres. Environ. Entomol. 13:1527–1530.

Lindeman, W.C., and G.E. Ham. 1979. Soybean plant growth, nodulation, and nitrogen fixation as affected by root temperature. Soil Sci. Soc. Am. J. 43:1134–1137.

Löhnis, M.P. 1951. Manganese toxicity in field and market garden crops. Plant Soil 3:193–222.

Lund, Z.F. 1970. The effect of calcium and its relation to several cations in soybean root growth. Soil Sci. Soc. Am. Proc. 34:456–459.

Lunt, O.R., and A.M. Kofranek. 1970. Manganese and aluminum tolerance of azalea (cv. Sweetheart Supreme). p. 559–573. *In* R.M. Samish (ed.) Recent advances in plant nutrition, Vol. 2. Plant analysis and fertilizer problems. Gordon and Breach, New York.

Lyons, J.M., D. Graham, and J.K. Raison (ed.) 1979. Low temperature stress in crop plants. Academic Press, New York.

MacRobbie, E.A.C. 1971. Fluxes and compartmentalization in plant cells. Annu. Rev. Plant Physiol. 22:75–96.

Mann, J.D., and E.G. Jaworski. 1970. Comparison of stresses which may limit soybean yield. Crop Sci. 10:620–642.

Markhart, A.H., III, E.L. Fiscus, A.W. Naylor, and P.J. Kramer. 1979. The effect of temperature on water and ion transport in soybean and broccoli root systems. Plant Physiol. 64:83–87.

----, M.M. Peet, N. Sionit, and P.J. Kramer. 1980. Low temperature acclimation of root fatty acid composition, leaf water potential, gas exchange, and growth of soybean seedlings. Plant Cell Environ. 3:435–441.

Marschner, H., and V. Römheld. 1983. *In vivo* measurement of root-induced pH changes at the soil-root interface: Effect of plant species and nitrogen source. Z. Pflanzenphysiol. 111:241–252.

Martin, C.A., D.K. Cassel, and E.J. Kamprath. 1979. Irrigation and tillage effects on soybean yield in a coastal plain soil. Agron. J. 71:592–594.

Matsumoto, H., and S. Morimura. 1980. Repressed template activity of the chromatin of pea roots treated by aluminum. Plant Cell Physiol. 21:951–959.

----, ----, and E. Takahashi. 1977. Less involvement of pectin in the precipitation of aluminum in pea root. Plant Cell Physiol. 18:325–335.

Mauney, J.R., K.E. Fry, and G. Guinn. 1978. Relationship of photosynthetic rate to growth and fruiting of cotton, soybean, sorghum, and sunflower. Crop Sci. 18:259–263.

----, G. Guinn, K.E. Fry, and J.D. Hesketh. 1979. Correlation of photosynthetic carbon dioxide uptake and carbohydrate accumulation in cotton, soybean, sunflower and sorghum. Photosynthetica 13:260–266.

McWilliam, J.R. 1983. Physiological basis for chilling stress and the consequences for crop production. p. 113–132. *In* C.D. Raper, Jr. and P.J. Kramer (ed.) Crop reactions to water and temperature stresses in humid, temperate climates. Westview Press, Boulder, CO.

Mederski, H.J. 1983. Effects of water and temperature stress on soybean plant growth and yield in humid, temperate climates. p. 35–48. *In* C.D. Raper, Jr. and P.J. Kramer (ed.) Crop reactions to water and temperature stresses in humid, temperate climates. Westview Press, Boulder, CO.

----, and D.L. Jeffers. 1973. Yield response of soybean varieties grown at two soil moisture stress levels. Agron. J. 65:410–412.

Minchin, F.R., and J.S. Pate. 1973. The carbon balance of a legume and the functional economy of its root nodules. J. Exp. Bot. 24:259–271.

Mohanty, P., and J.S. Boyer. 1976. Chloroplast response to low leaf water potential. IV. Quantum yield is reduced. Plant Physiol. 57:704–709.

Munns, D.N., L. Jacobson, and C.M. Johnson. 1963. Uptake and distribution of manganese in oat plants. II. A kinetic model. Plant Soil 19:193–204.

Murfet, I.C. 1977. Environmental interaction and the genetics of flowering. Annu. Rev. Plant Physiol. 28:253–278.

Musser, R.L., P.J. Kramer, and S.A. Thomas. 1983a. Physiological effects of root versus shoot chilling of soybean. Plant Physiol. 73:778–783.

----, S.A. Thomas, J.F. Thomas, and P.J. Kramer. 1983b. Effects of shoot chilling during floral initiation and development of 'Ransom' soybeans. Plant Physiol. Suppl. 72:43.

----, ----, R.R. Wise. T.C. Peeler, and A.W. Naylor. 1984. Chloroplast ultrastructure, chlorophyll fluorescence, and pigment composition in chilling-stressed soybeans. Plant Physiol. 74:749–754.

Nelson, O.E. 1983. Genetics and plant breeding in relation to stress tolerance. p. 351–357. *In* C.D. Raper, Jr. and P.J. Kramer (ed.) Crop reactions to water and temperature stresses in humid, temperate climates. Westview Press, Boulder, CO.

Newman, E.I. 1976. Water movement through root systems. Phil. Trans. Roy. Soc. London Bull. 273:463–478.

Nielsen, C.S. 1942. Effects of photoperiod on microsporogenesis in Biloxi soybean. Bot Gaz. 104:99–106.

Nishiyama, I., and T. Sataka. 1979. Male sterility caused by cooling treatment at the young microspore stage in rice plants. XIX. The difference in susceptibility to coolness among spikelets on a panicle. Jpn. J. Crop Sci. 48:181–186.

Ojima, M., J. Fukui, and I. Watanabe. 1965. Studies on the seed production of soybean. II. Effect of three major nutrient elements supply and leaf age on the photosynthetic activity and diurnal changes in photosynthesis of soybean under constant temperature and light intensity. Crop Sci. Soc. Jpn. Proc. 33:437–442.

O'Neill, S., and A.C. Leopold. 1982. An assessment of phase transitions in soybean membranes. Plant Physiol. 70:1405–1409.

Orians, G.H., and O.T. Solbrig. 1977. A cost-income model of leaves and roots with special reference to arid and semi-arid areas. Am. Nat. 111:677–690.

Ormrod, D.P. 1964. Net carbon dioxide exchange rates in *Phaseolus vulgaris* (L.) as influenced by temperature, light intensity, leaf area index, and age of plants. Can. J. Bot. 42:393–401.

Osmond, C.B., O. Björkman, and D.J. Anderson. 1980. Physiological processes in plant ecology. Springer-Verlag New York, New York.

Osmond, D.L., and C.D. Raper, Jr. 1982. Root development of field-grown flue-cured tobacco. Agron. J. 74:541–546.

----, R.F. Wilson, and C.D. Raper, Jr. 1982. Fatty acid composition and nitrate uptake of soybean roots during acclimation to low temperature. Plant Physiol. 70:1689–1693.

Parker, M.W., S.B. Hendricks, H.A. Borthwick, and F.W. Went. 1949. Spectral sensitivities for leaf and stem growth of etiolated pea seedlings. Am. J. Bot. 36:194–204.

Parkhurst, C.E., and J.I. Sprent. 1975. Effects of water stress on the respiratory and nitrogen fixing activity of soybean root nodules. J. Exp. Bot. 26:287–304.

Pate, J.S., and P. Farrington. 1981. Fruit set in *Lupinus angustifolius* cv. Unicrop. II. Assimilate flow during flowering and early fruiting. Aust. J. Plant Physiol. 8:307–318.

Patterson, D.T., and E.P. Flint. 1980. Potential effects of global atmospheric CO_2 enrichment on the growth and competitiveness of C_3 and C_4 weed and crop plants. Weed Sci. 28:71–75.

----, M.M. Peet, and J.A. Bunce. 1977. Effect of photoperiod and size at flowering on vegetative growth and seed yield of soybean. Agron. J. 69:631–635.

Patterson, R.P., C.D. Raper, Jr., and H.D. Gross. 1979. Growth and specific nodule activity of soybean during application and recovery of a leaf moisture stress. Plant Physiol. 64:551–556.

Peet, M.M. 1984. CO_2 enrichment of soybeans. Effects of leaf/pod ratio. Physiol. Plant. 60:38–42.

----, and P.J. Kramer. 1980. Effects of decreasing source/sink ratio in soybeans on photosynthesis, photorespiration, transpiration and yield. Plant Cell Environ. 3:201–206.

Phillips, D.A., K.D. Newell, S.A. Hassell, and C.E. Felling. 1976. The effect of CO_2 enrichment on root nodule development and symbiotic N_2 reduction in *Pisum sativum* L. Am. J. Bot. 63:356–362.

Plant, W. 1956. The effects of molybdenum deficiency and mineral toxicities on crops in acid soils. J. Hortic. Sci. 31:163–176.

Potter, J.R., and J.W. Jones. 1977. Leaf area partitioning as an important factor in growth. Plant Physiol. 59:10–14.

Raper, C.D., Jr., and S.A. Barber. 1970. Rooting systems of soybeans. I. Differences in root morphology. Agron. J. 62:581–584.

----, and R.J. Downs. 1973. Factors affecting the development of flue-cured tobacco grown in artificial environments. IV. Effects of carbon dioxide depletion and light intensity. Agron. J. 65:247–252.

----, L.R. Parson, D.T. Patterson, and P.J. Kramer. 1977. Relationship between growth and nutrient accumulation for vegetative cotton and soybean plants. Bot. Gaz. 138:129–137.

----, and R.P. Patterson. 1980. Environmental sensitivity of acetylene reduction activity in prediction of N fixation in soybeans. Agron. J. 72:717–719.

----, ----, M.L. List, R.L. Obendorf, and R.J. Downs. 1984. Photoperiod effects on growth rate of in vitro cultured soybean embryos. Bot. Gaz. 145:157–162.

----, and G.F. Peedin. 1978. Photosynthetic rate during steady-state growth as influenced by carbon-dioxide concentration. Bot. Gaz. 139:147–149.

----, and J. F. Thomas. 1978. Photoperiodic alteration of dry matter partitioning and seed yield in soybeans. Crop Sci. 18:654–656.

----, W.W. Weeks, R.J. Downs, and W.H. Johnson. 1973. Chemical properties of tobacco leaves as affected by carbon dioxide depletion and light intensity. Agron. J. 65:988–992.

Reitz, L.P. 1974. Breeding for more efficient water use—is it real or a mirage? Agric. Meterol. 14:3–11.

Ritchie, J.T. 1983. Reduction of stresses related to soil water deficit. p. 329–340. *In* C.D. Raper, Jr. and P.J. Kramer (ed.) Crop reactions to water and temperature stresses in humid, temperate climates. Westview Press, Boulder, CO.

Rogers, H.H., N. Sionit, J.D. Cure, J.M. Smith, and G.E. Bingham. 1984. Influence of elevated carbon dioxide on water relations of soybeans. Plant Physiol. 74:233–238.

----, J.F. Thomas, and G.E. Bingham. 1983. Response of agronomic and forest species to elevated atmospheric carbon dioxide. Science 220:428–429.

Rorison, I.H. 1965. The effect of aluminum on uptake and incorporation of phosphate by excised sanfoin roots. New Phytol. 63:23–27.

Rufty, T.W., Jr., P.S. Kerr, and S.C. Huber. 1983. Characterization of diurnal changes in activities of enzymes involved in sucrose biosynthesis. Plant Physiol. 73:428–433.

----, G.S. Miner, and C.D. Raper, Jr. 1979. Temperature effects on growth and manganese tolerance in tobacco. Agron. J. 71:638–644.

----, C.D. Raper, Jr., and S.C. Huber. 1984. Alterations in internal partitioning of carbon in soybean plants in response to nitrogen stress. Can J. Bot. 62:501–508.

----, ----, and W.A. Jackson. 1981. Nitrogen assimilation, root growth, and whole plant responses of soybeans to root temperature and carbon dioxide and light in the aerial environment. New Phytol. 88:607–619.

Runge, E.C.A., and R.T. Odell. 1960. The relation between precipitation, temperature, and yield of soybeans on the Agronomy South Farm, Urbana, Illinois. Agron. J. 52:245–247.

Ryle, G.J.A., R.A. Arnott, C.E. Powell, and A.J. Gordon. 1983. Comparisons of the respiratory effluxes of nodules and roots in six temperate legumes. Ann. Bot. 52:469–477.

----, C.E. Powell, and A.J. Gordon. 1979a. The respiratory costs of nitrogen fixation in soyabean, cowpea, and white clover. I. Nitrogen fixation and the respiration of the nodulated root. J. Exp. Bot. 30:135–144.

----, ----, and ----. 1979b. The respiratory costs of nitrogen fixation in soyabean, cowpea, and white clover. II. Comparisons of the cost of nitrogen fixation and the utilization of combined nitrogen. J. Exp. Bot. 30:145–153.

Sakamoto, C.M., and R.H. Shaw. 1967a. Light distribution in field soybean canopies. Agron. J. 59:7–9.

----, and ----. 1967b. Apparent photosynthesis in field soybean communities. Agron. J. 59:73–75.

Sartain, J.B., and E.J. Kamprath. 1975. Effect of liming a highly Al-saturated soil on the top and root growth and soybean nodulation. Agron. J. 67:507–510.

----, and ----. 1978. Aluminum tolerance of soybean cultivars based on root elongation in solution culture compared with growth in acid soil. Agron. J. 70:17–20.

Sato, K., and T. Ikeda. 1979. The growth responses of soybean plant to photoperiod and temperature. IV. The effect of temperature during the ripening period on the yield and character of seeds. Jpn. J. Crop Sci. 48:283–290.

Schmid, J., and E.R. Keller. 1980. The behavior of three cold-tolerant and a standard soybean variety in relation to the level and duration of a cold stress. Can. J. Plant Sci. 60:821–829.

Seddigh, M., and G.D. Jolliff. 1984. Effects of night temperature on dry matter partitioning and seed growth of indeterminate field-grown soybean. Crop Sci. 24:704–710.

Sharp, R.E., and W.J. Davies. 1979. Solute regulation and growth by roots and shoots of water-stressed maize plants. Planta 147:43–49.

Shaw, R.H. 1983. Estimates of yield reductions in corn caused by water and temperature stresses. p. 49–65. *In* C.D. Raper, Jr. and P.J. Kramer (ed.) Crop reactions to water and temperature stresses in humid, temperate climates. Westview Press, Boulder, CO.

Shepard, J.F., D. Bidney, T. Barsby, and R. Kemble. 1983. Genetic transfer in plants through interspecific protoplast fusion. Science 219:683–688.

Shibles, R. 1980. Adaptation of soybeans to different seasonal durations. p. 279–285. *In* R.J. Summerfield and A.H. Bunting (ed.) Advances in legume science. Royal Botanic Gardens, Kew, UK.

----, and C.R. Weber. 1965. Leaf area, solar radiation interception and dry matter production by soybeans. Crop Sci. 5:575–577.

----, and ----. 1966. Interception of solar radiation and dry matter production by various soybean planting patterns. Crop Sci. 6:55–59.

Silvius, J.E., R.R. Johnson, and D.B. Peters. 1977. Effect of water stress on carbon assimilation and distribution in soybean plants at different stages of development. Crop Sci. 17:713–716.

Simon, E.W. 1979. Seed germination at low temperatures. p. 37–45. *In* J.M. Lyons et al. (ed.) Low temperature stress in crop plants: The role of the membrane. Academic Press, New York.

Sinclair, T.R. 1980. Leaf CER from post-flowering to senescence of field-grown soybean cultivars. Crop Sci. 20:196–200.

----, and C.T. de Wit. 1975. Photosynthate and nitrogen requirements for seed production by various crops. Science 189:565–567.

Sionit, N. 1983. Response of soybeans to two levels of mineral nutrition in CO_2-enriched atmosphere. Crop Sci. 23:329–333.

----, H. Hellmers, and B.R. Strain. 1980. Growth and yield of wheat under CO_2 enrichment and water stress. Crop Sci. 20:687–690.

----, ----, and ----. 1982. Interaction of atmospheric CO_2 enrichment and irradiance on plant growth. Agron. J. 74:721–725.

----, and P.J. Kramer. 1977. Effect of water stress during different stages of growth of soybeans. Agron. J. 69:274–277.

----, H.H. Rogers, G.E. Bingham, and B.R. Strain. 1984. Photosynthesis and stomatal conductance with CO_2-enrichment of container- and field-grown soybeans. Agron. J. 76:447–451.

----, B.R. Strain, H. Hellmers, and P.J. Kramer. 1981. Effects of atmospheric CO_2 concentration and water stress on water relations of wheat. Bot. Gaz. 142:191–196.

Soldati, A., and E.R. Keller. 1977. Abklarung von Komponenten des Ertragsaufbaues bei der Sojabohne (*Glycine max* (L.) Merr.) unter verschiedenen klimatischen Bedingungen in der Schwiz. Schweiz Landwirts. Forsch. 16:257–258.

Sprent, J.I. 1971. The effect of water stress on nitrogen fixing root nodules. I. Effect on the physiology of detached soybean nodules. New Phytol. 70:9–17.

Stuckey, D.J. 1976. Effect of planting depth, temperature, and cultivar on emergence and yield of double cropped soybeans. Agron. J. 68:291–294.

Sullivan, C.Y., and W.M. Ross. 1979. Selecting for drought and heat resistance in grain sorghum. p. 263–281. *In* H. Mussell and R.C. Staples (ed.) Stress physiology in crop plants. John Wiley and Sons, New York.

Tanner, J.W., and D.J. Hume. 1978. Management and production. p. 158–217. *In* A.G. Norman (ed.) Soybean physiology, agronomy, and utilization. Academic Press, New York.

Taylor, A.O., and J.A. Rowley. 1971. Plants under climate stress. I. Low temperature, high light effects on photosynthesis. Plant Physiol. 47:719–725.

Taylor, H.M. 1980. Modifying root systems of cotton and soybeans to increase water absorption. p. 75–84. *In* N.C. Turner and P.J. Kramer (ed.) Adaptation of plants to water and high temperature stress. John Wiley and Sons, New York.

Terry, D.L., and C.B. McCants. 1968. The leaching of ions in soils. N.C. Agric. Exp. Stn. Tech. Bull. 184.

----, and ----. 1970. Quantitative prediction of leaching in field soils. Soil Sci. Soc. Am. Proc. 34:271–276.

Terry, N., P.S. Evans, and D.E. Thomas. 1975. Manganese toxicity effects on leaf cell multiplication and expansion and on dry matter yield of sugar beets. Crop Sci. 15:205–208.

Thaine, R., S.L. Overden, and J.S. Turner. 1959. Translocation of labelled assimilates in the soybean. Aust. J. Biol. Sci. 12:349–372.

Thawornwong, N., and A. Van Diest. 1974. Influences of high acidity and aluminum on the growth of lowland rice. Plant Soil 41:141–159.

Thomas, J.F., and C.D. Raper, Jr. 1975. Differences in progeny of tobacco due to temperature treatment of the mother plant. Tobacco Sci. 19:37–41.

----, and ----. 1976. Photoperiodic control of seed filling for soybeans. Crop Sci. 16:667–672.

----, and ----,. 1977. Morphological response of soybeans as governed by photoperiod, temperature, and age at treatment. Bot. Gaz. 138:321–328.

----, and ----. 1978. Effect of day and night temperature during floral induction on morphology of soybeans. Agron. J. 70:893–898.

----, and ----. 1981. Day and night temperature effects on carpel initiation and development soybeans. Bot. Gaz. 142:183–187.

----, and ----. 1983a. Photoperiod and temperature regulation of floral initiation and anthesis in soya bean. Ann. Bot. 51:481–489.

----, and ----. 1983b. Photoperiod effects on soybean growth during the onset of reproductive development under various temperature regimes. Bot. Gaz. 144:471–476.

----, and ----. 1984. Photoperiod regulation of floral initiation for soybeans of different ages. Crop Sci. 24:611–614.

----, ----, C.E. Anderson, and R.J. Downs. 1975. Growth of young tobacco plants as affected by carbon dioxide and nutrient variables. Agron. J. 67:685–689.

----, ----, and W.W. Weeks. 1981. Day and night temperature effects on nitrogen and soluble carbohydrate allocation during early reproductive growth of soybeans. Agron. J. 73:577–582.

Thompson, L.M. 1970. Weather and technology in the production of soybeans in the central United States. Agron. J. 62:232–236.

Thorne, J.H. 1981. Morphology and ultrastructure of maternal seed tissues of soybean in relation to the import of photosynthate. Plant Physiol. 67:1016–1025.

----. 1982. Characterization of the active sucrose transport system of immature soybean embryos. Plant Physiol. 70:953–958.

Thrower, S.L. 1962. Translocation of labelled assimilates in the soybean. II. The pattern of translocation in intact and defoliolated plants. Aust. J. Biol. Sci. 15:629–649.

Trang, K.M., and J. Giddens. 1980. Shading and temperature as environmental factors affecting growth, nodulation, and symbiotic N_2 fixation. Agron. J. 72:305–308.

Trim, A.R. 1959. Metal ions as precipitants for nucleic acids and their use in the isolation of polynucleotides from leaves. Biochem. J. 73:298–304.

Turner, N.C., and M.M. Jones. 1980. Turgor maintenance by osmotic adjustment: A review and evaluation. p. 87–103. *In* N.C. Turner and P.J. Kramer (ed.) Adaptation of plants to water and high temperature stress. John Wiley and Sons, New York.

van den Driesche, R., D.J. Connor, and B.R. Tunstall. 1971. Photosynthetic response of brigalow to irradiance, temperature and water potential. Photosynthetica 5:210–217.

Van Volkenburgh, E., and W.J. Davies. 1977. Leaf anatomy and water relations of plants grown in controlled environments and in the field. Crop Sci. 17:353–358.

Van Wambeke, A. 1976. Formation, distribution and consequences of acid soils in agricultural development. p. 15–24. *In* M.J. Wright (ed.) Plant adaptation to mineral stress in problem soils. Cornell Univ. Agric. Exp. Stn., Ithaca, NY.

Vieira da Silva, J. 1983. Environmental conditions and crop characteristics in relation to stress tolerance. p. 341–346. *In* C.D. Raper, Jr. and P.J Kramer (ed.) Crop reactions to water and temperature stresses in humid, temperate climates. Westview Press, Boulder, CO.

----, A.W. Naylor, and P.J. Kramer. 1974. Some ultrastructural and enzymatic effects of water stress in cotton (*Gossypium hirsutum* L.) leaves. Proc. Natl. Acad. Sci. 71:3243–3247.

Volk, G.M., and C.E. Bell. 1945. Some major factors in the leaching of calcium, potassium, sulfur, and nitrogen from sandy soil. Fla. Agric. Exp. Stn. Bull. 416.

Wallace, S.U., and I.C. Anderson. 1984. Aluminum toxicity and DNA synthesis in wheat roots. Agron. J. 76:5–8.

Wann, M., and C.D. Raper, Jr. 1979. A dynamic model for plant growth: Adaptation for vegetative growth of soybeans. Crop Sci. 19:461–467.

----, and ----. 1984. A dynamic model for plant growth: Validation study under changing temperatures. Ann. Bot. 53:45–52.

Wardlaw, I.F. 1974. Phloem transport: Physical chemical or impossible. Annu. Rev. Plant Physiol. 25:515–539.

----, and D. Bagnell. 1981. Phloem transport and the regulation of growth of *Sorghum bicolor* (Moench) at low temperature. Plant Physiol. 68:411–414.

Warrington, I.J., M. Peet, D.T. Patterson, J. Bunce, R.M. Hazlemore, and H. Hellmers. 1977. Growth and physiological responses of soybean under various thermoperiods. Aust. J. Plant Physiol. 4:371–380.

Watson, D.J. 1963. Climate, weather and yield. p. 337–350. *In* L.T. Evans (ed.) Environmental control of plant growth. Academic Press, New York.

Wells, R., L.L. Schulze, D.A. Ashley, H.R. Boerma, and R.H. Brown. 1982. Cultivar differences in canopy apparent photosynthesis and the relationship to seed yield in soybean. Crop Sci. 22:886–890.

Wenkert, W., E.R. Lemon, and T.R. Sinclair. 1978. Water content-potential relationship in soybean: Changes in component potentials for mature and immature leaves under field conditions. Ann. Bot. 42:295–307.

Went, F.W. 1953. The effect of temperature on plant growth. Annu. Rev. Plant Physiol. 4:347–362.

Williams, L.E., T.M. DeJong, and D.A. Phillips. 1981. Carbon and nitrogen limitations on soybean seedling development. Plant Physiol. 68:1206–1209.

----, ----, and ----. 1982. Effect of changes in shoot carbon-exchange rate on soybean root nodule activity. Plant Physiol. 69:432–436.

Willing, R.P., and A.C. Leopold. 1983. Cellular expansion at low temperature as a cause of membrane lesions. Plant Physiol. 71:118–121.

Wilson, J.M. 1983. Interaction of chilling and water stress. p. 133–147. *In* C.D. Raper, Jr. and P.J. Kramer (ed.) Crop reactions to water and temperature stresses in humid, temperate climates. Westview Press, Boulder, CO.

Wise, R.R., J.R. McWilliam, and A.W. Naylor. 1983. A comparative study of low-temperature-induced ultrastructural alterations of three species with differing chilling sensitivities. Plant Cell Environ. 6:525–535.

Wolfe, J. 1978. Chilling injury in plants—the role of membrane lipid fluidity. Plant Cell Environ. 1:241–247.

Wong, S.C. 1979. Elevated atmospheric partial pressure of CO_2 and plant growth. I. Interactions of nitrogen nutrition and photosynthetic capacity in C_3 and C_4 plants. Oecologia 44:68–74.

Woodward, R.G., and H.M. Rawson. 1976. Photosynthesis and transpiration in dicotyledonous plants. II. Expanding and senescing leaves of soybean. Aust. J. Plant Physiol. 3:257–267.

Wright, K.E., and B.A. Donahue. 1953. Aluminum toxicity studies with radio-active phosphorus. Plant Physiol. 28:674–680.

16 Seed Metabolism

R. F. Wilson
USDA-ARS and
North Carolina State University
Raleigh, North Carolina

The term *metabolism* describes a continuum of chemical and physical processes in living organisms and cells where simple or elemental compounds are systematically assembled to form complex constituents with a net conservation of energy or where complex biological products are converted to simpler substances with the release of energy to sustain life function. One can but wonder at the complexity of these anabolic and catabolic reactions over the time span from fertilization of ovules to the germination of seed. Since the publication of the first edition of the Soybean monograph, much has been learned about the biological basis for the associate interactions that collectively constitute the life cycle of a soybean [*Glycine max* (L.) Merr.] seed. Basic understanding of metabolic processes in soybean has evolved from the initial observation stage to a level of knowledge sufficient to explain the detailed workings of various interactive reactions. Progress is being made through more directed means to gain control over specific reactions or systems and to apply basic information in a directed manner to enhance both quality and quantity of the primary seed constituents. This chapter seeks to cover the fundamental biochemical, genetic, and physiological events that contribute to the current level of understanding and to provide a platform from which focus may be brought to bear upon future research opportunities.

16-1 PROPERTIES OF SOYBEAN SEED DEVELOPMENT

The anatomical diagram shown in Fig. 16-1 outlines the logistical avenues by which sucrose, water, and intermediary metabolites may enter the developing soybean seed (Thorne, 1981). The developing seed structure is attached to two large vascular bundles in the pod placenta by the funiculus. The vascular bundles extend through the funiculus and enter the seed coat at the chalazal end of the hilum. Each bundle then branches below the tracheid bar to form two lateral bundles that are positioned along the length of the hilum near the inner surface of the seed coat.

Soybeans: Improvement, Production, and Uses, 2nd ed.—Agronomy Monograph no. 16.

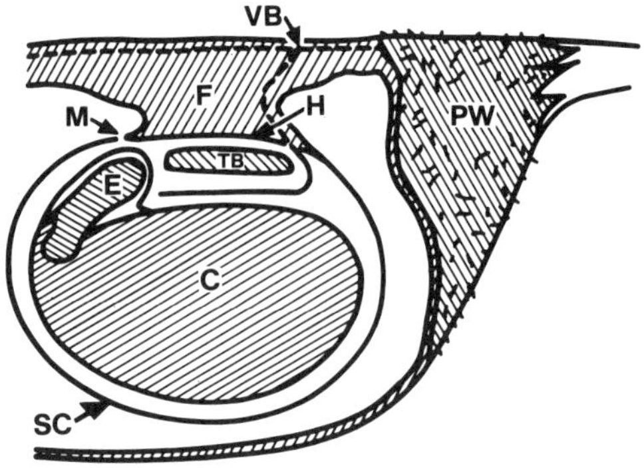

Fig. 16–1. Anatomical sketch of a soybean seed. Abbreviations: C—cotyledon, EA—embryonic axis, F—funiculus, H—hilum, M—micropyle, PW—pod wall, SC—seed coat; TB—tracheid bar; VB—vascular bundle. Adapted from Thorne (1981).

Although numerous smaller bundles branch from the lateral bundles to form an extensive network within the seed coat, there is no vascular connection between the seed coat and cotyledons of the developing embryo. Both structures are separated by an embryonic sac. Because of the physical appearance of the embryonic sac, it is highly probable that photosynthate and other intermediary metabolites diffuse through the seed coat to the surface of the enclosed embryo via an apoplastic route of intercellular transport without entering the embryo sac symplasm (Thorne, 1981). The rate of solute uptake by the embryo then may depend upon solute concentrations in the vascular system. Depending upon maturity group, genotype, and growth conditions, the total growth period for soybean may range from 108 to 144 days (Gay et al., 1980) and the duration of the seed fill, from anthesis to physiological maturity may range from 18 to 70 days (Reicosky et al., 1982). Within that period, seed development progresses through at least five discernible stages (Bils and Howell, 1963). Based upon a 60-day time frame, cell division is essentially complete in cotyledons at 15 days after flowering (DAF). Electron micrographs of seed at 15 DAF reveal cells with limited organelle inclusions; albeit nucleus, ribosomes, immature plastids with starch grains, and mitochondria may be observed (Adams et al., 1983b). A period of greatly increased metabolic activity and dry weight accumulation follows during the 3rd week after anthesis (Ohmura and Howell, 1962). Between 26 and 36 DAF there is a proliferation of protein and lipid bodies. Approximately 50% of the oil in mature seed is accumulated during this period, and starch content of seeds also peaks at approximately 11 to 12% of seed dry weight (Wilson et al., 1978). From 36 to 52 DAF protein and lipid bodies increase in size to proclude observation of most other subcellular structures in thin transmission electron microscopic sections. Approximately 50% of the protein in mature seed is accumulated during this period; lipid accumulation plateaus; and starch content declines with a concomitant increase in soluble sugar concentration of which 20 to 50% may be oligosaccharides such as raffinose and stachyose (Yazdi-Samadi et al., 1977). In the final week maximum dry weight is achieved, starch concentration declines to 1 to 3% of seed dry weight, chlorophyll is degraded in most soybean genotypes, total P_1 levels decline, and certain soluble proteins are modified as the seeds dehydrate in the final stage of seed maturation (Hsu and Obendorf, 1982).

The mechanism by which the seed maturation process is initiated has not been aptly elucidated. Germinable seed may be obtained when maturation is induced in immature seed through carefully controlled dehydration (Adams et al., 1983a). Hence, the process of maturation is not exclusive to a particular stage of seed development, and may be induced prematurely during seed growth by environmental conditions (water stress, temperature stress, and photoperiodism) or physical damage (mechanical, insect, or disease injury). Maturation effectively terminates anabolic mechanisms. Because the metabolic events associated with oil, protein, and starch deposition are not synchronized during seed development, a certain period of time is required before the deposition of each of these seed constituents comes to fruition. Thus, the length or duration of the seed-fill period may have a significant impact upon seed composition.

The period of seed growth prior to maturation may be defined as the seed-fill period. Based upon several observations extension of the seed-fill period beyond an average value for a given genotype (delayed maturation) should result in lower oil and starch concentrations, higher protein concentration, and greater seed size (Sato and Ikeda, 1979; Egli et al., 1978). The opposite trends in seed composition may be expected when for any reason the seed-fill period falls below an accepted norm (hastened maturation) for a given genotype. Under average conditions, however, the seed-fill period within cultivars does not fluctuate greatly. Although duration of the seed-fill period may differ among cultivars, genotypes with similar seed-fill periods may also exhibit differences in seed growth rate. Seed growth rate is highly correlated with seed size (Egli et al., 1978; Gent, 1983; Zeiher et al., 1982). Thus, differences in seed size among cultivars may not be exclusively attributed to seed-fill duration, but to different rates of seed dry matter accumulation (Beever and Cooper, 1982; Egli and Leggett, 1973).

It has been shown that genotypic differences in seed growth rate are not a function of short-term photosynthate production (Egli and Leggett, 1976), which implies that the supply of assimilates to the developing cotyledon is not a limiting factor. Soybean, however, may be stimulated artificially to produce about 30% more seed with a 25% increase in yield over control treatments at normal reproductive load (Streeter and Jeffers, 1979). In such cases, an increase in the number of pods $(node)^{-1}$ and/or seed $(pod)^{-1}$ generally results in a decline in seed size and probably a slower seed growth rate (Kollman et al., 1974; Openshaw et al., 1979). Hence, in response to a greater reproductive load, long-term photosynthate supply could become a factor which may limit the rate of seed growth. Another hypothesis for genotypic differences in seed growth rate also has been suggested by Egli et al. (1981). In that work, it was observed that genotypes with high seed growth rates had a greater number of cells $(seed)^{-1}$. For example, the cv. Essex produced mature seed weighing 113 mg $(seed)^{-1}$ at a growth rate of 4.3 mg $(seed)^{-1}$ $(day)^{-1}$ and contained about 6.6×10^6 cells $(seed)^{-1}$; whereas the cv. Emerald produced mature seed weighing 262 mg $(seed)^{-1}$ at a growth rate of 9.6 $(seed)^{-1}$ $(day)^{-1}$ and contained about 10.2×10^6 cells $(seed)^{-1}$. The seed-fill period for both cultivars was not significantly different. At this time, more information is needed to show the genetic basis for cell number in soybean seed. In addition the effects of this trait upon yielding ability (potential), the concentration for primary constituents (protein, oil, and carbohydrate) in soybean seed and seed metabolism have not yet been demonstrated.

16–2 PRIMARY CONSTITUENTS OF SOYBEAN SEED

16.2.1 Protein

16–2.1.1 Classification

A generalized classification system is often evoked to categorize plant proteins on the basis of solubility in various solvents (Smith and Grierson, 1982). Four classical types of protein have been established: albumin, globulin, prolamin, and glutenin. Albumins, a group of relatively simple proteins 12 to 100 kD in size, are found in nearly all living tissues, are soluble in water, yield mostly amino acids upon cleavage by enzymes or acids, and may be coagulated (denatured) by heat. Proteins of the albumin-type generally precipitate from solution only at high ammonium sulfate concentrations between 70 to 100% of saturation. Globulin-type proteins are relatively more complex as indicated by the range of molecular weight (18 to 360 kD) and are soluble in salt solutions (0.5 to 1.0 *M* NaCl). Globulins typically contain high levels of glutamine, asparagine, and arginine, but relatively low concentrations of S-containing amino acids such as methionine and cysteine. Prolamins are a low-molecular-weight (16 to 60 kD) class of proteins soluble in 70% aqueous-alcohol. These proteins are typically enriched in glutamine and proline but relatively defficient in arginine, histidine, tryptophan, and lysine. Glutenin-type proteins (30 to 50 kD) are practically insoluble in water; partly soluble in alcohol or dilute acid; and soluble in alkaline solvents (Smith and Grierson, 1982).

The predominant type of protein in seed crops is highly species dependent. Cereal grains such as corn (*Zea mays* L.) and wheat (*Triticum aestivum* L.) contain prolamin and glutenin-type proteins, respectively; whereas the major protein type of grain-legumes is globulin (Osbourne and Campbell, 1898). In soybean, globulin-type proteins account for about 90% of the seed protein (Muller, 1983) and about 36% of the seed dry weight (Hughes and Murphy, 1983). Soybean globulin is the predominant source of vegetable protein in the USA. When cooked, 83 to 89.7% of the protein is highly digestable (Eggum and Beames, 1983). Although soybean contains a higher concentration of essential amino acids than other pulse-legumes, the combined level of methionine, cysteine, and threonine in soy-protein is still lower than the minimum FAO nutritional recommendation of 2.5 g of essential amino acid $(g\ N)^{-1}$. As a result, supplemental addition of these amino acids may be required in the preparation of certain soy-protein food products.

16–2.1.2 Globulin Composition

Originally the globulin fraction from soybean was thought to consist of two components identified as legumin and vicilin. The distinction of two different fractions within globulin-type proteins was made on the basis of physical and biochemical characteristics. Legumin-like proteins were less soluble in dilute salt solutions than vicilin-like proteins (Derbyshire et al., 1976). Furthermore, legumin was not denatured at 95 °C and contained slightly higher percentages of N and S than vicilin. With the advent of more sophisticated ultracentrifugation technology, pioneered independently by Svedberg and Danielson, soybean proteins could be fractionated into two bands with sedimentation densities of 7S and 11S which roughly corresponded to the original legumin and vicilin fractions (Muller, 1983). As additional protein bands with distinct biochemical differences were identified among various legume species, the nomenclature glycinin and β-conglycinin was adopted to specify the 11S and 7S proteins found in the species *Glycine* (Koshiyama, 1983). The major sedimentation equilibrium bands now identified from soybean globulin have been listed in Table 16–1. Various two-dimensional electrophoretic systems (isoelectric focusing, nonequilibrium pH gradient in one direction followed by SDS-electrophoresis into the adjacent quadrant) may be used to separate over 450 subunits of the various globulin proteins from soybean seed (Mei-Guey et al., 1983). Of that total, only 35 nonidentical subunits have been associated with specific globulin fractions: glycinin (24 subunits), β-conglycinin (6 subunits), lectin (4 subunits), and trypsin-inhibitor (1 subunit). Consequently, there appears to

Table 16–1. Globulin-type protein composition of soybean seed. Adapted from Koshiyama (1983) and Hill and Breidenbach (1974).

S_{w20}	Nomenclature	Total soybean globulin	Molecular weight	Amino acid	
				Methionine	Cysteine
		%	kD	%	
2S	α-Conglycinin	12.7	18–33	0.5	4.8
7S	γ-Conglycinin	3.0	104	1.4	2.2
7S	β-Conglycinin	35.0	141–171	1.4	2.2
11S	Glycinin	41.8	317–360	1.5	3.0
15S	Poly-glycinin	7.5	ND†	ND	ND

†Not determined.

be significantly greater information than knowledge concerning soybean protein composition. Yet, great strides have been taken to improve that level of understanding.

16–2.1.3 Globulin Structure

16–2.1.3.1 11S Proteins—Basic research has indicated that the secondary structural configuration of 11S proteins is a mixture of 5% α-helix, 35% β-sheath, and 60% random-coil (Koshiyama, 1983). Hence, glycinin is a compactly folded molecule. As previously mentioned, a number of nonidentical subunits comprise the tertiary structure of 11S proteins where a 40 kD acidic (A) and a 20 kD basic (B) polypeptide are linked by disulfide bonds to form a dimer. Genetic diversity for the number of A and B polypeptide types in 11S proteins has been shown among several soybean cultivars. At this time, five groups have been established in regard to the acidic and basic polypeptide array, consisting of: I, 7-A plus 8-B; II, 7-A plus 7-B; III, 6-A plus 7-B; IV, 6-A plus 5-B; V, 6-A plus 3-B (Koshiyama, 1983). The association of a specific acidic polypeptide with a basic polypeptide, however, is nonrandom as indicated by the proposed subunit composition for a cultivar with 6-A and 5-B polypeptides: $A_{1a}B_2$, $A_{1b}B_{1b}$, A_2B_{1a}, A_3B_4, A_4B_3, and A_5B_3 (Staswick and Nielsen, 1983). The basis for the polymorphic composition of 11S subunits is probably governed by specific genetic traits. At least for the case mentioned above, the presence of A_4B_3 and A_5B_3 subunits appears to be controlled by a single gene. The recessive trait (*gl_1gl_1*) expressed in the cv. Raider prevents the formation of these particular subunits (Staswick and Nielsen, 1983; Harada et al., 1983). The immediate significance of that observation is found in the total number of methionine and cysteine residues per subunit. From amino acid sequencing the $A_{1a}B_2$ and A_2B_{1a} subunits have 12.3 to 14.1 methionine plus cysteine residues, whereas the $A_{1b}B_{1b}$ and A_3B_4 subunits have 8.1 to 8.8, and the A_4B_3 and A_5B_3 subunits contain only 1.3 to 2.3 residues of these essential amino acids (Koshiyama, 1983). Hence, genetic variability for 11S subunit composition purports a possible means to enhance both methionine and cysteine levels in glycinin. The degree of variability for such traits in soybean germplasm is virtually unknown albeit significant differences in glycinin composition have been recognized in many *Glycine max* cultivars (Hughes and Murphy, 1983) and in the accessions of *Glycine* subgenus *Glycine* (Staswick et al., 1983). The quarternary structure of glycinin is thought to form an annular hexagonal structure consisting of three dimers (Koshiyama, 1983). Gaining an understanding of the complexity of the structure is the first critical step in the formulation of molecular genetic strategies for the possible modulation of glycinin composition.

16–2.1.3.2 7S Proteins—The complexity of 7S proteins is equally intriguing. In deference to glycinin, the 7S constituents are considered to be glycoproteins. Beta-conglycinin exists as a hexameric molecule where each of the six subunits are trimers formed by various combinations of three nonidentical polypeptides (Hill and Breidenbach, 1974a). Beta-conglycinin may be separated into three molecular weight classes in accordance with isomeric configuration (Table 16–2). The major polypeptides α, α^1 (both 57 kD), and β (42 kD) also differ in methionine content and impart that characteristic to the various known isomers. Four additional molecular species ($\alpha^1{}_2\beta$, $\alpha_2\alpha^1$, $\alpha^1{}_3$, and β_3) should also be found if the polypeptides were randomly associated. The formation of the latter trimers, however, has not been shown by in vitro experimentation (Koshiyama, 1983). Thus, structural hindrance rather than specific genetic variability may preclude the natural occurrence of the latter four species. Nevertheless, it should be noted that genetic elimination of the β polypeptide could significantly elevate the level of methionine in β-conglycinin without impairing trimer formation. At this time, however, there are no known attempts to identify or induce genetic resouces with such a mutation.

16–2.1.3.3 2S Proteins—Considerably more work has been conducted on the genetics and biochemistry of the 2S globulins, even though these constituents of soybean are not technically storage proteins (Muller, 1983). In general, the 2S proteins exhibit antinutritional characteristics attributed principally to the antigenic properties of trypsin and α-chymotrypsin proteinase inhibitors. Two types of trypsin inhibitors are found in soybeans, the Kunitz (21.5 kD) and the Bowman-Birk (8.0 kD) proteins (Stahlhut and Hymowitz, 1983).

16–2.1.3.3.1 Kunitz Proteinase Inhibitor

The Kunitz protein is the major trypsin inhibitor found in soybeans. It is primarily associated with cell walls, protein bodies, and nucleii, but not usually found in the cytoplasm (Horisberger and Tacchini-Vonlanthen, 1983). Ingestion in nondenatured form may cause pancreatic hypertrophy and growth inhibition in monogastric animals and humans, possibly in response to an unbalance of methionine and cysteine in the pancreas. For that reason it is necessary to cook raw soybeans or to heat process the extracted protein concentrate during the preparation of soyfood products. The primary structure of the Kunitz inhibitor consists of 181 amino acids with the active site at the arginine^{63}-isoleucine^{64} location

Table 16–2. Subunit composition of β-conglycinin from soybean seed. Adapted from Koshiyama (1983).

Class	Isomer	Polypeptide molecular species	Methionine residues
kD			
141	β_1	$\alpha^1\beta_2$	3.1
141	β_2	$\alpha\beta_2$	2.8
156	β_3	$\alpha\alpha^1\beta$	4.7
156	β_4	$\alpha_2\beta$	4.4
171	β_5	$\alpha_2\alpha^1$	6.6
171	β_6	α_3	6.0

within the molecule (Horisberger and Tacchini-Vonlanthen, 1983). The search for a biological alternative to alleviate the nutritional problems presented by an active Kunitz protein has identified four basic variants for the *Ti* trait among 3000 accessions of the Soybean Germplasm Collection. Most U.S. cultivars and 88.7% of the Collection may be classified as having the *Ti*(A) gene; a second gene, *Ti*(B), is found in about 11% of the entries of the Collection; a third gene *Ti*(C) is found in about 0.3% of the Collection entries; and the fourth type of variant is devoid of dominant genes encoding the Kunitz trypsin-inhibitor protein. Germplasm resources representative of the fourth group are: PI 157440 and PI 196168 (Orf and Hymowitz, 1979). Even with the Ti trait in the recessive state, trypsin inhibitor activity may only be reduced by 30 to 50% because of the presence of other types of proteinase inhibitors. Inheritance studies have shown that F_2 seed from a *Ti*(A) × *Ti*(B) × *Ti*(C) cross yields a 1:2:1 phenotypic ratio. Thus, the *Ti* trait is probably inherited as codominant alleles in a multi-allelic system at a single locus (Orf and Hymowitz, 1979). In addition, the *Ti* trait is not linked with flower color, determinancy, seed-coat peroxidase, or seed lectin.

16–2.1.3.3.2 Bowman-Birk Proteinase Inhibitor

Bowman-Birk proteinase inhibitors have the distinctive structural feature of two independent active sites located on opposite sides of the molecule. Each active site reacts specifically with either trypsin or α-chymotrypsin (Stahlhut and Hymowitz, 1983). The primary structure contains 71 amino acids of which 21% are methionine (one residue) and cysteine (14 residues), however, the protein contains no or little glycine and tryptophan (Horisberger and Tacchini-Vonlanthen, 1983). The protein may be hydrolyzed to yield a 4.2 kD subunit with the trypsin site in the vicinity of arginine^{15}-serine^{16} and a 3.1 kD subunit with the α-chymotrypsin site at leucine^{43}-serine^{44} (Odani and Ikenaka, 1978a). Apparently, all active sites of doubleheaded proteinase inhibitors are occupied by serine residues. Substitution of glycine for serine at the reactive centers renders the protein inactive. Thus, an α-methyl or methylene group in the active site is absolutely necessary for proteinase binding (Odani and Ikenaka, 1978b). Five isomers of the Bowman-Birk inhibitor have been reported in soybeans. Each isomer has slightly different amino acid composition, inhibitory activity, and isoelectric point. Three phenotypes have been reported from the soybean Germplasm Collection that exhibit variants of the five isomers; specifically isomer-III, which exhibits the greatest reactivity with trypsin and α-chymotrypsin and generally accounts for 35% of the isomeric composition, and isomer-V which generally accounts for 10% of the isomeric composition and exhibits the lowest reactivity with both proteinases (Stahlhut and Hymowitz, 1983). Examples of soybean cultivars that lack isomer-III are: 'Bavender Special C' and 'Roanoke'; Those that lack isomer V are: 'Cutler', 'Disoy', and 'Hardee'; and cultivars having both isomers are: 'Amsoy 71', 'Bienville', and 'Williams'.

Bowman-Birk-type proteins are found in nucleii and protein bodies but not usually in cytoplasm, lipid bodies, or cell walls of the cotyledons (Horisberger and Tacchini-Vonlanthen, 1983); yet, there is little information on the genetic mechanisms governing their synthesis. Because these molecules may react with a wide range of proteinases, one may speculate that these double-headed proteinase inhibitors evolved as a defense mechanism against insects or animals. Although it may seem desirable to human nutrition to develop germplasm devoid of the Bowman-Birk type proteins, these proteins contain relatively high levels of cysteine, an essential amino acid in relatively low concentration in soybean. Thus, the elimination of Bowman-Birk proteinase inhibitors by genetic manipulation becomes a value judgement. Nevertheless, valuable information from basic research on the genetic regulation of Bowman-Birk proteinase inhibitors may eventually yield the means to biologically inactivate the proteins by modifying the primary structure at the reactive sites and to govern the concentration of these proteins in soybean seed. Until experimentation is initiated, it will be difficult to determine the function of Bowman-Birk proteins in developing seed or the appropriate action that should be taken toward the development of new germplasm resources with modified 2S protein composition (Kritsman et al., 1975).

16-2.1.4 Phytate

Soybean are rich in natural chelating agents that may decrease mineral nutrient availability of soy-protein food products. The most prominent chelating agent in soybeans is phytic acid or inositol hexaphosphoric acid (Prattley and Stanley, 1982; Graf, 1983). Significant amounts of phytic acid are localized in the crystalline globoid inclusions within protein bodies. The chelation activity, however, is pH dependent (Prattley et al., 1982a). At acid pH, phytic acid forms an insoluble complex with only 7S proteins. The most probable binding region within the primary 7S protein structure is in the vicinity of lysine-arginine or lysine-histidine bonds. The binding constant for these interactions with basic amino acids at low pH is estimated to be 2.3×10^5 (Prattley et al., 1982a). Hence, the complex is not easily disrupted by heat treatment. Because the isoelectric point at which 7S proteins precipitate from solution is near pH 4.5, significant quantities of phytic acid may accompany the protein isolate during commercial extraction procedures (Prattley et al., 1982b). As the pH of the concentrate is increased in subsequent food-processing steps phytin (Ca_5Mg-phytate) may be formed. Although a phytin-protein complex is relatively more soluble in water and less stable to heat, affinity for minerals and divalent cations is increased. Phytic acid may be dissociated from the protein in vitro above pH 10 at high ionic strength or in vivo in the alkaline environment of the intestinal tract. Free phytic acid, however, retains chelating properties at high pH levels forming insoluble, hydrolysis-resistant salts with Cu, Zn, Ni, Mn, Fe, and Ca (Cosrove, 1966). These cations must be in free-solution within the duodenum to interact with carrier-proteins that facilitate mineral absorption. Thus, salts of phytic acid reduce mineral bioavailability and may significantly compromise nutritional requirements especially in the dietary needs of infants. To abate such problems, the best available method to regulate the phytic acid concentration in soybean protein is through postharvest processing. The basic technology involves ultracentrifugation with either the addition of cations at low pH to inhibit protein interaction with phytic acid, or the addition of sequestering agents at high pH. The expense incurred to remove phytic acid is not publicized; however, a potential cost-effective alternative would be the development of soybean germplasm resources that are inherently low in phytic acid. Yet from assessments of the biological importance of phytic acid in seed metabolism, there may be reason to argue against attempts to lower its level in soybean through genetic manipulation. Phytic acid may serve an essential role in seed biochemistry, and certainly has beneficial industrial and medical application (Graf, 1983).

Although there has been little basic research on the biochemistry, function or genetic variability for phytic acid in soybean seed, information from other plant species has indicated that phytic acid is an end product of P and C metabolism (Cosgrove, 1966; Mandal and Biswas, 1970; Roberts and Loewus, 1968). As a storage product, phytic acid may account for as much as 90% of the total P in developing seed tissues. The capacity of phytic acid to sequester inorganic P may have significant impact upon C partitioning into either starch or sucrose. It is also conceivable that phytic acid may modulate phospholipase activity in developing seed by chelating Ca ions.

Phytase, myo-inositol hexaphosphate phosphohydrolase (EC 3.1.3.8), also is present in developing and germinating seed. During germination phytase activity peaks near 60 h after inhibition and catalyses the hydrolysis of phytic acid to myo-inositol and inorganic P (Mandal and Biswas, 1970). The myo-inositol released by the reaction is thought to be a precursor of phosphatidylinositol synthesis or may be metabolized to form uronic acid and pentose units of the acidic polysaccharides utilized in the cell wall formation (Roberts et al., 1968). In addition to the hydrolysis product, the phytase reaction releases free energy of about 60 kcal (phytate molecule)$^{-1}$. Hence phytic acid probably serves several vital metabolic functions in seeds, especially during germination. It is not known whether phytic acid concentration can be lowered or if phytase activity can be increased in developing seed without causing undesirable biological consequences. The crux of the problem with phytic acid in soybean is the interaction of phytic acid with the 7S proteins, which complicates the removal of phytic acid from extracted proteins. Perhaps then it would be more appropriate to apply molecular and conventional genetic information toward elimination or lowering the concentration of 7S proteins in soybean instead of altering phytic acid levels. Because glycinin generally contains greater methionine and cysteine content than β-conglycinin and also has properties vital for gelatination in the preparation of various soyfood products (Hughes and Murphy, 1983), lowering the concentration of 7S proteins relative to 11S proteins could make a significant improvement of soybean protein quality as well. Beachy et al. (1983a) have recently reported that the soybean cv. Keburi lacks the α' subunit of the 7S seed proteins. As understanding of the molecular genetics of soybean protein synthesis improves such innovations may indeed become reality.

16–2.1.5 Protein Synthesis

16–2.1.5.1 Genetic Control—Eucaryotic plants contain three different protein synthetic mechanisms localized respectively in the cytoplasm, chloroplast, and mitochondria (Ciferri et al., 1976). Each system has different tRNA, rRNA, and aminoacyl-tRNA synthetases which are codified by the genetic information within each organelle. Among the three systems, considerably more information is available on the genetic information contained in chloroplasts. Apparently, the structure and size of chloroplast-DNA is highly conserved among all land plants. In soybean, chloroplast DNA is circular and consists of 151 k bases with two inverted repeats approximately 21 to 23 k base pairs long (Palmer et al., 1983). Genes partially encoding rRNA are found within the inverted repeat regions (Spielmann et al., 1983). The molecular weight of chloroplast DNA is about 100 mD which is theoretically large enough to code for about 100 50-kD proteins (Ciferri et al., 1976). In comparison mitochondria DNA may be capable of coding for only 25 50-kD proteins. Currently, about 10% of the chloroplast genome (plastome) has been characterized with regard to known protein products. Storage proteins, however, are thought to be encoded by the nuclear genome in the cytoplasm. Although there is little information on the structure and expression of genes encoding 11S proteins, Beachy et al., (1983b) have extensively characterized the DNA fragments and complementary mRNAs which govern synthesis of the subunits for the 7S soybean seed proteins.

Protein synthesis is a complex energy requiring multi-enzyme process which takes place on polysomes consisting of an mRNA template and several rRNA ribosomes (Boulter, 1981). In corn, polysomes are bound to membranes closely associated with protein bodies, whereas in soybean, polysomes are not associated with protein bodies. In vitro protein synthesis in soybean, however, may be demonstrated with either free or membrane-bound polysomes (Beachy et al., 1978). Ribosomes in chloroplasts and mitochondria are 70S and require f-met tRNA to initiate protein synthesis. Cytoplasmic ribosomes in eucaryotic organisms are 80S and require met-tRNA for initiation (Ciferri et al., 1976). Chloramphenicol specifically inhibits protein synthesis by 70S ribosomal systems, 80S ribosomal systems are inhibited specifically by cycloheximide. Although various translation factors for both types of protein synthetic systems (RNA polymerase, ribosomal protein, aminoacyl-tRNA synthetases, transformylation, and elongation factors) are coded for by the nuclear genome, the respective complement of factors for the 70S and 80S systems are not interchangeable and function specifically with only one system (Ciferri et al., 1976).

In developing seed, DNA and RNA synthesis precedes storage protein synthesis and deposition, measurable after the termination of cell division (Boulter, 1980). The appearance of globulin proteins and subunits from 12 to 34 DAF follows the order 2S, 7S (α, α^1), 11S (A, B.), 7S (β), 11S (A_4); (Hill and Breidenbach, 1974b; Meinke et al., 1981). After an initial lag phase and exponential increase period, the rate of protein synthesis becomes constant, eliciting a linear increase in globulin $(\text{cell})^{-1}$ until physiological maturity (Boulter, 1980). During the linear increase, the proportion of rRNA to tRNA remains constant; however, the total amount of the translation factors may increase via endopolyploidy.

Globulin synthesis is localized in the rough endoplasmic reticulum (80S ribosomal system). The 11S proteins generally contain $< 1\%$ carbohydrate. The 7S proteins (glycoproteins) may contain 2 to 6% mannose, N-acetylglucosamine, or mannose (Boulter, 1981). Pulse-chase experiments with radiolabeled amino acids show 50% of the incorporated radioactivity in the rough endoplasmic reticulum (RER) at 45 min after removal of labeled substrate. After an additional 20 to 30 min, significant amounts of radioactivity are found in protein bodies (Chrispeels et al., 1982). Similar experiments with radioactive glucosamine or mannose show that glycosylation of 7S proteins may occur in the RER or in dictyosomes. Thus, it is highly probable that proteins synthesized in the RER are transported to the membrane-bound protein bodies via golgi vesicles (Boulter, 1981).

Pulse-chase experiments with radioactive glycine indicate that the 38 to 44 kD acidic subunits of 11S proteins are not synthesized independently from the 19 to 22 kD basic subunits. Significant levels or radioactivity initially appear in 59 to 62 kD proteins prior to the simultaneous increase in specific-radioactivity in both A and B subunits (Barton et al., 1982). Observations of that nature have led to the theory that the A- and B-type subunits are derived from a 60 kD precursor protein molecule (Croy et al., 1982). Evidence in support of this theory includes the characterization of the mRNAs that encode 7S and 11S proteins. In rather elegant experiments, double-stranded DNA complimentary to poly $(A)^+$mRNA from developing soybean cotyledons was inserted into the plasmid vector pBR322. After restriction-nuclease digestion, six clones were prepared that selected mRNA molecules encoding the 47 and 50 kD subunits of 7S proteins, and two clones selected mRNA that codes for a 60 kD protein. The 3′ to 5′ sequences of the latter two mRNAs were identical except that one of the mRNA molecules contained 150 additional codons at the 5′ end. The average size of these mRNA molecules (19S, 710 kD, and 2.2 k bases) was deemed sufficiently large to accommodate the transcription of one A- and one B-type subunit. Furthermore, after detailed sequencing it was determined that the genetic code for a 20 kD basic polypeptide was initiated at the 3′ end of the mRNA for the proposed 11S precursor molecule. The base sequence was unbroken upstream beyond the N-terminus of the basic subunit for at least 30 amino acids approaching the C-terminus of an acidic subunit. No termination or initiation codons were found in the pre-N-terminus sequence of the A subunit (Croy et al., 1982). Apparently then the A- and B-type molecular species arise from post-translational modification of the 60 kD precursor molecule. Hence, the single 11S protein precursor theory was supported. In addition, these data also demonstrated that at least four single-copy legumin genes per haploid genome were present in nuclear DNA.

A similar mechanism is suspected for the synthesis of 7S protein subunits. A 7S precursor protein is coded by high molecular weight mRNA (Croy et al., 1982). Greater than eight ribosomes may be associated with the mRNA for the 7S precursor (Beachy et al., 1978). From hybridization experiments, it is estimated that the mRNAs for 7S proteins are derived from genes present in single or low copy number in the nuclear genome (Croy et al., 1978).

16–2.1.5.2 Molecular Mechanism—Although different protein synthetic systems are present in soybean seed, the sequence of biochemical events from initiation to peptide-chain termination are similar (Boulter, 1978). Protein synthesis is initiated by the dissociation of monomer ribosomes, in the presence of appropriate factors, into native subunits. The native subunits of a 70S ribosome are 30S and 50S, and 40S and 60S for an 80S ribosome. In either case, the small subunit associates with mRNA and contains two active sites for amino-acyl (A) and peptidyl (P) tRNA binding. Formation of an initiation complex requires additional factors, GTP, and the attachment of f-met tRNA (70S system) or met-tRNA (80S system) at the P-binding site. A functional ribosome is then formed with reassociation of native subunits, which in effect sandwich the mRNA molecule, and the release of the specialized factors. Peptide-chain elongation proceeds by selection of aminoacyl-tRNA molecules compatible with the next mRNA codon and attachment of these molecules to the 'A'-binding site on the complete ribosome. A peptidyltransferase then catalyses the hydrolysis and transesterification of the aminoacyl moiety from the tRNA at the P-binding site to the aminoacyl tRNA at the A-binding site. The GTP and various transfer factors facilitate displacement of the t-RNA molecule remaining at the P-binding site and the translocation of the newly formed peptidyl-tRNA to that site. This sequence is repeated as the ribosome moves to the next mRNA codon reading the genetic code from the 5′ to the 3′ end of the mRNA molecule. Eventually, a peptide-chain termination sequence on the mRNA is encountered where various release factors and a specific peptidyl-tRNA hydrolase are evoked to free the peptide from the ribosomal complex (Weissbach and Pestka, 1977).

The nature and timing of the protein synthesis in developing soybean seed is dictated primarily by the level of transcribable mRNA (Smith and

Grierson, 1982); however, the functional link between the genetic information encoded by mRNA and protein synthesis is tRNA. Although total amounts of tRNA and aminoacyl-tRNA synthetases may not change during seed ontogeny, selective alteration within the tRNA population may have significant implications with the regulation of protein synthesis during seed differentiation and aging.

As cells pass through different physical states from development to germination, both quantitative and qualitative changes are shown in certain isoaccepting tRNAs and aminoacyl-tRNA synthetases (Sinclair and Pillay, 1981). As mentioned earlier, different tRNA molecular species are characteristic of the protein synthetic mechanisms in the chloroplast, mitochondria, and cytoplasm. With respect to aging, or more appropriately senescence of the tissue, changes are noted in certain tRNA species specifically associated with the chloroplast (Sinclair and Pillay, 1981). In particular the level of leucyl-tRNA is affected. The symptoms associated with the maturation of seed tissues appears to be highly negatively correlated with leucyl-tRNA levels. Furthermore, the mechanism of cytokinin action in regarding the symptoms of senescence is manifested by an increased acylation of leucyl-tRNA. The specific utilization of leucyl-tRNA in this process, however, is unknown, yet the implication remains that regulation of specific tRNA and aminoacyl-tRNA synthetases may control not only the timing of protein synthesis with respect to seed age but also may ultimately determine cell longevity.

The dynamics of protein synthesis and cell viability, in response to sudden alteration of the growth environment, however, cannot be exclusively attributed to tRNA metabolism. The formation of heat-shock (hs) proteins serves as an appropriate example. The hs proteins include 15 or more types of polypeptides, 15 to 23 kD in size, that appear in soybean as well as other plant and animal tissues within 3 to 4 min after a shift in growth temperature from 28 to 42 °C (Schoffl and Key, 1982). The sudden elevation in temperature apparently causes a rapid change in protein synthesis where the mechanisms at normal temperature are shut down and a new mechanism is initiated. The abrupt nature of the response tends to indicate that temperature-induced regulation of gene expression operates at the level of transcription or translation. Although absolute levels of poly(A)$^+$ mRNA(cell)$^{-1}$ remain constant, hybridization studies between poly(A)$^+$ mRNA and corresponding cDNA show the abundant presence of a new class of mRNA in soybean tissues after heat shock at 40 to 42 °C that is not present in tissues maintained at 28 °C (Schoffl and Key, 1982). The new class of mRNA may account for 20% of the total mRNA at 40 °C and collectively the concentration may be 10-fold greater than the average level of the most abundant mRNA species at 28 °C. Approximately 13 different hs-mRNA species have been identified with an average length of 850 nucleotides and about 19 000 copies (cell)$^{-1}$. Of these mRNA species, one clone contains genetic information for at least 13 proteins at 15 to 18 kD; another clone encodes four proteins at 21 to 23 kD; and two clones select a major hs-protein at 18 kD (Schoffl and Key, 1982). Other higher-molecular weight proteins may also exist, however, as yet, the function of the known hs-proteins has not been determined.

A striking feature of hs-protein metabolism is the rapidity at which these proteins disappear upon sudden shift of growth temperature from 40 °C to ambient levels. Hence, the purpose served by these proteins at high temperature is short lived upon return to normal growth conditions. Although the exact mode of genetic regulation has not been ascertained it is of interest to note that the base sequence of the hs-mRNA is highly homologous with poly(A)$^+$ mRNA from tissues grown at 28 °C (Schoffl and Key, 1982). Thus, the genetic information for hs-proteins may exist in regions of the mRNA molecules that are not encoded at normal growth temperatures. While as yet unproven, that theory may provide a plausible explanation for the extremely short generation time for hs-proteins. If the modulation of gene expression for hs-protein metabolism by growth conditions is achieved by subtle modification of existing transcription or translation factors, the implications of such a finding could have significant impact upon understanding the molecular basis for genotype × environmental interactions affecting proteins and enzymes for other traits.

The rapidity of hs-protein disappearance with a sudden temperature shift to ambient conditions also serves to illustrate the dynamics of protein turnover. Protein-turnover is an important means by which biological systems may alter general or specific protein constituents to maintain vital cell functions in response to environmental conditions or as needed during different stages of seed ontogeny. Protein-turnover is defined as the flux of amino acids through a protein and is a function of the balance between synthetic and degradative metabolic processes (Huffaker and Peterson, 1974). Because of inherent difficulties in measuring protein-turnover there has been little information on relation of this process to the regulation of biological function in developing soybean seed. Considerable research, however, has documented the developmental biochemistry of protein metabolism in germinating soybean cotyledons.

16–2.1.6 Germination

16–2.1.6.1 Protein Synthesis—Germination of the embryonic axis proceeds in three phases: (i) imbibition with rapid escalation of metabolic activity, (ii) a lag phase of metabolic activity, and (iii) a period of between 2 and 4 days after imbibition (DAI) of increased cell division, cell expansion, and DNA synthesis (Smith, 1982). All components necessary for protein synthesis, a high level of ribosomes and performed mRNA are present in the tissue at imbibition. New poly (A)$^+$ mRNA is formed within 6 h after imbibition initiating specific changes in cytoplasmic protein synthesis which precedes an increase in plastidic and mitochondria protein synthesis (Krul, 1974). The most actively transcribed denovo proteins from 6 to 48 h after imbibition are primarily hydrolases, such as: isocitratelyase, endopeptidase, phosphatease, α-amylase, proteases, sucrose phosphate synthetase, malate synthetase, peroxidase, and α-glycosidase (Smith, 1982). After 4 DAI, approximately 470 different denovo polypeptides may be found in the germinating tissues. The rate of synthesis for 56% of these translation products remains constant during the 22- to 26-day period before abscision of the cotyledons (Skadsen and Cherry, 1983). Synthetic rates for 28% of the denovo proteins increase with age, while the synthetic rate for the remaining 16% decline with age (Skadsen and Cherry, 1983). The total protein synthetic activity during the first 10 DAI, however, declines exponentially (Pillay, 1977). Genetic regulation of protein synthesis is primarily related to the availability of mRNA. Between 4 and 12 DAI mRNA levels may decline by 70%. Other factors which may contribute to the loss of translational capacity with increased age are: shortening of poly $(A)^+$ mRNA length, a decline in polysome content, and inactivation of ribosomes and translation factors (Pillay, 1977). No change is found, however, in DNA levels (Krul, 1974). The decline in protein synthetic activity with increased age is first perceived by dysfunction of the chloroplast followed by the mitochondrial and cytoplasmic mechanisms. This is evidenced in the 66% reduction of chloroplast proteins at 12 DAI (Skadsen and Cherry, 1983). Removal of the epicotyl prior to 18 DAI reverses the senescence process. The tissues regreen and protein synthetic activity is restored to the rates at 4 DAI. With prevention of epicotyl regeneration the succulent condition may be maintained for as long as 45 days (Skadsen and Cherry, 1983). Apparently, the cotyledons may lose as much as 90% of the total nucleic acid content and 80% of total protein before epicotyl removal is ineffectual in reversing senescence (Krul, 1974).

16–2.1.6.2 Protein Hydrolysis—Concurrent with and in support of protein synthetic activity in germinated soybean cotyledons, storage proteins sequestered within the protein-bodies are digested by proteolytic enzymes. Several forms of endopeptidases are the major proteases in soybeans. In general, endopeptides cleave peptide linkages with α-carboxyl groups of aspartic and glutamic acid residues and release the acyl portion of acidic amino acids (Bond and Bowles, 1983). Other proteases classified as exopeptidases complete the peptide digestion. Little is known about the exopeptidases in soybean other than the fact that three types (amino-, carboxy-, and di-peptidases) exist and respectively react with different peptide substrates. Both endo-and exo-peptidases are found in germinated and ungerminated seed (Ashton, 1976).

It may be somewhat surprising that protein synthesis occurs in germinated seed simultaneously with intense proteolytic activity. Apparently these antigonistic processes, however, are compartmentalized to some extent during germination. Hydrolysis of the constituents of protein bodies in soybean seed proceeds via internal digestion rather than at the periphery of the organelles. Between 0 and 2 DAI protein bodies begin to swell forming an angular mass as the limit-membrane is disrupted. The activities of proteases, acid phosphatase, phosphodiesterase, α-mannosidase, phospholipases, and ribonuclease increase significantly between 2 and 5 DAI (Smith, 1982). From 4 to 7 DAI the protein masses coagulate; and as the digestion process continues vacuoles formed by the vacancy fuse to yield cells with large central vacuoles (Ashton, 1976). By 9 DAI most of the 7S globulin is digested and at 16 DAI the 11S proteins are virtually totally dissembled (Catsimpoolas et al., 1968).

16-2.1.7 Future Prospects

Soybean is the major source of vegetable-protein in the USA and perhaps the world (Boulter, 1980). Future demand for soy-protein, however, may well hinge upon the development of unique soybean germplasm resources for specialized purposes. With the present base of information plus the integration of molecular and classical genetic methodologies, great potential exists for achieving improvement in both the quality and quantity of soy-protein.

The capabilities of qualitative and quantitative genetic technology have been aptly described in previous chapters of this monograph. With regard to selection procedures, considerable emphasis has been placed upon increasing protein concentration. In most cases, however, the evidence has shown a negative correlation between percent protein and yield, and percent protein and percent oil (Brim and Burton, 1979). Although genetic correlations of that nature should not be considered as casual relations, the genetic basis for yield depression when percent protein is the selection criterion is unknown. Selection for yield per se does not affect the concentration of protein or oil (Kenworthy and Brim, 1979). In a rare case, however, crosses between two highly adapted lines with high protein concentrations followed by six cycles of recurrent-selection produced populations in which both yield and percent protein increased (Brim and Burton, 1979; Burton and Brim, 1983). Hence, with appropriate genetic-recombination the simultaneous enhancement of both traits may be possible.

Improvement of protein quality in soybean depends first of all upon the perception of a particular problem. For human nutrition, the essential amino acid concentration of soy-protein is a significant issue. Methionine and cysteine collectively account for about 4.6% of the amino acids in soybean protein. To meet minimum FAO recommendation the concentration of methionine alone should be increased from 1.6 to 3.0% of the total amino acid content (Burton et al., 1982). Even though the relative difference between actual and desired levels is small, methionine concentration, by virtue of nutritional importance is considered as the most significant limiting amino acid followed by cysteine and threonine (Boulter, 1980). At this time, no soybean breeding programs or other efforts to alter essential amino acid composition are known. Limited genetic variability, time and cost of assay are the major impediments to such programs. Selection for increased protein concentration in soybean has not increased percent methionine (Burton et al., 1982). Hence, there may be no indirect method by which elevated concentrations of methionine or other essential amino acids may be achieved. The fact that methionine and cysteine levels among soybean globulins are greater in 2S (primarily proteinase inhibitors) > 11S (acidic subunits) > 11S (basic subunits) > 7S (α^1 -subunits) > 7S (α-subunits) > 7S (β-subunits), however, may suggest several new approaches for increasing the essential amino acid composition of soy-protein. These approaches may include:

1. Elimination of the β-subunit in 7S proteins to elevate methionine levels in β-conglycinin-type globulins.
2. Elimination of 7S proteins.
3. Substitution of methionine for alanine or valine in basic subunits of 11S proteins via modification of specific mRNA codon composition.
4. Selection for a greater 11S/7S ratio.
5. Substitution of glycine for serine at the active site of Bowman-Birk proteinase inhibitors.

Until proven otherwise, all things are possible. The challenge to test these and other ideas may prove to be vital to the continued prosperity of the soy-protein industry.

16–2.2 Carbohydrate

16–2.2.1 Composition

Starch and soluble sugars are the major constituents of the total nonstructural carbohydrate (TNC) fraction in soybean seed. Although the TNC concentration remains relatively constant during seed development, averaging about 12.4% of seed dry weight (Wilson et al., 1978), changes do occur in TNC composition. In general, starch levels peak at about 11 to 12% of seed dry weight near mid-pod fill, whereas total soluble sugar levels increase steadily throughout development (Yazdi-Samadi et al., 1977). At maturity, starch may account for only 1 to 3% of seed dry weight and the concentration of total soluble sugars may range from 6.2 to 16.5% of seed dry weight (Yazdi-Samadi et al., 1977; Konno, 1979). The major components of the soluble carbohydrate fraction in mature soybean seed are: sucrose (41.3 to 67.5%), stachyose (12.1 to 35.2%), and raffinose (5.2 to 15.8%). Lesser constituents are: pinitol (1-D-3-0-methylchiro-inositol), myo-inositol, galactopinitol, manninotriose, fructose, glucose, and galactose (Schweizer et al., 1978). Of these lesser compounds pinitol is the most prominent monosaccharide in soybean seed (Phillips et al., 1982).

16–2.2.2 Biosynthesis

The accumulation of the oligosaccharides, raffinose, and stachyose begins at mid-pod fill concomitant with the decline in seed starch levels (Amuti and Pollard, 1977; Konno, 1979). Accumulation of raffinose appears before significant deposition of stachyose is observed and is thought to be the immediate metabolic precursor for stachyose synthesis (Yazdi-

Samadi et al., 1977). The pathway of raffinose synthesis requires sucrose and galactinol or possibly galactopinitol as intermediates (Schweizer et al., 1978). Galactinol synthetase activity has been demonstrated in developing soybean seed (Handley et al., 1983). In addition, pulse-chase experiments with $^{14}CO_2$ (Thorne, 1980) have shown that 95 to 98% of the photosynthate translocated to soybean seed at 35 to 40 DAF was [^{14}C]-sucrose. At 1 h after the exposure to $^{14}CO_2$, < 10% of the radioactivity found in the seed was associated with oligosaccharides. Although raffinose and stachyose are translocatable sugars (Kandler and Hopf, 1980), these data suggest that both of these sugars were synthesized by the seed in response to the decline in starch synthesis. The predominant pathway for starch synthesis is via ADPG-starch synthetase (EC 1.4.1.21) which is localized within chloroplasts/plastids (Pavlinova and Turkina, 1978). Hence, the timing of starch depletion and oligosaccharide increase in developing soybean seed may be related to a decline in plastid integrity as the seed matures.

16–2.2.3 Genetic Control—Because the human digestive tract lacks the enzyme α-galactosidase, raffinose, and stachyose are not metabolized to constituent sugars when ingested (Hymowitz et al., 1972a). Hence, these oligosaccharides become fermentation substrates in the intestinal trait. The resultant products of oligosaccharide fermentation contribute significantly to the socially unacceptable condition known as flatulence. Although there is genetic variability among soybean cultivars for oligosaccharide content (Hymowitz et al., 1972a), few breeding programs have attempted to modify the sugar content of soybean seed. Genetic studies have shown, however, a high degree of heritability for total sugar content based upon F_3 progeny means (Openshaw and Hadley, 1981). Apparently, the trait is maternally inherited and governed by additive gene action (Openshaw and Hadley, 1978). No significant genotype $\times$ environment interaction for sugar content in soybean seed has been found (Hymowitz et al., 1972b; Openshaw and Hadley, 1981). Although selection for total sugar content does not affect the relative concentration of oligosaccharides (Openshaw and Hadley, 1978), total sugar content is positively correlated with percent oil and negatively correlated with percent protein and yield (Hymowitz et al., 1972a; Openshaw and Hadley, 1981). Comparisons between high oil (cv. Hawkeye) and low oil (PI 86002) lines also have shown that oligosaccharide content was positively correlated with percent oil even though there was no significant difference between these genotypes in the relative concentration of raffinose or stachyose (Hsu et al., 1973). During germination a decline in oligosaccharide levels can be measured within 15 h after imbibition (Smith, 1982). The degradation of both stachyose and raffinose is essentially completed by 13 DAI (Hsu et al., 1973).

16–2.2.4 Starch Degradation

The primary starch hydrolytic enzymes known in plant tissues are α- and β-amylase, and starch phosphorylase. Although the latter enzyme has not been found in soybean seed, both α-amylase and β-amylase activity may be detected in developing, mature, and germinated tissue (Adams et al., 1981a). Intracellular localization studies have suggested that α-amylase is found within chloroplasts whereas β-amylase is located outside the chloroplast (Preiss and Levi, 1980). In developing seed, α-amylase activity is highest near mid-pod fill and parallels the changes in starch accumulation in chloroplasts (Adams et al., 1981a). Beta-amylase activity, however, increased during the latter stages of seed development and was the predominant source of amylolytic activity in mature seed (Adams et al., 1981a, 1981b). Although starch reserves in soybean seed at the onset of germination were low, both α- and β-amylase activities were detectable with peak activity at 5 DAI. Genetic variability for β-amylase activity has been shown in the cvs. Chestnut and Altona (low activity), and Wells (high activity); (Adams et al., 1981b). The significance of low β-amylase activity is not clear. Beta-amylase catalyses the liberation of β-anomeric maltose from the nonreducing ends of α-1,4 glucans, but does not attack raw starch grains (Mikami and Morita, 1983). Apparently, starch grain hydrolysis is initiated by α-amylase, and β-amylase reacts with the resultant starch fragments. Hence, α-amylase is probably the key hydrolytic enzyme for starch degradation in soybean seed (Adams et al., 1981a).

16–2.3 Oil

16–2.3.1 Classification

The lipid constituents of soybean seed comprise a diverse group of compounds. The most abundant types of lipid in soybean oil are classified as *glycerolipid* which contain fatty acids esterified with glycerol. A standardized system has been adopted to describe the stereochemical structure and generic nomenclature for the different types of fatty acid and glycerolipid (IUPAC-IUB, 1977). By definition the term *fatty acid* designates an aliphatic monocarboxylic acid that may be liberated by hydrolysis from glycerolipids. In soybean seed, the major fatty acids are palmitic (16:0), stearic (18:0), oleic (18:1), linoleic (18:2), and linolenic (18:3). The numerical abbreviation for each fatty acid is a widely used convention that denotes the number of carbon atoms and the number of unsaturated C/C bonds in the molecular structure. Palmitic acid (16:0) contains 16 carbons and zero unsaturated bonds. Both 16:0 and 18:0 are fully saturated fatty acids, whereas 18:2 and 18:3 are polyunsaturated fatty acids. In healthy biological tissues fatty acids are seldom in free solution, but are found as ester derivatives of glycerol. Glycerolipids may be subdivided into groups that differ in polarity. Neutral glycerolipids (NL) are the mono-, di, and tri-esters of glycerol with fatty acids. Triacylglycerol (TG) is a storage lipid in developing seed and is the major component of soybean oil. More polar glycerolipids such as phospholipid (PL) and glycoglycerolipid (GL) are metabolically active in developing and germinating seed and have a prime function as constituents of membranes. The term *phospholipid* may be used for any lipoidal compound that contains a phosphoric acid moiety esterified to monoacylglycerol (MG) or diacylglycerol (DG). The term *lyso* precedes the generic name for all polar glycerolipids esterified with only one fatty acid. The simplest diacyl-PL structure, then is phosphatidic acid (PA). Common derivatives of PA in soybean seed are: phosphatidylcholine (PC), phosphatidylethanolamine (PE), phosphatidylinositol (PI), phosphatidylglycerol (PG), diphosphatidylglycerol or cardiolipin (DPG), and N-acylphosphatidylethanolamine (NPE). In order to designate the configuration of all glycerolipids, the carbon atoms of glycerol are numbered sterospecifically in accordance with the Fischer projection. In that context, the prefix *sn* (sterospecific numbering) is used to identify each of the three carbon atoms in glycerol. With regard to the phospholipids the phosphoryl moiety is esterified with glycerol at *sn*-3 carbon position. A similar positioning of the polar group occurs in glycoglycerolipids. The polar group with glycoglycerolipids, however, consists of one or more monosaccharide residues (usually galactose) joined by a glycosyl linkage to the *sn*-3 position of MG or DG. Mono- (MGDG), and di-(DGDG) galactosyldiacylglycerol are the major glycoglycerolipids in soybean seed. Both of these lipids are almost exclusively associated with plastids or chloroplasts, but comprise a relatively low percentage of the total glycerolipid in developing soybean seed.

To recapitulate, there are at least 12 major forms of glycerolipid (discounting a minor fraction of lyso-derivatives) in crude soybean oil. The average molecular weight of these compounds is < 0.1 kD. Although lipid molecules are relatively small in comparison with the polypeptides of storage proteins, the composition of a specific glycerolipid is made more complex by the number of isomers that may exist with different fatty acids. Using TG as an example, there are potentially five different fatty acids that may be esterified at only three positions on the glycerol moiety. Hence, with random assortment at all *sn*-positions and equal concentration of each of the five fatty acids there should be 125 different TG molecular species. By the same token there may be 25 molecular species for each type of diacyl-PL and GL. It should be pointed out, however, that the esterification of the different fatty acids with glycerol is nonrandom and distinct differences exist in concentration among the fatty acids in soybean seed. The evidence for a nonrandom assortment of fatty acids with glycerolipids is shown by a stereospecific positional acyl analysis of TG (Fatemi and Hammond, 1977a). As shown with 10 soybean cultivars and plant introductions, five lines selected for altered fatty acid composition, and in two accessions of *Glycine* subgenus *Glycine*, the fatty acid distribution at each *sn*-position is not proportional to the total concentration of the respective fatty acids in mature seed. Irrespective of genotype, about 97.4% of the total concentration of 16:0 plus 18:0 in TG occurs at the *sn*-1 and *sn*-3 positions, indicating virtually no saturated fatty acid at the *sn*-2 position. In addition, over all genotypes tested about 39% of the total concentration of 18:1 occurs at *sn*-3, about 44.3% of the total concentration of 18.2 occurs at *sn*-2, and about 37.2% of the total concentration of 18:3 occurs at *sn*-1 (Fatemi and Hammond,

1977a). If fatty acids had been esterified with glycerol by random assortment, the concentration of the respective fatty acids at each *sn*-position should equal 33% of the total concentration for that acid in TG. The biochemical basis for these findings and the impact upon molecular species composition will be discussed later.

16-2.3.2 Composition in Developing Seed

As shown in Table 16-3 the glycerolipid composition of soybean oil changes during seed development. With increased seed age the percentage of PL and DG declines while TG increases to 94.7% of the total glycerolipid in mature seed (75 DAF). The accumulation pattern of total glycerolipid is sigmoidal during seed development (Konno, 1979). Approximately 50% of the total glycerolipid in mature seed is deposited between 30 and 45 DAF. The changes in glycerolipid composition during seed ontogeny also are accompanied by changes in the concentration of individual fatty acids. In general the concentration of 16:0 and 18:0 (S, saturated fatty acid) remains constant, the concentration of 18:1 (M, monoene fatty acid) peaks between 30 and 45 DAF then declines, the concentration of 18:2 (D, diene fatty acid) increases linearly throughout seed development, and the concentration of 18:3 (T, triene fatty acid) declines until about 60 DAF then increases slightly during the final stages of seed maturation (Simmons and Quackenbush, 1954; Hirayama and Hujii, 1965). The fatty acid composition for mature seed from a typical cv. Dare is shown later in this section. The nature of these changes in fatty acid composition during seed development reflects and is directly related to changes in concentration among the various glycerolipid molecular species. At any given stage of seed development, 12 to 15 types of TG molecular species (not distinguishing the possible isomeric forms of each species) may be found (Marai et al., 1983). The predominant TG molecular species in most soybean cultivars are designated as: SD_2, MD_2, and D_3; where each alphabetic symbol (identified above) represents a particular type of fatty acid esterified with glycerol. The combined concentration of these three TG species increases from about 20% (35 DAF) to about 44% (maturity) of the total TG composition (Roehm and Privett,

Table 16-3. Glycerolipid composition during the ontogeny of soybean seed. From R. F. Wilson, unpublished; cv. Dare.

DAF†	Glycerolipid‡			Total glycerolipid
	PL	DG	TG	
	mol %			μmol (seed)$^{-1}$
30	13.7	2.0	84.3	8.1
45	10.8	1.6	87.6	28.2
60	9.6	1.6	88.8	39.8
75	3.9	1.4	94.7	40.9
$LSD_{0.05}$	4.2	0.4	4.6	13.0

†Days after flowering.
‡PL, phospholipid; DG, diacylglycerol; TG, triacylglycerol.

1970). Concomitantly, there are various changes in the concentration of other TG species. In particular the collective concentration of the SDT, D_2T, ST_2, MT_2, and DT_2 species declines from about 40% (35 DAF) to about 13% (maturity) of the total TG composition (Roehm and Privett, 1970). From these observations and others (Fatemi and Hammond, 1977b), one may predict the effect of biological alterations in the concentration of given fatty acids upon the level of related TG molecular species. With an increased percentage of 18:1 and lower 18:2, for example, one may expect to find high levels of M_2D, M_3, SM_2, and lower levels of MD_2, D_3, and SD_2 in the mature seed. Such trends have been confirmed by experimental observations (Wilson et al., 1976; Fatemi and Hammond, 1977b; Wilson, 1984).

16-2.3.3 Biosynthesis

16-2.3.3.1 Fatty Acid—Although fatty acid composition of various organs of soybean plants and cultured cells from those organs may be different (Ezzat and Pearce, 1980), the biochemical mechanism for fatty acid synthesis is highly conserved in plant tissues. Fatty acid synthetase is a multienzyme complex that catalyses the addition of two-carbon fragments from malonyl-CoA to an initial molecule of acetyl-CoA (Stumpf, 1981). Thus, the carbons from acetyl-CoA form the omega terminus of the fatty acid molecule. Only one molecule of acetyl-CoA is directly utilized by the enzyme complex for each fatty acid produced. To initiate the reaction, however, the CoA-derivatives are converted to ACP (acyl-carrier protein) derivatives which are bound to the synthetase complex. The acyl-ACP chain is then repeatedly elongated while attached to the complex. The final product of the reaction is 16:0-ACP. There is now evidence that at least three fatty acid synthetic mechanisms exist in higher plants: 16:0-ACP synthetase which produces 16:0-ACP, 16:0-ACP elongase which produces 18:0-ACP, and 18:0-ACP desaturase which produces 18:1-ACP. Current information shows that exogenous 16:0 is not metabolized to 18:0 and exogenous 18:0 is not converted to 18:1 (Rinne and Canvin, 1971; Stumpf, 1980; Stumpf, 1977; Wilson, 1981). Apparently, there is no mechanism by which an exogenous long chain fatty acid may be converted to an ACP-derivative associated with the fatty acid synthetase complex. Endogenous 16:0-ACP, 18:0-ACP, and 18:1-ACP are readily hydrolyzed to the free acid by thiolases. There acids are then converted to CoA-derivatives before esterification with glycerol.

The subcellular location of the 16:0, 18:0, and 18:1 synthetic mechanisms may be exclusively in plastids or chloroplasts (Stumpf, 1980). The enzyme activities for the synthesis of these three fatty acids has been shown to band in sucrose-gradients at a density of about 1.23 $g(cc)^{-1}$ with marker enzymes endemic to plastids (Beevers, 1982). In addition ACP which is essential for fatty acid synthetase activity (Stumpf, 1981), is exclusively found in intact plastids (Beevers, 1982; Stumpf, 1980). Proplastids isolated from nongreen soybean cell culture suspensions have been shown to synthesize 16:0, 18:0, and 18:1, from acetate, malonate, or pyruvate (Nothelfer et al., 1977). Also fatty acid synthesis in preparations of disrupted chloroplasts from soybean cell culture was stimulated by the addition of ACP (Nothelfer and Spener, 1979). Thus, these data fortify the argument that denovo fatty acid synthesis occurs in plastids. The primary product of these mechanisms, 18:1 (Stumpf, 1981), is the first and most rapidly labeled fatty acid found in incubations with radioactive acetate (Porra and Stumpf, 1976; Rinne and Canvin, 1971).

Oleic acid (18:1) is thought to be exported from plastids and desaturated to 18:2 and 18:3 in the cytoplasm, most likely in the ER (Stumpf, 1980). One of the first indications that 18:1 was the precursor of 18:2 and 18:3 by sequential desaturation in developing soybean seed came from pulse-chase studies with $^{14}CO_2$ (Dutton and Mounts, 1966). More definitive evidence has appeared in recent years that supports the sequential desaturation hypothesis (Stumpf, 1977). The mechanism by which the desaturation of 18:1 and 18:2 occurs, however, is still debated, and the enzymes catalyzing the reactions have not been characterized in any plant tissue. The debate centers principally upon whether 18:1-CoA or 18:1 esterified to PL is the substrate for the desaturase enzymes. The prevalent opinion now is that 18:1-CoA is transferred to PC by an *sn*-3 glycerolphosphate acyltranferase system or by acyl-exchange, and that 18:1-PC is the primary substrate for the desaturase reaction (Stymne and Glad, 1981).

The degree of polyunsaturated fatty acid synthesis in soybean seed is modified by environmental interaction. Temperature is thought to be the major environmental factor that influences fatty acid composition (Howell and Cartter, 1953; Howell and Collins, 1957). Under controlled environmental conditions, prolonged exposure to low day/night temperatures during the reproductive period elicits a higher concentration of starch, 18:2 and 18:3; a lower concentration of total oil and 18:1, but has no effect upon percent protein, sugar, or amino acids in the seed (Sato and Ikeda, 1979; Wolf et al., 1982). Alteration of fatty acid composition in response to temperature shifts, however, may be relatively rapid. In seed of soybean plants grown at 22.5 °C the ratio of 18:1 to 18:2 plus 18:3 in PC was 0.6. Within 24 h after the plants are exposed to 13 °C the ratio in seed PC dropped to 0.1. In the same time period the ratio in seed-PC from plants exposed to 32 °C increased to 2.1 (Slack and Roughan, 1978). No change, however, was noted in TG fatty acid composition. Hence, the initial response to temperature occurs in membrane lipids. Presumably, increased polyunsaturation of the membrane glycerolipids increases membrane fluidity and may enhance cell function at low temperature (Graham and Patterson, 1982). A causal relation between polyunsaturated fatty acid concentration and tolerance to chilling sensitivity, however, has not yet been established.

The biological basis for the response of plant lipid composition to temperature was thought to be related to molecular oxygen availability. Harris and James (1969) first proposed that molecular oxygen availability was the major rate-limiting factor for desaturase activity. The proof of the hypothesis rested upon two observations: (i) the desaturase reactions required O_2, and (ii) the solubility of O_2 in water (cell sap) decreased at higher temperatures. Hence, it seemed that O_2 availability at different temperatures could alter the rate of 18:2 and 18:3 synthesis. More recent investigations have shown (with safflower [*Carthamis tinctorius* L.] seed) that a decline in 18:2 formation could be induced at O_2 concentrations in the cell sap below 86 μM (Browse and Slack, 1983). The actual O_2 concentration in the cells equilibrated at 13 °C was 330 and 210μM in cells maintained at 37 °C. Although the characteristic response in fatty acid composition to temperature was observed in these treatments, the O_2 concentrations were well above critical levels. At a constant temperature, work with seed from other plant species has shown that an increased concentration of O_2 in cell sap had no effect upon unsaturated fatty acid composition (Mazliak, 1979). Hence, considerable doubt was cast upon the O_2-limitation hypothesis. As an alternative explanation for the temperature related change in fatty acid composition, it was noted that the activation energy for the 18:1-desaturase system was low, 21 to 27 kJ $(mol)^{-1}$, compared to 67 kJ $(mol)^{-1}$ for the rate-limiting step in fatty acid synthesis (Browse and Slack, 1983). Thus, the 18:1-desaturase could exhibit greater activity at lower temperature compared to the overall rate of fatty acid synthesis. Mazliak (1979), however, showed that the 18:0-ACP desaturase exhibited a temperature optimum at 17°C. Similar temperature optima have been reported for the synthesis of 18:1, 18:2, and 18:3 in developing soybean seed (Rinne and Canvin, 1971). Although the activities of these enzymes at different temperatures could be mediated by membrane fluidity, it is not known whether changes in enzyme conformation or changes in the tissue level of these enzymes are responsible for the observed responses.

16–2.3.3–2 Glycerolipid—Although fatty acids account for > 95% of the total radioactivity in lipids from developing soybean seed incubated with [^{14}C]acetate, nearly all of that radioactivity is associated with PL, DG, and TG. In typical acetate incorporation studies, PL will contain the majority of the radioactivity (Stearns and Morton, 1973). Because of the rapid accumulation of radioactive fatty acid in PL, the logistics of fatty acid synthesis and PL biosynthesis must be highly coordinated. An excellent review of the enzymatic reactions catalyzing the synthesis of the respective phospholipids has been prepared by Mudd (1980). There is convincing evidence that many of the described PL synthetic enzymes are localized in the ER (Beevers, 1982). The proposed rate-limiting step for all of these reactions in the enzyme *sn*-3-glycerolphosphate acyltransferase, which catalyses the formation of lyso PA and PA (Porra, 1979). The PA is then hydrolyzed to DG, and DG is utilized in the synthesis of all other PL molecules. The subcellular localization of *sn*-3-glycerolphosphate acyltransferase, however, is reported to be the plastids of developing seed (Porra, 1979). The enzyme also was shown to exhibit specificity for acyl substrates, where greater activities were attained with 16:0>18:0>18:1>18:2. In view of these data, it is conceivable that the products of fatty acid synthetases in the plastid may be exported to the ER and other organelles as derivatives of PA or DG rather than as free acids or acyl-CoA.

Evidence for complex PL synthesis in the plastid is limited. It is presumed that PL such as PC and PE, formed in the ER, are transported to plastids and other subcellular structures and incorporated into the respective membranes. The means by which such transport might occur is not well characterized; however, it is suspected that specific exchange proteins mediate the transfer (Yamada et al., 1980; Julienne et al., 1981). The distinction of whether PL is synthesized in the plastid or is imported from other subcellular locations may be moot, yet, understanding the process by which complex PL occur and are metabolized in the plastid may have significant impact upon TG biosynthesis and TG composition.

In addition to the structural function of PL in membranes, PL also may be metabolized to DG which may be subsequently utilized in TG biosynthesis (Wilson et al., 1980). Turnover of PC in developing soybean seed has been coupled with TG synthesis in other studies as well (Stymne and Appelqvist, 1980). The metabolism of PL molecular species radiolabeled with 16:0, 18:0, 18:1, 18:2, or 18:3 are highly correlated with the formation of related TG molecular species (Wilson, 1981). As an example, SD and DD were the predominant PL molecular species labeled with [^{14}C]18:2. In the same tissue the majority of radioactivity among TG molecular species occurred in D_3, SD_2, and D_2T. Thus, PL composition and metabolism at the site of TG biosynthesis may have a significant impact upon the TG composition of the seed.

The TG biosynthesis is catalyzed by the enzyme diacylglycerol acyltransferase (DGAT). Virtually all of the published information concerning DGAT has come from animal tissues. The first enzymatic characterization of DGAT from any plant tissue was performed with safflower seed (Ichihara and Noda, 1982) and in spinach (*Spinacia oleracea* L.) leaves (Martin and Wilson, 1983). The pH optimum for the enzyme was about pH 8.0 and DGAT apparently exhibited broad specificity for DG and acyl-CoA substrates. It was generally assumed that DGAT is localized in the ER which also is presumed to be involved in oleosome (oil body) biogenesis. The justification for these assumptions has been entirely based upon electron micrographic (EM) evidence. The EM photographs of plant cells have been used to show that invaginations of the ER form vessicles (spherosomes) which develop into oil bodies (Frey-Wyssling et al., 1963). It was implied then that DGAT was associated with ER membranes that form the limiting membrane bounding the spherosomes. Although oil bodies might synthesize TG (Harwood et al., 1971; Gurr et al., 1974), there has been no direct biochemical proof that spherosomes develop into oleosomes. In fact, Gurr (1980) and Ichihara and Noda (1981) contended that the origin of oleosomes was independent of the ER and spherosomes in oil-seeds. In support of the latter contention, DGAT has been localized among subcellular fractions of spinach leaves (Martin and Wilson, 1984). In that study DGAT activity was not found in the ER, but sedimented in a region of sucrose-gradients enriched with intact chloroplasts. Additional characterization with purified chloroplast preparations revealed that DGAT was not a stromal enzyme nor could it be associated directly with thylakoid membranes. The DGAT activity, however, was strongly associated with the chloroplast envelope membranes. The TG synthesis from [^{14}C]acetate was previously shown in purified intact chloroplasts from pea (*Pisum sativum* L.) and spinach leaves (Panter and Boardman, 1973; Dubacq et al., 1983). In developing seed, the activity of *sn*-3-glycerolphosphate acyltransferase in plastids closely paralleled TG accumulation during seed development (Porra, 1979). Hence, these data collectively suggested that chloroplasts or plastids possess both the means and the method to synthesize TG.

A link between chloroplastic DGAT activity and TG deposition in oleosomes has not yet been made. It is interesting to note, however, that osmiophillic bodies increase in size and number within chloroplasts of senescing leaf tissue. Where once the content of the osmiophilic bodies was thought to be exclusively tocopherol (Barr and Arntzen, 1969), it has now been demonstrated that these inclusions contain TG (Burke et al., 1984). An increase in osmiophilic bodies also has been observed in chlorophyll-deficient mutants (Palmer et al., 1979). Soybean genotypes containing the recessive Y_{11} gene had a threefold greater number of osmiophillic bodies than sibling genotypes with the normal dominant trait. In addition, both biotypes contained the same number of mitochondria, chloroplasts, and total cells. At least 10 other single recessive genes have been identified in soybean that express the chlorophyll-deficient trait (Nissly et al., 1981). Although leaf tissues contain low levels of TG compared to seed, similarities might exist in lipid metabolism between leaf chloroplasts from chlorophyll-deficient mutants and plastids from soybean seed. If such a relation does exist, the degree of plastidic TG synthesis may then be governed by the functional status (source or sink) of the tissue.

16-2.3.4 Quality

Aside from the physical properties of triacylglycerol crystalline structure (Lutton, 1972), discussions on soybean oil quality generally center upon polyunsaturated fatty acid content. Soybean, like most of the major edible vegetable oil crops, contains a high level of 18:2. Unlike the other edible vegetable oils, however, soybean also has a high level of 18:3. Both of these fatty acids have significant effect upon oil quality. A high concentration of 18:2 is positively correlated with nutritional quality, but 18:3 is negatively correlated with flavor quality. To clarify these points, it is necessary to briefly describe the biological properties of 18:2 and 18:3 that pertain to the major aspects of soybean oil nutritional and flavor quality.

16–2.3.4.1 Nutrition—With regard to dietary nutritional quality, 18:2 is an important metabolic precursor, in man and other mammalian species, for longer chain fatty acids that play a critical role in maintaining good health (Tinoco et al., 1979). It is somewhat ironic that mammalian species possess the biochemical mechanisms to metabolize 18:2, but lack the ability to synthesis 18:2. Thus, 18:2 is recognized as an essential dietary fatty acid (Pryde, 1980). In humans, it is recommended that 18:2 should contribute about 1% of the daily dietary caloric intake to satisfy the average minimum requirement for essential fatty acid (Holman, 1978). As the caloric content and composition of the diet varies, however, the absolute amount of 18:2 needed also will change. To bring the absolute requirement of 18:2 into perspective, comparisons have been made between a number of different vegetable oils supplied in equal amounts proportional with an average caloric intake. It is estimated that 40% of the calories in the average daily diet in the USA are derived from vegetable oil. If olive oil (which contains about 6% 18:2) was supplied at that level, 18:2 would contribute 1% of the total calories or the minimum daily requirement. Likewise if soybean oil were the exclusive vegetable oil used in the diet, 18:2 could contribute about 20% of the total caloric intake (Holman, 1978). Hence for practical purposes, the average adult person in the USA ingests a considerable amount of 18:2 in excess of minimum standards.

The essentiality of 18:3 in the human diet has recently been questioned. Although there is virtually no direct evidence to suggest that 18:3 is an essential fatty acid, the WHO-FAO and ANRC have recommended that 18:3 and 18:2 should be included in human diets (Tinoco et al., 1979). Basic research is needed to determine whether 18:3 is fully effective as an essential fatty acid and furthermore to ascertain any potential beneficial aspects of 18:3 in human health. In view of the rather low levels of 18:2 that are required for good health, however, one might expect that nearly any diet could provide sufficient levels of 18:3.

16–2.3.4.2 Flavor—The major problem with vegetable oils that contain high concentrations of 18:2, and 18:3 in particular, is flavor stability. Both of these fatty acids may be oxidized by various chemical or enzymatic mechanisms. The secondary reaction products from oxidized polyunsaturated fatty acids generally exhibit characteristic flavors that have a deleterious effect upon oil flavor quality, especially in liquid cooking oil (Chang et al., 1983). To abate this problem, soybean may be heat treated to denature proteins such as lipoxygenase; a variety of antioxidants may be added to inhibit peroxidation of polyunsaturated fatty acids by free radicles, molecular oxygen, and lipoxygenase (Peterman and Siedow, 1983); and vegetable oils may be partially hydrogenated to lower the level of 18:2 and 18:3 in the refined product (Pryde, 1980; Carpenter et al., 1976). In addition to these preventative measures, advances have also been made toward the development of soybean germplasm that is inherently low in 18:2 and 18:3, and in lipoxygenase activity.

16–2.3.4.2.1 Lipoxygenase

Lipoxygenase (EC 1.13.11.12), a protein that is essentially ubiquitous in all plant species and plant organs, catalyses the hydroperoxidation by molecular oxygen of lipids, such as 18:2 and 18:3, that contain a methylene interrupted *cis-, cis-*1,4 pentadiene moiety within the acyl structure (Andrawis et al., 1982). Lipoxygenase is thought to be a single polypeptide with a molecular weight of about 100 kD (Vliegenthart et al., 1979). Immunochemical analysis has shown that three lipoxygenase isozymes (designated as L-1, L-2, and L-3) occur in the seed of most soybean cultivars (Yabuuchi et al., 1982). In purified form each isozyme exhibits different pH optimum and substrate specificity. Optimal lipoxygenase activity is achieved at pH 9.0 with L-1, at pH 6.5 with L-2, and at pH 4.5 with L-3 (Roza and Francke, 1973). Although each isozyme may be assayed with either 18:2 or 18:3 (free fatty acid), L-1 is reported to react with polyunsaturated acyl groups esterified to PC (Eskola and Laakso, 1983) and L-3 may react with certain TG molecular species (Hildebrand and Hymowitz, 1981). At the respective pH optima, the reaction rate with free-18:2 is typically greater for L-1 $>$ L-2 $>$ L-3. The reaction of L-2 with 18:2, however, may be stimulated four-fold by the addition of 0.5 m*M* Ca^{2+} (Andrawis et al., 1982). With methyl-18:2 L-2 and L-3 exhibit greater reactivity than L-1. In addition, L-1 is 36 times more heat stable than L-2 at 69 °C (Hildebrand and Hymowitz, 1981).

The products of the lipoxygenase reaction are hydroperoxy-acids, such as L-13-hydroperoxy-18:2 and L-13-hydroperoxy *cis*-9, 15-trans-11-18:3 (Verhagen et al., 1978; Vick and Zimmerman, 1979a). These products may be generated in lieu of molecular oxygen by autooxidation of 18:2 and 18:3 initiated by free radicles. Furthermore, the lipoxygenase reaction may be demonstrated under anaerobic conditions with the addition of the hydroperoxy-acid reaction product (Verhagen et al., 1978). Each lipoxygenase molecule contains one nonheme Fe atom which is essential for enzyme activity (Galliard and Chan, 1980). Apparently, compounds such as L-13-hydroperoxy-18:2 stimulate lipoxygenase activity by red-ox reaction whereby the Fe atom is activated from the native (Fe II) state to the high spin (Fe III) state (Cheesbrough and Axelrod, 1983). In this process, the hydroperoxy-acyl moiety is oxidized and converted to various oxo-dienoic acids (Verhagen et al., 1978; Streckert and Stan, 1975). One of the products, trihydroxyoctadecenoic acid is associated with the bitter taste of oxidized soybean oil (Moll et al., 1979). Various other objectionable flavors also result as by-products from this reaction (Sessa, 1979; Chang et al., 1983). Many of these flavors are unique to soybean oil by virtue of the oxidation by-products from 18:3 (Vick and Zimmerman, 1979b). In addition to flavor, compounds derived from 18:3 such as 12-oxo *cis*-10,15-phytodienoic acid have structural similarities with prostagladins in mammals and could be a precursor of jasmonic acid (Vick and Zimmerman, 1979b; Zimmerman and Feng, 1978).

During soybean seed development, the activity of each lipoxygenase isozyme increases slowly until about 10 days before maturation (Hildebrand and Hymowitz, 1983). It is not known whether the escalation of lipoxygenase activity has a causal effect or is a result of the senescence process. It is known, however, that free radical generation increases with senescence and that free radicals stimulate lipoxygenase activity (Leshem et al., 1981). The application of cytokinin to the senescent tissue reverses the effect and concomitantly free radical generation and lipoxygenase activity subside. In addition, allopurinol, a specific inhibitor of xanthine oxidase which yields superoxide free radicals, also elicits similar effects. Thus, incipient prevention of free radical formation may be an effective means to regulate the senescence process in plant tissues (Leshem and Barness, 1982). In germinating seed, L-1 activity is high initially and declines to basal levels by 7 DAI. The activity of L-2 and L-3, however, peak at 5 DAI (Hildebrand and Hymowitz, 1983). Nearly all lipoxygenase activity is absent from germinated seed when the cotyledons turn yellow (Vernooy-Gerritsen et al., 1983). Thus, the relation to senescene in germinating seed is not clearly defined.

Evidence from a wide range of soybean genotypes, that varied significantly in maturity group, oil content, and percent protein, indicates that lipoxygenase activity is not affected by environmental interaction (Chapman et al., 1976). Genetic regulation of lipoxygenase activity in soybean cultivars, however, may soon become a reality. Several genotypes which lack specific lipoxygenase isozymes have been selected from nearly 6500 accessions of the World Soybean Collection. Examples of entries that lack L-1 are: PI 133226 and PI 408251 (Hildebrand and Hymowitz, 1981). Entries which lack L-3 are: PI 417458 and PI 205085 (Kitamura et al., 1983). Inheritance studies have shown that the L-1 and L-3 traits segregate independently (Kitamura et al., 1983), with no maternal or cytoplasmic effects (Hildebrand and Hymowitz, 1982). Thus, it appears that the lipoxygenase isozymes are governed by single-independent genes (Lx_1, Lx_2, and Lx_3) which are recessive in the selected germplasm. Germplasm has recently been developed with the lx_1lx_3 trait (Kitamura et al., 1983). It is also believed that germplasm resources containing the lx_2 gene have been discovered (N.C. Nielsen, 1984, personal communication). Although it has been assumed that the total elimination of lipoxygenase activity from soybean seed will not critically affect seed survival, it will be interesting to see if the rate and timing of seed maturation is delayed in lx_1 lx_2 lx_3 lines which should be produced in the near future.

16–2.3.4.2.2 Genetic Alteration of Oil Content and Composition

It is surprising to note that the current information on the genetic control of oil content and fatty acid composition of soybean is extremely superficial. In no case has there been a documentation of specific genes or the number of genes that may be involved in soybean lipid-synthetic processes. It is known, however, that soybean oil content and fatty acid composition are not simply inherited traits (Brim et al., 1968; Howell et al., 1972). Although these quantitative traits are heritable (White et al., 1961), the estimates for the heritability of percent oil, $h^2 = 0.28 \pm 0.03$ (Burton and Brim, 1981) and percent 18:1, $h^2 = 0.21 \pm 0.06$ (Burton et al., 1983) are low. Thus these estimates of heritability indicate that environmental interaction, primarily temperature effects (Collins and Howell, 1957; Gupta and Dhindsa, 1982), have a significant influence upon the variability of the traits. With such a large component of the variation due to environmental factors, selection progress is slow or perhaps impossible with conventional selection methodology. Because of these problems, adaptations of existing (Hammond and Fehr, 1975; Hammond et al., 1972) and recently developed breeding methods (Burton et al., 1983; Burton and Brim, 1981) have been implemented to improve the effectiveness of selection programs. The use of recurrent selection procedures has elicited increases in percent oil of 0.35% $(\text{cycle})^{-1}$, (Burton and Brim, 1981); and gains of 1.6% $(\text{cycle})^{-1}$ when 18:1 was the selected trait (Burton et al., 1983). In the latter case, it was determined that the effect of environmental interaction upon 18:1 concentration did not change the ranking of selected genotypes at different locations. Hence, population improvement could be expected even with significant environmental variation. The degree of progress made in the selection for high 18:1 concentration with recurrent selection has been highly significant and has resulted in a decline in the concentrations of 18:2 and 18:3 (Wilson et al., 1976, 1981; Wilson, 1984). The use of physically induced genetic mutations also has been highly successful in altering fatty acid composition in soybean seed. In addition to the independent research of Wilcox et al. (1984) at Purdue University, researchers at Iowa State University have released two experimental germplasm lines, A5 and A6, which are respectively distinguished by low percent 18:3 and high percent 18:0 (Hammond and Fehr, 1983a, 1983b). The fatty acid composition of several selected soybean germplasm resources is presented in Table 16–4. Examples of genotypes exhibiting either the high or low percent oil trait are listed in Table 16–5.

Table 16–4. Fatty acid composition of selected soybean germplasm.

Germplasm†	Fatty acid					Maturity group
	16:0	18:0	18:1	18:2	18:3	
	mol%					
Dare	12.3	2.9	21.3	56.1	7.4	V
N79-2077	5.0	3.6	48.8	36.5	6.1	IV
N78-2245	11.1	4.4	46.7	33.6	4.2	V
A5	9.3	3.9	39.8	42.9	4.1	0
PI 123440	9.7	3.5	27.8	54.6	4.4	V
A6	8.0	28.1	19.8	35.5	6.6	0
PI 171441	12.6	3.1	15.4	55.9	13.0	VI

†N79-2077 and N78-2245, developed at Raleigh, NC; A5 and A6 developed by W.R. Fehr at Iowa State Univ.

Table 16–5. Oil content of selected soybean germplasm. Adapted from Bernard (1956, 1957, 1966, 1970) and Hartwig and Edwards (1975, 1980).

Maturity group	High oil		Low oil	
	Germplasm	Oil	Germplasm	Oil
0	PI 290145	24.0	PI 181571	14.6
	PI 297513	24.9	PI 153297	15.1
I	PI 189896	22.6	PI 205085	13.7
	PI 189907	22.5	PI 81765	13.8
II	PI 92598	23.4	PI 81767	13.5
	PI 79885	23.5	PI 81773	14.2
IV	PI 88349	23.5	PI 181550	15.5
	PI 88310	22.4	PI 200499	15.5
V	PI 371611	27.6	PI 408042	13.3
	PI 235347	25.6	PI 407876	14.5
VI	PI 374221	26.7	PI 212605	13.9
	PI 360839	24.4	PI 175189	14.5
VII	PI 330633	25.5	PI 417496	15.2
	PI 285093	25.4	PI 123439	16.3
VIII	PI 221716	26.6	PI 323579	13.6
	PI 200832	25.4	PI 209833	13.8

With the identification of genetic differences among soybean genotypes for the concentration of specific fatty acids or percent oil, more detailed efforts can be made to determine the number and subcellular location of genes that affect these phenotypic traits. At present, it is thought that percent oil and fatty acid composition are governed by maternal influence (Brim et al., 1968; Singh and Hadley, 1968; Martin et al., 1983). Although these traits appear to be quantitatively inherited, presumably governed by major structural genes with smaller additive gene effects upon phenotypic expression, dominance or epistatis cannot be totally discounted. The heterogeneity of lipid synthetic mechanisms associated with various subcellular organelles in soybean seed strongly implies that different gene loci are involved in each system. Hence, if dominant or recessive genes do govern these respective activities, considerable effort will be needed to distinguish loci with specific mutations and demonstrate their function independent of gene products emanating from other lipid synthetic systems in soybean seed. With regard to oil synthesis, attempts are being made to determine the degree of interaction between the nuclear genome and plastome in the synthesis of the primary subunits of diacylglycerol acyltransferase (R.F. Wilson, unpublished). The results of these studies may provide basic information needed to define the biological basis for genetic variability in soybean oil content. Similar investigations should be initiated also with the gene products that catalyse the synthesis of specific fatty acids. Information of this nature may benefit future efforts to alter the oil content and fatty acid composition of soybean oil through genetic manipulation.

16–2.3.5 Germinating Seed

16–2.3.5.1 Synthesis—Although the primary lipid metabolic function in germinating seed is essentially degradative in nature, the fact that germinated soybean cotyledons actively synthesize fatty acid and glycerolipids (including TG) should not be overlooked (Negishi, 1976; Yoshida and Kajimoto, 1981). All enzymes necessary for fatty acid synthesis are present and active in cotyledons during the first 24 h after imbibition (Harwood, 1975). Inhibitors of protein synthesis have no effect upon that fatty acid synthetic activity. Peak activity for fatty acid synthesis from [^{14}C] acetate is observed between 3 and 10 DAI (Place et al., 1979; Porra and Stumpf, 1976). As in developing seed, the primary [^{14}C]acyl products are 16:0, 18:0, and 18:1. The activities of acetyl-CoA synthetase, acetyl-CoA carboxylase, and fatty acid synthetase are contained in plastids (Place et al., 1979). Hence, fatty acid synthesis parallels plastid or chloroplast development in the cotyledons during seed germination.

During the first 48 h of germination, total lipid content and dry weight remains relatively constant; however, significant changes are noted in the concentration of individual glycerolipids where TG concentration declines and percent PL increases (Harwood, 1975). The increase in PL concentration during this period is attributed to PA, PG, and PC. Changes in PL molecular species composition also occur (Harwood, 1976; Nishihara and Kito, 1978; Kahn et al., 1960). Peak synthetic activity for *sn*-3-glycerolphosphate acyltransferase is observed at 3 DAG (Porra, 1979). At 6 DAG, 65% of the radioactivity from [^{14}C]acetate accrues in PL (PI, PC, PE, PG) with the remainder in TG (Negishi, 1976). The PI synthetic activity peaks during this period as does phytase activity (Robinson and Carman, 1982; Yoshida and Kajimoto, 1978). Although total polar-glycerolipid content may plateau, there is a decline in the ratio of PL to GL. Peak synthetic activities for PC, PE, and MGDG, however, occur between 12 to 16 DAI which is contemporal with visible symptoms of tissue senescence (Dykes et al., 1976; Huber and Newman, 1976; Place et al., 1979). At 15 to 22 DAI the cotyledons yellow, the internal structural integrity of chloroplast membranes becomes disrupted, starch grains disappear, fatty acid synthesis declines, and there is a significant increase in osmiophilic globules within the chloroplasts (Sato et al., 1983; Hudak, 1981; Place et al., 1979; Bricker and Newman, 1980; Huber and Newman, 1976). In the senescing cotyledons the entire plastid volume may be filled with osmiophilic globules, which may contain TG. If the epicotyl is removed before the senescence process becomes irreversible, the cotyledons regreen, accumulate starch, and osmiophilic globule content declines (Place et al., 1979; Huber and Newman, 1976). Typical changes in fatty acid composition and content during these phases of cotyledon ontogeny are shown by Wetterau et al. (1978) and Yoshida and Kajimoto (1981). In retrospect to the previous discussion of DGAT localization, oleosome formation, and tissue function these data add to the speculation that the origin of oleosomes in developing seed may be linked with plastids rather than spherosomes from the ER.

16–2.3.5.2 Triacylglycerol Hydrolysis—The TG content in cotyledons declines linearly from 0 to 13 DAI (Kahn et al., 1960). The rate of

Table 16–6. Beta-oxidation of palmitic acid. Adapted from Galliard and Chan (1980).

Step	Substrate	Enzyme	Product
1	16:0	Thiokinase + ATP	16:0-CoA
2	16:0-CoA	Acyl dehydrogenase + FAD	16:1-CoA (2t) + FADH
3	16:1-CoA (2t)	Enoyl hydrase + H_2O	3-Hydroxy-16:1-CoA (2t)
4	3-Hydroxy-16:1-CoA (2t)	β-Hydroxyacyl-dehydrogenase + NAD	3-Keto-16:1-CoA (2t) + $NADH_2$
5	3-Keto-16:1-CoA (2t)	β-Ketoacylthiolase + CoA	14:0-CoA + Acetyl-CoA
6	14:0-CoA	Repeats steps 2 to 5 (6 times)	7 Acetyl-CoA + 6 FADH + $NADH_2$

TG hydrolysis is greater in seedlings grown in the light than in etiolated cotyledons (Yoshida and Kajimoto, 1978). There is no change in TG fatty acid composition, however, between 0 and 6 DAI (Singh et al., 1968; Toshida and Kajimoto, 1977). Hence, during that period the hydrolysis of TG molecular species appears to be random. The hydrolysis of TG to fatty acid, MG, and DG is catalyzed by lipase enzymes which exhibit different pH optima. Some controversy exists, however, concerning the subcellular localization of these lipases and whether they are present or inactive in ungerminated seed (Lin et al., 1983; Beevers, 1982; Lin et al., 1982). During germination, neutral lipase (pH 7.0) activity peaks at 1 DAI and is thought to be associated with spherosomes (Lin et al., 1982). The neutral lipase apparently will hydrolyze MG and DG to yield constituent fatty acids, but will not attack TG. Acid lipase (pH 4.5) activity appears with oleosomes but peaks (3 DAI) before significant decline in TG content is observed (Beevers, 1982). The acid lipase, however, does accept TG as a substrate. A third type of lipase, alkaline (pH 7.5 to 9), exhibits peak activity between 5 and 7 DAI. The activity of this enzyme parallels catalase activity and is thought to be located in glyoxysomes (Lin et al., 1982; Lin et al., 1983). Although alkaline lipase may be the most significant lipase in terms of TG hydrolytic activity, it appears that this enzyme reacts only with TG containing highly unsaturated acyl groups (Beevers, 1982; Lin et al., 1983). Hence, additional information is needed to understand more fully the nature of biochemical and genetic regulation to TG hydrolysis in germinating seed.

The metabolic fate of the fatty acids liberated from TG has been fairly well characterized. Fatty acids are rendered to acetyl-CoA, FADH, and $NADH_2$ by β-oxidation. The enzymes associated with the β-oxidation cycle are localized in glyoxysomes (Beevers, 1982). The multienzyme sequence of steps in the β-oxidation of 16:0, 18:1, and 18:2 is shown in Tables 16–6 and 16–7. Hydrolysis of one molecule of ATP per molecule fatty acid is necessary to initiate the reactions. As shown, complete β-oxidation of 16:0 yields 8 acetyl-CoA, 7 FADH, and 7 $NADH_2$; 18:1 yields 9 acetyl-CoA, 7 FADH, and 8 $NADH_2$; 18:2 yields 9 acetyl-CoA, 6 FADH, and 8 $NADH_2$. If all of the FADH and $NADH_2$ molecules

Table 16–7. Beta-oxidation of oleic acid (18:1) and linoleic acid (18:2). Adapted from Galliard and Chan (1980).

Step	Substrate	Enzyme	Product
1	18:1	Thiokinase + ATP	18:1-CoA (9c)
2	18:1-CoA (9c)	3 Cycles β-oxidation	12:1-CoA (3c) + 3 Acetyl-CoA + 3 FADH + 3 $NADH_2$
3	12:1-CoA (3c)	Isomerase	12:1-CoA (2t)
4	12:1-CoA (2t)	4 Cycles β-oxidation	6 Acetyl-CoA + 4 FADH + 5 $NADH_2$
1	18:2	Thiokinase + ATP	18:2-CoA (9,12c)
2	18:2-CoA (9,12c)	3 Cycles β-oxidation	12:2-CoA (3,6c) + 3 Acetyl-CoA + 3 FADH + 3 $NADH_2$
3	12:2-CoA (3,6c)	Isomerase	12:2-CoA (2t,6c)
4	12:2-CoA (2t,6c)	2 Cycles β-oxidation	8:1-CoA (2t) + 2 Acetyl-CoA + 1 FADH + 2 $NADH_2$
5	8:1-CoA (2t)	*Cis*Δ2-enoyl hydratase	D(−)3-Hydroxy-8:0-CoA
6	D(−)3-Hydroxy-8:0-CoA	3 Cycles β-oxidation	4 Acetyl-CoA + 2 FADH + 3 $NADH_2$

produced by these reactions were utilized in mitochondrial oxidative phosphorylation, a net total of 34 ATP (from 16:0), 37 ATP (from 18:1), and 35 ATP (from 18:2) could potentially be generated from the respective fatty acids. If the free energy of the reaction: ADP + $H_3PO_4 \rightarrow$ ATP + H_2O, equals + 8 kcal $(mol)^{-1}$; then the potential amount of free energy released by the β-oxidation of 18:1 could equal 296 kcal $(mol)^{-1}$. Hence, the catabolism of fatty acid is an important source of metabolic energy for tissues of germinating seed.

Acetyl-CoA generated from the fatty acids may be sequentially converted to succinate and oxaloacetate (in glyoxysomes) and to phosphoenolpyruvate (in the cytoplasm). In plant tissues (but not in mammalian systems), phosphoenolpyruvate may be utilized in sucrose biosynthesis by reversal of the glycolytic pathway (gluconeogenisis). Empirically, one molecule of 16:0 should yield two molecules of sucrose if all of the acetyl-CoA liberated in β-oxidation was metabolized via gluconeogenesis. Thus, a metabolic circle is completed. Sucrose is the primary form of C imported by developing seed. Carbon from sucrose is eventually transformed to TG during seed ontogeny. In germinated seed, C from TG may be utilized in sucrose biosynthesis to support the life function of developing tissues.

16–2.3.6 Future Prospects—In the recent past, major basic research thrusts have been undertaken in photosynthesis, N_2 fixation, plant growth regulation, and molecular biology. At present, the number of investigative efforts in those areas by far eclipses the amount of work and resources invested in plant lipid research. The opportunities in plant lipid research are unparalleled among the major research thrusts currently in vogue. These opportunities exist not only in the ability to gain biological control over specific processes such as the lipoxygenase reaction, fatty acid composition of glycerolipids, and TG content, but also extend to the regulation of generalized physiological and morphological phenomenon like senescence, plant maturation, and environmentally induced responses of many genotypic traits. Future advances in lipid research will undoubtedly improve understanding of the molecular biology of seed metabolism.

16-3 SYNOPSIS

It is realized that many important aspects of seed metabolism concerning seed development, primary seed constituents, and other biological processes may have been overlooked in this chapter. Such information was not discounted intentionally. Hopefully, this chapter has adequately sampled the vast amount of information that has been compiled since 1973. By no means, however, has our understanding of the metabolic processes in the soybean seed plateaued. As more information is gathered and implemented in the benefit of our Society, more innovative research accomplishments will be expected. The current status of soybean as a major world food resource may depend upon the degree to which these research objectives are pursued and realized.

REFERENCES

Adams, C.A., T.H. Broman, S.W. Norby, and R.W. Rinne. 1981a. Occurrence of multiple forms of α-amylase and absence of starch phosphorylase in soybean seeds. Ann. Bot. 48:895–903.

----, ----, and R.W. Rinne. 1981b. Starch metabolism in developing and germinating soya beans is independent of β-amylase activity. Ann. Bot. 48:433–439.

----, M.C. Fjerstad, and R.W. Rinne. 1983a. Characteristics of soybean seed maturation: Necessity for slow dehydration. Crop Sci. 23:265–267.

----, S.W. Norby, and R.W. Rinne. 1983b. Ontogeny of lipid bodies in developing soybean seeds. Crop Sci. 23:757–759.

Amuti, K.S., and C.J. Pollard. 1977. Soluble carbohydrates of dry and developing seeds. Phytochemistry 16:529–532.

Andrawis, A., A. Pinsky, and S. Grossman. 1982. Isolation of soybean lipoxygenase-2 by affinity chromatography. Phytochemistry 21:1523–1525.

Ashton, F.M. 1976. Mobilization of storage proteins of seeds. Ann. Rev. Plant Physiol. 27:95–117.

Barr, R., and C.J. Arntzen. 1969. The occurrence of δ-tocopheryl quinone in higher plants and its relation to senescence. Plant Physiol. 44:591–598.

Barton, K.A., J.F. Thompson, J.T. Madison, R. Rosenthal, N.P. Jarvis, and R.N. Beachy. 1982. The biosynthesis and processing of high molecular weight precursors of soybean glycinin subunits. J. Biol. Chem. 257:6089–6095.

Beachy, R.N., J.F. Thompson, and J.T. Madison. 1978. Isolation of polyribosomes and messenger RNA active in *in vitro* synthesis of soybean seed proteins. Plant Physiol. 61:139–144.

----, J. Bryant, J.J. Doyle, K. Kitamura, and B.F. Ladin. 1983a. Molecular characterization of a soybean variety lacking a subunit of the 7S seed storage protein. p. 413–422. *In* Plant molecular biology. Alan R. Liss, New York.

----, J.J. Doyle, B.F. Ladin, and M.A. Schuler. 1983b. Structure and expression of genes encoding the soybean 7S seed storage proteins. NATO Adv. Sci. Inst. Ser. 63:101–112.

Beever, J.S., and R.L. Cooper. 1982. Dry matter accumulation patterns and seed yield components of two indeterminate soybean cultivars. Agron. J. 74:380–383.

Beevers, H. 1982. Fat metabolism in seeds: Role of organelles. p. 223–235. *In* J.L.G.M. Wintermans and P.J.C. Kuiper (ed.) Biochemistry and metabolism of plant lipids. Elsevier Biomedical Press, Amsterdam.

Bernard, R.L. 1956. Agronomic evaluation of soybean plant introductions: Group O Maturity. USDA-RSLM 186. University of Illinois, Urbana.

----. 1957. Agronomic evaluation of soybean plant introductions: Group IV Maturity. USDA-RSLM 193. University of Illinois, Urbana.

----. 1966. Evaluation of maturity groups I and II of the USDA Soybean Collection. USDA-RSLM 230. University of Illinois, Urbana.

----. 1970. Evaluation of recent additions to maturity groups 00 to IV of the USDA Soybean Collection. USDA-RSLM 242. University of Illinois, Urbana.

Bils, R.F., and R.W. Howell. 1963. Biochemical and cytological changes in developing soybean cotyledons. Crop Sci. 3:304–308.

Bond, H.M., and D.J. Bowles. 1983. Characterization of soybean endopeptidase activity using exogenous and endogenous substrates. Plant Physiol. 72:345–350.

Boulter, D. 1976. Biochemistry of protein synthesis in seeds. p. 231–250. *In* Proc. Workshop Genetic Importance of Seed Proteins, Washington, DC. 18–20 Mar. 1974. National Academy of Science (USA), Washington, DC.

----. 1980. Ontogeny and development of biochemical and nutritional attributes in legume seeds. p. 127–134. *In* R.J. Summerfield and A.H. Bunting (ed.) Advanced legume science Proc. International legume conference. Royal Botany Garden Publishing Kew, UK.

----. 1981. Biochemistry of storage protein synthesis and deposition in the developing legume seed. p. 1–31. *In* Advances Botanical Research. Academic Press, New York.

Bricker, T.M., and D.W. Newman. 1980. Quantitative changes in the chloroplast thylakoid polypeptide complement during senescence. Z. Pflanzenphysiol. 98:339–346.

Brim, C.A., and J.W. Burton. 1979. Recurrent selection in soybeans. II. Selection for increased percent protein in seeds. Crop Sci. 19:494–498.

----, W.M. Schutz, and F.I. Collins, 1968. Maternal effect on fatty acid composition and oil content of soybeans, *Glycine max* (L.). Merr. Crop Sci. 8:517–518.

Browse, J., and C.R. Slack. 1983. The effects of temperature and oxygen on the rates of fatty acid synthesis and oleate desaturation in safflower (*Carthamis tinctorius*) seed. Biochim. Biophys. Acta 753:145–152.

Burke, J.J., W. Kalt-Torres, J.R. Swafford, J.W. Burton, and R.F. Wilson. 1984. Studies on genetic male-sterile soybeans. III. The initiation of monocarpic senescence. Plant Physiol. 75:1058–1063.

Burton, J.W., and C.A. Brim. 1981. Recurrent selection in soybeans. III. Selection for increased percent oil in seeds. Crop Sci. 21:31–34.

----, and ----. 1983. Registration of a soybean germplasm population. Crop Sci. 23:191.

----, A.E. Purcell, and W.M. Walter, Jr., 1982. Methionine concentration in soybean protein from populations selected for increased percent protein. Crop Sci. 22:430–432.

----, R.F. Wilson, and C.A. Brim. 1983. Recurrent selection in soybeans. IV. Selection for increased oleic acid percentage in seed oil. Crop Sci. 23:744–747.

Carpenter, D.L., J. Lehmann, B.S. Mason, and H.T. Slover. 1976. Lipid composition of selected vegetable oils. J. Am. Oil Chem. Soc. 53:713–718.

Catsimpoolas, N., T.G. Campbell, and E.W. Meyer. 1968. Immunochemical study of changes in reserve proteins of germinating soybean seeds. Plant Physiol. 43:799–805.

Chang, S.S., G-H. Shen, J. Tang, Q.Z. Jin, H. Shi, J.T. Carlin, and C.T. Ho. 1983. Isolation and identification of 2-pentenyl-furans in the reversion flavor of soybean oil. J. Am. Oil Chem. Soc. 60:553–557.

Chapman, G.W., Jr., J.A. Robertson, D. Burdick, and M.B. Parker. 1976. Chemical composition and lipoxygenase activity in soybeans as affected by genotype and environment. J. Am. Oil Chem. Soc. 53:54–56.

Cheesbrough, T.M., and B. Axelrod. 1983. Determination of the spin state of iron in native and activated soybean lipoxygenase-1 by paramagnetic susceptibility. Biochemistry 22:3837–3840.

Chrispeels, M.J., T.J.V. Higgins, S. Craig, and D. Spencer. 1982. Role of the endoplasmic reticulum in the synthesis of reserve proteins and the kinetics of their transport to protein bodies in developing pea cotyledons. J. Cell Biol. 93:5–14.

Ciferri, O., D. Tiboni, M.L. Munoz-Calvo, and G. Camerino. 1976. Protein synthesis in plants: Specificity and role of the cytoplasmic and organellar systems. NATO Adv. Study Inst. Ser. Life Sci. 12:155–166.

Collins, F.I., and R.W. Howell. 1957. Variability of linolenic and linoleic acids in soybean oil. J. Am. Oil Chem. Soc. 34:491–493.

Cosgrove, D.J. 1966. Chemistry and biochemistry of inositol polyphosphates. Rev. Pure Appl. Chem. 16:209–224.

Croy, R.R.D., G.W. Lycett, J.A. Gatehouse, J.N. Yarwood, and D. Boulter. 1982. Cloning and analysis of cDNA's encoding plant storage protein precursors. Nature (London) 295:76–79.

Derbyshire, E., D.J. Wright, and D. Boulter. 1976. Legumin and vicilin storage proteins of legume seeds. Phytochemistry 15:3–24.

Dubacq, J-P, D. Drapier, and A. Tremolieres. 1983. Polyunsaturated fatty acid synthesis by a mixture of chloroplasts and microsomes from spinach leaves: Evidence for two distinct pathways of the biosynthesis of trienoic acids. Plant Cell Physiol. 24:1–9.

Dutton, H.J., and T.L. Mounts. 1966. Desaturation of fatty acids in seeds of higher plants. J. Lipid Res. 7:221–225.

Dykes, C.W., J. Kay, and J.L. Harwood. 1976. Incorporation of choline and ethanolamine into phospholipids in germinating soyabean. Biochem. J. 158:575–581.

Egli, D.B., J. Fraser, J.E. Leggett, and C.G. Poneleit. 1981. Control of seed growth in soyabeans. (*Glycine max* (L.) Merr.) Ann. Bot. 48:171–176.

----, and J.E. Leggett. 1973. Dry matter accumulation patterns in determinate and indeterminate soybeans. Crop Sci. 13:220–222.

----, and ----. 1976. Rate of dry matter accumulation in soybean seeds with varying source-sink ratios. Agron. J. 68:371–374.

----, ----, and J.M. Wood. 1978. Influence of soybean seed size and position on the rate and duration of filling. Agron. J. 7:127–130.

Eggum, B.O., and R.M. Beames. 1983. The nutritive value of seed proteins. p. 499–531. *In* W. Gottschalk and H.P. Muller (ed.) Seed proteins: Biochemistry, genetics, nutritive value. Nijhoff Junk, The Hague.

Eskola, J., and S. Laakso. 1983. Bile salt-dependent oxygenation of polyunsaturated phosphatidylcholines by soybean lipoxygenase-1. Biochim. Biophys. Acta 751:305–311.

Ezzat, K.S., and R.S. Pearce. 1980. Fatty acids of lipids from cultured soybean and rape cells. Phytochemistry 19:1375–1378.

Fatemi, S.H., and E.G. Hammond. 1977a. Glyceride structure variation in soybean varieties: I. Stereospecific analysis. Lipids 12:1032–1036.

----, and ----. 1977b. Glyceride structure variation in soybean varieties: II. Silver ion chromatographic analysis. Lipids 12:1037–1042.

Frey-Wyssling, A., E. Grieshaber, and K. Muhlethaler. 1963. Origin of sphereosomes in plant cells. J. Ultrastruct. Res. 8:506–516.

Galliard, T., and H. W-S. Chan. 1980. Lipoxygenases. p. 132–162. *In* P.K. Stumpf and E.E. Conn (ed.) The biochemistry of plants, Vol. 4. Academic Press, New York.

Gay, S., D.B. Egli, and D.A. Reicosky. 1980. Physiological aspects of yield improvement in soybeans. Agron. J. 72:387–391.

Gent, M.P.N. 1983. Rate of increase in size and dry weight of individual pods of field grown soya bean plants. Ann. Bot. 51:317–329.

Graf, E. 1983. Applications of phytic acid. J. Am. Oil Chem. Soc. 60:1861–1867.

Graham, D., and B.D. Patterson. 1982. Response of plants to low, non-freezing temperatures: Proteins, metabolism, and acclimation. Ann. Rev. Plant Physiol. 33:347–372.

Gupta, K., and K.S. Dhindsa. 1982. Fatty acid composition of oil of different varieties of soybean. J. Food Sci. Technol (Tokyo) 19:248–250.

Gurr, M.I. 1980. The biosynthesis of triacylglycerols. p. 205–249. *In* P.K. Stumpf and E.E. Conn (ed.) The biochemistry of plants, Vol. 4. Academic Press, New York.

----, J. Blades, R.S. Appleby, C.G. Smith, M.P. Robinson, and B.W. Nichols. 1974. Studies on seed oil triglycerides: Triacylglyceride biosynthesis and storage in whole seeds and oil bodies of *Crambe* abyssinica. Eur. J. Biochem. 43:281–290.

Hammond, E.G., and W.R. Fehr. 1975. Oil quality improvement in soybeans, *Glycine max* (L.) Merr. Fette Seifen. Anstrichm. 77:97–101.

----, and ----. 1983a. Registration of A5 germplasm line of soybean. Crop Sci. 23:192.

----, and ----. 1983b. Registration of A6 germplasm line of soybean. Crop Sci. 23:192–193.

----, ----, and H.E. Snyder. 1972. Improving soybean quality by plant breeding. J. Am. Oil Chem. Soc. 49:33–35.

Handley, L.W., D.M. Pharr, and R.F. McFeeters. 1983. Relationship between galactinol synthase activity and sugar composition of leaves and seeds of several crop species. J. Am. Soc. Hortic. Sci. 108:600–605.

Harada, K., Y. Toyokawa, and K. Kitamura. 1983. Genetic analysis of the most acidic 11S globulin subunit and related characters in soybean seeds. Jpn. J. Breed. 33:23–30.

Harris, P., and A.T. James. 1969. Effect of low temperature on fatty acid biosynthesis in seeds. Biochim. Biophys. Acta 187:13–18.

Hartwig, E.E., and C.J. Edwards, Jr. 1975. Evaluation of soybean germplasm. U.S. Department of Agriculture, Stoneville, MS.

----, and ----. 1980. Evaluation of soybean germplasm. U.S. Department of Agriculture, Stoneville, MS.

Harwood, J.L. 1975. Lipid synthesis by germinating soyabean. Phytochemistry 14:1985–1990.

----, 1976. Synthesis of molecular species of phosphatidycholine and phosphatidylethanolamine by germinating soyabean. Phytochemistry 15:1459–1463.

----, A. Sodja, P.K. Stumpf, and A.R. Stumpf. 1971. On the origin of oil droplets in maturing castor bean seeds, *Ricinus communis*. Lipids 6:851–854.

Hildebrand, D.F., and T. Hymowitz. 1981. Two soybean genotypes lacking lipoxygenase-1. J. Am. Oil Chem. Soc. 58:583–586.

----, and ----. 1982. Inheritance of lipoxygenase-1 activity in soybean seeds. Crop Sci. 22:851–853.

----, and ----. 1983. Lipoxygenase activities in developing and germinating soybean seeds with and without lipoxygenase-1. Bot. Gaz. 144:212–216.

Hill, J.E., and R.W. Breidenbach. 1974a. Proteins of soybean seeds. I. Isolation and characterization of the major components. Plant Physiol. 53:742–746.

----, and ----. 1974b. Proteins of soybean seeds. II. Accumulation of the major protein components during seed development and maturation. Plant Physiol. 53:747–751.

Hirayama, O., and K. Hujii. 1965. Glyceride structure and biosynthesis of natural fats. III. Biosynthetic process of triglycerides in maturing soybean seed. Agric. Biol. Chem. 29:1–6.

Holman, R.T. 1978. How essential are essential fatty acids? J. Am. Oil Chem. Soc. 55:744A–781A.

Horisberger, M., and M. Tacchini-Vonlanthen. 1983. Ultrastructural localization of Bowman-Birk inhibitor on thin sections of *Glycine max* (L.) Merr. cv. Maple Arrow by the gold method. Histochemistry 77:313–321.

Howell, R.W., C.A. Brim, and R.W. Rinne. 1972. The plant geneticist's contribution toward changing lipid and amino acid composition of soybeans. J. Am. Oil Chem. Soc. 49:30–32.

----, and J.L. Cartter. 1953. Physiological factors affecting composition of soybeans. I. Correlation of temperature during certain portions of the pod-filling stage with oil percentage in mature seed. Agron. J. 45:526–528.

----, and F.I. Collins. 1957. Factors affecting linolenic and linoleic acid content of soybean oil. Agron. J. 49:593–597.

Hsu, F.C., and R.L. Obendorf. 1982. Compositional analysis of *in vitro* matured soybean seeds. Plant Sci. Lett. 27:129–135.

Hsu, S.H., H.H. Hadley, and T. Hymowitz. 1973. Changes in carbohydrate contents of germinating soybean seeds. Crop Sci. 13:407–410.

Huber, D.J., and D.W. Newman. 1976. Relationships between lipid changes and plastid ultrastructural changes in senescing and regreening soybean cotyledons. J. Exp. Bot. 27:490–511.

Hudak, J. 1981. Plastid senescence. I. Changes of chloroplast structure during natural senescence in cotyledons of *Sinapis alba* (L.). Photosynthetica 15:174–178.

Huffaker, R.C., and L.W. Peterson. 1974. Protein turnover in plants and possible means of its regulation. Ann. Rev. Plant Physiol. 25:363–392.

Hughes, S.A., and P.A. Murphy. 1983. Varietal influence on the quality of glycinin in soybeans. J. Agric. Food Chem. 31:376–379.

Hymowitz, T., F.I. Collins, J. Panczner, and W.M. Walker. 1972a. Relationship between the content of oil, protein, and sugar in soybean seed. Agron. J. 64:613–616.

----, W.M. Walker. F.I. Collins, and J. Panczner. 1972b. Stability of sugar content in soybean strains. Commun. Soil Sci. Plant Anal. 3:367–373.

Ichihara, K., and M. Noda. 1981. Triacylglycerol synthesis by subcellular fractions of maturing safflower seeds. Phytochemistry 20:1245–1249.

----, and ----. 1982. Some properties of diacylglycerol acyltransferase in a particulate fraction from maturing safflower seeds. Phytochemistry 21:1895–1901.

International Union of Pure and Applied Chemistry—International Union of Biochemistry. Commission on Biochemical Nomenclature. 1977. The nomenclature of lipids. Lipids 12:455–468.

Julienne, M., C. Vergnolle, and J-C. Kader. 1981. Activity of phosphatidylcholine—transfer protein from spinach (*Spinacia oleracea*) leaves with mitochondria and chloroplasts. Biochem. J. 197:763–766.

Kahn, V., R.W. Howell, and J.B. Hanson. 1960. Fat metabolism in germinating soybeans. I. Physiology of native fat. Plant Physiol. 35:854–860.

Kandler, O., and H. Hopf. 1980. Occurrence, metabolism, and function of oligosaccharides. p. 221–270. *In* P.K. Stumpf and E.E. Conn (ed.) The biochemistry of plants, Vol. 3. Academic Press, New York.

Kenworthy, W.J., and C.A. Brim. 1979. Recurrent selection in soybeans. I. Seed yield. Crop Sci. 19:315–318.

Kitamura, K., C.S. Davies, N. Kaizuma, and N.C. Nielsen. 1983. Genetic analysis of a nullallele for lipoxygenase-3 in soybean seeds. Crop Sci. 23:924–927.

Kollman, G.E., J.G. Streeter, D.L. Jeffers, and R.B. Curry. 1974. Accumulation and distribution of mineral nutrients, carbohydrate, and dry matter in soybean plants as influenced by reproductive sink size. Agron. J. 66:549–554.

Konno, S. 1979. Changes in chemical composition of soybean seeds during ripening. Jpn. Agric. Res. Q. 13:186–194.

Koshiyama, I. 1983. Storage proteins of soybean. p. 427–450. *In* W. Gottschalk and H.P. Muller (ed.) Seed proteins: Biochemistry, genetics, nutritive value. Nijhoff/Junk, The Hague.

Kritsman, M.G., A.S. Konikova, and R.N. Korotkina. 1975. Effect of inhibitors or proteolysis on incorporation of labeled amino acids into isolated proteins. Biokhimiya 40:1131–1134.

Krul, W.R. 1974. Nucleic acid and protein metabolism of senescing and regenerating soybean cotyledons. Plant Physiol. 54:36–40.

Leshem, Y.Y., and G. Barness. 1982. Lipoxygenase as effected by free radical metabolism: Senescence retardation by the xanthine oxidase inhibitor allopurinol. p. 275–278. *In* J.F.G.M. Wintermans and P.J.C. Kuiper (ed.) Biochemistry and metabolism of plant lipids. Elsevier Biomedical Press, Amsterdam.

----, J. Wurzburger, S. Grossman, and A.A. Frimer. 1981. Cytokinin interaction with free radical metabolism and senescence: Effects on endogenous lipoxygenase and purine oxidation. Physiol. Plant. 53:9–12.

Lin, Y-H., R.A. Moreau, and A.H.C. Huang. 1982. Involvement of glyoxysomal lipase in the hydrolysis of storage triacylglycerols in the cotyledons of soybean seedlings. Plant Physiol. 70:108–112.

----, L.T. Wimer, and A.H.C. Huang. 1983. Lipase in the lipid bodies of corn scutella during seedling growth. Plant Physiol. 73:460–463.

Lutton, E.S. 1972. Lipid structures. J. Am. Oil Chem. Soc. 49:1–9.

Mandal, N.C., and B.B. Biswas. 1970. Metabolism of inositol phosphates. I. Phytase synthesis during germination in cotyledons of mung beans, *Phaseolus aureus*. Plant Physiol. 45:4–7.

Marai, L., J.J. Myher, and A. Kuksis. 1983. Analysis of triacylglycerols by reversed-phase high pressure liquid chromatography with direct liquid inlet mass spectroscopy. Can. J. Biochem. Cell. Biol. 61:840–849.

Martin, B.A., B.F. Carver, J.W. Burton, and R.F. Wilson. 1983. Inheritance of fatty acid composition in soybean seed oil. Soybean Genet. Newsl. 10:89–92.

----, and R.F. Wilson. 1983. Properties of diacylglycerol acyltransferase from spinach leaves. Lipids 18:1–6.

----, and ----. 1984. Subcellular localization of triacylglycerol synthesis in spinach leaves. Lipids 19:117–121.

Mazliak, P. 1979. Temperature regulation of plant fatty acyl desaturases. p. 391–409. *In* J.M. Lyons et al., (ed.) Low temperature stress in crop plants. Academic Press, New York.

Mei-Guey, L., D. Tyrell, R. Bassette, and G.R Reeck. 1983. Two-dimensional electrophoretic analysis of soybean proteins. J. Agric. Food Chem. 31:963–968.

Meinke, D.W., J. Chen, and R.N. Beachy. 1981. Expression of storage-protein genes during soybean seed development. Planta 153:130–139.

Mikami, B., and Y. Morita. 1983. Location of SH-groups along the polypeptide chain of soybean β-amylase J. Biochem. 93:777–786.

Moll, C., U. Biermann, and W. Grosch. 1979. Occurrence and formation of bitter-tasting trihydroxy-fatty acids in soybeans. J. Agric. Food Chem. 27:239–243.

Mudd, J.B. 1980. Phospholipid biosynthesis. p. 250–282. *In* P.K. Stumpf and E.E. Conn (ed.) The biochemistry of plants, Vol. 4. Academic Press, New York.

Muller, H.P. 1983. The genetic control of seed protein production in legumes. p. 309–354. *In* W. Gottschalk and H.P. Muller (ed.) Seed proteins: Biochemistry, genetics, nutritive value. Nijhoff/Junk, The Hague.

Negishi, T. 1976. Incorporation of [1-^{14}C]acetate into phospholipids by soybean seedlings. J. Am. Oil Chem. Soc. 53:77–79.

Nishihara, M., and M. Kito. 1978. Changes in the phospholipid molecular species composition of soybean hypocotyl and cotyledon after dedifferentiation. Biochim. Biophys. Acta. 531:25–31.

Nissly, C.R., R.L. Bernard, and C.N. Hittle. 1981. Inheritance of two chlorophyll-deficient mutants in soybeans. J. Hered. 72:141–142.

Nothelfer, H.G., R.H. Barackhaus, and F. Spener. 1977. Localization and characterization of the fatty acid synthesizing system in cells of *Glycine max* (L.) Merr. suspension cultures. Biochim. Biophys. Acta 489:370–380.

----, and F. Spener. 1979. Stimulation of fatty acid biosynthesis in *Glycine max* suspension cultures by acyl carrier protein. Plant Sci. Lett. 16:361–365.

Odani, S., and T. Ikenaka. 1978a. Studies on soybean trypsin inhibitors. XIII. Preparation and characterization of active fragments from Bowman-Birk proteinase inhibitor. J. Biochem. 83:747–753.

Odani, S., and T. Ikenaka. 1978b. Studies on soybean trypsin inhibitors. XIV. Change of the inhibitory activity of Bowman-Birk inhibitor upon replacements of the α-chymotrypsin reactive site serine residue by other amino acids. J. Biochem. 84:1–9.

Ohmura, T., and R.W. Howell. 1962. Respiration of developing and germinating soybean seeds. Physiol. Plant. 15:341–350.

Openshaw, S.J., and H.H. Hadley. 1978. Maternal effects on sugar content in soybean seeds. Crop Sci. 18:581–584.

----, and ----. 1981. Selection to modify sugar content of soybean seeds. Crop Sci. 21:805–808.

----, ----, and C.E. Brokoski. 1979. Effects of pod removal upon seeds of nodulating and nonnodulating soybean lines. Crop Sci. 19:289–290.

Orf, J.H., and T. Hymowitz. 1979. Genetics of the Kunitz trypsin inhibitor: An antinutritional factor in soybeans. J. Am. Oil Chem. Soc. 56:722–726.

Osbourne, T.B., and G.F. Campbell. 1898. Proteids of the soybean (*Glycine lipida*). J. Am. Chem. Soc. 20:419–428.

Palmer, J.D., G.P. Singh, and D.T.N. Pillay. 1983. Structure and sequence evolution of three legume chloroplast DNAs. Mol. Gen. Genet. 190:13–19.

Palmer, R.G., M.A. Sheridan, and M.A. Tabatabai. 1979. Effects of genotype, temperature, and illuminance on chloroplast ultrastructure of a chlorophyll mutant in soybeans. Cytologia 44:881–891.

Panter, R.A., and N.K. Boardman. 1973. Lipid biosynthesis by isolated plastids from greening pea, *Pisum sativum*. J. Lipid Res. 14:664–671.

Pavlinova, O.A., and M.V. Turkina. 1978. Biosynthesis and the physiological role of sucrose in the plant. Sov. Plant Physiol. 25:815–828.

Peterman, T.K., and J.N. Siedow. 1983. Structural features required for inhibition of soybean lipoxygenase-2 by propyl gallate. Plant Physiol. 71:55–58.

Phillips, D.V., D.E. Dougherty, and A.E. Smith. 1982. Cyclitols in soybean. J. Agric. Food Chem. 30:456–458.

Pillay, D.T.N. 1977. Protein synthesis in aging soybean cotyledons loss in translational capacity. Biochem. Biophys. Res. Commun. 79:796–804.

Place, M.A., M.S. Morgan, A. Rutkoski, D.W. Newman, and J.G. Jaworski. 1979. Fatty acid metabolism in senescing and regreening soybean cotyledons. Planta 147:246–250.

Porra, R.J. 1979. Formation of phospholipids from *sn*-glycerol-3-phosphate and free fatty acids or their derivatives by homogenates of soybean cotyledons. Phytochemistry 18:1651–1656.

----, and P.K. Stumpf. 1976. Lipid biosynthesis in developing and germinating soybean cotyledons. Arch. Biochem. Biophys. 176:53–62.

Prattley, C.A., and D.W. Stanley. 1982. Protein-phytate interactions in soybeans. I. Localization of phytate in protein bodies and globoids. J. Food Biochem. 6:243–253.

----, ---- and F.R. Van De Voort. 1982a. Protein-phytate interactions in soybeans. II. Mechanisms of protein-phytate binding as affected by calcium. J. Food Biochem. 6:255–271.

----, ----, T.K. Smith, and F.R. Van De Voort. 1982b. Protein-phytate interactions in soybeans. III. The effect of protein-phytate complexes on zinc bioavailability. J. Food Biochem. 6:273–282.

Preiss, J., and C. Levi. 1980. Starch biosynthesis and degradation. p. 371–423. *In* P.K. Stumpf and E.E. Conn (ed.) The biochemistry of plants, Vol. 3. Academic Press, New York.

Pryde, E.H. 1980. Composition of soy oil. p. 13–31. *In* D.R. Erikson et al. (ed.) Handbook of soybean oil processing and utilization. American Soybean Association and American Oil Chemist's Society, St. Louis.

Reicosky, D.A., J.H. Orf, and C. Poneleit. 1982. Soybean germplasm evaluation for length of the seed filling period. Crop Sci. 22:319–322.

Rinne, R.W., and D.T. Canvin. 1971. Fatty acid biosynthesis from acetate and CO_2 in the developing soybean cotyledon. Plant Cell Physiol. 12:387–393.

Roberts, R.M., J. Deshusses, and F. Loewus. 1968. Inositol metabolism in plants. V. Conversion of myo-inositol to uronic acid and pentose units of acidic polysaccharides in root tips of *Zea mays.* Plant Physiol. 43:979–989.

----, and F. Loewus. 1968. Inositol metabolism in plants. VI. Conversion of myo-inositol to phytic acid in *Wolffiella floridana.* Plant Physiol. 43:1710–1716.

Robinson, M.L., and G.M. Carman. 1982. Solubilization of microsomal-associated phosphatidylinositol synthetase from germinating soybeans. Plant Physiol. 69:146–149.

Roehm, J.N., and O.S. Privett. 1970. Changes in the structure of soybean triglycerides during maturation. Lipids 5:353–358.

Roza, M., and A. Francke. 1973. Product specificity of soyabean lipoxygenases. Biochim. Biophys. Acta. 316:76–82.

Sato, K., and T. Ikeda. 1979. The growth responses of soybean to photoperiod and temperature. IV. The effect of temperature during the ripening period on the yield and characters of seeds. Jpn. J. Crop Sci. 48:283–290.

----, ----, and S. Minagawa. 1983. Changes in the fine structure of soybean cotyledons during seed ripening, germination, and emergence. Jpn. J. Crop Sci. 52:65–72.

Schoffl, F., and J.L. Key. 1982. An analysis of mRNAs for a group of heat shock proteins of soybean using cloned cDNAs. J. Mol. Appl Genet. 1:301–314.

Schweizer, T.F., I. Horman, and P. Wursch. 1978. Low molecular weight carbohydrates from leguminous seeds; a new disaccharide: galactopinitol. J. Sci. Food Agric. 29:148–154.

Sessa, D.J. 1979. Biochemical aspects of lipid-derived flavors in legumes. J. Agric. Food Chem. 27:234–239.

Simmons, R.O., and F.W. Quackenbush. 1954. Comparative rates of formation of fatty acids in the soybean seed during its development. J. Am. Oil Chem. Soc. 31:601–603.

Sinclair, D., and D.T. Pillay. 1981. Localization of tRNAs and aminoacyl-tRNA synthesis in cytoplasm, chloroplasts, and mitochondria of *Glycine max* (L.) Z. Pflanzenphysiol. 104:299–301.

Singh, B.B., and H.H. Hadley. 1968. Maternal control of oil synthesis in soybeans, *Glycine max* (L.) Merr. Crop Sci. 8:622–625.

----, ----, and F.I. Collins. 1968. Distribution of fatty acids in germinating soybean seed. Crop Sci. 8:171–175.

Skadsen, R.W., and J.H. Cherry. 1983. Quantitative changes in *In vitro* and *In vivo* protein synthesis in aging and rejuvenated soybean cotyledons. Plant Physiol. 71:861–868.

Slack, C.R., and P.G. Roughan. 1978. Rapid temperature-induced changes in fatty acid composition of certain lipids in developing linseed and soybean cotyledons. Biochem. J. 170:437–439.

Smith, H. 1982. Nucleic acid and protein synthesis during germination. p. 337–361. *In* H. Smith and D. Grierson (ed.) The molecular biology of plant development, Vol. 18. Botanical Monograph. University of California Press, Berkeley, CA.

----, and D. Grierson. 1982. Seed maturation and deposition of storage proteins. p. 306–336. *In* H. Smith and D. Grierson (ed.) The molecular biology of plant development, Vol. 18. Botanical Monograph. University of California Press, Berkeley.

Spielmann, A., W. Ortiz, and E. Stutz. 1983. The soybean chloroplast genome: Construction of a circular restriction site map and location of DNA regions encoding the genes for rRNAs, the large subunit of ribulose-1,5-bisphosphate carboxylase, and the 32 kD protein of the photosystem II reaction center. Mol. Gen. Genet. 190:5–12.

Stahlhut, R.W., and T. Hymowitz. 1983. Variation in the low molecular weight proteinase inhibitors of soybeans. Crop Sci. 23:766–769.

Staswick, P.E., P. Brove, and N.C. Nielsen. 1983. Glycinin composition of several perennial species related to soybean. Plant Physiol. 72:1114–1118.

----, and N.C. Nielsen. 1983. Characterization of a soybean cultivar lacking certain glycinin subunits. Arch. Biochem. Biophys. 223:1–8.

Stearns, E.M., and W.T. Morton. 1973. Incorporation of acetate into fatty acids and complex lipids of soybean cotyledon slices under aerobic and anaerobic conditions. Lipids 8:668–674.

Streckert, G., and H-J. Stan. 1975. Conversion of linoleic acid hydroperoxide by soybean lipoxygenase in the presence of guaiacol: Identification of the reaction products. Lipids 10:847–854.

Streeter, J.G., and D.L. Jeffers. 1979. Distribution of total nonstructural carbohydrates in soybean plants having increased reproductive load. Crop Sci. 19:729–734.

Stumpf, P.K. 1977. Lipid biosynthesis in developing seeds. p. 75–84. *In* M. Tevini and H.K. Lichtenthaler (ed.) Lipids and lipid polymers in higher plants. Springer-Verlag New York, New York.

----. 1980. Biosynthesis of saturated and unsaturated fatty acids. p. 177–204. *In* P.K. Stumpf and E.E. Conn (ed.) The biochemistry of plants, Vol. 4. Academic Press, New York.

----. 1981. Plants, fatty acids, compartments. Trends Biochem. Sci. 6:173–176.

Stymne, S., and L-A. Appelqvist. 1980. The biosynthesis of linoleate and α-linolenate in homogenates from developing soybean cotyledons. Plant Sci. Lett. 17:287–294.

----, and G. Glad. 1981. Acyl exchange between oleoyl-CoA and phosphatidylcholine in microsomes of developing soyabean cotyledons and its role in fatty acid desaturation. Lipids 16:298–305.

Thorne, J.H. 1980. Kinetics of ^{14}C-photosynthate uptake by developing soybean fruit. Plant Physiol. 65:975–979.

----. 1981. Morphology and ultrastructure of maternal seed tissues of soybean in relation to the import of photosynthate. Plant Physiol. 67:1016–1025.

Tinoco, J., R. Babcock, I. Hincenbergs, B. Medwadowski, P. Miljanich, and M.A. Williams. 1979. Linolenic acid deficiency. Lipids 14:166–173.

Verhagen, J., G.A. Veldink, M.R. Egmond, J.F.G. Vliegenthart, J. Boldingh, and J. Vander Star. 1978. Steady-state kinetics of the anaerobic reaction of soybean lipoxygenase-1 with linoleic acid and 13-L-hydroperoxylinoleic acid. Biochim. Biophys. Acta 529:369–379.

Vernooy-Gerritsen, M., A.L.M. Bos, G.A. Veldink, and J.F.G. Vliegenthart. 1983. Localization of lipoxygenases-1 and 2 in germinating soybean seeds by an indirect immunofluorescence technique. Plant Physiol. 73:262–267.

Vick, B.A., and D.C. Zimmerman. 1979a. Substrate specificity for the synthesis of cyclic fatty acids by a flaxseed extract. Plant Physiol. 63:490–494.

----, and ----. 1979b. Distribution of a fatty acid cyclase enzyme system in plants. Plant Physiol. 64:203–205.

Vliegenthart, J.F.G., G.A. Veldink, and J. Boldingh. 1979. Recent progress in the study on the mechanism of action of soybean lipoxygenase. J. Agric. Food Chem. 27:623–626.

Weissbach, H., and S. Pestka. 1977. Molecular mechanisms of protein biosynthesis. Academic Press, New York.

Wetterau, J.R., D.W. Newman, and J.G. Jaworski. 1978. Quantitative changes of fatty acids in soybean cotyledons during senescence and regreening. Phytochemistry 17:1265–1268.

White, H.B., Jr., F.W. Quackenbush, and A.H. Probst. 1961. Occurrence and inheritance of linolenic and linoleic acids in soybean seed. J. Am. Oil Chem. Soc. 38:113–117.

Wilcox, J.R., J.F. Cavins, and N.C. Nielsen. 1984. Genetic alteration of soybean oil composition by a chemical mutagen. J. Am. Oil Chem. Soc. 61:97–100.

Wilson, R.F. 1981. Aspects of glycerolipid metabolism in developing soybean cotyledons. Crop Sci. 21:519–524.

----. 1984. Effect of genetic selection upon lipid metabolism in soybeans. p. 77–88. *In* C. Ratledge et al. (ed.) Biotechnology of fats and health. Monograph 2. American Oil Chemist's Society, Champaign, IL.

----, J.W. Burton, and C.A. Brim. 1981. Progress in the selection for altered fatty acid composition in soybeans. Crop Sci. 21:788–791.

----, ----, J.A. Buck, and C.A. Brim. 1978. Studies on genetic male-sterile soybeans. I. Distribution of plant carbohydrate and nitrogen during development. Plant Physiol. 61:838–841.

----, R.W. Rinne, and C.A. Brim. 1976. Alteration of soybean oil composition by plant breeding. J. Am. Oil Chem. Soc. 53:595–597.

----, H.H. Weissinger, J.A. Buck, and G.D. Faulkner. 1980. Involvement of phospholipids in polyunsaturated fatty acid synthesis in developing soybean cotyledons. Plant Physiol. 66:545–549.

Wolf, R.B., J.F. Cavins, R. Kleiman, and L.T. Black. 1982. Effect of temperature on soybean seed constituents: Oil, protein, moisture, fatty acids, amino acids, and sugars. J. Am. Oil Chem. Soc. 59:230–232.

Yabuuchi, S., R.M. Lister, B. Axelrod, J.R. Wilcox, and N.C. Nielsen. 1982. Enzyme-linked immunosorbent assay for the determination of lipoxygenase isozymes in soybean. Crop Sci. 22:333–337.

Yamada, M., T. Tanaka, and J-I. Ohnishi. 1980. Phospholipid exchange protein from higher plants. p. 161–168. *In* P. Mazliak et al. (ed.) Biogenesis and function of plant lipids. Elsevier Biomedical Press, Amsterdam.

Yazdi-Samadi, B., R.W. Rinne, and R.D. Seif. 1977. Components of developing soybean seeds: Oil, protein, sugars, starch, organic acids, and amino acids. Agron. J. 69:481–486.

Yoshida, H., and G. Kajimoto. 1977. Changes in lipid component and fatty acid composition of axis and root of germinating soybeans. Agric. Biol. Chem. 41:2431–2436.

----, and ----. 1978. Fatty acid distribution in glycolipids and phospholipids in cotyledons of germinating soybeans. Agric. Biol. Chem. 42:1323–1330.

----, and ----. 1981. Changes in composition of non-polar lipids of the axis and root of germinating soybeans. Agric. Biol. Chem. 45:1187–1199.

Zeiher, C., D.B. Egli, J.E. Leggett, and D.A. Reicosky. 1982. Cultivar differences in N redistribution in soybeans. Agron. J. 74:375–379.

Zimmerman, D.C., and P. Feng. 1978. Characterization of a prostaglandin-like metabolite of linolenic acid produced by a flax seed extract. Lipids 13:313–316.

17 Fungal Diseases

Kirk L. Athow
Purdue University
West Lafayette, Indiana

A large number of fungi parasitize soybean [*Glycine max* (L.) Merr.] and the diseases caused by the fungi vary greatly in importance. Disease severity and prevalence are closely associated with environmental conditions and host susceptibility in the presence of a disease-producing fungus. Therefore, some soybean diseases are geographically limited, whereas others are distributed widely.

The disease-producing fungus may be restricted to a specific plant part-root, stem, leaf, or seed—or it may attack several or all parts of the plant. In the following discussion, the diseases are grouped according to the plant part most typically affected and the important diseases are discussed more completely.

Some general control measures are applicable to most diseases. These are mostly cultural practices that include removal or destruction of residue, proper soil preparation and drainage, good fertility, crop rotation, and use of pathogen-free seed. Cultural methods of control are not mentioned under each individual disease, but are emphasized when they are an important or the only method of control.

17-1 LEAF DISEASES

17-1.1 Brown Spot

Brown spot was first observed in the USA in 1922 in North Carolina (Wolf, 1923), then in 1923 in Delaware (Wolf and Lehman, 1926). It has since been reported from all the soybean-growing areas of the USA and is the most prevalent disease of soybean in the North Central states. Brown spot may cause severe defoliation, and yield losses of 8 to 34% have been attributed to the disease (Backman et al., 1979; Lim, 1980, 1983; Pataky and Lim, 1981; Williams and Nyvall, 1980; Young and Ross, 1978). Brown spot has been reported from the cooler areas of soybean adaptation such as Argentina, Brazil, Canada, Peoples Republic of China, Germany, Italy, Japan, Korea, Manchuria, Taiwan, the USSR,

Soybeans: Improvement, Production, and Uses, 2nd ed.—Agronomy Monograph no. 16.

and Yugoslovia. The disease apparently rarely develops in the warmer regions of the world.

17–1.1.1 Symptoms

Brown spot is primarily a leafspot disease. The first symptom of the disease is the appearance of irregular, dark-brown patches on the cotyledons. Next, reddish brown angular spots 1 to 5 mm in diameter develop on the unifoliate leaves, these quickly turn yellow and drop. The disease progresses upward on the plant from the lower leaves. Lesions frequently coalesce so that it is difficult to distinguish them individually.

The reddish brown color of the spots distinguishes this disease from bacterial blight which results in shiny black lesions, and the rapid and almost complete yellowing of the leaves is not characteristic of other soybean diseases (Plate 17–1). A longer disease cycle from infections that take place near the time of flowering accounts for the apparent reduction in brown spot during mid-season (Young and Ross, 1979). The disease may appear on the stems and pods of plants approaching maturity as indefinitely margined, dark lesions ranging in size from tiny flecks to several centimeters in length. The lesions on these parts are not distinct enough to be diagnostic.

17–1.1.2 Pathogen

The disease is caused by the fungus *Septoria glycines* Hemmi, first described in Japan in 1915 (Hemmi, 1915). Wolf and Lehman (1926) gave a more complete description of the fungus and the disease. Brown pycnidia form in mature lesions. The pycnidia are immersed and open to the surface by a large pore mostly on the upper side of the leaves. Pycnidiospores are hyaline, filiform, curved, and indistinctly one- to three-septate. They range from 1.4 to 2.1 μm × 21 to 50 μm.

Pycnidiospores overwinter on diseased leaves and stems. The fungus enters the leaf through stomata and grows intercellularly. It is also seed borne, but the fungus is not frequently isolated from seed. It invades the seed either directly by stomatal penetration of the pod or systemically through placental and funicular tissue (MacNeill and Zalasky, 1957).

17–1.1.3 Control

The disease is most prevalent when soybean crops are grown on the same land in consecutive years because inoculum for initial infection arises mainly from infected debris. Differences in susceptibility of cultivars have been observed in the early growth stages but these differences are not as readily detectable at later stages of growth (Lim, 1979). Treating seed with fungicides has not given control but spraying with foliar fungicides from bloom to pod fill has been effective in seasons favorable for development of the disease.

17–1.2 Frogeye Leafspot

Frogeye leafspot was first reported on soybean in the USA in 1924 by Moore (Melchers, 1925) from South Carolina. Subsequently, the disease has been observed in Albama, Arkansas, Delaware, Florida, Georgia, Illinois, Indiana, Iowa, Louisiana, Maryland, Mississippi, Missouri, New York, North Carolina, Ohio, Oklahoma, and Virginia. It occurs most commonly in the southern part of the USA where warm, humid conditions prevail. The disease has been recorded in Australia, Brazil, Canada, Peoples Republic of China, Columbia, Germany, Guatemala, India, Japan, Manchuria, the USSR, Taiwan, and Venezuela.

17–1.2.1 Symptoms

Frogeye is primarily a disease of the leaves, but stems, pods, and seeds may be infected. Lesions on the leaves vary in size and shape and have a gray or light tan central area with a narrow, reddish brown to purple border (Plate 17–2). The absence of yellowing around the spot is a distinguishing symptom. When the lesions are numerous, the leaf becomes dry and drops. Lesions on stems and pod are not distinctive. Infected seeds have a conspicuous gray to brown discolored area, ranging from minute specks to large zonate spots, usually with some cracking or flaking of the seed coat. Laviolette et al. (1970), using closely related susceptible and resistant lines, showed that moderately heavy infection reduced seed yield 12 to 15%.

17–1.2.2 Pathogen

The fungus causing frogeye was described as *Cercospora sojina* Hara in Japan in 1915 (Hara, 1915). Miura (1921), apparently unaware of Hara's paper, described the same fungus as *C. daizu*. Lehman (1928) gave a complete description of the disease and the causal organism, but listed the fungus as *C. diazu*. The accepted name is *C. sojina* Hara. The fungus overwinters in diseased leaves, stems, and seeds. Hyaline spores 5 to 7 μm $\times$ 39 to 70 μm with zero to six septations, produced on diseased residue, are a source of primary inoculum. Infected seed germinates poorly and the resulting seedlings are generally weak (Sherwin and Kreitlow, 1952), but spores produced on the cotyledons of infected seedlings are a prime source of inoculum and infected seeds are a means of distant dissemination of the fungus. Race 2 did not appear to infect seed which may have contributed to its short existence in the Midwest.

Physiologic specialization has been reported in the fungus. Athow and Probst (1952) and Probst et al. (1965) described two physiologic races in Indiana. Ross (1968) reported races 3 and 4 from North Carolina, and Phillips and Boerma (1981) reported race 5 from Georgia. Sinclair (1982) listed new isolates from the USA which Yorinori (1980) had separated into 8 races, and 10 groups of isolates from Brazil which showed differential reactions on some of the cultivars on which they had been tested.

These isolates and the five described races have not been evaluated on a uniform set of differential cultivars, so it is not certain that each is a valid race. However, it does indicate the diversity in the fungus.

17- 1.2.3 Control

Frogeye has been effectively controlled by resistant cultivars. Athow and Probst (1952) and Probst et al. (1965) showed that resistance to races 1 and 2 was controlled by two separate and independent, dominant genes designated Rcs_1 and Rcs_2, respectively. All cultivars resistant to race 2 were also resistant to race 1. Most of these derived their resistance from the cv. Ogden through their progenitor 'Kent' which is resistant to races 1, 2, 3, and 5. The reaction of Ogden or Kent to race 4 has not been determined. Kent has the gene Rcs_2 (Probst et al., 1965). The cv. Davis is resistant to races 1 to 5 and all the new USA and Brazilian isolates. Phillips and Boerma (1981, 1982) reported that resistance in Davis and 'Lincoln' to race 5 was controlled by independent dominant genes in each. Lincoln has the gene Rcs_1 (Athow and Probst, 1952). Boerma and Phillips (1983) reported that the Rcs_2 gene in Kent for resistance to race 2 did not condition resistance to race 5. On the other hand, the gene in Davis for resistance to race 5 also conditioned resistance to race 2. Furthermore, the gene in Davis and Rcs_2 in Kent were at different loci. They suggested the symbol Rcs_3 for the single dominant gene in Davis for resistance to races 2 and 5. The relation of the gene Rcs_2 in Kent to the gene for resistance in Davis has not been determined. In the meantime, Davis and several other cultivars provide sources of resistance to all known races.

Foliar fungicides such as benomyl, triphenyltin hydroxide, and thiabendazole applied from bloom to early pod set were effective in controlling frogeye leafspot according to Backman et al. (1979) and Horn et al. (1975).

17–1.3 Downy Mildew

This disease has ben found wherever soybean plants are grown in the USA. It has also been reported from 21 other countries and probably is coexistent with the crop. Tests have shown that downy mildew can reduce yield as much as 8%.

17-1.3.1 Symptoms

Downy mildew of soybean in its early stages produces yellowish green areas of indenfinite size and shape on the upper surface of the leaves. Diseased areas later become grayish brown to dark-brown necrotic lesions generally surrounded by chlorotic yellowish green margins (Plate 17–3). When the chlorotic margin is absent, the outer edge of the lesion appears darker than the center. A fluffy, gray, or faintly purple mold-like growth develops in the lesion on the undersurface of the leaf. This is the best diagnostic character of the disease and readily differentiates it from brown spot and frogeye leafspot which it may resemble at times. Severely diseased leaves become dry, curl at the edge, and fall prematurely.

Some plants are systemically infected. These plants remain small and each set of leaves that open are smaller than normal and curl downward at the edges and have a gray-green mottled appearance with the undersurface covered with the gray, moldy growth (Plate 17–5).

17–1.3.2 Pathogen

Miura (1921) briefly described downy mildew on soybean in Manchuria and identified the causal organism as *Peronospora trifoliorum* de Bary var. *manshurica* Naoumoff (1914). Downy mildew on soybean was reported for the first time in the USA by Haskell and Wood (1923). Gaumann (1923) raised *P. trifoliorum* var. *manshurica* to specific rank. The downy mildew pathogen then became *P. manshurica* (Naoum.) Syd. ex Gaum. Lehman and Wolf (1924) compared mildew-infected soybean plants from North Carolina with similarly infected specimens from Manchuria and found the fungus to be identical on both. The fungus sporulates abundantly on conidiophores that make up the fluffy growth on the underside of the leaf. The ovoid to subglobose dilutely gray conidiospores, 24 × 20 μm in size, are the chief source of infection in an epidemic development of the fungus in the field. Light brown smooth-walled, globose oospores are 20 to 23 μm in diameter.

Physiologic specialization in *P. manshurica* was first demonstrated by Geeseman (1950), who designated races 1, 2, and 3 on three differential cultivars. Lehman (1953) reported race 4 from North Carolina and later (1958) races 3A, 5, 5A, and 6, using 10 differential cultivars. Grabe and Dunleavy (1959) reported races 7 and 8 using a set of 14 cultivars as differentials. Dunleavy (1971) added races 9 through 23, and in 1977 races 24 through 32 using the following cultivars as differentials: 'Pridesoy', 'Norchief', 'Mukden', 'Richland', 'Roanoke', 'Illini', 'S100', 'Palmetto', 'Dorman', 'Kabott', and Ogden . Only 25% of the 80 isolates collected in 1977 were of races previously described and included four races: 14 (10%), 18 (5%), 20 (2%), and 22 (8%). This suggests a rapid change in the biotypes of the fungus probably influenced by the resistance or susceptibility of the currently predominant cultivars.

17–1.3.3 Control

The fungus overwinters as oospores on diseased crop residue and on the seed. A small percentage of the infected seed produces systemically infected plants that serve as centers of infection. Hildebrand and Koch (1951) demonstrated that chemical seed treatment virtually eliminated the incidence of systemically infected plants.

Resistance to most races has been reported in several genotypes (Dunleavy, 1970; Dunleavy and Hartwig, 1970). The cvs. Mendota, Kanrich, and Pine Dell Perfection appear to be resistant to all known races.

The resistance of Kanrich and Pine Dell Perfection was shown by Bernard and Cremeens (1971) to be controlled by the single gene *Rpm*. The commercial cv. Union presently grown in the southern Midwest has this gene for resistance. The fact that other resistant cultivars have not been developed suggests that downy mildew is not a limiting factor in soybean production.

17–1.4 Target Spot

Target spot was first reported in the USA in 1945 (Olive et al., 1945) and is generally distributed throughout the south. The disease has not been reported in other soybean areas of the USA. It does occur in Brazil, Cambodia, China, Japan, and Nicaragua.

Hartwig (1959) measured yield losses from target spot of 18 to 32% in 5 of 10 yrs in the Delta area of Mississippi. During the other 5 yrs, rainfall in August and September was below normal and disease did not develop sufficiently to cause defoliation.

17–1.4.1 Symptoms

Target spot is a disease that occurs primarily on the leaves, but may be found on the petioles, stems, pods, and seeds. Leaf lesions are reddish brown, circular to irregular, and vary from pinpoint size when young to 10 to 15 mm or more when mature (Plate 17–4). The large lesions sometimes possess concentric rings of dead tissue, which suggested the name *target spot*. The lesions are frequently surrounded by a dull-green or yellowish green halo. Presence of narrow, elongated lesions along the veins on the lower leaf surface is a good diagnosic character. Premature defoliation may follow lesion development.

Spots on petioles and stems are dark brown and range from specks to elongated spindle-shaped lesions. Pod lesions are generally circular, about 2 mm in diameter, with slightly depressed, purple-black centers and brown margins. The fungus penetrates the pod wall in some cases and causes a small, blackish brown spot on the seed.

17–1.4.2 Pathogen

Target spot is caused by the fungus *Corynespora cassiicola* (Berk. & Curt.) Wei. [syn. *Cercospora melonis* (Cooke) Lindau, *C. vignicola* Kawamura, *Helminthosporium vignae* L. Olive, and *H. vignicola* (Kawamura) L. Olive]. The fungus attacks plants of several genera and there appears to be some pathogenic variability among isolates. Jones (1961) found that certain isolates from cotton (*Gossypium hirsutum* L.), soybean, sesame (*Sesamum indicum* L.), and cowpea [*Vigna sinensis* (Torner) Savi] did not differ, but Olive et al. (1945) and Spencer and Walters (1969) reported two distinct races based on the differential response of cowpea and soybean. Morphologically, all isolates are similar, having conidia ranging from 7.4 to 12.4 μm in width and 39.5 to 234.8 μm in length (average 7.7 × 134.7 μm). No differential response has been noted on other hosts, which include velvetleaf (*Abutilon theophrastii* Medic.), pepper (*Capsicum frutescens* L.), sickelpod (*Cassia tora* L.), watermelon (*Citrullus vulgaris* Schrad.), cucumber (*Cucumis sativus* L.), guar [*Cyamopsis tetrogonoloba* (L.) Taub.] okra (*Hibiscus esculentus* L.), blue lupine (*Lupinus angustifolius* L.), yellow lupine (*L. luteus*), Florida velvetleaf [*Mucuna deiringianum* (Bort.) Merr.], mungbean (*Phaseolus aureus* Roxb.), lima bean (*P. limensis* Macf.), common bean (*P. vulgaris* L.), and castorbean (*Ricinus communis* L.).

Corynespora cassiicola has been reported as a cause of a root and stem rot of soybean in Nebraska (Boosalis and Hamilton, 1957) and Canada (Seaman et al. [1965]). Symptoms are dark, reddish brown lesions, which turn dark violet-brown with age, on the roots and stems of soybean seedlings. This fungus does not cause disease of the foliage and the foliar pathogen found in the southern USA does not attack the roots. Because of pathological and morphological differences, the fungus causing root and stem rot may be a species of *Corynespora* other than *C. cassiicola.*

17-1.4.3 Control

Short rotations with other crops appear to have little or no effect on the severity of target spot. Most of the cultivars grown in southeastern USA, where the disease is a potential threat, are resistant to the fungus. The exact mode of inheritance of resistance has not been reported, but the character is highly heritable. In the absence of resistant cultivars, the foliar fungicides suggested for control of frogeye leafspot should be effective for the control of target spot.

17-1.5 Powdery Mildew

Powdery mildew was reported on soybean in North Carolina in 1947 by Lehman (1947), and the disease has been more of a problem in the southern USA and on plants grown in greenhouses. The disease was rarely observed in the Midwest prior to 1973, but since then it has developed to damaging proportions in some fields during cooler than normal seasons. Dunleavy (1980) measured yield losses of 0 to 26%, with an average loss of 13%, during the 3 yrs 1976 through 1978. Frequently, the disease develops too late in the season to cause yield loss. The increase in powdery mildew coincided with a shift in acreage to the susceptible cvs. Harosoy, Amsoy, and Corsoy from the resistant cvs. Hawkeye, Lincoln, Ford, Adams, and Lindarin. Powdery mildew has been reported from Brazil, Canada, China, Germany, India, Peru, Puerto Rico, and South Africa.

17-1.5.1 Symptoms

White, powdery patches composed of mycelium and conidia develop on cotyledons, stems, pods, and, particularly, on the upper leaf surface.

The lesions enlarge and coalesce until the entire surface of infected plant parts are covered. Most common symptoms on the leaves are green and yellow islands, interveinal necrosis, necrotic specks, and crinkling of the leaf blade followed by defoliation. These symptoms may be almost absent when mycelial growth is abundant. Some cultivars have adult plant resistance (Mignucci and Lim, 1980) but are susceptible as seedlings. Chlorotic spots and veinal necrosis are associated with the resistant reaction.

17–1.5.2 Pathogen

The causal fungus is the obligate parasite *Microsphaera diffusa* Cke. & Pk. (syn. *Erysiphe polygoni* D.C., *E. glycines* Tai). It was incorrectly identified as *E. polygoni* because immature cleistothecia lack the characteristic appendages dichotomously branched three to five times at the tip. The fungus produces abundant conidia 2.8 to 5.4 μm long and 1.7 to 2.1 μm wide borne in chains on short, simple conidiophores. Conidia are barrel-shaped with flattened ends. Cleistothecia are hemispherical, light yellow to tan when immature, turning rusty brown and then black at maturity with 4 to 50 appendages. The pyriform asci are borne on short stalks and contain up to six yellow, ovoid ascospores measuring 9 $\times$ 18 μm.

17–1.5.3 Control

A high degree of resistance to powery mildew has been reported by Demski and Phillips (1974) in southern cultivars and by Grau and Laurence (1975) in northern cultivars. The latter showed that resistance was controlled by a single, dominant gene.

The fungicides benomyl, [methyl 1-(butylcarbamoyl)-2-benzimidazolecarbamate], thiophanate [1,2-bis(3-ethoxycarbonyl-2-thioureido)benzene], and thiabendazole [2-(4′-thiazolyl)-benzimidazole] used as a foliar spray are effective in controlling the disease. The cost-benefit ratio will govern its use.

17–1.6 Alternaria Leafspot

Alternaria leafspot is a common disease of soybean in many parts of the USA, and also has been reported from Argentina, Australia, Brazil, China, India, Kenya, Poland, South Africa, Taiwan, the USSR, and Yugoslovia. Young plants are occasionally infected, but it is generally a leaf disease of plants nearing maturity and occurs too late to cause damage. The spots are brown with concentric rings and range from 0.5 to 2.5 cm in diameter (Fig. 17–1). The spots often coalesce to form large dead areas. Alternaria leafspot resembles target spot, but lacks the reddish brown color and the greenish yellow halo that characterize target spot.

The fungus associated with alternaria leafspot generally has been reported only as *Alternaria* spp. *Alternaria tenuissima* (Needs ex Fr.) Wilts was reported on soybean in Kenya (Nattrass, 1961). Gibson (1922)

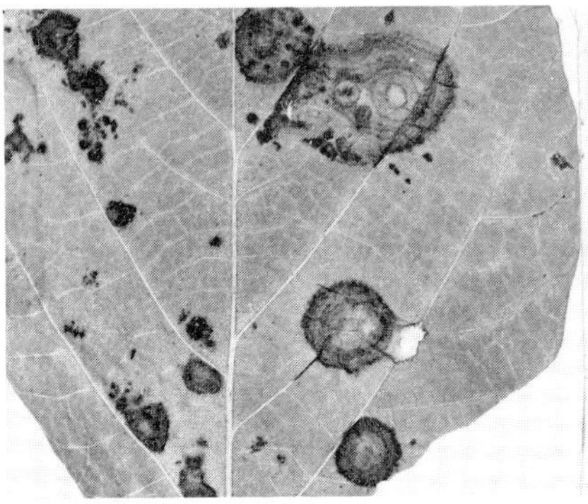

Fig. 17–1. Concentric lesions of Alternaria leafspot.

Fig. 17–2. Phyllosticta leafspot on young soybean leaves.

described *A. atrans* Gibson as a weak parasite infecting leaves of soybean in Arizona through aphid and sunburn injuries.I have observed a black, sooty growth of *Alternaria* on pods injured by grasshopper feeding, and species of *Alternaria* are frequently isolated from normal-appearing seed in Indiana. The incidence of *Alternaria* on seed generally increases accompanying decreases in *Cercospora kikuchii*, *C. sojina*, and *Diaporthe phaseolorum* var. *sojae* following foliar applications of the systemic fungicide benomyl. Membrs of the genus *Alternaria* apparently are not vigorous parasites of soybean.

17–1.7 Phyllosticta Leafspot

Phyllosticta leafspot is characterized by round to oval or irregular shaped spots, 1 to 2 cm in diameter on the younger leaves. The spots form from the margin of the leaf inward (Fig. 17–2). At first the spots are dull gray but later turn gray to tan with a dark-brown border. Small, black specks—the fruiting structures of the fungus—are found abundantly in the older leafspots. The disease usually does not persist long after the third or fourth leaf stage of growth of the soybean. However, it has caused severe defoliation on the Eastern Shore of Maryland (Jehle et al., 1952), southeastern Missouri (Crall, 1948), and Iowa (Crall, 1950). The disease can also cause lesions on stems, petioles, and pods. Walters and Martin (1981) reported spots on the pods in Arkansas as large as 8 mm in diameter with dark purplish red borders surrounding lighter, brownish centers on which were numerous dark pycnidia. Lesions on the stems and petioles are superficial, 1 to 4 mm wide by 1 to 12 mm long, and are gray to almost white with narrow purplish to brown borders. The disease also has been reported from Brazil, Bulgaria, Canada, China, Estonia, Germany, Italy, Japan, and the USSR.

The causal fungus was originally described by Massalongo (1900) in Italy as *Phyllosticta sojaecola* Massal. Tehon and Daniels (1927) reported *P. glycineum* on soybean in Illinois. Olive and Walker (1944) found that the fungus causing Phyllosticta leafspot in Maryland was similar to the one described by Massalongo. Although there are slight differences in spore size in the several reports, the name *P. sojaecola* Massal. generally is accepted. Pycnidia are 65 to 170 μm in diameter. Conidia are 1.8 to 3.2 μm $\times$ 5.2 to 7 μm, one-celled, narrowly elliptical with rounded ends, and hyaline in color. The ascigerous stage, designated *Pleosphaerulina sojaecola* (Massal.) Miura, has been reported in the Orient but not in the USA.

17-2 ROOT AND STEM DISEASES

Diseases of roots and stems generally are the most destructive because they may result in the death of the plants at various stages of development, particularly after completion of vegetative growth during which the plant exerts its maximum competition. Diagnosis of these diseases is often difficult because more than one pathogen may contribute to the syndrome. Also, some pathogens attack both roots and stems, some attack roots only, and others attack only stems. As a group, the fungi causing these diseases are mostly facultative saprophytes; i.e., they are parasites that can grow also on dead organic matter. They are often less specific and may attack other plant species or survive in the soil for a long time in the absence of a host plant. This tends to make control more difficult. Root and stem diseases are also easily confused with or influenced by herbicide damage and unfavorable soil or environmental conditions for growth.

17-2.1 Brown Stem Rot

Brown stem rot was first discovered in 1944 in central Illinois (Allington, 1946). It is now one of the most frequently occurring soybean diseases in the North Central states and Canada. Outside of these areas, it has been reported only for southeastern USA (Ross and Smith, 1963; Phillips, 1970), Mexico (Morgan and Dunleavy, 1966), and Egypt (Kamal-Abo-el-Dahab, 1968).

Physiological age of the plant and air temperature have been implicated as factors in the development of the disease. Allington and Chamberlain (1948) found no disease development at 21°C or above, whereas Phillips (1971) reported that development was extensive and unaffected by air temperature from 15 to 27°C. Gray (1974) found vascular browning more extensive at 22 than 28°C, but temperature within this range was not limiting. Chamberlain and McAlister (1954), and Gray (1974) showed that disease developed more rapidly in older plants. Temperature may influence, but does not restrict, the geographic distribution. Greater susceptibility of older plants probably is a factor in the time of disease appearance. The disease is rarely observed until late July or early August when plants are normally in the R4 to R5 growth stage as defined by Fehr et al. (1971).

17-2.1.1 Symptoms

The first symptom is a brown discoloration of the pith and vascular elements extending upward from the roots or base of the stem (Fig. Plate 17-6). There is no external evidence of the disease at this time and signs of early infection generally go unnoticed unless the stems are cut open and examined. Under favorable conditions for the disease, the browning becomes continuous throughout the stem, and the lower portion of the stem may exhibit a dull, brown external discoloration. During periods of hot, dry weather following infection, the leaves may wilt or develop interveinal necrosis as a result of shortage of water. Chamberlain (1961) demonstrated that a metabolite of the causal fungus as well as physical plugging was responsible for the reduced water flow in the vessels of infected stems. Some isolates of the causal fungus cause premature defoliation whereas others do not (Gray, 1971, 1974). The foliage symptoms have at times been singularly associated with brown stem rot, but other types of injury to the roots and stem that impede the movement of water may produce the same effect.

The effect of brown stem rot on yield depends greatly upon the environmental conditions. The largest losses occur in seasons with a cool period in early August followed by hot, dry weather. Dunleavy and Weber (1967) showed that heavily infected plants yielded only 44% of normal. Two-thirds of the yield loss was due to a reduction in seed number and one-third to a reduction in seed size. Heavily infected plants lodge severely causing difficulties in harvesting which may result in the greatest reduction in yield.

17-2.1.2 Pathogen

Presley and Allington (1947) identified the causal organism as a member of the genus *Cephalosporium*. In 1948, Allington and Chamberlain (1948) described the fungus as a new species, *C. gregatum* Allington and Chamberlain. This later was reduced to synonymy as *Phialophora gregata* (Allington and Chamberl.) W. Gams (syn. *Cephalosporium gregatum* Allington and Chamberl.). The fungus produces hyaline conidia in irregular aerial or decumbent conidial heads. The conidia are ovoid to elliptical and 1.7 to 4.3 μm $\times$ 3.4 to 9.4 μm in size. Variation in virulence among isolates have been observed (Phillips, 1971; Gray, 1971). The latter (1974) reported that only some isolates of the fungus caused defoliation and reduced yield. The fungus may persist in the soil several years even in the absence of the soybean host.

17-2.1.3 Control

The disease increases in severity with continuous growing of soybean. Rotation was recognized early as a means of control (Allington, 1946).

Five years of corn between crops of soybean has been shown to reduce brown stem rot infection from 100 to 6% of the plants (Dunleavy and Weber, 1967). A rotation with 3 yrs between soybean crops appears adequate to prevent economic loss.

Some differences in susceptibility exist among cultivars, but none are immune. Plant introduction 84946-2, a strain of unknown origin selected from an introduction from Korea, shows a high proportion of disease-free plants. Segregating generations from crosses of this strain with commercial cultivars have averaged more healthy plants that the commercial cultivars. The resistant cvs. BSR 201 (Maturity Group II) and BSR 302 (Maturity Group III) developed in Iowa have this source of resistance. Plant introduction 86150 may have a somewhat higher degree of resistance. Gray (1974) suggested selecting for resistance to defoliation as well as internal stem browining. Early maturing cultivars tend to escape severe infection but generally yield less than later-maturing cultivars.

17–2.2 Phytophthora Root Rot

Phytophthora root rot is one of the most destructive diseases of soybean. The disease was first observed in northwestern Ohio and northeastern Indiana in the late 1940s although the cause was not known at that time. The first published report of the disease was by Suhovecky and Schmitthenner (1955). Since then, the disease has been reported in most of the soybean-producing areas of the USA, and in Canada, Australia, Hungary, Japan, and New Zealand. The disease is more severe on heavy clay soils with small pore size than on lighter soils. It is particularly prevalent on low, poorly drained areas, but it also appears on higher ground if the soil remains wet for several days. Soil compaction, which reduces the pore size of the soil and aeration of the roots, increases the disease. Kittle and Gray (1979) reported that a soil temperature of 15°C which is 10 to 15°C below the optimum for root development was most favorable for root infection. I have observed that disease development is more rapid in the field or greenhouse with an air temperature of 25°C or above even though a lower temperature favors infection.

The disease may cause seed rot, pre-, or postemergence damping-off, or a more gradual killing or reduction of plant vigor throughout the season (Plate 17–7). Young plants, however, are most susceptible and die most quickly. The disease is primarily a root rot, although stem rot (Kaufmann and Gerdemann, 1958) and infection of leaves and stems, when splashed by contaminated soil (Morgan, 1963), have been reported.

17–2.2.1 Symptoms

The first noticeable symptoms of the disease, except for a poor stand, often are yellowing and wilting of the leaves (Plate 17–8). The lateral roots are almost completely destroyed and the main root becomes dark brown (Plate 17–9). The dark discoloration generally extends up the stem several centimeters and occasionally to the second or third node. The main root and stem usually remain firm and there is no definite lesion other than the dark discoloration. Some cultivars are not readily killed but shown yellowing and poor vigor similar to that associated with wet soil and N deficiency. The condition probably is a result of N deficiency because as the roots are destroyed there are fewer functional nodules.

17-2.2.2 Pathogen

Suhovecky and Schmitthenner (1955) first reported the disease to be caused by a species of *Phytophthora*. Herr (1957) referred to the causal organism as *P. cactorum* (Leb. & Cohn) Schroet. Kaufmann and Gerdemann (1958) considered the organism distinct from previously described species of *Phytophthora* suggesting the name *P. sojae*. Hildebrand (1959) found that isolates of the soybean fungus from Illinois, North Carolina, and Ontario, Canada were morphologically similar to *P. megasperma* but because of host specificity suggested the trinomial *P. megasperma* Drechs. var. *sojae* A. A. Hildeb. Kuan and Erwin (1980) suggested the name *P. megasperma* Drechs. f. sp. *glycinea* Kuan & Erwin because host specificity was more correctly designated by *formae speciales*. They replaced *sojae* with *glycinea* to indicate the genus of the host.

The optimum temperature for growth of the fungus is 25 to 28°C. It produces amphigynous and paragynous antheridia and thin-walled spherical or subspherical oogonia 29 to 59 μm in diameter. Oospores develop when an antheridium fertilizes an oogonium. The oospores have thick, smooth walls which protect them from desiccation and provide a means of survival during unfavorable soil conditions. The oospores germinate and produce a germ tube which develops into hyphae or a sporangium. The sporangia are 42 to 65 μm $\times$ 23 to 53 μm in size and obpyriform in shape. The sporangia produce hundreds of minute zoospores which are released into the soil water that fills the pore spaces between the soil particles. The zoospores swim in the free water by means of two flagella—one anterior and one posterior—, are attracted to the soybean roots where they attach themselves, encyst, germinate, and infect the roots directly through the epidermis. The zoospores are considered the main source of root infection. After infection, fungal development is limited to the cortex and stele of lateral roots in resistant cultivars whereas ramification to tap root and hypocotyl of the susceptible cultivars occurs (Beagle-Ristaino and Rissler, 1983). The fungus produces new oospores, sporangia, and zoospores in diseased tissue.

The fungus can survive in the soil for long periods of time in the absence of soybean, and, undoubtedly, has been present in some areas for many years. The disease did not become important until the susceptible cvs. Hawkeye, Harosoy, Lindarin, Ogden, Jackson, and others became widely grown.

Isolates of the fungus differ in virulence (Hildebrand, 1959; Hilty and Schmitthenner, 1962; Averre and Athow, 1964). Physiologic specialization was reported by Morgan and Hartwig (1965). Race 3 was

Table 17–1. Reaction of differential soybean cultivars to physiologic races 1 through 23 of *Phytophthora megasperma* f. sp. *glycinea.*

Differential cultivar	Reaction† to physiologic race																						
	1	2	3	4	5	6	7	8	9	10	11	12	13	14	15	16	17	18	19	20	21	22	23
Harosoy	S	S	S	S	S	S	S	S	S	S	S	R	S	S	S	R	S	R	R	S	S	S	S
Harosoy 63	R	R	S	S	S	S	S	S	S	R	R	R	R	S‡	R	R	R	R	R	S	S	S	S
Sanga	R	S	R	R	R	R	R	R	R	S	S	S	R	R	R	S	R	R	S	S	R	R	S
Mack	R	R	R	S	S	R	R	R	R	R	R	S	R	S	R	S	R	S	S	S	R	S	R
Altona	R	R	R	R	S	S	S	S	S	R	S	R	S	R	R	R	S	R	R	R	R	S	S
PI 171442	R	R	R	R	R	S	S	R	R	S	R	S	R	R	S	R	S	R	S	S	S	S	R
PI 103091	R	R	R	R	R	S	R	S	R	R	R	S§	R	R	R	R	S	R	S	R	R	R	R

†Abbreviations: R = Resistant and S = Susceptible.

‡Keeling (1982) reported Harosoy 63 resistant to race 14. Moots et al. (1983) reported the majority of Harosoy 63 plants had killing lesions with race 14. I have found Harosoy 63 to be uniformly susceptible to an isolate of race 14 from Keeling. Races 4 and 14 then are the same.

§Athow and Laviolette (1981, unpublished data), and Buzzell and Anderson (1984, personal communication) have found that PI 103091 is susceptible to race 12. Races 12 and 19 are then the same.

SOYBEANS:
Improvement, Production, and Uses

Second Edition

Chapter 17—Fungal Diseases

Kirk L. Athow

Plates 17-1, 17-2, 17-3, 17-4, 17-5, 17-6, 17-7, 17-8, 17-9, 17-10, 17-11, 17-12, 17-13, 17-14, 17-15, and 17-16

Chapter 19—Nematodes

R. D. Riggs and D. P. Schmitt

Plates 19-1, 19-2, 19-3, and 19-4

Chapter 20—Integrated Control of Insect Pests

Sam G. Turnipseed and Marcos Kogan

Plates 20-1, 20-2, 20-3, 20-4, and 20-5

Plate 17–1. Brown spot showing yellowing of the leaf.

Plate 17–2. Frogeye leafspot.

Plate 17–3. Downy mildew on lower and upper side of soybean leaf.

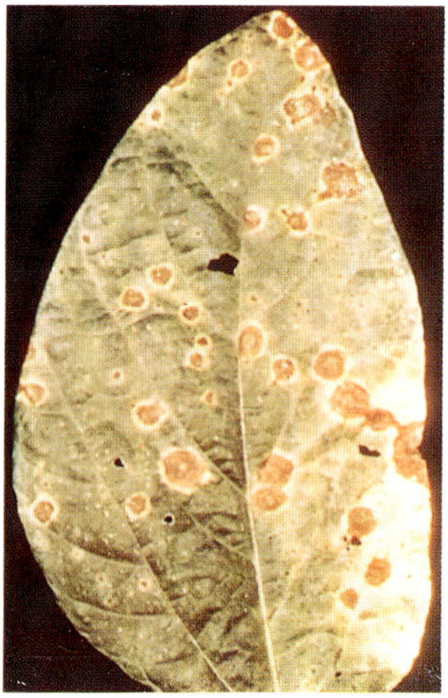

Plate 17–4. Lesions caused by the target spot fungus on soybean leaf.

Plate 17–5. Soybean plant systemically infected with *Peronospora manshurica* (*right*); healthy plant (*left*).

Plate 17–6. Internal stem discoloration caused by brown stem rot fungus (*left*); healthy (*right*).

Plate 17–7. Phytophthora root rot in a commercial soybean field in Indiana. Green area in foreground is resistant variety.

Plate 17–8. Dead and dying plants infected with *Phytophthora megasperma* f. sp. *glycinea*.

Plate 17–9. Young soybean plant killed by *Phytophthora* (*right*) showing decayed roots and discolored stem; healthy plant (*left*).

Plate 17–10. Stem canker lesion on lower part of soybean stem.

Plate 17–11. Root decay and lesions caused by *Rhizoctonia solani*.

Plate 17–12. Rhizoctonia root rot showing dead and wilting plants.

◀ Plate 17–13. Soybean stem with the cottony, mycelial growth of the sclerotium blight fungus.

Plate 17–14. Purple seed stain.

Plate 17–15. Soybean leaf infected with the purple seed stain fungus, *Cercospora kikuchii.*

Plate 17–16. Pod and stem blight showing pynidia on stems and pods (*left*) and infected seed (*lower right*).

Plate 19–1. Aboveground symptoms of *H. glycines* infection. (A) Race 1 in North Carolina causes general chlorosis and stunting; (B) Race 3 in Arkansas causes severe stunting but less chlorosis.

Plate 19–2. Soybean field showing damage by *M. incognita*. (A) Root-knot nematodes alone, foreground untreated, background treated with 1, 2–dibromo–3–chloropropane; right-resistant cultivar: (B) Root-knot nematodes + southern blight, two rows of a *M. incognita*–resistant cultivar.

Plate 19–3. *Rotylenchulus reniformis* females and egg masses (*arrows*) on roots of soybean. Courtesy, W. Birchfield.

Plate 19–4. Necrosis of soybean roots infected with *Pratylenchus scribneri*. Courtesy, R. B. Malek.

Plate 20–1. Lepidopterous pests: (a) green cloverworm, (b) soybean looper, (c) velvetbean caterpillar, and (d) corn earworm. Courtesy, Merle Shepard, Clemson University.

a

b

c

d

e

Plate 20–2. Pests other than Lepidoptera: (a) Mexican bean beetle adult, (b) Mexican bean beetle larva, (c) bean leaf beetle, (d) southern green stink bug, and (e) threecornered alfalfa hopper.

Plate 20–3. Diseases of lepidopterous pests: (a) nuclear polyhedrosis virus of the corn earworm, (b) *Entomophthora gammae* infecting soybean looper, (c) *Entomophthora* spp. infecting small larvae of corn earworm, (d) *Nomuraea rileyi* infecting soybean looper. Courtesy, G. R. Carner, Clemson University.

Plate 20–4. Predators of insect pests. A predaceous beetle, *Calleida decora* (*top*) and predaceous bug, *Nabis roseipennis* (*bottom*) both preying upon small lepidopterous larvae. Courtesy, Merle Shepard, Clemson University.

Plate 20–5. Parasites of lepidopterous larvae. *Microplitis demolitor* imported from Australia, parasitizing a small corn earworm larva (*top*) and *Apanteles ruficrus* imported from Australia, parasitizing a small soybean looper larva (*bottom*). Courtesy, Merle Shepard, Clemson University.

reported by Schmitthenner from Ohio in 1972, race 4 from Kansas by Schwenk and Sim (1974), races 5 and 6 from Ontario, Canada by Haas and Buzzell (1976), races 7, 8, and 9 by Laviolette and Athow from Indiana (1977), races 10 to 16 from Mississippi by Keeling (1979), races 17 to 20 from Mississippi and Arkansas by Keeling (1982a), races 21 and 22 from Indiana by Laviolette and Athow (1983), and race 23 from-Nebraska by White et al. (1983). The 23 physiologic races, as distinguished by their reaction on seven differential cultivars, are presented in Table 17–1.

A single, dominant gene for resistance has been reported in each of the differential cultivars except Harosoy and PI 103091. The gene symbol assigned to Harosoy is *rps* to indicate the recessive allele that conditions for susceptibility. The gene in 'Harosoy 63' has been designated Rps_1 (Bernard et al., 1975; Laviolette et al., 1979). 'Sanga', PI 84637 (Mueller et al., 1978), and D60-9647 (data unpublished from intercrossing Sanga, PI 84637, and D60-9647) have the gene Rps_1^b. 'Mack' and PI 54615-1 (Mueller et al., 1978) have the gene Rps_1^c. The gene Rps_3 has been reported in PI 171442 or PI 86972-1 (Mueller et al., 1978) and occurs in combination with Rps_1^b in the cv. Tracy (Athow et al., 1979); 'Altona' has the gene Rps_6 (Athow and Laviolette, 1982). In addition, Rps_2 occurs in the cv. CNS and in derived strains based upon results of root inoculation with races 1 and 2 (Kilen et al., 1974); Rps_1^k was reported by Bernard and Cremeens (1981) from the cv. Kingwa; Rps_4 was reported from PI 86050 in combination with Rps_1^c by Athow et al. (1980); and Rps_5 was reported by Buzzel and Anderson (1981) from Harosoy[6] × T240 (type collection strain). Rps_1, Rps_1^b, Rps_1^c, Rps_1^k are an allelomorphic series at the Rps_1 locus. Recently, Layton et al. (Purdue Univ., 1985, unpublished) have found four allelic genes at the Rps_3 locus which tentatively have been designated Rps_3^b, Rps_3^c, Rps_3^d, and Rps_3^e. Each of these genes gives resistance to several physiologic races (Table 17–2) and is the basis for disease resistance. The relationship between Rps_2 and the other genes is not clear. Rps_2 is characterized by resistance to root infection but susceptibility to hypocotyl infection, whereas all the other genes are equally effective in root and hypocotyl.

17–2.2.3 Control

Cultural practices, systemic fungicides, and resistant cultivars are the control methods that have been suggested or effectively used. Improved soil tilth and drainage would do much to reduce phyophthora root rot. However, the trends in soybean culture which emphasize larger equipment, earlier planting, reduced tillage, increased use of herbicides, and less rotation all enhance disease development (Duncan and Paxton, 1981; Kittle and Gray, 1979; Tachibana and Van Diest, 1983; Abney et al., 1979).

Resistant cultivars have provided the most effective control even with the large number of physiologic races of the fungus. The gene Rps_1

Table 17–2. Genes for resistance to specific races of *Phytophthora megasperma* f. sp. *glycinea* and their source.

Gene	Source	Reaction† to physiologic race																						
		1	2	3	4	5	6	7	8	9	10	11	12	13	14	15	16	17	18	19	20	21	22	23
rps	Harosoy	S	S	S	S	S	S	S	S	S	S	S	R	S	S	S	R	S	R	R	S	S	S	S
Rps_1	Harosoy 63	R	R	S	S	S	S	S	S	S	R	R	R	R	S	R	R	R	R	R	S	S	S	S
Rps_{1b}	Sanga	R	S	R	R	R	R	R	R	R	S	S	S	R	R	R	S	R	R	S	S	R	R	S
Rps_{1c}	Mack	R	R	R	S	S	R	R	R	R	R	R	S	R	S	R	S	R	S	S	S	R	S	R
Rps_{1k}	Kingwa	R	R	R	R	R	R	R	R	R	R	R	S	R	R	R	S	R	R	S	S	R	R	ND
Rps_3	PI 86972–1	R	R	R	R	R	S	S	R	R	S	R	S	R	R	S	R	S	R	S	S	S	S	R
Rps_4	PI 86050	R	R	R	R	S	S	S	S	S	R	S	R	R	R	R	R	ND	ND	ND	ND	ND	ND	ND
Rps_5	L62–904	R	R	R	R	R	S	S	R	R	S	R	S	R	R	S	R	ND	ND	ND	ND	ND	ND	ND
Rps_6	Altona	R	R	R	R	S	S	S	S	S	R	S	R	S	R	R	R	S	R	R	R	R	S	S

†Abbreviations: R = Resistant, S = Susceptible and ND = Not determined.

provided resistance to races 1 and 2 in the cvs. Clark 63, Harosoy 63, Hawkeye 63, Lindarin 63, Chippewa 64, Beeson, Calland, Protana, Amsoy 71, Cutler 71, Bonus, Harwood, Century, Oakland, Pella, Wells, Hardin, Evans, Hodgson, and others. The cultivars with the Rps_1 gene are susceptible to race 3 and many of the races found since 1972. The cvs. Vickery, Corosoy 79, Williams 79, Wells II, Beeson 80, Vinton 81, Mack, Lee 68, Pickett 71, Davis, and Centennial have the gene Rps_1^c and are resistant to all the races that have been found in the heavy production areas except for race 4. 'Williams 82' and 'Century 84' have the gene Rps_1^k and are resistant to all races found in the Midwest. The new cvs. Keller and Miami have the two genes Rps_1^c Rps_3, and 'Winchester' has the two genes Rps_1^b Rps_3. Tracy was the first cultivar with the two gene combination Rps_1^b Rps_3 (Athow et al., 1979), but it is sensitive to the herbicide metribuzin [4-amino-6-(1,1-dimethylethyl)-3-(methylthio)-1,2,4-triazin-5(4*H*)-one]. Kilen and Barrentine (1983) reported linkage between the recessive gene *hm* for sensitivity to metribuzin and the genes at the Rps_1 locus for resitance to *P. megasperma* f. sp. *glycinea.* Tracy was replaced by 'Tracy M' which was selected for nonsensitivity to metribuzin.

Although there are several two or three-gene combinations which give resistance to all races of the fungus, there has been a reluctance by some breeders to use specific gene resistance for fear that selection pressure that it might exert on the fungus population would hasten the evolution of additional races. Several of the proponents have advocated the use of so-called *tolerant* cultivars (Schmitthenner and Walker, 1979; Buzzell and Anderson, 1982; Tooley and Grau, 1982; Walker and Schmitthenner, 1982). Schmitthenner and Walker (1979) defined tolerance as the ability of susceptible plants to survive infection without showing severe symptom development such as death, stunting, or yield loss. In reality, these tolerant plants actually have varying degrees of resistance to root infection even though they are susceptible to pre- or postemergence hypocotyl infection. This led Jimenez and Lockwood (1980) to suggest that *field resistance* would be a less ambiguous term, and Tooley and Grau (1982) to refer to it as *rate reducing resistance.* Even though the term *tolerance* has been badly misused, resistance to root infection is a useful type of resistance. This resistance apparently is nonrace specific and polygenically controlled in nature and could be expected to be more durable because of less selection pressure on the pathogen. Walker and Schmitthenner (1982) reported that heritability of field resistance was high and controlled by relatively few genes which were independent of the specific *Rps* genes for resistance. Therefore, the two types of resistance together should compliment each other. Buzzell and Anderson (1982) suggested that selection for tolerance based on low plant loss, followed by backcrossing to include race-specific resistance, should provide effective long-term disease control.

The systemic fungicide metalaxyl has activity against *Phytophthora, Pythium,* and downy mildew fungi. The seed treatment formulation offers protection against the seedling phase of the soybean phytophthora. It has been suggested (Schmitthenner, 1983) for use with those cultivars with a high degree of field resistance. A granular formulation is intended for in-furrow, band, or broadcast application. Varying degrees of effectiveness have been reported by Papavizas et al. (1979), Vaartaja et al. (1979), Morton et al. (1982), Anderson and Buzzell (1982), Schmitthenner (1983), and others, but it may give season-long protection depending upon the rate used, the amount of rainfall, and the soybean cultivar. The cost-benefit relationship may be the deciding factor.

17–2.3 Stem Canker

Stem canker is a destructive disease because the causal fungus is capable of killing plants from midseason until maturity. The disease has occurred throughout the north central USA since the late 1940s and was reported from Ontario, Canada in 1952. Since 1975, the disease has become a serious threat to soybean production in the eight southeastern states of Alabama, Florida, Georgia, Louisiana, Mississippi, North Carolina, South Carolina, and Tennessee. There are no reports of the disease outside the USA and Canada. The history and the distribution of stem canker are difficult to trace because for a long time its symptoms were included in the disease complex known as pod and stem blight. The term *stem canker* was first suggested for this disease by Crall (1950). Personal communications suggest that this disease has been present for many years and probably occurs in some other countries.

The disease became prevalent in the Midwest with the release of two very susceptible cvs.—Hawkeye in 1948 and Blackhawk in 1951. They were sister lines from the cross of Mukden × Richland, but neither parent was as susceptible as its progeny. Estimates of up to 80% infected plants in some fields have been reported (Crall, 1951; Dunleavy, 1954; Hildebrand, 1952). Yield losses due to stem canker have been estimated as high as 50% (Andrews, 1950). In Indiana, we measured yield losses of 20% in fields with approximately 35% plants killed and we have observed much heavier infection. Some field yield losses in the southeastern USA have been as high as 50 to 100% (Bachman et al., 1981). The earlier the plants are killed the greater the loss in seed yield. Uninfected plants do not compensate for the loss of adjacent plants because infection rarely occurs before early August when the plants have reached full height and the pods are beginning to develop. I have never observed infection on plants less than 62 days of age, and the earliest infection generally is noted on plants 70- to 80-days old, irrespective of time of planting (Athow, 1957). Later plantings invariably have less infection probably because of a shorter period of susceptibility.

17–2.3.1 Symptoms

Dead plants with dried leaves attached may be the first indication of stem canker. However, the first symptom of the disease is a small, reddish brown superficial spot at the base of a branch or leaf petiole on the lower part of the stem—most often at one of the first eight nodes. The spot is first observable in the leaf scar after the petiole has fallen. The spot rapidly enlarges into a slightly sunken lesion that girdles the stem and kills the plant (Plate 17–10). At the time the plant is killed, green stem tissue above and below the reddish brown lesion is diagnostic and distinguishes stem canker from other soybean root and stem diseases. The stem becomes brittle and is easily broken at the lesion. Hobbs et al. (1981) reported a top dieback caused by the stem canker fungus. The upper five to six internodes are darker brown than the lower internodes on the same plant. As the tip dies, it is sometimes overgrown with a dark growth of *Alternaria*. This symptom of the disease has also been observed in Michigan and Indiana. In Indiana, it appears to be associated with late-season lodging caused by wind and rain.

17–2.3.2 Pathogen

Welch and Gilman (1948) were the first to recognize that the organism causing stem canker differed in morphology and pathogenicity from the pod and stem blight fungus, *Diaporthe phaseolorum* (Cke. & Ell.) var. *sojae* (Lehman) Wehm. They considered the stem canker fungus a strain of *D. phaseolorum* var. *batatatis* (Harter and Field) Wehm. Athow and Caldwell (1954) described the fungus as *Diaporthe phaseolorum* (Cke. & Ell.) Sacc. var. *caulivora* Athow and Caldwell. This trinomial generally has been accepted but there has been some dissension (Threinen et al., 1959; Whitehead, 1966).

The fungus can overwinter on diseased stems and in diseased seed. In the spring, infected stems are covered with perithecia that contain numerous eight-spored asci 29.8 to 40.2 μm $\times$ 4.5 to 7.0 μm in size. The hyaline, one-septate ascospores are elongate-ellipsoid, and measure 8.6 to 11.8 μm $\times$ 3.0 to 3.9 μm. The presence of diseased crop residue or the planting of infected seed has not been shown to influence the amount of the disease. The infection process is not thoroughly understood, although Crall (1956) implicated the blade and petiolar portions of the leaves and Athow (1957) showed that removal of the first six trifoliolate leaves practically eliminated the disease. All available evidence suggests that infected seed does not result in systemic infection of the developing plant although the cotyledons are infected prior to falling to the ground. Sporulation on the fallen cotyledons may be a source of inoculum. Infected seed also may be a means of long-range dissemination of the fungus. Roy and Miller (1983) found that isolates of *Diaporthe* and *Phomopsis* from cotton produced lesions on soybean similar to *Diaporthe phaseolorum* var. *caulivora* from soybean. This suggested that they represented a single species, or, at most, a few closely related species composed of perithecial and nonperithecial strains.

17–2.3.3 Control

In the Midwest, the disease has been controlled partially by the use of less susceptible cultivars. Harosoy 63, Lindarin 63, and 'Mandarin

Ottawa' are among the least susceptible. None of the cultivars now commonly grown in the North Central states are highly susceptible, and it is unusual to find more than 5 to 10% infection. 'Mandarin' is the common parent in most of these cultivars. Higby and Tachibana (1982) reported that the cv. Midwest had a high degree of resistance. In the southern USA, the disease has not been reported on the cvs. CNS, Tracy, Centennial, and Braxton (Keeling, 1982b; Weaver et al., 1984). Cultivar CNS is the suspected source of resistance. On the other hand, the cvs. Bragg, Coker 237, Forrest, Hutton, Jeff, Lee 74, Mack, Pickett 71, Davis, and Essex, and the breeding line J77-339 are highly or moderately susceptible (Keeling, 1982a; Krausz and Fortnum, 1983). Cultivars Tokyo or Peking, used for resistance to cyst nematode, are the common ancestors in most of the cultivars susceptible to stem canker. With the exception of J77-339, most of the cultivars had been grown for several years before the disease appeared. New virulent biotypes have been implicated in the sudden appearance of the disease in a rather extensive area. However, highly susceptible cultivars may be chiefly responsible as were the highly susceptible cvs. Hawkeye and Blackhawk in the Midwest 30 yrs ago.

Disease-free seed, plowing under of diseased residue, and rotation with some crop other than cotton should help reduce the disease. Foliar fungicides may be effective if adequate coverage can be obtained in the late vegetative stage prior to infection to the stem.

17–2.4 Rhizoctonia Root Rot and Aerial Web Blight

Rhizoctonia root rot is a problem of increasing importance in the USA and several other countries, particularly Brazil and Argentina. Intensive monoculture of soybean and the increasing use of herbicides are two reasons for this disease increase. Rhizoctonia root rot occurs early in the season when the soil is unusually wet. It may cause preemergence damping-off, a hypocotyl rot, or root rot. Young infected soybean plants develop a reddish brown decay of the outer cortical layer of the root and basal stem. Frequently, this decay causes a sunken, reddish lesion that girdles the stem at or just above the soil line and the plant dies (Plate 17–11). In extremely wet seasons, the disease may persist until the plants have made full growth and are setting pods. The lower part of the tap root and the secondary root system are usually killed. As the soil becomes dry, infected plants wilt and die or show wilting during the hot part of the day. Such plants may develop new roots just below the soil line, but these shallow roots do not provide the plant with sufficient water when the soil dries out. Occasionally, most of the plants in a fairly large spot are killed but more often individual plants or groups of plants scattered through the field are killed (Plate 17–12). Tachibana et al. (1971) reportd 42 and 48% yield reduction in two severely infected fields in Iowa. The causal fungus is *Rhizoctonia solani* Kuehn [syn. *Pellicularia filamentosa* (Pat.) Rogers]. It is a widespread soil-inhabiting fungus that attacks a large number of crop plants. High temperature (26–32°C), high moisture-holding capacity of the soil (>70%), and a soil pH > 6.6 favor the disease (Lewis and Papavizas, 1977). They also reported variability among isolates and differences in susceptibility among soybean cultivars.

Rhizoctonia leaf rot and aerial or web blight caused by *R. solani* Kuehn has been reported from many parts of the world, particularly in the humid, tropical, and semitropical regions (Hepperly et al., 1982). Severely infected leaves collapse, with the petioles remaining attached to the stems. Petioles of infected leaves remain green after the leaflets become necrotic. Dark brown sclerotia between 0.25 and 0.05 mm in length develop on petioles and leaves. Irregularly shaped, tan, sunken lesions are found on pods at the point where dead, blighted leaf tissues come in contact with pods. Infected seed has a tan discoloration, with mycelium growing on the seed surface. Foliar sprays with benomyl reduce the incidence of the foliage phase of the disease.

17–2.5 Pythium Rot

Several species of *Pythium* can attack soybean, causing seed decay and pre- or postemergence killing of seedlings. Infected seedlings that emerge often have nearly normal cotyledons but defective or dead growing points. This condition has been termed *baldhead* and may result in the eventual death of the seedlings or a weak plant with a secondary stem development from the cotyledonary node. Infected seedlings frequently have swollen hypocotyls and lesions at the junction of hypocotyl and primary root. Other symptoms reported by Schlub and Lockwood (1981) include a curling growth habit and reddish brown lesion on the hypocotyl and cotyledons. While pythium rot is primarily a seedling disease, it may occur on soybean until midseason. In the early stages of infection of older plants, the first symptoms is a water-soaked appearance of the outer tissues of the root and lower stem. Wilting of the plant follows. The smaller roots are decayed and the outer tissues of the larger roots and the lower stem become soft, moist, brown, and slough off easily. The disease may occur in large or small areas but generally only individual plants or small groups of plants are killed. The disease is not obviously related to topography although it is favored by high soil moisture.

Pythium debaryanum Hesse was first reported on soybean by Lehman and Wolf (1926a). In 1952, Hildebrand and Koch reported *P. ultimum* Trow as the causal organism of a root rot in Ontario, Canada. Morgan and Hartwig (1964) in Mississippi first reported *P. aphanidermatum* (Edson) Fitz. on soybean. Klag et al. (1978) reported that *P. ultimum*, *P. irregulare*, *P. spinosum* Saw, and *P. mamillatum* Meurs. were isolated from early to midseason and were most pathogenic at 20°C. *Pythium myriotylum* Drechs. was isolated later in the season and was more pathogenic at 30°C. Southern et al. (1976) found that *P. myriotylum* was prevalent on soybean in Florida where it apparently was favored by warmer temperature. *Pythium ultimum*, *P. debaryanum*, *P. irregulare*, *P. spinosum*, and *P. mamillatum* grow better than *P. aphanidermatum* and

P. myriotylum at cooler temperatures and are the cause of seed decay and preemergence killing of seedlings when soybean seed is planted in cold soil. *Pythium aphanidermatum* does not cause much seed rot even at higher temperature more favorable for its growth (Thomson et al., 1971). Under warm, moist conditions it and *P. myriotylum* are capable of causing the root rot phase of the disease which is identical in appearance to the disease caused by the other species in cool, wet conditions. Because of the differences in the temperature requirements the various species are more prevalent in certain regions. All of the species may occur in the intermediate area but here the conditions generally are not highly conducive to the disease.

All of the above species of *Pythium* attack a wide range of crop species. Only slight differences in cultivar susceptibility have been noted in soybean. Seed treatment with metalaxyl will provide control of damping-off if stand establishment is a problem. Morton et al. (1982) reported 0.30 g active ingredient (a.i.) ha was superior to 0.15 g a.i. ha under heavy disease pressure.

17–2.6 Fusarium Wilt and Root Rot

Several species of *Fusarium* have been reported to cause root and stem diseases of soybean. Cromwell (1917) reported a fusarium blight or wilt which he believed to be caused by *F. tracheiphilum* Smith. It has been shown that a similar disease of soybean in southeastern USA can be caused by race 1 of the cowpea wilt *Fusarium, F. oxysporum* Schecht. f. *tracheiphilum* Amst. & Amst. (Armstrong and Armstrong, 1950), race 2 of the cotton wilt *Fusarium, F. oxysporum* Schlecht. f. *vasinfectum* Amst. & Amst. (Armstrong and Armstrong, 1958), and an isolate from soybean, *F. oxysporum* Schlecht. f. *glycines* Amst. & Amst. (Armstrong and Armstrong, 1965). The diseases caused by these organisms are similar and are chiefly blights on plants that have passed the early succulent stages, but flaccid leaves and stem tips are sometimes observed on younger plants. Vascular discoloration is characteristic at all stages of the disease. Root rotting is of minor importance. Infection with these forms of *Fusarium* occurs at 28°C or above.

Ferrant and Carroll (1981) and Carroll and Leath (1983) reported that *F. oxysporum* Schlect. emend. Synd. & Hans. was recovered from 84% of the plants with symptoms of fusarium wilt in Delaware. First evidence of the disease was poor emergence. Symptoms consisted of randomly scattered round to elongated patches of yellow to brown plants with interspersed green plants. Soybean was killed at varoius stages of development. Individual plants showed stunting, chlorosis, sudden wilting, epinasty, and extensive defoliation. Pinkish spore masses were present on the surface of basal stems of older plants. Roots show brown to black cortical decay with vascular discoloration extending into the stem. On some plants, reddening of the pith was followed by tissue disintegration. Losses as high as 59% of average yield have been reported by Ferrent and Carroll (1981).

In Iowa (Dunleavy, 1961) and in Minnesota (French, 1962, 1963; and French and Kennedy, 1963), a root rot phase caused by *F. orthoceras* Appel. & Wr. has been reported. The fungus is usually confined to roots or the lower portion of the stems. Wilting sometimes occurs from severely rotted roots when soil moisture is low. This phase of the disease is most frequently observed on seedlings or young plants, and entire fields can be affected. Infected plants may survive but are weak. Wilting of older plants is much less common but is occasionally observed. In these cases, the tap root and secondary roots are almost completely rotted. In contrast with the high temperature requirements of the forms mentioned above (28°C or higher), French (1963) reported infection with the root rotting form only from 14 to 23°C, with most infection at the lower temperature.

Corriveau and Carroll (1983) found that plants infected with *F. oxysporum* had significantly fewer nodules per plant than noninfected plants. Rhizophere population of *F. oxysporum* were greater from plants with no-tillage as compared to conventional tillage (Carroll and Leath, 1983). Datnoff and Sinclair (1983) reported no apparent interaction between *F. oxysporum* and *Rhizocotonia solani.* Armstrong and Armstrong (1965) and Leath and Carroll (1980) found that cultivars differ in susceptibility, and the latter described a rapid screening method to detect resistant plants in the growth chamber.

17–2.7 Charcoal Rot

Charcoal rot is a disease of the roots and lower stem caused by the soil- and root-inhabiting fungus *Macrophomina phaseolina* (Tassi) Goid. [syn. *M. phaseoli* (Maubl.) Ashby, *R. bataticola* (Taub.) Butler]. Disease development is favored by high tempertaure (30 to 35°C) and moisture stress (Meyer et al., 1973) and is frequently more prevalent on sandy soils. The mycelium of the fungus penetrates and colonizes root tissue of young seedlings (Ammon et al., 1974) and subsequently invades cortical stem tissue; later, intracellular invasion occurs followed by the formation of sclerotia. The disease causes black streaks in the woody portion of the roots and stems. When the bark is peeled from the root and the base of the stem, small, black sclerotia may be seen (Fig. 17–3). Sclerotia, the propogating bodies of the causal fungus, are the principal means of survival. They are not abundant in root or stem tissue until the onset of moribundity suggesting that sclerotia are indicative of host cell death. They are often abundant enough to give a grayish black color to the tissue beneath the bark. Short et al. (1978) suggested that the differences in numbers of propagules in diseased tissues reflect differences in compatibility between selected soybean cultivars and *M. phaseolina.*

The fungus is widely distributed in soils in the warmer section of the USA and has been reported from most of the other countries where soybean is grown. It also attacks over 400 species of cultivated plants

Fig. 17–3. Soybean stem base and root with small, black sclerotia of the charcoal rot fungus.

Fig. 17–4. Irregular black sclerotia of sclerotinia stem rot fungus on and in soybean stem. Courtesy, D. W. Chamberlain.

and weeds. Sclerotia of the fungus may survive free in the soil or embedded in host residue in dry soils for long periods. They cannot survive more than 7 or 8 weeks in wet soil. No special control effort has been made because the disease rarely occurs on vigorously growing plants. The fungicides benomyl and thiophanate-methyl have been reported by Ilyas et al. (1976) to be effective in reducing survival of sclerotia in soil. The number of sclerotia increase with successive susceptible crops. Improved soil fertility and water relations will reduce the disease severity.

17–2.8 Sclerotium Blight

Sclerotium blight causes a rot at the base of the stem, but differs from charcoal rot in that the fungal sclerotia are larger (1 to 2 mm in diameter) and rounder, and are brown instead of black. They are produced on a cottony, mycelial growth on the outside of the stem rather than under the bark (Plate 17–13). Infected plants die prematurely, sometimes before the seed has formed. The disease is found in the sandy-soil areas of the South where high temperatures occur; the name *southern blight* is sometimes applied to the disease. Losses may be as high as 20 to 30% of the plants in some fields, but it is more common to see small, scattered areas of killed plants.

Sclerotium blight is caused by the fungus *Sclerotium rolfsii* Sacc. (perfect stage *Corticium rolfsii* Curzi). The fungus has broad hyphae (5–9 μm wide) with regular clamp connections which along with shape, size, and color of sclerotia distinguish *S. rolfsii* from *Sclerotinia sclerotiorum.* The fungus attacks a wide variety of plants, including practically all of the summer legumes adapted to the South (Epps et al., 1951; Higgins, 1927). Agarwal and Kotasthane (1971) evaluated 25 cultivars in India and reported seedling mortality from *S. rolfsii* ranged from 1.6 to 65%. 'Improved Pelican', 'Shelby', Palmetto , 'Hood', and Jackson were classified resistant with < 10% mortality. Bragg, Kent, Hampton, and Davis were the most susceptible (35 to 65% seedling mortality), whereas 'Hill', 'Scott', Dare, 'Bienville', Pickett, Clark 63, 'Semmes', Harosoy, Lee, 'Dorman', and Hardee were intermediate or moderately resistant according to their classification.

17–2.9 Sclerotinia Stem Rot

Sclerotinia stem rot or blight is sometimes called *white mold* because of the white, cottony growth on the stems of infected plants. Symptoms are not usually observed until late flowering and disease development is minimal until the beginning of pod development. Stem lesions originate at leaf axils where flowers are positioned (Abawi and Grogan, 1975; Abawi et al., 1975; Cline and Jacobsen, 1983; Grau et al., 1982) and advance up and down the stem. Lesions most frequently are on the main stem 15 to 40 cm above the soil surface. Lateral branches are less frequently diseased and symtpoms are rarely observed on the main stem at the soil line. Lesions are initially gray but later appear tan and water-soaked. The lesions on lateral branches also originate at nodes. A white fluffy mycelium develops on infected tissue, and large, black sclerotia of various shapes are formed externally, partially covered by the white fungus growth and internally in the stem pith and pods (Fig. 17–4). Pod infection without infection of the main stem occasionally occurs. The sclerotia are much larger than those of charcoal rot. They are black and irregularly shaped, thus differing from the round, brown sclerotia of sclerotium blight. Plants are often killed by the girdling of the stem, or the plants may wilt and become chlorotic with poor pod development.

Sclerotinia root rot was first reported in the USA in 1924 but has been a disease of minor importance except in widely scattered localized areas. The disease has also been reported from Argentina, Brazil, Canada, Hungary, Nepal, and South Africa. The causal fungus, *Sclerotinia sclerotiorum* (Lib.) Korf & Dumont, is the cause of a serious disease of green bean (*Phaseolus vulgaris* L.), cabbage (*Brassica oleracea* L.), sunflower (*Helianthus annuus* L.), peanut (*Arachis hypogea* L.), and a number of other cultivated crops. The disease in soybean frequently has been associated with a previous crop of one of these susceptible crops (Phipps and Porter, 1982; Grau et al., 1982; Cline and Jacobsen, 1983).The fungus survives in the soil for long periods in the form of sclerotia. With favorable temperature (20 to 30°C), and moisture, an abundance of spores are produced in cup-shaped fruiting structures, apothecia, that develop from the sclerotia on or near the soil surface. Cultivars with dense foliage and narrow rows which produce a tight canopy are conducive to sporulation and infection (Phipps, 1983). Trifluralin (*a,a,a,* trifluro-2,6-dinitro-*N,N*-dipropyl-f-toluidine) and metribuzin increase germination of sclerotia and numer of apothecia/sclerotium according to Radke and Grau (1982). Cultivars differ in their susceptibility (Grau et al., 1982; Cline and Jacobsen, 1983). Corsoy, Hodgson, 'Hodgson 78', Williams, and Union were the least susceptible.

17–2.10 Anthracnose

Anthracnose has been found in all soybean-producing areas of the USA and has been reported from Argentina, Brazil, China, India, Korea, and Thailand. The disease has long been considered a minor disease of soybean, but its relative importance has increased greatly in recent years. This is particularly true in the warmer, humid soybean-growing regions. Backman et al. (1979, 1982) reported that anthracnose had the greatest potential to reduce yield (an average of 20%) of any one disease in Alabama. Yield losses of 30 to 50% have been reported in Thailand and 100% in India (Sinclair, 1982). Seed infection in some parts of Brazil renders the seed worthless for planting the following year.

Soybean plants are subject to anthracnose at all stages of growth. When infected seeds are planted, many of the germinating seeds are killed before emergence. Dark brown, sunken cankers often develop on the cotyledons of young seedlings. The fungus may completely destroy one or both cotyledons or grow from them into the tissues of the young stem. There it produces numerous small, shallow, elongated, reddish brown lesions or large, deep-seated, dark brown lesions that kill the young plants. The causal fungus is capable of growing within stem, leaf, and pod tissues of soybean plants without giving external evidence of its presence until conditions are favorable for its sporulation on the surface of the invaded tissue. On older plants, anthracnose reaches its most destructive development stage during rainy periods in late summer. Lower branches and shaded leaves are killed by the fungus. Young pods also may be attacked. Indefinitely shaped brown areas develop and coalesce to cover the entire surface of the diseased stems and pods, and numerous, black-fruiting bodies (acervuli) are formed on the diseased area. As these structures develop, several short, dark spines (setae) emerge from each one, giving a dark, stubble-beard appearance to the diseased area. Seeds in diseased pods may be shriveled and moldy or infected seeds may show no outward sign of disease. Anthracnose is caused by several species of fungi from two closely related genera. *Glomerella glycines* (Hori) Lehman and Wolf was first reported as the causal agent of anthracnose in America (Lehman and Wolf, 1926b). The fungus *Colletotrichum dematium* (Pers. ex Fr.) Grove var. *truncatum* (Schw.) Arx. (syn. *C. dematium* (Fr.) Grove var.

truncata (Schw.) Arx., *C. truncatum* (Schw.) Andrus & Moore) is now considered the most common cause of anthracnose of soybean. Roy (1982) reported that *C. dematium* var. *truncatum, C. gloeosporioides* Penz., *C. graminicola* (Ces.) Wilson, *Glomerella cingulata* (Stonem.) Spauld & v. Schr., and *G. glycines* reduced seedling emergence but only *C. dematium* var. *truncatum, C. gloeosporioides*, and *G. cingulata* caused severe hypocotyl and cotyledonary infection and significant stunting of seedlings. All of these fungi survive from one growing season to the next in diseased crop residue left in the field. Most, if not all, overwinter in infected seed which serve as a source of primary inoculum as well as spreading the fungi to new locations.

Some differences in cultivar susceptibility have been noted but none is highly resistant. Fentin hydroxide (triphenyltin hydroxide) and benomyl applied as a foliar spray at early pod stage (R3) gave effective control of anthracnose (Backman et al., 1979, 1982; Horn et al., 1979). A second application was necessary if there were 2 wet days during the 10 days following the first application. Muchovey et al. (1980) found that on low pH soils, liming with calcium carbonate ($CaCO_3$), calcium hydroxide ($Ca[OH]_2$), or calcium sulfate ($CaSO_4$) reduced anthracnose seedling blight. Fungicidal seed protectants will improve the stand from infected seed.

17–3 SEED DISEASES

A number of fungi may be found associated with soybean seed. Some of these have been mentioned already; *Cercospora sojina, Peronospora manshurica, Cornespora cassicola, Colletotrichum* spp., *Diaporthe phaseolorum* var. *caulivora,* etc., and others that have not been directly implicated. This is an area of plant pathology that has received increased attention since 1970.

17–3.1 Purple Seed Stain

Purple seed stain is one of the most prevalent and widely distributed diseases of soybean seed. It was observed on soybean in the USA for the first time in 1924 (Gardner, 1924) in Indiana. The disease has since been observed in most states and foreign countries where soybean is grown.

17–3.1.1 Symptoms

The discoloration on the seed varies from pink or light purple to dark purple, and ranges in size from a small spot to the entire area of the seed coat (Plate 17–14). For this reason, the disease has been referred to as purple speck or purple spot as well as the more commonly accepted name of purple stain. Cracks often occur in the discolored areas, giving the seed a rough, dull appearance. Although the symptoms of purple stain are most conspicuous on the seed, the causal fungus also attacks leaves, stems, and pods (Murakishi, 1951; Walters, 1980). The foliar symptoms become most evident during the latter part of the growing season as the plants approach maturity. The first symptoms appear during late R5 (beginning seed) and early R6 (full seed) growth stages. Upper leaves exposed to the sun have a light purple, leathery appearance (Plate 17-15). Reddish purple, irregular lesions later occur on both upper and lower leaf surfaces. Lesions vary from pinpoint spots to irregular patches up to 1 cm in diameter and may coalesce to form large necrotic areas. Veinal necrosis may also be observed. Numerous infections cause rapid chlorosis and necrosis of leaf tissue, resulting in defoliation starting with the young upper leaves. The most obvious symptom is the premature blighting of the younger upper leaves over large areas, even entire fields. Lesiosn on petioles and stems are slightly sunken, reddish purple areas several millimeters long. Infection of petioles increases defoliation. On more susceptible cultivars, round, reddish purple lesions, which later become purplish black, occur on pods. Heavily infected stems have a dull gray to dark brown appearance and dry up 7 to 10 days prematurely. Estimated yield losses from Cercospora leaf blight in southern USA have been as high as 3.5 million t in some years.

The amount of purple discoloration in seed lots is important from the grading standpoint, and the value of the seed may be lowered by excessive amounts of purpling. The U.S. grading standards do not allow more than 5% purple stain in no. 1 yellow soybean. Weather conditions during the time of flowering and plant maturity apparently have a pronounced influence on the percentage of discolored seed that develop, since this varies widely from year to year and location to location for a given cultivar. More than 50% of the seed of some cultivars may be discolored. The discoloration is not detrimental to soybean for processing because the color disappears when the seed is heated.

17-3.1.2 Pathogen

Matsumoto and Tomoyasii (1925) described the causal fungus as *Cercosporina kikuchii* Mat. and Tomoy. In 1953, Chupp changed the name of the causal fungus to *Cercospora kikuchii* (Mat. & Tomoy.) Chupp, which is now generally accepted. However, Jones (1959) demonstrated that 10 other species of *Cercospora* isolated from other hosts are capable of causing purple discoloration of soybean seed when the fungi are injected into developing pods. Whether these other 10 species of *Cercospora* are capable of causing purple stain of soybean seed under natural conditions has not been determined.

The causal fungus overwinters on diseased leaves and stems as well as in infected seed. Seed infected with *C. kikuchii* is slightly lower in germination than noninfected seed (Lehman, 1950; Wilcox and Abney, 1973; Athow and Laviolette, unpublished). The incidence of the disease in harvested seed did not differ appreciably whether planted with purple stained or purple stain-free seed (Wilcox and Abney, 1973). However, when diseased seeds are planted, the fungus grows from the seed coat into the cotyledons and from these into the stem of a small percentage of the seedlings. The fungus produces abundant filiform, hyaline conidiophores, 0 to 22 septate and 70 to 164 μm $\times$ 4 to 5 μm in size on the diseased seedlings. The spores are blown by wind and splashed by rain to other leaves. Leaf spots develop and the fungus produces a second crop of spores that cause infections on other leaves, stems, and pods. Laviolette and Athow (1972) found that most seed infection took place during full bloom. Roy and Abney (1976) also reported the highest incidence of purple stained seed from inoculations at full-bloom or early pod stage, but also found a high percentage of purple stained seed from inoculatons at the green bean stage (R5).

17-3.1.3 Control

Cultivars differ in susceptibility and there are cultivars with moderate resistance. On the other hand, some susceptible cultivars have been released. Plant introduction 80837 appears to be almost immune to purple stain and is being used to develop cultivars with higher degrees of resistance. Cultivar CNS, the original source of resistance to bacterial pustule, has good resistance to purple stain and has contributed to lower susceptibility of many of the southern cultivars. Wilcox et al. (1975) reported heritabilities were high for disease incidence in the F_2 and F_3 generation of a cross of Harosoy (susceptible) $\times$ PI 80837 (resistant). Athow and Laviolette (1978, unpublished) using the same cross found that the seeds from F_1 plants were almost as susceptible as the susceptible parent, but segregation in the F_2 and F_3 generations indicated that one or possibly two recessive genes conditioned resistance in PI 80837.

Significant reduction in the amount of purple-stained seed has been reported with benomyl applied at full-bloom to early pod stage (Abney et al., 1975; Ensminger and Abney, 1978; Tenne and Sinclair, 1978; Miller and Roy, 1982). A second application may be necessary in 14 days if 2 days of rain occur during the interval since the first application. Benomyl applied in the R5 to R6 stages has been reported to increase yield by reducing foliar infection when conditions were favorable for its development.

17-3.2 Pod and Stem Blight

Pod and stem blight is found in all the soybean-producing areas in the USA. The disease also has been reported from Argentina, Brazil, Canada, China, Egypt, Guyana, India, Japan, Korea, Senegal, Taiwan, Thailand, and the USSR. It probably has a much wider geographic distribution.

17-3.2.1 Symptoms

The most noticeable symptom of the disease is the linear rows of dark pycnidia on stems, petioles, and pods (Plate 17-16). Leaf blade infection has been reported by Lehman (1923) and Luttrell (1947). The disease first appears on the petioles of lower leaves and upon broken lower branches. Pycnidia develop on the main stem and upper branches after the death of the entire plant following maturity or death from other causes. Dead stems may be covered with the pycnidia or the pycnidia may be limited to small patches, usually near the nodes. Pycnidia frequently are found on dry, poorly developed pods. Proof is lacking that the fungus is the cause of failure of the pods to develop normally. The most important aspect of the disease is its effect on the seed. Infected seed may exhibit various degrees of seed-coat cracking and shrivelling, and the seed frequently is partially or completely covered with a white mold (Plate 17–16). Infected seeds may be smaller than healthy ones, but the fungus is sometimes present in seeds that are normal in size and appearance. Infected seeds usually do not germinate, or if they do, may produce weak seedlings. Cotyledonary lesions are the most diagnostic symptoms of the disease in seedlings arising from infected seeds. These lesions range from pinpoint necrotic areas to more extensive ones that may involve the whole cotyledon. Lesions vary from almost colorless to brown or bright red. The seed coat frequently remains attached to diseased cotyledons after emergence. Wallen and Seaman (1963) reported that the percentage of seeds infected decreased greatly during 2 yrs in storage and germinability rose proportionally in the same priod. We have noted similar results with infected seed that was not severely shrivelled or moldy, indicating the fungus is relatively short lived in the seed and that the infected seed is not killed or permanently injured by the pathogen at the time of infection or during dormancy. The seed germination phase is obviously the critical stage for disease development.

17–3.2.2 Pathogen

The disease was first reported by Lehman (1923) in North Carolina in 1920. He named the causal fungus *Diaporthe sojae.* Wehmeyer (1933), in his monograph of the genus *Diaporthe*, changed the name to *D. phaseolorum* (Cke. & Ell.) Sacc. var. *sojae* (Lehman) Wehm. Pycnidia on stems and petioles develop in compacted groups in a black stromata beneath the epidermis. The pycnidia contain conidia borne on short, simple, unicellular conidiophores. The conidia are unicellular, hyaline, and fusiform-elliptical and measure 4.9 to 9.8 μm $\times$ 1.8 to 3.2 μm. The perithecial stage is found much less frequently than the pycnidial on overwintered stems. The perithecia, borne singly in a black stroma in the cortical tissue, produce eight-spored asci 38.0 to 51.2 μm $\times$ 3.3 to 5.6 μm. Welch and Gilman (1948) reported that the organism was heterothallic but Athow and Caldwell (1954) failed to induce perithecial formation by pairing single ascospore isolates in culture.

Kmetz et al. (1974) reported three distinct *Phomopsis* type fungi occurred on the aboveground parts of soybean and infected seed: a form which produced only pycnidia and readily rotted seed, designated as an unknown species of *Phomopsis*; a form which produced pycnidia and perithecia characteristic of *D. phaseolorum* var. *sojae* and which slightly inhibited seed germination and caused limited seed rot; and a form which produced pycnidia and perithecia characteristics of *D. phaseolorum* var. *caulivora* and which was capable of inducing stem canker symptoms and was moderately virulent on seed. *Diaporthe phaseolorum* var. *caulivora* probably should not be grouped with the other two forms because it does not have a naturally occurring *Phomopsis* (imperfect) stage, and produces the distinct stem canker disease. The other two forms have either primarily or exclusively, the *Phomopsis* stage, and both are intimately associated with the entire pod and stem blight disease complex. The differences between them, aside from the infrequent occurrence of perithecia with *D. phaseolorum* var. *sojae*, were small, scattered stroma in the medium, single short-beaked pycnidia with alpha and beta spores for *D. phaseolorum* var. *sojae* vs. extensive, large-spreading stroma, multichambered, long-beaked pycnidia with alpha spores only. The latter type, referred to as *Phomopsis* spp., was more virulent and caused seed rotting. These differences are not discrete and may represent a continuous variation among isolates of the pod and stem blight fungus. Further research is necessary to determine if one or more species is involved in the pod and stem blight disease. The name pod and stem blight is not descriptive of the seed decay that is now recognized as the most important aspect of the disease, and sometimes has been referred to as Phomopsis seed decay.

No relationship has been shown between the amount of seed-borne inoculum and the severity of the disease on the resulting plants (Garzonio and McGee, 1983). The major source of inoculum is infected soybean debris from the current or previous season (Athow and Caldwell, 1954; Hildebrand, 1954; Kmetz et al., 1979). Garzonio and McGee (1983) found most severe infection following continuous soybean, less with a corn-soybean rotation, and least following continuous corn. The causal fungus was isolated from immature, symptomless stems, pods, and occasionally green seeds. Dark blotches and pycnidia develop on infected plants that ripen 7 to 10 days prematurely. The pathogen is not systemic in the host, but remains semidormant and close to the point of infection until the plant begins to mature (Hill et al., 1981). The sudden appearance of pycnidia late in the season probably results from multiple localized infections.

There is a high positive correlation between disease incidence and rainfall during pod fill (Shortt et al., 1981). Most seed infection, however, takes place after R7 stage of growth (Athow and Laviolette, 1973; Kmetz et al., 1978; McGee and Brandt, 1979).The highest incidence of seed infection occurs with delay in harvest after the seed is ripe, particularly if accompanied by moist, warm weather. High humidity-high temperature result in the most pod and stem blight infection (Spilker et al., 1981). The amount of pod and stem blight in soybean seed increased as much as fivefold in plants infected with bean pod mottle virus (Stuckey et al.,

1982). Soybean mosaic and bean yellow mosaic did not affect the amount of pod and stem blight.

17-3.2.3 Control

Thorough plowing under diseased crop residue and rotation will give reasonably good control in most years because they reduce the amount of primary inoculum. A fungicidal seed protectant will increase emergence and reduce the number of diseased seedlings from severely infected seed (Wallen and Seaman, 1963). This practice is of commercial value only with seed in which the incidence of the pathogen is high.

Foliar spray with benomyl has significantly reduced the incidence of seed infection in some years. The first application should be applied at the R4 growth stage if 2 wet days have occurred since bloom. A second application is generally needed if 2 additional wet days occur during the 10-day period following the first application (Ellis et al., 1974; Ross, 1975; Abney, 1978; Tenne and Sinclair, 1978; Backman et al., 1979; McGee and Brandt, 1979; Hill et al., 1981; Miller and Roy, 1982). During dry years, seed infection is so low as to make foliar sprays unnecessary.

Cultivars differ in susceptibility but the trait is not highly heritable. Plant introduction 80837 has low seed infection in the Midwest, but becomes heavily infected when grown in the South out of its range of best adaptation. Cultivars Delmar and Ransom have a high degree of resistance which appears to be heritable.

17-4 FUNGI ASSOCIATED WITH OTHER DISEASES

The bibliographies of soybean diseases by Ling (1951), Chamberlain and Lipscomb (1967), and Sinclair and Dhingra (1975) list a number of other fungi associated with soybean diseases. Some of these fungi cause serious soybean diseases in other parts of the world, but have not been reported or are of minor importance in the USA. Others may be questionable pathogens of soybean or are the cause of diseases of rare occurrence. The literature on these diseases in most cases is not cited and the reader is directed to the above bibliographies and the *Compendium of Soybean Diseases* (Sinclair, 1982) for the original literature citations and additional information.

Soybean rust caused by *Phakosphora pachyrhizi* Syd. is common in Japan, Taiwan, China, and Formosa. It also occurs in Australia, Brazil, Java, Sarawak, Okinawa, Ryukyu Islands, Cambodia, Malaya, India, Puerto Rico, and the Phillipines.

Soybean blast or soybean-sleeping blight and soybean scab or Sphaceloma scab have been reported only from Japan. The causal fungus of blast is *Septogloeum sojae* Yoskii and Nishizawa. It produces superficial brown obscure spots or streaks on leaves, petioles, stems, and pods. The causal organism of scab is *Sphaceloma glycines* Kurata and Kuribayaski which attacks leaves, stems, and pods.

Stemphylium leaf blight caused by the fungus *Stemphylium botryosum* Wallr. occurs in India. Symptoms are small, circular necrotic spots with dark brown margins and gray centers on leaves of all ages. A dark green fungal growth is often observed on the surface of the lesions. Later the leaves dry up and severe defoliation results.

Neocosmospora vasinfecta E. F. Smith has been reported to cause a stem rot in the USA and Nigeria. Symptoms are similar to those of brown stem rot. Plants infected with *Phialophora gregata* have mycelium in both pith and xylem vessels, whereas plants infected with *N. vasinfecta* have mycelium in pith tissues only.

Murasaki mopa disease, also called *violet rot,* is a serious root rot in Japan, Formosa, Korea, Manchuria, and southern Rhodesia. The causal organism is *Helicobasidium mompa* Tanaka.

Yeast spot of soybean seed caused by *Nematospora coryli* Pegl. has been reported from Brazil, South Africa, Zaire, and the USA. It is frequently associated with stinkbug damage.

Trotteria venturioides Sacc. has been reported as the causal fungus of black leaf mildew in the Philippines.

Species of *Ascochyta* have been reported to cause leafspots—*A. sojae* Miura in China and Japan; *A. phaseolorum* Sacc. in Japan and Tanzia, and *Ascochyta* spp. in Germany, France, Italy, the USSR, Zaire, and the Netherlands.

Botrytis cinerea Pers. has been reported to cause a seedling disease and seed decay in Poland.

In Madagascar, *Coniothyrium sojae* Bourequet was found to be the fungus responsible for lesions at the stem base.

Chaetoseptoria wellmanii Stev. has ben reported as the causal organism of a leafspot in Costa Rica.

Corticium sasakii (Shira) Matsumoto attacks leaves, petioles, and stems in Japan, China, Formosa, whereas *C. centrifugum* (Lev.) Bres. has been reported to cause stem canker in China, Japan, Korea, and the USSR.

Dactuliophora glycines Leakey was identified as the causal fungus that produced large leafspots in northern Rhodesia.

Isariopis griseola Sacc. was reported to cause angular brown spots on leaves in the USSR.

A pod rot in China and Japan was attributed to *Macrophoma mame* Hara.

Several species of *Mycosphaerella* have been reported as the causal organisms producing leafspots of soybean. *Mycosphaerella cruenta* (Sacc.) Lan. has been reported from Cambodia and the USA; *M. phaseolorum* Siemszko from Nyasaland and the USSR; and *M. soja* Hori from China, Japan, and Korea.

Ophionectria soja Hara was reported as the causal fungus of a disease that results in browning and shrivelling of the stem near the ground level and eventual death of plants on the dikes of paddy fields in China and Japan.

Pleosphaerulina glycines Saw. has been reported attacking soybean leaves in Taiwan, Bermuda, Canada, China, Estonia, Germany, Italy, Japan, Korea, the USSR, and the USA.

The causal fungus of a leafspot in Ethiopia and northern Rhodesia was identified as *Pyrenochaeta glycines* R D Stewart.

Synchytrium dolichi (Cooke) Gaum. has been reported to cause brown, raised scabs on stems and pods in Kenya.

Verticillium albo-atrum Reinke and Berthold has been identified as the fungus causing a basal stem and root rot in Germany. This organism has been associated with a similar disease in Indiana and Iowa.

REFERENCES

Abawi, G.S., and R.S. Grogan. 1975. Sources of primary inoculum and effects of temperature and moisture on infection of *Whetzelinia sclerotiorum* in beans. Phytopathology 65:300–309.

----, F.J. Polach, and W.T. Molin. 1975. Infection of bean by ascospores of *Whetzelinia sclerotiorium*. Phytopathology 65:673–678.

Abney, T.S. 1978. Effectiveness of foliar fungicide combinations for control of *Diaporthe phaseolorum* var. *sojae* seed infection. Phytopathol. News 12:204.

----, T.L. Richards, and K.W. Roy. 1975. Effect of benomyl foliar applications on purple stain of soybeans. Proc. Am. Phytopathol. Soc. 2:82.

----, D.H. Scott, and G.B. Bergeson. 1979. Carbofuran increases Phytophthora damage in soybean. *In* Proc. 9th Int. Congr. Plant Protection, Washington, DC. August, 1979. American Phytopathological Society, St Paul.

Agarwal, S.C., and S.R. Kotasthane. 1971. Resistance in some soybean varieties against *Sclerotium rolfsii* Sacc. Indian Phytopathol. 24:401–403.

Allington, W.B. 1946. Brown stem rot of soybean caused by an unidentified fungus. Phytopathology 36:394.

----, and D.W. Chamberlain. 1948. Brown stem rot of soybeans. Phytopathology 38:793–802.

Ammon, V.T., T.D. Wyllie, and M.F. Brown, Jr. 1974. An ultrastructural investigation on pathological alterations induced by *Macrophomina phaseolina* (Tassi) Goid in seedlings of soybean, *Glycine max* (L.) Merrill. Physiol. Plant Pathol. 4:1–4.

Anderson, T.R., and R.I. Buzzell. 1982. Efficacy of metalaxyl in controlling Phytophthora root and stalk rot of soybean cultivars differing in field tolerance. Plant Dis. 66:1144–1145.

Andrews, E.A. 1950. Stem blight of soybeans in Michigan. Plant Dis. Rep. 34:214.

Armstrong, G.M., and J.K. Armstrong. 1950. Biological races of the Fusarium causing wilt of cowpea and soybeans. Phytopathology 40:181–193.

----, and ----. 1965. A wilt of soybean caused by a new form of *Fusarium oxysporum*. Phytopathology 55:237–239.

Armstrong, J.K., and G.M. Armstrong. 1958. A race of the cotton wilt Fusarium causing wilt of Yelredo soybean and flue-cured tobacco. Plant Dis. Rep. 42:147–151.

Athow, K.L. 1957. Studies of sobyean infection by the stem canker fungus. Phytopathology 47:2.

----, and R.M. Caldwell. 1954. A comparative study of Diaporthe stem canker and pod and stem blight of soybean. Phytopathology 44:319–325.

----, and F.A. Laviolette. 1973. Pod protection effects on soybean seed germination and infection with *Diaporthe phaseolorum* var. *sojae* and other microorganisms. Phytopathology 63:1021– 1023.

----, and ----. 1982. *Rps*$_6$, a major gene for resistance to *Phytophthora megsperma* f. sp. *glycinea* in soybean. Phytopathology 72:1564–1567.

----, ----, and J.R. Wilcox. 1979. The genetics of resistance to physiologic races of *Phytophthora megasperma* var. *sojae* in the soybean cultivar Tracy. Phytopathology 69:641–642.

----, ----, E.H. Mueller, and J.R. Wilcox. 1980. A new major gene for resistance to *Phytophthora megasperma* var. *sojae* in soybeans. Phytopathology 70:977–980.

----, and A.H. Probst. 1952. The inheritance of resistance to frogeye leaf spot of soybeans. Phytopathology 42:660–662.

Averre, C.W., and K.L. Athow. 1964. Host-parasite interaction between *Glycine max* and *Phytopthora megasperma* var. *sojae*. Phytopathology 55:886.

Backman, P.A., M.A. Crawford, J. White, D.L. Thurlow, and L.A. Smith. 1981. Soybean stem canker: A serious disease in Alabama. Highlights Agric. Res. 28:6.

----, R. Rodriguez-Kabana, J.M. Hammond, and D.L. Thurlow. 1979. Cultivar, environmental, and fungicide effects on foliar disease losses in soybeans. Phytopathology 69:562–564.

----, J.C. Williams, and M.A. Crawford. 1982. Yield losses in soybeans from anthracnose caused by *Colletotrichum truncatum*. Plant Dis. 66:1032–1034.

Beagle-Ristaino, J.E., and J.F. Rissler. 1983. Histopathology of susceptible and resistant soybean roots inoculated with zoospores of *Phytophthora megasperma* f. sp. *glycinea*. Phytopathology 73:590–595.

Bernard, R.L., and C.R. Cremeens. 1971. A gene for general resistance to downy mildew of soybeans. J. Hered. 62:359–362.

----, and ----. 1981. An allele at the rps_1 locus from the variety 'Kingwa'. Soybean Genet. Newsl. 8:40–42.

----, P.E. Smith, M.J. Kaufmann, and A.F. Schmitthenner. 1975. Inheritance of resistance to Phytophthora root and stem rot in the soybean. Agron. J. 49:391.

Boerma, H.R., and D.V. Phillips. 1983. Genetic implications of the susceptibility of Kent soybean to *Cercospora sojina*. Phytopathology 74:1666–1668.

Boosalis, M.G., and R.I. Hamilton. 1957. Root and stem rot of soybean caused by *Corynespora cassiicola* (Berk. & Curt.) Wei. Plant Dis. Rep. 41:696–698.

Buzzell, R.I., and T.R. Anderson. 1981. Another major gene for resistance to *Phytophthora megasperma* var. *sojae* in soybean. Soybean Genet. Newsl. 8:30–33.

----, and ----. 1982. Plant loss response of soybean cultivars to *Phytophthora megasperma* f. sp. *glycinea* under field conditions. Plant Dis. 66:1146–1148.

Carroll, R.B., and S. Leath. 1983. Effect of tillage practices on Fusarium wilt of soybean. Phytopathology 73:811.

Chamberlain, D.W. 1961. Reduction in water flow in soybean stems by a metabolite of *Cephalosporium gregatum*. Phytopathology 51:863–865.

----, and D.F. McAlister. 1954. Factors affecting the development of brown stem rot of soybean. Phytopathology 44:4–6.

----, and B.R. Lipscomb. 1967. Bibliography of soybean diseases. USDA Mimeo. CR-50-67 (July). Crops Res. Div., Oilseeds and Industrial Crops and Crops Protection Research Branch. U.S. Department of Agriculture, Washington, DC.

Cline, M.N., and B.J. Jacobsen. 1983. Methods for evaluating soybean cultivars for resistance to *Sclerotinia sclerotiorum*. Plant Dis. 67:784–786.

Corriveau, J.L., and R.B. Carroll. 1983. Comparison of nodulation and nitrogen fixation of till and no-till soybeans infected with *Fusarium oxysporum*. Phytopathology 73:828.

Crall, J.M. 1948. Defoliation of soybeans in southeastern Missouri caused by *Phyllosticta glycineum*. Plant Dis. Rep. 32:184–186.

----. 1950. Soybean diseases in Iowa in 1949. Plant Dis. Rep. 34:96–97.

----. 1951. Soybean diseases in Iowa in 1950. Plant Dis. Rep. 35:320–321.

----. 1956. Observations on the occurrence of soybean stem canker. Phytopathology 46:10.

Cromwell, R.O. 1917. Fusarium-blight or wilt disease of soybean. J. Agric. Res. 8:421–439.

Datnoff, L.E., and J.B. Sinclair. 1983. Field interaction between *Rhizoctonia solani* and *Fusarium oxysporum* on soybeans. Phytopathology 73:843.

Demski, J.W., and D.V. Phillips. 1974. Reaction of soybean cultivars to powdery mildew. Plant Dis. Rep. 58:723–726.

Duncan, D.R., and J.D. Paxton. 1981. Trifluralin enhancement of Phytophthora root rot of soybean. Plant Dis. 65:435–436.

Dunleavy, J. 1954. Soybean diseases in Iowa in 1953. Plant Dis. Rep. 38:89–90.

----. 1961. Fusarium blight of soybeans. Iowa Acad. Sci. Proc. 68:106–113.

----. 1970. Sources of immunity and susceptibility to downy mildew of soybean. Crop Sci. 10:507–509.

----. 1971. Races of *Peronospora manshurica* in the United States. Am. J. Bot. 58:209–211.

----. 1977. Nine new races of *Peronospora manshurica* found on soybeans in the Midwest. Plant Dis. Rep. 61:661– 663.

----. 1980. Yield loses in soybeans induced by powdery mildew. Plant Dis. 64:291–292.

----, and E.E. Hartwig. 1970. Sources of immunity from and resistance to nine races of the soybean downy mildew fungus. Plant Dis. Rep. 54:901–902.

----, and C.R. Weber. 1967. Control of brown stem rot of soybeans with corn-soybean rotations. Phytopathology 57:114–117.

Ellis, M.A., M.B. Ilyas, F.D. Tenne, J.B. Sinclair, and H.L. Palm. 1974. Effect of foliar applications of benomyl on internally seedborne fungi and pod and stem blight in soybean. Plant Dis. Rep. 58:760–763.

Ensminger, M.P., and T.S. Abney. 1978. Effect of foliar fungicides on fungal colonization and seed quality of soybeans in the Midwest. Phytopathol. News. Am. Phytopathol. Soc. 12:144.

Epps, W.M., J.C. Patterson, and I.E. Freeman. 1951. Physiology and parasitism of *Sclerotium rolfsii.* Phytopathology 41:245–256.

Fehr, W.R., C.E. Caviness, D.T. Burmood, and J. S. Pennington. 1971. Stages of development descriptions for soybeans, *Glycine max* (L.) Merrill. Crop Sci. 11:929–931.

Ferrant, N.P., and R.B. Carroll. 1981. Fusarium wilt of soybean in Delaware. Plant Dis. 65:596–599.

French, E.R. 1962. Fusarium root rot of soybean. Phytopathology 52:732.

----. 1963. Effect of soil temperture and moisture on the development of Fusarium root rot of soybean. Phytopathology 53:875.

----, and B.W. Kennedy. 1963. The role of *Fusarium* in the root rot complex of soybean in Minnesota. Plant Dis. Rep. 47:672–676.

Gardner, M.W. 1924. Indiana plant diseases, 1924. Proc. Indiana Acad. Sci. 35:237–257.

Garzonio, D.M., and D.C. McGee. 1983. Comparison of seeds and crop residue as sources of inoculum for pod and stem blight of soybeans. Plant Dis. 67:1374–1376.

Gaumann, E. 1923. Beitrage du einer monographie der gattung Peronospora Corda Beitr. (German monograph) Cryptogramen Scheiz. Dd. 5, Heft 4.

Geeseman, G.E. 1950. Physiologic races of *Peronospora manshurica* on soybeans. Agron. J. 42:257–258.

Gibson, F. 1922. Sunburn and aphid injury of soybean and cowpea. Ariz. Agric. Exp. Stn. Tech. Bull. 2.

Grabe, D.F., and J. Dunleavy. 1959. Physiologic specialization in *Peronospora manshurica.* Phytopathology 49:791–793.

Grau, C.R., and J.A. Laurence. 1975. Observations on resistance and heritability of resistance to powdery mildew of soybean. Plant Dis. Rep. 59:458–460.

----, V.L. Radke, and F.L. Gillespie. 1982. Resistance of soybean cultivars to *Sclerotinia sclerotiorum.* Plant Dis. 66:506–508.

Gray, L.E. 1971. Variation in pathogenicity of *Cephalosporium gregatum* isolates. Phytopathology 61:1410–1411.

----. 1974. Role of temperature, plant age, and fungas isolate in the development of brown stem rot of soybean. Phytopathology 64:94–96.

Haas, J.H., and R.I. Buzzell. 1976. New race 5 and 6 of *Phytophthora megasperma* var. *sojae* and differential reactions of soybean cultivars for race 1 to 6. Phytopathology 66:1361–1362.

Hara, K. 1915. Spot disease of soybean. (In Japanese.) Agric. Country 9:28.

Hartwig, E.E. 1959. Effect of target spot on yield of soybeans. Plant Dis. Rep. 43:504–505.

Haskell, R.J., and J.I. Wood. 1923. Diseases of cereal and forage crops in the United States in 1922. Plant Dis. Rep. Suppl. 27:164–265.

Hemmi, T. 1915. A new brown-spot disease of the leaf of *Glycine Hispida* Maxim. caused by *Septoria glycines* sp. N. Sapporo Nat. Hist. Soc. Trans. 6:12–17.

Hepperly, P.R., J.S. Mignucci, J.B. Sinclair, R.S. Smith, and W.H. Judy. 1982. Rhizoctonia web blight of soybean in Puerto Rico. Plant Dis. 66:256–257.

Herr, L.J. 1957. Factors affecting a root rot of soybean incited by *Phytophthora cactorum.* Phytopathology 47:15–16.

Higby, P.M., and H. Tachibana. 1982. Resistance to stem canker of soybeans. Phytopathology 72:1136.

Higgins, B.B. 1927. Physiology and parasitism of *Sclerotium rolfsii.* Phytopathology 17:417–448.

Hildebrand, A.A. 1952. Stem canker, a disease of increasing importance on soybeans in Ontario. Soybean Dig. 12:12–15.

----. 1954. Observation on the occurrence of the stem canker and pod and stem blight fungus on mature stems of soybeans. Plant Dis. Rep. 38:640–646.

----. 1959. A root and stalk rot of soybean caused by *Phytophthora megasperma* Drechsler var. *sojae* var. nov. Can. J. Bot. 37:927–957.

----, and L.W. Koch. 1951. A study of the systemic infection by downy mildew of soybean with special reference to symptomatology, economic significance and control. Sci. Agric. 31:505–518.

----, and ----. 1952. Observations on a root and stem rot of soybeans new to Ontario, caused by *Pythium ultimum* Trow. Sci. Agric. 32:574–580.

Hill, H.C., N.L. Horn, and W.L. Steffan. 1981. Mycelial development and control of *Phomopsis sojae* in artificially inoculated soybean stems. Plant Dis. 65:132–134.

Hilty, J.W., and A.F. Schmitthenner. 1962. Phytopathogenic and cultural variability of single zoospore isolates of *Phytophthora megasperma* var. *sojae.* Phytopathology 52:859–862.

Hobbs, T.W., A.F. Schmitthenner, C.W. Ellett, and R.E. Hite. 1981. Top dieback of soybean caused by *Diaporthe phaseolorum* var. *caulivora.* Plant Dis. 65:618–620.

Horn, N.L., F.N. Lee, and R.B. Carver. 1975. Effects of fungicides and pathogens on yield of soybeans. Plant Dis. Rep. 59:724–728.

Ilyas, M.B., M.A. Ellis, and J.B. Sinclair. 1976. Effect of soil fungicides on *Macrophomina phaseolina* sclerotium viability in soil and in soybean stem pieces. Phytopathology 66:355–359.

Jehle, R.A., A.E. Jenkins, K.W. Kreitlow, and H.S. Sherwin. 1952. an outbreak of phyllosticta canker and leaf spot of soybeans in Maryland. Plant Dis. Rep. 36:155–158.

Jimenez, B., and J.L. Lockwood. 1980. Laboratory method for assessing field tolerance of soybean seedlings to *Phytophthora megasperma* var. *sojae.* Plant Dis. 64:775–778.

Jones, J.P. 1959. Purple stain of soybean seeds incited by several *Cercospora* species. Phytopathology 49:430–432.

----. 1961. A leaf spot of cotton caused by *Corynespora cassiicola.* Phytopathology 51:305–308.

Kamal-Abo-el-Dahab, M. 1968. Occurrence of brown stem rot disease of soybean in Egypt (U.A.R). Phytopathol. Mediterr. 7:28–33.

Kaufmann, M.J., and J.W. Gerdemann. 1958. Root and stem rot of soybean caused by *Phytophthora sojae* N. sp. Phytopathology 48:201–208.

Keeling, B.L. 1979. Research on Phytophthora root and stem rot: Isolation, testing procedures, and seven new physiologic races. p. 367–370. F.T. Corbin (ed.) World soybean research conference II: Proceedings Westview Press, Boulder, CO.

----. 1982a. Four new physiologic races of *Phytophthora megasperma* f. sp. *glycinea.* Plant Dis. 66:334–335.

----. 1982b. A seedling test for resistance to soybean stem canker caused by *Diaporthe phaseolorum* var. *caulivora.* Phytopathology 72:807–809.

Kilen, T.C., and W.L. Barrentine. 1983. Linkage relationship in soybean between genes controlling reactions to Phytophthora root rot and metribuzin. Crop Sci. 23:894–896.

----, E.E. Hartwig, and B.L. Keeling. 1974. Inheritance of a second major gene for resistnace to Phytophthora rot in soybean. Crop Sci. 14:260–262.

Kittle, D.R., and L.E. Gray. 1979. The influence of soil temperature, moisture, porosity, and bulk density on the pathogenicity of *Phytophthora megasperma* var. *sojae.* Plant Dis. Rep. 63:231–234.

Klag, N.G., G.C. Papavizas, G.A. Bean, and J.G. Kantzes. 1978. Root rot of soybeans in Maryland. Plant Dis. Rep. 62:235–239.

Kmetz, K., C.W. Ellett, and A.F. Schmitthenner. 1974. Isolation of seedborne *Diaporthe phaseolorum* and *Phomopsis* from immature soybean plants. Plant Dis. Rep. 58:978–982.

----, and ----. 1979. Soybean seed decay: Source of inoculum and nature of infection. Phytopathology 69:798–801.

----, A.F. Schmitthenner, and C.W. Ellett. 1978. Soybean seed decay: Prevalence of infection and symptom expression caused by *Phomopsis* sp., *Diaporthe phaseolorum* var. *sojae,* and *D. phaseolorum* var. *caulivora.* Phytopathology 68:838–840.

Krausz, J.P., and B.A.Fortnum. 1983. An epiphtotic of Diaporthe stem canker of soybean in South Carolina. Plant Dis. 67:1128–1129.

Kuan, Ta-Li, and D.C. Erwin. 1980. Formae speciales differentiation of *Phytophthora megasperma* isolates from soybean and alfalfa. Phytopathology 70:333–338.

Laviolette, F.A., and K.L. Athow. 1972. *Cercospora kikuchii* infection of soybean as affected by stage of plant development. Phytopathology 62:71.

----, and ----. 1977. Three new physiologic races of *Phytophthora megasperma* var. *sojae.* Phytopathology 67:267–268.

----, and ----. 1983. Two new physiologic races of *Phytophthora megasperma* f. sp. *glycinea.* Plant Dis. 67:497–498.

----, ----, E.H. Mueller, and J.R. Wilcox. 1979. Inheritance of resistance in soybeans to physiologic races 5, 6, 7, 8, and 9 of *Phytophthora megasperma* var. *sojae.* Phytopathology 69:270– 271.

----, ----, A.H. Probst, J.R. Wilcox, and T.S. Abney. 1970. Effect of bacterial pustule and frogeye leafspot on yield of Clark soybean. Crop Sci. 10:418–419.

Leath, S., and R.B. Carroll. 1980. Screening for resistance to *Fusarium oxysporum* in soybean. Plant Dis. 66:1140–1143.

Lehman, S.G. 1923. Pod and stem blight of soybean. Ann. Mo. Bot. Gard. 10:111–178.

----. 1928. Frogeye leaf spot of soybean caused by *Cercospora diazu* Miura. J. Agric. Res. 36:811–833.

----. 1947. Powdery mildew of soybean. Phytopathology 37:434.

----. 1950. Purple stain of soybean seeds. N.C. Agric. Exp. Stn. Bull. 369.

----. 1953. Race 4 of the soybean downy mildew fungus. Phytopathology 43:460– 461.

----. 1958. Physiologic races of the downy mildew fungus on soybean in North Carolina. Phytopathology 48:83–86.

----, and F.A. Wolf. 1924. A new downy mildew on soybean. J. Elisha Mitchell Sci. Soc. 39:164–169.

----, and ----. 1926a. Pythium root rot of soybean. J. Agric. Res. 33:375–380.

----, and ----. 1926b. Soybean anthracnose. J. Agric. Res. 33:381–390.

Lewis, J.A., and G.C. Papavizas. 1977. Factors affecting *Rhizoctonia solani* infection of soybeans in the greenhouse. Plant Dis. Rep. 61:196–200.

Lim, S.M. 1979. Evaluation of soybean for resistance to Septoria brown spot. Plant Dis. Rep. 63:242–245.

----. 1980. Brown spot severity and yield reduction in soybean. Phytopathology 70:974–977.

----. 1983. Response to *Septoria glycines* of soybean nearly isogenic except for seed color. Phytopathology 73:719–722.

Ling, L. 1951. Bibliography of soybean diseases. Plant Dis. Rep. Suppl. 204:109–173.

Luttrell, E.S. 1947. *Diaporthe phaseolorum* var. *sojae* on crop plants. Phytopathology 37:445–465.

MacNeill, B.H., and H. Zalasky. 1957. Histological study of host-parasite relationships between *Septoria glycines* Hemmi and soybean leaves and pods. Can. J. Bot. 35:501–505.

Massalongo, C. 1900. De nonullis speciebus novis micromycetum agri veronensis. Atti R. 1st Veneto Sci. Lett. Arti. 59:683–690.

Matsumoto, T., and R. Tomoyasii. 1925. Studies on purple speck of soybean seed. Ann. Phytopathol. Soc. Jpn. 1:1–14.

McGee, D.C., and C.L. Brandt. 1979. Effect of foliar application of benomyl on infection of soybean seeds by *Phomopsis* in relation to time of inoculation. Plant Dis. Rep. 63:675–677.

Melchers, L.E. 1925. Diseases of cereal and forage crops in the United States in 1924. Plant Dis. Rep. Suppl. 40:186.

Meyer, W.B., J.B. Sinclair, and M.N. Khare. 1973. Biology of *Macrophomina phaseoli* in soil studied with selective media. Phytopathology 63:613–620.

Mignucci, J.S., and S.M. Lim. 1980. Powdery mildew development on soybean with adult plant resistance. Phytopathology 70:919–921.

Miller, W.A., and K.W. Roy. 1982. Effect of benomyl on the colonization of soybean leaves, pods, and seeds by fungi. Plant Dis. 66:918–920.

Miura, M. 1921. Diseases of the main agricultural crops of Manchuria. South Manchuria Railway Co. Agric. Exp. Stn. Bull. 11 (In Japanese.) (English Abstr.) Jpn. J. Bot. 1:(1922)9.

Moots, C.K., C.D. Nickell, L.E. Gray, and S.M. Lim. 1983. Reaction of soybean cultivars to 14 races of *Phytophthora megasperma* f. sp. *glycinea*. Plant Dis. 67:764–767.

Morgan, F.L. 1963. Soybean leaf and stem infection by *Phytophthora megasperma* var. *sojae*. Plant Dis. Rep. 47:880–882.

----, and J. Dunleavy. 1966. Brown stem rot of soybean in Mexico. Plant Dis. Rep. 50:598–599.

----, and E.E. Hartwig. 1964. *Pythium aphanidermatum*, a virulent soybean pathogen. Phytopathology 54:901.

----, and ----. 1965. Physiologic specialization in *Phytophthora megasperma* var. *sojae*. Phytopathology 55:1277–1279.

Morton, H.V., C.L. Kern, and T.D. Taylor. 1982. Control of Pythium damping off and Phytophthora root rot of soybeans with metalaxyl. Phytopathology 72:971.

Muchovey, J.J., R.M.C. Muchovey, O.D. Dhingra, and L.A. Maffia. 1980. Suppression of anthracnose of soybean by calcium. Plant Dis. 64:1088–1089.

Mueller, E.H., K.L. Athow, and F.A. Laviolette. 1978. Inheritance of resistance to four physiologic races of *Phytophthora megasperma* var. *sojae*. Phytopathology 68:1318–1322.

Murakishi, H.H. 1951. Purple seed stain of soybean. Phytopathology 41:305–318.

Naoumoff, N. 1914. Materiaux pour la flore mycologique de la Russie. Fungi ressurienses I. Bull. Soc. Mycol. Fr. 30:64–83.

Nattrass, R.M. 1961. Host list of Kenya fungi and bacteria. Mycol. Paper 81. Commonwealth Mycological Institute, Kew, UK.

Olive, L.S., D.C. Bain, and C.L. Lefebvre. 1945. A leaf spot of cowpea and soybean caused by an undescribed species of *Helminthosporium*. Phytopathology 35:822–831.

----, and E.A. Walker. 1944. A severe leafspot of soybean caused by *Phyllosticta sojaceola*. Plant Dis. Rep. 28:1122–1123.

Papavizas, G.C., F.W. Schwenk, J.C. Locke, and J.A. Lewis. 1979. Systemic fungicides for controlling Phytophthora root rot and damping-off of soybean. Plant Dis. Rep. 63:708–712.

Pataky, J.K., and S.M. Lim. 1981. Efficacy of benomyl for controlling Septoria brown spot of soybeans. Phytopathology 71:438–442.

Phillips, D.V. 1970. Incidence of brown stem rot of soybean in Georgia. Plant Dis. Rep. 54:987–988.

----. 1971. Influence of air temperature on brown stem rot of soybean. Phytopathology 61:1205–1208.

----, and H.R. Boerma. 1981. *Cercospora sojina* race 5: A threat to soybeans in the southeastern United States. Phytopathology 71:334–336.

----, and ----. 1982. Two genes for resistance to race 5 of *Cercospora sojina* in soybeans. Phytopathology 72:764–766.

Phipps, P.M. 1983. Incidence of Sclerotinia blight of soybean. Phytopathology 73:968.

----, and D.M. Porter. 1982. Sclerotinia blight of soybean caused by *Sclerotinia minor* and *Sclerotinia sclerotiorum*. Plant Dis. 66:163–165.

Presley, J.T., and W.B. Allington. 1947. Brown stem rot of soybean caused by a *Cephalosporium*. Phytopathology 37:681.

Probst, A.H., K.L. Athow, and F.A. Laviolette. 1965. Inheritance of resistance to race 2 of *Cercospora sojina* in soybeans. Crop Sci. 5:332.

Radke, V.L., and C.R. Grau. 1982. Effect of herbicides on carpogenic germination of sclerotia of *Sclerotinia sclerotiorum*. Phytopathology 72:1139.

Ross, J.P. 1968. Additional physiologic races of *Cercospora sojina* on soybeans in North Carolina. Phytopathology 58:708–709.

----. 1975. Effect of overhead irrigation and benomyl sprays on late season foliar diseases, seed infection, and yields of soybean. Plant Dis. Rep. 59:809–813.

----, and T.J. Smith. 1963. Brown stem rot of soybean in North Carolina and Virginia. Plant Dis. Rep. 47:329.

Roy, K.W. 1982. Seedling diseases caused in soybean by species of *Colletotrichum* and *Glomerella*. Phytopathology 72:1093–1096.

----, and T.S. Abney. 1976. Purple seed stain of soybean. Phytopathology 66:1045–1049.

----, and W.A. Miller. 1983. Soybean stem canker incited by isolates of *Diaporthe* and *Phomopsis* spp. from cotton in Mississippi. Plant Dis. 67:135–137.

Schlub, R.L., and J.L. Lockwood. 1981. Etiology and epidemiology of seedling rot of soybean by *Pythium ultimum*. Phytopathology 71:134–138.

Schmitthenner, A.F. 1972. Evidence for a new race of *Phytophthora megasperma* var. *sojae* pathogenic to soybean. Plant Dis. Rep. 56:536–539.

----. 1983. Relative efficacy of metalaxyl seed and soil treatments for control of Phytophthora root rot of soybean. Phytopathology 73:801.

----, and A.K. Walker. 1979. Tolerance versus resistance for control of Phytophthora root rot of soybeans. p. 35–44. *In* Proc. 9th Soybean Seed Res. Conf., Chicago, IL. December. American Seed Trade Association, Washington, DC.

Schwenk, F.W., and T. Sim. 1974. Race 4 of *Phytophthora megasperma* var. *sojae* from soybean proposed. Plant Dis. Rep. 58:352–354.

Seaman, W.L., R.A. Shoemaker, and E.A. Peterson. 1965. Pathogenicity of *Corynespora cassiicola* on soybean. Can. J. Bot. 43:1461–1469.

Sherwin, H.S., and K.W. Kreitlow. 1952. Discoloration of soybean seeds by the frogeye fungus, *Cercospora sojina*. Phytopathology 42:568–572.

Short, G.E., T.D. Wyllie, and V.D. Ammon. 1978. Quantitative enumeration of *Macrophomina phaseolina* in soybean tissue. Phytopathology 68:736–741.

Shortt, B.J., A.P. Grybauskas, F.D. Tenne, and J.B. Sinclair. 1981. Epidemiology of Phomopsis seed decay of soybean in Illinois. Plant Dis. 65:62–64.

Sinclair, J.B. (ed.) 1982. Compendium of soybean diseases. 2nd ed. American Phytopathological Society, St. Paul.

----, and O.D. Dhingra. 1975. An annotated bibliography of soybean diseases. INSOY Ser. 7, University of Illinois, Urbana.

Southern, J.W., N.C. Schenck, and D.J. Mitchell. 1976. Comparative pathogenicity of *Pythium myriotylum* and *P. irregulare* to the soybean cultivar Bragg. Phytopathology 66:1380–1385.

Spencer, J.A., and H.J. Walters. 1969. Variations in certain isolates of *Corynespora cassiicola*. Phytopathology 59:58–60.

Spilker, D.A., A.F. Schmitthenner, and C.W. Ellett. 1981. Effects of humidity, temperature, fertility and cultivar on the reduction of soybean seed quality by *Phomopsis* sp. Phytopathology 71:1027–1029.

Stuckey, R.E., S.A. Ghabrial, and D.A. Reicosky. 1982. Increased incidence of *Phomopsis* sp. in seed from soybeans infected with bean pod mottle virus. Plant Dis. 66:826–829.

Suhovecky, A.J., and A.F. Schmitthenner. 1955. Soybeans affectd by early root rot. Ohio Farm Home Res. 40:85–86.

Tachibana, H., D. Jowett, and W.R. Fehr. 1971. Determination of losses in soybeans caused by *Rhizoctonia solani*. Phytopathology 61:1444–1446.

----, and A. Van Diest. 1983. Association of Phytophthora root rot of soybean with conservation tillage. Phytopathology 73:844.

Tehon, L.R., and E.Y. Daniels. 1927. Notes on the parasitic fungi of Illinois III. Mycologia 19:100–129.

Tenne, F.D., and J.B. Sinclair. 1978. Control of internally seedborne microorganisms of soybean with foliar fungicides in Puerto Rico. Plant Dis. Rep. 62:459–463.

Thomson, T.B., K.L. Athow, and F.A. Laviolette, 1971. Effect of temperature on the pathogenicity of *Pythium aphanidermatum*, *P. debaryanum*, and *P. ultimum* on soybean. Phytopathology 61:933–935.

Threinen, J.T., T. Kommedahl, and R.J. Klug. 1959. Hybridization between radiation-induced mutants of two varieties of *Diaporthe phaseolorum*. Phytopathology 49:797–801.

Tooley, P.W., and C.R. Grau. 1982. Identification and quantitative characterization of rate-reducing resistance to *Phtophthora megasperma* f. sp. *glycinea* in soybean seedlings. Phytopathology 72:727–733.

Vaartaja, O., R.E. Pitblado, R.R. Buzzell, and L.G. Crawford. 1979. Chemical and biological control of Phytophthora root and stalk rot of soybean. Can. J. Plant Sci. 59:307–311.

Walker, A.K., and A.F. Schmitthenner. 1982. Developing soybean varieties tolerant to Phytophthora rot. p. 67–78. *In* Proc. 12th Soybean Seed Res. Conf., Chicago, IL. 7–8 December. American Seed Trade Association, Washington, DC.

Wallen, V.R., and W.L. Seaman. 1963. Seed infection of soybean by *Diaporthe phaseolorum* and its influence on host development. Can. J. Bot. 41:13–21.

Walters, H.J. 1980. Soybean leaf blight caused by *Cercospora kikuchii*. Plant Dis. 64:961–962.

----, and K.F. Martin. 1981. *Phyllosticta sojaecola* on pods of soybeans in Arkansas. Plant Dis. 65:161–162.

Weaver, D.B., B.H. Cosper, P.A. Backman, and M.A. Crawford. 1984. Cultivar resistance to field infestations to soybean stem canker. Plant Dis. 68:877–879.

Wehmeyer, L.E. 1933. The genus *Diaporthe* Nitschke and its segregates. University of Michigan Press, Ann Arbor, MI.

Welch, A.W., and J.C. Gilman. 1948. Hetero- and homo-thallic types of *Diaporthe* on soybeans. Phytopathology 38:628–637.

White, D.M., J.E. Partridge, and J.H. Williams. 1983. Races of *Phytophthora megasperma* f. sp. *glycinea* on soybean in eastern Nebraska. Plant Dis. 67:1281–1284.

Whitehead, M.C. 1966. Stem canker and blight of birdsfoot-trefoil and soybean incited by *Diaporthe phaseolorum* var. *sojae*. Phytopathology 56:396–400.

Wilcox, J.R., and T.S. Abney. 1973. Effects of *Cercospora kikuchii* on soybeans. Phytopathology 63:796–797.

----, F.A. Laviolette, and R.J. Martin. 1975. Heritability of purple seed stain resistance in soybean. Crop Sci. 15:525- 526.

Williams, D.J., and R.F. Nyvall. 1980. Leaf infection and yield losses caused by brown spot and bacterial blight diseases of soybean. Phytopathology 70:900–902.

Wolf, F.A. 1923. Report of the division of plant pathology. N.C. Agric. Exp. Stn. Ann. Rep. 46:92.

----, and S.G. Lehman. 1926. Brown spot disease of soybeans. J. Agric. Res. 33:365–374.

Yorinori, .T. 1980. *Cercospora sojina*: Pathogenicity, new races and seed transmission in soybean. Ph.D. thesis. Univeristy of Illinois, Urbana.

Young, L.D., and J.P. Ross. 1978. Resistance evaluation and inheritance of a nonchlorotic response to brown spot of soybean. Crop Sci. 18:1075–1077.

----, and ----. 1979. Brown spot development and yield response of soybean inoculated with *Septoria glycines* at various growth stages. Phytopathology 69:8–11.

18 Viral and Bacterial Diseases

J. P. Ross
USDA-ARS and North Carolina State University
Raleigh, North Carolina

Soybean (*Glycine max* (L.) Merr.] diseases caused by viruses are probably found in all soybean-producing areas. Whereas certain local units of production may be free of virus diseases, the transmission of virus through seed to the seedling is not uncommon and is probably the main reason for the wide distribution of some soybean viruses. Factors which determine the establishment of a virus in an area where it has been introduced are the presence of a suitable vector and the cultivation of susceptible soybean cultivars. The presence of natural wild hosts, particularly perennial plants also enhance the probability of establishment, especially if the virus is not transmitted through soybean seed. Since environmental conditions can markedly affect insect population levels and alternate host survival, the occurrence of soybean viral epiphytotics may be sporadic.

Although many viruses will infect soybean, this chapter will deal only with those virus diseases that have been described as naturally occurring on soybean and appear sufficiently significant to be a problem in soybean production in some area of the world. Table 18–1 presents some characteristics of these viruses.

18–1 RECENT DEVELOPMENTS IN VIROLOGY

The discipline of plant virology has made rapid advances since 1975. Viruses have been more fully characterized and classified into groups based not only on properties such as inclusion bodies within host cells, host range, and transmission characteristics but also on properties of the virus particle, such as the molecular weight of the nucleic acid, protein molecular weight, sedimentation coefficients, morphology, and antigenic relationships. Employment of more sophisticated characteristics than those of the dilution end point, thermal inactivation point and longevity in vitro has become more practical due to recent advances in technology (Francki, 1980).

Since 1965, virologists have been attempting to form a standardized acceptable system of nomenclature for viruses through the work of the

Soybeans: Improvement, Production, and Uses, 2nd ed.—Agronomy Monograph no. 16.

Table 18–1a. Characteristics of viruses that occur naturally on soybean.

Virus group	Virus (Abbr.)	Particle morphology, size	Transmission†	Natural host range
Alfalfa mosaic virus	Alfalfa mosaic (AMV)	Bacilliform 58×18 nm	Aphid (NP), M, seed	Wide
Bromovirus	Cowpea chlorotic mottle (CCMV)	Isometric 26 nm	Beetle, M	Fabaceae Cucurbitaceae
Carlavirus (?)	Cowpea mild mottle (CMMV)?	Straight to flexuous rod 13×650 nm	Whitefly (SP), M. seed	Wide
Comovirus	Bean pod mottle (BPMV)	Isometric 30 nm	Beetle, M, seed	Fabaceae
	Cowpea mosaic (CMV)	Isometric 20–24 nm	Beetle, M	Fabaceae
	Cowpea severe mosaic (CSMV)	Isometric 25nm	Beetle, M	Fabaceae
Cucumovirus	Soybean stunt (SSV)	Isometric 20–28 nm	Aphid (NP), M, seed	Wide
	Peanut stunt (PSV)	Isometric 30 nm	Aphid (NP), M, seed	Wide
Geminivirus (?)	Soybean crinkle leaf (SCLV)		Whitefly (P)	Wide
	Soybean yellow mosaic (SYMV)		Whitefly (P)	Fabaceae
Ilarvirus	Tobacco streak (TSV)	Polyhedral 28–32 nm	M, seed	Wide
Luteovirus	Soybean dwarf (SDV)	Isometric 25 nm	Aphid (C)	Fabaceae
	Indonesian (ISDV)	Isometric 26 nm	Aphid (C)	Soybean
Nepovirus	Tobacco ringspot (TRV) (Bud blight)	Isometric 29 nm	Nematode, thrip, grasshopper, M, seed	Wide
Potyvirus	Soybean mosaic (SMV)	Flexuous rod 12×750 nm	Aphid (NP), M, seed	Fabaceae
	Bean yellow mosaic (BYMV)	Flexuous rod 15×750 nm	Aphid (NP), M	Wide
	Peanut mottle (PMV)	Flexuous rod 12×750 nm	Aphid (NP), M	Fabaceae

†NP = Nonpersistent, M = mechanical, P = persistent, C = circulative, and SP = semipersistent.

Table 18-1b. Characteristics of viruses that occur naturally on soybean.

Virus group	Virus (Abbr.)	Virus distribution	Distinguishing symptoms	Immunity in soybean	Strains
Alfalfa mosaic virus	Alfalfa mosaic (AMV)	Worldwide	Mild mottle	—	Several
Bromovirus	Cowpea chlorotic mottle (CCMV)	USA-GA	—	Single dom. gene	—
Carlavirus(?)	Cowpea mild mottle (CCMV)?	Ivory Coast, Thailand	Leaf malformation, mosaic, top necrosis	—	—
Comovirus	Bean pod mottle (BPMV)	USA	Mottle of young leaves	None	—
	Cowpea mosaic (CMV)	USA, Africa, Asia	—	—	—
	Cowpea severe mosaic (CSMV)	Western hemisphere	Severe stunting	—	—
Cucumovirus	Soybean stunt (SSV)	Japan, USA, S. Africa, Indonesia	Mottled seed, concentric or broken lines	—	Four strains
	Peanut stunt (PSV)	USA, Japan	—	Present	—
Geminivirus(?)	Soybean crinkle Leaf (SCLV)	Thailand	Leaf malformation	—	—
	Soybean yellow mosaic (SYMV)	India	Bright yellow areas	Present	—
Ilarvirus	Tobacco streak (TSV)	Brazil, USA	Bud necrosis	—	Present
Luteovirus	Soybean dwarf (SDV)	Japan (Hokkaido)	Stunting, leaf curling, rugose	None	Dwarfing or chlorotic
	Indonesian (ISDV)	Indonesia	Dwarfing, dark green leaves	—	—
Nepovirus	Tobacco ringspot (TRV) (Bud Blight)	USA, Canada, Asia, Australia	Terminal necrosis	None	Many
Potyvirus	Soybean mosaic (SMV)	Worldwide	Mottled seed	One dom. gene, allelic series; one rec. gene	Many
	Bean yellow mosaic (BYMV)	USA, Japan	Yellow mottling	Present	Many
	Peanut mottle (PMV)	Worldwide	—	Single dom. gene	Many

†NP = Nonpersistent, M = mechanical, P = persistent, C = circulative, and SP = semipersistent.

International Committee of Taxonomy of Viruses (ICTV) of the International Association of Microbiology Societies (Mathews, 1979). At present there are 23 groups and two families which contain plant viruses (Francki, 1981). Ten of these groups contain viruses which affect soybean and will be discussed in this chapter (Table 18–1). Viruses within each group have basic similar properties, such as the number of pieces containing genetic information, but may differ in some distinguishable characteristic such as serological relationships, host range, and transmission characteristics. The names of the groups are siglas which may utilize part of the common name of the group type virus as the prefix (*pot*ato virus *Y* is the type virus of the *poty*virus group) or may use characteristics of the viruses as a prefix (*ne*matode-transmitted, with *po*lyhedral particles, hence, *nepo*virus group).

18–2 SYMPTOMATOLOGY

The range of symptoms exhibited by virus-infected soybean plants vary from those that are barely detectable to those that cause severe stunting or death. Certain viruses may produce typically unique symptoms which can be of great aid in virus identification. Plant reaction to virus infection, however, may vary greatly depending on soybean cultivar, environment, and stage of plant development at time of inoculation. Therefore, the use of symptoms as an aid in identifying the virus should be tempered with caution; knowledge of the symptoms elicited by the virus on various cultivars can be of aid. A single virus strain may produce local lesions on inoculated leaves, a systemic necrosis, or a mild mosaic depending on the soybean genotype (Ross, 1969a). The existence of an array of virus strains, each producing somewhat different symptoms, may lead to confusion in virus identity; plants infected with soybean mosaic and bean pod mottle viruses may manifest symptoms unlike those produced by either virus alone (Ross, 1968). Generally, viruses that cause severe symptoms, such as severe stunting, plant barrenness, or death are not endemic but rather occur sporadically following a specific sequence of events.

18–2.1 Foliage Symptoms

The variety of foliar symptoms expressed by virus-infected soybean are valuable as an indication that the plants are infected with a virus. However, the use of foliar symptoms to identify the virus is of limited value unless the scientist is familiar with the range of symptoms caused by many viruses. Foliar symptoms caused by other factors, such as herbicides, may appear similar to those caused by virus.

In some instances, foliar symptoms may fade and become masked. For example, foliar symptoms on SMV-infected field-grown plants may disappear when daily maximum temperatures frequently exceed 30 °C.

Some viruses cause their most noticeable symptoms in the younger leaves (BPMV), whereas others, such as SMV, normally cause the older leaves to become distorted and crinkled. Foliar symptoms on greenhouse-grown plants may be vastly different from those that develop on the same cultivars grown in the field.

Initial symptoms of many virus infections are often uniquely different and more severe than those that develop later. Initial symptoms of TRSV may include recurving of the terminal growing point to form a crook which is brittle and often dies; subsequent growth from axillary buds is stunted. Young leaves developing soon after inoculation of many viruses manifest a yellow netting associated with the veins, later however, this symptom does not appear in the young leaves.

Some of the more common foliar symptoms are: mosaic or mottling (various sized areas of yellow and green), distortion in leaf shape (cupping), crinkling or raised areas (rugosity), enations from veins, chlorotic or necrotic spots, general chlorosis, dwarfing of leaf size, and bronzing of older leaves.

18–2.2 Stem Symptoms

Symptoms of virus infection on stems usually result from altered growth habits. In many cases the younger the plant at time of infection, the more pronounced the alteration in plant growth. Stunting of plant growth is one of the most common stem symptoms, and some viruses (TRSV, TSV, SDV, and SSV) cause pronounced height reduction.

Necrosis on uneven growth from the apical meristem may create a markedly distorted plant; if the apical meristem loses dominance and axillary buds initiate growth from short internodes, the plant may develop a witches broom-like appearance (TRSV). Plants infected by both SMV and BPMV may develop a stem curvature soon after infection, but subsequent stem growth is straight although the plants may be stunted (Lee and Ross, 1972).

Height reduction caused by virus infection may mainly result from the initial reaction following infection. The percent height reduction of BPMV-infected plants measured at plant maturity were similar to those measured 30 days after inoculation of plants at the second trifoliolate leaf stage. Hence, the chronic phase of the disease caused no growth reduction (Windham and Ross, 1985).

A symptom called *soybean green stem* was attributed by Schwenk and Nickell (1980) to BPMV infection. This condition is observed after plant maturity and consists of the stems remaining green after all leaves have abscissed and pods dried; it can reduce the threshing efficiency by increasing seed loss during combining.

The stem pith of plants infected with TRSV may have a brown discoloration particularly at the nodes.

18–2.3 Pod and Seed Symptoms

Symptoms of pods of TRSV-infected plants may consist of distortion and blotching, poorly filled pods, and a high percentage of one-seeded pods. The barrenness of plants infected with this virus prior to flowering is largely attributed to failure of anthers to dehisce and poor pollen germination; many seed set on such plants are the result of cross-pollination by insects (Brim et al., 1964). Plants infected with TRSV prior to flowering often possess clusters of "dudded" flowers (undeveloped pods). These plants fail to mature normally and retain their leaves after noninfected plants have matured.

Pods of SMV-infected plants may lose their pubescence and appear glabrous, and when plants are infected with SMV and BPVM, pods may show dark streaks and be distorted.

Other than a reduction in seed size, viruses can cause unique patterns of discoloraton (mottling) on the seed coat. The color of the mottling is apparently controlled by the allelic series of genes governing seed hilum pigmentation and is under the modification by virus infection of the *I* and i^i alleles (Wilcox and Laviolette, 1968). Some seed from SMV-infected plants display a mottling radiating out from the hilum whereas SSV (Japan) may cause a concentric ring or broken line pattern (Koshimizu and Iizuka, 1963); SSV (Indonesia) caused both rings, streaks, and mottle, and the mottled seed transmitted the virus to progeny at a high rate (Roechan et al., 1975). The seed mottling caused by SMV was found to increase as temperatures during early pod set were decreased from 30 to 20 °C (Ross, 1970). However, certain cultivars, such as Forrest, tend to have mottled seed in the apparent absence of SMV, therefore, seed-coat mottling is not conclusive evidence that the parent plant was SMV-infected. Hill et al. (1980) also obtained evidence that indicated mottled seed is not a reliable indication that the mother plant was SMV-infected. Since soybean genotypes, temperature at early pod set (Ross, 1970), stage of plant growth at inoculation (Ross, 1969b), and strain of SMV (Ross, 1968) can affect the amount of seed mottling caused by SMV, and since Lin and Hill (1983) reported that BPMV may cause seed mottling, many factors should be considered in drawing conclusions about the condition of a crop based on whether or not its seed were mottled.

18–2.4 Root Symptoms

The effect of virus infection on root symptoms is less noticeable than that on the aboveground plant parts. Keeling (1982) reported that under controlled conditions of greenhouse or growth chamber the root volume and root dry weight of SMV-infected plants were reduced up to 61 and 38%, respectively, compared to these measurements of uninfected plants 30 days after virus inoculation. Orellana et al. (1978) reported that nodules on TRSV-infected plants were drastically reduced or eliminated until plants were 40-days old. Since root functions are dependent on the energy related processes of plant tops, any detrimental effect on physiological processes such as photosynthesis, respiration, or other essential metabolic conversions will be reflected in energy-requiring processes of the root such as root growth, nodule formation, and dinitrogen (N_2) fixation.

18–3 TRANSMISSION

Soybean viruses are known to be transmitted by aphids, whiteflies, beetles, grasshoppers, thrips, nematodes, through seed, and by mechanical means. The transmission characteristics of the viruses are of great importance in determining the economic impact of the diseases they cause. Although soybean plants are hosts to over 50 viruses and virus strains (Sinclair, 1982), less than 20 are probably of economic importance; this probably is due in large part to their transmission relationships.

18–3.1 Aphids

The aphid-transmitted viruses that affect soybean belong to four virus groups: potyviruses, cucumoviruses, alfalfa mosaic virus, and luteoviruses. The first three groups are stylet-borne and can be acquired and transmitted in short periods (some < 1 min), and the ability to transmit the virus is rapidly lost (1–4 h). Luteoviruses require a latent period between virus acquisition by the aphid and its ability to transmit the virus (about 24 h); the aphid can transmit virus for weeks.

18–3.1.1 Potyviruses

Soybean mosaic is probably the most widespread soybean virus disease because it is seed borne, transmitted by over 30 aphid species (Abney et al., 1976), and usually does not cause severe affects on the host. These characteristics and its diverse pathogenic strains make SMV adaptable to worldwide distribution. Research with other members of the potyvirus groups have indicated that certain virus strains can be transmitted by certain aphid species, but other strains may not be transmitted by the same species (Hollings and Brunt, 1981). Hence, virus-aphid vector specificity is important. The seven SMV strains discerned by Cho and Goodman (1979) were found as seed transmitted viruses mainly in the soybean germplasm collection grown either in Illinois or Mississippi; all transmissions were done by mechanically rubbing leaves with sap from the infected source plants. Some of their isolates could infect cultivars considered SMV-resistant in the USA. One of the reasons these pathogenic isolates are not prevalent in the USA probably relates to a vector-virus strain specificity, which may play a major role in preventing outbreaks of mosaic epidemics in the USA.

Hill et al. (1980) concluded that most spread of SMV within soybean fields in Iowa was from initial infection foci of seedlings infected from seed transmission rather than from sources outside the field. The PMV was also spread within peanut (*Arachis hypogaea* L.) fields from seed-borne infections of peanut seedlings, and the amount of spread was dependent on aphid population levels (Paguio and Kuhn, 1974). The PMV apparently is not seed transmitted in soybean and was found to be transmitted into soybean fields from nearby virus-infected peanut fields (Demski, 1975). This study found no correlation between PMV spread and winged aphid populations which suggests that perhaps some other vector beside aphids may be involved.

18-3.1.2 Cucumoviruses

Although peanut stunt virus (PSV) and soybean stunt virus (SSV) are stylet borne like the potyviruses, there is not the specificity between virus strains and aphid vectors that exists for the potyviruses. The four strains of SSV described by Takahashi et al. (1980) are all readily transmitted by *Myzus persicae*, *Aphis glycines*, and *Macrosiphum solani*.

18-3.1.3 Alfalfa Mosaic Virus

The AMV is transmitted by many aphid species, however, recent reports have been limited to Japan (Takahashi et al., 1980).

18-3.1.4 Luteoviruses

Soybean dwarf virus (SDV) is transmitted in a circulative or persistent manner. Extensive research has been done on its transmission characteristics in Japan (Tamada, 1975). This virus is only transmitted after a latent period of 15 to 27 h following virus acquisition by its only known aphid vector *Aulacorthum solani*. Other characteristics of the transmission of SDV are that the aphids retain the virus after molting, there is a loss of transmission ability with time, and the latent period is reduced and the retention period of the virus by the aphid is increased when the quantity of virus acquired during feeding is increased.

The recently identified Indonesian soybean dwarf virus (ISDV) is distinct from SDV (Japan) (Iwaki et al., 1980). It is transmitted by *Aphis glycines*, but not by *A. cracacivora* or *Aulacorthum solani*, after a minimum of a 6-h acquisition feeding period. The aphids were able to transmit the virus for 9 days after acquisition and also after molting.

18–3.2 Beetles

Some members of the families Chrysomelidae and Coccinelidae, order Coleoptera, are efficient vectors of soybean viruses (Fulton et al., 1980). These insects are leaf-feeding beetles as adults and complete their life cycle in the soil as larvae and pupae which often feed on the soybean roots. The viruses are relatively stable and may be readily transmitted by mechanical means. However, all mechanically transmitted soybean viruses are not transmitted by beetles and recent studies have revealed mechanisms of vector-virus specificity which account for beetle transmission of some viruses but not others (Gergerich et al., 1983).

18-3.2.1 Comoviruses

The comoviruses, bean pod mottle virus (BPMV) and cowpea mosaic virus (CMV), are transmitted by various chrysomelid beetles. *Cerotoma trifurcata* Forst., the bean leaf beetle, is an excellent vector of BPMV (Ross, 1963; Walters, 1964; Walters and Lee, 1969) and is found in high populations in soybean fields in some areas of Arkansas, Louisiana, and North Carolina. Other beetle vectors of BPMV include *Diabrotica undecimpunctata howardii* Barber from N. C. (Ross, 1963), *Epicauta vittata* from Mississippi (Patel and Petri, 1971) and *D. balteata* Le Conte, *Colaspis flavida* Say, and *C. lata* Schiff from Louisiana.

A yellow strain of CMV was transmitted in Nigeria by two species of thrips, two species of chrysomelid beetles, a curculionid beetle, and two grasshopper species (Whitney and Gilmer, 1974). Jansen and Staples (1970) found that both *D. undecimpunctata howardii* and *C. trifucata* can transmit CMV better from cowpea than from soybean; this suggests that the virus would be more of a problem in cowpea than in soybean. McLaughlin et al. (1977) found that a severe isolate of CMV was transmitted by *C. trifurcata* from *Desmodium canescens* to cowpea but not to soybean, and from cowpea to cowpea but not to soybean. Since the comoviruses can be transmitted by many insects, the transmission efficiency of each vector on the various virus hosts and its period of activity assumes great importance in determining the relevance of the virus in soybean production. *Ceratoma trifurcata* transmits BPMV from soybean to soybean immediately after feeding and the longer it feeds on infected plants the greater the probability of transmission. Soybean fields in North Carolina have been found 100% infected with BPMV especially when soybean seeds are planted in late June, followng small grain, adjacent to earlier-planted soybean fields which served as a reservoir of both virus and beetle vectors. Bean leaf beetles were found to transmit BPMV from naturally infected *D. paniculatum* to soybean, and in this role they can initiate the virus infection into a field of soybean from a perennial weed host (Walters and Lee, 1969).

18-3.2.2 Bromovirus

The bromovirus, cowpea chlorotic mottle, is transmitted by *Cerotoma ruficornis* and *C. trifurcata*; *Diabrotica balteata*, *D. undecimpunctata*, and *D. adelpha*; and *Epilachna varivestis* (Fulton et al., 1980). This virus was found in Georgia to be transmitted by *D. undecimpunctata howardii* to soybean and cowpea mainly from sources outside the fields rather than from infected plants within the fields (Demski and Chalkley, 1979). This may explain why this virus disease has not become a serious problem in soybeans.

18–3.3 Whiteflies

Soybean viruses transmitted by species of the genus *Bemisia* are mainly limited to tropical and subtropical regions. The yellow mosaic virus of soybean (SYMV) is transmitted by *B. tabaci* and has caused considerable damage to soybean crops in Uttar Pradesh in India (Suteri, 1974). This virus is acquired by the whitefly vector in 2 to 3 h, can be transmitted in 24 h, and is retained by the vector for at least 3 weeks; the vector-virus relationship is considered persistent. The virus is not mechanically transmitted. Many of the yellow mosaic diseases of legumes of southeast Asia with causal agents which are whitefly-transmitted may be caused by the same virus or related viruses (Muniyappa et al., 1976). These viruses are not as well known as are other viruses which are mechanically transmitted, occur in high concentrations in the host, and hence are more easily purified.

Another white fly-transmitted agent, causing disease in soybean, has recently been reported from Thailand, and is thought to be a virus (Iwaki et al., 1983). Soybean leaf crinkle (SCL), as the name implies, causes leaf malformation. Whiteflies were able to transmit the agent for 9 days after acquisition. Although both SCL and SYMV are thought to be in the geminivirus group, their identities have not been firmly established.

Cowpea mild mottle virus (CMMV), found infecting soybean in Thailand (Iwaki et al., 1982), is transmitted by white flies in a semipersistent manner; transmission was successful after only a 20-min acquisition feeding period, but the authors believe no latent period is necessary after acquisition. The vector retained the virus for only 1 day or less. This virus was found in soybean in the Ivory Coast in 1978, however, no vector was reported (Thouvenel et al., 1982). The CMMV is different from other whitefly transmitted soybean viruses since it is readily mechanically transmitted and is seed borne.

18–3.4 Thrips

Although the occurrence of thrips on soybean is not uncommon, they have not been firmly established as a vector in the field of a soybean virus. Messieha (1969) reported that under laboratory conditions, nymphs of *Thrips tabaci* could transmit TRSV from soybean to soybean. *Sericothrips variabilis* and *Franklinella tritici* did not serve as vectors. *Thrips tabaci* required an 8-h acquisition feeding period to become viruliferous and a 24-h feeding period to transmit the virus. Viruliferous thrips could transmit the virus for at least 2 weeks.

Whitney and Gilmer (1974) reported that a yellow strain of cowpea mosaic virus (CMV) could be transmitted by two genera of thrips. Since this virus infects soybean, these insects may play a role in CMV transmission in soybean.

18–3.5 Nematodes

The only soybean virus known to be transmitted by nematodes is TRSV. *Xiphinema americanum* Cobb, the dagger nematode, has been shown to be capable of transmission at a low level of efficiency, and the importance of this vector in soybean has not been established (McGuire and Douthit, 1978).

18–3.6 Grasshoppers

Grasshoppers, *Melanoplus differentialis* (Thomas), were shown to transmit TRSV but its importance is questionable since if the insect remained on the test feeding plant for over 30 s, no transmission occurred; if only one bite was taken, transmission was 2.4% (Dunleavy, 1957).

The acridid grasshoppers *Cantantops spissus* and *Zonocerus variegatus* were found to transmit a yellow strain of cowpea mosaic in Nigeria from cowpea to cowpea at the rate of 18 to 21% with single insects (Whitney and Gilmer, 1974).

18–3.7 Seed

At least eight viruses are transmitted by soybean seed, the most efficient means of virus distribution. Rates of virus transmission by seed as low as 0.01% (1 in 10 000) may be economically significant but difficult to detect. Hence, reports that certain viruses are not seed transmitted or certain lines do not transmit a virus in its seed should be accepted with an understanding of how the conclusion was reached. With most, if not all, seed-borne viruses, the younger the soybean plant when infection occurs, the greater the percentage of seed that will transmit the virus. However, many other factors can also affect virus transmission in soybean seed. Iizuka (1973) found that infections that occur after flowering often produce little or no seed transmission. Early infections with SMV, SSV, and TRSV gave 34, 95, and 100% transmission, respectively. The AMV and BYMV were not seed transmitted. Demski and Harris (1974) reported that if the planting date was delayed until 21 August in Georgia, no seed transmission of TRSV occurred from plants inoculated on their primary leaves. Seed transmission of TRSV from earlier planting dates was 40 to 60%.

Strains of viruses may be seed transmitted at different rates, and also soybean genotypes may transmit virus at different rates in their seed. Ross (1968), working with two strains of SMV, found SMV-1 was transmitted almost twice as frequently (11.1%) in 'Lee' as was SMV-2 (5.8%); in 'Hill' infected with SMV-1, only 1.3% transmission occurred, and SMV-2 was not seed transmitted in Hill. Ganekar and Schwenk (1974) found that a strain of TSV isolated from tobacco (*Nicotiana tabacum* L.), unlike a strain of TSV isolated from soybean, was not seed transmitted in soybean. Bowers and Goodman (1982) reported a number of soybean plant introductions and cultivars that are resistant to SMV seed transmission.

Whereas SMV and TRSV have long been known to be seed transmitted, BPMV has just recently been shown to be transmitted in seed from Nebraska at the rate of 0.1% (Lin and Hill, 1983). Prior reports had indicated that it was not seed transmitted. A 0.1% rate of transmission of BPMV is adequate to initiate damaging virus epidemics in the presence of high early season vector (bean leaf beetle) populations.

Viruses may remain viable in seed as long as the seed is viable. Hence, TRSV was found to survive in seed and be transmitted to seedlings after 5 yrs of storage with little decrease in the rate of transmission (Laviolette and Athow, 1971).

Studies by Iizuka (1973) have shown that SMV is usually transmitted only when the embryo is infected, and that as plants mature, virus in the seed tissues is inactivated; both pollen and ovules may transmit the virus to the embryo. Bowers and Goodman (1979) found that in seed of SMV-infected plants of 'Merit', which does not transmit SMV in seed, the virus was present in the embryos of immature seed but no infectious virus was present in embryos of mature seed.

18–3.8 Mechanical

Spread of a virus throughout a soybean crop by machinery requires a readily transmissible virus and a tractor operation in which plant contact is made by machinery throughout the field. Bean pod mottle virus was spread throughout an experimental soybean field of the author when a tractor sprayed an insecticide at weekly intervals to control bean leaf beetles, the virus vector; the sprayer boom brushed across the top leaves of young plants and almost all plants became infected in the virtual absence of the insect vector. Since CMV is also readily transmitted mechanically and is in the same virus group (comovirus) as BPMV, it may also be transmitted by machinery.

18–4 CONTROL

In a practical sense, control of virus diseases of soybean can be accomplished by planting resistant cultivars (if available), or can be based on the transmission properties of the virus either in the vector or in the seed. Since problems caused by virus diseases are limited to specific areas where conditions are favorable for adequate virus inoculum and large vector populations, virus problems may be seasonal.

18–4.1 Resistance

Resistance to soybean virus diseases can be exhibited by failure of the virus to replicate and move systemically in the host, by the host reacting to infection with mild symptoms, or by the host failing to transmit a normally seed-transmitted virus in its seed. Resistance to SMV probably has received more attention than that of other viruses. Kiihl and Hartwig (1979) reported that the resistance in PI 96 986 (Ross, 1969a) was governed by a dominant allelomorphic series at a single locus (*rps*). Plants heterozygous at this locus often produced a necrotic reaction to the virus. Kwon and Oh (1980), on the other hand, reported that resistance to a strain of SMV that caused necrosis was controlled by a single recessive gene. Cho and Goodman (1979) classified isolates of SMV found in the U.S. germplasm collection into seven strains, some of which caused symptoms on cultivars previously reported resistant. From these results and that of others, resistance in soybean to SMV is very complex because of the pathogenic diversity of the virus. The natural appearance of a new SMV strain capable of causing considerable damage to a previously SMV-resistant cultivar in Korea apparently occurred after the release of a highly SMV-resistant cultivar (Cho et al., 1977).

Bowers and Goodman (1982) reported 12 germplasm lines that were resistant to seed transmission of SMV. This was based on planting 1000 seed of each line from SMV-infected plants.

Bock (1973), working in east Africa, found no PMV resistance in soybean; however, Demski and Kuhn (1975), working in Georgia, found 14 of 70 soybean cultivars (old and new) that were highly resistant to infection. As plants of the susceptible cultivars aged, they became more difficult to infect, and after 5 to 6 weeks growth in the greenhouse, infections were limited to inoculated leaves, and no systemic movement was detected. Hence, soybean apparently possess a resistance to initial infection of PMV, and susceptible plants also develop resistance to systemic movement. Resistance to PMV infection found in 'Dorman' and 'CNS' is conditioned by an allele at a single locus (*Rpv*) which is completely dominant (Boerma and Kuhn, 1976). Shipe et al. (1979) found 145 plant introductions out of 2161 lines in Maturity Groups II, III, and IV were resistant to PMV. Since a second gene for resistance (rpv_2) was found to be recessive in 'Peking' (Shipe et al., 1979) it is not known which resistant genes are present in the 145 lines in Maturity Groups II, III, and IV.

Provvidenti (1975) reported that 'Corsoy', 'Cutler 71', 'Swift', and 'Williams' were resistant to BYMV in field and greenhouse experiments.

Cooperative trials by several pathologists in the early 1960s revealed no resistance to TRSV in the then existant germplasm collection. Orellana (1981) evaluated 630 plant introductions in maturity Groups 00 to VII of *Glycine soja* Sieb and Zucc. (= *G. ussuriensis* Reg & Maack) for resistance to TRSV. Plant introduction 407 287 reacted with local chlorotic and necrotic local lesions, and no systemic infection ensued. This species has the same chromosome number as *G. max*.

Immunity to SSV, reported from Indonesia, was found in 'Taichung', 'Bonus', and 'No. 1592' (Roechan et al., 1975); and Tamada (1977) reported eight cultivars resistant to five strains of SSV in Japan. Immunity to PSV was found in 'Bragg', Corsoy, 'Hark', 'Harosoy', 'Ogden', and 'Wayne' by Milbrath and Tolin (1977).

Resistance to a soybean strain of CCMV was found in 18 soybean cultivars which developed necrotic local lesions on the inoculated leaves (Harris and Kuhn, 1971). This resistance in 'Lee', Bragg, and Hill was later determined to be governed by a completely dominant allele at a single locus (*Rcv*) (Boerma et al., 1975).

In plants with field resistance or tolerance, the virus moves systemically throughout the plant but either the symptoms are mild (field resistance) or the symptoms and virus concentrations are similar to that of nontolerant lines but the effect of the disease on yield is less. Data from a study using closely related pairs of SMV-susceptible and resistant lines showed that one pair (recurrent parent 'Dare') yielded essentially the same in the presence of the virus whereas in other pairs, resistant lines outyielded their mosaic-susceptible siblings. Whether this type of resistance found in Dare-related lines could be conscientiously selected in a breeding program is not known. However, breeders' nurseries frequently have a high incidence of SMV, and selection for high yield under these conditions may result in a covert selection for field resistance to SMV.

Evaluations for immunity in soybean to BPMV have been conducted by several workers including the author. No immunity has been found. Scott et al. (1974) evaluated various plant introductions of several species of *Glycine* by mechanically inoculating the leaves of seedlings. All plants of *G. clandestina*, *G. gracilis*, *G. koidzumii*, and *G. ussuriensis* were susceptible, but plants of *G. falcata, G. javanica*, and *G. tomentella* were immune. Most accessions of *G. wightii* were immune. Studies with BPMV have shown that 'Davis' manifests only a 3% yield reduction when inoculated in the V2-3 growth stage, whereas Forrest and 'Centennial' sustain 10 to 13% and 9% yield reductions, respectively, (Windham and Ross, 1985). Since foliar symptoms of Davis were noticeably milder than those of Forrest and Centennial, Davis could be considered field resistant.

18–4.2 Vector Control by Chemicals

The spread of virus into and throughout an annual crop, such as soybean, by invertebrate vectors can be important in considering virus disease control. Generally, if the virus has not spread throughout the crop by flowering, little or no yield loss will occur. Delaying inoculation of soybean with BPMV for 2 weeks (V2 vs. V6) resulted in a 50% reduction in yield loss. (Windham and Ross, 1985).

The feeding habit of migratory aphids and the rapidity that a stylet-borne virus can be acquired and transmitted renders the control of these viruses impractical by insecticide applications to the soybean crop; furthermore, in most localities soybean is not a preferred host for indigenous aphids.

Experiments on control of the luteovirus, SDV, which is transmitted in a circulative manner by aphids, were conducted in Japan using granulated insecticides applied to the soil (Tamada, 1975). Although the insecticides were ineffective in preventing the migratory viruliferous alatae from transmitting the virus into the crop, subsequent transmission within the crop was reduced.

Eliminating early season spread of BPMV by insecticide applications to control the bean leaf beetle probably would be an effective means of reducing yield loss caused by this virus. Windham and Ross (1985) found that virus infection prior to the V6 growth stage was essential for significant yield loss to occur.

Control of whitefly populations with either two soil applications of aldicarb (Singh et al., 1971) or with four spray applications of insecticide (Rataul and Singh, 1974) was highly effective in preventing SYMV spread in India. Since soybean grown in tropical or subtropical areas serves as a host for whitefly and SYMV is transmitted in a persistent or semipersistent manner, insecticides applied to the soybean crop can kill or incapacitate the vector before transmission can occur.

18–4.3 Weed Control

Any virus-infected host plant can serve as an inoculum reservoir and/or a mechanism of virus survival during times when soybean are not grown. In areas with no killing frost, volunteer soybean may also play a role in perpetuating the virus. The SMV was found in volunteer soybean in southern Brazil during the winter and could be an important source of the virus for the next soybean crop (Fett, 1978).

Other hosts, particularly perennials, can also play a vital role in maintaining the virus during winter months. On Hokkaido, Japan, white (*Trifolium repens* L.) and red clover (*T. pratense* L.) are symptomless hosts of SDV, and distribution of the virus in soybean is closely associated with SDV in clover (Tamada, 1975). The virus is transmitted to soybean from white and red clover by the foxglove aphid, *Aulacorthum solani*, and the abundance of this aphid and these overwintering hosts make this soybean disease one of the most severe on Hokkaido. No evidence was presented on the use of methods to control the disease by eliminating the clover.

Naturally infected *Desmodium paniculatum* was found to be an efficient source of BPMV for the bean leaf beetle (Walters and Lee, 1969). Since this species is a perennial host it could serve to initiate BPMV epidemics in soybean.

Elimination of virus-infected weed hosts near soybean fields to control a virus disease in soybean depends on the type of vegetation, availability of effective herbicide, and cost of application. If the weed host is killed, one or two applications may suffice to control the disease.

18–4.4 Control of Seed Transmission

Control of soybean virus diseases by imposing methods based on virus transmission properties in seed should consider the following: (i) percentage virus transmission by the seed planted; (ii) sources of virus other than soybean seed, e.g., nearby volunteer soybean, or other hosts; (iii) vector prevalence. In production areas where vectors are absent, virus transmission in seed at low rates would have no impact on crop yield since no in-crop spread would occur. Since soybean is planted in high densities (250 000–350 000 plants ha^{-1}), seed transmission rates as low as 0.1 to 0.05% could initiate high virus incidence if high vector populations existed during early plant growth. These low rates of seed transmission may go undetected. Only recently was BPMV reported to be transmitted by seed at the rate of 0.1% after several reports of it not being seed transmitted (Lin and Hill, 1983).

Elimination of virus-infected plants in the seed crop by roguing would be time consuming and probably impractical. Very sensitive techniques have recently been employed to detect virus in seed (Lister, 1978; Bryant et al., 1983); as little as 25 to 50 ng of purified SMV was detected in extracts of 100 homogenized seed. A disadvantage of this method is that the ratio between the percentage of seed-producing infected seedlings and the concentration of virus in the seed extract varies with the cultivar, and at low levels of virus infected seed (<5%) the experimental error may overshadow a positive reading (Bryant et al., 1983). In spite of this, these sensitive methods may find application in the future in controlling seed-borne soybean viruses. Since low rates of virus transmission in seedlings can be important, sensitive methods are needed.

18–5 LOSSES AND YIELD REDUCTIONS

The magnitude of yield loss caused by a virus disease depends in large part on the following factors: (i) growth stage at which most infection occurred (Ross, 1969b); (ii) percentage of plants that become infected (Ross, 1983a; Thongmeearkom et al., 1978; Horn et al., 1973); (iii) distribution of infected plants (Hill et al., 1980); (iv) degree of susceptibility or sensitivity of the cultivar to the virus (Ross, 1977, 1983a); (v) virulence of the virus strain(s) present (Ross, 1968); (vi) environmental conditions (Bryant et al., 1982; Ross, 1970); (vii) the degree of yield compensation manifested by noninfected plants (Ross, 1983a). The variations and interactions of these factors on soybean growth and yield make estimations of yield losses caused by virus diseases generally imprecise.

The loss of crop yield value caused by virus may be a direct reduction in total yield, and may be reflected by a reduction in seed size (Ross, 1968; Tamada, 1977). Other ways in which virus diseases cause a loss in crop value are by decreasing seed quality by causing seedcoat discoloration (SMV and SSV), predisposing seed to increased infection by fungi (SMV, BPMV) (Ross, 1977; Stuckey et al., 1982; Windham and Ross, 1986), and causing cracked seedcoats (Windham and Ross, 1986). The percent oil and protein composition of seed may be also altered by virus infection (Demski and Jellum, 1975).

One of the main difficulties in determining the effect of virus on soybean yields under natural conditions is the spread of virus into the healthy control plots; insecticide treatments to prevent style-borne aphid transmission virus is ineffective. To avoid this problem, mosaic-resistant and susceptible sibling lines were used by Ross (1977) that were produced by backcrossing the SMV resistance from PI 96 983 or PI 170 893 into 'Semmes', Dare, 'Pickett 71', 'Lee 68', and 'Ransom.' Rows artificially inoculated with SMV were arranged throughout the field to serve as an inoculum source for aphid transmission into the plots. Yields of some susceptible lines were 30% less than their sibling resistant line. Hartwig and Keeling (1982) used three pairs of closely related SMV resistant and susceptible lines that were inoculated artificially in the V5 growth stage. Yields were reduced up to 12%, and stage of growth at time of inoculation did not effect the yield loss. Bryant et al. (1982) reported that during one growing season plants inoculated with SMV at growth stage V-1 sustained a greater yield loss than plants inoculated at stage R-2, but the next season there was no difference in yield loss between the two inoculation treatments. Davis artifically inoculated at 3 weeks of age with CCMV yielded 23 to 31% less than noninoculated controls (Harris and Kuhn, 1971).

When certain viruses in combination infect the same plant, yield reductions may be greater than the sum of the losses caused by each virus alone. Infection by SMV and BPMV caused such a synergistic yield reduction and reduced yields up to 80% (Ross, 1968). Demski and Jellum (1975) inoculated two cultivars with either one or two of the following viruses at the V-1 growth stage: SMV, TRSV, and CCMV. All combinations of viruses caused greater yield reductions than either virus alone, and SMV in combination with TRSV or CCMV caused the greatest yield reductions.

The yield loss caused by a virus disease may be greatly affected by the ability of the noninfected plants to compensate for the stunting of adjacent infected plants. If plants are killed or severely stunted early in the season and the virus fails to spread, little or no yield loss may occur (Thongmeearkom et al., 1978). However, if infected plants are not severely stunted, the infected plants may not allow adjacent noninfected plants a significant competitive yield advantage, and yields may be affected by subtle compensative abilities (Ross, 1983b). If distribution of infected plants within a field is clustered or nonrandom because of the nature of virus spread, compensative effects would be less important than if distribution was thoroughly random (Hill et al., 1980; Windham and Ross, 1985).

18–6 VIRUS IDENTIFICATION

Identification of the causal agent of a disease is often one of the first steps in controlling a plant disease. Identification of a plant virus may be a complex task because of its submicroscopic size and fastidious reproductive habits. Hence, knowledge of its host range, mode of transmission, and symptomatology on various hosts is important in virus identification and is usually the first type of information obtained. Table 18–2 is a guide to the identification of soybean viruses. All soybean viruses (in Guide), except SDV (Japanese and Indonesian), SCLV, and SYMV, are mechanically transmissible, and therefore the reaction of various hosts (symptomatology) is attainable with a minimum of equipment. Information on transmission characteristics and insect vectors (aphids, beetles, and whitefly) are also important; stylet-borne (nonpersistent) viruses, however, can lose their ability to be aphid transmitted after being successively mechanically transferred. Evidence has been obtained that indicates a helper component is produced in infected plants that is necessary for stylet-borne virus transmission; with repeated mechanical transmission, the helper component is lost, and aphids no longer can transmit the virus (Govier and Kassanis, 1974).

Since both nonpersistent and circulative viruses can have specificity with their vector and even different strains of the same virus may be transmitted by different aphid species (Harris and Maramorsch, 1977), these characteristics may aid in virus identification.

Table 18–2. Guide for identification of 15 soybean viruses based on transmission, host range, and symptoms.

- A. Transmitted by aphids
  - B. Transmission persistent, not mechanically transmitted—Soybean dwarf and Indonesian soybean dwarf
  - BB. Transmission nonpersistent, mechanically transmitted
    - C. Soybean seedcoat mottled
      - D. Mottling as streaks, saddle, or blotches—Soybean mosaic
      - DD. Mottling pattern of concentric rings or broken lines—Soybean stunt
    - CC. Soybean seed coat not mottled
      - DD. Host range narrow (mainly Fabaceae)
        - E. Systemic infection in *Chenopodium amaranticolor*—Bean yellow mosaic
        - EE. Local infection in *C. amaranticolor*—Peanut mottle
      - DD. Host range wide
        - E. Systemic chlorosis and necrotic flecking in *C. amaranticolor*—Alfalfa mosaic
        - EE. Local infection in *C. amaranticolor*—Peanut stunt
- AA. Not transmitted by aphids
  - B. Transmitted by beetles
    - C. Infects *Cucumis sativus*—Cowpea chlorotic mottle
    - CC. Does not infect *C. sativus*
      - D. Does not produce symptoms in *Vigna unguiculata*—Bean pod mottle
      - DD. Produces systemic symptoms in *V. unguiculata*
        - E. Produces systemic symptoms in *C. amaranticolor*—Cowpea mosaic
        - EE. Local infection only in *C. amaranticolor*—Cowpea severe mosaic
  - BB. Not transmitted by beetles
    - C. Transmitted by whitefly (*Bemisia tabaci*)
      - D. Mechanically transmitted, infects *Arachis hypogaea*—Cowpea mild mottle
      - DD. Not mechanically transmitted, does not infect *A. hypogaea*
        - E. Host range wide (Solanaceae)—Soybean crinkle leaf
        - EE. Host range narrow—(Fabaceae)—Soybean yellow mosaic
    - CC. Not transmitted by *B. tabaci*
      - D. Transmitted by nematode *Xiphenema americanum*—Tobacco ringspot
      - DD. Not transmitted by *X. americanum*—Tobacco streak

The morphology of the virus particle is also of great aid in virus identification, and observations of virus-infected leaf dips under the electron microscope, can reveal long flexuous rods of viruses belonging to the potyvirus group; isometric particles are difficult to discern. By observing virus-infected leaf tissue under the light microscope, virus-induced inclusions may be observed (McWhorter, 1965; Edwardson and Christie, 1978). Many viruses produce typical inclusions which are uniquely characteristic of the virus group to which they belong and hence can be useful in identifying the virus.

After the identity of the virus has been narrowed to a few possible choices by host range, symptomatology, transmission, and particle morphology, serological techniques are useful in the final identification. Many serological techniques are used for plant virus identification. Besides the older microprecipitin and agar diffusion tests, newer more sensitive techniques include the enzyme-linked immuno-sorbent assay (ELISA). Antisera to many soybean viruses are available from the American Type Culture Collection, Rockville, MD.

18-7 VIRUS STRAINS

Pathogenic variation among plant viruses makes them similar to other plant pathogens. Different isolates of a virus may elicit a range of reactions when infecting various soybean cultivars. Of the soybean viruses, SMV has been investigated most with regard to pathogenic strains. The importance of different pathogenic virus strains becomes economically significant when one is found causing significant yield loss on a previuosly resistant cultivar. In Korea, in 1974 and 1975, a severe outbreak of a necrotic disease was found to be caused by a strain of SMV on cultivars previously considered resistant. (Cho et al., 1977). Ross (1969a) grouped seven SMV isolates into three strains based on the reaction of 24 soybean lines; later a new strain was detected that caused severe symptoms on previously resistant cultivars (Ross, 1975). This severe strain was also detected in Mississippi (Kiihl and Hartwig, 1979). Probably the most extensive investigation on SMV strains was reported in Japan (Takahashi et al., 1980); 102 isolates were classified into five strains (A,B,C,D, and E) based on symptomatology and pathogenicity on four soybean cultivars (Harosoy, Shiromame, Ou 13, and Tokatinugaba). Strains A, B, and C caused normal mosaic symptoms, were seed transmitted in six cultivars at about the same percentage (36–41%), and were readily aphid transmitted; strains D and E produced severe symptoms of crinkling, stunting, chlorosis, and top necrosis, were difficult to transmit with aphids, and were transmitted in only 3% of the seed. Cho and Goodman (1979) classified 98 isolates of SMV (from field collections and plant introductions from the U. S. germplasm collection) into seven strains based on the reaction of eight soybean cultivars.

The abundance of such pathogenic variability among SMV isolates is difficult to explain. However, the appearance of a severe or necrotic strain does not mean that it necessarily will pose problems for soybean production. Properties such as efficiency of aphid transmisssion by indigenous aphid species and percentage of seed transmission by the cultivars in production play as great a role in determining the yield loss caused by an SMV strain as does the pathogenicity of the virus isolate.

Pathogenic variation of other soybean viruses has also been reported. Isolates of SDV have been separated on their ability to cause either chlorosis or dwarfing (Tamada, 1975) or dwarfing, rugosity or both (Takahashi et al., 1980). Isolates of SSV and AMV were separated into four and three strains, respectively, based on pathogenicity and symptoms on soybean cultivars (Takahashi et al., 1980). The CMV had been separated into two subgroups, severe and yellow, but was subsequently separated into two separate viruses; the yellow strain remained CMV and the severe strain became cowpea severe mosaic virus; both infect soybean (McLaughlin et al., 1977; Thongmeearkom et al., 1978). Viruses such as TRSV often exist in variants which are adapted to certain hosts. Hence, a strain of TRSV from tobacco may infect soybean poorly if at all, whereas, other variants infect soybean and may not infect tobacco.

Moore and Scott (1971) described a strain of BPMV, J-10, that caused systemic yellow mottling in *Chenopodium quinoa* whereas their other BPMV isolates caused only local infection; variety Hill gave more severe symptoms with J-10 than with other isolates.

Much information concerning plant viruses that affect soybean is available in the Descriptions of Plant Viruses published by the Commonwealth Mycological Institute, Ferry Lane, Kew, Surrey, UK.

18–8 BACTERIAL DISEASES

The major bacterial disease of soybean is bacterial blight caused by *Pseudomonas syringae* pv. *glycinea.* This pathogen is seedborne, distributed worldwide, and in some parts of the world the disease is an important factor in reducing soybean yields. Although other bacterial diseases have been described, at present they are apparently of minor significance since little research on them has been reported since the first edition of this monograph in 1973.

In 1980, the Committee on Taxonomy of Phytopathogenic Bacteria of the International Society for Plant Pathology proposed the naming of pathovars of phytopathogenic bacteria and listed the proposed names and strains of pathovars (Dye et al., 1980). This proposal was accepted and an explanation of the new bacterial nomenclature has been published (Bradbury, 1983). The old and revised names of bacterial pathovars of soybean are listed in Table 18–3.

Bacterial pustule, caused by *Xanthomonas campestris* pv. *phaseoli*, occurred mainly in most warm temperate areas of soybean production such as the southern soybean-growing region of the USA and some areas of Brazil. Today, however, pustule-resistant cultivars, utilizing a single

Table 18–3. Bacterial diseases of soybean.

Disease name	Causal organism	
	New name	Previous name
Blight	*Pseudomonas syringae* pv. *glycinea* (Coerper) Young, Dye, Wilke)	*Pseudomonas glycinea*
Pustule	*Xanthomonas campestris* pv. *phaseoli* (Smith) Dye	*Xanthomonas phaseoli* var. *sojensis*
Wildfire	*Pseudomonas syringae* pv. *tabaci* (Wolf and Forster) Young, Dye, and Wilkie	*Pseudomonas tabaci*
Tan spot	*Corynebacterium flaccumfaciens* (Hedges) Dowson pv. *flaccumfaciens*	*Corynebacterium flaccumfaciens*

recessive gene from CNS, makes the disease virtually nonexistent in these production areas. The disease was only occasionally observed by Fett (1979) and by Lehman et al. (1976) in the southern areas of Brazil where mostly pustule-resistant cultivars are grown.

Wildfire, caused by *P. syringae* pv. *tabaci*, usually occurs only in the presence of bacterial pustule, and since pustule has been controlled by growing resistant cultivars, wildfire is rarely found in areas where it had previously occurred.

The foliar disease tan spot was observed in 1975 in Iowa (Dunleavy, 1983). The causal organism, *Corynebacterium flaccumfaciens* pv. *flaccumfaciens*, also induces wilting and a leafspot on *Phaseolus vulgaris*. Tan spot of soybean was rarely found in Iowa from 1975 through 1981 (16 counties) and is considered of little economic importance. Initial symptoms consist of a chlorotic area, frequently beginning at the leaf perimeter and progressing toward the midrib. After several days the chlorotic area becomes tan in color; these areas commonly fall out during high winds and leaves may appear ragged.

18–9 BACTERIAL BLIGHT

Bacterial blight is generally most severe in the more cool temperate soybean-producing regions of the world. The disease is quite prevalent in the soybean-producing states of the north central USA. In Brazil, for instance, blight was found in well over half of the commercial soybean fields in the southern states of Rio Grande-do-Sul and Santa Cantaria which are between 36 and 32° S Lat (Lehman et al., 1976). Yield losses caused by bacterial blight on Corsoy in Iowa were near 18% (Williams and Nyvall, 1980). In these field tests, plants were inoculated five times at 10- to 15-day intervals starting in full bloom.

Pathogenic bacteria are transmitted on and in seed, and seedlings become infected at an early stage of growth. Laurence and Kennedy (1974) found that populations of *P. syringae* pv. *glycinea* increased on the seed during germination regardless of whether seed were from resistant or susceptible cultivars, and Daft and Leben (1972) reported that increased seedling infection via injured cotyledons was obtained by increasing the percentage of abrasive sand particles in the soil.

The pathogen may survive in plant debris during seasons in which soybean is not grown, however, this type of survival depends on several factors. Daft and Leben (1973) found that bacteria overwintered in Ohio in leaf debris better than in stem debris, and Kennedy (1969) reported that in Minnesota the bacterium could survive until the following spring in leaves left on the ground surface but died when leaves were placed underground. Fett (1979), working in the state of Paraná, Brazil, found that the bacterium did not survive in debris from infected soybean from one growing season to the next. Apparently, the blight organisms will survive in leaf debris on the surface of the ground from one crop to the next if temperatures are similar to those of Ohio and Minnesota (40–44° N Lat), but in warmer climates, such as in Paraná (Latitude 32–34° S Lat), the organism succumbs.

The main mode of survival of the bacterium from one season to the next is in seed. Leben (1975) investigated controlling the disease by eliminating bacteria from seed and found that treating the seed with certain antagonistic bacteria would reduce seedling infection; treatment of seed with oxytetracyline hydrochloride reduced seedling infection from 50 to 90% to 0 to 2%. Leben also suggested that production of disease-free seed, such as that obtained in the greenhouse, could break the seed-borne cycle. Parashar and Leben (1972) developed a sensitive method of detecting *P. syringae* pv. *glycinea* in lots of seed by soaking the seed in sterile water for 2 h, shaking them in sand to injure cotyledons, and then germinating the seed in water-saturated vermiculite. Some seed in lots of 1500 each, with infection rates of $< 10\%$, consistently developed lesions on their cotyledons when exposed to this treatment; if the treatment was omitted prior to planting, no lesions developed.

After seedlings become infected, the bacterium can survive and multiply on the leaf surface (Mew and Kennedy, 1971, 1982; Kennedy and Ercolani, 1978), in buds, or within various parts of the plant without producing typical water-soaked lesions. For copious foliar infections to occur, wind-driven rains are considered necessary. Hence, when Daft and Leben (1972) produced conditions of driving rain, which forced bacteria into stomata, lesions developed. Typical early leaf symptoms of bacterial blight are angular water-soaked lesions surrounded by a chlorotic halo. The lesions later become black or brown, and on the underside of the leaf they glisten when wet; leaves typically appear tattered (Kennedy and Tachibana, 1973).

Heavy infection of young leaves may produce general chlorosis. This symptom and the yellow halos surrounding the lesions have been investigated by several workers. Mitchell and Young (1978) identified a chlorosis-inducing toxin from cultures of the blight organism as either coronatine or its sterioisomer. Gnanamanikam et al. (1982) found that different strains of *P. syringae* pv. *glycinea* differed widely in their production of coronatine and that some chlorosis-producing strains did not produce significant amounts of the toxin. These researchers believe that since coronatine is found in infected leaf tissue, it is responsible for some of the chlorotic symptoms of bacterial blight; however, other materials are also probably responsible for this chromatic symptom. Gulya and Dunleavy (1979) found a toxin produced by *Pseudomonas syringae* pv. *glycinea* in culture which appeared to inhibit synthesis of chlorophyll and aminolevulinic acid, a precursor to chlorophyll synthesis.

Cross et al. (1966) differentiated seven pathogenic races of *P. syringae* pv. *glycinea* by inoculating the unifoliolate leaves of seven soybean cultivars (Acme, Chippewa, Flambeau, Harosoy, Lindarin, Merit, Norchief). Five to 7 days after inoculation the leaves were classified as susceptible, intermediate, or resistant. Thomas and Leary (1980) described a new race (no. 8) from Australia which elicited an intermediate response on all the seven test cultivars except Lindarin. Fett and Sequeria (1981) further characterized the physiological races of the pathogen; in their tests the reaction of some host-pathotype combinations differed from those previously obtained by Cross et al. (1966). The age of the unifoliolate leaf when inoculated was found to influence the symptom expression. These workers differentiated a new pathogenic race (no. 9).

The hypersensitive resistant reaction of cultivars inoculated with incompatible races was found to be associated with accumulation of hydroxyphaseolin and other isoflavoid compounds in host tissue (Keen and Kennedy, 1974). Races of *P. syringae* pv. *glycinea* compatible with the cultivar did not cause the production of these compounds by the host.

REFERENCES

Abney, T.S., J.O. Sillings, T.L. Richards, and D.B.Broersma. 1976. Aphids and other insects as vectors of soybean mosaic virus. J. Econ. Entomol. 69:254–256.

Bock, K.R. 1973. Peanut mottle virus in east Africa. Ann. Appl. Biol. 74:171–179.

Boerma, H.R., and C.W. Kuhn. 1976. Inheritance of resistance to peanut mottle virus in soybean. Crop Sci. 16:533–534.

----, ----, and H.B. Harris. 1975. Inheritance of resistance to cowpea chlorotic mottle virus (soybean strain) in soybeans. Crop Sci. 15:849–850.

Bowers, G.R., and R.M. Goodman. 1979. Soybean mosaic virus: Infection of soybean seed parts and seed transmission. Phytopathology 69:569–572.

----, and ----. 1982. New sources of resistance to seed transmission of soybean mosaic virus in soybeans. Crop Sci. 22:155–156.

Bradbury, J.F. 1983. The new bacterial nomenclature—What to do. Phytopathology 73:1349–1350.

Brim, C.A., K.L. Athow, and J.P. Ross. 1964. The effect of tobacco ringspot virus on natural hydridization in soybean. Agron. Abstr. American Society of Agronomy, Madison, WI, p. 62.

Bryant, G.R., D.P. Durand, and J.H. Hill. 1983. Development of a solid phase radioimmunoassay for detection of soybean mosaic virus. Phytopathology 73:623–629.

----, J.H. Hill, T.B. Bailey, H. Tachibana, D.P. Durand, and H.I. Benner. 1982. Detection of soybean mosaic virus in seed by solid-phase radioimmunoassay. Plant Dis. 66:693–695.

Cho, E.K., B.J. Chung, and S.H. Lee. 1977. Studies on identification and classification of soybean virus diseases in Korea. II. Etiology of a necrotic disease of *Glycine max*. Plant Dis. Rep. 61:313–317.

----, and R.M. Goodman. 1979. Strains of soybean mosaic virus: Classification based on virulence in resistant soybean cultivars. Phytopathology 69:467–470.

Cross, J.E., B.W. Kennedy, J.W. Lambert, and R.L. Cooper. 1966. Pathogenic races of the bacterial blight pathogen of soybeans, *Pseudomonas glycinea*. Plant Dis. Rep. 50:557–560.

Daft, A.C., and C. Leben. 1972. Bacterial blight of soybeans: Epidemiology of blight out breaks. Phytopathology 62:57–62.

----, and ----. 1972. Bacterial blight of soybeans: Seedling infection during and after emergence. Phytopathology 62:1167–1170.

----, and ----. 1973. Bacterial blight of soybeans: Field overwintered *Pseudomonas glycinea* as possible primary inoculum. Plant Dis. Rep. 57:156–157.

Demski, J.W. 1975. Source and spread of peanut mottle virus in soybean and peanut. Phytopathology 65:917–920.

----, and J. Chalkley. 1979. Non-movement of cowpea chlorotic mottle virus from cowpea and soybean. Plant Dis. Rep. 63:761–764.

----, and H.B. Harris. 1974. Seed transmission of viruses in soybean. Crop Sci. 14:888–890.

----, and M.D. Jellum. 1975. Single and double virus infection of soybean: Plant characteristics and chemical composition. Phytopathology 65:1154–1156.

----, and C.W. Kuhn. 1975. Resistant and susceptible reaction of soybeans to peanut mottle virus. Phytopathology 65:95–99.

Dunleavy, J.M. 1957. The grasshopper as a vector of tobacco ringspot virus in soybean. Phytopathology 47:681–682.

----. 1983. Bacterial tan spot, a new foliar disease of soybean. Crop Sci. 23:473–476.

Dye, D.W., J.F. Bradbury, M. Groto, A.C. Hayward, R.A. Elliott, and M.N. Schroth. 1980. International standards for naming pathovars of phytopathogenic bacteria and a list of pathovar names and pathotype strains. Rev. Plant Pathol. 59:153–168.

Edwardson, J.R., and R.G. Christie. 1978. Use of virus-induced inclusions in classification and diagnosis. Ann. Rev. Phytopathol. 16:31–35.

Fett, W.F. 1978. Volunteer soybean: Survival sites for soybean pathogens between seasons in southern Brazil. Plant Dis. Rep. 62:1013–1016.

----. 1979. Survival of *Pseudomonas glycinea* and *Xanthomonas phaseoli* var. *sojensis* in leaf debris and soybean seed in Brazil. Plant Dis. Rep. 63:79–83.

----, and Sequeria, L. 1981. Further characterization of the physiologic races of *Pseudomonas glycinea*. Can. J. Bot. 59:283–287.

Francki, R.I.B. 1980. Limited value of the thermal inactivation point, longevity *in vitro* and dilution and point as criteria for the characterization, identification and classification of plant viruses. Intervirology 13:91–98.

----. 1981. Plant virus taxonomy. p. 3–16. *In* E. Kustok (ed.) Handbook of plant virus infections—Comparative diagnosis. Elsevier/North Holland Biomedical Press, Amsterdam.

Fulton, J.P., H.A. Scott, and R. Gamez. 1980. Beetles. p. 115–132. *In* K.F. Harris and K. Maramorsch (ed.) Vectors of plant pathogens. Academic Press, New York.

Ganekar, A.M., and A.W. Schwenk. 1974. Seed transmission and distribution of tobacco streak virus in six cultivars of soybeans. Phytopathology 64:112–114.

Gergerich, R.C., H.A. Scott, and J.P. Fulton. 1983. Regurgitant as a determinant of specificity in the transmission of plant viruses by beetles. Phytopathology 73:936–938.

Gnanamanikam, S.S., A.N. Starratt, and E.W.B. Ward. 1982. Coronatine production in vitro and in vivo and its relation to symptom development in bacterial blight of soybean. Can. J. Bot. 60:645–650.

Govier, D.A., and B. Kassanis. 1974. Evidence that a component other than the virus particle is needed for aphid transmission of potato virus Y. Virology 57:285– 286.

Gulya, T., and J.M. Dunleavy. 1979. Inhibition of chlorophyll synthesis by *Pseudomonas glycinea*. Crop Sci. 19:261–264.

Harris, H.B., and C.W. Kuhn. 1971. Influence of cowpea chlorotic mottle virus (soybean strain) on agronomic performance of soybeans. Crop Sci. 11:71–73.

Harris, K.F., and K. Maramorosch (ed.) 1977. Aphids as virus vectors. Academic Press, New York.

Hartwig, E.E., and B.L. Keeling. 1982. Soybean mosaic virus investigations with susceptible and resistant soybeans. Crop Sci. 22:955–957.

Hill, J.H., B.S. Lucas, H.I. Benner, H. Tachibana, R.B. Hammond, and L.P. Pedigo. 1980. Factors associated with the epidemiology of soybean mosaic virus in Iowa. Phytopathology 70:536–540.

Hollings, M., and A.A. Brunt. 1981. Potyviruses. p. 731–807. *In* E. Kurstak (ed.) Handbook of plant virus infections-comparative diagnosis. Elsvier/North Holland Biomedical Press, Amsterdam.

Horn, N.L., L.D. Newsom, and R.L. Jansen. 1973. Economic injury thresholds of bean pod mottle and tobacco ringspot virus infections of soybeans. Plant Dis. Rep. 57:811–813.

Iizuka, N. 1973. Seed transmission of viruses in soybean. Bull. Tohoku Natl. Agric. Exp. Stn. 46:131–141.

Iwaki, M., M. Roechan, H. Hibino, H. Tochihara, and D.M. Tantera. 1980. A persistant aphidborne virus of soybean, Indonesian soybean dwarf virus. Plant Dis. 64:1027–1030.

----, P. Thongmeearkom, M. Prommin, Y. Honda, and T.Hibi. 1982. Whitefly transmission and some properties of cowpea mild mottle virus on soybean in Thailand. Plant Dis. 66:365–368.

----, ----, Y. Honda, and N. Deema. 1983. Soybean leaf crinkle: a new whitefly-borne disease of soybean. Plant Dis. 67:546–548.

Jansen, W.P., and R. Staples. 1970. Effect of cowpeas and soybeans as source or test plants of cowpea mosaic virus on vector efficiency and retention of infectivity of the bean leaf beetle and the spotted cucumber beetle. Plant Dis. Rep. 54:1053–1054.

Keeling, B.L. 1982. Effect of soybean mosaic virus on root volume and dry weight of soybean plants. Crop Sci. 22:629–630.

Keen, N.T., and B.W. Kennedy. 1974. Hydroxyphaseollin and related isoflavanoids in the hypersensitive resistance reaction of soybeans to *Pseudomonas glycinea.* Physiol. Plant Pathol. 4:173–185.

Kennedy, B.W. 1969. Detection and distribution of *Pseudomonas glycinea* in soybean. Phytopathology 59:1618–1619.

----, and G.L. Ercolani. 1978. Soybean primary leaves as a site for epiphytic multiplication of *Pseudomonas glycinea.* Phytopathology 68:1196–1201.

----, and H.Tachibana. 1973. Bacterial diseases. *In* B.E. Caldwell (ed.) Soybeans: Improvement, production, and uses. Agronomy 16:491–526.

Kiihl, R.A.S., and E.E. Hartwig. 1979. Inheritance of reaction of soybean mosaic virus in soybeans. Crop Sci. 19:372–375.

Koshimizu, S., and Iizuka, T. 1963. Studies on soybean mosaic virus. Diseases in Japan. Bull. Tohoku Natl. Agric. Exp. Stn. 27:1–104.

Kwon, S.H., and J.H. Oh. 1980. Resistance to a necrotic strain of soybean mosaic virus in soybean. Crop Sci. 20:403–404.

Laurence, J.A., and B.W. Kennedy. 1974. Population changes of *Pseudomonas glycinea* on germinating soybean seeds. Phytopathology 64:1470–1471.

Laviolette, F.A., and K.L. Athow. 1971. Longevity of tobacco ringspot virus in soybean seed. Phytopathology 61:755.

Leben, C. 1975. Bacterial blight of soybean: Seedling disease control. Phytopathology 64:844–847.

Lee, Y.S., and J.P. Ross. 1972. Top necrosis and cellular changes in soybean doubly infected by soyban mosaic and bean pod mottle viruses. Phytopathology 62:839–845.

Lehman, P.S., C.C. Machado, and M.T. Tarrago. 1976. Frequency and severity of soybean diseases in the states of Rio-Grande-do-Sul and Santa-Cantarina Brazil. Fitopatol. Bras. 1:183–194.

Lin, M.T., and J.H. Hill. 1983. Bean pod mottle virus: Occurrence in Nebraska and seed transmission in soybeans. Plant Dis. 67:230–233.

Lister, R.M. 1978. Application of the enzyme-linked immunosorbent assay for detecting viruses in soybean seed and plants. Phytopathology 68:1393–1400.

Mathews, R.E.F. 1979. Classification and nomenclature of viruses. Third report of the International Committee on Taxonomy of viruses. Intervirology 12:129–296.

McGuire, J.M., and L.B. Douthit. 1978. Host effect on acquisition and transmission of tobacco ringspot virus by *Xiphenema americanum.* Phytopathology 68:457–459.

McLaughlin, M.R., P. Thongmeearkon, G.M. Milbrath, and R. Goodman. 1977. Isolation and some properties of a yellow subgroup member of cowpea mosaic virus from Illinois. Phytopathology 67:844–847.

McWhorter, F.P. 1965. Plant virus inclusions. Ann. Rev. Phytopathol. 3:287–312.

Messieha, M. 1969. Transmission of tobacco ringspot virus by thrips. Phytopathology 59:943–945.

Mew, T.W., and B.W. Kennedy. 1971. Growth of *Pseudomonas glycinea* on the surface of soybean leaves. Phytopathology 61:715–716.

----, and ----. 1982. Seasonal variation in populations of pathogenic pseudomonads on soybean leaves. Phytopathology 72:103–105.

Milbrath, G.M., and S.A. Tolin. 1977. Identification, host range and serology of peanut stunt virus isolated from soybean. Plant Dis. Rep. 61:637–640.

Mitchell, R.E., and H. Young. 1978. Identification of a chlorosis-inducing toxin of *Pseudomonas glycines* as caronatine. Phytochemistry 17:2028–2029.

Moore, B.J., and H.A. Scott. 1971. Properties of a strain of bean pod mottle virus. Phytopathology 61:831–833.

Muniyappa, V., H.R. Reddy, and G. Shivashanka. 1976. Studies on the yellow mosaic disease of horsegram (*Dolichos biflorus*). II. Host range studies. Mysorc. J. Agric. Sci. 10:611–614.

Orellana, R.G. 1981. Resistance to bud blight in introductions from the germplasm of wild soybean. Plant Dis. 65:594–595.

----, F. Fan, and C. Sloger. 1978. Tobacco ringspot virus and Rhizobium interactions in soybean: Impairment of leghemoglobin accumulation and nitrogen fixation. Phytopathology 68:577–582.

Paguio, O.R., and C.W. Kuhn. 1974. Incidence and source of inoculum of peanut mottle virus and its effect on peanut. Phytopathology 64:60–64.

Parashar, R.D., and C. Leben. 1972. Detection of *Pseudomonas glycinea* in soybean seed lots. Phytopathology 62:1075–1077.

Patel, V.C., and H.N. Petri. 1971. Transmission of bean pod mottle virus to soybean by the striped blister beetle, *Epicauta vittata*. Plant Dis. Rep. 55:628–629.

Provvidenti, R. 1975. Resistance in *Glycine max* to isolates of bean yellow mosaic virus in New York State. Plant Dis. Rep. 59:917–919.

Rataul, H.S., and L. Singh. 1974. Control of soybean yellow mosaic virus in soybean *Glycine max* L. by controlling the vector whitefly *Bemisia tabaci* Genn in Punjab. Punjab Agric. Univ. J. Res. 11:73–76.

Roechan, M., M. Iwaki, and D.M. Tantera. 1975. Virus diseases of legume plants in Indonesia. Contrib. Cent. Res. Inst. Agric. (Bogor, Indones.) 15:1–16.

Ross, J.P. 1963. Transmission of bean pod mottle virus by beetles. Plant Dis. Rep. 47:1049–1050.

----. 1968. Effect of single and double infections of soybean mosaic and bean pod mottle viruses on soybean yield and seed characters. Plant Dis. Rep. 52:344–348.

----. 1969a. Pathogenic variation among isolates of soybean mosaic virus. Phytopathology 59:829–832.

----. 1969b. Effect of time and sequence of inoculation of soybeans with soybean mosaic and bean pod mottle viruses on yields and seed characters. Phytopathology 59:1404–1408.

----. 1970. Effect of temperature on mottling of soybean seed caused by soybean mosaic virus. Phytopathology 60:1798–1800.

----. 1975. A newly recognized strain of soybean mosaic virus. Plant Dis. Rep. 59:806–808.

----. 1977. Effect of aphid-transmitted soybean mosaic virus on yields of closely related resistant and susceptible soybean lines. Crop Sci. 17:869–872.

----. 1983a. Effect of soybean mosaic on component yields from blends of mosaic resistance and susceptible soybeans. Crop Sci. 23:343–346.

----. 1983b. Compensatory yield response in soybean: A subtle but important expression of disease resistance. Phytopathology 83:844 (Abstr.)

Schwenk, F.W., and C.D. Nickell. 1980. Soybean green stem caused by bean pod mottle virus. Plant Dis. 64:863–865.

Scott, H.H., J.V. Van Scyoc, and C.E. Van Scyoc. 1974. Reactions of *Glycine* spp. to bean pod mottle virus. Plant Dis. Rep. 58:191–192.

Shipe, E.R., G.R. Buss, and C.W. Roane. 1979. Resistance to peanut mottle virus (PMV) in soybean (*Glycine max*) plant introductions. Plant Dis. Rep. 63:757–760.

----, ----, and E.A. Tolin. 1979. A second gene for resistance to peanut mottle virus in soybean. Crop Sci. 19:656–658.

Sinclair, J.B. (ed.) 1982. Compendium of soybean diseases. American Phytopathology Society, St. Paul.

Singh, H., G.S. Sandhu, and G.S. Mavi. 1971. Control of yellow mosaic virus in soybean. *Glycine max* (L.) Merrill by the use of granular insecticides. Indian J. Entomol. 33:272–278.

Stuckey, R.E., S.A. Ghabrial, and D.A. Reicosky. 1982. Increased incidence of *Phomopsis* sp. in seeds of soybeans infected with bean pod mottle virus. Plant Dis. 66:826–829.

Suteri, B.D. 1974. Occurrence of soybean yellow mosaic in Uttar Pradaesh. Curr. Sci. 43:689–690.

Takahashi, K., T. Tanaka, W. Iida, and Y. Tsuda. 1980. Studies on virus diseases and causal viruses of soybean in Japan. Bull. Tohoku Natl. Agric. Exp. Stn. 62:1–130.

Tamada, T. 1975. Studies on the soybean dwarf disease. Rep. Hokkaido Natl. Agric. Exp. Stn. 25:1–144.

----. 1977. The virus diseases of soybean in Japan. Food Fert. Tech. Ctr. Tech. Bull. 33:1–11.

Thomas, M.D., and J.V. Leary. 1980. A new race of *Pseudomonas glycinea*. Phytopathology 70:310–312.

Thongmeearkom, P., E.H. Paschal, and R.M. Goodman. 1978. Yield reduction in soybeans infected with cowpea mosaic virus. Phytopathology 68:1549–1551.

Thouvenel, J.C., Monsarrat, A., and Fauquet, C. 1982. Isolation of cowpea mild mottle virus from diseased soybeans in the Ivory Coast. Plant Dis. 66:336–337.

Walters, H.J. 1964. Transmission of beanpod mottle virus by bean leaf beetles. Phytopathology 54:240.

----, and Lee, F.N. 1969. Transmission of bean pod mottle virus from *Desmodium paniculatum* to soybean by the bean leaf beetle. Plant Dis. Rep. 53:411.

Whitney, W.K., and R.M. Gilmer. 1974. Insect vectors of cowpea mosaic virus in Nigeria. Ann. Appl. Biol. 77:17–21.

Wilcox, J.R., and F.A. Laviolette. 1968. Seedcoat mottling response of soybean genotypes to infection with soybean mosaic virus. Phytopathology 58:1446–1447.

Williams, D.J., and R.F. Nyvall. 1980. Leaf infection and yield losses caused by brown spot and bacterial blight diseases of soybeans. Phytopathology 70:900–902.

Windham, M.T., and J.P. Ross. 1985. Phenotypic response of six soybean cultivars to bean pod mottle virus infection. Phytopathology 75:305–309.

----, and ----. 1986. Effect of beanpod mottle virus on soybean seed quality and mycoflora. Plant Dis. (In press.)

19 Nematodes

R. D. Riggs
University of Arkansas
Fayetteville, Arkansas

D. P. Schmitt
North Carolina State University
Raleigh, North Carolina

Nematode damage to soybean [*Glycine max* (L.) Merr.] has been reported from most areas of the world where this crop is grown. In some areas only one or two nematode species are responsible for most of the problems; however, many nematode species parasitize soybean. In the USSR, 84 species have been recovered from soil in which soybean was grown (Shavrov, 1968) and over 100 species (Schmitt and Noel, 1984) have been associated with soybean in the USA.

The first published report of a nematode problem on soybean was in 1915 (Hori, 1916); and occurred as moon night or yellow dwarf caused by soybean cyst nematode (*Heterodera glycines* Ichinohe). The problem had been known in Japan for a number of years prior to the report, and although it was not reported from China until 1938 (Nakata and Asuyana, 1938), soybean researchers in China say it was present before 1900. China is considered to be the area where the soybean evolved and it has been cultivated there for over 5000 yrs. It is likely that numerous nematode problems have occurred through the years, but the only nematodes that have been reported from Eastern Asia are soybean cyst, root-knot [*Meloidogyne incognita* (Kofoid & White) Chitwood, *M. arenaria* (Neal) Chitwood, *M. hapla* Chitwood, and *M. javanica* (Treub) Chitwood,] (Raut and Sethi, 1980; Sasser, 1977) and lesion (*Pratylenchus* spp.) (Gotoh and Oshima, 1963) nematodes. In western Asia there are few reports of plant parasitic nematodes on soybean (Trabulsi et al., 1980).

In various areas of Africa *M. incognita* and *M. javanica* (Ibrahim et al., 1972; Javaid and Ashraf, 1978), *Scutellonema cavenessi* Sher (Germani, 1981), *Rotylenchulus reniformis* Linford & Oliveira (El Sherif et al., 1974), and *Tylenchorhynchus clarus* Allen (Aboul-Eid and Osman, 1981) have been associated with soybean. In Europe, *Pratylenchus neglectus* (Rensch) Filipjev & Stekhoven, *P. thornei* Sher & Allen, *Helico-*

Soybeans: Improvement, Production, and Uses, 2nd ed.—Agronomy Monograph no. 16.

tylenchus digonicus Perry, and *Tylenchorhynchus dubius* (Butschli) Filipjev have been found on soybean in Yugoslavia (Ivezic et al., 1980), but there are few other reports. No reports were found relative to soybean in Australia.

Many nematode species parasitize soybean in South America. *Heterodera glycines* has recently been found in Colombia (Norton, 1983). One to four species of root-knot nematodes have been reported from several areas but most of the surveys have been made in Brazil (Carvalho, 1958; Kloss, 1960; Lehman et al., 1976; Lordello, 1955) and Ecuador (Bridge, 1976) where 10 to 15 species of phytoparasitic nematodes have been identified as parasites of soybean.

Species of eight genera of nematodes have been identified on soybean in Canada: *Meloidogyne, Helicotylenchus, Tylenchorhynchus, Paratylenchus, Criconemella, Paratrichodorus,* and *Xiphinema* (Johnson, 1977). In the USA, *H. glycines* is probably the most widespread species parasitic on soybean (Riggs, 1977), but *M. incognita* is generally distributed and *M. arenaria* is becoming more common throughout soybean-growing areas of southern and southeastern USA. Other economically important nematodes in the USA are *R. reniformis, Hoplolaimus columbus* Sher, *Pratylenchus* spp., and *Belonolaimus* spp. Each may be the most damaging species in a particular area. In addition, many other species have been associated with soybean, sometimes in high numbers, but most have not been shown to be pathogenic. One factor that has contributed to a greater awareness of nematodes associated with soybean in the USA is the widespread occurrence of soybean cyst nematode at damaging levels. There have also been larger numbers of nematologists working on this important crop.

19-1 SOYBEAN CYST NEMATODE

Soybean cyst nematode (SCN) was first reported in Japan in 1915 (Hori, 1916) but it was probably there as early as 1881. Chinese scientists report (1981, personal communication) that it has been known there since before 1900, although the first published report was in 1938 (Nakata and Asuyana, 1938). This pest has been a serious problem in Korea since its discovery in 1936 (Yokoo, 1936). *Heterodera glycines* was found in North Carolina, in 1954 (Winstead et al., 1955), and was in 23 states by 1984 (Fig. 19-1). The announcement of SCN in Egypt (Diab, 1968) was later rescinded (Aboul-Eid and Ghorah, 1974). The latest discoveries of SCN infestations were in Colombia (Norton, 1983) and in Java (Nishazawa, 1984 personal communication). A list of soybean diseases in Taiwan (Hung, 1958) included SCN. There are several races of SCN which increases its seriousness. This nematode will probably eventually infest all acreage cropped intensively with soybean since it seems to be adapted to most or all environments where the crop is grown.

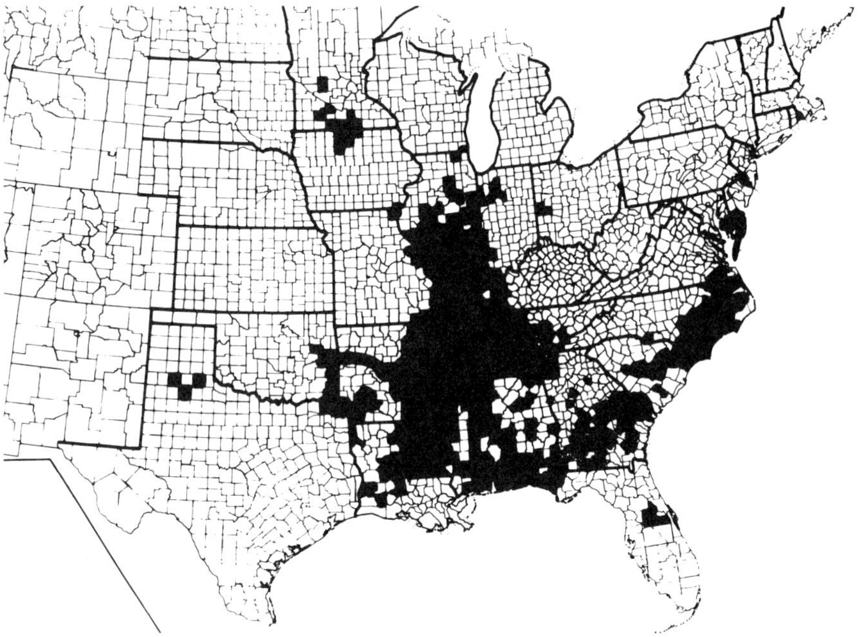

Fig. 19–1. Map of USA showing location of counties infested with *H. glycines* in 1984.

19–1.1 Life Cycle and Disease Cycle

The cyst of *H. glycines* is a brown, tough, lemon-shaped body which is filled with 100 or more eggs. It has a posterior gelatinous mass that may contain eggs. During a growing season, about 75% of the eggs in the soil will hatch or die (Slack et al., 1981). However, a given cyst may contain eggs that do not hatch readily. Price (1976) found that he could select SCN isolates in which eggs hatched readily and others in which eggs hatched poorly. Since the nematode can persist in the soil for as many as 7 yrs in the absence of a host (Slack et al., 1981), the failure of eggs to hatch readily may be a survival mechanism. In addition, a period of dormancy starting in late September or early October and continuing until sometime between January and May is documented in the USA (Fig. 19–2) (Ross, 1963; unpublished data) and doubtless exists elsewhere. Low winter temperatures prevent egg hatch during an extended period when a host is absent. This insures an abundant supply of infective juveniles when the next crop is planted. In most cases the eggs hatch readily in water, but there is some evidence that a hatching stimulant may be operative (Masamune et al., 1982; Okada, 1972).

Second stage juveniles (J2) of *H. glycines*, which hatch from the eggs, emerge from the cyst or egg mass, penetrate the roots, migrate to the stele and begin to feed (Endo, 1964). The number and rate of penetration by J2s is related to soil temperature (Hamblen et al., 1972). One day after inoculation four times as many J2s were found in 'Lee' soybean roots at

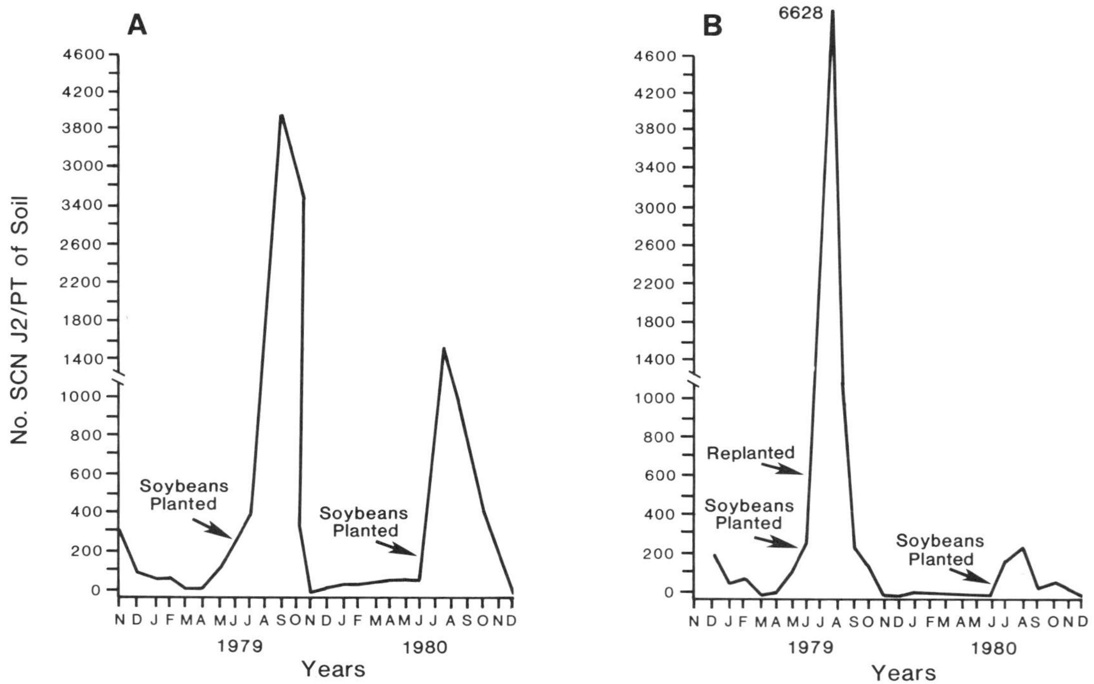

Fig. 19–2. *Heterodera glycine* second stage juvenile (J2) population curves based on monthly sampling of small plots. (A) 1979 and 1980 susceptible cultivars; (B) 1979 susceptible, 1980 resistant cultivar.

28 °C as at 22°C. At 14 days, 12 times as many J2s were found in roots at 28 °C as at any other temperature within a range of 14 to 32 °C.

Within 48 h after penetration of soybean roots, cell wall perforations occur at the feeding site (Gipson et al., 1971). This is the beginning of the syncytium which becomes a sink where nutrients accumulate (Gipson et al., 1971; Jones and Dropkin, 1975a, 1975b). The process of syncytial development involves a continuation of cell wall dissolution, and a breakup of the central vacuoles into many small vacuoles. Additional changes are occuring within the cytoplasm. It becomes dense, the rough endoplasmic reticulum-like material increases, and plastids decrease. The developing nematodes feed on the syncytia (Endo, 1964; Gipson et al., 1971). The females swell and become lemon-shaped. The J4 stage males become vermiform and adult males leave the roots. The time required for the nematodes to mature and for new J2s to emerge depends on the temperature (Hamblen et al., 1972) and the host (Schilling, 1982). Since one life cycle on soybean requires 21 to 25 days the potential for population increase in a single season is considerable.

19–1.2 Damage by Soybean Cyst Nematode

Loss of yield is the damage of most concern to soybean growers but it may not be the most obvious damage. Infected roots become greatly discolored and necrotic which suppresses shoot growth (Fig. 19–3). Severe chlorosis and stunting led the Japanese to call the disease caused by

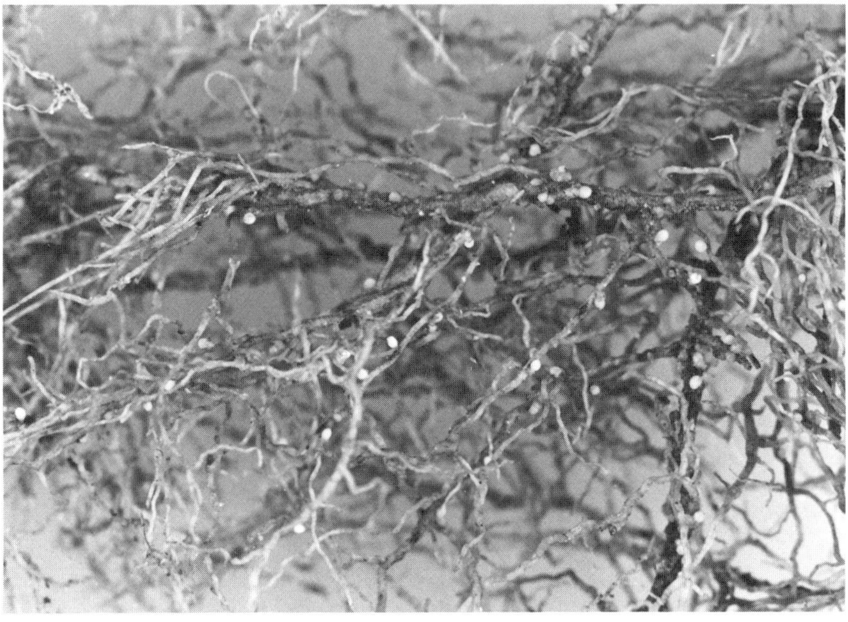

Fig. 19–3. Soybean root system with *H. glycines* females attached. Note root discoloration.

soybean cyst nematode *yellow dwarf.* Similar symptoms have also been observed on soybean in North Carolina (Plate 19–1A). However, in most areas of the USA, stunting occurs with little or no chlorosis (Plate 19–1B). The severeity of stunting is related to the initial population density. For example, number of eggs (Bonner, 1981), J2s (Bonner and Schmitt, 1985), or eggs + J2s (Noel et al., 1980; Noel and Stanger, 1982) per unit of soil at planting are negatively correlated with yield. However, the amount of chlorosis is associated with the race of the nematode. Races 1 and 2 in North Carolina produce the symptoms in Plate 19–1A which appear to be related to a reduction or inhibition of *Bradyrhizobium* nodulation (Barker et al., 1972; Lehman et al., 1971). There is less reduction in nodulation with races 3 and 4.

19–1.3 Variability in Soybean Cyst Nematode

Heterodera glycines possesses a high degree of genetic variability that is expressed phenotypically. In 1962, Ross (1962b) reported the occurrence of physiological variation within *H. glycines.* Following other reports of variation (Epps and Duclos, 1970; Miller, 1966, 1967; Riggs et al., 1968; Sugiyama et al., 1968) four races wre described (Golden et al., 1970) based on the amount of reproduction on four soybean lines and cultivars compared to a susceptible cultivar (Table 19–1). An additional race was described from Japan (Inagaki, 1979) (Table 19–1). The present system has been very useful, however, there is evidence that the variation may be much greater (Riggs et al., 1981). Therefore, a more comprehensive system for designating races of SCN should be considered, not to attempt to describe all the variability but to update it.

Great variability has been evident in the field as resistant cultivars are very effective for 3 yrs or more, but eventually a race of *H. glycines* builds up that is capable of damaging the resistant cultivar (Slack et al., 1981). There have been instances where damage to a resistant soybean cultivar occurred the 2nd yr it was planted. In some cases in Arkansas, cotton had been on the field for 10 to 15 yrs prior to the planting of the SCN-resistant cv. Forrest which sustained yield loss in the 2nd yr (R. D. Riggs, unpublished). This may indicate that the virtually undetectable

Table 19–1. Host differentiation for identification of races of *Heterodera glycines* (Golden et al, 1970; Inagaki, 1979).

	Reproduction on			
Race	Pickett	Peking	PI 88788	PI 90763
1	–†	–	+	–
2	+	+	+	–
3	–	–	–	–
4	+	+	+	+
5	+	–	+	–

† Lee is the susceptible host used for comparison to the differentials. Reproduction on Lee is converted to 100%; + = 10% or more of the number of mature females recovered from cv. Lee; – = Less than 10% of the number of mature females recovered from cv. Lee.

population remaining in the field after 10 or more yrs of a nonhost crop has a broader host capability than those juveniles that hatch readily. A genetic mechanism may inhibit hatch and provide for this broader host capability.

Resistance in soybean to *H. glycines* appears to be controlled by at least 10 genes (Caldwell et al., 1960; Hartwig and Epps, 1970; Matson and Williams, 1965; Sugiyama and Katsumi, 1966; Thomas, 1974). Some of the genes govern resistance to all races while others are necessary for resistance to only one or two of the races (Thomas, 1974). The only work on the inheritance of the parasitic capabilities of the nematode indicate that two or three alleles at a single locus are involved (Price, 1976).

19–2 ROOT-KNOT NEMATODES

The root-knot disease was first observed on soyean in the greenhouse by Franck in 1882 (Franck, 1882). There was little else reported about this disease on soybean until 1922 when McClintock (McClintock, 1922) reported on the resistance of certain soybean cultivars. By 1944 many soybean fields in North Carolina were known to be infested with *Meloidogyne* spp. (Atkinson, 1944). *Meloidogyne* spp. are major nematode problems in certain soybean production areas and probably rank second to SCN in importance on soybean worldwide.

Six species, *M. arenaria*, *M. bauruensis* Lordello, *M. hapla*, *M. incognita*, *M. inornata* Lordello and *M. javanica*, have been reported to damage soybean (Carvalho, 1954; Lordello, 1956a, 1956b). *Meloidogyne incognita* occurs worldwide (Sasser, 1977); whereas *M. arenaria*, *M. hapla* and *M. javanica*, although widely distributed (Sasser, 1977), are more limited by temperature requirements. *Meloidogyne hapla* is a cooler temperature species, whereas the other two are more common in warmer areas. *Meloidogyne bauruensis* and *M. inornata* (Lordello, 1956b) have been reported on soybean only in Brazil.

Root-knot nematodes generally are most important in sandy and sandy loam soils. They may occur in all soybean-growing areas in the USA but appear to be most severe in Florida, Georgia, Alabama, and Louisiana.

19–2.1 Life Cycle and Biology

Meloidogyne spp. penetrate soybean roots as second stage juveniles. Penetration occurs near the root tip in most cases and the juvenile migrates through the root tissue to the protoxylem poles. As the juveniles start to feed, the surrounding cells multiply by dividing rapidly. Cells immediately adjacent to the nematode lip region enlarge as the nematode feeds upon them (Dropkin and Nelson, 1960; Kaplan et al., 1979; Veech and Endo, 1970). The juveniles enlarge as they mature and as giant cells develop both from enlargement and from incorporation of neighboring cells. Eventually 5 to 9 multinucleate, thick-walled giant cells may be found grouped around a developing female. The life cycle in soybean is longer than in tomato (*Lycopersicon esculentum* Mill.), and may extend to 39 days or longer (Gommers and Dropkin, 1977). Males may or may not be present. *Meloidogyne* spp. reproduce parthogenetically, so males are not necessary. At maturity the females produce a gelatinous matrix, which is usually extruded at the root surface, in which 300 or more eggs are deposited. As many as four generations of root-knot nematodes may be produced on soybean in one growing season depending on the planting date and maturity group of the soybean cultivars. The nematode over winters as eggs or J2s (Kinloch, 1982).

Meloidogyne incognita has been divided into four races and *M. arenaria* into two races on the basis of host differentials (Eisenback et al., 1981). The different species have also been characterized cytologically (Eisenback et al., 1981) and there are intraspecific groupings. However, these groups have not been tested on a large number of soybean cultivars to determine the race reactions on different soybean cultivars. Five populations of *M. incognita* from West Tennessee were tested on a series of soybean cultivars and considerable variability was observed (Bernard, 1980).

19–2.2 Damage and Symptoms

Meloidogyne spp. can cause stunting and chlorosis of soybean (Plate 19–2). The galls on the roots are diagnostic and are quite distinct morphologically from the nodules produced by N_2-fixing bacteria (Fig. 19–4). Lack of N_2 fixation is probably not involved in the symptoms produced by *Meloidogyne* spp. (Baldwin et al., 1979) as it is with *H. glycines.*

Interaction with soil microflora may result in greater damage than is produced by the nematode alone. The galls are often invaded by various fungi which may result in root discoloration and decay (Melendez and Powell, 1967). Goswami and Agrawal (1978) found that *M. incognita* with either *Fusarium oxysporum* or *F. solani* caused soybean to yield less than soybean with either organism alone. On the other hand, Ross (1965) was unable to demonstrate an interaction between *M. incognita* and *F. oxysporum.* There are beneficial microflora, e.g., *Glomus macrocarpus,* which help reduce the yield loss due to *M. incognita* (Kellum and Schenck, 1980). However, such interactions vary with the cultivar (Schenck et al., 1975) and more research will be necessary before this mechanism can be manipulated in the field.

The number of nematode units (eggs or juveniles)/volume of soil necessary to cause damage varies due to environmental influence. For example, soybean is more tolerant of *M. incognita* at low temperatures (18–22 °C) and less tolerant at high temperature (30 °C) than at 26 °C, which is more nearly optimum for nematode reproduction (Nardacci and Barker, 1979). In addition, adequate moisture, particularly at pod-filling time, may reduce the amount of yield loss caused by a given level of

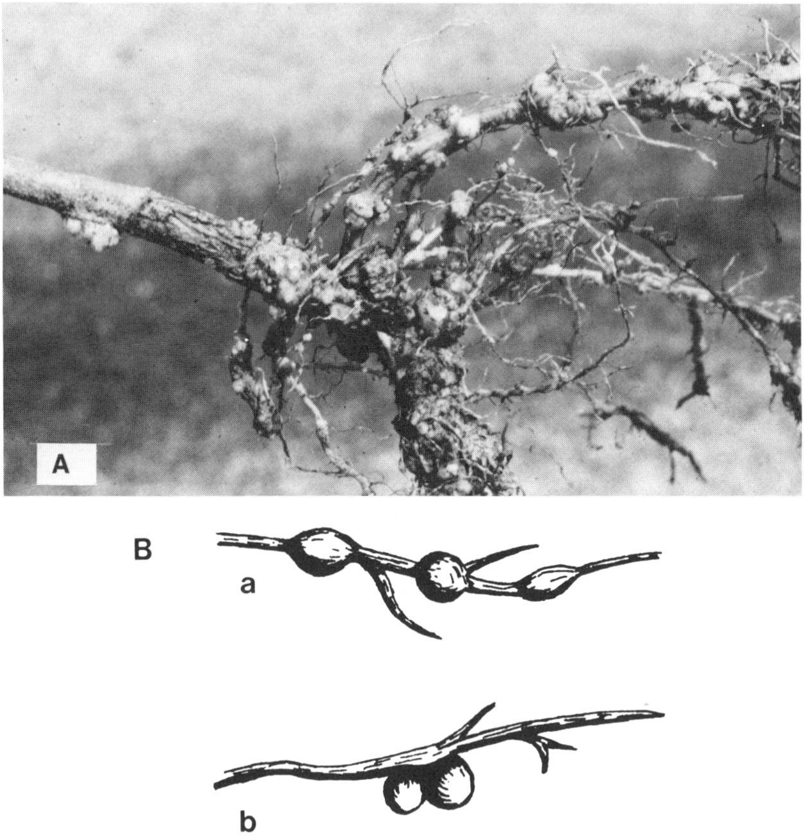

Fig. 19–4. (A) Soybean root with galls caused by *Meloidogyne* spp. (B) Root-knot nematode galls (a) and *Rhizobium* spp. nodules (b).

infection (Barker, 1982). However, the growth and yield of soybean is generally inversely proportional to initial population density (Kinloch, 1982; Nardacci and Barker, 1979). In Florida, predictive sampling was best done from November through January and the yield suppression was equivalent to 4.36 kg ha^{-1} for each juvenile/10 cm^3 of soil (Kinloch, 1982).

19–3 RENIFORM NEMATODE

Rotylenchulus reniformis was described by Linford and Oliveira (Linford and Oliveira, 1940) from cowpea roots in Hawaii in 1940. Reniform nematodes have been reported from at least 38 countries, mostly from tropical or subtropical climates (Heald and Thames, 1982). It has a broad host range and was first reported on soybean in 1958 in Brazil (Carvalho, 1954). *Rotylenchulus macrodoratus* Dasgupta, Raski & Sher, also has been reported to parasitize soybean (Cohn and Mordechai, 1977), but *R. reniformis* is the more important species on this crop.

19–3.1 Life Cycle and Biology

Second stage juveniles of the reniform nematode emerge from the eggs and become mature males and females in 9 to 10 days without feeding (Birchfield, 1962). The females partially penetrate the root, leaving their posterior ends outside the tissue (Plate 19–3). The posterior portion of the nematode enlarges and a gelatinous matrix is produced. Rebois (1973) reported an average of 59 eggs/female parasitizing soybean cultivars that were good hosts and the life cycle was completed in 19 days. Even the best host cultivars had only 72 eggs/egg mass (Lim and Castillo, 1979). Males and females can be distinguished beginning at J3, but reproduction may be parthenogenetic (Sivakumar and Seshadri, 1971). Nakasono (1978) demonstrated that, of six isolates from Japan and the USA, three reproduced by amphimixis and three by parthenogenesis.

19–3.2 Symptoms and Damage

When the infective female reniform nematode penetrates the cortical tissue, the ruptured cortical cells become disorganized and devoid of cytoplasm (Rebois, 1973; Rebois et al., 1975). The nematode begins to feed on an endodermal cell, which becomes the first syncytial cell. As feeding continues, the walls of surrounding parenchyma cells become part of the syncytium and eventually 100 to 200 cells may be incorporated. On the other hand, the giant cell formed as a result of parasitism of soybean by *R. macrodoratus* has only one nucleus and is probably formed from a single cell (Cohn and Mordechai, 1977). As a result of the damage to the soybean root system, plants become stunted and chlorotic. The root system also may be necrotic and stunted. Singh (1975) demonstrated that plant height and root and shoot weights were reduced by *R. reniformis* parasitism. In addition, there was a reduction of as much as 33% in seed yield (Rebois, 1971). The most diagnostic sign of the presence of reniform nematode is the soil-covered egg masses on the roots (Plate 19-3).

19–4 LESION NEMATODES

Several species of *Pratylenchus* spp. are associated with soybean. These include *P. brachyurus* (Godfrey) Filipjev & Stekhoven, *P. coffeae* (Zimmerman) Filipjev & Stekhoven, *P. hexincisus* Taylor & Jenkins, *P. neglectus*, *P. crenatus* Loff, *P. alleni* Ferris, *P. scribneri* Steiner, *P. agilis* Thorne & Malek, and *P. zeae* Graham. This large group of species is distributed over a wide geographical range (Schmitt and Noel, 1984), probably due to their wide host range.

19–4.1 Life Cycle and Biology

Some *Pratylenchus* spp. are parthenogenetic and others are amphimictic. The female deposits eggs in the root and probably a few in the soil. The first- and second-stage juveniles are contained within the egg. The second stage hatches and goes through three more molts to become an adult. All stages outside the egg are infective. Lesion nematodes penetrate into host roots and migrate within the tissue as they feed on the cells.

19–4.2 Symptoms and Damage

Pratylenchus spp. typically cause lesions in the cortical tissue of the root (Acosta and Malek, 1981) (Plate 19–4). Some thickening of intact cells adjacent to the nematode-affected cells occurs (Lindsey and Cairns, 1971). There may be some toxic reaction associated with the nematode (Brooks and Perry, 1967). The damage to the root may result in severe chlorosis and stunting of the shoot tissue. In extremely severe cases, plants may be killed. Some *Pratylenchus* spp. are involved in disease complexes with fungi (Lindsey and Cairns, 1971). The correlation of numbers of lesion nematodes to crop response has been characterized for few species (Endo, 1967). The relationship between population densities of *P. brachyurus* and seed yield of a sensitive cultivar is negative and linear in sandy soils and curvilinear in sandy clay loam soils (Schmitt and Barker, 1981).

19–5 LANCE NEMATODES

Several species of *Hoplolaimus* are associated with soybean. These include *H. columbus* Sher, *H. galeatus* (Cobb) Thorne, and *H. magnistylus* Robbins. *Hoplolaimus columbus* is the most pathogenic of these species. Lance nematodes are predominantly endo- or semiendoparasites, although a small portion of the population is found external to the root, presumably feeding ectoparasitically. Crop hosts of *Hoplolaimus* spp. include cotton, soybean, corn, and wheat (S. Lewis, 1983, personal communication).

19–5.1 Life Cycle

Hoplolaimus columbus is parthenogenetic, whereas, *H. galeatus* and *H. magnistylus* are amphimictic. Eggs of *H. columbus* hatch 9 to 15 days after oviposition (Fassuliotis, 1975). The entire cycle requires 45 to 49 days at temperatures between 21 to 27 °C (Smith, 1969).

19–5.2 Symptoms and Damage

Hoplolaimus columbus causes root damage during its migration through the root. Cortical cells are ruptured and cells become brown in the areas where the nematode has fed. The cortical tissue is sloughed off and plants produce no secondary roots (Lewis et al., 1976). This root damage results in chlorosis, stunting, and reduced pod production.

The damage caused by *H. galeatus* usually is not economically significant. This nematode appears to be an efficient parasite that is able to feed without adversely affecting its host. The host-parasite relationship involving *H. magnistylus* has not been characterized.

19–6 STING NEMATODES

Belonolaimus longicaudatus Rau is the sting nematode of greatest economic importance in soybean production. However, this nematode may be a composite taxon involving two or more species (R. Robbins, 1983, personal communication). It is primarily distributed in the coastal plains of the southeastern USA in sandy soil (84 to 94% sand) and parasitizes many field crops.

There are a number of biotypes of *B. longicaudatus*. Three Florida populations of this nematode were differentiated as races on tomato, rough lemon [*Citrus limon* (L.) Burm. f.], peanut (*Arachis hypogaea* L.), and strawberry (*Fragaria* spp.) (Abu-Gharbeh and Perry, 1970). Populations from Georgia and North Carolina appear to be different from those in Florida (Robbins and Barker, 1973).

Belonolaimus nortoni Rau is also associated with soybean in the USA. It is found in several North Central and Southern states. It is also associated with sandy soils, mostly along inland rivers and streams rather than in coastal areas.

19–6.1 Life Cycle

Belonolaimus spp. are presumed to have the typical four juvenile stages and an adult stage. Populations of *B. longicaudatus* decline rapidly in the winter. The eggs begin to hatch in the spring, but the length of the life cycle has not been determined. Reproduction is by amphimixis.

19–6.2 Damage

The feeding relationship of *B. longicaudatus* to the root of *Phaseolus vulgaris* L. (Standifer, 1959) has been characterized and is presumed to be similar on soybean. The nematode feeds ectoparasitically at the apex of the root, causing a yellow area to form. Roots become swollen and curve into the form of a "j". Lesions form at the root apices and zone of maturation. A few individuals can induce severe root pruning, stunting, chlorosis, wilting, and death of plants (Sasser et al., 1972).

19–7 OTHER NEMATODES

Numerous other nematode species belonging to several genera have been found in the soybean rhizosphere. Spiral nematodes that have been recovered include *Helicotylenchus digonicus* (Yugoslavia), *H. pseudorobustus* (Steiner) Golden (Nigeria and USA), *H. dihystera* (Cobb) Sher (USA), *Rotylenchus* spp. (USA), *Scutellonema bradys* (Steiner & LeHew) Andrassy and *S. brachyurum* Steiner (Brazil, USA, Senegal). Stunt nematodes that have been found include *Tylenchorhynchus dubius* (Yugoslavia, USA), *T. claytoni* Steiner (USA), *T. ewingi* Hopper (USA), *T. goffarti* Sturhan (USA), and *T. martini* (USA), *Quinsulcius acutus* (Allen) Siddiqi (Ecuador, USA), *Merlinius* spp. (Ecuador) and others. Only *T. claytoni* has been shown to reduce soybean yields in microplots (Ross et al., 1967) but control in the field did not increase yields (Schmitt and Noel, 1984). *Criconemella* spp. (Ecuador, Canada, Brazil, USA) including *C. ornata* (Raski) Luc & Raski did not reduce yield (Barker et al., 1982) nor did *Hemicriconemoides mangifera* Siddiqui (Ecuador). *Paratylenchus* spp. including *P. projectus* Jenkins (USA) and *P. tenuicaudatus* Wu (USA) also have been reported. *Paratrichodorus* spp., *Xiphinema americanum* Cobb, and *Longidorus* spp. have been reported from widely separated areas, (Canada, Brazil, Ecuador, Mexico, and USA). Only *X. americanum* has been shown to significantly reduce soybean yields (Schmitt and Noel, 1984).

19–8 NEMATODE MANAGEMENT

Nematode management may be grouped in three major categories: chemicals, resistance, and cultural. The method used has been directly related to the value of the soybean crop. Prior to the 1960s, soybean prices were too low to justify chemical control. More recently, the rising market price of soybean stimulated an increase in the use of chemicals and/or cultivars with resistance to *M. incognita* and *H. glycines* along with a tendency for growers to move away from rotation. However, the increasing cost of chemicals and environmental problems associated with them, in addition to the development of races that readily damage resistant cultivars has again resulted in a need to consider different methods of nematode management.

The control tactics, even nematicides, are fairly species-specific. Determination of the kinds and quantities of nematodes in a given field is necessary in order to employ the best control strategy.

19–8.1 Chemical

The chemicals available for use on soybean (Epps et al., 1964; Minton and Parker, 1979; Minton et al., 1978) are fumigants and nonfumigants. The fumigants are volatile compounds which are toxic to nematodes. Examples of effective fumigants include 1,3-dichloropropene (1,3-D), and sodium-*N*-methyldithiocarbamate. These are effective against most nematodes but may be phytotoxic to soybean and are labeled for use 10 to 14 days, preferably longer if conditions are cool and wet, prior to planting.

The most practical application is on land that is bedded and the chemical can be applied when the beds are formed. Deep injection generally is necessary for greatest efficacy of these compound since they volatilize and move upward.

The nonfumigants are chemicals of low valatility that require water for activity. These may be systemic (e.g., aldicarb and fenamiphos) in the plant or contact (ethoprop and carbofuran) in the soil. At the concentrations normally used, nematodes are rarely killed; rather their hatch, penetration and/or development are affected. Certain of the nematicides may be more effective against one nematode than another; e.g., aldicarb has been more effective against *H. glycines* than *M. hapla* and carbofuran has been particularly effective against *Pratylenchus brachyurus*, but has variable affects on *H. glycines* (Schmitt, unpublished). Soil conditions, temperature, and rainfall may influence the effectiveness of the chemical (Schmitt, unpublished). Under certain circumstances, there may be interactions between nematicides and certain herbicides that decreases the activity of the nematicide or increases injury to the crop (Schmitt et al., 1983).

19–8.2 Resistant Cultivars

Soybean cultivars are available with resistance to cyst (Epps and Hartwig, 1972; Kinloch et al., 1982; Riggs and Hamblen, 1962, 1966; Saito et al., 1973) root-knot (Acosta and Negron, 1982; Crittendon, 1955; Ibrahim and El-Saedy, 1982; Kim et al., 1982; Kinloch, 1982; Kinloch and Hinson, 1974; Kinloch et al., 1982; Minton et al., 1978; Rodriguez-Kabana and Thurlow, 1980; Yoshii, 1977), reniform (Birchfield et al., 1971; Lim and Castillo, 1979), lesion (Acosta et al., 1979; Dickerson and Franz, 1974; Zirakparvar, 1982), and lance (Nyczepir and Lewis, 1979) nematodes and possibly other nematodes that have not been tested. Planting resistant cultivars is an effective method of reducing nematode damage and is probably the least expensive method available.

Cultivars having resistance to one to three species of root-knot (*M. incognita, M. arenaria,* and *M. javanica*) and/or SCN are available (Kinloch et al., 1982). Compilation of a complete list of cultivars with resistance to one or more nematode species is extremely difficult because there are so many sources of information, public, and private.

The main problem encountered in the use of resistant soybean cultivars has been the practice of planting them year after year. This has resulted in the selection of pathotypes of *H. glycines* or *M. incognita* (Bernard, 1980; Riggs, 1960, unpublished) that parasitize resistant cultivars. This has been demonstrated in pot tests (Riggs et al., 1977), and has been observed under field conditions (Slack et al., 1981; 1983, personal communication from several states) for SCN. In some cases the resistance is difficult to transfer, and sources of resistance to new pathotypes are not always available, factors that complicate the task of plant breeders.

19–8.3 Cultural Practices

Crop rotation is a practice that may be employed to control a number of soybean-parasitic nematodes (Crittendon, 1956; Epps et al., 1964; Ferris and Bernard, 1967; Johnson et al., 1975; Kinloch, 1974; Nishizawa, 1978; Ozaki and Asai, 1963; Ross, 1962a; Slack et al., 1981). On a soil infested with soybean cyst nematode, 2 yrs of a nonhost crop reduced the population of nematodes enough that a susceptible cultivar produced a normal crop the next year (Slack et al., 1981). A resistant soybean cultivar was as good as a non-host (Slack et al., 1981). No additional advantage was obtained by a 3rd yr in a nonhost. Grain sorghum is a good nonhost to use in rotations to control SCN but may not be good in a lesion nematode rotation and its effectiveness in a root-knot nematode rotation would depend on the species and location. *Meloidogyne incognita* can be a serious parasite on grain sorghum in some areas but not reproduce well on it in other areas (Riggs, unpublished). In planning a rotation system, all nematodes present must be considered.

19–8.4 Other Management Considerations

Biocontrol of nematodes may offer some promise in the future, but is not currently a feasible practice. However, without natural biological predation, antagonism, and competition, the major control practices probably would not be as effective as they are. Many organisms have been isolated from different species of nematodes, and some have proven pathogenic or have predaceous activity (Morgan-Jones and Rodriguez-Kabana, 1981; Morgan-Jones et al., 1981; Ownley Gintis et al., 1982a; 1982b). Mass production and application techniques have not been developed.

Any practice that reduces stress on the soybean plant will improve the capability of the plant to tolerate nematode parasitism. Irrigation to insure sufficient moisture during periods of low rainfall will help the soybean overcome nematode damage. Subsoiling to open up compacted layers will allow the soybean roots to grow deeper, if the subsoil is not too acidic, where the moisture supply is likely to be greater. Proper fertilization based on soil anlayses will aid the plant in obtaining needed nutrients in spite of nematode parasitism and damage. These practices may be less effective against root-knot nematodes, because secondary organisms often invade the roots and cause decay more than in other types of nematode parasitism.

19–9 PROSPECTS FOR FUTURE CONTROL OF SOYBEAN PARASITIC NEMATODES

Perhaps our greatest challenge in controlling plant parasitic nematodes on soybean is to improve education and extension programs. Traditional methods are still highly effective when used properly. Research, teaching, and extension nematologists or plant pathologists must become more effective in their education of growers. Farmers must know the species and population densities of nematodes in every field in order to devise effective management systems. This can be achieved by an assay program that provides not only genus identification, which is the most common practice today, but species and race identification also. There are opportunities and challenges for private laboratories to establish assay services for growers. Without grower acceptance of assay programs, the prospects for future effective management of nematodes will be diminished because proper strategy development is dependent on a good set of data.

We are entering an age where most farmers will own or have access to computers that will be of great assistance in planning control strategies for plant parasitic nematodes. Nematode researchers are developing damage-cost relationships. As growers develop case histories of nematode problems they will be able to input current assay data into their own, service organization, public or private computers. As more sophistication is attained in this area, farmers should be able to run programs on the computer to project results of selected management practices. They will be able to determine the cost and return from a control practice, and project the nematode numbers at the end of the season. Models can be used to test how long rotations must be to effectively reduce populations of pathogenic nematodes to levels where resistant cultivars or susceptible cultivars can be grown profitably.

Once a decision has been made that nematode population management is necessary and can be profitable, the traditional methods of management—resistance, rotation, and nematicides—are still effective. The future of any of these practices, however, is heavily dependent on wise use of each. Resistant cultivars become susceptible because of overuse. Nematicides are withdrawn because of health risks, especially from overuse or abuse. Indeed, these three management tactics are effective and can be effective for the foreseeable future if they are properly used. Grower education cannot be overemphasized. The development of novel approaches to nematode control would be timely. For example, instead of developing a new resistant cultivar each time a new race of nematode appears, effort could be directed at manipulating behavior of genes and stabilizing the gene pool of the nematode population.

Effective fumigants are being replaced by organophosphates and carbamates which, although less effective in controlling nematodes, can give fairly satisfactory yield increases, sometimes even comparable to the fumigants. With considerable emphasis being placed on biotechnology, the development of new classes of effective but safe nematicides seems feasible.

The prospects for future control of soybean-parasitic nematodes should be good, but will certainly require the efforts of many people.

These people must work together in a synergistic, energetic manner because the crop is so widely planted and the nematodes are ubiquitous.

REFERENCES

Aboul-Eid, A.H., and H.A. Osman. 1981. Control of *Tylenchorhynchus clarus* on soybean with systemic nematicides. Nematol. Medit. 9:105–107.

Aboul-Eid, H.Z., and A.I. Ghorah. 1974. Pathological effects of *Heterodera cajani* on cowpea. Plant Dis. Rep. 58:1130–1133.

Abu-Gharbieh, W.I., and V.G. Perry. 1970. Host differences among Florida populations on *Belonolaimus longicaudatus* Rau. J. Nematol. 2:209–216.

Acosta, N., and R.B. Malek. 1981. Symptomatology and histopathology of soybean roots infected by *Pratylenchus scribneri* and *P. alleni*. J. Nematol. 13:6–12.

----, ----, and D.I. Edwards. 1979. Susceptibility of soybean cultivars to *Pratylenchus scribneri*. J. Agric. Univ. P.R. 63:103–110.

----, and J.A. Negron. 1982. Susceptibility of six soybean cultivars to *Meloidogyne incognita* race 4. Nematropica 12:181–187.

Atkinson, R.E. 1944. Diseases of soybeans and peanuts in the Carolinas in 1943. Plant Dis. Rep. Suppl. 148:254–259.

Baldwin, J.G., K.R. Barker, and L.A. Nelson. 1979. Effects of *Meloidogyne incognita* on nitrogen fixation in soybean. J. Nematol. 11:156–161.

Barker, K.R. 1982. Influence of soil moisture, cultivar, and population density of *Meloidogyne incognita* on soybean yield in microplots. J. Nematol. 14:429 (Abstr.).

----, D. Huisingh, and S.A. Johnston. 1972. Antagonistic interaction between *Heterodera glycines* and *Rhizobium japonicum* on soybean. Phytopathology 62:1201–1205.

----, D.P. Schmitt, and V.P. Campos. 1982. Response of peanut, corn, tobacco and soybean to *Criconemella ornata*. J. Nematol. 14:576–581.

Bernard, E.C. 1980. Reassessment of resistance among soybeans to *Meloidogyne incognita*. J. Nematol. 12:215 (Abstr.).

Birchfield, W. 1962. Host-parasite relations of *Rotylenchulus reniformis* on *Gossypium hirsutum*. Phytopathology 52:862–865.

----, C. Williams, E.E. Hartwig, and L.R. Brister. 1971. Reniform nematode resistance in soybean. Plant Dis. Rep. 55:1043–1045.

Bonner, M.J. 1981. The relationship of field population densities of *Heterodera glycines* to soybean growth and yield. M.S. thesis. North Carolina State University, Raleigh.

----, and D.P. Schmitt. 1985. Population dynamics of *Heterodera glycines* life stages on soybean. J. Nematol. 17:153–158.

Bridge, J. 1976. Plant parasitic nematodes from the lowlands and highlands of Ecuador. Nematropica 6:18–22.

Brooks, T.L., and V.G. Perry. 1967. Pathogenicity of *Pratylenchus brachyurus* to citrus. Plant Dis. Rep. 51:569–573.

Caldwell, B.E., C.A. Brim, and J.P. Ross. 1960. Inheritance of resistance of soybeans to the cyst nematode, *Heterodera glycines*. Agron. J. 52:635–636.

Carvalho, J.C. 1954. A soja C. seus inimigos do solo. (In Portuguese with English summary.) Rev. Inst. Adolfo Lut 14:45–52.

----. 1958. *Rotylenchus elisensis* nova especie associada compraizes de soja. Rev. Inst. Adolfo Lutz 17:43–46.

Cohn, E., and M. Mordechai. 1977. Uninucleate giant cell induced in soybean by the nematode *Rotylenchulus macrodoratus*. Phytoparasitica 5:85–93.

Crittenden, H.W. 1955. Root-knot nematode resistance of soybeans. Phytopathology 45:347 (Abstr.).

----. 1956. Control of *Meloidogyne incognita acrita* by crop rotations. Plant Dis. Rep. 40:977–980.

Diab, K.A. 1968. Occurrence of *Heterodera glycines* from the Golden Island, Giza, U.A.R. Nematologica 14:148.

Dickerson, O.J., and T.J. Franz. 1974. Resistance to *Pratylenchus* spp. in dry-edible beans and soybeans. Proc. Am. Phytopathol. Soc. 1:125 (Abstr.).

Dropkin, V.H., and P.E. Nelson. 1960. The histopathology of root-knot nematode infections in soybeans. Phytopathology 50:442–447.

Eisenback, J.D., H. Hirschmann, J.N. Sasser, and A.C. Triantaphyllou. 1981. A guide to the four most common species of root-knot nematodes (*Meloidogyne* species) with a pictorial key. International *Meloidogyne* Project. North Carolina State University, Raleigh.

El Sherif, M.A., S.L. Hafiz, and B.A. Oteifa. 1974. Root phenolic content of resistant and susceptible soybean cultivars infected with *Rotylenchulus reniformis* nematodes. Ann. Agric. Sci. 1:245–250.

Endo, B.Y. 1964. Penetraton and development of *Heterodera glycines* in soybean roots and related anatomical changes. Phytopathology 54:79–88.

----. 1967. Comparative population increase of *Pratylenchus brachyurus* and *P. zeae* in corn and in soybean varieties Lee and Peking. Phytopathology 57:118–120.

Epps, J.M., and L.A. Duclos. 1970. Races of soybean cyst nematode in Missouri and Tennessee. Plant Dis. Rep. 54:319–320.

----, and E.E. Hartwig. 1972. Reaction of soybean varieties and strains to race 4 of the soybean cyst nematode. J. Nematol. 4:222 (Abstr.).

----, J.N. Sasser, and G. Uzzell, Jr. 1964. Lethal dosage concentrations of nematocides for the soybean cyst nematode and the effect of a nonhost crop in reducing the population. Phytopathology 54:1265–1268.

Fassuliotis, G. 1975. Feeding, egg laying, and embryology of the Columbia lance nematode, *Hoplolaimus columbus.* J. Nematol. 7:152–158.

Ferris, V.R., and R.L. Bernard. 1967. Population dynamics of nematodes in fields planted to soybeans and crops grown in rotation with soybeans. I. The genus *Pratylenchus* (Nemata:Tylenchida). J. Econ. Entomol. 60:405–410.

Franck, A.B. 1882. Gallen der *Anguillula radicicola* Greff and *Soja hispida, Medicago sativa, Lactuca sativa* and *Pirus communis.* Ver. Bot. Vereins. Der. Prov. Brandenburg 23:45–55.

Germani, G. 1981. Pathogenicity of the nematode *Scutellonema cavenessi* on peanut and soybean. Rev. Nematol. 4:203–208.

Gipson, I., K.S. Kim, and R.D. Riggs. 1971. An ultrastructural study of syncytium development in soybean roots infected with *Heterodera glycines.* Phytopathology 61:347–353.

Golden, A.M., J.M. Epps, R.D. Riggs, L.A. Duclos, J.A. Fox, and R.L. Bernard. 1970. Terminology and identity of infraspecific forms of the soybean cyst nematode (*Heterodera glycines.* Plant Dis. Rep. 54:544–546.

Gommers, F.J., and V.H. Dropkin. 1977. Quantitative histochemistry of nematode-induced transfer cells. Phytopathology 67:869–873.

Goswami, B.K., and D.K. Agrawal. 1978. Interrlationships between species of Fusarium and root-knot nematode, *Meloidogyne incognita,* in soybean. Nematol. Medit. 6:125–128.

Gotoh, A., and Y. Oshima. 1963. *Pratylenchus*-arten and ihre geographische verbreitung in Japan (Nematoda:Tylenchia). Jpn. J. Appl. Entomol. Zool. 7:187–199.

Hamblen, M.L., D.A. Slack, and R.D. Riggs. 1972. Temperature effects on penetration and reproduction of soybean-cyst nematode. Phytopathology 62:762 (Abstr.).

Hartwig, E.E., and J.M. Epps. 1970. An additional gene for resistance to the soybean cyst nematode, *Heterodera glycines.* Phytopathology 60:584 (Abstr.).

Heald, C.M., and W.H. Thames. 1982. The reniform nematode, *Rotylenchulus reniformis.* p. 139–143. *In* R.D. Riggs (ed.) Nematology in the Southern Region of the United States. Univ. of Arkansas South. Coop. Ser. Bull 276.

Hori, S. 1916. Phytopathological notes. 5. Sick soil of soybean caused by nematode. (In Japanese.) J. Plant Prot. (Suwon, Korea) 2:927–930.

Hung, Y. 1958. A preliminary report on the plant-parasitic nematodes of soybean crop of the Pintung District, Taiwan, China. Agric. Pest News 5(4):1–5.

Ibrahim, I.K.A., and M.A. El-Saedy. 1982. Resistance of 20 soybean cultivars to root-knot nematodes, in Egypt. J. Nematol. 14:447 (Abstr.).

----, I.A. Ibrahim. and S.I. Massoud. 1972. Induction of galling and lateral roots on five varieties of soybeans by *Meloidogyne javanica* and *Meloidogyne incognita.* Plant. Dis. Rep. 56:882–884.

Inagaki, H. 1979. Race status of five Japanese populations of *Heterodera glycines.* Jpn. J. Nematol. 9:1–4.

Ivezic, M., G. Pvar, M. Vrataric, D. Samata, and A. Djudjar. 1980. Parasitic nematodes of sunflower, soybean and oil rape in the territory of Osijck, Yugoslavia. Zb. Rad. Poljopr. Univ. Beogradu 10:33–42.

Javaid, I., and M. Ashraf. 1978. Some observations on soybeans diseases in Zambia and occurrence of *Pyrenochaeta glycines* on certain varieties. Plant Dis. Rep. 62:46–47.

Johnson, A.W., C.C. Dowler, and E.W. Hauser. 1975. Crop rotation and herbicide effects on population densities of plant-parasitic nematodes. J. Nematol. 7:158–168.

Johnson, P.W. 1977. Genera of plant parasitic nematodes associated with soybeans on the heavier textured soils of Essex, Kent and Lambton counties of Southeastern Ontario. Can. Plant Dis. Sur. 57(1–2):19–22.

Jones, M.G.K., and V.H. Dropkin. 1975a. Cellular alterations induced in soybean roots by three endoparasitic nematodes. Physiol. Plant Pathol. 5:119–124.

----, and ----. 1975b. Scanning electron microscopy of syncytial transfer cells induced in roots by cyst nematodes. Physiol. Plant Pathol. 7:259–263.

Kaplan, D.T., I.J. Thomason, and S.D. Van Gundy. 1979. Histological study of the compatible and incompatible interaction of soybeans and *Meloidogyne incognita*. J. Nematol. 11:338–343.

Kellam, M.K., and N.C. Schenck. 1980. Interactions between a vesicular-arbuscular mycorrhizal fungus and root-knot nematode on soybean. Phytopathology 70:293–296.

Kim, D.G., D.R. Choi, and Y.E. Choi. 1982. Resistance of soybean cultivars to root-knot nematode species *Meloidogyne incognita* and *M. hapla* in Korea. Korean J. Plant Prot. 21:34–37.

Kinloch, R.A. 1974. Nematode and crop response to short-term rotations of corn and soybean. Proc. Soil Crop Sci. Soc. Fla. 33:86–88.

----. 1982. The relationship between soil population of *Meloidogyne incognita* and yield reduction of soybean in the coastal plain. J. Nematol. 14:162–167.

----. 1982. The role of plant resistance in the management of soybean nematodes in Florida. Nematologica 28:156 (Abstr.).

----. and K. Hinson. 1974. Comparative resistance of soybeans to *Meloidogyne javanica*. Nematropica 4:17–18.

----, C.K. Hiebsch, and K. Hinson. 1982. Combined resistance in soybean breeding lines to *Meloidogyne arenaria*, *M. incognita* and *Heterodera glycines* (race 3). J. Nematol. 14:451 (Abstr.).

Kloss, G.R. 1960. Catalogo de nematoides fitopagos do Brasil. Boleti Fitossanitario III, no. 182.

Lehman, P.S., D. Huisingh, and K.R. Barker. 1971. The influence of races of *Heterodera glycines* on nodulation and nitrogen-fixing capacity of soybean. Phytopathology 61:1239–1244.

----, C.C. Machado, and M.T. Tarrago. 1976. Frequency and severity of soybean diseases in the states of Rio-Grande-do-Sol and Santa Catarina Brazil. Fitopatol. Bras. 1:183–194.

Lewis, S.A., F.H. Smith, and W.M. Powell. 1976. Host-Parasite relationships of *Hoplolaimus columbus* on cotton and soybean. J. Nematol. 8:141–145.

Lim, B.K., and M.B. Castillo. 1979. Screening soybeans for resistance to reniform nematode disease in the Phillipines. J. Nematol. 11:275–282.

Lindsey, D.W., and E.J. Cairns. 1971. Pathogenicity of the lesion nematode, *Pratylenchus brachyurus*, on six soybean cultivars. J. Nematol. 3:220–226.

Linford, M.B., and J.M. Oliveira. 1940. *Rotylenchulus reniformis*, nov. gen., n. sp., a nematode parasite of roots. Proc. Helm. Soc. Wash. 7:35–42.

Lordello, L.G.E. 1955. Nematodes attacking soybean in Brazil. Plant Dis. Rep. 39:310–311.

----. 1956a. Nematoides que parasitam a soja na regaio de Bauru. Bragantia 15(6):55–64.

----. 1956b. *Meloidogyne inornata* sp. n., a serious pest of soybean in the state of Sao Paulo, Brazil (Nematoda, Heteroderidae). Rev. Bras. Biol. 16:65–70.

Masamune, T., M. Anetai, M. Takasugi, and N. Katsui. 1982. Isolation of a natural hatching stimulus, glycinoeclepin A, for the soybean cyst nematode. Nature (London) 297:495–496.

Matson, A.L., and L.F. Williams. 1965. Evidence of a fourth gene for resistance to the soybean cyst nematode. Crop Sci. 5:477.

McClintock, J.A. 1922. Resistant plants for root-knot nematode control. Ga. Agric. Exp. Stn. Circ. 77:1.

Melendez, P.L., and N.T. Powell. 1967. Histological aspects of the Fusarium wilt-root-knot complex in flue-cured tobacco. Phytopathology 57:286–292.

Miller, L.I. 1966. Variation in development of two morphologically different isolates of *Heterodera glycines* obtained from the same field. Phytopathology 56:585 (Abstr.).

----. 1967. Development of 11 isolates of *Heterodera glycines* on six legumes. Phytopathology 57:647 (Abstr.).

Minton, N.A., and M.B. Parker. 1979. Effects on soybeans and nematode populations of three soil fumigants applied at several rates at time of planting. Nematropica 9:36–39.

----, ----, and B.G. Mullinix, Jr. 1978. Effects of cultivars, subsoiling, and fumigation on soybean yields and *Meloidogyne incognita* nematode populations. J. Nematol. 10:43–47.

Morgan-Jones, G., and R. Rodriguez-Kabana. 1981. Fungi associated with cysts of *Heterodera glycines* in an Alabama USA soil. Nematropica 11:69–74.

----, B. Ownley Gintis, and R. Rogriguez-Kabana. 1981. Fungal colonization of *Heterodera glycines* cysts in Arkansas, Florida, Mississippi and Missouri soils USA. Nematropica 11:155–164.

Nakasono, K. 1978. Sexual attraction of *Rotylenchulus reniformis* females and some differences in the attractiveness between amphimictic and parthenogenetic populations (Nematoda:Nacobbidae) Helminthol. Abstr. 47:31 (Abstr.).

Nakata, K., and H. Asuyana. 1938. Survey of the principal diseases of crops in Manchuria. (In Japanese.) Bur. Indus. Rep. 32:166.

Nardacci, J.F., and K.R. Barker. 1979. The influence of temperature on *Meloidogyne incognita* on soybean. J. Nematol. 11:62–70.

Nishizawa, T. 1978. Annual population changes of soil nematodes in the field with continous cropping or rotation. The Kasitart J. 12:3–33.

Noel, G.R., P.V. Bloor, R.F. Posdol, and D.I. Edwards. 1980. Influence of *Heterodera glycines* on soybean yield components and observations of economic injury levels. J. Nematol. 12:232–233 (Abstr.).

----, and B.A. Stanger. 1982. Estimation of *Heterodera glycines* populations and yield reduction of soybean. J. Nematol. 14:461.

Norton, D.C. 1983. *Heterodera glycines* on soybean in Colombia. Plant Dis. 67:1389.

Nyczepir, A.P., and S.A. Lewis. 1979. Relative tolerance of selected soybean cultivars to *Hoplolaimus columbus* and possible effects of soil temperature. J. Nematol. 11:27–31.

Okada, T. 1972. Hatching stimulant in the egg of the soybean cyst nematode, *Heterodera glycines* Ichinohe. Appl. Entomol. Zool. 7:234–237.

Ozaki, K., and K. Asai. 1963. Studies on the rotation systems. II. The relationships between crop sequence and the soybean cyst nematode population in the soil. Res. Bull. Hakkaido Natl. Agric. Exp.Stn. (In Japanese with English summary.) 81:11–21.

Ownley Gintis, B., G. Morgan-Jones, and R. Rodriguez-Kabana. 1982a. Fungal colonization of young cysts of *Heterodera glycines* in soybean field. J. Nematol. 14:464 (Abstr.).

----, ----, and ----. 1982b. Mycoflora of young cysts of *Heterodera glycines* in North Carolina USA. Nematropica 12:295–304.

Price, M. 1976. Variability of *Heterodera glycines* races and genetics of parasitic capabilities on soybean. Ph.D. diss. University of Arkansas, Fayetteville (Diss. Abstr. 77- 23392).

Raut, S.P., and C.L. Sethi. 1980. Studies on the pathogenicity of *Meloidogyne incognita* on soybean. Indian J. Nematol. 10:166–174.

Rebois, R.V. 1971. The effect of *Rotylenchulus reniformis* inoculum levels on yield, nitrogen, potassium, phosphorus and amino acids of seeds of resistant and susceptible soybean (*Glycine max*). J. Nematol. 3:326–327 (Abstr.).

----. 1973. Effect of soil temperature on infectivity and development of *Rotylenchulus reniformis* on resistant and susceptible soybeans, *Glycines max*. J. Nematol. 5:10–13.

----, P.A. Madden, and B.J. Eldridge. 1975. Some ultrastructural changes induced in resistant and susceptible soybean roots following infection by *Rotylenchulus reniformis*. J. Nematol. 7:122–139.

Riggs, R.D. 1977. Worldwide distribution of soybean-cyst nematode and its economic importance. J. Nematol. 9:34–39.

----, and M.L. Hamblen. 1962. Soybean-cyst nematode host studies in the family Leguminosae. Ark. Agric. Exp. Stn. Rep. Ser. 110:1–18.

----, and ----. 1966. Further studies on the host range of the soybean cyst nematode. Ark. Agric. Exp. Stn. Bull. 718:1- 19.

----, ----, and L. Rakes. 1977. Development of *Heterodera glycines* pathotypes as affected by soybean cultivars. J. Nematol. 9:312–318.

----, ----, and ----. 1981. Infra-species variation in reactions to hosts in *Heterodera glycines* populations. J. Nematol. 13:171–179.

----, D.A. Slack, and M.L. Hamblen. 1968. New biotype of soybean cyst nematode. Ark. Farm Res 17(5):11.

Robbins, R.T., and K.R. Barker. 1973. Comparisons of host range and reproduction among populations of *Belonolaimus longicaudatus* from North Carolina and Georgia. Plant Dis. Rep. 57:750–754.

Rodriguez-Kabana, R., and D.L. Thurlow. 1980. Evaluation of selected soybean cultivars in a field infested with *Meloidogyne arenaria* and *Heterodera glycines*. Nematropica 10:50–55.

Ross, J.P. 1962a. Crop rotation effects on the soybean cyst nematode population and soybean yields. Phytopathology 52:815–818.

----. 1962b. Physiological strains of *Heterodera glycines*. Plant Dis. Rep. 46:766–769.

----. 1963. Seasonal variation of larval emergence from cysts of the soybean cyst nematode, *Heterodera glycines*. Phytopathology 53:608–609.

----. 1965. Predisposition of soybeans to Fusarium wilt by *Heterodera glycines* and *Meloidogyne incognita*. Phytopathology 55:361–364.

----, C.J. Nusbaum, and H. Hirschmann. 1967. Soybean yield reduction by lesion, stunt, and spiral nematodes. Phytopathology 57:463–464 (Abstr.).

Saito, M., K. Suvada, and S. Sakai. 1973. Breeding of soybean varieties resistant to *Heterodera glycines* in Hokkaido. Agric. Hortic. 48:69–74.

Sasser, J.N. 1977. Worldwide dissemination and importance of the root-knot nematodes *Meloidogyne* spp. J. Nematol. 9:26–29.

----, L.A. Nelson, and K.R. Barker. 1972. Effects of *Trichodorus christiei* and *Belonolaimus longicaudatus* on growth and yield of soybean following chemical soil treatment. J. Nematol. 4:233–234.

Schenck, N.C., R.A. Kinloch, and D.W. Dickson. 1975. Interaction of endomycorrhizal fungi and root-knot nematode on soybean. p. 607–617. *In* F.E. Sanders (ed) Endomycorrhizas. Academic Press, New York.

Schilling, K. 1982. Life cycle studies of variants of soybean-cyst nematode, *Heterodera glycines* Ichinohe. J. Nematol. 14:470 (Abstr.).

Schmitt, D.P., and K.R. Barker. 1981. Damage and reproductive potentials of *Pratylenchus brachyurus* and *P. penetrans* on soybean. J. Nematol. 13:327–332.

----, F.T. Corbin, and L.A. Nelson. 1983. Population dynamics of *Heterodera glycines* and soybean response in soil treated with selected nematicides and herbicides. J. Nematol. 15:432–437.

----, and G.R. Noel. 1984. Nematode parasites of soybeans. *In* W.R. Nickle (ed.) Plant and insect parasitic nematodes. Academic Press, New York.

Shavrov, G.N. 1968. Study of the plant nematode fauna of soybeans in the Primorskiy Kray. Soobshch. dal' nevost. Fil. V.L. Komorova sib Otd. Akad. Nauk. SSSR 26:141–144.

Singh, N.D. 1975. Studies on selected hosts of *Rotylenchulus reniformis* and its pathogenicity to soybean (*Glycine max*). Nematropica 5:46–51.

Sivakumar, C.V., and A.R. Seshadri. 1971. Life history of the reniform nematode, *Rotylenchulus reniformis*, Linford and Oliveira, 1940. Indian J. Nematol. 1:7–20.

Slack, D.A., R.D. Riggs, and M.L. Hamblen. 1981. Nematode control in soybeans: Rotation and population dynamics of soybean cyst and other nematodes. Ark. Agric. Exp. Stn. Rep. Ser. 263:1–36.

Smith, F.H. 1969. Host-parasite and life history studies of the lance nematode (*Hoplolaimus columbus*, Sher, 1963) on soybean (*Glycine max* (L.) Merrill). Ph.D. diss. Univ. of Georgia, Athens, GA (Diss. Abstr. 72-3783).

Standifer, M.S. 1959. The pathologic histology of bean roots injured by sting nematodes. Plant Dis. Rep. 43:983–986.

Sugiyama, S., K. Hiruma, T. Miyahara, and K. Kohuhun. 1968. Studies on the resistance of soybean varieties to soybean cyst nematode. II. Differences of physiological strains of the nematode from Kerawana and Kikojogahara. Jpn. J. Breed. 18:206–221.

----, and H. Katsumi. 1966. (A resistant gene of soybeans to the soybean cyst nematode observed from a cross between Peking and Japanese varieties). (Translation from Japanese.) Jpn. J. Breed. 16:83–86.

Thomas, J.D. 1974. Genetics of resistance to races of the soybean-cyst nematode. M.S. thesis. Univ. of Arkansas, Fayetteville, AR.

Trabulsi, I.Y., M.A. Ali, and M.E. Abd-Elsamea. 1980. Response of soybean cultivars to infection by *Meloidogyne incognita* and *Rhizobium japonicum* alone and in combination. Nematol. Medit. 8:171–175.

Veech, J.A., and B.Y. Endo. 1970. Comparative morphology and enzyme histochemistry of root-knot resistant and susceptible soybean. Phytopathology 60:896–902.

Winstead, N.N., C.B. Skotland, and J.N. Sasser. 1955. Soybean cyst nematode in North Carolina. Plant Dis. Rep. 39:9–11.

Yokoo, T. 1936. Host plants of *Heterodera schachtii* Schmidt and some instructions. Korean Agric. Exp. Stn. Bull. 8(43):47–174.

Yoshii, K. 1977. Reaction of soybean varieties to the root-knot nematode (*Meloidogyne incognita*). Fitopatlogia 12:35–38.

Zirakparvar, M.E. 1982. Susceptibility of soybean cultivars and lines to *Pratylenchus hexincisus*. J. Nematol. 14:217–220.

20 Integrated Control of Insect Pests

Sam G. Turnipseed
Clemson University
Blackville, South Carolina

Marcos Kogan
University of Illinois and
Illinois Natural History Survey
Champaign, Illinois

The exact impact of insect pests on soybean [*Glycine max* (L.) Merr.] in the USA is difficult to estimate. In most years, insects are not a limiting factor in production except perhaps to relatively small areas in northern Florida. Insects, however, represent an important economic factor as both yield and seed quality may be substantially reduced if treatments are not applied on time when populations tend to exceed economic injury levels. An indirect indication of the economic importance of insects in soybean production are records of insecticide use in the various regions of the USA. A survey conducted by the USDA-ERS in 1980 (Hanthorn et al., 1982) showed that 11% of the total hectares planted to soybean received insecticide treatments. Treated hectareage varied from 2% in the North Central states to 47% in the Southeast. These summaries, however, do not portray the complexity of yearly fluctuations. For instance, in 1983 over 1.2 million ha of soybean were sprayed with insecticides and miticides in Illinois, representing about one-third of the state's hectareage (Colwell, 1984).

Much variation in economic impact of insect pests results from the heterogeneity of the agroecological conditions under which U.S. soybean is grown. Production is highly concentrated in two major natural regions: the coastal plains and the interior plains, with smaller areas in production also found in the interior and the Appalachian highlands. These regions extend from the Atlantic and Gulf Coasts, westward and northward between 28 and 46° N Lat. Little is planted west of the 98° meridian. These natural regions have a variety of soil types, temperature, and precipitation regimes; and soybean is grown in association with several other crops

Soybeans: Improvement, Production, and Uses, 2nd ed.—Agronomy Monograph no. 16.

forming characteristic agroecosystems. It is this diversity of growing conditions and the frequent aberrations in weather patterns that explain the vast fluctuations in insect impact from year to year and from region to region.

Various surveys have been performed over the years in different states (Balduf, 1923; Blickenstaff and Huggans, 1962; Tugwell et al., 1973; Deitz et al., 1976; Kogan, 1980). These various sources have identified over 700 species of phytophagous insects collected in soybean fields. Not all of these species are soybean feeders and the actual role in the soybean community of many of them is not known.

Insect colonizers of soybean fields in the USA are species that fall into three categories (Kogan, 1981): (i) polyphagous species that have expanded their host range to include soybean as an accepted host (e.g., grasshoppers, cutworms, armyworms, and spider mites); (ii) oligophagous species feeding on native legumes that have adapted secondarily to soybean (e.g., bean leaf beetle [*Cerotoma trifurcata* (Forster)], Mexican bean beetle (*Epilachna varivestis* Mulsant), and the leaf miner *Odontota horni* Smith); and (iii) nonlegume oligophagous feeders that shifted hosts to incorporate soybean in their diet (example is *Dectes* stem borer). Species in a fourth potentially damaging category—that of immigrants from other regions of the world—have failed so far to find their way into the USA. An exception perhaps is the southern green stink bug (*Nezara viridula* L.) of probable southeast Asian origin. Its immigration into the USA preceded by many years the expansion of soybean production in the New World.

Despite the relative richness of the insect fauna associated with soybean, Kogan (1980) found, based on assessments of entomologists in 18 states, that 14 species accounted for 98.6% of the total insect damage potential to the crop in the USA. These species are discussed in this chapter in greater detail.

20–1 PLANT RESPONSE TO DAMAGE

Insects are capable of injuring every vegetative and reproductive organ of a soybean plant (Fig. 20–1). The effect of injury on stand establishment and on growth and yield of the plant depends on timing and intensity of injury and on growing conditions following time of injury. Soybean is known for its tolerance to injury, but seed yield and quality will fall if injury exceeds the tolerance level. Plants respond differently to different types of injury so it is necessary to recognize the various types of injury and the effects they have on the plant. Poston et al. (1983) defined three principal types of plant responses to insect injury: (i) susceptible response, (ii) tolerant response, and (iii) overcompensatory response (Fig. 20–2). Soybean displays a susceptible response to insects feeding on seeds and pods after the R4 and R5 stages. An overcompensatory response occurs when plants are moderately (20–30%) defoliated

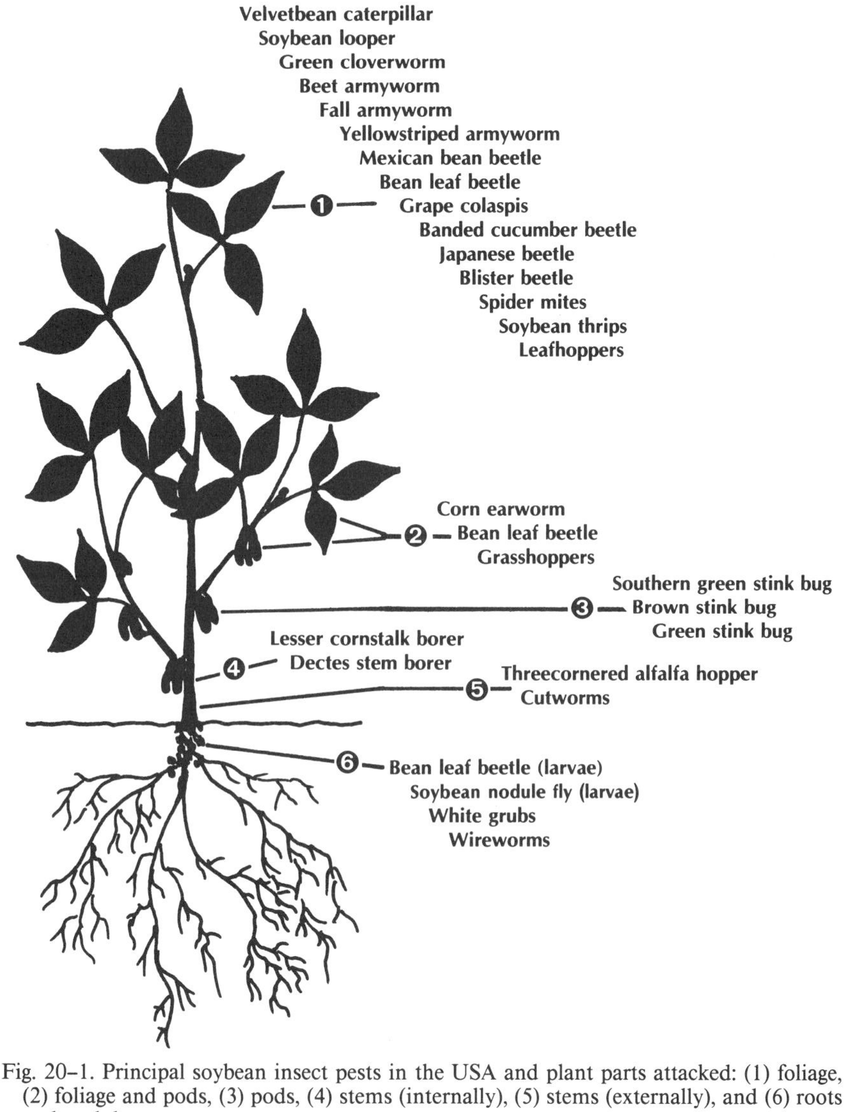

Fig. 20–1. Principal soybean insect pests in the USA and plant parts attacked: (1) foliage, (2) foliage and pods, (3) pods, (4) stems (internally), (5) stems (externally), and (6) roots and nodules.

before the R1 stage (see below). For levels of deblossoming below as much as 50% prior to R3 soybean usually has a tolerant response.

20–1.1 Types of Injury and Plant Response

Insects injure plants mainly as a result of their feeding activities. Feeding injury, however, often exposes plants to secondary infections by plant pathogens, and in some instances disease transmission is the main

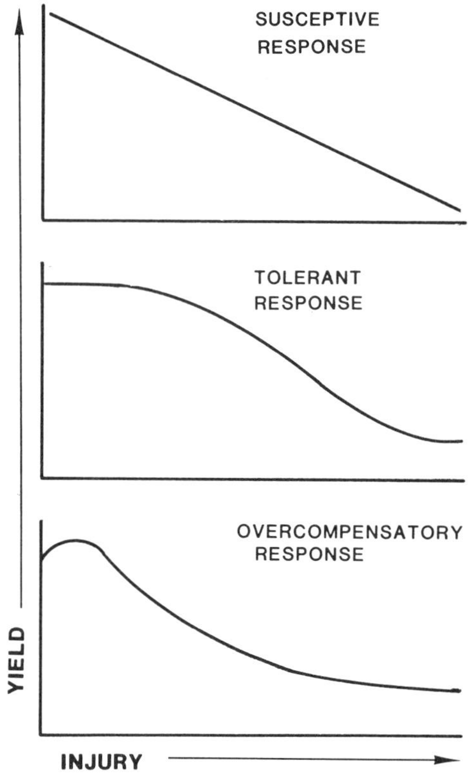

Fig. 20–2. Major categories of plant response to injury. After Poston et al. (1983).

consequence of feeding. This latter type of insect related injury is discussed in chapters 16 and 17 in this book. According to the plant parts that are predominantly affected, one may recognize two major groups of insect pests: (i) those that feed on vegetative organs (roots, stems, and foliage) and (ii) those that feed on reproductive organs (flowers, pods, and seeds).

20–1.1.1 Injury to Vegetative Organs

Insects with mandibulate mouthparts (e.g., lepidopterous caterpillars, beetles, and grasshoppers) remove parts of plant tissue with each bite. The consequence of sustained feeding depends on what plant part is attacked. Insects feeding on stems, externally or internally (as borers) weaken stems, and cause breakage or interruption of sap flow within the plant. Early season injury by stem feeding insects results in poor stand establishment or delay in plant growth. Optimal plant density varies with soil and varietal characteristics, and there is a certain density that is usually recommended for different row spacings. Thus, Pendleton and Hartwig (1973) recommended 30 plants m^{-1} of row at 102-cm row width, 24 plants m^{-1} at 76 cm, and 18 plants m^{-1} at 50 cm. If insect injury grossly reduces plant population below these levels a control action or replanting may be necessary. Insect injury, however, is often concentrated in patches resulting in irregular gaps in the stand. It seems that the length of gaps has a greater effect on yield than a comparable reduction in stand that is evenly distributed. Gaps of 25, 50, and 100 cm were artificially opened in plots of 20-m long rows to simulate cutworm injury to 'Williams' soybean. In each plot, total length of the gaps was 3 m, obtained with 12- × 25-cm gaps, 6- × 50-cm gaps, or 3- × 100-cm gaps. The 100-cm gaps resulted in 19% yield reduction. There was, however, no statistical difference in yields of check plots and in plots with 25- and 50-cm gaps, although both treatments tended to reduce yields (M. Jeffords et al., University of Illinois, unpublished data).

The most obvious type of insect injury to soybean is defoliation (Fig. 20–3). The effect of defoliation on yield has been extensively studied using simulated (hand) defoliation effected at various stages of plant growth. Turnipseed (1973) and Kogan and Turnipseed (1980) have reviewed the early literature on simulated injury. Most studies have shown that soybean has highest tolerance to defoliation occurring up to the R3 stage. In general, prior to R3, plants recover from up to 30% leaf area loss without noticeable effect on yields. Although one may expect that recovery from defoliation depends on growing conditions during the rest of the season, studies by Turnipseed (1972) indicated that the relationship of defoliation to yield was similar in irrigated and in nonirrigated, and in early and in late-planted soybean, despite considerable difference in mean yields among main treatments. Results on the combined effect of drought and defoliation on soybean yields by Caviness and Thomas (1980) confirmed Turnipseed's data. No statistical differences in yield were observed in interactions of irrigation and defoliation, although irrigation increased yields by 38 to 51% in very dry years.

Fig. 20–3. Soybean field heavily injured by Mexican bean beetle adults and larvae in Clark County, IL, 1982.

There is no apparent difference in response between determinate and indeterminate cultivars (Fehr et al., 1977; Gazzoni and Minor, 1979), except at the highest levels of defoliation (100%). Studies in Africa (Tayo, 1980) indicated that early apical bud removal caused a 36% yield increase in 'Hampton', and 5.5 to 17% yield decrease in 'Bossier.' The author implied that differences were due to divergent growth habits—indeterminate in Hampton, and determinate in Bossier. In the USA, however, both cultivars have determinate growth.

More recent research has concentrated on a critical assessment of the methodology and on the refinement of measurements of physiological effects of defoliation. For instance, Poston et al. (1976) compared artificial (handmade) defoliation with insect-induced defoliation. Other studies were conducted to determine the effect of sequential defoliation at successive growth stages (Thomas et al., 1978), and the use of hole punchers to inflict defoliation at a rate comparable to feeding rates of two species of caterpillars (Hammond and Pedigo, 1982). These studies have helped refine the methodology, but in general they did not change much the picture that has been delineated since the early experiments by Kalton et al. (1949). This set of data (as analyzed by Stone and Pedigo, 1972), and data from Thomas et al. (1978) and more recently Saito et al. (1983) have allowed computation of second degree equations describing the relationship between percent defoliation and yield at various stages of growth (Fig. 20–4). These equations have been useful in computing economic injury levels for soybean defoliators (Stone and Pedigo, 1972; Kogan, 1976). In summary, under most experimental conditions and at all growth stages, yield reductions are not expected to occur below 20% defoliation. Before R2 and after R6 there is no significant yield reduction at defoliation levels below 30%. Figure 20–4 (redrawn from Saito et al., 1983) illustrates these relationships for the Japanese cv. Bonminori.

In addition to the effect on yield, defoliation may affect seed quality. In general, protein concentration is not greatly affected by defoliation,

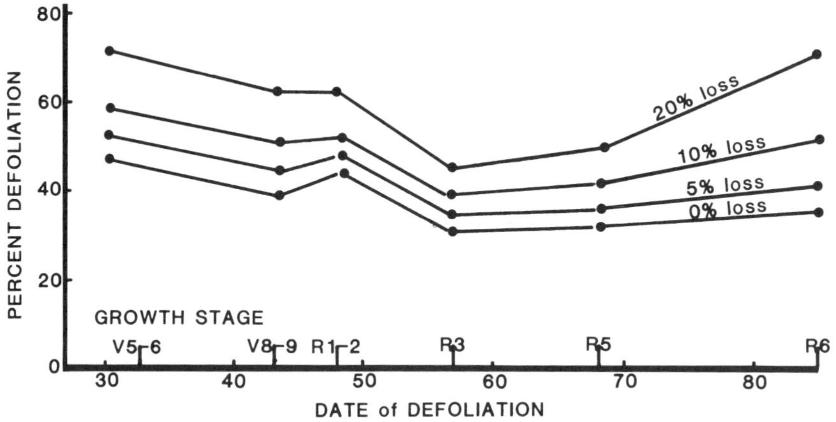

Fig. 20–4. Relationship between date of defoliation (days after planting) and defoliation level, expressed as isolines of percent yield reduction. Redrawn from Saito et al. (1983).

although there is a tendency for protein to decrease and for oil to increase with the reduction of seed weight resulting from defoliation (Turnipseed, 1972; Thomas et al., 1978).

20-1.1.2 Injury to Reproductive Organs

When insects feed on the reproductive organs of soybean the result may be shedding of blossoms or young pods, abortion of seeds, seed malformation, production of unfertile seed, changes in the oil/protein ratio and transmission or enhancement of penetration of disease organisms (Todd and Herzog, 1980; Shortt et al., 1982).

Soybean normally sheds from 20 to 80% of all flowers that are produced (Carlson, 1973). Chapin and Sullivan (Clemson University, 1979, unpublished data) removed 50 or 100% of all blossoms in 'Bragg' soybean and repeated treatments with a second removal a week after and a third removal 14 days after the initial deblossoming. Only the highest levels of deblossoming (50% 3 ×, 100% 2 ×, and 100% 3 ×), caused reduced yields. All other treatments had no significant effect on yields. This experiment also showed that plants compensated for reduced pod numbers by an increase in seed size.

Symptoms of injury to pods and seeds differ between insects with chewing or sucking mouth parts. Mandibulate pod feeders produce scars on the pods that may facilitate infections by secondary fungal diseases (Shortt et al., 1982). Other pod feeders remove portions or destroy entire seeds within pods. Certain pod borers eat preferentially on the germ thus reducing grain quality for seed (Kobayashi and Oku, 1980). Stink bugs, besides directly damaging seeds, also transmit *Nematospora coryli* Peglion, the causal agent of the yeast spot disease.

Direct injury to pods and seeds may result in yield losses and decreases in seed quality although there is considerable compensation, particularly from mandibulate feeders, if injury occurs in the pod set stage (R3). Estimates of the effect of pod removal on yield have been obtained by artificial depodding (Turnipseed, 1973; Thomas et al., 1974). Yields of the determinate cv. Hampton were not reduced after removal of pods 1.3 cm long and longer, when the first set of pods were full length (R4-R5); later removals caused yield losses (Turnipseed, 1973). Mueller and Engroff (1980) reported that levels of corn earworm [*Heliothis zea* (Boddie)] of up to 46 larvae per meter of row did not produce detectable differences in yields when infestations occurred from stages R3 and up to R6 in both irrigated and nonirrigated soybean. They attributed the lack of a detectable effect on the ability of soybean to compensate through additional enlargement and complete development of remaining seeds. Similar results were reported by McWilliams (1983) who used both *H. zea* and *H. virescens* (F.). In this study, it was shown that feeding on young pods resulted in an average destruction of 50 seeds per larva, whereas larvae feeding on fully grown seeds completed development on only five seeds. These data provide valuable information for further refinements of economic injury levels for pod-feeding insects.

Stink bugs feed directly on the seeds but their damage potential cannot be measured exclusively on the number of seeds that they pierce, because the extent of injury depends also on what part of the seed is pierced during seed development. Germination may be reduced by a single stink bug puncture in the radicle-hypocotyl axis, whereas several punctures on cotyledons may not affect germination (Jensen and Newsom, 1972). Other effects of stink bug injury are: abortion of seeds and pods, foliar retention if pod abortion is extensive, changes in oil and protein content of seeds, and transmission of the yeast spot disease agent (Todd and Herzog, 1980). The feeding behavior of stink bugs is much more complex than that of pod borers or other mandibulate pod feeders. Therefore, injury to yield loss relation is more difficult to define and, in general, economic injury levels for stink bugs are much lower than those for other pod feeders (see below).

20–2 ECONOMICS OF INTEGRATED CONTROL

Sound economic principles must be adhered to in the development of tactics for pest control whether it be for insects, nematodes, plant pathogens, or weeds. An assessment of the potential risk from a pest should be made before control action is taken. Successful soybean producers cannot afford to adopt such practices as: (i) applying nematicides as insurance without soil sampling to determine infestation levels, (ii) using a $24 postemergence herbicide combination when an $8 one would control their specific weed problems, or (iii) applying an insecticide without knowing what insects and how many are present in their fields. For an insect pest, we generally work within the context of an economic injury level which has been defined as the level of pest population capable of producing damage that if prevented would at least offset the cost of control used to suppress the population (Newsom et al., 1980). Before examining more closely economic injury levels for single insect pests and insect pest complexes, we will outline briefly a few factors involved in integrated control of insect pests.

Integrated control of soybean insect pests employs a few uncomplicated components. These include: (i) monitoring of crop growth and insect development, (ii) assessing presence and effectiveness of natural control agents, (iii) applying control measures on the basis of economic damage thresholds, and (iv) using chemicals at minimum effective dosages that control pests while having the least adverse effect on natural enemies (Turnipseed and Kogan, 1983).

In the South, where risk of economic loss is high, soybean insect pests usually are monitored on a weekly basis from mid-July through crop maturity. Scouting data, which are used in making treatment decisions, usually include plant growth stage, plant injury (defoliation and pod damage), levels of pest species and impact of natural enemies on pest populations. In the Midwest, scouting is done less frequently and research has been directed toward predicting pest population development to appropriately allocate sampling resources (Moffit et al., 1986).

Treatment decisions in soybean are based on economic thresholds, which often include plant injury estimates in addition to numbers of insects (population level) per unit of area (Turnipseed and Kogan, 1983). Two basic reasons for these injury/population thresholds are that: (i) soybean compensates remarkably for injury to foliage, blooms, pods, and even growing tips of plants, and (ii) soybean usually is attacked by more than one pest species at the same time, necessitating some combination of numerical threshold values for separate species. Thus, a treatment decision adjusted for stage of crop growth may be based on: (i) percentage defoliation, (ii) combined numbers of defoliating species present, and (iii) an evaluation of the level of natural enemy activity against the pest(s) involved.

Natural enemies and insecticides are two important regulators of insect pest populations. Compatibility of insecticides and natural enemies is vital to integrated control programs. Economic considerations are obvious in the choice of insecticides for insect suppression. If a chemical is used that selectively kills the pest but spares natural enemies, then only the cost of application is involved. However, in many cases where a nonselective chemical is used that destroys natural enemies, pests resurge to even higher levels within 14 to 21 days necessitating yet another application.

20–2.1 Economic Injury Levels for Single Pests

We have indicated that insects attack almost all stages of soybean but the most important economic consequences occur from foliage and pod feeders in mid- or late season. Development of thresholds for single species is important where pest complexes are not involved. The green cloverworm [*Plathypena scabra* (F.)] is a foliage feeder and an insect pest of significance in many midwestern production areas. Stone and Pedigo (1972) developed economic injury levels for green cloverworm in Iowa which included larval foliage consumption, plant growth stage, costs of insecticidal applications, and value of soybean. They concluded that population levels for which treatment is recommended should be revised upward. A somewhat similar situation exists in much of Florida where the velvetbean caterpillar (*Anticarsia gemmatalis* Hubner) is the only foliage-feeding insect of significance. Population levels of 18 to 24 insects m^{-1} of row were designated as requiring treatment to prevent economic loss (Strayer, 1973; Reid, 1975). A static model has been used to permit computation of economic injury levels for single foliage or multiple defoliating species. This model is useful for quick calculations of changes in market value of soybean and cost of sprays on economic injury level (Kogan, 1976; Ruesink, 1982).

Economic thresholds have also been developed for pod-feeding insects. In certain cases in the South, corn earworm will feed on young pods in the absence of other pod or foliage insects. Smith and Bass (1972) caged corn earworm on soybean in Alabama when young pods were developing. They concluded that nine medium-sized corn earworms per meter of row would justify control measures. This is quite probably a conservative economic threshold since Newsom et al. (1980) indicated that work in Arkansas during similar growth stages with 15 and 20 corn earworms per meter of row resulted in no yield loss or adverse effects on quality. However, based on these studies thresholds in Arkansas have been raised only from 6.5 to 10 larvae per meter of row (Mueller and Engroff, 1980) demonstrating the cautious approach that entomologists take in this matter.

The southern green stink bug feeds on seed within pods in southern production regions. Their outbreaks often occur in the absence of other major insect pests. Caged soybean plants were infested at various population densities to determine effects of feeding damage on soybean yield and quality (Todd and Turnipseed, 1974). Germination, emergence, and seedling survival as well as yield and quality were all reduced by as few as three stink bugs per meter of row.

In situations where an insect pest acts independently to reduce soybean yield or quality, then single pest thresholds are helpful. This is, however, most often not the way insects feed on the crop.

20–2.2 Economic Injury Levels for Multiple Pests

Under most field situations, more than one species often occur simultaneously on plants. Green cloverworm, velvetbean caterpillar, corn earworm, and soybean looper [*Pseudoplusia includens* (Walker)] may all feed on foliage *at the same time* in fields in the southern USA. Thus, thresholds are needed not just for single species acting alone but for complexes as they appear on the crop. Economic damage thresholds (Turnipseed and Kogan, 1976) require quick and reasonably accurate assessments of both defoliation and population estimates of defoliating species present. Although untrained personnel tend to grossly overestimate defoliation, efficient estimates are made by practiced surveyors (Kogan and Turnipseed, 1980). Economic damage thresholds are quite useful where defoliating complexes prevail.

What does a grower do when developing soybean pods are being attacked by corn earworm or stink bugs and foliage is being consumed at the same time, but both at just under threshold levels? About the best available advice, which is inadequate, is that treatment should be made when either corn earworm or stink bugs or the defoliating complex approaches their threshold levels. At this point, it is evident that additional research is needed even though the described situation is not nearly as complex as in many actual field occurrences. In many production regions, soybean may have foliage and pod feeders at near economic levels plus at the same time, (i) larvae of *Dectes texanus texanus* Leconte boring in main stems, (ii) threecornered alfalfa hopper [*Spissistilus festinus* (Say)] girdling leaf petioles, (iii) larvae of the bean leaf beetle feeding on roots and nodules, and (iv) larvae of the soybean nodule fly [*Rivellia quadrifasciata* (Macquart)] feeding on nodules. This insect situation is indeed complex and needs much more attention than it is now receiving. The unraveling of the complexities of this situation and the development of combined thresholds for all major insect pests is admirable but also inadequate unless considered in the context of the impact of other pest classes such as nematodes, plant diseases, and weeds (see section 20–8).

20–3 LEAF-FEEDING INSECTS

A majority of phytophagous insects on soybean feed primarily on leaves. Leaf-feeders range in size from tiny whiteflies and thrips to large beetles and caterpillars. Most have chewing mouthparts and cause losses by defoliation, and others suck plant juices causing losses by a general decline in vigor.

20–3.1 Larvae of Lepidoptera (Moths)

Lepidopterous larvae, as a group, are perhaps the most conspicuous and, often, the most damaging insect pests of soybean. Three species of this group will be discussed in detail and several others of lesser importance will be mentioned.

20–3.1 Green Cloverworm

The green cloverworm [*Plathypena scabra* (F.)] is one of the few insect species reported to reach economic injury levels throughout most soybean production areas of the USA (Sherman, 1920; Balduf, 1923; Stone and Pedigo, 1972). It is the most important pest species in many midwestern states. However, in the Southeast, green cloverworm is not important as a phytophagous species and, in fact, is considered beneficial in soybean insect management. This is because larvae that are present during July serve as hosts to numerous natural enemies (predators, parasites, and diseases) that help regulate important pests occurring later in the season.

Larvae of the green cloverworm may feed on soybean foliage in July, but under normal conditions outbreaks of economic significance do not occur until mid-August. Larvae reach a length of about 3.2 cm when fully grown and are green with faint white stripes laterally (Plate 20–1.1). After completion of larval development, pupation occurs on or just under the soil surface or under plant refuse. Pupae are sometimes attached by a few silken strands to plant parts. During the summer, the pupal stage lasts approximately 10 days. The slate gray to brown moths then emerge, mate, and after approximately 7 days begin to lay eggs. Moths have a wing span of slightly more than 2.5 cm and mouthparts project from the undersurface of the head to give a snout-like appearance. Eggs are deposited on either surface of soybean leaves and begin to hatch in about 5 days. Winter is passed in adult or pupal stages.

There may be up to four generations per year in the South, depending on climate and season. In the Midwest, there are possibly two generations per year. The first eggs in the spring are deposited on leguminous plants such as clover and alfalfa. Eggs of later generations are deposited on soybean. Additional biological information is found in Pedigo (1980).

20–3.2 Soybean Looper

Several species of loopers may infest soybean, but well over 90% of total larval numbers have been identified as the soybean looper [*Pseudoplusia includens* (Walker)] (Hensley et al., 1964; Canerday and Arant, 1966; Herzog, 1980). Loopers may cause extensive foliage loss and occasional pod damage in southern states, but seldom reach economic injury levels north of Arkansas, Tennessee, or North Carolina. Soybean looper is an annual immigrant into soybean production areas from overwintering locations such as southern Florida. Although loopers may be present in fields in July, peak infestations seldom occur before late August or early September. This insect eats large holes in leaves, and high populations may completely defoliate plants.

Infestations of soybean looper often increase to economic proportions due to two factors; the presence of cotton (*Gossypium hirsutum* L.) and the misuse of insecticides. Availability of cotton nectar as a carbohydrate food source enables adult females to produce substantially more eggs, leading to extremely high populations (Jensen et al., 1974). Also, the misuse of insecticides through unnecessary application or the use of those that are highly toxic will decimate natural enemies, thus causing loopers to reach economic levels where they ordinarily would not be important.

Larvae are somewhat larger when full grown than the green cloverworm, with a thick body tapering to the head (Plate 20–1.2). The color is green, with lighter stripes running the length of the body. When the insect is crawling, the middle of the body forms a characteristic hump or loop from which the common name of *looper* was derived. The larval stage is usually completed in 14 to 21 days, after which pupation occurs in a silken cocoon attached to the soybean plant. In summer months, moths emerge from cocoons in about 14 days and later begin depositing eggs. Moths have a wingspan of approximately 3.8 cm and are brownish with two white spots in the middle of the front wings. After overwintering in the pupal stage, moths emerge in early spring and lay eggs on alternate hosts such as vegetables and other legumes. Several generations occur each year. Herzog (1980) presented a review of the biological information and sampling procedures for loopers on soybean.

20–3.3 Velvetbean Caterpillar

The velvetbean caterpillar (*Anticarsia gemmatalis* Hübner) is the most serious defoliator in the USA, and is second only to the corn earworm in overall economic importance (Kogan, 1980; Turnipseed and Kogan, 1983). It survives the winter only in the tropics and possibly in southern Florida. As the season progresses, adult moths migrate northward with infestations on soybean seldom reaching economic injury levels north of a line from Arkansas through North Carolina (Herzog and Todd, 1980). Hinds and Osterberger (1931) indicated that this insect preferred soybean to velvetbeans and called it the *soybean caterpillar.* Serious infestations of soybean fields usually occur during September on late-maturing cultivars. Velvetbean caterpillars may completely strip foliage in just a few days and immediately afterward begin feeding on pods. The outer wall and nearly mature seed may be consumed, but of most importance in nearly mature fields is the clipping of pods from plants. As much as 240 to 300 kg of bean have been lost from clipping in areas of South Carolina (S.G. Turnipseed, 1983, personal observations).

Larvae are usually greenish and striped but may be brownish to almost black in the fall (Plate 20–1.3). These brownish to black stages are often misidentified as *armyworms* by untrained observers. When disturbed, these caterpillars wriggle rapidly and drop from the plant. After growing to a length of about 5 cm, mainly in upper portions of plants, larvae move downward and pupate just under the soil surface at the base of plants. Following a heavy larval infestation many pupae may be collected by lightly scratching beneath debris under plants. Moths emerge in about 10 days and begin depositing eggs. These adults are light brown with an oblique dark line across the front and back wings. Several generations may be produced annually, depending on geographical location and climate.

20–3.4 Other Lepidopterous Larvae

Soybean foliage is attacked by several other lepidopterous larvae including the corn earworm, armyworms (*Spodoptera* spp.), and woollybears (larvae of Arctiidae) which occasionally produce damaging infestations.

Foliage feeding by the corn earworm is often overlooked because this insect is primarily a pod feeder. However, in certain southern areas and occasionally on the East Coast, earworms may cause economic losses by feeding on foliage (Nickels, 1926).

Full-grown larvae of the corn earworm are about 3.8-cm long and vary from light green to almost black (Plate 20–1.4). Larvae are lighter on the underside and have light and dark lateral stripes that run the length of the body. Mature larvae crawl down or drop to the ground and burrow into the soil to a depth of 10 to 15 cm, where pupation occurs in a small earthen cell. Moths may emerge in a few days and begin depositing eggs of another generation. Usually, four generations develop in a season in the South and one generation from egg to adult requires about 30 days in midsummer. Overwintering is accomplished in the pupal stage. Moths emerge in the spring and early generations develop on other host plants such as corn. Moths have a wing span of about 3.8 cm and are light brown with irregular lines and dark areas near wing tips.

Representative of armyworms attacking soybean is the beet armyworm [*S. exigua* (Hubner)]. This insect overwinters as a pupa or an adult. Adults have a wing span of 3.2 cm, with front wings grayish brown and rear wings whitish to dark in front. Eggs are deposited in early spring in masses of about 80 and are covered with hairs and scales from the female's body. A cycle from egg to adult requires 25 to 30 days during warmer months. There are four generations a year in the South. This insect is more prevalent on soybean in the southern Mississippi Delta than in other soybean-growing areas. Other members of the armyworm group causing occasional damage are the fall armyworm [*S. frugiperda* (J.E. Smith)] and the yellowstriped armyworm (*S. ornithogalli* Guenee).

The yellow woollybear [*Spilosoma virginica* (F.)] is the most common species of the family Arctiidae attacking soybean in the USA. It occasionally reaches economic levels in the North but not in the South. The adult, a white-winged moth with a wing span of 3.7 to 5 cm, lays clusters of white round eggs on leaves. The extremely hairy larvae are white when young, but usually turn dark brown or reddish as they grow older. The species feeds on many cultivated and wild plants. Woollybears overwinter as pupae inside silken cocoons heavily covered with interwoven hairs from the body of the caterpillar. Usually there are two generations in the North. Heavy infestations commonly occur late in the season. If populations are high, it is common to see caterpillars crawling across roads as they move from one field to the next. In 1983, there were reports of infestations requiring insecticide sprays, but in most years, heavy populations are decimated by a fungus of the genus *Entomophthora.*

20–3.2 Coleoptera (Beetles)

Beetles as a group are generally less important as foliage feeders than lepidopterous larvae, but in certain areas members of this group cause significant economic losses. Beetles may be identified by the front wings or elytra, which appear as hardened shields and protect the hind wings that are folded underneath.

20–3.2.1 Mexican Bean Beetle

During the past 10 to 15 yrs, a dramatic shift has occurred with respect to geographical distribution of economic populations of the Mexican bean beetle (*Epilachna varivestis* Mulsant) on soybean. Prior to the 1970s losses in soybean were observed only in the Coastal Plains from Maryland and southward although the beetles' range extended into midwestern production regions. Currently, soybean damage is more extensive in southern Indiana and surrounding areas than in southeastern Coastal Plains.

Eggs, larvae, pupae, and adults occur on soybean through most of the season. However, foliage losses do not usually reach economic proportions until August, because soybean growth during June and July compensates for damage from beetle populations. Also, there is considerable early season predation by beneficial species such as geocorids and lady beetles (Coleoptera: Coccinellidae) which cause high mortality in the first beetle generation.

After overwintering in clumps of grass or in protected areas near soybean fields, the first emerging adults (Plate 20–2a) may begin feeding on alternate hosts before moving to young soybean. However, many beetles move directly from hibernation to early planted soybean where they feed for a few days before depositing masses of yellow eggs on the undersurface of lower leaves. After a week or longer, depending on temperatures, small yellow larvae hatch and begin feeding. Full-grown larvae are approximately 8-mm long, yellow, and covered with several rows of branching yellowish spines. In colder weather, in late season, spines may be completely black. When mature, larvae change into an orangish yellow pupal stage on the undersurface of lower leaves. Several pupae may occur on one leaflet. When first emerging from the pupal stage, adults are soft bodied and of a light lemon color. Soon eight small black spots appear on each wing cover and color darkens to yellow and, with advancing age they acquire a copper hue. Up to four generations may occur during a season, depending on latitude (Turnipseed and Shepard, 1980).

Larvae and adults feed on the undersurface of leaves between veins, leaving a lace-like appearance to leaves. After about 2 days remaining veins change from green to brown, which is characteristic of Mexican bean beetle feeding.

20–3.2.2 Bean Leaf Beetle

The bean leaf beetle [*Cerotoma trifurcata* (Forster)], in addition to feeding on all parts of the soybean plant, is also instrumental in transmission of a disease agent known as bean pod mottle virus (Ross, 1963; Walters, 1964). Adults usually feed on foliage and occasionally on flowers or pods. This insect seems to be most destructive to soybean from Louisiana north into Illinois and in Tennessee and North Carolina. From South Carolina south and west into Mississippi, the bean leaf beetle only occasionally causes economic losses, and numbers seem to be higher on soybean grown on heavier soils.

Damage to soybean foliage is in the form of small holes eaten in leaves. Damage in early season often goes unnoticed until leaves fully expand. Although some damage may occur on seedlings, beetles in numbers of economic importance do not usually appear until late August or September. Late-season adults may feed on pods and produce considerable injury through the secondary invasion of fungi (Shortt et al., 1982).

A summary of bean leaf beetle development is presented in Kogan et al. (1980). Adults overwinter in protected areas such as woodlots, near soybean fields. Adults that emerge in April and May are about 5-mm long and are yellowish, brownish, or reddish with or without black spots or markings (Plate 20–2b). This stage feeds on the undersurface of leaves and when disturbed, drops to the ground. Small spindle-shaped eggs are deposited in soil at the base of plants and hatch into slender white larvae that feed on roots and nodules. Pupation occurs in earthen cells in the soil and adults emerge a few days later. There may be from one to three generations each season.

20–3.2.3 Other Coleoptera

Several other adult Coleoptera occasionally feed on foliage of soybean. Of these insects, Japanese beetle, blister beetles, cucumber beetles, and *Colaspis* spp. are of some economic importance.

The Japanese beetle (*Popillia japonica* Newman) is not a serious problem of soybean in Southeastern states, but adults have been reported feeding on foliage in North Central states (Gould, 1963). Adults are about 1.3-cm long and are metallic green or greenish bronze. They skeletonize bean leaves in late July and early August. Larvae live in the soil and are similar to, but smaller than, white grubs. There is usually one generation a year, but in colder regions larvae may require 2 yrs to develop.

Feeding from adult blister beetles occasionally results in severe defoliation of border areas of soybean fields. These beetles are 2.0- to 2.5-cm long, and wing covers may be black with gray margins (margined blister beetle, *Epicauta pestifera* Werner) or yellow with dark brown stripes [striped blister beetle, *E. vittata* (F.)]. If one of these insects is accidentally mashed on the human skin, a blister may occur, caused by a substance called *cantharidin* within the beetle's body. Eggs are deposited in soil, where larvae and pupae develop. Larvae feed largely on grasshopper eggs. Most species have one generation per year.

Cucumber beetles are often found in soybean fields (Kretzschmar, 1948; Nettles et al., 1970) but seldom contribute to foliage losses of economic significance. There are three common species, all of which are about 6-mm long with a basic greenish yellow color. These species may be separated by the presence of black spots (spotted cucumber beetle, *Diabrotica undecimpunctata howardi* Barber), stripes [striped cucumber beetle, *Acalymma vittata* (F.)], or bands (banded cucumber beetle, *D. balteata* LeConte). Larvae feed on roots and nodules in soil. There are usually two generations per year, but up to four or five may occur in the deep South.

In the southern Mississippi Valley, *Colaspis* spp. may cause economic losses by feeding on foliage. The larvae also feed on roots.

20–3.3 Miscellaneous Leaf Feeders

Many species that are neither Lepidoptera nor Coleoptera sometimes feed on soybean leaves and some of these species may at times cause economic damage.

20-3.3.1 Thrips

Thrips are probably more numerous on soybean than any other insect group, with the most prevalent species being *Sericothrips variabilis* (Beach) (Blickenstaff and Huggans, 1962). Adults are about 1.3-mm long, very slender, and have alternate brown and white bands on the abdomen. Reports of thrips causing damage to soybean are numerous (Ratcliffe et al., 1960; Rodriguez and Ortega, 1962; Petty, 1967), but in most areas it is doubtful that thrips damage causes any economic loss. Observations in South Carolina (Turnipseed, 1973), Arkansas (Mueller and Luttrell, 1977), and Illinois (Irwin and Kuhlman, 1979) indicated complete recovery of soybean from high early season thrips populations. When thrips injury occurred in combination with another stress factor such as herbicide injury, however, seedlings began to drop leaves and die (Wedberg and Kuhlman, 1976).

Also, thrips may transmit tobacco ringspot virus to soybean to a limited extent (Bergeson et al., 1964). On the positive side, moderate infestations of thrips appear to act as prey for predators such as *Orius insidiosus* (Say), causing increased predator numbers which help control eggs and small larvae of several important pests (Irwin and Yeargan, 1980).

Feeding injury appears as a silvering and light streaking of leaflets along the veins. Damage is more severe during cool weather. In the South, adult populations appear by mid-May on the undersurface of leaflets of early planted soybean. Tiny whitish larvae hatch within a few days from eggs inserted in the leaf tissue. Later larval stages are yellow. There are several generations each year.

20-3.3.2 Leafhoppers

The potato leafhopper [*Empoasca fabae* (Harris)] is the most abundant species of leafhoppers found on soybean (Helm et al., 1980). Glabrous types of soybean are severely stunted by feeding of potato leafhoppers. Robbins and Daugherty (1960) found that both incidence of and oviposition by potato leafhopper were highest on glabrous soybean and lowest on lines with dense pubescence. Broersma et al. (1972) and Turnipseed (1977) indicated that length and orientation of hairs rather than density alone were most important in keeping leafhopper populations at low levels. Commercial cultivars of soybean possess normal pubescence, and this prevents economic populations from developing.

Different species of leafhoppers found on soybean are usually < 6-mm long, are green to brown, and jump when disturbed. Leafhoppers feed by inserting their needle-like mouthparts into leaves from which juices are removed. Eggs are usually deposited in the tender stems, buds, or leaves. Several active, wingless nymphal stages occur, which feed on soybean before changing into winged adults. From one to several generations may occur, depending on the species and climate.

20-3.3.3 Mites

Mites, (Acari: Tetranychidae), commonly called *red spiders* or *spider mites*, are tiny arthropods of the class Arachnida that under certain circumstances may devastate soybean fields in late season. If carbamate insecticides are used in mid-season for control of such pests as corn earworm and weather is dry following application, then conditions for mite outbreaks are optimal. During 1980 and 1982, such conditions occurred and locally severe mite outbreaks were observed in South Carolina (S.G. Turnipseed, personal observations). In 1983, drought conditions accounted for a major outbreak of spider mites in the Midwest. In Illinois alone, more than 80 000 ha were sprayed to control spider mites (Colwell, 1984). Kogan and Rodriguez (1977) suggested that drought not only provided optimal conditions for mite development but also changed the soybean physiology, making it more nutritious for the pests.

Mites are approximately 0.4-mm long and vary in color from white to greenish or red. Colonies are found on the undersurface of leaves in tiny webbing where nymphs and adults feed by sucking plant juices. Extensive feeding causes leaves to appear yellow or brown, shrivel, die, and eventually drop from the plant. All stages from eggs to adults usually occur together. There are many generations each season (Poe, 1980).

20-4 POD-FEEDING INSECTS

The corn earworm and stink bugs are probably the two most important insect pests of soybean. Other insects such as velvetbean caterpillars and bean leaf beetles feed on pods but significant monetary loss is less frequent.

20-4.1 Corn Earworm

Corn earworm (*Heliothis zea* Boddie) is probably the most important insect of row crops in the USA. It is a major pest on corn, cotton, peanut, and soybean. Kogan (1980), assessing the most important soybean insect pests over a 10-yr period, found that corn earworm had the greatest economic impact in crop production in the USA. Its primary damage is to developing pods before seed enlargement begins, affecting yield and seed quality (Biever et al., 1983). Such feeding also may cause significant losses of blossoms, foliage, and even growing tips of plants. Although the most serious damage to soybean occurs in a geographic band that includes Coastal Plain areas from southern Virginia to Alabama (Stinner et al., 1980) serious damage does occur in other southern production regions.

It is of interest to note that Stinner et al (1980) report that early planting of early maturing (Maturity Groups V and VI) cultivars avoids losses to corn earworm in North Carolina and that later planted cultivars of Maturity Group VII or VIII often suffer severe losses. Their observations were made during 1971–1975. Observations in the southern part of South Carolina during 1980 to 1983 indicate that exactly the opposite situation occurs; early planted Maturity Group V or VI soybean that bloom in July are often devastated by corn earworm, whereas later-planted Maturity Group VII or VIII soybean that bloom during mid- to late August seldom suffer economic loss. Natural enemies are much more important in late compared to early season. These above described differences indicate that drastic changes in an insect's economic impact can occur with changes in geography or time (1971–1975 vs. 1980–1983).

Details of the life and seasonal history and control of earworm were discussed in the previous section 20–3.

20–4.2 Stink Bugs

Three of the most important members of the stink bug complex are the green stink bug [*Acrosternum hilare* (Say)]; the southern green stink bug [*Nezara viridula* (L.)]; and the brown stink bug [*Euschistus servus* (Say)]. Southern green stink bug causes more economic loss in the USA than all other stink bug species combined (Kogan, 1980). It is found from Louisiana and southern Arkansas across the southeastern states. Green and brown stink bugs occur in the South and farther north in Missouri and some midwestern states. Other species in the genera *Euschistus* and *Thyanta* may damage soybean. Also, "broadheaded bugs" in the genus *Alydus* may cause damage that is indistinguishable from that of stink bugs (Daugherty and Jackson, 1967). A detailed summary of stink bug species found on soybean in the USA was presented by Todd and Herzog (1980).

Adult stink bugs are winged bright green or brown insects 1.3- to 2.0-cm long (Plate 20–2c). Adults and nymphs possess scent glands that give off a foul odor when the bugs are disturbed.

Major stink bug damage on soybean occurs when pods and seed are pierced by mouthparts of nymphs or adults feeding on plant juices. The nature of this damage has been described by several workers (Miner, 1961; Daugherty et al., 1964; Miner, 1966; Miner and Wilson 1966; Turner, 1967). Feeding on young pods may cause shriveled seed and even pod abortion. When developing seed are fed upon they become sunken and wrinkled in the feeding area, the seed coat is stained, and underlying cotyledonous tissue is streaked with white. Fully developed seed may be punctured after bean plants are mature, causing a slight staining and discoloration.

When high infestations of stink bugs occur early and are not detected or controlled, complete yield loss may occur. Although lower infestations may not cause yield losses, in many instances the presence of damaged seed may result in reduction in seed quality. Heavy pod feeding by stink bugs will cause plants to remain green throughout the season when non-damaged plants are mature.

A yeast-spot disease organism (*Nematospora coryli* Peglion) that contributes to the described damage is transmitted through the salivary glands of stink bugs into the soybean seed (Daugherty, 1967; Foster and Daugherty, 1969; Clarke and Wilde, 1970). Thus, damage is caused by mechanical feeding in combination with infection by the yeast spot organism and perhaps other disease organisms (Kilpatrick and Hartwig, 1955).

Winter is passed in the adult stage. The first generation develops on many wild and cultivated hosts (Todd and Herzog, 1980). Adults migrate to soybean during the summer, when infestations increase rapidly with pod formation and seed development (Miner, 1966). Peak populations in the South usually occur in late September and early October. Several generations may occur.

Prior to 1976, the southern green stink bug commonly caused economic damage throughout the South Carolina Coastal Plain. However, following the severely cold winter of 1975–1976 southern green stink bugs were not found in damaging numbers in the state and populations have failed to reach significant levels through the 1984 season (M.J. Sullivan, 1985, personal communication).

20–4.3 Miscellaneous Pod Feeders

When young pods are present, occasional damage may occur from velvetbean caterpillars, soybean loopers, green cloverworms, or perhaps other insects that are normally foliage feeders.

Bean leaf beetles often cause pod damage and pod clipping by the velvetbean caterpillar was described in the previous section 20–3. The bean leaf beetle feeds with its chewing mouthparts on the outer pod, exposing the seed, which becomes water-soaked and discolored. This damage may superficially resemble that of stink bugs. Disease organisms may enter through this feeding area and reduce the crop value (Shortt et al., 1982).

20–5 STEM-FEEDING INSECTS

Insects feeding in or on soybean stems seldom cause losses of economic significance. Damage usually occurs in the seedling stage, resulting in reduced stands or skips in rows. Soybean, however, compensates for such damage with little adverse effect on yields, except when feeding is severe. Controls are seldom necessary.

20–5.1 Threecornered Alfalfa Hopper

Soybean stems are weakened by feeding of adults and nymphs of the threecornered alfalfa hopper [*Spissistilus festinus* (Say)]. On seedling soybean, these insects inject the needle-like mouthparts into the stem at the base of plants and cause girdling with a series of feeding punctures. As the main stem hardens the insect moves up, damaging the upper main stem and leaf petioles (Mueller, 1980). Adults are about 6.4-mm long, green, triangular-shaped, blunt anteriorly, and pointed posteriorly (Plate 20–2d). Nymphs are similar in shape and color, but lack wings and have numerous spiny projections over the body. This insect is found throughout the southern states and is not found in the Midwest north of Missouri (Mueller, 1980). Appropriate survey methods for threecornered alfalfa hopper were determined by Boyer (1967) in Arkansas.

In simulating threecornered alfalfa hopper injury, Caviness and Miner (1962) showed that stand reductions of 45% 2 weeks before blooming caused no significant yield loss. Observations by Tugwell and Miner (1967) indicated that girdling of up to 68% of soybean stems by this insect failed to reduce yields. Mueller and Jones (1983) detected yield differences only at the highest level of infestation (81–96%). At these levels yield losses ranged from 16 to 49%, but this high threshold resulted from infestation at early vegetative stages. Sparks and Newson (1984) reported that the most important injury to soybean occurs after the beginning of flowering. Feeding of adults and nymphs on leaf petioles and the peduncles and pedicels of racemes seriously disrupts nutrient source-sink relationships. Yield losses may occur if high winds late in the season cause breakage of hopper-weakened stems.

Winter is passed in the adult stage on young pine trees (Newsom et al., 1984). Emerging adults feed on a variety of plants and move to seedling soybean in May or June where eggs are deposited in stems. Eggs hatch in about 10 days and nymphs feed on soybean, passing through several molts before reaching the adult stage. There may be several generations each season.

20–5.2 Lesser Cornstalk Borer

Damage by lesser cornstalk borer [*Elasmopalpus lignosellus* (Zeller)] occurs most frequently in sandy soils, during dry weather, and on late-planted soybean. This insect does not cause damage outside southern soybean-growing areas.

Larvae tunnel into stems of seedlings at the soil surface, leaving a characteristic webbing with adhering soil particles at the entrance hole. These larvae are yellowish green with reddish brown cross bands and may be found in the tunnel-like webbing or in stems. Smaller plants usually die, and larger plants that are damaged may be broken off by high winds.

Winter is usually passed in the larval or pupal stage, with adult moths appearing in early spring. Moths have a wingspan of slightly < 2.5 cm and color is brownish with females having darker forewings than males. First eggs are laid on corn [*Zea mays* (L.)], johnsongrass [*Sorghum halepense* (L.) Pers.] or other hosts. Larvae begin feeding on leaves or roots and later burrow into stems. Larvae may be present at planting time, in which case soybean seedlings are attacked as soon as they emerge. In other instances, eggs may be deposited on young soybean, with developing larvae causing damage. A generation from egg to adult requires about 4 weeks under optimum conditions. Several such generations occur each year.

20–5.3 Other Stem Feeders

A new pest of soybean is a weed borer (*Dectes texanus texanus* LeConte) which burrows down the center of soybean stems (Daugherty and Jackson, 1969; Campbell, 1980). This borer is a beetle larva that normally feeds on weeds, but was reported infesting soybean in 1968 in Missouri, and later in North Carolina, Arkansas, and Texas. Eggs are laid most commonly in petioles (Campbell, 1980). Larvae tunnel down the leaf petioles, into and down the stem and girdle the plant from within, causing breakage about 5 cm above the soil surface. Below the girdled area the stem is sealed with frass, wherein the larva overwinters, pupating in late spring. There is one generation per year. In one Missouri field 90% of plants were infested with borers, about 60% of stalks were girdled, and 16% were broken off.

Cutworms may reduce stands, particularly when soybean is planted after early vegetables. In South Carolina, the granulate cutworm [*Feltia subterranea* (F.)] was observed destroying numerous seedlings in fields planted after an early spring tomato (*Lycopersicon esculentum* Mill.) crop (Turnipseed, unpublished observations). In the North, the black cutworm [*Agrotis ipsilon* (Hufnagel)] may occasionally affect plant stand and require control actions (Kogan and Kuhlman, 1982). Cutworms are thick-bodied larvae of Lepidoptera that may attain a length of 4 cm. These larvae may be light brown to black, are usually sluggish, and roll into a ball when disturbed. Feeding occurs at night with daylight hours spent beneath the soil in cracks, or under clods or plant refuse.

20–6 SEED-, ROOT-, OR NODULE-FEEDING INSECTS

Our knowledge concerning the economic consequences of insects that attack soybean below the soil surface is severely limited. In a recent treatment of undergound soybean insects, Eastman (1980) discussed their identification and life histories. Newsom et al. (1978) indicated that nodule-feeding may be more important than previously realized because of effects of injury on N_2 fixation. Although insect-feeding alone may limit production in certain cases, undoubtedly of more importance is their interactive impact with nematodes and soil diseases. Substantial increases in research efforts will be necessary before we can begin to make meaningful progress in understanding how the various pests and pest classes function beneath the soil surface.

20–6.1 Soybean Nodule Fly

The soybean nodule fly [*Rivellia quadrifasciata* (Macquart)] is a recently detected pest of soybean in the USA that is likely as important as bean leaf beetle larvae in attacking underground portions of the plants. Adults are among the most common Diptera collected from soybean, but larval damage to nodules in the USA, where the fly occurs throughout most soybean-producing regions (Eastman, 1980), was not determined until 1975 (Eastman and Wuensche, 1977). Eggs are similar to rice (*Oryza sativa* L.) grains in shape, 1-mm long, and a chalky to creamy white. The full-grown larva is about 8-mm long and creamy to tannish white. Pupation occurs in russet to reddish yellow puparia in the soil. The adult is about 5-mm long with a russet head, black thorax, and reddish yellow abdomen. Wings are marked with four black bands.

Larvae of the soybean nodule fly damage soybean nodules by boring a hole into the nodule and consuming the contents with high populations reducing N_2 fixation, plant growth, and seed yield (Newsom et al., 1978). Apparently, early planted soybean attracts and supports the largest populations of soybean nodule fly.

20–6.2 Bean Leaf Beetle

While adults of the bean leaf beetle (*C. trifurcata*) attack aboveground parts, larvae feed on roots and, more importantly, nodules of soybean (Isely, 1930; Kogan et al., 1980). There are three larval stages, and the older ones seem to disperse away from the tap root, probably in search of nodules, their prime source of nourishment (Levinson et al., 1979). Jackson (1967) reported that larval feeding on nodules in northern Missouri caused N deficiency.

Life and seasonal histories of bean leaf beetle were discussed previously.

20–6.3 Miscellaneous Underground Pests

Other below ground feeders are: (i) larvae of the seedcorn maggot [*Hylemya platura* (Meigen)] on planted seed in northern growing areas (Funderburk et al., 1984); (ii) white grubs, which are larvae of *Phyllophaga* spp., on roots in midwestern states; and (iii) larvae of cucumber beetles (*Diabrotica* spp.) on roots and nodules in most production areas. Occasional damage to seed, roots, or nodules may be caused by larvae of the grape colaspis (*C. brunnea*) and by several species of wireworms, flea beetles, and other soil-inhabiting insects (Turnipseed, 1973).

20–7 COMPONENTS OF INTEGRATED CONTROL

The central idea of integrated pest management (IPM) is to take maximum advantage of available control methods in a harmonious way. The IPM programs usually blend a combination of methods or tactics into regionally suitable control strategies. At present, soybean insect control programs rely primarily on the temporary suppression of insect outbreaks that approach or exceed the economic injury level. This is accomplished through the remedial application of selective insecticides. These programs require adequate measurements of pest population levels present in the field at various stages of the crop cycle and also assessments of the level of injury already effected by those pests. This information is obtained through scouting procedures. Second, on the basis of scouting results, crop managers use established thresholds and decision charts to determine the necessity of an insecticidal application. Insecticides must be applied at minimum effective rates and only when necessary to avoid economic loss. In this section we discuss scouting procedures, the use of forecasts, and plans to help in the decision-making process in soybean insect control. We then present a summary of the current status of methods used in managing soybean insect pests.

20–7.1 Scouting

Insect pests invade soybean fields at different stages of the crops' growth cycle. To optimize the effort invested in scouting, timing of the scouting program should be based on adequate knowledge of the phenology of major pests. For instance, if the main problem in a region is stink bugs there is no reason to start scouting at early stages of vegetative growth, because stink bugs will not invade fields until the beginning of pod development.

Scouting should provide information on: (i) the population level of target pests and status of their natural control agents (parasites, predators, and disease); and (ii) data on the status of the crop and current level of injury.

20–7.1.1 Assessment of Pest Populations

Various methods are used for sampling insect populations. No particular method is equally efficient to sample all insect species at all stages of plant growth and under every possible growing condition. There is vast literature on sampling methods for soybean insects (Kogan and Herzog, 1980) but for most practical purposes the following methods are usually adopted in IPM programs. Direct observations determine pest populations and levels of injury from seed emergence up to about V4-V5 growth stage. After that, either a sweep net or a ground cloth are recommended sampling procedures for most scouting needs. Efforts have been made to calibrate both sweep net and ground cloth to permit accurate population measurements. For example, Marston et al. (1979) have developed calibration ratios to convert counts of soybean caterpillars to absolute populations. They introduced known amounts of caterpillars on plants at various stages of growth and used the ground cloth method to determine recovery. Recovery was about 50 to 60% depending on the stage of soybean growth and development. Use of published ratios is dangerous, however, because efficiency of sampling varies considerably with growing conditions and individual differences among surveyors.

Quite often in the decision-making plan there is need to provide information on the presence of natural enemies and, in particular, incidence of diseases. Consequently, an effort should also be made upon scouting soybean fields to assess the presence of these control agents. This may be particularly difficult because of the need to identify some of the natural enemies either in the field or in the laboratory. A wealth of information on natural enemies of soybean pests including sampling procedures can be found in Pitre (1983).

20–7.1.2 Measurements Taken on the Crop

For decision plans on insect control, the following crop measurements are useful: (i) row width, (ii) plant density, (iii) plant height, (iv) phenological stage of growth (usually using Fehr and Caviness, 1977, V & R system), and (v) the current level of injury observed in the field. This latter type of measurement is one most likely to pose difficulties for scouts. It is particularly difficult to assess defoliation. The most practical method is the use of visual estimates. This method, however, varies enormously among individual scouts. Scouts must be thoroughly trained in the calibration of defoliation estimates. To do this, plants or individual leaflets with increasing levels of defoliation are carefully premeasured and used in visual comparisons with random field samples of leaves (Kogan and Kuhlman, 1982). This technique has been used in the establishment of a management program for soybean insects in Brazil (Kogan et al., 1977).

Pod injury is rated visually in the field. Seed damage is usually rated in the laboratory by establishing classes of injury or by measuring deviations from the normal mean weight of 100 seeds. Differential flotation has also been used to measure seed injury (Kogan and Turnipseed, 1980).

20–7.2 Forecasting

Scouting is essential in managing insects on soybean. However, final adoption of management programs by soybean farmers will depend on the ability of making scouting and other operations efficient and simple to accomplish. In vast production areas of the USA, insect pests are sporadic in time and space and in most years there is no need to apply insecticides. Improvement of a forecasting capability to provide growers, early in the season, with information on the probability of a pest outbreak would greatly enhance adoption of scouting procedures and insect management in general. Forecasts are based on predictive models, but the initialization of these models requires collection of data on regional insect movements and populations. At present, several predictive models have been developed for soybean insect pests, for example, the Mexican bean beetle (Waddill et al., 1976; Reichelderfer and Bender, 1979), the southern green stink bug (Marsolan and Rudd, 1976), *Heliothis* spp. (Stinner et al., 1974; Hartstack et al., 1976), the velvetbean caterpillar (*Anticarsia gemmatalis* Hübner) (Menke, 1973; Menke and Greene, 1976; University of Florida, *Consortium for Integrated Pest Management Modeling Component*) and the bean leaf beetle, *Cerotoma trifurcata* (Forster) (Zavaleta and Dixon, 1982). These models are at various stages of development and they are only as good as the biological and behavioral information available for their development. Currently, few practical programs exist that operate on a forecasting mode for soybean. In South Carolina, growers are advised weekly as to specific cultivars that will be at risk from *H. zea* damage through use of a pheromone and light trap monitoring system with information published in a newspaper through a terminal (J.W. Chapin, 1984, personal communication). The program emphasizes the relatively long response time between initiation of moth flights and economic loss and has proven useful in allocating scouting resources. Moffitt et al. (1986) have shown that adoption of an insect management program based on forecasting will require a better than 90% accuracy for widespread acceptance by farmers. Given the fact that all these models operate on assumptions of certain weather patterns it is difficult to achieve this level of accuracy. However, improvements are constantly being made in the models and it is expected that programs will increase the use of forecasting systems as subsidiaries in the decision-making phase.

20–7.3 Decision Making

With the available scouting information, it is then possible to establish criteria for determining the optimal strategy for control of a given pest situation. Most currently established insect management programs use economic injury levels and remedial applications of minimal rates of insecticides via the so-called *static decision models* usually consisting of charts easily accessed by users (e.g., Kogan and Kuhlman, 1982). Kogan and Helm (1984) present one such chart.

Modern agriculture, however, requires more precise definition of the parameters necessary for correct economic decisions. Such definition may become possible with the adoption of dynamic models in which not only the population dynamics of the pest is modeled with an acceptable degree of reliability, but also crop growth is modeled to respond to contemporary weather conditions which enter the model in real time. Preliminary models to describe crop phenology have been successfully deployed and used to help in irrigation decisions in soybean (Swaney et al., 1983). This model employs the crop simulation model SOYGROW (Wilkerson et al., 1983). Using the SOYGROW model, Wilkerson et al. have developed a decision-making plan for the control of soybean insect pests within a given season. These models consider the combined effect of insecticides, diseases, and natural enemies. As more information becomes available, models will become more realistic and will help considerably in decision-making plans. Zavaleta and Kogan (1984) have used the bean leaf beetle model to estimate the economic benefits of breeding host plant resistance in soybean. Such an approach helps define not only priorities in research invested to develop a given control procedure, but also the level of control that is necessary to maintain populations below the economic injury level.

Soybean insect control is primarily concerned with the temporary suppression of insect outbreaks that approach or exceed economic thresholds. To be effective, this type of control requires integration of various components that deal with insect populations. Several of the more important of these components are discussed below.

20–7.4 Chemical Insecticides

The use of conventional chemical insecticides constitutes the only presently available tool that affords consistent, economical, and effective suppression of insect outbreaks on soybean. Such insecticides, when used safely in a sensible manner in accordance with extension service recommendations and label restrictions, do not pose threats to users or to the environment. Chemical insecticides form the backbone of integrated control systems for soybean insect pests. This is the situation now, it has been in the past, and will continue to be for years to come as other methods of control are developed.

Even though the overwhelming majority of chemical insecticides are used properly and safely, it is their misuse that causes problems and creates adverse publicity. The public's attention is called to the infrequent poisoning of a person or animals, but problems that result in ineffective control are still too common in many farming situations. Broad spectrum chemicals are sometimes applied when unnecessary and at unnecessarily high rates. This causes destruction of natural enemies such as predators, parasites, and insect diseases and often leads to later resurgences of pests in greater numbers than existed when insecticides were initially applied (Bartlett, 1964; Newsom, 1967; Shepard et al., 1977; Morrison et al., 1979; Van Duyn and McLeod, 1983; Turnipseed, 1984). Thus, as a consequence of initial misuse, a producer is forced to make a second application a few weeks later. Research has shown that low rates of certain insecticides afford adequate pest control (Turnipseed et al., 1974) and allow survival of beneficial species (Turnipseed et al., 1975).

Since chemical insecticides are so important in integrated control of soybean insects, scientists in private companies and the public sector must increase efforts to develop new and effective materials. Newsom (1970) cautioned that the current furor over pesticides may discourage vitally needed research on insecticides. Insecticidal research on soybean must be encouraged and should include studies on new types and methods of use of chemicals in integrated programs designed to conserve natural enemies of pest species and protect the integrity of the environment.

20–7.5 Insect Diseases and Microbial Insecticides

Insect diseases constitute a key component of the natural enemy complex which also includes predators and parasites. Together these endemic natural enemies are so important that they keep most pests at subeconomic levels most of the time (Pitre, 1983; Marston et al., 1984). Insect diseases are particularly effective against major lepidopterous pests (Carner, 1980).

Fungi are the most obvious and widespread insect pathogens found in soybean fields, and *Nomuraea rileyi* (Farlow) Samson is the predominant species of this group (Plate 20–3d). It occurs in all the major soybean-producing areas of the USA (Carner, 1980) and in South America, Australia, and other parts of the world. This organism infects and causes disease epizootics in larvae of such pests as green cloverworm, soybean looper, corn earworm, and velvetbean caterpillar. Cadavers which are first white and then turn greenish with sporulation, are often found on plants or soil under conditions of adequate humidity. Another important fungus is *Entomophthora gammae* Weiser infecting larvae of the soybean looper (Plate 20–3.2 and 3.3). There are two forms of infection by *E. gammae*, one being the conidial spore form in which cacavers are discolored and shriveled and the other the resting spore form that results in black and turgid cadavers.

Although there are no reports of naturally occurring bacterial pathogens in soybean insect pests, *Bacillus thuringiensis* Berliner is effective as a microbial insecticide against most lepidopterous pests and is recommended for control of green cloverworms, soybean loopers, and velvetbean caterpillars. Microbial insecticides are applied to control pest outbreaks just as chemical ones are used. They generally, however, require 2 or more days for mortality to occur and for this reason soybean growers seldom use them.

Several nuclear polyhedrosis viruses (NPVs) are effective against different lepidopterous species. These include: (1) soybean looper-NPV, which, imported from Central America, is effective in the field as a microbial insecticide, and has become established as a permanent natural control agent (Livingston et al., 1980; Carner, 1980); (2) velvetbean caterpillar NPV, imported from South America, is very effective as a microbial insecticide (Moscardi and Corso, 1982), but it has not been established in the USA as a permanent natural control agent (Carner and Turnipseed, 1977; Allen and Knell, 1977); and (3) *Heliothis* NPV (Plate 20–3a) which has been marketed commercially as a microbial insecticide and also recurs annually in some production regions as a natural mortality factor (Carner, 1980).

Most insect pathogens are effective as endemic natural control agents and some as microbial insecticides. There has been some success in the importation of pathogens for establishment as permanent natural control agents and as microbial insecticides. Insect pathogens are safe to use and they do not usually adversely affect other natural enemies such as predators and parasites. Strong research efforts are needed to fully exploit the potential of insect pathogens in integrated control programs for soybean insect pests.

20–7.6 Predators and Parasites

The importance of the native natural enemy complex, which includes predators and parasites as well as diseases (see section 20–7.5), cannot be overemphasized in the integrated control of soybean insect pests. Without these natural enemies, multiple applications of insecticides would be required during the growing season to keep pest species below economic injury levels. Currently, insecticidal applications are infrequent because the importance of natural enemies was carefully considered during the development of recommendations for pest species. In evaluation programs for insecticides in South Carolina, a short-lived broad spectrum insecticide is applied to soybean in early July to disrupt natural enemies. This practice *insures* that high populations of such pests as corn earworm and soybean looper will be available to work with in late July and early August (S.G. Turnipseed, personal observations).

A complex of polyphagous predators in soybean is quite adaptable to various pest insects (Newsom et al., 1980) and this complex is particularly important in preying upon eggs and small larvae before damage occurs. Although the predator complex (Plate 20–4) varies somewhat in different production regions, its major components are reported to consist of nabids, geocorids, and spiders (Turnipseed, 1973; McCarty et al., 1980; Pitre, 1983). Reed et al. (1984) found that 25% of corn earworm eggs were consumed by predators after placement on foliage in a soybean field for 24 h during August. Major components of the predator complex during this study were adult and immature geocorids and nabids, adults of the beetles *Notoxus monodon* (F.) and *Lebia analis* Dejean, and immature lady beetles. An average of 10.5 total predators m^{-1} was present. With similar mortality during the entire egg stage (ca. 3 days) and additional predation on small larvae, it is simple to discern the importance of this group of natural enemies.

In South Carolina up to 20% of the lepidopterous complex in soybean is parasitized (Plate 20–5), depending upon the season, with the major parasites being: *Apanteles marginiventris* (Cresson) on green cloverworm, corn earworm, soybean looper, and velvetbean caterpillar; and *Meteorus autographae* (Muesebeck) on soybean looper, corn earworm, and velvetbean caterpillar (McCutcheon and Turnipseed, 1981). These parasites apparently do not kill as many lepidopterous pests as do predators and insect pathogens, but they do so when larvae are small and before a significant amount of feeding is done. Native parasites of other pests such as stink bugs may be more important than parasites of Lepidoptera. *Telenomus basalis* (Wollaston) is quite effective in destroying stink bug egg masses, and *Trichopoda pennipes* (F.) has killed 60 to 80% of southern green stink bug adults (Harper et al., 1983).

Two concerted efforts to import and establish exotic parasites from abroad against soybean insect pests have been successful (Jones et al., 1983). *Pediobius foveolatus* (Crawford) was imported from India and annual releases provide some control of Mexican bean beetles in the Atlantic

Coastal Plain. Also *Euplectrus puttleri* Gordh, imported from South America, has become established in southern Florida where it kills substantial numbers of the velvetbean caterpillar each year.

Thus, both native and imported predators and parasites, when considered in conjunction with insect diseases, are extremely important components in the integrated control of soybean insect pests. Attempts to control soybean insect pests that do not consider the role of natural enemies are often doomed to failure.

20–7.7 Cultural Practices

Various cultural practices such as planting date and maturity, row spacing and overall crop mix with soybean can be important in managing pest species. There has been growing interest in the effect of reduced tillage on arthropods (Sloderbeck and Edwards, 1979; House and Stinner, 1983; Sloderbeck and Yeargan, 1983). The early planting of early maturity (Maturity Group V) soybean was reported to avoid damage from corn earworm in North Carolina (Stinner et al., 1980). In South Carolina, however, exactly the opposite situation occurs; early planted Maturity Group V or VI soybean is often heavily infested with corn earworm, whereas cultivars that bloom later seldom require treatment (see section 20–4). Row spacing is also important in managing corn earworm because adults preferentially oviposit in soybean fields with open canopies (Johnson et al., 1975). Thus, soybean planted in narrow rows usually has fewer problems with corn earworm because of early canopy closure. Yet another cultural factor is important in the development of economic populations of corn earworm in soybean: the presence of corn in the agroecosystems involved. Earlier generations build up on corn before soybean is susceptible to attack by corn earworm. In the North, however, sweet corn is readily infested by corn earworms, but soybean growing nearby has never been observed with corn earworm injury. The soybean looper usually is not a problem in areas where cotton is absent but can be devastating where the crop is grown because moths use cotton nectaries for maximum oviposition in nearby soybean fields (see section 20–3.2). Reduced tillage and double cropping can impact on the population dynamics of Mexican bean beetle, redlegged grasshopper (Sloderbeck and Edwards, 1979), and of other pest species (Turnipseed, 1973).

Small acreage trap crops of early maturing cultivars effectively attract and hold populations of the southern green stink bug where they can be controlled with an insecticide before they move to the main part of the soybean crop (Newsom et al., 1980). Such trap crops can also be used for control of the bean leaf beetle and an important disease of soybean, bean pod mottle virus, for which it is the only important vector.

In addition to pest species, cultural practices can also be important in the development of natural enemies. An excellent review is available on this subject (Mayse et al., 1983).

20–7.8 Resistant Cultivars

The greatest potential tool for effective management of soybean insect pests is available through development of resistant cultivars. Examples and mechanisms of insect resistance in plants were recently reviewed by Maxwell and Jennings (1980).

The role of soybean pubescence in resistance to the potato leafhopper, *Empoasca fabae* (Harris), has been one of the first types of resistance identified in soybean (Hollowell and Johnson, 1934). Pubescence seems to interfere with oviposition and with both feeding and attachment mechanisms of leafhopper nymphs, thus reducing survival on the pubescent types (Lee, 1983). Pubescence interferes also with normal development and behavior of other small arthropods on soybean (Turnipseed, 1977).

Interest in insect resistance in soybean has increased enormously during the late 1960s and early 1970s. Active breeding programs to introduce insect resistance into commercial cultivars exist in at least 10 states through either university or USDA programs or a combination of both (North Carolina, South Carolina, Mississippi, Georgia, Florida, Louisiana, Illinois, Indiana, Ohio, and Maryland) and several private seed companies are now engaged in breeding for insect resistance in soybean. Much of this emphasis was generated by the identification of three plant introductions highly resistant to the Mexican bean beetle by Van Duyn et al., (1971). The three plant introductions were PI 229358, PI 227687, and PI 171451. Elden et al. (1974) screened Maturity Groups III through V for resistance to the same insect and reported PI 90481, PI 96089, and PI 157413 as the most resistant. Other researchers have determined that the plant introductions reported by Van Duyn et al. were also resistant to other insect species: corn earworm (Clark et al., 1972; Beland and Hatchett, 1976; Hatchett et al., 1976; Smith and Brim, 1979); the bean leaf beetle and striped blister beetle (Clark et al., 1972); the soybean looper (Kilen et al., 1977); cabbage looper [*Trichoplusia ni* (Hubner)] (Luedders and Dickerson, 1977); stink bugs (Jones and Sullivan, 1979; Miranda et al., 1979; Panizzi et al., 1981); and the sweet potato whitefly [*Bemisia tabaci* (Gen.)] (Rossetto et al., 1977).

Interest in the identification of new sources of resistance particularly for stink bugs, has led to extensive screening of the germplasm. Results are, however, rather ambiguous because of the strong influence of crop phenology in the expression of resistance to stink bugs. Thus, direct comparison of results obtained by various researchers may not be possible because of the difficulties in clearly stating the stage of pod and seed development at which evaluations were made.

20–8 INTERACTIVE IMPACT OF INSECTS, WEEDS, NEMATODES, AND DISEASES

Entomologists are often confronted with an insect species feeding on foliage, another on pods, and still another on roots and nodules, perhaps *all at the same time.* This is frustrating because adequate answers cannot be given concerning their total impact on soybean yield and quality. Research programs need to be developed to provide these answers.

But consider the plight of soybean producers in managing pests on their crop. They are confronted with these insects *and simultaneously* with weeds, nematodes, and plant diseases, any one or all of which may cause economic loss. Scientists in the separate pest disciplines have developed economic thresholds for single pests or, in a few instances, for several similar pests that act in the same manner. These thresholds are certainly better than none at all and they are probably reasonably accurate in certain situations. Unfortunately few, if any, pest species or classes impinge on production of the crop in the absence of other species, classes, or environmental factors that concurrently stress the plant. The first efforts to analyze the combined effects of insects and weeds are just beginning to reveal the complexities of the problem (Higgins et al., 1983). Pedigo et al. (1981) made an effort to compile information on all major soybean pests, but the impact of weeds, diseases, nematodes, and insects is treated separately as research to support real integration is still lacking.

Newsom (1982), addressing primarily pest-induced stress stated, "It does not appear likely that development of more efficient plants will revolutionize yields of soybean until the barriers to increase yields of presently available cultivars are better understood. Undoubtedly, plant breeders and genetic engineers can develop improved potentials for yields but these cannot be fully realized without protection from pests. Adequate protection from pests cannot be provided until the interactions of pest complexes are better understood. Research toward establishing a better understanding and effective management of pest interactions, as well as others that occur among inputs to production systems, will probably contribute more to significant increases in soybean yields in the near future than any other research effort."

The problem of adequately addressing stress from pest complexes may appear overwhelming, but not necessarily. Soybean is currently produced successfully by farmers in many areas of the world. These farmers address multiple stress factors to the best of their ability. As responsible scientists we can and must successfully address the problem of stress from pest complexes. How can we accomplish this? The answer is obviously through strong interdisciplinary efforts in research and extension. We must develop an understanding of the physiological processes involved in determining crop yield and quality. Then we must understand how pest complexes interfere with these processes and, thus, adversely affect yield and quality. Such efforts should be conducted within the same framework in which the farmers operate, in individual soybean fields and production systems that they use.

ACKNOWLEDGMENT

We are grateful for the assistance provided by the following persons in the preparation of this chapter: Merle Shepard and G. R. Carner,

Clemson University, for color slides used in Plates 20–1 to 5; C. R. Edwards, Purdue University, for reviewing the manuscript; C. G. Helm, Illinois Natural History Survey, for editorial assistance; Jenny Kogan, Soybean Insect Research Information Center, Univ. of Illinois, for assisting in information retrieval; and Sandra McGary, Illinois Natural History Survey, for word processing the manuscript.

This chapter is a contribution of the South Carolina Agric. Exp. Stn., Clemson University, the Illinois Natural History Survey, and the Illinois Agric. Exp. Stn., Univ. of Illinois at Urbana-Champaign. It was supported in part by Experiment Stations Project S-157, and the USDA through the *Consortium for Integrated Pest Management*, grant no. 82-CRSR-2-1000, to Texas A&M University. The opinions expressed herein are those of the authors and not of the USDA or the funding and supporting institutions.

REFERENCES

Allen, G.E., and J.D. Knell. 1977. A nuclear polyhedrosis virus of *Anticarsia gemmatalis*. I. Ultrastructure, replication, and pathogenicity. Fla. Entomol. 60:233–240.

Balduf, W.V. 1923. The insects of the soybean in Ohio. Ohio Agric. Exp. Stn. Res. Bull. 366:144–181.

Bartlett, B.R. 1964. Integration of chemical and biological control. P. 489–511. *In* P. Debach (ed.) Biological control of pests and weeds. Reinhold Publishing Co., New York.

Beland, G.L., and J.H. Hatchett. 1976. Expression of antibiosis to the bollworm in two soybean genotypes. J. Econ. Entomol. 69:557–560.

Bergeson, G.B., K.L. Athow, F.A. Laviolette, and Sister Mary Thomasine. 1964. Transmission, movement, and vector relationships of tobacco ringspot virus in soybeans. Phytopathology 54:723–728.

Biever, K.D., G.D. Thomas, P.E. Boldt, and C.M. Ignoffo. 1983. Effects of *Heliothis zea* (Lepidoptera: Noctuidae) on soybean yield and quality. J. Econ. Entomol. 76:762–765.

Blickenstaff, C.C., and J.L. Huggans. 1962. Soybean insects and related arthropods in Missouri. Mo. Agric. Exp. Stn. Res. Bull. 803.

Boyer, W.P. 1967. Survey methods for threecornered alfalfa hopper (*Spissistilus festinus*) in soybeans in Arkansas. USDA Coop. Econ. Insect Rep. 17:324–325.

Broersma, D.B., R.L. Bernard, and W.H. Luckmann. 1972. Some effects of soybean pubescence on populations of the potato leafhopper. J. Econ. Entomol. 65:78–82.

Campbell, W.V. 1980. Sampling coleopterous stem borers in soybean. P. 357–373. *In* M. Kogan and D.C. Herzog (ed.) Sampling methods in soybean entomology. Springer-Verlag, New York, New York.

Canerday, T.D., and F.S. Arant. 1966. The looper complex in Alabama (Lepidoptera, Plusiinae). J. Econ. Entomol. 59:742–743.

Carlson, J.B. 1973. Morphology. *In* B.E. Caldwell (ed.) Soybeans: improvement, production, and uses. Agronomy 16:17–95.

Carner, G.R. 1980. Sampling pathogens of soybean insect pests. P. 559–574. *In* M. Kogan and D.C. Herzog (ed.) Sampling methods in soybean entomology. Springer-Verlag, New York, New York.

----, and S.G. Turnipseed. 1977. Potential of a nuclear polyhedrosis virus for control of the velvetbean caterpillar in soybean. J. Econ. Entomol. 70:608–610.

Caviness, C.E., and F.D. Miner. 1962. Effects of stand reduction in soybeans simulating threecornered alfalfa hopper injury. Agron. J. 54:300–302.

----, and J.D. Thomas. 1980. Yield reduction from defoliation of irrigated and non-irrigated soybeans. Agron. J. 72:977–980.

Clark, W.J., F.A. Harris, F.G. Maxwell, and E.E. Hartwig. 1972. Resistance of certain soybean cultivars to bean leaf beetle, striped blister beetle and bollworm. J. Econ. Entomol. 65:1669–1672.

Clarke, R.G., and G.E. Wilde. 1970. Association of the green stink bug and the yeast spot disease organism of soybeans. 1. Length of retention, effect of molting, isolation from feces and saliva. J. Econ. Entomol. 63:200–204.

Colwell, C.E. 1984. Insect situation and insecticide use in 1983. p. 110–125. *In* Thirty-sixth Illinois custom spray operators training school manual. Illinois Cooperative Extension Service, University of Illinois, Urbana.

Daugherty, D.M. 1967. Pentatomidae as vectors of yeast-spot disease of soybeans. J. Econ. Entomol. 60:147–152.

----, and R.D. Jackson. 1967. Damage to soybeans by the broadheaded bug, *Alydus pilosulus.* Entomol. Soc. Am. N. Central Stn. Branch Proc. 22:14–15.

----, and ----. 1969. Economic damage to soybeans caused by a Cerambycid beetle. Entomol. Soc. Am. N. Cent. Stn. Branch Proc. 24:36.

----, M.H. Neustadt, C.W. Gehrike, L.E. Cavanah, L.F. Williams, and D.E. Green. 1964. An evaluation of damage to soybeans by brown and green stink bugs. J. Econ. Entomol. 57:719–722.

Deitz, L.L., J. Van Duyn, J.R. Bradley, R.L. Rabb, W.M. Brooks, and R.E. Stinner. 1976. A guide to the identification and biology of soybean arthropods in North Carolina. N. C. Agric. Exp. Stn. Tech. Bull. 238.

Eastman, C.E. 1980. Sampling phytophagous underground soybean arthropods. p. 327–354. *In* M. Kogan and D.C. Herzog (ed.) Sampling methods in soybean entomology. Springer-Verlag New York, New York.

----, and A.L. Wuensche. 1977. A new insect damaging nodules of soybean: *Rivellia quadrifasciata* (Macquart). J. Ga. Entomol. Soc. 12:190–199.

Elden, T.C., J.A. Schillinger, and A.L. Steinhauer. 1974. Field and laboratory selection for resistance in soybeans to the Mexican bean beetle. Environ. Entomol. 3:785–788.

Fehr, W.R., and C.E. Caviness. 1977. Stages of soybean development. Iowa Coop. Ext. Serv. Spec. Rep. 80.

----, ----, and J.J. Vorst. 1977. Response of indeterminate and determinate soybean cultivars to defoliation and half-plant cut-off. Crop Sci. 17:913–917.

Foster, J.E., and D.M. Daugherty. 1969. Isolation of the organism causing yeast-spot disease from the salivary system of the green stink bug. J. Econ. Entomol. 62:424–427.

Funderburk, J.E., L.G. Higley, and L.P. Pedigo. 1984. Seedcorn maggot (Diptera: Anthomyiidae) phenology in central Iowa and examination of a thermal-unit system to predict development under field conditions. Environ. Entomol. 13:105–109.

Gazzoni, D.L., and H.C. Minor. 1979. Efeito do desfolhamento artificial em soja sobre o rendimento e seus componentes [Effect of artificial defoliation of soybean on yield and yield components]. Anais do I Seminario Nacional de Pesquisa de Soja. (Londrina, PR 24 a 30 de Setembro de 1978). 2:47–57 EMBRAPA, Centro Nacional de Pesquisa de Soja, Londrina, Brazil.

Gould, G.E. 1963. Japanese beetle damage to soybeans and corn. J. Econ. Entomol. 56:776–781.

Hammond, R.B., and L.P. Pedigo. 1982. Determination of yield-loss relationships for two soybean defoliators by using simulated insect-defoliation techniques. J. Econ. Entomol. 75:102–107.

Hanthorn, M., C. Osteen, R. McDowell, and L. Roberson. 1982. 1980 pesticide use on soybeans in the major producing states. USDA-ERS, Staff Rep. AGES 820106, Washington, DC.

Harper, J.D., R.M. McPherson, and M. Shepard. 1983. Geographical and seasonal occurrence of parasites, predators and entomopathogens. p. 7–19. *In* H.N. Pitre (ed.) Natural enemies of arthropod pests in soybean. Southern Coop. Ser. Bull. 285.

Hartstack, A.W., J.A. Witz, J.P. Hollingsworth, R.L. Ridgway, and J.D. Lopez. 1976. MOTHZV-2: A computer simulation of *Heliothis zea* and *Heliothis virescens* population dynamics, users manual. USDA-ARS S-127. U.S. Government Printing Office, Washington, DC.

Hatchett, J.H., G.L. Beland, and E.E. Hartwig. 1976. Leaf-feeding resistance to bollworm and tobacco budworm in three soybean plant introductions. Crop Sci. 16:277–280.

Helm, C.G., M. Kogan, and B.G. Hill. 1980. Sampling leafhoppers on soybean. p. 260–282. *In* M. Kogan and D.C. Herzog (ed.) Sampling methods in soybean entomology. Springer-Verlag New York, New York.

Hensley, S.D., L.D. Newsom, and J. Chapin. 1964. Observations on the looper complex of the noctuid subfamily Plusiinae. J. Econ. Entomol. 57:1006–1007.

Herzog, D.C. 1980. Sampling soybean looper on soybean. p. 141–168. *In* M. Kogan and D.C. Herzog (ed.) Sampling methods in soybean entomology. Springer-Verlag New York, New York.

----, and J.W. Todd. 1980. Sampling velvetbean caterpillar on soybean. p. 107–140. *In* M. Kogan and D.C. Herzog (ed.) Sampling methods in soybean entomology. Springer-Verlag New York, New York.

Higgins, R.A., L.P. Pedigo, and D.W. Staniforth. 1983. Selected preharvest morphological characteristics of soybeans stressed by simulated green cloverworm (Lepidoptera: Noctuidae) defoliation and velvetleaf competition. J. Econ. Entomol. 76:484–491.

Hinds, W.E., and B.A. Osterberger. 1931. The soybean caterpillar in Louisiana. J. Econ. Entomol. 24:1168–1173.

Hollowell, E.A., and H.W. Johnson. 1934. Correlation between rough-hairy pubescence in soybean and freedom from injury by *Empoasca fabae*. Phytopathology 24:12.

House, G., and J. Stinner. 1983. Arthropods in no-tillage soybean agroecosystems; Community composition and ecosystem interactions. Environ. Manage. 7:23–28.

Irwin, M.E., and D.E. Kuhlman. 1979. Relationships among *Sericothrips variabilis* (Beach), systemic insecticides and soybean yield. J. Ga. Entomol. Soc. 14:148–154.

----, and K.V. Yeargan. 1980. Sampling phytophagous thrips on soybean. p. 283–304. *In* M. Kogan and D.C. Herzog (ed.) Sampling methods in soybean entomology. Springer-Verlag New York, New York.

Isely, D. 1930. The biology of the bean leaf beetle. Ark. Agric. Exp. Stn. Bull. 249.

Jackson, R.D. 1967. Soybean insect problems. Soybean Dig. 27(11):16–18.

Jensen, R.L., and L.D. Newsom. 1972. Effect of stink bug damaged soybean seeds on germination, emergence, and yield. J. Econ. Entomol. 65:261–264.

----, ----, and J. Gibbens. 1974. The soybean looper: Effects of adult nutrition on oviposition, mating frequency, and longevity. J. Econ. Entomol. 67:467–470.

Johnson, M.W., R.E. Stinner, and R.L. Rabb. 1975. Ovipositional response of *Heliothis zea* (Boddie) to its major hosts in North Carolina. Environ. Entomol. 4:291–297.

Jones, W.A., Jr., and M.J. Sullivan. 1979. Soybean resistance to the southern green stink bug *Nezara viridula*. J. Econ. Entomol. 72:628–632.

----, S.Y. Young, M. Shepard, and W.H. Whitcomb. 1983. Use of imported natural enemies against insect pests of soybean. p. 63–77. *In* H.N. Pitre (ed.) Natural enemies of arthropod pests in soybean. Southern Coop. Ser. Bull. 285.

Kalton, R.C., C.R. Weber, and J.C. Eldredge. 1949. The effect of injury simulating hail damage to soybeans. Iowa Agric. Home Econ. Exp. Stn. Res. Bull. 359:736–796.

Kilen, T.C., J.H. Hatchett, and E.E. Hartwig. 1977. Evaluation of early generation soybeans for resistance to soybean looper. Crop Sci. 17:397–398.

Kilpatrick, R.A., and E.E. Hartwig. 1955. Fungus infection of soybean seed as influenced by stink bug injury. Plant Dis. Rep. 39:177–180.

Kobayashi, T., and T. Oku. 1980. Predator-prey interactions in soybean communities: Implications of pesticide perturbations. Abstr. Int. Congr. Entomol. 16:256.

Kogan, M. 1976. Evaluation of economic injury levels for soybean insect pests. p. 515–533. *In* L.D. Hill (ed.) World soybean research conference proceedings. The Interstate Printers and Publishers, Danville, IL.

----. 1980. Insect problems of soybeans in the United States. p. 303–325. *In* F.T. Corbin (ed.) World Soybean research conference II: Proceedings. Westview Press, Boulder, CO.

----. 1981. Dynamics of insect adaptations to soybean: Impact of integrated pest management. Environ. Entomol. 10:363–371.

----, and C.G. Helm. 1984. Soybean insects in Illinois: Lessons of the 1983 season. p. 110–120. *In* Tenth Annual Illinois Crop Protection Workshop. Champaign, IL. 6–8 March. Illinois Cooperative Extension Service, University of Illinois, Urbana.

----, and D.C. Herzog (ed.) 1980. Sampling methods in soybean entomology. Springer-Verlag, New York, New York.

----, and D.E. Kuhlman. 1982. Soybean insects: Identification and management in Illinois. Ill. Agric. Exp. Stn. Bull. 773.

----, and J.G. Rodriguez. 1977. Outbreaks of spider mites on soybean: Possible result of plant nutritional quality. Proc. N. Central Br. Entomol. Soc. Am. 32:22–23.

----, and S.G. Turnipseed. 1980. Soybean growth and assessment of damage by arthropods. p. 3–29. *In* M. Kogan and D.C. Herzog (ed.) Sampling methods in soybean entomology. Springer-Verlag New York, New York.

----, ----, M. Shepard, E.B. de Oliveira, and A. Borgo. 1977. Pilot insect pest management program for soybean in southern Brazil. J. Econ. Entomol. 70:659–663.

----, G.P. Waldbauer, G. Boiteau, and C.E. Eastman. 1980. Sampling bean leaf beetle on soybean. p. 201–236. *In* M. Kogan and D.C. Herzog (ed.) Sampling methods in soybean entomology. Springer-Verlag New York, New York.

Kretzschmar, G.P. 1948. Soybean insects in Minnesota with special reference to sampling techniques. J. Econ. Entomol. 41:586–591.

Lee, Y.I. 1983. The potato leafhopper, *Empoasca fabae*, soybean pubescence and hopper-burn resistance. Ph.D. diss. Univ. of Illinois, Urbana. (Diss. Abstr. Int. 36B:2057).

Levinson, G.A., G.P. Waldbauer, and M. Kogan. 1979. Distribution of bean leaf beetle eggs, larvae and pupae in relation to soybean plants: Determination by emergence cages and soil sampling techniques. Environ. Entomol. 8:1055–1058.

Livingston, J.M., P.J. McLeod, W.C. Yearian, and S.Y. Young III. 1980. Laboratory and field evaluation of a nuclear polyhedrosis virus of the soybean looper, *Pseudoplusia includens*. J. Ga. Entomol. Soc. 15:194–199.

Luedders, V.D., and W.A. Dickerson. 1977. Resistance of selected soybean genotypes and segregating populations to cabbage looper feeding. Crop Sci. 17:395–397.

Marsolan, N.F., and W.G. Rudd. 1976. Modeling and optimal control of insect pest populations. Math. Biosci. 30:231–244.

----, W.A. Dickerson, W.W. Ponder, and G.D. Booth. 1979. Calibration ratios for sampling soybean lepidoptera: Effect of larval species, larval size, plant growth stage and individual sampler. J. Econ. Entomol. 72:110–114.

----, D.L. Hostetter, R.E. Pinnell, W.A. Dickerson, and D.B. Smith. 1984. Natural mortality of lepidopteran eggs and larvae in Missouri soybeans. Ann. Entomol. Soc. Am. 77:21–28.

Maxwell, F.G., and P.R. Jennings (ed.) 1980. Breeding plants resistant to insects. John Wiley and Sons, New York.

Mayse, M., H.N. Pitre, and W.H. Whitcomb. 1983. Effects of cultural practices on natural enemies. p. 49–55. *In* H.N. Pitre (ed.) Natural enemies of arthropod pests in soybean. Southern Coop. Ser. Bull. 285.

McCarty, M.T., M. Shepard, and S.G. Turnipseed. 1980. Identification of predaceous arthropods in soybeans by using autoradiography. Environ. Entomol. 9:199–203.

McCutcheon, G.S., and S.G. Turnipseed. 1981. Parasites of lepidopterous larvae in insect resistant and susceptible soybeans in South Carolina. Environ. Entomol. 10:69–74.

McWilliams, J.M. 1983. Relationship of soybean pod development to bollworm and tobacco budworm damage. J. Econ. Entomol. 76:502–506.

Menke, W.W. 1973. A computer simulation model: The velvetbean caterpillar in the soybean agroecosystem. Fla. Entomol. 56:99–102.

----, and G.L. Greene. 1976. Experimental validation of a pest management model. Fla. Entomol. 59:135–142.

Miner, F.D. 1961. Stink bug damage to soybeans. Ark. Farm Res. 10:12.

----. 1966. Biology and control of stink bugs on soybeans. Ark. Exp. Stn. Bull. 708.

----, and T.H. Wilson. 1966. Quality of stored soybeans as affected by stink bug damage. Ark. Farm Res. 15:2.

Miranda, M.A.C., de, C.J. Rossetto, D. Rossetto, N.R. Braga, H.A.A. Mascarenhas, J.P.F. Teixeira, and A. Massariol. 1979. Resistencia de soja a *Nezara viridula* e *Piezodorus guildinii* em condicoes de campo [Soybean resistance to *Nezara viridula* and *Piezodorus guildinii* under field conditions]. Bragantia 38:181–188.

Moffitt, L.J., R.L. Farnsworth, L. Zavaleta, and M. Kogan. 1986. Economic impact of public pest information: Soybean insect forecasts in Illinois. Am. J. Agric. Econ. 68: 274–279.

Morrison, D.E., J.R. Bradley, Jr., and J.W. Van Duyn. 1979. Populations of corn earworm and associated predators after applications of certain soil-applied pesticides to soybeans. J. Econ. Entomol. 72:97–100.

Moscardi, F., and I.C. Corso. 1982. "Projeto piloto" para utilizacao do *Baculovirus anticarsia*, a nivel de agricultor, no controle de *Anticarsia gemmatalis* en soja. [Pilot project for use of *Baculovirus anticarsia*, at the farm level, for the control of *Anticarsia gemmatalis* in soybean]. p. 266–270. *In* EMBRAPA-Centro Nacional de Pesquisa de Soja, Resultados de Pesquisa de Soja. Londrina, Parana, Brazil.

Mueller, A.J. 1980. Sampling threecornered alfalfa hopper on soybean. p. 382–393. *In* M. Kogan and D.C. Herzog (ed.) Sampling methods in soybean entomology. Springer-Verlag New York, New York.

----, and B.W. Engroff. 1980. Effects of infestation levels of *Heliothis zea* on soybean. J. Econ. Entomol. 73:271–275.

----, and J.W. Jones. 1983. Effects of main-stem girdling of early vegetative stages of soybean plants by threecornered alfalfa hoppers (Homoptera: Membracidae). J. Econ. Entomol. 76:920–922.

----, and R.G. Luttrell. 1977. Thrips on soybeans. Ark. Farm. Res. 1977. (July–August):7.

Nettles, W.C., F.H. Smith, and C.A. Thomas. 1970. Soybean insects and diseases. S.C. (Clemson Univ.) Ext. Circ. 504.

Newsom, L.D. 1967. Consequences of insecticide use on nontarget organisms. Annu. Rev. Entomol. 12:157–186.

----. 1970. The end of an era and future prospects for insect control. Tall Timbers Conf. Ecol. Anim. Control. Habitat Manage. Proc. 2:117–136.

----. 1982. Current status of plant protection for soybean in the United States. *In* Soybean Research in China and the United States. Proc. of the 1st China/USA Soybean Symp., Urbana, IL. 25–27 July. INTSOY Ser. 25. Urbana, IL.

----, E.P. Dunigan, C.E. Eastman, R.L. Hutchinson, and R.M. McPherson. 1978. Insect injury reduces nitrogen fixation in soybeans. La. Agric. 21:15–16.

----, M. Kogan, F.D. Miner, R.L. Rabb, S.G. Turnipseed, and W.H. Whitcomb. 1980. General accomplishments toward better pest control in soybean. p. 51–98. *In* C.B. Huffaker (ed.) New technology of pest control. John Wiley and Sons, New York.

----, P.L. Mitchell, and N.N. Troxclair, Jr. 1984. Overwintering of threecornered alfalfa hopper in Louisiana. J. Econ. Entomol. 76:1298–1302.

Nickels, C.B. 1926. An important outbreak of insects infesting soybeans in lower South Carolina. J. Econ. Entomol. 19:614–618.

Panizzi, M.C., I.A. Bays, R.A.S. Kiihl, and M.P. Porto. 1981. Identificação de genotipos fontes de resistencia a percevejos-pragas da soja [Identification of genotypes as resistance sources to stink bugs on soybeans]. Pesqui. Agropecu. Bras. 16:33–37.

Pedigo, L.P. 1980. Sampling green cloverworm on soybean. p. 169–186. *In* M. Kogan and D.C. Herzog (ed.) Sampling methods in soybean entomology. Springer-Verlag New York, New York.

----, R.A. Higgins, R.B. Hammond, and E.J. Bechinski. 1981. Soybean pest management. p. 417–537. *In* D. Pimentel (ed.) CRC Handbook of pest management in agriculture, Vol. 3. CRC Press, Boca Raton, FL.

Pendleton, J.W., and E.E. Hartwig. 1973. Management. *In* B.E. Caldwell (ed.) Soybeans: Improvement, production, and uses. Agronomy 16:211–237.

Petty, H.B. 1967. How to control soybean insects. Farm Technol. 23:42–47.

Pitre, H.M. (ed.) 1983. Natural enemies of arthropod pests in soybean. Southern Coop. Ser. Bull. 285.

Poe, S.L. 1980. Sampling mites on soybean. p. 312–324. *In* M. Kogan and D.C. Herzog (ed.) Sampling methods in soybean entomology. Springer-Verlag New York, New York.

Poston, F.L., L.P. Pedigo, R.B. Pearce, and R.B. Hammond. 1976. Effects of artificial and insect defoliation on soybean net photosynthesis. J. Econ. Entomol. 69:109–112.

----, ----, and S.M. Welch. 1983. Economic injury levels: Reality and practicality. Bull. Entomol. Soc. Am. 29:49–53.

Ratcliffe, R.H., T.L. Bissel, and W.E. Bickley. 1960. Observations on soybean insects in Maryland. J. Econ. Entomol. 53:131–133.

Reed, T.D., M. Shepard, and S.G. Turnipseed. 1984. Assessment of the impact of arthropod predators on noctuid larvae in cages in soybean fields. Environ. Entomol. 13:954–961.

Reichelderfer, K.H., and F.E. Bender. 1979. Application of a simulative approach to evaluating alternative methods for the control of agricultural pests. Am. J. Agric. Econ. 61:258–267.

Reid, J.C. 1975. Larval development and consumption of soybean foliage by the velvetbean caterpillar, *Anticarsia gemmatalis* (Hubner) (Lepidoptera: Noctuidae), in the laboratory. Ph.D. diss. Univ. of Florida, Gainesville (Diss. Abstr. Int. 36B:2057).

Robbins, J.C., and D.M. Daugherty. 1969. Incidence and oviposition of potato leafhopper on soybeans of different pubescence types. Entomol. Soc. Am. N. Cent. Stn. Branch. Proc. 24:35–36.

Rodriguez, J., and A. Ortega. 1962. Biologia y combate del trips de la soya en el Valle del Yaqui. Agric. Tech. Mexico 2:29–33.

Ross, J.P. 1963. Transmission of bean pod mottle virus in soybeans by beetles. Plant Dis. Rep. 47:1049–1050.

Rossetto, D., A.S. Costa, M.A.C. Miranda, V. Nagai, and E. Abramides. 1977. Diferenças na oviposição de *Bemisia tabaci* em variedades de soja [Variation in the oviposition of *Bemisia tabaci* in different soybean varieties]. An. Soc. Entomol. Bras. 6:256–263.

Ruesink, W.G. 1982. Analysis and modeling in pest management. p. 353–373. *In* R.L. Metcalf and W.H. Luckmann (ed.) Introduction to insect pest management. John Wiley and Sons, New York.

Saito, T., H. Kawamoto, and K. Kiritani. 1983. Effect of artificial defoliation on growth and yield of soybean; Development of dynamic economic injury level and control threshold. Jpn. J. Appl. Entomol. Zool. 27:203–210.

Shepard, M., G.R. Carner, and S.G. Turnipseed. 1977. Colonization and resurgence of insect pests of soybean in response to insecticides and field isolation. Environ. Entomol. 6:501–506.

Sherman, F. 1920. The green cloverworm (*Plathypena scabra* Fabr.) as a pest on soybeans. J. Econ. Entomol. 13:295–303.

Shortt, B.J., J.B. Sinclair, C.G. Helm, M.R. Jeffords, and M. Kogan. 1982. Soybean seed quality losses associated with bean leaf beetles and *Alternaria tenuissima*. Phytopathology 72:615–618.

Sloderbeck, P.E., and C.R. Edwards. 1979. Effects of soybean cropping practices on Mexican bean beetle and redlegged grasshopper populations. J. Econ. Entomol. 72:850–853.

----, and K.V. Yeargan. 1983. Green cloverworm (Lepidoptera: Noctuidae) populations in conventional and double-crop, no-till soybeans. J. Econ. Entomol. 76:785–791.

Smith, C.M., and C.A. Brim. 1979. Field and laboratory evaluations of soybean lines for resistance to corn earworm leaf feeding. J. Econ. Entomol. 72:78–80.

Smith, R.H., and M.H. Bass. 1972. Soybean response to various levels of pod-worm damage. J. Econ. Entomol. 65:193–195.

Sparks, Jr., A.N., and L.D. Newsom. 1984. Evaluations of the pest status of the threecornered alfalfa hopper (Homoptera: Membracidae) on soybean in Louisiana. J. Econ. Entomol. 77:1553–1558.

Stinner, R.E., J.R. Bradley, Jr., and J.W. Van Duyn. 1980. Sampling *Heliothis* spp. on soybean. p. 407–421. *In* M. Kogan and D.C. Herzog (ed.) Sampling methods in soybean entomology. Springer-Verlag New York, New York.

----, A.P. Gutierrez, and G.D. Butler, Jr. 1974. An algorithm for temperature dependent growth rate simulation. Can. Entomol. 106:519–524.

Stone, J.D., and L.P. Pedigo. 1972. Development and economic-injury level of the green cloverworm on soybean in Iowa. J. Econ. Entomol. 65:197–201.

Strayer, J.R. 1973. Economic threshold studies and sequential sampling for management of the velvetbean caterpillar, *Anticarsia gemmatalis* (Hubner), on soybean. Ph.D. diss. Clemson University, Clemson, SC (Diss. Abstr. Int. 35B:301).

Swaney, D.P., J.W. Mishoe, J.W. Jones, and W.G. Boyers. 1983. Using crop models for management: impact of weather characteristics on irrigation decisions in soybeans. Trans. ASAE 26:1808–1814.

Tayo, T.D. 1980. The response of two soya-bean varieties to the loss of apical dominance at the vegetative stage of growth. J. Agric. Sci. 95:409–416.

Thomas, G.D., C.M. Ignoffo, K.D. Biever, and D.B. Smith. 1974. Influence of defoliation and depodding on yield of soybeans. J. Econ. Entomol. 67:683–685.

----, ----, D.B. Smith, and C.E. Morgan. 1978. Effects of single and sequential defoliation on yield and quality of soybeans. J. Econ. Entomol. 71:871–874.

Todd, J.W., and D.C. Herzog. 1980. Sampling phytophagous Pentatomidae on soybean. p. 438–478. *In* M. Kogan and D.C. Herzog (ed.) Sampling methods in soybean entomology. Springer-Verlag New York, New York.

----, and S.G. Turnipseed. 1974. Effects of southern green stink bug damage on yield and quality of soybeans. J. Econ. Entomol. 67:421–26.

Tugwell, P., and F.D. Miner. 1967. Soybean injury caused by the threecornered alfalfa hopper. Ark. Farm Res. 16:12.

----, E.P. Rouse, and R.G. Thompson. 1973. Insects in soybeans and a weed host (*Desmodium* sp.). Ark. Agric. Exp. Stn. Rep. Ser. 214.

Turner, J.W. 1967. The nature of damage by *Nezara viridula* (L.) to soybean seed. Queensl. J. Agric. Anim. Sci. 24:105–107.

Turnipseed, S.G. 1972. Response of soybeans to foliage losses in South Carolina. J. Econ. Entomol. 65:224–229.

----. 1973. Insects. *In* B.E. Caldwell (ed.) Soybeans: Improvement, production and uses. Agronomy 16:545–572.

----. 1977. Influence of trichome variations on populations of small phytophagous insects in soybeans. Environ. Entomol. 6:815–817.

----. 1984. Insecticide use and selectivity in soybean. p. 227–237. *In* P.C. Matteson (ed.) Int. Workshop in Integrated Pest Control for Grain Legumes, Proc., Goiania, Goias, Brazil. April 1983. Brasilia, DF, Brazil. Dep. Difus. Tecnol. EMBRAPA, Brasilia, DF, Brazil.

----, and M. Kogan. 1976. Soybean entomology. Annu. Rev. Entomol. 21:247–282.

----, and ----. 1983. Soybean pests and indigenous natural enemies. p. 1–6. *In* H.N. Pitre (ed.) Natural enemies of arthropod pests in soybean. Southern Coop. Ser. Bull. 285.

----, and M. Shepard. 1980. Sampling Mexican bean beetle on soybean. p. 189–200. *In* M. Kogan and D.C. Herzog (ed.) Sampling methods in soybean entomology. Springer-Verlag New York, New York.

----, J.W. Todd, G.L. Greene, and M.H. Bass. 1974. Minimum rates of insecticides on soybeans: Mexican bean beetle, green cloverworm, corn earworm, and velvetbean caterpillar. J. Econ. Entomol. 67:287–291.

----, ----, and W.V. Campbell. 1975. Field activity of selected foliar insecticides against geocorids, nabids and spiders on soybeans. J. Ga. Entomol. Soc. 10:272–277.

Van Duyn, J.W., and P.J. McLeod. 1983. Pesticide effects on natural enemies. p. 56–62. *In* H.N. Pitre, (ed.) Natural enemies of arthropod pests in soybean. Southern Coop. Ser. Bull. 285.

----, S.G. Turnipseed, and J.D. Maxwell. 1971. Resistance in soybeans to the Mexican bean beetle. I. Sources of resistance. Crop. Sci. 11:572–573.

Waddill, V.H., B.M. Shepard, J.R. Lambert, G.R. Carner, and D.N. Baker. 1976. A computer simulation model for populations of Mexican bean beetles on soybeans. S.C. Agric. Exp. Stn. Bull. Clemson, SC. 590.

Walters, H.J. 1964. Transmission of bean pod mottle virus by bean leaf beetles. Phytopathology 54:240.

Wedberg, J.L., and D.E. Kuhlman. 1976. Thrips problems in soybeans. p. 17–20. *In* Twenty-eighth Illinois custom spray operators training school. Illinois Cooperative Extension Service, University of Illinois, Urbana.

Wilkerson, G.G., J.W. Jones, K.J. Boote, I.I. Ingram, and J.W. Mishoe. 1983. Modeling soybean growth for crop management. Trans. ASAE 26:63–73.

Zavaleta, L., and M. Kogan. 1984. Economic benefits of breeding host plant resistance: A simulation approach. N. Cent. J. Agric. Econ. 6:28–35.

Zavaleta, L.R., and B.L. Dixon. 1982. Economic benefits of Kalman filtering for insect pest management. J. Econ. Entomol. 75:982–988.

21 Processing and Utilization

T. L. Mounts
Northern Regional Research Center
Peoria, Illinois

W. J. Wolf
Northern Regional Research Center
Peoria, Illinois

W. H. Martinez
USDA-ARS
Beltsville, Maryland

The soybean [*Glycine max* (L.)] has long been recognized as a valuable component of medicine, food, and feed in ancient China. It was first cultivated in the USA in 1766 to provide the necessary ingredients to make soy sauce and vermicelli for the English market (Hymowitz and Harlan, 1983). This initial activity ceased with the demise of its promoter, Samuel Bowen; and interest in the soybean lay essentially dormant for a century.

Several events—the Perry Expedition (1853–1854), development of the State Agricultural Experiment Station and Extension systems in the USA, and the Chinese-Japanese War (1894–1895) which enhanced trade and access to soybean seed—were critical to the expanded production and utilization of the soybean in the USA (Probst and Judd, 1973). The first USDA bulletin devoted entirely to soybean and their use as a forage crop was issued at the turn of the century (Williams, 1899). From an estimated production of 20 000 ha in 1907, soybean hectarage for hay and silage steadily increased to 1.9 million ha by 1940.

Other contemporary and subsequent events shifted the mode of utilization into new and different forms. Soybean seeds imported from Manchuria by European countries (1900–1910) were successfully used as a source of oil in soap manufacture; World War II greatly increased demand for fats and oils; and the effect of the boll weevil on cottonseed production created a need for an alternate oilseed crop. By the mid-1930s, defatted soybean meal had also become an accepted protein ingredient in livestock and poultry production. These events reversed the U.S. position of net importer of oil, meal, and bean in the period 1900 to 1918, to one of net exporter of soybean meal in the late 1930s. By 1941, soybean hectarage

Soybeans: Improvement, Production, and Uses, 2nd ed.—Agronomy Monograph no. 16.

harvested for processing into oil and meal (2.4 million ha) exceeded the hectarage providing forage silage and seed (2.2 million ha). Currently, the majority of the more than 28 million ha of soybean crops cultivated in the USA are harvested for domestic or foreign processing into meal and oil products (Table 21–1).

21–1 SOYBEAN OIL

Initially, growth of the U.S. soybean industry was influenced more by the shortage of oil and its relatively high price than the need for protein. In the 1920s, the oil was used mostly in soaps, paints, and varnishes. Use of soybean oil in food was restricted for a considerable period because of flavor stability problems (Dutton, 1981).

Deficiencies in the performance of soybean oil as an industrial drying oil combined with strong competition from synthetic resins and detergents materially limited the size of the industrial market (Smith and Circle, 1972). Efforts to develop new industrial uses began at the U.S. Regional Soybean Industrial Products Laboratory, organized in 1936, and continued later at the Northern Regional Research Laboratory (1940). Cost of the raw material relative to petroleum continued to be a dominant factor. Despite the development of new end uses such as speciality nylons, high-pressure lubricants, plasticizers and coatings, domestic use of soybean oil for industrial products has remained essentially constant at about 2% of supply or 90 700 t since 1960 (USDA, 1984a).

As interest in industrial use declined, demand for edible fats and oils rose, stimulating interest in improving soybean oil quality. Though the

Table 21–1. Soybean supply and use (USDA, 1950–1982).

Year	Supply: For all purposes	Supply: Harvested for hay	Supply: Harvested for bean	Utilization: Total production for bean†	Utilization: Processed for oil and meal	Utilization: Exports
	10^3 ha			10^6 t		
1929	983	–‡	287	0.2	0.1	–
1939	3 874	1 859	1 748	2.4	1.6	0.3
1942	5 547	1 062	4 007	5.1	3.6	<0.1
1947	5 286	523	4 622	5.1	4.4	0.1
1952	6 463	439	5 846	8.1	6.4	0.9
1957	8 520	180	8 447	13.1	9.6	2.4
1962	11 509	184§	11 181	18.2	12.9	4.9
1967	16 514	–	16 106	26.6	15.7	7.3
1972	18 981	–	18 502	34.6	19.7	13.0
1977	23 886	–	23 421	48.1	25.2	19.1
1982	29 226¶	–	28 667	60.7	30.2	24.6

†Difference between "total" and sum of "processed" and "exports" accounted for as seed, feed, and residual.
‡Data not available.
§Minimal hectarage reported as part of "other hay" category after 1964.
¶Preliminary data.

exigencies of World War II forced increased use of soybean oil in margarine, in certain instances as much as 30%, quality remained a problem. Through a unique collaboration begun in 1946 of government, academic, and industrial research organizations fostered by the National Soybean Processors Association, technologies were devised to make soybean oil the leading edible oil in the USA (Dutton, 1981).

The relative market share held by soybean oil in the most recent decades is indicated in Tables 21–2, 21–3, and 21–4. Since 1960, soybean oil's share of the edible oil products market has increased from 54 to 76% (Table 21–2). This increase occurred primarily at the expense of cottonseed oil and lard for which price, availability, improved processing technology, and nutritional considerations (cholesterol and polyunsaturation) were contributing factors.

Within the major classes of edible products (Table 21–3), use of soybean oil in baking and frying fats, salad and cooking oils, and margarine has changed from a 4-3-3 distribution in 1960 to a 3-5-2 distribution, respectively, in 1982–1983.

As documented in Table 21–4, soybean oil held over 80% of the margarine market by 1960. But the 40% growth in this market since 1960 must be considered moderate when compared to a 90% increase in utilization of baking fats and oils, and 320% increase for salad and cooking

Table 21–2. Market share of selected fats and oils in edible oil products in the USA. (USDA, 1984a)

Item	1960	1965	1970	1975	1980–1981†	1982–1983†
	%					
Soybean oil	54	56	66	66	72	76
Cottonseed oil	22	19	10	6	5	5
Corn oil	5	‡	‡	5	5	5
Peanut oil	1	1	2	1	1	1
Lard	9	7	6	2	3	2
Edible tallow	5	5	6	6	6	5
Coconut oil	3	3	4	1	3	3
Palm oil	‡	‡	‡	7	2	2

†Census started reporting annual tables on a marketing year (October–September) basis beginning in 1978/1979.
‡Census data withheld to avoid disclosing figures for individual companies.

Table 21–3. Distribution of soybean oil as used in various edible oil products in the USA. (USDA, 1984a)

1960	1965	1970	1975	1980–1981†	1982–1983†	
	%					
Baking and frying fats	37	35	36	30	31	32
Salad and cooking oils	28	37	40	46	49	50
Margarine	34	27	23	24	19	17
Other edible products	1	1	1	‡	1	1

†Census started reporting annual tables on a crop year (October–September) basis in 1978
‡Less than 0.5%.

Table 21–4. Soybean oil consumed in end products and percent of total fats and oils used in the USA (USDA, 1984a).

Item	1950		1960		1965		1970		1975		1980–1981†		1981–1982†	
	10^3 t	%	10^3 t	%	10^3 t	%	10^3 t	%	10^3 t	%	10^3 t	%	10^3 t	%
Baking and frying fats	361	48	530	51	667	53	990	61	918	54	1213	63	1357	67
Salad and cooking oils	156	27	402	46	709	56	1120	73	1375	76	1917	80	1981	80
Margarine	208	50	501	81	504	72	639	79	711	82	756	82	781	86
Other edible products	—‡	—	12	10	16	9	17	8	10	5	20	11	23	13
Edible total	725	23	1446	54	1896	56	2766	66	3014	66	3906	72	4142	74
Inedible	140	7	99	5	116	5	101	4	83	4	92	4	92	4
Total	865	17	1545	34	2012	36	2867	44	3097	46	3998	50	4234	53

†Census started reporting annual tables on a crop year (October–September) basis in 1978.
‡Included in "salad and cooking oils."

oils by 1981 to 1982 (USDA, 1984a). The increased utilization of salad and cooking oils, coupled with a rapid increase in soybean oils's share of this market (46%, 1960; 80%, 1981/1982), produced an exceptional rate of increase in tonnage of soybean oil consumed by this market area since 1960.

Future rate of growth for soybean oil in domestic edible markets will be moderated by diminished population growth. Maintenance of current market share will depend upon nutritional concerns with total calorie consumption. Export demand will be subject to pressure from the anticipated rapid expansion in availability of palm and palm kernel oils and the growing ability of other nations to meet and exceed their domestic food needs. Future increase in soybean oil utilization may well depend upon expanded industrial use in areas such as grain dust suppressants, pesticide application, and the ability to cost-effectively convert this renewable resource to chemical feedstocks.

21–2 SOYBEAN PROTEIN

Soybean meal plays a key role as a protein ingredient in feeds in the USA. About 16.2 Tg were forecast to be available for consumption as feeds in 1984; consumption by animal classes was forecast as follows (USDA, 1984b):

Class	Tg	Percent
Poultry	7.2	46
Swine	4.4	32
Beef	1.9	9
Dairy	1.7	9
Others	1.0	4
Total	16.2	100

Poultry feeds are the largest outlet for meal followed by swine (*Sus* spp.) feeds; the two account for 78% of total usage. Soybean meal has been estimated to make up more than 90% of the oilseed meals consumed in poultry feeds. Smaller quantities of soybean meal are fed to beef and dairy cattle (*Bos* spp.). Pet foods are another market for soybean meal.

Production of edible protein products is small as compared to soybean meal for feed uses. No official statistics for soy flour, concentrate and isolate production are kept, and individual producers consider their own production figures as proprietary information. Estimates obtained from industry sources for 1982 are shown in Table 21–5. There has been no significant change in recent years. Flours and grits were produced in the largest quantities estimated to be 159 000 t; concentrates, isolates, and textured flours were manufactured at 36 000 to 43 000 t each; textured concentrates represented only 4000 t. Selling prices for the protein products are also included in Table 21–5. Prices of concentrates and isolates

Table 21-5. Production estimates for edible soybean protein products in the USA for 1982.

Product	Minimum protein content†	Annual production	Selling prices‡
	—— % ——	10^3 t	$/kg
Defatted flours and grits	50	159	0.22–0.24
Concentrates	70	36	0.92–0.95
Isolates	90	41	2.20–2.42
Textured flours	50	43	0.66
Textured concentrates	70	4	1.21

†Dry basis.
‡January 1985 prices.

relative to flours and grits reflect the added costs of processing involved in their manufacture.

21-3 SOYBEAN PROCESSING

Soybean is processed to yield oil and meal by a hexane extraction process (Becker, 1978). In recent years, there has been a trend towards increasing the capacity of soybean processing plants. In 1978, 50% of the processing plants had < 900 t day^{-1} capacity (Mogush, 1980), while by 1982, 76% had capacities of > 900 t day^{-1} (American Soybean Association, 1983). On the average, about 80% of the U.S. capacity has been used annually (Mogush, 1980).

21-3.1 Traditional Extraction Technology

Although large quantities of soybean are received at the extraction plants at harvest time, purchase and receipt of bean continue throughout the year. The soybean are cleaned and dried if above about 13% moisture (w/w) and stored in large concrete silos (Fig. 21-1); most processors prefer to "age" freshly harvested soybean for about 21 days to facilitate processing. Belt conveyors and bucket elevators bring the soybean into the plant where they are collected in storage bins. Prior to dehulling, the soybean pods are dried to 9.5 to 10.5% moisture (w/w) and then tempered for several days. Before being fed into the cracking rolls, the soybean are cleaned and passed over a magnet to remove stray metal. A shaker screen follows the cracking rolls, where the fines drop through and the hulls are removed by fan aspiration. The cracked bean is conditioned to 10 to 11% moisture and 74 to 79°C (165–175°F) in a steam-jacketed cooker, a vertical-stack type or a rotary steam-tube dryer type. Moisture as live steam or water spray is added to the conditioner if necessary. Flaking of the bean is accomplished in a roll stand consisting of a pair of smooth-surface rolls regulated to produce flakes about 0.254-mm (0.010 inch) thick. The

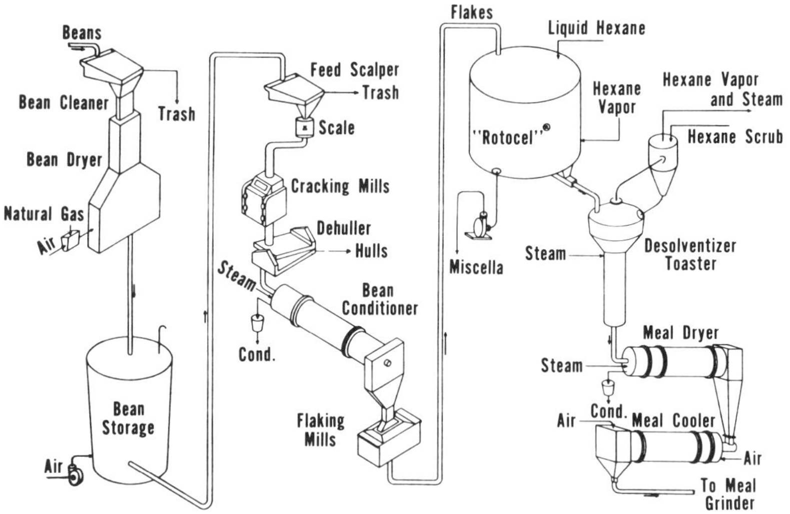

Fig. 21–1. Flow chart for the solvent extraction of soybean. Courtesy, Dravo Corporation.

flakes are then transported to the extractor by mass-flow type enclosed conveyors designed to minimize flake breakage.

Percolation type extractors are used for the hexane extraction of soybean. The liquid solvent or miscella (solvent containing dissolved oil) is pumped over a bed of flakes, percolates down through the bed, and leaves at the bottom through a perforated plate, mesh screen, or wedge wire screen bar system countercurrent to the movement of the flakes. Some of the types of percolation extractors are rotary cells, chain and basket, perforated belt, chain conveyor, and filter. With one rotary type the cells rotate in a horizontal plane and fresh solvent is pumped onto flakes at the end of the extraction cycle; as extraction proceeds the miscella becomes richer in oil. The extracted flakes are allowed to drain before being dropped into the discharge hopper.

The extracted flakes contain about 30 to 35% hexane, 7 to 8% water, and 0.5 to 1.0% oil. For processing into livestock meal, solvent is reclaimed from the flakes in the desolventizer-toaster (D-T) unit. Description of the D-T operation and further processing of the defatted flakes into edible soy protein proteins is discussed later (section 21–7).

The miscella leaving the extractor contains about 25 to 30% oil; it is first filtered to remove suspended fines, and then enters a series of evaporator stages for reclamation of the solvent. The first stage evaporator is heated with vapors from the DT and gives a miscella containing 65 to 78% oil; the second stage evaporator yields 90 to 95% oil content. Vapors from the two evaporator stages pass to condensers, and the recovered hexane is recycled to the extractor. Final solvent removal is accomplished in the oil stripper, a steel cylindrical vacuum column in which live steam flows upward countercurrent to the flow of oil. The oil, essentially free of solvent, is cooled to ambient temperature and pumped to storage.

21–3.2 New Developments

The introduction of fluidbed driers for soybean drying, dehulling, and conditioning is estimated to have reduced heat requirements to 50% of the traditional soybean drying and preparation process (Florin and Bartesch, 1983). The soybean are taken from storage, cleaned, and introduced into the fluidbed drier. Hot air is used as the fluidization gas and the soybean are quickly heated to a surface temperature of 75 to 92°C. With this heat treatment the hulls are loosened from the cotyledons and cracked open. The product stream proceeds through cracking rolls and hammer mills, to halve the bean and free the detached hulls by discharge air and the halved soybean are then conditioned. On leaving the fluidbed conditioner, the bean halves continue through cracking rolls and flakers to the extraction plant. In this process the soybean are only heated once in the drier and retain heat throughout to the extractor. The drying rate can be varied by adjustment of the temperature of the fluidbed and this flexibility allows direct dehulling of freshly harvested soybean. Use of fluidbed technology can reduce the drying and tempering time from 48 to 72 h to 10 to 20 min.

Another system for the drying and preparation of soybean is currently under study and is known as MIVAC drying (microwave vacuum) (Moore, 1983). This system uses electrical energy for drying and conditioning; cracking and dehulling proceed according to conventional methods and the cracked, dehulled soybean go directly to the flaker. The air used in the dehulling process is heated by the waste heat generated during production of the microwave energy.

Pretreatment of both whole soybean and flakes has been proposed for the inactivation of enzymes prior to oil extraction. The goal of such pretreatments is to improve crude oil quality to facilitate physical refining of the oil as an alternative to the traditional alkali refining process. Infrared (IR) heat treatment of whole soybean at 104°C for 5 min prior to bean preparation for oil extraction was adequate to achieve improved oil quality compared to untreated bean (Kouzek Kanani et al., 1984). Lipoxygenase activity was reduced to 0.5% of that found in untreated bean. The peroxide value determined for crude oil extracted from treated soybean was 0.6 cmol kg^{-1} compared to 2.8 cmol kg^{-1} of crude oil extracted from untreated soybean. This diminished oxidative deterioration was reflected in fewer off-flavors in treated-soybean oil compared to oil from untreated bean. Water-degummed oil from IR-treated soybean had a much lower P content (24 mg kg^{-1}) than that from raw soybean (185 mg kg^{-1}). This indicates an inactivation of phospholipase D by IR-treatment resulting in less formation of nonhydratable phospholipids during preparation and extraction of beans. The P content of oil from IR-treated soybean was reduced to 4 mg kg^{-1} by treating with 1% bleaching earth and was below the maximum (5 mg kg^{-1}) specified for steam refining. These laboratory tests suggest that IR-treatment of whole soybean may be successfully employed for enzyme inactivation prior to processing.

Moist-heat treatment of soybean flakes to inactivate phospholipase D has been commercialized as the ALCON process, an additional stage between conventional bean preparation and the extractor (Kock, 1983). In this process, the flakes are conditioned to moisture levels of 15 to 20% and a temperature of 95 to 110°C. The conditioned flakes are stirred while being tempered for a period of about 15 min. Stirring within the tempering equipment causes flake agglomeration, which gives the flakes a completely different granulated type structure having an increased bulk density. Finally, the flakes are dried and cooled to appropriate extraction conditions. Crude oil extracted from soybean flakes given this moist-heat treatment had a decreased amount of nonhydratable phospholipids which was confirmed by the low P content after water degumming.

The hexane used for oil extraction of soybean is a hydrocarbon fraction derived from petroleum having a boiling range of 63 to 69°C. It is an excellent solvent for vegetable oils and is essentially free of N-or S-containing compounds and unsaturated hydrocarbons (Mustakas, 1980). Interest in alternative solvents has increased due to disadvantages in the use of hexane. Being a petroleum product, hexane may be of limited availability in the future; it is extremely flammable and forms explosive mixtures with air. Hexane vapors are toxic, hence maximum concentrations in the workplace must be controlled. Recovery of hexane from oil and meal is energy intensive and incomplete, requiring that some of the solvent be continuously replaced.

Recently, Shell Development Company (Sullivan et al., 1982) and the Northern Regional Research Center (Baker and Sullivan, 1983) have conducted pilot-plant studies using isopropanol (IPA) as a solvent for extracting soybean. An IPA/water (88:12) azeotrope was employed near the boiling point (80.2 °C) and on cooling the miscella a phase separation occurs which allows recovery of the oil by nondistillation techniques in a phase separator. Residual IPA in the oil-rich phase is recovered in an oil stripper and a semi-refined oil is obtained that does not require degumming and produces smaller amounts of soapstock than hexane-extracted oil. The oil produced by IPA extraction has a lower phosphatide content than hexane extracted oil but is otherwise similar (Baker and Sullivan, 1983).

The IPA-extracted flakes leave the extractor with a higher solvent content than hexane extracted flakes. Mechanical screw pressing was used to reduce the IPA content prior to desolventizing and toasting. The resultant meal appears comparable to hexane processed meal for animal nutrition.

Workers at Texas A&M University have evaluated water as a solvent for oil recovery from soybean (Lawhon et al., 1981; Johnson and Lusas, 1983). In this process, finely ground soybean are mixed with water and the resulting slurry is centrifuged to yield an oil phase, an aqueous phase (containing protein, soluble sugars, and some oil), and an insoluble residue fraction (cell wall polysaccharides). The protein can be recovered by ultrafiltration followed by spray drying. This process has some disadvantages such as: low oil yields; the presence of oil in the protein fraction which may impart poor flavor stability; and a need to operate under more stringent sanitary conditions than are needed with hexane due to the potential for microbiological contamination.

Supercritical fluid (SCF) technology may offer a viable alternative to present extraction methods. Supercritical carbon dioxide (SC-CO_2) as a solvent for recovery of soybean oil has been evaluated recently on a laboratory scale (Friedrich et al., 1982). The SC-CO_2 extracted oil contains lower amounts of phosphatides and other nontriglyceride components than does hexane-extracted oil.

Defatted soybean meals with protein solubilities > 70% and with improved flavors were produced by SC-CO_2 extraction. Optimum extraction conditions were 82.8 MPa, 85°C, and 10.5 to 11.5% moisture (Eldridge et al., 1986). The usual grassy-beany and bitter flavors of hexane-defatted soybean flours were only minimally detected in optimally SC-CO_2 extracted materials. Meal characteristics were: Nitrogen Solubility Index, 62: Flavor Score, 7.2 (on a scale of 1–10, i.e., 1=strong, 10=bland); Lipoxygenase Units, 3 μm O_2 consumed min^{-1} mg^{-1} protein; trypsin inhibitor activity, 24.8 g kg^{-1}; and urease activity (pH increase), 2.1. Heat treatment of the defatted meal is required to inactivate urease and lower the trypsin-inhibitor activity. Disadvantages of the SC-CO_2 extraction technique are the expensive high pressure equipment and sophisticated technology required for implementation. If engineering problems can be overcome, however, this process may find application in the oilseed processing industry.

While several solvents have been examined, none has been developed to the point where it is used as a commercial replacement for hexane.

21–4 SOYBEAN OIL PROCESSING

Processing of soybean oil is designed to convert the crude oil as extracted from the bean into a finished product free of impurities and amenable to a variety of food formulations. Compositional analyses for crude and refined soybean oils are shown below (Pryde, 1980).

Composition	Crude oil	Refined deodorized oil
Triglycerides (%)	95–97	> 99
Phosphatides (%)	1.5–2.5	0.003–0.006
Unsaponifiable matter (%)	0.6	0.3
Plant sterols (%)	0.33	0.13
Tocopherols (%)	0.15–0.21	0.05–0.10

(continued on next page)

Continued.

Composition	Crude oil	Refined deodorized oil
Hydrocarbons (squalene) %	0.014	0.01
Free fatty acids (%)	0.3–0.7	< 0.05
Trace metals		
Iron (mg kg^{-1})	1–3	0.1–0.3
Copper (mg kg^{-1})	0.03–0.05	0.02–0.06

Traditionally, oils have been processed through stages of degumming, alkali refining, bleaching, and deodorization. Hydrogenation of the oil is practiced to improve stability and modify the physical nature of the oil from a liquid to a solid.

21–4.1 Degumming

The principal method of degumming employed in the USA is a batch treatment of the oil with 1 to 3% of water, based on oil volume (Carr, 1978). The mixture is agitated for 30 to 60 min at 70 to 80°C. The hydrated phosphatides and gums are separated by centrifuging. About 90% of the phosphatides are removed from the oil by water degumming (Myers, 1957). Although most of the remaining phosphatides are removed during alkali refining, vegetable oils often contain some phosphatides that are not removed by hydration. Beal et al. (1956) concluded that the residual P content of a satisfactorily refined oil should be between 2 and 20 mg kg^{-1} (0.06–0.6% phosphatide). Oils with P contents above this range were found to have decreased oxidative stability. Most commercially produced oils will contain < 5 mg kg^{-1} P.

21–4.2 Refining/Physical Refining

Refining of vegetable oils is practiced to remove free fatty acids, phosphatides and gums, prooxidant metals, coloring matter, insoluble matter, settlings, and miscellaneous unsaponifiable materials. The treatment has little effect on the triglycerides of the oil. The first step in the conventional process is called *caustic refining*. If a crude oil is to be caustic refined, it is usually treated with 300 to 1000 mg kg^{-1} of food-grade, 75% phosphoric acid at ambient temperature at least 4 h prior to the refining step, to increase the efficiency of phosphatide removal during caustic refining (Carr, 1978). Oil refining is usually a continuous process. A 15 to 20% sodium hydroxide (NaOH) solution based on the free fatty acid content of the oil plus 0.10 to 0.13% excess, is proportioned into the crude oil, mixed in a high shear in-line mixer, and then held for 3 to 15 min in a slow speed mixer. The soap-oil mixture is heated to 75 to 80°C and then separated into refined oil and soapstock by centrifuging. Refined oil is washed once or twice with 10 to 20% (w/w) of soft water at 90°C. Of all the unit processes, caustic refining has the most significant effect on oil quality and, if the oil is not properly refined, subsequent processing operations such as bleaching, hydrogenation, and deodorization will be impaired, and finished products will fail to meet quality standards. Protection of the oil from exposure to air is important during refining to minimize oxidative deterioration. Caustic refining is preferably conducted in hermetic disk-type, self-cleaning centrifuges, and completely closed systems where the oil has no contact with air from the moment it is pumped into the system until it leaves as a neutralized and dried oil. (Braae, 1976).

Crude oils processed from normal soybean refine satisfactorily and, after bleaching and deodorization, give high-quality soybean salad oils. Such oils are bland, or have midly beany, buttery, or nutty flavors and will receive flavor scores in the 7 to 8 range on a 10-point flavor intensity scoring scale (1 = extreme, 10 = bland). A flavor score of 6 is generally considered the breakpoint between satisfactory and unsatisfactory oils.

List et al. (1977) evaluated refining methods for improvement of flavor quality of oils extracted from field- and storage-damaged bean. They found that the strength or excess of alkali used had no significant effect on the flavor score of the oils, which were considered unsatisfactory as salad oils. Double-refining of phosphoric acid-degummed damaged bean oil showed some improvement over single-refining, but the flavor scores were merely borderline satisfactory and high refinery losses were anticipated in plant-scale operations.

Physical refining is an alternate process to caustic refining for removing the free fatty acids present in crude oil. Recent developments in the physical refining of edible oils, i.e., the removal of free fatty acids by steam distillation in a simultaneous deacidification-deodorization step, have required changes in degumming procedures. Degumming of the oil with phosphoric acid is an important pretreatment for physical refining (Sullivan, 1976). List et al. (1978b) stirred crude soybean oil with 0.2% (w/w) 85% phosphoric acid (H_3PO_4) for 15 min at 60°C, and then used centrifugation and decantation to separate gums. The degummed oil was washed twice with water (20% by weight), bleached with 0.5% earth at 105°C under vacuum, filtered, and then deodorized at 260°C, 1 mm of Hg for 1 h. The phosphoric acid-degummed oil was compared to a water-degummed oil, after both were steam refined, in organoleptic evaluations conducted as described by Moser et al. (1947, 1950, 1965). Evaluations of the freshly deodorized oils showed that the phosphoric acid degummed oil had significantly less intense flavors and a better flavor score than the water-degummed oil. The improvement in flavor score is attributed to enhanced Fe removal effected by the phosphoric acid pretreatment (List et al., 1978a). After water washing and bleaching, the phosphoric acid pretreated oil had an Fe content of 0.1 mg kg^{-1}, whereas the water-degummed oil contained 0.5 mg kg^{-1} of Fe. Iron is an active catalyst for oxidative deterioration of soybean oil, especially at deodorization temperatures (Beal et al., 1956).

21–4.3 Bleaching

Bleaching of alkali-refined oils removes entrained soaps and reduces color bodies in the oil; it is more appropriately referred to as adsorption treatment. Batch or continuous vacuum bleaching is generally practiced; this consists of agitation of the oil with 0.5 to 1.5% acid-activated earth at 90 to 95°C for 15 to 30 min at a high vacuum (3–10 mm of Hg absolute pressure) followed by filtration to give a clean, clear oil. Cowan (1966) reported the flavor evaluation of two deodorized soybean oils prepared in the same commercial plant. The bleaching step was included for one sample but omitted for the other. The bleached oil showed a significantly higher flavor score both initially and after accelerated storage. It has been suggested that bleaching removes peroxides and secondary oxidation products and that this added function gives the observed improvement in flavor scores (Wiedermann, 1981). Such removal is by processes involving chemisorption and subsequent chemical reaction on the surface of activated clays, i.e., decomposition and dehydration or pseudoneutralization of peroxides. Careful selection of the type of bleaching clay was shown to give a dramatic improvement in terms of peroxide reduction. Bleached oil having a peroxide value of 0.0 can be obtained for subsequent deodorization.

21–4.4 Deodorization

Deodorization is the last process step used to improve the taste, odor, color, and stability of the oil by removal of undesirable substances. All commercial deodorization, whether in continuous, semicontinuous, or batch units, is essentially a steam stripping of the oil for removal of free fatty acids and other volatile materials. Deodorization is conducted at 1 to 6 mm Hg and 210 to 274°C for 3 to 8 h in batch or 15 to 120 min in continuous or semicontinuous units, with 5 to 15%, or 1 to 5% stripping steam, respectively (Zehnder, 1976). The goal of deodorization is to produce a finished oil that has a bland flavor, a maximum free fatty acid content of 0.05%, and a zero peroxide value. The deodorization process will not produce a good quality finished oil unless the previous treatment of that oil was correct. In addition to the removal of free fatty acids and volatile odor compounds and the decomposition of peroxides, deodorization also reduces the tocopherol content of soybean oil by about one third. Tocopherols are natural antioxidants found in vegetable oils and contribute significantly to oxidative stability. Frankel et al. (1959) observed that at high concentrations of tocopherol, the synergistic effect between citric acid and tocopherol was decreased. They suggested that the residual tocopherol content in deodorized soybean oil was close to the optimum concentration for maximum oxidative stability. Citric acid (0.005 to 0.01% weight of oil) is added to the oil at the cooling stage in the deodorizer, to protect the oil against oxidation. Citric acid acts as a metal chelating agent and reduces the activity of prooxidant metals. Miyakoshi and Komoda (1978) evaluated the flavor of freshly deodorized oils, with varying concentrations of citric acid (8.5, 22.1, and 49 mg kg^{-1} added at 130°C during cooling. Oils were also evaluated after storage at 60°C and after 10-h exposure to fluorescent light. They found that the efficiency of citric acid is independent of concentration, and that as little as 10 mg kg^{-1} citric acid is effective in stabilizing soybean oil against flavor and oxidative deterioration.

21–4.5 Hydrogenation

When soybean oil is to be used as a cooking oil or in the formulation of margarines and shortenings, it must be partially hydrogenated to improve high-temperature stability and to physically harden the oil (Allen, 1978). Most hydrogenations are performed as batch processes; however, continuous systems have been patented and are in use (Coombes et al., 1974). Conditions of hydrogenation range from 0.01 to 0.1% Ni catalyst, at 140 to 225°C and at pressures of 0.05 to 0.6 MPa gauge. Soybean oil is generally hydrogenated to an iodine value of 110 to 115 for use as a cooking oil, 80 to 90 for margarine base stock oil, and 60 to 70 for shortenings (Weihrauch et al., 1977).

Partial hydrogenation of edible oils is practiced to increase stability by the selective reduction of linolenic acid. Commercially, a dual purpose salad/cooking oil is prepared by hydrogenation of soybean oil with Ni catalysts under selective conditions, such as 35 to 97 $\times$ 10^3 Pa, 0.05% catalyst at 177°C. The oil is hydrogenated to an iodine value of 110 to 115 and must be winterized to meet the requirements of the standard American Oil Chemists' Society cold test (Link, 1975). This test calls for the oil to remain clear for a minimum of 5.5 h at 0°C. Stearine, high melting glycerides, and palmitic and stearic acid fractions are removed by the winterization process (Numenz, 1978). Oil is chilled slowly to about 6°C during a 24-h period; at this point cooling is stopped and the oil/crystal mixture is allowed to stand for 6 to 8 h. The yield of liquid oil is approximately 75 to 85%. By-product stearine is generally used for shortening manufacture. Mounts et al. (1978) showed that although hydrogenation of soybean oil to linolenic acid contents of 3.3 and 0.4%, gave a significant improvement in the oxidative stability, as measured by the active oxygen method (AOM), it did not give a significant improvement in the flavor stability of the oil during accelerated storage tests up to 8 days at 60°C. The oils in these tests were all treated with citric acid on the cooling side of deodorization. Frankel (1980) attributes the improvement of the oxidative stability of soybean oil to the conversion of linolenic acid into isolinoleic and monoenoic acids, which are more difficult to oxidize. Some of these compounds have double bonds between positions 14 and 16 in the C chain and may produce flavor compounds similar to those of linolenic acid upon oxidation, which may explain why hydrogenation does not have a significant impact on flavor stability during storage. Lowering of the linolenic acid content does provide increased stability during use of soybean oil as a cooking oil. It is still, however, only a partial solution to the problem, and thermal oxidation will eventually produce objectionable odors and flavors.

After deodorization, all soybean oils, whether or not hydrogenated, are subject to oxidation when the oil is in contact with air. Areas of treatment of edible oils with N_2 to ensure proper protection are summarized in Table 21–6.

21–4.6 Energy Conservation

A major concern of the edible oil industry in the 1970s and 1980s has been energy conservation. All of the processing steps just discussed consume energy to heat and cool the oil and to generate the required vacuum levels. The bleaching, deodorization, and hydrogenation steps offer the greatest opportunity for conservation (Gavin, 1983) by means of various energy recovery techniques.

21–5 FOOD USES OF SOYBEAN OIL

21–5.1 Salad and Cooking Oils

As noted earlier, salad and cooking oils constitute the largest volume usage of soybean oil. Partial selective hydrogenation of soybean oil to lower its linolenic acid content to about 3% considerably improves the oil's flavor and oxidative stability. Blending soybean oil with an oil containing little or no linolenic acid is also done to lower the linolenic acid content, i.e., with cottonseed oil (Carpenter et al., 1976). Some dual purpose salad/frying oils contain an antifoam agent, usually a silicone compound (Freeman et al., 1973; Lorenz, 1978).

Soybean oil is the major oil used in mayonnaise and prepared salad dressings. Most mayonnaises contain 77 to 83% oil (Newkirk et al., 1978). A thicker product containing 80 to 84% oil is often used in institutions. While oil is the major component, mayonnaise is an oil-in-water emulsion, that is, the oil is the dispersed rather than the continuous phase. Prepared salad dressings use mayonnaise as the base and include one or more of the following ingredients: minced onions, minced green peppers, chopped stuffed olives, hard-cooked eggs, and chili sauce (Anonymous, 1963). Imitation mayonnaises contain considerably more water than do regular mayonnaises and only 14 to 40% oil. Spoonable salad dressings must contain at least 30% oil. Both these products use a starch paste as a thickener (Newkirk et al., 1978; Weiss, 1983). Most commercial pourable salad dressings contain 55 to 65% oil while low calorie dressings have a low oil content, i.e., 4 to 14% (Watt and Merrill, 1963).

21–5.2 Margarines

Margarine, originally developed as a butter substitute, is recognized as a high-quality, nutritious product available in several forms for table

Table 21–6. Summary of treatment areas with N_2 to protect edible oils.

Area of treatment	Method of introducing N_2	Classification	Approximate usage
Manufacturing-pumping from deodorizer	In-line between cooler and storage tank	Sparging	Approx. 0.025 ft^3/gal (3.7×10^{-4} m^3 L^{-1})
Bulk-oil storage	Nitrogen in-line sparger directly into headspace	Blanketing	Enough to maintain positive pressure
Filling of tank cars	In-line between storage and tank car	Sparging	1000 ft^3 (28 m^3) per 8000–gal (30 240 L) car
Tank car or truck	Into headspace of car after filling	Blanketing	Undetermined
Customer's plant-pumping from tank car to oil storage	Sparging in-line during pumping	Sparging	Approx. 0.025 ft^3/gal (3.7×10^{-4} m^3 L^{-1})
Storage in tanks	Directly into headspace plus N_2 from in-line sparger	Blanketing	Enough to maintain positive pressure
Pumping from storage to filler or header	Sparging in-line during pumping	Sparging	Approx. 0.025 ft^3/gal (3.7×10^{-4} m^3 L^{-1})
Filler bowl or header	Entrance into closed filler bowl or header	Blanketing	Maintain light pressure
Closing or capping machine	Shroud or purge technique	Blanketing or purging	Undetermined

use and is no longer considered a substitute (Massiello, 1978). Regular margarine contains 80% fat and about 16 to 18% aqueous phase. Cow's milk, pure water, or water plus some edible protein, such as nonfat dry milk solids or soybean protein may constitute the aqueous phase (Code of Federal Regulations, 1977a, 1977b). Other ingredients are 2 to 3% salt (except for salt-free margarine); emulsifiers, such as mono- and diglycerides and/or lecithin; preservatives; flavoring; coloring, usually β-carotene; fortifiers, i.e., 15 000 U.S. Pharmacopeia (USP) units of vitamin A and sometimes 2 000 USP units of vitamin D; and optional ingredients such as butter, nutritive sweetners and fat antioxidants.

Until 1955, stick or brick-type margarine was the only table-grade type available. By 1976, this type constituted only 22% of the margarine market and 10 types of margarine and manufactured spreads were available. Currently, products for the consumer retail market are the principal margarine production (Table 21–7).

Basic steps involved in margarine manufacture include: (i) formulating the margarine oil blend, (ii) preparing the aqueous phase, (iii) preparing the emulsion, (iv) solidifying the emulsion and controlling its plasticity, (v) packaging the margarine, and (vi) tempering the packaged margarine, if necessary.

The oil formulation, the manner and degree of super-cooling the emulsion, and the extent of mechanically "working" the supercooled emulsion during the crystallizing stage are all used to control margarine plasticity. If the fat composition is rapidly supercooled and allowed to solidify without agitation, the margarine will become quite firm and have a narrow plastic range. The plastic range is extended by mechanically

Table 21–7. U.S. margarine production, 1985.†

Type of margarine	Quantity	
	10^3 t	%
Consumer retail products		
¼ lb (0.11 kg) Sticks	503.8	
1 lb (0.45 kg) Soft tubs	134.8	
Country patties	10.4	
Solids or rolls	64.4	
Spreads	220.4	
Diet and imitation, all sizes	43.5	
Subtotal	977.3	82.8
Food service products		
1 lb (0.45 kg) Solids or rolls	72.6	
Individual servings	14.9	
Subtotal	87.5	7.4
Bakery and industrial products		
Bulk sizes more than 1 lb (0.45 kg)	96.7	8.2
Total production	1180.6‡	100.0

†National Association of Margarine Manufacturers based on data reported by U.S. Bureau of Census (1986, personal communication.)

‡The sum of margarine packaged by package sizes does not agree with the total production because some margarine is not packaged during the same month in which it is produced.

working the emulsion while the fat is crystallizing from the supercooled state (Wiedermann, 1978). The melting point, fatty acid composition, and structure of the individual triglycerides influence the structural, nutritional and lubricity aspects of the margarine. Stick margarines usually are made from a blend of two or three intermediate fats, or of a low IV fat and a liquid oil. Soft margarines contain a high proportion of liquid oil; fluid margarines have the highest proportion.

The fats and oils selected for the margarine oil blend are mixed in the proper proportions and heated to 38°C (100°F) or more, and the oil-soluble ingredients are added. These additives include emulsifiers, usually mono- and diglycerides and sometimes lecithin, plus specified amounts of vitamins and desired flavoring and coloring ingredients.

The aqueous phase is prepared separately. A milk phase is prepared by adding dried protein such as whey or nonfat dry milk solids to water, then pasteurizing and cooling the mixture. Water-soluble ingredients, usually salt and preservatives, are also added (Wiedermann, 1978).

In the batch-continuous process, the oil and aqueous phases are prepared batch-wise in separate tanks, then the two phases are blended; the resultant emulsion is solidified on a continuous basis, as will be described later. Batch-wise blending of the two phases is done in a premix tank that often is referred to as a churn. Holding time in the churn is kept to a minimum because the temperature (30°C) is ideal for bacterial growth. If automatic proportioning equipment is used, the two phases can be prepared on a continuous rather than batchwise basis if proper temperature control is employed. Most margarine manufacturers in the USA prefer to weigh the major ingredients into the churn (Moustafa, 1979).

The oil and aqueous phases are emulsified to the proper degree as, or just before, the blend enters the first of three heat exchange chilling tubes of the continuous processing unit. Water in the emulsion is dispersed as droplets of about 5-μm diam and then kept from coalescing. Mono- and diglycerides help create and maintain an emulsion containing small droplets (Moustafa, 1979). The emulsion is cooled in a matter of seconds (5–10) by ammonia refrigerant vaporizing in the jackets of the heat exchangers. Scrapers continually remove solidified fat from the inner wall of the heat exchangers (often referred to as the A units) to promote rapid heat transfer. Small crystals form as the emulsion is supercooled (e.g., to 45–50°F or 7–10°C).

For regular (stick type) margarine, the supercooled, still-fluid melt is piped to either of twin crystallizers, referred to as B units. These are enlarged, empty cylinders wherein the emulsion remains stationary (typically, ca. 2 min) until crystal development proceeds to a point where the product is sufficiently firm to withstand the forces applied in extruding, shaping, and wrapping the margarine in high-speed automatic machinery. By use of twin crystallizers, the emulsion can remain stationary in one while the other is being filled. The temperature increases several degrees in the B unit because of the heat of crystallization.

Whipped margarines are produced by incorporating N_2 into the margarine. The gas is introduced ahead of one of the heat exchangers or at the suction side of the feed pump. This type margarine requires vigorous mixing in the B unit (as defined above) to limit the extent of crystal development; otherwise, the margarine becomes too firm. For this process, the B unit is a worker-type crystallizer. It has radial pins on a rotating central shaft, and these pins intermesh with stationary pins protruding from the cylinder wall. The agitation is controlled to allow crystal growth and yet prevent the formation of a firm crystal lattice. The soft, semifluid mass is packaged immediately in specially designed print-forming (i.e., stick-forming) machines that squeeze little or no N from the margarine. The gas increases the margarine's volume by 50% and makes the margarine softer and easier to spread at refrigerator temperatures.

Soft-tub margarines containing as much as 70 to 80% liquid oil are produced by chilling the emulsion in a single A unit and then mixing the chilled mass in a large, agitated crystallizer. Working the chilled emulsion to a limited degree prevents the fat crystals from growing into a firm network. The soft, fluid margarine is packaged in plastic tubs by a liquid-filling machine. Crystal development continues in the tub to give a soft, semisolid product. After packaging, soft tub margarines are tempered by holding 24 h or more at about 7°C (45°F), so that the crystal structure can become fully developed and stabilized (Moustafa, 1979).

21–5.3 Shortenings

Soybean oil constitutes about 64% of the fat used in the manufacture of shortening, while the other major fats (24%) are edible tallow and lard (Anonymous, 1982). Shortenings are used in the preparation of many foods and impart a tender quality to baked goods. In addition, in baked goods, shortenings enhance the aeration of leavened products; add to the flavor; promote a desirable grain and flavor; assist in the development of flakiness in products such as pie crusts, Danish and puffed pastry products; modify the wheat gluten, particularly in the development of yeast-raised doughs; and act as emulsifiers for the retention of liquids (Baldwin et al., 1972). Shortenings perform two chief functions in baked goods: (i) a leavening and creaming action, and (ii) a lubricating function. In icings and fillings, shortenings are used to entrain large volumes of air bubbles and thus produce a fine delicate structure. Large quantities of shortenings are used in the preparation of fried foods both by deep fat and by pan and grill frying. Fats play a dual role by aiding in the transfer of heat to the food being fried and by being partially absorbed by the food, contributing to nutritive value and to flavor.

Prior to 1961, shortenings were formulated from highly hydrogenated oils having substantially reduced polyunsaturated fatty acid (PUFA) contents ranging from 5 to 12%. Since 1961, shortenings have been produced with higher levels of polyunsaturated fatty acids, typically containing 22 to 32% PUFA. This change has been in response to research which suggests that intake of polyunsaturates plays a beneficial role in reducing blood cholesterol levels.

The basic steps in shortening manufacture are: (i) preparation of the individual basestocks and hardfats, (ii) formulation of the fat blend and other ingredients, (iii) solidifying and plasticizing the fat blend, (iv) packaging, and (v) tempering the shortening, when necessary. Tempering generally is limited to the plastic shortenings and margarines used for baking.

Tempering consists of holding the shortening for 24 to 72 h in a room maintained at a constant temperature, usually somewhere between 27 and 32°C (80–90°F). The holding time is determined by factors such as container size and type of shortening. Crystallization continues slowly during the tempering step. Tempering stabilizes the crystal structure against changes that might otherwise occur during subsequent temperature variations encountered in normal handling and storage.

Plastic shortenings generally are solidified and plasticized in scraped surface, heat exchanger equipment similar to that used in the production of whipped or soft margarines described earlier. The melted fat plus optional ingredients are chilled rapidly from 46–49°C (115–120°F) to 16–18°C (60–65°F) in one or more A units to produce numerous nuclei for crystal formation. The supercooled melt then is piped to a worker-type B unit wherein the shortening is agitated as the crystals grow and the mass partially solidifies. From 10 to 20% air or an inert gas is added and dispersed in the shortening as small bubbles. In this form, the air improves the whiteness of the product and sometimes contributes to its creaming ability. The plastic mass is pumped through a homogenizing valve and then to package fillers. The packages vary in size from 0.45-kg (1-lb) tins that are sealed after filling to open-end drums 172-kg (380-lb) having removable covers.

21–5.4 Emulsifiers

Emulsifiers have a number of functions in both plastic and pourable shortenings as well as in other lipid-containing food products. The functions include: stabilizing emulsions as in margarines and salad dressings; antispattering in margarines; texture control in bread and cakes; dough conditioning and antistaling in bread; aerating in cakes, toppings, and icings; plasticizing cake icings; and wetting in coffee whiteners and instant foods. Food emulsifiers probably are used in greater quantities than any other food additive (Nash and Brickman, 1972). A number of factors influence emulsifier selection, including: ingredient formulation, flavor, type of homogenizing and heating equipment, product preparation technique, finished product form (e.g., liquid, powder, and plastic solid), storage requirements, costs, and legal aspects.

There now are seven, legally sanctioned emulsifiers that are widely used in yeast-raised bakery products. Each is fat-derived, and they include (Landfried, 1977): (i) mono- and diglycerides (40–50% content), (ii) distilled monoglycerides (90%), (iii) succinylated monoglycerides, (iv) ethoxylated mono- and diglycerides, (v) Polysorbate 60, (vi) calcium stearoyl-2-lactylate, and (vii) sodium stearoyl-2-lactylate. Several of the above, as well as others, are used in cakes and other chemically leavened baked goods. The mono- and diglyceride-type emulsifiers will vary in hardness with the degree to which the fatty acid portion has been hydrogenated.

21–5.5 Nutritional Considerations

The disappearance of visible animal and vegetable oils and fats in the USA was estimated at 5.8 Tg for 1984. Based on 1980 USDA statistical data, Rizek et al. (1983) have estimated average total fat consumption at 13.5 Tg or 169 g (0.37 lb.) per person per day for 1980. However, actual consumption is considered to be about 23% (40 g) less due to waste and/or losses. As shown in Table 21–8, of the 169 g of total edible fat consumed, 72 g (42.6%), was visible fat. About 13.3% of the visible fat is estimated to be supplied by unhydrogenated soybean oil (SBO) and 39.6% by hydrogenated soybean (HSBO) plus small amounts of other hydrogenated vegetable oils such as corn (*Zea mays* L.), peanut, (*Arachis hypogaea* L.), cotton (*Arachis hypogaea* L.), and sunflower (*Helianthus annuus* L.) (Table 21–8).

Since hydrogenated soybean oil alone contributes more than 23% of the total fat intake and about 10% of the total calorie intake, questions have been raised concerning its nutritional value. However, these concerns do not involve the actual hydrogenation reaction that converts alkene bonds in unsaturated fatty acids to saturated or alkane bonds. The questions are instead based on the isomerization reaction, which is a secondary or side reaction that occurs during hydrogenation and is the reaction which produces the new fatty acid structures or isomers that are not present in unhydrogenated SBO. As a result, a variety of mono- and polyunsaturated isomers are formed which consists of both *cis* and *trans* positional isomers. Metabolic and other studies have shown that, with minor differences, monounsaturated *trans* fatty acids are metabolized and utilized similar to the corresponding *cis* acids. Numerous studies in which high levels of *trans* fatty acids were fed to animals for long periods of time with adequate dietary essential fatty acids, have shown no adverse effects. These studies as well as others with humans have been reviewed recently by Emken (1985). Recent studies (Ohlrogge et al., 1982) have indicated that monounsaturated positional and geometric isomers in partially hydrogenated soybean oils are metabolized and do not accumulate abnormally in lipids extracted from human tissues.

21–5.6 Lecithin

An important by-product of soybean oil processing is lecithin which is obtained during the degumming of the oil. The sludge which is obtained from the degumming centrifuges contains about 40 to 50% water. The sludge is dried to a moisture content of about 1% in continuous, agitated-

Table 21–8. Dietary sources of energy for the U.S. population (Emken, 1985.)

Energy Source	Total calories	Daily per capita fat consumption					
		Visible		Nonvisible		Total	
	%	g	%	g	%	g	%
Butter and animal	24.5	21.2	29.4	77.4	79.8	98.6	58.3
Soybean oil	3.3	9.6	13.3	3.7	3.8	13.3	7.9
Hydrogenated oil	9.8	28.5	39.6	11.0	11.8	39.5	23.4
Vegetable oils (other)†	4.4	12.7	17.7	4.9	5.1	17.6	10.4
Total fat	42.0	72.0	42.6	97.0	57.4	169	100
Total carbohydrate	46.0						
Total protein	12.0						

†Includes corn, peanut, cottonseed, and sunflower oils.

film evaporators, operating on either a vertical or horizontal axis. The horizontal axis type is preferred since the tendency for the lecithin film to break is reduced (Van Nieuwenhuyzen, 1976). The commercial soybean lecithin produced contains about 35% soybean oil and 65% phosphatides, carbohydrates, and moisture (Brekke, 1980). While trading rules specify only six grades of lecithin (NSPA, 1984–1985), a more definitive classification of lecithin types has been proposed as outlined in Table 21–9 (Flider, 1985). For a detailed discussion of the manufacture of these types the reader is referred to the recent monograph *Lecithins* (Szuhaj and List, 1985). Lecithins are used in many food applications in which advantage is taken of their surface-active effects. Lecithin products have both a lipophilic and a hydrophilic group in the same molecule and therefore act at the boundary between immiscible materials (Szuhaj, 1980). The major functions of lecithin in food uses are in collodial dispersion, wetting, lubrication and release, crystallization control and starch complexing. These same functions find application in nonfood uses such as coatings manufacture and finishing, glass and ceramics processing, and metal processing.

21–6 NONFOOD USES OF SOYBEAN OIL

Because of its low cost and availability, soybean oil is the most important vegetable oil used for industrial products. In 1983, 5.8% of the 4.5 million t of soybean oil used in the USA was applied to industrial purposes (USDA, 1984c). The overall consumption of fats and oils in inedible products has declined in recent years as shown in Table 21–10.

Table 21-9. Lecithin types (Flider, 1985).

- I. Natural
  - A. Plastic
    1. Unbleached
    2. Single-bleached
    3. Double-bleached
  - B. Fluid
    1. Unbleached
    2. Single-bleached
    3. Double-bleached
- II. Refined
  - A. Custom-blended natural
  - B. Oil-free phosphatides
    1. As is
    2. Custom-blended
  - C. Fractionated oil-free phosphatides
    1. Alcohol-soluble
       - a. As is
       - b. Custom-blended
    2. Alcohol-insoluble
       - a. As is
       - b. Custom-blended
- III. Modified chemically

Table 21-10. Nonfood (industrial) consumption of fats and oils

Inedible products	1978†	1983‡
	10^3 t	
Soap	396	368
Paint and varnish	114	66
Fatty acids	1 035	844
Feed	630	670
Resins and plastics	62	82
Lubricants	75	42
Other	333	277
Total	2 645	2 349

†Bureau of the Census (1979).
‡USDA (1984c).

Use of soybean oil for inedible products, however, has grown by 13% since 1976 to a total of 260 000 t in 1983. The primary markets for soybean oil are in alkyd paints and in an epoxidized derivative as a plasticizer/stabilizer for vinyl plastics (Pryde, 1983). Modified soy fatty acids and amines have found application as surfactants and antierosion agents; brominated soybean oil is added to fruit-based soft drinks to improve cloud stability and reduce ring deposits. Soapstock, a soybean oil processing by-product, is used as a feed additive and as a chemical feedstock in the fatty acid industry. New products from soybean and other vegetable oils and new applications have been the object of research in the areas described below.

21-6.1 Coatings and Plastics

High-solids and water-dispersible coatings have been developed. Derivatization of soybean oil by treatment with ozone yielded high solid resins suitable for baked coatings on metal surfaces (Thomas and Gast, 1979). Water-dispersible resins, containing 70 to 80% vegetable oil-derived material and 20 to 30% of petrochemical-derived materials gave films which dried rapidly at room temperatures to form flexible, adherent coatings (Schneider and Gast, 1978, 1979).

Engineering thermoplastics are exemplified by those nylons that are cast and formed into gears, gear housing, and the like. One experimental type of nylon can be made from soybean oil which absorbs less moisture than nylon used for clothing and, consequently, has better dimensional stability and dielectric properties in moist environments (Perkins et al., 1975).

21-6.2 Lubricants

Building blocks used in the petrochemical industry, such as synthesis gas (a mixture of carbon monoxide and hydrogen), can be used with vegetable oils to make lubricants. One such derivative made from soybean oil could be used at extraordinarily low temperatures (to almost

−70°C or −90°F) (Dufek et al., 1974). Other derivatives of soybean oil were prepared that could serve as substitutes for sperm whale oil and as lubricants for continuous casting of steel (Bell et al., 1977). Investigations on the sulfurization of soybean oil have given some important leads to the development of high-pressure lubricant additives for use in automative automatic transmission fluids (Schwab et al., 1978).

21–6.3 Diesel Fuels

Vegetable oils for diesel fuel have a number of advantages. They are liquid fuels from renewable resources and have a favorable energy input/output ratio, unless produced on irrigated land. They would permit crop production even in a petroleum shut-off and have potential for making marginal lands productive. They consume less energy than does alcohol production and have higher energy content than alcohol. They have cleaner emissions and simpler technology than alcohol production. One disadvantage is that vegetable oils for diesel fuels as yet are not economically feasible. Further research and development are needed.

In comparing the properties of soybean oil with diesel oil as shown below, it can be noted that the cetane number and the heat content are similar.

Property	No. 2 diesel oil	Soybean oil
Density, kg L^{-1}	0.852	0.925
Cetane number	49	37
Heat content, 10^6 J kg^{-1}	42.4	39.1
Viscosity, CTS at 100°F	1.9–4.1	36
Volatility		
°C	220–355	nonvolatile
°F	430–675	nonvolatile
Flash point		
°C	> 52	300
°F	> 125	570

However, soybean oil is far more viscous than diesel oil and has low volatility. Research has shown that vegetable oils can be used successfully in a naturally aspirated, air-cooled, *indirect*-injection diesel engine; they cannot be used neat in *direct*-injection engines. Since the great majority of farm tractors in the USA have direct-injection engines for greater fuel efficiency, some kind of modification to vegetable oils appears to be necessary before they can be used in this type of equipment.

Scientists at Purdue University have tested blends of reclaimed cooking oil with diesel oil in campus buses with direct injection engines (Engelman et al., 1978). The optimum blend was 20% in diesel oil which gave the lowest smoke emission and fuel consumption at full load. The incorporation of aqueous ethanol into vegetable oil to form a microemulsion serves not only to extend diesel fuel supplies but possibly also to improve combustion properties. Vegetable oils, when injected into the combustion chamber of a diesel engine cylinder, do not form the atomized spray typical of no. 2 diesel oil. As a consequence, combustion is incomplete, and injector coking, ring sticking, and lubricant contamination are major problems. It is hoped that the microemulsions will form a better spray pattern and give superior engine performance because of the lower viscosities of the microemulsions compared to the original oil. Vegetable oil esters, formed by the transesterfication reaction with simple alcohols, are also being evaluated for fuel use. For a more complete update on this area, see the proceedings of a recent conference on this topic (USDA, 1983).

21–6.4 Ag-Chem Uses

Interest in using soybean oil as carriers for pesticides or as spray adjuvants has resulted in extensive research to develop formulations, to define application and dispersal equipment, and to evaluate the efficiency of the new formulations in field trials. For a complete review of current research in this area see the Proceedings of the Ag-Chem Uses of Soybean Oil Workshop (American Soybean Association, 1984). Pesticides must be applied as liquid sprays in most cases. The most extensively used carrier to dissolve or disperse the solid or viscous organic pesticides has been a special phytobland grade of mineral oil. Soybean oil is a less expensive alternative for this purpose. Research has been directed to confirm that soybean oil fulfills the requirements for use as a solvent or carrier, i.e., the agricultural chemical must be soluble or dispersable, and it must be phytobland or nonphytotoxic. With normal application rates the oil/pesticide solution is diluted with water with an added surfactant or emulsifier. Through use of ultra-low volume (ULV) application systems, oil-soluble pesticides can be applied directly without further dilution with water. As a spray adjuvant, soybean oil is a competitive alternative to mineral oil for addition to water-based solutions, emulsions or suspensions of herbicides to alter the spray characteristics. Adjuvants provide adhesive effects for dry residues remaining from particulate sprays or reduce the surface tension of the spray droplets, which causes them to readily spread over the surface of the plant. Tests have demonstrated that soybean oil is a viable alternative to mineral oil in these applications in that most pesticides are sufficiently soluble or dispersible for use; soybean oil is less volatile and improves droplet stability important to control of wind drift in ULV applications; and soybean oil is not phytotoxic to plants.

Another significant potential market for soybean oil is its use as a dust suppressant in grain handling. Recent research (Lai et al., 1981) has demonstrated that dust concentration during handling is reduced by more than 90% by application of soybean oil to wheat and corn at a rate of 0.03%. The dust suppressant effect was retained throughout the 1-yr period of the test. Considerable testing is required before this technique of dust suppressant is adopted including determination of threshold limits of odor detection of soybean oil and combinations of oil and lecithin, final determination of optimum treatment levels, assessing the quality of end products from treated grain, and effect of oil treatment on handling machinery.

21–7 DEFATTED SOYBEAN PROTEIN PROCESSING

21–7.1 Toasting for Feed Use

The bulk of defatted meal produced in the USA is used in animal feeds. Meal intended for feeds is prepared by processing hexane-laden flakes coming from the extractor in conventional D-T units (Fig. 21–1). Such units consist of several compartmentalized stages in which the flakes are agitated and moved downward through each stage. In the first stages, the hexane is stripped out by injecting live steam which condenses and raises the moisture content to about 20%. In later stages, the flakes are toasted and partially dried. After leaving the D-T unit, the flakes go to a drier where moisture content is reduced to 10 to 12%. Flakes from the drier, after cooling, are ground in a hammer or other type mill, screened to the desired grit size and stored as finished meal. An alternative to the conventional DT is the desolventizer-toaster dryer cooler (D-TDC) where the entire operation is carried out in one piece of equipment (Lebrun et al., 1985). The D-TDC consists of four stages: (i) predesolventizing; (ii) desolventizing-toasting; (iii) drying; and (iv) cooling.

Toasting is necessary to obtain optimal growth when soy meal is fed to animals. Enhancement of growth by toasting is caused by inactivation of a number of antinutritional factors including trypsin inhibitors and hemagglutinins plus denaturation of the storage proteins to make them more digestible (Liener, 1981). The antinutritional factors are discussed later (section 21–8).

21–7.2 Conversion into Edible Products

In the USA, edible soybean proteins fall into three major classes: (i) flours and grits; (ii) concentrates; and (iii) isolates. All three types are made from defatted soybean flakes which are prepared by hexane extraction essentially as described earlier (section 21–3) except that greater attention is paid to sanitation, and desolventization is performed in a flash desolventizer or vapor desolventizer-deodorizer instead of a D-T unit (Becker, 1978; Mustakas et al., 1980). Both desolventizing processes use superheated hexane to flash evaporate the hexane from the flakes with a minimum of protein denaturation which is desirable if the flakes are intended for preparation of isolates. If toasted or partially toasted, defatted flakes are required, the toasting operation is carried out in a flake stripper-cooker or deodorizer (Milligan and Suriano, 1974; Becker, 1978).

This operation also removes residual hexane and volatile flavor components.

21-7.2.1 Flours and Grits

These products containing a minimum of 50% protein are made by grinding and sifting defatted flakes. Grits are coarse-ground flakes and are graded according to particle size. Flours are ground finer than grits and standards for flours require that 97% of the product passes through a No. 100 mesh screen. Flours with smaller particle sizes are also available.

21-7.2.2 Protein Concentrates

Concentrates are defined as products containing a minimum of 70% protein on a dry basis. They are manufactured from defatted flakes or flours by extracting with aqueous ethanol (Mustakas et al., 1962) or with a dilute acid at pH 4.5 (Sair, 1959) to remove soluble sugars (sucrose, raffinose, and stachyose) and minor constituents. The acid-leached concentrate is neutralized before spray drying to make the proteins more soluble in food systems. The alcohol process is most widely used. Yields of concentrates are 60 to 70% of the weight of starting flakes or flour. Removal of raffinose and stachyose is preferred for some food applications because these oligosaccharides are not hydrolyzed and absorbed but rather pass intact into the lower digestive tract where they are fermented to produce flatus (Rackis, 1981).

21-7.2.3 Isolates

These are the most highly refined class of soybean proteins available and contain a minimum of 90% protein on a moisture-free basis. Defatted flakes prepared with a minimum of heat treatment are extracted with dilute alkali ($pH < 9$) and centrifuged. The extract is then adjusted to pH 4.5 to precipitate the proteins. The resulting protein curd is recovered by centrifuging, washed, and then usually neutralized with food grade alkali before spray drying to yield the proteinate form of isolate. Unneutralized (isoelectric type) isolates are also available but are less frequently used because of their insolubility. Yields of isolates are about 30% of the weight of the starting flakes or about 60% of the protein in the flakes. Because of the lower yields and the cost of processing, isolates are more expensive than flours or concentrates.

21-7.2.4 Textured Protein Products

Flours, concentrates, and isolates are also processed to give them meat-like textures. Flours and concentrates are textured by thermoplastic extrusion (Harper, 1978). The flours or concentrates are mixed with water and additives to form a dough, which is then fed into an extruder where the dough is subjected to high temperatures, pressure, and mechanical stresses. The dough is squeezed through a die that puffs it and gives it a fibrous texture similar to certain meat products. Bits, ribbons, or sheets of products can be obtained by using an appropriate die in the extruder.

Textured isolates can be made by a spinning process analogous to spinning textile fibers. Isolate is dissolved in sodium hydroxide (pH 10–11) to form a spinning dope which is then pumped through a spinnerette into a coagulating bath containing acid and salts. The protein coagulates to form continuous filaments which are gathered into bundles or "tows". The tows are then washed, stretched, blended with fats, colors, flavors, and other additives for fabrication into a variety of sizes and shapes resembling meat products, such as chicken, beef, bacon, and seafood (Thulin and Kuramoto, 1967). Several companies in the USA and Europe have used the spinning process in the past, but only two American companies are spinning fibers at present. The process is complex and expensive and the meat analogs have not been widely accepted by consumers.

An alternative to spinning is the use of jet cooking to form fibers (Hoer, 1972). A slurry of isolate is pumped through a heat exchanger under high pressure at 116 to 157°C and expelled through a slot-like or small circular nozzle. The isolate is coagulated by the high temperature to form fibers which are dropped through ambient air for cooling, collected in a vessel, and finally centrifuged to remove excess water. The fibrous isolate is sold frozen.

21–7.3 Product Range and Composition

The protein products derived from soybean are available with a range of characteristics depending upon intended use.

21–7.3.1 Feed Products

Defatted soybean meal used for feeds is available in two forms—with hulls and dehulled—that differ primarily in protein and crude fiber contents. Table 21–11 shows compositions of the two types of meal plus edible grade protein products; typical data for soybean are included for comparative purposes. Meal containing 44% protein is processed by omit-

Table 21–11. Composition of soybean, defatted soybean meals, flours, concentrates, and isolates.

Product	Moisture	Protein	Fat	Fiber	Ash
			%		
Soybean†	11.0	37.9	17.8	4.7	4.5
Defatted meal, with hulls‡	10.4	44.0	0.5	7.0	6.0
Defated meal, dehulled‡	10.7	47.5	0.5	3.5	6.0
Full-fat soy flour†	5.0	44.3	21.0	2.0	4.9
Defatted grits and soy flour†	7.0	54.9	0.8	2.4	6.0
Lecithinated soy flour†	5.5	49.9	15.5	2.1	5.0
Protein concentrate†	7.5	66.6	—	3.5	5.5
Protein isolate†	5.0	93.1	—	0.2	4.0

†Smith and Circle (1972). ‡Allen (1984).

ting the dehulling step or by dehulling, grinding the hulls, and then adding them back to the meal after the oil has been extracted. The presence of the hulls decreases the protein content and increases the fiber content. Dehulled meal contains 47.5% minimum protein and only 3.5% fiber.

21-7.3.2 Edible Products

Defatted flours and grits are identical in composition (Table 21-11) but differ in particle size. Grits are available in three particle sizes as measured with U.S. Standard Screens: coarse (No. 10–20 screen); medium (No. 20–40 screen); and fine (No. 40–80 screen). Flours are likewise available in various particle sizes; most flours are 100 mesh, but 200 mesh flours are also manufactured. In addition to differences in particle sizes, grits and flours are also available with varying heat treatments ranging from raw (minimum of moist heat treatment) to fully cooked or toasted. Cooking is used to remove the grassy-beany flavor of raw soybean flakes, to alter functional properties such as protein solubility and water absorption and to improve nutritional value.

Concentrates are available in two basic forms depending on the method of preparation as outlined earlier. Concentrates prepared by alcohol extraction have a lower content of water-soluble protein than those obtained by the dilute acid leach process because alcohol causes extensive denaturation and insolubilization of the proteins. The insoluble cell wall polysaccharides make up most of the nonprotein constituents not accounted for in Table 21-1.

Isolates are available in two basic types as described earlier: (i) isoelectric isolates and (ii) proteinates. Two U.S. manufacturers also provide proteinates with a range of functional properties including gelation, emulsification, water binding, and combatability with acidic foods. Enzymatic hydrolysates of isolates are prepared commercially and are used primarily as whipping agents in confectionary items (Gunther, 1979).

As discussed earlier, flours, concentrates, and isolates are also available in textured forms that have compositions essentially the same as those of the untextured forms (Table 21-11).

21-7.4 By-products

Hulls are a major by-product obtained in the preparation of dehulled, defatted soybean flakes used for 47.5% protein meal and for edible protein products. Hulls constitute 7 to 8% of the soybean and are of low nutritional value because of their high fiber content. They are used mainly in ruminant feeds and find some specialty applications such as vitamin carriers in feeds.

Processing of defatted flakes into concentrates yields by-products consisting of the materials extractable with aqueous ethanol or dilute acid. The major extractables are the sugars, sucrose, raffinose, and stachyose. Some proteins are also extracted, especially by the dilute acid extraction process. The solubles are either discarded or, in the case of the alcohol process, concentrated in the recovery of the alcohol. The recovered soluble sugars are concentrated into a molasses-like syrup (referred to as condensed soybean solubles when concentrated to 60% solids) or dried. The solubles are used as an additive in feeds; feeding studies with rats (*Rattus* spp.) and pigs (*Sus scrofra domesticus*) have been reported (Cline et al., 1976).

Manufacture of isolates yields two by-products: (i) the spent flakes remaining after extraction of the proteins and (ii) the pH 4.5 solubles (also referred to as whey) remaining after precipitation of the proteins at pH 4.5. The spent flakes are disposed of by incorporation into feeds. The whey is discarded or, as practiced by a Brazilian company, sprayed onto soybean flakes before they are desolventized in a D-T unit, thereby incorporating them into soybean meal used for feed.

21–8 UTILIZATION OF DEFATTED SOYBEAN PROTEIN PRODUCTS

As pointed out in the "Introduction," defatted soybean meal is used primarily for feeds. Only small amounts of defatted soybeans are processed further into edible grade proteins (Table 21–5) that are utilized as ingredients in a variety of food products.

21–8.1 Feed Uses

Because of its high nutritional value, wide availability, low cost and consistency of composition, soybean meal has displaced most of the animal proteins (meat meal, tankage, and fish meal) and has become a vital component of poultry and swine feeds in the USA and Canada. Extent of useage of soybean meal is affected by its price relative to that of other oilseed meals. Uniformity in composition is supported by analyses of 21 413 dehulled soybean meal samples in 1976 to 1982. The yearly means of protein content (on an as-is basis) varied only from 48.50 to 49.31% (Jones, 1984).

For 1984, it was estimated that the amounts of oilseed meals fed in processed feeds in the USA would be as follows (USDA, 1984b):

Meal	Tg	Percent
Soybean	16.2	90
Cottonseed	1.4	8
Sunflower	0.4	2
Linseed	0.1	—
Total	18.1	100

Soybean meal is clearly the dominant oilseed meal available as a high protein ingredient for animal feeding in the USA. In contrast, the amount of animal proteins (tankage and meat meals, fish meals, and milk products) available for feeding was only 1.8 Tg or 10% of the quantity of oilseed meals.

Meals available for livestock and poultry feeding and their specifications as set by the National Soybean Processors' Association are compared with the approximate composition of soybean and their seed parts in Table 21–12. The 44% protein meal contains the hulls and is therefore high in fiber; its use is primarily for older pig rations and for cattle feeding. Dehulled meal (47.5–49.0% protein) contains only 3.3 to 3.5% fiber and is mainly utilized in poultry feeds, especially in broiler rations. Mill feed, containing 13% protein, consists of hulls removed by air aspiration of undehulled soybean flakes after hexane extraction; it contains some soybean meal and flour. Soybean mill run contains only 11% protein and consists of soybean hulls plus such meal as adheres to them when they are separated from full-fat flakes prior to the hexane extraction process. Because of their high fiber contents, mill feed and mill run are fed mainly to ruminants.

21-8.1.1 Poultry

It is estimated that about 25% of the poultry diets fed in the USA consists of defatted soybean meal (McNaughton, 1981). Composition of a practical chicken (*Gallus gallus domesticus*) starter ration containing 21% protein is shown in Table 21–13. This ration contains 29% of dehulled, extracted soybean meal which supplies 68% of the total protein. Dehulled soybean meal is preferred for poultry rations because it is lower in crude fiber and higher in metabolizable energy than undehulled meal and the other oilseed meals commonly used in feeds. Cottonseed and peanut meals generally are used in poultry rations only if their price per unit of protein is significantly lower than soybean meal. Methionine is the first limiting amino acid of soybean meal when it is fed to poultry,

Table 21–12. Approximate composition of soybean, seed parts, and meal products.†

Component	Protein	Fat	Carbohydrate	Ash
	%			
Whole beans	40	21	34	4.9
Cotyledon	43	23	29	5.0
Hull	8	1	86	4.3
Hypocotyl	41	11	43	4.4
Meal products	Min.	Min.	Max. (fiber)	
Flakes and meal (undehulled, extracted)	44	0.5	7.0	
Flakes and meal (dehulled, extracted)	47.5–49.0	0.5	3.3–3.5	
Mill feed‡	13		32	
Mill run‡	11		35	

†Data for soybean and soybean parts from Kawamura (1967) on a dry basis. Data for meal products from NSPA (1984–1985) on a 12% moisture basis.
‡Typical analysis on a 13% moisture basis.

Table 21–13. Composition of practical chicken starter rations. Data from Powell and Gehle (1976).

Ingredient	Content
	%
Yellow corn	57.25
Soybean meal (49% protein)	29.00
Fish solubles	0.65
Wheat middlings	2.50
Delactosed whey	1.50
Coastal bermudagrass, dehydrated	5.00
Minerals	3.50
Vitamins	0.25
Animal fat	0.25
DL-Methionine	0.10
Choline chloride	0.10

hence synthetic methionine is often added to starter rations. The need for protein in chicken rations decreases as the birds mature, but soybean meal is still the primary protein source used in rations for finishing broilers and for laying hens.

Soybean meal proteins are highly digestible by chickens. A recent collaborative study (Engster et al., 1985) performed with roosters indicated a protein digestibility of 92.0% for soybean meal which was comparable to corn (94.0%) but higher than for a meat meal (88.0%) or wheat middlings (89.0%).

Turkey (*Meleagris gallopavo*) rations are also a substantial outlet for soybean meal; about 18% of the 7.2 Tg of soybean meal estimated to be used in 1984 for poultry was expected to be fed to turkeys (USDA, 1984b). Starter rations for turkeys require more protein than chick starters, consequently higher soybean meal levels are used in turkey starters than in chick starters. Diets recommended for turkey poults 0 to 4 weeks old call for protein contents of 28% (1200 Kcal/lb of diet) to 32% (1400 Kcal/lb of diet); for finishing rations the protein content is reduced to 13.0 to 15.5% of the diet (Scott, 1984).

21–8.1.2 Swine

Because of its excellent amino acid profile, dependable supply and competitive price, soybean meal is the major source of supplemental protein used in swine diets. For 1984, predicted useage of soybean meal for swine feeding was 4.4 Tg (USDA, 1984b).

Like poultry, swine require high levels of protein in their diets and the major amino acids of concern for swine feeding are lysine, methionine plus cystine, and tryptophan. Protein levels recommended for various production stages are as follows (Jensen, 1984):

Production Stage	Percentage protein in diet
Starter, 10–30 lb	20
Grower, 30–120 lb	16

(continued on next page)

Continued.

Production Stage	Percentage protein in diet
Finisher, 120 lb to market wt	14
Breeder	
Gestation	12
Lactation	14

Typically, corn-soybean meal blends are used to provide these protein contents. Beneficial effects of moist heating on soybean meal used for feeding swine have long been recognized. Commercially processed meals give N and amino acid digestibilities (Rudolph et al., 1983) that are comparable to optimal responses obtained with experimentally heated meals (Vandergrift et al., 1983).

Although not ruminants, growing swine apparently can tolerate wide ranges of crude fiber in the diet. Addition of 6% soybean hulls to a corn-dehulled soybean meal diet gave an increase in daily gain without an increase in feed per gain (Kornegay, 1978). Typical swine diets include 23% (preweaner diet) to 7% (finisher) of undehulled soybean meal (Sievert, 1972).

21-8.1.3 Cattle

Because of rising prices of milk proteins since 1975, there has been an increased interest in the use of soybean proteins in calf (*Bos* spp.) milk replacers. There is ample literature indicating that use of defatted soy flour in milk replacers results in poor performance of calves (see review by Barr, 1981). The causes for poor performance with defatted soy flour are not clearly established but factors such as poor digestibility of nutrients, presence of trypsin inhibitors, and antigenicity have been proposed as responsible. A variety of chemical treatments of soy flour have been evaluated to improve soy flour for use in calf starters. Alcohol extraction or merely contacting with alcohol is reported beneficial and one U.S. company sells a calf milk replacer that contains alcohol-treated soy flour as a partial replacer for milk protein. Barr (1981) estimated that about 70% of the dairy herd replacement calves are fed milk replacers and that 60 to 65% of the replacers contain some soy proteins. About three to five million calves were believed to consume milk replacers containing chemically processed soy flour or soybean protein concentrate. Most of these products have 50% or less of the milk proteins replaced by soy proteins.

Soybean meal is, however, an excellent protein source for more mature cattle. Total useage of soybean meal in 1984 was estimated to be 3.6 Tg with about an equal distribution between dairy and beef cattle (USDA, 1984b). For ruminant feeding, soybean meal must compete with urea, by-product meals and high-fiber oilseed meals (cottonseed, peanut, and sunflower meals).

A variety of treatments including heat and addition of formaldehyde and tannins have been applied to soybean meal and other protein sources to make the proteins more resistant to degradation in the rumen but available for digestion in the small intestine (Broderick, 1975). For example, 0.3% formaldehyde substantially reduced ruminal degradation of soybean meal proteins but had little effect on apparent digestibility of crude protein in steers. Higher formaldehyde levels provided little additional protection against ruminal protein degradation but lowered digestibility of the protein, presumably because of excessive cross-linking of the proteins (Spears et al., 1985). If commercialized, the production of protected proteins could provide additional outlets for soybean meal because it has a good balance of essential amino acids.

21–8.2 Food Uses

Edible soybean proteins are used primarily as ingredients in a variety of processed foods (Table 21–14). Defatted flours are the least expensive form of soybean proteins and level of useage varies from 1 to 2% (as functional ingredient such as moisture retainer) to nearly 100% (meat analogs where oil, salt, flavors, and colors are only other ingredients). Protein concentrates sell for three to four times as much as flours but have the advantage of being low in raffinose and stachyose which are largely removed during processing. These oligosaccharides are believed to cause flatulence. Isolates are the most expensive and inefficient form of soybean proteins to use (only 60 to 70% of the protein in defatted soybean flakes is recovered in the production of isolates). An important

Table 21–14. Food uses of soybean proteins.

Protein form	Uses
Defatted flours and grits	Baked goods (breads, crackers, and sweet goods)
	Ground meat extenders
	Meat analogs
	Nonfat dry milk replacers
	Breakfast cereals
	Infant foods
	Diet foods
	Soup mixes
	Confections
Concentrates	Processed meats
	Frozen meat dinners
	Breakfast foods
	Infant foods
Isolates	Whole milk replacers
	Coffee whiteners
	Cake mixes
	Beverage products
	Confections
	Processed meats
	Meat analogs
	Infant formulas

use is in formulas for infants that are allergic to cow's milk; in this application they may be the sole source of protein during the 1st months of life.

Edible grades of soybean protein, especially defatted flour, are generally less expensive than animal proteins but they are less widely used than those of animal origin because of limitations in nutritional and functional properties. Current status of these problems is discussed here.

21-8.2.1 Nutrient-Antinutrient Properties

It has been known since 1917 that soybean meal needs to be cooked in order to support normal growth of rats (Osborne and Mendel, 1917). Numerous studies have been conducted to identify the factor(s) responsible for the poor growth-promoting activity of raw soybean or meal (Liener, 1981). Several heat-liable, as well as heat-stable, antinutrients are now known to occur in soybean.

The most extensively studied heat-labile antinutritional factors in raw soybean meal are the trypsin inhibitors. They cause hypertrophy of the pancreas and increase secretion of enzymes by this organ. The enzymes secreted by the hyperactive pancreas are rich in methionine and cystine, but the enzymes are excreted in the feces, hence there is loss of these S-containing amino acids to the animal. Because S amino acids are first limiting in soybean proteins, their loss in the feces has been postulated to be responsible for the poor growth of rats when they are fed raw soybean meal (Lyman, 1957). The trypsin inhibitors account for about 40% of the growth inhibition of raw soybean plants. Poor digestibility of the undenatured proteins is believed to account for much of the remaining growth inhibition of raw soybean (Kakade et al., 1973).

Recent studies showed that continuous ingestion of soy flour with varying levels of trypsin inhibitor activity results in the formation of pancreatic lesions (McGuinness et al., 1980; Liener et al., 1985). At 6 months, hypertrophy (enlargement caused by increased cell size) and hyperplasia (enlargement caused by increased number of cells) were noted and gross inspection of the pancreas revealed nodules after 15 months of exposure to the trypsin inhibitors (Liener et al., 1985). Detailed histological examination of the pancreas indicated nodular hyperplasia at 6 months of feeding. There was a positive relationship between incidence of the lesion and both time of exposure and level of dietary trypsin inhibitor. After 18 months of ingesting the inhibitor-containing diets, acinar adenoma was observed; adenoma was most prevalent in rats consuming the highest level of trypsin inhibitor in the soy flour diets (Liener et al., 1985). In a related study, protein isolates as well as soy flours were fed for 2 yrs and histological changes in the pancreas were evaluated. Nodular hyperplasia and acinar adenoma were the major pathological changes noted in the pancreas and the development of the lesions appeared to be related only to the level of trypsin inhibitor irrespective of the source (Gumbmann et al., 1985).

The significance of these effects on the pancreas in rats in relation to practical feed rations and human consumption of soybean foods is unknown. It must be remembered that soybean products are always used in foods which are ultimately baked or cooked before eating. These final preparation steps probably significantly reduce the antinutritional factors. Trypsin inhibitors, also, are not unique to soybeans; they occur in many foods that are staples in Western diets. The British diet is estimated to contribute about 330 mg of trypsin inhibitor activity per person per day. Eggs (28%), milk (17%), and potatoes (13%) (*Solanum tuberosum* L.) contribute almost 60% of the total trypsin inhibitor intake (Doell et al., 1981).

Other heat-labile antinutrients in soybean are lectins (hemagglutinins), goitrogens, and antivitamins (Liener, 1981). Lectins are readily inactivated by moist heat and do not appear to have major effects on the nutritional quality of soybean proteins (Turner and Liener, 1975). The effects of the goitrogens in soybean are relatively weak and can be overcome by adding potassium iodide to the diet; they can also be partially eliminated by heat (Block et al., 1961). The antivitamins are likewise inactivated by heat treatment and their effects can be counteracted by supplementation with vitamins and minerals (Liener, 1981).

Heat-stable factors affecting nutritional quality of soybean protein products include phytates, flatulence factors, and allergens (Liener, 1981). Phytates occur in soybean meal at a level of about 1.5% and have long been implicated in interfering with bioavailability of minerals such as Ca, Mg, and Zn (Cheryan, 1980; Erdman, 1979). Recent studies with humans confirmed earlier reports with laboratory animals. For example, addition of sodium phytate to a basal diet containing egg albumin as the protein source decreased dietary Zn absorption in young men from 34 to 18% (Turnlund et al., 1984). Likewise, addition of phytate to a cow's milk-based infant formula (phytate concentration similar to that of soy-based formula) reduced Zn absorption in adult men from 32 to 16% as compared to 14% absorption for a soy isolate-based formula (Lonnerdal et al., 1984). A soy flour-based formula gave a Zn absorption of only 8%. Phytic acid in soybean flakes is hydrolyzed only slowly by autoclaving and hence is stable to moist heat during processing of soybean into oil and meal. Mineral supplementation is, therefore, used to counteract the low absorption caused by phytate.

Flatulence caused by ingestion of soybean products is attributed mainly to raffinose and stachyose and is a problem only with full-fat and defatted flours (Steggerda et al., 1966; Rackis, 1981). Processing of defatted flakes into concentrates and isolates removes the oligosaccharides and eliminates flatulence. Tofu does not cause flatulence and fermentation procedures such as preparation of tempeh also reduce the flatulence activity of soybean (Calloway et al., 1971).

Allergic responses to ingestion of soybean protein by humans are comparatively rare. The medical literature documents occasional cases of allergy to soybean dust in workers in soybean processing plants or food plants using soy flour (Bush and Cohen, 1977), but the most common allergic response is to eating soybean foods such as soy-based infant formulas (Halpern et al., 1973). The use of soy flours in milk replacers has long been known to cause gastrointestinal disturbances in calves. Recent work has attributed the disturbances to allergic responses caused by β-conglycinin and glycinin, the two major storage proteins (Kilshaw and Sissons, 1979). Heat treatment used in the normal preparation of soybean meal apparently is insufficient to eliminate the antigenicity of these proteins, whereas hot alcohol reportedly inactivates them (Sissons et al., 1982).

21-8.2.2 Structure-Function Relationships

Soy proteins in the form of flours, concentrates, and isolates are used as ingredients in the processing of a variety of foods ranging from baked goods to dairy analogs and processed meats. In some foods, they are major contributors of dietary protein (e.g., textured proteins used as meat extenders, meat analogs, and infant formulas), but most applications are for functional purposes where low levels of soy protein are often used.

Proteins in soybean are a complex mixture and commercial soy protein preparations reflect this complexity. Soy flours contain the naturally occurring mixture, whereas some of the minor proteins are removed in the preparation of concentrates and isolates. A major difference between the various soy proteins is molecular size, as revealed by ultracentrifugation (Table 21-15). The 7S and 11S fractions make up about 70% of the total protein. Six isomers of β-conglycinin are the major proteins found in the 7S fraction; they have a quaternary structure made up of three subunits (Thanh and Shibasaki, 1978) and are glycoproteins. The 11S fraction consists of 11S globulins or glycinins which likewise have a quaternary structure but consist of six subunits instead of three. Each glycinin subunit, in turn, contains an acidic (molecular wt. 37 000) and a basic (molecular wt. 20 000) polypeptide chain that are linked together by a disulfide bond. The complete amino acid sequence of one of the subunits is known (Staswick et al., 1984).

The quaternary structures of β-conglycinin and glycinin are sensitive to heat, pH, ionic strength, and organic solvents such as alcohols. Manipulation of these and other parameters can change the various functional properties of soy proteins that are summarized in Table 21-16. For example, depending upon conditions, heating can decrease solubility (Wolf and Tamura, 1969), increase foaming properties (Eldridge et al., 1963a) or induce gelation of soy proteins (Circle et al., 1964). Because of their differences in structure, β-conglycinin and glycinin also have other properties that set them apart from each other. Glycinin has solubility properties that are sensitive to temperature at low ionic strength and can be separated from β-conglycinin and the other proteins by cryoprecipitation (Wolf and Sly, 1967). The two proteins also differ in solubility as a function of pH. Glycinin has a minimum solubility at pH 6.7 whereas

Table 21–15. Approximate distribution of ultracentrifugal fractions of water-extractable soybean proteins (Wolf and Cowan, 1975).

Fraction	Amount of total	Components	Molecular wt	Other characteristics
	%			
2S	22	Trypsin inhibitors	8 000–21 500	Globulins
		Cytochrome C	12 000	Heme protein
7S	37	Hemagglutinins	11 000	Glycoproteins, albumins
		Lipoxygenases	102 000	Contain Fe
		β-Amylase	61 700	—
		β-Conglycinins	180 000–210 000	Glycoproteins, globulins, three subunits
11S	31	Glycinins	350 000	Globulins, six subunits
15S	10		>350 000	Polymer form of other proteins?

Table 21–16. Summary of functional properties of soy proteins important in food applications.†

Property	Functional criteria
Organoleptic/kinesthetic	Color, flavor, odor, texture, mouthfeel, smoothness, grittiness, and turbidity.
Hydration	Solubility, wettability, water absorption, swelling, thickening, and gelling syneresis.
Surface	Emulsification, foaming (aeration, whipping), protein-lipid, film formation, lipid-binding, and flavor-binding.
Structural/rheological	Elasticity, grittiness, cohesiveness, chewiness, viscosity, adhesion, network-crossbinding, aggregation, stickiness, gelation, dough formation, texturizability, fiber formation, and extrudability.
Other	Compatability with additives, enzymatic, and antioxidant.

†These properties vary with pH, temperature, protein concentration, protein fraction, prior treatment, ionic strength, and dielectric constant of the medium. They are also affected by other treatments, interactions with other macromolecules in the medium, by processing treatments and modification, by physical, chemical, or enzymatic methods (Kinsella, 1979).

β-conglycinin has its minimum in solubility at pH 5.0. These differences in solubility are used in conjunction with cryoprecipitation to separate β-conglycinin and glycinin from each other (Thanh and Shibasaki, 1976). Both proteins form gels on heating but the resulting gels differ in properties. Gels made from crude glycinin are higher in tensile and shear strength and have greater water-holding capacity than gels make from crude β-conglycinin (Saio et al., 1974).

Beta-Conglycinin and glycinin are not separated and available to the food industry on a commercial scale. However, the ratio of β-conglycinin to glycinin in different soybean varieties varies so that the ratio of the two proteins in isolates may be variable. Changes in the β-conglycinin: glycinin ratio result in differences in the physical properties of tofu. Tofu, made from soybean high in β-conglycinin content, is softer than tofu from soybean having a high content of glycinin (Saio et al., 1969).

Because of the variety of foods that soy proteins are used in, they must have a range of functional properties (Table 21–16). A given protein does not possess optimal functional properties for all uses. Processing is, therefore, tailored to modify functional properties to meet specific needs. Soy flours are processed with a range of moist heat treatments; for example, a light heat treatment gives a flour that is best suited for use in bread. Isolates, likewise, are modified by proprietary processes and are provided by the industry in different forms to meet specific applications. The chemistry of protein modification is still poorly understood, but a large literature on physical and chemical properties is now available (Smith and Circle, 1972; Kinsella, 1979).

21–8.2.2.1 Flavor and Color—Raw soybean has strong grassy-beany and bitter flavors and processing is necessary to remove or reduce them to acceptable levels. A large number of compounds has been isolated from soybean proteins; major compounds having flavor characteristics of soybeans are tabulated in Table 21–17. Except for the phenolic compounds (formed by decarboxylation of the corresponding cinnamic acids), all of these compounds are derived from lipids through enzymatic or chemical oxidation. These compounds are present at low concentrations (mg kg^{-1}) but interact strongly with the proteins making it difficult to remove them. Moist heat treatment is commonly used to remove or modify the flavor compounds in defatted soy flakes and flours, but bound forms of flavors persist and may be released later when the soy product is incorporated into a food and during mastication when the food is eaten. A combination of alcohol extraction and heat is used in the preparation of concentrates. Extraction of defatted flakes with a mixture of hexane/ethanol followed by toasting gives good results on a laboratory scale (Honig et al., 1976) and has been patented commercially for preparation of protein concentrates (Hayes and Simms, 1973), but may be too expensive to be competitive with regular moist heat treatment. A recent survey of commercial soy protein products indicated that residual flavors are still present (Warner et al., 1983).

Recognition that many of the flavors found in soybean products, particularly soy milk, may be generated enzymatically by lipoxygenase, resulted in development of the hot-grind procedure where lipoxygenase is inactivated during the initial grinding step which is carried out at 80°C or higher (Wilkens et al., 1967). This principle has been adopted by the Japanese soy milk industry although vacuum pan stripping is also included in processing to remove residual volatile flavor compounds (Shurtleff and Aoyagi, 1984).

Color of soybean protein products is less of a problem than flavor. Soy flours are cream to light yellow in color; increased heat treatment causes darkening probably because of browning reactions. Concentrates and isolates are cream colored. Extruded soy flours and concentrates are light to dark tan. The compounds responsible for the yellow color of soybean have not been identified. They are partially extractable with alcohol but concentrates prepared with alcohol still retain some yellow color. Alcohol washing of isolates removes some of the pigments (Eldridge et al., 1963b). Residual pigments appear to be firmly bound. Concentration of tan-brown pigments has been observed in the portion of isolates

Table 21–17. Compounds with flavor characteristics of soybean (Kinsella, 1979; Sessa et al., 1976).

Type	Specific compounds	Flavor characteristics
Alcohols	Isopentanol, hexanol, heptanol, octenol	Grassy, moldy, mushroom, musty
Aldehydes	Hexanal, heptenal, hexenal, decadienal	Grassy, potato chip-like
Ketones	Hexanone, ethyl vinyl ketone	Grassy-beany
Furans	2-Pentyl furan	Grassy-beany
Phenols	4-Vinylguaicol, 4-Vinylphenol	Cooked off-flavor
Phosphatides	Oxidized phosphatidylcholines	Bitter

that is irreversibly insolubilized by precipitation at pH 4.6, the so-called *acid-sensitive fraction* (Anderson and Warner, 1976).

21-9 FULL-FAT SOYBEAN PRODUCTS

Full-fat soy flour was manufactured for many years in the USA but was discontinued several years ago. It is, however, manufactured in Europe and some is imported into the USA. In England, enzyme active full-fat soy flour is added to virtually all bread (Pringle, 1974).

Soy milk has long been a dietary staple in China. With the introduction of modern processing techniques that have resulted in improved flavor of the product, however, soy milk has become a popular beverage in other parts of the Far East, and was imported into the USA in the early 1980s. Several companies were planning plants for the USA and conducting marketing studies in 1984 to 1985. In Japan, where soy milk has not been a traditional part of the diet, soy milk markets grew rapidly in the late 1970s and early 1980s. In 1983 to 1984, soy milk held about 5% of the Japanese market occupied by cow's milk (Haumann, 1984; Shurtleff and Aoyagi, 1984). Japanese consumers are buying soy milk primarily for health reasons—the desire to maintain a high proportion of plant foods in the diet and to avoid sources of cholesterol such as meat, eggs, and cow's milk. Four large food companies make about 80% of the soy milk produced in Japan.

Tofu became a popular item in the USA in the late 1970s. It is produced in small shops and several small factories and is available in many supermarkets. Long a staple in China and Japan, it became westernized and a wide range of recipes for its use has been developed (Shurtleff and Aoyagi, 1975).

Two fermented soybean foods, miso and tempeh, have also been introduced in the USA since the late 1970s. Miso, a paste-like product resembling peanut butter in consistency, is made by fermenting cooked soybean with salt with or without rice (*Oryza sativa* L.) or barley (*Hordeum vulgare* L.). In Japan, miso serves as a soup base but other uses for Western cuisine have been developed (Shurtleff and Aoyagi, 1976). Tempeh is made by inoculating cooked soybean with the mold, *Rhizopus oligosporus*, and fermenting for about 24 h. The mold mycellium permeates the bean mass and binds it together. When sliced and deep-fat fried, tempeh is crisp and golden brown; it is a traditional food of Indonesia (Shurtleff and Aoyagi, 1979).

REFERENCES

Allen, R.D. 1984. Feedstuffs ingredient analysis table: 1984 edition. Feedstuffs 56(30):25–30.

Allen, R.R. 1978. Principles and catalysts for hydrogenation of fats and oils. J. Am. Oil Chem. Soc. 55:792–795.

American Soybean Association. 1983. Soya bluebook. American Soybean Association, St. Louis.

----. 1984 Proceedings of the Ag-Chem Uses of soybean oil. American Soybean Association, St. Louis.

Anderson, R.L., and K. Warner. 1976. Acid-sensitive soy proteins affect flavor. J. Food Sci. 41:293–296.

Anonymous. 1963. Prepared salad dressings. Good Housekeeping 156(June):151.

----. 1982. Food fats and oils. Institute of Shortening and Edible Oils, Washington, DC.

Baker, E.C., and D.A. Sullivan. 1983. Development of a pilot-plant process for the extraction of soy flakes with aqueous isopropyl alcohol. J. Am. Oil Chem. Soc. 60:1271–1277.

Baldwin, R.R., R.P. Baldrey, and R.G. Johansen. 1972. Fat systems for bakery products. J. Am. Oil Chem. Soc. 49:473–477.

Barr, G. 1981. Soybean meal in calf milk replacers. J. Am. Oil Chem. Soc. 58:313–320.

Beal, R.E., E.B. Lancaster, and O.L. Brekke. 1956. The phosphorus content of refined soybean oil as a criterion of quality. J. Am. Oil Chem. Soc. 33:619–624.

Becker, K.W. 1978. Solvent extraction of soybeans. J. Am. Oil Chem. Soc. 55:754–761.

Bell, E.W., L.E. Gast, F.L. Thomas, and R.E. Koos. 1977. Sperm oil replacements: Synthetic wax esters from selectively hydrogenated soybean and linseed oils. J. Am. Oil Chem. Soc. 54:259–263.

Block, R.J., R.H. Mandl, H.W. Howard, C.D. Bauer, and D.W. Anderson. 1961. The curative action of iodine on soybean goiter and the changes in the distribution of iodoamino acids in the serum and in thyroid gland digests. Arch. Biochem. Biophys. 93:15–24.

Braae, B. 1976. Degumming and refining practices in Europe. J. Am. Oil Chem. Soc. 53:353–357.

Brekke, O.L. 1980. Oil degumming and soybean lecithin. p. 71–88. *In* D.R. Erickson et al. (ed.) Handbook of soy oil processing and utilization. American Soybean Association, St. Louis and American Oil Chemists' Society, Champaign, IL.

Broderick, G.A. 1975. Factors affecting ruminant responses to protected amino acids and proteins. p. 211–259. *In* M. Friedman (ed.) Protein nutritional quality of foods and feeds, Part 2. Marcel Dekker, New York.

Bureau of the Census. 1979. Current Industrial Report M 20K(79)–13.

Bush, R.K., and M. Cohen. 1977. Immediate and late onset asthma from occupational exposure to soybean dust. Clin. Allergy 7:369–373.

Calloway, D.H., C.A. Hickey, and E.L. Murphy. 1971. Reduction of intestinal gas-forming properties of legumes by traditional and experimental food processing methods. J. Food Sci. 36:251–255.

Carpenter, D.L., J. Lehmann, B.S. Mason, and H.T. Slover. 1976. Lipid composition of selected vegetable oils. J. Am. Oil Chem. Soc. 53:713–718.

Carr, R.A. 1978. Refining and degumming systems for edible fats and oils. J. Am. Oil Chem. Soc. 55:765–771.

Cheryan, M. 1980. Phytic acid interactions in food systems. CRC Crit. Rev. Food Sci. Nutr. 13:297–335.

Circle, S.J., E.W. Meyer, and R.W. Whitney. 1964. Rheology of soy protein dispersions. Effect of heat and other factors on gelation. Cereal Chem. 41:157–172.

Cline, T.R., D.L. Jones, and M.P. Plumlee. 1976. Use of condensed soybean solubles in nonruminant diets. J. Anim. Sci. 43:1015–1018.

Code of Federal Regulations. 1977a. Title 9:319.700 (Animal and Plant Health Inspection Service, USDA). U.S. Government Printing Office, Washington, DC.

----. 1977b. Title 21:166 (Food and Drugs). U.S. Government Printing Office, Washington, DC.

Coombes, W.A., R.A. Zavada, J.E. Hansen, W.A. Singleton, and R.R. King. 1974. Continuous hydrogenation of fatty materials. U.S. Patent 3 792 067. Date issued: 12 February.

Cowan, J.C. 1966. Key factors and recent advances in the flavor stability of soybean oil. J. Am. Oil Chem. Soc. 43:300A, 302A, 318A-321A.

Doell, B.H., C.J. Ebden, and C.A. Smith. 1981. Trypsin inhibitor activity of conventional foods which are a part of the British diet and some soya products. Qual. Plant. Plant Foods Hum. Nutr. 31:139–150.

Dufek, E.J., W.E. Parker, and R.E. Koos. 1974. Some esters of mono-, di-, and tricarboxysteric acid as lubricants: Preparation and evaluation. J. Am. Oil Chem. Soc. 51:351–355.

Dutton, H.J. 1981. History of the development of soy oil for edible uses. J. Am. Oil Chem. Soc. 58:234–236.

Eldridge, A.C., P.K. Hall, and W.J. Wolf. 1963a. Stable foams from unhydrolyzed soybean protein. Food Technol. 17:1592–1595.

----, J.P. Friedrich, K. Warner, and W.F. Kwolek. 1986. Preparation and evaluation of supercritical carbon dioxide defatted soybean flakes. J. of Food Sci. 51:584–587.

----, W.J. Wolf, A.M. Nash, and A.K. Smith. 1963b. Alcohol washing of soybean protein. J. Agric. Food Chem. 11:323–328.

Emken, E.A. 1985. Nutritional considerations in soybean oil usage. p. 242–250. *In* Richard Shibles (ed.) World soybean research conference III: Proceedings. Westview Press, Boulder, CO.

Engelman, H.W., D.A. Guenther, and T.W. Silvis. 1978. Vegetable oil as a diesel fuel. p. 1–8. *In* Am. Soc. Mech. Eng., Proc. Energy Technol. Conf. and Exhib., Houston, TX. 5–9 Nov. 1978.

Engster, H.M., N.A. Cave, H. Likuski, J.M. McNab, C.A. Parsons, and F.E. Pfaff. 1985. A collaborative study to evaluate a precision-fed rooster assay for true amino acid availability in feed ingredients. Poult. Sci. 64:487–498.

Erdman, J.W., Jr. 1979. Oilseed phytates: Nutritional implications. J. Am. Oil Chem. Soc. 56:736–741.

Flider, F.J. 1985. The manufacture of soybean lecithins. p. 21–37. *In* B.F. Szuhaj and G.R. List (ed.) Lecithins. American Oil Chemical Society, Champaign, IL.

Florin, G., and H.R. Bartesch. 1983. Processing of oilseeds using fluidbed technology. J. Am. Oil Chem. Soc. 60:193–197.

Frankel, E.N. 1980. Soybean oil flavor stability. p. 229–244. *In* D.R. Erickson et al. (ed.) Handbook of soy oil processing and utilization. American Soybean Association, St. Louis, and American Oil Chemists' Society, Champaign, IL.

----, P.M. Cooney, H.A. Moser, J.C. Cowan, and C.D. Evans. 1959. Effect of antioxidants and metal inactivators in tocopherol-free soybean oil. Fette Seifen Anstrichm. 61:1036–1039.

Freeman, I.P., F.B. Padley, and W.L. Sheppard. 1973. Use of silicones in frying oils. J. Am. Oil Chem. Soc. 50:101–103.

Friedrich, J.P., G.R. List, and A.J. Heakin, 1982. Petroleum-free extraction of oil from soybeans with supercritical CO_2. J. Am. Oil Chem. Soc. 59:288–292.

Gavin, A.M. 1983. Energy conservation in edible oil processing (U.S. view). J. Am. Oil Chem. Soc. 60:420–426.

Gumbmann, M.R., W.L. Spangler, G.M. Dugan, J.J. Rackis, and I.E. Liener. 1985. The USDA trypsin inhibitor study. IV. The chronic effects of soy flour and soy protein isolate on the pancreas in rats after two years. Qual. Plant. Plant Foods Hum. Nutr. 35:275–314.

Gunther, R.C. 1979. Chemistry and characteristics of enzyme-modified whipping proteins. J. Am. Oil Chem. Soc. 56:345–349.

Halpern, S.R., W.A. Sellars, R.B. Johnson, D.W. Anderson, S. Saperstein, and J.S. Relsch. 1973. Development of childhood allergy in infants fed breast, soy, or cow milk. J. Allergy Clin. Immunol. 51:139–151.

Harper, J.M. 1978. Food extrusion. CRC Crit. Rev. Food Sci. Nutr. 11:155–215.

Haumann, B.F. 1984. Soymilk—New processing, packaging expand markets. J. Am. Oil Chem. Soc. 61:1784–1786, 1788, 1790–1793.

Hayes, L.P., and R.P. Simms. 1973. Defatted soybean fractionation by solvent extraction. U.S. Patent 3 734 901. Date issued: 22 May.

Hoer, R.A. 1972. Protein fiber forming. U.S. Patent 3 662 672. Date issued: 16 May.

Honig, D.H., K. Warner, and J.J. Rackis. 1976. Toasting and hexane:ethanol extraction of defatted soy flakes. Flavor of flours, concentrates and isolates. J. Food Sci. 41:642–646.

Hymowitz, T., and J.R. Harlan. 1983. Introduction of soybean to North America by Samuel Bowen in 1765. Econ. Bot. 37:371–379.

Jensen, A.H. 1984. Dietary nutrient allowances for swine. Feedstuffs 56(30): 38, 40–41, 43.

Johnson, L.A., and E.W. Lusas. 1983. Comparison of alternative solvents for oils extraction. J. Am. Oil Chem. Soc. 60:229–242.

Jones, F.T. 1984. A survey of soybean meal used in poultry feeds 1976 to 1982. Poult. Sci. 63:1462–1463.

Kakade, M.L., D.E. Hoffa, and I.E. Liener. 1973. Contribution of trypsin inhibitors to the deleterious effects of unheated soybeans fed to rats. J. Nutr. 103:1772–1778.

Kawamura, S. 1967. Quantitative paper chromatography of sugars of the cotyledon, hull, and hypocotyl of soybeans of selected varieties. Tech. Bull. Fac. Agric. Kagawa Univ. 18:117–131.

Kilshaw, P.J., and J.W. Sissons. 1979. Gastrointestinal allergy to soyabean protein in preruminant calves. Allergenic constituents of soyabean products. Res. Vet. Sci. 27:366–371.

Kinsella, J.E. 1979. Functional properties of soy proteins. J. Am. Oil Chem. Soc. 56:242–258.

Kock, M. 1983. Oilseed pretreatment in connection with physical refining. J. Am. Oil Chem. Soc. 60:198–202.

Kornegay, E.T. 1978. Feeding value and digestibility of soybean hulls for swine. J. Anim. Sci. 47:1272–1280.

Kouzeh Kanani, M., D.J. van Zuilichem, J.P. Roozen, and W. Pilnik. 1984. A modified procedure for low temperature infrared radiation of soybeans: III. Pretreatment of whole beans in relation to oil quality and yield. Lebensm. Wiss. Technol. 17:39–41.

Lai, F.S., Y. Pomeranz, B.S. Miller, C.R. Martin, D.F. Aldis, and C.S. Chang. 1981. Status of research on grain dust. Adv. Cereal Sci. Technol. 4:237–337.

Landfried, B.W. 1977. Surfactants used by bread makers. Cereal Foods World 22:338–340.

Lawhon, J.T., L.J. Manak, K.C. Rhee, K.S. Rhee, and E.W. Lusas. 1981. Combining aqueous extraction and membrane isolation techniques to recover protein and oil from soybeans. J. Food Sci. 46:912–916, 919.

Lebrun, A., M. Knott, and J. Beheray. 1985. Meal desolventizing and finishing. J. Am. Oil Chem. Soc. 62:793–799.

Liener, I.E. 1981. Factors affecting the nutritional quality of soya products. J. Am. Oil Chem. Soc. 58:406–415.

----, Z. Nitsan, C. Srisangnam, J.J. Rackis, and M.R. Gumbmann. 1985. The USDA trypsin inhibitor study. II. Time related biochemical changes in the pancreas of rats. Qual. Plant. Plant Foods Hum. Nutr. 35:243–257.

Link, W.E. (ed). 1975. Official and tentative methods. 3rd ed. American Oil Chemists' Society, Champaign, IL.

List, G.R., C.D. Evans, K. Warner, R.E. Beal, W.F. Kwolek, L.T. Black, and K.J. Moulton. 1977. Quality of oil from damaged soybeans. J. Am. Oil Chem. Soc. 54:8–14.

----, T.L. Mounts, and A.J. Heakin. 1978a. Steam-refined soybean oil: II. Effect of degumming methods on removal of prooxidants and phospholipids. J. Am. Oil Chem. Soc. 55:280–284.

----, ----, K. Warner, and A.J. Heakin. 1978b. Steam-refined soybean oil: I. Effect of refining and degumming methods on oil quality. J. Am. Oil Chem. Soc. 55:277–279.

Lonnerdal, B., A. Cederblad, L. Davidsson, and B. Sandstrom. 1984. The effect of individual components of soy formula and cows' milk formula on zinc bioavailability. Am. J. Clin. Nutr. 40:1064–1070.

Lorenz, K. 1978. Dimethyl polysiloxanes in baking and frying fats and oils. Baker's Dig. 52(2):36–40, 69.

Lyman, R.L. 1957. The effect of raw soybean meal and trypsin inhibitor diets on the intestinal and pancreatic nitrogen in the rat. J. Nutr. 62:285–294.

Massiello, F.J. 1978. Changing trends in consumer margarines. J. Am. Oil Chem. Soc. 55:262–265.

McGuinness, E.S., R.G.H. Morgan, D.A. Levison, D.L. Frape, G. Hopwood, and K.G. Wormsley. 1980. The effects of long-term feeding of soya flour on the rat pancreas. Scand. J. Gastroenterol. 15:497–502.

McNaughton, J.L. 1981. Color, trypsin inhibitor and urease activity as it affects growth of broilers. J. Am. Oil Chem. Soc. 58:321–324.

Milligan, E.D., and J.F. Suriano. 1974. System for production of high and low protein dispersibility index edible extracted soybean flakes. J. Am. Oil Chem. Soc. 51:158–161.

Miyakoshi, K., and M. Komoda. 1978. Effect of deodorizing and heating conditions on the content of citric acid and its decomposed products in edible oils. Yukagaku 27:381–384.

Mogush, J.J. 1980. Soybean processing industry—U.S. and foreign. p. 889–896. *In* F.T. Corbin (ed) World soybean research conference II: Proceedings. Westview Press, Boulder, CO.

Moore, N.H. 1983. Oilseed handling and preparation prior to solvent extraction. J. Am. Oil Chem. Soc. 60:189–192.

Moser, H.A., H.J. Dutton, C.D. Evans, and J.C. Cowan. 1950. Conducting a taste panel for the evaluation of edible oils. Food Technol. 4:105–109.

----, C.D. Evans, J.C. Cowan, and W.F. Kwolek. 1965. A light test to measure stability of edible oils. J. Am. Oil Chem. Soc. 42:30–33.

----, C.M. Jaeger, J.C. Cowan, and H.J. Dutton. 1947. The flavor problem of soybean oil: II. Organoleptic evaluation. J. Am. Oil Chem. Soc. 24:291–296.

Mounts, T.L., K.A. Warner, G.R. List, J.P. Friedrich, and S. Koritala. 1978. Flavor and oxidative stability of hydrogenated and unhydrogenated soybean oils: Effects of antioxidants. J. Am. Oil Chem. Soc. 55:345–349.

Moustafa, A. 1979. Soft margarine in the U.S.A. Report prepared for the American Soybean Association, St. Louis.

Mustakas, G.C. 1980. Recovery of oil from soybeans. p. 49–65. *In* D.R. Erickson et al. (ed.) Handbook of soy oil processing and utilization. American Soybean Association, St. Louis and American Oil Chemists Society, Champaign, IL.

----, L.D. Kirk, and E.L. Griffin, Jr. 1962. Flash desolventizing defatted soybean meals washed with aqueous alcohols to yield a high-protein product. J. Am. Oil Chem. Soc. 39:222–226.

----, E.D. Milligan, J. Taborga A., and D.A. Fellers. 1980. Conversion of soybean extraction plant in Bolivia to production of flours for human consumption: Feasibility study. J. Am. Oil Chem. Soc. 57:55–58.

Myers, N.W. 1957. Design and operation of a commercial soybean-oil refining plant using acetic anhydride as a degumming reagent. J. Am. Oil Chem. Soc. 34:93–96.

Nash, N.H., and L.M. Brickman. 1972. Food emulsifiers—science and art. J. Am. Oil Chem. Soc. 49:457–461.

National Soybean Processors Association. 1984–1985. Yearbook and trading rules. National Soybean Processors Association, Washington, DC.

Newkirk, D.R., A.J. Shephard, and W.D. Hubbard. 1978. Comparison of total fat, fatty acids, cholesterol, and other sterols in mayonnaise and imitation mayonnaise. J. Am. Oil Chem. Soc. 55:548A-549A.

Numenz, G.M. 1978. Old and new in winterizing. J. Am. Oil Chem. Soc. 55:396A-398A.

Ohlrogge, J.B., R.M. Gulley, and E.A. Emken. 1982. Occurrence of octadecenoic fatty acid isomers from hydrogenated fats in human tissue lipid classes. Lipids 17:551–557.

Osborne, T.B., and L.B. Mendel. 1917. The use of soy bean as food. J. Biol. Chem. 32:369–387.

Perkins, R.B., Jr., J.J. Roden, III, and E.H. Pryde. 1975. Nylon-9 from unsaturated fatty derivatives: Preparation and characterization. J. Am. Oil Chem. Soc. 52:473–477.

Powell, T.S., and M.H. Gehle. 1976. Effect of various pullet restriction methods on performance of broiler breeders. Poult. Sci. 55:502–509.

Pringle, W. 1974. Full-fat soy flour. J. Am. Oil Chem. Soc. 51:74A-76A.

Probst, A.H., and R.W. Judd. 1973. Origin, U.S. history and development, and world distribution. *In* B.E. Caldwell (ed.) Soybeans: improvement, production, and uses. 16:1–15.

Pryde, E.H. 1980. Composition of soybean oil. p. 13–31. *In* D.R. Erickson et al. (ed.) Handbook of soy oil processing and utilization. American Soybean Association, St. Louis, and American Oil Chemists Society, Champaign, IL.

----. 1983. Utilization of commercial oilseed crops. Econ. Bot. 37:459–477.

Rackis, J.J. 1981. Flatulence caused by soya and its control through processing. J. Am. Oil Chem. Soc. 58:503–509.

Rizek, R.L., S.O. Welsh, R.M. Marston, and E.M. Jackson. 1983. Levels and sources of fat in the U.S. food supply and in diets of individuals. p. 13–43. *In* E.G. Perkins and W.H. Visek (ed.) Dietary fats and health. American Oil Chemists Society, Champaign, IL.

Rudolph, B.C., L.S. Boggs, D.A. Knabe, T.D. Tanksley, Jr., and S.A. Anderson, 1983. Digestibility of nitrogen and amino acids in soybean products for pigs. J. Anim. Sci. 57:373–386.

Saio, K., M. Kamiya, and T. Watanabe. 1969. Food processing characteristics of soybean 11S and 7S proteins. Part I. Effect of difference of protein components among soybean varieties on formation of tofu-gel. Agric. Biol. Chem. 33:1301–1308.

----, I. Sato, and T. Watanabe. 1974. Food use of soybean 7S and 11S proteins. High temperature expansion characteristics of gels. J. Food Sci. 39:777–782.

Sair, L. 1959. Proteinaceous soy composition and method of preparing. U.S. Patent 2 881 076. Date issued: 7 April.

Schneider, W.J., and L.E. Gast. 1978. Water-dispersible urethane polyesteramide coatings from linseed oil. J. Coat. Technol. 50(646):76–81.

----, and ----. 1979. Poly(ester-amide-urethane) water dispersible and emulsifiable resins. J. Coat. Technol. 51(654):53–57.

Schwab, A.W., W.K. Rohwedder, and L.E. Gast. 1978. Hydrogen sulfide adducts of methyl trans,trans-9,11-octadecadienoate. J. Am. Oil Chem. Soc. 55:860–864.

Scott, M.L. 1984. Dietary nutrient allowances for chickens, turkeys. Feedstuffs 56(30): 64, 66.

Sessa, D.J., K. Warner, and J.J. Rackis. 1976. Oxidized phosphatidylcholines from defatted soybean flakes taste bitter. J. Agric. Food Chem. 24:16–21.

Shurtleff, W., and A. Aoyagi. 1975. The book of tofu, Vol. 1. Autumn Press, Hayama-shi, Kanagawa-ken, Japan.

----, and ----. 1976. The book of miso. Autumn Press, Brookline, MA.

----, and ----. 1979. The book of tempeh. Harper and Row Publishing, New York.

----, and ----. 1984. Soymilk. Industry and market. The Soyfoods Center, Lafayette, CA.

Sievert, C.W. 1972. Formulating quality feeds. p. 103–136. *In* Feed industry red book. Communications Marketing, Edina, MN.

Sissons, J.W., A. Nyrup, P.J. Kilshaw, and R.H. Smith. 1982. Ethanol denaturation of soya bean protein antigens. J. Sci. Food Agric. 33:706–710.

Smith, A.K., and S.J. Circle. 1972. Soybeans: Chemistry and technology, Vol. I. Proteins. AVI Publishing Co., Westport, CT.

Spears, J.W., J.H. Clark, and E.E. Hatfield. 1985. Nitrogen utilization and ruminal fermentation in steers fed soybean meal treated with formaldehyde. J. Anim. Sci. 60:1072–1080.

Staswick, P.E., M.A. Hermodson, and N.C. Nielsen. 1984. The amino acid sequence of the A_2B_{1a} subunit of glycinin. J. Biol. Chem. 259:13424–13430.

Steggerda, F.R., E.A. Richards, and J.J. Rackis. 1966. Effects of various soybean products on flatulence in the adult man. Proc. Soc. Exp. Biol. Med. 121:1235–1239.

Sullivan, D.A., B.D. Campbell, M.F. Conway, and F.N. Grimsby. 1982. Isopropanol extraction of oilseeds. Oil Mill Gazet. 86(10):24–27.

Sullivan, F.E. 1976. Steam refining. J. Am. Oil Chem. Soc. 53:358–360.

Szuhaj, B.F. 1980. Food and industrial uses of soybean lecithin. p. 681–691. *In* F.T. Corbin (ed.) World soybean research conference II: Proceedings. Westview Press, Boulder, CO.

----, and G.R. List (ed.) 1985. Lecithins. American Oil Chemists Society, Champaign, IL.

Thanh, V.H., and K. Shibasaki. 1976. Major proteins of soybean seeds. A straightforward fractionation and their characterization. J. Agric. Food Chem. 24:1117–1121.

----, and ----. 1978. Major proteins of soybean seeds. Subunit structure of β-conglycinin. J. Agric. Food Chem. 26:692–695.

Thomas, F.L., and L.E. Gast. 1979. New solventless polymeric protective coatings from fatty acid derivatives. J. Coat. Technol. 51(657):51–59.

Thulin, W.W., and S. Kuramoto. 1967. "Bontrae"—A new meat-like ingredient for convenience foods. Food Technol. 21:168–171.

Turner, R.H., and I.E. Liener. 1975. The effect of the selective removal of hemagglutinins on the nutritive value of soybeans. J. Agric. Food Chem. 23:484–487.

Turnlund, J.R., J.C. King, W.R. Keyes, B. Gong, and M.C. Michel. 1984. A stable isotope study of zinc absorption in young men: Effects of phytate and α-cellulose. Am. J. Clin. Nutr. 40:1071–1077.

U.S. Department of Agriculture. 1950–1982. Agricultural statistics. USDA-ERS, U.S. Government Printing Office, Washington, DC.

----. 1983. Proceedings of vegetable oil as diesel fuel—Seminar III. USDA-ARS, ARM-NC28, Oct. 19–20. U.S. Government Printing Office, Washington, DC.

----. 1984a. U.S. soybean industry. ERS, Agric. Econ. Rep.

----. 1984b. Feed. Outlook and situation report. p. 37–39. USDA-ERS FdS-295. U.S. Government Printing Office, Washington, DC.

----. 1984c. Oil crops. Outlook and situation report. USDA-ERS OCS-6, U.S. Government Printing Office, Washington, DC.

Vandergrift, W.L., D.A. Knabe, T.D. Tanksely, Jr., and S.A. Anderson. 1983. Digestibility of nutrients in raw and heated soyflakes for pigs. J. Anim. Sci. 57:1215–1224.

Van Nieuwenhuyzen, W. 1976. Lecithin production and properties. J. Am. Oil Chem. Soc. 53:425–427.

Warner, K., T.L. Mounts, J.J. Rackis, and W.J. Wolf. 1983. Relationships of sensory characteristics and gas chromatographic profiles of soybean protein products. Cereal Chem. 60:102–106.

Watt, B.K., and A.L. Merrill. 1963. Composition of foods. USDA Agric. Handb. 8. U.S. Government Printing Office, Washington, DC.

Weihrauch, J.L., C.A. Brignoli, J.B. Reeves, III, and J.L. Iverson. 1977. Fatty acid composition of margarines, processed fats and oils: A new compilation of data for tables of food composition. Food Technol. 31(2):80–85, 91.

Weiss, T.J. 1983. Mayonnaise and salad dressings. p. 211–246. *In* Food oils and their uses. 2d ed. AVI Publishing Co., Westport, CT.

Wiedermann, L.H. 1978. Margarine and margarine oil, formulation and control. J. Am. Oil Chem. Soc. 55:823–829.

----. 1981. Degumming, refining and bleaching soybean oil. J. Am. Oil Chem. Soc. 58:159–166.

Wilkins, W.F., L.R. Mattick, and D.B. Hand. 1967. Effect of processing method on oxidative off-flavors of soybean milk. Food Technol. 21:1630–1633.

Williams, T.A. 1899. The soybean as a forage crop. USDA Farmers' Bull. 58.

Wolf, W.J., and J.C. Cowan. 1975. Soybeans as a food source. CRC Press, Cleveland.

----, and D.A. Sly. 1967. Cryoprecipitation of soybean 11S protein. Cereal Chem. 4:653–668.

----, and T. Tamura. 1969. Heat denaturation of soybean 11S protein. Cereal Chem. 46:331–344.

Zehnder, C.T. 1976. Deodorization 1975. J. Am. Oil Chem. Soc. 53:364–369.

SUBJECT INDEX